CW01642898

Building Performance Simulation for Design and Operation

When used appropriately, building performance simulation has the potential to reduce the environmental impact of the built environment, to improve indoor quality and productivity, as well as to facilitate future innovation and technological progress in construction. Since publication of the first edition of *Building Performance Simulation for Design and Operation*, the discussion has shifted from a focus on software features to a new agenda, which centres on the effectiveness of building performance simulation in building life cycle processes.

This new edition provides a unique and comprehensive overview of building performance simulation for the complete building life cycle from conception to demolition, and from a single building to district level. It contains new chapters on building information modelling, occupant behaviour modelling, urban physics modelling, urban building energy modelling and renewable energy systems modelling. This new edition keeps the same chapter structure throughout including learning objectives, chapter summaries and assignments. Moreover, the book:

- Provides unique insights into the techniques of building performance modelling and simulation and their application to performance-based design and operation of buildings and the systems which service them.
- Provides readers with the essential concepts of computational support of performance-based design and operation.
- Provides examples of how to use building simulation techniques for practical design, management and operation, their limitations and future direction.

It is primarily intended for building and systems designers and operators, and postgraduate architectural, environmental or mechanical engineering students.

Jan L.M. Hensen is Professor in Building Performance Simulation at Eindhoven University of Technology. His teaching and research focuses on development and application of computational modelling and simulation for high-performance buildings while considering building physics, indoor environmental quality and building energy systems. He is, amongst other roles, Co-Editor of the first edition of *Building Performance Simulation for Design and Operation* and the founding Co-Editor of the *Journal of Building Performance Simulation.*

Roberto Lamberts is Professor in Construction at the Department of Civil Engineering of the Federal University of Santa Catarina. He is also currently a board member of the International Building Performance Simulation Association (IBPSA) and counsellor of the Brazilian Council for Sustainable Buildings. He is, amongst other roles, Co-Editor of the first edition of *Building Performance Simulation for Design and Operation* and editorial board member of the *Journal of Building Performance Simulation.*

Building Performance Simulation for Design and Operation

Expanded Second Edition

Edited by Jan L.M. Hensen and Roberto Lamberts

Routledge
Taylor & Francis Group
LONDON AND NEW YORK

Second edition published 2019
by Routledge
2 Park Square, Milton Park, Abingdon, Oxon, OX14 4RN

and by Routledge
52 Vanderbilt Avenue, New York, NY 10017

Routledge is an imprint of the Taylor & Francis Group, an informa business

© 2019 selection and editorial matter, Jan L.M. Hensen and Roberto Lamberts; individual chapters, the contributors

The right of Jan L.M. Hensen and Roberto Lamberts to be identified as the authors of the editorial material, and of the authors for their individual chapters, has been asserted in accordance with sections 77 and 78 of the Copyright, Designs and Patents Act 1988.

All rights reserved. No part of this book may be reprinted or reproduced or utilised in any form or by any electronic, mechanical, or other means, now known or hereafter invented, including photocopying and recording, or in any information storage or retrieval system, without permission in writing from the publishers.

Trademark notice: Product or corporate names may be trademarks or registered trademarks, and are used only for identification and explanation without intent to infringe.

First edition published by Spon Press 2011

British Library Cataloguing-in-Publication Data
A catalogue record for this book is available from the British Library

Library of Congress Cataloging-in-Publication Data
Names: Hensen, Jan, editor. | Lamberts, Roberto, editor.
Title: Building performance simulation for design and operation / [edited by] Jan L.M. Hensen and Roberto Lamberts.
Description: Second edition. | Abingdon, Oxon ; New York, NY : Routledge, 2019. | Includes bibliographical references and index.
Identifiers: LCCN 2018051032 | ISBN 9781138392199 (hbk : alk. paper) | ISBN 9780429402296 (ebk : alk. paper)
Subjects: LCSH: Buildings—Performance—Computer simulation.
Classification: LCC TH453 .B875 2019 | DDC 690/.22—dc23
LC record available at https://lccn.loc.gov/2018051032

ISBN: 978-1-138-39219-9 (hbk)
ISBN: 978-0-429-40229-6 (ebk)

Typeset in Times New Roman
by Apex CoVantage, LLC

The International Building Performance Simulation Association (IBPSA) makes every effort to ensure the accuracy of all the information contained in publications that it endorses. However, IBPSA, our agents and our licensors make no representations or warranties whatsoever as to the accuracy, completeness or suitability for any purpose of the content. Any opinions or views expressed in this publication are the opinions and views of the authors and are not the views of IBPSA. The accuracy of the content should not be relied upon and should be independently verified with primary sources of information. IBPSA shall not be liable for any losses, actions, claims, proceedings, demands, costs, expenses, damages and other liabilities whatsoever or howsoever caused arising directly or indirectly in connection with, in relation to or arising out of the use of this content.

MIX
Paper from responsible sources
FSC www.fsc.org FSC™ C013985

Printed in the United Kingdom
by Henry Ling Limited

Contents

List of contributors viii
Foreword to the second edition by Joe Clarke xiii
Foreword to the first edition by Joe Clarke xiv
Preface to the second edition xvi
Preface to the first edition xvii

1 **Building performance simulation – challenges and opportunities** 1
JAN L.M. HENSEN AND ROBERTO LAMBERTS

PART I
Fundamentals 11

2 **Thermal load and energy performance prediction** 13
JEFFREY D. SPITLER

3 **Ventilation performance prediction** 76
JELENA SREBRIC

4 **People in building performance simulation** 117
ARDESHIR MAHDAVI AND FARHANG TAHMASEBI

5 **Indoor thermal quality performance prediction** 146
CHRISTOPH VAN TREECK AND DANIEL WÖLKI

6 **Weather and climate in building performance simulation** 191
DRURY B. CRAWLEY AND CHARLES S. BARNABY

7 **Daylight performance predictions** 221
CHRISTOPH REINHART

8 **Moisture modeling and durability assessment of building envelopes: recent advances** 270
AYTAÇ KUBILAY, XIAOHAI ZHOU, DOMINIQUE DEROME AND JAN CARMELIET

9 **Computational modeling in architectural acoustics** 315
ARDESHIR MAHDAVI

PART II
Design 341

10 **The role of simulation in performance-based building** 343
GODFRIED AUGENBROE

11 **BIM and BPS: a case study of integration cost metrics and design options** 374
TIMOTHY HEMSATH, MATTHEW GOLDSBERRY AND JOEL YOW

12 **Building simulation for policy support** 384
DRURY B. CRAWLEY

PART III
Operational optimization 397

13 **Building simulation for practical operational optimization** 399
DAVID E. CLARIDGE AND MITCHELL T. PAULUS

14 **Modeling and simulation in building automation systems** 454
GREGOR P. HENZE

PART IV
Systems 501

15 **HVAC systems performance prediction** 503
JONATHAN WRIGHT

16 **Micro-cogeneration system performance prediction** 534
IAN BEAUSOLEIL-MORRISON

17 **Modeling in building-to-grid integration** 559
SEN HUANG, THOMAS SEVILLA AND WANGDA ZUO

18 **Modelling HVAC and renewable energy plant and control** 586
CHRIS UNDERWOOD AND SIMON REES

19 **A view on future building system modeling and simulation** 631
MICHAEL WETTER

PART V
Urban 657

20 Integrated resource flow modelling of the urban built environment 659
DARREN ROBINSON

21 Urban building energy modeling 696
CHRISTOPH REINHART AND CARLOS CEREZO DAVILA

22 Urban physics simulation for climate change adaptation of buildings and urban areas 723
BERT BLOCKEN, TWAN VAN HOOFF, YASIN TOPARLAR, CHRISTOF GROMKE, HAMID MONTAZERI AND WENDY JANSSEN

Index 767

Contributors

Godfried Augenbroe is Professor in the School of Architecture at the Georgia Institute of Technology. He directs the High Performance Building Lab with a thriving masters and PhD program. Augenbroe has published over 200 refereed papers and several books. He has been the main advisor of 35 PhD dissertations in areas of building performance, simulation, knowledge management, building pathology, sustainability, urban energy systems, uncertainty and risk analysis. In 2015 he received the Lifetime Distinguished Achievement award from IBPSA International. In 2017 he was honoured by Georgia Tech with the Outstanding Doctoral Thesis Advisor award.

Charles S. Barnaby is an independent consultant specializing in building modelling software development. He has served as President of the International Building Performance Simulation Association (IBPSA). He is a Fellow of IBPSA and ASHRAE. His interests include standards for building information exchange, weather data and peak load calculation methods.

Ian Beausoleil-Morrison is Professor in the Faculty of Engineering and Design at Carleton University in Ottawa. He has served on ASHRAE Technical Committees and as an Operating Agent for the International Energy Agency's Energy Conservation in Buildings Implementing Agreement. He is an IBPSA Fellow and served as IBPSA President from 2010 to 2015. He is the Co-Founder and has been Co-Editor of the *Journal of Building Performance Simulation* since its establishment in 2008.

Bert Blocken is Professor and Chair of Building Physics at the Department of the Built Environment at Eindhoven University of Technology and is a part-time professor at the Department of Civil Engineering at KU Leuven. His research areas are urban physics, wind engineering and sports aerodynamics.

Jan Carmeliet is Professor at the Chair of Building Physics, Department of Mechanical and Process Engineering at ETH Zurich since 2008. His research interests concern physical processes in multiscale porous materials, urban climate and urban energy systems. He obtained in PhD from Katholieke Universiteit Leuven in 1992, where he was a professor.

David E. Claridge (MS and PhD, Stanford University) is Professor of Mechanical Engineering at Texas A&M University and is Director of the Energy Systems Laboratory. He is a Fellow of ASME and ASHRAE and recipient of the E.K. Campbell Award from ASHRAE, received a Faculty Distinguished Achievement in Research Award from Texas A&M University and has received three Best Paper Awards from ASME.

Joe Clarke is Professor and Director of the Energy Systems Research Unit at the University of Strathclyde, Glasgow. His main focus is the role that energy systems simulation can play in helping to reduce energy demand, accelerate the take-up of renewable energy technologies, mitigate environmental impacts and improve human wellbeing. A major aspect of his work involves the development and dissemination of software tools for energy systems simulation and support for the application of these tools in design, research, teaching and policy-making contexts. He is the progenitor of the ESP-r building simulation program and founding member and Past President of the International Building Performance Simulation Association, which promotes the technology worldwide.

Drury B. Crawley is Bentley Fellow and Director, Building Industry Development at Bentley Systems, Inc., where he focuses on building performance, BIM, net-zero-energy buildings, interoperability, resilience and sustainability. He is currently President of IBPSA-USA and Vice President of IBPSA. Prior to being made a Bentley Fellow in 2014, he led the development of Bentley's suite of building performance software. Before joining Bentley in 2010, he developed and managed EnergyPlus and the USDOE's Commercial Buildings Initiative promoting net-zero-energy buildings. He co-founded Climate.OneBuilding, a repository of free weather data for building performance simulation (http://climate.onebuilding.org/) and is a Fellow of IBPSA and ASHRAE.

Carlos Cerezo Davila is Research Scientist at MIT and the Environmental Design Director for KPF New York. His work focuses on the development of workflows and tools to incorporate building performance simulation in design at all scales. At MIT, his research centres on the application of energy simulation and uncertainty analysis at an urban scale, in collaboration with municipalities such as Boston, Chicago, Lisbon and Kuwait. He holds a masters from Harvard University (2013) and a PhD from MIT (2017).

Dominique Derome is Senior Scientist at the Laboratory for Multiscale Studies in Building Physics, Empa, where she has worked since 2008. Previously, she was Associate Professor in the Department of Building, Civil and Environmental Engineering of Concordia University, Montreal, where she obtained her PhD. She has a multi-disciplinary background in architecture, building engineering and soft matter physics.

Matthew Goldsberry is Computational Design Principal with HDR Architecture, where he oversees the applied research and implementation of advanced computational design workflows. He is responsible for developing new computational tools and workflows to facilitate design exploration, automated analysis and advanced data management.

Christof Gromke is Researcher at the Institute for Hydromechanics at Karlsruhe Institute of Technology (KIT). His research interests comprise building and environmental aerodynamics and urban meteorology with special focus on the effects of vegetation on flow and dispersion processes.

Timothy Hemsath is the Sustainability Design Lead in HDR's Omaha office and has worked professionally in Omaha and Philadelphia as a project architect. Previously, he was an Associate Professor at the University of Nebraska-Lincoln's College of Architecture, where he provided leadership in the realm of digital fabrication and energy modelling.

Jan L.M. Hensen is Professor in Building Performance at Eindhoven University of Technology and at the Czech Technical University in Prague. His teaching and research focuses on development and application of building performance simulation with the objective to improve energy efficiency and indoor environmental quality of buildings. He is a Fellow of IBPSA, ASHRAE and REHVA.

Gregor P. Henze is Professor of Architectural Engineering and endowed C.V. Schelke Chair at the University of Colorado Boulder. His teaching focuses on building systems engineering and his research emphasizes model predictive control and reinforcement learning control for building energy systems and building thermal mass, fault detection and diagnosis, urban energy systems, energy analytics and decision analysis as well as building-to-grid integration. He is Associate Director of the Renewable and Sustainable Energy Institute, Joint Professor at the National Renewable Energy Laboratory in Golden, Colorado, as well as Co-Founder and Chief Scientist of QCoefficient, Inc.

Twan van Hooff is Senior Research Fellow at the Department of Civil Engineering at KU Leuven and part-time Assistant Professor at the Department of the Built Environment at Eindhoven University of Technology. His areas of expertise are building ventilation, building energy simulation and urban physics.

Sen Huang is a building control engineer with the Optimization and Control Group at Pacific Northwest National Laboratory (PNNL). At PNNL, he primarily works on developing transactive control strategies for building-to-grid integration and applying system modelling in the design/evaluation process of the building controls. His research interests include modelling of cyber-physical systems and large-scale building simulations.

Wendy Janssen graduated with an MSc in Building Physics and Services at Eindhoven University of Technology. Since 2016, she has been Advisor Hydro and Wind and Data Scientist at the Port of Rotterdam. Before 2016, she was a researcher on climate adaptation of buildings and cities at Eindhoven University of Technology.

Aytaç Kubilay is a scientist at the Laboratory for Multiscale Studies in Building Physics at Empa (Swiss Federal Laboratories for Materials Science and Technology). His research interests include numerical modelling of microclimate, wind flow in urban areas, wind-driven rain, porous media and heat and mass transfer. He completed his PhD in 2014 at the Chair of Building Physics at ETH Zurich, where he worked on the modelling of wind-driven rain in urban areas using numerical and experimental methods.

Roberto Lamberts is Professor in Building Construction at the Department of Civil Engineering of the Federal University of Santa Catarina. His teaching and research focuses on thermal comfort, thermal performance and energy efficiency in buildings and building simulation. He is a member of IBPSA and ASHRAE.

Ardeshir Mahdavi is Director of the Department of Building Physics and Building Ecology as well as the Head of the Institute of Architectural Sciences at TU Wien. His research is concerned with building physics, building performance simulation, building controls and diagnostics, building ecology and human ecology. He has authored over 700 publications and supervised 70 doctoral students. Professor Mahdavi is the recipient of the IBPSA Distinguished Achievements Award.

Hamid Montazeri is Senior Research Fellow at the Department of Civil Engineering at KU Leuven and part-time Assistant Professor at the Department of the Built Environment at Eindhoven University of Technology. His areas of expertise are urban physics and wind engineering.

Mitchell T. Paulus is Energy Engineer at Command Commissioning. He currently develops data analytics and software to implement into all phases of the existing building commissioning process. He earned his PhD from Texas A&M University in 2017 with a focus on remote HVAC system optimization utilizing typical building sensor data.

Simon Rees is Professor of Building Energy Systems in the School of Civil Engineering at the University of Leeds and leads the teaching of building physics in the school's Architectural Engineering programme. His research has included the performance of low-energy cooling systems and numerical modelling of ventilation and environmental heat transfer. His work has recently focused on geothermal heat exchange as well as assessment of microgeneration technologies and thermal energy networks.

Christoph Reinhart is Professor in Building Technology at MIT where he is leading the Sustainable Design Lab, an interdisciplinary group with a grounding in architecture that develops design workflows, planning tools and metrics to evaluate the environmental performance of buildings and neighbourhoods. He is also the head of Solemma, a technology company and Harvard University spinoff.

Darren Robinson is Professor of Architectural and Urban Sciences and Director of Research in the School of Architecture at the University of Sheffield. He has many research interests, but he is mainly recognized for his expertise in the energy simulation of buildings in their urban context that are occupied by people whose behaviours are stochastic in nature. He has over 100 refereed scientific publications to his credit, including the book *Computer Modelling for Sustainable Urban Design*. He sits on the editorial advisory boards of the *Building and Environment* (BAE) journal and the on-line journal *Sustainability*. A Fellow of IBPSA, he is a recipient of the CIBSE Napier-Shaw Medal, the JPBS Best Paper Prize and the BAE Best Paper Award (twice).

Thomas Sevilla leads the research and development team for smart city IoT sensors for 3Z Telecom in Miramar, Florida. Additionally, he is a PhD student in the Department of Civil, Architectural and Environmental Engineering at the University of Miami. His PhD research, which he started in 2016, mainly focuses on the design of control for distributed energy resources and their complex interconnections with the power grid.

Jeffrey D. Spitler is Regents Professor and OG&E Energy Technology Chair in the School of Mechanical and Aerospace Engineering at Oklahoma State University, where he teaches classes and performs research in the areas of heat transfer, thermal systems, building simulation, design cooling load calculations, HVAC systems and ground source heat pump systems. He is a Fellow of IBPSA and ASHRAE.

Jelena Srebric is Professor of Mechanical Engineering and the founding Director of the Cluster for Sustainability in the Built Environment at the University of Maryland (CITY@UMD). She conducts research and teaches in the fields of building energy consumption, indoor air quality and associated respiratory health of occupants. She is a Fellow of the International Society of Indoor Air Quality and Climate and an International Member of the Serbian National Academy of Engineering.

Farhang Tahmasebi is Lecturer at UCL Institute for Environmental Design and Engineering in London. His research focuses on building performance simulation, high-performance buildings and occupant behaviour modelling. Prior to his appointment at UCL, he acted as a researcher and educator at TU Wien Institute of Architectural Sciences. He is currently a subtask leader at IEA-EBC Annex 79, which focuses on applying occupant behaviour models in performance-based design process.

Yasin Toparlar is a guest researcher at the Chair of Building Physics at Eindhoven University of Technology. He earned his PhD in 2018 from the same university with the thesis "A multiscale analysis of the urban heat island effect". His research focus is mainly on urban climatology, urban sustainability, building energy simulation and climate change adaptation.

Christoph van Treeck is Professor for Energy Efficiency and Sustainable Building at RWTH Aachen University. He specializes in the fields of building and community energy systems performance modelling and simulation, building information modelling, computational fluid mechanics and thermal ergonomics of the indoor environment.

Chris Underwood is Emeritus Professor of Energy Modelling for the Built Environment at Northumbria University. His expertise lies in the modelling of HVAC and urban renewable plant and control. His most recent work has been on domestic scale heat pumps, thermal storage and the citywide electrification of heat.

Michael Wetter is Computational Staff Scientist at Lawrence Berkeley National Laboratory (LBNL). His research includes the development of new generation computing tools for the design and operation of building and district energy systems. He is a Fellow of IBPSA, a recipient of the IBPSA Outstanding Young Contributor Award, a Board Member of IBPSA, Co-Operating Agent of IBPSA Project 1 and a Past President of IBPSA-USA.

Daniel Wölki is Group Leader at the Institute of Energy Efficiency and Sustainable Building (E3D) at RWTH Aachen University. His research and teaching focuses on thermal comfort, human thermophysiology, building physics, system modelling and cyber-physical systems.

Jonathan Wright is Professor of Building Optimization at Loughborough University and is an IBPSA Fellow. He has over 30 years of research experience in the field of model-based optimization and its application to the optimum design and operation of buildings. His publications span the core elements of building optimization studies: model development, the development of optimization methods and the application of model-based optimization to building design. He has wide experience in teaching building energy and mechanical systems topics to both university undergraduates and postgraduates.

Joel Yow worked at HDR Architecture as Computational Design Principal, where he focused on the integration of data, advanced analysis and design excellence. He is now leading his own data-focused practice, where he works with other firms, owners and builders to develop computational tools, implement data-focused strategies and refine digital design workflows.

Xiaohai Zhou is Research Scientist at the Laboratory for Multiscale Studies in Building Physics at Empa and is Lecturer at the Chair of Building Physics at ETH Zurich. He is working on building envelope retrofitting, heat and moisture transport in porous building materials and durability of building materials.

Wangda Zuo is Associate Professor and Lewis-Worcester Faculty Fellow in the Department of Civil, Environmental and Architectural Engineering at the University of Colorado Boulder. His research focuses on modelling and simulation of smart, sustainable and resilient buildings and cities. He is the recipient of the IBPSA-USA Emerging Professional Award and the Eliahu I. Jury Early Career Research Award.

Foreword to the second edition

No research field stands still, and this is particularly true of building performance prediction in light of the myriad developments underpinning the clean energy transition agenda being pursued in many parts of the world. Such developments are extending both the scale and the scope of the energy utilization challenge at local, regional and national levels. Never has there been a greater need to consider the implications of design intentions in a holistic manner in order to maintain balance between conflicting aspects – for example, between energy saving and well-being, between renewables integration and utility costs, or between heightened complexity and operational resilience.

In this second edition, distinguished researchers contribute to these challenges by updating the reader on developments relating to the traditional targets of building performance simulation: ventilation systems, thermal comfort, daylight utilization, moisture flow, HVAC systems operation and automated control. It also introduces new materials that address domains that have hitherto been missing: the modelling of occupant behaviour, a dominant cause of the performance gap observed in practice; building-integrated renewable energy systems, seen as one way of attaining net-zero-energy designs; electrical grid interaction, forming the basis of future smart grids; micro-generation, providing a means to reduce the overall load on the electricity supply network; and urban scale considerations, facilitating community energy schemes and providing the link to the wider issue of resource utilization.

Additional new material then completes the state-of-the-art review by considering emerging concepts relating to building information management and modern approaches to software engineering. Such developments will surely help to integrate the design community by enabling model exchange and faster response to new functionality requests from users.

The book remains essential reading to those practitioners and researchers who seek to understand and apply building simulation in a professional manner.

Joe Clarke
August 2018

Foreword to the first edition

The fossil fuels are entering their tertiary stage and steps are being taken in many countries to kick-start the transition to an alternative energy infrastructure. A pressing question is how this transition can best be managed, negative impacts mitigated, and the various technology options blended over time: fossil fuel de-carbonisation and sequestration in the short term, the deployment of energy efficiency measures, the switch to new and renewable source of energy, and the removal of barriers confronting new nuclear plant. A key aspect of any future energy infrastructure will be real-time demand management to facilitate the matching of demand with supply – especially where the latter comprises significant inputs from stochastic, distributed renewable energy sources. Because a large portion of a country's energy demand is associated with the built environment, it is here that productive action can be taken to reduce energy consumption whilst ensuring that expectations relating to human comfort/health and environmental protection are met.

The built environment is inherently complex and as a consequence conflicts abound, proffered solutions are often polarised and consensus is difficult to attain. This situation gives rise to three fundamental engineering challenges: how to consider energy systems in a holistic manner in order to address the inherent complexity; how to include environmental and social considerations in the assessment of cost-performance in order to ensure sustainable solutions; and how to embrace inter-disciplinary working in order to derive benefit from the innovative approaches to be found at the interface between the disciplines. In short, energy systems require an integrated approach to design: will the widespread deployment of micro-CHP within the urban environment be acceptable if the global carbon emission reduction to result is attained at the expense of reduced local air quality and increased maintenance cost?

Integrated building performance simulation has emerged as an apt means of addressing the above challenges while allowing collaborating practitioners to identify the action combinations that will be most effective in providing acceptable overall performance as a function of the unique climate, design and operational parameters defining specific buildings and communities, planned or existing. IBPS does this by modelling the heat, air, moisture, light, electricity, pollutant and control signal flows within building/plant systems and, thereby, nurturing performance improvement by design. The benefits of the power and universal applicability of the approach comes at a price, however: application requires an understanding of design hypothesis abstraction, computer model building, multiple domain simulation, performance trade-offs, and the translation of outcomes to design evolution.

This book presents the complementary views of distinguished researchers in the field, arranged in a progressive format that covers the myriad issues underpinning the application of modelling and simulation when used to support decision-making relating to building performance and operation. In addition to the wide scope of topics covered, the book provides useful examples of the practical application of building simulation to formulate design and operational solutions that are acceptable in terms of performance criteria relating to indoor air quality, thermal/visual/acoustic comfort, operational/embodied energy, carbon emissions, and capital/running cost. A unique feature of the book is the balance between theory and practice on the one hand, and between the issues at the individual building and community level on the other. The book is essential reading to those practitioners and researchers who seek to understand and apply building simulation in a professional manner.

Joe Clarke
Energy Systems Research Unit
University of Strathclyde
Glasgow
December 2009

Preface to the second edition

The first edition of this book was published in 2011. It was very well received, judging from the number of copies sold and the positive feedback of many readers from various fields. Since then, several years have passed and many new developments have taken place in the field of building performance simulation. All of the first edition chapter authors felt that – given recent developments in their particular areas – after eight years, their chapter could do with revisions and updates, and in several cases also expansions. Therefore, the collective authors agreed that it was high time for a new edition of the book.[1]

We have also received numerous requests to publish an expanded second edition with more information on building information modelling, occupant behaviour modelling, urban physics modelling, urban building energy modelling and renewable energy systems modelling. In response, this second edition has five additional chapters.

For improved readability, in this second edition, the order of the chapters has changed. They are now organized according to the themes: fundamentals, design, operational optimization, systems and urban aspects.

As before, we would like to express our sincere gratitude to all co-authors, reviewers and others for the time and effort they have devoted to this book project.

Jan L.M. Hensen
Roberto Lamberts
September 2018

Note

1 In this context we also agreed that it would be a good idea for each chapter author to give a webinar presenting a summary of his/her chapter and to provide updates on recent developments. Feel free to watch these webinars, which were broadcasted in 2016/2017; they are still accessible from www.ibpsa.org/?page_id=695.

Preface to the first edition

The rise of technology over the past decades has been something of a mixed blessing. On the one hand it has increased our freedom to move and communicate and has provided us with more comfort. On the other hand, it is widely understood that the energy use currently required to drive our modern way of living has led to critical environmental problems. These problems have been highlighted to such a degree through the explosion of research and news coverage over recent years, that it is now common knowledge that our lifestyle is unsustainable. In modern terminology, to slow down and hopefully reverse the manmade damage, we need to develop a sustainable and net-zero-energy built environment. This will involve not only the design of net energy producing new 'green' buildings, but also the optimization of energy use of existing buildings.

In line with the rise of technology, buildings and the systems within them have become exponentially more complex in recent times. The modern built environment is populated by a variety of building types with highly demanding performance and user requirements. The difficulties involved in optimizing energy use in buildings have been recognized for quite some time. The complexity of the task arises from the number of variables from a wide range of fields that must be considered. Many professionals and researchers, including ourselves, concluded that solving such a complex problem requires two things: interdisciplinary research involving a wide variety of disciplines, and well-developed technological tools to make the problem manageable.

In 1986 a group of like-minded individuals established the International Building Performance Simulation Association, IBPSA (www.ibpsa.org), a non-profit society of building performance simulation researchers, developers and practitioners dedicated to improving the built environment. IBPSA provides a forum for researchers, developers and practitioners to review building model developments, facilitate evaluation, encourage the use of software programs, address standardization, and accelerate integration and technology transfer.

IBPSA covers broad areas of building environmental and building services engineering. Typical topics include building physics (including heat, air and moisture flow, electric and day lighting, acoustics, smoke transport); heating, ventilation and air-conditioning systems; energy supply systems (including renewable energy systems, thermal storage systems, district heating and cooling, combined heating and power systems); human factors (including health, productivity, thermal comfort, visual comfort, acoustical comfort, indoor air quality); building services; and advancements and developments in modelling and simulation, such as coupling with CAD, product modelling, software interoperability, user interface issues, validation and calibration techniques. All these topics may be addressed at different levels of resolution (from microscopic to the urban scale), and for different stages in the building life cycle (from early sketch design, via detailed design to construction, commissioning, operation, control and maintenance) of new and existing buildings worldwide.

In essence, IBPSA has two key objectives: to use computer simulation to (a) provide better support for the design of buildings; and (b) provide better support for building operation and management in the use phase of buildings. These two objectives have informed our own research over the last decades. This book aims to give the reader a thorough understanding of the recent progress made in building simulation and the key challenges that still need to be overcome.

The main motivation for developing this book was that at the time of writing no comprehensive textbook on the subject was available, even though building performance simulation has become an essential technology for architectural and engineering design and consultancy practices which aim to provide innovative solutions for their clients.

This book sets out to fill this gap by providing unique insight into the techniques of building performance modelling and simulation and their application to performance-based design and operation of buildings and the systems which service them. It provides readers with the essential concepts of computational support of performance-based design and operation – all in one book. It provides examples of how to use building simulation techniques for practical design, management and operation, and highlights their limitations and suggests future research directions.

This book provides a comprehensive overview of building performance simulation for the complete building life-cycle from conception to demolition. It addresses theory, development, quality assurance and use in practice of building performance simulation. The book is therefore both theoretical and practical, and as such will be of interest to those concerned with modelling issues (universities, research organizations and government agencies) and real world applications (architects, engineers, control bodies, building operators). The book is primarily intended for (future) building and systems designers and operators of a postgraduate level. However, due to the interdisciplinary nature of research into the built environment, the book should also prove useful for a variety of connected fields.

The interdisciplinary nature of the research becomes clear when it is understood that building performance simulation draws its underlying theories from many disciplines including physics; mathematics; material science; biophysics; human behavioral, environmental and computational sciences. The book would lend itself to adaption for multidisciplinary courses, for example, AEC related university courses which address building performance prediction and operational issues. Other courses might include it on their recommended reading list, especially at the postgraduate level.

The book begins by introducing and describing the key features of building performance simulation and sets the scene for the rest of the book. The concepts of performance indicators and targets are discussed, followed by a discussion of the current and future role of building simulation in performance-based building design and operation. This will lay the foundations for in-depth discussions of performance prediction for key aspects such as energy demand, indoor environmental quality (including thermal, visual, indoor air quality and moisture phenomena), HVAC and renewable system performance, urban level modelling, building operational optimization and automation. The book ends with a discussion of future directions for building performance simulation research and applications in practice. The book aims to show that when used appropriately, building performance simulation is a very powerful technique capable of helping us achieve a sustainable built environment, and at the same time improving indoor quality and productivity, as well as stimulating future innovation and technological progress in the architecture, engineering and construction (AEC) industry.

We believe this book to be long overdue. We have been contemplating the idea of writing it for many years. However, due to the interdisciplinary nature of the subject, writing such a book required the cooperation of many individuals. Toward the end of 2006 the idea became more concrete, and in early 2007 the co-authors and the publisher enthusiastically joined the adventure. Despite our busy schedules, in 2008 a symposium was organized in Brazil to bring co-authors together and allow them to present the content of their chapters. We would like to thank Eletrobrás and Conselho Nacional de Desenvolvimento Científico e Tecnológico (CNPq) for their financial support of this event. The event proved to be an important catalyst that allowed the book to progress to its finished state, even if it took a fair bit longer to finish the book than originally anticipated. In the end, we are really very pleased with the results, and hope you will enjoy reading the book too.

This book is the result of cooperation and dedication of many individuals, in particular of course all co-authors. We would like to take the opportunity to also acknowledge the support of our universities: Eindhoven University of Technology, the Netherlands, and Universidade Federal de Santa Catarina, Brazil. Last but not least, we wish to express our gratitude to Duncan Harkness, Roel Loonen, Ana Paula Melo, Martin Ordenes Mizgier, Jikke Reinten and Marija Trcka for their editorial and practical support.

Jan L.M. Hensen
Roberto Lamberts
March 2010

1 Building performance simulation – challenges and opportunities

Jan L.M. Hensen and Roberto Lamberts

Scope

The aim of this chapter is to provide a general view of the context and current state of building performance simulation as an introduction to the following chapters in this book.

Learning objectives

To appreciate the context, background, current state, challenges and potential of building performance simulation.

Key words

Context/state-of-the-art/future needs/quality assurance

Introduction

Building performance covers a wide variety of aspects. In the past and for the immediate future much attention is rightfully devoted to energy efficiency of buildings and communities. However, it is vitally important to realize that the primary purpose of buildings is to protect against undesirable outside conditions, and in the case of commercial buildings and homes, to provide healthy, comfortable and also, in many cases, productive indoor environments. Awareness of the significant potential benefits of providing these features is rapidly increasing; see Figure 1.1. As such, it can be expected that indoor environmental quality for health and well-being will arguably become more important than energy-related performance in the not-so-distant future.

From the perspective outlined above, building energy performance should not be addressed as an objective in itself, but rather as a cost function. Other important challenges that the built environment must manage include:

- The range of scale from tiny to enormous buildings;
- The range of construction from "do it yourself" to industrial;
- The idiosyncrasy of bespoke buildings that present unique challenges in the design, building and operational phases;
- The wide variety of stakeholders in the built environment with varying and often conflicting interests;
- The necessity for building designs to compromise or optimize depending on available resources and (conflicting) stakeholder "wishes";
- The expected longevity of buildings and their future uses;
- The increasing complexity of buildings and integrated building systems;

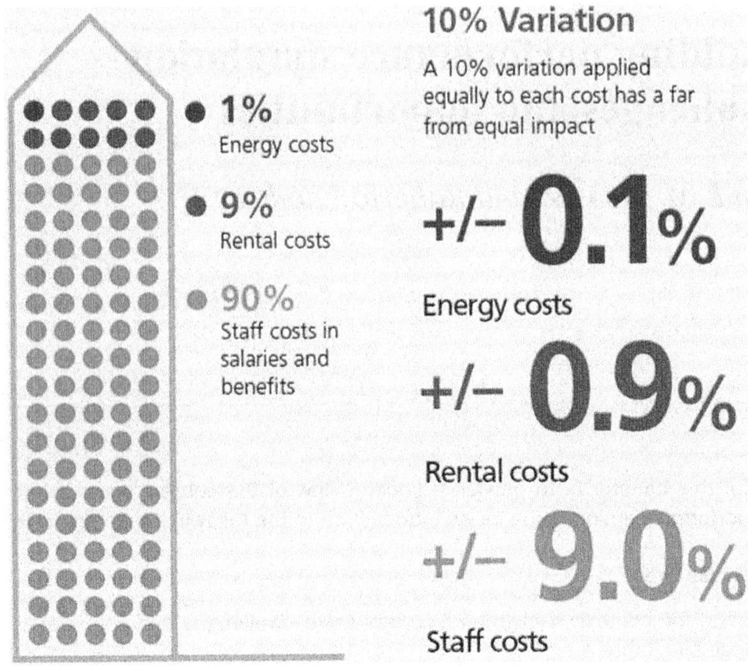

Figure 1.1 Typical business operating costs and the effect of a 10% variation applied to each cost

Source: WGBC 2014

- The increasing interconnectedness of buildings at the local and district levels;
- The change from energy consumer to energy prosumer;
- And more.

Despite the above discussion, immediate legislative pressures have placed decarbonization of the energy supply for residential and tertiary sectors high on the agenda. At present, fossil fuel consumption of buildings is responsible for a considerable portion of the total CO_2 emissions. The EU low-carbon economy roadmap (EU 2011) suggests that by 2050, the EU should cut greenhouse gas emissions to 80% below 1990 levels. Important milestones on the way to achieving this target are 40% emissions cuts by 2030 and 60% by 2040. As indicated in Figure 1.2, the current policy will not have the desired effect, and thus all sectors will need to contribute and collaborate where necessary.

For the building sector, this implies the need to contribute many innovative building solutions such as building-integrated electricity production, energy storage systems, adaptive building skins, switchable glazing, super-insulation, demand response, etc. Furthermore, these innovative solutions must be thoroughly tested to understand how they can be integrated into existing buildings or combined and optimized in new designs, and to determine how robust they may be to future scenarios.

Thus, the ultimate challenge for the building sector is to achieve a zero-carbon built environment by designing and developing high-performance buildings with indoor environments optimized for health, comfort and/or productivity. To face such a complex challenge, expertise is required from many different technical and non-technical disciplines. This expertise should be integrated and managed in a multi-actor, multi-scale, multi-physics and performance-based approach.

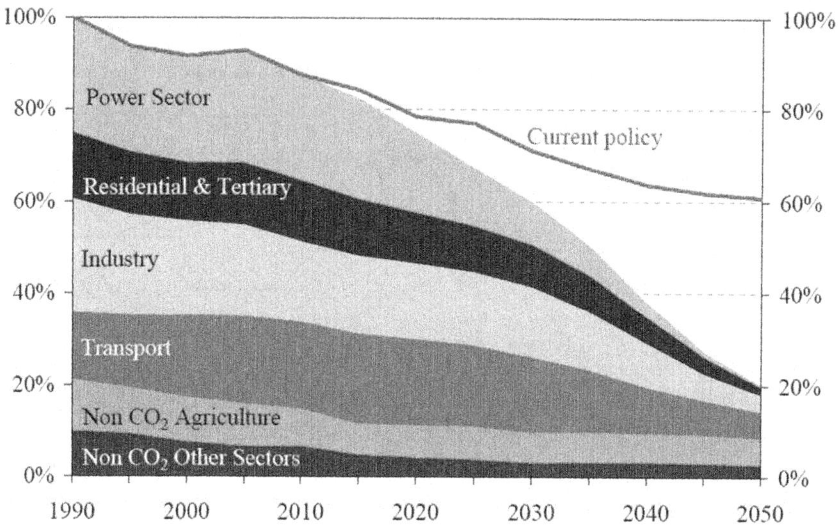

Figure 1.2 EU GHG emissions towards an 80% reduction within the EU

Source: EU 2011

We believe that computational building performance modelling and simulation potentially offers huge opportunities in this endeavour.

Building performance simulation

Computational modelling (i.e. creating a computer-based representation of a real system at a particular level of abstraction) and simulation (i.e. enacting, implementing or instantiating a model to predict the (future) behaviour of a real system) is one of the most powerful and widely used analysis techniques, with applications ranging from economic and climate forecasting to the testing of games and engineering problems. Nevertheless, it is very important to recognize that (1) simulation does not directly generate solutions or answers – its main purpose is to increase understanding, and (2) most of the time it is difficult to ensure the quality of simulation results.

Computational building performance modelling and simulation primarily covers the aspects sketched in Figure 1.3. Research, development and application of building performance simulation started in the 1960s. For reviews of the history and current state of building performance simulation, see e.g. Kusuda (1999), Spitler (2006), Clarke and Hensen (2015) and Wang and Zhai (2016). Over time, numerous building energy software tools have been developed; see BEST (2018) for an overview of currently available tools.

Both the power and the complexity of building performance modelling and simulation arise from its use of many underlying theories from diverse disciplines, mainly from physics, mathematics, material science, biophysics and human behavioural, environmental and computational sciences. The fidelity and robustness of simulation models have improved significantly and are set to continue.

The proliferation of activities and publications of the International Building Performance Simulation Association (IBPSA – www.ibpsa.org) provides ample evidence that the discipline of building performance simulation is continually evolving and maturing.

Whereas in the early days of simulation the main agenda focused on modelling and software features, the discussion has moved in recent decades to the effectiveness of building performance simulation in building life cycle processes. The development of powerful and

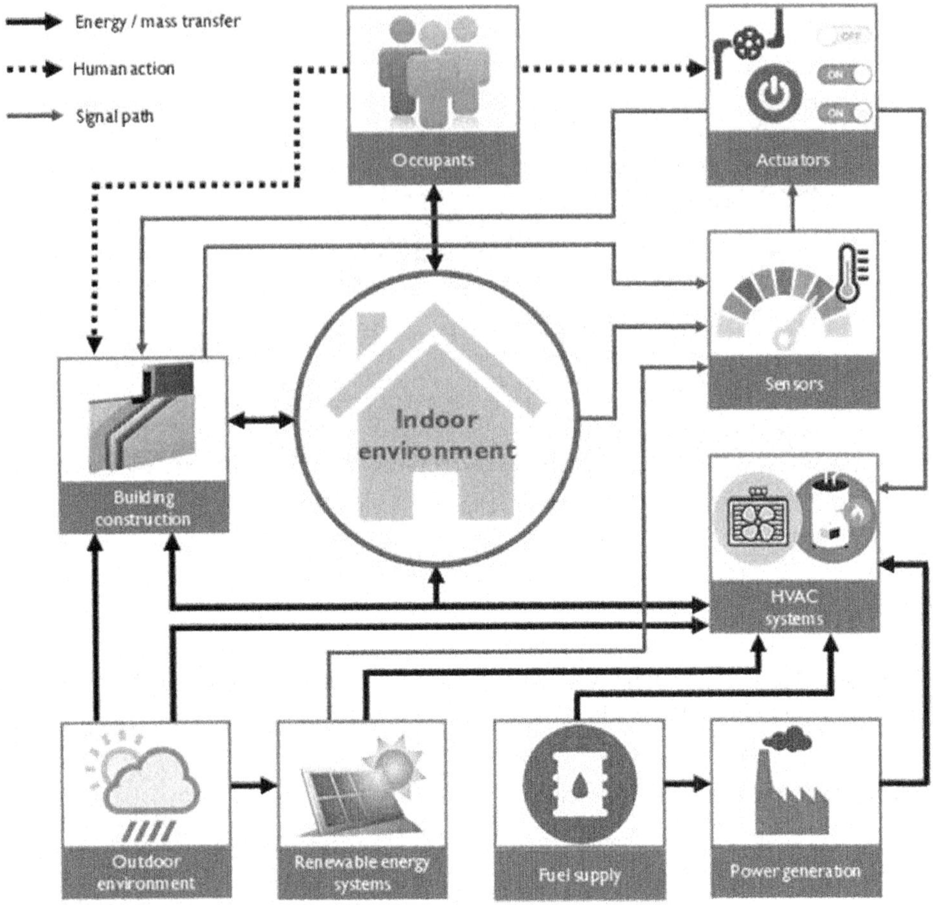

Figure 1.3 Dynamic interactions of (continuously changing) sub-systems in buildings

user-friendly interfaces has resulted in the wide use of building performance simulation. Nevertheless, many theoretical and practical challenges still need to be overcome before the full potential of building performance modelling and simulation can be realized.

One of the aims of this book is to demonstrate that building performance simulation can be used for many different applications. At present, there is a growing acceptance that modelling and simulation have an important role to play in the development of new or renovated high-performance sustainable buildings.

Other application areas are not directly related to design or operation of a particular building or district but serve more general objectives, such as policy support (see Chapter 12). Similarly, modelling and simulation have an important role in the development of novel building systems and components (see e.g. Loonen et al. 2017).

There also exists a considerable and rapidly increasing interest – in practice and research – in the use of simulation for post-construction activities such as commissioning, operation and management. The uptake in current practice is still limited, but we expect that the next decade will see strong growth in the application of building performance simulation for such activities. The two main reasons for this are (1) the current (considerable) discrepancy between predicted and actual energy consumption in buildings (see e.g. De Wilde 2014), and (2) the emergence of new business models driven by whole lifetime building (energy) performance.

Another area that has gained considerable interest recently is that of urban modelling and simulation, which is why this second edition book now includes no less than three chapters on the topic.

Quality assurance of simulation-based decisions

Quality assurance of simulation-based decisions – not only for design, but for all sorts of applications – is a very important and ongoing issue addressed throughout this book. So, in this introductory chapter we limit ourselves to some general aspects.

Firstly, the quality of simulation results depends, of course, on the physical correctness of the model. Although Robinson (1999) is not related to building simulation, the conclusion that it is not possible to validate a model and its results, but only to increase the level of confidence that is placed in them, seems to be equally true for our domain. Secondly – and likely even more importantly – quality assurance of results for simulation-based decisions depends on much more than only the physical correctness of the model; the quality of the end result (i.e. the results to be communicated to decision makers) can only be "assured" when it is based on quality assurance during all steps of a simulation study.

In other words, while the correctness of the model may be thought of as having the greatest influence on the credibility of simulation results, the accuracy of the problem formulation is also a greatly influential factor that must be properly considered. Accordingly, it is crucial to begin employing validation, verification and testing (VV&T) techniques in this first phase and continue to employ them over the full life cycle of the simulation study, which culminates in the presentation of the results. Such a VV&T procedure is very eloquently described by Balci (1994).

The difference between verification, validation and testing is as follows. Model verification aims to identify any inaccuracies that may arise and discover any errors in the model when it undergoes transformation from one form to another. Model verification deals with *building the model right*. Model validation aims to substantiate that the behaviour of the model is accurate, consistent and in line with the objectives of the study in the chosen target domain. In essence, model validation governs the process of *building the right model*.

Model testing is conducted to identify any inaccuracies that may exist and to reveal errors within the model. Testing of the model involves the input of test data or test cases, which is then evaluated to see if the model functions correctly and as expected. If this testing returns the result of "test failed", this implies failure in the model and not in the test. Validation and verification are the dual reasons that such testing is performed. Testing that aims to validate is primarily concerned with evaluating the model's behavioural accuracy, while verification tests are focused on determining the accuracy of the transformation of the model from one form to another. VV&T is used as a term to capture the entirety of these testing processes.

Model VV&T serves to safeguard against three key error types associated with simulation applications (Balci 1990). The Type I error occurs when the credibility of the model is incorrectly deemed unacceptable. In contrast, the Type 2 error occurs when the credibility of the model is incorrectly deemed acceptable. The Type 3 error occurs when a model solves a problem other than the problem of interest. The terms 'model builder's risk' and 'model user's risk' are used to refer to the risk of committing Type 1 and Type 2 errors, respectively. The result of committing Type 1 errors is that the model development will increase in terms of cost. However, the results of Type 2 and Type 3 errors may well render the simulation study useless, and decisions based on its results may have major negative consequences.

The life cycle of a simulation study is presented in Figure 1.4. The phases are depicted by the shaded ovals. The relations between the phases are represented by the dashed arrows. The credibility assessment stages are depicted by the solid arrows. It should be noted that the dashed arrows do not imply that the life cycle can be seen as a rigidly linear or sequential process.

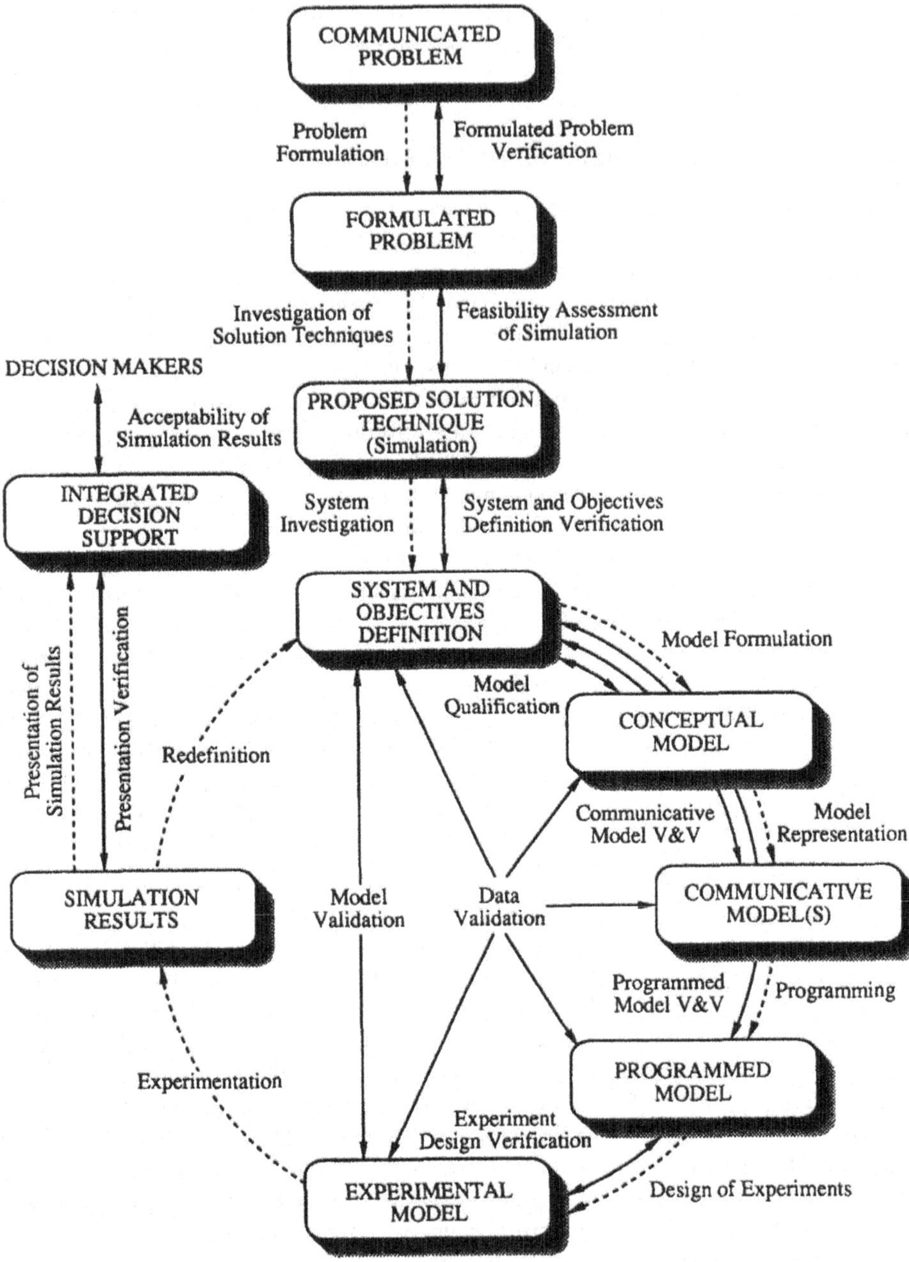

Figure 1.4 Verification, validation and testing throughout the life cycle of a simulation study

Source: Balci 1994

Rather, these dashed arrows simply indicate the general direction of travel through the development life cycle. Another noteworthy point is that there is an associated VV&T activity for each life cycle phase. Negative results from any of these activities may lead to the need to revisit an earlier phase or even to begin the process afresh. VV&T should not be seen as a set of discrete phases or steps within the life cycle; it should be seen as a continuous set of actions that take place during the entirety of the life cycle phases.

The phases can be grouped in three categories as follows. The problem definition phases deal with the communicated problem, formulated problem, proposed solution technique and system and objectives definition. The model development phases consist of all the phases depicted in the circle in Figure 1.4. The decision support phases consist of the simulation results and integrated decision support phases.

Conducting VV&T activities across the full life cycle is done with the purpose of identifying shortcomings in quality in any phase of the study. These activities, then, act as safeguards from the inception of the project, when the problem is communicated, until project completion, when the results of the implemented simulation are presented.

Although most studies are initiated by a decision maker in response to a recognized problem, this problem is rarely defined with sufficient clarity and specificity to enable an analyst (problem-solver/consultant/research group) to immediately begin working towards a solution. Generally, an essential first step is to carefully formulate the problem as precisely and accurately as possible. This problem formulation should function to allow the analyst to proceed by conceiving of specific and appropriate research endeavours.

A clear problem formulation allows the analyst to identify alternative techniques capable of addressing the formulated problem. These techniques should then be evaluated, and those that are determined to cost too much in terms of time or money or those that are deemed to be unable to yield sufficient benefits in terms of the given objectives should be discarded. The analyst should then weigh the remaining techniques against each other to determine which will yield the greatest benefit-cost ratio. This selection procedure of available techniques underlines the point that it is unwise at best and invalid at worst to rely on the aphorism "when all else fails, use simulation". If used, simulation must be shown to the decision maker to be a credible solution to the problem at hand.

Here, it is worth to consider the "ten reasons why not to simulate" by Banks and Gibson (1997), which can be summarized as:

Don't simulate when:

1. The problem can be solved using "common sense analysis";
2. The problem can be solved analytically (using a closed form);
3. It's easier to change or perform direct experiments on the real thing;
4. The cost of the simulation exceeds possible savings;
5. There aren't proper resources available for the project;
6. There isn't enough time for the model results to be useful;
7. There is no data – not even estimates;
8. The model can't be verified or validated;
9. Project expectations can't be met;
10. System behaviour is too complex, or can't be defined.

In the conceptual and communicative model phases, one has to decide about the modelling complexity, which includes the scope (how much of the real world is represented) and resolution (number of variables in the model and their spatial and temporal precision or granularity). In this context, it is advisable to adhere to Occam's razor or Einstein's principle, which imply that a model should be as simple as possible but not simpler. Thus, as Robinson (2008) points out:

> All simulation models are simplifications of reality. . . . The issue in conceptual modeling is to abstract an appropriate simplification of reality. . . . The overarching requirement is the need to avoid the development of an overly complex model. In general, the aim should be: to keep the model as simple as possible to meet the objectives of the simulation study.

The implication of the above is that for the same physical artefact (e.g. a building, a façade or an HVAC component) a different modelling approach is to be preferred depending on the objective of the simulation. Hensen (2004) elaborates this for building airflow-related performance studies, and Gaetani et al. (2017) for occupant behaviour modelling.

It is interesting to note that the majority of validation and verification efforts in our field are restricted to the programmed model phase. Very few publications address VV&T during the other phases of a simulation study. However, it can be expected that in our field, like in many others, numerous issues could arise from poor/absent VV&T in the other phases – starting with the communicated problem and ending with the decision support phase.

It is unfortunate that it is still common practice not to report confidence levels for simulation results, because it is well known that, for example, real and predicted energy consumption of low-energy buildings are very sensitive for uncertainties in occupant behaviour and weather, to mention just two aspects.

Further explanation and details of the life cycle application of VV&T can be found in Balci (1994). Here it suffices to state that conducting VV&T activities is of paramount importance in guiding complex, large-scale simulation applications to an efficacious end. It is crucial that the importance of VV&T is understood by all stakeholders in the process, including those sponsoring the study and those conducting it. Even though its importance is clear, implementing VV&T across all project phases costs a great amount of time and effort. In real applications in which the simulation study is often limited by the time available, documentation of VV&T activities is sometimes necessarily jettisoned. In order to remedy this less than ideal situation, further research attention is needed to determine how best to provide computer-aided assistance so that VV&T techniques can be automated within applications of building performance simulation.

From the above it is hopefully also clear that the primary and paramount requirement for quality assurance is sufficient domain knowledge of the user. According to Becker and Parker (2009), we should nevertheless appreciate the distinction between being a subject matter expert in something, being able to describe that thing so it can be simulated, and actually implementing and testing the simulation.

Discussion

As a (future) simulator, please note that simulation is a skill that needs to be learned. As implied above, the first step is to acquire sufficient domain knowledge, and then skills and knowledge relating to principles, assumptions and limitations of modelling and simulation. Only with this combined domain and simulation knowledge will it be possible to assure the quality and have confidence in the eventual results. When learning about modelling and simulation, it is important to do that step by step, thus moving from less to more complex tools, as explained in the "Black belt energy modeling matrix" by Franconi (2010).

In the context of user aspects of quality control, it is very encouraging to see that professional organizations such as the American Society of Heating, Refrigerating and Air Conditioning Engineers (ASHRAE) and the Chartered Institution of Building Services Engineers (CIBSE) are collaborating with the International Building Performance Simulation Association (IBPSA) to bring this to the attention of professionals. Both ASHRAE and CIBSE have many publications related to theoretical and practical aspects of building performance simulation. ASHRAE now also has a Building Energy Modeling Professional certification scheme. The purpose of this certification is to certify an individual's ability to evaluate, choose, use, calibrate and interpret the results of energy modelling software when applied to building and systems energy performance and economics and to certify individuals' competence to model new and existing buildings and systems with their full range of physics.

Synopsis

This chapter demonstrates that building performance simulation has the potential to deliver, directly or indirectly, substantial benefits to building stakeholders and to the environment. It also explains that the building simulation community faces many present and future challenges. These challenges can be categorized in two main groupings:

- Provide better decision support. Issues here include product development support, policy support, early phase design support, multi-scale approaches (from construction detail to district level), uncertainty and sensitivity analysis, robustness analysis (employing use and environmental change scenarios), optimization under uncertainty, inverse approach (to address "how to" instead of being able to answer "what if" questions), multi-physics (particularly inclusion of electrical power flow modelling) and integration in the construction process (using building information modelling (BIM), process modelling, etc.);
- Provide support for building operation and management, with issues such as accurate in-use energy consumption prediction, whole building (total energy) performance analysis and model predictive (supervisory multi-input multi-output control).

Many (but not all) of these issues are addressed in the following chapters.

References

Balci, O. (1990) Guidelines for successful simulation studies. In Balci, O., Sadowski, R.P. and Nance, R.E. (eds), *Proceedings 1990 Winter Simulation Conferences*. Piscataway, NJ: IEEE, pp. 25–32.

Balci, O. (1994) Validation, verification, and testing techniques throughout the life cycle of a simulation study. *Annals of Operations Research* 53: 121–173.

Banks, J. and Gibson, R.R. (1997) *Don't Simulate When: Ten Rules for Determining When Simulation Is Not Appropriate*. Institute of Industrial Engineers, IIE Solutions.

Becker, K. and Parker, J.R. (2009) A simulation primer. In Gibson, D. and Baek, Y. (eds), *Digital Simulations for Improving Education: Learning through Artificial Teaching Environments*. Hershey, PA: IGI Global.

BEST. (2018) Building Energy Software Tools Directory. www.buildingenergysoftwaretools.com/.

Clarke, J.A. and Hensen, J.L.M. (2015) Integrated building performance simulation: Progress, prospects and requirements. *Building and Environment* 91: 294–306.

EU. (2011) *A Roadmap for Moving to a Competitive Low Carbon Economy in 2050*. Brussels: European Commission.

Franconi, E. (2010) *Black Belt Energy Modeling Matrix*. https://d231jw5ce53gcq.cloudfront.net/wp-content/uploads/2017/04/Pathways-to-Zero_Black-Belt-Modeling_Matrix_2010.pdf.

Gaetani, I., Hoes, P. and Hensen, J.L.M. (2017) On the sensitivity to different aspects of occupant behaviour for selecting the appropriate modelling complexity in building performance predictions. *Journal of Building Performance Simulation* 10: 601–611.

Hensen, J.L.M. (2004) Integrated building airflow simulation. In Malkawi, A. and Augenbroe, G. (eds), *Advanced Building Simulation*. New York: Spon Press, pp. 87–118.

Kusuda, T. (1999) Early history and future prospects of building system simulation. In *Proceedings of the 6th International IBPSA Conference*. Kyoto, Japan, pp. 3–15.

Loonen, R.C.G.M., Favoino, F., Hensen, J.L.M. and Overend, M. (2017) Review of current status, requirements and opportunities for building performance simulation of adaptive facades. *Journal of Building Performance Simulation* 10(2): 205–223.

Robinson, S. (1999) Simulation verification, validation and confidence: A tutorial. *Transactions of the Society for Computer Simulation International* 16(2): 63–69.

Robinson, S. (2008) Conceptual modelling for simulation part I: Definition and requirements. *Journal of the Operational Research Society* 59(3): 278–290.

Spitler, J.D. (2006) Building performance simulation: The now and the not yet. *HVAC&R Research* 12(3a): 549–551.

Wang, H. and Zhai, J.Z. (2016) Advances in building simulation and computational techniques: A review between 1987 and 2014. *Energy and Buildings* 128: 319–335.
WGBC. (2014) *Health, Wellbeing and Productivity in Offices, the Next Chapter for Green Building.* World Green Building Council.
Wilde, P. de. (2014) The gap between predicted and measured energy performance of buildings: A framework for investigation. *Automation in Construction* 41: 40–49.

Recommended reading

We recommend that you read the rest of this book, but not necessarily from the beginning to the end. Read the chapters according to your personal interest. This book aims to provide a comprehensive and in-depth overview of various aspects of building performance modeling and simulation, such as the role of simulation in design, outdoor and indoor boundary conditions, thermal modeling, airflow modeling, thermal comfort, acoustics, daylight, moisture, HVAC systems, micro-cogeneration systems, building simulation in operational optimization and in building automation, urban-level modeling and simulation, building simulation for policy support, and finally a view on future building system modeling and simulation.

The structure of each chapter is similar. They start with identifying the chapter's scope and learning objectives. This may be followed by an introduction and sections on scientific foundation, computational methods, application examples and a discussion and/or synopsis. As we are limited by space constraints, most chapters include recommendations for further reading. In order to practice what has been learned, most chapters finish with some suggested activities or assignments.

Activities

1 If you are not yet an IBPSA member, we strongly advise you to check out www.ibpsa.org. There is a good chance that there is already a group of like-minded people in your region. In case you are interested in regular updates of what is going on in the building performance simulation field, make sure that you sign up for IBPSA News.
2 If you are interested in leading edge research in this field, we recommend looking up the *Journal of Building Performance Simulation*, which aims to make a substantial and lasting contribution to the international building community by supporting authors and the high-quality, original research they submit. This journal also offers a forum for original review papers and researched case studies.
3 We are very keen to know your opinion about this book. So, after reading (parts of) this book, please email your comments, suggestions or other feedback to j.hensen@tue.nl and roberto.lamberts@ufsc.br.

Part I

Fundamentals

2 Thermal load and energy performance prediction

Jeffrey D. Spitler

Scope

This chapter covers heat transfer to, from and within the building envelope; calculation of building heating and cooling loads; and energy calculations related to the building envelope and simulation of rooms and zones to predict temperatures, loads and required heat transfer to and from the building system.

Learning objectives

After studying this chapter, the student should be able to do the following:

1 Understand how fundamental knowledge of heat transfer is applied to building simulation.
2 Understand the basis for simulation of rooms and zones within buildings.
3 Apply the heat balance concept to building surfaces and zone air.

Key words

Load calculations/energy calculations/convection/radiation/conduction/heat transfer/infiltration

Introduction

Prediction of the building thermal load and thermal response is fundamental to building performance simulation. Strictly speaking, the building thermal load is defined as the amount of heat that must be removed (cooling load) or added (heating load) to maintain a constant room air temperature. This quantity is usually used to determine design loads for sizing ducts and equipment. In practice, a constant room air temperature is seldom maintained due to a variety of factors, e.g. setback thermostatic control of space temperature or use of natural ventilation for cooling. Accordingly, for prediction of energy performance, the actual heat extraction/addition rates to the space are of interest.

The building is generally divided into "zones" – portions of the building that can be considered to have an approximately uniform air temperature. This is most commonly a single room or group of rooms with similar thermal characteristics served by the same system, like a row of similar offices along the south side of the building. Each zone forms a control volume over which heat transfer into and out of the zone is analyzed.

This requires analysis of all three modes of heat transfer – conduction, convection, and radiation – within the building envelope and between the building envelope and its surroundings. (Here, the term building envelope refers to the walls, roofs, floors, and fenestrations that make up the building.)

The three modes of heat transfer all occur simultaneously, and it is the simultaneous solution of all three modes of heat transfer that complicates the analysis. In practice, this simultaneous solution is done with a building simulation program, though the very simplest forms of cooling and heating load calculation can be done by hand, relying on tabulated factors that were computed with a building simulation program.

Before concerning ourselves with the simultaneous solution, we will first consider each of the three modes independently. Then, after considering the three modes of heat transfer, various means for solving the entire problem will be considered. Each of the zone models utilizes a different approach for the simultaneous solution.

Conduction heat transfer

Conduction heat transfer through the building envelope, as well as conduction heat transfer into and out of the building envelope, is an important contribution to the building load and the thermal response of the zone.

Steady-state conduction

Heat transfer through building walls and roofs is generally treated as a pure conduction heat transfer process, even though, for example, convection and radiation may be important in an internal air gap[1] in the wall. Conduction is the transfer of heat through a solid[2] via random atomic and molecular motion in response to a temperature gradient. Elements of the building envelope, such as thermal bridges and corners, distort the temperature gradients so that the heat flows in directions other than purely perpendicular to the envelope surfaces. Such heat flow is said to be multi-dimensional, and while it is possible to analyze multi-dimensional conduction heat transfer, for building load calculations, it is generally approximated as being one-dimensional; however, the approximations do take into account the impact of thermal bridges. Heat loss from foundation elements is also multi-dimensional, but, again, approximations are made that simplify the calculation procedure.

One-dimensional steady-state heat conduction is described by the Fourier equation:

$$\dot{q} = -kA\frac{dT}{dx} \tag{2.1}$$

where:

$\dot{q}$ = heat transfer rate, W;
k = thermal conductivity, W/(m K);
A = area normal to the heat flow, m²;
$\frac{dT}{dx}$ = temperature gradient, °C/m or K/m, which are equivalent.

For a single layer of a wall, with heat flow in the x direction, and with conductivity k and surface temperatures T_2 and T_1, as shown in Figure 2.1, Equation 2.1 can be integrated to give:

$$\dot{q} = -kA\frac{(T_2 - T_1)}{(x_2 - x_1)} \tag{2.2}$$

A convenient form of Equation 2.2 involves definition of the unit thermal resistance, R:

$$R = \frac{(x_2 - x_1)}{k} = \frac{\Delta x}{k} \tag{2.3}$$

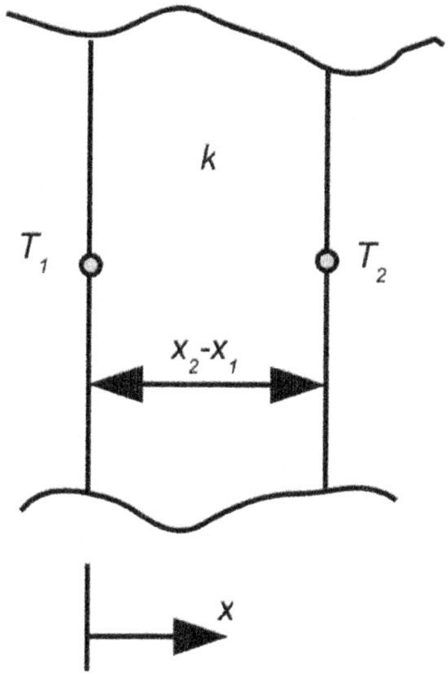

Figure 2.1 A single-layer plane wall

where:

R = unit thermal resistance, m²K/W;
Δx = thickness of the layer; units are consistent with k: m.

Then, the conduction heat transfer rate through the wall is given by:

$$\dot{q} = A\frac{(T_1 - T_2)}{R} \tag{2.4}$$

Example 2.1

A 100 mm thick uninsulated wall of area 10 m² is made out of concrete with conductivity of 1.2 W/mK. What is the *R*-value of the wall? Under steady-state conditions, with the exterior surface temperature of the wall at −20 °C and the interior surface temperature of the wall at 15 °C, what is the total heat loss through the wall?

Solution:

The *R*-value is determined by dividing the thickness by the conductivity:

$$R = 0.1 \text{ m}/1.2 \text{ W/mK} = 0.083 \text{ m}^2\text{K/W}$$

The conduction heat transfer rate through the wall is given by Equation 2.4:

$$\dot{q} = 10\text{m}^2 \frac{\left(15\,^\circ\text{C} - (-20\,^\circ\text{C})\right)}{0.083\dfrac{\text{m}^2\text{K}}{\text{W}}} = 4{,}200 \text{ W}$$

End of solution.

A more common situation is that the wall has multiple layers, as shown in Figure 2.2. Here, the wall is made up of three layers with two exterior surfaces (1 and 4) and two interior interfaces

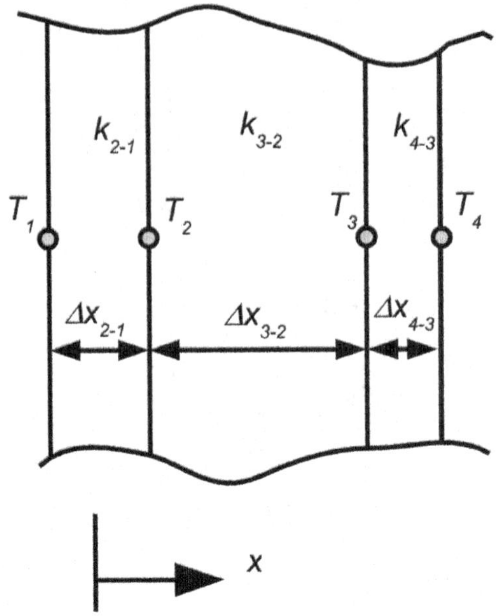

Figure 2.2 Multi-layer wall

(2 and 3). Each of the three layers is defined by a conductivity (e.g. k_{2-1} for the leftmost layer) and thickness (e.g. Δx_{2-1} for the leftmost layer).

Individual unit thermal resistances are determined for each layer:

$$R_{2-1} = \frac{\Delta x_{2-1}}{k_{2-1}} \tag{2.5a}$$

$$R_{3-2} = \frac{\Delta x_{3-2}}{k_{3-2}} \tag{2.5b}$$

$$R_{4-3} = \frac{\Delta x_{4-3}}{k_{4-3}} \tag{2.5c}$$

There is an analogy between conduction heat transfer and electricity, as illustrated in Figure 2.3, where the unit thermal resistances (analogous to electrical resistances[3] per unit area) are shown connecting the surface temperatures (analogous to voltages). As a consequence, the total unit thermal resistance of the wall may be determined by simply adding the three individual unit thermal resistances (in series):

$$R_{4-1} = R_{2-1} + R_{3-2} + R_{4-3} \tag{2.6}$$

Then, the heat flux (analogous to current per unit area) may be determined as:

$$q'' = \frac{(T_1 - T_4)}{R_{4-1}} \tag{2.7}$$

where:

q'' = heat flux rate (heat transfer rate per unit area), W/m^2.

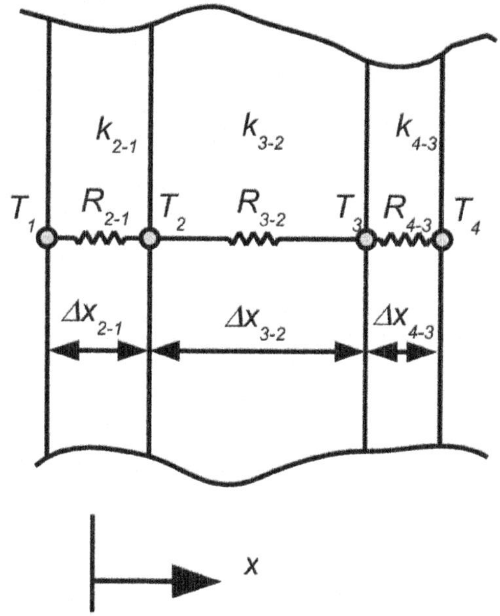

Figure 2.3 Multi-layer wall analysis based on electrical analogy

Example 2.2

The concrete wall of Example 2.1 is improved by adding a 50 mm thick layer of expanded polystyrene, with an *R*-value of 1.8, and a layer of 12.7 mm thick drywall, with an *R*-value of 0.079. With these additional layers, what is the *R*-value of the wall? Under steady-state conditions, with the exterior surface temperature of the wall at −20 °C and the interior surface temperature of the wall at 15 °C, what is the total heat loss through the wall?

Solution:

Because the three layers are in series with each other, the total *R*-value is determined by adding the resistances of the three layers, as modeled in Equation 2.6:

$$R = 0.083 + 1.8 + 0.079 = 1.96\ \text{m}^2\cdot\text{K/W}$$

The conduction heat transfer rate through the wall is given by Equation 2.4:

$$\dot{q} = 10\text{m}^2 \frac{\left(15\,^\circ\text{C} - (-20\,^\circ\text{C})\right)}{1.96\dfrac{\text{m}^2\text{K}}{\text{W}}} = 179\ \text{W}$$

A dramatic reduction in heat loss can generally be had when insulation is added to uninsulated walls.

End of solution.

Generally, heating load calculations are performed for conditions that are assumed to be approximately steady state: that is nighttime, hours after the building occupants have gone home and the sun has gone down. In this case, the conduction heat transfer through the above-grade portion of the building envelope is determined by first calculating the overall thermal resistance and conductance (U-factor) of each surface, multiplying the U-factors by the surface areas, summing the UA values for each room, and multiplying the total UA for each room by the design temperature difference. For slab-on-grade floors, basement walls, and basement floors, a simple multi-dimensional analysis is used such that the solution is given in one-dimensional form.

Transient conduction

The discussion of conduction heat transfer in the previous section treated it as a steady-state phenomenon. While for heating load calculations this is generally sufficient, strong daily variations in incident solar radiation and outdoor air temperatures occurring under energy analysis and cooling load design conditions cannot be ignored in most cases. This is particularly true as the thermal capacitance of the wall or roof increases. Thermal capacitance of a wall or roof element may be defined as the amount of energy required to raise the temperature by one degree:

$$C_{th} = Mc_p = \rho V c_p \tag{2.8}$$

where:

C_{th} = thermal capacitance of a wall or roof layer, J/K;
M = mass of the wall or roof layer, kg;
c_p = specific heat of the wall or roof layer material, J/kg·K;
ρ = density of the wall or roof layer material, lb_m/ft^3 or kg/m^3;
V = volume of the wall or roof layer, m^3.

The energy required to raise the temperature of the layer is then given by:

$$Q = C_{th}\Delta T = Mc_p\Delta T = \rho V c_p \Delta T \tag{2.9}$$

where:

ΔT = temperature increase, °C.

Example 2.3 Thermal storage

Consider the 100 mm thick layer of concrete in the 10 m^2 wall in Example 2.1. During a warm summer day, it is warmed from a temperature of 20 °C to a temperature of 30 °C. If the concrete has a density of 1600 kg/m^3 and a specific heat of 900 J/kgK, how much energy is absorbed in warming the concrete layer?

Solution:

The total volume of the concrete is 1 m^3; multiplying by the density gives the total mass of the concrete as 1600 kg. The required energy input is the mass multiplied by the specific heat multiplied by the rise in temperature:

$$Q = 1600 \text{ kg} \times 900 \text{ J/kgK} \times 10 \text{ °C} = 14{,}400 \text{ kJ}$$

End of solution.

Comparing the amount of energy required to raise the temperature of the layer by 10 °C to the expected heat transmission rate under steady-state conditions suggests that this heat storage effect will not be unimportant. However, because the thermal capacitance is distributed throughout each layer, there is no simple way to calculate the transient behavior. The distributed thermal capacitance is sometimes represented as shown in Figure 2.4, with the two parallel lines below each resistance representing the distributed capacitance.

In order to further illustrate the effects of thermal storage in the wall, consider what would happen if the three-layer (100 mm concrete, *R*-1.8 insulation, 12.7 mm drywall) wall described in Example 2.2 was placed in an office building in Atlanta, facing southwest. Under clear sky, high-temperature conditions, the heat transmission through the wall may be calculated using two methods:

- A method that ignores the thermal mass in the wall, computing the conduction heat gain with Equation 2.4 at each hour of the day. In this case, the exterior temperature changing throughout the day gives conduction heat gains that change over the day. This method may be referred to as "Quasi-steady-state" because it treats each hour of the day as if the wall comes immediately to steady-state conditions (labeled "Quasi-SS" in Figure 2.5).

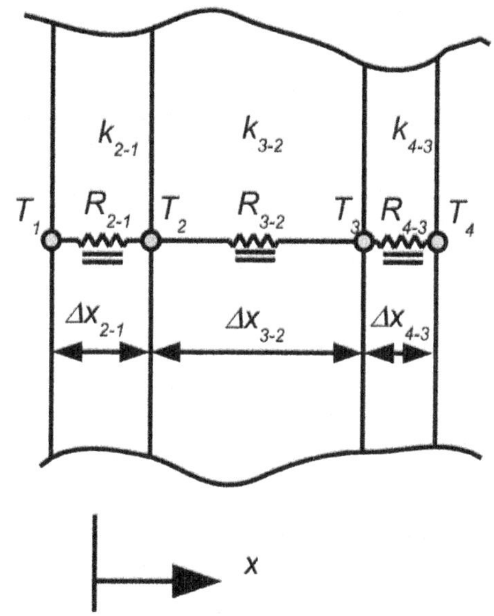

Figure 2.4 Electrical analogy with distributed thermal capacitance

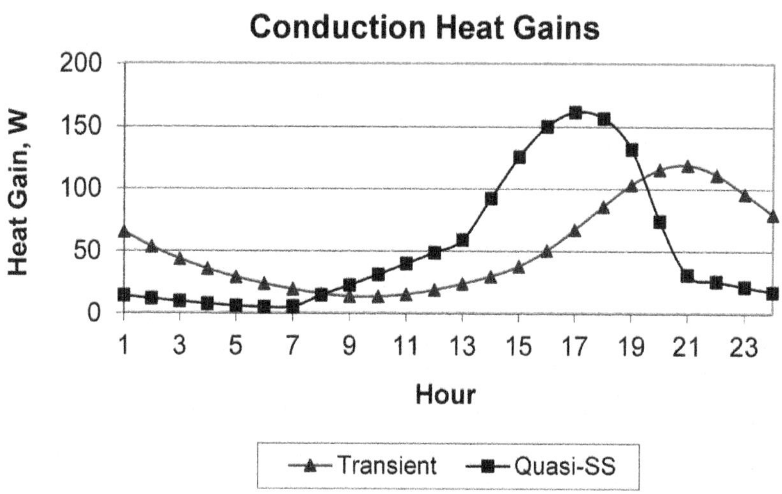

Figure 2.5 Comparison of transient and quasi-steady-state conduction heat gain calculations

- A method, described later in this chapter, which takes into account the thermal capacitance of each layer (labeled “Transient” in Figure 2.5).

Conduction heat gains computed with each method are shown in Figure 2.5. In Figure 2.5, two effects of the thermal mass may be observed:

- The first may be referred to as “Time Delay” – the peak heat gain, when calculated with the quasi-steady-state method, occurs at 5 p.m. (hour 17). When the thermal capacitance is accounted for in the calculation, we can see that the peak heat gain actually occurs at 9 p.m. (hour 21). This four-hour delay in the peak heat gain may be important, especially if the peak heat gain is delayed such that it occurs after the occupants have gone home.
- The second effect may be referred to as “Damping” – the peak heat gain, when calculated with the transient method, is about 30% lower than that calculated with the quasi-steady-state method.

Both effects are important, and this is why transient conduction calculations are typically performed as part of cooling load calculation procedures and energy analysis.

Transient one-dimensional conduction in a homogenous slab is described by the heat diffusion equation:

$$\frac{k}{\rho c_p}\frac{\partial^2 T}{\partial x^2} = \frac{\partial T}{\partial t} \tag{2.10}$$

where:

k = thermal conductivity, W/mK;
ρ = density, kg/m^3;
c_p = specific heat, J/kgK;
T = temperature at any point in the slab, °C;
x = distance, m;
t = time, s.

Non-linear, time-dependent boundary conditions and walls and roofs that are not homogeneous preclude obtaining a simple solution of Equation 2.10. However, calculation of transient conduction heat transfer through walls and roofs may be performed with a number of different methods. These methods include:

1 Z-transform methods – methods based on Z-transform theory, including “response factors” and “conduction transfer functions.” Z-transform methods have a high degree of accuracy and are computationally efficient. This class of methods is the most widely used for load calculations and building energy analysis.
2 Numerical methods – finite difference and finite element methods. Numerical methods also have a high degree of accuracy but typically are less computationally efficient than Z-transform methods. However, they offer significant increased flexibility for treating such phenomena as radiation across air gaps and phase-change materials.
3 Lumped parameter methods – treating walls and roofs as discrete resistances and lumped capacitances. These methods are typically used in conjunction with lumped capacitance zone models.

Each of the methods is discussed later.

Transfer function and response factor methods

Due to their computational efficiency and accuracy, Z-transform methods are widely used in both design load calculations and building energy analysis applications. Z-transform methods have been utilized in building performance in one of three formulations, all of which utilize a fixed time step, with the most common time step being one hour:

- Response factors.
- Conduction transfer functions (CTF).
- Periodic response factors (PRF).

Response factors may be thought of as time series coefficients relating the current heat flux to past and present values of interior and exterior temperatures. (The interior and exterior temperatures may be air temperatures, sol-air temperatures, or surface temperatures, depending on the application.) Particularly for thermally massive constructions, large numbers of response factors may be required. Conduction transfer functions replace much of the required temperature history with heat flux history. In other words, many of the response factors are replaced with coefficients that multiply past values of heat flux. Periodic response factors are used for design day load calculations where the design day is assumed to repeat.

The conduction transfer function formulation for the inside heat flux is:

$$q_i''(\theta) = -Z_o T_{i,\theta} - \sum_{j=1}^{nz} Z_j T_{i,\theta-j\delta} + Y_o T_{o,\theta} + \sum_{j=1}^{nz} Y_j T_{o,\theta-j\delta} + \sum_{j=1}^{nq} \Phi_j q''_{ki,i,\theta-j\delta} \tag{2.11}$$

And, for the outside heat flux:

$$q_o''(t) = -Y_o T_{i,\theta} - \sum_{j=1}^{nz} Y_j T_{i,\theta-j\delta} + X_o T_{o,\theta} + \sum_{j=1}^{nz} X_j T_{o,\theta-j\delta} + \sum_{j=1}^{nq} \Phi_j q''_{ko,i,\theta-j\delta} \tag{2.12}$$

where:

X_j = outside CTF coefficient, $j = 0,1, \ldots, nz.$;
Y_j = cross CTF coefficient, $j = 0,1, \ldots, nz.$;
Z_j = inside CTF coefficient, $j = 0,1, \ldots, nz.$;
Φ_j = flux CTF coefficient, $j = 0,1, \ldots, nq.$;
T_i = inside surface temperature, °C;
T_o = outside surface temperature, °C;
q_o'' = conduction heat flux into wall at outside face, W/m²;
q_i'' = conduction heat flux out of wall at inside face, W/m²;
θ = current time, seconds or hours;
δ = length of time step, seconds or hours.

The subscript on each of the CTF terms indicates the time, relative to the current time, in terms of the number of time steps to which the term applies; e.g. X_1 is applied to the temperature one time step earlier, which is given in the temperature subscript as $\theta - 1\delta$. The first terms in the series (those with subscript "*0*") apply to the current time and have been extracted from the summation to facilitate solving for the current temperature in the solution scheme.

The two summation limits, *nz* and *nq*, depend on the wall construction and method used to calculate the CTFs. Generally, walls that are more thermally massive will require more terms. Response factors have a similar formulation, for which $nq = 0$. Theoretically, *nz* is then infinite, but in practice, a large number of terms, perhaps 100 or more, may be sufficient. The CTF formulation is computationally more efficient.

Although the use of either response factors or conduction transfer functions is relatively straightforward, a more difficult task is determining the response factors or conduction transfer function coefficients. A detailed explanation of an analytical procedure for determining response factors and conduction transfer function coefficients of multilayer slabs is given by Hittle (1981). Spitler (1996) cites a number of other methods, of which the most common are the Laplace method (Mitalas and Arseneault 1970; Maestre et al. 2010) and the state-space method (Seem et al. 1989). More recently, time-domain (Davies 1996) and frequency-domain (Wang and Chen 2003; Urbikain and Davies 2012; Sanza Pérez et al. 2017) methods have been developed. Although not usually done as a precursor to simulation, it is also possible to derive response factors or CTF coefficients experimentally, as demonstrated by Brown and Stephenson (1993).

Periodic response factors are used for design load calculations, where heat gains for a single day, assumed to be preceded by one or more identical days, are desired. Since periodic response factors are used in conjunction with simplifications to both the exterior and interior heat balances, heat gain is calculated based on an equivalent exterior temperature, the sol-air temperature, and the interior air temperature, which is presumed constant. Conductive heat gain with periodic response factors is calculated as:

$$q_\theta = A\sum_{j=0}^{23} Y_{Pj}\left(T_{e.\theta-j\delta} - T_{rc}\right) \tag{2.13}$$

where:

q_θ = hourly conductive heat gain for the surface, W;
A = surface area, m^2;
Y_{Pj} = jth periodic factor;
$T_{e,\theta-j\theta}$ = sol-air temperature, j hours ago, °C;
T_{rc} = presumed constant room air temperature, °C;
θ = the current hour;
δ = the time step (one hour).

Periodic response factors can be computed with the same methods as conduction transfer functions or directly from the values of the conduction transfer function coefficients, as described by Spitler and Fisher (1999).

Building simulation tools that use conduction transfer functions and periodic response factors will generally have internal routines for calculating the values based on the user's layer-by-layer description of the surface construction. Other options include the ASHRAE *Toolkit for Building Load Calculations* (Pedersen et al. 2001) and the software that accompanies the ASHRAE *Load Calculations Application Manual* (Spitler 2014).

Example 2.4

A wall is made up of five layers, as shown in Table 2.1, listed from outside to inside. Conduction transfer function coefficients for this wall have been calculated with a cooling load calculation program that uses the Laplace transform method; these values are shown in Table 2.2 (Mitalas and Arseneault 1970). For a day where the exterior and interior surface temperatures are as shown in Table 2.3, determine the interior surface heat flux. Assume the previous days had identical temperatures.

Solution:

Based on the provided CTF coefficients, $nz = 5$ and $nq = 5$. Because the calculation relies on past values of q'' to start the calculation, we must assume some values of q'' to get started. After starting the calculation, we will compute the same day over and over until a steady periodic

Table 2.1 Layer descriptions for Example 2.4

Layer	*Thickness, mm*	*Density, kg/m3*	*Conductivity, W/mK*	*Specific Heat, J/kg K*
Face brick	100	1920	0.87	800
Air gap	20	1	0.11	1005
Insulation	50	91	0.04	840
Solid concrete block	100	2096	1.63	920
Plaster	13	720	0.16	840

Table 2.2 Conduction transfer function coefficients for Example 2.4

J	X_j	Y_j	Z_j	Φ_j
0	2.171573E+01	3.375092E−05	1.033658E+01	
1	−4.514066E+01	5.617864E−03	−1.794393E+01	1.463783E+00
2	2.844604E+01	2.266381E−02	8.642595E+00	−5.568298E−01
3	−5.162262E+00	1.133217E−02	−1.021217E+00	2.488959E−02
4	1.826630E−01	8.227039E−04	2.664688E−02	−2.266001E−04
5	−1.023128E−03	8.775257E−06	−1.882663E−04	

Table 2.3 Exterior and interior surface temperatures for Example 2.4

Time	$T_{o,t}$	$T_{i,t}$
1	26.42	22.09
2	25.44	22.08
3	24.67	22.07
4	24.08	22.05
5	23.89	22.02
6	24.28	22.00
7	25.25	21.97
8	27.00	21.93
9	29.53	21.90
10	32.44	21.88
11	35.75	21.86
12	38.86	21.84
13	41.19	21.84
14	42.75	21.84
15	43.33	21.86
16	42.75	21.88
17	41.39	21.91
18	39.25	21.95
19	36.72	21.99
20	34.19	22.02
21	32.06	22.05
22	30.11	22.08
23	28.56	22.09
24	27.39	22.10

solution is reached. We are assuming that values of zero will work, though with this very heavy wall, it may take a number of days to converge. Therefore, an improved estimate, which we will make use of here, is simply to find the steady-state heat transfer rate based on the average exterior and interior temperature. The U-value for the wall is 0.592 W/m^{2}K and the average difference between the exterior and interior surface temperatures is 10.42 °C, yielding an estimate for the average heat flux of 6.16 W/m^2. Then, applying Equation 2.11 for the first hour, we get:

$$q_i''(1) = -(10.336575 \times 22.09 - 17.943933 \times 22.10 + 8.642595 \times 22.09 - 1.021217 \times 22.08 + 0.026647 \times 22.05 - 0.000188 \times 22.02) + 0.000034 \times 26.42 + 0.005618 \times 27.39 + 0.022664 \times 28.56 + 0.011332 \times 30.11 + 0.000823 \times 32.06 + 0.000009 \times 34.19 + 1.463783 \times 6.16 - 0.556830 \times 6.16 + 0.024890 \times 6.16 - 0.000227 \times 6.16 = 6.07 \text{ W/m}^2$$

Applying it again for the second hour, we get:

$$q_i''(2) = -(10.336575 \times 22.08 - 17.943933 \times 22.09 + 8.642595 \times 22.10 - 1.021217 \times 22.09 + 0.026647 \times 22.08 - 0.000188 \times 22.05) + 0.000034 \times 25.44 + 0.005618 \times 26.42 + 0.022664 \times 27.39 + 0.011332 \times 28.56 + 0.000823 \times 30.11 + 0.000009 \times 32.06 + 1.463783 \times 6.07 - 0.556830 \times 6.16 + 0.024890 \times 06.16 - 0.000227 \times 6.16 = 5.90 \text{ W/m}^2$$

Equation 2.11 is applied for each hour of the day. Then, on the second day, the first day is used as history. For example, for the first hour of the second day:

$$q_i''(1) = -(10.336575 \times 22.09 - 17.943933 \times 22.10 + 8.642595 \times 22.09 - 1.021217 \times 22.08 + 0.026647 \times 22.05 - 0.000188 \times 22.02) + 0.000034 \times 26.42 + 0.005618 \times 27.39 + 0.022664 \times 28.56 + 0.011332 \times 30.11 + 0.000823 \times 32.06 + 0.000009 \times 34.19 + 1.463783 \times 7.64 - 0.556830 \times 7.88 + 0.024890 \times 8.02 - 0.000227 \times 08.03 = 7.32 \text{ W/m}^2$$

Doing this for four days gives the results shown in Table 2.4. By the fourth day, the results are more or less converged.

Note that in an annual energy simulation, it is common to start with a short pre-conditioning period of one or a few days so as to develop reasonable historical values for past values of the heat fluxes prior to January 1.

End of solution.

Numerical methods

The literature on finite difference, finite volume, and fine element methods applied to conduction heat transfer problems is voluminous. For a discussion of numerical methods applied to transient heat conduction in buildings, in the context of building simulation, see Clarke (2001); Underwood and Yik (2004); Waters and Wright (1985); Lewis and Alexander (1990); and Walton (1993). During the early years of building performance simulation, numerical methods were often considered too computationally demanding for practical application in hourly energy calculations. Increasing computational speed has rendered this a moot point for

Table 2.4 Heat fluxes at the interior surface

Hour	*Day*			
	1	*2*	*3*	*4*
1	6.07	7.32	7.34	7.34
2	5.90	6.96	6.97	6.97
3	5.69	6.57	6.58	6.58
4	5.44	6.17	6.18	6.18
5	5.16	5.77	5.78	5.78
6	4.88	5.38	5.39	5.39
7	4.60	5.02	5.02	5.02
8	4.36	4.70	4.71	4.71
9	4.17	4.46	4.46	4.46
10	4.07	4.31	4.31	4.31
11	4.09	4.28	4.28	4.28
12	4.24	4.40	4.40	4.40
13	4.53	4.66	4.67	4.67
14	4.96	5.07	5.07	5.07
15	5.48	5.57	5.57	5.57
16	6.06	6.13	6.13	6.13
17	6.64	6.70	6.70	6.70
18	7.17	7.22	7.22	7.22
19	7.59	7.64	7.64	7.64
20	7.89	7.92	7.92	7.92
21	8.03	8.06	8.06	8.06
22	8.02	8.05	8.05	8.05
23	7.88	7.90	7.90	7.90
24	7.64	7.66	7.66	7.66

many applications, and numerical methods are now used in a number of building simulation programs. However, space precludes a detailed treatment here, and the reader should consult the works cited above for more information.

Numerical methods treat the building envelope surface as being made up of discrete capacitances and resistances; this can be seen most clearly in the derivation of the finite volume method. For building walls modeled in one dimension, the ESP-r program uses a default value of three discrete capacitances per layer. This gives reasonable computational speed for annual energy analysis. However, there are some applications for which further simplification achieved by using even fewer discrete capacitances is highly desirable. Methods that use only one to three discrete capacitances to model a wall are often called lumped parameter methods.

Lumped parameter methods

Lumped parameter methods treat the individual building envelope surfaces as being made up of a small number of discrete capacitances and resistances, typically only one or two. They are of particular interest for very short time-step (on the order of a few seconds) simulation of building systems. Transfer function models are unsuitable for such time steps because the required number of CTF coefficients becomes very large as the time steps become short. Numerical methods can be used, but they will typically require too much computational time at such short time steps. Furthermore, lumped parameter methods can be conveniently implemented within simulation environments, such as MATLAB®, that are commonly used for simulation of control systems. The history of lumped parameter methods is described by Gouda et al. (2002).

Typically, the total of the discrete resistance would be equivalent to the total wall resistance, and the total of the discrete capacitances would be equal to the total wall capacitance. However, an approximate distribution of the discrete resistances and capacitances must be made. Underwood and Yik (2004) give suggested distributions of resistances and capacitances for a range of wall and floor types. Underwood (2014) describes an optimization-based method for tuning the parameters of a second-order (two discrete capacitances) model and validates it against a conduction transfer function-based model.

Deviations from one-dimensional conduction-only heat transfer

For the most part, the models used for load calculations and energy calculations treat the wall heat transfer problem as a one-dimensional, transient conduction-only problem. Walls often include thermal bridges, for which modeling approaches are reviewed by Quinten and Feldheim (2016). Seem et al. (1989) and Carpenter et al. (2003) describe development of CTF coefficients for two- and three-dimensional surfaces. Karambakkam et al. (2005) describe a simple procedure for incorporating the effects of thermal bridges within the one-dimensional representation.

Heat transfer from the building foundation to and from the ground is also a multi-dimensional problem that is necessarily treated in a simplified manner. For purposes of design cooling load calculations, heat loss to the foundation is often neglected completely. For purposes of design heating load calculations, it is necessary to include the heat lost through the foundation, and a number of approaches have been proposed. A simple approach for on-grade slabs, which utilizes a perimeter heat loss factor, is given by ASHRAE (2017). For basements, the same reference also gives a simplified approach based on the original work of Latta and Boileau (1969). For energy calculations, more general approaches are described by Beausoleil-Morrison and Mitalas (1997) and Krarti and Choi (1996). Beyond these approaches, numerical methods may be used, but in most cases they will require more computational time than the rest of the building simulation, perhaps by orders of magnitude.

Likewise, moisture transport into and out of the wall (see Chapter 8) is most often neglected in load and energy calculations, despite its clear importance under some conditions. Mendes et al. (2003) give an overview of candidates for modeling moisture transport within a building simulation program. They also performed a study using the candidates to compute transient heat and mass conduction through four different walls in three climates while retaining an existing building simulation program to predict inside humidities. The effects of moisture transport on the conduction heat transfer were surprisingly high; one case showed a 200% overprediction in conduction heat gain if moisture transport was not taken into account. Barbosa and Mendes (2008) describe a more comprehensive study of a single building in Brazil. Compared to a detailed whole-building hygrothermal model, ignoring moisture effects resulted in a 13% overprediction of the peak cooling load and a 4% underprediction of annual energy consumption. Computationally efficient treatment of combined heat and mass transport (Berger et al. 2017) is a topic of ongoing research.

Other deviations from the standard one-dimensional transient conduction-only model include internal radiation (Goss and Miller 1989), diffuse infiltration (Liu and Claridge 1992, Virtanen et al. 1992), and internal convection (Shipp 1983; Park et al. 1987). Strand and Pedersen (1997) describe development of conduction transfer functions for walls with internal heat gain, e.g. radiant heating systems.

Convection heat transfer

Surface convection heat transfer

Thermal convection is the transport of energy in a fluid or gas by mixing in addition to conduction. For load calculations, we are primarily interested in convection between an envelope surface (wall, roof, etc.) and the indoor or outdoor air. The rate of convection heat transfer depends on the temperature difference, whether the flow is laminar or turbulent, and whether it is buoyancy driven or driven by an external flow. The convection heat transfer rate is usually expressed[4] as:

$$\dot{q} = hA(T - T_s) \tag{2.14}$$

where:

h = convection coefficient, W/m²K;
T = bulk temperature of the air, °C;
T_s = surface temperature, °C.

The convection coefficient, often referred to as the film coefficient, may be estimated with a convection correlation, or, for design conditions, may be read from a table.

Example 2.5 Convection

A 10 m² wall is heated by the sun to a temperature of 60 °C at a time when the outdoor air is at 20 °C, and a light breeze results in a convection coefficient of 10 W/m²K. What is the heat lost by the surface due to convection?

Solution:

Equation 2.14 gives the heat transferred to the surface by convection. The heat transferred away from the surface by convection would be given by:

$$\dot{q} = hA(t_s - t) = 10\frac{\text{W}}{\text{m}^2\text{K}}\left(10\ \text{m}^2\right)\left(60°\text{C} - 20°\text{C}\right) = 4{,}000\ \text{W}$$

Note that the convection coefficient is typically given in units of W/m²K, but it could just as well be given in units of W/m² °C; h is always applied to temperature differences and a temperature difference is the same in either Kelvin or degrees Celsius; 40 K = 40 °C when we are referring to a temperature difference.

End of solution.

Exterior surface convection

Convection to exterior surfaces may be represented with a range of models, all of which involve the use of a convection coefficient:

$$q''_{convection,\, out,\, j,\, \theta} = h_c(T_o - T_{os,\, j,\, \theta}) \tag{2.15}$$

where:

h_c is the convection coefficient, W/m²K.

Mirsadeghi et al. (2013) review exterior convection correlations applied in building simulation and recommend correlations suitable for two categories of buildings: single story and high-rise (6–8 stories). Previous review papers include Palyvos (2008), McClellan and Pedersen (1997) and Cole and Sturrock (1977). Given the very complex wind-driven and buoyancy-driven airflows around a building, a convective heat transfer model might be very complex and difficult to use. A correlation developed by Yazdanian and Klems (1994) and recommended by Mirsadeghi et al. (2013) seems to strike a reasonable balance between accuracy and ease of use for low-rise buildings. The correlation takes the form:

$$h_c = \sqrt{\left[C_t(\Delta t)^{1/3}\right]^2 + \left[aV_o^b\right]^2} \tag{2.16}$$

where:

C_t = turbulent natural convection constant, given in Table 2.5;
ΔT = temperature difference between the exterior surface and the outside air, °C;
a, b = constants given in Table 2.1;
V_o = wind speed at standard conditions, m/s.

For high-rise buildings, Taki and Loveday (1996) recommend the correlation:

$$h_c = CV_s^{0.5} \tag{2.17}$$

where:

$C = 16.7\dfrac{W}{m^2K}\left(\dfrac{m}{s}\right)^{-0.5}$;
V_s = wind speed near surface, m/s.

The correlation was based on windspeeds between 0.2 m/s and 4 m/s. Loveday and Taki do not make a recommendation for windspeeds below 0.2 m/s, but a minimum convection coefficient of 7.5 W/m^{2}K might be inferred from their measurements.

Mirsadeghi et al. (2013) note there are gaps in available correlations; e.g. experimentally based correlations for building roofs are not available. Development of correlations based on experimental measurements (Liu and Harris 2007; Shao et al. 2009) and CFD modeling (Blocken et al. 2009; Defraeye et al. 2011; Karava et al. 2011; Hu et al. 2018) remain active areas of research.

Interior surface convection

Interior convective heat transfer to and from room surfaces occurs under a wide range of regimes varying from natural convection to forced convection and from laminar flow to turbulent flow. At this time, there are no satisfactory models that cover the entire range of conditions.

Table 2.5 Convection correlation coefficients for the Yazdanian and Klems (1994) model

Wind direction	C_t	*a*	*b*
	W/m²K⁴ᐟ³	*W/m²K(m/s)*	–
Windward	0.84	2.38	0.89
Leeward	0.84	2.86	0.617

Table 2.6 Interior surface convection coefficients for use with the heat balance model

Position of surface	*Direction of heat flow*	$h_c, \frac{W}{m^2 - K}$
Horizontal	Upward	4.15
Sloping – 45 degrees	Upward	3.98
Vertical	Horizontal	3.18
Sloping – 45 degrees	Downward	2.39
Horizontal	Downward	1.02

However, Beausoleil-Morrison (2001) has developed a model for rooms with ceiling diffusers that incorporates correlations (Alamdari and Hammond 1983; Fisher and Pedersen 1997) from a range of different flow regimes. Goethals et al. (2011) and (Peeters et al. 2011) review available correlations. Subsequent to these two review papers, correlations for rooms with floor registers (Clark and Novoselac 2013), windows with venetian blinds (Clark et al. 2013), and rooms with active chilled beams and radiant walls (Le Dréau et al. 2015) have been reported.

For many buildings with conventional mechanical cooling systems, the cooling loads are only modestly sensitive to the interior convection coefficients. Buildings that are highly glazed are a notable exception. For buildings that rely on nighttime ventilative cooling and thermal storage, the predicted number of hours for which thermal comfort criteria are exceeded is sensitive (Leenknegt et al. 2012) to the interior convective heat transfer correlation.

A relatively simple model, strictly applicable for natural convection conditions, utilizes fixed convection coefficients extracted from the surface unit conductances (ASHRAE 2017). The surface unit conductances, which are combined convection–radiation coefficients, have a radiative component of about 5.1 W/(m^{2}K). By subtracting the radiative component, we obtain the convective coefficients shown in Table 2.6. Once the convective coefficient is obtained, the convective heat flux from the wall to the zone air is

$$q''_{convection,\, in,\, j,\theta} = h_c(T_{is,\, j,\theta} - T_i) \tag{2.18}$$

Bulk convection – infiltration and ventilation

Infiltration and ventilation result in the introduction of outdoor air into the space. Methods used to estimate the quantity of infiltration and/or ventilation air are discussed in Chapter 7. Introduction of outdoor air results in both sensible and latent heat gain, which will be computed as follows:

$$\dot{q}_{infiltration,\,\theta} = \dot{m}_a c_p (T_o - T_i) = \frac{\dot{Q} c_p}{v_o}(T_o - T_i) \tag{2.19}$$

$$\dot{q}_{infiltration,\, latent,\,\theta} = \dot{m}_a (W_o - W_i) i_{fg} = \frac{\dot{Q}}{v_o}(W_o - W_i) i_{fg} \tag{2.20}$$

where:

$\dot{q}_{infiltration,\theta}$ = sensible heat gain due to infiltration at time θ, W;
$\dot{q}_{infiltration,latent,\theta}$ = latent heat gain due to infiltration at time θ, W;
$\dot{m}_a$ = mass flow rate of infiltration on a dry-air basis, kg/s;
c_p = specific heat of the infiltration air, J/kg K;
T_o = outside air temperature, °C;
T_i = inside air temperature, °C;

$\dot{Q}$ = volume flow rate of air, m^3/s;
v_o = specific volume of the outside air, m^3/kg;
W_o = outside air humidity ratio, kg H_2O vapor/kg dry air;
W_i = inside air humidity ratio, kg H_2O vapor/kg dry air;
i_{fg} = phase-change enthalpy, J/kg.

Bulk convection – air system

For rooms where heating or cooling is provided by an air system, the heat transfer to or from the space is affected by bulk convection. The air temperature and/or flow rate to the zone may be thermostatically controlled. It is possible to formulate the effects of the system in several different ways:

1 If the system is modeled simultaneously, and the mass flow rate and/or temperature of air entering the space is determined by the system simulation, it is possible to determine the system heat transfer rate as:

$$\dot{q}_{system,\,\theta} = \dot{m}_a c_p (T_{sys} - T_i) \tag{2.21}$$

2 If a design load calculation is being performed, a desired room air temperature can be set and the heat balance can be arranged to solve for the required system heat transfer rate.
3 If the system is not being simulated or it is being simulated sequentially, it is often convenient to represent the system heat transfer as a piecewise linear function of zone temperature, called a control profile:

$$\dot{q}_{system,\,\theta} = a + bT_i \tag{2.22}$$

where a and b are equation coefficients that apply over a specific range of temperatures.

Radiation heat transfer

Thermal radiation is the transfer of energy by electromagnetic waves. In building load calculations, thermal radiation is generally thought of as being a surface-to-surface phenomenon either between surfaces within the building or between the surface of the sun and the building surfaces. Gases, aerosols, and particulates also emit and absorb radiation between surfaces. However, for analysis of radiation heat transfer between surfaces within a building, the radiation path lengths are short enough that emission and absorption of the indoor air can be neglected. For radiation heat transfer between the sun and the building surfaces, emission and absorption in the earth's atmosphere is accounted for in the models used to determine incident solar irradiation.

Thermal radiation emitted by any surface will have a range of wavelengths, depending upon the temperature of the emitting surface. The amount of thermal radiation absorbed, reflected, or emitted by any surface will depend on the wavelengths and direction in which the radiation is incident or emitted relative to the surface. Dependency on the wavelength is referred to as "spectral"; surfaces for which the properties are effectively independent of the wavelength are referred to as "gray" surfaces. Dependency on the direction is referred to as "specular"; surfaces for which the properties are effectively independent of the direction are referred to as "diffuse" surfaces. Properties of interest include:

- Absorptance,[5] α, the ratio of radiation absorbed by a surface to that incident on the surface.
- Emittance, ε, the ratio of radiation emitted by a surface to that emitted by an ideal "black" surface at the same temperature.

- Reflectance, ρ, the ratio of radiation reflected by a surface to that incident on the surface.
- Transmittance, τ, the ratio of radiation transmitted by a translucent surface to that incident on the surface.

Analysis of thermal radiation is greatly simplified when the surfaces may be treated as gray and diffuse. With two exceptions, surfaces in buildings are treated as both gray and diffuse in load calculation procedures.

The first exception is based on the fact that thermal radiation wavelength distributions prevalent in buildings may be approximately lumped into two categories – short wavelength radiation (solar radiation or visible radiation emitted by lighting) and long wavelength radiation (radiation emitted by surfaces, people, equipment, etc., that is all at relatively low temperatures compared to the sun). Treatment of the two wavelength distributions separately in building load calculations might be referred to as a "two-band" model, but in practice this only means that surfaces may have different absorptances for short wavelength radiation and long wavelength radiation. As an example, consider that a surface painted white may have a short wavelength absorptance of 0.4 and a long wavelength absorptance of 0.9.

The second exception is for analysis of windows. Solar radiation is typically divided into specular ("direct" or "beam") and diffuse components. Since the window transmittance, absorptance, and reflectance tend to be moderately strong functions of the incidence angle, these properties are generally calculated for the specific incidence angle each hour.

A notable feature of thermal radiation is that the emission is proportional to the fourth power of the absolute temperature (i.e. Kelvin or Rankine). As an example, consider a case with only two surfaces separated by a non-participating medium. The radiation heat transfer rate between surfaces 1 and 2 is given by:

$$\dot{q}_{1-2} = \frac{\sigma\left(T_1^4 - T_2^4\right)}{\dfrac{1-\varepsilon_1}{A_1\varepsilon_1} + \dfrac{1}{A_1F_{1-2}} + \dfrac{1-\varepsilon_2}{A_2\varepsilon_2}} \tag{2.23}$$

where:

σ	=	Stefan-Boltzmann constant, 5.673×10^{-8} W/m²K⁴;
T_1 and T_2	=	surface temperatures of surfaces 1 and 2, K;
ε_1 and ε_2	=	emittances of surfaces 1 and 2;
F_{1-2}	=	view factor from surface 1 to surface 2.

Example 2.6 Radiation

A 10 m² wall has an air gap separating a layer of brick and a layer of concrete block. At a time when the temperature of the brick surface adjacent to the air gap is 0 °C and the temperature of the concrete block surface adjacent to the air gap is 10 °C, what is the radiation heat transfer rate across the air gap?

Solution:

The radiation heat transfer rate can be calculated with Equation 2.23. In order to use Equation 2.23, there are a few terms that must first be evaluated. First, the temperatures T_1 and T_2 are needed in absolute temperature, Kelvin. Taking the concrete block as surface 1 and the brick as surface 2:

$$T_1 = 10 + 273.15 = 283.15 \text{ K}$$
$$T_2 = 0 + 273.15 = 273.15 \text{ K}$$

The surface emissivities are not given, but, in general for building materials that are not polished metals, a value of 0.9 is a reasonable assumption.

The view factor between two parallel surfaces facing each other is approximately one. Equation 2.23 can then be evaluated as:

$$\dot{q}_{1-2} = \frac{\sigma\left(283.15^4 - 273.15^4\right)}{\frac{1-0.9}{10\cdot 0.9} + \frac{1}{10\cdot 1} + \frac{1-0.9}{10\cdot 0.9}} = 329 \text{ W}$$

If a low emissivity coating, e.g. a layer of aluminum foil, was applied to the concrete block layer and not compromised by being splattered with mortar or coated with dust during the bricklaying, the emittance might be as low as 0.05. (In practice, this would be nearly impossible, but just for purposes of example, we'll assume that it might be done.) The radiation heat transfer could then be calculated as:

$$\dot{q}_{1-2} = \frac{\sigma\left(283.15^4 - 273.15^4\right)}{\frac{1-0.05}{10\cdot 0.05} + \frac{1}{10\cdot 1} + \frac{1-0.9}{10\cdot 0.9}} = 20 \text{ W}$$

End of solution.

In practice, rooms in buildings have more than two surfaces, and, in general, every surface exchanges radiant heat with every other surface. While it is possible to analyze the complete radiation network, building performance simulations typically adopt a simpler approximation. Several approximations are discussed in the section below entitled "Longwave radiation – exterior."

Shortwave radiation – exterior

Determination of incident solar radiation is covered in Chapter 6. For opaque surfaces, the next step is to determine the portion of the solar radiation that is absorbed. Although the absorptance of a surface to solar radiation will depend on the incidence angle, absorptance is often nearly constant until the incidence angle approaches 90°, at which point the solar irradiation is very low. Accordingly, the solar absorptance is reasonably approximated as a constant value and the absorbed solar irradiation is given by:

$$q''_{solar,\, out,\, j,\, \theta} = \alpha G_t \tag{2.24}$$

where:

α = solar absorptivity of the surface, dimensionless;
G_t = total solar irradiation incident on the surface, W/m^2.

Sample values of solar absorptances can be found in Table 2.7.

Shortwave radiation – windows

For transparent and semi-transparent surfaces, e.g. windows, the transmission of solar radiation through the window is of particular interest, as is the amount of solar radiation that is absorbed and then transmitted into the space in the form of longwave radiation or convection. For simple

Table 2.7 Solar absorptances

Surface	*Absorptance*
Brick, red (Purdue)[a]	0.63
Paint, cardinal red[b]	0.63
Paint, matte black[b]	0.94
Paint, sandstone[b]	0.50
Paint, white acrylic[a]	0.26
Sheet metal, galvanized, new[a]	0.65
Sheet metal, galvanized, weathered[a]	0.80
Shingles, Aspen gray[b]	0.82
Shingles, autumn brown[b]	0.91
Shingles, onyx black[b]	0.97
Shingles, generic white[b]	0.75

[a] F. P. Incropera and D. P. DeWitt (1990) *Fundamentals of Heat and Mass Transfer, 3rd Edition*. New York: Wiley.
[b] D. S. Parker, J.E.R. McIlvaine, S. F. Barkaszi, D. J. Beal and M. T. Anello (2000) Laboratory Testing of the Reflectance Properties of Roofing Material. FSEC-CR670–00. Florida Solar Energy Center, Cocoa, FL.

window systems, consisting of one, two, or three panes of homogeneous glass, analytical calculations based on ray tracing may be used. See Duffie and Beckman (1991) for an example. In current practice, there are a number of features that complicate the analysis:

- Modern windows often incorporate thin coatings of metallic oxides or other materials that are spectrally selective. Short of an analysis that breaks down the spectrum into small wavelength bins, some approximation is needed.
- Furthermore, the properties of the thin coatings are often considered proprietary by window manufacturers, in which case only overall window properties are given rather than detailed layer-by-layer properties.
- Internal shading layers, such as curtains and venetian blinds, are difficult to analyze, and needed data are rarely available. Actual positions of curtains and venetian blinds are controlled by occupants and seldom known accurately. A more detailed discussion of this subject can be found in Chapter 7.

Detailed model

A range of different approaches may be taken. A detailed model of such a window requires a method for determining the layer-by-layer absorption of solar radiation and transmission of solar radiation through all layers into the space. Direct (beam) solar radiation and indirect (diffuse) solar radiation are necessarily treated separately, and the transmission and absorption of direct solar radiation is a function of the incidence angle. Models for this analysis are described by Klems (2002) and Wright and Kotey (2006). As noted above, data (optical properties: transmittance, absorptance, and reflectance) supporting the models can be difficult to find; the WINDOW 7.6 program (LBNL 2017) developed by Lawrence Berkeley National Laboratory comes with data for thousands of different window layers, many of which are commercially available glazing systems (i.e. if the window manufacturer discloses exactly which window layers were utilized in a specific window/glazing system, all data needed for a detailed analysis would then be available. However, since that information is often incomplete, the utility of the large data set of window layers is limited for building simulation practitioners).

Data that are actually provided by window manufacturers typically include:

- Number of layers – is the window single pane, double pane, or triple pane?
- Some description of the glass type – coloration and whether it has a low-E coating, for example.
- *U*-factor (NFRC 2017b).
- Solar heat gain coefficient (SHGC) at normal incidence angle (NFRC 2017a).
- Visual transmittance (NFRC 2017a).

With this information and the WINDOW 7.6 program, it may be possible, though tedious, to develop a layer-by-layer description that matches the overall performance. An alternative would be to consult a source such as Chapter 15 of the 2017 *ASHRAE Handbook – Fundamentals*, where a table of angle-dependent properties for typical windows can be found. Using the manufacturer's description of the window, including the SHGC at normal incidence angle, a similar window can be sought from the table.

Regardless of how the optical properties are obtained, for any given window, an overall solar transmittance (τ_{sol}) and absorptances (α_1 for outer pane, α_2 for inner pane) for each pane are the result, as shown in Table 2.8. These results, for a single pane window and two double pane windows, give each of the properties of interest at intervals of 10 degrees. Properties for beam radiation would be found from the table, interpolating with the actual incidence angle. Properties for diffuse radiation would be taken from the "Hemispherical" column.

With these properties in hand, a detailed model of the window would then balance the heat at each pane, as described below in the section titled "Surface heat balances – transparent and semi-transparent surfaces."

Table 2.8 Window optical properties and SHGC

Angle	*0*	*10*	*20*	*30*	*40*	*50*	*60*	*70*	*80*	*90*	*Hemispherical*
	Single pane, 3 mm thick, clear										
τ_{sol}	0.834	0.833	0.831	0.827	0.818	0.797	0.749	0.637	0.389	0	0.753
α_1	0.091	0.092	0.094	0.096	0.100	0.104	0.108	0.110	0.105	0	0.101
SHGC	0.859	0.859	0.857	0.854	0.845	0.825	0.779	0.667	0.418	0	0.781
	Double pane, outer pane 3.2 mm thick with low-E coating; inner pane 5.7 mm thick										
τ_{sol}	0.408	0.410	0.404	0.395	0.383	0.362	0.316	0.230	0.106	0	0.338
α_1	0.177	0.180	0.188	0.193	0.195	0.201	0.218	0.239	0.210	0	0.201
α_2	0.060	0.060	0.061	0.061	0.063	0.063	0.061	0.053	0.038	0	0.059
SHGC	0.469	0.472	0.466	0.459	0.448	0.428	0.382	0.291	0.152	0	0.400
	Double pane, both panes 5.7 mm thick, clear										
τ_{sol}	0.607	0.606	0.601	0.593	0.577	0.546	0.483	0.362	0.165	0	0.510
α_1	0.167	0.168	0.170	0.175	0.182	0.190	0.200	0.209	0.202	0	0.185
α_2	0.113	0.113	0.115	0.116	0.118	0.119	0.115	0.101	0.067	0	0.111
SHGC	0.701	0.701	0.698	0.691	0.678	0.648	0.585	0.456	0.237	0	0.606

Note: Sample data generated with the WINDOW 5.2 program.

Simple model

Table 2.8 also shows, for each window, solar heat gain coefficients, also as a function of incidence angle. A simpler model, appropriate for zone models that do not directly utilize a heat balance, e.g. the transfer function method, can utilize these values along with the manufacturer-specified *U*-factor to compute the heat gains due to fenestration. See (Spitler 2014) for more in-depth coverage; a brief treatment follows here.

In this simple model, conduction heat gain is computed separately from the transmitted and absorbed solar heat gain. Because the thermal mass of the glass is very low, the conduction is approximately steady state. Accordingly, for each hour, the conduction heat gain may be calculated as:

$$q_\theta = UA(T_{o,\theta} - T_{rc}) \tag{2.25}$$

where:

q_θ = hourly conductive heat gain, W, for the window;
U = overall heat transfer coefficient for the window, W/m^2K, as specified by the window manufacturer;
A = window area – including frame, m^2;
$T_{o,\theta}$ = outdoor air temperature, °C;
T_{rc} = presumed constant room air temperature, °C;
θ = the current hour.

Transmitted and absorbed solar heat gains are calculated for each hour with a step-by-step procedure, as follows:

1. Compute incident angle, surface azimuth angle, incident direct (beam) irradiation, and diffuse irradiation on window, as described in Chapter 3.
2. If exterior shading exists, determine sunlit area and shaded area, as described in Chapter 3.
3. For windows without interior shading, the beam, diffuse, and total transmitted and absorbed solar heat gains are given by:

$$q_{SHG,D} = E_D A_{sunlit} SHGC(\theta) \tag{2.26}$$

$$q_{SHG,d} = (E_d + E_r) A\ SHGC_{diffuse} \tag{2.27}$$

$$q_{SHG} = q_{SHG,D} + q_{SHG,d} \tag{2.28}$$

where:

$q_{SHG,D}$ = direct (beam) solar heat gain, W/m^2;
$q_{SHG,d}$ = diffuse solar heat gain, W/m^2;
q_{SHG} = total solar heat gain, W/m^2;
E_D = incident direct (beam) irradiation, W/m^2;
E_d = incident diffuse irradiation from sky, W/m^2;
E_r = incident diffuse reflected irradiation, W/m^2;
$SHGC(\theta)$ = angle dependent *SHGC* interpolated from Table 2.8;
$SHGC_{diffuse}$ = the *SHGC* for diffuse irradiation, taken from the "Hemispherical" column of Table 2.8;
A_{sunlit} = the unshaded area of the window, m^2;
A = the total area of the window, including the frame, m^2.

If there is no interior shading, the window heat gain calculation is complete.

Interior shading

The detailed and simple models described above are both adequate for cases where the window unit is made up of one or more layers of glass or similar glazing material. The use of interior shading devices, such as venetian blinds or draperies, through which air may pass, further complicates the analysis. Likewise, insect screening can similarly complicate the analysis. For the detailed method, see Wright (2008) and Wright et al. (2009) for the most comprehensive treatment of the problem. For the simple model, an approximate method utilizing interior attenuation coefficients (IAC) is described by ASHRAE (2017) and Spitler (2014).

Example 2.7

On July 21 in Seattle, Washington (47.45° N latitude and 122.3° W longitude), the outdoor air temperature is 28.1 °C, and the room air temperature is 22 °C at 3 p.m. Pacific Daylight Savings Time. The southwest-facing double pane with low-E coating window shown in Table 2.8 is 1.2 m high and 1.2 m wide; the exterior surface is flush with the outside of the building. The rated *U*-factor for the window is 1.66 W/m²K. At that time,

- the incident solar beam radiation is 486.1 W/m²,
- the incidence angle is 55.88°, and
- the incident solar diffuse radiation is 193.5 W/m² (106.8 W/m² diffuse from sky and 86.7 W/m² reflected).

Using the simple model, compute the solar heat gain for the window. Using the detailed model, determine the transmitted solar radiation and the solar radiation absorbed by both panes.

Solution:

First, with the simple model, the conduction heat gain may be determined. Then, the window properties, including appropriate values of SHGC, are determined. Lastly, the solar heat gains with and without internal shading are calculated.

Simple model – conduction heat gain

The *U*-factor is given by the manufacturer as 1.66 W/m²K. The area of the window is 1.44 m². The air temperature at noon is 28.1 °C. From Equation 2.4, noting that $U = 1/R$, the conduction heat gain at noon is:

$$q_{cond15} = 1.66 \times (1.44) \times (28.1 - 22.0) = 14.6 \text{ W}$$

Simple model – solar heat gain calculation

The incidence angle is 55.88°; interpolating in Table 2.8 gives a value of:

$$SHGC\,(55.88) = 0.335$$

Then, $q_{SHG,D} = E_D A_{sunlit} SHGC(\theta) = 486.1 \times 1.44 \times 0.401 = 280.7 \text{ W}$

$$q_{SHG,d} = (E_d + E_r)\,A\;SHGC_{diffuse} = 193.5 \times 1.44 \times 0.4 = 111.5 \text{ W}$$

$$q_{SHG} = q_{SHG,D} + q_{SHG,d} = 392 \text{ W}$$

Combined with conduction, the total heat gain from the window is then 407 W.

Detailed model

With the detailed model, values of transmittance and absorptance for beam radiation are determined by interpolating in Table 2.8:

$$\tau_{sol}(55.88°) = 0.335$$
$$a_1(55.88°) = 0.211$$
$$a_2(55.88°) = 0.062$$

values for diffuse radiation are read from the last column:

$$\tau_{sol,d} = 0.338$$
$$a_{1,d} = 0.201$$
$$a_{2,d} = 0.059$$

Then, the transmitted beam solar radiation is determined by multiplying the beam radiation by the beam transmittance and area:

$$q_{trans,D} = E_D A_{sunlit} \tau_{sol} = 486.1 \times 1.44 \times 0.335 = 234.5 \text{ W}$$

The transmitted diffuse solar radiation is determined by multiplying the diffuse radiation by the diffuse transmittance and area:

$$q_{trans,d} = (E_d + E_r) A\, \tau_{sol,d} = 193.5 \times 1.44 \times 0.338 = 94.2 \text{ W}$$

And the total transmitted solar radiation is:

$$q_{trans} = q_{trans,D} + q_{trans,d} = 329 \text{ W}$$

In a similar manner, the absorbed solar radiation in each pane is determined.

$$q_{abs,1} = q_{abs,D,1} + q_{abs,d,1} = E_D A_{sunlit} \alpha_1(55.88^o) + E_d A \alpha_{1,d}$$
$$= 486.1 \times 1.44 \times 0.211 + 193.5 \times 1.44 \times 0.201 = 203.7 \text{W}$$

$$q_{abs,2} = q_{abs,D,2} + q_{abs,d,2} = E_D A_{sunlit} \alpha_2(55.88^o) + E_d A \alpha_{2,d}$$
$$= 486.1 \times 1.44 \times 0.062 + 193.5 \times 1.44 \times 0.059 = 59.7 \text{W}$$

Comparing the results from the simple model and the detailed model, we see that the simple model predicts a total heat gain through the window of 407 W. The detailed model predicts 329 W of transmitted solar radiation and a total absorbed radiation of 263 W. Of the 263 W, only a fraction will flow into the room. The quantity that flows into the room will depend on the other outside and inside conditions and will be determined using a heat balance, described in the section titled "Surface heat balances – transparent and semi-transparent surfaces." Depending on those conditions, the detailed model may predict more or less heat gain to the space than the simple model.

End of solution.

Shortwave radiation – interior

Treatment of shortwave radiation, once it has been transmitted through the window, depends on the zone model described in the "Zone models" section below. For transfer-function type models, a pre-computed transfer function or response function is applied to the solar heat gains determined with Equations 2.26–2.28. These heat gains include both transmitted and inward-flowing absorbed solar radiation.

For heat balance models and thermal network models, it is necessary to make an estimate of how the transmitted direct and diffuse solar radiation are distributed. In other words, the amount of transmitted solar radiation absorbed by each surface in the room must be determined. This can be (and sometimes is) analyzed in a very detailed manner, accounting for exactly where the radiation strikes each room surface, and then accounting for each reflection until it is all absorbed.

However, this level of detail is difficult to justify for most load calculation and energy calculation programs, as the amount of effort required to obtain the requisite input data is disproportionate to the user's ability to predict future configurations of furniture and window treatments. Therefore, a simpler model is often employed, assuming that some fraction of the transmitted beam radiation is absorbed by the floor. That fraction not absorbed by the floor is then assumed to be diffusely reflected and, along with the transmitted diffuse radiation, uniformly absorbed by all room surfaces.

If the total transmitted diffuse radiation and the reflected direct radiation (from the floor) are divided by the total interior surface area of the zone and distributed uniformly, then for all surfaces except the floor, the absorbed solar radiation flux is given by:

$$q''_{solar,\,in,\,j,\theta} = \frac{\sum_{i=1}^{M} \dot{q}_{trans,\,d,\,i} + (1-\alpha_{floor})\sum_{i=1}^{M} \dot{q}_{trans,\,D,\,i}}{\sum_{j=1}^{N} A_j} \tag{2.29}$$

where:

$q''_{solar,\,in,\,j,\theta}$ = solar radiation absorbed by interior surfaces other than the floor, W/m^2;
$\dot{q}_{trans,\,d,\,i}$ = diffuse solar radiation transmitted by the ith window, W;
$\dot{q}_{trans,\,D,\,i}$ = direct solar radiation transmitted by the ith window, W;
α_{floor} = the solar absorptance of the floor;
A_j = the area of the jth interior surface, m^2;
M = the number of windows;
N = the number of interior surfaces.

Then, the absorbed solar radiation flux (W/m^2) for the floor is given by:

$$q''_{solar,\,in,\,floor,\,\theta} = \frac{\sum_{i=1}^{M} \dot{q}_{trans,\,d,\,i} + (1-\alpha_{floor})\sum_{i=1}^{M} \dot{q}_{trans,\,D,\,i}}{\sum_{j=1}^{N} A_j} + \frac{\alpha_{floor}\sum_{i=1}^{M} \dot{q}_{trans,\,D,\,i}}{A_{floor}} \tag{2.30}$$

This is a fairly simple model for distribution of transmitted solar heat gain. A number of improvements might be made, including determining which interior surfaces are actually sunlit by the direct solar radiation and allowing for additional reflection of the beam radiation. Beyond that, more sophisticated algorithms are used for analysis of daylighting as described in Chapter 7 and might be adapted for cooling load calculation use.

Longwave radiation – exterior

Long wavelength (thermal) radiation to and from exterior surfaces is also a very complex phenomenon. The exterior surfaces radiate to and from the surrounding ground, vegetation, parking lots, sidewalks, other buildings, and the sky. In order to make the problem tractable, a number of assumptions are usually made:

- Each surface is assumed to be opaque, diffuse, and isothermal and to have uniform radiosity and irradiation.
- Each surface is assumed to be gray, having a single value of absorptivity and emissivity that applies over the thermal radiation spectrum. (The surface may have a different value of absorptivity that applies in the solar radiation spectrum.)
- Radiation to the sky, where the atmosphere is actually a participating medium, may be modeled as heat transfer to a surface with an effective sky temperature.
- Lacking any more detailed information regarding surrounding buildings, it is usually assumed[6] that the building sits on a flat, featureless plane, so that a vertical wall has a view factor between the wall and the ground of 0.5 and between the wall and the sky of 0.5.
- Without a detailed model of the surrounding ground, it is usually assumed to have the same temperature as the air. Obviously, for a wall with a significant view to an asphalt parking lot, the ground temperature would be somewhat higher.

With these assumptions, the net long wavelength radiation into the surface is given by

$$q''_{\text{radiation},out,j,\theta} = \varepsilon\sigma\left[F_{s-g}\left(T_g^4 - T_{os,j,\theta}^4\right) + F_{s-sky}\left(T_{sky}^4 - T_{os,j,\theta}^4\right)\right] \tag{2.31}$$

where:

ε = surface long wavelength emissivity;
σ = Stefan-Boltzmann constant, 5.67×10^{-8} W/m^{2}K^4;
F_{s-g} = view factor from the surface to the ground;
F_{s-sky} = view factor from the surface to the sky;
T_g = ground temperature, K;
T_{sky} = effective sky temperature, K;
$T_{os,j,\theta}$ = surface temperature, K.

Since it is usually assumed that the building sits on a featureless plain, the view factors are easy to determine:

$$F_{s-g} = \frac{1 - \cos\alpha}{2} \tag{2.32}$$

$$F_{s-sky} = \frac{1 + \cos\alpha}{2} \tag{2.33}$$

where α is the tilt angle of the surface from horizontal.

It is often convenient to linearize this equation by introducing radiation heat transfer coefficients:

$$h_{r,g} = \varepsilon\sigma\left[\frac{F_{s-g}(T_g^4 - T_{os,j,\theta}^4)}{T_g - T_{os,j,\theta}}\right] \tag{2.34}$$

$$h_{r,\,sky} = \varepsilon\sigma\left[\frac{F_{s-sky}\left(T_{sky}^{4} - T_{os,\,j,\,\theta}^{4}\right)}{T_{sky} - T_{os,\,j,\,\theta}}\right] \tag{2.35}$$

Then Equation 2.31 reduces to

$$q''_{radiation,\,out,\,j,\,\theta} = h_{r,\,g}\left(T_{g} - T_{os,\,j,\,\theta}\right) + h_{r-sky}\left(T_{sky} - T_{os,\,j,\,\theta}\right) \tag{2.36}$$

If the radiation coefficients are determined simultaneously with the surface temperature, Equation 2.36 will give identical results to Equation 2.37.

A number of models are available (Cole 1976; Algarni and Nutter 2015) for estimating the effective sky temperature seen by a horizontal surface under clear sky conditions. Perhaps the simplest is that used by the BLAST program (McClellan and Pedersen 1997), which simply assumes that the effective sky temperature is the outdoor dry bulb temperature minus 6 K.

For surfaces that are not horizontal, the effective sky temperature will be affected by the path length through the atmosphere. An approximate expression based on Walton's heuristic model (1983) is

$$T_{sky,\,\alpha} = \left[\cos\frac{\alpha}{2}\right]T_{sky} + \left[1 - \cos\frac{\alpha}{2}\right]T_{o} \tag{2.37}$$

where:

$T_{sky,\alpha}$ = effective sky temperature for a tilted surface, °C or K;
T_{sky} = effective sky temperature for a horizontal surface, °C or K;
T_{o} = outdoor air dry bulb temperature, °C or K.

Longwave radiation – interior

From a building load calculation perspective, radiation is often divided into two categories – short wavelength and long wavelength. Short wavelength radiation is radiation with wavelengths near the visible spectrum, such as solar radiation and a portion of the radiation given off by lights. Long wavelength radiation is given off by sources near room temperature, such as people, equipment, walls, etc.

This section describes models for distribution of long wavelength radiation given off by surfaces and heat gain sources in the room. The exchange of thermal radiation between surfaces in an enclosure is fairly well understood and is covered in any elementary heat transfer textbook. However, the solution procedure is generally considered to require too much computational time for an hourly load calculation.

Furthermore, rooms are seldom empty, and describing all of the interior surfaces and furnishings in detail is likely to be burdensome to the designer and to have little point, as the arrangement of the furnishings is not likely to remain constant over the life of the building. Therefore, simpler methods (Carroll 1980a; Walton 1980; Davies 1988; Liesen and Pedersen 1997) are often used for estimating radiation heat transfer. Many of these are based on defining a single temperature to which each surface radiates. Two additional simplifications are usually made when analyzing radiation heat transfer inside a room:

- Furnishings (e.g. desks, chairs, tables, shelves) are usually lumped into a single surface, sometimes called "internal mass."
- Radiation from equipment, lights, and people is usually treated separately. See the next section.

A reasonably simple model with acceptable accuracy is Carroll's *mean radiant temperature method* (Carroll 1980a). The model represents all of the surfaces in the room as a single fictitious surface, the so-called *mean radiant temperature* (MRT).

First, Carroll defines an MRT view factor, which is an approximate view factor that compensates for the self-weighting in the mean radiant temperature:

$$F_i = 1 / (1 - A_i F_i / \sum_{i=1}^{N} A_i F_i) \tag{2.38}$$

where:

N = the number of surfaces in the zone.

Since the MRT view factors (F_i) appear on both sides of the equation, iteration is required to determine the values. Carroll (1980) suggests setting F_i to one as an initial guess. The MRT view factors can be calculated once at the beginning of the simulation during the initialization phase. This procedure for calculating the MRT view factors is suitable for most zones. In the case of geometries with large but significantly different coplanar surfaces, Carroll suggests adjusting the procedure for estimating the MRT view factors.

Also, before the simulation begins, a first estimate of the radiation coefficients for each surface to the mean radiant temperature should be made as follows:

$$h_{rad,\,ref,\,i} = \frac{4\sigma \bar{T}_{ref}^{3}}{1/F_i + (1-\varepsilon_i)/\varepsilon_i} \tag{2.39}$$

where:

$h_{rad,\,ref,\,i}$ = radiation coefficient for *i*th surface at reference temperature, W/m²K;
$\bar{T}_{ref}$ = mean reference temperature, taken to be 300 K;
F_i = the MRT view factor for *i*th surface, (–);
ε_i = the longwave emissivity of *i*th surface, (–).

Then, at each simulation time step and/or at each simulation iteration where radiation heat transfer is to be calculated, the estimates of the radiation coefficients are improved with a two-step process. First, given the actual surface temperatures, new estimates of the radiation coefficients ($h_{rad,1,j}$ – the 1 in the subscript refers to the first update) are calculated with this expression:

$$h_{rad,1,i} = (0.865 + T_i / 200) h_{rad,\,ref,\,i} \tag{2.40}$$

Then the first estimate of the mean radiant temperature, $T_{MR,1}$, is calculated as the average surface temperature, weighted by the estimated radiative coefficients and surface areas:

$$T_{MR,1} = \sum_{i=1}^{N} A_i h_{rad,1,i} T_i / \sum_{i=1}^{N} A_i h_{rad,1,i} \tag{2.41}$$

A second, improved estimate of each radiation coefficient is made using the first estimate of the mean radiant temperature:

$$h_{rad,2,i} = (0.865 + T_{MR,1} / 200) h_{rad,1,i} \tag{2.42}$$

Then, a second improved estimate of the mean radiant temperature is made:

$$T_{MR,2} = \sum_{i=1}^{N} A_i h_{rad,1,i} T_i / \sum_{i=1}^{N} A_i h_{rad,2,i} \tag{2.43}$$

The net radiation flux in W/m² leaving each surface is then given by:

$$q''_{rad} = h_{rad,2,i}(T_i - T_{MR,2}) \tag{2.44}$$

And the net radiation in W leaving each surface is:

$$q_{rad} = h_{rad,2,i}A_i(T_i - T_{MR,2}) \tag{2.45}$$

Example 2.8

A room with a south-facing window is shown in Figure 2.6. The window is 5 m wide and 1.5 m high; its lower edge is 1 m above the floor and it is centered in the wall from east to west.

The surfaces are summarized in Table 2.9. On a cold winter's night, they have the temperatures shown in the last column.

Determine the net radiation leaving each of the seven surfaces.

Solution:

Based on the areas shown in Table 2.9, the MRT view factors are computed iteratively with Equation 2.38 and shown in the second column of Table 2.10. With the MRT view factors known, the reference values of the radiation coefficient are calculated with Equation 2.39. Using the known surface temperatures, Equation 2.40 is used to make the first estimate of the radiation coefficient, shown in the fourth column.

The first estimate of the mean radiant temperature, $T_{MR,1}$, made with Equation 2.41, is 15.91 °C. Using this estimate of mean radiant temperature, the second estimates of the radiation

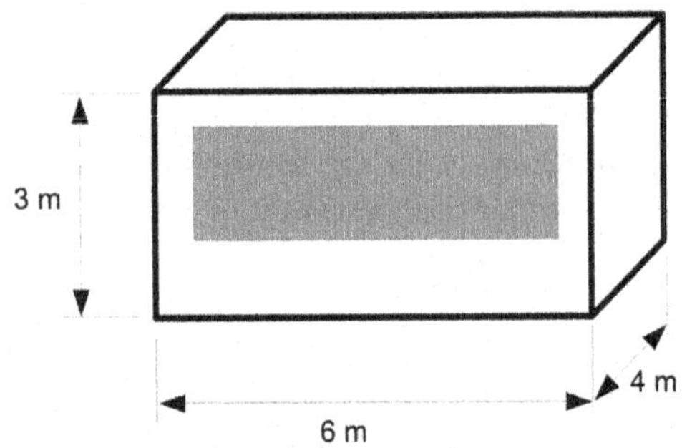

Figure 2.6 South-facing zone

Table 2.9 Surfaces of room in radiation example

Number	*Name*	*Area (m²)*	*Emissivity*	*Temperature (°C)*
1	Floor	24	0.9	14
2	North wall	18	0.9	17
3	East wall	12	0.9	17
4	West wall	12	0.9	17
5	Ceiling	24	0.9	18
6	Window	7.5	0.9	15
7	South wall (surrounding window)	10.5	0.9	9

Table 2.10 Radiation example intermediate results

Number	F_i	$h_{rad,ref,i}$	$h_{rad,1,i}$	$h_{rad,2,i}$	q''_{rad} *(W/m²)*	q_{rad} *(W)*
1	1.318	7.04	6.58	6.22	−11.85	−284.48
2	1.197	6.47	6.14	5.80	6.35	114.27
3	1.114	6.07	5.76	5.44	5.96	71.47
4	1.114	6.07	5.76	5.44	5.96	71.47
5	1.318	7.04	6.72	6.35	13.30	319.17
6	1.065	5.42	4.93	4.66	−32.18	−241.36
7	1.096	5.98	5.63	5.31	−4.81	−50.55

coefficients, calculated with Equation 2.42, are given in the fifth column of Table 2.10. A second estimate of the mean radiant temperature, $T_{MR,2}$, made with Equation 2.43, is also 15.91 C. Using Equations 2.44 and 2.45, the net radiative flux and net radiation leaving each surface are given in the last two columns of Table 2.10.

End of solution.

Zone models

To this point, all of the mechanisms for heat transfer into and out of the zone have been considered. In order to determine what actually happens in the zone, the individual mechanisms need to be analyzed simultaneously. A fundamental way of doing this is known as the heat balance model, though many versions of the model exist, so it might be more accurately referred to as a class of models. Heat balance models ensure that all energy flows in each zone are balanced and involve the solution of a set of energy balance equations for the zone air and the interior and exterior surfaces of each wall, roof, and floor. These energy balance equations are combined with equations for transient conduction heat transfer through walls and roofs and algorithms or data for weather conditions including outdoor air dry bulb temperature, wet bulb temperature, solar radiation, and so on.

A second class of models, known as thermal network models, are a form of heat balance model characterized by discretization of the building into a network of nodes, with interconnecting paths, through which energy flows; the use of numerical sub-models for conduction heat transfer (e.g. finite difference or finite volume); and provision for multiple zone air temperatures.

A third class of models uses a heat balance model to precalculate zone responses to heat gains. This class of models utilizes transfer functions or response factors to relate current values of the cooling load or heating load to past values of heat gains, and, in most cases, past values of the cooling load or heating load.

The heat balance and thermal network models will balance all of the energy flows every hour, while the transfer function models may not. The heat balance method is also more flexible, able to incorporate such features as variable convection coefficients. However, the transfer function method generally runs faster than the heat balance method.

Thermal network models are even more flexible than most heat balance models, while also requiring the most computation time.

The next three sections describe the three different models, with emphasis on the heat balance model.

Zone models: heat balance models

The heat balance model may be implemented with a wide variety of sub-models, e.g. different exterior convection models, different transient conduction models, and different interior radiation models. What is described in this section is one implementation of the heat balance model made up of a particular combination of sub-models. Here, the sub-models are chosen primarily for simplicity in illustrating the model.

The heat balance model utilizes the heat balance concept at all interior and exterior zone surfaces, as well as applying a heat balance to the zone air.

Surface heat balances – opaque surfaces

To illustrate the heat balance model, consider a simple zone with seven surfaces: four walls, a window in one of the walls, a roof, and a floor. The zone has solar energy coming through the window; heat conducted through the exterior walls and roof; and internal heat gains due to lights, equipment, and occupants. The heat balances on both the interior and exterior surfaces of a single wall or roof element are illustrated in Figure 2.7. The heat balance on the *j*th exterior surface at time θ is represented conceptually by:

$$q''_{conduction,\, out,\, j,\theta} = q''_{solar,\, out,\, j,\theta} + q''_{convection,\, out,\, j,\theta} + q''_{radiation,\, out,\, j,\theta} \tag{2.46}$$

where:

$q''_{conduction,\, out,\, j,\theta}$ = conduction heat flux, W/m²;
$q''_{solar,\, out,\, j,\theta}$ = absorbed solar heat flux, W/m²;
$q''_{convection,\, out,\, j,\theta}$ = convection heat flux, W/m²;
$q''_{radiation,\, out,\, j,\theta}$ = thermal radiation heat flux, W/m².

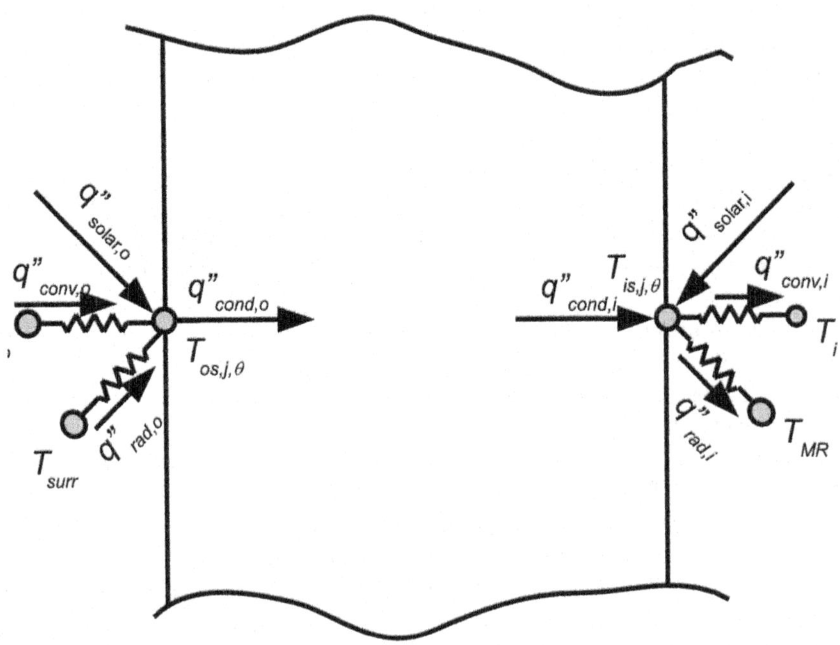

Figure 2.7 Opaque surface heat balance

A few features of Figure 2.7 that should be noted are:

- $q''_{conduction,out,j,\theta}$ is not equal to $q''_{conduction,in,j,\theta}$ unless steady-state heat transfer conditions prevail. This would be unusual for cooling load calculations.
- For the exterior surface, the wall is likely to radiate to both the sky and surroundings, and perhaps to other buildings. For this figure, only one interchange is shown.
- For the interior surface, the wall actually radiates to all other surfaces. But, as described above, this is simplified by computing a mean radiant temperature for the zone.
- The solar radiation incident on the interior surface will have been transmitted through the window.

The interior surface heat balance on the *j*th surface at time θ may be represented conceptually as:

$$q''_{conduction,in,j,\theta} + q''_{solar,in,j,\theta} = q''_{convection,in,j,\theta} + q''_{radiation,in,j,\theta} \tag{2.47}$$

where:

$q''_{conduction,in,j,\theta}$ = conduction heat flux, W/m²;
$q''_{solar,in,j,\theta}$ = absorbed solar heat flux, W/m²;
$q''_{convection,in,j,\theta}$ = convection heat flux, W/m²;
$q''_{radiation,in,j,\theta}$ = thermal radiation heat flux, W/m².

Exterior surface heat balance

We will first consider the exterior surface heat balance. Before deriving the full transient exterior surface heat balance equation, it may be helpful to consider a simpler case, based on a *steady-state* heat balance. Consider the exterior surface of a wall, as shown in Figure 2.8.

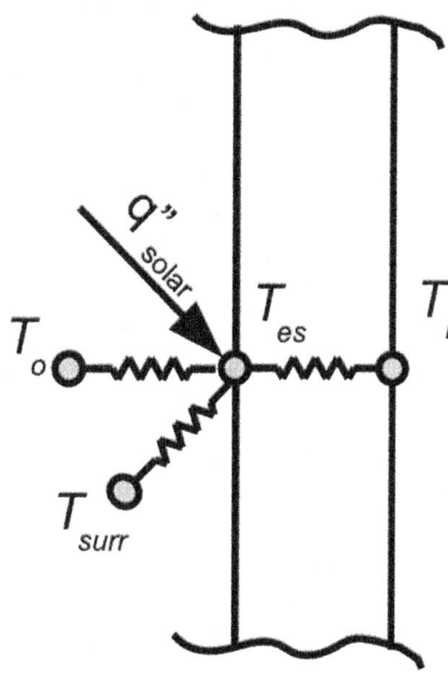

Figure 2.8 Exterior surface steady-state heat balance

where:

t_{es} = exterior surface temperature, °C;
t_{is} = interior surface temperature, °C;
t_o = outdoor air temperature, °C;
q''_{solar} = absorbed solar heat flux, W/m²;
t_{surr} = temperature of the surroundings, °C.

Neglecting any transient effects, a heat balance on the exterior surface might be written as:

$$q''_{conduction} = q''_{solar} + q''_{convection} + q''_{radiation} \tag{2.48}$$

With each component of the heat transfer defined as follows:

$$q''_{conduction} = U_{s-s}\left(T_{es} - T_{is}\right) \tag{2.49}$$

$$q''_{solar} = \alpha G_t \tag{2.50}$$

$$q''_{convection} = h_c\left(T_o - T_{es}\right) \tag{2.51}$$

$$q''_{radiation} = h_r\left(T_{surr} - T_{es}\right) \tag{2.52}$$

where:

U_{s-s} = surface-to-surface conductance, W/m²;
α = absorptance of the surface to solar radiation;
G_t = total incident solar radiation flux, W/m²;
h_c = exterior surface convection coefficient, W/m²K;
h_r = exterior surface radiation coefficient, W/m²K.

Here, the radiation heat transfer has been linearized, though in the heat balance solution procedure, the radiation heat transfer coefficient will be recalculated on an iterative basis. Substituting Equations 2.49–2.52 into 2.48, the exterior surface temperature can then be determined:

$$T_{es} = \frac{\alpha G_t + U_{s-s}T_{is} + h_c T_o + h_r T_{surr}}{U_{s-s} + h_c + h_r} \tag{2.53}$$

A similar procedure for the interior surface yields a similar equation for the interior surface temperature. The interior and exterior surface temperatures would then be solved simultaneously. When all interior surface temperatures are known, the cooling load can be calculated as the sum of the convective heat gains from the interior surfaces, lighting, equipment, people, and infiltration.

Example 2.9

For a very lightweight wall, a steady-state surface heat balance will give an approximate value of the exterior surface temperature and the conduction heat flux into the surface. Consider such a wall on a low-rise building with a surface-to-surface *U*-factor of 0.3 W/m²K, a solar absorptivity of 0.8, and an emissivity of 0.9. The weather conditions include incident solar radiation is 680 W/m², the outdoor dry bulb temperature is 28.1 °C, the sky temperature is 22.1 °C and the wind speed is 3.6 m/s with the surface on the windward side of the building. At this time, the interior surface temperature of the wall is 21 °C.

Estimate the exterior surface temperature and the conduction heat flux into the surface.

Solution:

The first step is to establish the boundary conditions, where they are not explicitly stated. The sky temperature is given as 22.1 °C; because the wall does not face the sky (i.e. it is oriented vertically, not horizontally), the effective sky temperature should be calculated with Equation 2.37. The tilt angle, α, is 90° and the effective sky temperature is

$$T_{sky,\alpha} = \left[\cos\frac{90°}{2}\right]22.1 + \left[1 - \cos\frac{90°}{2}\right]28.1 = 23.9°C$$

The temperature of the ground will be assumed to be equal to the air temperature (22.1 °C), and view factors from the wall to the ground and from the wall to the sky will both be taken to be 0.5, as discussed in the "Longwave radiation – exterior" section.

For low-rise buildings, the Yazdanian and Klems (1994) convection correlation is appropriate. The convection coefficient is a function of wind speed and temperature difference. Similarly, the radiation coefficients, $h_{r,g}$ and $h_{r,sky}$, are functions of the surface temperature and sky temperature. Since the surface temperature is not known at the beginning, it is necessary to make an initial guess, then iterate until a solution is achieved.

Making an initial guess of t_{os} = 50 °C gives the following values for convection and radiation coefficients:

$$h_c = 7.76 \text{ W/m}^2\text{K, from Equation 2.16,}$$
$$h_{r,sky} = 3.05 \text{ W/m}^2\text{K, from Equation 2.34,}$$
$$h_{r,grd} = 3.11 \text{ W/m}^2\text{K, from Equation 2.35.}$$

Using these values in the surface heat balance, Equation 2.53,

$$T_{es} = \frac{\alpha G_t + U_{s-s}T_{is} + h_c T_o + h_{r,sky}t_{sky} + h_{r,g}T_g}{U_{s-s} + h_c + h_{r,sky} + h_{r,g}}$$

$$T_{es} = \frac{0.8(680) + 0.3(21) + 7.76(28.1) + 3.05(23.9) + 3.11(28.1)}{0.3 + 7.76 + 3.05 + 3.11} = 65.3°C$$

A second iteration with t_{os} = 65.3 °C, gives the following values for convection and radiation coefficients:

$$h_c = 7.91 \text{ W/m}^2\text{K, from Equation 2.16,}$$
$$h_{r,sky} = 3.29 \text{ W/m}^2\text{K, from Equation 2.34,}$$
$$h_{r,grd} = 3.35 \text{ W/m}^2\text{K, from Equation 2.35.}$$

Using these values with the heat balance equation gives a new estimate of the surface temperature as 63.7 °C. A few additional iterations give slightly different values of the coefficients and a final estimate of the outside surface temperature as 63.81 °C. With this value of the outside surface temperature, the conductive heat flux, from Equation 2.9, is 12.8 W/m^2.

A few additional calculations illustrate the heat balance with the final values of the coefficients:

- The absorbed solar radiation is 0.8 × 680 = 544 W/m^2;
- The heat leaving by convection is 7.90 × (63.81 – 28.1) = 282.0 W/m^2;

- The heat radiating to the sky is 3.26 × (63.81 – 23.9) = 130.4 W/m²;
- The heat radiating to the ground is 3.33 × (63.81 – 28.1) = 118.8 W/m²;
- The heat conducted through the wall is 12.8 W/m².

A quick check of the results confirms that the heat transferred to the surface (544 W/m²) is equal to the sum of the heat being transferred away from the surface (282.0 + 130.4 + 118.8 + 12.8 = 544 W/m²).

End of solution.

In practice, these calculations are further complicated by the transient conduction heat transfer and the need to analyze all modes of heat transfer simultaneously for all surfaces in a room.

The transient surface heat balance requires use of the conduction transfer function described above. For any given hour in the simulation, past values of the exterior surface temperature and conduction heat flux will be known or assumed. Therefore, all the historical terms from Equation 2.12 may be gathered into a single term, $H_{out,j,\theta}$, which can be calculated at the beginning of the time step.

$$H_{out,j,\theta} = -\sum_{n=1}^{N_y} Y_n T_{is,j,\theta-n\delta} + \sum_{n=1}^{N_x} X_n T_{os,j,\theta-n\delta} + \sum_{n=1}^{N_q} \Phi_n q''_{conduction,out,j,\theta-n\delta} \quad (2.54)$$

and Equation 2.12 may be represented as:

$$q''_{conduction,\,j,\theta} = -Y_o T_{is,j,\theta} + X_o T_{os,j,\theta} + H_{out,j,6} \quad (2.55)$$

Then, by substituting the expressions for conduction heat flux (Equation 2.12), absorbed solar heat gain flux (Equation 2.24), convection heat flux (Equation 2.15), and radiation heat flux (Equations 2.34–2.36) into the exterior surface heat balance equation (Equation 2.46) and recasting the equation to solve for the exterior surface temperature, the following expression results:

$$T_{os,j,\theta} = \frac{Y_o T_{is,j,\theta} - H_{out,j,\theta} + \alpha G_t + h_c T_o + h_{r-g} T_g + h_{r-sky} T_{sky}}{X_o + h_c + h_{r-g} + h_{r-sky}} \quad (2.56)$$

Note that h_c, h_{r-g}, and h_{r-sky} all depend on the exterior surface temperature. While Equation 2.56 might be solved simultaneously with Equations 2.12, 2.34, 2.35, and 2.36 in a number of different ways, it is usually convenient to solve them by successive substitution. This involves assuming an initial value of the exterior surface temperature; then computing h_c, h_{r-g}, and h_{r-sky} with the assumed value; then solving Equation 2.56 for the exterior surface temperature; then computing h_c, h_{r-g}, and h_{r-sky} with the updated value of the exterior surface temperature; and so on until the value of the exterior surface temperature converges.

Also, the current value of the interior surface temperature appears in Equation 2.56. For thermally massive walls, Y_o will usually be zero. In this case, the exterior surface heat balance may be solved independently of the current hour's interior surface temperature. For thermally non-massive walls, the exterior surface heat balance must usually be solved simultaneously with the interior surface heat balance.

Example 2.10

For a day in September, the wall in Example 2.4 on a low-rise building is exposed to the weather conditions shown in the last four columns of Table 2.11. For hours 1–16, the surface temperatures and heat fluxes have already been calculated. At hour 17, the interior surface temperature has already been calculated and it is 21.91 °C.

Table 2.11 Weather conditions and previous hour

Time	$T_{os,\theta}$ (°C)	$T_{is,\theta}$ (°C)	q_o'' (W/m²)	q_i'' (W/m²)	*Incident solar (W/m²)*	$T_{o,\theta}$ (°C)	*Wind speed (m/s)*	*Sky temperature (°C)*
1	27.39	22.09	−83.55	11.59	0.00	21.69	3.6	15.69
2	25.97	22.08	−71.15	10.86	0.00	21.22	3.6	15.22
3	24.75	22.07	−61.16	10.06	0.00	20.76	3.6	14.76
4	23.70	22.05	−52.59	9.23	0.00	20.38	3.6	14.38
5	22.83	22.02	−44.90	8.42	0.00	20.10	3.6	14.10
6	22.14	22.00	−37.46	7.62	0.00	20.00	3.6	14.00
7	21.67	21.97	−29.21	6.86	0.00	20.18	3.6	14.18
8	22.36	21.93	−0.30	6.13	40.00	20.63	3.6	14.63
9	23.51	21.90	21.58	5.45	72.59	21.47	3.6	15.47
10	24.97	21.88	39.18	4.83	99.05	22.66	3.6	16.66
11	28.42	21.86	91.80	4.29	198.22	24.06	3.6	18.06
12	35.65	21.84	196.58	3.86	423.81	25.63	3.6	19.63
13	43.92	21.84	264.88	3.61	627.37	27.14	3.6	21.14
14	51.75	21.84	292.69	3.68	784.36	28.25	3.6	22.25
15	58.04	21.86	280.05	4.19	873.26	29.01	3.6	23.01
16	61.82	21.88	226.59	5.17	873.85	29.30	3.6	23.30
17		21.91			760.91	29.04	3.6	23.04
18					475.74	28.38	3.6	22.38
19					0.06	27.38	3.6	21.38
20					0.00	26.16	3.6	20.16
21					0.00	24.96	3.6	18.96
22					0.00	23.92	3.6	17.92
23					0.00	23.00	3.6	17.00
24					0.00	22.24	3.6	16.24

The wall has an exterior solar absorptivity, $\alpha = 0.8$, and an exterior thermal emissivity, $\varepsilon = 0.9$. The surface is on the windward side of the building.

Determine the exterior surface temperature and exterior conductive heat flux for hour 17.

Solution:

Like Example 2.9, the boundary conditions and appropriate correlations must be established first. The effective sky temperature for hour 17, calculated with Equation 2.37, is 24.79 °C. The Yazdanian and Klems correlation will again be used for the convection coefficient. The radiation coefficients are calculated in the same way as in Example 2.9.

Based on the previous hours, the history term for the conduction can be calculated with Equation 2.54:

$$\begin{aligned} H_{out,j,17} = & -(0.005618 \times 21.88 + 0.022664 \times 21.86 + 0.011332 \times 21.84 + 0.000823 \\ & \times 21.84 + 0.000009 \times 21.84) - 45.140663 \times 61.82 + 28.446037 \times 58.04 \\ & - 5.162262 \times 51.75 + 0.182663 \times 43.92 - 0.001023 \times 35.65 + 1.463783 \\ & \times 226.59 - 0.556830 \times 280.05 + 0.024890 \times 292.69 - 0.000227 \\ & \times 264.88 = -1216.33 \end{aligned}$$

An initial guess for the hour 17 surface temperature must be made; the previous hour's surface temperature of 61.82 °C is a reasonable starting point. Using that value gives the following values for convection and radiation coefficients:

$$h_c = 7.91 \text{ W/m}^2\text{K, from Equation 2.16,}$$
$$h_{r,sky} = 3.25 \text{ W/m}^2\text{K, from Equation 2.34,}$$
$$h_{r,grd} = 2.95 \text{ W/m}^2\text{K, from Equation 2.35.}$$

Then, applying the heat balance, Equation 2.56:

$$T_{os,j,\theta} = \frac{Y_o T_{is,j,\theta} - H_{out,j,\theta} + \alpha G_t + h_c T_o + h_{r-g} T_g + h_{r-sky} T_{sky}}{X_o + h_c + h_{r-g} + h_{r-sky}}$$

$$T_{os,j,\theta} = \frac{3.375092 \cdot 10^{-5}(21.91) - (-1216.33) + 0.8(760.9) + 7.91(29) + 2.95(29) + 3.25(24.79)}{21.7157 + 7.91 + 2.95 + 3.25}$$

$$= 62.00°\text{C}$$

The new estimate of surface temperature is so close to the original estimate that, to three significant figures, the convection and radiation coefficients do not change. Further iteration gives a final estimate of the surface temperature as 61.99 °C and the conduction flux can be calculated with Equation 2.55:

$$q''_{conduction,j,\theta} = -Y_o T_{is,j,\theta} + X_o T_{os,j,\theta} + H_{out,j,\theta} = 3.375092 \cdot 10^{-5}(21.91) + 21.7157(62.0) - 1216.33 = 129.9 \text{ W/m}^2.$$

End of solution.

Interior surface heat balance

Much like the outside surface heat balance, the inside surface heat balance insures that the heat transfer due to absorbed solar heat gain, convection, and long wavelength radiation is balanced by the conduction heat transfer. Again, this comes about by solving for the surface temperature that results in a heat balance being achieved. This section will discuss how each heat transfer mechanism is modeled. For each mechanism, there are a number of possible models that could be used, ranging from very simple to very complex. We will again follow the approach of selecting a reasonably simple model for each heat transfer mechanism and referring the reader to other sources for more sophisticated and accurate models.

Like the exterior surface heat balance, the interior surface heat balance may be formulated to solve for a specific surface temperature. First, a history term that contains all of the historical terms for the interior CTF equation should be defined:

$$H_{in,j,\theta} = -\sum_{n=1}^{N_z} Z_n T_{is,j,\theta-n\delta} + \sum_{n=1}^{N_y} Y_n T_{os,j,\theta-n\delta} + \sum_{n=1}^{N_q} \Phi q''_{conduction,in,j,\theta-n\delta} \quad (2.57)$$

and then Equation 2.11 may be represented as:

$$q''_{conduction,in,j,\theta} = -Z_o T_{is,j,\theta} + Y_o T_{os,j,\theta} + H_{in,j,\theta} \quad (2.58)$$

Furthermore, the net radiation leaving the surface is the surface-to-surface radiation minus the radiation due to internal heat gains:

$$q''_{radiation,in,j,\theta} = q''_{radiation-surf,in,j,\theta} - q''_{radiation,ihg,j,\theta} \tag{2.59}$$

Then, by substituting the expressions for conduction heat flux (Equation 2.11), convection heat flux (Equation 2.18), radiation heat flux (Equations 2.38–2.45), and absorbed solar heat gain (Equation 2.29 or 2.30) into the interior surface heat balance (Equation 2.47), and solving for the interior surface temperature, we obtain:

$$T_{is,j,\theta} = \frac{q''_{solar,in,j,\theta} + Y_o T_{os,j,\theta} + H_{in,j,\theta} + h_c T_i + h_{r,j} T_{f,j} + q''_{balance} + q''_{radiation-ihg,in,j,\theta}}{Z_o + h_c + h_{r,j}} \tag{2.60}$$

Note that $h_{r,j}$, $T_{f,j}$, and $q''_{balance}$ all depend on the other surface temperatures as well as on $T_{is,j,\theta}$. With a more sophisticated convection model, h_c might also depend on the surface temperature. As in the exterior heat balance, it is convenient to solve the equations iteratively with successive substitution.

Surface heat balances – transparent and semi-transparent surfaces

The heat balance on windows must be treated differently than the heat balances on walls and roofs. The primary reason for this is that solar radiation may be absorbed throughout the window rather than just at the interior and exterior surfaces. This can lead to some rather arduous calculations, so we will make some simplifying assumptions:

- A window contains very little thermal mass, so we will assume that it behaves in a quasi-steady-state mode.
- Most of the overall thermal resistance of a window comes from the convective and radiative resistances at the interior and exterior surfaces and (if a multiple-pane window) between the panes. The conductive resistance of the glass or other glazing materials is quite small in comparison. Therefore, we will neglect the conductive resistance of the glass itself.
- Neglecting the conductive resistance causes each layer to have a uniform temperature. Therefore, there will be a single heat balance equation for each layer rather than an interior and an exterior surface heat balance equation for the entire window system.
- Layer-by-layer absorptance data may not generally be available for actual windows. If not, the engineer will have to make an educated guess as to the distribution of absorbed solar radiation in each layer.

Consider the thermal network for a double pane window shown in Figure 2.9. It has incident solar radiation from the outside, $q''_{solar,out,j,\theta}$ and solar radiation incident from the inside, $q''_{solar,in,j,\theta}$. The solar radiation incident from the inside was transmitted through a window and possibly reflected before striking the inside surface of the window. For both solar radiation fluxes, a certain amount is absorbed by both panes.

In addition to the heat transfer modes addressed earlier in this chapter, Figure 2.9 also shows radiation and convection heat transfer between the panes. The radiative resistance may be estimated from a simple radiation network analysis as:

$$R_R = \frac{\left[\frac{1-\varepsilon_1}{\varepsilon_1} + 1 + \frac{1-\varepsilon_2}{\varepsilon_2}\right]}{4\sigma T_m^3} \tag{2.61}$$

where:

$\varepsilon_1, \varepsilon_2$ = emissivities of the two window surfaces, dimensionless;
σ = Stefan-Boltzmann constant, 5.67×10^{-8} W/m²K⁴;
T_m = mean temperature of the two window surfaces, K.

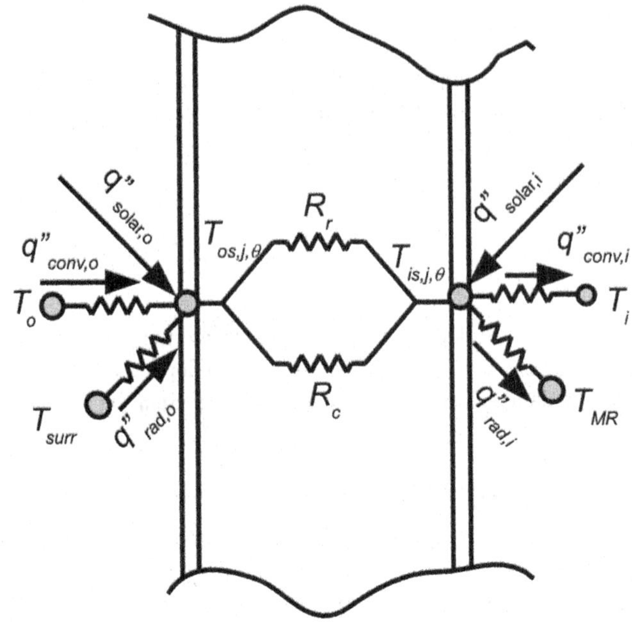

Figure 2.9 Heat balance for a double pane window

The convection resistance may be determined with a suitable correlation, such as that given by Wright (1996). Wright's correlation was developed for cavity height-to-width ratios greater than 40 and Rayleigh numbers less than 10^6, but it should be usable for any practical window configuration. The Nusselt number is defined based on the cavity width:

$$\mathrm{Nu} = h_c \cdot \frac{l}{k} \quad \text{or} \quad h_c = \mathrm{Nu} \cdot \frac{k}{l} \tag{2.62}$$

where:

l = cavity width or spacing between the panes of glass, m;
k = thermal conductivity of the fill gas, W/mK.

Rayleigh number is also defined based on the cavity width:

$$\mathrm{Ra} = \frac{\rho^2 l^3 g c_p \Delta T}{\mu k T_m} \tag{2.63}$$

where:

ρ = fill-gas density, kg/m^3;
c_p = fill-gas specific heat, J/kgK;
μ = fill-gas dynamic viscosity Ns/m^2;
T_m = mean temperature of fill gas, K;
g = acceleration due to gravity, m/s^2.

Then, the Nusselt number is correlated with the following equations, depending on the value of the Rayleigh number:

$$\mathrm{Nu} = 0.0673838 \cdot \mathrm{Ra}^{(1/3)} \qquad \mathrm{Ra} > 5 \times 10^4 \tag{2.64a}$$

$$\text{Nu} = 0.028154 \cdot \text{Ra}^{0.4134} \qquad 10^4 < \text{Ra} \le 5 \times 10^4 \tag{2.64b}$$

$$\text{Nu} = 1 + 1.75967 \times 10^{-10} \cdot \text{Ra}^{2.2984755} \qquad \text{Ra} \le 10^4 \tag{2.64c}$$

Once the convection coefficient is found, the convective resistance is:

$$R_C = \frac{1}{h_c} \tag{2.65}$$

Both the radiative resistance and convective resistance may be adjusted according to the surface temperatures. A *U*-factor may be defined:

$$U_{airspace} = \frac{1}{R_R} + \frac{1}{R_C} \tag{2.66}$$

As done previously with convection and resistance coefficients, the value of $U_{airspace}$ may be updated between iterations.

The heat balance for each pane may then be defined in a manner analogous to the heat balances previously developed for opaque exterior surfaces and opaque interior surfaces. The heat balance for each pane may then be cast in a form to solve for the pane temperatures:

$$T_{os,j,\theta} = \frac{q''_{absorbed,outer,j,\theta} + U_{airspace} T_{is,j,\theta} + h_{c,o} T_o + h_{r-g} T_g + h_{r-sky} T_{sky}}{U_{airspace} + h_{c,o} + h_{r-g} + h_{r-sky}} \tag{2.67}$$

$$T_{is,j,\theta} = \frac{q''_{absorbed,inner,j,\theta} + U_{airspace} T_{os,j,\theta} + h_{c,i} T_i + h_{rad,2,j} T_{MR,2} + q''_{radiation-ihg,in,j,\theta}}{U_{airspace} + h_{c,i} + h_{rad,2,j}} \tag{2.68}$$

A window with more panes would be analyzed in the same manner, but there would be an additional heat balance equation for each additional pane. Likewise, for a single pane window, there would only be a single heat balance equation:

$$T_{os,j,\theta} = T_{is,j,\theta} = \frac{q''_{absorbed,j,\theta} + h_{c,o} T_o + h_{r-g} T_g + h_{r-sky} t_{sky} + h_{c,i} T_i + h_{rad,2,j} T_{MR,2} + q''_{radiation-ihg,in,j,\theta}}{h_{c,o} + h_{r-g} + h_{r-sky} + h_{c,i} + h_{rad,2,j}} \tag{2.69}$$

Example 2.11

The 1.2 m high and 1.2 m wide window in Example 2.7, at 3 p.m. PDST on July 21 in Seattle, absorbed 141.5 W/m² in the outer pane and 41.5 W/m² in the inner pane. The rated *U*-factor for the window is 1.66 W/m²K. At that time, the following conditions may be assumed to apply:

- outdoor air temperature and ground temperature is 28.1 °C,
- the effective sky temperature is 23.9 °C,
- the wind speed is 3.6 m/s,
- mean radiant temperature of zone is 21.8 °C,
- the air temperature of the zone is 21.5 °C,
- all of the window surfaces have emissivities of 0.84, except the inside of the outer pane, which has a low-E coating and an emissivity of 0.033,
- there are no internal heat gains (e.g. lighting, equipment, occupants),

- the interior convection coefficient from Table 2.6 is 3.18 W/m²K, and
- the interior radiation coefficient is 6.35 W/m²K.

Note that several of the quantities would, in an actual simulation, be determined simultaneously with the window surface temperatures and heat flux.

Find the interior and exterior surface temperatures and convective and radiative heat gains from the window.

Solution:

Equations 2.67 and 2.68 represent the heat balance equations for the outside and inside panes, respectively. By stipulating values for the mean radiant temperature and air temperature in the zone, as well as the interior radiative and convective coefficients, the problem has been simplified so that these two heat balances can be considered independent of other surfaces and the zone air. As in most heat balance analyses, initial guesses must be made for surface temperatures; arbitrarily, we will choose 50 °C for the outer pane and 25 °C for the inner pane. Using these guesses gives the following values for convection and radiation coefficients:

$$h_c = 7.80 \text{ W/m}^2\text{K, from Equation 2.16,}$$
$$h_{r,sky} = 3.05 \text{ W/m}^2\text{K, from Equation 2.34,}$$
$$h_{r,grd} = 3.11 \text{ W/m}^2\text{K, from Equation 2.35,}$$
$$U_{airspace} = 2.44 \text{ W/m}^2\text{K, from Equation 2.66.}$$

Then, Equation 2.67 gives the outer pane temperature:

$$T_{os,j,\theta} = \frac{q''_{absorbed,outer,j,\theta} + U_{airspace}T_{is,j,\theta} + h_{c,o}T_o + h_{r-g}T_g + h_{r-sky}T_{sky}}{U_{airspace} + h_{c,o} + h_{r-g} + h_{r-sky}}$$

$$= \frac{141.46 + 2.44(25) + 7.8(28.1) + 3.11(28.1) + 2.44(23.9)}{2.44 + 7.8 + 3.11 + 2.44} = 35.47\ ^\circ\text{C}$$

and Equation 2.68 gives the inner pane temperature:

$$T_{is,j,\theta} = \frac{q''_{absorbed,inner,j,\theta} + U_{airspace}T_{os,j,\theta} + h_{c,i}T_i + h_{rad,2,j}T_{MR,2} + q''_{radiation-ihg,in,j,\theta}}{U_{airspace} + h_{c,i} + h_{rad,2,j}}$$

$$= \frac{41.46 + 2.44(50) + 3.18(21.5) + 6.35(21.8) + 0}{2.44 + 3.18 + 6.35} = 30.94\ ^\circ\text{C}$$

With the new estimates of the pane temperatures, the radiation and convection coefficients are:

$$h_c = 7.62 \text{ W/m}^2\text{K,}$$
$$h_{r,sky} = 2.83 \text{ W/m}^2\text{K,}$$
$$h_{r,grd} = 2.89 \text{ W/m}^2\text{K,}$$
$$U_{airspace} = 2.26 \text{ W/m}^2\text{K.}$$

Solving Equations 2.67 and 2.68 again gives an outer pane temperature of 36.80 °C and an inner pane temperature of 27.86 °C. Then, with a few more iterations, the coefficients converge to:

$h_c = 7.63\ W/m^2K$,

$h_{r,sky} = 2.85\ W/m^2K$,

$h_{r,grd} = 2.91\ W/m^2K$,

$U_{airspace} = 2.26\ W/m^2K$.

Final values of the outer and inner pane temperatures are 36.36 °C and 28.03 °C, respectively. A final check on the window heat balance shows the following:

- The incident solar radiation is 680 W/m^2.
- Of that, 228.5 W/m^2 are transmitted, 182.9 W/m^2 are absorbed by the two panes, and 268.7 W/m^2 are reflected back outside.
- The 182.9 W/m^2 absorbed is distributed as follows:
  - convected to the outside: 63.0 W/m^2,
  - radiated to the sky: 35.6 W/m^2,
  - radiated to the ground: 24.0 W/m^2,
  - convected to the zone air: 20.8 W/m^2,
  - radiated to the other zone surfaces: 39.5 W/m^2, and
  - $63+35.6+24+20.8+39.5 = 182.9\ W/m^2$.

End of solution.

Zone air heat balance

Finally, with the assumption that the zone air has negligible thermal storage capacity, a heat balance on the zone air may be represented conceptually as

$$\sum_{j=1}^{N} A_j q''_{convection,in,j,\theta} + q_{infiltration,\theta} + q_{system,\theta} + q_{internal,conv,\theta} = 0 \qquad (2.70)$$

where:

A_j = area of the jth surface, m^2;
$q_{infiltration,\theta}$ = heat gain due to infiltration, W;
$q_{system,\theta}$ = heat gain due to the heating/cooling system, W;
$q_{internal,conv,\theta}$ = convective portion of internal heat gains due to people, lights, or equipment, W.

The heat balance may be cast in several forms:

- Solving for the required system capacity to maintain a fixed zone air temperature.
- Solving for the zone temperature when the system is off.
- Solving for the zone temperature and system capacity with a system that does not maintain a fixed zone air temperature.

For the purposes of design cooling load calculations, the first formulation is usually of the most interest. The second formulation may be useful when modeling setback conditions or to help determine thermal comfort for naturally cooled buildings. The third formulation is the most general – with a fairly simple model of the system, it is possible to model the first condition (by specifying a system with a very large capacity) or to model the second condition (by specifying a system with zero capacity). Also, while the first formulation is suitable for determining required system airflow rates and cooling coil capacities, it may be desirable to base the central plant equipment sizes on actual heat extraction rates. A more detailed discussion of the subject of HVAC system performance prediction can be found in Chapter 15.

Before each formulation is covered, each of the heat transfer components will be briefly discussed.

Convection from surfaces

Convection from individual surfaces is described with Equation 2.18. The total convection heat transfer rate to the zone air is found by summing the contribution from each of the N surfaces:

$$\dot{q}_{convection,in,j,\theta} = \sum_{j=1}^{N} A_j q''_{convection,in,j,\theta} = \sum_{j=1}^{N} A_j h_{c,i,j}(T_{is,j,\theta} - T_i) \quad (2.71)$$

Convection from internal heat gains

Convection from internal heat gains is found by summing the convective portion of each individual internal heat gain:

$$\dot{q}_{internal,conv,\theta} = \sum_{j=1}^{M} \dot{q}_{j,\theta} F_{con,j} \quad (2.72)$$

where:

$\dot{q}_{internal,conv,\theta}$ = convective heat transfer to the zone air from internal heat gains, W;
$\dot{q}_{j,\theta}$ = heat gain for the jth internal heat gain element, W;
$F_{con,j}$ = convective fraction for the jth internal heat gain element.

Although the convective fraction for any heat gain element will vary with its surface temperature, the radiative environment, local airflow, and other factors, it is often treated as a constant for specific heat gains. Typical values are tabulated by Spitler (2014) and ASHRAE (2017).

Heat gain from infiltration

Methods used to estimate the quantity of infiltration air are discussed in Chapter 3. The same methods apply to cooling load calculations. Both a sensible and latent heat gain will result and are computed as follows:

$$\dot{q}_{infiltration,\theta} = \dot{m}_{a,infiltration,\theta} c_p (T_o - T_i) = \frac{\dot{Q} c_p}{v_o}(T_o - T_i) \quad (2.73)$$

$$\dot{q}_{infiltration,latent,\theta} = \dot{m}_{a,infiltration,\theta}(W_o - W_i) i_{fg} = \frac{\dot{Q}}{v_o}(W_o - W_i) i_{fg} \quad (2.74)$$

Zone air heat balance formulations

The simplest formulation of the zone air heat balance is to determine the cooling load – i.e. for a fixed zone air temperature, determine the required system heat transfer. In this case, Equations 2.71, 2.72, and 2.73 can be substituted into Equation 2.70 to give:

$$\dot{q}_{system,\theta} = -\sum_{j=1}^{N} A_j h_{c,i,j} \left(T_{is,j,\theta} - T_i\right) - \dot{m}_{a,infiltration} c_p \left(T_o - T_i\right) - \dot{q}_{internal,conv,\theta} \tag{2.75}$$

Likewise, the zone air heat balance can be formulated to determine the instantaneous zone temperature when there is no system heat transfer. Setting the system heat transfer rate in Equation 2.75 equal to zero and solving for the zone air temperature gives:

$$T_i = \frac{\sum_{j=1}^{N} A_j h_{c,i,j} (T_{is,j,\theta}) + \dot{m}_{a,infiltration} c_p T_o + \dot{q}_{internal,conv,\theta}}{\sum_{j=1}^{N} A_j h_{c,i,j} + \dot{m}_{a,infiltration} c_p} \tag{2.76}$$

Finally, the zone air heat balance can be formulated to determine the zone temperature when there is system heat transfer. Substituting the piecewise linear expression for system capacity in Equation 2.22 into the zone air heat balance (Equation 2.70) and solving for the zone air temperature gives

$$T_i = \frac{a + \sum_{j=1}^{N} A_j h_{c,i,j} \left(T_{is,j,\theta}\right) + \dot{m}_{a,infiltration} c_p T_o + \dot{q}_{internal,conv,\theta}}{-b + \sum_{j=1}^{N} A_j h_{c,i,j} + \dot{m}_{a,infiltration} c_p} \tag{2.77}$$

Note that the control profile coefficients depend on the value of the zone air temperature. Therefore, it is usually necessary to choose a and b based on an intelligent guess of the zone air temperature. Then, using those values of a and b, solve Equation 2.77 for T_i. If the value of T_i is not within the range for which a and b were chosen, then another iteration must be made.

Example 2.12

The zone shown in Figure 2.6 is conditioned by a variable air volume (VAV) system. The VAV system has a maximum flow of 140 L/s of air at 15 °C at zone air temperatures of 25 °C and higher. The minimum fraction the VAV terminal unit is 0.3, which occurs at zone air temperatures of 22 °C and lower. Between 22 °C and 25 °C, the airflow is modulated linearly. Below 20 °C, reheat is provided. At a certain time, the outdoor air temperature is 28.1 °C, the infiltration rate is 6 L/s, the surface temperatures and convection coefficients are as shown in Table 2.12, and other heat gains to the space are summarized in Table 2.13. Find the zone air temperature and cooling provided by the system.

Table 2.12 Zone surface temperatures and convection coefficients

Surface number	*Surface name*	*Area (m²)*	*Temperature (°C)*	*h_c (W/m²K)*
1	Floor	24.0	22.3	4.15
2	North wall	18.0	24.1	3.18
3	East wall	12.0	23.7	3.18
4	West wall	12.0	23.8	3.18
5	Ceiling	24.0	25.3	1.02
6	Window	7.5	30.9	3.18
7	South wall (surrounding window)	10.5	24.3	3.18

Table 2.13 Other convective heat gains

Internal heat gain type	*Convective portion (W)*
Lighting	57
Equipment	207
People	117
$\dot{q}_{internal,conv,\theta}$	381

Solution:

First, it might be noted that the surface temperatures in Table 2.12 are not generally known in advance, but rather are determined simultaneously with the zone air temperature. Then, the first step in solving the problem is to develop a piecewise linear profile for the system heat transfer rate, in the form shown in Equation 2.22. The zone air heat balance, Equation 2.70, is developed with the convention that heat transfer from the system to the zone air is positive when heat is being added to the zone.

The system heat transfer rate can be computed for three ranges:

$T_i > 25$ °C, where the VAV terminal unit will be fully open.
25 °C $\geq T_i \geq$ 22 °C, where the VAV terminal unit is modulated between full open and 30% of full flow.
22 °C $> T_i \geq$ 20 °C, where the VAV terminal unit is at the minimum fraction.

As the supply air temperature is assumed to be held constant at 15 °C, regardless of zone air temperature, the system heat transfer may be computed at two points within each of the above three ranges, and then the coefficients of the linear equation giving system heat transfer rate as a function of zone air temperature may be determined. For each point, the heat transfer rate is determined with Equation 2.78:

$$\dot{q}_{system,\theta} = \dot{m}_a c_p (T_{SA} - T_i) \tag{2.78}$$

where T_{SA} is the supply air temperature, °C.

With the air density taken as 1.22 kg/m³ and the specific heat of the air taken as 1006 J/kgK, the heat transfer rates shown in Table 2.14 can be computed.

Fitting a line for each of the temperature ranges gives the set of a and b coefficients for the piecewise linear representation of the system heat transfer, often referred to as the system control profile, shown in Table 2.15.

Before performing the zone air heat balance, the infiltration mass flow rate and specific heat must be determined. Based on the outdoor air temperature, 28.1 °C, the density may be taken as 1.22 kg/m³, giving a mass flow rate of 0.007 kg/s, and the specific heat is 1006 J/kgK.

The zone air heat balance, as formulated in Equation 2.77, has two summations that can be computed first. Both summations utilize values taken from Table 2.12. The first summation is:

$$\sum_{j=1}^{N} A_j h_{c,i,j} T_{is,j,\theta} = (24 \times 4.15 \times 22.3) + (18 \times 3.18 \times 24.1) + \ldots = 7580.9$$

The second summation is similar to the first, but without the surface temperatures:

$$\sum_{j=1}^{N} A_j h_{c,i,j} = (24 \times 4.15) + (18 \times 3.18) + \ldots = 314.9$$

Table 2.14 System heat transfer rate

System state	T_i *(°C)*	*Flow (L/s)*	$\dot{q}_{system,\theta}$, *W*
VAV TU fully open	30	140	−2579
VAV TU fully open	25	140	−1719
VAV TU min. frac	22	42	−361
Reheat comes on	20	42	−258

Table 2.15 System control profile

Temperature range	*a*	*b*
$T_i > 25$ °C	2579.12	−171.9411
25 °C $\geq T_i \geq$ 22 °C	9600.04	−452.7782
22 °C $> T_i \geq$ 20 °C	773.73	−51.5823

$$T_i = \frac{a + \sum_{j=1}^{N} A_j h_{c,i,j}\left(T_{is,j,\theta}\right) + \dot{m}_{a,infiltration} c_p T_o + \dot{q}_{internal,conv,\theta}}{-b + \sum_{j=1}^{N} A_j h_{c,i,j} + \dot{m}_{a,infiltration} c_p}$$

With numerical values determined for all of the terms in the zone air heat balance, as formulated in Equation 2.77, can now be applied to determine the zone air temperature. It is necessary to first guess which zone temperature range applies – an initial guess that the temperature is between 22 °C and 25 °C will be used. The zone air heat balance is then:

$$T_i = \frac{a + \sum_{j=1}^{N} A_j h_{c,i,j}\left(T_{is,j,\theta}\right) + \dot{m}_{a,infiltration} c_p T_o + \dot{q}_{internal,conv,\theta}}{-b + \sum_{j=1}^{N} A_j h_{c,i,j} + \dot{m}_{a,infiltration} c_p}$$

$$= \frac{9600.04 + 7580.9 + (0.007 \times 1006 \times 28.1) + 381}{-(-452.7782) + 314.9 + (0.007 \times 1006)}$$

= 22.92 °C. So, our initial guess was correct; if it had not been, we would have tried the range suggested by the resulting temperature.

With a temperature of 22.92 °C, the zone load can be computed from Equation 2.70, or more simply from Equation 2.22:

$$\dot{q}_{system,\theta} = a + bT_i = 9600.04 - 452.7782(22.92) = -778 \text{ W}$$

The system will provide 778 W of cooling or remove 778 W from the space.

End of solution.

Zone models: thermal network models

Thermal network models (Lewis and Alexander 1990; Sowell 1990; Walton 1993; Clarke 2001) discretize the building into a network of nodes with interconnecting energy flow paths.

The energy flow paths may include conduction, convection, radiation, and airflow. Thermal network models may be thought of as heat balance models with finer granularity. Where heat balance models generally have one node representing the zone air, a thermal network model may have several. Where heat balance models generally have a single exterior node and a single interior node per wall, thermal network models may have additional nodes. Where heat balance models generally distribute radiation from lights in a simple manner, thermal network models may model the lamp, ballast, and luminaire housing separately. Thermal network models are the most flexible of the three zone models discussed here. However, the added flexibility requires more computational time and, often, more user effort.

Furthermore, at the present time, most implementations of thermal network models are primarily research tools. One program that is more widely used is ESP-r (ESRU 2017), which has been under development by Professor Joe Clarke and colleagues at the University of Strathclyde since the 1970s. A few of the features made possible by the thermal network approach include 2-D and 3-D conduction heat transfer (Strachan et al. 1995), integrated computational fluid dynamics analysis (Clarke et al. 1995), and phase-change materials (Heim and Clarke 2003).

Implementation of thermal network models in advanced simulation environments (Riederer 2005; Sasic Kalagasidis et al. 2007; Zuo et al. 2016; Wetter and van Treeck 2017) remains an active area of ongoing research and development. As this was written, IBPSA (2017) was undertaking an international collaborative research project to expand building and HVAC models in Modelica.

Zone models: transfer function models

An alternative to the heat balance models and thermal network models are the transfer function models. Within this category, we may include models that use room transfer functions or room response factors to determine cooling loads from the heat gains. Examples of implementations include the DOE2 (York and Cappiello 1981) program, the ASHRAE transfer function method (TFM) (McQuiston and Spitler 1992), and the radiant time series method (RTSM) (Spitler et al. 1997; Spitler 2014) cooling load calculation procedures.

All of these models make certain simplifications to the heat balance model. These include:

- There is no exterior heat balance. Instead of modeling convection to the outdoor air, radiation to the ground and sky, and solar radiation separately, they are modeled as a single heat transfer between an "equivalent" temperature, known as the sol-air temperature, and the surface temperature. This allows the resistance between the sol-air temperature and the surface temperature to be included as a resistance in the transient conduction analysis, and it allows the exterior driving temperature for the transient conduction analysis to be determined prior to the load calculation. This has the limitation that a single fixed combined convection and radiation coefficient must be used, independent of the surface temperature, sky temperature, air temperature, wind speed, etc.
- There is no interior surface heat balance. Instead, for radiation purposes, it is assumed that the other surfaces in the zone are effectively at the zone air temperature. Then, a single, fixed value of the surface conductance is used and folded into the transient conduction analysis.
- There is no zone air heat balance. Cooling loads are determined directly, but the zone air temperature is assumed to be constant. A second set of transfer functions can be used to estimate the heat extraction rate.
- The storage and release of energy by the walls, roofs, floors, and internal thermal mass are approximated with a predetermined zone response. Unlike the heat balance model, this

phenomenon is considered independently of the through-the-wall conduction heat transfer. This has implications for the accuracy of the calculation. For cooling load calculations, it usually leads to a small overprediction of the cooling load. For zones with large quantities of high-conductance surfaces (e.g. glass), it results in a significant overprediction, as described by Rees et al. (1998) and Spitler and Nigusse (2010) for the RTSM. The latter reference describes a correction procedure that ameliorates this problem.

This also has implications for presentation of results, as it makes it possible to identify the individual heat gains and cooling loads on a component-by-component basis. For example, see Figure 2.10, where lighting heat gains for a conference room in an office building are plotted alongside the resulting cooling loads. The cooling loads show the delay and damping effects caused by some of the lighting heat gain being absorbed in the room structure and later released.

Each of the simplifications lead to common features of transfer function models – combined convection and radiation heat transfer, sol-air temperatures, and room transfer functions or response factors. Each of these features are discussed later, followed by a brief discussion of space air transfer functions, which allow the effects of system capacity and control to be included in a prediction of the actual room air temperature and actual heat extraction rate of the system.

Combined convection and radiation heat transfer

An important feature of transfer function models is that convection and radiation analysis are often combined into a single surface conductance. This requires a linear approximation of the radiation heat transfer. Equation 2.23 quantifies the radiation heat transfer between two surfaces that only view each other. Considering that surface 2 in Equation 2.23 could represent, in aggregate, the surroundings, then approximately $A_2 >> A_1$ and Equation 2.23 can be simplified to:

$$\dot{q}_{1-2} = A_1 \varepsilon_1 \sigma \left(T_1^4 - T_2^4 \right) = A_1 h_r (T_1 - T_2) \tag{2.79}$$

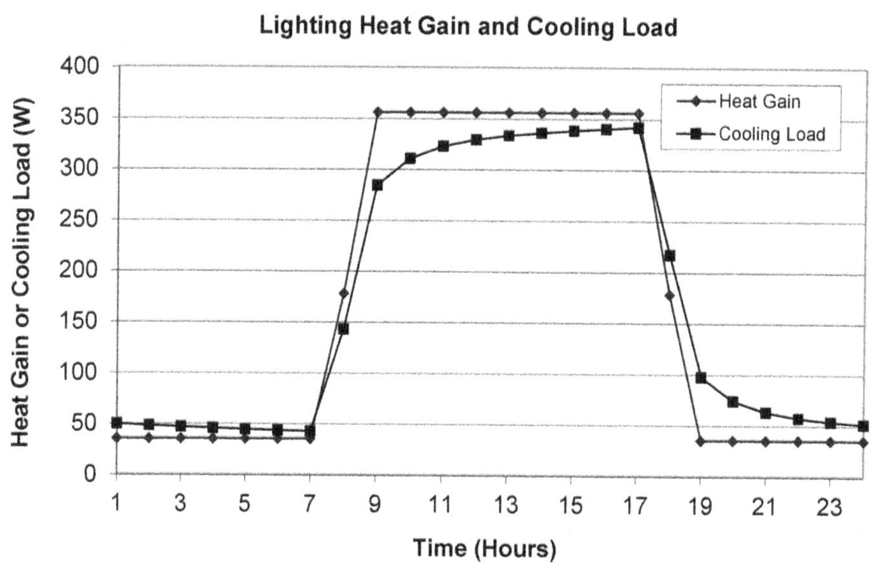

Figure 2.10 Lighting heat gain and the resulting cooling load for a conference room

$$h_r = \varepsilon_1\sigma(T_1^2 + T_2^2)(T_1 + T_2) \approx 4\varepsilon_1\sigma\bar{T}^3 \tag{2.80}$$

where:

h_r = radiation coefficient, W/m²K.

Note that while the linearized radiation coefficient is calculated with absolute temperatures, it is applied to a temperature difference, which is the same whether it is taken in Kelvin or degrees Celsius:

$$\dot{q}_r = h_r A(T - T_w) \tag{2.81}$$

A surface conductance, which combines the radiation and convection coefficient, can then be defined as:

$$h_o = h_c + h_r \tag{2.82}$$

This has the advantage of allowing the combined convection and radiation heat transfer rate to be expressed as:

$$\dot{q} = h_o A(T - T_w) \tag{2.83}$$

Or, the inverse of the surface conductance can be treated as a conductive resistance, simplifying the analysis so that it appears as simply a conduction analysis. For the interior conductance, the indoor air temperature is taken as a suitable approximation to the actual effective temperature with which convection and radiation are exchanged. For the exterior conductance, an equivalent temperature, the sol-air temperature, is utilized, as described in the next section.

Sol-air temperature

In the case of the exterior conductance, the problem is further complicated by the presence of solar radiation. For simplified load calculation procedures, an equivalent air temperature, the sol-air temperature, is defined that gives approximately the same heat flux to the surface as the combined effects of solar radiation, convection, and radiation to the surroundings. It is given by:

$$T_e = T_o + \alpha G_t / h_o - \varepsilon\delta R / h_o \tag{2.84}$$

where:

T_e = sol-air temperature, °C;
T_o = outdoor air temperature, °C;
α = absorptance of the surface to solar radiation;
G_t = total incident solar radiation flux, W/m²;
h_o = exterior surface conductance, W/m²K;
ε = exterior surface emittance;
δR = difference between the thermal radiation incident on the surface from the sky or surroundings and the radiation emitted by a blackbody at the outdoor air temperature, W/m²K.

The last term, $\varepsilon\delta R/h_o$, is often taken as 4 °C for horizontal surfaces (i.e. facing the sky) and 0 °C for vertical surfaces.

Room transfer functions and response factors

Room transfer functions and response factors relate the hourly heating and cooling loads due to individual types of heat gains to previous values of that type of heat gain and previous values of the cooling load due to that type of heat gain.

For example, consider a particular zone with equipment turned on for a certain fraction of the day. The equipment heat gain is partially radiant and partially convective. The convective portion immediately becomes space cooling load; the radiant portion is absorbed by thermal mass in the zone and later convected to the room air. If this zone is modeled with a set of heat balance equations and pulsed with a one-hour unit heat gain, the space cooling load each hour will be a fractional number indicating how much of the original pulse will be convected into the room air each hour. This set of fractional numbers is a set of thermal response factors (Mitalas and Stephenson 1967).

Room thermal response factors can be computed with a heat balance program. However, a more computationally efficient form is often desired, and this involves computation of a set of room transfer functions, sometimes called weighting factors. One procedure for computation of is described by the *DOE-2 Engineer's Manual* (York and Cappiello 1981). For any zone, there are many possible combinations of heat gain terms and cooling load terms that may be used. A typical form is:

$$Q_\theta = v_0 q_\theta + v_1 q_{\theta-\delta} + v_2 q_{\theta-2\delta} - w_1 Q_{\theta-\delta} - w_2 Q_{\theta-2\delta} \tag{2.85}$$

where:

Q_θ = cooling or heating load at hour θ, W;
q_θ = heat gain at hour θ, W;
δ = time interval, typically one hour.

The terms v_0, v_1 . . . , w_1, w_2 . . . , are the coefficients of the room transfer function:

$$K(z) = \frac{v_0 + v_1 z^{-1} + v_2 z^{-2}}{1 + w_1 z^{-1} + w_2 z^{-2}} \tag{2.86}$$

which relates the transform of the corresponding parts of the cooling load and of the heat gain. These coefficients depend of (1) the time interval of δ, (2) the nature of the heat gain (fraction radiant and location in room), and (3) the thermal characteristics of the room.

The RTSM cooling load calculation method is similar in many respects to the TFM, but it assumes that the design day cooling load calculation is for a single day, with the previous days having the same conditions. In other words, any energy stored in the building overnight will be consistent with the previous days having been identical in weather and internal heat gains. In the RTSM, the radiant portion of the heat gains is converted to cooling loads using radiant time factors, a 24-term periodic response factor series.

The radiant time series for a particular zone gives the time-dependent response of the zone to a single steady periodic pulse of radiant energy. The series shows the portion of the radiant pulse that is convected to the zone air for each hour. Thus, r_0 represents the fraction of the radiant pulse convected to the zone air in the current hour, r_1 in the last hour, and so on. The radiant time series thus generated is used to convert the radiant portion of hourly heat gains to hourly cooling loads according to Equation 2.87.

$$Q_\theta = r_0 q_\theta + r_1 q_{\theta-\delta} + r_2 q_{\theta-2\delta} + r_3 q_{\theta-3\delta} + \ldots + r_{23} q_{\theta-23\delta} \tag{2.87}$$

where:

Q_θ	=	cooling load (Q) for the current hour (θ);
q_θ	=	heat gain for the current hour;
$q_{\theta-v\delta}$	=	heat gain n hours ago;
r_0, r_1, etc.	=	radiant time factors.

Space air transfer functions

To this point, the transfer function models have only considered determination of heating load or cooling load, i.e. the heat transfer rate required to maintain a constant room temperature. However, room temperature might vary for a number of reasons, and it is desirable to have a way of accounting for variations in room temperature, as well as predicting room temperature in some cases.

The heat extraction characteristics of the system are approximated by a simple linear expression, similar to a single segment of the control profile described for the heat balance model:

$$ER_\theta = W_\theta + \left(ST_{r\theta}\right) \tag{2.88}$$

where:

ER_θ	=	rate of heat removal from the space at time θ, W;
$T_{r\theta}$	=	actual air temperature in the space at time θ, °C;
W	=	intercept characterizing the performance of cooling equipment, W;
S	=	slope characterizing the performance of cooling equipment, W/°C.

This linear relationship is expected to hold over the throttling range of the control system.

The heat extraction rate and the room air temperature are related by the space air transfer function:

$$\sum_{i=0}^{1} p_i\left(ER_{\theta-i\delta} - Q_{\theta-i\delta}\right) = \sum_{i=0}^{2} g_i\left(T_{rc} - T_{r\theta-i\delta}\right) \tag{2.89}$$

where:

g_i and p_i	=	coefficients of the space air transfer function;
Q_θ	=	calculated cooling load for room at time θ, based on an assumed constant room temperature of T_{rc}, W.

Previous equations for heat extraction rate (2.88) and the space air transfer function (2.89) can be solved simultaneously to yield:

$$ER_\theta = \left(W_\theta g_0 + I_\theta S\right) / \left(S + g_0\right) \tag{2.90}$$

where:

$$I_\theta = T_{rc}\sum_{i=0}^{2} g_{i,\theta} - \sum_{i=1}^{2} g_{i,\theta}T_{r\theta-i\delta} + \sum_{i=0}^{1} p_i Q_{r\theta-i\delta} - \sum_{i=0}^{1} p_i ER_{\theta-i\delta} \tag{2.91}$$

If the value of ER_θ calculated by Equation 2.90 results in a value greater than $ER_{\max}$, ER_θ is set equal to $ER_{\max}$. If the value of ER_θ calculated by results in a value less than $ER_{\min}$, ER_θ is set equal to $ER_{\min}$. Then $T_{r\theta}$ is calculated from the expression:

$$T_{r\theta} = \left(I_\theta - ER_\theta\right) / g_{0,\theta} \tag{2.92}$$

Note that Equation 2.90 requires some prior values of ER_θ and $T_{r\theta}$, which must be assumed to begin the computation process. The computation is then repeated until the results for successive days are the same. At that time, the results are independent of the values assumed initially.

Synopsis

This chapter has covered methods for determining zone loads for design load calculation or energy calculation purposes. The heat balance model has been covered in some detail, although for each of the needed sub-models (e.g. exterior convection on walls, transient conduction, interior radiation, etc.), only one of the simpler possibilities is covered. The number of possible variations on sub-models is enormous. The thermal network model, which might be characterized as a heat balance model, but with much more detailed sub-models for many of the phenomenon, has only been briefly described, and the reader should consult the references given in the section for additional detail. Finally, the chapter has described the underlying basis for transfer function models, which, though limited in some respects, are widely used today for both energy calculations and load calculations.

Despite this chapter being by far the longest in the book, numerous topics have been omitted or given cursory treatment. The "Recommended reading" section below gives some initial directions for further information on methods for modeling zones. Beyond those resources, the reader will find that ongoing developments are reported in the academic literature and, in particular, the conference proceedings of the International Building Performance Simulation Association (www.ibpsa.org) serve as an archive for much of the work done since the mid-1980s.

The field continues to evolve for several reasons, including:

- In the nature of the case, buildings have essentially infinite variation in design. Phenomena that are unimportant for most buildings may be very important for some buildings. Therefore, methods are augmented to more accurately model a wider range of buildings.
- As building designs for common building types evolve to use less energy, the relative importance of heat transfer mechanisms shifts and methods are refined to improve accuracy.
- New developments in building components such as fenestration, lighting, and internal equipment require continuous updates to data and at least intermittent improvements in methods.
- Moore's Law improvements in computing hardware and corresponding developments in simulation environments and simulation methodologies used in the broader world continue to facilitate new possibilities for more accurate and flexible zone models.
- Experimental research continues to better inform models, both by experimental validation and by investigation of specific phenomena such as convection heat transfer leading to improved sub-models.

Finally, some reflection on experimental research and the current state of the art does suggest that zone models will be undergoing improvement for many years in the future. Prediction of infiltration rates and accounting for the effects of imperfections introduced in the real-world construction process continue to be problematic. As another example, consider that, after many years of investigation of interior convective heat transfer with an eye towards improving building simulation, most load calculation design procedures still use convection coefficients that were experimentally measured in the 1930s. Though progress has been made in this area, an interior convection algorithm that covers all or most room conditions and room ventilation configurations is still not available. Similar limitations in modeling of other phenomena exist, indicating that further developments in zone modeling will continue indefinitely.

Notes

1 Even though heat transfer in an air gap is due to convection and radiation, it is approximated as being a conduction process, with a fixed thermal resistance that is independent of the temperatures of the gap surfaces.
2 Technically, conduction also occurs in liquids and gases, too, but here we are only concerned with conduction in solids.
3 The analogy is to discrete electrical resistors when expressed in thermal resistances ($R' = \Delta x/kA$) and heat transfer rates. For one-dimensional planar walls, it is often convenient to consider the analogy on a per unit area basis, in which case it may be expressed in terms of unit thermal resistances and heat fluxes (heat transfer rates per unit area).
4 Also known as Newton's law of cooling.
5 Despite some attempts to reserve the endings "-ivity" for optically pure surfaces and "-tance" for real-world surfaces, the terms absorptivity, emissivity, reflectivity, and transmissivity are often used interchangeably for absorptance, emittance, reflectance, and transmittance.
6 More detailed urban-scale models that explicitly represent the surrounding buildings and account for radiative interchange between buildings have been described in the literature. See Robinson et al. (2007), Kämpf (2009), Allegrini et al. (2016), Robinson et al. (2007), Kämpf (2009), and EPFL (2017).

References

Alamdari, F. and Hammond, G.P. (1983) Improved data correlations for buoyancy-driven convection in rooms. *Building Services Engineering Research and Technology* 4(3): 106–112.

Algarni, S. and Nutter, D. (2015) Survey of sky effective temperature models applicable to building envelope radiant heat transfer. In *2015 ASHRAE Annual Conference*, June 27, 2015–July 1, 2015. Atlanta, GA: American Society of Heating, Refrigerating and Air-Conditioning Engineers Inc., pp. 351–363.

Allegrini, J., Dorer, V. and Carmeliet, J. (2016) Impact of radiation exchange between buildings in urban street canyons on space cooling demands of buildings. *Energy and Buildings* 127(Supplement C): 1074–1084.

ASHRAE (2017) *ASHRAE Handbook – Fundamentals Volume*. Atlanta, GA: ASHRAE, Inc.

Barbosa, R.M. and Mendes, N. (2008) Combined simulation of central HVAC systems with a whole-building hygrothermal model. *Energy and Buildings* 40(3): 276–288.

Beausoleil-Morrison, I. and Mitalas, G. (1997) BASESIMP: A residential-foundation heat-loss algorithm for incorporating into whole-building energy-analysis programs. *Building Simulation '97*, Prague, pp. 1–8.

Beausoleil-Morrison, I. (2001) An algorithm for calculating convection coefficients for internal building surface for the case of mixed flow in rooms. *Energy and Buildings* 33: 351–361.

Berger, J., Mendes, N., Guernouti, S., Woloszyn, M. and Chinesta, F. (2017) Review of reduced order models for heat and moisture transfer in building physics with emphasis in PGD approaches. *Archives of Computational Methods in Engineering* 24(3): 655–667.

Blocken, B., Defraeye, T., Derome, D. and Carmeliet, J. (2009) High-resolution CFD simulations for forced convective heat transfer coefficients at the facade of a low-rise building. *Building and Environment* 44(12): 2396–2412.

Brown, W.C. and Stephenson, D.G. (1993) A guarded hot box procedure for determining the dynamic response of full-scale wall specimens – part 1. *ASHRAE Transactions* 99(Pt.1): 632–642.

Carpenter, S.C., Kosny, J. and Kossecka, E. (2003) Modeling transient performance of two-dimensional and three-dimensional building assemblies. *ASHRAE Transactions* 109(1): 561–566.

Carroll, J.A. (1980) An "MRT Method" of computing radiant energy exchange in rooms. In *Systems Simulation and Economic Analysis Conference*. San Diego, pp. 343–348.

Clark, J. and Novoselac, A. (2013) Flow and mixed convection heat transfer in buoyant jets from floor registers. *Energy and Buildings* 61(Supplement C): 140–145.

Clark, J., Peeters, L. and Novoselac, A. (2013) Experimental study of convective heat transfer from windows with venetian blinds. *Building and Environment* 59(Supplement C): 690–700.

Clarke, J.A. (2001) *Energy Simulation in Building Design*. Oxford: Butterworth-Heinemann.

Clarke, J.A., Dempster, W.M. and Negrao, C. (1995) The implementation of a computational fluid dynamics algorithm within the ESP-R system. *Proceedings of Building Simulation* 1995: 166–175.

Cole, R.J. (1976) The longwave radiative environment around buildings. *Building and Environment* 11: 3–13.

Cole, R.J. and Sturrock, N.S. (1977) The convective heat exchange at the external surface of buildings. *Building and Environment* 12: 207–214.

Davies, M.G. (1988) Design models to handle radiative and convective exchange in a room. *ASHRAE Transactions* 94(Pt. 2): 173–195.

Davies, M.G. (1996) A time-domain estimation of wall conduction transfer function coefficients. *ASHRAE Transactions* 102(1): 328–343.

Davies, M.G. (2004) *Building Heat Transfer*. Chichester, West Sussex: Wiley.

Defraeye, T., Blocken, B. and Carmeliet, J. (2011) Convective heat transfer coefficients for exterior building surfaces: Existing correlations and CFD modelling. *Energy Conversion and Management* 52(1): 512–522.

Duffie, J.A. and Beckman, W.A. (1991) *Solar Engineering of Thermal Processes, 2nd edition*. New York: John Wiley & Sons.

EPFL. (2017) *CitySim Software | LESO-PB*. https://citysim.epfl.ch/.

ESRU. (2017) ESRU Home Page. www.strath.ac.uk/research/energysystemsresearchunit/applications/esp-r/.

Fisher, D.E. and Pedersen, C.O. (1997) Convective heat transfer in building energy and thermal load calculations. *ASHRAE Transactions* 103(2): 137–148.

Goethals, K., Breesch, H. and Janssens, A. (2011) Sensitivity analysis of predicted night cooling performance to internal convective heat transfer modelling. *Energy and Buildings* 43(9): 2429–2441.

Goss, W.P. and Miller, R.G. (1989) Literature review of measurement and predictions of reflective building insulation system performance: 1900–1989. *ASHRAE Transactions* 95(Pt. 2): 651–664.

Gouda, M.M., Danaher, S. and Underwood, C.P. (2002) Building thermal model reduction using nonlinear constrained optimization. *Building and Environment* 37(12): 1255–1265.

Heim, D. and Clarke, J.A. (2003) Numerical modelling and thermal simulation of phase change materials with ESP-R. In *Eighth International IBPSA Conference, Eindhoven, Netherlands, Building Simulation 2003*, pp. 459–466.

Hittle, D.C. (1981) Calculating building heating and cooling loads using the frequency response of Multilayered Slabs. PhD, University of Illinois at Urbana-Champaign.

Hu, Z.-X., Cui, G.X. and Zhang, Z.-S. (2018) Numerical study of mixed convective heat transfer coefficients for building cluster. *Journal of Wind Engineering and Industrial Aerodynamics* 172: 170–180.

IBPSA. (2017) *IBPSA Project 1 – BIM/GIS and Modelica Framework for Building and Community Energy System Design and Operation*. https://ibpsa.github.io/project1/.

Kämpf, J.H. (2009) On the modelling and optimisation of urban energy fluxes. Doctoral Thesis, EPFL.

Karambakkam, B.K., Nigusse, B. and Spitler, J.D. (2005) A one-dimensional approximation for transient multi-dimensional conduction heat transfer in building envelopes. In *The 7th Symposium on Building Physics in the Nordic Countries*. Reykjavik, Iceland: The Icelandic Building Research Institute, pp. 340–347.

Karava, P., Jubayer, C.M. and Savory, E. (2011) Numerical modelling of forced convective heat transfer from the inclined windward roof of an isolated low-rise building with application to photovoltaic/thermal systems. *Applied Thermal Engineering* 31(11): 1950–1963.

Klems, J.H. (2002) Solar heat hain through fenestration systems containing shading: Summary of procedures for estimating performance from minimal data. *ASHRAE Transactions* 108(1): 512–524.

Krarti, M. and Choi, S. (1996) Simplified method for foundation heat loss calculation. *ASHRAE Transactions* 102(1): 140–152.

Latta, J.K. and Boileau, G.G. (1969) Heat losses from house basements. *Canadian Building* 19(10): 39.

LBNL (2017) *WINDOW | Windows and Daylighting*. https://windows.lbl.gov/software/window.

Le Dréau, J., Heiselberg, P. and Jensen, R.L. (2015) Experimental investigation of the influence of the air jet trajectory on convective heat transfer in buildings equipped with air-based and radiant cooling systems. *Journal of Building Performance Simulation* 8(5): 312–325.

Leenknegt, S., Wagemakers, R., Bosschaerts, W. and Saelens, D. (2012) Numerical sensitivity study of transient surface convection during night cooling. *Energy and Buildings* 53(Supplement C): 85–95.

Lewis, P.T. and Alexander, D.K. (1990) HTB2: A flexible model for dynamic building simulation. *Building and Environment* 25(1): 7–16.

Liesen, R.J. and Pedersen, C.O. (1997) An evaluation of inside surface heat balance models for cooling load calculations. *ASHRAE Transactions* 103(2): 485–502.

Liu, M. and Claridge, D.E. (1992) The energy impact of combined solar radiation/infiltration/conduction effects in walls and attics. In *Thermal Performance of the Exterior Envelopes of Buildings V*, Clearwater Beach, FL, pp. 39–47.

Liu, Y. and Harris, D.J. (2007) Full-scale measurements of convective coefficient on external surface of a low-rise building in sheltered conditions. *Building and Environment* 42(7): 2718–2736.

Maestre, I.R., Cubillas, P.R. and Pérez-Lombard, L. (2010) Transient heat conduction in multi-layer walls: An efficient strategy for Laplace's method. *Energy and Buildings* 42(4): 541–546.

McClellan, T.M. and Pedersen, C.O. (1997) Investigation of outside heat balance models for use in a heat balance cooling load calculation procedure. *ASHRAE Transactions* 103(2): 469–484.

McQuiston, F.C. and Spitler, J.D. (1992) *Cooling and Heating Load Calculation Manual*. Atlanta, GA: ASHRAE, Inc.

Mendes, N., Winkelmann, F.C., Lamberts, R. and Philippi, P.C. (2003) Moisture effects on conduction loads. *Energy and Buildings* 35(7): 631–644.

Mirsadeghi, M., Cóstola, D., Blocken, B. and Hensen, J.L.M. (2013) Review of external convective heat transfer coefficient models in building energy simulation programs: Implementation and uncertainty. *Applied Thermal Engineering* 56(1): 134–151.

Mitalas, G.P. and Arseneault, J.G. (1970) Fortran IV program to calculate z-transfer functions for the calculation of transient heat transfer through walls and roofs. In *Use Of Computers For Environmental Engineering Related To Buildings*, Gaithersburg, MD, pp. 633–668.

Mitalas, G.P. and Stephenson, D.G. (1967) Room thermal response factors. *ASHRAE Transactions* 3(Pt. 1): 2.1–2.10.

NFRC (2017a) Procedure for determining fenestration product solar heat gain coefficient and visible transmittance at normal incidence. ANSI/NFRC 200–2017.

NFRC (2017b) Procedure for determining fenestration product U-factors. ANSI/NFRC 100–2017. Greenbelt.

Palyvos, J.A. (2008) A survey of wind convection coefficient correlations for building envelope energy systems' modeling. *Applied Thermal Engineering* 28: 801–808.

Park, J.E., Kirkpatrick, J.R., Tunstall, J.N. and Childs, K.W. (1987) Building wall heat flux calculations. *ASHRAE Transactions* 93 (Pt. 2): 120–136.

Pedersen, C.O., Liesen, R.J., Fisher, D.E. and Strand, R.K. (2001) *Toolkit for Building Load Calculations (RP-987)*. Atlanta, GA: ASHRAE.

Peeters, L., Beausoleil-Morrison, I. and Novoselac, A. (2011) Internal convective heat transfer modeling: Critical review and discussion of experimentally derived correlations. *Energy and Buildings* 43(9): 2227–2239.

Quinten, J. and Feldheim, V. (2016) Dynamic modelling of multidimensional thermal bridges in building envelopes: Review of existing methods, application and new mixed method. *Energy and Buildings* 110(Supplement C): 284–293.

Rees, S.J., Spitler, J.D. and Haves, P. (1998) Quantitative comparison of North American and U.K. cooling load calculation procedures – results. *ASHRAE Transactions* 104(2): 47–61.

Riederer, P. (2005) Matlab/Simulink for building and HVAC simulation – state of the art. In *Building Simulation 2005*. Montreal, Canada: IBPSA, pp. 1019–1026.

Robinson, D., Campbell, N., Gaiser, W., Kabel, K., Le-Mouel, A., Morel, N., Page, J., Stankovic, S. and Stone, A. (2007) SUNtool – a new modelling paradigm for simulating and optimising urban sustainability. *Solar Energy* 81(9): 1196–1211.

Sanza Pérez, J., Andrés Chicote, M., Varela Díez, F. and Velasco Gómez, E. (2017) A new method for calculating conduction response factors for multilayer constructions based on frequency – Domain spline interpolation (FDSI) and asymptotic analysis. *Energy and Buildings* 148(Supplement C): 280–297.

Sasic Kalagasidis, A., Weitzmann, P., Nielsen, T.R., Peuhkuri, R., Hagentoft, C.E. and Rode, C. (2007) The international building physics toolbox in Simulink. *Energy and Buildings* 39: 665–674.

Seem, J.E., Klein, S.A., Beckman, W.A. and Mitchell, J.W. (1989) Transfer functions for efficient calculation of multidimensional transient heat transfer. *Journal of Heat Transfer* 111(February): 5–12.

Shao, J., Liu, J., Zhao, J., Zhang, W., Sun, D. and Fu, Z. (2009) A novel method for full-scale measurement of the external convective heat transfer coefficient for building horizontal roof. *Energy and Buildings* 41(8): 840–847.

Shipp, P.H. (1983) Natural convection within masonry block basement walls. *ASHRAE Transactions* 89 (1):405–419.

Sowell, E.F. (1990) Lights: A numerical lighting/HVAC test cell. *ASHRAE Transactions* 96 (Pt. 2): 780–786.

Spitler, J.D. (1996) *Annotated Guide to Load Calculation Models and Algorithms*. Atlanta, GA: ASHRAE, Inc.

Spitler, J.D. (2014) *Load Calculation Applications Manual, 2nd edition (SI)*. Atlanta, GA: ASHRAE.

Spitler, J.D. and Fisher, D.E. (1999) On the relationship between the radiant time series and transfer function methods for design cooling load calculations. *HVAC&R Research* 5(2): 125–138.

Spitler, J.D., Fisher, D.E. and Pedersen, C.O. (1997) The radiant time series cooling load calculation procedure. *ASHRAE Transactions* 103(2): 503–515.

Spitler, J.D. and Nigusse, B.A. (2010) Refinements and improvements to the radiant time series method. *ASHRAE Transactions* 116(1): 542–549.

Strachan, P., Nakhi, A. and Sanders, C. (1995) Thermal bridge assessments. *Building Simulation* 95: 563–570.

Strand, R.K. and Pedersen, C.O. (1997) Implementation of a radiant heating and cooling model into an integrated building energy analysis program. *ASHRAE Transactions* 103(1): 949–958.

Taki, A.H. and Loveday, D.L. (1996) External convection coefficients for framed rectangular elements on building facades. *Energy and Buildings* 24: 147–154.

Underwood, C.P. (2014) An improved lumped parameter method for building thermal modelling. *Energy and Buildings* 79(Supplement C): 191–201.

Underwood, C.P. and Yik, F.W.H. (2004) *Modelling Methods for Energy in Buildings*. Oxford: Blackwell Science.

Urbikain, M.K. and Davies, M.G. (2012) A frequency domain estimation of wall conduction transfer function coefficients. *Energy and Buildings* 51(Supplement C): 191–202.

Virtanen, M., Heimonen, I. and Kohonen, R. (1992) Application of the transfer function approach in the thermal analysis of dynamic wall structures. In *Thermal Performance of the Exterior Envelopes of Buildings V*, Clearwater Beach, FL, ASHRAE, pp. 240–251.

Walton, G. (1983) *Thermal Analysis Research Program Reference Manual*. National Bureau of Standards.

Walton, G.N. (1980) A new algorithm for radiant interchange in room loads calculations. *ASHRAE Transactions* 86 (Pt. 2): 190–208.

Walton, G.N. (1993) *Computer Programs for Simulation of Lighting/HVAC Interactions*. National Institute of Standards and Technology, NISTIR 5322.

Wang, S. and Chen, Y. (2003) Transient heat flow calculation for multilayer constructions using a frequency-domain regression method. *Building and Environment* 38: 45–61.

Waters, J.R. and Wright, A.J. (1985) Criteria for the distribution of nodes in multilayer walls in finite-difference thermal modelling. *Building and Environment* 20(3): 151–162.

Wetter, M. and van Treeck, C. (2017) IEA EBC Annex 60 Final Report, New Generation Computational Tools for Building & Community Energy Systems. ISBN 978-0-692-89748-5. September 2017. http://www.iea-annex60.org/pubs.html

Wright, J.L. (1996) A correlation to quantify convective heat transfer between vertical window glazings. *ASHRAE Transactions* 102(1): 940–946.

Wright, J.L. (2008) Calculating center-glass performance indices of glazing systems with shading devices. *ASHRAE Transactions* 113(2): 199–209.

Wright, J.L., Collins, M.R., Kotey, N.A. and Barnaby, C.S. (2009) Improved cooling load calculations for fenestration with shading devices. Final Report (ASHRAE 1311-RP). Atlanta.

Wright, J.L. and Kotey, N.A. (2006) Solar absorption by each element in a glazing/shading layer array. *ASHRAE Transactions* 112(2): 3–12.

Yazdanian, M. and Klems, J. (1994) Measurement of the exterior convective film coefficient for windows in low-rise buildings. *ASHRAE Transactions* 100(1): 1087–1096.

York, D.A. and Cappiello, C.C. (1981) DOE-2 Engineers Manual (Version 2.1A), Lawrence Berkeley Laboratory and Los Alamos National Laboratory.
Zuo, W., Wetter, M., Tian, W., Li, D., Jin, M. and Chen, Q. (2016) Coupling indoor airflow, HVAC, control and building envelope heat transfer in the Modelica Buildings library. *Journal of Building Performance Simulation* 9(4): 366–381.

Recommended reading

The following references treat various aspects of room and zone simulation. A couple of general works, not specific to a particular approach, are Davies (2004), which gives a thorough treatment of the theory of building heat transfer, mainly from an analytical perspective, and Spitler (1996), which is an annotated bibliography of load calculation models and algorithms.

For the heat balance model, the most comprehensive treatment is probably the Pedersen, et al. (2001) work, called a "toolkit" – it consists of a complete set of Fortran routines and a complete set of documentation, which, taken together, form the basis of a heat balance model. Emphasis is given to clarity rather than speed. The heat balance model is also treated in some detail by McQuiston et al. (2005). Clarke (2001) gives an overview of building simulation, mostly informed by development of the ESP-r program and so emphasizing the thermal network model approach. The text by Underwood and Yik (2004) treats several topics given scant coverage here, in particular numerical methods and lumped parameter methods for transient conduction.

For transfer function models, the York and Cappiello (1981) manual for the DOE-2 program is fairly comprehensive, and Spitler (2014) describes the radiant time series method in detail.

Clarke, J.A. (2001) *Energy Simulation in Building Design*. Oxford: Butterworth-Heinemann.
Davies, M.G. (2004) *Building Heat Transfer*. Chichester, West Sussex: Wiley.
McQuiston, F.C., Parker, J.D. and Spitler, J.D. (2005) *Heating, Ventilating, and Air Conditioning Analysis and Design, 6th edition*. New York: John Wiley and Sons.
Pedersen, C.O., Liesen, R.J., Fisher, D.E. and Strand, R.K. (2001) *Toolkit for Building Load Calculations (RP-987)*. Atlanta, GA: ASHRAE.
Spitler, J.D. (1996) *Annotated Guide to Load Calculation Models and Algorithms*. Atlanta, GA: ASHRAE.
Spitler, J.D. (2014) *Load Calculation Applications Manual, 2nd edition (SI)*. Atlanta, GA: ASHRAE.
Underwood, C.P. and Yik, F.W.H. (2004) *Modelling Methods for Energy in Buildings*. Oxford: Blackwell Science.
York, D.A. and Cappiello, C.C. (1981) *DOE-2 Engineers Manual*. Berkeley: University of California.

Activities

1 A lightweight wall made with an exterior insulation finish system (EIFS) has the layers shown in Table 2.16. The resulting conduction transfer function coefficients are shown in Table 2.17. If the exterior wall surface temperatures are those given in Table 2.18 for Atlanta, and the interior surface temperature is held constant at 22 °C, compute the conduction heat flux at the interior surface for each hour of the day. Assume the day repeats.

2 A lightweight wall made with an exterior insulation finish system has the layers shown in Table 2.16. The resulting conduction transfer function coefficients are shown in Table 2.17. If the exterior wall surface temperatures are those given in Table 2.18 for Seattle, and the interior surface temperature is held constant at 22 °C, compute the conduction heat flux at the interior surface for each hour of the day. Assume the day repeats.

3 A lightweight wall made with an exterior insulation finish system has the layers shown in Table 2.16. The resulting conduction transfer function coefficients are shown in Table 2.17. If the exterior wall surface temperatures are those given in Table 2.18 for Denver, and the

Table 2.16 Layer descriptions for EIFS wall

Layer	*Thickness, mm*	*Density, kg/m³*	*Conductivity, W/mK*	*Specific Heat, J/kg K*
Stucco	12.7	1856	0.69	840
Extruded polystyrene	50.0	32	0.03	1500
Equivalent homogeneous layer – steel studs and insulation	89.0	65	0.09	550
Plaster board and skim	12.7	800	0.43	420

Table 2.17 Conduction transfer function coefficients for EIFS wall

j	X_j	Y_j	Z_j	Φ_j
0	6.18619E+00	1.25614E−01	2.17049E+00	
1	−6.332701E+00	2.043432E−01	−1.949532E+00	8.691221E−02
2	4.842522E−01	7.784391E−03	1.167875E−01	

Table 2.18 Wall surface temperatures, °C

Local time	*Atlanta, July 21, SE-facing*	*Seattle, September 21, SW-facing*	*Denver, December 21, S-facing*
1	29.3	20.0	6.5
2	28.7	19.5	5.9
3	28.2	19.1	5.4
4	27.8	18.7	5.0
5	27.4	18.4	4.9
6	27.1	18.3	5.2
7	27.7	18.5	5.8
8	44.6	21.0	19.2
9	55.2	23.6	40.7
10	60.8	26.2	53.7
11	62.5	32.8	61.9
12	60.8	46.3	65.9
13	56.3	58.6	65.2
14	49.4	68.0	60.0
15	44.4	73.5	49.0
16	44.0	73.8	28.0
17	43.2	67.5	16.4
18	41.6	51.8	15.0
19	39.2	25.7	13.3
20	35.8	24.5	11.6
21	33.1	23.3	10.2
22	31.9	22.2	9.0
23	30.9	21.3	7.9
24	30.0	20.5	7.2

interior surface temperature is held constant at 22 °C, compute the conduction heat flux at the interior surface for each hour of the day. Assume the day repeats.

4 A heavyweight brick-and-block wall is made up of the layers shown in Table 2.1. The resulting conduction transfer function coefficients are shown in Table 2.2. If the exterior wall surface temperatures are those given in Table 2.18 for Atlanta, and the interior surface temperature is held constant at 22 °C, compute the conduction heat flux at the interior surface for each hour of the day. Assume the day repeats.

5 A heavyweight brick-and-block wall is made up of the layers shown in Table 2.1. The resulting conduction transfer function coefficients are shown in Table 2.2. If the exterior wall surface temperatures are those given in Table 2.18 for Seattle, and the interior surface temperature is held constant at 22 °C, compute the conduction heat flux at the interior surface for each hour of the day. Assume the day repeats.

6 A heavyweight brick-and-block wall is made up of the layers shown in Table 2.1. The resulting conduction transfer function coefficients are shown in Table 2.2. If the exterior wall surface temperatures are those given in Table 2.18 for Denver, and the interior surface temperature is held constant at 22 °C, compute the conduction heat flux at the interior surface for each hour of the day. Assume the day repeats.

7 For Problems 1 and 4, compute the surface-to-surface U-factors and then compute the interior conduction heat flux at each hour. For each wall, quantify the delay and damping effects of the thermal mass.

8 Consider the zone shown in Figure 2.6, used for Example 2.8. If the south wall and window were coated with aluminum foil, the emissivity might be reduced to 0.1. If this were done, determine the radiative heat flux leaving each surface. How much of a reduction in heat loss might one expect, based solely on the inter-surface radiation?

9 Consider the zone shown in Figure 2.6, used for Example 2.8. If under hot summer conditions, the surface temperatures for surfaces 1–7 are 23.0, 23.2, 23.7, 22.5, 25.9, 30.1, and 28.2, respectively, what will the radiative heat fluxes leaving each surface be?

10 For a very lightweight wall, with negligible thermal mass, a steady-state surface heat balance will give an approximate value of the exterior surface temperature and the conduction heat flux into the surface. Consider the wall in Table 2.1 if it is painted such that the solar absorptivity is 0.9 and the emissivity is 0.9. The weather conditions for Atlanta are shown in Tables 2.19 (incident solar radiation on the wall) and 2.20 (hourly air temperature). Assume the sky temperature is 6 °C below the outdoor air temperature, the wind speed is a constant 4 m/s with the surface on the windward side of the building, and the building is a low-rise building. Also assume the interior surface temperature of the wall is constant at 22 °C. Find the exterior surface temperature and conductive heat flux at hour 15.

11 Find the exterior surface temperature and conduction heat flux at hour 15 for the situation described in Problem 10, except for the Seattle weather conditions.

12 Find the exterior surface temperature and conduction heat flux at hour 15 for the situation described in Problem 10, except for the Denver weather conditions.

13 Write a computer program or develop a spreadsheet to find the exterior surface temperature and conduction heat flux at all hours for the situation described in Problem 10.

14 Write a computer program or develop a spreadsheet to find the exterior surface temperature and conduction heat flux at all hours for the situation described in Problem 11.

15 Write a computer program or develop a spreadsheet to find the exterior surface temperature and conduction heat flux at all hours for the situation described in Problem 12.

16 Table 2.21 gives beam and diffuse solar radiation incident on a southeast-facing vertical window in Atlanta on July 21 along with the corresponding incidence angle. For each hour of the day, find the transmitted and absorbed solar radiation in both panes of a double pane window with a low-E coating – the second window in Table 2.8.

Table 2.19 Incident solar radiation on vertical walls, W/m²

Local time	*Atlanta, July 21, SE-facing*	*Seattle, September 21, SW-facing*	*Denver, December 21, S-facing*
1	0.0	0.0	0.0
2	0.0	0.0	0.0
3	0.0	0.0	0.0
4	0.0	0.0	0.0
5	0.0	0.0	0.0
6	0.0	0.0	0.0
7	10.7	0.0	0.0
8	326.4	40.0	230.8
9	514.7	72.6	606.8
10	602.0	99.0	815.0
11	609.7	198.2	930.0
12	549.4	423.8	966.2
13	432.6	627.4	925.8
14	273.8	784.4	806.3
15	158.6	873.3	592.0
16	139.1	873.9	201.8
17	122.1	760.9	0.0
18	99.3	475.7	0.0
19	69.5	0.1	0.0
20	25.9	0.0	0.0
21	0.0	0.0	0.0
22	0.0	0.0	0.0
23	0.0	0.0	0.0
24	0.0	0.0	0.0

Table 2.20 Outdoor air temperatures, °C

Local time	*Atlanta, July 21*	*Seattle, September 21*	*Denver, December 21*
1	29.3	22.3	6.5
2	28.7	21.8	5.9
3	28.2	21.3	5.4
4	27.8	20.9	5.0
5	27.4	20.6	4.9
6	27.1	20.4	5.2
7	27.1	20.5	5.8
8	27.4	21.0	7.0
9	28.0	21.8	8.7
10	29.0	23.0	10.6
11	30.3	24.5	12.8
12	31.8	26.2	14.8
13	33.5	27.9	16.3
14	34.9	29.2	17.3
15	36.0	30.1	17.7
16	36.7	30.5	17.3

(*Continued*)

Table 2.20 (Continued)

Local time	*Atlanta, July 21*	*Seattle, September 21*	*Denver, December 21*
17	36.8	30.3	16.4
18	36.4	29.7	15.0
19	35.6	28.6	13.3
20	34.4	27.3	11.6
21	33.1	26.0	10.2
22	31.9	24.8	9.0
23	30.9	23.8	7.9
24	30.0	23.0	7.2

Table 2.21 Solar radiation incident on a southeast-facing surface in Atlanta, July 21

Hour	*Beam (W/m^2)*	*Diffuse (W/m^2)*	*Incidence angle (°)*
1	0.0	0.0	133.85
2	0.0	0.0	128.31
3	0.0	0.0	109.73
4	0.0	0.0	98.06
5	0.0	0.0	86.98
6	0.0	0.0	76.84
7	8.0	2.7	68.09
8	247.1	79.3	61.37
9	385.8	128.8	57.43
10	441.4	160.6	56.86
11	430.5	179.1	59.76
12	363.0	186.4	65.66
13	248.6	184.0	73.82
14	99.8	173.9	83.58
15	0.0	158.6	94.40
16	0.0	139.1	105.90
17	0.0	122.1	117.83
18	0.0	99.3	129.93
19	0.0	69.5	141.94
20	0.0	25.9	153.29
21	0.0	0.0	162.27
22	0.0	0.0	163.92
23	0.0	0.0	156.62
24	0.0	0.0	145.73

17 Table 2.21 gives beam and diffuse solar radiation incident on a southeast-facing vertical window in Atlanta on July 21 along with the corresponding incidence angle. For each hour of the day, find the transmitted and absorbed solar radiation in both panes of a double pane window with clear panes – the third window in Table 2.8.

18 For the window of Problem 16, determine the outer pane and inner pane surface temperatures for each hour and the convective flux and radiative flux into the room for

each hour. The outdoor air temperatures are those shown for Atlanta in Table 2.20. For purposes of solving this problem, you may assume that the distance between the panes is 13 mm; the sky temperature is 6 °C below the outdoor air temperature; the building is low-rise, so the Yazdanian and Klems (1994) model is applicable for exterior convection; the interior zone air temperature is fixed at 21.5 °C and the zone MRT is fixed at 21.8 °C; the interior convection coefficient is 4.18 W/m^2K; and all window panes have an emissivity of 0.84 except the low-E coated pane, which has an emissivity of 0.033 on the inside of the outer pane.

19 Work Problem 18, except use the window from Problem 17. In this case, you may assume that all window panes have an emissivity of 0.84.

Answers to selected problems

1 At hour 12, conduction heat flux is 14.71 W/m^2; at hour 18, conduction heat flux is 7.67 W/m^2.

2 At hour 12, conduction heat flux is 5.50 W/m^2; at hour 18, conduction heat flux is 15.05 W/m^2.

3 At hour 12, conduction heat flux is 14.98 W/m^2; at hour 18, conduction heat flux is −1.86 W/m^2.

4 At hour 12, conduction heat flux is 8.34 W/m^2; at hour 18, conduction heat flux is 15.06 W/m^2.

5 At hour 12, conduction heat flux is 0.54 W/m^2; at hour 18, conduction heat flux is 11.93 W/m^2.

6 At hour 12, conduction heat flux is −3.75 W/m^2; at hour 18, conduction heat flux is 11.06 W/m^2.

8 The radiation heat flux leaving the north wall is 3.60 W/m^2.

9 The radiation heat flux leaving the north wall is −8.44 W/m^2.

10 At hour 15, the exterior surface temperature is 43.30 °C and the conduction heat flux is 7.88 W/m^2.

11 At hour 15, the exterior surface temperature is 72.68 °C and the conduction heat flux is 18.75 W/m^2.

12 At hour 15, the exterior surface temperature is 48.99 °C and the conduction heat flux is 9.99 W/m^2.

16 At hour 11, the transmitted solar radiation is 197.05 W/m^2, the absorbed solar radiation in the outer pane is 129.67 W/m^2, and the absorbed solar radiation in the inner pane is 36.85 W/m^2.

18 At hour 11, the temperature of the outer pane is 37.4 °C and the temperature of the inner pane is 27.8 °C. The inward longwave radiative heat flux is 38.38 W/m^2 and the inward convective heat flux is 20.17 W/m^2.

3 Ventilation performance prediction

Jelena Srebric

Scope

This chapter first presents an introduction to the state of the art in modeling and simulation of airflow, air temperature, and contaminant concentrations in buildings. Following the introduction of fundamental modeling concepts, the chapter continues with simulation examples for different ventilation systems, such as mechanical, natural, and hybrid ventilation. Overall, the entire chapter has outlined specific recommendations for potential new users to produce reliable ventilation performance predictions.

Learning objectives

To develop an understanding of the principles of ventilation performance predictions required for thermal comfort and indoor air quality studies.

Key words

Ventilation performance/air distribution/computational fluid dynamics/multi-zone airflow network models

Introduction

Ventilation systems are responsible for delivering air in air-conditioned spaces to accomplish the ventilation as well as heating/cooling tasks. As a result, ventilation systems have significant impact on thermal comfort and indoor air quality (IAQ). For example, increased ventilation is specified as one of the criteria in the LEED (Leadership in Energy and Environmental Design) certification process under indoor environmental quality credits (USGBC 2009). In the same group of credits, thermal comfort is also included, and ventilation provides thermal comfort to support the well-being of occupants. In the past decade, it has been found that the ventilation systems are associated with occupational health, perceived air quality including thermal comfort, and occupants' productivity (NIOSH 1997; Fang et al. 1998; Wargocki et al. 2002; National Research Council 2006; Zhao et al. 2015). This central role of ventilation systems in achieving high indoor environmental quality requires reliable predictions of ventilation system performance. State-of-the-art prediction approaches include multi-zone airflow network and computational fluid dynamics (CFD) simulation models. Both of these models are capable of calculating indoor air parameters such as airflow rates, temperature, and contaminant concentrations in buildings with different ventilation systems. Nevertheless, important fundamental differences between the two simulation techniques exist and define their accuracy and reliability in predicting ventilation performance. Knowledge of the strengths and weaknesses of both multi-zone airflow network and CFD models is the key for successful use of these simulation techniques for design and operation of ventilation systems.

Modeling approaches

Before simulation models, such as multi-zone airflow network and CFD, became widely accepted tools for ventilation performance predictions, semi-empirical equations were the standard design approach. These equations played an important historic role, and the most successful ones are still in use. For example, equations for estimating infiltration flow rates are included in the multi-zone airflow network models. The use of semi-empirical equations, multi-zone airflow network, or CFD models are complementary to each other, because each of these models resolves the airflow patterns with a different level of detail. More detailed models are always more time consuming, and, therefore, appropriate models have to be chosen based on the particular ventilation design problem.

Overview of semi-empirical equations

The simplest models for ventilation performance prediction are the semi-empirical equations, which are also called single-zone models. These equations represent a class of analytical models calibrated with empirical coefficients to predict bulk ventilation airflow rates through different types of openings. For openings at the building envelope, such as windows or doors, the airflow is driven by the pressure difference between the indoor and outdoor environments. This pressure difference is caused by the wind airflow around a building as well as the temperature difference between indoor and outdoor environments, which is called the *stack effect*. It is important to notice that this assumption of combined *wind* and *stack effect* is strictly true only in the absence of heating, ventilating, and air-conditioning (HVAC) systems, which cause indoor pressure changes depending on the system balancing and maintenance. As a result, this assumption is reliable for ventilation performance predictions through building openings in a natural ventilation regime.

Assuming the airflow through openings is proportional to the square root of the pressure differential across the openings, the total airflow rate due to the combined wind and stack effect can be approximated as following:

$$\dot{Q}_{tot} = \sqrt{\dot{Q}_w^2 + \dot{Q}_s^2}, \tag{3.1}$$

where $\dot{Q}_{tot}$ is the total volumetric flow rate [m^3/s], $\dot{Q}_w$ and $\dot{Q}_s$ are volumetric flow rates due to the wind and stack effect respectively [m^3/s]. For single-sided ventilation that represents natural airflow through a single opening for a single space in a building, the following simple correlations can be used (Allard 1998; CIBSE 2005):

$$\dot{Q}_w = 0.05AU_H, \tag{3.2}$$

$$\dot{Q}_s = 0.2A\sqrt{\frac{gh\Delta T}{T_{avg}}}, \tag{3.3}$$

where A is the opening area [m^2], U_H is the wind velocity at the building height H [m/s], g is the gravitational acceleration [m/s^2], h is the opening height [m], ΔT is the inside-outside temperature difference [K], and T_{avg} is the average of inside and outside temperature [K]. These simple equations are good first estimates for a single space (single zone) with bi-directional flow through a single opening. Nevertheless, natural ventilation scenarios for multiple spaces result in much more complex airflow patterns that require much more complex equations or use of tools such as multi-zone airflow network or CFD models.

Another airflow pattern that is often predicted with semi-empirical equations is building infiltration. One of the first widely used infiltration models was developed at the Lawrence Berkley National Laboratory (LBNL), and it is called the LBNL model (Sherman and Grimsrud 1980).

This model was similar to other formulations at that time, because it used a power law relationship between the airflow rate and a pressure difference across the building envelope. Nevertheless, the model was unique because it used the effective leakage area for building openings, which could be obtained from fan pressurization tests. The modern version of this model is known as the ASHRAE basic model (American Society of Heating, Refrigerating and Air-Conditioning Engineers [ASHRAE] 2017, chapter 16):

$$\dot{Q}_{inf} = \frac{A_e}{1000}\sqrt{C_s \Delta T + C_w U^2}, \tag{3.4}$$

where $\dot{Q}_{inf}$ is the total volumetric flow rate of infiltration [m^3/s], A_e is the effective air leakage area [cm^2], ΔT is the average indoor-outdoor temperature difference for time interval of calculation [K], U is the average wind speed measured at local weather station for time interval of calculation [m/s], C_s is the stack coefficient [(L/s)^2/(cm^4·K)], and C_w is the wind coefficient [(L/s)^2/(cm^4 (m/s)^2)]. The stack and wind coefficients are available in tables (ASHRAE 2017, chapter 16). The simplicity of this model has made it very popular, and important model assumptions are often overlooked. Among these assumptions are that the neutral pressure level is half of the building height and the effects of wind directionality are averaged over the time period of interest. In real buildings, the neutral pressure level, where the indoor and outdoor pressures are equal, is often at different elevations depending on indoor and outdoor environmental conditions. In addition, the time averaging of wind directionality neglects the changes in leakage area available for infiltration as the wind impinges on different building surfaces.

The pressure created on the exterior surface of a building by wind has a very complex distribution that depends on many factors, such as wind direction, wind velocity, air density, building geometry, surface orientation, and nature of the surrounding environment (Etheridge and Sandberg 1996). Many of these factors are very hard to predict because of the stochastic nature of the wind itself. Usually, the wind pressures are positive on the windward surface and negative on the leeward surface of a building. The time-averaged static pressures over the building surface are proportional to the velocity or dynamic pressure of undisturbed air stream. The pressure due to wind at a reference height (H) is calculated by the following expression:

$$p_v = \frac{1}{2} \cdot \rho_a \cdot U_H^2, \tag{3.5}$$

where U_H is the reference wind velocity at reference height from the ground [m/s], ρ_a is the ambient air density [kg/m^3], and p_v is often referred to as the stagnation air pressure [Pa]. The reference height is typically 10 m above the ground. The difference between the pressure on a building surface and the atmospheric air pressure, at the same height for an otherwise undisturbed wind approaching the building, is defined as following:

$$p_w = C_p \cdot p_v = C_p \cdot \frac{\rho_a \cdot v^2}{2}, \tag{3.6}$$

where p_w is the equivalent static air pressure on the building surface [Pa], and C_p is the wind pressure coefficient [–]. This coefficient defines what portion of the wind kinetic energy is transformed to pressure energy on the vertical building surface. An overview of wall-averaged wind pressure coefficients for different types of buildings is available in the form of databases (Blevis 1984; Hagentoft 2001). These databases, with appropriate figures describing each case, resulted from extensive literature reviews. As an example, Table 3.1 shows averaged values of the wind pressure coefficients for different surfaces of low-rise buildings. In this example, the incident wind angle, θ, at the windward surface 1, changes from 0° to 45°, and results in different wind pressure coefficients.

Table 3.1 Wind pressure coefficients C_p for low-rise buildings up to three stories in height

Wall			*Surface nomenclature*
Location	Wind angle: 0°	Wind angle: 45°	
Surface 1	0.4	0.1	
Surface 2	−0.2	−0.35	
Surface 3	−0.3	0.1	
Surface 4	−0.3	−0.35	
ROOF, pitch angle – less than 10°			
Location	Wind angle: 0°	Wind angle: 45°	
Front	−0.6	−0.5	
Rear	−0.6	−0.5	
ROOF, pitch angle – between 10° and 30°			
Location	Wind angle: 0°	Wind angle: 45°	
Front	−0.35	−0.45	
Rear	−0.35	−0.45	
ROOF, pitch angle – greater than 30°			
Location	Wind angle: 0°	Wind angle: 45°	
Front	0.3	−0.5	
Rear	−0.5	−0.5	

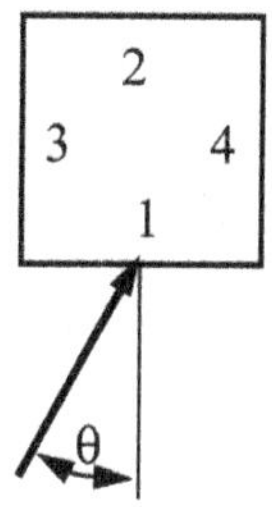

Source: Hagentoft 2001

In addition to databases, wind pressure coefficients are also represented with correlations resulting from wind tunnel experiments. The literature on wind pressure coefficients is extensive, and the simplest correlations can be found in the *ASHRAE Handbook – Fundamentals* (2017, chapter 24). These correlations are typically represented as a function of wind incident angle, such as the following equation for low-rise buildings (Swami and Chandra 1988):

$$C_p = \ln\left(\begin{array}{l} 1.248 - 0.703\sin\left(\frac{\theta}{2}\right) - 1.175\sin^2\theta + 0.131\sin^3(2\theta G) + \\ +0.769\cos\left(\frac{\theta}{2}\right) + 0.07G^2\sin^2\left(\frac{\theta}{2}\right) + 0.717\cos^2\left(\frac{\theta}{2}\right) \end{array}\right), \quad (3.7)$$

where θ is the incident wind angle [rad], and G is a shape factor [−], calculated by taking the natural logarithm of the ratio of the windward wall length to the length of the adjacent wall. This and other wind pressure coefficient equations can be used to calculate wind pressures on the building surfaces as input parameters for a multi-zone airflow network model.

Overall, even though the semi-empirical equation models incorporate numerous assumptions, they can provide good estimates of bulk airflow properties when the domain of interest such as a building can be treated as a single zone. For more complex situations, the semi-empirical models can be used for specifying boundary conditions, such as infiltration rates, for the multi-zone airflow network models.

Overview of multi-zone airflow network modeling approaches

The simplest mathematical models for calculating the distributions of indoor airflows, temperature, and contaminant concentrations are probably the multi-zone airflow network models. These models solve the conservation equation for mass, energy, and species concentration based on a number of approximations. The models divide a given space or a building into a number of volumes (zones). Then, the multi-zone airflow network models solve the mass, energy, and concentration equations for each zone to obtain a solution for specific boundary conditions and the perfect mixing assumption. Since the multi-zone airflow network models are simple and easy to understand, they are preferred models for design problems as long as modeling results are reliable.

A popular type of multi-zone airflow network models determines airflow within a space using Bernoulli's equation (Walton 1989). This multi-zone airflow network approach is effective for the simultaneous analyses of HVAC system, infiltration, and multi-room airflow problems. At present, the two most widely used multi-zone airflow network models are CONTAM and COMIS, which are developed by the National Institute of Standards and Technology (NIST) and Lawrence Berkeley National Laboratory, respectively (Dols and Walton 2002; Feustel and Smith 1997). Both of these simulation tools are available on a shareware basis and have user-friendly interfaces accompanied with detailed user manuals. The comprehensive fundamentals are available in the manuals, and the airflow mass balance applied to each zone in a building for a transient problem is given in Equation 3.8:

$$\frac{\partial m_i}{\partial t} = \rho_i \frac{\partial V_i}{\partial t} + V_i \frac{\partial \rho}{\partial t} = \sum_j F_{j,i} + F_i, \tag{3.8}$$

where m_i is the mass of air in zone i [kg], t is time [s], ρ is the density [kg/m^3], V_i is the volume of zone i [m^3], $F_{j,i}$ is the airflow rate [kg/s] between zone j and zone i; a positive value indicates flow from j to i and a negative value indicates flow from i to j, and F_i is the airflow rate [kg/s] for non-flow processes that could add or remove significant quantities of air from the zone, usually sources and sinks specified as boundary conditions. Furthermore, flow paths (fluid flow components) are the connections between zones that allow airflow in one direction for small openings and in both directions for large openings. The flow path equations represent fluid flow components such as doors, windows, and structural leakages that connect zones and allow airflow movements.

Figure 3.1(a) shows an office building that has a detailed multi-zone airflow network model. All four building floors are modeled, and Figure 3.1(b) presents the first floor in the graphical interface of a multi-zone airflow network model. Each separate space is a zone with appropriate flow paths, so the model is similar to the floor plan. The lines in the model define zone boundaries, while each dot represents either a flow path that connects zones or mechanical system air supplies and exhausts.

The airflow through a flow path, also known as a fluid flow component, is usually driven by pressure difference between connected zones. Flow path equations, based on one-dimensional fluid mechanics laws, are applied to quantify the airflow movement through the flow paths. For the pressure-driven flow paths, the airflow rate from zone j to zone i, $F_{j,i}$ [kg/s], is a function of the pressure difference along the flow path, ΔP:

$$F_{j,i} = f(P_j - P_i) = f(\Delta P), \tag{3.9}$$

where P_j, P_i are the pressures at zone j and i, [Pa], which include static, dynamic, and stack pressures for each zone. The following power-law equation is widely used to present the function $f(\Delta P)$ in Equation 3.9:

$$F_{j,i} = C \cdot (\Delta P)^n \tag{3.10}$$

where C is the flow coefficient (discharge coefficient) of the flow path, and n is the flow exponent of the flow path, usually 0.5 for large openings, 1.0 for narrow openings, and 0.6–0.7 for typical infiltration openings (Dols and Walton 2002).

(a)

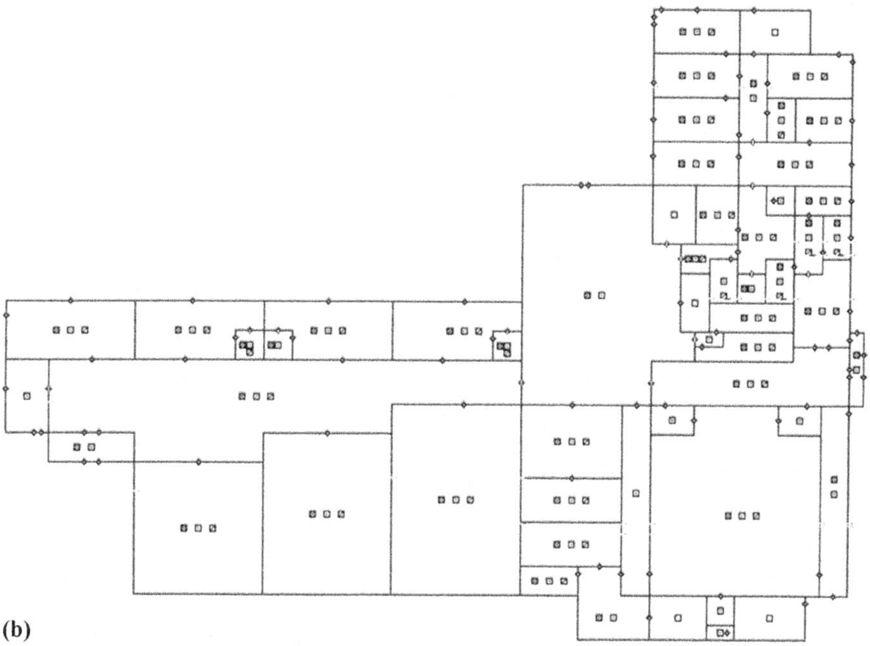

(b)

Figure 3.1 (a) Four-story office building and (b) a representation of the first floor in a multi-zone airflow network simulation tool

The remaining key problems in effective use of multi-zone airflow network models are related to the perfect mixing assumption and flow coefficients of large openings. A recent experimental study has shown that the perfect mixing assumption for the temperature field is acceptable as long as the temperature gradient is relatively small, which is defined in the literature (Wang and Chen 2008). However, the method is unreliable if applied to a single space containing several different sub-volumes with non-uniform indoor parameter distribution, such as temperature field with displacement ventilation or contaminant concentration in a space with a point contaminant source (Srebric et al. 2008). Under these conditions, a multi-zone airflow network model can easily result in inaccurate predictions of contaminant flow rate intensity and direction from a point source.

Furthermore, for large openings, such as doors and windows, the airflow is typically bi-directional, and the fundamental nature of bi-directional airflow is based on two opposing flow forces creating unstable balance that changes dynamically. As a result, Equation 3.9 cannot accurately account for airflow through large openings because it relies on flow coefficients that do not have a general equation. Application of multi-zone airflow network models to a new complex geometry requires a more sophisticated simulation or experiment to determine new flow coefficients. Experimental and numerical methods are emerging to bridge the problems of multi-zone airflow network models, because they are not only computationally inexpensive but also relatively easy to set up when compared to CFD models (Srebric et al. 2008).

Hybridization of multi-zone airflow network models by coupling them with energy simulations and CFD is an important trend not only to improve the multi-zone airflow network model accuracy, but also to provide the necessary boundary conditions to energy simulation and CFD models (Hensen 1999; Clarke 2001; Novoselac 2004). The most widely used hybrid simulation tools are ESP-r and EnergyPlus developed by the University of Strathclyde, Glasgow, and the U.S. Department of Energy, respectively (Clarke 2001; DOE 2008). EnergyPlus is available on the shareware basis and can be obtained for implementation of custom simulation modules upon signing a developer's license agreement. Each of these hybrid simulation tools has included energy, multi-zone airflow network, and CFD simulation modules.

The use of hybrid models is still not a standard practice, but it appears to be a clear trend for building design and operation. For example, a study demonstrated that the hybrid simulations with coupled simulation tools increase simulation accuracy, while using minimal computer resources for practical design applications (Djunaedy et al. 2004). The study also proposed a methodology for selecting an appropriate simulation strategy, which, in principle, first uses the simplest models and then increases the complexity of simulations to resolve building design problems. The main suggestion is to start with energy simulation models that assume uniform indoor parameters, then include multi-zone airflow network models, if necessary. Only apply CFD as a last resort, because it is the most computationally intensive simulation model. As the available computational power steadily increases, the problems associated with intense CFD computations will disappear. CFD will still require much more intense work on input and model development; therefore, multi-zone airflow network models will still be an important simulation option. Nevertheless, problems in determining reliable flow path equations for the multi-zone airflow network models stimulated the development of CFD simulation models for ventilation performance predictions.

Overview of CFD modeling approaches

CFD models can provide detailed indoor parameters of design interests such as velocities, temperatures, and contaminant concentrations in a reliable and fast manner. These simulation models are recommended as an inexpensive approach compared to taking measurements and a more detailed approach compared to the multi-zone airflow network models. In principle, a CFD model subdivides the interior space into a number of cells. For each cell, the conservation of mass is satisfied so that the sum of mass flows into or out of a cell from all its neighbors is balanced to zero. Similarly the exchange of momentum from the flow into or out of a cell must be balanced in each direction with pressure, gravity, viscous shear, and energy transport by turbulent eddies. For transitional and fully turbulent flows, which normally exist in room flow, the challenge is to properly describe the transport of momentum and energy. The nature of turbulence is complex and not yet fully understood. The turbulent flow is three-dimensional and random with many vortices (turbulent eddies) that enhance mixing in the flow field. The mixing decreases velocity gradients and dissipates kinetic energy of the fluid stream. The dissipation is irreversible, and kinetic energy is transformed to the internal energy of the fluid. To study the impact of turbulent mixing on thermal comfort and indoor air quality, partial differential equations that govern turbulent flow must be solved.

Indoor airflow is typically modeled as turbulent incompressible flow of Newtonian fluid, and the following transport equations describe the flow in the tensor notation.

Mass continuity:

$$\frac{\partial \rho}{\partial t} + \frac{\partial \rho \tilde{u}_i}{\partial x_i} = 0, \tag{3.11}$$

where ρ is the air density [kg/m^3], $\tilde{u}_i$ is the instantaneous velocity component in the x_i direction [m/s], x_i is the coordinate (for $i = 1, 2, 3$, x_i corresponds to three perpendicular axes) [m], and t is the time [s].

Momentum conservation:

$$\frac{\partial \rho \tilde{u}_i}{\partial t} + \frac{\partial \rho \tilde{u}_i \tilde{u}_j}{\partial x_j} = -\frac{\partial \tilde{p}}{\partial x_i} + \frac{\partial}{\partial x_j}\left[\mu\left(\frac{\partial \tilde{u}_i}{\partial x_j} + \frac{\partial \tilde{u}_j}{\partial x_i}\right)\right] + \rho\beta\left(T_o - \tilde{T}\right)g_i, \tag{3.12}$$

where $\tilde{u}_j$ is the instantaneous velocity component in the x_j direction [m/s], $\tilde{p}$ is the instantaneous pressure [Pa], μ is the molecular viscosity [Pa s], β is the thermal expansion coefficient of air [1/K], T_o is the temperature of a reference point [K], $\tilde{T}$ is the instantaneous temperature [K], and g_i is the gravity acceleration in the i direction [m/s^2].

Energy conservation:

$$\frac{\partial \rho \tilde{T}}{\partial t} + \frac{\partial \rho \tilde{u}_j \tilde{T}}{\partial x_j} = \frac{\partial}{\partial x_j}\left(\frac{k}{c_p}\frac{\partial \tilde{T}}{\partial x_j}\right) + S_T, \tag{3.14}$$

where k is the thermal conductivity of air [W/(m K)], S_T is the thermal source, where S_T = (heat flux for the source/c_p) [W (kg K)/J], and c_p is the specific heat at constant pressure [J/(kg K)]. In addition, $\frac{k}{c_p} = \frac{1}{k/c_p} = \frac{\mu}{\mu k/c_p} = \frac{\mu}{Pr}$, where Pr is the *Prandtl number*.

Species concentration conservation:

$$\frac{\partial \rho \tilde{c}}{\partial t} + \frac{\partial \rho \tilde{u}_j \tilde{c}}{\partial x_j} = \frac{\partial}{\partial x_j}\left(\rho D \frac{\partial \tilde{c}}{\partial x_j}\right) + S, \tag{3.15}$$

where $\tilde{c}$ is the instantaneous species concentration [m_c^3/m_{air}^3], D is the molecular diffusion coefficient for the species [m_c^2/s], and S is the species source [m_c^3/s]. In addition, $\rho D = \frac{1}{1/\rho D} = \frac{\mu}{\mu/\rho D} = \frac{\mu}{Sc}$, where Sc is the *Schmidt number*.

Ventilation performance predictions with CFD use the Boussinesq approximation for thermal buoyancy as shown in Equation 3.12. This approximation makes air density a constant in all the governing equations and considers the buoyancy influence on air movement by the temperature difference. The conservation equations (Equations 3.11–3.15) must be solved numerically. The solution provides the field distributions of pressure, air temperature, velocity, and contaminant concentrations. Depending on the numerical solution procedure of the conservation equations, there are three CFD methods: (1) direct numerical simulation (DNS), (2) large-eddy simulation (LES), and (3) *Reynolds averaged Navier-Stokes* (RANS) equations with turbulence models. DNS solves the highly reliable Navier-Stokes equations without approximations. Therefore, DNS requires a grid resolution as fine as the *Kolmogorov micro-scale* of small eddies for which the inertia and viscous effects are of equal magnitude. Based on dimensional analyses, the ratio of smallest to largest length scales in turbulent flows is in order of $Re^{3/4}$, where Re is the *Reynolds number*. For three-dimensional flows, such as indoor airflow, the approximate number of cells or grid points becomes $Re^{9/4}$ (Versteeg and Malalasekera 2007). Since the Reynolds number for a typical indoor airflow pattern is approximately 10^5, the required grid number for solving the three-dimensional airflow pattern is approximately 10^{11} to 10^{12}. Current supercomputers have an approximate grid resolution as fine as 10^8, and the new PetaFlops supercomputers are expected to allow 10^9 and 10^{12} grid resolution in the near future (Nakahashi 2008). In addition, the DNS method requires very small time steps, where the ratio of smallest to largest time scales is approximated as $Re^{1/2}$ (Versteeg and Malalasekera 2007). Overall, current computer capacity is still far too small to solve ventilation airflows with the DNS method in design practice.

LES was developed in the early 1970s by Deardorff (1970) for meteorological applications. He separated turbulent motion into large eddies and small eddies. The separation of the two did not significantly affect the evolution of large eddies. LES solves the large-eddy motion by a set of filtered equations governing three-dimensional, time-dependent motion. Turbulent transport approximations are used for small eddies, and the small eddies are modeled independently from the flow geometry. The success of LES stems from the fact that the main contribution to turbulent transport comes from large-eddy motion. LES is also a more realistic application of computational effort for building performance simulations than DNS is, because LES can be performed on a large and fast workstation or computer clusters. Nevertheless, LES is still too time consuming because it calculates time-dependent flow, although the time and space steps can be larger than those for DNS.

The RANS method is the fastest but potentially least accurate method. RANS solves ensemble-averaged Navier-Stokes equations by using turbulence modeling. In RANS, all unsteadiness is averaged and regarded as part of the turbulence (Ferziger and Peric 2001). The grid number used for the simulation is normally much less than that for LES or DNS. Most importantly, steady flow can be solved as time-independent; therefore, computing costs are less than those for DNS and LES are. However, turbulence modeling introduces an additional error into the calculation. Nevertheless, calculations based on RANS are popular and widely used in many building design solutions, because they give reasonably good-quality results with an affordable computing effort. Even with more detailed calculations such as LES and DNS, the results should be averaged for design assessments of ventilation performance, because the current design guidelines employ the steady-state assumption.

The first step in averaging the conservation equations is to decompose instantaneous variables into their mean and fluctuating parts, which is called *Reynolds decomposition*. The mean parts are denoted with capital letters, and the fluctuating parts are denoted with a "′" superscript:

$$\tilde{u}_i = U_i + u_i' \qquad \tilde{p} = P + p' \qquad \tilde{T} = T + T' \qquad \tilde{c} = C + c'. \tag{3.16}$$

To illustrate typical mean and fluctuating parts of indoor airflow variables, Figure 3.2 shows instantaneous values of velocity and temperature measured in a full-scale environmental chamber with a square mixing diffuser at the ceiling (Srebric 2000). The results are for a period of 4.5 minutes sampled at 5 Hz. Two measurement points are selected to show how different airflow variables can easily exist within the same space. Both measurement points are at the

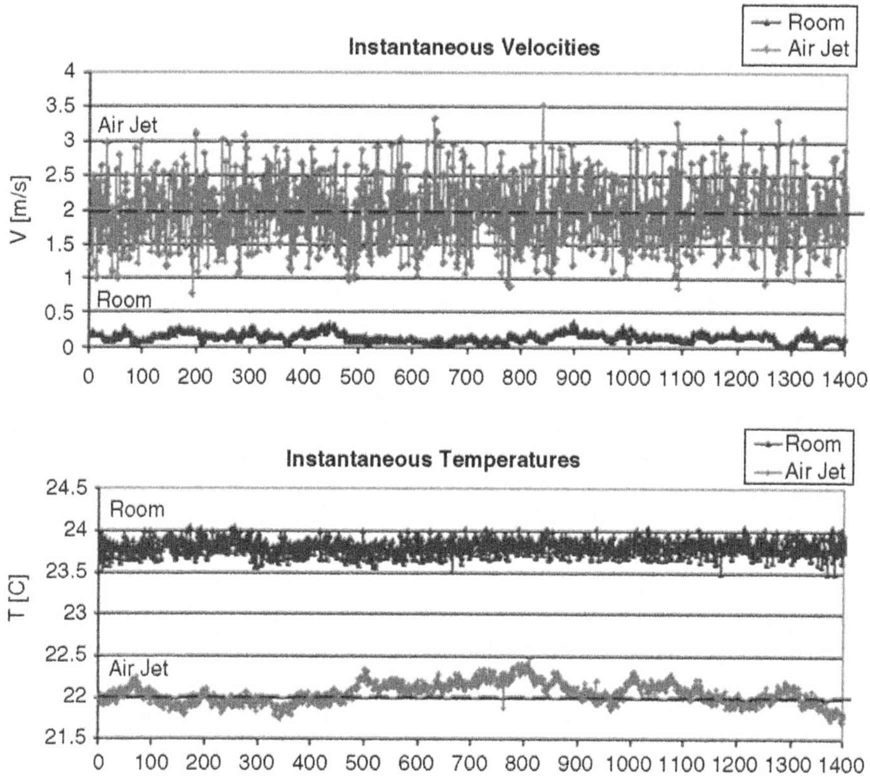

Figure 3.2 Instantaneous characteristics of air velocities (a) and temperatures (b) in a room with a square ceiling diffuser measured at two different heights (for room $h = 1.10$ m, and for jet $h = 2.40$ m)

same vertical line, but at different heights (h). The first measurement point is in the air jet region close to the ceiling ($h = 2.40$ m), and the second one is in the occupied zone ($h = 1.10$ m) in the room. The results show that the fluctuation and unsteadiness are much higher in the jet region than are those in the occupied zone.

Since any instantaneous variable can be represented as the sum of the average value and fluctuation around the average, a general variable can be introduced:

$$\tilde{\phi}(x_i,t) = \phi(x_i) + \phi'(x_i,t)\cdot \tag{3.17}$$

The averaging for statistically steady flow gives the mean value of the variables and has the following integral form:

$$\phi(x_i) = \lim_{T\to\infty}\frac{1}{T}\int_0^T \tilde{\phi}(x_i,t)\,dt. \tag{3.18}$$

The averaging time, T, has to be much larger than the time scale of turbulence to obtain results independent of the integration interval. To evaluate the mean flow parameters for unsteady turbulent flow, ensemble averaging is used:

$$\phi(x_i,t) = \lim_{N\to\infty}\frac{1}{N}\sum_{n=1}^{N}\tilde{\phi}(x_i,t), \tag{3.19}$$

where N is the number of independent samples. A measurement with the same initial and boundary conditions has to be repeated N times, where N has to be large enough to eliminate the influence of fluctuation on the results. The averaging introduces an additional term $\rho \overline{u_i' u_j'}$ called *Reynolds stresses* in Equation 3.12, and additional terms $\rho \overline{u_i' \phi'}$ called the scalar fluxes in Equations 3.13 and 3.14. These non-linear terms in the original equations are grouped with viscous diffusion terms to enable their modeling with eddy-viscosity turbulence approximation (Versteeg and Malalasekera 2007). Turbulence should be considered as an engineering approximation rather than scientific law (Ferziger and Peric 2001). Hence, the evaluation of the turbulence model's performance and justification of its assumptions is performed through its application in solving design problems.

Turbulence modeling

The modeling of eddy viscosity and diffusivity is based on an analogy with laminar flow and molecular transport. Turbulent eddies are represented as molecules, which on a microscopic scale collide and exchange energy. Although the analogy is empirical and questionable, the performance of the model is good for many practical applications (Wilcox 2006). The performance depends particularly on the equation or equations used to define the distribution of eddy viscosity, which are called turbulence models.

The simplest eddy-viscosity models are zero-equation turbulence models. These models have one algebraic equation for turbulent viscosity, μ_t, and no (zero) additional partial differential equations to resolve turbulent properties. The basis for zero-equation models is *Prandtl*'s mixing-length hypothesis (Wilcox 2006). Forming an analogy with the molecular transport of momentum, *Prandtl* postulated the mixing-length hypothesis in the following form:

$$\mu_t = \rho l_{\mathrm{mix}}^2 \left| \frac{dU}{dx_j} \right|. \tag{3.20}$$

where μ_t is the turbulent viscosity [Pa s], l_{mix} is the mixing length [m], ρ is the fluid density [kg/m^3], and U is the average velocity [m/s].

The mixing length is an empirical variable defined as a free path of small eddies (fluid lumps) over which their original momentum is preserved. The eddy viscosity and mixing length are not physical properties of the fluid, but they are properties of the flow field. The product of the mixing length and the velocity gradient is called mixing velocity, which is also the property of the turbulent flow. Calibration of the mixing length is always needed for complex flows such as indoor airflows. A zero-equation turbulence model was developed and validated for indoor airflow CFD simulations (Srebric et al. 1999). Nerveless, more typical turbulence modeling requires one or two additional partial differential equations to calculate parameters for the turbulent viscosity. The standard k-ε model is the most widely used two-equation model in practice (Launder and Spalding 1974). The two additional transport equations include the turbulent kinetic energy and turbulent energy dissipation rate as:

$$\frac{\partial \rho k}{\partial t} + \frac{\partial \rho U_j k}{\partial x_j} = \frac{\partial}{\partial x_j}\left(\frac{\mu_t}{\sigma_k} \frac{\partial k}{\partial x_j} \right) + 2\mu_t S_{ij} \cdot S_{ji} - \rho\varepsilon, \tag{3.21}$$

$$\frac{\partial \rho \varepsilon}{\partial t} + \frac{\partial \rho U_j \varepsilon}{\partial x_j} = \frac{\partial}{\partial x_j}\left(\frac{\mu_t}{\sigma_\varepsilon} \frac{\partial \varepsilon}{\partial x_j} \right) + C_{1\varepsilon} \frac{\varepsilon}{k} 2\mu_t S_{ij} \cdot S_{ji} - C_{2\varepsilon} \rho \frac{\varepsilon^2}{k}, \tag{3.22}$$

where k is the turbulent kinetic energy [J/kg], ε is the turbulent energy dissipation rate [J/(kg s)], S_{ij} is the strain rate tensor ($S_{ij} = \frac{1}{2}\left(\frac{\partial U_i}{\partial x_j} + \frac{\partial U_j}{\partial x_i}\right)$) [1/s], and the empirical model constants are $\sigma_k = 1.0$, $\sigma_\varepsilon = 1.3$, $C_{1\varepsilon} = 1.44$, and $C_{2\varepsilon} = 1.92$.

The turbulent viscosity is then calculated based on k and ε without assumptions for length scale, which is difficult to estimate:

$$\mu_t = \rho C_\mu \frac{k^2}{\varepsilon}, \tag{3.23}$$

where $C_\mu = 0.09$ is the empirical constant.

With an eddy-viscosity turbulence model, such as the standard k-ε model, the transport equations for all indoor airflow variables, including turbulence properties, can be written in terms of the general variable ϕ, as the following time-averaged Navier-Stokes equation:

$$\frac{\partial \rho \phi}{\partial t} + \frac{\partial \rho U_j \phi}{\partial x_j} = \frac{\partial}{\partial x_j}\left(\Gamma_{\phi,\text{eff}} \frac{\partial \phi}{\partial x_j}\right) + S_\phi, \tag{3.24}$$

where ϕ can be 1, u_i, T, c, k, and ε to represent the mass, momentum, energy, species concentration, turbulent kinetic energy, and turbulent energy dissipation rate conservation laws respectively; $\Gamma_{\phi,\text{eff}}$ is the effective diffusion coefficient in each conservation law; and S_ϕ is the source for a specific variable ϕ. Table 3.2 shows different variables with corresponding effective diffusion coefficients and source terms for Equation 3.24. Overall, Equation 3.24 has transient $\frac{\partial \rho \phi}{\partial t}$, convection $\frac{\partial \rho U_j \phi}{\partial x_j}$, diffusion $\frac{\partial}{\partial x_j}\left(\Gamma_{\varphi,\text{eff}} \frac{\partial \phi}{\partial x_j}\right)$, and source S_ϕ terms as defined by different conservation laws. For ventilation performance predictions, a system of partial differential equations will typically have the mass conservation, momentum conservation for three velocity components, energy conservation, and two equations for turbulent properties, k and ε, which is a system of seven partial differential equations for each cell (grid point).

Table 3.2 The k-ε turbulence model parameters

Equation	ϕ	$\Gamma_{\phi,eff}$	S_ϕ
Continuity	1	0	0
Momentum	U_i	$\mu + \mu_t$	$-\frac{\partial p}{\partial x_i} + \rho\beta(T_o - T)g_i$
Turbulent kinetic energy	k	$\mu + \frac{\mu_t}{\sigma_k}$	$2\mu_t S_{ij} \cdot S_{ji} - \rho\varepsilon$
Dissipation rate of kinetic energy	ε	$\mu + \frac{\mu_t}{\sigma_\varepsilon}$	$C_{1\varepsilon}\frac{\varepsilon}{k}2\mu_t S_{ij} \cdot S_{ji} - C_{2\varepsilon}\rho\frac{\varepsilon^2}{k}$
Temperature	T	$\frac{\mu}{Pr} + \mu_t$	S_T
Concentration	C	$\frac{\mu}{Sc} + \mu_t$	S

The standard k-ε model introduces two additional differential equations for the turbulent transport processes. To solve the two additional equations requires considerable computing effort because of the highly non-linear characteristics of the equations. For example, the prediction of ventilation performance in an office requires several hours of computing time on a PC. If a designer wants to use the CFD method to vary the design, such as to change HVAC equipment location, the additional computing time renders this process inefficient, especially for conceptual designs. Therefore, a zero-equation model, referring to an algebraic turbulence model, is more suitable for design, when the simplicity of the model does not lead to incorrect solutions of the transport flow equations (Srebric et al. 1999).

Due to the complexity of turbulent flow, it is difficult to obtain a universal turbulence model for all types of ventilation performance predictions. A general conclusion is that the turbulent models perform differently from one case to another, although all simulated flows are indoor airflows. In fact, for indoor airflow simulations, no turbulent models perform significantly better than the standard k-ε model. The standard k-ε model is stable, easily implemented, widely validated, and reasonably accurate for many indoor airflow applications. Nevertheless, the standard k-ε model is not universal and has provided erroneous results in certain indoor airflow regimes. For example, the standard k-ε model has had problems in predicting strong buoyancy, separated flows, axisymmetric jets, swirl flows, and heat transfer from a wall. Therefore, many modified k-ε models, such as the RNG k-ε model (Yokhot and Orszag 1986), have emerged to improve performance for certain classes of problems. Compared to the standard k-ε model, the RNG k-ε model has an additional source term, R, in the equation for the turbulent energy dissipation rate, ε. This term improves flow prediction for regions with large strain rates, and the term is negligible for small strain rates. Therefore, the model can better predict separate flows, which are commonly present in indoor airflow. Several studies have shown that the RNG k-ε model has better performance than the standard k-ε model for indoor airflow simulations (Chen 1995; Loomans 1998). However, the RNG model assumes high Re numbers for the flow field, which is one of the assumptions that causes problems in modeling with the standard k-ε model.

CFD boundary conditions

Boundary conditions are necessary for the mathematical solution of the system of partial differential equations representing the conservation laws. It is important to notice that CFD takes boundary conditions as an additive source term on the left-hand side of Equation 3.24. This is a rather crude but quite practical representation of ventilation transport process. As a result, CFD models should be a simplified version or real detailed geometry, such as the example office space with six cubicles in Figure 3.3(a). The entire office was reduced to floor, ceiling, four walls, cubicle partitions, tables, computers, lamps, and displacement ventilation (DV) diffuser as shown in Figure 3.3(b). All of these items represent CFD boundary conditions with defined size, location, and source term intensity for the conservation laws. This is an appropriate level of details for CFD models used for ventilation performance predictions, and additional items in the office can be neglected without sacrificing accuracy of CFD results. In principle, only important boundary conditions have to be introduced in CFD models. There are three types of boundaries of practical importance:

- free boundary,
- symmetry surface, and
- conventional boundary.

Free boundary is a boundary surface that is adjacent to an inviscid stream. Examples are the air supply and exhaust. For a supply, the boundary conditions are:

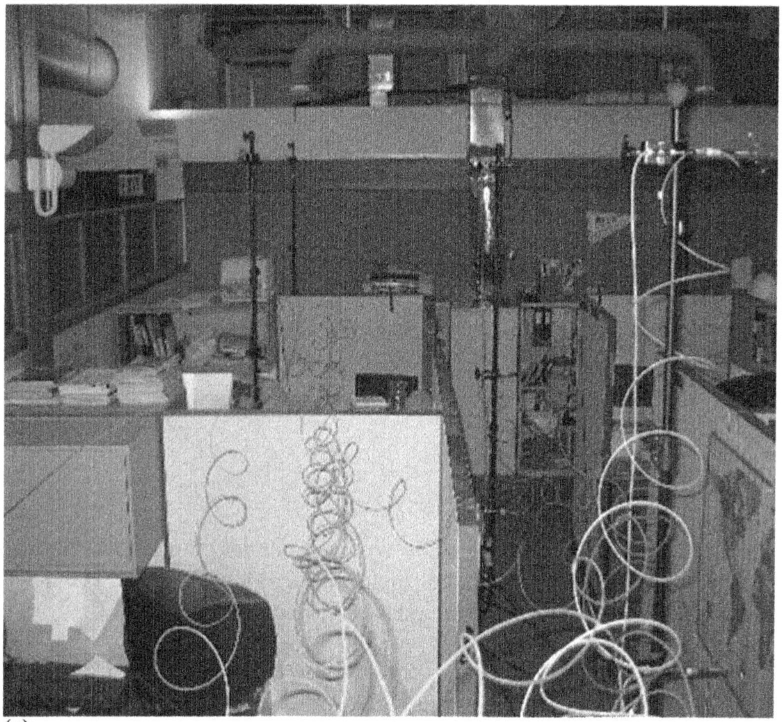

(a)

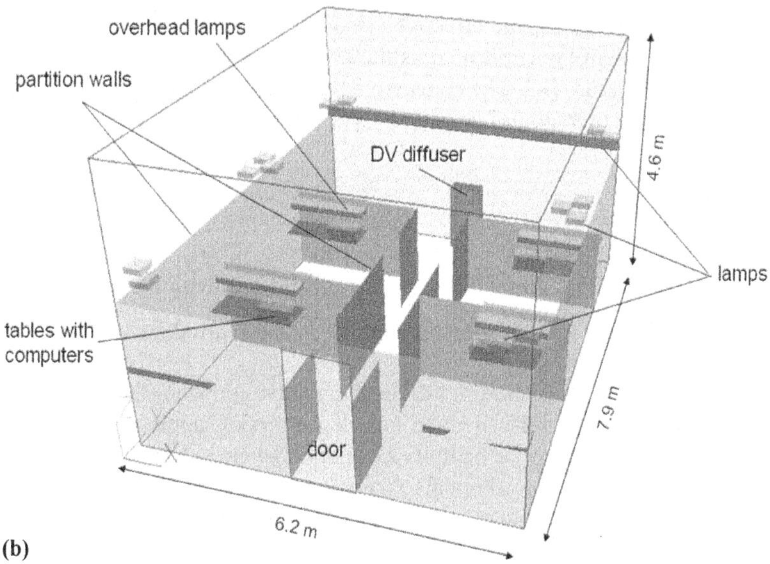

(b)

Figure 3.3 (a) An example office and (b) its simplified model for a CFD model

Source: Srebric et al. 2008

$$u_i = u_{supply},\ T = T_{supply},\ k = k_{supply},\ \varepsilon = \varepsilon_{supply},\ and\ c = c_{supply}, \tag{3.25}$$

where subscripts "supply" are the parameter values at the supply opening. These equations apply only for a slot opening, while supply openings with diffusers require more complex specifications of boundary conditions, which are different for different type of ventilation methods discussed later in this chapter.

Pressure is normally given for an exhaust opening, and zero gradients normal to the surface are assumed for the other parameters:

$$p = p_{exhaust}, \frac{\partial U_i}{\partial x_i} = 0, \quad \frac{\partial T}{\partial x_i} = 0, \quad \frac{\partial k}{\partial x_i} = 0, \quad \frac{\partial \varepsilon}{\partial x_i} = 0, \quad \frac{\partial C}{\partial x_i} = 0\,, \tag{3.26}$$

where $p_{exhaust}$ is the pressure at a return [Pa], and x_i is the coordinate normal to the surface of the exhaust opening or diffuser [m].

Symmetry surface boundaries are used to reduce the size of simulation domains that are symmetric. If the x_i coordinate is normal to the symmetry surface, the following equations describe the boundary conditions of the symmetry surface:

$$\frac{\partial U_i}{\partial x_i} = 0, \quad \frac{\partial T}{\partial x_i} = 0, \quad \frac{\partial k}{\partial x_i} = 0, \quad \frac{\partial \varepsilon}{\partial x_i} = 0, \quad \frac{\partial C}{\partial x_i} = 0\,, \tag{3.27}$$

Conventional boundary is the most common type of boundary conditions that include wall, ceiling, and floor surfaces and the surfaces of furniture, appliances, and occupants. If the x_i coordinate is parallel to the surface, the boundary conditions are:

$$\tau = \mu_{eff} \frac{\partial U_i}{\partial x_j} \quad \dot{q} = h(T_w - T), \tag{3.28}$$

where τ is shear stress [Pa], μ_{eff} is the effective viscosity, $\mu_{eff} = \mu + \mu_t$ [Pa s], h is the convective heat transfer coefficient [W/(m²K)], $\dot{q}$ is the convective heat transfer rate [W], and T_w is the wall (surface) temperature [K]. The convective heat transfer coefficient is determined from the following equation, which is similar to the Reynolds analogy:

$$h = \frac{\mu_{eff}}{Pr_{eff}} \frac{c_p}{\Delta x_j}, \tag{3.29}$$

where Pr_{eff} is the effective Prandtl number, and the effective diffusion coefficient for temperature equation, $\frac{\mu_{eff}}{Pr_{eff}} = \frac{\mu}{Pr} + \frac{\mu_t}{Pr_t}$, is based on laminar, Pr, and turbulent, Pr_t, Prandtl numbers.

Equations 3.28 and 3.29 can be used as wall boundary conditions for the laminar flows as well as for the turbulent flows. The zero-equation turbulence models apply directly to the specified equations. However, the standard *k-ε* or RNG *k-ε* model developed for high *Re* numbers needs adjustment for the near wall region where the *Re* number is very small. The *k-ε* models use additional equations for the near-wall region, called wall functions (Tennekes and Lumley 1972).

The wall functions are dimensionless velocity, temperature, and concentration profiles in the wall boundary region. The wall boundary layer has three regions, the laminar sub-layer ($0 < y^+ \leq 5$), buffer zone ($5 < y^+ \leq 30$), and inertial sub-layer ($30 < y^+ \leq 130$), where y^+ is the dimensionless distance to the wall. The wall has a damping effect on the turbulent flow, and the layer closest to the wall has predominant viscous forces. In the buffer zone, viscous forces and Reynolds stresses have the same order of magnitude and neither can be neglected. Finally, the inertial sub-layer is predominately turbulent. The wall functions give expressions for laminar and turbulent sub-layers, while the buffer zone is neglected. A near wall flow is considered laminar if $y^+ \leq 11.63$, and if $y^+ > 11.63$ the flow is turbulent. The wall function equations for velocity and temperature are:

$$u^+ = y^+,\ T^+ = Pr\,y^+,\ \text{for } y^+ \le 11.63 \tag{3.30}$$

$$u^+ = \frac{1}{\kappa}\ln\left(E\,y^+\right),\ T^+ = Pr_t\left[u^+ + E\left(\frac{Pr}{Pr_t} - 1\right)\left(\frac{Pr_t}{Pr}\right)^{1/4}\right],\ \text{for } y^+ > 11.63 \tag{3.31}$$

$$u^+ = \frac{U_P}{u_\tau},\quad u_\tau = \sqrt{\frac{\tau_w}{\rho}},\quad y^+ = \frac{y_P u_\tau \rho}{\mu},\quad T^+ = \frac{(T_w - T_P)u_\tau}{q} \tag{3.32}$$

where u^+ is the dimensionless mean velocity [−], U_P is the velocity parallel to the wall [m/s], u_τ is the friction velocity [m/s], τ_w is the wall shear stress [Pa], y^+ is the dimensionless wall distance [−], y_P is the distance to the wall [m], κ is the von Karman constant (κ = 0.41) [−], E is the roughness parameter (E = 9 for hydraulically smooth walls with constant shear stress) [−], T_w is the wall temperature [K], T_P is the air temperature at location denoted as P [K], and $\dot{q}$ is the convective heat transfer rate [W].

In the viscous sub-layer, the mean velocity profile is approximated as the linear function of wall distance as shown in Figure 3.4. At the wall surface (y = 0), air velocity is zero ("no-slip" conditions) and air temperature is equal to the wall surface temperature. For the turbulent sub-layer, the velocity profile is the log-law function.

Assumptions used to derive the wall functions restrict the application of these formulae to a certain class of flows. The assumptions are the *Prandtl mixing hypothesis*, the *Boussinesq eddy-viscosity assumption*, a fully developed flow, and no pressure gradients or other momentum sources (constant shear stresses). For typical ventilation performance predictions, these assumptions are acceptable and wall functions are widely used. However, predictions of the heat transfer in the near-wall region tend to be incorrect and dependent on the distance to the wall y_P. Improved calculation of heat transfer is possible with improved temperature profile equations or use of Equation 3.28with a prescribed empirical value for the convective heat transfer coefficient h (Novoselac 2004). This temperature profile or heat flux can be obtained from coupling CFD with the energy simulation tools such as ESP-r or EnergyPlus (Hensen 1999). Overall, a correct prediction of the wall boundary conditions is crucial for accurate ventilation performance predictions.

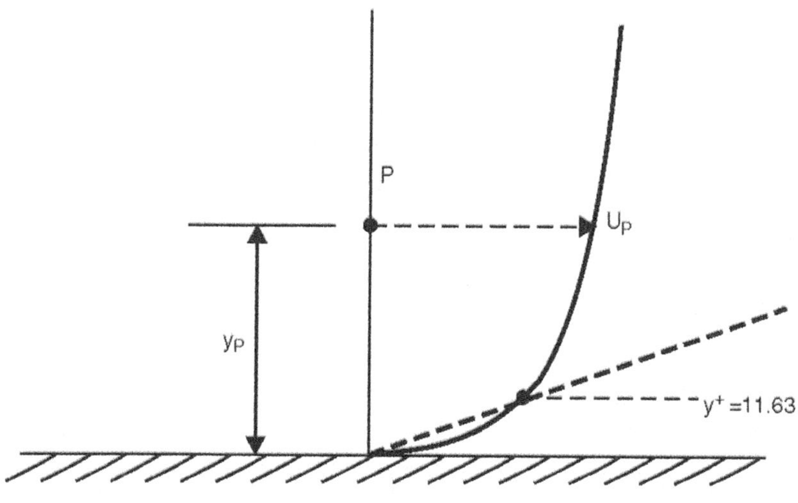

Figure 3.4 Velocity distribution at the near wall region

CFD model performance

Professional organizations such as ASHRAE provide guidelines for quality assessment of CFD model predictions (Chen and Srebric 2002; Sorensen and Nielsen 2003; ASHRAE 2017). The quality assessment has three steps that include (1) verification, (2) validation, and (3) reporting of results. The verification step determines whether a CFD model has capabilities to represent important physical phenomena in support of ventilation design decisions. The validation step establishes a CFD model and user capabilities to accurately simulate physical phenomena. Finally, the reporting of results provides instructions to enable quality assessment of CFD results with sufficient details describing verification and validation of ventilation performance predictions. These steps represent guidelines for best practices in CFD use for ventilation performance predictions and cannot be extended into a standard design procedure, because CFD models are still under intense development (AIAA 2002).

Verification of a CFD model starts with the assessment of necessary models to realistically represent ventilation transport processes and finishes with the evaluation of numerical simulation methods. These verifications assessments include selection of:

- airflow and heat transfer models,
- turbulence model, and
- numerical solution methods.

For ventilation design, it is important to achieve thermal comfort, so both velocity and temperature fields need to be modeled. The airflow and heat transfer models are connecting convection, diffusion, and/or radiation in the simulation domain. Therefore, mass, momentum, and energy conservation laws have to be solved simultaneously with turbulence model. It is important to notice that most ventilation airflows are weakly turbulent because of the high *Rayleigh number*, *Ra*, and sometimes high Reynolds number, *Re*. Typical *Ra* ranges from 10^9 to 10^{12}, and *Re* from 10^4 to 10^7, while turbulence occurs when $Ra > 10^9$ and/or $Re > 10^4$ (ASHRAE 2017). For a new CFD user, discussion on different turbulence models is useful, but the starting point in gaining modeling experience should be a CFD model with the standard k-ε model (Equations 3.21 and 3.22). This model assumes fully developed turbulent flow, and calculation errors can occur in the simulation regions where this assumption is incorrect such as near wall boundaries. Other more complex turbulence models are recommended only for more experienced users because they can introduce additional partial differential equations that are hard to solve.

As discussed, with the standard k-ε model seven partial differential equations have to be solved simultaneously for each grid point in the simulation domain. The equations are discretized and solved with numerical solution procedures that start with an initial guess for calculated variable distributions. The solution procedure for the system of discretized transport equations is iterative because the equations are coupled and the convection terms in momentum equations are non-linear. The discretization introduces an error that can be reduced by refining the grid size. A commonly accepted method for grid refinement is to double the grid size until the two different grid resolutions give similar CFD results (Wilcox 2006). In this way, the discretization error is minimized, and other numerical and physical approximations, such as turbulence models, can be explored as potential sources of simulation errors.

Due to the non-linearity of convection terms in the transport equations, solution procedures can be unstable. The variables must be solved simultaneously, and the progress towards a solution has to be simultaneous. To control the rate of change for the calculated variables, relaxation factors are used. Relaxation factors serve to control the change in variable calculations and, hence, decrease the numerical instability. A general problem with relaxation factors is that they depend on a solved case, and therefore their values have to be estimated and corrected during calculations. The relaxation only influences the calculation procedure and the final solution of a problem is independent of relaxation factors. Overall, the numerical solution should be independent of grid distributions, initial variable distributions, and relaxation factors. The criterion for ending the numerical procedure is called the convergence criterion. Usually, the residual of the continuity equation must be smaller than a certain predefined value, such as:

$$\text{Residual for mass} = \frac{\text{Sum of absolute residuals in each cell}}{\text{Total mass inflow}} < 0.1\%, \tag{3.33}$$

$$\text{Residual for energy} = \frac{\text{Sum of absolute residuals in each cell}}{\text{Total heat gain}} < 1\%. \tag{3.34}$$

In the end, converged and grid-independent CFD results represent approximate performance of ventilation system in a building. The level of accuracy depends on the airflow and heat transfer models, turbulence model, and numerical solution methods selected in the verification step. The next step is the validation to quantitatively address the model accuracy.

Validation includes quality assessment of both CFD tool and user in modeling a ventilation system in a building. This step is necessary for every new user who plans to use CFD as a design tool. Example simulation cases for ventilation systems are available in the literature or in the end of this chapter. It is important that the validation data are detailed and complete. Validation data have to include (1) detailed geometry and boundary conditions of simulation domain, (2) detailed measurements of flow properties such as distributions of velocity, pressure, and temperature, and (3) complementary overall measurements such as total mass flow rate and heat fluxes (Versteeg and Malalasekera 2007).

It is important to notice that measured data also contain uncertainties and errors introduced by acquisition equipment, experimental setup, and data collection process. For example, accuracy of low-velocity measurement probes is relatively low when collecting data in the occupied zone, where velocities are typically below 0.2 m/s and many times even below 0.05 m/s as shown in Figure 3.2(a) (Melikov et al. 2007). Depending on probe type and calibration, the inaccuracy of velocity measurements can be the same order of magnitude as the measured velocity. On the contrary, temperature probes can be highly accurate, and depending on the manufacturer and price, a typical accuracy is in the range from ±0.1 °C to ±0.5 °C. As a result, errors in velocity or turbulence predictions can point to measurement as well as model uncertainties. Nevertheless, errors in temperature distribution, if not resulting from erroneous velocity distributions, can point to modeling problems. Overall, high accuracy of CFD results when compared to measured data is desirable, but not always feasible. The validation criteria should reflect CFD application requirements (ASHRAE 2017). Luckily, for ventilation design process, model changes are typically incremental, and comparative analyses of results can be sufficient.

When creating a CFD model for either validation of design process, the best practice is to start from very simple geometry and boundary conditions and only gradually increase model complexity. This approach is recommended for all users, experienced and inexperienced, when starting a new type of modeling exercise. The first CFD model should probably include just the simulation domain with supply and exhaust diffusers to calculate simple mass balance in the domain. The subsequent modeling efforts should gradually include other important objects, such as partitions, computers, and occupants. Finally, the CFD model should also include temperature field and non-isothermal boundary conditions. The gradual introduction of complexity in the simulation domain minimizes probability of simple input errors, but also helps determine sources of potential convergence or instability problems. Overall, a successful validation process confirms that appropriate decisions were made with respect to airflow and heat transfer models, turbulence model, and numerical solution methods, but it also confirms the user's knowledge and capabilities.

Reporting is the last step in the quality assurance process for use of CFD models for ventilation performance predictions. A report should contain all of the verification and validation steps outlined with sufficient details that another user can reproduce CFD results (Chen and Srebric 2002). If the simulation case is not new, the report can reference another study with detailed case geometry, boundary conditions, and measurement uncertainties/errors and proceed to list the details of numerical models. Both quantitative and qualitative evaluations have to be included. For example, simple qualitative statements such as "bad" or "good" agreement between CFD results and measured data can be misleading if the comparison criteria are not specific. Finally, the report has to draw conclusions that can benefit the broader design community interested in this type of CFD simulation for ventilation performance predictions.

Multi-zone airflow network and CFD model selection

Both multi-zone airflow network and CFD models are appropriate for evaluations of the ventilation system performance. As outlined in the model overview, multi-zone airflow network is applicable to evaluations where perfect mixing assumption holds, and only bulk flow properties are needed. On the contrary, CFD is suitable for applications with known detailed boundary conditions such as locations, size and intensity of mass, momentum, and heat and contaminant sources. Hybridization of these two models by their coupling can take advantage of both model strengths, while mitigating some of the weaknesses. In order to compare different models, experimental measurements were conducted in an interior cubicle office with displacement ventilation system as shown in Figure 3.3. Detailed measured and simulation data are available in the literature (Srebric et al. 2008). In this chapter, the focus is on practical aspects of efforts necessary for model setup, execution, and accuracy.

Three different models were developed for this cubicle office: CFD, coupled CFD-multi-zone airflow network, and multi-zone airflow network. Each model simulated the temperature field and contaminant distribution. A point source of SF_6 gas was placed in front of displacement diffuser, and the source was slightly shifted towards one side of the diffuser. The simplest model, which is the multi-zone airflow network model, included nine zones that covered the lower level and another nine zones that covered the upper level of the office. A total of 18 zones were sufficient to simulate this office with the multi-zone airflow network model. For the coupled model, CFD handled the area close to the diffuser, which was called the source zone with approximately 100,000 grid points. An additional 12 zones covered the space outside of the source zone with the multi-zone airflow network model. Finally, a stand-alone CFD model covered the entire space and used approximately 200,000 grid points. The computational time was t_{mz} (couple of seconds) for multi-zone airflow network, $4500 \times t_{mz}$ for coupled model, and $16{,}000 \times t_{mz}$ for CFD model.

Considering the computational time of each method, the coupled model is three and a half times faster than the CFD model and significantly slower than the multi-zone airflow network model. In addition, the time required to set up a coupled model is much shorter than the time needed to set up the CFD model, because the computation domain covered by the multi-zone simulation required only associated source/sink intensities, rather than a detailed geometry, detailed space layout, and detailed source/sink locations. Further interesting observations come from the evaluation of simulation result accuracy as shown in Figure 3.5. The asymmetry of the concentration in this case was nicely captured by the coupled model, while the multi-zone airflow network model failed to predict the asymmetry due to its inherent assumption of the uniform concentrations in each simulation zone. Nevertheless, the multi-zone model performed well outside of the immediate vicinity of the source zone. Even the buoyancy-driven transport of contaminants with displacement ventilation can be captured with the multi-zone simulations as long as the temperature field is correctly specified or calculated.

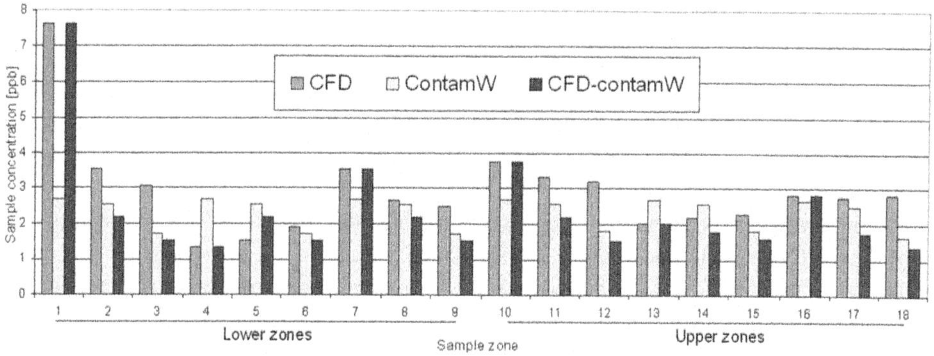

Figure 3.5 The comparison of CFD, multi-zone airflow network (ContamW), and coupled (CFD-contamW) model results

Source: Srebric et al. 2008

Overall, the justification for using either of the three models, CFD, multi-zone airflow network, or coupled models, is case dependent. At present, users of the simulation models are responsible for making this judgment based on the tradeoffs between required accuracy and available computational time. These tradeoffs become quite obvious for simulation domains that are much larger than a single cubicle office, such as an entire building as shown in Figure 3.1. This building has four stories plus basement with an approximate area of 2000 m^2 per floor. A multi-zone airflow network model was created for the entire building with a total of 250 zones, where each floor had roughly 50 zones. For each zone, we had at least four flow elements that included a window, door, air supply, and exhaust. This resulted in at least 1000 flow elements. This model setup took more than a week to finish, while the simulations were still very fast (measured in seconds). It was relatively easy to decide that the development and execution of CFD model for building shown in Figure 3.1 was too time consuming from both model setup and computational time requirements. The assessments of ventilation performance in this building had to be accomplished with the multi-zone airflow network model.

Ventilation system performance

Ventilation systems in buildings are designed and operated to accomplish two primary functions: (1) to deliver fresh air to occupants while removing internally generated contaminants and (2) to provide acceptable thermal comfort in the vicinity of occupants. During the operation of a ventilation system, occupants' focus is on the thermal comfort, because perception of thermal comfort is immediate and thermal discomfort is intolerable. However, poor air quality is more difficult to notice, and, therefore, occupants' response time is longer. In addition, designers traditionally pay more attention to the thermal comfort than to the air quality, because thermal comfort is better defined by standards. For the ventilation system design, the thermal comfort is typically defined by the ANSI/ASHRAE Standard 55 (ASHRAE 2013), and the indoor air quality requirements are specified in ANSI/ASHRAE Standard 62.1 (ASHRAE 2016). Chapter 5 discusses in more detail thermal comfort parameters, while the current chapter discussions will focus on indoor air quality parameters. It is important to notice that simulations of thermal comfort and indoor air quality parameters are based on the same system of equations, such as the equations in Table 3.2. Calculated airflow rates, velocities, turbulence parameters, temperatures, and contaminant concentrations are inputs into the thermal comfort and indoor air quality indicators defined by these two standards.

Indoor air quality (IAQ) indicators

To evaluate performance of ventilation systems, a large number of indoor air quality indicators, such as air exchange rates, air exchange efficiency, ventilation effectiveness, and contaminant removal effectiveness, are in use today. Each of these indicators is useful for a certain type of ventilation system analysis. However, the air exchange efficiency (ε_a) and the contaminant removal effectiveness (ε) appear to be the most suitable for use in comparative analysis of ventilation strategies, because they are general and almost all other indicators are extensions of these two (Novoselac and Srebric 2003). These indicators are applicable to all types of ventilation strategies and airflow patterns. Also, they can be easily measured in the field or the laboratory, which qualify them for extensive use in design and standards.

The first developed IAQ indicator is the contaminant removal effectiveness ε (Yaglou and Witheridge 1937), which is based on the room/space average contaminant concentration $\langle C \rangle$, the contaminant concentration at the supply C_s, and the concentration at the exhaust C_e:

$$\varepsilon = \frac{C_e - C_s}{\langle C \rangle - C_s}. \tag{3.35}$$

On the contrary, air exchange efficiency (ε_a) is based on a local value of age of air (τ), and it was developed in chemical engineering. Later, ε_a was experimentally and theoretically evaluated for use in the indoor air quality field (Etheridge and Sandberg 1996). The contaminant removal effectiveness represents the ratio between shortest possible time needed for replacing the air in the room (τ_n) and the average time for air exchange (τ_{exe}):

$$\varepsilon_a = \frac{\tau_n}{\tau_{exe}} = \frac{\tau_n}{2\langle \tau \rangle}. \tag{3.36}$$

The average time for air exchange can be calculated as $\tau_{exe} = 2\cdot\langle \tau \rangle$, where $\langle \tau \rangle$ represents the average local values of age of air [s]. Also, the shortest possible time needed for replacing the air in the room (τ_n) can be calculated as a reciprocal value of the number of air changes per hour (ACH) in the room (τ_n = 1/ACH, and ACH = volume flow rate/room volume [1/hour]).

It is important to keep in mind that ε_a and ε are only qualitative indicators of indoor air quality and that additional parameters such as flow rates and supply contaminant concentration or local contaminant distribution should be taken into account in the design process. Detailed distributions of IAQ or thermal comfort indicators can be calculated with CFD-based tools for different types of ventilation systems (see Chapter 5 as well).

As an example using CFD, the contaminant removal effectiveness (ε) and the air exchange efficiency (ε_a) were calculated in the same room, and results are presented in the middle vertical cross section, shown in Figure 3.6 (Novoselac and Srebric 2003). More specifically, Figure 3.6(a) presents two-dimensional airflow velocities with a bypass and a recirculation in the occupied zone. Figure 3.6(b) shows corresponding age of air normalized by the shortest time needed for the replacement of air (τ/τ_n). Figures 3.6(c) and 3.6(d) represent contaminant concentration normalized with the contaminant concentration at the exhaust (C/C_n) for two different positions of a contaminant source; in the room center (Figure 3.6(c)) and within the air jet (Figure 3.6(d)). The comparison of Figure 3.6(b) with 3.6(c) and 3.6(d) show that, depending on the position of the contaminant source, the mean age of air and the contaminant concentration field might have a similar or completely different distribution. Consequently, the air exchange efficiency (ε_a) and the contaminant removal effectiveness (ε) might have the same or conflicting values for perceived indoor air quality in the same space. Nevertheless, these two parameters are the general single number IAQ indicators actually evaluating performance of the air distribution system.

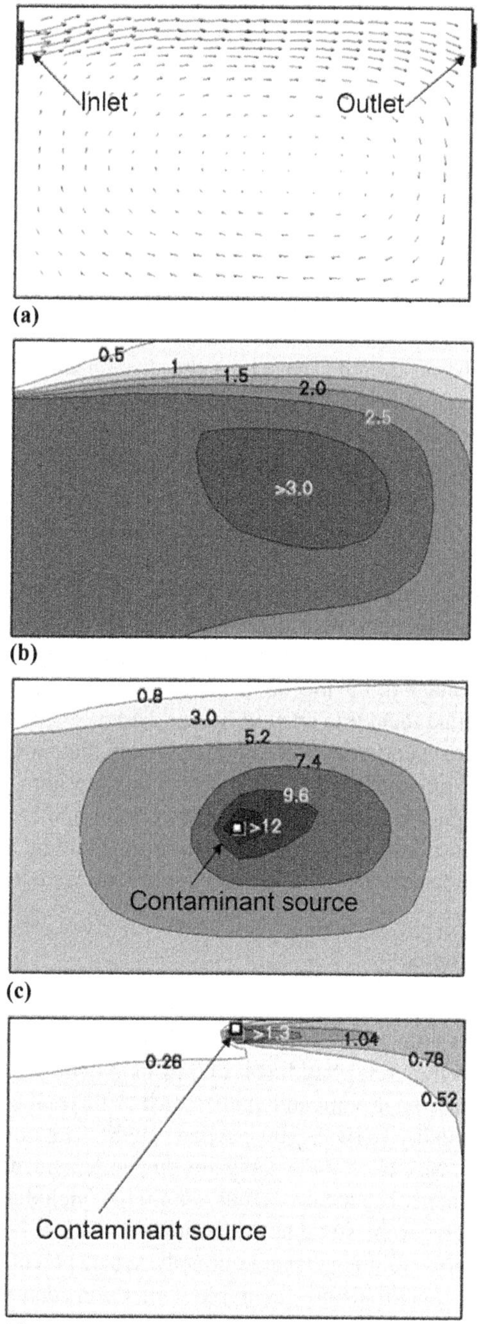

Figure 3.6 Normalized local age of air and contaminant concentration distributions for the same airflow field and different positions of the contaminant source (a) airflow pattern, (b) normalized age of air τ/τ_n ($\varepsilon_a = 0.21$ for this airflow pattern), (c) normalized concentration C/C_e ($\varepsilon_a = 0.21$ and $\varepsilon = 0.19$ for contaminant source at room center), and (d) normalized concentration C/C_e ($\varepsilon_a = 0.21$ and $\varepsilon = 2.20$ for contaminant source in the supply air jet)

Source: Novoselac and Srebric 2003

Overall, air exchange efficiency (ε_a) quantifies the quality of the airflow pattern. The airflow pattern depends on the inlet and outlet positions, the temperature difference between the supply jet and room air, the jet inertial force, and the intensity of buoyant flow created by cooling/heating loads. On the contrary, contaminant removal effectiveness (ε) is the indicator of the contamination level in a room. It depends not only on the airflow pattern, but also on the intensity, size, position, and type of contaminant sources relative to this pattern (Novoselac and Srebric 2003). Overall, different ventilations systems result in different air quality levels that can be assessed with CFD.

Ventilation examples

A classification of ventilation systems based on the driving mechanism for air circulation includes (1) mechanical, (2) hybrid, and (3) natural ventilation. Mechanical ventilation uses fans, ducts, and diffusers, called air distribution systems, to move the air through building spaces. As an alternative to mechanical ventilation, natural ventilation takes advantage of pressure difference between indoor and outdoor environments due to wind and stack effect to accomplish the same ventilation task. As a result, natural ventilation is one of the passive technologies that offer significant energy saving opportunities for buildings in mild climates. Nevertheless, natural ventilation is not as reliable as mechanical ventilation, because it depends on the building geometry, surrounding terrain, and stochastic nature of outdoor air temperature, humidity, and wind velocity. In recent years, many building design projects explored opportunities to take advantage of the mechanical ventilation reliability and natural ventilation energy savings by combining them into one hybrid ventilation system. This hybrid system uses both natural and mechanical ventilation via a control system that switches between or complements the two operational modes based on an optimization algorithm designed to minimize the HVAC energy consumption while satisfying thermal comfort and IAQ requirements. Overall, the mechanical ventilation is still the dominant system for building ventilation, while natural ventilation is becoming more widely used for its potential to save energy.

Natural and hybrid ventilation

Natural ventilation is the oldest form of ventilation, and it has been used as long as buildings have been constructed. At present, natural ventilation is an interesting alternative to mechanical ventilation even in commercial buildings, due to its potential to reduce building energy consumption and to increase environmental quality such as thermal comfort, ventilation effectiveness, and indoor air quality (CIBSE 2005; Allard 1998). The potential benefits of natural ventilation are not always easy to take advantage of due to dependence on weather conditions and surrounding environment. The local weather conditions, including wind velocity, air temperature, and humidity, have to be suitable for natural ventilation. In addition, in urban environments, local air and noise pollution can dramatically reduce potential for natural ventilation, even if the local weather conditions are favorable. Other considerations in design of natural ventilation have to include fire and safety regulations as well as security issues (Allard 1998). Furthermore, Chapter 20 covers urban-level performance predictions that can be directly used in design considerations of natural ventilation.

The design process of natural ventilation can be based on empirical correlations, multi-zone airflow network modeling, and/or CFD simulations. At present, there is no standard design procedure for natural ventilation, such as the established design procedures for mechanical ventilation. In principle, design of natural ventilation has to take into account different considerations. From the perspective of ventilation, design considerations have to include (1) climate analysis and site planning, (2) layout and sizing of window openings at building envelope, and (3) floor planning and layout of internal partitions. These three steps are not independent of each other, and, as a result, the design process is typically iterative. As the final outcome, well-designed natural ventilation should be able to satisfy requirements for thermal comfort and indoor air quality as specified in the standards (ASHRAE 2013; ASHRAE 2016).

The first step in natural ventilation design, climate analysis and site planning, relies on assessment of wind conditions in the building's local microenvironment. Figure 3.7 shows an example of site planning for a residential neighborhood in Thailand to promote natural ventilation for all of the residences (Tantasavasdi et al. 2001). It was demonstrated that even in this hot and humid climate, certain parts of the year could be suitable for natural ventilation, provided that front row houses did not block the incoming wind completely. Local weather data, described in Chapter 6, should be used to determine how many hours/days in a month/season a building would be able to take advantage of natural ventilation. Typically, outdoor air temperature and humidity are compared to the thermal comfort zone (ASHRAE 2017), and associated wind velocities are assessed for their potential to provide sufficient air change rate per hour (ACH) for a building. In principle, ACH can be correlated to the incoming wind velocity and temperature difference between indoor and outdoor environments, because they are the two driving forces that create the pressure difference to establish airflow rates through open windows (Awbi 1991; Clarke 2001; Mahdavi and Proglhof 2008). The calibrated correlations can be found in the form of equations (Clarke 2001; Mahdavi and Proglhof 2008) or as graphically represented functions (Awbi 1991) with the following general form:

$$ACH = \text{Function}\,(\textit{wind velocity, temperature difference}). \tag{3.37}$$

This simple equation can be used only for rough (but useful) estimates, especially if a particular building has an available calibrated function. The calculation of the actual airflow rates for different outdoor weather conditions is not simple, because the actual airflow rates depend not only on the combined effect of wind and stack, but also on window distribution and size. In the design process for predicting these airflow rates, the starting point is determining wind pressure on the external building surface. Semi-empirical equations (ASHRAE 2017, chapter 24) or CFD can be used to calculate pressure coefficients for different site plans, and wind pressures can be estimated for different building shapes and different locations of building openings.

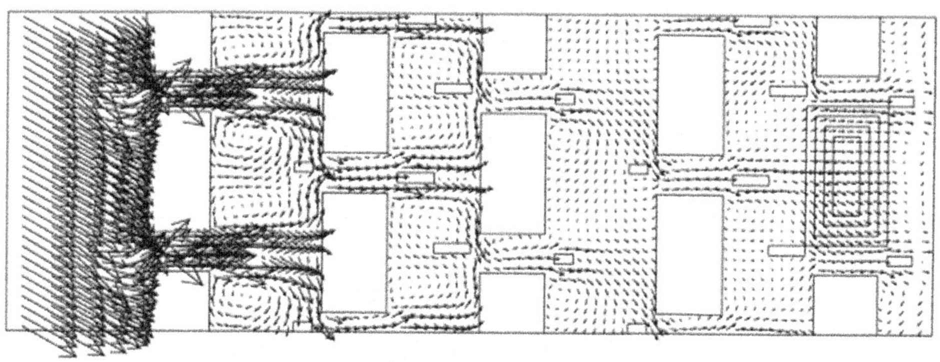

Figure 3.7 An example of CFD calculated wind velocities to support site planning and promote natural ventilation for all residences in an urban neighborhood

Source: Tantasavasdi et al. 2001

These CFD simulations need to incorporate several rows of buildings, and incoming wind profile is typically defined as the power law wind velocity profile (Awbi 1991; Allard 1998):

$$U(z) = K\,U_H\,z^a, \tag{3.38}$$

where U_H is the reference wind velocity [m/s], z is the distance from the ground [m], while the coefficient K and the exponent a are constants dependent on the terrain surface roughness [–]. The values for these coefficients are given in Table 3.3. The reference wind velocity U_H is measured at weather stations, and it is defined as the hourly averaged wind speed, i.e. the wind speed that is exceeded 50% of the time at a particular site (see Chapter 6).

Once the wind velocities and pressure distribution or pressure coefficients are established, the design process of natural ventilation focuses on the internal environment through calculations of airflow at the building openings and the internal environment. Empirical correlations for airflow through different types of openings are available in the literature, not only for the simple single-sided ventilation, but also for much more complex natural ventilation cases (Allard 1998; CIBSE 2005). In addition, CFD can be used to study the flow through building openings as well as the internal environment to address the design issue of floor planning and layout of internal partitions. Figure 3.8 shows the calculated velocities with CFD simulations

Table 3.3 Typical values for terrain dependent parameters

Terrain types	*Power-law coefficients*	
	K	*a*
Open flat country	0.68	0.17
Country with scattered wind breaks	0.52	0.20
Urban (industrial or forest areas)	0.35	0.25
Center of large city	0.21	0.33

Source: Allard 1998

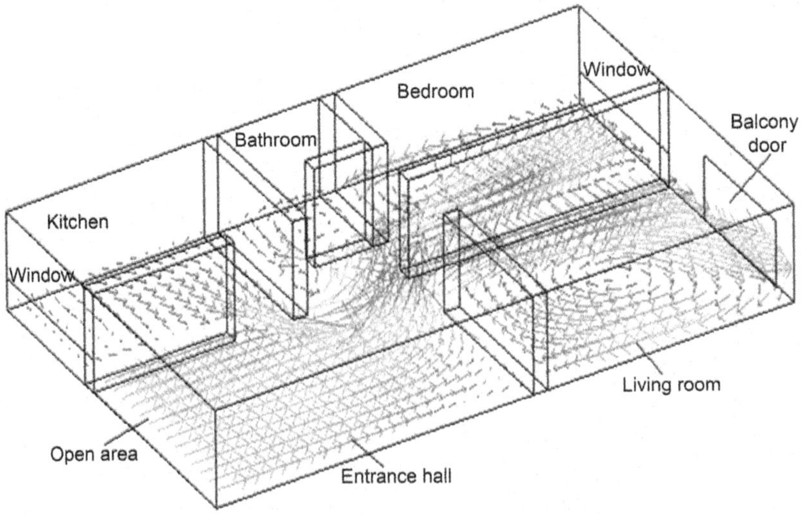

Figure 3.8 A CFD example showing air velocities used in floor planning and layout of internal partitions for a naturally ventilated apartment

that can be used in design decisions of floor planning. In particular, the partitions are selected to promote natural ventilation by minimizing obstructions in the cross-ventilation airflow pattern.

For buildings that use both natural and mechanical ventilation, in the form of hybrid ventilation, the last step is the integration of the two ventilation modes. This integration is achieved through the building management system that provides different control options depending on the climate and season (Lomas et al. 2007). In particular, Equation 3.37 can be calibrated with experimental or multi-zone airflow network model data, which are used in the development of optimal control strategy (Mahdavi and Proglhof 2008). This model-based control of HVAC systems is an important trend for integrated building design (Hensen and Augenbroe 2004) that will dynamically optimize not only the ventilations systems, but also other components of HVAC operation, as also discussed in Chapters 13 and 14.

Mechanical ventilation

For mechanically ventilated rooms, thermal comfort and IAQ strongly depend on the performance of air supply diffusers. The prediction of diffuser flow characteristics is challenging, because the jet flow from a diffuser is complicated due to the complex diffuser geometry and the effects of confined space. Jet formulae are commonly used to describe the flow and thermal characteristics of supply diffusers. The jet formulae are empirical expressions for jet velocity and temperature decay, profiles, and trajectory for inclined jets and separation distance for cold ceiling jets. Researchers have spent numerous decades developing the jet formulae, and today the approach is the most widely used to design mechanical ventilation systems.

Diffuser air jets are classified according to the most important parameters for jet diffusion, such as jet temperature and distance to the nearest enclosure surface. A jet can be isothermal if the jet temperature is the same as the room air temperature or non-isothermal if the temperatures are different. With respect to the distance from enclosure surfaces, a jet can be free or attached. When an air diffuser is installed close to the side walls, floor, or ceiling, the jet attaches to the surface due to the *Coanda effect*. The attachment occurs because the jet has a restricted entrainment on the wall side, and, as a result, a zone of lower air pressure forms and drags the jet towards the wall. On the contrary, a free jet has unrestricted momentum diffusion and room air entrainment. In practice, if the enclosure is big enough to have a minimal impact on the jet trajectory and spread, a jet can be treated as a free jet.

An air jet is normally divided into four zones: core, transitional, main, and terminal, as shown in Figure 3.9 (Awbi 1991). The notation in the figure includes the air supply velocity, U_o, the effective area of the supply opening, A_o, the centerline velocity, U_m, and the distance from the supply opening, x. In the core zone, the centerline velocity is equal to the initial jet velocity. The core region usually extends approximately four diameters or widths of the supply opening. The transitional region can be neglected for circular jets or plane jets when the aspect ratio (width/height of square opening) is smaller than 13.5 (Awbi 1991). For most commonly used supply diffusers, the first two zones are relatively small compared to the third or main jet zone. In the main jet zone, the flow is fully developed, and the velocity profiles are self-similar and can be described with non-dimensional jet profiles.

A fully developed jet flow is very important for ventilation design because diffuser jets usually enter the occupied space within their main zone (third zone). Therefore, the main zone is a subject of study by many researchers (Li et al. 1993). Actually, not only jet velocity profiles, but also jet temperature and concentration profiles tend to be self-similar in the main zone. Within a jet, heat and mass transfer are more intense than momentum transfer. Consequently, for the temperature and concentration profiles, the core region fades faster. The main zone for temperature and concentration develops faster than that for momentum (Grimitlyn and Pozin 1993).

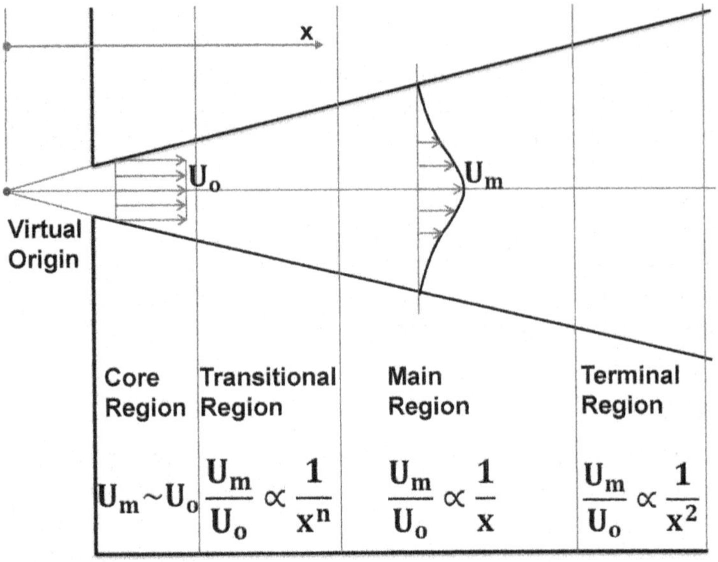

Figure 3.9 Jet zones and characteristic velocity profiles in log-log plot

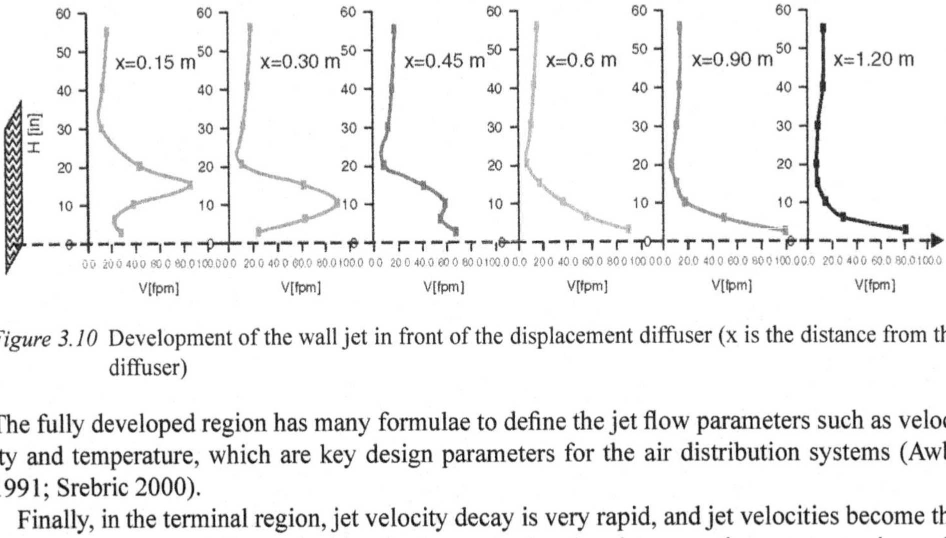

Figure 3.10 Development of the wall jet in front of the displacement diffuser (x is the distance from the diffuser)

The fully developed region has many formulae to define the jet flow parameters such as velocity and temperature, which are key design parameters for the air distribution systems (Awbi 1991; Srebric 2000).

Finally, in the terminal region, jet velocity decay is very rapid, and jet velocities become the same order as room airflow velocities. In the terminal region, jet separation may occur in a cold air distribution system, which can cause draft problems. The theory of the terminal jet region is still under development, and no formulae are available to describe jet degradation.

Some diffuser jets can be successfully predicted with the jet formulae, but not all of them. For example, Figure 3.10 shows the development of jet flow in front of the displacement diffuser (Srebric 2000). The jet drops immediately to the floor in the front of the diffuser because of the low air supply velocity and buoyancy effect. The jet then spreads over the floor and reaches the opposite wall. In front of the diffuser, the jet velocity profile changes along its trajectory. Close to the diffuser, no jet formulae can be used since the jet is in a transition region. Only after 0.9 m does the jet form an attached jet. This result was obtained experimentally, but it can also be simulated with CFD.

CFD approach is very popular due to its flexibility in testing different diffuser layouts and supply air jet parameters. Nevertheless, special attention has to be paid to boundary condition, as the supply diffusers are usually dominant sources of momentum that create airflow pattern responsible for temperature and concentration distributions. The supply boundary condition may be specified as a supply velocity, pressure, or airflow rate. The airflow inlet condition requires specifying the mass flow rate and the temperature and contaminant concentration at the supply diffuser. The velocities and the pressure are calculated, by extrapolation, at the supply diffuser boundaries. In order to have an accurate CFD simulation for a turbulent flow, experimentally measured values of the turbulence parameters are also required for supply diffusers. For the k - ε turbulence models, these values are turbulent kinetic energy k and turbulent dissipation rate ε. When these values are not available from experimental data, they can be estimated from the following equations:

$$k = \frac{3}{2}(U_{ref} TI)^2, \tag{3.39}$$

$$\varepsilon = C_\mu^{3/4} \cdot \frac{k^{3/2}}{\ell}, \tag{3.40}$$

$$\ell = 0.07L, \tag{3.41}$$

where U_{ref} is the mean stream velocity [m/s], TI is the turbulence intensity [−], ℓ is the turbulence length scale [m], C_μ is the $k\text{–}\varepsilon$ turbulence model constant, and L is the characteristic length of the supply diffuser (for a duct L is the equivalent radius) [m].

For ventilation performance predictions, the inlet boundary is especially important due to the use of supply diffusers with potentially complex geometry designed to produce particular performance characteristics. Detailed diffuser modeling is possible for limited regions in the vicinity of the diffuser, but is not very practical in room-flow simulations, because including the small geometric details of the diffuser in the CFD model of the room would result in a grid distribution that requires enormous computational time. Therefore, most room airflow simulations should use simplified diffuser modeling that replicates the diffuser performance without explicitly modeling the fine geometrics details of the diffuser. The most obvious simplified method is to replace the actual diffuser with less complicated diffuser geometry, such as slot opening, that supplies the same flow momentum and airflow rate to the space as the actual diffuser does (Nielsen 1992; Srebric and Chen 2002). The simplified methods are classified as jet momentum modeling at the air supply devices and jet momentum modeling in front of the air supply devices (Fan 1995). The modeling at the supply device has several variations such as basic, slot, and momentum model, while the methods for the modeling in front of the diffuser are prescribed velocity, box, and diffuser specification model (Srebric and Chen 2002). The most widely used methods are:

- momentum,
- box, and
- prescribed velocity.

The *momentum method* decouples the momentum and mass boundary conditions for the diffuser in CFD simulations (Chen and Moser 1991). The diffuser is represented with an opening that has the same gross area, mass flux, and momentum flux as a real supply diffuser. This model enables specification of the source terms in the conservation equations over the real diffuser area. The air supply velocity, U_o, for the momentum source term is calculated from the mass flow rate $\dot{m}$ and the diffuser effective area A_o:

$$U_0 = \frac{\dot{m}}{\rho A_0} \tag{3.42}$$

The specification of boundary conditions with the momentum method is very simple, but the method might not work well for certain types of diffusers (Srebric 2000; Srebric and Chen 2002).

The *box method* is based on the wall jet flow generated close to the diffuser (Nielsen 1992; Srebric and Chen 2002). The details of the flow in the immediate vicinity of the supply opening are ignored, and the supplied jet is described along imaginary box surfaces in front of the supply diffuser. The data for velocity distribution in a wall jet (or a free jet) generated by different commercial diffusers can be obtained from diffuser catalogs or design guide books as *ASHRAE Fundamentals* and from other textbooks (Awbi 1991; Rajaratnam 1976; Etheridge and Sandberg 1996).

The *prescribed velocity method* has also been successfully used in the numerical prediction of room air movement. The inlet profiles are given as boundary conditions of a simplified slot diffuser, and they are represented only by a few grid points. All the variables, except the velocities, are predicted in a volume close to the diffuser as well as in the rest of the room. The velocities are prescribed in the volume in front of the diffuser as the fixed analytical values obtained for a wall jet from the diffuser, or they are given as measured values in front of the diffuser (Nielsen 1992).

More detailed description of available simplified methods and their applicability to commonly used supply diffusers can be found in the literature (Chen and Srebric 2002). Figure 3.11 shows how real diffusers can be simplified by the momentum method and box method in CFD simulations (Srebric 2000).

Mixing and displacement ventilation

Depending on the type of the supply diffusers, mechanical ventilation is classified as either mixing or displacement ventilation. Mixing ventilation relies on the forced airflow in the occupied space to remove air contaminants, bring in fresh air for breathing, and preserve thermal comfort. These tasks are accomplished by completely mixing the air in the occupied zone with the supply air before it is exhausted. Different versions of this system exist and include supply with different ceiling diffusers, such as square or round ceiling diffusers. In contrast, displacement ventilation relies on stratified flow with minimal or no mixing, and several different variations of this system exist.

In displacement ventilation systems for cooling applications, conditioned air with a temperature slightly lower than the desired room air temperature in the occupied zone is supplied. The air supply diffusers are located at or near the floor level, and the supply air is directly introduced to the occupied zone. Returns are located at or close to the ceiling through which the warm room air is exhausted from the room as shown in Figure 3.12. The supply air is spread over the

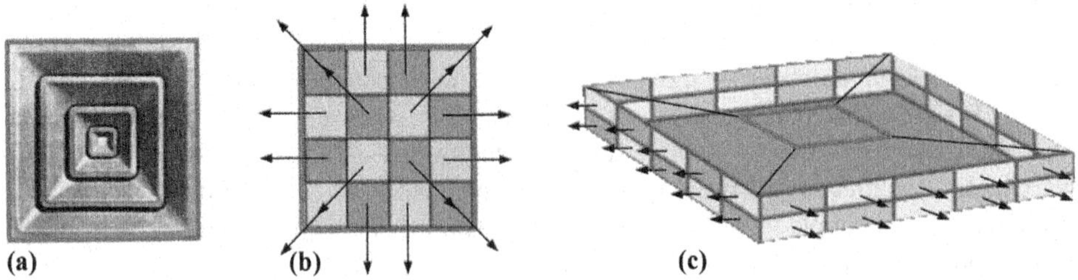

Figure 3.11 An example of simplified boundary condition for (a) supply (diffuser) modeling for square diffuser, (b) diffuser modeling by momentum method, and (c) diffuser modeling by box method

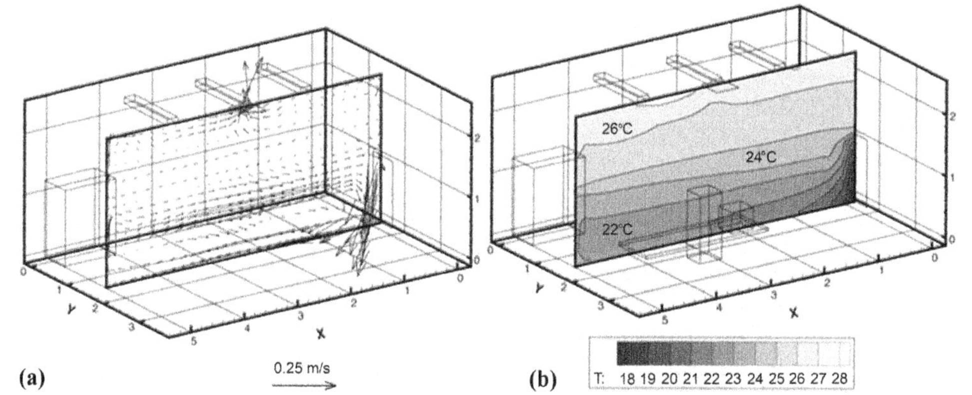

Figure 3.12 Computational fluid dynamics (CFD) simulation results for the displacement ventilation (a) velocities and (b) temperatures in the vertical middle section

floor and then rises as it is heated by the heat sources in the occupied zone. Heat sources, such as people and computers, in the occupied zone create upward convective flows in the form of thermal plumes. These plumes remove heat and contaminants that are less dense than air from the surrounding occupied zone.

Design example for mixing and displacement ventilation

An example of a practical CFD use in air distribution system design includes a decision on type of air distribution system with respect to targeted indoor air quality, thermal comfort, and building energy consumption. The objective of a practical study was to evaluate air distribution systems in a painting studio based on airflow pattern, air velocities, air temperatures, and contaminant concentrations. A properly designed ventilation system should provide better indoor air quality, such as lower contaminant concentrations, while maintaining good thermal comfort. Three different air distribution systems were evaluated, and one was selected based on the results of this study. The studied ventilation systems include existing mixing ventilation, modified mixing ventilation, and displacement ventilation, and one of the three systems was recommended for installation.

The mixing ventilation generates a strong jet flow, while the jet from displacement diffuser is weak. As a result, the mixing ventilation "steers" the air in the entire space and creates airflow with mostly uniform air parameters. On the contrary, the displacement diffuser makes a stratified flow for all indoor parameters. Nevertheless, these general descriptions are not sufficient to understand capabilities and deficiencies of a particular ventilation system for a particular space.

To select the best strategy, designers should account for use of the space and multiple different solutions, not only for the ventilations system type, but also for occupant activities and space layout. Figure 3.13 shows the layout of the studied space during art classes when occupants were creating paintings and afterwards drying them in the storage space. The study was performed with CFD simulations, which showed that existing mixing ventilation is as good as the modified mixing ventilation and better than the displacement ventilation. The comparison was based on the contaminant concentrations that need to be removed from the painting studio. Figure 3.14 shows the contaminant concentration levels for two different mixing ventilation systems. As the final result of the study, the selected system was the existing mixing system with several modifications. The recommended modifications included decreasing the flow rate on the supply diffuser above the storage space and adding an exhaust opening in the corner of

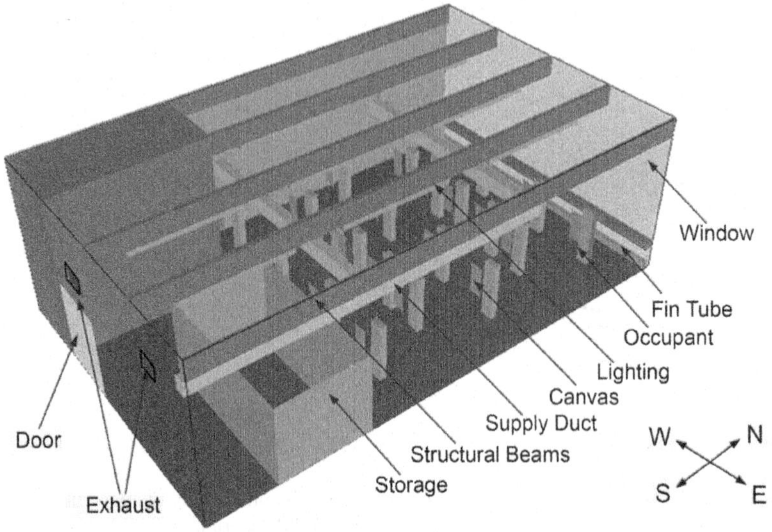

Figure 3.13 Layout of the painting studio during art classes

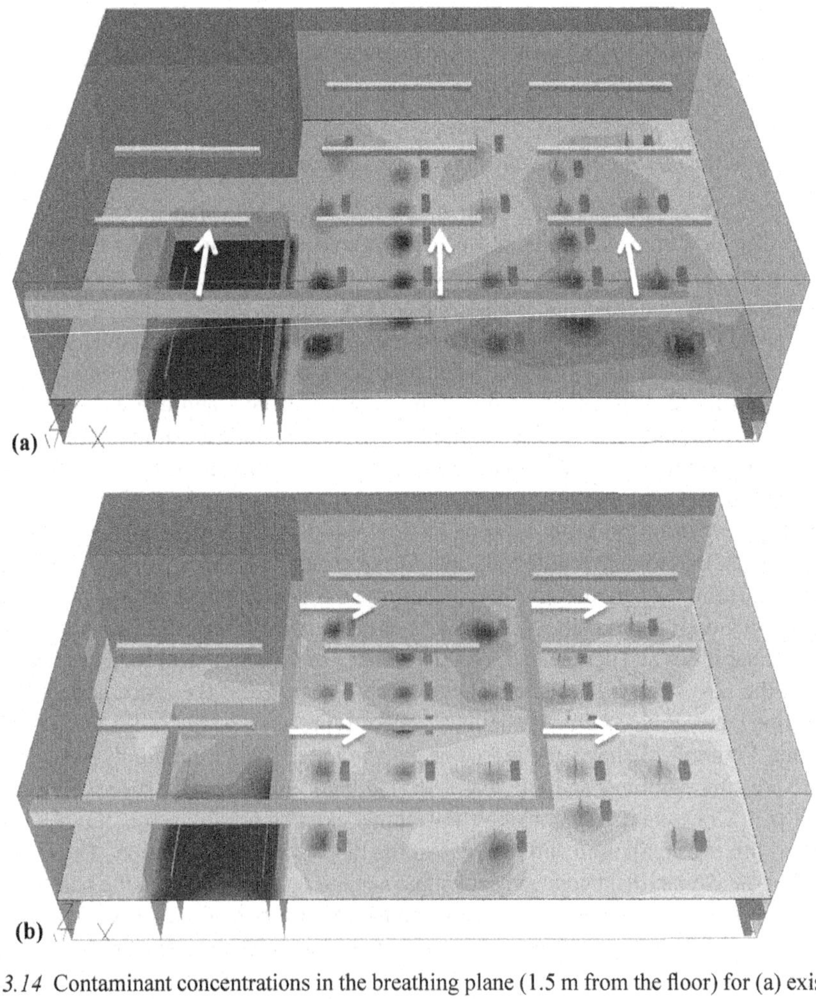

Figure 3.14 Contaminant concentrations in the breathing plane (1.5 m from the floor) for (a) existing and (b) modified mixing ventilation

the room close to the storage area. The study concluded that these minor modifications are more effective on the improvement of the indoor air quality than the change of the air distribution system. In this way, costly and unnecessary modifications were avoided, while the goal of better ventilation was achieved.

The example of the ventilation in a painting studio is a case where only mechanical ventilation can be used to address removal of contaminants from internal sources. Nevertheless, there are many other practical applications where natural ventilation can be used by itself or as hybrid ventilation to include both mechanical and natural ventilation modes.

Discussion

This chapter briefly introduced different simulation tools available for performance prediction of mechanical and natural ventilation systems. The tools range from semi-empirical correlations to simple multi-zone airflow network simulations and complex CFD simulations. Each of the described methods has a set of implied assumptions such as perfect mixing and bulk flow for simpler models, and even much more complex models are based on numerous assumptions such as the eddy viscosity or the Boussinesq approximation. A responsible user of different models has to be aware of all of these assumptions, so that over a period of extensive use it becomes clear how these approximations make models either effective or ineffective in treating specific design and operation questions. It is easy to become excited about opportunities for simulation tools, but they have severe limitations that should be understood and addressed through future research efforts.

A good starting point in learning specific simulation tools for ventilation performance prediction is to reproduce an existing study with detailed published data. In particular, it is important to pick a study that has both simulation and experimental data to provide a new user with an opportunity to understand the accuracy of simulation tools as described in Chapter 1. For example, when comparing the simulation results to measured data, users can understand the instrumentation accuracy, and the difference between computational and experimental results can be put into perspective. If the simulation results and experimental data appear not to "disagree," but the difference is still smaller than the accuracy of the instruments used for measurements, then the agreement is actually "good," even though the graphical representation appears to present discrepancies. Two case studies are presented as assignments in the end of this chapter, so that a new user can start building experience with multi-zone airflow network and CFD simulation tools.

The accuracy of simulation tools strongly depend on the appropriate setting of boundary conditions and numerical simulation parameters. As discussed, every CFD user who intends to apply the results in design and operational practice is responsible for verification of the model capabilities, validation of calculated data, and appropriate reporting of not only the results, but also the simulation parameters. For new users, the verification process includes searching for a similar previously published study as well as understanding the required physical models. The validation process does not have to be conducted on every single simulated case. In principle, the user should find an existing case as close as possible to the new design case of interest, and first perform simulations and validation for the existing case. In this way, the user builds necessary experience and can make an informed evaluation of new simulation results. Last, but not least, a detailed report of CFD results that includes important simulation parameters provides a quality assessment of the CFD study conducted.

Synopsis

This chapter described a range of different analytical and simulation tools to predict ventilation performance in design and operation of buildings. The performance predictions used examples of mechanical, natural, and hybrid ventilation to illustrate capabilities of different simulation tools for ventilation design. Simulation tool users are strongly advised to expand the briefly introduced concepts by exploring the recommended readings. A clear future trend will be represented by the coupled simulation tools that have available empirical equations, multi-zone airflow network models, CFD models, and energy simulations in a single platform. The users of coupled tools will be responsible to use the simplest, most applicable available tools in order to accurately resolve design and operational problems for ventilation systems.

References

AIAA. (2002) Guide for the verification and validation of computational fluid dynamics simulations. American Institute of Aeronautics and Astronautics (AIAA) *Standard* G-077–1998.

Allard, F. (1998) *Natural Ventilation in Buildings: A Design Handbook*. London: James & James Science Publishers Ltd, 356 pp.

ASHRAE (2013) *ANSI/ASHRAE Standard 55–2013: Thermal Environmental Conditions for Human Occupancy*. American Society of Heating, Refrigerating and Air-Conditioning Engineers (ASHRAE), Inc.

ASHRAE (2016) *ANSI/ASHRAE Standard 62.1–2016: Ventilation for Acceptable Indoor Ari Quality (IAQ)*. American Society of Heating, Refrigerating and Air-Conditioning Engineers (ASHRAE), Inc.

ASHRAE (2017) *Handbook of Fundamentals*. Chapters 13, 16, 20, and 24. Atlanta, GA: American Society of Heating, Refrigerating and Air-Conditioning Engineers (ASHRAE).

Awbi, H.B. (1991) *Ventilation of Buildings*. London: Chapman & Hall.

Blevis, R.D. (1984) *Applied Fluid Dynamics Handbook*. New York: Van Nostrand Reinhold Company Inc.

Chen, Q. (1995) Comparison of different k-ε models for indoor airflow computation. *Numerical Heat Transfer, Part B, Fundamentals* 28: 353–369.

Chen, Q. and Moser, A. (1991) Simulation of a multiple-nozzle diffuser. *Proceeding of 12th AIVC Conference* 2: 1–14.

Chen, Q. and Srebric, J. (2002) A procedure for verification, validation, and reporting of indoor environment CFD analyses. *International Journal of HVAC&R Research* 8(2): 201–216.

CIBSE (2005) *Natural Ventilation in Non-domestic Buildings*. London: The Chartered Institution of Building Service Engineers (CIBSE).

Deardorff, J.W. (1970) A numerical study of three-dimensional turbulent channel flow at large Reynolds numbers. *Journal of Fluid Mechanics* 42: 453–480.

Djunaedy, E., Hensen, J.L.M. and Loomans, M.G.L.C. (2004) Selecting an appropriate tool for airflow simulation in buildings. *Building Services Engineering Research and Technology* 25(3): 269–278.

DOE. (2008) User guides: Getting started and overview. Department of Energy, November 2008, 75 pages.

Dols, W.S. and Walton, G.N. (2002) CONTAMW2.0 User Manual. National Institute of Standards and Technology (NIST), NISTIR 6921, November 2002, 162 pages.

Etheridge, D. and Sandberg, M. (1996) *Building Ventilation – Theory and Measurement*. West Sussex: John Wiley & Sons.

Fan, Y. (1995) CFD modeling of the air and contaminant distribution in rooms. *Energy and Buildings* 1995(23): 33–39.

Fang, L., Clausen, G. and Fanger, P.O. (1998) Impact of temperature and humidity on the perception of indoor air quality. *Indoor Air* 8: 80–90.

Ferziger, J.H. and Peric, M. (2001) *Computational Methods for Fluid Dynamics*. New York: Springer-Verlag.

Feustel, H.E. and Smith, B.V. (1997) *COMIS 3.0 – User's Guide*. Lawrence Berkeley National Laboratory, August 1997, 151 pp.

Grimitlyn, M.I. and Pozin, G.M. (1993) Fundamentals of optimizing air distribution in ventilated spaces. *ASHRAE Transactions* 99(1): 1128–1138.

Hagentoft, C.E. (2001) *Introduction to Building Physics*. Lund, Sweden: Studentlitteratur.

Hensen, J. (1999) Comparison of coupled and decoupled solutions for temperature and air flow in a building. *ASHRAE Transactions* 105: 962–969.

Hensen, J. and Augenbroe, G. (2004) Performance simulation for better building design. *Energy and Buildings* 36(8): 735–736.

Launder, B.E. and Spalding, D.B. (1974) The numerical computation of turbulent flows. *Computer Methods in Applied Mechanics and Energy* 3: 269–289.

Li, Z.H., Zhang, J.S., Zhivov, A.M. and Christianson, L.L. (1993) Characteristics of diffuser air jets and airflow in the occupied regions of mechanically ventilated rooms – a literature review. *ASHRAE Transactions* 99(1): 1119–1127.

Lomas, K.J., Cook, M.J. and Fiala, D. (2007) Low energy architecture for a sever U.S. climate: Design and evaluation of a hybrid ventilation system. *Energy and Buildings* 39: 32–44.

Loomans, M. (1998) The measurement and simulation of indoor air flow. PhD Dissertation, Eindhoven University of Technology, The Netherlands.

Mahdavi, A. and Proglhof, C. (2008) A model-based approach to natural ventilation. *Building and Environment* 43(4), April 2008, *Building Performance Simulation*: 620–627.

Melikov, A.K., Popiolek, Z., Silva, M.C.G., Care, I. and Sefker, T. (2007) Accuracy limitations for low-velocity measurements and draft assessment in rooms. *HVAC and R Research* 13(6): 971–986.

Nakahashi, K. (2008) Building-cube method: A CFD approach for near-future PetaFlops computers. In 8th World Congress on Computational Mechanics (WCCM8), June 30–July 5, 2008, Venice, Italy.

National Research Council. (2006) Review and assessment of the health and productivity benefits of green schools: An interim report committee to review and assess the health and productivity benefits of green schools. National Research Council of the National Academies.

Nielsen, P.V. (1992) Description of supply openings in numerical models for room air distribution. *ASHRAE Transactions* 98(1): 963–971.

NIOSH. (1997) *NIOSH facts: Indoor Environmental Quality (IEQ). National Institute for Occupational Safety and Health* (NIOSH). www.cdc.gov/niosh/topics/indoorenv/.

Novoselac, A. (2004) Combined airflow and energy simulation program for building mechanical system design. PhD Thesis, Department of Architectural Engineering, The Pennsylvania State University, October 2004, 276 pp.

Novoselac, A. and Srebric, J. (2003) Comparison of air exchange efficiency and contaminant removal effectiveness as IAQ indices. *ASHRAE Transactions* 109(2): 339–349.

Rajaratnam, N. (1976) *Turbulent Jets*. Amsterdam: Elsevier Scientific Publishing Co.

Sherman, M. and Grimsrud, D.T. (1980) Infiltration-pressurization correlation: Simplified physical modeling. *ASHRAE Transactions* 86(2): 778–807.

Sorensen, D.N. and Nielsen, P.V. (2003) Quality control of computational fluid dynamics in indoor environments. *Indoor Air* 13: 2–17.

Srebric, J. (2000) Simplified methodology for indoor environment design. PhD Thesis, 250 pages, Department of Architecture, Massachusetts Institute of Technology, Cambridge, MA.

Srebric, J. and Chen, Q. (2002) Simplified numerical models for complex air supply diffusers. *International Journal of HVAC&R Research* 8(3): 277–294.

Srebric, J., Chen, Q. and Glicksman, L.R. (1999) Validation of a zero-equation turbulence model for complex indoor airflows. *ASHRAE Transactions* 105(2): 414–427.

Srebric, J., Yuan, J. and Novoselac, A. (2008) In-situ experimental validation of a coupled multi-zone and CFD model for building contaminant transport simulations. *ASHRAE Transactions* 114(1): 273–281.

Swami, M.V. and Chandra, S. (1988) Correlations for pressure distribution on buildings and calculation of natural-ventilation airflow. *ASHRAE Transactions* 94(1): 243–266.

Tantasavasdi, C., Srebric, J. and Chen, Q. (2001) Natural ventilation design for houses in Thailand. *Energy and Buildings* 33(8): 815–824.

Tennekes, H. and Lumley, J.L. (1972) *A First Course in Turbulence*. Cambridge, MA: MIT Press.

USGBC (2009) LEED for New Construction & Major Renovations. U.S. Green Building Council (USGBC), Approved November 2008 and Updated July 2012, 105 pages.

Versteeg, H.K. and Malalasekera, W. (2007) *An Introduction to Computational Fluid Dynamics*. Harlow: Longman.

Vukovic, V., Tabares-Velasco, P.C. and Srebric, J. (2009) Real-time identification of indoor pollutant source positions based on neural network locator of contaminant sources (LOCS) and optimized sensor networks. Accepted by *Journal of the Air & Waste Management Association*.

Walton, G.N. (1989) Airflow network models for element-based building airflow modeling. *ASHRAE Transactions* 95(2): 611–620.
Wang, L. and Chen, Q. (2008) Evaluation of some assumptions used in multi-zone airflow network models. *Building and Environment* 43(10): 1671–1677.
Wargocki, P., Sundell, J., Bischof, J., Brundrett, G., Fanger, P.O., Gyntelberg, F., Hanssen, O., Harrison, P., Pickering, A., Seppanen, O.A. and Wouters, P. (2002) Ventilation and health in non-industrial indoor environments: Report from European Multidisciplinary Scientific Consensus Meeting (EUROVEN). *Indoor Air* 12: 113–128.
Wilcox, D.C. (2006) *Turbulence Modeling for CFD*. La Cañada, CA: DCW Industries, Inc.
Yaglou, C.P. and Witheridge, W.N. (1937) Ventilation requirements. *ASHRAE Transactions* 42(Pt. 2): 423–436.
Yokhot, V. and Orszag, S.A. (1986) Renormalization group analysis of turbulence – I – basic theory. *Journal of Scientific Computing* 1(1): 3–51.
Yoshino, H., Yun, Z., Kobayashi, H. and Utsumi, Y. (1995) Simulation and measurement of air infiltration and pollutant transport using a passive solar test house. *ASHRAE Transactions* 101(Pt. 1): 1091–1099.
Zhao, M., Kim, Y.S. and Srebric, J. (2015) Occupant perceptions and a health outcome in retail stores. *Building and Environment* 93(2): 385–394.

Recommended reading

Clarke, J.A. (2001) *Energy Simulation in Building Design, 2nd edition*. Oxford: Butterworth-Heinemann.
Etheridge, D. and Sandberg, M. (1996) *Building Ventilation – Theory and Measurement*. West Sussex: John Wiley & Sons.

Assignments and model answers

Two assignments are included to enable practice of multi-zone airflow network and CFD modeling. Both assignments are specified in such a way that users can conduct them with any of the available simulation tools. The first assignment is to model a house with a multi-zone airflow network tool, and the second assignment is to model an office with a CFD tool.

1. Multi-zone airflow network assignment

The objective of this assignment is to simulate contaminant dispersion due to infiltration in a small house. To correctly predict dispersion, the prediction of infiltration and airflow pattern has to be accurate. This particular case is selected because it has detailed measured data published in the literature (Yoshino et al. 1995). This study conducted measurements in a three-zone single-story building as shown in Figure 3.15(a). Rooms 1, 2, and 3 had contaminant concentration sensors and were completely isolated from the unmarked room on the right-hand side in Figure 3.15(a). Each of the three rooms (zones) had a burst source releasing different tracer gases: $CHCl_2F$, N_2O, and SF_6. This setup allowed simultaneous tracer gas releases and concentration measurements within the three rooms (zones).

This assignment is focused on simulation of SF_6 with a point source in room 3. Figure 3.15(b) presents constructed multi-zone airflow network model for three relevant rooms, defining the contaminant source location in room 3 (zone 3). The following model parameters were selected based on the experimental study (Yoshino et al. 1995, Vukovic et al. 2009):

- The height of the floor is set to 2.8 m with zero elevation from the ground.
- Table 3.4 specifies detailed properties of the three building zones.
- For all flow paths between the zones, one-way flow elements are selected. The choice is based on a multi-zone airflow network tool that allows users to define flow exponent

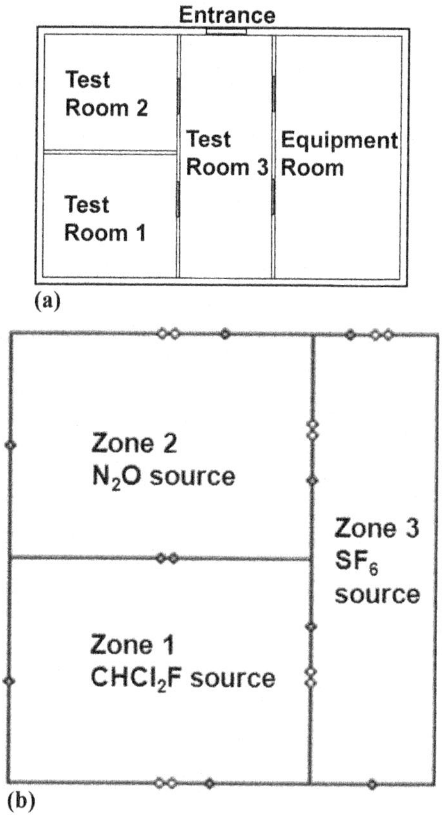

Figure 3.15 (a) The floor plan of the test house and (b) multi-zone airflow network model (◊-flow paths symbol)

Sources: (a) Yoshino et al. 1995; (b) Vukovic et al. 2009

Table 3.4 The summary of the zone properties

Zone number	*Zone name*	*Floor area* [m^2]	*No. of flow paths*	*Temperature* [°C]
1	Room 1	7.425	9	33.4
2	Room 2	7.425	9	26.0
3	Room 3	9.735	10	28.2

values only for one-way flow elements. Consequently, to take into account two-way flows, two differently elevated elements are placed for each real flow path.

- Additional flow elements define infiltration due to background leakage. Table 3.5 specifies the properties of flow path elements between the zones. Relative elevation of the flow elements defines the position of the center of the leakage area of the element with respect to the floor level. All selected flow elements have discharge coefficient 0.6 and pressure drop of 10 Pa during the measurements.
- Outdoor conditions are specified as following: southwest wind of 1.06 m/s, normal atmospheric pressure, and outdoor temperature of 25.3 °C.

Table 3.5 The summary of airflow path properties

Connected zones	*Relative elevation [m]*	*Leakage area [cm²]*
1–2	0.625	86
1–2	1.875	86
1–3	0.625	124
1–3	1.875	124
1–3	1.4	3
1-ext_s	0.85	26
1-ext_s	1.95	26
2–3	0.625	150
2–3	1.875	150
2–3	1.4	3
2-ext_n	1.725	21
2-ext_n	2.175	21
2-ext_n	1.4	6
2-ext_w	1.4	6
3-ext_n	0.625	141
3-ext_n	1.875	141
3-ext_n	1.4	172
3-ext_s	1.4	172
Legend: ext – exterior, n – north, s – south, w – west		

- The initial uniform concentration of SF_6 in room 3 is 1 ppm, which is a normalized value, so that the results are independent of source strength.

Model validation: All measured concentration data showed non-zero background contaminant concentrations. Background concentration levels also differ with respect to contaminant species. In order to compare calculated SF_6 to measured data, these background concentrations are subtracted from all measured concentration values. This procedure enables the assignment modeling task to be independent of the background concentration levels. Figure 3.16 shows a validation that compares simulation results to measured data for SF_6 concentrations in each of the three rooms. Overall, even though the model is simple, the exercise is challenging because it includes the validation.

2. CFD assignment

This assignment requires CFD modeling of a displacement ventilation case shown in Figure 3.17. In addition, Table 3.6 and Table 3.7 provide the most important thermal and flow boundary conditions to be used in any CFD simulation tool, and additional detailed geometry data are available in the literature (Srebric 2000). This particular setup has also available detailed measured data in the occupied zone, so it presents an opportunity for users to validate and verify their simulation technique (Chen and Srebric 2002). Figure 3.12 shows the simulated velocities and temperatures that are plotted in the vertical middle section. The simulated results are in very good agreement with the observed flow pattern and measured velocities and temperatures (Srebric 2000). To substantiate this statement, validation results are presented in Figure 3.17.

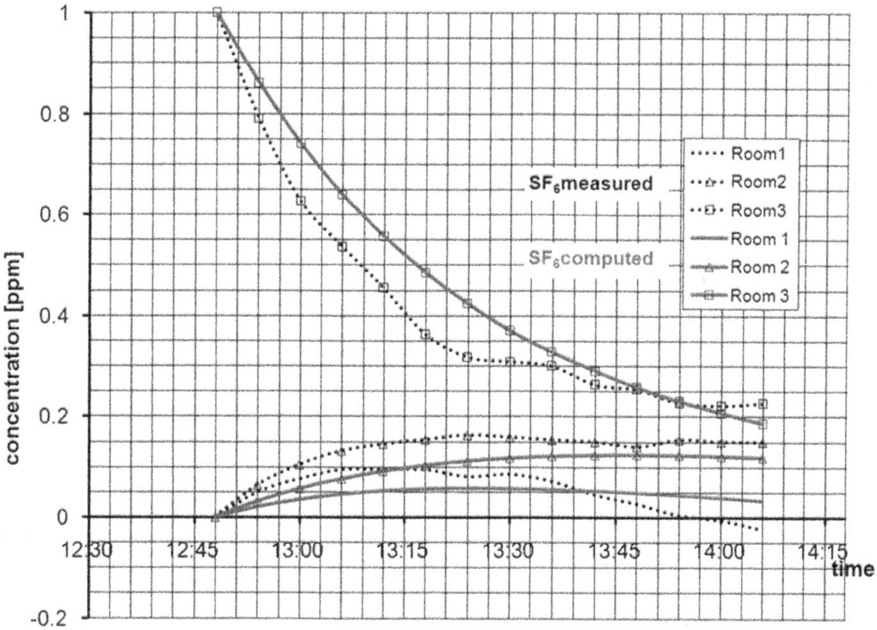

Figure 3.16 Measured vs. computed SF_6 concentrations

Source: Vukovic et al. 2009

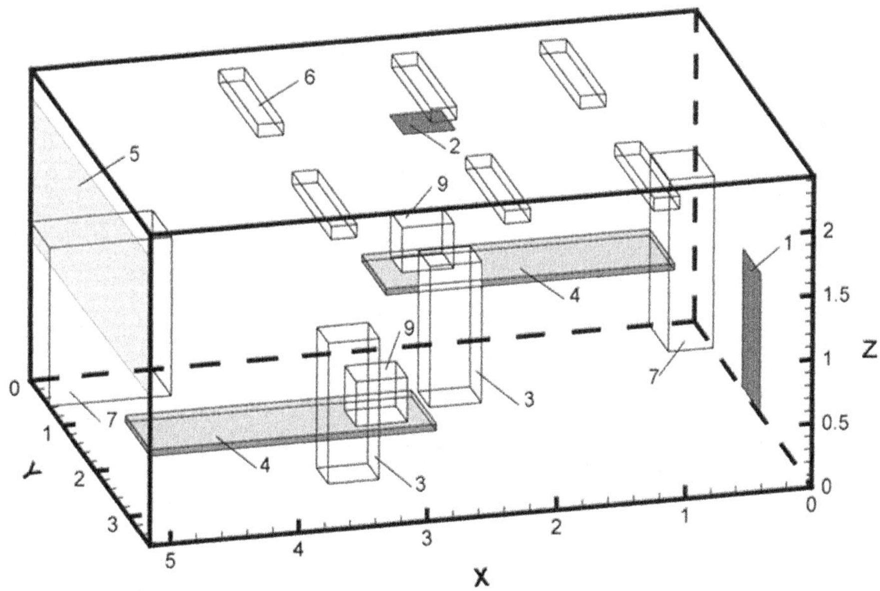

Figure 3.17 The layout of the displacement ventilation simulated and tested in the chamber (inlet -1, outlet -2, person -3, table -4, window -5, fluorescent lamps -6, cabinet -7, computer -9)

Table 3.6 The geometrical, thermal, and flow boundary conditions for the diffuser and window

Displacement ventilation case			
Supply diffuser	Size: 0.53 m × 1.1 m	Temperature: 17.0 °C	Velocity: 0.086 m/s
Window	Size: 3.65 m × 1.16 m	Temperature: 27.7 °C	–

Table 3.7 The size and capacity of the heat sources

Heat source	*Size*	*Power*
Baseboard heater	$1.2 \times 0.1 \times 0.2$ m^3	1500 W
Person	$0.4 \times 0.35 \times 1.1$m^3	75 W
Computer 1	$0.4 \times 0.4 \times 0.4$m^3	108 W
Computer 2	$0.4 \times 0.4 \times 0.4$m^3	173 W
Overhead lighting	$0.2 \times 1.2 \times 0.15$	34 W

Model validation: The detailed comparison between measured and calculated velocity and air temperature is given in Figures 3.18(a) and (b), respectively. The right bottom pictures in Figure 3.18 shows the floor plan of the room where the measurements were carried out at nine positions. In these figures, the computed results are compared with the experimental data in different vertical sections in the room. Overall, Figure 3.18 shows a very good agreement between the computed and measured results.

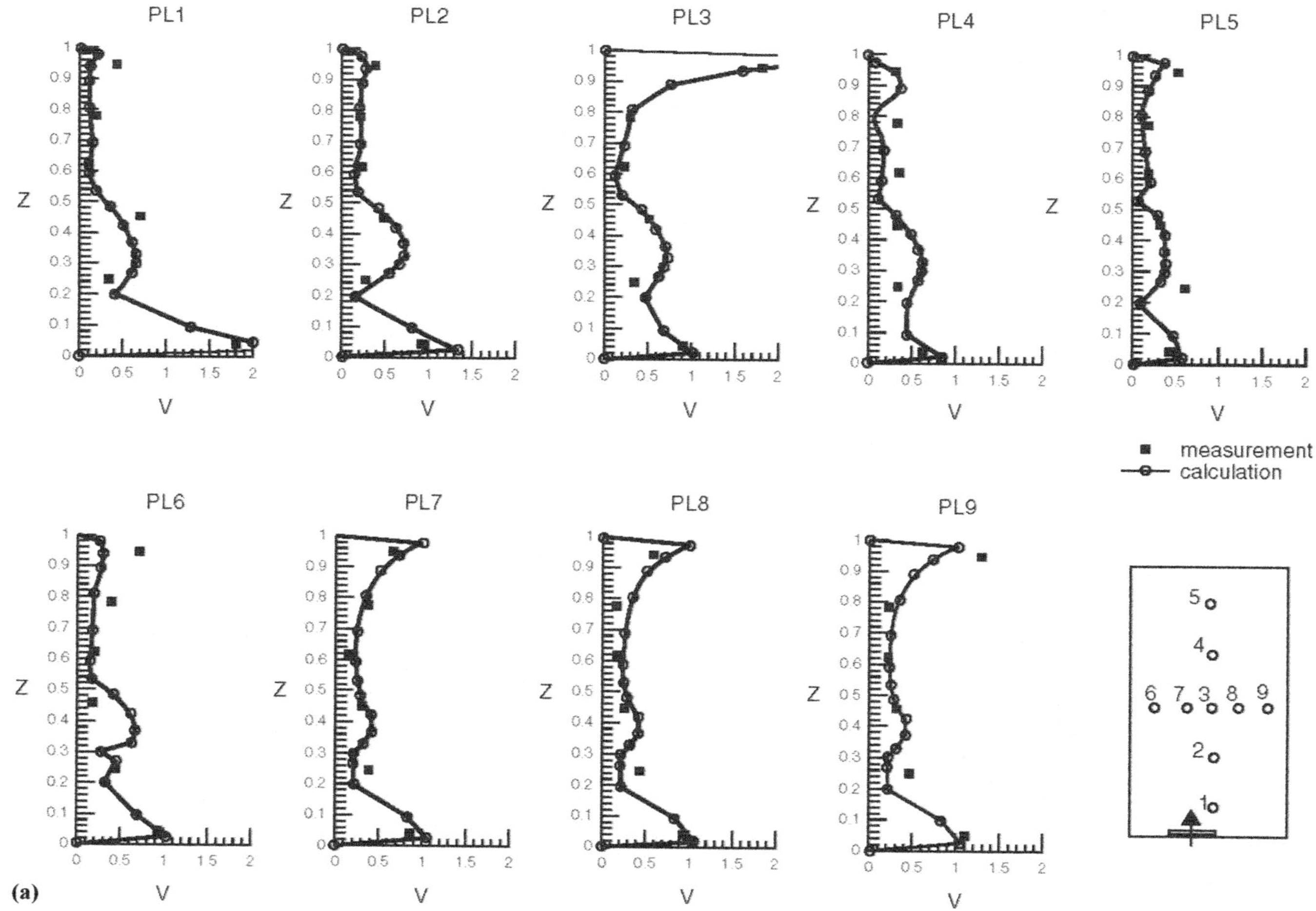

Figure 3.18 Comparison of (a) the velocity profiles and (b) the temperature profiles at nine positions in the room between the simulated and measured data for the displacement ventilation case

Legend: Z = height/total room height (H), V = velocity/inlet velocity (V_{in}), $\theta = (T - T_{in}/T_{out} - T_{in})$, H = 2.43 m, V_{in} = 0.086 m/s, T_{in} = 17.0 °C, and T_{out} = 26.7 °C

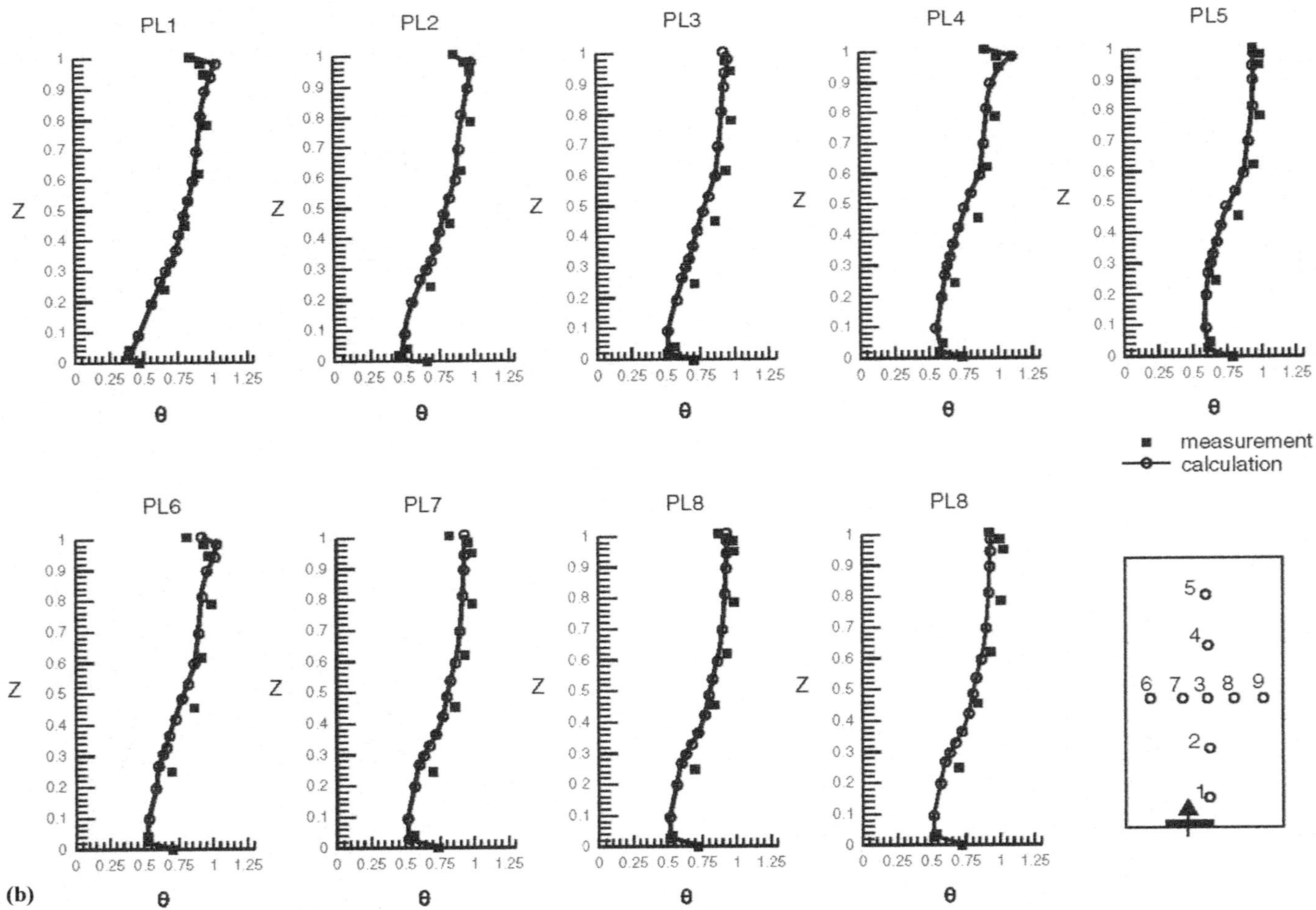

Figure 3.18 (Continued)

4 People in building performance simulation

Ardeshir Mahdavi and Farhang Tahmasebi

Scope

This chapter addresses the representation of people's presence and behavior in building performance simulation.

Learning objectives

The main objective of this chapter is to provide a conceptual framework to systematically situate people in the context of building performance simulation. Toward this end, the motivation behind and the methodological options for the representation of inhabitants' presence and behavior in buildings are discussed.

Key words

Buildings/performance/simulation/inhabitants/presence/behavior

Introduction and overview

The subject of people in building performance simulation can be approached from a number of distinct vantage points (Mahdavi 2011). Simulation tools are of course developed and used by people. Moreover, people's requirements concerning health and comfort inform relevant input assumptions in simulation models, for instance, when heating and cooling set-points are specified for the operation of buildings' environmental control systems. Likewise, people's preferences and needs – for example, with regard to thermal, visual, and acoustical comfort conditions – provide the context to evaluate and interpret simulation results. The present treatment, however, focuses on the representation of people's presence and activities in buildings. These are considered in simulation, simply because they have an impact on the performance of buildings. For instance, buildings' thermal and indoor air quality performance is not only influenced by inhabitants' presence as a source of (sensible and latent) heat and pollutants, but also due to their actions, including use of water, equipment, and appliances, as well as operation of control devices for heating, cooling, shading, ventilation, and illumination. Similarly, the sound field and thus the acoustical performance of rooms are influenced by the presence of people in a room and the associated sound absorption. Also, buildings' safety requirements (pertaining, for example, to means of egress) cannot be evaluated without considering the behavior of people in critical situations such as when a fire alarm sounds or systems malfunction.

Building performance simulation has made substantial advances in the last decades. Computational routines to capture the relevant physical (energy and mass transfer) phenomena have become increasingly efficient. Progress has been also made in the representation of building geometry and construction. Furthermore, empirically grounded high-resolution representations of contextual factors (specifically, weather conditions) have become broadly available.

In comparison, representations of inhabitants in simulation models are less developed. Due, in part, to a relatively short history of pertinent research, there is still a lack of shared understanding of the essential questions and challenges in the representation of people in the building performance simulation. To address this state of affairs, the present chapter focuses on a number of key questions:

1 How do inhabitants influence building performance?
2 What kinds of occupancy-related input information do simulation tools need?
3 What kinds of modeling techniques have been applied toward representation of people in building performance simulation?
4 Do different types of simulation-supported performance queries require different kinds of occupancy-related models, for instance with regard to representational method and informational resolution?
5 What options are available to obtain empirical information for the understanding and modeling of people's presence and behavior in buildings?
6 What are the key issues in the evaluation (validation) of the reliability, fidelity, and predictive capabilities of occupancy-related models?
7 What kinds of approaches have been pursued to implement occupancy-related models in building performance simulation?

Note that, in the framework of the present chapter, we focus predominantly on building simulation applications pertaining to thermal and visual performance domains. Hence, the emphasis is on those aspects of human behavior in buildings that are directly and naturally relevant to the buildings' energy performance, thermal and visual comfort, and indoor air quality. Consequently, modeling challenges pertaining to other domains (e.g. safety and security applications) are not explicitly discussed.

People's impact on building performance

Both empirical and simulation-aided studies have shown the large impact of occupant behavior on buildings' energy use. For instance, Azar and Menassa (2012) observed that energy models of office buildings in different climatic zones in the United States are highly sensitive to occupancy-related behavioral parameters. More specifically, Yang and Becerik-Gerber (2014) showed that application of HVAC schedules that use observation-based personalized occupancy profiles in an office building could save up to 9% energy compared to the conventional schedules. As another example, a post-occupancy evaluation of 26 UK "EcoHomes" revealed that a total of 51%, 37%, and 11% of the variation in heat, electrical, and water consumption, respectively, could be explained by occupant behavior (Gill et al. 2010). Likewise, using post-occupancy monitored data to evaluate 18 design-phase building energy models in Canada, Samuelson et al. (2016) showed that revising the initial assumptions pertaining to occupancy, plug loads, and lighting density – on average – can result in up to 32% reduction in Energy Use Index estimation error.

A useful high-level distinction can be made between "passive" and "active" effects of inhabitants on building performance:

Passive effects result from the sheer presence of people in buildings. Consider, for instance, performance simulation results pertaining to buildings' energy and indoor environmental (thermal, visual, acoustical) performance: Depending on their activity, people release not only sensible and latent heat, but also water vapor, carbon dioxide, and other execrations and odorous substances. Similarly, in the building and room acoustics domain, the presence of people in a space has an effect on the sound field via the introduction of additional sound absorption. To account for the passive effects of people's presence in buildings, simulation specialists typically rely on documented results of empirical studies, such as people's activity-dependent metabolic and respiration rates (see Figure 4.1). Provided such information is available, the modeling process is as such straightforward, barring a possible complexity: The aforementioned passive effects such as heat emission may depend on the context (e.g. thermal conditions in the occupied space). This interdependence suggests that – at least in case of highly detailed numeric simulation models – a dynamic (runtime) coupling between "model inhabitant" and "model environment" may be necessary.

Active effects pertain to the circumstance that, in most buildings, inhabitants have varying degrees of freedom to directly operate the actuators of control devices such as windows, shades, luminaires, radiators, and fans. They may also manipulate control systems' settings, such as heating and cooling temperature set-points and associated schedules. Such control actions modulate energy and mass transfer rates between outdoor and indoor environments and within different spaces in a building. As such, people's active effects on building performance and corresponding simulation-based predictions can be substantial. Systematic simulation-based building performance assessments should thus address control-oriented user behavior. General information about building type (residential, commercial) and environmental systems (free-running, air-conditioned), as well as organizational and administrative information (e.g. working hours), can provide at best a rough idea of the scope and frequencies of inhabitants' control actions and their performance implications. Developing more dependable models of inhabitants' actions, however, represents a major challenge. The current state of domain knowledge provides insufficient grounds for formulating general-purpose white-box behavioral models. Black-box models, on the other hand, require massive amounts of observational data from different building types and locations.

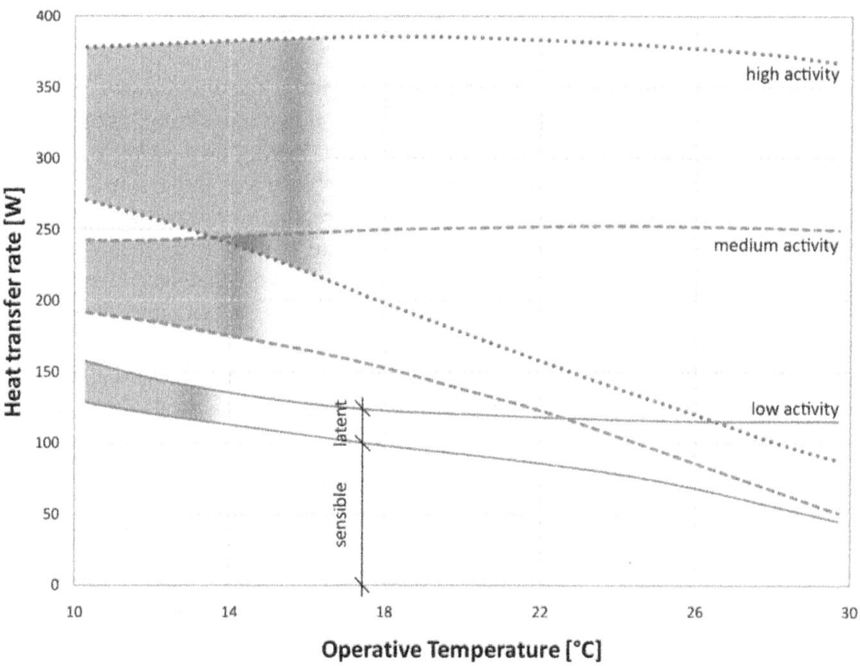

Figure 4.1 Heat transfer rate of human body as a function of activity and the ambient (operative) temperature

Source: Adapted from Rietschel and Raiß 1968

General informational requirements of simulation models

The answer to the question of simulation models' occupancy-related informational requirements can be deceptively simple: Simulation models must contain assumptions regarding inhabitants' state, location, and performance-relevant behavior. Toward this end, multiple theoretical and practical challenges must be addressed. For one thing, where should the necessary information come from? And how should it be framed in a simulation model? This section specifically deals with the latter question.

Simulation models of mass and energy transfer phenomena in buildings frequently deploy numeric solutions of mathematical – typically differential – equations. Depending on the targeted resolution, equation systems may be solved for very different scales in terms of spatial (nodal) networks or discretized time steps. High spatial and temporal resolution may be advantageous in view of capturing a system's behavior realistically. It comes, however, with higher informational and computational costs. In praxis, model resolution levels and commensurate tools are selected depending on the nature of the projects, queries, and available expertise and computational resources. In the process, compromises must be made between desired resolution and available resources. This naturally has implications for the choices pertaining to the representation of inhabitants in the simulation models. Ideally, the granularity of occupancy-related information should match the spatial and temporal granularity of the simulation domain.

To pursue this question systematically, the following basic conceptual reflection would be useful. Consider the temporal representation of occupancy-related circumstances in performance modeling. The time aspects can be accommodated via two complementary alternatives, namely events and states (see basic illustrative examples of Table 4.1). Occupancy-related processes can be captured in terms of events occurring at specific points in time. As such, events can be associated with inhabitants' state and location as well as their actions. Arrival in a room (or occupying a workstation) is a presence-related event, as is leaving a room (or vacating a workstation). Opening a window (or closing one) is an event that signifies a control-oriented action.

The relationship between events and states can be simply stated as follows: Events bring about changes in states. The event of an inhabitant's arrival in a room brings about a change in the room's state of occupancy (from vacant to occupied). Putting on a jacket is an event that brings about a change in the thermal resistance of the inhabitant's clothing. Similarly, the event of operating a device or manipulating its settings (e.g. opening a window, changing a thermostat's position) brings about a change in the device's state or its settings (e.g. from closed to open window, from higher to lower temperature set-points). Note that state changes need not be Boolean (as is the case in the simple instances of Table 4.1): In principle, the position of a window, blind, or valve can vary continuously.

Ideally, representations of buildings' inhabitants in simulation models should rely on comprehensive empirically based assumptions concerning all events associated with or triggered by

Table 4.1 Basic instances of events and states pertaining to inhabitants' presence (at a workstation) and control actions (operation of a window)

Domain	*Subject*	*Event*	*State*	
			Prior to the event	*After the event*
Presence	Workstation	Arrival	Vacant	Occupied
		Departure	Occupied	Vacant
Control action	Window	Opening	Closed	Open
		Closing	Open	Closed

the inhabitants. As also highlighted in a recent international survey (O'Brien et al. 2017), realizing this in praxis is, however, a major challenge due to multiple circumstances. For one thing, there is the previously mentioned paucity of comprehensive observational data. Moreover, the information (including recommended standard-based input data) that is available has typically a low level of granularity in terms of both inter-inhabitant diversity and temporal resolution.

Theoretically, event patterns can be generated synthetically to emulate the stochastic appearance of real patterns with arbitrary levels of granularity. Aside from the issue of the empirical grounding of such an approach, event patterns should also address input requirements of typical simulation applications. In many instances, such applications deploy state-based representations of external boundary conditions and internal processes in accordance with the temporal discretization of the simulation domain (in terms of computation time steps). This implies the need for the translation of event-based information into a state-based representation. Thereby, rules must be explicated to assign states to intervals based on the events' time stamps. As the very basic illustration of Figure 4.2 implies, the event-to-state mapping might result in effectively different distributions of states over time as a function of both the interval size and the deployed assignment rule.

Given this background, we can now discuss one of the central challenges of the dynamic representation of occupancy-related phenomena in building performance simulation, namely the feedback problem. To understand this problem, a comparison with the representation of weather conditions in thermal performance simulation is useful. In conventional (thermal) simulation, weather conditions are obviously assumed to influence a building's performance, but not the other way around. Even though non-zero, the feedback effect of buildings' operation on the surrounding conditions is typically ignored, because of its presumed small magnitude. Consequently, time series of the values of relevant micro-climatic variables can be processed in advance and simply deployed throughout the simulation process as a pre-packaged representation of the external boundary conditions (i.e. a weather file).

Occupancy-related variables are different, in that their values may be dependent on simulation results (i.e. pertinent indoor environmental conditions). This applies not only to inhabitants' passive effects (see, for example, the dependency of people's heat transfer rates on indoor – operative – temperature, as depicted in Figure 4.1), but also to active effects. For instance, operating a window may change indoor environmental conditions, which in turn may influence inhabitants' disposition for further control actions. Likewise, adaptive changes to inhabitants' states (activity, clothing) and even their location may be influenced dynamically by indoor environmental factors. For instance, a drop in indoor air temperature may encourage an adaptive action in terms of increased clothing or even trigger an incentive to move from

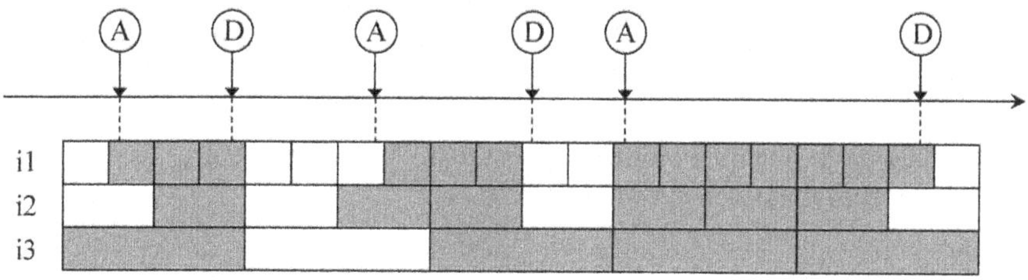

Figure 4.2 Mapping of a theoretical chain of arrival (A) and departure (D) events to presence states (shaded cells) for three different interval sizes (i1, i2, i3). The resulting fraction of presence over the observation period is different for different interval sizes (65% for i1, 60% for i2, and 80% for i3). Note that, in this illustrative instance, an interval's state is declared occupied if, according to the event time stamp information, the state of presence can be assumed to have lasted more than 50% of the interval duration.

one area of a space or building to another. This implies that pre-packaged descriptions (tables and schedules) of inhabitants' states and especially their actions do not establish a principally adequate solution for certain types of detailed performance queries. Rather, in such cases, there is a need for dynamic run-time coupling of building-related energy and mass transfer processes with occupancy-related behavioral phenomena. The mechanics of integrating behavioral models in building simulation are further discussed in the following sections.

Modeling methods

Overview

As stated previously, occupancy-related models for building performance simulation mainly address inhabitants' presence and control actions. The models can target different entities, such as workstations, windows, and shades. The presence and behavior models can be non-probabilistic or probabilistic and may deploy different mathematical methods (e.g. linear regression, logistic regression, and hidden Markov models). The models may only consider the state of devices or also the dynamics of occupants' operation of devices. What follows is a brief overview of modeling approaches for representing inhabitants' presence and behavior in building performance simulation.

Presence models

Schedules

Frequently, presence patterns in building performance models are represented via typical occupancy schedules. These schedules provide the occupancy ratio or the probability of presence as a function of the time of day and the day of week. Commonly three types of days (namely, weekdays, Saturdays, and Sundays and public holidays) are distinguished and treated separately. For building designs, such occupancy schedules can be obtained from relevant standards or previous empirical studies. In case of existing buildings, interviews, questionnaires, or monitored data can be deployed to infer typical occupancy patterns. A widely used example of occupancy schedules for different types of buildings has been provided in ASHRAE Standard 90.1 (ASHRAE 2013). However, for example in the case of office buildings, a number of studies have documented noticeably lower occupancy rates compared to those included in this standard (see, for example, Duarte et al. 2013).

Recently, the focus of the scientific community has shifted towards development of probabilistic presence models. However, this has not removed the need for reliable building occupancy profiles for different types of buildings in different contexts. Specifically, recent studies suggest that the viability of typical building-level simulation results (such as annual heating and cooling demands) is primarily dependent on the availability of reliable estimations of actual occupancy, rather than the probabilistic or non-probabilistic nature of the deployed representation of presence patterns (Tahmasebi and Mahdavi 2017).

Non-probabilistic models

Specific building simulation use cases may require individuals' presence states as input information, while they do not necessarily require randomly generated non-repeating occupancy state profiles. In such cases, non-probabilistic approaches can provide simple solutions. For example, a probability threshold may be defined to derive the occupancy states based on the presence probabilities in the occupancy schedules. That is, each time interval is assumed to be occupied if the corresponding presence probability is higher than the probability threshold. Specifically, for the purpose of simulation-based building systems predictive control, Mahdavi and Tahmasebi (2015) developed this approach by identifying the threshold probability, such that the area under the resulting binary occupancy profile is as close as possible to the area under the profile of presence probability derived from past monitored data.

Probabilistic models

In general, probabilistic presence models generate non-repeating daily profiles of occupancy states while maintaining the available information or assumptions regarding the presence patterns' traits. Thereby, the probabilistic occupancy models provide the occupancy states for each occupant, which can be used as the background for probabilistic action models. In addition, provided that a proper approach to represent diversity is adopted, probabilistic occupancy models can emulate the inter-occupant diversity (see, for example, O'Brien et al. 2016; Mahdavi et al. 2017). As such, for the purpose of an annual building performance simulation, the probabilistic occupancy model must provide 365 daily presence profiles for each occupant in the model. It should be noted that the random nature of daily occupancy patterns implies that slight differences could emerge, if the process would be repeated.

To model presence patterns in a probabilistic manner, Markov chains have been commonly used (e.g. Page et al. 2008; Richardson et al. 2008; Wang et al. 2011). A Markov chain is a memoryless process, in which future states are determined based on the current state and the state changing probabilities. Thus, in case of a presence model, the Markov chain includes the transition probabilities between occupied and vacant states in each time step to determine an arrival when occupants are absent or a departure when occupants are present. The Markov chain is called homogeneous, if the transition probabilities are not time-dependent; otherwise, it is called inhomogeneous (Wolf et al. 2018).

The transition probabilities can be obtained directly from the monitored data or derived based on available typical presence patterns. As an example of the latter possibility, the model developed by Page et al. (2008) uses as input a profile of presence probability and parameter of mobility (defined as the ratio of state change probability to state persistence probability) and returns random non-repeating daily profiles of occupancy states (Figure 4.3). In a number of modeling efforts, survival analysis has been used in combination with the Markov model to obtain the duration of intermediate vacancy or occupancy periods (see, for example, Reinhart 2001; Wilke et al. 2013).

For the cases in which the occupancy level cannot be directly measured, hidden Markov models have been deployed for occupancy prediction based on measured indoor environmental parameters such as CO_2 and sound level (see, for example, Dong et al. 2010). A hidden Markov model consists of a Markov chain with unobserved states and a series of observations, whose values are influenced by the state of the Markov chain. Thus, the model allows obtaining information about the occupancy as an unobserved parameter based on other observations (Wolf et al. 2018).

Behavior models

Schedules

In building performance models, non-adaptive occupant actions (such as use of electric equipment) can be simply represented by schedules. Thereby, the assumption is that the time of the day, day of the week, or the month of the year can be used to make assumptions regarding occupant behavior. However, as these behavioral patterns strongly depend on occupant presence

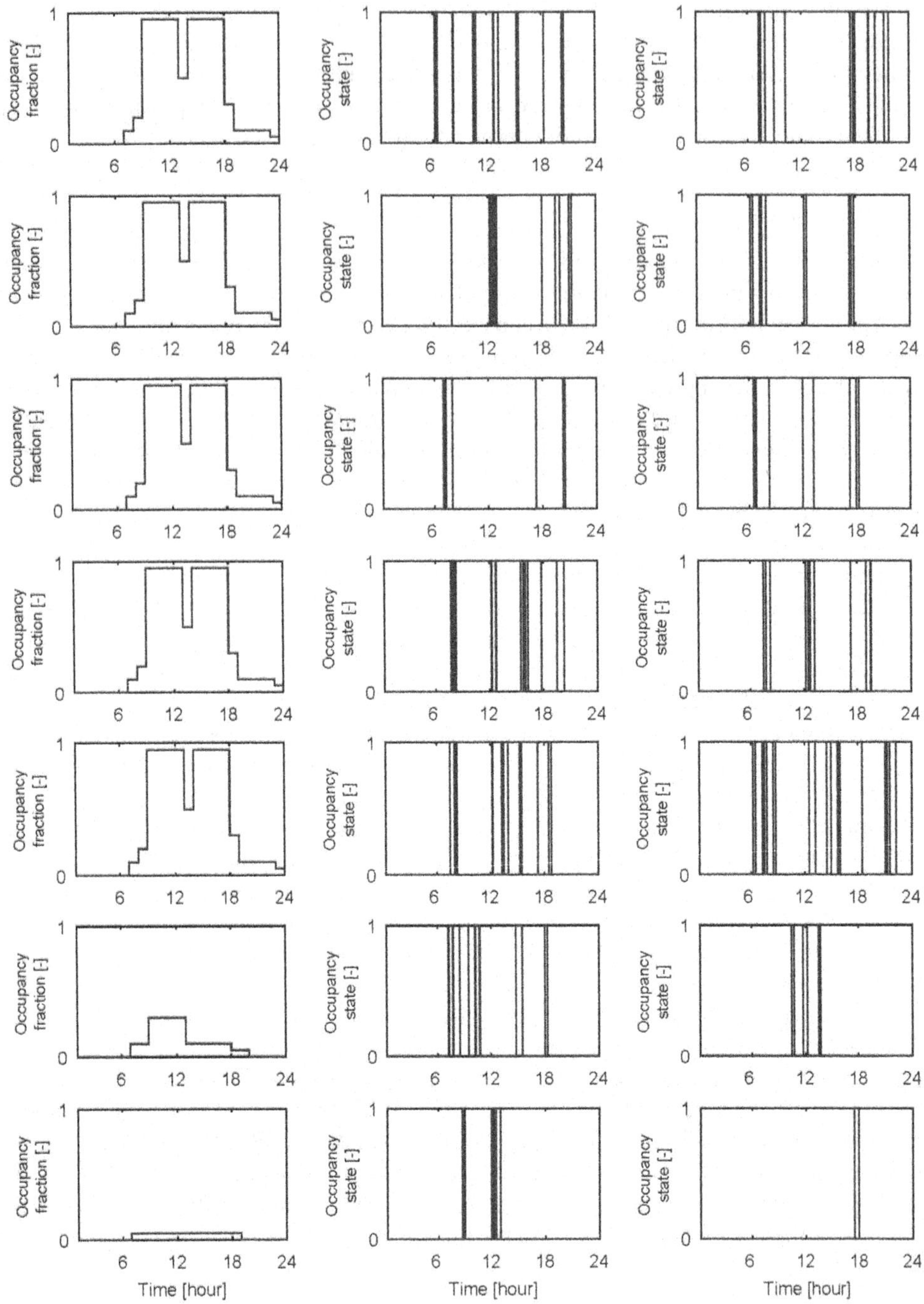

Figure 4.3 An example of random non-repeating occupancy profiles generated by a probabilistic occupancy model (Page et al. 2008). The left column represents presence schedules for weekdays, Saturday, and Sunday based on ASHRAE 90.1 (ASHRAE 2013) as the model input; the middle and right columns represent two weeks of Boolean occupancy profiles generated by the model.

patterns, the schedules for the use of lighting and equipment should be defined in accordance with the assumptions regarding occupancy schedules.

Exclusive reliance on schedules for representing adaptive occupant behavior (such as operation of windows and shades), which is arguably much more dependent on the environmental conditions rather than time of the day or the year, is not recommended. In building performance simulation, if the adaptive occupant behavior is not linked to any indoor environmental parameter, the operation of building devices remains indifferent to alternative design and operation options. In certain cases, this approach may be adopted to study the performance of design alternatives regardless of occupant behavior. However, it arguably does not capture the dynamics of real interactions between inhabitants and building systems.

Non-probabilistic models

Traditionally, rule-based models have been used to represent the state of buildings' environmental control devices. Thereby, one or multiple threshold values for the assumed pertinent variables are set by the modeler for the operation of the devices. For example, a set-point value for incident solar radiation can be used to operate the shading devices. Thus, if incident solar radiation on the window exceeds the threshold value, the shade is closed. Otherwise, the shade is open. Also, the shades can be deployed if the solar radiation and outdoor temperature both exceed specific set-points.

There are simplifications inherent to this modeling approach, which the building performance modeler should be aware of. First, in the course of simulation, multiple devices might be operated in an unrealistically simultaneous manner, if a set-point is assigned to large spatial zones. Second, if no deadband is considered for a set-point and the model entails a feedback, an unrealistically large number of successive state changes can occur. As such, one can argue that these models actually simulate automated control of building devices, rather than inhabitants' behavior. However, given their simplicity and the clarity of assumptions they make, these models can be still very useful for specific queries. For instance, they can be applied to establish a kind of benchmark "reasonable behavior" case, suggesting, for instance, that the inhabitants would behave in a well-informed, energy-conscious manner.

Furthermore, especially in case of non-adaptive actions, occupancy schedule information can also be used within non-probabilistic behavioral models. For example, Mahdavi et al. (2016a) developed a linear regression model for estimating plug loads in office buildings using the presence probability at each time interval. Given the fact that such models rely on average patterns of occupancy (included within an occupancy schedule), they tend to provide better estimations of the aggregate long-term (e.g. annual) performance indicator values in comparison with interval-specific indicator values (e.g. peak values).

Probabilistic models

In general, probabilistic behavior models estimate the probability of inhabitants operating environmental control devices (or the probability of specific device states) based on the previous state of devices and a number of independent variables pertaining to time, occupancy states and events, as well as indoor and outdoor environment. The estimated probabilities of actions or states are commonly used in the framework of the inverse transform sampling method (involving comparison with random numbers) to determine the occurrence of an action or the state of devices in multiple runs of the building model. Prediction of actions and states can be also combined in a behavior model to address a situation in which occurrence of an action (for example, lowering of a shade) is not enough to determine the resulting state of the device (for example, a shade that can be deployed from 0 to 100%).

In this context, one of the most common modeling methods is logistic regression as a form of a generalized linear model (GLM), in which the dependent variable follows a Bernoulli distribution (probability distribution of a single yes/no experiment). Logistic regression makes it possible to use the existing observational data to estimate the probability of occupant control-oriented actions or systems' on/off states based on a combination of independent variables pertaining to – for example – indoor and outdoor environment, as well as occupancy traits. In logistic regression, the logit function (representing the natural logarithm of the odds of having an action or a specific state) is deployed as a link to map the linear combination of independent variables (with an unbounded domain) onto the Bernoulli distribution (with a domain bounded between 0 and 1). To this end, the linear function of the independent variables is set equivalent to the logit function:

$$logit(p) = \ln\left(\frac{P}{1-p}\right) = \beta_0 + \beta_1 x_1 + \cdots + \beta_n x_n \tag{4.1}$$

where P is the probability of an action or state, x_1 to x_n are the independent variables, and β_1 to β_n are the regression coefficients. Subsequently, the inverse logit function gives the probability of occurrence of an action or observing a specific state based on the estimated regression coefficients as follows:

$$P = logit^{-1}\left(\beta_0 + \beta_1 x_1 + \cdots + \beta_n x_n\right) = \frac{e^{\beta_0 + \beta_1 x_1 + \cdots + \beta_n x_n}}{1 + e^{\beta_0 + \beta_1 x_1 + \cdots + \beta_n x_n}} \tag{4.2}$$

The logistic regression models for the occupant control-oriented actions are normally used within a Markov process to calculate the probabilities of possible state transitions (e.g. Herkel et al. 2008; Haldi and Robinson 2009; Andersen et al. 2013; Gunay et al. 2017). Additionally, in a number of probabilistic behavior models, survival models have been used to estimate the duration of time until an event happens (e.g. Reinhart 2004; Gunay et al. 2016a). Figure 4.4 illustrates an example of a logistic regression model, where indoor temperature has been used to estimate the probability of opening windows in an office space, both in terms of individual inhabitants (W1 to W7) and in aggregate (all).

As an extension to the generalized linear models for incorporation of random effects in the model, a generalized linear mixed model (GLMM) can be also deployed to integrate occupant behavior diversity in modeling efforts (for details, see Madsen and Thyregod 2011; Haldi et al. 2017).

The significance of deployment scenarios

The reliability and appropriateness of a specific behavioral model cannot be discussed in isolation from the specific circumstances of its deployment in the simulation-assisted building performance evaluation workflow. Consequently, it would be misguided to assume that a specific modeling approach or technique can be appropriately applied to all kinds of use cases (see Gaetani et al. 2016; Mahdavi and Tahmasebi 2016a).

Performance simulation models can assume different levels of resolution with regard to the representation of the underlying (physical) phenomena, required (input) information, and produced results (output). Generally speaking, the choice of a specific level of resolution in these aspects is not independent of the types of queries, for which the simulation model is expected to provide answers. This also applies to choices in the type and resolution of representations of people's presence and behavior in building performance simulation models. The relationship between these choices and the purpose of the simulation-assisted analyses is not well

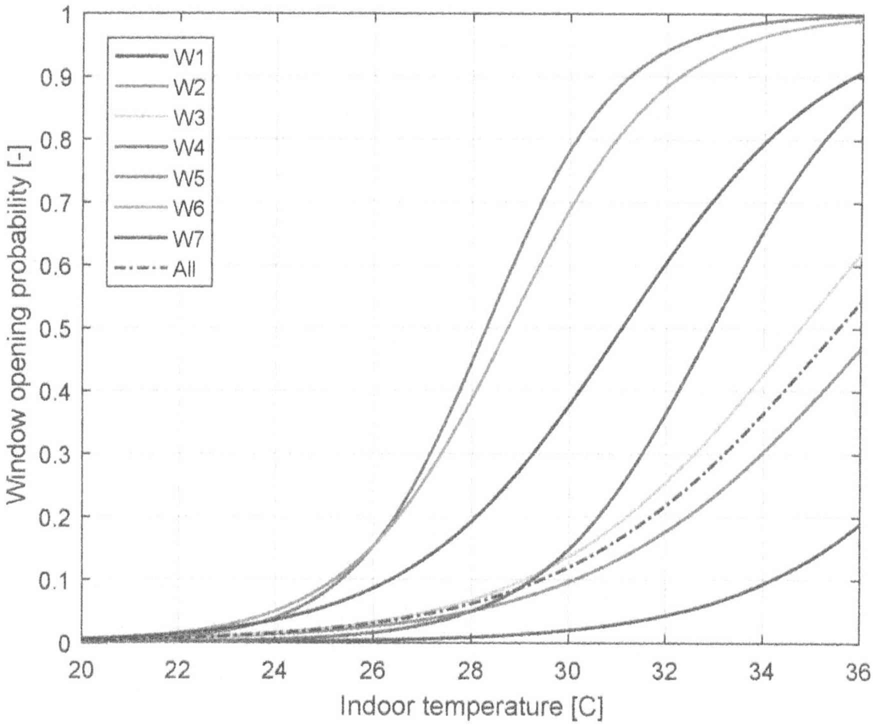

Figure 4.4 An example of estimating window-opening probability based on indoor temperature using logistic regression

understood. This represents a practical problem, as it implies that adopted methods in capturing people's presence and behavior in a simulation process may in fact be incongruous with regard to the specific simulation use scenario at hand. It can also be argued that the criteria for the evaluation of the representational fidelity of people's presence and behavior in buildings are not independent of the types of the studies undertaken in the course of simulation tool deployment.

Few studies have explicitly addressed the fitness of occupancy-related models with regard to different simulation queries. An earlier proposal in a different context (Gupta and Mahdavi 2004) suggested viewing and structuring performance queries in terms of a multi-dimensional query space. The classification of the queries was intended to render them more suitable for analysis, resulting in enhanced responses through selection and execution of appropriate computational tools and techniques. Specific to the deployment of occupancy models, sensitivity analysis has been used in the past to arrive at the minimal required user model resolution with regard to a number of building performance indicators and design parameters (Hoes et al. 2009; Gaetani et al. 2017). For example, when a performance indicator is determined to be more sensitive to the occupancy-related assumptions, the simulation effort should start with a more sophisticated model of occupancy (and if the performance indicator still does not fall within the required target value range, a higher resolution level should be applied). However, the focus in these cases was on the design stage. As such, they did not involve empirical data to confirm the conjecture that using more sophisticated models would necessarily provide more accurate results.

To work toward a systematic understanding of query types and model attributes, consider the following simple observations. Simulation-supported building performance analyses require input models in multiple categories, such as building fabric, building systems, external conditions, and use patterns. However, queries are not always interested in simultaneous assessment of all these categories. This implies that, in a specific instance, detailed variations pertaining to just one category (e.g. building fabric) may be of interest. In such an instance, input models regarding other categories may be standardized, entailing explicit default assumptions without considering uncertainty and diversity. There is as such nothing wrong with exploring strategic parametric uncertainty in a selected model input category, while maintaining persistent representations of others. The important point for the proper interpretation of any simulation result is the transparent and detailed description of all underlying assumptions. As such, simulated data resulting from opaque and inscrutable model input assumptions are of little use. Moreover, attaching uncertainty ranges to simulation outcome does not make it more meaningful, if the pertinent distribution functions of the input variables are either unknown or arbitrary.

Given the multiplicity of building performance simulation deployment scenarios (involving different users, different phases of the building delivery process, different queries, etc.), a corresponding well-structured conceptual framework in terms of a multi-dimensional simulation deployment space could be useful (Mahdavi and Tahmasebi 2016a). Such a framework would not only represent a prerequisite for evaluating the suitability of alternative modeling techniques and resolutions with regard to people's presence and behavior in buildings, but could also contribute to substantiating the evaluation process of such modeling techniques.

Table 4.2 briefly outlines nine dimensions that are suggested to capture the space of possible queries in building performance simulation, namely, (I) phase in the building delivery process, (II) purpose of the study, (III) relevant domain, (IV) building type, (V) indoor climate control strategy, (VI) physical subject, (VII) spatial resolution, (VIII) performance indicator, and (IX) temporal resolution.

The relevance of this conceptual construct for the choice of occupancy-related models can be clearly illustrated, once specific performance queries are projected onto this space. Consider, for instance, an early design, low-resolution (whole building) query pertaining to the generation of a thermal quality certification of a building's fabric, versus a detailed, high-resolution (micro-zone-level) HVAC systems configuration query. Whereas in the former case the representation of inhabitants may be adequately treated as a normalized boundary condition

Table 4.2 Dimensions of the simulation deployment space

	Dimension	*Remarks, examples, instances*
I	Phase in the building delivery process	Early design, detail design, HVAC systems design, building operation
II	Purpose (or nature) of the study	Parametric study of design options, generation of building quality and/or energy performance compliance documents, HVAC system sizing, HVAC controls
III	Domain (discipline)	Energy, thermal comfort, lighting, acoustics, fire safety
IV	Building type	Function of the building (residential, commercial, educational, mixed use)
V	Indoor climate-control strategy	Passive, hybrid (mixed mode), fully air-conditioned
VI	Physical subject	Building details, whole buildings, campus, district, urban
VII	Spatial resolution	Whole building, individual floors, rooms, micro-zones
VIII	Performance indicator (results)	Annual heating/cooling demand, peak heating/cooling loads, thermal comfort indices
IX	Temporal resolution (horizon)	Entire life cycle, annual, monthly, daily, hourly, sub-hourly

(realized in terms of presence diversity profiles and rule-based device operation routines), in the latter case a more realistic representation (realized in terms of probabilistic patterns of presence and control-oriented actions together with the consideration of inhabitants' diversity) could be more appropriate.

A previous case study (Mahdavi and Tahmasebi 2016a) demonstrated the desirability and usability of the above framework in more detail, whereby probabilistic and non-probabilistic occupancy models were considered as a case in point. The findings suggested that a particular modeling technique for generating occupancy-related input information for performance simulation cannot be declared *a priori* to be superior to others. Rather, the circumstances pertaining to the nature of application scenarios, such as time horizon of predictions or granularity of performance indicators, must be carefully considered. In other words, there are good reasons to suggest that the choice of an appropriate occupancy-related model (and perhaps even the criteria for evaluating its performance) depends on the position of the relevant simulation-based query within the proposed application space. However, this does not mean that by situating individual queries in the proposed multi-dimensional (see Table 4.2) space of all queries, they can be mechanistically mapped onto specific kinds of occupancy-related models. At the present stage, what the proposed framework can offer is to encourage deeper reflection and explicit reasoning with regard to the choice and deployment of appropriate occupancy-related models.

The empirical basis

Overview

Given the current state of occupancy-related representations in building performance simulation, systematic observational data regarding people's presence and actions is both necessary and critical. Specifically, the reliability of occupant behavior models for building performance simulation largely depends on the quantity and quality of the underlying empirical data. Most theories regarding subjective evaluation processes of environmental conditions as well as causal theories of occupants' control-oriented behavior involve indoor or outdoor environmental parameters as independent variables. Consequently, collecting such data is a precondition of any empirical study aiming to explain occupants' operation of control systems and devices (Mahdavi and Taheri 2016).

In recent years, a large number of occupant behavior-monitoring campaigns have been conducted (although commonly with relatively small sample sizes) to obtain data for understanding and representing occupants' presence patterns and their interactions with building environmental control systems. As such, building inhabitants and their control-oriented behavior in buildings are typically studied using either field (*in situ*) or controlled (laboratory) monitoring studies. These are briefly discussed in the following sections. Note that in both instances, considerations pertaining to research methodology and ethical issues are of essential importance. Observational studies need to address, among other challenges, the implications of the Hawthorne effect (Mahdavi 2011; O'Brien et al. 2018). The provision of detailed information regarding the nature of an observational study to the study's participants can influence inhabitants' behavioral patterns. Providing little or false information may counteract such influence but may not be an acceptable option from the ethical and legal standpoints. The design of such studies and the interpretation of associated results must thus carefully follow the applicable requirements and regulations (Chen et al. 2018).

A further crucial challenge faced by most past observational studies may be identified in terms of the absence of an explicit systematic ontology for building-related data monitoring systems and campaigns. The collection, storage, and processing of behaviorally relevant data is most useful if it is situated in a broad strategic approach. Toward this end, the formulation of a comprehensive and consistent ontology represents an indispensable requirement. To address this gap, a recent effort proposed a systematic ontology for building-related data monitoring and processing applications (Mahdavi and Taheri 2016). Specifically, in order to construct a well-formed schema for building monitoring, Mahdavi and Taheri (2016) identified a number of fundamental data categories (see Figure 4.5). These include (1) occupants, (2) indoor environmental conditions, (3) external environmental conditions, (4) control systems and devices, (5) equipment, and (6) energy flows. Thereby, the authors suggested that the data obtained via sensors, meters, and other sources (e.g. simulation-powered virtual sensors, human agents) in the above six categories could be captured in terms of a structure consisting of values, associated sources, and possible actors (for details, see Mahdavi and Taheri 2016; Mahdavi et al. 2018, 2016c).

Field studies

A large number of field studies have been conducted in the past decades to understand how building occupants interact with buildings' environmental control systems such as windows, blinds, and luminaires (see, for example, Boyce 1980 Fritsch et al. 1990; Herkel et al. 2005; Lindelöf and Morel 2006; Love 1998; Mahdavi 2011; Mahdavi et al. 2008; Rijal et al. 2007; Andersen et al. 2009). As such, there are significant differences between individual studies in this set in terms of size and type of buildings observed, relevant control devices (luminaires, shades, windows, etc.), duration of observation, and measured environmental factors, as well as measurements' resolution and precision. Nonetheless, most of these studies share a common

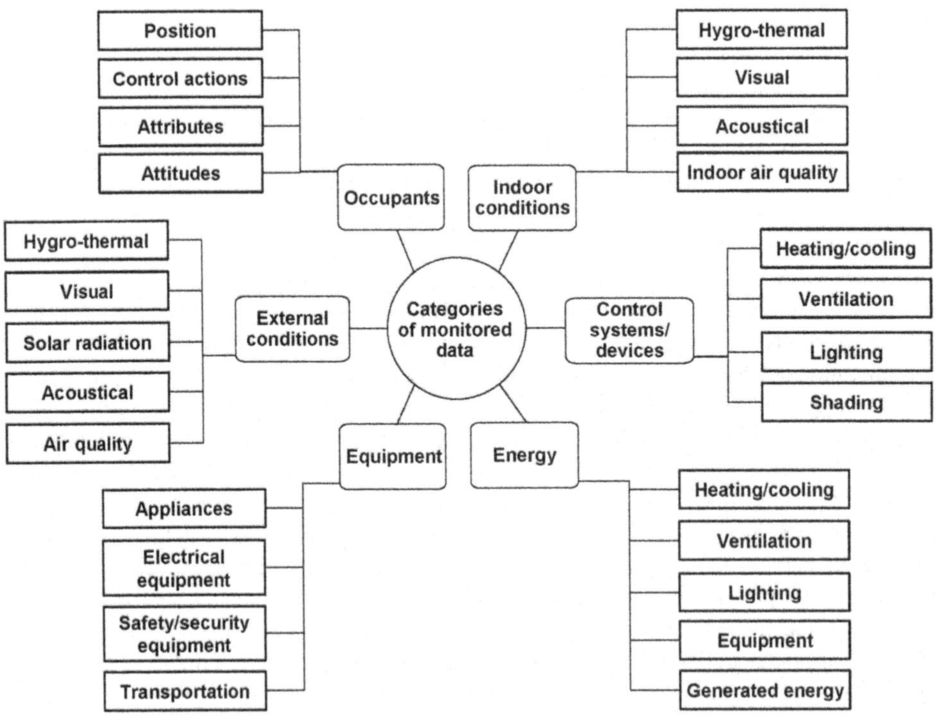

Figure 4.5 Fundamental elements of a building monitoring ontology including pertinent data categories and sub-categories

Source: Mahdavi and Taheri 2016

feature in that they attempt to establish a link between user control actions (or the state of user-controlled devices) and measurable indoor or outdoor environmental parameters.

Field studies can provide realistic data on inhabitants (obtained from their actual environment) and offer the potential for a large sample size and long-term data collection. However, extensive *in situ* monitoring studies involve labor-intensive and potentially invasive deployment, adjustment, replacement, and retrieval of the sensors (Gilani and O'Brien 2016; O'Brien et al. 2018). In addition, as shown in previous studies (e.g. O'Brien and Gunay 2014; Day et al. 2012), relying alone on physical sensor data cannot fully reveal potential causal factors underlying occupant behavior. Therefore, a field study needs also to obtain direct information from the inhabitants (e.g. via interviews and surveys).

Controlled studies

Controlled studies (typically in laboratory settings) have been used for a long time in research pertaining to thermal comfort and indoor air quality (Fanger 1970; Parsons 2003; Wu and Mahdavi 2014). A key advantage of such studies is the possibility to strategically manipulate a variety of parameters relevant to hypothesis testing and model evaluation. Such parameters include specifically suspected independent variables pertaining to indoor environment together with control device states and settings. Laboratory settings also enable the researchers to obtain relevant individual physiological data such as skin and body temperature, skin wittedness, and hormone levels. More recently, laboratory settings have been also utilized to study inhabitants' behavior under controlled circumstances (see, for example, Schweiker et al. 2016).

Controlled laboratory studies face, of course, a number of challenges. They are typically expensive and time-consuming. Moreover, they need to provide a realistic and familiar surrounding for the participants. Nonetheless, they have the potential to provide the researchers with the opportunity to design a test environment tailor-made for investigating detailed research questions. Laboratory studies also offer more freedom in terms of selecting participants, the experimental procedure, as well as monitoring, diagnostics, and control equipment (Wagner et al. 2018a).

Model validation

General considerations

If performance simulation is to effectively support decision making, it must be based on reliable models of the relevant processes, including inhabitants' behavior. Given the recent increase in the number of behavioral models proposed for incorporation in building performance simulation applications, it is important to understand the necessary conditions for the systematic evaluation of models of inhabitants' presence and actions in buildings (Mahdavi 2015). The current lack of general procedures and guidelines for the evaluation of proposed user-related behavioral models may be due, in part, to the following three circumstances:

- First, systematic occupancy-related studies in the context of the built environment belong to a relatively young field of inquiry. By comparison, the thermal comfort research has a longer tradition and is arguably better established. Nonetheless, even in that domain many open research questions and challenges persist (Schweiker and Wagner 2016; Shipworth et al. 2016).
- Second, a persistent problem for both model development and model evaluation lies in the rather limited availability of large-scale observational data. Consequently, the demographic basis of the majority of proposed behavioral models is often very small (Mahdavi

and Tahmasebi 2016b). As data is hard to come by, models are often developed and disseminated with insufficient empirical backing.

- Third, behavioral models should be ideally derived based on the concurrent consideration of multiple parameters of physical, physiological, psychological, and socio-cultural nature. The multifariousness of potential influencing and contributing factors to behavior actions creates a kind of background "noise." Against this background, it is often difficult to discern the typically low-strength "signal" of causal factors hypothesized to be behind behavioral manifestations (Mahdavi et al. 2016b).

Nonetheless, avoiding basic misconceptions and fallacies could improve the quality of development and evaluation in behavioral modeling (Mahdavi 2015):

- It is critical not to confuse simulation (computational, typically dynamic representation of a system's behavior) with prediction. Long-term exact predictions of buildings' energy and thermal performance are unrealistic, even if occupancy-related processes could be accurately modeled. The long-term unpredictability of external weather conditions alone is sufficient to falsify overarching predictive accuracy claims. A more reasoned view of performance simulation appears to lie in its utility toward complex system analysis, rather than accurate long-term predictions. As such, the frequent mismatch between simulation-based predictions and respective observations (the so-called performance gap) can be the result of multiple sources of uncertainty, pertaining not only to internal (occupancy-related) processes, but also to external (weather) conditions, building fabric, and building systems.
- In model comparison and evaluation discourse, the term "deterministic," which has a weighty philosophical baggage (Mahdavi 2015), is often used in a potentially misleading manner to characterize fixed diversity profiles (e.g. assumed fixed schedules of occupants' presence) and rule-based behavioral models. Subsequently, the inference is made that simulation results would be necessarily more "accurate" if occupancy-related diversity profiles and rule-based assumptions are simply replaced with more detailed probabilistic ones (see, for example, D'Oca et al. 2014). However, there is no conclusive empirical evidence suggesting that specific modeling formalisms automatically result in more accurate building performance simulations (Mahdavi and Tahmasebi 2016a).
- The construction of black-box models of inhabitants' control actions and the use of data-driven probabilistic methods toward generation of realistic behavioral patterns may be very useful. However, they do not replace fundamental studies to understand the motivational (and potentially causally effective) field behind behavioral manifestations shaped by physiological, psychological, and social factors. Such studies could facilitate the development of gray- (or even white-) box behavioral models that could be more illuminating and effective in the long run (Mahdavi et al. 2016b).
- Any statements about the validity of specific behavioral models can be assessed only based on properly and meticulously prepared documents of the model development and evaluation procedures (research design, empirical basis, hypotheses and assumed causal factors, limitations, etc.). Thus, any independent instance could retrace, comprehend, and reappraise such procedures. Moreover, behavioral models cannot be declared to be "validated" based on a limited set of observational data. Specifically, data sets for model development and model evaluation should not be conflated. Paucity of empirical information does not justify testing a model based on the same data set, which was used for its development. Specifically, it is important not to extrapolate from a single limited behavioral study to all kinds of populations, building types, locations, and climates.
- Similar to other domains where model evaluation is critical, in the behavioral modeling field too we must safeguard against bias. As such, internal evaluation by model developers

does not provide conclusive evidence for a model's general reliability. While not easy to conduct, external evaluation procedures, double blind studies, and round robin tests are undoubtedly in a better position to convincingly support the evaluation of a model's performance.

- It is essential to exercise care while incorporating insufficiently documented and rudimentarily tested behavioral models in broadly used simulation applications, lest potential users are misled into assuming such models necessarily capture the "reality" of inhabitants' presence and behavior in buildings.

Given this background, a number of relevant challenges with regard to the evaluation of occupancy-related models are further addressed in the following. These include model performance indicators, the standing of coefficients in models' mathematical formulations, the problem of models' feedback, and the model performance comparison challenge.

Evaluation statistics

One of the fundamental challenges of evaluation procedures pertaining to behavioral models of building inhabitants is the paucity of systematically classified model performance metrics. Whereas an ultimate ontology of fit-for-purpose metrics for behavioral model evaluation cannot be offered here, the previously discussed distinction between predictions of "states" and "events" (or "actions") can provide a suitable starting point. The central step in model evaluation can be framed in terms of the comparison of predicted and monitored events and states. Specific aspects of model performance can be also evaluated by comparing the predicted action probabilities with the monitored actions (Fabi et al. 2015; Tahmasebi and Mahdavi 2018). From the large number of indicators, which have been used in previous evaluation studies of behavioral models, two broad categories can be inferred: the indicators addressing aggregate aspects of models' predictions and those addressing the interval-by-interval congruence between predictions and measurements. In other words, whereas the first category "vertically" aggregates observations and predictions independently before the overall comparison, the second category compares first "horizontally" time series data pairs, which can then be further processed statistically (Mahdavi and Tahmasebi 2016b). Illustrative listings of these two types of indicators are provided in Figure 4.6. Note that in this framework, the indicators, which address aggregate traits of the predictions (such as total number of actions, median state durations, etc.) are grouped along with those, which address the proximity of predicted probability distributions to those of the measured ones (such as Jensen-Shannon divergence).

It can be argued that, in general, while a superior performance in terms of aggregate indicators is specifically desired in simulation studies geared at performance levels over longer periods of time (such as conventional use of building performance simulation models for estimation of annual energy demands), the indicators resulting from interval-by-interval contrast of predictions and measurements are of more interest in studies, in which short-term performance predictions play a central role (e.g. predictive building systems control).

However, for any specific use case, the selection of a proper set of indicators and the adequate interpretation of their values require deep domain knowledge and experience. In many instances, listings of indicators' numeric values must be augmented with appropriate data visualization and analysis techniques to provide an effective basis for the evaluation of the predictive performance of behavioral models. To illustrate this point, consider the case of an external evaluation study of a probabilistic shade operation model. In the course of this study, model predictions of shade positions were compared with corresponding observations concerning an office building in Hartberg, Austria (Mahdavi et al. 2008). The studied model appeared to

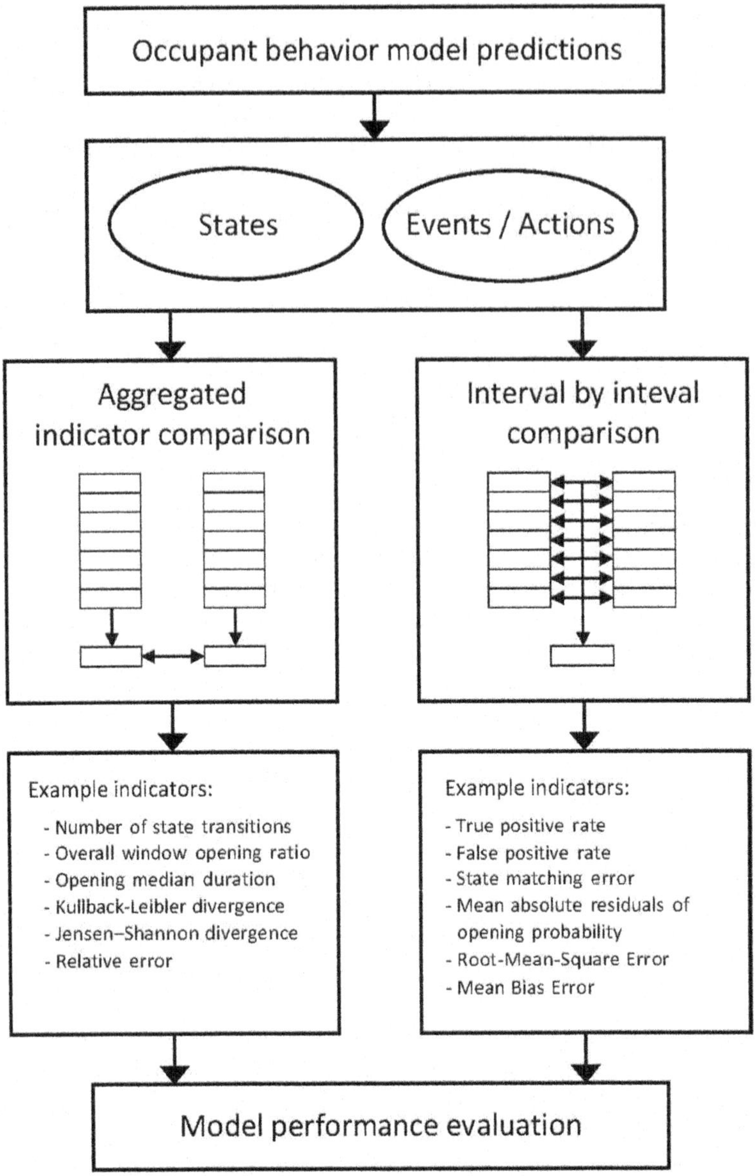

Figure 4.6 Categories, aggregation structures, and example indicators for occupant behavior model evaluation

Source: Mahdavi and Tahmasebi 2016b

provide relatively reasonable shade-state predictions in terms of aggregate and interval-based indicators. However, the time series plot of the observed and predicted opening and closing actions (Figure 4.7) reveals a different picture: With an ample overestimation of closing actions in the cold season, the model drastically fails to capture inhabitants' use of shade devices in this period.

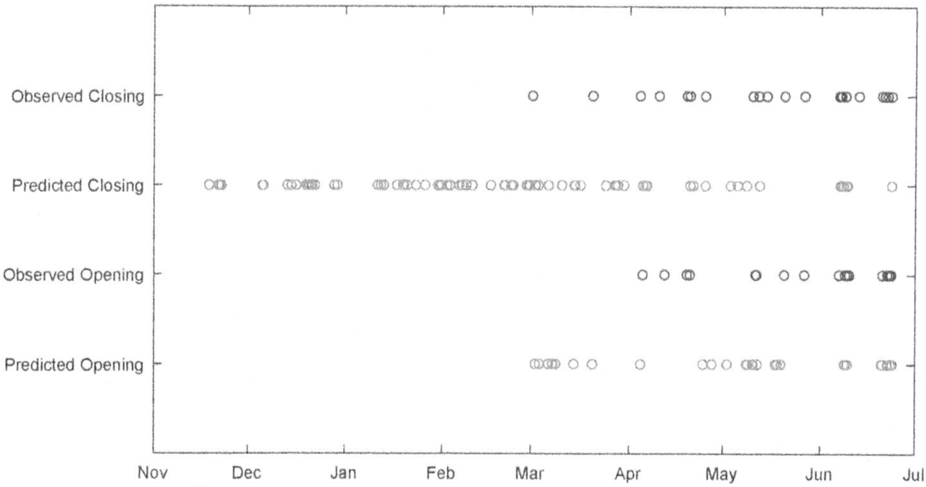

Figure 4.7 Time series plot of observed openings and closings of external shades in an office building together with those predicted by a probabilistic shade operation model (dots in this graph signify opening and closing events)

The original and calibrated model coefficients

The original coefficients of occupant behavior models are generally the result of available data obtained from a limited number of buildings (at times even just one building). Hence, when such models are used with their original coefficients for other buildings, their predictions can largely deviate from observations. Experience suggests that the predictive performance of the occupant behavior models can be largely improved by refitting the models to the data obtained from the building under study.

To better appreciate this point, consider a recent comparative study of two probabilistic window operation models (see Figure 4.8). In the course of this study, the predictions made by these two models (referred to as A and B in Figure 4.8) were compared with actual observations from an office building in Vienna, Austria (Tahmasebi and Mahdavi 2016). Thereby, not only the prediction results by models with their original coefficients were considered, but also those made by calibrated model versions, whose coefficients were adjusted based on observational data. As Figure 4.8 documents, the original models' errors are considerably reduced due to the calibration process. However, note that the use of on-site monitored data for calibration of occupant behavior models is by no means feasible in the majority of use cases in building performance simulation (i.e. design decision support). Consequently, the general use of occupant behavior models with their original coefficients involves a substantial degree of uncertainty.

Given the variety of potential settings, in which occupant behavior models can be deployed, ongoing, and future, more extensive model development and cross-sectional investigations are needed. The resulting broader empirical foundation could be expected to foster the development of more representative and inclusive models in the future. Presumably, such future models would provide guidelines for the adequate manipulation of the models' default coefficients to account for the specific characteristics of a given context. Toward this end, it would be preferable to develop occupant behavior models without excessive overfitting to a specific data set,

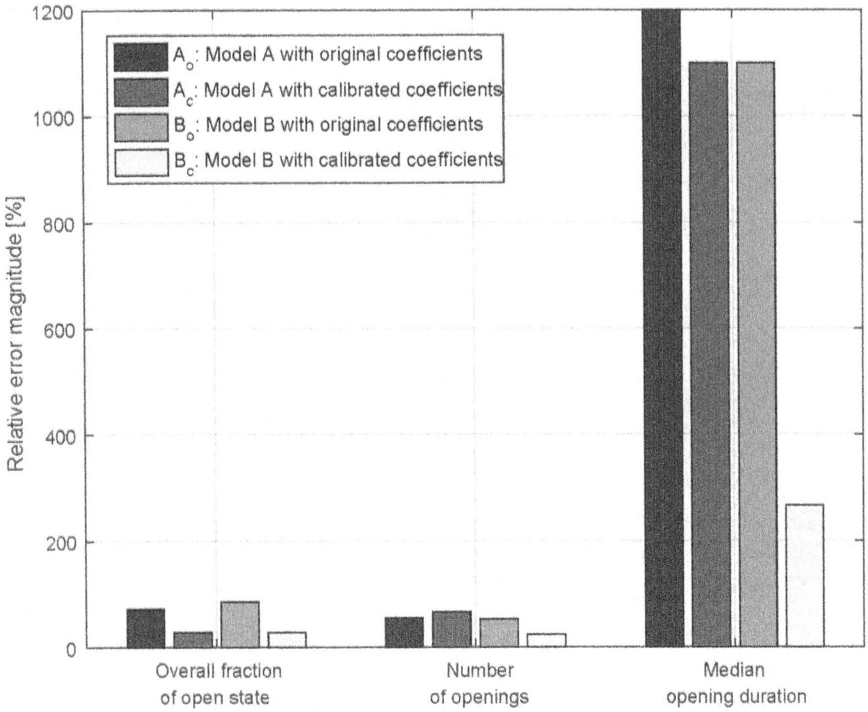

Figure 4.8 Comparison of prediction errors by two probabilistic window operation models (A and B) with original and calibrated coefficients in heating season. Errors are displayed for three model performance indicators (overall fraction of open states, number of opening actions, and median opening duration).

as it may involve the exclusion and inclusion of independent variables due to special context-specific conditions.

Addressing the model feedback in the evaluation process

The inherent feedback in occupant behavior models with independent variables related to indoor environment poses a challenge for model evaluation efforts: The behavioral models' output (for instance, window states) has an impact on the models' input (for instance, indoor temperature). Empirical data matching every possible sequence of actions predicted by behavioral models cannot be obtained. In this regard, different approaches have been adopted in previous efforts to evaluate the predicative performance of occupant behavior models.

A number of model evaluation studies (e.g. Schweiker et al. 2012; Fabi et al. 2015) have neglected the feedback in occupant behavior models. Hence, models' predictions at each time interval do not have any impact on the indoor environmental factors for the next interval. This omission can render the efforts toward the validation of behavioral models inconclusive. For example, in evaluating a number of probabilistic and non-probabilistic window operation models, Tahmasebi and Mahdavi (2016) showed that running two probabilistic models without considering feedback led to a large overestimation of the fraction of open state and opening duration: In real circumstances, opening of the windows resulted in room temperature drop, which in turn triggered window closing actions. In other words, without considering the models' feedback, opening of the window does not reduce the indoor air temperature and is therefore not followed by a closing action. In the same study, the disregard of models' feedback also countered the tendency of single-threshold non-probabilistic models to predict an unrealistically large number of actions. As such, windows are operated according to these models as soon as the temperature falls below or rises above a certain threshold, which, in the realistic scenario (including feedback) would result in a large number of opening and closing actions.

Given these circumstances, the use of calibrated building performance models may represent an alternative for evaluating occupant behavior models (Tahmasebi and Mahdavi 2012). Thereby, the calibrated building model serves as a virtual representation of the building, which has the potential to reliably emulate the building's response to the predicted occupant behaviors. Needless to say, the development of highly accurate calibrated building models requires extensive input data pertaining to outdoor environment as well as the physical and operational characteristics of the building.

There are also other approaches to occupant behavior model evaluation, which do not require the inclusion of feedback but instead focus on specific aspects of models' predictions. One method relies on discontinuous model runs. That is, as opposed to the aforementioned two approaches, the predicted state of the environmental control device at each time interval is not inherited by the next time step. Thus, at each time interval the model is fed anew with the monitored state of the device in addition to the explanatory environmental parameters. The predicted state of the device is then compared with the measured state in the next time interval. In this model run scenario, there is no need to include the model feedback. However, it cannot capture the intended continuous run of the models, where state prediction for a time step has implications for the subsequent intervals. Consequently, this method can only capture the model potential in correctly predicting the device states in specific intervals and not the overall performance of the model in timeframes such as a day or a season.

Another model evaluation approach that does not require the consideration of feedback is the use of predicted action probabilities instead of the resulting states or actions. For example, in a recent effort, Tahmasebi and Mahdavi (2018) used the coefficient of discrimination (Tjur 2009) to evaluate window operation models in view of their performance in discriminating opening and closing actions. Thereby, the model performance is derived in terms of the difference between the average of estimated action probabilities for the so-called successes (intervals, in which an opening or a closing action has been observed) and the failures (intervals, in which an opening or a closing action has not been observed). This approach does not require Monte Carlo simulation of the model and inclusion of model feedback. However, the method does not yield specific insights about model performance issues such as state duration of devices or frequency of actions in specific periods.

Performance comparison of different occupant behavior models

Recent studies on occupant behavior modeling suggest that one cannot simply declare *a priori* that a particular modeling technique for generating occupancy-related input information for performance simulation is superior to others. Rather, one needs to carefully consider the circumstances pertaining to the nature of an application scenario, such as time horizon of predictions or granularity of performance indicators.

As an example with respect to presence models, Mahdavi and Tahmasebi (2015) concluded that, for the purpose of short-term prediction of occupant presence patterns (i.e. a 24-hour future time horizon used in a model-predictive control scenario), a non-probabilistic presence model outperformed two widely used probabilistic models. However, in using presence models for long-term simulations (namely, annual model runs), the probabilistic models could provide better representations of the occupancy patterns' distributions and peak (Tahmasebi and

Mahdavi 2017). Nonetheless, this better performance did not translate into more accurate estimations of building-level annual and peak heating and cooling demands.

In a further study (Tahmasebi and Mahdavi 2016), the performance of a number of probabilistic and simple non-probabilistic window operation models were explored with regard to the prediction of window operation in an office building. The probabilistic models entailed the widely used Markov chain based logistic regression models that estimated the probability of window opening and closing actions based on the previous window state and a number of occupancy-related and environmental independent variables. The non-probabilistic models embodied simple rules for operating windows based on indoor and outdoor temperature. The study results suggested that, while probabilistic models could emulate the seemingly random character of occupant behavior and provided probabilistic distributions of performance indicators, their use did not guarantee more reliable predictions. Leaving aside the large errors that resulted from using such models without necessary adjustments, probabilistic window operation models overestimated the occupants' operation of windows in the heating season and thus the annual and peak heating demands (see Figures 4.9 and 4.10). However, as compared with rule-based models, the probabilistic models displayed a better performance in predicting window operations and assessing thermal comfort in the free-running season.

The comparison of model predictions with observed data in the case of two plug load models (Mahdavi et al. 2016a) also revealed that the simple non-probabilistic model could provide fairly reasonable predictions of annual energy use associated with plug loads. However, the

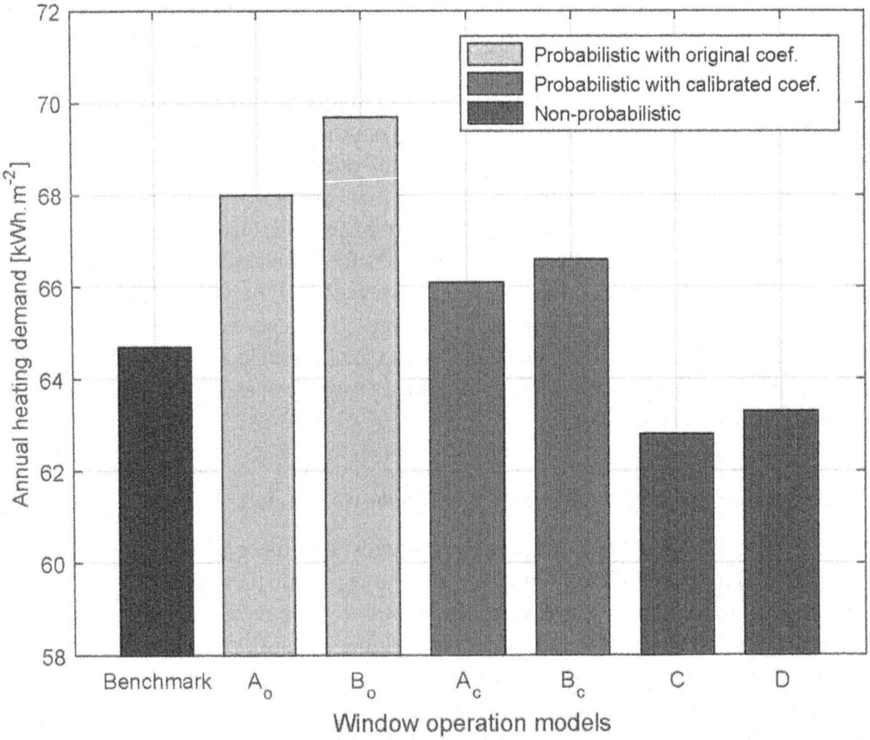

Figure 4.9 Annual heating demand for an office area computed under different window operation assumptions: A_o, B_o, A_c, and B_c denote the use of two probabilistic window operation models with original and calibrated coefficients, C and D denote two non-probabilistic models, and Benchmark denotes the use of observational data

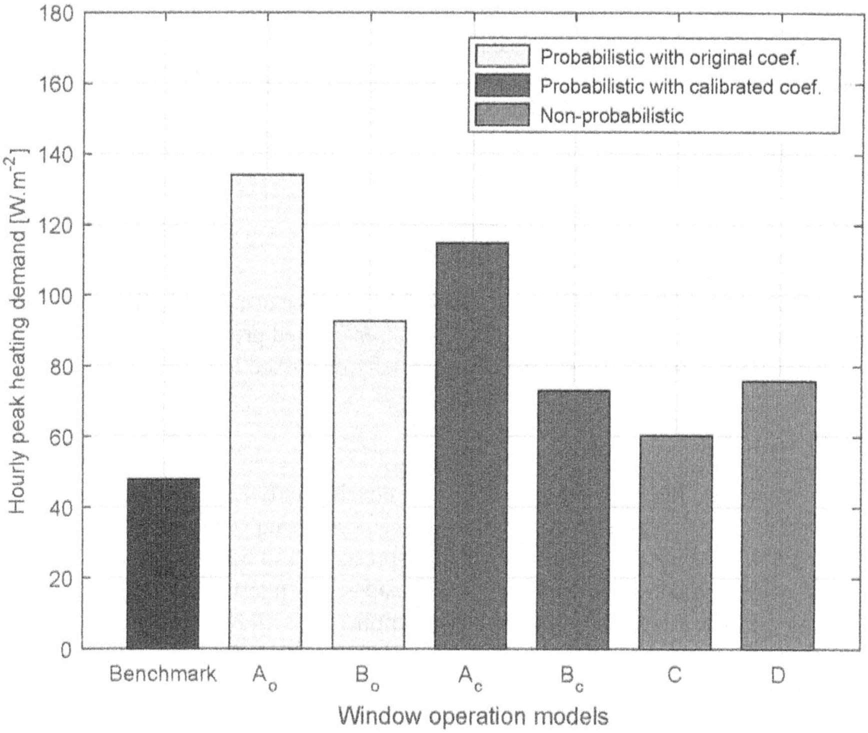

Figure 4.10 Hourly peak heating demands for an office area computed under different window operation assumptions: A_o, B_o, A_c, and B_c denote the use of two probabilistic window operation models with original and calibrated coefficients, C and D denote two non-probabilistic models, and Benchmark denotes the use of observational data

probabilistic plug load model outperformed the simplified model in terms of peak load and the distribution of predictions.

The above observations and further similar studies underline the importance of reliable information and transparent assumptions regarding the actual patterns of presence and behavior in buildings. In many instances pertaining to predicting buildings' future performance, the availability and fidelity of such basic information appears to be more important than the differences between predictions of alternative occupancy-related models.

Concluding reflections on the problem of model evaluation

A critical review of the efforts on understanding and modeling occupant behavior underlines a number of challenges in the evaluation process of the models. These include the paucity of underlying empirical information with sufficiently high quality and of representative nature, shortcomings in model documentation (especially concerning the applicability of the proposed models with regard to building type, location, climate, and deployment scenarios), model input requirements that cannot be met in realistic model deployment situations, problems associated with model coefficients and their calibration, the lack of a set of comprehensive, adequate, and universally accepted model performance metrics, and – last but not least – the problem of feedback, i.e. the inclusion of the predicted actions' impact on environmentally relevant model input variables.

All in all, the definition and pursuit of rigorous model validation procedures in the behavioral modeling field may be arguably characterized as work in progress. As a consequence, both model developers and potential users would be well advised to be cautious in promoting insufficiently tested behavioral models. Specifically, statements concerning models' validity and overall applicability in the building delivery process would be of little credibility without comprehensive empirical backing and carefully executed model-testing procedures.

Implementation in simulation environments

In general, three approaches are commonly adopted to include occupant presence and behavior models in building performance simulation tools: user-defined profiles and rules, user custom codes, and co-simulation. These approaches are briefly discussed later.

User-defined profiles and rules

To provide users with the simplest workflow, tool developers typically include, in the simulation applications, the functionalities to define occupancy-related profiles and non-probabilistic rules. As such, to execute these types of occupant presence and behavior models, the modeler only needs to fill the required fields within the simulation tool interface (or directly in the model input text file). For example, building simulation applications allow the modelers to define the schedules for occupant presence that include the occupancy ratio or the probability of presence at each simulation time interval throughout the year. Similarly, the state of environmental control devices can be controlled by schedules or simple non-probabilistic rules.

User custom codes

Given the limited modeling possibilities offered by the incorporated functionalities in building performance simulation tools, more elaborate occupant behavior models can be implemented in the programming environments provided within these tools. Such modeling modules manage the run-time data exchange between the building and occupant behavior models. For example, the EnergyPlus Energy Management System (DOE 2017) gives the building modelers the possibility to develop custom control and modeling routines. This approach is widely used by advanced simulation users (see, for example, Gunay et al. 2016b; Tahmasebi and Mahdavi 2016) due to the relative maturity of the internal modules of the simulation tools as compared to third-party tools with a short development history. In addition, given the integrated nature of the solution, it does not necessitate learning a new simulation interface and it works unaffected by problems associated with coupling different simulation environments. However, most building simulation applications lack comprehensive debugging capabilities, which impedes the effectiveness of the user custom codes (Yan et al. 2015) and makes the modeling process more error-prone.

Co-simulation

As another alternative with a higher degree of freedom, the occupancy-related models can be implemented in external programming and simulation environments to be used with the building simulation model in a co-simulation platform. Thereby, individual code components can be instantiated by different tools, and run-time data exchange can be conducted at discrete communication intervals (Yan et al. 2015). One of the recent examples of using a co-simulation approach in occupant behavior modeling is OB-FMU (Hong et al. 2016), which follows the FMI standard (FMI 2017). In addition, some researchers use the external interfaces of the simulation tools to couple the building model with the occupant behavior models implemented in programming environments such as Matlab (see, for example, Langevin et al. 2015).

Concluding remarks

In this chapter, we discussed the processes through which inhabitants can influence building performance. We then explored the kinds of occupancy-related input information required for thermal and visual building performance simulation applications, as well as the empirical sources of such information. We further argued that different types of simulation-supported building performance queries might require different kinds of occupancy-related models. Moreover, we not only addressed different kinds of modeling techniques and formalisms toward the representation of people in building performance simulation, but also discussed the key challenges in the validation of such formalisms. Finally, we briefly reviewed the approaches toward implementing occupancy-related models in building performance simulation.

As previously alluded to, research and development pertaining to the understanding and modeling of people's behavior in buildings has by no means resulted in a fully established and mature field of scientific and professional activity. Despite the recent increase in the related efforts, much work remains to be done. Thereby, significant further progress is more likely to be achieved if the most urgent research questions are approached in an interdisciplinary manner, involving collaboration between representatives from such diverse domains as architecture, building construction, physics, systems science, mathematics, statistics, computer science, and, most importantly, human and social sciences (physiology, psychology, sociology).

References

Andersen, R., Fabi, V., Toftum, J., Corgnati, S.P. and Olesen, B.W. (2013) Window opening behaviour modelled from measurements in Danish dwellings. *Building and Environment* 69: 101–113.

Andersen, R., Toftum, J., Andersen, K.K. and Olesen, B.W. (2009) Survey of occupant behaviour and control of indoor environment in Danish dwellings. *Energy and Buildings* 41(1): 11–16.

ASHRAE. (2013) *ASHRAE 90.1–2013: Energy Standard for Buildings Except Low-Rise Residential Buildings*. Atlanta, GA: American Society of Heating, Refrigerating and Air-Conditioning Engineers, Inc.

Azar, E. and Menassa, C.C. (2012) A comprehensive analysis of the impact of occupancy parameters in energy simulation of office buildings. *Energy and Buildings* 55: 841–853.

Boyce, P. (1980) Observations of the manual switching of lighting. *Lighting Research & Technology* 12(4): 195–205.

Chen, C., Schweiker, M. and Day, J.K. (2018) Ethics and privacy. In Wagner, A., O'Brien, W. and Dong, B. (eds), *Exploring Occupant Behavior in Buildings*. Cham: Springer.

Day, J., Theodorson, J. and Wymelenberg, K.V.D. (2012) Understanding controls, behaviors and satisfaction in the daylit perimeter office: A daylight design case study. *Journal of Interior Design* 37(1): 17–34.

D'Oca, S., Fabi, V., Corgnati, S.P. and Andersen, R.K. (2014) Effect of thermostat and window opening occupant behavior models on energy use in homes. *BUILD SIMUL* 7: 683–694.

DOE. (2017) *EnergyPlus Version 8.7 Documentation, Application Guide for EMS*. Washington, DC: U.S. Department of Energy.

Dong, B., Andrews, B., Lam, K.P., Höynck, M., Zhang, R., Chiou, Y.-S. and Benitez, D. (2010) An information technology enabled sustainability test-bed (ITEST) for occupancy detection through an environmental sensing network. *Energy and Buildings* 42(7): 1038–1046.

Duarte, C., Wymelenberg, K.V.D. and Rieger, C. (2013) Revealing occupancy patterns in an office building through the use of occupancy sensor data. *Energy and Buildings* 67: 587–595. doi:10.1016/j.enbuild.2013.08.062.

Fabi, V., Andersen, R.K. and Corgnati, S. (2015) Verification of stochastic behavioural models of occupants' interactions with windows in residential buildings. *Building and Environment* 94(1): 371–383. doi:10.1016/j.buildenv.2015.08.016.

Fanger, P.O. (1970) *Thermal Comfort*. Copenhagen: Danish Technical Press.

FMI. (2017) *Functional Mock-up Interface*. www.fmi-standard.org/.

Fritsch, R., Kohler, A., Nygard-Ferguson, M. and Scartezzini, J.-L. (1990) A stochastic model of user behaviour regarding ventilation *Building and Environment* 25(2): 173–181.

Gaetani, I., Hoes, P. and Hensen, J.L. (2016) Occupant behavior in building energy simulation: Towards a fit-for-purpose modeling strategy. *Energy and Buildings* 121: 188–204.

Gaetani, I., Hoes, P. and Hensen, J.L. (2017) On the sensitivity to different aspects of occupant behaviour for selecting the appropriate modelling complexity in building performance predictions. *Journal of Building Performance Simulation* 10(5–6): 601–611.

Gilani, S. and O'Brien, W. (2016) Review of current methods, opportunities, and challenges for in-situ monitoring to support occupant modelling in office spaces. *Journal of Building Performance Simulation* 10(5–6): 444–470.

Gill, Z.M., Tierney, M.J., Pegg, I.M. and Allan, N. (2010) Low- energy dwellings: The contribution of behaviors to actual performance. *Building Research & Information* 38(5): 491–508. doi:10.1080/09613218.2010.505371.

Gunay, H.B., O'Brien, W., Beausoleil-Morrison, I. and Gilani, S. (2016a) Modelling plug-in equipment load patterns in private office spaces. *Energy and Buildings* 121: 234–249.

Gunay, H.B., O'Brien, W. and Beausoleil-Morrison, I. (2016b) Implementation and comparison of existing occupant behaviour models in EnergyPlus. *Journal of Building Performance Simulation* 9(6): 567–588.

Gunay, H.B., O'Brien, W., Beausoleil-Morrison, I. and Gilani, S. (2017) Development and implementation of an adaptive lighting and blinds control algorithm. *Building and Environment* 113: 185–199.

Gupta, S. and Mahdavi, A. (2004) Exploring performance query space. In *SimBuild 2004, IBPSA-USA National Conference*. Boulder, CO.

Haldi, F., Calì, D., Andersen, R.K., Wesseling, M. and Müller, D. (2017) Modelling diversity in building occupant behaviour: A novel statistical approach. *Journal of Building Performance Simulation* 10(5–6): 527–544. doi: 10.1080/19401493.2016.1269245.

Haldi, F. and Robinson, D. (2009) Interactions with window openings by office occupants. *Building and Environment* 44: 2378–2395.

Herkel, S., Knapp, U. and Pfafferott, J. (2005) A preliminary model of user behavior regarding the manual control of windows in office buildings. In *Proceedings of The Ninth International IBPSA Conference, Building Simulation 2005*. Montréal, Canada, pp. 403–410.

Herkel, S., Knapp, U. and Pfafferott, J. (2008) Towards a model of user behaviour regarding the manual control of windows in office buildings. *Building and Environment* 43(4): 588–600.

Hoes, P., Hensen, J.L.M., Loomans, M.G.L.C., de Vries, B and Bourgeois, D. (2009) User behavior in whole building simulation. *Energy and Buildings* 41: 295–302.

Hong, T., Sun, H., Chen, Y., Taylor-Lange, S.C. and Yan, D. (2016) An occupant behavior modeling tool for co-simulation. *Energy and Buildings* 117: 272–281.

IEA. (2017) *IEA-EBC Annex 66. Occupants Behavior Research Bibliography*. www.annex66.org/?q=biblio, last visited 2017–04–24.

Langevin, J., Wen, J. and Gurian, P.L. (2015) Simulating the human-building interaction: Development and validation of an agent-based model of office occupant behaviors. *Building and Environment* 88: 27–45.

Lindelöf, D. and Morel, N. (2006) A field investigation of the intermediate light switching by users. *Energy and Buildings* 38: 790–801.

Love, J.A. (1998) Manual switching patterns observed in private offices. *Lighting Research & Technology* 30(1): 45–50.

Madsen, H. and Thyregod, P. (2011) *Introduction to General and Generalized Linear Models*. Boca Raton, FL: CRC Press.

Mahdavi, A. (2011) The Human Dimension of Building Performance Simulation. Driving better design through simulation (Keynote). In Soebarto, V., Bennetts, H., Bannister, P., Thomas, P.C. and Leach, D. (eds), *Proceedings of the 12th Conference of The International Building Performance Simulation Association*. Sydney, Australia, ISBN: 978-0-646-56510-1, pp. K16–K33.

Mahdavi, A. (2015) Common fallacies in representation of occupants in building performance simulation. In *Proceedings of Building Simulation Applications 2015–2nd IBPSA-Italy Conference*. Bozen-Bolzano University Press, ISBN: 978-88-6046-074-5, pp. 1–7.

Mahdavi, A., Del Bolgia, M. and Tahmasebi, F. (2016b) Understanding the user-driven natural ventilation in Buildings: Can we benefit from the operationalization of high-level human-ecological concepts? In *Proceedings of 9th Windsor Conference*. Cumberland Lodge, Windsor, UK, April 7–10, 2016.

Mahdavi, A., Glawischnig, S., Schuss, M., Tahmasebi, F. and Heiderer, A. (2016c) Structured building monitoring; Ontologies and platform. In Mahdavi, A., Martens, B. and Scherer, R. (eds), *eWork and eBusiness in Architecture, Engineering and Construction: ECPPM 2016*. London, UK: CRC Press, pp. 651–658.

Mahdavi, A., Mohammadi, A., Kabir, E. and Lambeva, L. (2008) Occupants' operation of lighting and shading systems in office buildings. *Journal of Building Performance Simulation* 1(1): 57–65.

Mahdavi, A. and. Taheri, M. (2016) An ontology for building monitoring. *Journal of Building Performance Simulation* 10(5–6): 499–508.

Mahdavi, A., Taheri, M., Schuss, M., Tahmasebi, F. and Glawischnig, S. (2018) Structured building data management: Ontologies, queries, and plat-forms. In Wagner, A., O'Brien, W. and Dong, B. (eds), *Exploring Occupant Behavior in Buildings*. Cham: Springer.

Mahdavi, A. and Tahmasebi, F. (2015) Predicting people's presence in buildings: An empirically based model performance analysis. *Energy and Buildings* 86: 349–355.

Mahdavi, A. and Tahmasebi, F. (2016a) The deployment-dependence of occupancy-related models in building performance simulation. *Energy and Buildings* 177: 313–320.

Mahdavi, A. and Tahmasebi, F. (2016b) On the quality evaluation of behavioural models for building performance applications. *Journal of Building Performance Simulation* 10(5–6): 554–564.

Mahdavi, A., Tahmasebi, F. and Kayalar, M. (2016a). Prediction of plug loads in office buildings: Simplified and probabilistic methods. *Energy and Buildings* 129: 322–329.

Mahdavi, A., Tahmasebi, F., O'Brien, W. and Burak Gunay, H. (2017) The diversity challenge in models of occupants' presence in buildings. In Baratieri, M., Corrado, V., Gasparella, A. and Natuzzi, F. (eds), *Building Simulation Applications Proceedings – 3rd IBPSA-Italy Conference*. Bolzano, Italy. Free University of Bozen – Bolzano, 3. ISSN: 2531–6702, Paper-No. 70.

O'Brien, W., Gaetani, I., Gilani, S. and Carlucci, S. (2017) International survey on current occupant modelling approaches in building performance simulation. *Journal of Building Performance Simulation* 10(5–6): 653–671.

O'Brien, W., Gilani, S. and Gunay, H.B. (2018) In situ approaches to studying occupants. In Wagner, A., O'Brien, W. and Dong, B. (eds), *Exploring Occupant Behavior in Buildings*. Cham: Springer.

O'Brien, W. and Gunay, H.B. (2014) The contextual factors contributing to occupants' adaptive comfort behaviors in offices – A review and proposed modeling framework. *Building and Environment* 77: 77–87.

O'Brien, W., Gunay, H.B., Tahmasebi, F. and Mahdavi, A. (2016) A preliminary study of representing the inter-occupant diversity in occupant modelling. *Journal of Building Performance Simulation* 10(5–6): 509–526.

Page, J., Robinson, D., Morel, N. and Scartezzini, J.L. (2008) A generalized stochastic model for the simulation of occupant presence. *Energy and Buildings* 40(2): 83–98.

Parsons, K. (2003) *Human Thermal Environments*. Taylor & Francis Group, CRC Press. ISBN 0-415-23793-9.

Reinhart, C.F. (2001) Daylight Availability and Manual Lighting Control in Office Buildings Simulation Studies and Analysis of Measurements. PhD Thesis, Technical University of Karlsruhe, Germany, October 2001.

Reinhart, C.F. (2004) Lightswitch-2002: A model for manual and automated control of electric lighting and blinds. *Solar Energy* 77(1): 15–28.

Richardson, I., Thomson, M. and Infield, D. (2008) A high-resolution domestic building occupancy model for energy demand simulations. *Energy and Buildings* 40(8): 1560–1566. doi:10.1016/j.enbuild.2008.02.006.

Rietschel, H. and Raiß, W. (1968) *Lehrbuch der Heiz- und Lüftungstechnik* (15. Aufl.) Berlin/Göttingen/Heidelberg: Springer-Verlag.

Rijal, H.B., Tuohy, P., Humphreys, M.A., Nicol, J.F., Samuel, A. and Clarke, J. (2007) Using results from field surveys to predict the effect of open windows on thermal comfort and energy use in buildings. *Energy and Buildings* 39: 823–836.

Samuelson, H.W., Ghorayshi, A. and Reinhart, C.F. (2016) Analysis of a simplified calibration procedure for 18 design-phase building energy models. *Journal of Building Performance Simulation* 9. doi:10.1080/19401493.2014.988752.

Schweiker, M., Brasche, S., Hawighorst, M., Bischof, W. and Wagner, A. (2016) Presenting LOBSTER, an innovative climate chamber, and the analysis of the effect of a ceiling fan on the thermal sensation and performance under summer conditions in an office-like setting. In *Proceedings of 9th Windsor Conference*. Cumberland Lodge, Windsor, UK, April 7–10, 2016.

Schweiker, M., Haldi, F., Shukuya, M. and Robinson, D. (2012) Verification of stochastic models of window opening behaviour for residential buildings. *Journal of Building Performance Simulation* 5(1): 55–74.

Schweiker, M. and Wagner, A. (2016) Exploring potentials and limitations of the adaptive thermal heat balance framework. In *Proceedings of 9thWindsor Conference*. Cumberland Lodge, Windsor, UK, April 7–10, 2016.

Shipworth, D., Huebner, G., Schweiker, M. and Kingma, B.R.M. (2016) Diversity in Thermal Sensation: Drivers of variance and methodological artefacts. In *Proceedings of 9thWindsor Conference*. Cumberland Lodge, Windsor, UK, April 7–10, 2016.

Tahmasebi, F. and Mahdavi, A. (2012) Optimization-based simulation model calibration using sensitivity analysis. In *7th Conference of IBPSA-CZ*. Brno, Czech Republic.

Tahmasebi, F. and Mahdavi, A. (2016) An inquiry into the reliability of window operation models in building performance simulation. *Building and Environment* 105: 343–357.

Tahmasebi, F. and Mahdavi, A. (2017) The sensitivity of building performance simulation results to the choice of occupants' presence models: A case study. *Journal of Building Performance Simulation* 10(5–6), 625–635.

Tahmasebi, F. and Mahdavi, A. (2018) On the utility of occupants' behavioural diversity information for building performance simulation: An exploratory case study. *Energy and Buildings* 176(2018), 380–389.

Tjur, T. (2009) Coefficients of determination in logistic regression models – A new proposal: The coefficient of discrimination. *The American Statistician* 63(4), 366–372.

Wagner, A., Anderson, R., Zhang, H., de Dear, R., Schweiker, M.,. Goh, E., Lichtenbelt, W.V.M., Streblow, R., Goia, F., Park, S. and Kleber, M. (2018a) Laboratory approach to studying occupant behavior. In Wagner, A., O'Brien, W. and Dong, B. (eds), *Exploring Occupant Behavior in Buildings*. Cham: Springer.

Wagner, A., O'Brien, W. and Dong, B., ed. (2018b) *Exploring Occupant Behavior in Buildings*. Cham: Springer.

Wang, C., Yan, D. and Jiang, Y. (2011) A novel approach for building occupancy simulation. *Building Simulation* 4(2): 149–167.

Wilke, U., Haldi, F., Scartezzini, J.-L. and Robinson, D. (2013) A bottom-up stochastic model to predict building occupants' time-dependent activities. *Building and Environment* 60: 254–264.

Wolf, S., Andersen, R.K., Barthelmes, V., Gunay, B., Møller, J.K., Madsen, H., O'Brien, W., Schweiker, M. and Yilmaz, S. (2018) Chapter 5 in IEA-EBC Annex 66 report.

Wu, Y. and Mahdavi, A. (2014) Assessment of thermal comfort under transitional conditions. *Building and Environment* 76: 30–36.

Yan, D., O'Brien, W., Hong, T., Feng, X., Gunay, H.B., Tahmasebi, F. and Mahdavi, A. (2015) Occupant behavior modeling for building performance simulation: Current state and future challenges. *Energy and Buildings* 107: 264–278.

Yang, Z. and Becerik-Gerber, B. (2014) The coupled effects of personalized occupancy profile based HVAC schedules and room reassignment on building energy use. *Energy and Buildings* 78: 113–122.

Recommended reading

Systematic treatment of occupants' presence and actions in buildings and related models represent an emerging field of inquiry, resulting in a paucity of standard textbooks. In this context, the book titled *Exploring Occupant Behavior in Buildings* represents a useful resource (Wagner et al. 2018b). Moreover, as a result of the activities of IEA Annex 66, a database of relevant publications has been created (IEA 2017). Likewise, the list of references at the end of this chapter includes a number of important and valuable contributions pertaining to the representation of inhabitants in building performance simulation.

Assignments

Explicit assignments are not included for this chapter, given the specific character of the material covered. However, various activities could deepen the appreciation of the challenges associated with systematic modeling of people's presence and actions in buildings. For example, readers might generate reports of their own presence and daily interactions with a building's environmental systems. Thereby, the general scheme captured in Table 4.1 can provide a conceptual framework. Similarly, systematically collected data from specific buildings can be structured and used as input data for energy performance simulation runs. The results could then be compared to those obtained by using standard occupancy-related input models.

5 Indoor thermal quality performance prediction

Christoph van Treeck and Daniel Wölki

Scope

This chapter provides background information on the human thermoregulation, thermal sensation, and comfort perception. Furthermore, various influencing factors and performance indicators which are of practical relevance for the assessment of an occupant's thermal comfort are addressed. Corresponding prediction methods in the simulation environment are emphasized based on a description of the necessary conditions for thermal comfort in uniform and non-uniform environments. In this regard, the various levels of detail of available simulation models and their accuracy are distinguished, depending on the conceptual stages of a building during the planning process. The chapter explains various thermal comfort indices and summarizes international codes and standards for thermal quality performance prediction. The chapter also gives application examples that address the simulation of local thermal comfort involving detailed numerical human model, draught risk analysis in conjunction with a coupled CFD approach, as well as typical results obtained by the use of a thermal multi-zone model.

Learning objectives

The intention of this chapter is to give a concise overview of the topic of indoor thermal quality performance prediction with special emphasis on (but not restricted to) building energy simulations. This includes an introduction to the dynamic reactions of the human body to thermal stimuli as well as the aspects of thermal sensation and thermal comfort perception. The case studies presented here are designed to make readers aware of the different levels of detail in simulation and the physical resolution of the various methods, to enable them to select an appropriate assessment method in simulation practice and to introduce them to current codes and standards available.

Key words

Thermal comfort/influence factors/comfort indices/calculation methods and algorithms/relevant industrial areas/design stage-related levels of detail/uncertainty/temperature sensation/comfort perception/human thermoregulation/codes and standards/case studies/thermal imaging/comfort tracking

Introduction

From the human body to thermal quality performance prediction in buildings

It is well known from experiments that the sensation of temperature and the perception of thermal comfort are related to the thermal state of the human body, as detected by thermoreceptors, that they depend on skin and core temperatures, respectively, and their variation over time (Fanger 1970; Gagge 1973; Stolwijk 1971). The efforts for regulating body temperature and moisture sensation on the skin are also of significance (Gagge 1973). The thermal state of the body is the integral value of several thermophysical and thermoregulatory processes with varying impacts on individual body parts. The body becomes acclimatized due to physiological adaptations. With these remarkable abilities, the human body maintains its core temperature at a fairly constant level over a wide range of ambient conditions.

Thermal sensation and comfort can be expressed by the use of diverse assessment scales and are usually regarded in terms of steady-state, uniform conditions for the body as a whole and in relation to the indoor climate. The latter is a rational approach for a large number of simulation applications in practice (Fanger 1970). Recently, so-called adaptive models also take into account the behavioural adaptation of people in naturally ventilated buildings and relate the acceptability of thermal indoor conditions to the change in outdoor climate over time.

The ergonomic requirements of the local thermal environment can be obtained by taking objective measurements, using a system, as given in Figure 5.1. A non-uniform environment is then correlated to a uniform environment using the concept of equivalent temperatures at the level of individual body parts. Performing this type of experiment is, however, costly and time-consuming.

Building performance simulation offers a cost-effective tool for a thorough investigation of the aforementioned effects. Models are available that enable a statistical correlation of performance indicators with subjective evaluations of thermal sensation and comfort, thus reflecting the impact of the surrounding physical ambience on a person's well-being. Here, commonly used simulation approaches range from steady-state monthly and annual balances (rough level) over dynamic multi-zone models up to computational methods that model fluid flow and radiative heat transfer in detail (fine level). Combined approaches exist in order to bridge the gap between the multi-scale and multi-level problem (Beausoleil-Morrison 2000; van Treeck 2004).

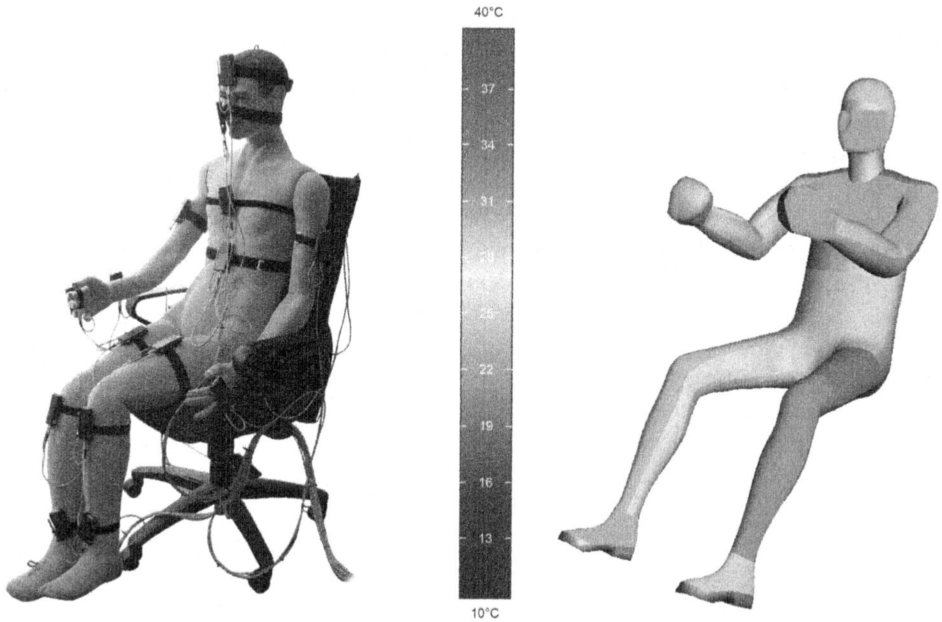

Figure 5.1 Manikin equipped with flat, heated sensors that measure the dry heat exchange between a person and his/her surroundings. The colour bar expresses the appropriate local resultant surface temperatures (RST) in °C.

To assess the indoor thermal quality of a building in terms of thermal comfort, descriptive integral quantities are required, such as the frequency of temperatures exceeding a certain temperature limit during a pre-defined period, or an even more sophisticated model for the prediction of the overheating risk. Relevant prediction methods are introduced in this chapter. The required data can be acquired by the use of dynamic simulation models that are based on information regarding a building's geometry, its structure, the materials of the corresponding components, the site-specific conditions, and the selected energy concept.

Researchers increasingly tend to concentrate on both highly refined and combined numerical models for the assessment of human comfort. For such an integrated study, a flow solver, a thermal radiation solver, and a numerical human model that simulates the heat exchange between the human body surface and the physical environment are needed. The latter also includes the detailed replication of the corresponding human thermoregulatory actions. In addition, such a model is required to relate the computed skin surface temperatures to local thermal sensation votes.

When predicting performance, a distinction is generally made between the aspects of indoor thermal quality and indoor air quality. The latter is described in more detail in Chapter 3 ("Ventilation Performance Prediction") of this volume. With regard to health criteria, engineers are confronted with complex integrated relationships that involve thermal, physical, chemical, and microbiological phenomena such as the sick building syndrome (de Dear 1998; Kröling 1998; Stolwijk 1991) or the tight building syndrome (Recknagel et al. 2007). The thermal indoor climate and air quality influence both health and productivity. Therefore, a well-designed indoor thermal environment helps to increase the mental performance capabilities of persons and simultaneously reduces staff absence.

Engineers in the building sector are faced with the problem that the energy concept, including the HVAC system, is usually designed according to heating and cooling demands in order to maintain operative room temperatures at a reasonable level – and not primarily according to local thermal effects influencing thermal comfort. The latter temperature level can be determined by a comfort zone. Such local effects are usually not considered during the early design stages of a building. It is particularly important to investigate the interaction between the energy system, the building envelope, and the (thermally active) building components, especially in the context of energy-efficient building design (Pfafferott et al. 2007). This, however, is important as modifications of the design become more expensive with the advancing planning process.

People dissatisfied with their thermal environment are likely to try to thwart the situation and negatively influence a building's energy use (de Dear and Brager 2002; van der Linden et al. 2002). Consequently, the indoor thermal quality influences energy consumption of the building due to varying cooling and heating demands. As explained later in this chapter, several standards give recommendations for a reasonable choice of parameters that affect the thermal environment in buildings in terms of both thermal comfort and system layout. Here, the most significant innovation in existing standards over the past years is the consideration of user behaviour in naturally ventilated spaces. The latter is expressed as conscious actions that aim to improve people's thermal situation.

Other relevant industrial areas

Besides the building and the HVAC sector, the assessment of thermal comfort is likewise relevant for a number of other industrial areas, such as the automotive industry, the aircraft industry, and the railway/coach industries, as indicated in Figures 5.2 and 5.3. In vehicles, for example, many effects interfere with each other in a small space and affect the overall thermal state of the human body. Such effects include, for example, short-wave radiation, which is locally absorbed

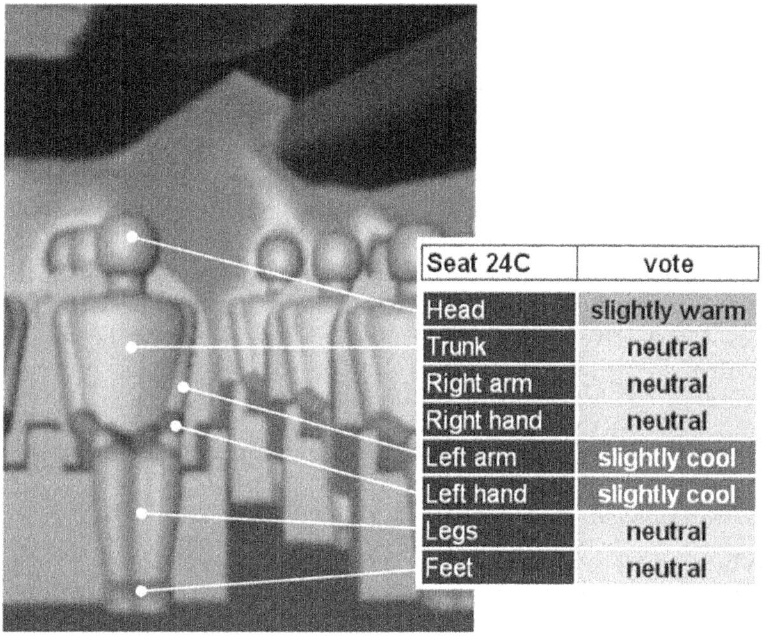

Figure 5.2 Example of a thermal flow analysis for an airplane passenger cabin using a lattice Boltzmann type CFD solver and equivalent temperatures for thermal comfort assessment. The cross-section shows the thermal plumes of the convective flow.

Source: van Treeck et al. 2007b

Figure 5.3 Flow analysis inside an open train carriage. The picture shows the average velocity field.

Source: van Treeck et al. 2007a

by parts of the body, or streams of cold air, which are produced by the air-conditioning system pointing to local body parts. In combination with skin areas covered with sweat due to solid contact with fabrics (seat surface, clothing, etc.), local discomfort may arise. A single effect may accordingly dominate the perception of overall individual comfort.

In addition to the aspect of thermal quality, health, occupational safety, and risk management may also become important issues with respect to the design. For example, operative temperatures in spaces equipped with machinery must be limited to moderate levels in order to ensure safe working conditions. Special performance indicators that address this issue are called heat stress indices. An extensive summary on the most important indices can be found in Havenith and Fiala (2016).

Human thermoregulation, energy balance, and temperature sensation

Human thermoregulation

The human body is able to maintain its core temperature at approximately 37 °C over a wide range of ambient conditions. Here, the thermal state of the body is the result of several thermophysiological processes that even include local autonomous thermoregulation within different tissue materials of diverse body parts. The corresponding heat loss to the environment due to convective and radiative heat transfer, heat conduction caused by contacting surfaces, evaporation, diffusion, and respiration is compensated by internal heat production due to conscious or unconscious muscle activity as well as the basal metabolic rate. In combination, they form the total metabolic rate.

In general, the human thermoregulation involves four essential autonomic mechanisms that control the heat exchange between human beings and their physical environment. In this regard, the skin blood flow can be controlled by vasoconstriction (narrowing of blood vessels) and vasodilatation (widening of blood vessels). Evaporative heat loss is the result of sweating, while shivering increases metabolic heat production because of the increased muscle activity. Conscious reactions, such as changing clothes, performing exercises, or changing one's position inside an indoor space can actively aid the thermoregulation process.

Human thermoregulation is triggered by various stimuli from the indoor environment. Skin temperature is detected by cutaneous thermoreceptors which are inhomogeneously distributed across the human body surface and cause body-part-specific sensitivities. Warm and cold receptors operate within different but overlapping ranges, and cold receptors are more densely distributed across the skin surface than are warm receptors (Persson 2010); it is important to distinguish between receptors sensing warmth, cold, and pain. The behaviour of the warm and cold receptors makes it possible to use a so-called set-point concept within thermoregulation models, where set-point temperatures represent local or global skin surface temperatures that arise under thermal-neutral ambient conditions. Both types show a dynamic overshoot in their response frequency as soon as skin temperatures change over time (Hensel 1982). Concerning the body core, thermosensitive neurons are mainly concentrated within the hypothalamic region. The hypothalamus can also be stimulated by local warming or cooling of individual body locations, which causes vasodilatation and sweating or vasoconstriction and shivering, respectively. It plays a central role in processing information from thermal stimulations.

To summarize, thermoregulatory reactions of the central nervous system are the result of multiple functions of signals from the body core and its periphery. Moreover, local changes of the skin temperature of body parts also cause local thermoregulatory reactions such as modifications of the perspiration rate or local blood flow rates. Significant indicators for the activity of the human thermoregulatory system are the mean skin temperature, its variation over time, and the hypothalamus temperature. Calorimetric values only exert a slight influence; temperature values are the main indicators. The corresponding indicators can be correlated with the prior mentioned autonomic thermoregulatory responses in order to form a detailed thermoregulation model (Fiala et al. 1999; Fiala et al. 2001; Gagge 1973; Gordon 1974; Holmér 2004; Huizenga et al. 2001; Stolwijk 1971; Tanabe et al. 2002; Wissler 1985).

Heat exchange with the environment

In the simulation context, heat exchange between the human body and the environment can be computed by numerically approximating the governing partial differential equations or by the use of simplified algebraic expressions based on linearized heat transfer coefficients (Bejan 1993; Modest 2003). A general form of the corresponding heat balance equation of the human body is given by Equation 5.1. It is usually related to the outer clothing or the naked skin surface, at which the sensible and latent heat losses can be expressed as a function of the skin temperature or the skin wetness. However, both parameters are linked to environmental parameters such as air temperature, air velocity, average radiant temperature, and water vapour pressure. For a detailed description of the single terms of Equation 5.1, the reader is referred to Fanger (1970), Fiala (1998), Fiala et al. (1999), Stolwijk (1971), or ASHRAE Standard 55.

$$\underbrace{M \quad W}_{\substack{\text{net heat}\\ \text{production}}} = \underbrace{\left(Q_{conv} + Q_{cond} + Q_{rad} + E_{skin}\right)}_{\substack{\text{skin}\\ \text{heat losses}}} + \underbrace{\left(C_{resp} + E_{resp}\right)}_{\substack{\text{respirative}\\ \text{heat losses}}} + \underbrace{S}_{\substack{\text{heat}\\ \text{storage}}} \tag{5.1}$$

The left-hand side of Equation 5.1 starts with the net heat production ($M-W$), which comprises the total metabolic rate M that is reduced by the portion of energy converted into external mechanical work W, depending upon the mechanical efficiency of muscle tissue. Metabolic activity is described by the metabolic equivalent in [met], which represents the heat production per unit of skin area where 1 met = 58 W/m^2 for a seated adult at rest.[1] Tables that contain the metabolic heat production for different levels of activity have been summarized in standards such as DIN EN ISO 8996, EN ISO 7730, and ASHRAE Standard 55. According to Havenith et al. (2002), it is important to remember that most of the tables in literature that contain estimates for the human metabolism, expressed as [W/m^2], refer to an average male with a body surface area of 1.8 m^2 and 70 kg body weight. Consequently, they require correction as soon as they are applied to specific sub-populations or individuals.

On the right-hand side of Equation 5.1, the terms for heat loss and storage are accumulated. The sensible heat losses from the skin surface involve convective and radiative heat transfer Q_{conv} and Q_{rad}, heat conduction through contact Q_{cond} (often neglected by using adiabatic boundary conditions in simulation), and latent heat loss due to sweat evaporation and moisture diffusion E_{skin}. The respiratory system participates by means of sensible C_{resp} and latent heat flux E_{resp}. The heat storage term S, i.e. the change of the body internal heat content, can be described in terms of the body's thermal capacity and its rate of change in connection with transient ambient conditions.

Mathematical human models and possible ways of classification

As mentioned before, the heat exchange between a human being and his/her surroundings can be generally described by Equation 5.1, which represents the basic equation for most of the existing mathematical human models available in the current literature. However, individual heat exchange expressions can be different. In the context of building performance simulations or for the general assessment of the thermal quality of indoor spaces, a multitude of mathematical human models have evolved during the past 10 decades. Their level of detail has advanced from reduced order models, such as empirical equations or one-node models, to highly sophisticated multi-element models.

In current literature, the characterization of those models is performed based on the number of body compartments, their spatial subdivision into sectors and tissue layers, and the type of discretization method used to numerically solve the partial differential equations that describe the heat transfer phenomena occurring inside the human body (finite differences, finite volumes, finite elements). In this regard, a common distinction is made between the following model categories:

- one-node or empirical models,
- two-node models,
- multi-node models, and
- multi-element models.

Here, one-node or empirical models approximate the human being by a single heat balance equation or an empirical equation. Probably the most well-known representatives of this category are the models of Burton (1934), Fanger (1970), and Pandolf et al. (1986).

Two-node models represent the next step in model evolution. They are often based on the so-called core and shell principle according to Machle and Hatch (1947) or Aschoff and Wever (1958). Here, a human being is idealized as two concentric cylinders, each representing a single node with a uniform nodal temperature. A classical representative of this model category is the two-node Pierce model (Gagge 1973; Gagge et al. 1986), which is also given as part of *ASHRAE Fundamentals*.

Multi-node models can be regarded as an expansion of the so-called two-node models. They are characterized by an increased number of tissue layers as well as an increased number of geometrical elements used to approximate the human anatomy. The most prominent model of this category is the model of Stolwijk (1971), which can be regarded as a milestone in the context of thermophysiological modelling.

The highest level of discretization, however, is represented by the multi-element models. Those models show high structural analogies to the prior mentioned multi-node models, which often makes it hard to distinguish between the two. However, the main difference between the two types of models is that multi-element models are able to consider spatial asymmetries due to the subdivision of the body into additional compartments (Tanabe et al. 2002; Streblow 2011) and sectors (Fiala et al. 1999, 2001).

A historical overview of well-known mathematical human models from the current literature with respect to their classification as well as to their field of application is given in Table 5.1.

A short review of thermal sensation versus comfort

It is important to distinguish between the terms skin temperature, thermal sensation, and thermal comfort (acceptability). As cutaneous thermoreceptors show an inhomogeneous distribution across the entire skin surface, they cause different regional sensitivities (Guyton 2002; Hensel 1982), which needs to be considered by formulae used for the calculation of the average skin temperature of persons. Details are given in Fiala (1998), McIntyre (1980), and Schmidt (2016). Fiala also summarizes the interpretation of the various thermal stimuli from the body core and its peripheral areas in terms of autonomous thermoregulatory responses.

Here, the autonomous thermoregulatory responses are correlated with the body core temperature and the skin temperature as well as its variation over time. The latter is known to be partly responsible for the effect that people feel colder when their skin temperature is actively decreasing compared to when the temperature remains at a constant but cold level (*ASHRAE Handbook – Fundamentals* (2017)). The moisture sensation on the skin is likewise significant, because skin moisture highly correlates with warm discomfort (Gagge 1973, 1979) as well as

Table 5.1 Historical overview of mathematical human models, their classification and field of application

Source	*Classification*	*Field of application*
Burton (1934)	empirical	uniform environments
Wissler (1964)	multi-element	non-uniform environments
Fanger (1970)	empirical	uniform environments
Stolwijk (1971)	multi-node	non-uniform environments
Azer and Hsu (1977)	two-node	uniform environments
Gagge et al. (1986)	two-node	uniform environments
Pandolf et al. (1986)	empirical	uniform environments
Smith (1991)	multi-element	non-uniform environments
Jones and Ogawa (1992)	two-node	uniform environments
Yokoyama et al. (1997)	multi-element	non-uniform environments
Fiala et al. (1999, 2001)	multi-element	non-uniform environments
Havenith (2001)	two-node	uniform environments
Huizenga et al. (2001)	multi-element	non-uniform environments
Kohri and Mochida (2002)	multi-element	non-uniform environments
Tanabe et al. (2002)	multi-element	non-uniform environments
Paulke and Wagner (2007)	multi-element	non-uniform environments
Foda and Sirén (2010)	multi-element	non-uniform environments
Streblow (2011)	multi-element	non-uniform environments
Holopainen (2012)	multi-element	non-uniform environments
Kingma (2012)	multi-element	non-uniform environments
Wölki (2017)	multi-element	non-uniform environments

with the effort of the human body to control its internal temperature level (Berglund and Cunningham, 1986).

The two-node Pierce model of Gagge (1973), for example, is used to derive two quantities that are dedicated to thermal sensation and thermal discomfort. The corresponding indices were obtained from (i) the modified effective temperature (*ET**) and (ii) the skin wetness. They are based on experimental data collected at uniform and steady-state conditions (cf. Fanger 1970; Nevins et al. 1966; Rohles and Wallis 1979). Provided the operative temperature is within the comfort range, humidity fluctuations tend not to have a distinct effect on human thermal sensation for values of relative humidity between 20 and 70 percent (Gonzalez and Gagge 1973; Gonzalez and Berglund 1979; Nevins et al. 1975).

For uniform steady-state conditions (Nevins et al. 1966; Rohles and Wallis 1979) near thermal neutrality, Fanger (1970) observed that the level of thermal comfort can be related to the disequilibrium in the heat balance of persons. For practical reasons, it might therefore be advisable to use reduced order models that allow the direct correlation of physical environmental parameters with a single thermal comfort index that is valid for the body as a whole. However, many applications require a much more detailed resolution of the local effects on human thermal comfort perception. This is especially the case for asymmetric and/or transient conditions. In this regard, the prior-mentioned approach is no longer applicable.

Mower (1976) showed that local thermal sensation is governed by skin temperatures, irrespective of the general thermal state of the body (Fiala 1998). However, local (dis)comfort was found to depend on overall comfort perception as well (Cabanac and Hardy 1969; Issing and Hensel 1981). Zhang (2003) showed that temperature sensation and comfort perception depend on the overall thermal state of the body, as captured by thermoreceptors and quantified by the weighted influence of the individual body parts. Figure 5.4(a) shows the relationship between the deviation of local skin temperatures from their set points at thermal neutrality and the local thermal sensation. In this regard, negative values for thermal sensation correspond to the sense of cold and positive values to the sense of warmth. The graphs indicate an asymmetric behaviour of the local thermal sensation and the local thermal comfort when related to higher and lower ambient temperatures as well as to the thermal state of the entire body (Zhang 2003).

In Figure 5.4(b), the assessed local thermal sensation is plotted against the comfort perception in terms of its acceptability. It can be seen that the corresponding local outcomes are influenced by the overall thermal sensation of a person. Therefore, the perceived sensation of optimum comfort shifts according to the overall feeling. If, for example, the core temperature

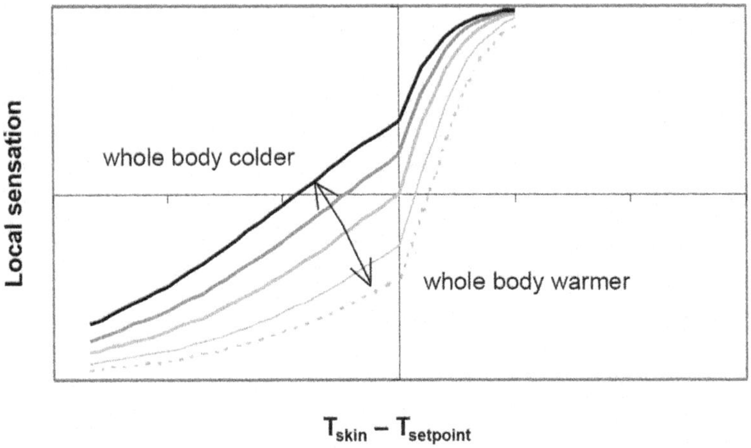

Figure 5.4(a) Local thermal sensation versus local skin temperature deviations from a predefined set point at thermal neutrality and the thermal state of the entire body

Source: Zhang 2003; Zhang et al. 2004. Reprinted with the author's permission

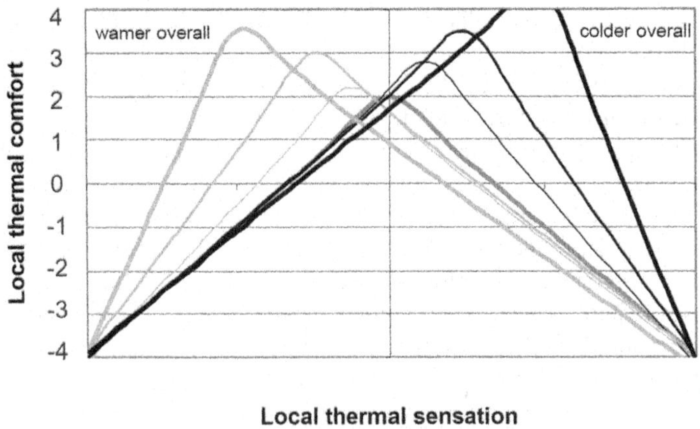

Figure 5.4(b) Local thermal comfort modified by the overall thermal sensation

Source: Zhang 2003; Zhang et al. 2004. Reprinted with the author's permission is high, cold stimuli are perceived as more pleasant than warm ones because they improve the thermal state of the body (Zhang 2003; Zhang et al. 2004; Zhang et al. 2010a, 2010b, 2010c).

In asymmetrical environments, parts of the body may therefore feel comfortable, even if the body as a whole does not. It is impossible, therefore, to apply the concept of neutral thermal sensation without modification. This effect is even stronger in transient situations. In order to account for dynamic responses to transient ambient conditions, for instance, a time derivative of the skin temperature needs to be included in a model that aims to consider the transient thermal sensation. Such a signal can be used to represent the increased fire rates of local thermoreceptors, which can be 10 times their basal level (Hensel 1982; Ring and de Dear 1991).

Performance indicators and conditions for thermal comfort

This section summarizes the environmental and personal parameters that affect the human thermal comfort perception. It explains the idea of operative and equivalent temperatures and introduces thermal sensation scales as well as basic environmental indices. The concept of the predicted mean vote (*PMV*) is introduced and discussed in the context of the behavioural adaptation of persons. The subsequent section details the issue of thermal comfort prediction in buildings, with a special focus on comfort envelopes and corresponding guidelines.

Definition of thermal comfort

Benzinger (1979) defines thermal comfort from a physiological point of view as "the absence of driving impulses from cutaneous and hypothalamic receptors causing the body to counteract with physiological adaptations". A resting and unclothed individual perceives ideal thermal comfort at thermoneutral conditions. The latter can be assumed to be at approximately 30 °C ambient temperature where conscious actions are not required to control the thermal state (Gagge et al. 1967). ASHRAE Standard 55 associates thermal comfort with the "condition of mind, which expresses satisfaction with the thermal environment".

Parameters affecting thermal comfort

The four main environmental parameters that influence the thermal sensation and comfort of persons are

- air temperature (McNall and Schlegel 1968),
- (relative) air velocity (Rohles et al. 1974),
- mean radiant temperature of the surrounding surfaces, and
- ambient water vapour pressure (Rohles and Nevins 1971).

Their influence, however, is only sensible if local climatic effects such as asymmetrical radiation, draught, or vertical temperature gradients are present, which affect a person's thermal state. Consequently, the environmental parameters mentioned can be used as indicators for the description of thermal discomfort (Fanger et al. 1985). The thermal comfort perception of human beings from different climatic regions and cultural backgrounds appears to be statistically uniform if similar environmental and personal parameters are taken into account (*ASHRAE Handbook – Fundamentals* 2017, Busch 1992; de Dear and Brager 2002; Fanger 1972).

The air velocity has a significant influence if, for example, a cold air stream is applied to local body parts at which it causes local asymmetries. Furthermore, the degree of turbulence is relevant, i.e. if fluctuations perpendicular to the main flow direction occur (*ASHRAE Handbook – Fundamentals* 2017). In summer, a high room air temperature may be compensated by an increase in the ventilation rate (EN ISO 7730). The influence of thermal radiation is particularly noticeable if the surrounding surfaces are colder than the environment (Hensen 1990); the effect of locally absorbed short-wave radiation has hardly been investigated to date.

Air humidity plays an important role too, because it influences the evaporation intensity of sweat and is significant if perspiration occurs at high indoor temperatures, where sensible heat loss is limited or non-existent and if the thermal comfort limit has already been attained (Recknagel et al. 2007). While discomfort in cold environments is governed by the average skin temperature, discomfort in hot environments is related to skin wetness as well as to skin and core temperatures (ASHRAE 2009; Berglund and Cunningham; Fanger 1970; Gagge 1979; Gonzalez and Gagge 1973; Nevins et al. 1975). Skin wetness expresses the fraction of skin covered with sweat and is a strong indicator of the level of thermal discomfort (Gagge 1979; Gonzalez and Gagge 1973). In winter, however, respiratory diseases frequently occur when humidity levels are low, causing dry mucous membranes and dust accumulation in rooms, which is further accelerated by the dry air (*ASHRAE Handbook – Fundamentals* 2017, Recknagel et al. 2007).

The human thermoregulatory response is further influenced by behavioural actions that refer to the adjustment of personal factors such as

- the activity level, and
- the clothing combination worn (insulation, moisture permeability).

The thermal resistance of clothing is usually expressed in units of [clo] where $1\,\mathrm{clo} = 0.155(\mathrm{m^2K})/\mathrm{W}$. Values of the insulation and vapour permeability of clothing ensembles can be obtained from measurements and are summarized in standards (EN ISO 9920, ASHRAE Standard 55). Similarly, metabolic rates for different levels of activity are given in standards in tabular form (EN ISO 7730, EN ISO 8996).

Environmental indices

Environmental indices are frequently used in a scope-dependent manner to express the physiological strain a thermal environment imposes. A distinction is made between rational and empirical indices (*ASHRAE Handbook – Fundamentals* 2017, EN ISO 11399). For hot ambient conditions, which often appear in industrial environments, the *wet bulb globe temperature index* (*WBGT*) and the *sweat rate* are also used to predict comfort diagnostically or analytically (ISO 7243, ISO 7933). Another formulation is the *heat stress index* (*HSI*) which, according to the *ASHRAE Handbook – Fundamentals*, expresses the ratio between the total evaporative heat loss required to reach a thermal equilibrium and the maximum possible evaporative loss for specific ambient conditions. Under cold conditions, for example, the *wind chill index* (*WCI*) and the *required clothing insulation* (*IREQ*) are defined in the *ASHRAE Handbook – Fundamentals* or ISO/TR 11079. EN ISO 11399 summarizes calculation procedures for the various situations. An extensive overview of various heat stress indices can be found in Havenith and Fiala (2016).

Equivalent homogeneous temperatures (Bohm et al. 1990; Wyon et al. 1989) or equal thermal environments (Mayer and Schwab 1993) represent concepts of parameter reduction. In this context, a set of physical quantities that belong to different heat transfer mechanisms is combined into a single value. This makes it possible to transfer inhomogeneous environments to uniform ones and enables the correlation between the thermal conditions and the thermal sensation of subjects. The corresponding subjective responses are assumed to be similar, irrespective of the combination of the heat transfer mechanisms.

The *operative temperature* (T_{OP}) represents "the uniform temperature of an imaginary black enclosure and the air within it in which an occupant would exchange the same amount of heat by radiation plus convection as in the actual non-uniform environment" (ASHRAE Standard 55). The operative temperature considers the average radiant temperature and the air temperature at actual velocity and can accordingly be approximated as the average between both temperatures if both values are close to each other and if the air velocity is small.

Other regional expressions exist for assessing humidity, such as the *humid operative temperature*, which is the temperature of an imaginary isothermal environment with a relative humidity of 100 percent that yields the same total heat loss from the skin surface at the same skin temperature and skin wetness as in the actual environment. The *effective temperature* (*ET**) by Gagge et al. (1971) is the same, except that it is defined for 50 percent relative humidity (*ASHRAE Handbook – Fundamentals* 2017). Both indices are defined in terms of operative temperatures and are useful indices for taking skin wetness into account. The *standard effective temperature* (*SET**) is equivalent to *ET**, however it is related to standardized clothing a person would wear at a given activity level.

Alongside the operative temperature, the *equivalent temperature* (T_{EQ}) also takes into account the dry heat loss from the clothed human body surface. The idea is to translate the non-uniform convective and radiative heat transfer into a uniform environment. ISO 14505–2 defines the equivalent temperature as "the uniform temperature of the imaginary enclosure with air velocity equal to zero in which a person will exchange the same dry heat by radiation and convection as in the actual non-uniform environment". The standard explains experimental methods and calculation procedures for T_{EQ} with respect to body segments and for the body as whole. For a historical review on this, the reader is referred to Nilsson (2004).

T_{OP}, T_{EQ}, *ET**, and the predicted mean vote *PMV* (see section below) are frequently used measures that are used to specify so-called comfort zones that relate the physical indoor quantities to a range of acceptable thermal conditions.

Subjective evaluation scales

The level of thermal comfort can be expressed by the use of well-defined assessment scales. However, thermal comfort is a subjective measure, and the wording for applying these scales is frequently not very precise. We distinguish between the following definitions:

The level of (dis)comfort can be expressed by the sequence {*comfortable, slightly uncomfortable, uncomfortably, very uncomfortable*}. Gagge et al. (1967) introduced the unpleasantness scale with the four terms {*pleasant, indifferent, slightly unpleasant, unpleasant*}. The so-called Bedford scale (Bedford 1936) uses mixed expressions in order to combine the prediction of thermal sensation and comfort. This seven-point scale reads {*much too warm, too warm, comfortably warm, comfortable, comfortably cool, too cool, much to cool*} with ratings from one to seven. Bedford showed a linear relationship between comfort assessment and equivalent temperatures in his scaling. In addition, he discovered that the influence of humidity is insignificant for temperatures below 24 °C.

Thermal sensation as a rational experience was suggested by Rohles and Nevins (1971) with the seven-point ASHRAE scale

$$\{cold, cool, slightly\ cool, neutral, slightly\ warm, warm, hot\}$$

which is numerically correlated with the range {*−3, −2, −1, 0, +1, +2, +3*}. Here, zero represents neutrality (ASHRAE Standard 55). In a large field study, Rohles and Nevins (1971) developed linear correlations between comfort level, dry bulb temperature, humidity, gender, and exposure time, which are summarized in the *ASHRAE Handbook – Fundamentals* (2017).

Another scale is the mean thermal vote (MTV), which is similar to the Bedford scale but counting from −3 to +3 for {*much too cold, too cold, cold but comfortable, neutral, hot but comfortable, too hot, much too hot*}, having again zero as the neutral value, like ASHRAE's seven-point scale (Nilsson et al. 1997; Wyon et al. 1989). Unlike the ASHRAE scale, however, it clearly distinguishes between acceptable, i.e. comfortable, ratings (−1, 0, +1), and ratings which are unacceptable.

It should be noted that differences in language and in subjective observations also cause misinterpretations of experimental data. The data transfer between experiments performed by different authors must accordingly be handled with care, even if the same scale has been used. An overview of existing scales, votes, and interpretations is given Table 5.2. Numbers in brackets represent possible scale extensions.

Predicted mean vote and predicted percentage of dissatisfied

Most engineering applications are related to moderate climates. In this context, a common method for steady-state and uniform conditions is to predict the average thermal sensation (the predicted mean vote, *PMV*) of persons according to the well-known Fanger model (Fanger 1970).

The idea of the model is that, close to thermal neutrality, the skin temperature and perspiration rate are the main parameters influencing the human body heat balance (*ASHRAE Handbook – Fundamentals* 2017). Fanger reduced the correlations discovered during his experiments to a single equation. The empirical *PMV* model thus expresses the disequilibrium between the energy balance equation for the body as a whole and the perceived level of comfort

Table 5.2 Overview of different subjective assessment scales

ASHRAE-scale[1,2,3,4] *Thermal sensation*		*COMFORT-scale* *Unipolar*[2,3]		*COMFORT-scale* *Bipolar*[3]		*BEDFORD-scale*[5]	
vote	attribute	vote	attribute	vote	attribute	vote	attribute
(−5)	unbearably cold						
(−4)	very cold			−4	extremely uncomfortable		
−3	cold			−3	very uncomfortable	−3	much too cold
−2	cool			−2	uncomfortable	−2	too cold
−1	slightly cool			−1	slightly uncomfortable	−1	cold but comfortable
0	neutral	0	comfortable			0	neutral
1	slightly warm	1	slightly uncomfortable	1	slightly comfortable	1	hot but comfortable
2	warm	2	uncomfortable	2	comfortable	2	too hot
3	hot	3	very uncomfortable	3	very comfortable	3	much too hot
(4)	very hot	(4)	extremely uncomfortable	4	extremely comfortable		
(5)	unbearably hot	(5)	unbearable				

Source: Redrawn from Schmidt 2016

[1] ASHRAE Standard 55
[2] DIN EN ISO 10551
[3] Zhang 2003
[4] *ASHRAE Handbook – Fundamentals* 2017
[5] Nilsson 2004

on the seven-point ASHRAE scale for a large group of persons exposed to the same thermal environment. Such a disequilibrium (i.e. the difference between the left- and right-hand side of the balance equation defined in the former section) occurs if, for a specific activity and clothing, the actual heat flow differs from the flow required for an optimum level of comfort (*PMV* = 0).

Since the obtained *predicted mean vote index* (*PMV*) is a statistical measure, it can be further related to a *predicted percentage of dissatisfied* (*PPD*) people. Figure 5.5 shows different correlations for estimating the *PPD* related to neutral assessments for the *PMV*. Fanger (1970) defines neutral conditions (bold line) for *PMV* = 0. EN ISO 7730, for example, contains clo-met diagrams indicating isolines of optimum operative temperatures for *PMV* = 0. In Fanger's correlation, *PMV* = 0 means that still five percent of the group may be dissatisfied, primarily stemming from differences in physiology and perception of occupants. Mayer (1997), however, argues that the neutral position of *PMV* must be shifted to the warmer side of the graph (*PMV* = +0.4, *PPD* < 16%). Corresponding differences are the result of the different classification methods the researchers used to characterize their *PMV* results. As it was unusual during former times to directly ask subjects about their thermal acceptance (e.g. *PPD*), the researchers clustered their *PMV* results in different manners. Fanger, for example, characterized *PMV* values ≥1 or ≤−1 as uncomfortable, whereas Mayer classified *PMV* values unequal to zero as uncomfortable. Consequently, the resulting correlations differ (Schmidt 2016). Further correlations have been published by Araújo and Araújo (1999), de Paula Xavier and Roberto (2000), and Yoon et al. (1999). They are summarized in van Hoof (2008).

Applying this approach to building simulation is straightforward. The calculation procedure, however, requires the six personal and environmental parameters, i.e. average air temperature, average radiant temperature, average air velocity, relative humidity, and level of activity and clothing, as corresponding input data, respectively. The calculation procedure is explained in detail in EN ISO 7730 and ASHRAE Standard 55. Equations for calculation of the surface temperature of clothing and the convective heat transfer coefficient have to be solved iteratively. Clothing insulation parameters can be estimated from EN ISO 9920, and metabolic rates for different activities are detailed in EN ISO 8996 and ASHRAE Standard 55, for example, and may accordingly be approximated.

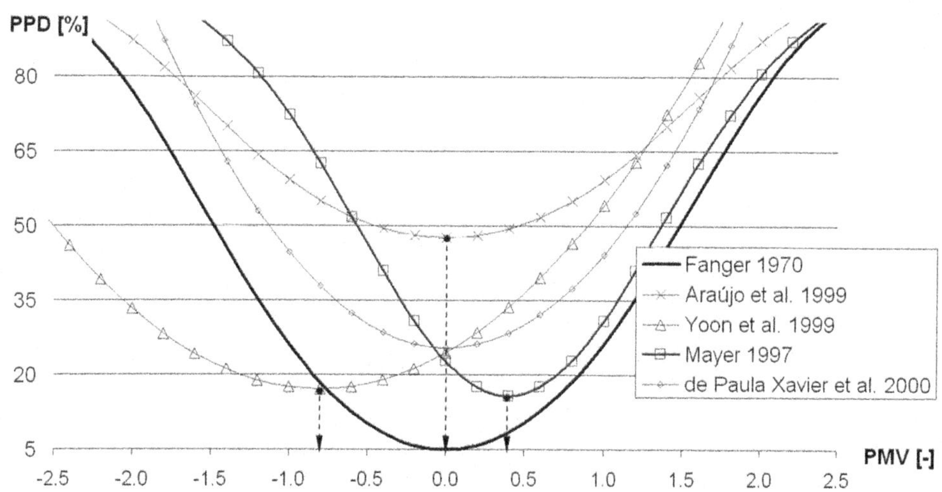

Figure 5.5 Different relationships for *PMV* and *PPD*

Source: Redrawn from the formulae summarized in van Hoof 2008

In practice, comfort envelopes, which are introduced in the next section, are often defined in terms of *PMV* ranges such as "$-0.5 < PMV < +0.5$", assuming typical combinations of clothing and activity, typical relative humidity, low local air velocities, and a reasonable range of operative temperatures. Consequently, the *PMV* model can be classified as a comfort criterion for the body as a whole.

In other words, setting the *PPD* to a certain value, for example six percent, causes a fixed *PMV* range ($-0.2 < PMV < +0.2$). Selecting the said parameters in a case-dependent manner, for instance assuming persons are performing sedentary work (1.2 met) and wearing winter (1.0 clo) or summer (0.5 clo) clothing, a valid range of operative temperatures can be obtained. In this case, the minimum required operative temperature for the heating period is 21 °C. The corresponding maximum operative temperature is 25.5 °C during the cooling period.

The accuracy of a comfort prediction strongly depends on the resolution of the simulation model. With a thermal multi-zone model including a detailed radiation model, most of the information is available. However, a detailed assessment requires the coupling with a CFD solver in order to replace average quantities with field values of the air temperature and the local flow velocity.

Local thermal comfort

The *PMV* model is formulated for the body as a whole, expressing an average assessment which is statistically representative for a group of people. At the same time, local thermal conditions such as the draught risk (Fanger 1964, 1988; Heiselberg 1994; Mayer and Schwab 1990), asymmetrical radiation (Fanger 1977, Fanger et al. 1985; Glück 1994), warm and cold floor surfaces (Olesen 2002, 2008), and vertical temperature gradients (Olesen et al. 1979), all of which can be a significant cause for discomfort (Fanger 1977), must be taken into account. Local effects are especially noticeable by persons that perform light, sedentary activities (*ASHRAE Handbook – Fundamentals* 2017).

For environmental conditions close to neutrality, the standards EN ISO 7730 and ASHRAE Standard 55 provide additional equations and diagrams to assess those issues in terms of a percentage of dissatisfied (PD). The criteria should be applied collectively for the whole-body assessment using the *PMV* model to identify local deficiencies.

- *Draught risk.* Local air velocity fluctuations cause a feeling of draught. This mainly occurs if parts of the body are cooled by a locally increased convective airflow. The model is especially applicable for the evaluation of thermal comfort for the neck, head, shoulder, and back region. For other body parts, such as arms and legs, the model might overestimate the draught rate. For activity levels > 1.2 met (EN ISO 7730), the influence of the draught rate decreases. Additionally, the local degree of turbulence significantly contributes to draught discomfort (Fanger et al. 1988). The standards provide an algebraic expression to assess the draught risk in terms of PD.

  In order to use the draught risk model within building performance simulations, it is necessary to know the airflow and temperature fields. Both can be obtained by CFD. Fanger (1964), Heiselberg (1994), and Nielsen (1981) provide empirical equations for assessing the risk of down draught from cold window surfaces. The methods are summarized in (Recknagel et al. 2007).
- *Compensation of high indoor temperatures.* Increased airflow rates can be used to compensate for high ambient temperatures. This effect is used in terms of the behavioural adaptation of occupants, e.g. the opening of windows to make use of natural ventilation. In its appendix, EN ISO 7730 quantifies the impact of increased air velocities with respect to

increasing indoor temperatures that exceed the respective comfort temperatures. However, for naturally ventilated buildings, it is suggested to use the adaptive comfort guidelines.

- *Asymmetrical radiant temperature.* Hot ceilings or cold vertical surfaces, such as cold windows, may lead to uncomfortable situations (Fanger 1977). Both standards summarize PD relations with respect to hot and cold vertical and horizontal surfaces. Related effects can be investigated via thermal multi-zone models if a geometry-based long-wave radiation model is available.
- *Warm and cold floors.* The standards further show the relation between PD and cold or warm floor surfaces, respectively. A minimum PD of six percent is obtained at approximately 23–25 °C for standing and sedentary persons, respectively. Here, multi-zone models are once more appropriate, if short- and long-wave radiation is to be considered in detail. A corresponding discussion on radiant floor heating and cooling is given by Olesen (2002, 2008).
- *Vertical temperature gradients.* EN ISO 7730 and ASHRAE Standard 55 provide a diagram to assess the PD caused by a vertical temperature gradient between head (1.1 m above the floor) and ankles (0.1 m above the floor). Assuming a well-distributed air temperature within a thermal zone, empirical equations that model the temperature stratification or natural stack effect may be applied where applicable. More recent investigations on the effect of air-displacement ventilation show that the temperature limits defined in the current standards are much too conservative, thus causing an exclusion of this technology so far. For more information on this topic, the reader is referred to Möhlenkamp et al. (2016). However, a full picture of the temperature distribution can thereby only be obtained with CFD or with measurements.

Inhomogeneous and transient conditions

The *PMV* model in combination with the introduced local comfort criteria is a straightforward, useful way of predicting discomfort by the use of simulation methods. This is particularly useful for the design of HVAC components. In many situations of practical interest, however, it is necessary to evaluate the human thermoregulatory responses towards inhomogeneous steady state or transient conditions, which increases the assessment tremendously.

The former sections have already set out the basic correlations between local thermal sensation and comfort perception, which both depend on the overall thermal state of the body. Information on human responses to asymmetrical (Fanger 1977) or transient uniform conditions (Gagge et al. 1967; Griffiths and McIntyre 1974; Hensen 1990) or non-uniform environments are available in the literature. Only a limited number of publications address the human thermoregulation under both non-uniform and transient conditions at the same time (Zhang et al. 2004).

Local ergonomic requirements of the thermal environment can be obtained by measurements with thermal manikins and by interviewing individuals exposed to the same ambient conditions. The real non-uniform environment is then correlated to a uniform environment using equivalent temperatures. The procedure is repeated for each segment of the body (EN ISO 14505–2 2004; Wyon et al. 1989). Corresponding regression analysis usually tends to show linear relationships between the local thermal sensation votes and the local equivalent temperatures. The correlations that are obtained in this way are accordingly limited to (i) the range of investigated ambient conditions and (ii) to the specific experimental settings such as clothing insulation and activity level. Experiences gained with the measurement and modelling of the human thermoregulatory responses as well as vital parameters under various climatic conditions are provided in Bohm et al. (1990), Han et al. (2001), Holmer (1995), Mayer (2000), Nilsson (2004), Schmidt et al. (2012), Strøm-Tejsen et al. (2007), and Zhang et al. (2004).

The comfort range obtained by the regression can be displayed in a chart with respect to the equivalent temperatures. The method is demonstrated in Figure 5.6, which has been adapted from the findings of Bohm et al. (1990) and Han et al. (2001) for a car cabin. The limiting values shown in the figure are determined by the selected percentage of (dis)satisfied people, in this case for 90 percent acceptability. Measured or simulated local equivalent temperatures (not shown in the figure) can then be visually compared with the comfort zone.

In order to address transient and non-uniform conditions, current research focuses on the integration of highly detailed numerical human models in numerical simulation tools as well as on their application in practice (Frijns et al. 2006; Kato et al. 2007; Kok et al. 2006; McGuffin et al. 2002; Nielsen et al. 2003; Nilsson 2004; Niu and Gao 2008; Paulke and Wagner 2007; Rugh et al. 2004; Sano et al. 2008; Streblow et al. 2008; van Treeck et al. 2007a; Wölki et al. 2016, 2017; Zhang and Yang 2008; Zhu et al. 2007, 2008). A detailed numerical analysis calls for the coupling between a CFD solver, a radiation solver, and a numerical human model, which is accordingly fairly complicated with respect to the mesh generation and the implementation of the numerical coupling procedure. An example of such a holistic numerical analysis is introduced in a later section that deals with the detailed numerical human model of Fiala et al. (1999, 2001).

In this regard, the dynamic thermal sensation (DTS) model proposed by Fiala in 1998 was used and compared to the static *PMV* model of Fanger (1970). The corresponding DTS index is suitable for the body as a whole and applicable under transient, dynamic, and asymmetric

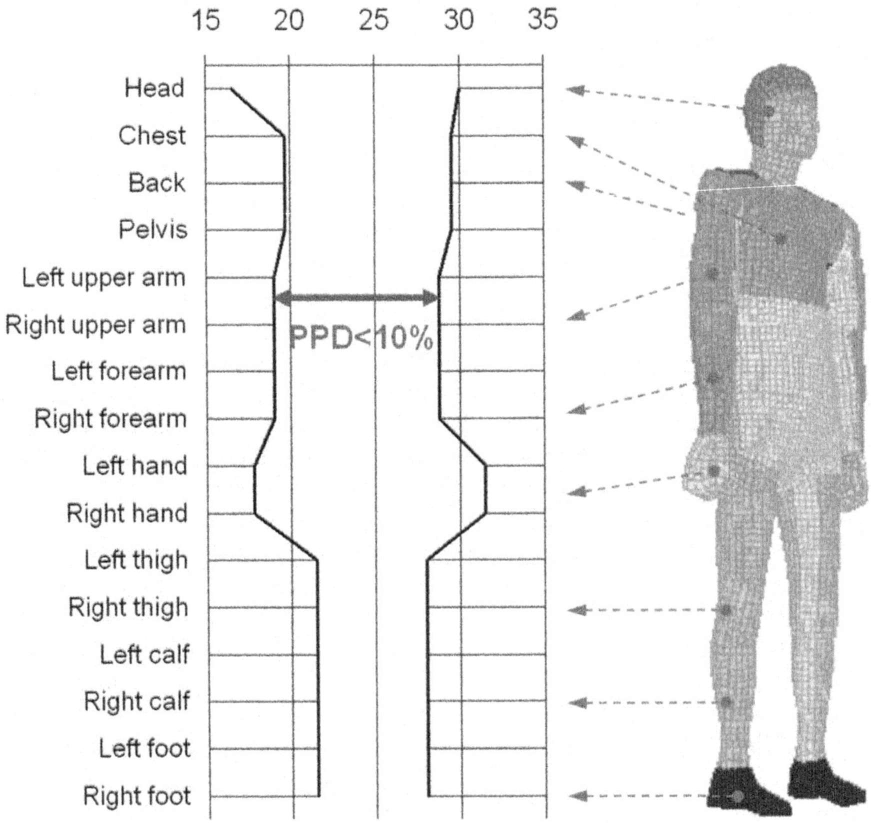

Figure 5.6 Example of a comfort zone for 90 percent acceptability in terms of equivalent temperatures

Source: Redrawn from the values presented in Bohm et al. (1990) and Han et al. (2001) for a car cabin ambient conditions. Because the model is based on physiological input parameters such as mean skin temperature, its derivative with respect to time, and the hypothalamus temperature (body core temperature), the DTS can be called a physiological comfort model. Here, the influence of asymmetrical environments is reflected in the calculation of the mean skin temperature, which is the weighted integral value of temperature information from local body locations. The transient nature of thermal sensation is considered in a so-called remembering function, whose influence vanishes with progressing time. The dynamic nature of the DTS index is shown in Figure 5.7 for a human model exposed to changing boundary conditions and subject to changes in the clothing insulation over time. It has to be emphasized that the nature of DTS differs from the static *PMV*, as the latter is a measure of the ambient climatic conditions and does not account for the real thermoregulatory responses of the human body as the DTS does. The skin and core temperatures for the above-mentioned experiment were determined by the use of Fiala's thermoregulation model.

From a series of experiments addressing the local and global temperature sensation and comfort perception of persons, Zhang et al. (2010a, 2010b, 2010c) derived a mathematical model for non-uniform, transient, homogenous, and steady-state conditions. The researchers developed local sensation and comfort expressions for 19 individual body segments as well as a global sensation and comfort index which takes the temperature sensitivity (weights) of the different body parts into account. Similar to Fiala's DTS, this model is formulated in terms of skin and core temperatures and considers their rate of change as well. Regression coefficients are detailed in Zhang et al. (2004) and Zhang et al. (2010a, 2010b, 2010c).

Some of the correlations between thermal sensation and comfort perception have already been mentioned in the former sections. Figure 5.4 shows the asymmetric nature of the static local sensation model, which is further extended by a term accounting for the rate of change of the skin temperature under transient conditions (Zhang et al. 2010a, 2010b, 2010c). The highest accuracy of the model's thermal sensation and comfort predictions is reached in warm

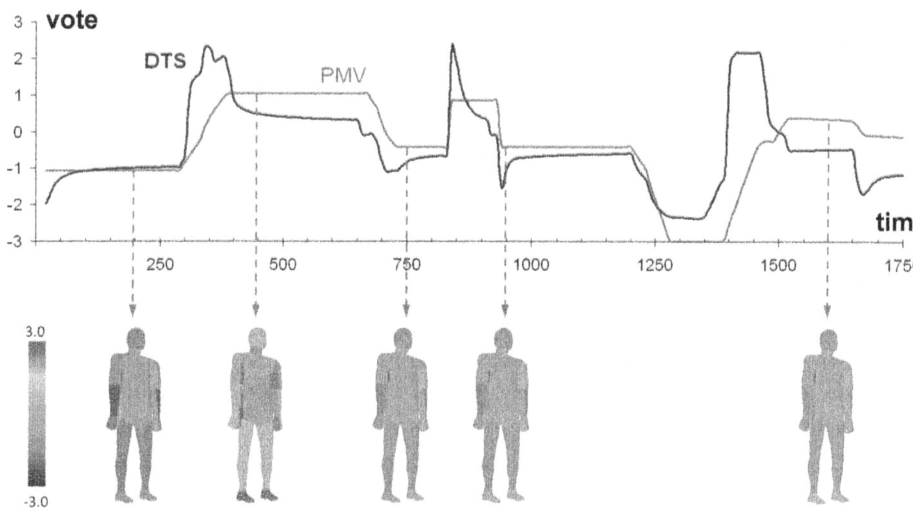

Figure 5.7 Numerical human model exposed to transient conditions and subject to changes of the ambient conditions and clothing insulation. The graph shows the corresponding *PMV* and DTS (Fiala 1998) indices together with the local thermal sensation vote on the surface of the human model.

Source: Modified from van Treeck and Frisch 2009

environments. As stated by Zhang, the validation was mainly performed in warm environments, in which local body parts were cooled (Zhang 2003; Zhang et al. 2010a). According to Schmidt (2016), additional validations for the cool comfort area are still subject to the current research.

Schmidt (2016) implemented the so-called balance comfort model (BCM), which represents a hybrid between Fanger's *PMV* approach and the model of Zhang et al. (2010a, 2010b, 2010c). Here, the global heat balance equation of Fanger is applied to local body parts. The researcher extended the original set of equations by a local heat conduction equation, which considers the influence of local temperature asymmetries, caused by contacting surfaces, on a person's thermal comfort. The calculation of the resulting local and global sensation and comfort values is done in a calorimetric manner.

Secondary factors affecting comfort

Besides the main environmental and personal parameters mentioned before, the indoor air quality (air age, chemical and microbiological issues, contamination) and other effects also influence comfort perception, health, productivity, and mental performance capability. The influence of acoustic effects, daylight and the use of blinds, electro-magnetic fields, and other secondary effects are summarized in the *ASHRAE Handbook – Fundamentals* (2017) and Recknagel et al. (2007).

Olesen and Hellwig (Recknagel et al. 2007) conclude that secondary psychosocial effects are not yet completely understood. Day-to-day variations in the perception are negligible according to Fanger (1972). Differences in the age-related comfort perception are mainly due to a lower activity level of elderly persons. Men and women show almost the same preferences, but women tend to have a lower metabolic heat production (Schmidt 2016).

For building performance simulations, it is important to note that a person's subjective thermal comfort perception seems to be statistically uniform if similar environmental conditions and personal aspects are taken into consideration (*ASHRAE Handbook – Fundamentals* 2017). In large experiments, Fanger (1982) discovered that persons cannot adapt their thermal preference. This, however, would mean that the same comfort conditions apply to all persons, independent of any cultural diversity (Fanger and Toftum 2002). However, several researchers showed that factors beyond the fields of physics and physiology influence the occupants' expectations and thermal preferences in buildings (de Dear and Brager 2002). Taking into account psychological effects, clothing preferences, and conscious behavioural actions leads to the formulation of adaptive comfort standards (see this section and the section "Prediction of thermal comfort in buildings").

Acceptability, satisfaction, and adaptation

Thermal sensation and comfort perception in terms of acceptability or satisfaction with the thermal environment is not only influenced by the aforementioned six environmental and personal factors. A person's behavioural adaptation and expectations are of comparable relevance (de Dear and Brager 2002).

Humphreys and Nicol (1998) state that in conditions of thermal discomfort, people immediately try to decrease their physiological strain, thus tending to restore their thermal comfort (see also Auliciems 1983). De Dear and Brager used extensive field studies (1998a) to show that people evaluate the indoor thermal performance differently if the building design makes provisions for the individual control of the indoor climate. Such behavioural actions include the opening of windows, wearing appropriate clothes (if allowed by the dress code), adjusting the scheduling, closing roller blinds, installing local fans, and taking refreshments and drinks. In this connection, De Dear and Brager (2002) indicate that persons living and working in naturally ventilated buildings accept a wider range of operative indoor temperatures than normally predicted by the traditional static thermal comfort envelopes (EN ISO 7730, ASHRAE Standard 55).

Figure 5.8 depicts the regression functions introduced by de Dear and Brager (2002). Both graphs show the observed and predicted comfortable indoor temperatures, which is plotted against the average outdoor temperatures. The solid lines represent the corresponding dependencies determined by the lab-based *PMV* model of Fanger (1970). Figure 5.8(a) depicts the results for HVAC buildings and Figure 5.8(b) for naturally ventilated (NV) buildings. It can be seen that the regression model for NV buildings, which was developed on

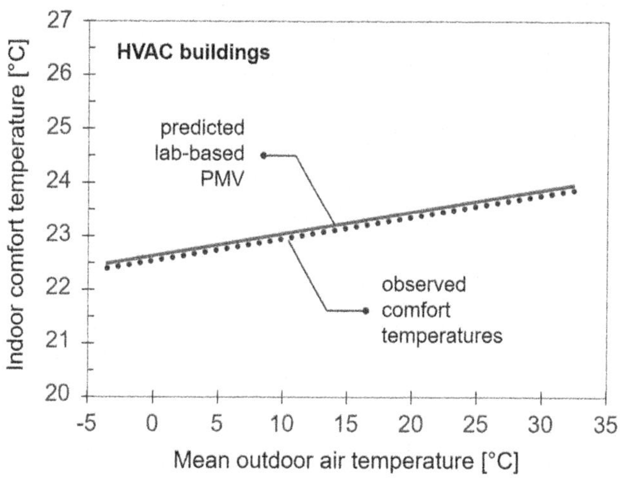

Figure 5.8(a) Qualitative behaviour of the observed and predicted comfortable operative indoor temperatures for HVAC buildings

Source: Redrawn from de Dear and Brager 2002

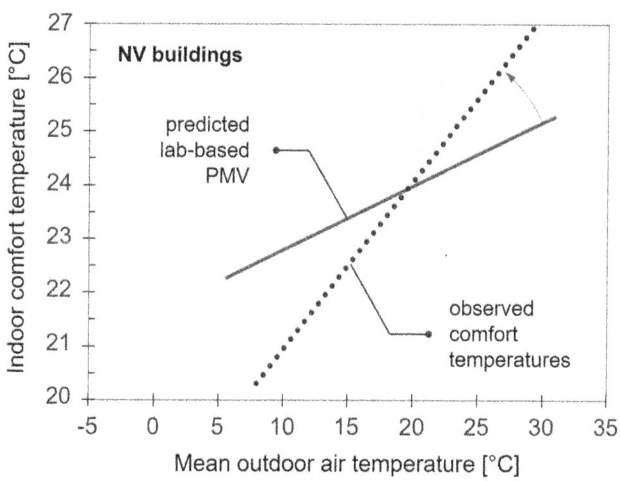

Figure 5.8(b) Qualitative behaviour of the observed and predicted comfortable operative indoor temperatures for naturally ventilated (NV) buildings

Source: Redrawn from de Dear and Brager 2002

subject experiments, shows a much steeper gradient than the corresponding lab-based *PMV* model. This indicates that the occupants of NV buildings accept a wider range of ambient conditions than do the occupants of HVAC buildings. Nevertheless, the *PMV* model predicts comfortable indoor operative temperatures for HVAC buildings very accurately (de Dear and Brager 2002).

Another important factor alongside behavioural adaptation is what people expect of the indoor climate. The latter refers to the ambient thermal conditions and the history of the preceding days (psychological adaptation). The following section discusses adaptive comfort guidelines which are mainly formulated as a function of the running average outdoor temperature (van der Linden et al. 2006).

The *PMV* model does, to some extent, include behavioural adaptations, as the model accounts for personal parameters such as clothing and activity. Fanger and Toftum (2002) enhanced the *PMV* model by multiplying *PMV* by a so-called expectancy factor, which is a function of region, season, and indoor environment. The expectancy factor can be regarded as a correction factor that extends the thermal sensation assessment range of people in non-air-conditioned buildings in warm climates (van Hoof and Hensen 2007; van Hoof 2008).

Prediction of thermal comfort in buildings

Comfort assessment in buildings

The assessment of thermal comfort in buildings can generally be done by designing the HVAC system layout in such a way that it meets the heating and cooling demand required for maintaining operative temperatures within pre-defined limits. This can be achieved by taking quantitative measurements or conducting surveys in existing buildings during their operation phase. An additional option is to use building performance simulations. As previously demonstrated in this book, simulation is a highly efficient method and provides details of a building's performance and its expected indoor thermal quality in advance. For example, a dynamic simulation can provide hourly indoor air temperatures during the summer season in order to assess the thermal quality of a naturally ventilated building zone. If the air temperature is controlled by an HVAC system, the level of comfort is mainly determined by the respective set-point temperature of the control device.

Professional HVAC designers often apply an annual or monthly static balancing approach to the system layout. If this approach is used, similar comfort categories should be applied for both the HVAC system layout and for the corresponding calculated energy consumption. In cases where the system is designed on the base of a dynamic simulation, EN 15251 recommends temperature ranges for heating and cooling with respect to a selected comfort category, type of clothing, and level of activity. However, it becomes clear that the level of the indoor thermal quality strongly influences the heating and cooling demands of a building. The next section summarizes the respective comfort envelopes.

When designing HVAC systems, it is important to distinguish between mechanically ventilated (i.e. heated and/or cooled) buildings, naturally ventilated buildings, and "mixed-mode" buildings. The evaluation methods differ, as occupants tend to show different expectations for naturally ventilated buildings. The latter assume that people adapt themselves to warmer conditions by changing clothes or by opening windows and doors (de Dear and Brager 2002). Adaptive comfort guidelines account for these effects by relating the individual expectation and local adaptation to a gradual change in the ambient climate over time.

Predicting the indoor thermal performance is also subject to a number of uncertainties (de Wit 2001). For a detailed review on the subject of occupant behaviour, the reader is referred to Chapter 4 of this volume.

Comfort envelopes

A number of standards define comfort envelopes or categories from the point of view of the indoor thermal quality performance. The comfort criteria of these categories are related to a certain range of *PMV* or *PPD* with respect to a selected clothing insulation level, level of activity, and climate zone in terms of humidity. Some formulations make use of a psychrometric chart to express satisfaction with the thermal environment in terms of operative temperature and humidity.

For moderate climates, EN 15251 defines four categories for (I) high, (II) normal, (III) moderate, and (IV) low expectations. For category II, for example, which is suggested for new or renovated buildings, it is recommended to keep $PPD < 10\%$ and thus to allow for $-0.5 < PMV < +0.5$ for mechanically ventilated buildings. For people working in an open plan office with sedentary, light activity (1.2 met), this would imply the following conditions: During the heating period, the minimum operative temperature (1 clo) should be maintained above 20 °C, and during the cooling period (0.5 clo) below 26 °C. Other configurations and settings are defined in diverse tables of EN 15251. Local thermal comfort criteria, such as draught risk or radiation asymmetries, have to be satisfied simultaneously in accordance with EN ISO 7730.

Standard EN ISO 7730 and the ASHRAE Standard 55 define "ergonomics of the thermal environment" and "thermal environmental conditions for human occupancy", respectively. EN ISO 7730 proposes three categories – A, B, and C – for limiting the *PPD* (and accordingly the range of *PMV*), each with respect to the body as a whole and to the four major local comfort criteria for steady-state conditions (cyclical temperature changes restricted to 1 K and drifts or ramps limited to 2 K per hour). The standard also summarizes equations and tables required to calculate the *PMV* and to assess the local effects, such as the draught risk (DR), in terms of PD. As indicated before, these criteria enable one to derive the permissible range of operative temperatures from estimated values for clothing and activity, humidity, and local air velocity. Table 5.3 summarizes some of these ISO categories.

Similarly, the ASHRAE Standard 55 is based on the energy balance method for the entire body and provides equations and calculation procedures, even for local conditions. Compared to the EN ISO 7730, the ASHRAE Standard 55 uses a similar methodology but draws the comfort zone in terms of the range of operative temperatures for 80 percent acceptability. It also accounts for wet bulb temperature limits. A psychrometric chart is used to define comfort envelopes for a summer zone (0.5 clo) and a winter zone (1.0 clo), respectively. The corresponding dependencies are shown in Figure 5.9, in which the operative indoor temperatures are correlated with the humidity ratio (or the dew point temperature). The limiting values for the operative temperatures are found by selecting $PPD < 10\%$ for the general thermal comfort and adds an additional 10 percent for local effects

Table 5.3 Categories for indoor thermal performance characterization according to EN ISO 7730. The standard also specifies local conditions with respect to vertical air temperature gradients, warm or cold floor surfaces, and asymmetric radiation fields.

EN ISO 7730 category	*Discomfort level for the body as a whole*		*Local discomfort*
	Predicted percentage of dissatisfied	*Predicted mean vote*	*Percentage of dissatisfied in terms of draught risk*
	PPD [%]	*PMV*	*DR [%]*
A	< 6	$-0.2 < PMV < +0.2$	< 10
B	< 10	$-0.5 < PMV < +0.5$	< 20
C	< 15	$-0.7 < PMV < +0.7$	< 30

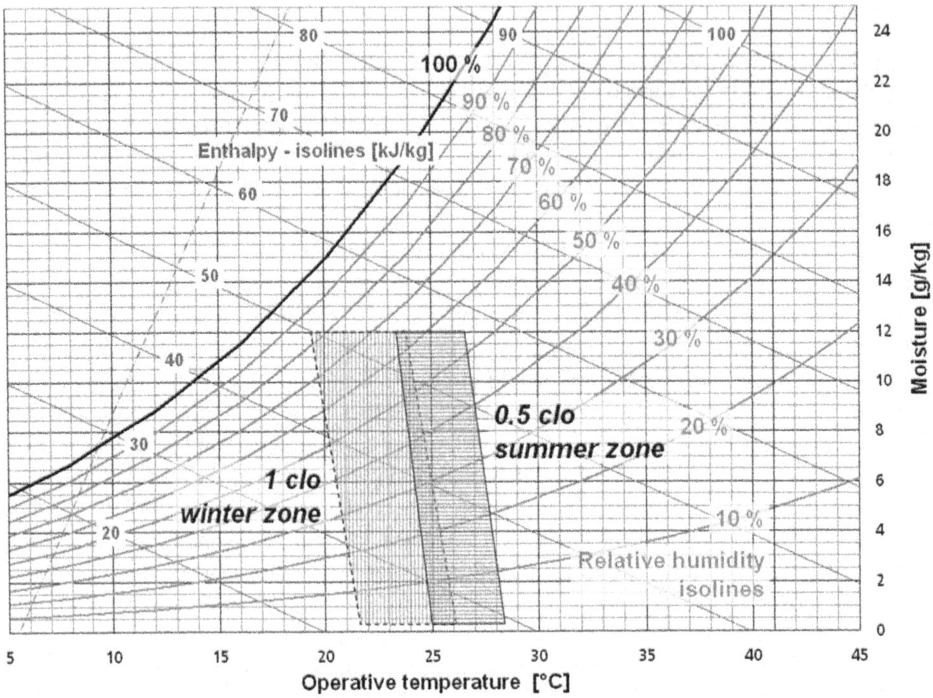

Figure 5.9 Psychrometric chart (Mollier chart) showing summer and winter comfort envelopes within ± 0.5 *PMV* limits. The comfort zones have been adapted from the specifications of the ASHRAE Standard 55.

which determine the range of *PMV*, provided the activity levels (1 to 1.3 met) and the air speed (< 0.2 m/s) have already been chosen. The humidity is limited by an upper and lower bound, as a high relative humidity causes discomfort due to skin wetness and increased skin friction and low humidity is accompanied by dry skin, throat, and eyes and may lead to respiratory disorders (*ASHRAE Handbook – Fundamentals* 2017).

In contrast to EN ISO 7730, the ASHRAE Standard 55 contains the dependence between the occupant-specific comfort temperatures, which is correlated with the outside weather conditions for free-running or naturally ventilated buildings. This issue is discussed in the following section.

Adaptive comfort guidelines

Adaptive thermal comfort (ATC) guidelines define comfort envelopes for free-running, non-air-conditioned buildings in terms of acceptable indoor temperatures which vary as a function of the running outdoor temperatures (Nicol and Pagliano 2007). The general assumption of the adaptive standard is that occupants constantly adapt themselves to the actual climatic situation to which they are exposed, thus accepting a wider temperature range. Possible adaptation strategies are changes related to one's clothing, the operation on windows, the use of blinds or fans as well as psychological and physiological adaptations (short-term and long-term).

Concerning the calculation of the running outdoor temperatures, several methods have been proposed. Humphreys (1978), for example, suggests exponentially weighted running average values over a seven-day period, van der Linden et al. (2006) a four-day period, Humphreys and Nicol (1998) and de Dear and Brager (2002) take monthly average values of the dry bulb outdoor temperature instead. Here, the adaptive comfort model of Humphreys and Nicol is based on extensive data from a wide range of climates, buildings, and cultures. The ATC model of de Dear and Brager, however, is suggested to be applicable to buildings for which cooling and central heating are not required (*ASHRAE Handbook – Fundamentals* 2009). Regarding this matter, it has to be remarked that the *ASHRAE Handbook – Fundamentals* contains the equation suggested by de Dear and Brager (1998a) to calculate the comfortable operative indoor temperatures for the ATC model.

Further standards that deal with the ATC model are EN ISO Standard 15251, the Dutch ISSO Standard 74, as well as ASHRAE Standard 55. As the operative indoor and the outdoor temperatures are the only inputs, the consideration of the ATC model in simulation and practice is straightforward. The application is applicable for buildings that provide a high level of occupant control in which persons mainly pursue light, sedentary activities (1.0 to 1–3 met). In addition, the permitted variation of clothing insulation is restricted to values between 0.5 to 1.0 clo (ASHRAE Standard 55). DIN EN 15251, however, remarks that the range in which the clothing insulation can be adapted by the occupants should be as wide as possible, in order to guarantee the maximum adaptation option.

The range of acceptable operative indoor temperatures is centred on the comfort temperature, which is defined by the ATC regression equation. The corresponding comfort envelope is based on *PMV* ranges with respect to certain categories of acceptability. De Dear and Brager (2002) propose the equation $T_{comf} = 0.31\ T_{a,out} + 17.8$ for the calculation of the comfortable operative indoor temperature T_{comf} in connection with the average outdoor temperature $T_{a,out}$ and suggest a bandwidth of 5 K to 7 K for 90 and 80 percent acceptability, respectively. The corresponding correlation is depicted in Figure 5.10(a), which is also part of ASHRAE Standard 55.

EN ISO 15251 (2007) suggests three categories (see Figure 5.10(b)) for naturally ventilated buildings. The comfort temperature regression equation in this case is $T_{comf} = 0.33\ T_{rm} + 18.8$ using the running average of an exponentially weighted, daily outdoor temperature T_{rm} (EN ISO 15251 2007). The temperature range for the three depicted comfort classification categories I, II, and III are 4 K, 6 K, and 8 K, respectively. The settings for mechanically ventilated buildings are applied outside the intervals indicated in the diagram.

Similarly, the 90, 80, and 65 percent limits (class A, B, or C) suggested by the Dutch ISSO Standard 74 (2014) are derived from research work of de Dear and Brager (1998a). The standard distinguishes between type Alpha buildings, which offer a high degree of occupant control, and type Beta buildings that consider centrally controlled HVAC systems. During the heating period, the same operative temperature limits are used for both types of buildings at outdoor temperatures below 10–12 °C. For further details, the reader is referred to ISSO Standard 74 (2014).

Van der Linden et al. (2008) quantify the differences between the ATC and *PMV* approaches for moderate climates. In this regard, the researchers investigate the range of input parameters for the *PMV* model in a moderate climate and compare it with the corresponding ATC method. Appropriate seasonal scenarios are selected, and correlations between average outdoor temperatures and clothing habits for naturally ventilated buildings are selected from De Carli et al. (2007). Different humidity values and air velocities are assumed for winter and summer months. These settings make it possible to obtain acceptable ranges of operative indoor temperatures, which are then compared to the Dutch ATC guideline. The results indicate that the adaptive ranges can be satisfactorily explained in terms of the *PMV* ranges for average outdoor temperatures up to 21 °C.

In other words, the *PMV* model in fact provides a flexible approach if parameters are adjusted according to the adaptations and expectations of persons. The ATC approach, however, defines a wider range of acceptable conditions, provided a naturally ventilated building is evaluated. The disregarding of psychological effects in connection with the heat balance method (de Dear

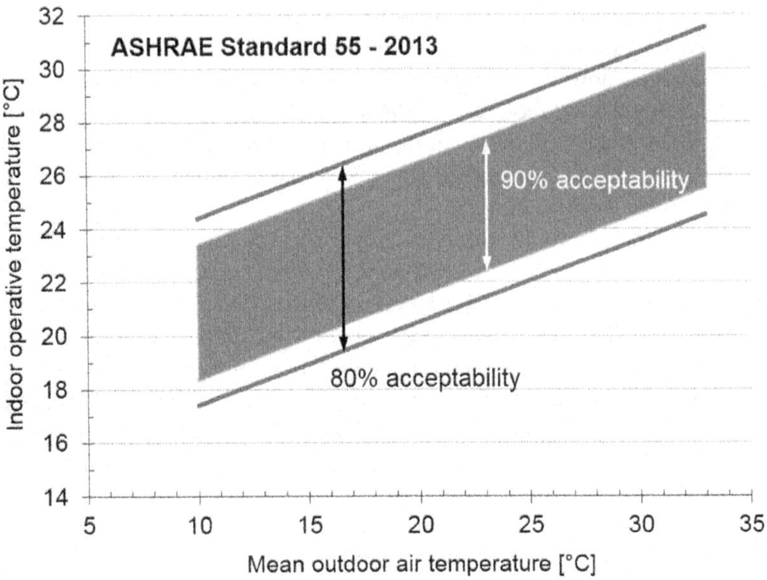

Figure 5.10(a) Acceptable conditions for naturally ventilated buildings

Source: Redrawn from de Dear and Brager 2002

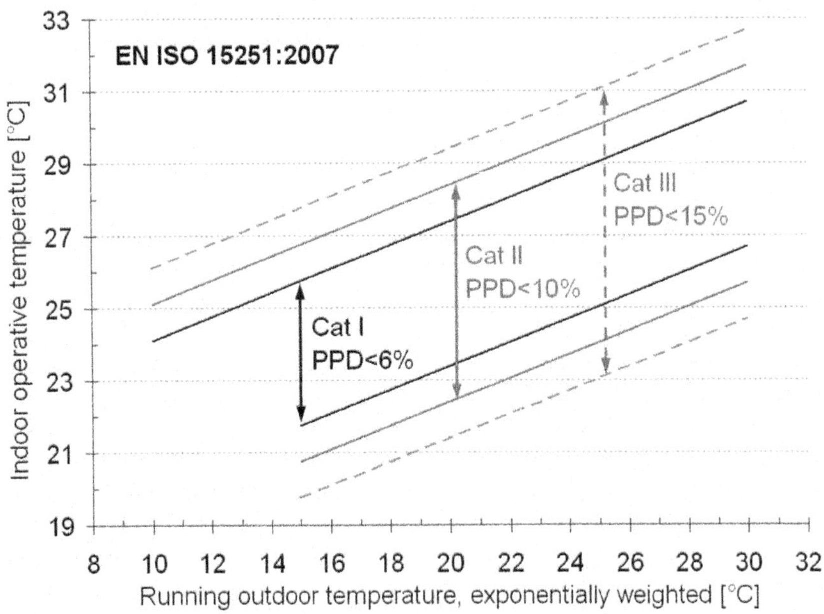

Figure 5.10(b) ATC design values for the three categories of EN ISO 15251 (2007)

and Brager 2002) can be compensated to some extent by applying the expectancy factor concept suggested by Fanger and Toftum (2002). Van Hoof and Hensen (2007) identified an energy reduction potential of the ATC method for type Alpha buildings of up to 10 percent.

Robinson and Haldi (2008) suggest that the adaptive effects can be interpreted as an aggregation of individual adaptive processes that may be further isolated in terms of successive adaptive temperature increments, which determine the slope of the adaptive curve of the envelope to some extent.

Recent approaches, such as the adaptive thermal heat balance model (ATHB) of Schweiker and Wagner (2015), attempt to combine the adaptive comfort approach of EN ISO 15251 (2007) with Fanger's well-known *PMV* model. In this regard, the researchers developed three equations, which they used to modify the clothing insulation level and the metabolic heat production of the *PMV* model. Their equations consider behavioural, physiological, and psychological adaptation processes of persons and were derived based on the ASHRAE RP884 database. The corresponding model evaluation was done by the use of an independent dataset the researchers gained from their own test facility. The prediction results for thermal sensation and comfort of the ATHB model have been demonstrated to show good agreement with experimental data and prediction results produced by the *PMV* and ATC model. ATHB itself is assumed to be applicable to naturally ventilated and air-conditioned buildings.

Long-term indicators

The discussion of the guidelines reveals that two types of indicators are needed to assess the simulation data in terms of an integrated measure for the annual thermal quality performance, as well as to capture the local thermal behaviour subject to the proper functioning of the HVAC system. An additional, local analysis calls for a refined model and is therefore applicable for selected critical periods of a reference year only.

Here, EN ISO 7730 and EN 15251 propose several long-term indicators that refer to the frequency at which the thermal quality of a building is not maintained within a desired comfort range. The corresponding outliers are subsequently weighted with corresponding penalty functions.

- *Hourly performance index*. The number of hours per year or the percentage of the time during which the predefined comfort criteria (e.g. *PMV* or operative temperature) cannot be satisfied is summarized and quantified.
- *Degree hours*. This performance indicator uses weighting factors that consider the time during which the operative temperatures exceed a predefined temperature interval (degree hours). It only considers the time during which the occupants are present. As a result, the corresponding weighting coefficients, $w = T_{op} - T_{op,limit}$, depend on the difference between the operative temperature T_{op} and its predefined upper and lower temperature limits $T_{op,limit}$. The product of the weighting factor w and the time is summed up for characteristic periods, i.e. it is distinguished between the heating ($T_{op} < T_{op,limit}$) and cooling period ($T_{op} > T_{op,limit}$). The unit of this performance index is degree hours.
- *PPD weighted performance index*. A similar procedure as for the degree hours method can be used to weight the *PPD* index. In this connection, the weighting factor w is determined as the ratio between the *PPD* value of the actual *PMV* and the reference *PPD* that corresponds to the *PMV* defined for the limits of the comfort range. It considers the time during which the *PMV* limit is exceeded. If the *PPD* is within the predefined range, the weight is zero.

  The *PPD*-based weighting method represents a stricter criterion for the long-term assessment of thermal performance of a building than does the method based on degree hours, because it results in a higher number of exceeding times. Corresponding examples are given in the appendix of EN 15251.

The introduced long-term indicators are subject to several discussions in the literature. Van der Linden et al. (2006) compare different assessment methods in connection with the Dutch TO/GTO methods (TO is based on exceeding temperatures and GTO means weighted excess hours) and the ATC method (ISSO Standard 74). The examples indicate that the ATC guideline is more rigorous than the degree hour method as a building's comfort category is already downgraded as soon as a single value drops out of a specified range of acceptability.

We therefore suggest using an integrated performance approach to provide a descriptive graphical evaluation of the simulation results. The latter should sketch the comfort envelopes as well as the weighted excess hours or similar integral quantities. However, the long-term indicators fail to accurately predict heat stress events that cause overheating.

Assessment of the overheating risk

Adaptive models assume that outdoor climatic conditions influence the behavioural and psychological adaptation of persons with respect to thermal comfort. Consequently, the acceptable thermal indoor conditions can be specified over a wider comfort range in naturally ventilated buildings (de Dear and Brager 2002). According to Baker and Standeven (1996), occupants only tolerate a limited number of heat stress events during which the indoor temperature exceeds a certain limit. It is, however, not clear if and how these stresses – if counted or weighted based on one of the above-described methods – accumulate and cause overheating.

In order to account for the accumulation of heat stimuli, Robinson and Haldi (2008) propose a mathematical model to predict the overheating risk in buildings for moderate climates. The model hypothesizes an analogy between the charging and discharging of a human's tolerance towards overheating stimuli and the charging of an electrical capacitor. It is assumed that during a warm period the overheating tolerance is discharged and during a series of cold days it is recharged. The tolerance is accordingly inverse to the probability for overheating. Robinson and Haldi (2008) formulate their overheating probability model in an analytical manner with empirical time constants for the (dis)charging rate of the tolerance level. This promising model was tested in a field survey, which revealed that overheating is caused by an accumulation of heat stress events rather than single events (Robinson and Haldi 2008).

As the adaptability is implicitly given in the (dis)charging time constants of the model, Robinson and Haldi (2008) also propose an explicit model which eliminates the adaptive increments, i.e. the elevations of the neutral temperature due to adaptation (Baker and Standeven 1996). Therefore, probability functions for the adaptive actions that are shown to correlate primarily with indoor temperatures are proposed. The elevation of the neutral temperature is further quantified by explicitly relating comfort temperatures and their increments to actions and their conjugates.

Certification

In order to assess the energy efficiency of an entire building, the above-described methods are required for the classification of the indoor thermal quality with respect to a single indicator. EN 15251 gives some recommendations according to the European Energy Performance of Buildings Directive (EPBD, 2003). It is useful to treat the issues of thermal quality and air quality separately, for example indicating the percentage of the periods when the indoor climate is within the limits prescribed by a certain comfort category (I, II, III, IV). For different energy labels and country-specific settings, the reader is referred to national standards.

Application: level of detail-dependent prediction in simulation

The introductory section already discussed the available range of simulation approaches. Based on the former section, the following introduces three typical examples for the prediction of the indoor thermal performance. It shall be noted that the resolution of simulation models with respect to space and time highly correlates with engineering costs. The latter, however, are mainly related to the time it takes to create or couple the respective models.

Comfort analysis with thermal multi-zone models

Dynamic building performance simulation (BPS) delivers most of the parameters that are relevant to assess indoor thermal comfort. In this connection, BPS usually assumes a well-distributed or stratified air temperature in space. The operative temperatures, i.e. the average air and radiant temperatures of the surrounding surfaces as well as the relative air humidity, are generally predicted at a high time resolution. The average air speed may be estimated from the thermodynamic conditions of the respective space. Quantities are usually evaluated on an hourly basis or less and mainly investigated over an entire year, i.e. 8760 time-lines or more are available for the subsequent post-processing (1 a = 365 d = 8760 h). For available tools that consider occupant behaviour, the reader is referred to Chapter 4.

This section deals with an example room (German VDI Standard 6020) of a naturally ventilated office building, which is exposed to a moderate climate. The room itself is modelled as a single rectangular space with a ground area of 17.5 m² and a large south-facing window that is equipped with additional shading devices. The building is of a heavy construction type. Thermal comfort is assessed via the static thermal comfort envelope of ISO 7730 and the two adaptive guidelines ISSO Standard 74 and EN ISO 15251.

Figures 5.11, 5.12, and 5.13 show the simulation results with respect to the comfort envelopes EN ISO 7730, EN ISO 15251, and ISSO 74, respectively. The corresponding temperature information on the different abscissae are based on the respective standards. For EN ISO 7730, the daily average ambient air temperature was used. As the standard can be regarded as static, different clothing insulation values were chosen for the summer (0.5 clo) and winter (1.0 clo) periods. In addition, information on local air speed and relative humidity were considered (see EN ISO 7730). For the sake of better readability, only the daily minimum and maximum values are depicted.

The frequency at which the criteria of the comfort zones are met or violated are summarized in Table 5.4 (for all time steps). In this case, the building would be classified as "worse than category C" according to EN ISO 7730, as "category III" for ISO 15251, and as "worse than category C" according to ISSO 74 with respect to their acceptability.

To be able to visualize the frequency of temperatures that exceed the specified daily maximum operative temperatures in connection to the point in time at which they occurred, a

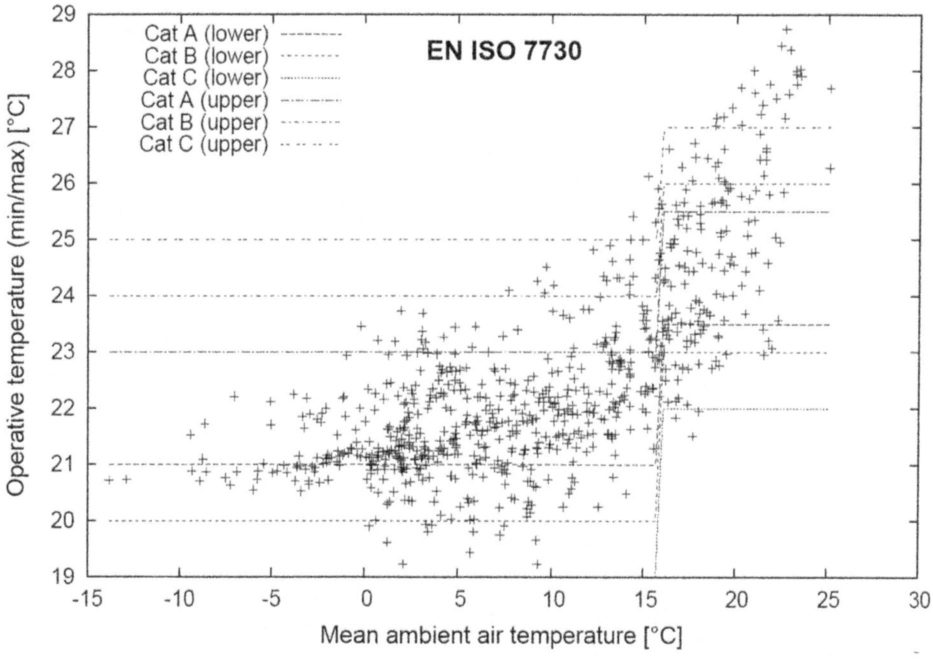

Figure 5.11 Daily minimum and maximum hourly operative room temperatures plotted against the daily average ambient air temperatures for the comfort envelope of EN ISO 7730

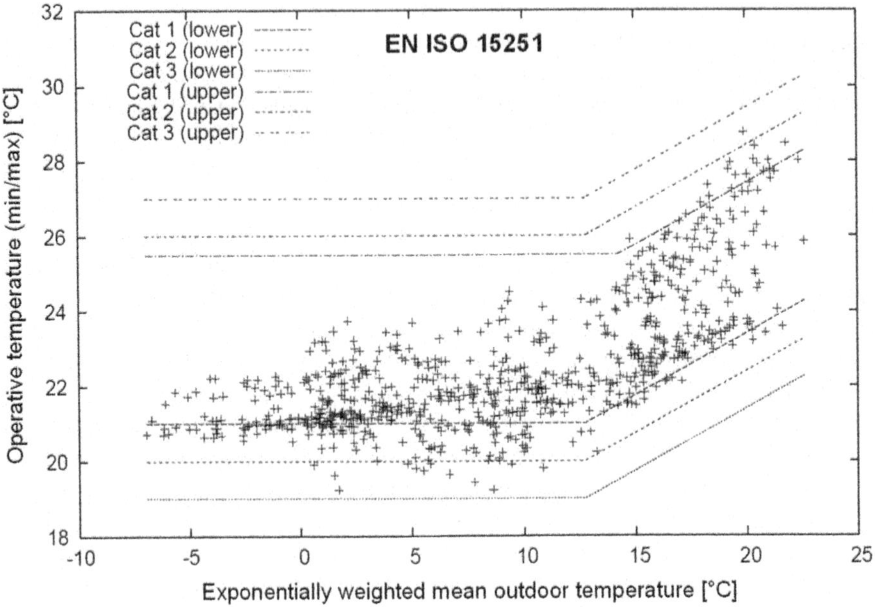

Figure 5.12 Daily minimum and maximum hourly operative room temperatures plotted against the exponentially weighted running average outdoor temperatures for the comfort envelope of the EN ISO 15251

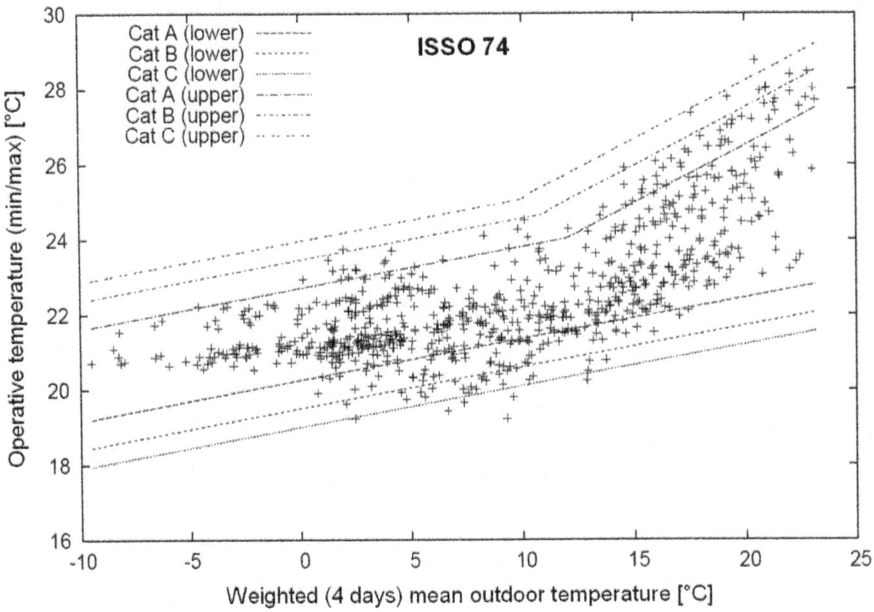

Figure 5.13 Daily minimum and maximum hourly operative room temperatures plotted against the weighted running average outdoor temperatures for the comfort envelope of the ISSO 74

calendar-type diagram is used as depicted in Figure 5.14. The figure evaluates the results in accordance with EN ISO 7730 for values within the upper band of category B. Such calendar-type layouts could be drawn for any kind or band of comfort envelope. This kind of visualization is useful for identifying local hot spots when combined with the diagrams shown in

Table 5.4 Comparison of the different comfort classification schemes (categories)

ISO 7730	*A*	*B*	*C*	*Else*
Acceptability	94%	90%	85%	< 85%
Frequency of occurrence	75.9%	16.1%	6.5%	1.6%

EN ISO 15251	*I*	*II*	*III*	*IV*
Acceptability	94%	90%	85%	< 85%
Frequency of occurrence	93.8%	6.0%	0.2%	0%

ISSO 74	*A*	*B*	*C*	*Else*
Acceptability	90%	80%	65%	< 65%
Frequency of occurrence	94.8%	4.4%	0.6%	0.1%

Figures 5.11–5.13. The latter method merely indicates whether temperatures were outside the permitted range but delivered no information when the event occurred.

Local draught risk analysis using CFD

The results depicted in the previous subsection were related to the body as a whole and expressed in terms of *PMV*/*PPD* and ATC. However, as already indicated, local thermal conditions are of high relevance, too. This section demonstrates the local draught risk analysis using the equations provided by EN ISO 7730 and ASHRAE Standard 55, respectively. For the local draught risk assessment, the average air temperature, the air velocity, as well as the degree of turbulence are required. Those values can be obtained by the use of computational fluid dynamics (CFD).

In CFD, the Navier-Stokes equations and the energy equation are solved numerically. This offers a detailed insight into the indoor airflow behaviour. The turbulent convective nature of the airflows calls for the application of turbulence models (Bejan 1993) and the refinement of the locations closest to the walls. A geometric model is required to set up a CFD simulation in order to create a surface and volume mesh. This procedure is very time-consuming with respect to the engineering work involved, thus it restricts the application of CFD in practice. Initial and boundary conditions, such as inflow and outflow conditions (velocity or pressure values), temperatures or heat fluxes at surfaces delimiting the flow domain are required. As shown by Beausoleil-Morrison (2000) or van Treeck (2004), realistic boundary conditions can be obtained from a preceding simulation or through coupling with a thermal multi-zone model.

The capabilities of such a high-resolution CFD analysis are demonstrated for the case of a cold-induced downdraught at a glass facade of an atrium situated inside a building of 26 m height. The south-facing atrium underwent a number of investigations that are detailed in Schälin et al. (1996). Van Treeck (2004) compared available measurement data with simulation results. In order to supply the CFD model with boundary conditions, the researcher implemented an algorithm for the coupling between a thermal solver and a CFD model.

Based on the computed velocity and temperature fields for a winter scenario with closed vents, the local draught risk is calculated and depicted in Figure 5.15 on a horizontal plane at 0.4 m above the ground. The figure clearly shows that the downdraught from the cold facade (15 °C air temperature) produces high local air velocities (up to 0.5 m/s). The inconvenience of this local situation at the floor area is expressed in terms of PD on a 2-D plane. It can be

Maximum values (ISO 7730 / CAT A-B)

	Jan	Feb	Mar	Apr	May	Jun	Jul	Aug	Sep	Oct	Nov	Dec
1	0.0	0.0	0.0	0.0	0.0	0.0	0.4	2.4	2.1	0.0	0.0	0.0
2	0.0	0.0	0.0	0.4	0.3	0.0	0.0	1.1	2.9	0.0	0.0	0.0
3	0.0	0.0	0.0	0.0	0.0	0.0	2.0	0.2	2.3	0.0	0.7	0.0
4	0.0	0.0	0.0	0.0	0.0	0.0	1.9	1.7	3.0	0.0	0.7	0.0
5	0.0	0.0	0.0	0.0	0.0	0.0	1.0	1.7	2.5	0.0	1.1	0.0
6	0.0	0.0	0.0	0.0	0.5	0.0	1.0	1.5	0.3	0.0	1.5	0.0
7	0.0	0.0	0.0	0.0	0.8	0.6	0.0	3.2	0.2	0.1	1.3	0.0
8	0.0	0.0	0.0	0.0	0.0	0.0	0.1	0.8	1.1	0.0	0.0	0.0
9	0.0	0.0	0.0	0.0	0.0	0.0	1.1	0.0	0.8	0.0	0.0	0.0
10	0.0	0.0	0.0	0.0	0.0	0.0	0.7	2.6	3.1	0.0	0.5	0.0
11	0.0	0.0	0.0	0.0	0.0	1.3	0.5	1.8	1.4	0.0	0.3	0.0
12	0.0	0.1	0.0	0.0	0.0	2.0	0.0	2.3	2.9	0.0	0.7	0.0
13	0.0	0.0	0.0	0.0	0.0	0.0	0.3	1.7	0.0	0.0	0.4	0.0
14	0.0	0.1	0.0	0.0	0.0	0.0	2.5	2.1	0.0	1.4	0.4	0.0
15	0.0	0.0	0.0	0.0	1.2	0.0	2.2	1.9	0.4	0.6	0.0	0.0
16	0.0	0.0	0.0	0.0	1.3	0.1	2.0	0.2	0.0	0.0	0.0	0.0
17	0.0	0.0	0.0	0.0	-0.1	0.0	2.2	0.1	0.8	0.0	0.0	0.0
18	0.0	0.0	0.0	0.3	-0.2	0.4	2.5	1.1	1.8	0.0	0.0	0.0
19	0.0	0.2	0.0	0.0	0.0	0.8	0.6	1.7	1.6	0.0	0.0	0.0
20	0.0	0.7	0.0	0.0	1.6	0.2	0.1	0.9	0.3	0.0	0.0	0.0
21	0.0	0.5	0.0	0.0	0.0	1.6	1.2	0.0	0.6	0.0	0.0	0.0
22	0.0	0.0	0.0	0.0	0.0	1.8	1.0	1.3	0.0	0.0	0.0	0.0
23	0.0	0.0	0.0	0.4	0.0	0.0	1.5	0.3	0.0	0.0	0.0	0.0
24	0.0	0.0	0.0	0.0	0.8	1.3	0.9	0.0	0.0	0.0	0.0	0.0
25	0.0	0.2	0.0	0.0	0.7	2.6	1.0	0.5	0.0	0.0	0.0	0.0
26	0.0	0.0	0.0	0.0	0.0	0.2	2.4	0.4	0.0	0.0	0.0	0.0
27	0.0	0.2	0.0	0.0	0.0	0.0	2.3	0.0	0.0	0.0	0.0	0.0
28	0.0	0.2	0.0	0.0	1.1	1.3	0.6	0.0	0.0	0.0	0.0	0.0
29	0.0		0.0	0.2	0.0	1.2	0.9	0.0	0.0	0.0	0.0	0.0
30	0.0		0.0	0.0	0.8	0.0	1.4	0.0	0.0	0.0	0.0	0.0
31	0.0		0.0		-0.3		2.4	0.3		0.0		0.0

Figure 5.14 Deviation of the daily maximum operative temperatures from the upper band of category B buildings specified in the comfort envelope of EN ISO 7730. The values are expressed in K. Dark colours indicate high deviations.

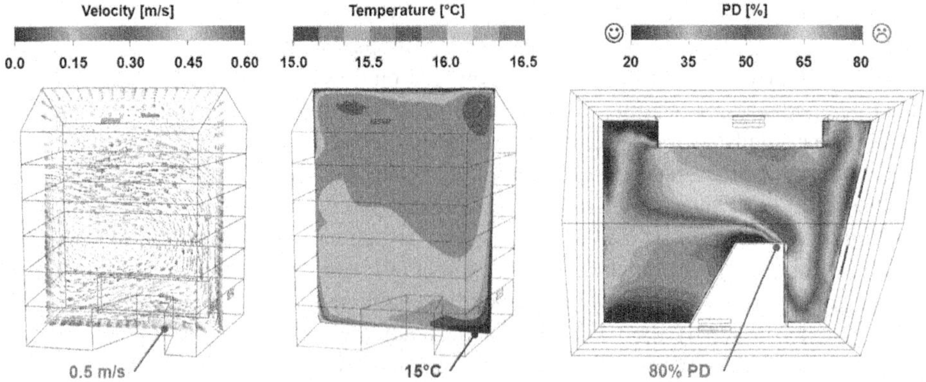

Figure 5.15 From left to right: averaged velocity field on a vertical plane, vertical temperature profile, and draught risk in terms of PD on a horizontal plane at a height of 0.4 m above the ground

Source: Modified from van Treeck 2004

seen that the corresponding acceptance is less than 20 percent. The situation can be remedied by installing local radiators near the facade or by improving the *U*-value of the glazing system.

Local and dynamic comfort analysis by the use of a numerical human model

The preceding sections have demonstrated possible ways of thermal comfort assessment for the entire human body and steady-state conditions during which the thermal body state was close to neutrality. This section addresses the application of a multi-segmented numerical human model in order to perform a detailed local and dynamic thermal comfort analysis. Surface, skin, and core temperatures are computed in order to express local thermal sensations for inhomogeneous conditions. The dynamic thermoregulatory responses of the numerical human model as well as its resulting thermal comfort indices are illustrated in Figures 5.16 to 5.18.

For a detailed simulation, both a flow and a thermal radiation solver are required in order to compute the convective and radiative heat transfer between the numerical model and the environment. To reduce the computational overhead, predefined view factors and heat transfer coefficients can be used (Fiala et al. 1999, 2001; van Treeck et al. 2008; van Treeck et al. 2009; Bolineni et al. 2015; Bolineni et al. 2016). With the help of a numerical human model, the body internal heat transfer mechanisms that occur inside the living tissue are modelled and summarized within a so-called passive system. The highly dynamic human thermoregulatory reactions such as shivering, sweating, and vasomotion are contained inside an active system part (Fiala et al. 1999, 2001; Stolwijk 1971; Tanabe et al. 2002). Amongst others, the simulation results of the numerical human model comprise surface, skin, and body core temperatures for individual body parts and their variation over time. Here, the surface temperatures can be converted into body part specific equivalent temperatures, which in turn can be related to experimentally obtained local thermal sensation votes.

The following section deals with the finite-element (FE) based equation solver FIALA-FE (Paulke and Wagner 2007), which is part of the commercial software tool THESEUS-FE. FIALA-FE itself is based on the mathematical framework of the finite-difference (FD) based model of Fiala et al. (1999, 2001). Within the corresponding software tool THESEUS-FE, the numerical human model is spatially discretized as sectors, using spherical and cylindrical elements

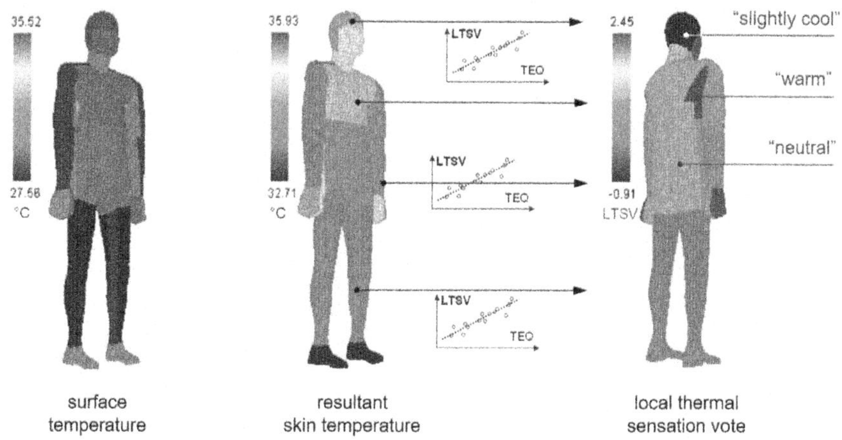

Figure 5.16 Simulated skin and surface temperatures using the numerical human model FIALA-FE (Paulke and Wagner 2007). Results are mapped to local thermal sensation votes via regression and displayed on the artificial skin surface of the shell model.

Source: Reprinted from van Treeck et al. 2009

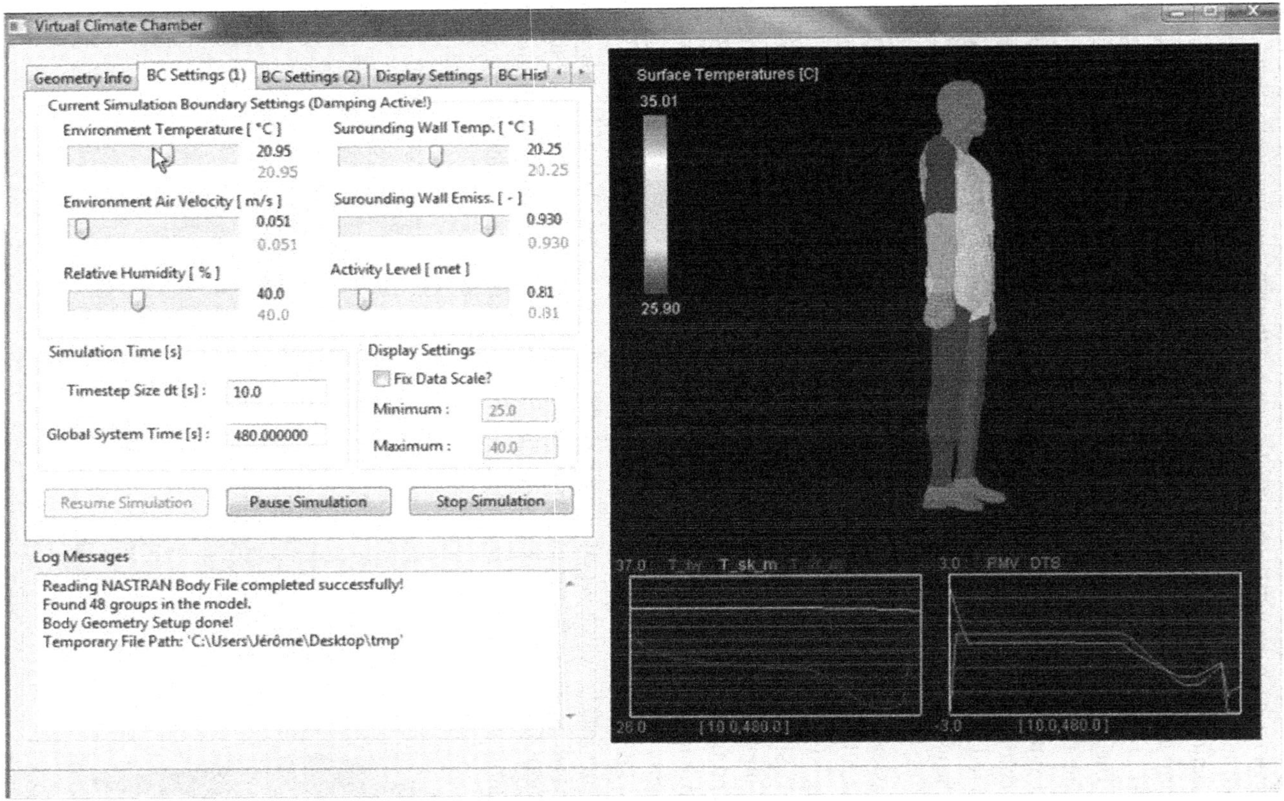

Figure 5.17 Graphical user interface for the visualization of the simulated thermal and physiological properties of the numerical human model FIALA-FE

Source: Frisch 2008

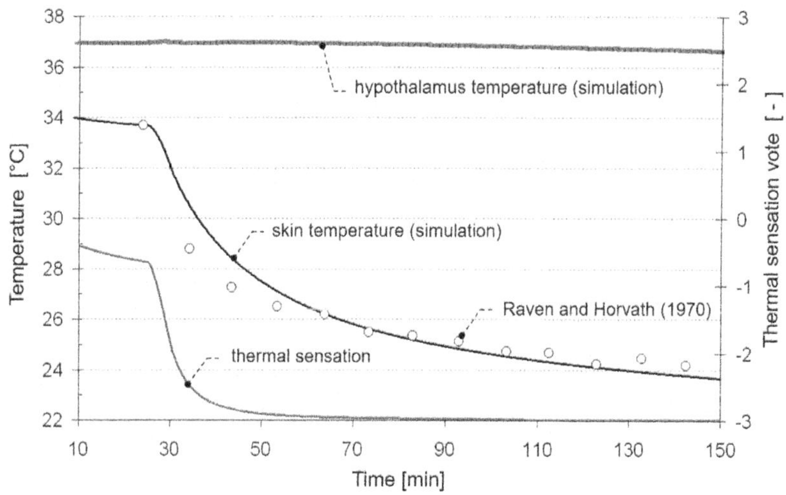

Figure 5.18(a) Comparison between the simulation and experimental results with respect to the dynamic thermoregulatory responses under transient ambient conditions (Fiala 1998; Paulke and Wagner 2007; Raven and Horvath 1970; van Treeck et al. 2009). Depiction of the temperature values.

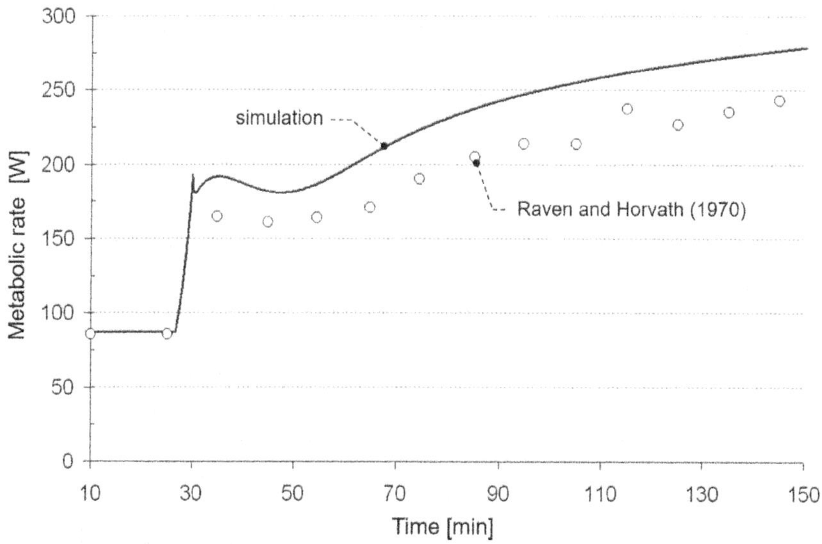

Figure 5.18(b) Depiction of the metabolic rate

that are linked to a corresponding shell model (THESEUS-FE 2014). In addition, the model is coupled to other solvers such as a parametric manikin model (van Treeck et al. 2009).

The skin and surface temperatures shown in Figure 5.16 are correlated with local thermal sensation votes (LTSV) that are based on body part specific equivalent temperatures. The corresponding link between the skin and equivalent temperatures can be established by linear regressions. The latter are the result of an experimental study in a joint research project described in van Treeck et al. (2009).

Figure 5.17 exemplarily shows a graphical user interface of an integrated simulation tool (a "virtual climate chamber"), which enables the monitoring of the simulated thermal state of the numerical human model. The tool can be used in a stand-alone manner or coupled with other external solvers (Frisch 2008; van Treeck et al. 2009).

The prediction of the dynamic human thermoregulatory responses for transient ambient conditions shall be demonstrated with the experiment of Raven and Horvath (1970), which also served as a benchmark for the model development of Fiala et al. (1999, 2001) and Paulke and Wagner (2007). The original experiment was carried out with 11 healthy unacclimatized male persons that were dressed only in athletic shorts (0.1 clo). The entire experiment consists of a 60-minute preconditioning phase at 28 °C and 45% relative humidity, followed by a two-hour cold exposure at 5 °C and 70% relative humidity. Physiological measurements were recorded by the researchers after 30 minutes of the preconditioning phase. During the experiment, diverse physiological and vital signals were monitored. For the following simulation, an air speed of 0.1 m/s and an activity level of 0.8 met were assumed for a naked humanoid.

Figure 5.18 compares the results of the average skin temperatures (a) and the metabolic rate (b) with the experimental results of Raven and Horvath (1970). It can be seen that the simulated temperature results of the skin are in good agreement with the experiment. After 30 minutes, the ambient conditions change and cause a dynamic change in metabolic rate, which is reasonably well predicted. Deviations can be attributed to the shivering rate, which is part of the increase in metabolic rate. The model, however, is calibrated to match the average thermoregulatory response of an average adult person. For a detailed discussion on the influence of different morphological, anthropometrical, and physiological parameters on the human metabolism, the reader is referred to Fiala et al. (1999, 2001) or Wölki (2017).

In terms of thermal comfort assessment, the dynamic thermal sensation (DTS) index of Fiala (1998) is shown in Figure 5.18(a). As described in the former sections, DTS considers the core, the average skin temperature, as well as its variation over time for the evaluation of global thermal comfort (Figure 5.7). It is a comfort index applicable in uniform, non-uniform, static, and transient conditions.

Discussion

Design stage-related level of detail of simulation

The former sections have demonstrated various examples for the prediction of the indoor thermal quality performance at different levels of resolution in space and time. In this regard, building performance simulation (BPS) can serve as an effective way to provide insight into the whole-year characteristics of a building's thermal quality. The resolution of spaces, however, is restricted because a uniform distribution of indoor air temperatures is assumed and no additional information on local airflow patterns is available. Integral measures are useful for identifying critical temperature periods in connection with the overheating risk. BPS is the recommended method for investigations on the indoor thermal quality performance, exploration of the thermal dynamics of a building and its single components, and validation of the overall conceptual building design.

In order to capture local effects, a detailed CFD analysis is required, which causes an increasing complexity for modelling and computation. The application of CFD is restricted to single spaces and selected critical periods, which may have been identified earlier via BPS. Hence, CFD can support building simulation to detect and investigate local situations, such as the position of the air inlets of an HVAC system. CFD results can be "translated" to comfort data and depicted in connection with the *PMV* or DR model, for example.

If the dynamic responses of the human body itself are also to be taken into consideration, computational codes need to be linked to a human thermoregulation model that provides detailed information of the skin and core temperatures as well as their variations over time. The application of such models in building performance simulation tools is still subject to current research. However, there is a growing tendency to use them in industrial applications and in the aircraft and vehicle industry.

In all cases, empirical knowledge is required to link environmental parameters (such as air velocity and temperature) or individual parameters (such as skin temperature and moisture accumulation) to thermal sensation and comfort of persons.

Figure 5.19 summarizes the different simulation approaches together with the applicable performance indicators. The "granularity" of a corresponding thermal comfort analysis is accordingly determined by the resolution of the balancing or simulation approach in space and time. In addition, the figure compares the data availability of the different building design stages to the resolution of the respective simulation approaches.

Hausladen et al. (2005, 2006), for example, proposed a "ClimaDesign" guideline for the prediction of the indoor climatic conditions and the energy efficiency of buildings during the very early design stages in the absence of detailed planning data. The researchers pointed out the interplay between individual design variables and their different sensitivities. Available are several other simplified tools which are applicable to early design stages, such as the Design Advisor of Urban and Glicksman (2006).

The visualization of simulation results is another important aspect of simulation practice. The expressiveness of results varies according to the level of detail. At a coarse level, for example, significant details may not be visible, while at a fine level the high information content needs to be condensed in order to extract and interpret relevant quantities.

Advanced local thermal comfort prediction based on thermal imaging and sensor data fusion

Energy-efficient control strategies for future HVAC systems follow an integrated and user-centred system approach. A human individual becomes consequently an explicit part of the control heuristics in order to deal with the available energy in a much more efficient and ecological

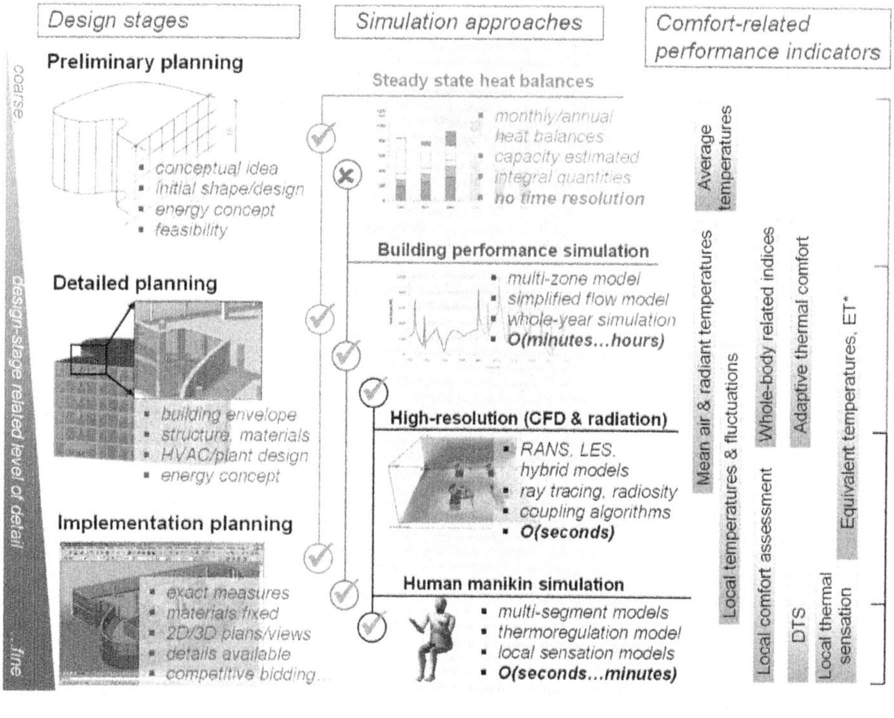

Figure 5.19 Simulation approaches and required resolution for relevant performance indicators with respect to the interplay between the different building design stages

way, while simultaneously keeping the indoor local and global thermal comfort at a maximum. Consequently, conventional air-conditioning strategies that are heating or cooling an entire air volume (macroclimate) are re-evaluated.

The latest research takes advantage of the progresses made in the fields of thermal imaging, computational image analysis, thermophysiological modelling, and thermal comfort assessment (Metzmacher et al. 2016, 2017; Schmidt 2016; Wölki 2017). Figure 5.20 depicts a first system implementation that comprises the full coupling between the thermophysiological model MORPHEUS, the balance comfort model (BCM), thermal imaging and person-tracking algorithms, as well as additional sensor hardware for the sensor data fusion and advanced assessment of the thermal states of individual body parts. The communication between the different system components is established by a central data server, which acts as a mediator platform for the data exchange between sensors, algorithms, and numerical models and enables the real-time assessment of the individual-specific thermal comfort perception. Subsequently, this information is used to control corresponding body part specific air-conditioning systems in a model-predictive manner.

Synopsis

The indoor thermal quality performance and the energy efficiency of a building, its components, and its HVAC system are inherently connected to each other. In the integrated building design, engineers are confronted with the balance of aesthetical, functional, and economical aspects, which interact with both the energy and the thermal quality performance. The conceptual planning stages are accompanied by different levels of detail with respect to the available simulation data. Consequently, the accuracy of the related simulations possess different levels of accuracy.

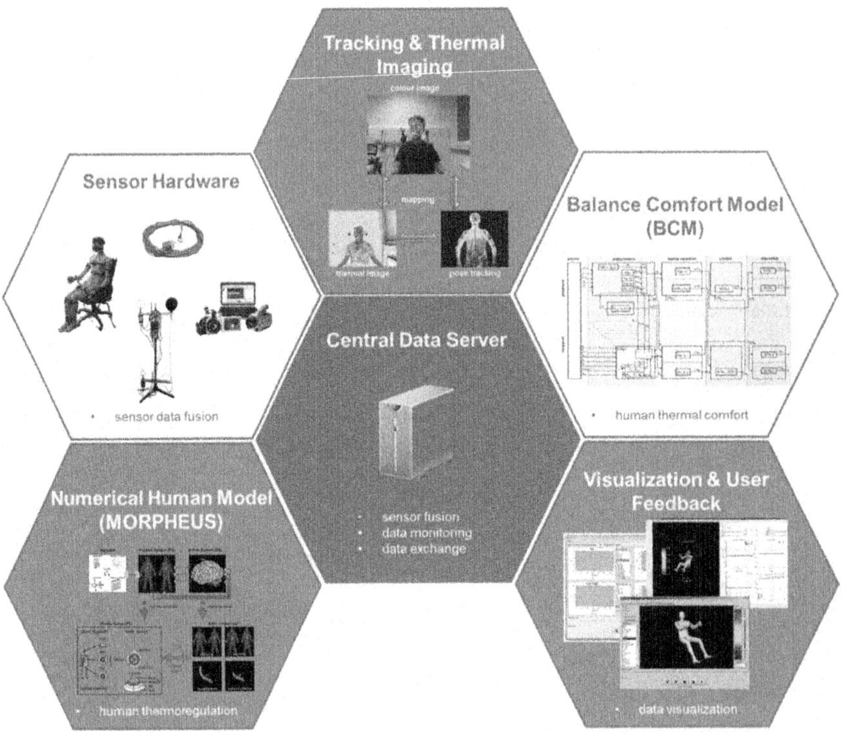

Figure 5.20 Advanced local thermal comfort prediction; overview of the system components

This chapter was written to summarize the relevant indicators and integral measures for the prediction of the thermal quality performance for buildings and various indoor spaces. Background information was provided on human thermoregulation, thermal sensation, as well as thermal comfort assessment methods applicable to uniform and non-uniform environments. The relevant codes and standards have been introduced as well as examples demonstrating the application of the various models and their relevant resolution.

While evaluating the thermal quality performance of buildings by means of simulation, special emphasis was set on the post-processing and interpretation of the obtained results. Here, descriptive integrated views, such as the calendar-type diagrams or the well-founded models that account for the accumulation of heat stress events, are required.

The *PMV*/*PPD* model, supported by the adaptive comfort guidelines, belongs to well-established and recommended methods in engineering practice. Their basic idea is to statistically relate the thermal sensation and acceptability of a group of people to environmental parameters and personal factors. The necessary data for this can be acquired by the use of building performance simulation, which makes the application of these models straightforward. However, in some situations, special attention needs to be paid to local effects. Some of these effects can be predicted with the help of empirical equations; others require the additional application of refined models such as CFD or even numerical human models. Prediction methods that account for asymmetrical and transient conditions are still subject to research and involve extensive experimental work with subjects.

Note

1 Differences exist, however, between ethnic and geographical groups (ASHRAE 2013).

References

Note: Codes and standards are summarized in the next section and are not repeated here.

Araújo, V.M.D. and Araújo, E.H.S. (1999) The applicability of ISO 7730 for the assessment of the thermal conditions of users of the buildings in Natal-Brazil. In Raw, G., Aizlewood, C. and Warren, P. (eds), *IndoorAir* 2: 148–153.

Aschoff, J. and Wever, R. (1958) Kern und Schale im Wärmehaushalt des Menschen. *Die Naturwissenschaften* 45(20): 477–485.

ASHRAE. (2009) *Handbook – Fundamentals (SI)*. American Society of Heating, Refrigerating and Air-Conditioning Engineers.

ASHRAE. (2017) *Handbook – Fundamentals (SI)*. American Society of Heating, Refrigerating and Air-Conditioning Engineers.

Auliciems, A. (1983) Psychophysical criteria for global thermal zones of building design. *International Journal of Biometeorology* 27: 69–86.

Azer, N.Z. and Hsu, S. (1977) The prediction of thermal sensation from a simple thermoregulatory model. *ASHRAE Transactions* 83 (Part 1).

Baker, N. and Standeven, M. (1996) Thermal comfort for free-running buildings. *Energy and Buildings* 23: 175–182.

Beausoleil-Morrison, I. (2000) The adaptive coupling of heat and air flow modeling within dynamic whole building simulation. PhD Thesis. University of Strathclyde, Glasgow.

Bedford, T. (1936) *The Warmth Factor in Comfort at Work*. MRC Industrial Health Board Report HMSO, vol. 76. London.

Bejan, A. (1993) *Convection Heat Transfer*. West Sussex: John Wiley & Sons, Inc.

Benzinger, T. (1979) The physiological basis for thermal comfort. In Fanger, P. and Valbjorn, O. (eds), *Indoor Climate*. Copenhagen: Danish Building Research Institute, pp. 441–476.

Berglund, L.G. and Cunningham, D.J. (1986) Parameters of human discomfort in warm environments. *ASHRAE Transactions* 92(2B): 732–746.

Bohm, M., Browen, M., Holmer, I., Nilsson, P.V. and Noren, O. (1990) *Evaluation of Vehicle Climate with a Thermal Manikind.* Swedish Institute of Agricultural Engineering.

Bolineni, S.R., Park, S., Nasyrov, V., Stratbücker, S., Grün, G., Schmidt, C. and van Treeck, C.A. (2016) Uniform segment-wise body parametric convection heat transfer coefficient model for sitting posture in vehicles. *Science and Technology for the Built Environment.* 5(23): 843–857. London: Taylor & Francis.

Bolineni, S.R., Stratbücker, S. and van Treeck, C. (2015) Development of reduced order flow responsive models of convective heat transfer for human body segments. *Building Simulation* 8(4): 393–404.

Brager, G.S. and de Dear, R.J. (1998) Thermal adaptation in the built environment: A literature review. *Energy and Buildings* 27(1): 83–96.

Burton, A.C. (1934) The application of the theory of heat flow to the study of energy metabolism. *The Journal of Nutrition* 7: 497–533.

Busch, J.F. (1992) A tale of two populations: Thermal comfort in air-conditioned and naturally ventilated offices in Thailand. *Energy and Buildings* 18: 235–249.

Cabanac, M. and Hardy, J.D. (1969) Réponses unitaires et thermorégulatrices lors de réchauffements et réfroidissements localisés de la région préoptique et du mésencéphale chez le lapin. *Journal of Physiology Paris* 61: 331–347.

de Carli, M., Olesen, B.W., Zarrella, A. and Zecchin, R. (2007) People's clothing behaviour according to external weather and indoor environment. *Building and Environment* 42: 3965–3973.

de Dear, R.J. (1998) Sick building syndrome and appropriate design. *Human Bioclimatology*, 87–109.

de Dear, R.J. and Brager, G.S. (1998a) Developing an adaptive model of thermal comfort and preference. *ASHRAE Transactions* 104: 1–18.

de Dear, R.J. and Brager, G.S. (1998b) Developing an adaptive model of thermal comfort and preference. *ASHRAE Technical Data Bulletin* 14(1): 27–49.

de Dear, R.J. and Brager, G.S. (2002) Thermal comfort in naturally ventilated buildings: Revisions to ASHRAE standard 55. *Energy and Buildings* 34: 549–561.

de Paula Xavier, A.A. and Roberto, L. (2000) Indices of thermal comfort developed from field survey in Brazil. *ASHRAE Transactions* 106: 45–58.

de Wit, S. (2001) Uncertainty in predictions of thermal comfort in buildings. PhD Thesis, TU Delft, The Netherlands.

DIN EN 15251 (2007) Eingangsparameter für das Raumklima zur Auslegung und Bewertung der Energieeffizienz von Gebäuden – Raumluftqualität, Temperatur, Licht und Akustik; Deutsche Fassung EN 15251:2007.

DIN EN ISO 8996 (2004) Ergonomics of the thermal environment. Determination of metabolic rate (ISO 8996:2004). German version EN ISO 8996:2004.

EPBD (2003) Directive 2002/91/EC of the European Parliament and of the Council of 16 December 2002 on the energy performance of buildings, European Commission. OJ L 1, 4.1.2003, pp. 65–71.

Fanger, P.O. (1964) Termiske luftströmninger langs vinduer og kolde vægge. Ingeniören 19.

Fanger, P.O. (1970) Thermal Comfort-Analysis and Applications in Environmental Engineering. PhD Technical University of Denmark and Laboratory of Heating and Air Conditioning, Copenhagen.

Fanger, P.O. (1972) *Thermal Comfort.* New York: McGraw-Hill.

Fanger, P.O. (1977) Thermal discomfort caused by radiant asymmetry, local air velocities, warm and cold floors, and vertical air temperature gradients. In Durrand, J. and Rayuand, J. (eds), *Thermal Comfort. Editions INSERM* 75: 145–152.

Fanger, P.O. (1982) *Thermal Comfort.* Malabar: Robert E. Krieger.

Fanger, P.O., Ipsen, B.N., Langkilde, G., Olesen, B.W., Christensen, N.K. and Tanabe, S. (1985) Comfort limits for asymmetric thermal radiation. *Energy and Buildings* 8: 225–236.

Fanger, P.P., Melikov, A.K., Hanzawa, H. and Ring, J. (1988) Air turbulence and sensation of draught. *Energy and Buildings* 12: 21–39.

Fanger, P.O. and Toftum, J. (2002) Extension of the PMV model to non-air-conditioned buildings in warm climates. *Energy and Buildings* 34: 533–536.

Fiala, D. (1998) Dynamic simulation of human thermoregulation and thermal comfort. PhD Thesis. De Montfort University Leicester, UK.

Fiala, D., Lomas, K. and Stohrer, M. (1999) A computer model of human thermoregulation for a wide range of environmental conditions: The passive system. *Journal of Applied Physiology* 87: 1957–1972.

Fiala, D., Lomas, K. and Stohrer, M. (2001) Computer prediction of human thermoregulatory and temperature responses to a wide range of environmental conditions. *International Journal of Biometeorology* 45: 143–159.

Foda, E. and Sirén, K. (2010) A new approach using the Pierce two-node model for different body parts. *International Journal of Biometeorology* 55: 519–532.

Frijns, A., Limpens-Neilen, D., Schoffelen, M. and van Leeuwen, G. (2006) Simulation of the Human Thermal Responses to Local Heating. In *Proceedings of IEA*, Maastricht, The Netherlands.

Frisch, J. (2008) Diploma thesis, TU München.

Gagge, A.P. (1973) Rational temp. indices of man's thermal env. and their use with a 2-node model of his temp. *Federation Proceedings* 32: 1572–1582.

Gagge, A.P. (1979) The role of humidity during warm discomfort. In Fanger, P.O. and Valbkorn, O. (eds), *Indoor Climate*. Copenhagen: Danish Building Research Institute, pp. 527–538.

Gagge, A.P., Fobelets, A.P. and Berglund, P.E. (1986) A standard predictive index of human response to the thermal environment. *ASHRAE Transactions* 92(2B): 709–731.

Gagge, A.P. and Gonzalez, R.R. (1973) Physiological bases of warm discomfort for sedentary man. *Archives des Sciences Physiologiques* 27: A409–A424.

Gagge, A.P., Stolwijk, J.A. and Hardy, J.D. (1967) Comfort and thermal sensation and associated physiological responses at various ambient temperatures. *Environmental Research* 1: 1–20.

Gagge, A.P., Stowijk, J.A.J and Nishi, Y. (1971) An effective temperature scale based on a simple model of human physiological regulatory response. *ASHRAE Transactions* 77(1): 247–262.

Glück, B. (1994) Zulässige Strahlungstemperatur-Asymmetrie. *Gesundheitsingenieur* 115(6): 285–344.

Gonzalez, R.R. and Berglund, L.G. (1979) Efficacy of temperature and humidity ramps in energy conservation. *ASHRAE Journal* 6: 34–41.

Gonzalez, R.R. and Gagge, A.P. (1973) Magnitude estimates of thermal discomfort during transients of humidity and operative temperature. *ASHRAE Transactions* 79(1): 88–96.

Gordon, R. (1974) *The Response of Human Thermoregulatory System in the Cold*. St. Barbara, CA: University of California.

Griffiths, I.D. and McIntyre, D.A. (1974) Sensitivity to temporal variations in thermal conditions. *Ergonomics* 17: 499–507.

Guyton, C. (2002) *Textbook of Medical Physiology*. Philadelphia, London: W.B. Saunders.

Han, T., Huang, L., Kelly, S., Huizenga, C. and Zhang, H. (2001) Virtual thermal comfort engineering. *SAE Technical Paper Series* 2001–01–0588.

Hausladen, G., Saldanha, M. and Liedl, P. (2006) *ClimaSkin: Konzepte für Gebäudehüllen, die mit weniger Energie mehr leisten*. München: Verlag Callwey.

Hausladen, G., Saldanha, M., Liedl, P. and Sager, C. (2005) *ClimaDesign: Lösungen für Gebäude, die mit weniger Technik mehr können*. München: Verlag Callwey.

Havenith, G. (2001) Individualized model of human thermoregulation for the simulation of heat stress response. *Journal of Applied Physiology* 90: 1943–1954.

Havenith, G. and Fiala, D. (2016) Thermal indices and thermophysiological modelling for heat stress. *Comprehensive Physiology* 6: 255–302.

Havenith, G., Holmér, I. and Parsons, K. (2002) Personal factors in thermal comfort assessment: Clothing properties and metabolic heat production. *Energy and Buildings* 34: 581–591.

Heiselberg, P. (1994) Stratified flow in rooms with a cold vertical wall. *ASHRAE Transactions* 100(1): 1011–1023.n

Hensel, H. (1982) *Thermal Sensation and Thermoreceptors in Man*. Springfield: Charles C. Thomas.

Hensen, J.L.M. (1990) Literature review on thermal comfort in transient conditions. *Building and Environment* 25(4): 309–316.

Herkel, S., Knapp, U. and Pfafferott, J. (2005) A preliminary model of user behaviour regarding the manual control of windows in office buildings. In *Building Simulation Conference*. Montreal, Canada.

Holmér, I. (1995) Heated manikins as a tool for evaluating clothing. *Annals of Occupational Hygiene* 39(6): 809–818.

Holmér, I. (2004) Thermal manikin history and applications. *European Journal of Applied Physiology* 92: 614–618.

Holopainen, R. (2012) A human model for improved thermal comfort. PhD, VTT science.

van Hoof, J. (2008) Forty years of Fanger's model of thermal comfort: Comfort for all? *IndoorAir* 18: 182–201.

van Hoof, J. and Hensen, J.L.M. (2007) Quantifying of relevance of adaptive thermal comfort models in moderate thermal climate zones. *Building and Environment* 42: 156–170.

Huizenga, C., Hui, Z. and Arens, E. (2001) A model of human physiology and comfort for assessing complex thermal environments. *Building and Environment* 36: 691–699.

Humphreys, M.A. (1978) Outdoor temperatures and comfort indoors. *Building Research and Practise* 6: 92–105.

Humphreys, M.A. and Nicol, J.F. (1998) Understanding the adaptive approach to thermal comfort. *ASHRAE Transactions* 104(1): 991–1004.

Issing, K. and Hensel, H. (1981) Static thermal sensation and thermal comfort. *Zeitschrift für Physikalische Medizin, Balneologie und Medizinische Klimatologie* 11: 354–365.

Jones, B.W. and Ogawa, Y. (1992) Transient interaction between the human body and the thermal environment. *ASHRAE Transactions* 98: 189–195.

Kato, S., Nagano, H. and Yang, L. (2007) *A CFD Manikin with a Thermo Physiology Model*. Helsinki: RoomVent.

Kingma, B.R.M. (2012) Human thermoregulation: A synergy between physiology and mathematical modelling. PhD Thesis, Universitaire Pers Maastricht, Maastricht.

Kohri, I. and Mochida, T. (2002) Evaluation method of thermal comfort in a vehicle with a dispersed two-node model. *Journal of the Human-Environmental System* 6(1): 19–29.

Kok, J., van Muijden, J., Burgers, S., Dol, H. and Spekreijse, S. (2006) Enhancement of aircraft cabin comfort studies by coupling of models for human thermoregulation, internal radiation, and turbulent flows. In Wesseling, P., Onate, E. and Periaux, J. (eds), *ECCOMAS CFD 06*. Egmond aan Zee, The Netherlands.

Kröling, P. (1998) Sick building syndrom: Ursachen und Prophylaxe gebäudebedingter Gesundheitsstörungen. *Allergologie* 21(5): 180–191.

Machle, W. and Hatch, T.F. (1947) Heat: Man's exchange and physiological responses. *Physiological Reviews Published* 27(2): 200–227.

Mayer, E. (1997) A new correlation between predicted mean votes (PMV) and predicted percentages of dissatisfied (PPD). In Woods, J.E., Grimsrud, D.T. and Boschi, N. (eds.), *Healthy Buildings International Conference*, Vol. 2: pp. 189–194.

Mayer, E. (2000) Manfitted measurement of thermal climate by a dummy representing suit for simulation of human heatloss (DRESSMAN). *Healthy Buildings* 2: 551–556.

Mayer, E. and Schwab, R. (1990) Untersuchungen der physikalischen Ursachen von Zugluft. *Gesundheitsingenieur* 111(1): 17–30.

Mayer, E. and Schwab, R. (1993) Evaluation of heat stress by an artificial skin. *IndoorAir* 6: 73–78.

McGuffin, R., Burke, R., Huizenga, C., Zhang, H., Vlahinos, A. and Fu, G. (2002) Human thermal comfort model and manikin. *SEA* 2002–01–1955.

McIntyre, D.A. (1980) *Indoor Climate*. London: Applied Science Publishers LTD.

McNall, P.E. and Schlegel, J.C. (1968) The relative effect of convective and radiation heat transfer on thermal comfort. *ASHRAE Transactions* 74: 131–143.

Metzmacher, H., Wölki, D., Schmidt, C., Frisch, J. and van Treeck, C. (2016) Energy-efficient climate control in electric vehicles through innovative sensor technology and novel methods for thermal comfort evaluation. In *ETA Conference*. Berlin, Germany, pp. 65–75.

Metzmacher, H., Wölki, D., Schmidt, C., Frisch, J. and van Treeck, C. (2017) Real-time assessment of human thermal comfort using image recognition in conjunction with a detailed numerical human model. In *Building Simulation 2017*. San Francisco, FL.

Modest, M.F. (2003) *Radiative Heat Transfer, 2nd edition*. New York: Academic Press.

Möhlenkamp, M., Wesseling, M., Wick, A., Gores, I. and Müller, D. (2016) Thermal comfort of displacement ventilation in environments with different mean room temperatures. In *Proceeding of 9th Windsor Conference: Making Comfort Relevant*. Cumberland Lodge, Windsor, UK, pp. 621–631.

Mower, G.D. (1976) Perceived intensity of peripheral thermal stimuli is dependent of internal body temperature. *Journal of Comparative and Physiological Psychology* 90: 1152–1155.

Nevins, R.G., Gonzalez, R.R., Nishi, Y. and Gagge, A.P. (1975) Effect of changes in ambient temperature and level of humidity on comfort and thermal sensations. *ASHRAE Transactions* 81(2): 169–182.

Nevins, R.G., Rohles, F.G., Springer, W. and Feyerherm, A.M. (1966) Temperature-humidity chart for thermal comfort of seated persons. *ASHRAE Transactions* 72: 283–291.

Nicol, F. and Pagliano, L. (2007) Allowing for thermal comfort in free-running buildings in the New European Standard EN15251.

Nielsen, P.V. (1981) Luftströmninger i ventilerede arbejdslokaler. SBI-rapport 128. Statens byggeforskningsinstitut.

Nielsen, P.V., Murakami, S., Kato, S., Topp, C. and Yang, J.H. (2003) *Benchmark Test for a Computer Simulated Person*. Aalborg University. www.cfd-benchmarks.com.

Nilsson, H.O. (2004) Comfort climate evaluation with thermal manikin methods and computer simulation models. PhD Thesis. Royal Institute of Technology, Sweden.

Nilsson, H.O., Holmér, I., Bohm, M. and Norén, O. (1997) Equivalent temperatures and thermal sensation, comparison with subjective responses. In *Conference in the Automotive Industry ATA 1*. Bologna, Italy, pp. 157–162.

Niu, J. and Gao, N. (2008) *Personalized Ventilation for Commercial Aircraft Cabins*. Copenhagen, Denmark: IndoorAir.

Olesen, B.W. (2002) Radiant floor heating in theory and practise. *ASHRAE Journal* 7: 19–24.

Olesen, B.W. (2008) Radiant floor cooling systems. *ASHRAE Journal* 9: 16–22.

Olesen, B.W., Schöler, M. and Fanger, P.O. (1979) Discomfort caused by vertical air temperature. In Fanger, P.O. and Valbjorn, O. (eds), *Indoor Climate*. Copenhagen: Danish Building Research Institute, pp. 561–579.

Pandolf, K.B., Stroschein, L.A., Drolet, L.L., Gonzalez, R.R. and Sawka, M.N. (1986) Prediction modelling of physiological responses and human performance in the heat. *Computers in Biology and Medicine* (16): 319–329.

Paulke, S. and Wagner, S. (2007) Some considerations on global and local thermal comfort based on Fiala's thermal manikin in THESEUS-FE. EUROSIM 2007.

Persson, P.B. (2010) Physiologie des Menschen – mit Pathophysiologie. Lehrbuch Kapitel 39. In *Energie- und Wärmehaushalt, Thermoregulation* 31. Auflage. Springer-Verlag, pp. 834–853.

Pfafferott, J., Herkel, S., Kalz, D. and Zeuscher, A. (2007) Comparison of low-energy office buildings in summer using different thermal comfort criteria. *Energy and Buildings* 39: 750–757.

Raven, P. and Horvath, S. (1970) Variability of physiological parameters of unacclimatized males during a two-hour cold stress of 5 °C. *International Journal of Biometeorology* 14: 309–320.

Recknagel, H., Sprenger, E. and Schramek, E.R. (2007) *Taschenbuch für Heizung, Klima und Technik, 73th edition*. München: Oldenbourg Industrieverlag.

Ring, J.W. and de Dear, R.J. (1991) Temperature transients: A model for heat diffusion through the skin, thermoreceptor response and thermal sensation. *Indoor Air* 1(4): 448–456.

Robinson, D. and Haldi, F. (2008) An integrated adaptive model for overheating risk prediction. *Journal of Building Performance Simulation* 1(1): 43–55.

Rohles, F.H. and Nevins, R.G. (1971) The nature of thermal comfort for sedentary man. *ASHRAE Transactions* 77(1): 239–246.

Rohles, F.H. and Wallis, S.B. (1979) Comfort criteria for air conditioned automotive vehicles. *SAE Technical Paper* 790122, 440–449.

Rohles, F.H., Woods, J.E. and Nevins, R.G. (1974) The effect of air movement and temperature on the thermal sensation of sedentary man. *ASHRAE Transactions* 80: 101–118.

Rugh, J.P., Farrington, R.B., Bharathan, D., Vlahinos, A., Burke, R., Huizenga, C. and Zhang, H. (2004) Predicting human thermal comfort in a transient nonuniform thermal environment. *European Journal of Applied Physiology* 92: 721–727.

Sano, J., Tanabe, S., Murakami, K., Nagayama, H., Oi, H. and Itami, T. (2008) *Evaluation of Thermal Environments in Vehicles by CFD Coupled with Numerical Thermoregulation-Model JOS*. Copenhagen, Denmark: IndoorAir.

Schälin, A., Moser, A., van der Maas, J. and Aiulfi, D. (1996) Application of air flow models as design tools for atria. In *Proceeding of Roomvent*. Yokohama, Japan.

Schmidt, C. (2016) Entwicklung eines Modellansatzes zur Bewertung der thermischen Behaglichkeit unter inhomogenen Klimabedingungen. PhD Thesis, RWTH Aachen University.

Schmidt, C., Wölki, D., Grün, G. and van Treeck, C. (2012) Accuracy of mean skin temperature calculations and measurements in thermal comfort-related assessment. In *The Changing Context of Comfort in an Unpredictable World*. Windsor: Cumberland Lodge.

Schweiker, M. and Wagner, A. (2015) A framework for an adaptive thermal heat balance model (ATHB). *Building and Environment* 94: 252–262.

Smith, C.E. (1991) A transient, three-dimensional model of the human thermal system. PhD Thesis, Kansas State University, Kanas.

Stolwijk, J. (1971) A mathematical model of physiological temperature regulation in man. CR-1855, NASA.

Stolwijk, J.A. (1991) Sick-building syndrome. *Environmental Health Perspectives* 95: 99–100.

Streblow, R. (2011) Thermal sensation and comfort model for inhomogeneous indoor environments. PhD Thesis, RWTH Aachen University, Fakultät für Maschinenwesen, Lehrstuhl für Gebäude- und Raumklimatechnik.

Streblow, R., Müller, D. and Gores, I. (2008) *A Coupled Simulation of the Thermal Environment and Thermal Comfort with an Adapted Tanabe Comfort Model*. Copenhagen, Denmark: IndoorAir.

Strøm-Tejsen, P., Zukowska, D., Jama, A. and Wyon, D. (2007) *Assessment of the Thermal Environment in a Simulated Aircraft Cabin using Thermal Manikin Exposure*. Helsinki: RoomVent.

Tanabe, S.I., Kobayashi, K., Nakano, J., Ozeki, Y. and Konishi, M. (2002) Evaluation of thermal comfort using combined multi-node thermoregulation (65MN) and radiation models and computational fluid dynamics (CFD). *Energy and Buildings* 34: 637–646.

THESEUS-FE. (2014) *Theory Manual*, release 4.4. www.theseus-fe.com.

van der Linden, A.C., Boerstra, A.C., Raue, A.K., Kurvers, S.R. and de Dear, R.J. (2006) Adaptive temperature limits: A new guideline in The Netherlands. *Energy and Buildings* 38: 8–17.

van der Linden, K., Boerstra, A.C., Raue, A.K. and Kurvers, S.R. (2002) Thermal indoor climate building performance characterized by human comfort response. *Energy and Buildings* 34: 737–744.

van der Linden, W., Loomans, M. and Hensen, J. (2008) *Adaptive Thermal Comfort Explained by PMV*. Copenhagen, Denmark: IndoorAir2008.

van Treeck, C. (2004) Gebäudemodell-basierte Simulation von Raumluftströmungen. PhD Thesis, Technische Universität München.

van Treeck, C. and Frisch, J. (2009) Model-adaptive analysis of indoor thermal comfort. In Strachan, P.A., Kelly, N.J. and Kummert, M. (eds.), *Building Simulation*. Glasgow: Building Simulation 2009 Organising Committee, pp. 1374–1381.

van Treeck, C., Frisch, J., Pfaffinger, M., Rank, E., Paulke, S., Schweinfurth, I., Schwab, R., Hellwig, R. and Holm, A. (2009) Integrated thermal comfort analysis using a parametric manikin model for interactive real-time simulation. *Journal of Building Performance Simulation*, in press.

van Treeck, C., Hillion, F., Cordier, A., Pfaffinger, M. and Rank, E. (2007b) Flow and thermal sensation analysis in an aeroplane passenger cabin using a lattice Boltzmann based approach. In *ICMMES 2007*, Germany.

van Treeck, C., Pfaffinger, M., Wenisch, P., Frisch, J., Yue, Z., Egger, M. and Rank, E. (2008) *Towards Computational Steering of Thermal Comfort Assessment*. Copenhagen, Denmark: Indoor Air.

van Treeck, C., Rank, E., Krafczyk, M., Tölke, J. and Nachtwey, B. (2006) Extension of a hybrid thermal LBE scheme for Large-Eddy simulations of turbulent convective flows. *Computers and Fluids* 35: 863–871.

van Treeck, C., Wenisch, P., Borrmann, A., Pfaffinger, M., Wenisch, O. and Rank, E. (2007a) ComfSim – Interaktive Simulation des thermischen Komforts in Innenräumen auf Höchstleistungsrechnern. *Bauphysik* 29(1): 2–7.

Urban, B.J. and Glicksman, L.R. (2006) The MIT Design Advisor – a fast, simple building design tool. In *Proceeding of IBPSA-Germany Conference BauSIM2006*, pp. 203–204.

Wissler, E. (1985) Mathematical simulation of human thermal behaviour using whole body models. In Shitzer, E. (ed.), *Heat Transfer in Medicine and Biology* 1(13): 325–373.

Wissler, E.H. (1964) A mathematical model of the human thermal system. *Mathematical Biophysics* 26: 147–166.

Wölki, D., Brüntjen, M.A., Schmidt, C. and van Treeck, C. (2016) A dynamic index for transient thermal comfort prediction. In *5th German-Austrian Conference of the International Building Performance Association, BauSIM 2014*. Aachen, Germany.

Wölki, D. (2017) MORPHEUS: Modelica-based implementation of a numerical human model involving individual human aspects. PhD Thesis, RWTH Aachen University, Fakultät für Bauingenieurwesen, Lehrstuhl für Energieeffizientes Bauen E3D.

Wyon, D.P., Larsson, S., Forsgren, B. and Lundgren, I. (1989) Standard procedures for assessing vehicle climate with a thermal manikin. *SAE Technical Paper 890049*, pp. 1–11.

Yokoyama, S., Kakuta, N. and Ochifuji, K. (1997) Development of a new algorithm for the heat transfer equation in the human body and its applications. *Applied Human Science* 16(4): 153–159.

Yoon, D.W., Sohn, J.Y. and Cho, K.H. (1999) The comparison on the thermal comfort sensation between the results of questionnaire survey and the calculation of the PMV values. *Indoor Air* 2: 137–141.
Zhang, H. (2003) Human thermal sensation and comfort in transient and non-uniform thermal environments. PhD Thesis, University of California, Berkeley.
Zhang, H., Arens, E., Huizenga, C. and Taeyoung, H. (2010a) Thermal sensation and comfort models for non-uniform and transient environments, part I: Local sensation of individual body parts. *Building and Environment* 45(2): 380–388.
Zhang, H., Arens, E., Huizenga, C. and Taeyoung, H. (2010b) Thermal sensation and comfort models for non-uniform and transient environments, part II: Local comfort of individual body parts. *Building and Environment* 45(2): 389–398.
Zhang, H., Arens, E., Huizenga, C. and Taeyoung, H. (2010c) Thermal sensation and comfort models for non-uniform and transient environments, part III: Whole-body sensation and comfort. *Building and Environment* 45(2): 399–410.
Zhang, H., Huizenga, C., Arens, E. and Wang, D. (2004) Thermal sensation and comfort in transient non-uniform thermal environments. *European Journal of Applied Physiology* 92: 728–733.
Zhang, Y. and Yang, T. (2008) *Simulation of Human Thermal Responses in a Confined Space*. Copenhagen, Denmark: IndoorAir.
Zhu, S.W., Kato, S., Ooka, R. and Sakoi, T. (2007) Development of a computational thermal manikin applicable in a non-uniform thermal environment (Part 1). *ASHRAE HVAC&R Research* 13(4): 661–679.
Zhu, S.W., Kato, S., Ooka, R., Sakoi, T. and Tsuzuki, K. (2008) Development of a computational thermal manikin applicable in a non-uniform thermal environment (Part 2). *ASHRAE HVAC&R Research* 14(4): 545–564.

International codes and standards

ASHRAE (2013) Standard 55: Thermal Environmental Conditions for Human Occupancy. Atlanta, GA: American Society of Heating, Refrigerating and Air-Conditioning Engineers
DIN V 18599 (2005) Energy efficiency of buildings. Calculation of the net, final and primary energy demand for heating, cooling, ventilation, domestic hot water and lighting. Beuth Verlag
DIN 33403–2 (2001) Klima am Arbeitsplatz und in der Arbeitsumgebung. Beuth Verlag
ISSO 74 (2014): Thermische behaaglijkheid; eisen voor de binnentemperatuur in gebouwen, publicatie 74, ISSO, Rotterdam
ISO 7243 (1989) Hot environments – Estimation of the heat stress on working man, based on the WBGT-index (wet bulb globe temperature)
ISO 7726 (1998) Ergonomics of the thermal environment – Instruments for measuring physical quantities
EN ISO 7730 (2005) Ergonomics of the thermal environment – Analytical determination and interpretation of thermal comfort using calculation of the PMV and PPD indices and local thermal comfort criteria
ISO 7933 (2004) Ergonomics of the thermal environment – Analytical determination and interpretation of heat stress using calculation of the predicted heat strain
EN ISO 8996 (2004) Ergonomics of the thermal environment – Determination of metabolic rate
EN ISO 9920 (2007) Ergonomics of the thermal environment – Estimation of thermal insulation and water vapour resistance of a clothing ensemble
ISO 10551 (1995) Ergonomics of the thermal environment – Assessment of the influence of the thermal environment using subjective judgement scales
ISO 11079 (2007) Ergonomics of the thermal environment – Determination and interpretation of cold stress when using required clothing insulation (IREQ) and local cooling effects
ISO 11399 (1995) Ergonomics of the thermal environment – Principles and application of relevant International Standards
EN ISO 13731 (2001) Ergonomics of the thermal environment – Vocabulary and symbols
ISO 14505–2 (2006) Ergonomics of the thermal environment – Evaluation of thermal environment in vehicles – Part 2: Determination of equivalent temperature
EN 15251 (2007) Indoor environmental input parameters for design and assessment of energy performance of buildings addressing indoor air quality, thermal environment, lighting and acoustics

VDI 6020 (2001) Requirements for methods of building performance simulation. VDI Verein Deutscher Ingenieure. Beuth Verlag

Activities and solution strategy

Choose a preferred building simulation tool. Create a suitable single-zone model of heavy weight type building, with an appropriate insulation standard and with a south-facing wall and window surface. All other walls are adiabatically connected with the rest of the building. Use weather data from a moderate climate zone and select typical internal heat sources.

Case A: natural ventilation

Set up a standard convective heating system and define a reasonable temperature threshold and a plant operation schedule for the winter period. Select a high infiltration rate during the day and omit mechanical cooling for the summer period.

- Compute the annual and monthly heating performance and validate your results using a spreadsheet tool.
- Compute the operative room temperatures for each hour of the year. Choose an adaptive thermal comfort guideline and compute the minimum, maximum, and average operative temperatures for each day. Plot these values against the weighted running average outdoor temperature as required by the standard. Indicate the comfort categories in your diagram.
- Classify the building in terms of the comfort category by counting the number of occurrences of the operative temperature values within each category (separately for the upper and lower bound).
- Locate critical periods in terms of overheating. Try to optimize your model by changing the design (night ventilation, increase/decrease the thermal capacity of the components, reduce internal heat gains, increase the ventilation rate during the day, etc.).
- Try to find a method to account for the accumulation of heat stress events that cause overheating.
- Add a radiant cooling system that covers the ceiling. Again, compute the operative room temperatures and check the thermal quality performance of the model. Evaluate the impact of the cooling device in terms of local thermal comfort. Use the methods described in the ASHRAE Standard 55 or EN ISO 7730.

Case B: mechanical ventilation

Use the scenario of case A and add a convective cooling system. Similarly, add a plant operation schedule for the summer period. Select a low infiltration rate during setback times.

- Again, compute the annual and monthly heating performance and validate your results with the help of a spreadsheet tool.
- Compute the operative room temperatures for each hour of the year. Choose a thermal comfort guideline based on the *PMV*/*PPD* model to evaluate the thermal quality performance.
- Design the cooling plant performance in such a way that the same comfort category is obtained from the point of view of *PPD* as with the natural ventilation case. Compare the annual energy consumption.

6 Weather and climate in building performance simulation

Drury B. Crawley and Charles S. Barnaby

Scope

This chapter addresses the environmental boundary conditions that drive building simulation models. The qualitative short- and long-term behavior of weather data is described. Practical representation of this information in hourly weather files and typical weather years is covered. The extensive role of modeled data is emphasized. Important trends caused by urbanization and climate change are considered.

Learning objectives

The reader will understand how highly varying weather phenomena are represented in practice for application in building performance simulation. The reader should understand the nature and limitations of commonly used typical year hourly weather files. Further, the reader should see that boundary conditions imposed on building models are often themselves model results. Determining the appropriateness and uncertainty of any building simulation results needs to include consideration of weather used in the study.

Key words

Weather data/typical weather year/solar models/climatic data

Introduction

Short-interval environmental data are required to drive building simulation models. These data provide the boundary conditions for component models that are combined in simulation applications. Ideally, environmental data should be observed exactly at the project site, subject to all local conditions. Further, the observations should be made at intervals consistent with the modeling being undertaken. These ideals are rarely realized. This chapter explores the strategies that have been developed to adapt available weather data sources to building simulation applications.

We should start with definitions of the two terms that are often used interchangeably but are not the same – weather and climate. The on-line dictionary at www.dictionary.com provides the following definitions:

- *Weather*: the state of the atmosphere with respect to wind, temperature, cloudiness, moisture, pressure, etc.
- *Climate*: the composite or generally prevailing weather conditions of a region, as temperature, air pressure, humidity, precipitation, sunshine, cloudiness, and winds, throughout the year, averaged over a series of years.

Thus, weather is what is happening now, while climate is the composite of weather over time. Building simulation can need both – the weather that occurred for a specific period or a representation of typical or extreme conditions.

Historically, large-scale (national) weather data observation programs have supported various activities such as general forecasting, aviation, and agriculture. Requirements for engineering applications have received minimal consideration during development of observation procedures. Thus, building simulation is often in the position of making do with what data are available. This situation is improving as technology lowers the cost and increases the sophistication of data observation, management, and modeling.

Hensen (1999) provides an overview of simulation-related weather data issues; most of that discussion remains applicable. An overriding point is that the type of simulation being conducted strongly determines weather data requirements. Traditionally, weather data for simulation have been represented in files of values for the 8760 hours of a year (or 8784 hours for a leap year). This form evolved for two reasons. First, until the 1990s, most weather observations were made manually; a one-hour measurement/reporting cycle was practical and sufficient for most applications. Second, a one-hour time step was (and still is) used in many building simulation models as a natural interval – short enough to capture behaviors of interest but long enough to allow practical execution times. Because of this history, it is common to think of weather data for building simulation as full-year hourly data. While hourly files remain important, we must consider other forms as simulation applications become more sophisticated.

Two attributes of modeling determine weather data needs. The nature of the process defines the items required (temperature, solar radiation, and so forth) while its time constant suggests the appropriate observation interval. Building structural components with substantial mass generally have relatively long time constants (hours); this accounts for the success of the hourly simulation approach. However, many building elements respond much more quickly – the air within the building, many furnishings, HVAC equipment, and daylighting controls, to name a few. A common approach in hourly models is to represent such elements with quasi steady state models that calculate average performance assuming current conditions persist indefinitely. To the extent that actual processes are non-linear, this simplification can be seriously in error. Shorter time steps are needed using sub-hourly weather values.

In addition to needing the right values at the right interval, building simulation applications require weather data that are complete and self-consistent (colloquially, "clean"). Applications can behave unpredictably or fail completely when confronted by unexpected values such as non-zero beam solar when the sun is below the horizon or wet-bulb temperature above the dry-bulb temperature. Typically, weather data, like any real measurements, contain erroneous or missing items. Thus, preprocessing, correction, and filling are important aspects of preparing weather data for simulation use.

ASHRAE Standard 209 recognizes the importance of weather data and climatic data throughout the building design, construction, and operation phases (ASHRAE 2018). Standard 209 describes 11 modeling cycles aligned with distinct building phases and requires a climate and site analysis to be performed prior to conceptual design modeling. Other modeling cycles have specific weather data needs such as those described in this chapter.

The sections below expand on this introductory background. Much of the material concerns primarily software developers – the issues discussed are addressed within simulation applications or weather data-processing utilities. However, the practitioner should generally understand how simulation weather boundary conditions are derived and the implications of those methods.

Characteristics of weather data

Selecting and applying weather data for building simulation should be informed by a qualitative understanding of the behavior of the driving phenomena. Figure 6.1 shows some example data observed at one-minute intervals for a typical day.

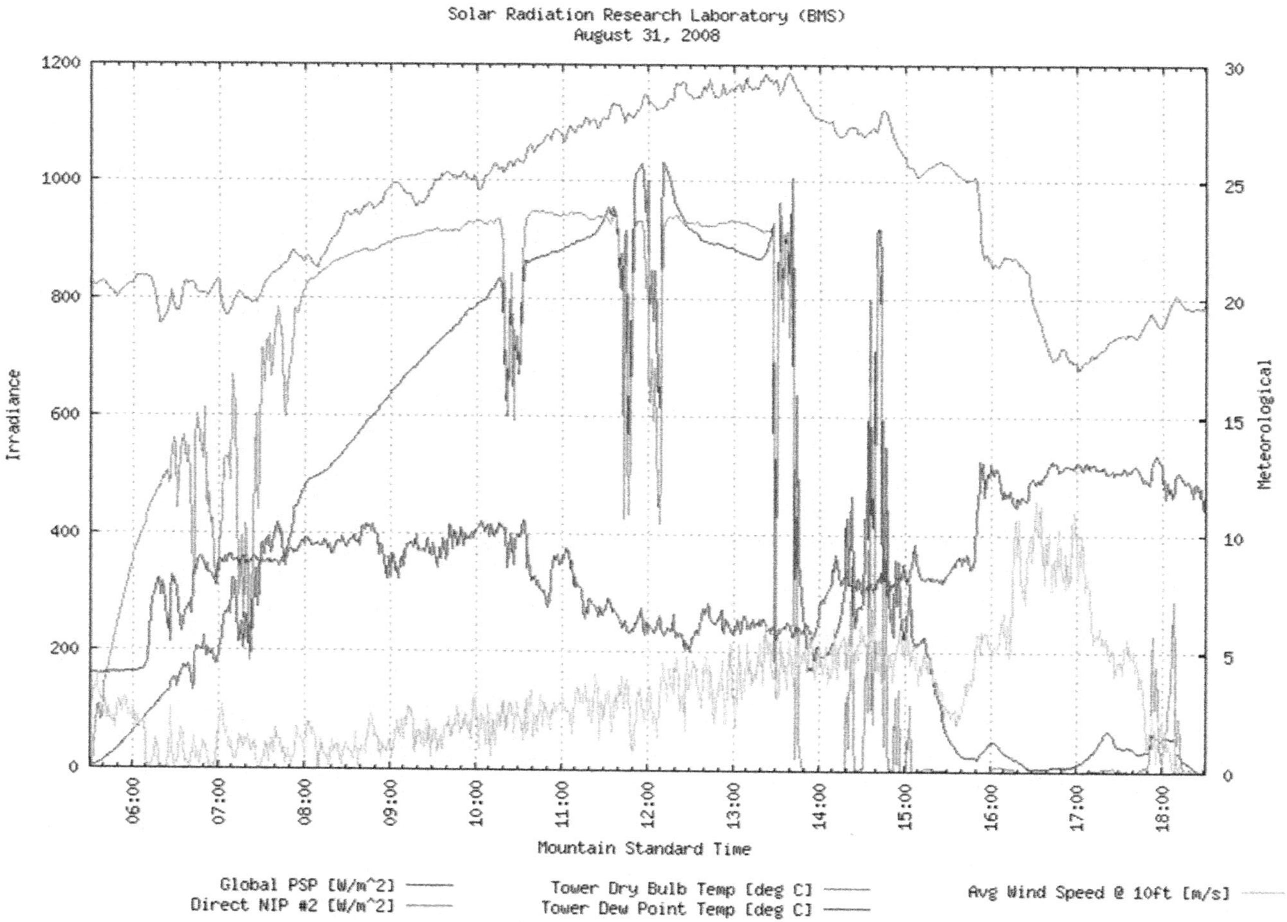

Figure 6.1 One-minute temperature, wind, and solar radiation observations for 31 August 2008 at National Renewable Energy Laboratory, Golden, Colorado, USA

By inspection, it is seen that this day begins with rising temperature and wind speed accompanied by occasional clouds. At about 16:00, there is a sudden drop in temperature and solar radiation along with a rise in dew point – probably due to a rain shower (confirmed by precipitation data not shown). The figure shows significant variability within single hours, with the following characteristics:

Temperature. Both dry-bulb and dew-point temperature vary relatively slowly, with only limited noise. Hour-to-hour changes are rarely more than a few degrees.
Solar. Large, rapid changes occur over a matter of minutes.
Wind. Wind velocity varies by a factor of two or more over time scales of a few minutes.

Representing this behavior with hourly values requires aggregation that necessarily omits a great deal of information. Conventional procedures for constructing hourly records use instantaneous values for meteorological data (such as temperatures, humidity, and wind speed) and totals over the prior hour for solar and illuminance data. Thus, for the Figure 6.1 example, the hourly record for 10:00 would show dry-bulb temperature of 25 °C and global solar radiation of about 705 Wh/m^2.

At the other end of the time scale, weather conditions also vary significantly from year to year. For many locations, there is no such thing as an average year – one or more anomalies occur during each year. Five- or ten-year simulations could be needed to assess long-term building performance, although representative single-year analysis is often possible using fabricated typical weather years, discussed later.

Simulation applications and weather data required

The following are some common uses of simulation and associated implications for weather data characteristics.

Energy design and compliance analysis of fully conditioned buildings. This is the "classic" simulation application. Energy savings information (generally annual) is derived by comparing results for design variants. In this application, the need for absolute accuracy is minimized – errors partially cancel, short-time inaccuracies are not significant, and the mechanical HVAC system will ensure actual building comfort. Similar baseline comparison procedures are used in energy code compliance and building labeling programs. Representative single-year hourly data are sufficient.

Performance of un- or semi-conditioned buildings. Overheating analysis, passive solar design, and other "floating space temperature" problems involve investigation of performance under non-average conditions. For example, a comfort study for a building without mechanical cooling might involve simulation using weather data from a particularly warm summer. Alternatively, multi-year studies are indicated. In any case, typical data are often not adequate.

Equipment sizing. Determination of maximum demand on HVAC components is typically based on design-day or short period calculations using near-extreme conditions. Designers need peak heating and cooling design conditions to perform load calculations (temperature, humidity, solar, and wind conditions). Most common data available are percentiles representing the average number of hours that a design condition will be exceeded.

Model calibration, building troubleshooting, control optimization, and actual savings estimation. Applications involving the performance of existing buildings require weather data observed during the study period at or near the building site. Historical data are generally of little use.

Engineering studies. Many engineering studies can be performed using simple weather information. For example, the annual number of hours when mechanical ventilation is feasible can be readily estimated using bin temperature data.

Natural ventilation design. Local wind conditions are highly variable; data from airport stations are notoriously unreliable for nearby sites. Great care is required when assembling wind data for study of buildings relying on wind-driven ventilation.

Daylighting studies. Although hourly illuminance data are sufficient to estimate electrical savings from sensor-control of lighting systems, detailed lighting studies involving visual comfort or control dynamics can require sub-hourly data.

Renewable energy systems. The output of solar electric systems depends on short-term variation and spectral make-up of incident radiation. Wind turbine output is proportional to the cube of wind velocity. Standard hourly data may produce unreliable results for systems that have non-linear characteristics such as these examples.

In short, the practitioner must consider the processes being simulated and take care to select appropriate weather data as boundary conditions. As can be seen from these examples, serious modeling errors can result from the use of inappropriate data. Conversely, there are situations where simple or representative data are sufficient; in these cases, seeking more detailed or "accurate" information has little value.

Climatic design condition data

To be able to size heating and cooling equipment and systems, engineers need to understand the peak heating and cooling conditions for the particular location. Data needed for load calculations include peak temperature, humidity, wind, and solar radiation. Chapter 14 in 2017 *ASHRAE Handbook – Fundamentals* (ASHRAE 2017a) covers the broad range of applications that require design conditions. The 2017 *Fundamentals* includes design conditions for 8118 locations throughout the world. What are the design conditions? Percentiles represent the number of hours that a particular design condition can be expected to be exceeded in a typical year (or month), based on the period of record. For example, a 99% annual heating dry-bulb temperature ≈ 88 hours (8,760 − (99.6% × 8760)) that typically you would see colder temperatures each year. For monthly cooling design conditions, the calculation is similar. A 5% monthly cooling dry-bulb temperature ≈ 36 hours (720 − (5% × 720)) in a 30-day month that typically you would see warmer temperatures. ASHRAE recommends the use of the 99% annual heating and the 5% monthly cooling. ASHRAE has a separate application, WDView, that allows further access to the electronic data and custom binning of temperature data (ASHRAE 2017b).

Data items required for simulation

Table 6.1 lists common weather data items used in building simulation. Several deserve additional discussion. As noted, dry-bulb temperature and wind are subject to significant local microclimate variation, so weather station observations (typically made at an open site such as an airport) often do not accurately represent nearby site conditions. As discussed later, dry-bulb temperature can be reliably modified with relatively simple methods. Local wind direction and velocity, however, cannot be modeled with any simple procedures. Developing data to drive simulations of wind-sensitive designs often requires on-site measurements, scale model wind tunnel studies, computational fluid dynamics (CFD) modeling, and/or use of conservative synthetic data. Regarding solar and illuminance data, note that these important values are normally generated using models, as is extensively discussed later.

Table 6.1 Weather data items of use for building simulation

Item	*Model use(s)*	*Availability and issues*
Dry-bulb air temperature	• Exterior surface convection • Infiltration/ventilation sensible heat transfer • Equipment (e.g. air-cooled condenser)	• Universally observed • Significant local effects (e.g. heat island)
Humidity (relative humidity, wet-bulb temperature, or dew-point temperature)	• Infiltration/ventilation latent heat transfer • Equipment (e.g. cooling tower)	• Commonly observed
Solar irradiance (direct and diffuse)	• Fenestration heat gain • Exterior surface heat balance • Solar thermal and photovoltaic systems	• Sparsely measured • If observed, often global only • Model sources widely used • Remote sensing opportunities
Solar illuminance (direct and diffuse)	• Daylight modeling	• Rarely measured (modeled from irradiance)
Sky temperature	• Exterior surface heat balance	• Rarely measured (modeled from temperature, humidity, and cloud cover)
Cloud cover/sky condition	• Sky models (e.g. for daylighting)	• Generally observed • Multiple data representation conventions • Evolution of automated instrumentation introduces uncertainties
Wind (velocity and direction)	• Exterior surface convection • Infiltration • Natural ventilation	• Generally observed • Local effects very significant for both velocity and direction • Low velocity observations unreliable
Ground temperature	• Below-grade heat transfer	• Measured for agricultural purposes, limited exploitation of observed values for building simulation
Ground surface albedo	• Reflected irradiance/ illuminance	• Inferred from presence of snow
Atmospheric pressure	• Psychrometric calculations	• Commonly observed
Precipitation	• Exterior surface wetting • Wind-driven rain	• Generally measured; inconsistent reporting formats
Weather conditions (e.g. type of precipitation, present weather conditions)	• Exterior surface wetting • Implied precipitation	• Generally reported; inconsistent reporting formats

Weather data formats

Representation and storage of short-term weather data have evolved along with the computer and information revolution of the last 50 years. Early automated data formats replicated punched card layouts formerly used for data entry by weather services. To save transcription costs, often data from only every third hour were made available. Weather data were expensive and were distributed on magnetic tape (one still occasionally hears the term "weather tape"). High cost, inconvenient formats, and omitted data made for difficult application to simulation.

Until recently, source files typically consisted of 8760 fixed format records (values located in specific columns of each data line). Custom programming was often required to decode the records, although the formats were generally simple and well documented. Most simulation applications resorted to using external weather utilities that could decode source files, generate missing data, and write "packed" binary files that could be efficiently read during model execution. Additionally, this strategy saved precious disk space – the packed files are typically 50–200 kilobytes as opposed to over a megabyte for the ASCII source.

Current formats, such as TMY3 (Wilcox and Marion 2008) and EPW (USDOE 2018b), are now more compatible with desktop spreadsheet tools. They are comma-separated and include metadata (such as site location) and column headings that make them somewhat self-documenting. In addition, many formats include field-level flags that indicate source and uncertainty. Figure 6.2 shows example data and illustrates that current files are straightforward in structure and admit of end-user manipulation.

The EPW format is particularly rich, allowing inclusion of design conditions, extreme period sequences, and sub-hourly data. That is, EPW is intended to support building simulation in general rather than being a format for use by a single simulation program. The EPW format has become a de facto standard that can be read or translated by more than 25 building simulation programs. Figure 6.3 shows some example data from an EPW file.

In general, managing weather data for simulation is becoming simpler. A helpful trend is toward free availability of weather files on the Internet (although files remain expensive in some European countries). For local studies, building automation systems now often record at least some weather parameters, often in an ad hoc (vendor specific) format. With spreadsheet manipulation, it is usually straightforward to adapt such data for simulation use. Thus, tasks that were formerly cumbersome, expensive, and technically demanding are now manageable with desktop computing tools.

Typical meteorological years

The most common building modeling application is full-year hourly simulation to calculate energy use. Extensive effort has been applied to finding representative or typical weather years that allow estimation of long-term (multi-year) performance from a single-year analysis.

Early work attempted to select a representative year from a multi-year set. For example, the test reference year (TRY) data set was developed for 60 locations in the United States (NCDC 1976) using a process in which years in the period of record (~1948–1975) that had months with extremely high or low mean temperatures were progressively eliminated until only one year remained. This results in selection of mild years that exclude extreme conditions.

The lack of extreme conditions and solar data in the TRY files led to development of the typical meteorological year (TMY) approach now in wide use. A TMY is a composite of typical months, not necessarily from the same year. A data record of *N* years contains *N* Januaries, *N* Februaries, and so forth. Various methods have been used to select the best representative for each of the 12. All rely on statistical measures of the similarity of the distributions of daily indices such as minimum, mean, and maximum dry-bulb temperature; minimum, mean, and maximum dew-point temperature; mean and maximum wind velocity; and daily total global and direct solar radiation. Weightings are used for the indices, influencing their importance. In some cases, final selections are made by hand, as illustrated in Figure 6.4. Wilcox and Marion (2008) provide a summary of methods used in the development of the TMY, TMY2, and TMY3 data sets developed for the United States.

	A	B	C	D	E	F	G	H	I	J	K	L	M	N	O	P	Q	R	S	T	U	V	W	X	Y	Z
1	722280	BIRMINGHAM MUNICIPAL AP	AL	-6	33.567	-86.75	189																			
2	Date (MM/DD/YYYY)	Time (HH:MM)	ETR (W/m^2)	ETRN (W/m^2)	GHI (W/m^2)	GHI source	GHI uncert (%)	DNI (W/m^2)	DNI source	DNI uncert (%)	DHI (W/m^2)	DHI source	DHI uncert (%)	GH illum (lx)	GH illum source	Global illum uncert (%)	DN illum (lx)	DN illum source	DN illum uncert (%)	DH illum (lx)	DH illum source	DH illum uncert (%)	Zenith lum (cd/m^2)	Zenith lum source	Zenith lum uncert (%)	TotCld (tenths)
3	1/1/1987	1:00	0	0	0	1	0	0	1	0	0	1	0	0	1	0	0	1	0	0	1	0	0	1	0	10
4	1/1/1987	2:00	0	0	0	1	0	0	1	0	0	1	0	0	1	0	0	1	0	0	1	0	0	1	0	10
5	1/1/1987	3:00	0	0	0	1	0	0	1	0	0	1	0	0	1	0	0	1	0	0	1	0	0	1	0	9
6	1/1/1987	4:00	0	0	0	1	0	0	1	0	0	1	0	0	1	0	0	1	0	0	1	0	0	1	0	5
7	1/1/1987	5:00	0	0	0	1	0	0	1	0	0	1	0	0	1	0	0	1	0	0	1	0	0	1	0	0
8	1/1/1987	6:00	0	0	0	1	0	0	1	0	0	1	0	0	1	0	0	1	0	0	1	0	0	1	0	0
9	1/1/1987	7:00	2	177	3	1	13	8	1	9	2	1	13	0	1	13	0	1	9	0	1	13	0	1	21	0
10	1/1/1987	8:00	153	1415	69	1	13	375	1	9	28	1	13	70	1	13	219	1	9	47	1	13	50	1	21	0
11	1/1/1987	9:00	389	1415	236	1	9	683	1	9	47	1	13	245	1	9	599	1	9	79	1	13	96	1	18	0
12	1/1/1987	10:00	580	1415	392	1	9	802	1	9	62	1	13	415	1	9	769	1	9	98	1	13	134	1	18	0
13	1/1/1987	11:00	711	1415	507	1	9	862	1	9	72	1	13	530	1	9	840	1	9	106	1	13	156	1	18	0
14	1/1/1987	12:00	773	1415	526	1	9	740	1	9	120	1	13	555	1	9	740	1	9	150	1	13	270	1	18	2
15	1/1/1987	13:00	761	1415	513	1	9	779	1	9	93	1	13	539	1	9	770	1	9	123	1	13	205	1	18	2
16	1/1/1987	14:00	677	1415	471	1	9	844	1	9	67	1	13	495	1	9	818	1	9	102	1	13	148	1	18	1
17	1/1/1987	15:00	526	1415	346	1	9	779	1	9	57	1	13	367	1	9	734	1	9	93	1	13	122	1	18	0
18	1/1/1987	16:00	319	1415	179	1	13	613	1	9	42	1	13	186	1	13	506	1	9	71	1	13	83	1	21	0
19	1/1/1987	17:00	80	1144	40	1	13	197	1	9	20	1	13	36	1	13	97	1	9	28	1	13	36	1	21	0
20	1/1/1987	18:00	0	0	0	1	0	0	1	0	0	1	0	0	1	0	0	1	0	0	1	0	0	1	0	4
21	1/1/1987	19:00	0	0	0	1	0	0	1	0	0	1	0	0	1	0	0	1	0	0	1	0	0	1	0	3
22	1/1/1987	20:00	0	0	0	1	0	0	1	0	0	1	0	0	1	0	0	1	0	0	1	0	0	1	0	10
23	1/1/1987	21:00	0	0	0	1	0	0	1	0	0	1	0	0	1	0	0	1	0	0	1	0	0	1	0	10
24	1/1/1987	22:00	0	0	0	1	0	0	1	0	0	1	0	0	1	0	0	1	0	0	1	0	0	1	0	10
25	1/1/1987	23:00	0	0	0	1	0	0	1	0	0	1	0	0	1	0	0	1	0	0	1	0	0	1	0	10
26	1/1/1987	24:00:00	0	0	0	1	0	0	1	0	0	1	0	0	1	0	0	1	0	0	1	0	0	1	0	10

Figure 6.2 Example TMY3 format data for Birmingham, AL, USA, displayed in a spreadsheet. Only “upper left” portion of file is shown; full dimensions are 68 columns by 8762 rows (two header rows and 8760 rows of hourly data).

	A	B	C	D	E	F	G	H	I	J	K	L	M	N	O	P	Q	R	S	T	U	V	W	X
1	LOCATION	Ahmedabad Intl AP	GJ	IND	ISD-TMYx	426470	23.077	72.635	5.5	57.6														
2	DESIGN CONDITIONS	1	2017 ASHRAE Handb		Heating	1	10.8	12.1	-2.5	3.1	24.9	0	3.8	24.6	7	22.6	6.2	22.8	1.2	50	Cooling	5	13.1	42.5
3	TYPICAL/EXTREME PERIODS	8	Wet Seas	Typical	17-Jul	23-Jul	Dry Seaso	Typical	8-Nov	14-Nov	Summer -	Extreme	27-May	2-Jun	Summer -	Typical	15-Apr	21-Apr	Winter - V	Extreme	24-Dec	30-Dec	Winter - V	Typical
4	GROUND TEMPERATURES	3	0.5				22.31	21.76	22.73	24.24	28.2	31.25	33.32	33.94	32.87	30.5	27.33	24.39	2				24.31	23.34
5	HOLIDAYS/DAYLIGHT SAVING	No	0	0	0																			
6	COMMENTS 1	NCEI ISD - #years=[15] Period of Record=2003-2017;Jan=2007;Feb=2003;Mar=2003;Apr=2016;May=2011;Jun=2007;Jul=2005;Aug=2010;Sep=2016;Oct=2008;Nov=2006;Dec=2012;																						
7	COMMENTS 2	Downloaded from Climate.Onebuilding.org -- Ground temps represent undisturbed earth temperatures - calculated from this weather data.																						
8	DATA PERIODS	1	1	Data	Sunday	1-Jan	31-Dec																	
9	2007	1	1	1	60	?9?9?9?9E	17	15	88	100640	0	0	373	0	0	0	0	0	0	0	158	0	9	9
10	2007	1	1	2	60	?9?9?9?9E	17	15	88	100640	0	0	373	0	0	0	0	0	0	0	80	0	9	9
11	2007	1	1	3	60	?9?9?9?9E	15.4	12.9	85	101009	0	0	355	0	0	0	0	0	0	0	64	0	8	8
12	2007	1	1	4	60	?9?9?9?9E	15	14	94	100636	0	0	372	0	0	0	0	0	0	0	80	0	10	10
13	2007	1	1	5	60	?9?9?9?9E	15	14	94	100636	0	0	372	0	0	0	0	0	0	0	53	0	10	10
14	2007	1	1	6	60	?9?9?9?9E	15.4	13.6	89	100860	0	0	364	0	0	0	0	0	0	0	339	0	9	9
15	2007	1	1	7	60	?9?9?9?9E	16	13.9	87	100915	0	0	367	0	0	0	0	0	0	0	310	0	9	9
16	2007	1	1	8	60	?9?9?9?9E	16.6	14.1	85	100969	27	124	362	5	0	5	517	0	521	197	285	0	8	8
17	2007	1	1	9	60	?9?9?9?9E	17.2	14.4	84	101023	322	1415	366	109	102	86	12039	3459	9518	2931	318	0	8	8
18	2007	1	1	10	60	?9?9?9?9E	22.1	14.7	63	100843	580	1415	378	351	467	159	39200	29905	17892	6053	121	0	5	5
19	2007	1	1	11	60	?9?9?9?9E	27	15	48	100663	783	1415	394	544	824	87	66110	43595	10607	4253	297	0	2	2
20	2007	1	1	12	60	?9?9?9?9E	25.4	16.7	59	101231	919	1415	394	644	819	112	78427	45461	13662	5625	198	0	4	4
21	2007	1	1	13	60	?9?9?9?9E	27	14	45	100663	977	1415	392	718	870	117	88002	50745	14370	5968	200	3.1	2	2
22	2007	1	1	14	60	?9?9?9?9E	28	16	48	100665	953	1415	400	693	795	157	82681	49121	18781	7634	200	3.1	2	2
23	2007	1	1	15	60	?9?9?9?9E	27	15.6	50	100936	850	1415	394	601	845	92	73459	45360	11303	4606	270	1.5	2	2
24	2007	1	1	16	60	?9?9?9?9E	29	15	43	100668	674	1415	399	450	781	77	54175	38682	9333	3636	290	3.1	1	1
25	2007	1	1	17	60	?9?9?9?9E	27	14	45	100663	438	1415	392	252	454	111	28031	22522	12432	4066	340	3.1	2	2
26	2007	1	1	18	60	?9?9?9?9E	26.4	16	53	100875	156	1415	396	54	131	40	5938	0	4370	1283	320	2.6	3	3
27	2007	1	1	19	60	?9?9?9?9E	23	16	65	100654	0	721	384	0	0	0	0	0	0	0	44	0	5	5
28	2007	1	1	20	60	?9?9?9?9E	22	16	69	100652	0	0	382	0	0	0	0	0	0	0	161	0	6	6
29	2007	1	1	21	60	?9?9?9?9E	20.8	16.9	78	101032	0	0	381	0	0	0	0	0	0	0	335	0	7	7
30	2007	1	1	22	60	?9?9?9?9E	20	17	83	100647	0	0	383	0	0	0	0	0	0	0	57	0	8	8
31	2007	1	1	23	60	?9?9?9?9E	19	17	88	100645	0	0	386	0	0	0	0	0	0	0	30	0	9	9
32	2007	1	1	24	60	?9?9?9?9E	18.6	15.9	84	101076	0	0	375	0	0	0	0	0	0	0	215	0	8	8

Figure 6.3 Example EPW format data for Ahmedabad, Gujarat, India, displayed in a spreadsheet. Only "upper left" portion of file is shown; full dimensions are 36 columns by 8768 rows (eight header rows and 8760 rows of hourly data).

The TMY procedure was used for the development of most of the current data sets widely available for building simulation. The "Data sources and tools" section (below) lists some of these. (Note that in some countries, current files are designated TRY [test reference year (Lund 1991)]; not to be confused with the older U.S. TRY [test reference year] data discussed earlier.)

ISO 15927–4 (ISO 2005) specifies the same method for constructing reference years (TRY) but does not use solar radiation as a selection parameter. Pernigotto et al. (2014a) proposed improvements and clarifications to the basic methods of ISO 15927–4 and demonstrated the impacts for 10 cities in Italy. Kalamees et al. (2012) looked at other weather parameters to get more accurate reference years for boreal areas of Finland. Pernigotto also provided extensive documentation of efforts to develop TMY3 or reference year data sets around the world. Recent work to create typical and design meteorological years includes Arima et al. (2017), Bulut (2004), Kalogirou (2003), Kim et al. (2017), Lee et al. (2010), Murphy (2017), Oko and Ogoloma (2011), Rahman and Dewsbury (2007), and Zhou et al. (2006), among many others. Hong et al. (2013) compare the resulting predicted energy performance using actual weather for a 30-year period to the TMY created. Standard 160 (ASHRAE 2016) specifies the development and use of moisture design reference years based on the 10th-percentile warmest and coldest years from a period of record.

Other work by Aguiar et al. (1999) investigate the use of statistical and stochastic models as alternatives to deriving TMYs from observed meteorological data. Chan (2016) proposes generating TMYs using genetic algorithms. Rastogi and Andersen (2015) similarly describe a method to create synthetic weather data for stochastic building simulation. Barreira et al. (2017) describe how a reference year for hygrothermal simulation could be constructed using ISO 15927–4. Most recently, Murano et al. (2018) recommend an improved weighting of variables when using ISO 15927–4 for improved assessment of residential buildings.

Studies have shown that TMY selection procedures produce weather years that yield average simulated energy use results for typical buildings, although predictions of peak energy demand may be less reliable. For example, Crawley and Lawrie (2015) found that office building end-use energy predictions made using typical years fall in the middle of the range of results from individual years (see Figure 6.5 for example results). Cui et al. (2017) found similar results comparing 55 years of meteorological data against typical years for 10 cities in China. Seo et al. (2009) showed that selection weightings have only a moderate effect on energy use predictions for typical buildings. Crawley and Lawrie (2015) and Narowski et al. (2013) discuss the issue of creating extreme or untypical meteorological years. Clarke (2001) discusses the formulation of a climate severity index (CSI) to describe the severity of weather impacts on building performance.

In summary, research studies and field experience have shown that the use of TMY-type single-year files is reasonably accurate and provides a practical route for estimating long-term building energy performance. However, since a single year cannot capture the full variability of the long-term record, peak energy predictions and uncertainty analyses based on TMYs are often not reliable.

Modeled data

Weather data generated using many types of models are used extensively in the building simulation field. As discussed earlier, the weather sequences used by simulation applications must contain all required items for all time steps. It is common for data sources to not include

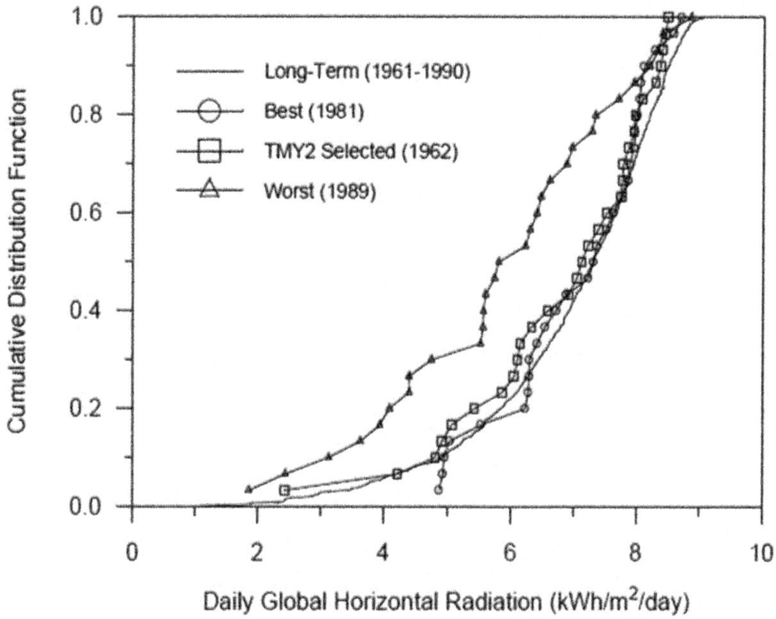

Figure 6.4 Cumulative distribution functions for June global horizontal solar radiation for Boulder, Colorado

Source: Wilcox and Marion 2008

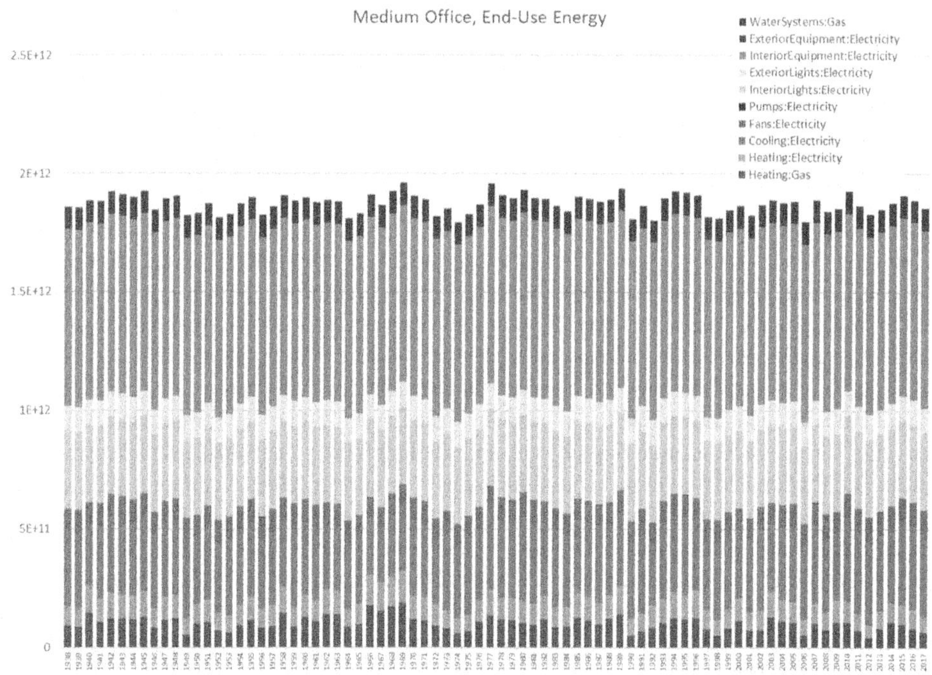

Figure 6.5 End-use energy predictions for 5000 m² office building in Washington, DC

Source: Crawley and Lawrie 2015, updated with data from 1938 to 2017 required items and/or have erroneous or missing values. The following are examples of how models are used in adapting available information to simulation requirements.

Solar models

Modeling of solar and illuminance data are particularly important, since these values are significant (in some cases, dominant) drivers of building performance but are sparsely observed. With a few exceptions, all of the weather file collections listed in this chapter include modeled solar and illuminance values. For example, in the TMY3 data set, less than 40 of 1020 sites include measured solar values.

In general, for each hour, solar models derive a theoretically based cloudless sky global horizontal irradiation (GHI) and then apply empirical modifiers that depend on cloud conditions. Davies and McKay (1989) describe and compare a number of first-generation procedures. Myers et al. (2005) analyze three additional models. Huang (2011) compared daily modeled and measured values. These and other studies rely on data sets of high-quality measured values for a range of locations. Model quality is assessed using the mean bias error (MBE) and the root mean square error (RMSE) calculated from monthly, daily, or hourly modeled and measured values. The MBE is zero when the average modeled value agrees with measured. RMSE represents the dispersion of the individual values and is zero only when every point is modeled perfectly. Current models produce very small MBEs (a few percent or less), while RMSEs are much larger (perhaps 6% of monthly, 20% for daily, and 30% for hourly). Figure 6.6 shows typical results for daily values.

Recently, models have been enhanced to use satellite data to augment other observations. These models offer improved geographical coverage and have similar or somewhat better accuracy than station-based models. Perez et al. (2004) demonstrate that for any site more than 20–25 km from a measurement station, satellite-derived values are the most accurate available estimates of hourly global horizontal irradiance. Several on-line services provide gridded, satellite-based, near real-time short-interval solar data for various regions of the world, including the Copernicus Atmospheric Monitoring Service (CAMS 2017), the National Solar Radiation Database (NSRDB) (NREL 2018; Habte et al. 2017), and the Amaterass Consortium (AMTR 2018).

Thus far, this discussion of solar models has focused on global horizontal irradiance. However, building simulation requires separate direct and diffuse radiation components incident on each exterior surface. If daylighting is being modeled, illuminance values are needed as well. Partitioning models are used to estimate the direct and diffuse portions of the horizontal irradiance. Daylight models calculate the luminous fraction of the total radiation. Transposition models are used to derive the intensity on an arbitrary surface. The transposition step is necessarily done within simulation programs, since surface-specific information is required. Surface incident values depend on simple beam geometry, the distribution of sky brightness, reflections from other surfaces and surrounding objects, and shading due to other surfaces and surrounding objects. Highly detailed modeling is possible using computer graphics techniques such as those presented by Jones et al. (2012). A common optimization is performing detailed solar geometry calculations for groups of days, as opposed to each day. Descriptions of many solar methods are found in Duffie and Beckman (2006), Clarke (2001), and USDOE (2018a). Muneer (2004) provides comprehensive treatment of both theory and calculation methods, including sample FORTRAN implementations. Underlying and ongoing research is documented in the technical literature; for example, see Hay and Davies (1978), Perez et al. (1990), and Gueymard (1987, 2009). Gueymard and Ruiz-Arias (2016) compared 140 direct/diffuse separation models to high-quality observations. Pernigotto et al. (2015) evaluated 22 horizontal diffuse irradiance models and 12 irradiance models for tilted surfaces and found that solar models can introduce

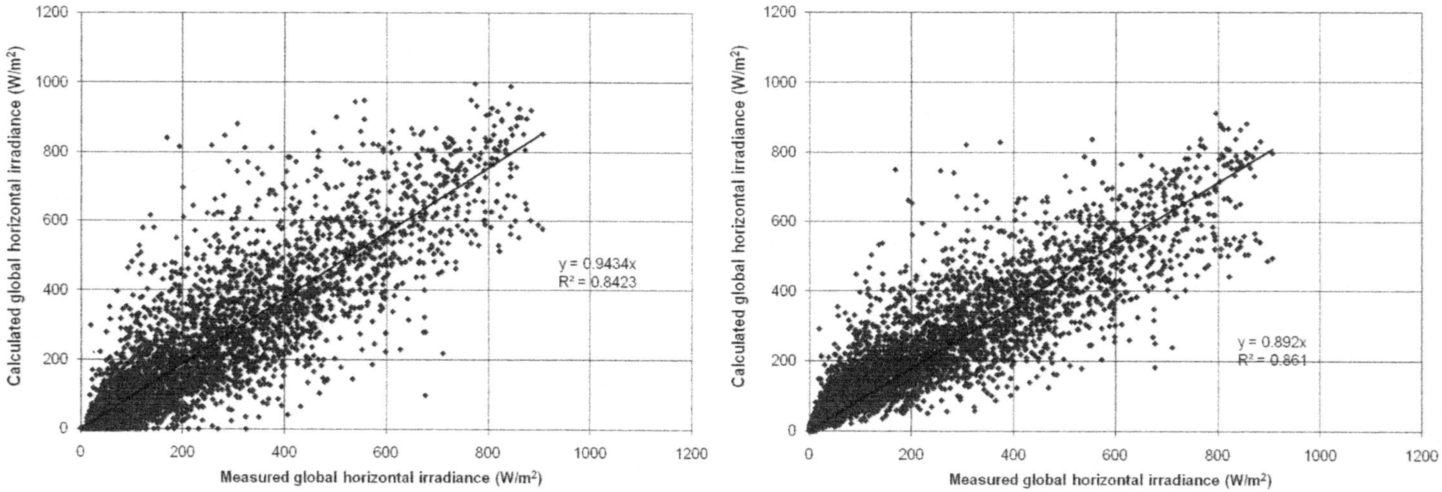

Figure 6.6 Measured daily global radiation vs. modeled for Beijing, China (left), and London, United Kingdom (right)

Source: Huang 2011, ©2011 ASHRAE, used with permission

as much as 10% uncertainty into building energy simulation results. Citations in these sources will lead the interested reader to numerous publications going back to the 1960s.

In summary, calculation of surface irradiation is among the most complex aspects of building performance simulation and introduces large uncertainty in short-term results. The process starts with values that have significant short-term uncertainty and chains through several steps, each of which produces representative results based on limited inputs. Accurate long-term averages can be expected, but realistic hour-to-hour variability is not captured. This shortcoming is most pronounced for partly cloudy conditions – inter-day differences are smaller for clear and cloudy days than for partly cloudy days. For example, one should not draw conclusions about daylighting rate of change using typical modeled illuminance data as input.

Other modeling applications

In addition to generation of solar data, models are used in many other situations that arise in preparation of weather data for simulation applications.

Generation of required items. Missing required data items must be inferred from other values. Estimation of irradiance and illuminance is discussed in the previous section. Long-wave radiant exchange models require sky temperature, typically estimated from dry-bulb temperature, humidity, and cloud cover. Temperature adjacent to underground surfaces can be approximated from annual temperature cycles and/or undisturbed ground surface temperature (Kusuda and Achenbach 1965). Preliminary investigations indicate that shallow (agricultural) ground temperature observations may have application in improving accuracy of model predictions for depths of interest for building simulation (Spitler 2009).

Filling gaps. As in any set of observations, weather data sources can be incomplete or include obviously erroneous values. For example, if within a single record the wet-bulb temperature is greater than the dry-bulb temperature, one or the other is incorrect (or both). Errors can originate from instrumentation malfunctions, data-processing bugs, or transcription mistakes (for pre-automation hand-entered data), to name a few possibilities. For short gaps up to a few hours, linear or cubic spline interpolation is often satisfactory. Baltazar and Claridge (2006) compared 17 algorithms and found the least bias using linear interpolation for filling dry-bulb and dew-point temperature gaps up to 6 hours. Methods for filling longer gaps are a topic of active current research. Exploitation of mesoscale models and remote sensing techniques (discussed later) should soon allow accurate repair of nearly all data gaps.

Reconstruction of short time step data. Simulation applications that utilize time steps shorter than an hour require corresponding short-interval weather data. Since hourly is the commonly available form, sub-hourly values must often be generated. As can be seen from Figure 6.1, linear interpolation can safely be used to estimate mid-hour temperature values. Reconstruction of plausible sub-hourly wind and especially solar data require more sophisticated models; for example, Meteonorm (Meteotest 2018) uses stochastic procedures to generate values for intervals as short as one minute. However, for general building energy estimation, simple linear methods are acceptable, since building mass dampens short-term variation.

Generation of weather sequences. A number of procedures have been developed that generate hourly weather sequences of up to a full year in length using longer term weather statistics as input; for example, see Meteotest (2018), Degelman (1991), Knight et al. (1991), Klein and Reindl (1998), Balcomb (1999), and David et al. (2010). Degelman (2007) found that synthetic weather yields simulation results very similar to those derived with observed data for a case study building. Rastogi (2016) developed a method for dealing with weather data sensitivity and uncertainty analyses in building simulation. These include a generalizable procedure for generating an ensemble of synthetic, but realistic, weather data.

Adjustment of weather sequences. A common requirement in practice is the need to adjust weather data to represent an alternative site, to account for urban effects, or to preview future climate conditions. Several methods and tools have been developed to modify data; among the earliest is Arens et al. (1980). A more recent tool is WeatherShift (Arup 2018). Additional methods are discussed in later sections ("Urban heat island" and "Climate change"). Mesoscale models (discussed later) have potential use here also.

Time-shifting. In some cases, weather data items must be manipulated to reconcile them with project site local time. This can be particularly important for solar values, since physically impossible conditions can arise if sun position is calculated relative to the project site and then applied to data observed elsewhere. Local time conventions, such as daylight savings, must also be correctly handled.

Predictive control strategies. Building control schemes that respond to expected (modeled) weather conditions have been proposed and studied. See Chapter 14.

Current developments

Ongoing developments in measurement and modeling of environmental data are finding application in building performance simulation. The following are some examples of the advances that are expected to yield improvements in the quality and availability of weather data for simulation.

Mesoscale models. Geographically aware regional models are now universally used for weather forecasting. It is possible to use these models to generate localized historical weather sequences by driving them with measured data from nearby stations. Mesoscale models remain too computationally intensive for routine use, but this may change given the rapid development of multi-processor computers. See Qiu et al. (2015) for recent work on mesoscale modeling. The Weather Research and Forecasting Model (WRF) is described by NCAR (2018a).

Reanalysis. Meteorological services are continuously improving historical records in support of climate studies. For example, the North American Regional Reanalysis (NARR 2018) makes available data sets of numerous variables at 3-hour intervals for the period 1979–present on a 32 km grid. Other projects include CFSR/CFSv2 (2018) and MERRA-2 (NASA GMAO 2018). Stackhouse et al. (2006) describe reanalysis data available through NASA. NCAR (2018b) provides a summary of reanalysis projects worldwide. Roth (2019) evaluates the utility of these sources for building-related applications.

Remote sensing. Many years of satellite observations now exist. These data are beginning to contribute to weather data used in practice, particularly for solar values (as discussed earlier). Satellite-based data offer uniform worldwide coverage, a major advantage. The 2017 ASHRAE clear sky model uses parameters derived from satellite-based aerosol estimates, see Roth (2016). Stackhouse et al. (2018) describe satellite-derived engineering data available from NASA.

Data sources and tools

A growing body of simulation and engineering weather data is available. This section lists and characterizes some of the sources. Additional Internet searching is recommended, since any published list is rapidly out of date. Links to data sets are provided in the references if available. Older, superseded data sets are included in a separate section below.

Several services provide consolidated access to many files:

- *EnergyPlus*. Free data for more than 2100 locations available in EPW weather format USDOE (2018c).
- *Climate.OneBuilding.org*. Free data for more than 6000 locations worldwide. Data include monthly design conditions, and EPW, DAYSIM/Radiance WEA, and ESP-r CLM weather formats. Also available are TMYx files derived from ISD (NCEI 2018) with hourly data through 2017 Climate.OneBuilding (2019).
- *White Box Technologies*. Weather data for energy calculations including historical and typical year for more than 10,000 locations worldwide. Includes ASHRAE IWEC2 (2011). Data provided in BINM (modified DOE-2) and EPW formats. Cost varies by data source and number of files White Box Technologies (2018).
- Other commercial services are readily found via Internet search; for example, Solargis (2018).

Regional and specialty data sets include:

- *Año Meteorológico Típico para Aplicaciones de Energía Solar (AMTUes)*. Data set created using the TMY methodology for five locations in Uruguay (Alonso-Suárez et al. 2016).
- *Argentina TMY (ArgTMY)*. EPW format, TMY data derived for 15 locations in northern Argentina using data from National Meteorological Service (SMN) of Argentina (Bre and Fachinotti 2016).
- *ASHRAE Design Weather Data*. WDView 6.0 provides design weather data for 8118 worldwide locations, including monthly design temperatures, bin data, and heating or cooling degree days to any base (hourly data are not provided). In addition, clear sky solar radiation can be generated for all locations (ASHRAE 2017a, 2017b).
- *Bundesinstitut für Bau-, Stadt- und Raumforschung (BBSR2018)*. Spatially continuous reference years using source data from weather stations and satellite and model data for the years 1995–2012 for current and future (2031–2060) variants. Heat island impacts are built into the spatially continuous data (Deutschen Wetterdienst (2018).
- *Bundesinstitut für Bau-, Stadt- und Raumforschung (BBSR2011)*. 15 locations with current and 2021–2050 data for *jahr* (year), *somm* (summer), and *wint* (winter) versions (Deutschen Wetterdienst 2014).
- *California Compliance Weather Data*. Data sets used in demonstrating compliance with Title 24 energy code (CEC 2016).
- *Canadian Weather for Energy Calculations 2016 (CWEC2016) and Canadian Weather Energy and Engineering Datasets (CWEEDS)*. CWEEDS includes 492 Canadian locations with at least 10 years of data for the period between 1998 and 2014. This 2016 version of the CWEEDS files uses satellite-derived solar estimates using methods developed at SUNY (State University of New York). The CWEC 2016 files were created by joining 12 typical meteorological months selected from the CWEEDS data set. The months are chosen using TMY methodology. The CWEEDS are available from Environment and Climate Change Canada (2016a, 2016b).
- *Chartered Institution of Building Services Engineers (CIBSE 2016)*. Test reference years and design summer years for 16 U.K. locations for use with building energy simulation software. Future weather data is available for a variety of emissions scenarios for 2020s, 2050s, and 2080s. The data sets are available in several formats, including EPW, Tas, and Excel (CIBSE 2016; Mylona 2012; Watkins et al. 2011).

- *Chinese Standard Weather Data (CSWD).* Developed for use in simulating building heating and air-conditioning loads and energy use, and for calculating renewable energy utilization, this set of 270 typical hourly data weather files were developed by Dr. Jiang Yi, Department of Building Science and Technology at Tsinghua University and China Meteorological Bureau. The source data include annual design data, typical year data, and extreme years for maximum enthalpy, and maximum and minimum temperature and solar radiation (China Meteorological Bureau et al. 2005).
- *Comitato Termotecnico Italiano Anno Climatici Tipo (CTIACT).* Typical reference year data for 110 localities throughout Italy Pernigotto et al. (2014b).
- *IMGW Weather Data Set for Poland.* The Polish Ministerstwo Infrastruktury prepared this set of 61 files for use in building energy calculations. The files are based on data from the Instytutu Meteorologii i Gospodarki Wodnej (IMGW 2009; Budzyński et al. 2004).
- *IMS Weather Data for Israel.* Weather data for Israeli locations developed by Faculty of Civil and Environmental Engineering, Technion – Israel Institute of Technology, Haifa, Israel, from data provided by the Israel Meteorological Service (IMS 2009).
- *Instituto Nacional de Meteorologia Arquivo por Brasil (INMET).* Derived from the INMET data set, typical weather files for 410 locations in Brazil and 1 in Antarctica. (Roriz 2012).
- *Indian Society of Heating, Refrigerating and Air-Conditioning Engineers (ISHRAE 2014).* Developed for use with building energy performance simulation programs, this set of 62 locations in India was created by the Indian Society of Heating, Refrigerating and Air-Conditioning Engineers (ISHRAE) in EPW and BIN formats (ISHRAE 2014).
- *Japanese Generated Meteorological Year (JGMY).* 157 locations in Japan synthetically generated (statistically correct) data. Developed by Degelman (1991).
- *New Zealand National Institute of Water and Atmospheric Research Ltd (NIWA).* The New Zealand Energy Efficiency and Conservation Authority (EECA) has developed a simulation-based Home Energy Rating Scheme (HERS) along with a supporting set of hourly weather data files for all climate zones of New Zealand. The files represent artificial years created from 12 representative months (Liley et al. 2007).
- *Passive House Institute Korea (PHIKO 2018).* Test reference year (TRY) was created, based on ISO 15927–4:2005 for assessing the annual energy use. The TRY is based on measured data from the years 2005 to 2014 sourced by the Korea Meteorological Administration. The preparation of the TRY was performed by Passive House Institute Korea (Kim et al. 2017).
- *RMY Australia Representative Meteorological Year Climate Files (RMY2012).* Developed for the Australia Department of Resources, Energy and Tourism for use in complying with the National House Energy Rating Scheme (NatHERS), these data are for 69 locations throughout Australia, covering the climate regions in the Building Code of Australia (Lily 2013). Three versions derived using solar radiation weights of 50% (A), 33% (B), and 17% (C).
- *Turkey TMY (TurTMY).* TMY data for the 81 Turkish provincial capital cities with source data from 1989 to 2006. Derived using the TMY2 methodology (Pusat, Ekmekçi, and Akkoyunlu 2015).
- *Typical Meteorological Year 3 (TMY3a).* The TMY3 set includes files for 1020 locations in the United States including Guam, Puerto Rico, and the U.S. Virgin Islands, derived from a 1991–2005 period of record. The files contain hourly values of solar radiation and meteorological elements for a one-year period. Their intended use is for computer simulations of solar energy conversion systems and building systems to facilitate performance comparisons of different system types, configurations, and locations in the United States and its territories. Because they represent typical rather than extreme conditions, they are not suited for designing systems to meet the worst-case conditions occurring at a location. The TMY3 were updated in 2015 to substantially improve the precipitation data. The files are available from the National Renewable Energy Laboratory (Wilcox and Marion 2008; Huang 2015).

Older data sources

- *Chinese Typical Year Weather (CTYW)*. 57 files for China, 1982–1997 period of record, Zhang and Huang (2004).
- *Canadian Weather for Energy Calculations (CWEC)*. 81 files for Canada, 1953–1995 period of record, Numerical Logics (1999).
- *Indian Weather Data from the Indian Society of Heating, Refrigerating and Air-Conditioning Engineers (ISHRAE)*. 58 locations in India, ISHRAE (2009).
- *International Weather for Energy Calculations (IWEC)*. 227 locations outside the USA and Canada, ASHRAE (2001).
- *Australia Representative Meteorological Year Climate Files (RMY)*. 66 locations throughout Australia, Energy Partners (2008).
- *Solar and Meteorological Surface Observation Network (SAMSON)*. 239 locations in the U.S., covering 30 individual years in 1961–1990. Used as source data for TMY2 (see later), NCDC (1993).
- *Solar and Wind Energy Resource Assessment (SWERA)*. 156 locations in 14 developing countries, SWERA (2009).
- *Test Reference Years (Brazil) (TRY)*. 17 major state capitals in Brazil, Carlo and Lamberts (2005).
- *Test Reference Year (EU) (TRY)*. Locations in major European cities, Lund (1991).
- *Test Reference Year (USA) (TRY)*. 60 U.S. locations, source period of 1948–1975, NCDC (1976).
- *Typical Meteorological Years (TMY)*. 234 U.S. locations. 26 locations with measured solar data, 1952–1975 period of record, NCDC (1981).
- *Typical Meteorological Years 2 (TMY2)*. 239 locations in U.S., derived from the SAMSON (see earlier) and the National Solar Radiation Database (NSRDB) with a period of 1961–1990; solar radiation values were calculated from cloud cover based on the Perez model (1990, 1991), NREL (1995).
- *Weather Year for Energy Calculations (WYEC)*. Crow (1970, 1980, 1983) created typical weather data for 51 U.S. and Canada locations, ASHRAE (1985).
- *Weather Year for Energy Calculations 2 (WYEC2)*. 77 locations in U.S. and Canada, new models for calculating solar radiation from cloud cover data (Perez et al. 1990; ASHRAE 1997; Stoffel and Rymes 1998).

Near real-time data

For some applications, notably those that use simulation models to forecast near-term building performance, very recent data are required. Commercial services provide gridded data over large regions. However, such data must be used with caution – Bhandari et al. (2012) evaluated several commercially available and public real-time weather data sets and found variation among the data to cause as much as a 90% variance in hourly peaks and a 7% variation in simulated annual energy consumption.

Example public sources that make such data available (although not necessarily in ready-to-use form) include:

Integrated Surface Database (ISD). The ISD contains global hourly and synoptic observations compiled from numerous sources for more than 35,000 stations worldwide. While some data begin in 1901, most are from the 1970s or later. Currently, data for more than 14,000 "active" stations are updated daily. ISD includes parameters such as wind speed and direction, wind gust, temperature, dew point, cloud data, sea level pressure, altimeter setting, station pressure, present weather, visibility, precipitation amounts for various time periods, snow depth, and various other elements as observed by each station. The data can be accessed on-line (NCEI 2018).

Oklahoma Mesonet. In the U.S., the state of Oklahoma operates a network of 120 stations with high-quality instruments (including pyranometers) recording observations at five-minute intervals. With about a one-hour lag, all data are available on-line (Mesonet 2018).

Weather data tools

Below are a few of the tools that can be used to edit, visualize, and evaluate weather data.

- *DView*. DView is a free tool that can be used to visualize and evaluate hourly simulation output including weather data. Hourly, daily, monthly, and annual summary graphs are possible. Version 3.1, NREL (2017).
- *Elements*. Tool for editing custom weather files for building energy modeling. Big Ladder Software (2016).
- *Energy Design Tools*. Series of tools for energy design from UCLA Department of Architecture and Urban Design. Climate Consultant is an easy-to-use tool to understand local climate. It reads weather files (EPW format) and displays psychrometric and more than 15 other visualizations. Latest update, V6.12, 2017. UCLA (2017).
- *Ladybug tools*. Ladybug imports standard EnergyPlus (EPW) weather files, performs detailed analysis of climate data, and provides interactive visualizations. Dragonfly models urban heat island, climate change, and local climate factors such as topography. Ladybug Tools (2018).
- *SunPath/Daylight*. Web-based tools for visualizing solar position and available daylight. Marsh (2018).

Selecting the right weather data

A problem often encountered in practice is selection of a suitable weather data source when no weather station is close to the project site under study. Although there is no generally accepted procedure, the first step is to identify some candidate sources with approximately the same latitude and elevation as the site. Then comparison of monthly statistics derived from candidate files (using a simulation program weather utility) to climatological summaries for the project location will generally allow selection of an acceptable match. If no similar source is found, data synthesis or adjustment procedures should be considered. Crawley (1998) recommended using weather files derived using a TMY-type methodology.

Urban heat island

Conditions at urban building sites are modified by the surrounding environment and can differ significantly from those at nearby weather stations (typically at airports). Howard (1833) first described the altered meteorological conditions caused by pollution in London as "city fog." Howard also measured the temperature differences between the urban center and the countryside for a number of years, publishing his initial findings in 1820. In a footnote to his table of mean monthly temperature differences, Howard wrote, "night is 3.70° *warmer* and day 0.34° *cooler* in the city than in the country" (emphasis in original), recognizing what today we call the urban heat island (UHI) effect.

More recently, Mitchell (1953, 1961) measured the extent and intensity of the heat island phenomena. Oke (1988) and Runnalls and Oke (2000) were the first to develop diagrams to explain the diurnal and seasonal patterns of heat islands. Their diagrams were confirmed by the temperature measurements by Streuker (2003) and Morris and Simmonds (2000). Specifically, Streuker's measurements reinforced Oke's findings (1973) that heat island intensity depends on urban concentration (population density), vegetation, and surface albedo. Oke (2006) presents a broad overview of urban influences on local weather conditions and guidance about how to make meteorological measurements at urban sites. The U.S. Environmental Protection Agency's Heat Island Reduction Initiative estimates that the heat island effect is in the range of 2–10 °F (1–5 °C) (USEPA 2007). Additional discussion and references relating to UHI are found in Chapter 12. This extensive body of knowledge has seen only limited application in the building simulation field. Erell and Williamson (2006) describe an urban canyon air temperature (CAT) model. Williamson et al. (2009) and Crawley (2008a, 2008b) offer two of the few studies to date on the impact on simulation results of using data modified to include urban influences. Erell et al. (2011) cover the challenge of simulating the urban microclimate – the spaces between buildings. Short et al. (2009) present the challenge of using passive stack ventilation to naturally ventilate buildings due to the London heat island. Nakano et al. (2015) describe a workflow to couple climate change predictions with UHI to capture local changes on a building site. Schneider dos Santos et al. (2016) provide a review of UHI monitoring, modeling, and remote sensing approaches and their application in building simulation. Zhao et al. (2018) evaluated the impacts of trees and their location on microclimates and human comfort within the urban context. Van der Heijden et al. (2013) describe a data-driven method based on a neural network to predict UHI intensity.There is a clearly detectable warming trend in long-term climate records, probably caused primarily by urbanization, although global warming may be at work as well (and it certainly will be in the relatively near future, as discussed later). For example, Roth (2016) analyzed building-related weather statistics for 8118 worldwide stations. Modest warming trends were found, somewhat more pronounced in winter than summer. Example trend distributions are shown in Figure 6.7.

Climate change

Over the past 30 years, the international scientific community, as organized through the Intergovernmental Panel on Climate Change (IPCC), has focused significant effort to characterize the potential impacts of greenhouse gas emissions from human activities on the global climate. IPCC Working Group I focused on creating atmosphere-ocean general circulation models (GCMs), similar to models used to predict the weather, in which the physics of atmospheric motion are used to predict the future climate (IPCC 2013). Four scenarios of human contribution to climate change, known as representative concentration pathways (RCP), are then simulated within the major GCMs to estimate the potential change in temperature and other climate variables. The IPCC released its fifth assessment report (AR5) in 2014 (IPCC 2014a, 2014b), in which it identifies buildings as the sector with the highest economic mitigation potential of any other energy sector. The potential change in predicted annual average global temperature during the coming century is 2–6 °C, as shown in Figure 6.8. It must be emphasized that the data shown are averages – some locations will experience much larger monthly or hourly changes.

Simulation methods will figure heavily in determining optimum mitigation strategies. To allow simulation of building behavior under projected conditions, techniques have been developed to adjust current weather files to reflect climate change scenarios. A "morphing" technique developed by Belcher et al. (2005) was used by Jentsch et al. (2008, 2013) in the publicly available tool CCWorldWeatherGen (Sustainable Energy Research Group 2013) and to produce weather years for 2020s, 2050s, and 2080s for 14 locations in the U.K. (CIBSE 2016). Eames et al. (2012) covered the aspects of the 5 km grid used in the CIBSE work. Moazami et al. (2017) reviewed two software tools for creating future weather data – CCWorldWeatherGen and WeatherShift (Arup 2018) – and recommend careful review of the resulting data

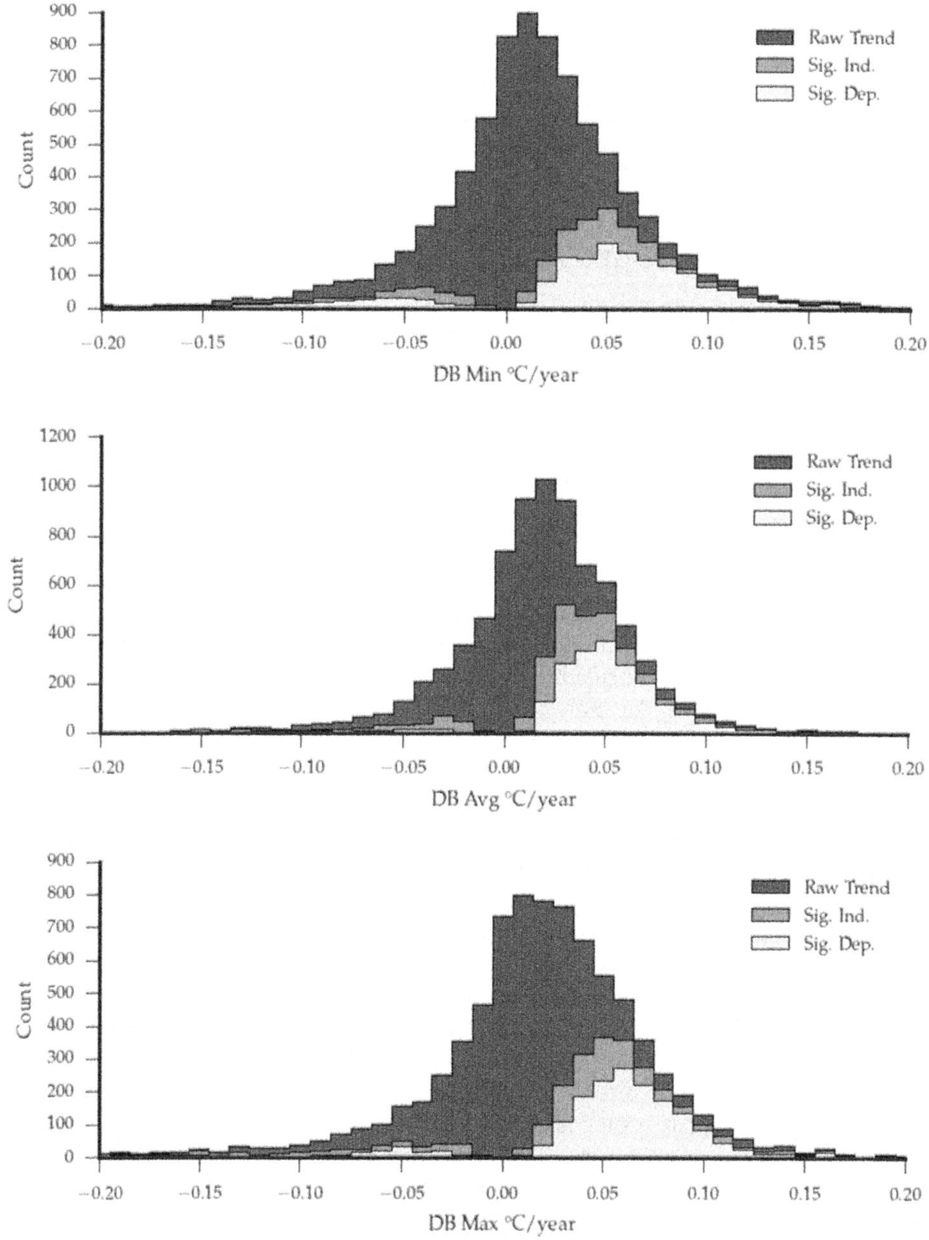

Figure 6.7 Histograms of the trends of monthly minimum, average, and maximum dry-bulb (DB) temperatures for the 8118 stations in the *ASHRAE Handbook – Fundamentals* (ASHRAE 2017a)

Note: "Raw Trend": all data; "Sig. Ind.": statistically significant trends assuming values are independent; "Sig. Dep.": statistically significant trends when corrected for serial dependence.
Source: Roth 2016, ©2016 ASHRAE, used with permission

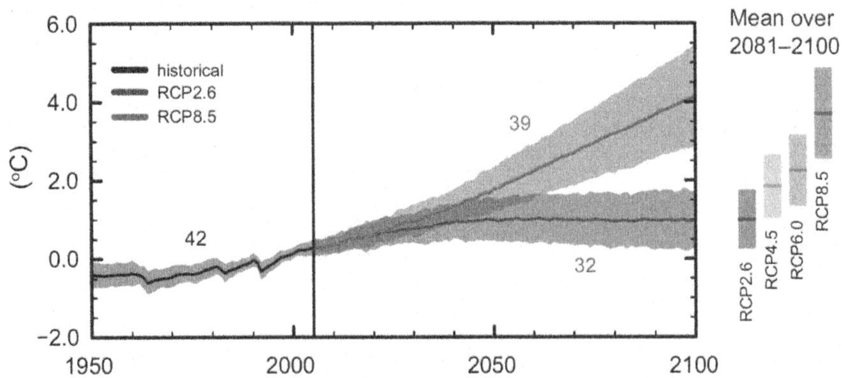

Figure 6.8 Change in global annual mean surface temperature relative to 1986–2005, uncertainty (shading) is shown for scenarios RCP2.6 (dark gray, lower range) and RCP8.5 (light gray, upper range). Black line (and shading) prior to 2005 are the modeled historical evolution. The mean and associated uncertainties averaged over 2081–2100 are given for all RCP scenarios as colored vertical bars. The numbers show the number of CMIP5 models used to calculate the multi-model mean.

Source: IPCC 2013, ©IPCC, used with permission

to ensure its usefulness. Crawley (2008a, 2008b) presents a methodology for generating predicted weather sequences for 2100. Guan (2009) characterizes proposed methods of preparing future weather data for simulation. Zhu et al. (2016) proposes deriving future TMYs from the hourly output of the general circulation models rather than morphing existing TMYs based on future monthly statistics. For study of building design robustness, Struck et al. (2009) and Crawley and Lawrie (2015) recommend using multi-year simulation with projected weather data, as opposed to single years projected from current reference years. DWD and BBSR (Deutschen Wetterdienst 2014, 2018) have created current and future test reference year data for 14 locations in Germany. These include typical, summer, and winter variants, along with similar future data for 2035. Rastogi (2016), and Rastogi and Andersen (2016) describe methods for incorporating climate change in analysis of weather-based uncertainty. Nik (2016) proposes a method to synthesize typical and extreme weather data sets from regional climate model results.

Several other tools are available for modeling and creating future climate and urban heat island files: Dragonfly (Ladybug Tools 2018), Urban Weather Generator (Nakano et al. 2015), and WeatherShift (Arup 2018).

In recent years, multi-day record-breaking high and low temperature events occurred in many regions. This emphasizes the need to study building performance under possible peak conditions. The techniques cited above for generation of future weather sequences focus on matching general trends and may not produce realistic extreme sequences. Wood and Eames (2017) and Sillmann et al. (2017) review modeling and prediction of short- and long-duration extreme events. Until such research converges on standard methods, practitioners should consider including simulation studies using recent-year weather data or shifted data based on recent years.

Synopsis

As can be seen from the summary presented in this chapter, an extensive body of knowledge and technology is employed in deriving the outdoor conditions that are used to simulate building performance. Some techniques are built into end-user simulation software, while others are used during the preparation and distribution of data sets.

Most of the details discussed are not of immediate concern to the practitioner. The following summarizes the application guidance found throughout the chapter.

- When doing *typical hourly simulation* for design analysis or code compliance, selecting and using a representative typical year file is generally sufficient. The most common practical difficulty is that there is no weather station near the project site. In this case, a more distant station perhaps can be used (with care) or data synthesis/adjustment techniques applied. Resorting to adjustment is becoming less necessary as more files become available.
- *Solar and illuminance data* are sparsely observed and thus nearly all commonly available information is derived with models. Care must be used when assessing short-term modeling results for any system that relies on these inputs, including photovoltaic systems and daylighting strategies.
- *Net-zero projects* and other low-energy designs should not rely on typical year simulations. Multi-year studies are indicated to ensure that the full range of potentially difficult conditions is evaluated.
- *Urban heat island effects* can be significant. Adjustment of airport data or use of specialized models may be required for sites in large urban centers.
- *Wind speed and direction* are heavily modified by local conditions. Observations at exposed weather stations are unreliable for sheltered project sites, even if nearby. Natural ventilation designs should be simulated using at-site measurements and/or conservative synthetic data.
- *Climate change* is beginning to be seen in local and regional extreme climate events. These trends will continue – causing significantly changed conditions in 50 to 100 years; this should be considered for long-life projects.
- *Model calibration* and other work with measured performance generally require contemporaneous weather data observed at the project site. Typical year data are not appropriate.

Because of the interaction of actual environmental conditions and shortcomings in observations (notably solar radiation), weather data remain a large source of uncertainty in building simulation predictions. This uncertainty may be trending lower, due to improvements such as remote sensing, enhanced models, and more observation stations. However, the practitioner must remain cognizant of his or her application and how model results are being used.

References

Aguiar, R., Camelo, S. and Gonçalves, H. (1999) Assessing the value of typical meteorological years built from observed and from synthetic data for building thermal simulation. In *Proceedings of Building Simulation 1999*. Kyoto, Japan, pp. 627–634.

Alonso-Suárez, R., Bidegain, M., Abal, G. and Modernell, P. (2016) *Año Meteorológico Típico para Aplicaciones de Energía Solar (AMTUes): Series típicas horarias para 5 sitios del Uruguay*. Memoria Técnica del LES/UdelaR, versión 2.4, junio 2016. http://les.edu.uy/productos/amtues-2/.

AMTR (2018) *Amaterass Consortium*. www.amaterass.org/.

Arens, E.A., Flynn, L.E., Nall, D.N. and Ruberg, K. (1980) *Geographical Extrapolation of Typical Hourly Weather Data for Energy Calculation in Buildings*. National Bureau of Standards Building Science Series 126. Gaithersburg: NIST.

Arima, Y., Ookab, R. and Kikumoto, H. (2017) Proposal of typical and design weather year for building energy simulation. *Energy and Buildings* 139: 517–524.

Arup (2018) *WeatherShift*. www.weather-shift.com.

ASHRAE (1985) *Weather Year for Energy Calculations*. Atlanta, GA: ASHRAE, Inc.
ASHRAE (1997) *WYEC2 Weather Year for Energy Calculations 2, Toolkit and Data*. Atlanta, GA: ASHRAE, Inc.
ASHRAE (2001) *International Weather for Energy Calculations* (IWEC Weather Files) User's Manual and CD. Atlanta, GA: ASHRAE, Inc.
ASHRAE (2011) *International Weather for Energy Calculations, Version 2* (IWEC Weather Files), User's Manual and DVD. Atlanta, GA: ASHRAE, Inc.
ASHRAE (2016) *ASHRAE Standard 160–2016 Criteria for Moisture-Control Design Analysis in Buildings*. Atlanta, GA: ASHRAE, Inc.
ASHRAE (2017a) *ASHRAE Handbook – Fundamentals*. Atlanta, GA: ASHRAE, Inc.
ASHRAE (2017b) *Weather Data Viewer, Version 6.0* (WDView 6.0) (CD). Atlanta, GA: ASHRAE, Inc.
ASHRAE (2018) *ASHRAE Standard 209–2018 Energy Simulated Aided Design for Buildings Except Low-Rise Residential Buildings*. Atlanta, GA: ASHRAE, Inc.
Balcomb, D. (1999) WeatherMaker: Weather file conversion and evaluation. In *Proceedings of the National Passive Solar Conference*. Portland, Maine, pp. 689–695.
Barreira, E., Simões, M.L., Delgado, J.M.P.Q. and Sousa, I. (2017) Procedures in the construction of a test reference year for Porto-Portugal and implications for hygrothermal simulation. *Sustainable Cities and Society* 32: 397–410.
Baltazar, J.C. and Claridge, D.E. (2006) Study of cubic splines and Fourier series as interpolation techniques for filling in short periods of missing building energy use and weather data. *Journal of Solar Energy Engineering* 128(2): 226–230.
Belcher, S.E., Hacker, J.N. and Powell, D.S. (2005) Constructing design weather data for future climates. *Building Services Engineering Research and Technology* 26(1): 49–61.
Bhandari, M., Shrestha, S. and New, J. (2012) New evaluation of weather datasets for building energy simulation. *Energy and Buildings* 49: 109–118.
Big Ladder Software (2016) *Elements*. https://bigladdersoftware.com/projects/elements/.
Bre, F. and Fachinotti, V.D. (2016) Generation of typical meteorological years for the Argentine Littoral Region. *Energy and Buildings* 129: 432–444.
Budzyński, K., Narowski, P.G. and Czechowicz, J. (2004) *Przygotowanie zbiorów zagregowanych danych klimatycznych dla potrzeb obliczeń energetycznych budynków*. Warsawa: Ministerstwo Infrastruktury.
Bulut, H. (2004) Typical solar radiation year for Southeastern Anatolia. *Renewable Energy* 29(9): 1477–1488.
CEC (2016) *2016 ACM Supporting Content*. Sacramento: California Energy Commission. www.energy.ca.gov/title24/2016standards/ACM_Supporting_Content/.
CAMS (2017) *User's Guide to the CAMS Radiation Service, Copernicus Atmosphere Monitoring Service*. https://atmosphere.copernicus.eu/.
Carlo, J.C. and Lamberts, R. (2005) *Processamento de Arquivos Climáticos para Simulação do Desempenho Energético de Edificações*, para Eletrobrás/PROCEL, AET N 02/04, Laboratório de Eficiência Energética em Edificações, Universidade Federal de Santa Catarina, Centro Technológico, Departamento de Engenharia Civil, Florianópolis, SC, Brasil. www.labeee.ufsc.br/downloads/arquivos-climaticos/formato-try-swera-csv-bin.
Chan, A.L.S. (2016) Generation of typical meteorological years using genetic algorithm for different energy systems. *Renewable Energy* 90: 1–13.
China Meteorological Bureau, Climate Information Center, Climate Data Office and Tsinghua University, Department of Building Science and Technology. (2005) *China Standard Weather Data for Analyzing Building Thermal Conditions*. Beijing: China Building Industry Publishing House.
CIBSE (2016) *Test Reference Years and Design Summer Years*. London: The Chartered Institution of Building Services Engineers. www.cibse.org/knowledge/cibse-weather-data-sets.
Clarke, J.A. (2001) *Energy Simulation in Building Design*. Oxford: Butterworth-Heinemann.
Climate.OneBuilding (2019) *Typical Weather and Design Condition Data*. http://climate.onebuilding.org/.
Crawley, D.B. (1998) Which weather data should you use for energy simulations of commercial buildings? *ASHRAE Transactions*, 104(2): 498–515. Atlanta: ASHRAE.
Crawley, D.B. (2008a) Estimating the impacts of climate change and urbanization on building performance. *Journal of Building Performance Simulation* 1(2): 91–115.

Crawley, D.B. (2008b) Building performance simulation: A tool for policymaking. PhD Thesis. University of Strathclyde, Glasgow.

Crawley, D.B. and Lawrie, L.K. (2015) Rethinking the TMY: Is the 'Typical' Meteorological Year Best for Building Performance Simulation? In *Proceedings of Building Simulation 2015: 14th Conference of International Building Performance Simulation Association*, December 7–9, 2015, Hyderabad, India, pp. 2655–2662.

Crow, L.W. (1970) Summary Description of Typical Year Weather Data, Chicago Midway Airport. *ASHRAE Research Project RP-100*. Atlanta, GA: ASHRAE, Inc.

Crow, L.W. (1980) Development of hourly data for Weather Year for Energy Calculations (WYEC), including solar data, at 21 weather stations throughout the United States. *ASHRAE Transactions* 87(1).

Crow, L.W. (1983) Development of hourly data for weather year for energy calculations (WYEC), including solar data, at 29 stations throughout the United States and 5 stations in Canada. *ASHRAE Research Project RP-364 Final Report*, November 1983. Atlanta, GA: ASHRAE, Inc.

Cui, Y., Yan, D., Hong, T., Chan, X., Lauo, X. and Zhang, Q. (2017) Comparison of typical year and multiyear building simulations using a 55-year actual weather data set from China. *Applied Energy* 195: 890–904.

David, M., Adelard, L., Lauret, P. and Garde, F. (2010) A method to generate typical meteorological years from raw hourly climatic databases. *Building and Environment* 45: 1722–1732.

Davies, J.A. and McKay, D.C. (1989) Evaluation of selected models for estimating solar radiation on horizontal surfaces. *Solar Energy* 43(3): 153–168.

Degelman, L.O. (1991) A statistically-based hourly weather data generator for driving energy simulation and equipment design software for buildings. In *Proceedings of Building Simulation 1991*. Nice, Sophia-Antipolis, France, pp. 592–599.

Degelman, L.O. (2007) Testing the accuracy of synthetically-generated weather data for driving building simulation models. In *Proceedings of Building Simulation 2007*. Beijing, China, pp. 1–8.

Deutschen Wetterdienst. (2014) *Testreferenzjahre von Deutschland für mittlere, extreme und zukünftige Witte-rungsverhältnisse*. https://www.bbsr-energieeinsparung.de/EnEVPortal/DE/Regelungen/Testreferenzjahre/TRY_node.html

Deutschen Wetterdienst. (2018) *Ortsgenaue Testreferenzjahre von Deutschland für mittlere, extreme und zukünftige Witterungsverhältnisse*. https://www.bbsr.bund.de/BBSR/DE/FP/ZB/Auftragsforschung/5EnergieKlimaBauen/2013/testreferenzjahre/01-start.html

Duffie, J.A. and Beckman, W.A. (2006) *Solar Engineering of Thermal Processes, 3rd edition*. New York: Wiley.

Eames, M., Kershawm, T. and Coley, D. (2012) The appropriate spatial resolution of future weather files for building simulation. *Journal of Building Performance Simulation* 5(6): 347–358.

Energy Partners (2008) *Australian Climate Data Bank: Weather Data Enhancement for Reference Meteorological Years*. Manuka, ACT, Australia: Energy Partners. www.nathers.gov.au/nathers-accredited-software/nathers-climate-zones-and-weather-files.

Environment and Climate Change Canada (2016a) *Canadian Weather Energy and Engineering Datasets (CWEEDS)*. Downsview, ON: Meteorological Service of Canada, ECCC. http://climate.weather.gc.ca/prods_servs/engineering_e.html.

Environment and Climate Change Canada (2016b) *Canadian Weather Year for Energy Calculations (CWEC)*. Downsview, ON: Meteorological Service of Canada, ECCC. http://climate.weather.gc.ca/prods_servs/engineering_e.html.

Erell, E., Pearlmutter, D. and Williamson, T. (2011) *Urban Microclimate – Designing the Spaces Between Buildings*. London: Routledge.

Erell, E. and Williamson, T. (2006) Simulating air temperature in an urban street canyon in all weather conditions using measured data from a reference meteorological station. *International Journal of Climatology* 26(12): 1671–1694.

Gueymard, C.A. (1987) An anisotropic solar irradiance model for tilted surfaces and its comparison with selected engineering algorithms. *Solar Energy* 38(5): 367–386 and Erratum. *Solar Energy* 40(2): 175.

Gueymard, C.A. (2009) Direct and indirect uncertainties in the prediction of tilted irradiance for solar engineering applications. *Solar Energy* 83(3): 432–444.

Gueymard, C.A. and Ruiz-Arias, J.A. (2016) Extensive worldwide validation and climate sensitivity analysis of direct irradiance predictions from 1-min global irradiance. *Solar Energy* 128: 1–30.

Guan, L. (2009) Preparation of future weather data to study the impact of climate change on buildings. *Building and Environment* 44(4): 793–800.

Habte, A., Sengupta, M. and Lopez, A. (2017) *Evaluation of the National Solar Radiation Database (NSRDB): 1998–2015*. Golden, CO: National Renewable Energy Laboratory.

Hay, J.E. and Davies, J.A. (1978) Calculation of the solar radiation incident on an inclined surface. In *Proceedings of the First Canadian Solar Radiation Data Workshop*, Toronto, Canada, pp. 59–72.

Hensen, J.L.M. (1999) Simulation of building energy and indoor environmental quality – Some weather data issues. In *Proceedings of the International Workshop on Climate Data and Their Applications in Engineering*. Prague: Czech Hydrometeorological Institute.

Hong, T., Chang, W.-K., Lin, H.-W. (2013) A sensitivity study of building performance using 30-year actual weather data In *Proceedings of Building Simulation 2013*, 13th Conference of International Building Performance Simulation Association. Chambéry, France, August 26–28, 2013, pp. 1398–1405.

Howard, L. (1833) *Climate of London Deduced from Meteorological Observations, 3rd edition*. London: Harvey & Darton.

Huang, J. (2011) *Development of 3,012 Typical Year Weather Files for International Locations*. ASHRAE Research Project 1477-RP Final Report. Atlanta, GA: ASHRAE.

Huang, J. (2015) *Improving the Reporting of Liquid Precipitation in the TMY3 Weather Files*. Moraga, CA: White Box Technologies.

IMGW (2009) *Typowe lata meteorologiczne i statystyczne dane klimatyczne do obliczeń energetycznych budynków*. www.miir.gov.pl/strony/zadania/budownictwo/charakterystyka-energetyczna-budynkow/dane-do-obliczen-energetycznych-budynkow-1/ (in Polish).

IMS (2009) Israeli Meteorological Service.

IPCC (2013) *Climate Change 2013, The Physical Science Basis*. Intergovernmental Panel on Climate Change. Cambridge: Cambridge University Press.

IPCC (2014a) *Climate Change 2014, Impacts, Adaptation and Vulnerability*. Intergovernmental Panel on Climate Change. Cambridge: Cambridge University Press.

IPCC (2014b) *Climate Change 2014: Synthesis Report*. Intergovernmental Panel on Climate Change. Cambridge: Cambridge University Press.

ISHRAE (2009) *Indian Weather Data*. New Delhi: ISHRAE.

ISHRAE (2014) *Indian Weather Data*. New Delhi: ISHRAE. www.ishrae.in.

ISO (2005) *ISO 15927–4:2005, Hygrothermal Performance of Buildings – Calculation and Presentation of Climatic Data – Part 4: Hourly Data for Assessing the Annual Energy Use for Heating and Cooling*. International Standards Organization. Geneva, Switzerland: ISO. www.iso.org/standard/41371.html.

Jentsch, M.F., Bahaj, S. and James, P.A.B. (2008) Climate change future proofing of buildings: Generation and assessment of building simulation weather files. *Energy and Buildings* 40(12): 2148–2168.

Jentsch, M.F., James, P.A.B., Bourikas, L. and Bahaj, A.S. (2013) Transforming existing weather data for worldwide locations to enable energy and building performance simulation under future climates. *Renewable Energy* 55: 514–524.

Jones, N.L., Greenberg, D.P. and Pratt, K.P. (2012) Fast computer graphics techniques for calculating direct solar radiation on complex building surfaces. *Journal of Building Performance Simulation* 5(5): 300–312.

Kalamees, T., Jylhä, K., Tietäväinen, H., Jokisalo, J., Ilomets, S., Hyvönen, R. and Saku, S. (2012) Development of weighting factors for climate variables for selecting the energy reference year according to the EN ISO 15927–4 standard. *Energy and Buildings* 47: 53–60.

Kalogirou, S.A. (2003) Generation of typical meteorological year (TMY-2) for Nicosia, Cyprus. *Renewable Energy* 28(15): 2317–2334.

Kim, S., Zirkelbach, D., Künzel, H.M., Lee, J.-H. and Choi, J. (2017) Development of test reference year using ISO 15927–4 and the influence of climatic parameters on building energy performance. *Building and Environment* 114: 374–386.

Klein, S.A. and Reindl, D.T. (1998) *Automated Generation of Hourly Design Sequences*, ASHRAE Research Project RP-962 Final Report. Atlanta, GA: ASHRAE, Inc.

Knight, K.M., Klein, S.A. and Duffie, J.A. (1991) A methodology for the synthesis of hourly weather data. *Solar Energy* 46(2): 109–120.

Kusuda, T. and Achenbach, P.R. (1965) *Earth Temperature and Thermal Diffusivity at Selected Stations in the United States*, National Bureau of Standards Report 8972. Gaithersburg: NIST.
Ladybug Tools (2018) *Ladybug and Dragonfly*. www.ladybug.tools/.
Lee, K., Yoo, H. and Levermore, G.J. (2010) Generation of typical weather data using the ISO Test Reference Year (TRY) method for major cities of South Korea. *Building and Environment* 45(4): 956–963.
Liley, J.B., Shiona, H., Sturman, J. and Wratt, D.S. (2007) *Typical Meteorological Years for the New Zealand Home Energy Rating Scheme*. Prepared for the Energy Efficiency and Conservation Authority. NIWA Client Report: LAU2007–02-JBL. Omakau, New Zealand: NIWA.
Lily, J.B. (2013) *Australian Climate Data and Reference Meteorological Years for NatHERS 2012*. Prepared for Australian Federal Government Department of Resources, Energy and Tourism, NIWA Client Report: LAU2013–03-JBL. Omakau. New Zealand: NIWA.
Lund, H. (1991) The design reference year. In *Proceedings of Building Simulation 1991*. Sophia Antipolis, France, pp. 600–606.
Marsh, A. (2018) *Sunpath* and *Daylight-Box*. http://andrewmarsh.com/apps/staging/sunpath3d.html, http://andrewmarsh.com/apps/staging/daylight-box.html.
Mesonet (2018) *Oklahoma Mesonet*. www.mesonet.org.
Meteotest (2018) *Meteonorm 7.3, A Global Meteorological Database for Engineers, Planners, and Education*. Bern, Switzerland: Meteotest. www.meteonorm.com.
Mitchell, J.M. (1953) On the causes of instrumentally observed secular temperature trends. *Meteorology* 10(4): 244–261.
Mitchell, J.M. (1961) The temperature of cities. *Weatherwise* 14: 224–229.
Moazami, A., Carlucci, S. and Geving, S. (2017) Critical analysis of software tools aimed at generating future weather files with a view to their use in building performance simulation. *Energy Procedia* 132: 640–645.
Morris, C.J.G. and Simmonds, I. (2000) Associations between varying magnitudes of the urban heat island and the synoptic climatology in Melbourne, Australia. *International Journal of Climatology* 20(15): 1931–1954.
Muneer, T. (2004) *Solar Radiation and Daylight Models, 2nd edition*. Oxford: Butterworth-Heinemann.
Murano, G., Dirutigliano, D. and Corrado, V. (2018) Improved procedure for the construction of a typical meteorological year for assessing the energy need of a residential building. *Journal of Building Performance Simulation* https://doi.org/10.1080/19401493.2018.1479774.
Murphy, S. (2017) The construction of a modified Typical Meteorological Year for photovoltaic modeling in India. *Renewable Energy* 111: 447–454.
Myers, D.R., Wilcox, S., Marion, W., George, R. and Anderberg, M. (2005) Broadband model performance for an updated national solar radiation database in the USA. *Presented at ISES 2005 Solar World Congress*. Orlando, FL. www.nrel.gov/docs/fy05osti/37699.pdf.
Mylona, A. (2012) The use of UKCP09 to produce weather files for building simulation. *Building Services Engineering Research and Technology* 33(1): 51–62.
Nakano, A., Bueno, B., Norford, L. and Reinhart, C.F. (2015) Urban weather generator – A novel workflow for integrating urban heat island effect within urban design process. In *Proceedings of Building Simulation 2015, 14th Conference of International Building Performance Simulation Association*. Hyderabad, India, December 7–9, 2015, pp. 1901–1908.
Narowski, P., Janicki, M. and Heim, D. (2013) Comparison of untypical meteorological years (UMY) and their influence on building energy performance simulations. In *Proceedings of Building Simulation 2013, 13th Conference of International Building Performance Simulation Association*. Chambéry, France, August 26–28, 2013, pp. 1414–1421.
NASA GMAO (2018) *Modern-era Retrospective analysis for Research and Applications, Version 2 (MERRA-2)*. Greenbelt, MD: NASA. https://gmao.gsfc.nasa.gov/reanalysis/MERRA-2/.
NARR (2018) *North American Regional Reanalysis/Environmental Modeling Center*. www.ncdc.noaa.gov/data-access/model-data/model-datasets/north-american-regional-reanalysis-narr.
NCAR (2018a) *The Weather Research and Forecasting Model (WRF)*. Boulder, CO: National Center for Atmospheric Research. www.mmm.ucar.edu/weather-research-and-forecasting-model.
NCAR (2018b) *Atmospheric Reanalysis: Overview & Comparison Tables*. Boulder, CO: National Center for Atmospheric Research. https://climatedataguide.ucar.edu/climate-data/atmospheric-reanalysis-overview-comparison-tables.

NCDC (1976) *Test Reference Year (TRY)*, Tape Reference Manual, TD-9706, September 1976. Asheville. North Carolina: National Climatic Data Center, U.S. Department of Commerce.
NCDC (1981) *Typical Meteorological Year User's Manual*, TD-9734, Hourly Solar Radiation – Surface Meteorological Observations, May 1981. Asheville, NC: National Climatic Data Center, U.S. Department of Commerce.
NCDC (1993) *Solar and Meteorological Surface Observation Network*, 1961–1990, Version 1.0, September 1993. Asheville, NC: National Climatic Data Center, U.S. Department of Commerce.
NCEI (2018) *Integrated Surface Database (ISD)*. Asheville, NC: National Centers for Environmental Information, U. S. Department of Commerce. www.ncdc.noaa.gov/isd.
Nik, V.M. (2016) Making energy simulation easier for future climate – Synthesizing typical and extreme weather data sets out of regional climate models (RCMs). *Applied Energy* 177: 204–226.
NREL (1995) *User's Manual for TMY2s (Typical Meteorological Years), NREL/SP-463–7668, and TMY2s, Typical Meteorological Years Derived from the 1961–1990 National Solar Radiation Data Base*, June 1995, CD-ROM. Golden, CO: National Renewable Energy Laboratory.
NREL (2017) *DView*, Version 3.1, July 2017.Golden, CO: National Renewable Energy Laboratory. https://beopt.nrel.gov/downloadDView.
NREL (2018) *National Solar Radiation Database*. https://nsrdb.nrel.gov/.
Numerical Logics (1999) *Canadian Weather for Energy Calculations*, Users Manual and CD-ROM. Downsview, ON: Environment Canada.
Oke, T.R. (1973) City size and the urban heat island. *Atmospheric Environment* 7(8): 769–779.
Oke, T.R. (1988) *Boundary Layer Climates*. New York: Routledge.
Oke, T.R. (2006) *Initial Guidance to Obtain Representative Meteorological Observations at Urban Sites*. IOM Report No.81, WMO/TD. No. 1250. Geneva, Switzerland: World Meteorological Organization.
Oko, Chika Ogbonna Chima and Ogoloma, O.B. (2011) Generation of a typical meteorological year for Port Harcourt zone. *Journal of Engineering Science and Technology* 6(2): 204–214.
PHIKO (2018) *PHIKO Test Reference Year*. Seoul, Korea: Passive House Institute Korea.
Perez, R., Ineichen, P., Seals, R., Michalsky, J. and Stewart, R. (1990) Modeling daylight availability and irradiance components from direct and global irradiance. *Solar Energy* 44(5): 271–289.
Perez, R., Kmiecik, M., Moore, K., Wilcox, S., George, R., Renne, D., Vignola, F. and Ineichen, P. (2004) Status of high resolution solar irradiance mapping from satellite data. In *Proceedings of the ASES Annual Meeting 2004*. Portland, OR, pp. 265–270.
Pernigotto, G., Prada, A., Baggio, P., Gasparella, A. and Mahdavi, A. (2015) Impact of solar irradiation models on simulated hourly energy performance of buildings. In *Proceedings of Building Simulation 2015*. Hyderabad, India, pp. 955–962.
Pernigotto, G., Prada, A., Gasparella, A. and Hensen, J.L.M. (2014a) Analysis and improvement of the representativeness of EN ISO 15927–4 reference years for building energy simulation *Journal of Building Performance Simulation* 7(6): 391–410.
Pernigotto, G., Prada, A., Cóstola, D., Gasparella, A. and Hensen, J.L.M. (2014b). Multi-year and reference year weather data for building energy labelling in north Italy climates. In *Energy and Buildings* 72: 62–72. CTIACT now incorporates 110 locations throughout Italy https://try.cti2000.it/.
Pusat, S., Ekmekçi, Ý. and Akkoyunlu, M.T. (2015) Generation of typical meteorological year for different climates of Turkey. *Renewable Energy* 75 (March 2015): 144–151.
Qiu, X., Yang, F., Corbett-Hains, H. and Roth, M. (2015) *Procedure to Adjust Observed Climatic Data for Regional or Mesoscale Climatic Variations*, ASHRAE Research Project 1561-RP Final Report. Atlanta, GA: ASHRAE, Inc.
Rahman, I.A. and Dewsbury, J. (2007) Selection of typical weather data (test reference years) for Subang, Malaysia. *Building and Environment* 42(10): 3636–3641.
Rastogi, P. (2016) On the sensitivity of buildings to climate: The interaction of weather and building envelopes in determining future building energy consumption. PhD Thesis, Lausanne, Switzerland: Ecole Polytechnique Fédérale de Lausanne.
Rastogi, P. and Andersen, M. (2015) Embedding stochasticity in building simulation through synthetic weather files. In *Proceedings of Building Simulation 2015*. Hyderabad, India, pp. 963–970.
Rastogi, P. and Andersen, M. (2016) Incorporating Climate Change Predictions in the Analysis of Weather-Based Uncertainty. In *Proceedings of SimBuild 2016*. Salt Lake City, Utah, pp. 181–188.

Roriz, M. (2012) *Arquivos Climáticos de Municípios Brasileiros*, Associação, São Carlos, SP, Brasil: Nacional de Tecnologia do Ambiente Construído. www.roriz.eng.br/epw_9.html.

Roth, M. (2016) *Updating the ASHRAE Climate Design Data for 2017*, ASHRAE Research Project 1699-RP Final Report. Atlanta, GA: ASHRAE, Inc.

Roth, M. (2019) *Evaluation of Climate Reanalysis Data for Use in ASHRAE Applications*. ASHRAE Research Project 1745-RP Final Report. Atlanta, GA: ASHRAE, Inc.

Runnalls, K.E. and Oke, T.R. (2000) Dynamics and controls of the near-surface heat island of Vancouver, British Columbia. *Physical Geography* 21(4): 283–304.

Schneider dos Santos, R., Taylor, J., Davies, M., Mavrogianni, A. and Symonds, P. (2016) Modelling and monitoring tools to evaluate the urban heat island's contribution to the risk of indoor overheating. In *Proceedings of the 3rd IBPSA-England Conference BSO 2016*. Great North Museum, Newcastle, UK, September 12th–14th, 2016.

Seo, D., Huang, J. and Krarti, M. (2009) Evaluation of typical weather year selection approaches for energy analysis of buildings. *ASHRAE Transactions* 115(2): 654–667.

Short, C.A., Cook, M.J. and Woods, A. (2009) Low energy ventilation and cooling within an urban heat island. *Renewable Energy* 34: 2022–2029.

Sillmann, J., Thorarinsdottir, T., Keenlyside, N., Schaller, N., Alexander, L.V., Hegerle, G., Seneviratne, S.I., Vautard, R., Zhang, X. and Zwiers, F.W. (2017) Understanding, modeling and predicting weather and climate extremes: Challenges and opportunities. *Weather and Climate Extremes* 18: 65–74.

Solargis (2018) *Solargis*. https://solargis.info.

Spitler, J.D. (2009) *Load Calculation Applications Manual*. Atlanta: ASHRAE.

Stackhouse, P.W., Whitlock, C.H., Chandler, W.S., Hoell, J.M., Zhang, T., Mikovitz, J.C., Leng, G.S. and Lilienthal, P. (2006) *Supporting Energy-Related Societal Applications Using NASA's Satellite and Modeling Data*. http://ntrs.nasa.gov.

Stackhouse, P.W., Zhang, T., Westberg, D., Barnett, A.J., Bristow, T., Macpherson, B. and Hoell, J.M. (2018) *POWER Release 8 (with GIS Applications) Methodology* (Data Parameters, Sources, & Validation), Langley, VA: NASA. https://power.larc.nasa.gov/documents/POWER_Data_v8_methodology.pdf.

Stoffel, T.L. and Rymes, M.D. (1998) Production of the weather year for energy calculations version 2 (WYEC2) data files. *ASHRAE Transactions* 104(2): 498–497.

Streuker, D.R. (2003) A study of the urban heat island of Houston, Texas. PhD Thesis. Houston: Rice University.

Struck, S., de Wilde, P., Evers, J., Hensen, J. and Plokker, W. (2009) On selecting weather data sets to estimate a building design's robustness to climate change. In *Proceedings of Building Simulation 2009*. Glasgow, Scotland, pp. 513–520.

Sustainable Energy Research Group (2013) *CCWorldWeatherGen*, Southampton: Sustainable Energy Research Group, University of Southampton. www.energy.soton.ac.uk/ccworldweathergen/.

SWERA (2009) *Solar and Wind Energy Resource Assessment*. https://openei.org/wiki/Solar_and_Wind_Energy_Resource_Assessment_(SWERA).

UCLA (2017) *Energy Design Tools*. www.energy-design-tools.aud.ucla.edu/.

USDOE (2018a) *Engineering Reference, EnergyPlus Version 8.9.0 Documentation*. Washington, DC: U.S. Department of Energy.https://energyplus.net/sites/all/modules/custom/nrel_custom/pdfs/pdfs_v8.9.0/EngineeringReference.pdf.

USDOE (2018b) *Auxiliary Programs, EnergyPlus Version 8.9.0 Documentation*. Washington, DC: U.S. Department of Energy. https://energyplus.net/sites/all/modules/custom/nrel_custom/pdfs/pdfs_v8.9.0/AuxiliaryPrograms.pdf.

USDOE (2018c) *Weather Data*. Washington, DC: U.S. Department of Energy. https://energyplus.net/weather.

USEPA (2007) *Heat Island Effect*. Washington, DC: U.S. Environmental Protection Agency. www.epa.gov/heatisland/.

Van der Heijden, M.G.M, Blocken, B. and Hensen, J.L.M. (2013) Towards the integration of the urban heat island in building energy simulations. In *Proceedings of Building Simulation 2013*, 13th Conference of International Building Performance Simulation Association. Chambéry, France, August 26–28, 2013, pp. 1006–1013.

Watkins, R., Levermore, G.J. and Parkinson, J.B. (2011) Constructing a future weather file for use in building simulation using UKCP09 projections. *Building Services Engineering Research and Technology* 32(3): 293–299.

White Box Technologies (2018) *Weather Data for Energy Calculations*. http://weather.whiteboxtechnologies.com/.

Wilcox, S. and Marion, W. (2008) *User's Manual for TMY3 Data Sets*, NREL/TP-581–43156. April 2008. Golden, CO: National Renewable Energy Laboratory. http://rredc.nrel.gov/solar/old_data/nsrdb/1991-2005/tmy3/.

Williamson, T., Erell, J.E. and Soebarto, V. (2009) Assessing the error from failure to account for urban microclimate in computer simulation of building energy performance. In *Proceedings of Building Simulation 2009*, Glasgow, Scotland, pp. 597–504.

Wood, M. and Eames, M. (2017) Efficient Summertime Overheating Analysis Using Decomposed Weather Files. In *Proceedings of Building Simulation 2017*. San Francisco, CA, pp. 355–362.

Zhang, Q. and Huang, J. (2004) *Chinese Typical Year Weather Data for Architectural Use* (in Chinese). Beijing: China Machine Press.

Zhao, Q., Sailor, D.J. and Wentz, E. (2018) Impact of tree locations and arrangements on outdoor microclimates and human thermal comfort in an urban residential environment. *Urban Forestry and Urban Greening* 32: 81–91.

Zhou, J., Yezheng, W. and Gang, Y. (2006) Generation of typical solar radiation year for China. *Renewable Energy* 31(12): 1972–1985.

Zhu, M., Pan, Y., Huang, Z. and Xu, P. (2016) An alternative method to predict future weather data for building energy demand simulation under global climate change. *Energy and Buildings* 113: 74–86.

Recommended reading

Chapter 14 Climatic Data, *ASHRAE 2017 Handbook – Fundamentals* (ASHRAE 2017a). www.ashrae.org.

Integrated Surface Database web site (NCEI 2018). www.ncdc.noaa.gov/isd.

IPCC 5th Assessment Reports (IPCC 2013, 2014a, 2014b). www.ipcc.ch/report/ar5/

ISO Standard 15927–4:2005. *Hygrothermal Performance of Buildings – Calculation and Presentation of Climatic Data – Part 4: Hourly Data for Assessing the Annual Energy Use for heating and Cooling* (ISO 2005). www.iso.org/standard/41371.html.

National Solar Radiation Database (NREL 2018). https://nsrdb.nrel.gov/.

User's Manual for TMY3 Data Sets (Wilcox and Marion 2008). http://rredc.nrel.gov/solar/old_data/nsrdb/1991-2005/tmy3/.

Activities

1 Download a weather file for a location near where you live. Open it in a spreadsheet application and familiarize yourself with the data format.

2 Make a weather file suitable for use with the simulation program of your choice.

3 Install DView (NREL 2017) and examine weather data you have downloaded or assembled. DView allows you to visualize solar radiation, wind, temperatures, relative humidity, pressure, and snow depth (where available). Compare a few hourly variables and look at heatmap and monthly profiles.

4 Do a climate and site analysis as described per ASHRAE Standard 209 (ASHRAE 2018).

7 Daylight performance predictions

Christoph Reinhart

Scope

This chapter provides an overview of the different steps required to use daylight simulations for building design. Following a brief introduction to the theory of daylight simulations including common simulation algorithms, sky models, and material descriptions, the debate turns towards the question of *what* to calculate, i.e. what daylighting performance metrics to use, how to address building occupant behavior and expectations, and how to interpret simulation results in order to help design teams make more informed design decisions. This chapter, however, does *not* provide a general introduction to the larger theory of daylighting, e.g. how to develop an overall building massing or how to design a static shading system.

Learning objectives

This chapter is concerned with the assessment of the luminous environment in daylit spaces via simulation. The reader will learn about the basic algorithms that are commonly used to model daylight and will be exposed to current and emerging metrics to evaluate the performance of daylit spaces.

Key words

Daylight simulations/daylighting metrics/energy savings from lighting controls/glare analysis

Introduction

Daylighting describes the use of natural light in buildings. The objectives for daylighting a building are manifold and range from questions of aesthetics and occupant health and comfort to energy savings for lighting and space conditioning. Daylight simulations can help design teams address these multiple aspects by allowing them to (a) predict the amount of light available inside or outside of buildings under selected sky conditions or over the course of a whole year and (b) interpret the results by converting them into meaningful performance metrics. Daylight simulations can be combined with simulations of electric light sources to ensure that both forms of space lighting complement each other. They can also be embedded within an integrated lighting-thermal simulation in order to assess the overall effect of a daylighting strategy on energy use for lighting, heating, and cooling (Winkelmann and Selkowitz 1985; Janak 1997; Reinhart and Wienold 2011). Within recent years, it has become increasingly evident that the success of a daylighting strategy in real buildings is closely related to how exactly the electric lighting *and* shading systems are being controlled, e.g. manually, automatically, or automatically with manual override. Consequently, a new branch of building performance simulation is emerging that specifically focuses on developing behavior models that mimic occupant use of personal controls in buildings, such as light switches, venetian blinds, thermostats, and operable windows (Newsham 1994; Reinhart 2004; Bourgeois et al. 2006; Mahdavi et al. 2008; Hoes et al. 2009; Gunay et al. 2014) (see also Chapter 4).

Before diving right into the theory of daylight simulations, it is worthwhile to explore the larger context within which these simulations are typically being carried out. Daylighting is an established design component within any building that claims to be 'sustainable' and/or 'green' and that houses occupants. The reason for the strong association of daylighting with sustainable design practice is that its benefits are well aligned with the larger goals of the green building movement, namely promoting resource efficiency of and improving occupant well-being in buildings. According to two recent surveys, daylight simulations are one among several 'tools' that are routinely being used to design for daylight in buildings (Reinhart and Fitz 2006; Galasiu and Reinhart 2007). Other popular tools are 'experience from previous work,' 'rules of thumb,' and (to a lesser degree) 'scale models.' Figure 7.1 shows that in a recent survey of close to 200 design practitioners with an interest in sustainable design, both 'designers' and 'engineers' rated daylight simulations as their most important design tool during design development, along with 'experience from previous work' in the case of the designers. While this is an impressive statistic for daylight simulations, the reader should note that the target survey audience were practitioners with an explicit interest in sustainable design (Galasiu and Reinhart 2007). The market penetration of daylight simulations in the design community at large is probably significantly lower at this point. Nevertheless, the number of design practitioners in North America using the US Green Building Council's (USGBC) LEED rating system is continuously growing: As of 2017, over 200,000 individuals had earned LEED credentials, such as being a 'LEED accredited professional,' and over 92,000 projects had used or were in the process of using LEED for certification (US Green Building Council 2017). As a complementary

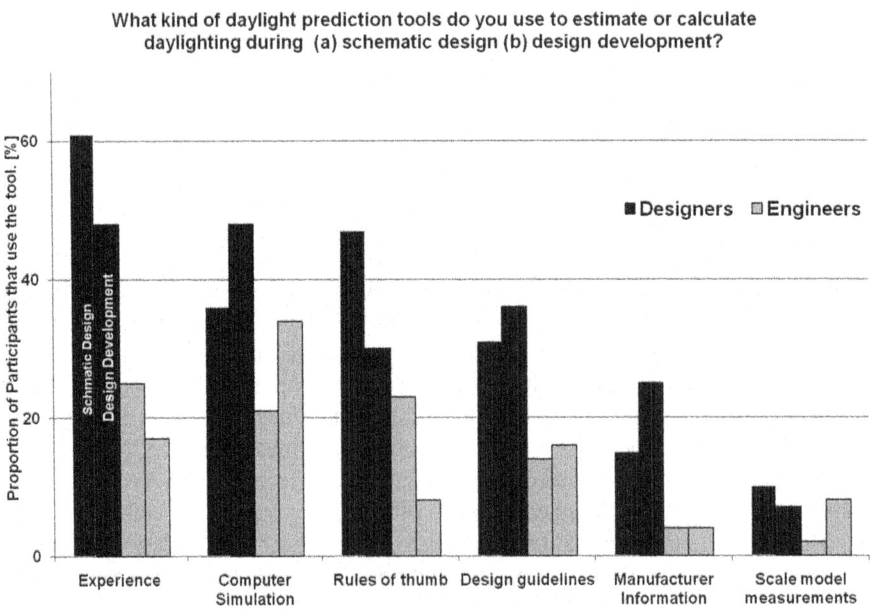

Figure 7.1 Results from a recent survey on the role of daylighting in sustainable design

Source: Galasiu and Reinhart 2007

effort, the American Institute of Architects (AIA) published specific energy modeling guidelines encouraging their members to integrate computer-based performance analysis into their design workflow (Amercian Institute of Architects 2012). Given that LEED requires/promotes the use of building performance simulations for energy and daylighting, the use of simulations is destined to continue to rise.

Apart from the earlier described 'push' towards the use of daylight simulations through the green building movement, the ubiquitous proliferation of three-dimensional computer-aided design (CAD) and building information modeling (BIM) software in building design practice and education is another factor that is helping to promote the use of daylight simulations. This is due to the fortuitous fact that the kind of three-dimensional building models that are commonly used for design presentation already contain a large amount of geometrical detail that can be reused for a computer-based daylighting analysis. These geometric models hence constitute a 'free' resource that greatly reduces the additional effort and costs required to carry out a daylighting analysis. Another factor that contributes to the increased use of environmental performance simulations by designers is that a number of daylighting and thermal analysis tools have been directly linked to designers' native working environments such a Rhinoceros 3D or SketchUp (Jakubiec and Reinhart 2011a; Guglielmetti et al. 2011; Sadeghipour et al. 2013). At this point, one could even argue that as certain analysis workflows – including shading studies and radiation maps (Figure 7.4) as well as daylight availability metrics (Figures 7.19 to 7.21) – have become so 'easy' to conduct, more attention has to be paid to educating the users of daylight simulation tools on how to interpret and use their results.

For the remainder of this chapter, it is assumed that the reader has decided to use daylight simulations for building design. The chapter accordingly provides an overview of the different steps required to do so. Following a brief introduction to the theory of daylight simulations, including common simulation algorithms, sky models, and material descriptions, the debate turns towards the question of *what* to calculate, i.e. what daylighting performance metrics to use, how to address building occupant behavior and expectations, and how to interpret simulation results in order to help design teams make more informed design decisions. In the spirit of the overall book, this chapter is based on the premise that an informed design decision is a better decision.

This chapter does *not* provide a general introduction to the larger theory of daylighting, e.g. how to develop an overall building massing or how to design a static shading system. To learn more about those topics, the reader may consult the "Recommended reading" section at the end of the chapter. While this chapter presents general concepts and avoids focusing on the limitations and capabilities of any one particular daylighting analysis tool, it should be noted that all of the examples presented are based on simulations using the Radiance light backward raytracer (Ward and Shakespeare 1998), a rigorously validated and flexible open source engine that is used by practitioners and researchers worldwide.

Computational models

This section deals with the various elements needed to simulate daylighting levels within a scene. While the study's scope may vary in scale from a façade detail to a whole neighborhood, the underlying simulation algorithms will often be identical. Figure 7.2 shows the basic elements required for a daylight performance simulation:

- The *scene* consists of a three-dimensional geometric model of the investigated daylit object(s), including optical material descriptions for all surfaces in the scene.
- The *sky model* quantifies the amount of direct sunlight and diffuse daylight coming from the different parts of the celestial hemisphere.

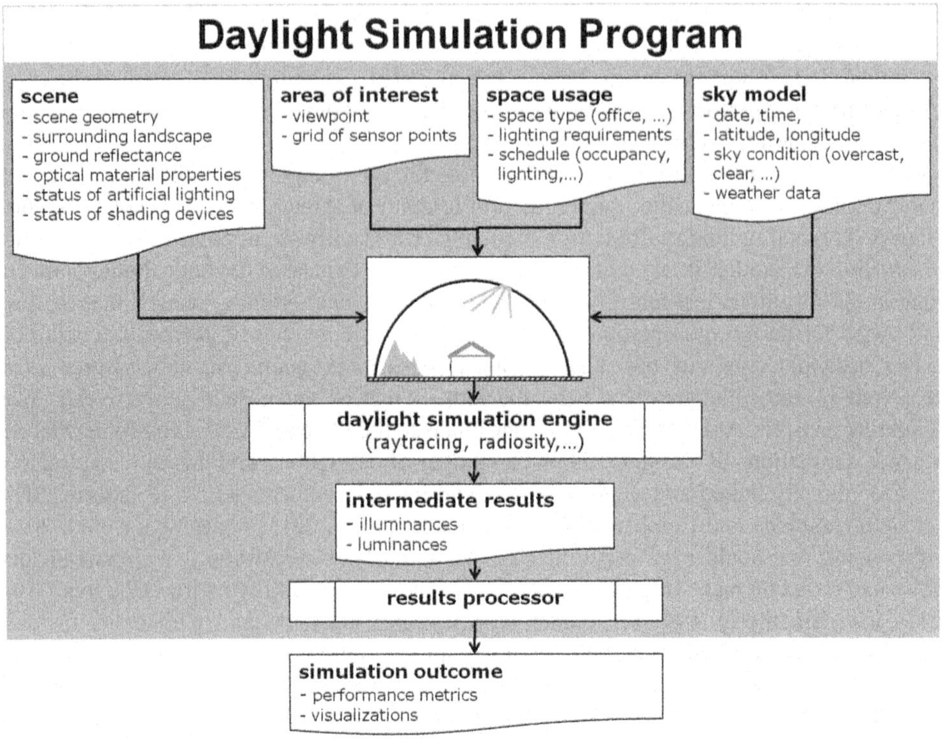

Figure 7.2 Elements needed for a daylight simulation

- *Areas of interest* in the scene can be select viewpoints and/or discrete sensors such as a grid of upward-facing illuminance sensors.
- *Space usage* information describes the type of space investigated (office, classroom, etc.), required lighting levels, and occupancy schedules.
- The *daylight simulation engine* combines the sky model with the scene and calculates illuminances and/or luminances within the scene.
- The *results processor* translates the raw simulation results into a format that can directly inform design decisions. Example formats are scene visualizations and false color maps of daylight availability metrics such as daylight factor and/or spatial daylight autonomy. These metrics will be discussed later.

While substantial amounts of time and resources in a daylighting analysis typically go towards building a scene and running the daylight simulations, the author would like to stress that the results processor is the most important, albeit frequently hastily performed, step within a daylight performance simulation. A daylighting modeler should, of course, pay close attention to working with a validated simulation engine and carefully review the scene model and simulation parameters. However, getting the lighting simulation right is just a prerequisite for a successful daylighting analysis. What determines *whether* the analysis is going to influence the design outcome is *what* metrics are being calculated. Following Steven Covey's second habit, to "begin with the end in mind" (Covey 1989), is particularly good advice for daylight simulationists, simply because the use of different daylighting metrics may lead to different, if not opposing, design conclusions.

Building the scene

A three-dimensional model is an obvious requirement for any daylight simulation. For simulation newcomers, however, it is usually less clear how detailed a daylighting model has to be at the different design phases to yield meaningful results and how the optical surface properties of all scene objects should be set.

The level of geometric complexity required for a daylight simulation depends on what is being calculated. From the earliest design stages onwards, it is advisable to model a site's immediate surroundings such as neighboring buildings and landscape that may affect the daylight availability on site. It is also good practice to explicitly model the ground plane and assign it a material that reflects onsite conditions rather than relying on the simulation program to treat ground reflectances automatically (Reinhart and Walkenhorst 2001). As shown in Figure 7.3, it generally suffices to represent buildings as simple blocks. While for direct shading studies the only required material distinction is whether a surface is transparent or opaque, it is generally good practice to make sure that all surfaces within a scene have been assigned meaningful optical properties. At this point, the reader should take note that even though a number of daylighting analysis programs currently come with default material libraries, the materials in these libraries are not necessarily populated with realistic values. The author therefore recommends that the reader carefully review the default materials used by his or her program of choice and – if necessary – prepare a custom library with common materials. Table 7.1 provides a list of common material descriptions. These materials should suffice for early design explorations. More opaque material definitions based on measurements in real buildings can be found on the *Lighting Materials for Simulation* website by Singapore University for Technology and Design (Jakubiec 2016). As a design evolves, product specific databases such as the *International Glazing Database* from the US National Fenestration Rating Council (LBNL 2017) provide optical surface properties of most commonly used glazing products in the US. The *Window 6.3 Research Version* further provides additional data on the optical properties of more Complex

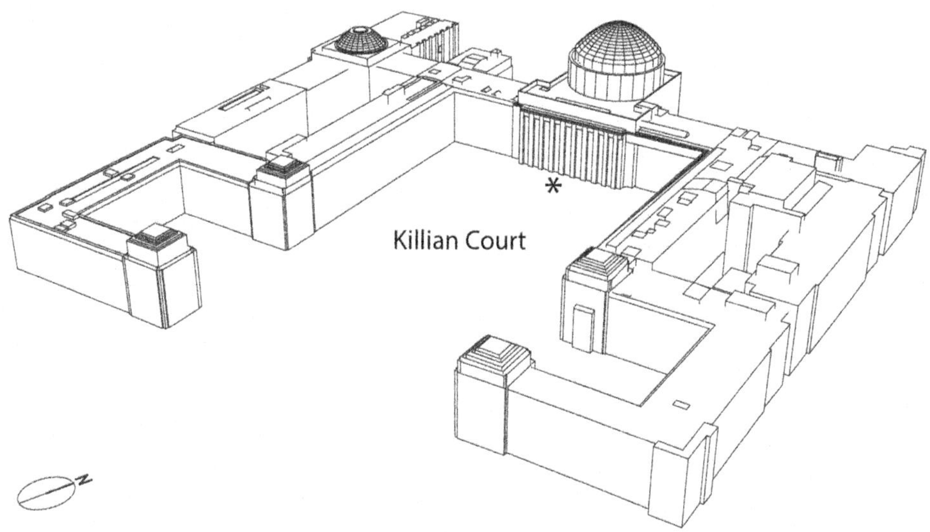

Figure 7.3 Massing model of select MIT buildings around Killian Court; * marks the position of the observer in Figure 7.7

Table 7.1 Optical properties of common material surfaces

Interior floor	20% diffuse reflectance[1]
Interior wall	50% diffuse reflectance[1]
Interior ceiling	80% diffuse reflectance[1]
Interior ceiling (high reflectance)	90% diffuse reflectance
Double glazing	72% direct visual transmittance
Single glazing	90% direct visual transmittance
Translucent glazing	20% diffuse-diffuse hemispherical transmittance[2]
Exterior building surfaces	44% diffuse reflectance[3]
Exterior ground	20% diffuse reflectance

[1] IESNA Handbook (IESNA 2011)
[2] Reinhart and Andersen 2006
[3] Leder et al. 2007

Fenestration System (CFS), such as window glazings combined with venetian blinds (LBNL 2017). A number of daylight simulation programs further allow the user to enter either measured or simulated bidirectional transmission and reflection distribution functions (BTDFs and BRDFs) for arbitrary CFSs. BTDFs/BRDFs describe the amount of light coming from a certain direction (characterized by two incident angles) that is transmitted through or reflected off a CFS in a certain direction and characterized by two outgoing angles. BTDFs/BRDFs are hence four-dimensional functions. The optical device that is used to measure or simulate these functions is called a goniophotometer (Andersen and deBoer 2006).

It is worth noting that a façade surface with a mostly diffuse reflectance corresponds to, for example, a brick wall with a moderate window-to-wall ratio. In contrast, a fully glazed curtain wall of a neighboring building may have a considerable specular reflective component that may lead to glare from direct solar reflections. Depending on the type of daylighting analysis at hand, it can therefore become necessary to explicitly model a specular neighboring façade that could have a decisive impact on the visual comfort conditions at the building site itself.[1] Another, often overlooked fact is that the surface property of the ground in immediate vicinity to a vertical window has a strong impact on the amount and quality of daylighting entering though the window, especially for interior points that do not 'see' the celestial hemisphere, such as the back of a room and/or a ceiling-mounted photocell-controlled dimming system.

Same as the design of a building evolves throughout the various stages of design, so should its daylighting model. During the initial programming and massing stage, simple block models suffice, e.g. for direct shading studies (shadow analysis onto the building envelope) and incident daily or seasonal irradiation patterns (Figure 7.4). At this stage, exposure of different façade sections to the celestial hemisphere can further be used to test the 'daylight feasibility' for different parts of the massing model (Reinhart and LoVerso 2010).

As the design analysis moves 'into' the building, façade openings, interior walls, and large pieces of furniture, especially partitions, need to be added to the model to develop a feeling of how deep the daylight might penetrate into the building. At this point, the thickness of exterior walls and roofs should be modeled along with specific material properties and dimensions of all elements around building envelope openings such as overhangs, light shelves, complex fenestration systems, and/or other light redirecting devices. Getting wall thicknesses, positions of building envelope openings, and interior dimensions right is crucial. Unfortunately, modeling window frames and mullions is time consuming and usually premature at the earlier design stages. In order to still yield meaningful simulation results of interior lighting conditions, a daylighting modeler may either assume that a fixed proportion of

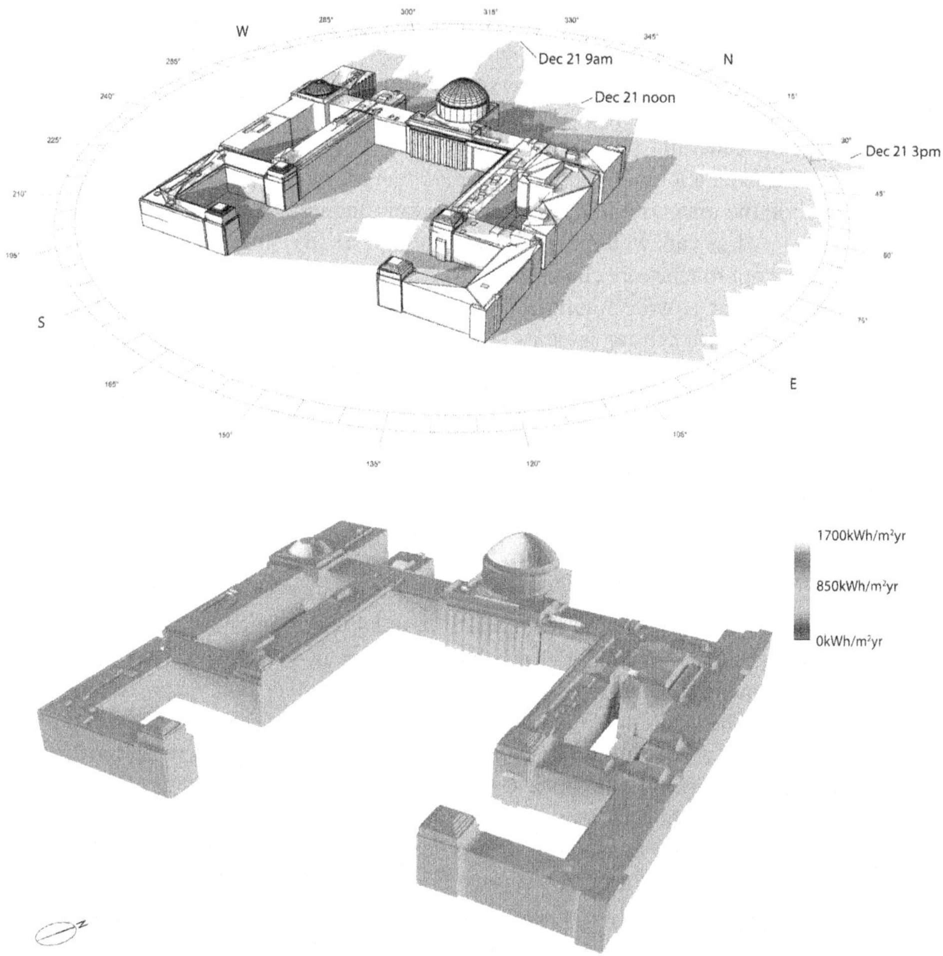

Figure 7.4 Two applications for the massing model from Figure 7.3: Top – Direct shading study on Dec. 21st at 9 am, noon, and 3 pm (generated with DIVA-for-Rhino; Solemma 2016); Bottom – Annual irradiation pattern (generated with Radiance and GenCumulativeSky; (Robinson and Stone 2004))

20% of window openings are used up by frame and mullions, which thus effectively reduce the visual transmittance of the window by about 20%, or use a generic framing system from a CAD library with roughly the same opaque proportion as expected for the final design. Given the myriad details required for a proper daylight simulation, it is worth remembering that – same as for other fields of building performance simulation – a relative comparison of different design options is typically easier to do and yields more reliable results than the use of absolute simulation results against benchmark levels and standards. In other words, it is easier to estimate the relative effect of increasing a window by 20% than to report the absolute illuminance level that one might expect to encounter in a room under specific sky conditions.

As the façade design becomes increasingly detailed, modeling movable shading devices – such as venetian blinds – might become a necessity. As explained below, adding this level of detail only makes sense for some performance metrics. Similarly, modeling the luminance coming from a computer screen is usually not necessary for metrics based on upward-facing illuminances but might become crucial for image-based visual comfort analysis procedures.

As a side note, if a daylighting analysis is to be combined with a thermal simulation for a fully integrated energy performance assessment, the model detail and settings required for the thermal calculation generally diverge from the daylighting/lighting model: The daylighting model may benefit from a lot more geometric detail than is typically provided or required for a thermal model. On the other hand, the division of a building into thermal zones whose boundaries may be defined as 'air walls' is a concept foreign to (day)lighting calculations. These different modeling approaches are challenging to resolve from a model maintenance standpoint (BIM approaches might provide solutions one day) and potentially explain the author's informal observation that architectural students – trained in considering geometric detail – tend to initially experience difficulties building thermal building models (Wasilowski and Reinhart 2009). As of 2018, the author recommended to his architecture students to use their architectural design model, which they use for representation and design development, for any daylighting analysis and to create a thermal model on a separate layering structure (organized by zone type, windows, and external shading objects) but within the same CAD file.

Modeling the sky

Once the scene has been modeled, it has to be combined with one or several sky conditions. One of the challenges of evaluating the daylighting performance of a space, building, or neighborhood is that ambient sky conditions are in permanent flux and depend on prevailing weather conditions as well as the position of the sun in the sky. Sky models are mathematical constructs, used by computer simulation tools, to describe the *sky luminous distribution*, i.e. the amount of light coming from different parts of the celestial hemisphere. Sky models differentiate between *direct sunlight*, i.e. light that is incident from the geometric position of the sun with respect to the site, and *diffuse daylight*, i.e. sunlight that is scattered off clouds, aerosols, air molecules, and water vapor before it hits the earth's surface (Figure 7.5). The ratio of direct sunlight to all daylight falling onto an unshaded horizontal surface over the course of a year depends on the local climate and may vary from over 70% for sunny climates such as Phoenix, AZ, USA, to around 40% for predominantly overcast climates such as London, UK.

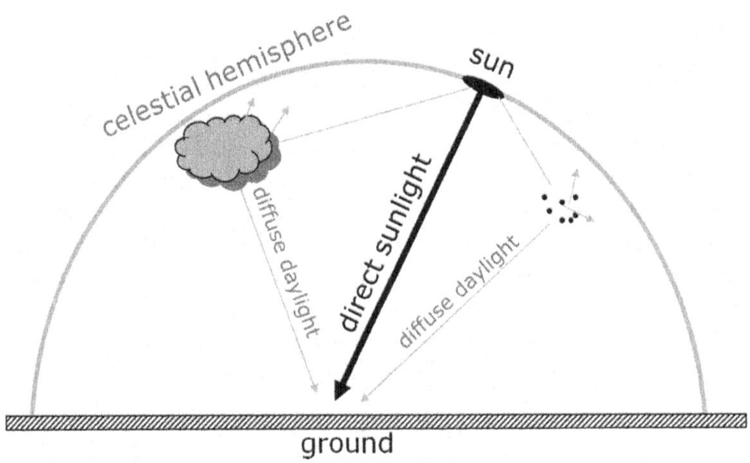

Figure 7.5 Daylight falling onto a site is usually divided into direct sunlight and diffuse daylight

Different sky models require different types of inputs. The simplest sky models only consider the position of the sun with respect to a site and require time of day, date, latitude, and longitude as input. The position of the sun in the sky is typically expressed by two angles, solar azimuth and solar altitude (Figure 7.6).

A detailed description of the trigonometric functions that are commonly used to calculate both angles for a particular site, date, and time can, for example, be found in the author's *Daylighting Handbook I* (Reinhart 2014, chapter 6). Direct sun models are used for shading studies (Figure 7.4, top) or sun charts. Sun charts are two-dimensional representations of the position of the sun with respect to a site over the course of a year. There exist a number of different projection modes for sun charts. In *orthographic* sun charts, the solar azimuth is plotted along the X axis with zero corresponding to due south and the solar altitude is plotted along the Y axis with 90 degrees corresponding to the zenith. Figure 7.7 shows an orthographic sun chart diagram for Boston, MA, USA (42° N, 71° W) with a shading mask from the main entrance to MIT's Kilian Court (marked with a '*' in Figure 7.3). The figure reveals when direct sunlight is incident onto the court entrance and can be used in combination with a local climate file to account for solar heat gain management and/or glare protection.

Historically, a series of increasingly refined, standardized sky luminance distributions have been presented over the past 100 years that consider diffuse daylight as well as direct sunlight. These distributions are mathematical functions of the *relative* luminous distribution of the celestial hemisphere with respect to the zenith luminance. Local radiation measurements combined with luminous efficacy models are required to translate a *relative* sky luminous distribution into *absolute* values.

The oldest, *uniform sky* model assumes an isotropic sky luminance distribution without any direct sunlight and initially served as the reference sky for the daylight factor metric, which will be introduced later (Waldram 1909). The *CIE overcast sky* with a 3:1 gradation of the sky luminance from the zenith to the horizon later replaced the uniform sky as the reference sky for the daylight factor (Moon and Spencer 1942). In 1973, the Commission Internationale de L'éclairage (CIE) published a standard sky luminous distribution for clear skies (CIE 1973), the *CIE clear sky*. The CIE overcast and clear skies served for several decades as two extreme

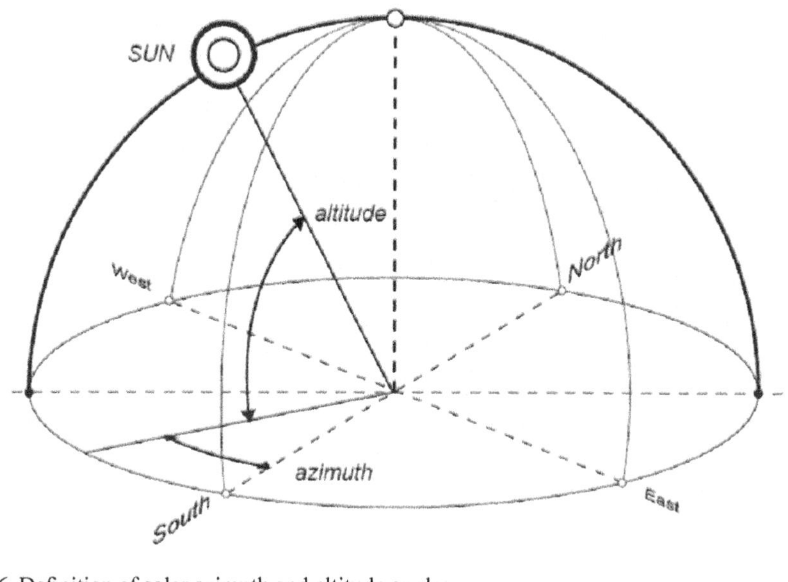

Figure 7.6 Definition of solar azimuth and altitude angles

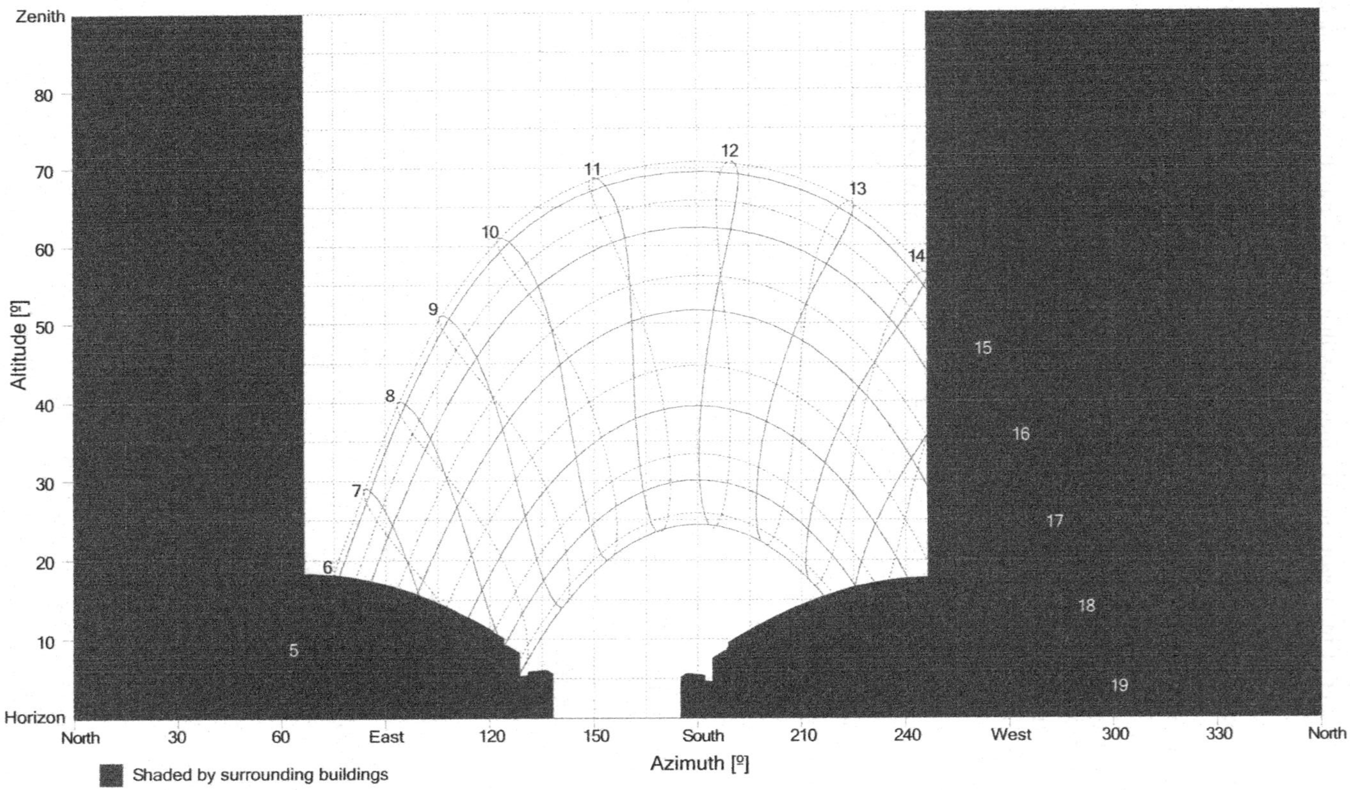

Figure 7.7 Sun chart diagram for Boston with a shading mask for the entrance to Killian Court marked as '*' in Figure 7.3 (generated using Autodesk Ecotect 5.6)

Source: Autodesk 2017

sky conditions to test building designs under. Then – during the mid-1990s – the *CIE standard general sky* (Kittler et al. 1997) as well as the *Perez sky* (Perez et al. 1993) were introduced. Both approaches attempted to "cover the spectrum of intermediate and cloudy skies between the two already standardized clear and overcast sky distributions" (Kittler et al. 1997; Kittler et al. 1998). The CIE standard general sky and Perez sky both rely on the same basic mathematical functions to describe the ratio of the luminance, L_a, of an arbitrary sky element to the zenith luminance, L_Z (Equation 7.1).

$$\frac{L_a}{L_Z} = \frac{f(\chi)\cdot\phi(Z)}{f(Z_S)\cdot\phi(0)} \quad \textit{with } \phi(Z) = 1 + a \cdot e^{\left(\frac{b}{\cos(Z)}\right)} \text{ and}$$
$$f(\chi) = 1 + c\cdot\left[e^{d\chi} - e^{d\frac{\pi}{2}}\right] + e\cdot\cos^2(\chi) \tag{7.1}$$

Parameters *a* to *e* are used to adopt the sky luminous distribution function to a range of clear, intermediate, or overcast sky conditions. As shown in Figure 7.8, χ is the shortest angular distance between a given sky element and the sun, *Z* is the angular distance between the sky element and the zenith, and Z_S is the angular distance between the sun and the zenith.

$\varphi(Z)$ is called the *luminance gradation* function. It defines the changes of luminance from horizon to zenith. For a uniform sky, this function corresponds to unity ($a = 0$). $f(\chi)$ is called the *scattering indicatrix*. It relates the changes of luminance of a sky segment to its angular distance from the sun. Its main purpose is to define the luminance in the *circumsolar region* surrounding the sun (CIE 2004).

The difference between the CIE general sky and the Perez sky is that the former defines 15 discrete parameter sets for variables *a* to *e*, whereas the Perez sky model continuously changes

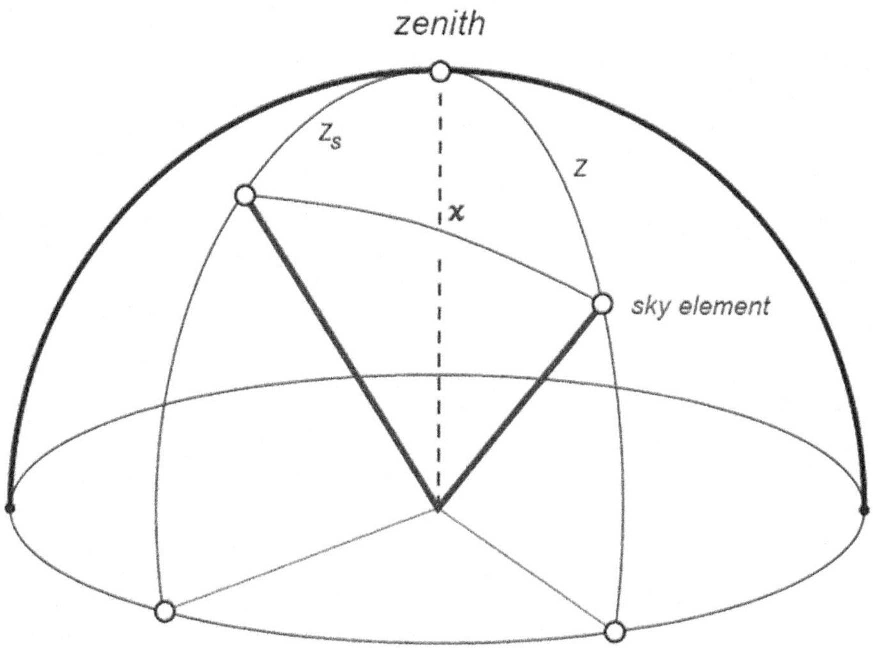

Figure 7.8 Angles defining the position of the sun and a sky element

Source: CIE 2004

those parameters as a function of direct and diffuse irradiances/illuminances. Both models vary between and include the earlier mentioned 'old' CIE overcast and clear skies. Same as Perez, the CIE general sky requires direct and diffuse radiation pairs along with building site, date, and time to identify the one of the 15 standardized general skies that best matches a particular sky condition for a given set of date, time, location, and irradiation data. While the CIE general sky is being promoted through the CIE and ISO standards, the building energy simulation community tends to use the Perez sky model with leading programs such as EnergyPlus, ESP-r, and TRNSYS all using it for modeling the celestial hemisphere for solar gains and daylighting predictions. CIE general sky and Perez have both been implemented into a number of commercial and open source daylight simulation programs, making them accessible to the larger design community. The author has thus far used the Perez model to generate annual indoor and outdoor illuminance time series for a number of reasons: (a) Perez can easily be used in combination with hourly annual climate files, (b) using the same sky model for both the daylighting and thermal simulations makes the two more consistent, and (c) the author has yielded convincing results comparing outside illuminance predictions based on Perez with measurements taken in Freiburg, Germany, and Ottawa, Canada (Reinhart and Walkenhorst 2001; Reinhart and Breton 2009). Another, more fundamental reason why the author conceptually prefers the continuous Perez sky model over the 15 discrete CIE general skies is that these skies seem to be viewed by their promoters to be reference skies under which a designer should test the daylighting performance of a particular building design. Rather than testing a design under a confusing number of intermediate skies, the author recommends to validate a design using visualizations under the CIE overcast sky and the CIE clear sky on solstice and equinox days to develop a general intuition of how the design performs under extreme skies. For a more representative performance analysis that takes the local climate into account, an annual daylight simulation is recommended.

Figure 7.9 visually compares the sky luminous distribution over Boston on April 2 at noon for the uniform, CIE overcast, CIE clear, and Perez skies. According to the typical meteorological year (TMY2) for Boston Logan Airport, April 2 is a day with intermediate sky conditions. The Perez model therefore lies between the extremes of the rotationally invariant, overcast CIE sky and the CIE clear sky distribution.

In order to carry out an annual daylighting or energy analysis of a building, annual climate data for a given site is required (see Chapter 6). This usually comprises hourly data for temperature, relative humidity, wind speed, and direction as well as direct and diffuse solar radiation. An ever-increasing number of climate files are available for download from the internet. As of 2017, annual typical meteorological years for over 2100 sites worldwide could be centrally downloaded free of charge from the EnergyPlus website (US-DOE 2017). The standardized format can also be directly imported into a number of daylight simulation programs (Crawley et al. 1999). A caveat, which a daylighting modeler should be aware of, is that these data sets are frequently based on local airport weather stations that typically measure total solar radiation using a pyranometer. If this is the case for a particular site, the direct and diffuse radiation and illuminance components that are reported in the climate files have been theoretically derived from total solar radiation using a separation model such as the Reindl (Reindl et al. 1990) and a luminous efficacy models such as Perez (Perez et al. 1990).

If the reader is interested in modeling the short time step behavior of daylight, e.g. for glare analysis or to predict energy savings from automated lighting controls (Clarke and Janak 1998), Olseth and Skartveit developed an autocorrelation model to generate sub-hourly time series from hourly global solar radiation (Olseth and Skartveit 1989). The model has been validated and implemented into the Radiance-based Daysim simulation software (Walkenhorst et al. 2002).

For cumulative light exposure levels (e.g. annual and seasonal), a number of authors have presented the concept of a cumulative sky (Mardaljevic and Rylatt 2003; Compagnon 2004;

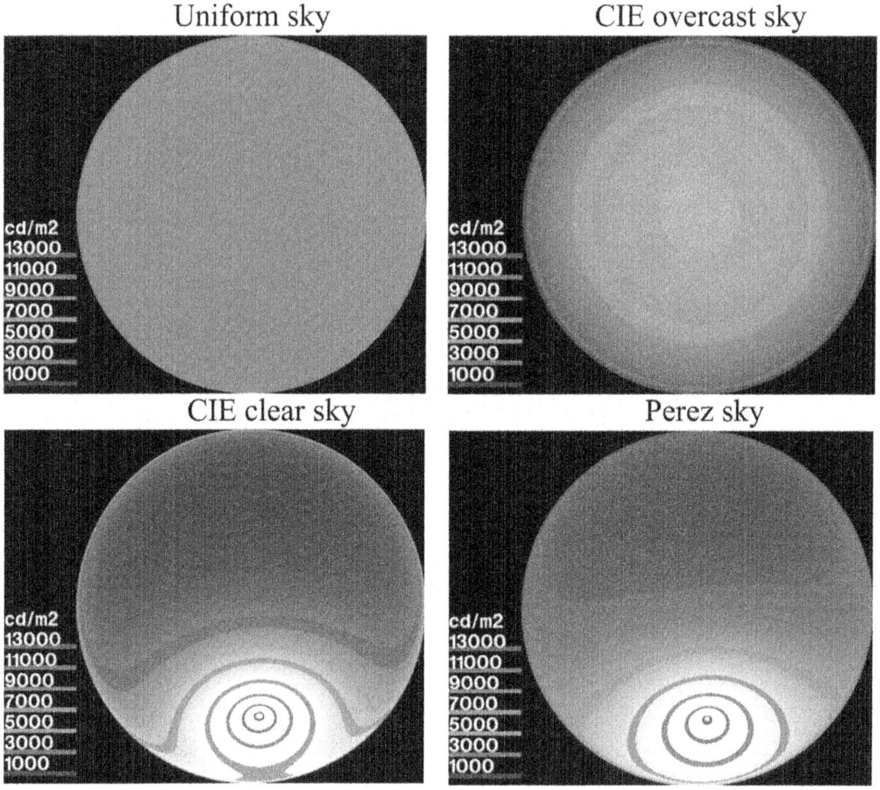

Figure 7.9 Sky luminous distribution over Boston on April 2 at noon according to the uniform, 'old' CIE overcast and CIE clear, as well as Perez sky models

Robinson and Stone 2004). The idea is to add a series of different sky conditions on top of each other and to then run a regular daylight simulation under this theoretical, very bright sky. The advantage of this method is that cumulated radiation and light exposure can be calculated in a single simulation run. An example annual radiation map generated with Radiance and 'GenCumulativeSky' (Robinson and Stone 2004) is shown in Figure 7.4, bottom.

Before moving on to the next section, a few emerging topics related to modeling the celestial hemisphere for daylight simulations are presented in the following.

Sky scanners, which measure the sky luminous distribution directly, have been commercially available for several decades. However, due to prohibitively high investment costs, their use has thus far been restricted to academic applications such as validation studies (Aizlewood 1993; Mardaljevic 1995) and for the development of sky models. With the recent advances in digital photography, a high-end digital camera in combination with a fisheye lens and a luminance meter now allow the collection of high dynamic range (HDR) photographs of the sky luminous distribution, making sky scanner data accessible to a larger number of researchers and design teams (Figure 7.10) (Inanici 2006). A few sites have started to collect hourly HDR sky images, which can then be mapped onto the celestial hemisphere and act as a light source in daylight simulations (Inanici 2009). It is likely that time series of HDR sky images will soon become available for a variety of sites worldwide, complementing the earlier mentioned climate files.

Another recent improvement to sky models is the addition of physically based color gradients, e.g. for modeling sunsets. A prominent, early example is the *Utah sky model* (Preetham et al. 1999). Initially, the resulting colored skies were mainly used to add realism to design

Figure 7.10 Calibrated high dynamic range photograph generated with a digital camera, a luminance meter, and Photosphere

visualizations. However – as the spectral effects of light on human health and alertness are becoming better understood (Figueiro et al. 2008) – several groups have started to model the changing spectral composition of daylight throughout the day and year in order to evaluate non-visual effects of light on humans (Pechacek et al. 2008; Inanici et al. 2015; Solemma 2018).

Areas of interest

Once a scene has been built and a sky condition has been selected, an 'area of interest' has to be defined within the scene. Commonly used formats are selected viewpoints (Figure 7.11), two-dimensional grids of illuminance sensors for false color representations (Figure 7.12), and/or a line of upward-facing illuminance sensors positioned along the central axis of a space. The choice of what area of interest a modeler chooses is intimately related to the purpose of the simulation and the performance metric used. Generally speaking, visualizations and false color distributions are more engaging than simple line plots revealing additional visual detail of a design. On the flipside, line plots allow the direct comparison of several design variants in a single diagram.

Space usage

Depending on the daylight performance metric used, different additional information regarding the intended use of a space is required, including target work plane illuminances, occupancy

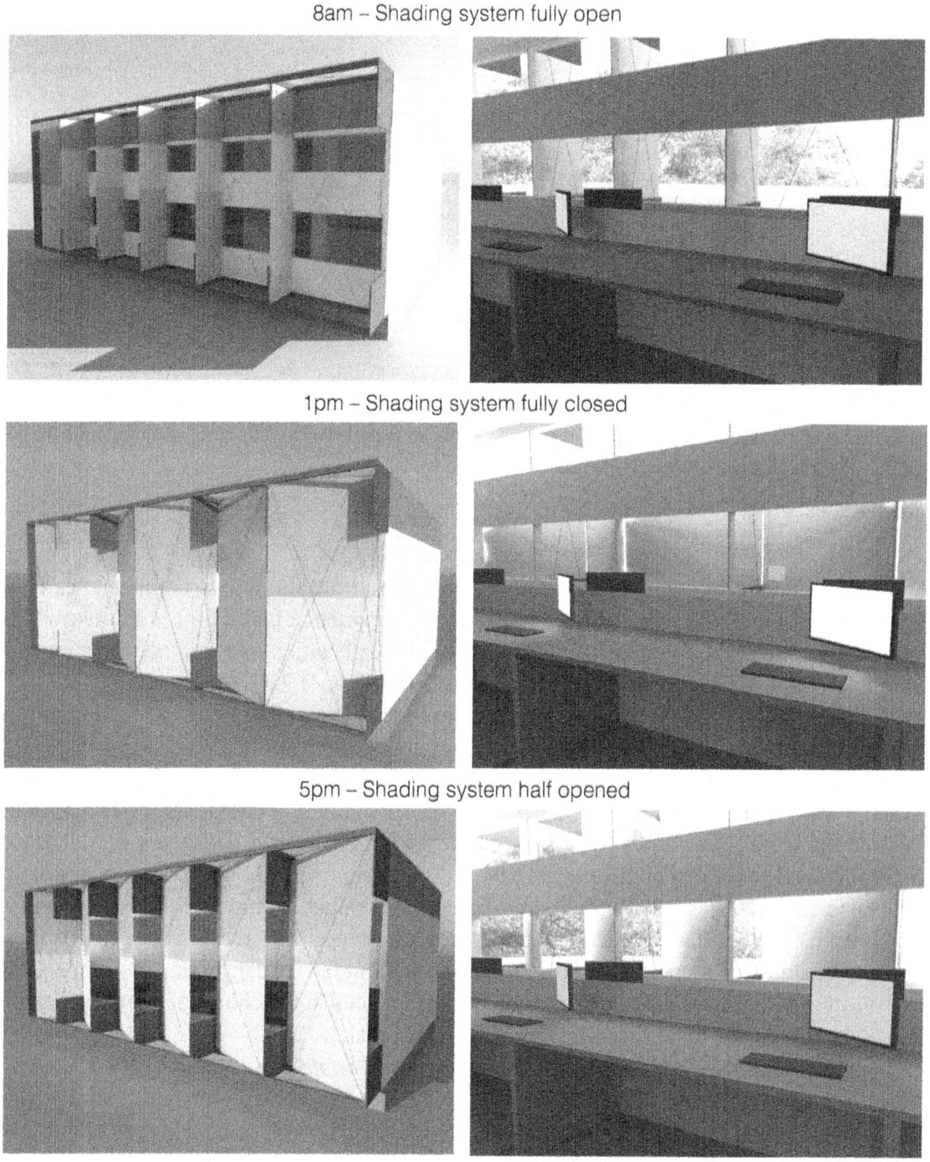

Figure 7.11 Radiance visualization study of a dynamic exterior shading system from the outside (left) and inside (right) at different times of the day; the surrounding landscape was mapped onto the sky dome to visualize how the building interior might 'feel'

Source: Simulation by Catherine De Wolf and Madeline Gradillas

schedules, etc. Suitable levels for these numbers can be found, for example, in the handbooks of the Illuminating Engineering Society of North America (IESNA 2011) and the American Society of Heating, Refrigerating and Air-Conditioning Engineers (ASHRAE 2017).

Daylight simulation algorithms

Once a complete daylighting model has been entered into a simulation program, the simulation engine needs to be invoked to carry out a global illumination calculation within a scene.

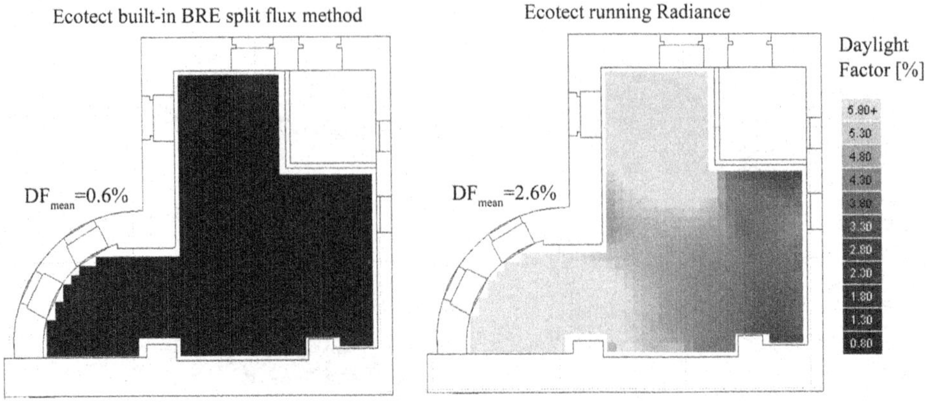

Figure 7.12 Daylight factor calculation in a sidelit space in Autodesk Ecotect v5.6 using the built-in BRE split flux method (left) and Radiance v3.8 (right)

Source: Ibarra and Reinhart 2009

A number of popular lighting programs now support multiple simulation engines to carry out the same type of analysis. It is hence of paramount importance that the program user understands *which* simulation engine is used for a particular calculation and *whether* the simulation engine is capable of reliably modeling the task at hand. Figure 7.12 shows an example of the exact same model of a sidelit space modeled in Autodesk Ecotect with the daylight factor distribution calculated using either Ecotect's internal daylighting engine (a variation of the BRE split flux method; Hopkinson, Longmore and Petherbridge, 1946) or the Radiance backward raytracer (Ward and Shakespeare 1998). The choice of simulation engine used has a profound impact on the simulation results, which vary in this example by *over 400%* between Radiance and BRE split flux (Ibarra and Reinhart 2009). The reason for this large discrepancy is that Radiance is capable of explicitly modeling light that is reflected multiple times, whereas the BRE split flux method bases its calculation solely on objects in direct view of a sensor (see below). Consequently, it underestimates the actual daylight factor distribution in the space. The objective of Figure 7.12 is to sensitize the reader to the fact that the choice of simulation engine used – even within the same simulation program – is crucial.

In 2017, the IBPSA-USA's *Building Energy Software Tools Directory* (IBPSA-USA 2017) listed 15 different programs that claimed to be able to 'model light.' Which one should a practitioner use? First of all, a reader should have some knowledge of the basic approaches commonly used to model the phenomena of (day)light in buildings. Three common methods are therefore briefly introduced in the following: BRE split flux, radiosity, and raytracing.

The BRE split flux or protractor method, developed at a research institution that is now called the British Building Research Establishment (BRE), only models diffuse sky conditions, e.g. for daylight factor calculations, and splits internal daylighting levels into three components: a sky component (SC), an externally reflected component (ERC), and an internally reflected component (IRC) (Figure 7.13) (Dufton 1946). Originally, the various components had to be geometrically derived using so-called protractor diagrams, which were placed on building floor plans and sections. This manual process, tedious in its execution, can nowadays be dramatically accelerated and streamlined using a three-dimensional computer model and a set of rays traced backwards from a sensor point (Figure 7.13). The BRE split flux method tends to yield meaningful results for scene locations that 'see' a lot of sky and for which the internally reflected component is small. For heavier obstructed spaces and points further away from any façade opening, the method becomes increasingly unreliable since the internally reflected component

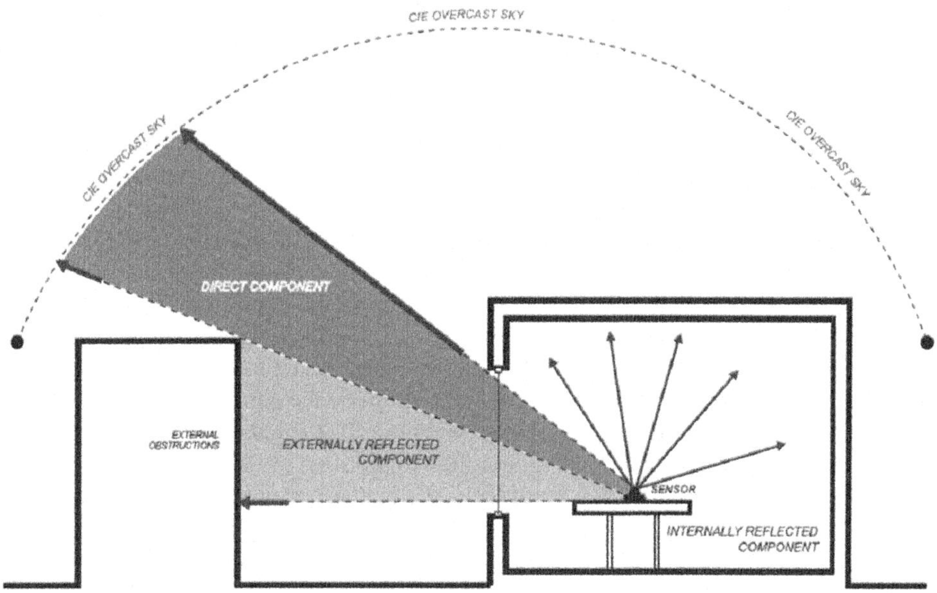

Figure 7.13 Principle of the BRE split flux method

is derived from static parameters that might have been derived in spaces that are significantly different from the space under investigation.

Two numerical approaches that are commonly used to simulate interior lighting conditions under varying sky conditions in arbitrarily shaped spaces are *radiosity* and *raytracing*.

Radiosity was originally developed to solve problems involving radiative heat transfer between surfaces based on form factors. Since the 1980s, radiosity has also been applied in computer graphics to calculate illuminance levels due to electric lighting or daylight. A form factor defines the fraction of radiative energy leaving a given surface to the energy that directly arrives at a second surface. In radiosity, each surface is treated as a perfectly diffuse reflector with a constant luminance so that the radiation exchange between two surfaces can be described by a single number, which depends on the reflective properties of the surfaces as well as the overall scene geometry. To calculate the indoor luminance distribution in a room due to daylight, the incoming luminous flux through all non-opaque parts of the building envelope is set equal to the available flux within the building. This assumption defines a set of equations that uniquely determine the luminances of all considered surfaces. The basic radiosity approach can be coupled with a finite element approach, which detects regions with a large luminance gradient between neighboring surface patches and subsequently subdivides the affected surfaces into sub-surfaces. A detailed description of radiosity methods can be found, for example, in Cohen and Wallace (1993).

The basic approach of *raytracing* is to simulate individual, representative light rays in a scene and to extrapolate the overall luminous distribution of the scene from the available lighting information. Forward raytracing (rays traveling *from* a light source) is generally used to analyze the interaction of light with individual components such as a luminaire or venetian blinds. The simulation output can be, for example, IESNA photometric files (ANSI/IESNA 2002) or bidirectional transmission and/or distribution functions (BTDF/BRDF). For more complex scenes – for example of a whole neighborhood – *light backward raytracing* is generally used. In backward raytracing, rays are emitted from the point of interest and traced backwards until they either hit a light source, such as the sun or celestial hemisphere, or another object. In the former case, the luminance distribution function of the light source determines the luminance contribution at the viewpoint. If a ray hits an object other than a light source, the luminance of that object needs to be calculated by secondary rays emitting from the object (Figure 7.14). The angular distribution under which secondary rays are *spawned* depends on the optical properties of the surface. Raytracing allows for arbitrarily complex optical surface properties, including purely specular surfaces such as a mirror, Lambertian surfaces such as conventional wall paint, transparent surfaces such as glazings, as well as arbitrary mixtures of these basic surface types. A ray path is usually aborted once a certain number of reflective bounces has been reached or if the relative weight of a ray falls below a given threshold value[2] (Ward and Shakespeare 1998).

An advantage of following a light ray 'backwards' from the point of measurement (viewpoint or virtual sensor) to a light source is efficiency, as the majority of rays that are emitted by the potentially many light sources within a scene never reach the particular viewpoint or sensor of interest (Ward and Shakespeare 1998). To illustrate the point, the reader should consider the interior visualization of the workspace in Figure 7.11 (right column), which is one of many similar workspaces in that building. For the visualization in Figure 7.11, the luminance distributions of objects in close vicinity to the viewpoint are of primary interest, and 'going backwards' the simulation naturally concentrates on those objects.

In scenes that involve both complex space geometries and complex fenestration systems in which the light is being redirected multiple times (the reader might think of a light pipe), regular backward raytracing becomes inefficient and (potentially) unreliable. Approaches that have been explored to tackle such scenes include replacing the CFS with a single, illuminating surface that has the optical properties of the CFS (Andersen and deBoer 2006) under a particular sky condition and using a hybrid forward and backward raytracing approach in which light is first forward traced from the light source in the scene and then collected from the areas of interest using backward raytracing (Schregle 2004; mental-images 2007). Another approach to deal with arbitrary CFS is the so-called three phase method, conceptualized by Peter Tregenza and implemented into Radiance by Greg Ward, that splits a daylight calculation into three components – interior raytracing, multiplication of interior results with a matrix that represents

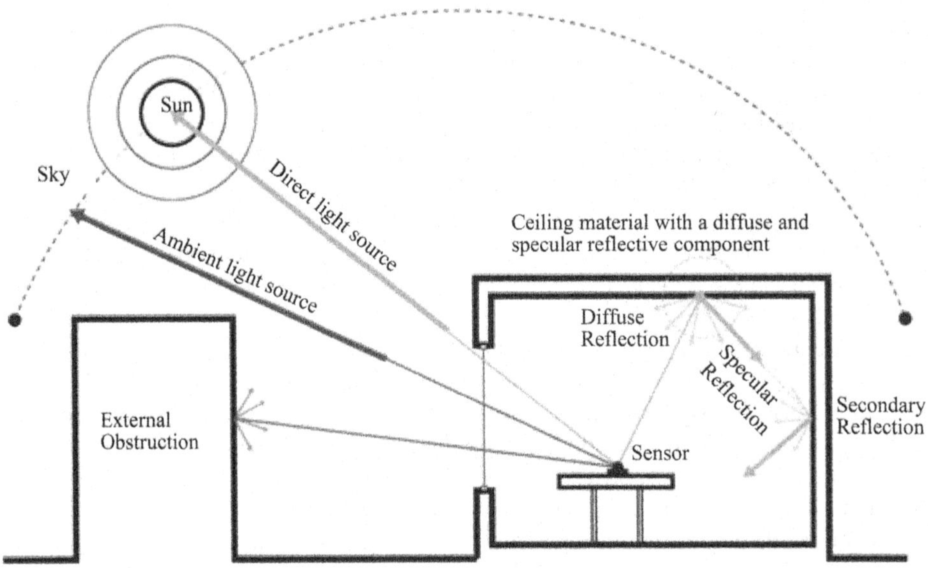

Figure 7.14 Principle of a light backward raytracer such as Radiance

any CFS layer, and another outside raytracing step to consider neighboring objects and the relationship between an envelope opening and the sky (Tregenza and Waters 1983; McNeil and Lee 2013).

Is radiosity or raytracing better suited to address a specific design question? The answer obviously depends on the question. A general advantage of radiosity compared to raytracing is that it requires less calculation time for straightforward geometries that do not contain too many surface elements. This advantage diminishes with rising model complexity as the calculation time in radiosity increases with the square of the number of considered elements, whereas in raytracing this relation is roughly linear (Tregenza 1983) or even sublinear (Ward 1994). A radiosity calculation yields the total luminance distribution in a room independent of the point of view of the spectator. Therefore, a scene walk-through can be realized a lot faster with radiosity than with raytracing, as each new viewpoint requires a new raytracing run. Radiosity-based walk-throughs also tend to look smoother. An advantage of raytracing over radiosity is that it is able to simulate specular and partly specular materials (Ward and Shakespeare 1998). This quality can be important in daylit spaces, especially for visual comfort analysis purposes, since window frames, shading devices, and furniture finishes often exhibit distinctly non-Lambertian surface properties.

In order to accomplish both, the ability to continuously explore a scene and to get physically accurate results for a variety of materials, Nathaniel Jones tested an alternative global illumination method to analyze daylit scenes, called *path tracing* (Jones and Reinhart 2016). Same as light backwards raytracing, path tracing follows rays from an observer's viewpoint backwards towards potential light sources. However, only a single ray is followed from each intersection before creating a scene visualization. The process is repeated iteratively and thus yields a succession of increasingly refined scene visualizations (Lafortune and Willems 1993). Initial results indicated that glare or daylight availability metrics derived from these intermediate images oftentimes converge quickly, providing designers with nearly immediate design feedback (Jones and Reinhart 2016).

Both radiosity and raytracing require the modeler to set several simulation parameters in accordance with the scene complexity and expected luminance gradient. A key simulation parameter for raytracing is the number of reflectances (bounces) considered before a ray is discarded. Setting the simulation parameters can be a complicated task – especially for simulation beginners – which is why graphical user interfaces are increasingly 'hiding' the actual parameters and asking the user to provide more qualitative descriptions of a scene, such as 'scene detail' or 'scene complexity.' These advances notwithstanding, setting the simulation parameters remains a delicate issue as selecting too-low values may lead to wrong results, whereas too-high values reduce the simulation uncertainty but also lead to unjustifiably long simulation runs.

A fundamental problem of raytracing is that it cannot provide a reliable scene-specific estimate of the remaining calculation errors. This can turn out to be a real impediment as simulation results may lie above *or* below the true illuminance levels. Simulation results are too low if, for example, a raytracer misses a small window or skylight in a room and therefore grossly underestimates the real illuminance. Results are too high if a raytracer interpolates between two bright luminance directions, e.g. from two neighboring windows, and ignores that a wall lies in the interpolated region.

Program validation

Once a modeler has developed a basic appreciation of the various daylight simulation algorithms that are commonly used, he or she can start searching for a simulation program that uses the desired simulation algorithm and calculates the daylight performance metrics the modeler is interested in (see the next section). Once a potential program has been identified, the modeler should also verify that the program has been validated. Validation may happen through three methods, (a) the comparison of simulation results to lighting scenarios that can also be solved analytically, (b) inter-program comparison, and (c) through comparison to measurements. In 2006 the International Commission on Illumination, CIE, released a document titled "CIE 171 – Test Cases to Assess the Accuracy of Lighting Computer Programs" (CIE 2006). The report proposed a series of analytical (electric lighting and daylight) and experimental (electric lighting only) test cases and had the objective to help "program users and developers to assess the accuracy of lighting simulation programs and to identify areas of strength and weakness within different programs." Similarly to the BESTEST/ASHRAE Standard 140 (ASHRAE 2007) cases for thermal simulation engines, the main focus of the CIE 171 test cases was to "test very isolated aspects of a simulation software such as the capability to model light sources including daylight, light transfer through openings and the ability to model inter-reflections" (CIE 2006). While the author agrees that the use of simple, idealized lighting scenarios as promoted by CIE 171 are very useful for basic software testing and 'bug finding,' the author also feels that a complementary set of more complex daylighting test cases that compare experimental measurements in 'real' daylit spaces with simulations is necessary to help design practitioners develop a feeling of how close a lighting simulation program, operated by a knowledgeable user, can model reality. This attitude seems to be shared by others, given that a 2006 survey of 185 daylighting modelers from 27 countries yielded a strong preference for Radiance: The survey participants named over 40 different software packages that they frequently used, but over 50% of all votes went to tools that are based on Radiance[3] (Reinhart and Fitz 2006).

Radiance is a mixed stochastic deterministic light backwards raytracer that had been validated based on illuminance measurements in full-scale spaces with a clear glazing and a light shelf (Mardaljevic 1995; Jarvis and Donn 1997), venetian blinds (Reinhart and Walkenhorst 2001), a translucent glazing (Reinhart and Andersen 2006), and laser-cut panels (McNeil and Lee 2013). A 2017 study further demonstrated that physically based renderings using Radiance and its GPU-based version AcceleRAD may be used in place of high dynamic range photographs to predict visual discomfort (Jones and Reinhart 2017).

For software developers interested in validating their tools against measurements, at least two data sets are currently available: The British Building Research Establishment (BRE) offers a very rich data set of indoor illuminances in a full-scale sidelit test room with a clear glazing and light shelf including sky scanner data (Aizlewood 1993). In 2009, the National Research Council Canada (NRC) in Ottawa, Canada, collected a new data set that consists of measured indoor and outdoor illuminances as well as direct and diffuse outdoor irradiances for five daylighting test cases of varying complexity (Reinhart and Breton 2009). The NRC data set has thus far been used to validate the mental ray raytracer within Autodesk 3ds Max Design 2009 and the Radiance-based Daysim 3.0 program. Given the rising interest in physically accurate daylight simulations, the author expects that other simulation programs will go through comparable experimental validation exercises using the BRE, NRC, or other experimental data sets. These validation studies should make it easier for beginning daylighting modelers to compare the validity and limitations of various simulation programs.

Dynamic simulations

The simulation methods discussed thus far are *static* in the sense that they were developed to simulate interior or exterior lighting levels under a particular sky condition, e.g. the CIE overcast sky. Since actual ambient daylighting conditions depend on the local climate and are constantly changing, *dynamic* or *annual* daylight simulations calculate time series of lighting levels due to daylight throughout the whole year. These time series are called *annual illuminance profiles* and typically have a time step varying from 1 hour down to a minute. Annual illuminance profiles can be used to predict electric lighting energy use in daylit spaces and/or combined with a thermal simulation program to predict overall energy use for heating, lighting, and cooling. Another application is to translate them into *climate-based* daylighting metrics that consider the quantity and character of daily and seasonal variations of daylight for a given building site (see the next section).

In principle, it is possible to carry out an annual simulation by simply repeating a static simulation multiple times under varying sky conditions. However, this procedure typically leads to exorbitant simulation times (Reinhart and Herkel 2000). A series of different dynamic daylight simulation methods were introduced in the past, but most researchers seem nowadays to use a *daylight coefficient* based approach (Tsangrassoulis et al. 1996; Janak 1997; Mardaljevic 2000; Reinhart and Walkenhorst 2001; McNeil and Lee 2013).

The concept of daylight coefficients, originally proposed by Tregenza and Waters (Tregenza and Waters 1983), is a numeric method to speed up an annual daylight simulation. The basic approach is to conceptually divide the celestial hemisphere into disjoint sky segments (Figure 7.15, left) and to calculate the contribution of each sky segment to the total illuminance at various sensor points in a building (Figure 7.15, right). These illuminance contributions, $E_\alpha(x)$, at a sensor point and orientation, x, for a sky segment, α, are then normalized with the luminance, L_α, and angular size, ΔS_α, of the sky segment (Equation 7.2). The resulting quantify is called a daylight coefficient, $DC_\alpha(x)$.

$$DC_\alpha(x) = \frac{E_\alpha(x)}{L_\alpha \Delta S_\alpha} \tag{7.2}$$

A complete set of daylight coefficients for all sky segments defines the relationship between a point within a scene and the celestial hemisphere. It takes into account scene geometry and material properties but is independent of the local climate. Daylight coefficients can hence be coupled with an arbitrary sky luminance distribution, L_α with $\alpha = 1. . . N$, by a simple linear superposition to calculate the total illuminance $E(x)$ at x (Equation 7.3):

$$E(x) = \sum_{\alpha=1}^{N} DC_\alpha(x) L_\alpha \Delta S_\alpha \tag{7.3}$$

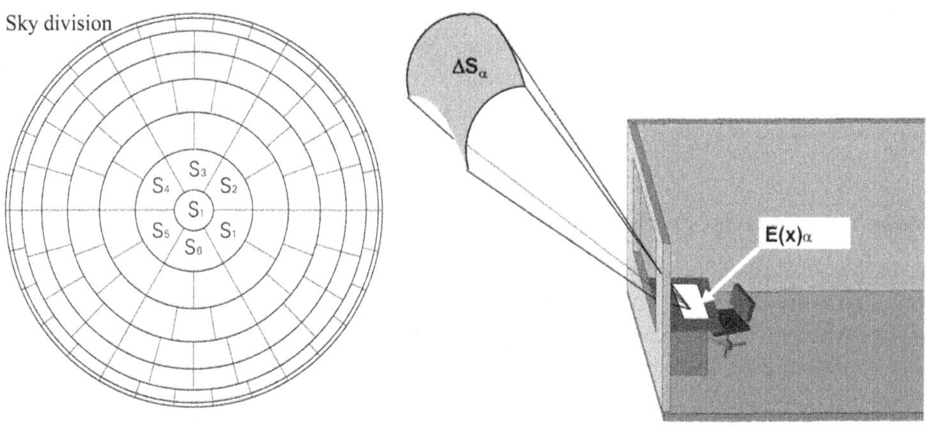

Figure 7.15 (Left) Division of the celestial hemisphere into disjoint sky segment; (right) Visual definition of a daylight coefficient

The time-consuming part of an annual daylight simulation based on daylight coefficients is the calculations of the coefficients, e.g. via raytracing. This step is hence carried out in an initial, pre-processing step before the annual simulation is invoked. Once a set of daylight coefficients is available, the illuminance calculation under any sky condition only requires milliseconds. The daylight coefficient concept can also be extended to luminance time series. If applied to all pixels in an image, the concept even allows the generation of annual time series of scene visualizations (Wienold 2009). Daylight coefficient approaches mainly differ in how they differentiate between contributions from diffuse daylight and direct sunlight. For diffuse daylight, most available methods split the celestial hemisphere into 145 sky segments according to the Tregenza division (Tregenza 1987) (Figure 7.15, left) as well as one or several ground segments. Contributions from direct sunlight can be divided into those that are directly incident onto a sensor and those that have been reflected at least once within the scene before hitting a sensor (Mardaljevic 2000; Bourgeois et al. 2008).

A number of simulation studies have yielded that dynamic daylight simulations using a daylight coefficient approach, backward raytracing, and the Perez all-weather and/or general sky models can simulate indoor illuminances under a range of overcast, intermediate, and clear sky conditions with a relative error below 25% compared to measurements (Mardaljevic 2000; Reinhart and Walkenhorst 2001; Reinhart and Breton 2009; McNeil and Lee 2013). This error might initially sound high, but the reader should consider that the human eye is a logarithmic sensor and that we barely notice illuminance changes over time in that order of magnitude. In other words, the resolution of daylight simulation programs roughly corresponds to the sensitivity of the human eye. On the other hand, indoor illuminance levels in daylit spaces may vary from tens to tens of thousands of lux, a range that can be captured by simulation programs. These facts suggest that the current generation of simulation programs provides sufficient accuracy to provide reliable design advice, at least for spaces of typical geometric complexity and with common façade elements such as clear and translucent glazings combined with overhangs, light shelves, standard venetian blinds, and roller shades. As mentioned above, accurately modeling more complex fenestration systems including mirror blinds and light wells can be tackled using photon mapping and/or matrix multiplications. The latter application remains an area of active research.

From a practical point of view, the additional effort required for carrying out a dynamic as opposed to a static daylight simulation is small. As far as model preparation is concerned, the only difference is that a local climate file has to be specified, which does often not require any serious preparation effort since climate data is nowadays available for many parts of the world in a standardized format (see Chapter 6 for more information). The annual simulation itself can be fully automated and, again, does not require any additional effort on behalf of the modeler. The main difference is that – in the case of raytracing – the simulation time is roughly five to eight times longer for a dynamic than for a static simulation (Reinhart and Walkenhorst 2001).

Simulation uncertainty, user errors, and quality control

The earlier discussed validation studies demonstrate how close daylight simulations can approach reality if they are carried out by an expert user and if all simulation input parameters are known. The critical reader might wonder how much of this information is typically available during the design of an actual building and how difficult it is to build a high-quality daylighting model of a space when one first starts using daylight simulations. These are valid concerns. A modeler is indeed commonly required to make a number of 'educated guesses' regarding material properties and building usage patterns, which can in turn have a decisive impact on how the building is going to perform. The effect of a modeler's level of expertise on model quality is something that we are only beginning to understand.

In 2005/2006, the author taught two full-semester lighting classes at McGill University, for which around 70 architecture students had to model the mean daylight in the classroom from Figure 7.12 using the Autodesk Ecotect version 5.6 graphical user interface (GUI) and Radiance as simulation engine (Ibarra and Reinhart 2009). Figure 7.16 shows the error distributions of the daylight factor calculations by the students for 2005 and 2006 vis-à-vis the best practice model from Figure 7.12. In both years, the majority of student models yielded significantly higher results than the best practice model with the average mean daylight factors in 2005 and 2006 being 8.3% and 5.3%, respectively, compared to 2.6% for the best practice model. A detailed analysis of the models showed that simulation novices made any number of modeling mistakes, ranging from incomplete and/or inaccurate geometries to not adjusting material properties or encountering software interoperability errors when importing geometries that had been prepared in another CAD tool. It is worth mentioning that the daylight factor calculation in Autodesk Ecotect was actually chosen for the study since it was a mature and visually intuitive workflow that the author had assumed could be easily followed by simulation novices. Chances are that modeling errors made using other simulation programs would have been comparable or worse. In response to these findings, Diego Ibarra and the author developed a series of simulation exercises as well as a simulation checklist to help simulation beginners build better models and develop a simulation 'mindset' (Ibarra and Reinhart 2013). This educational approach was formally tested in a 2012 class at MIT using the DIVA-for-Rhino GUI and Radiance as simulation engine. Figure 7.16 also shows the results for this class: Over 80%

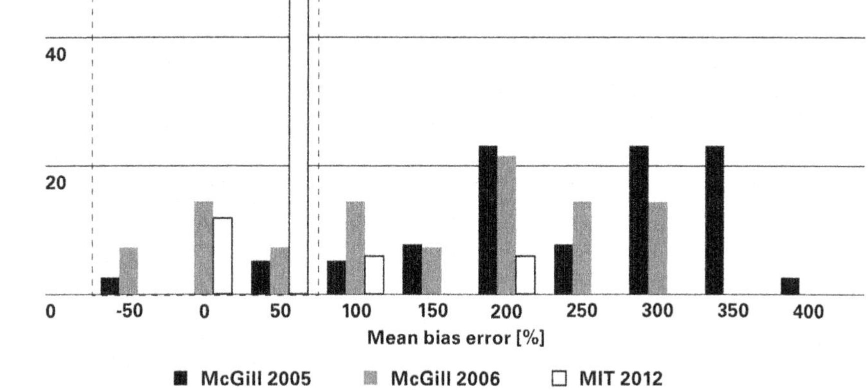

Figure 7.16 Frequency distribution for daylight factor prediction errors based on student models compared to best-practice models for three different classes held at McGill University and MIT

Source: Ibarra and Reinhart 2013

of students submitted results within a ±50% error range compared to the best-case model. Based on this finding, the author recommends the use of the simulation checklist in Table 7.2 to simulation novices. Experienced modelers might also find it to be a useful reference for model quality control.

Table 7.2 Daylight simulation checklist

Before you start	Did you decide which daylighting performance metrics to simulate and how to interpret the results?
	Do you have a general idea of what the results should look like? For example, a mean daylight factor in a standard sidelit space should lie between 2% and 5%; interior illuminance should lie between 100 lux and 3000 lux, and daylight autonomies should be above 50% in the well daylit portions of a space.
	Have you verified that the simulation program that you intend to use has been validated for the purpose that you intend to use it for, i.e. that the simulation engine produces reliable results *and* that the program supports the sky models related to your performance metric of choice? (An example would be the old CIE overcast sky for daylight factor calculations.)
	Have you secured credible climate data for your building site? (This is only required for climate-based daylighting performance metrics.)
Preparing the scene	Did you model all significant neighboring obstructions such as adjacent buildings and trees?
	Did you model the ground plane?
	Did you model wall thicknesses, interior partitions, hanging ceilings, and larger pieces of furniture? Try to model all space dimensions at least within a 5 cm tolerance. Façade details should be modeled with a 2 cm tolerance.
	Did you consider window frames and mullions by either modeling them geometrically or by using reduced visual transmittances for windows and skylights?
	Window glazings: – Did you check that all window glazings only consist of one surface? Several CAD tools model double/triple glazings as two/three closely spaced parallel surfaces, whereas daylight simulation programs tend to assign the optical properties of multiple glazings to a single surface. – Did you check that all windows are 'inserted' into the wall planes and not 'overlaid' on the wall surfaces? Several CAD tools suggest that you can create and visualize a window in many different ways, one simply being the placement of a window surface on top of a wall surface. If presented with two coplanar surfaces, a simulation program such as Radiance will either ignore the window or somehow 'guess' which surface to consider.
	Did you assign meaningful material properties to all scene components (see Table 7.1)?
	Did you model any movable shading devices such as venetian blinds? If yes, do the results make sense?
Setting up the simulation	Make sure that you set up your project files correctly. This may involve: – Checking that your project directory and file names do not contain any blanks (" "). – Verifying that all sensors have the correct orientation, i.e. work plane sensors are facing up and ceiling sensors are facing down. – Setting the resolution of the work plane to 0.5 m × 0.5 m or 1 ft × 1 ft and placing it around 0.85 m above the floor. – Selecting simulation parameters that correspond to the 'scene complexity.' To do so you should consult the technical manual of your simulation program. – Selecting the correct sky model (CIE, Perez, etc.).

In 1979, Joseph Lynes proposed the simple formula in Equation 7.4 to calculate the mean daylight factor in a room (Lynes 1979):

$$DF = \frac{A_{glazing}\tau_{vis}\theta}{A_{totlal}2(1 - R_{mean})} \tag{7.4}$$

Where DF is the mean daylight throughout the space; R_{mean} is the area-weighted mean surface reflectance; $A_{glazing}$ is the net glazing area; A_{total} is the total area of interior surfaces (including windows); θ is the sky angle, and τ_{vis} is the visible glazing transmittance. Figure 7.17 compares predictions of the Lynes formula to simulated mean daylight factors calculated in Radiance for over 2000 sidelit spaces. The figure shows a strong correlation ($R^2 = 0.9$) between the two prediction methods. While the daylight factor is not the premiere daylight metric that the author would recommend modelers to use from a design standpoint, the figure shows that crosschecking mean daylight factor predictions of a model against the Lynes formula provides for a quick and easy 'plausibility check.' If a simulation model predicts a significantly different mean daylight factor than the Lynes rule, chances are that the modeler has made a mistake somewhere along the way. In that case, the model should be carefully reviewed before the modeler tries to derive more advanced daylighting metrics from the it.

In closing, Table 7.3 lists three practical questions that a novice modeler should consider when initially choosing a simulation program. The first 'test' for a software to pass is whether it can actually calculate the set of metrics that the modeler is interested in. Next, the novice modeler should verify whether a potential software candidate has been validated and how (CIE test cases, measurements, etc.). This point becomes a key concern when the software is to be used in a consulting type environment, as the modeler has to prove that due diligence was used when choosing a software tool. Finally, the modeler should find out whether there is a support network available for the software. Such a network can have many forms: Having a group of users of the same tool within a company provides by far the best learning opportunities. One

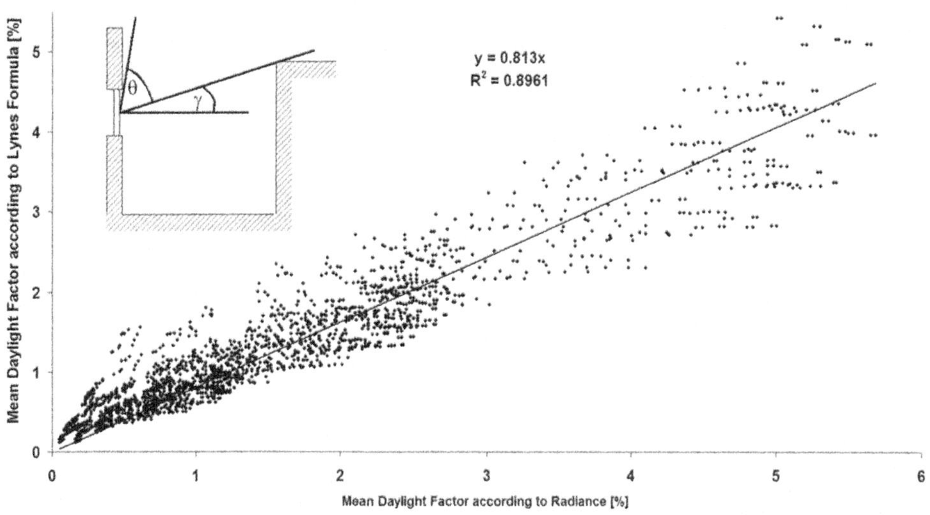

Figure 7.17 Mean daylight factor calculated with Radiance versus the mean daylight factor according to the Lynes formula (Equation 7.4) for 2304 sidelit spaces with varying levels of external obstructions; inset: graphical definition of the sky and obstruction angles

Source: Reinhart and LoVerso 2010

Table 7.3 Three questions to consider when choosing a daylight simulation tool

1. Which software can model the daylight performance metrics that I am interested in?
2. Has that software been validated (what are the supported sky model(s) and rendering algorithm(s))?
3. How easy is the software to learn? Is there any support network available?

can strategically build up such a group through hiring well-trained students. The next best support can be offered through active mailing lists, software vendor hotlines, and/or local simulation interest groups.

Daylight performance metrics

The previous section discussed how to calculate the global illumination in a daylit scene under a single or multiple sky conditions. This section now turns to the most important part of a simulation: the analysis of the results. The section aims to provide a balanced account of established and emerging *daylight performance metrics* together with a discussion of how these metrics can influence design decisions. A daylight performance metric is a quantitative evaluation of certain daylit qualities of a space, building, and/or neighborhood. Example metrics are the percentage of the total floor area of a building that is daylit or the number of times in a year when an occupant experiences visual discomfort. Multiple metrics can be bundled into an overall performance assessment, e.g. a space can be required to meet certain minimum standards regarding annual daylight availability, visual comfort, and energy usage. Performance metrics become especially useful design tools when they are combined with a target range so that a design iteration can be said to either 'satisfy' a metric or not. The author recommends that design teams initially formulate their project-specific definition of 'good daylight' along with a series of concrete metrics and target values. This exercise is useful, as good daylighting is interpreted by some to be the 'interaction of natural light and building form to achieve high quality workspaces,' whereas other think of it as being primarily an energy-efficiency design measure (Galasiu and Reinhart 2007). Both definitions are valid but may at times favor different design choices.

The remainder of this section is organized as follows. Initially, illuminance-based daylight availability metrics are being introduced followed by luminance-based visual comfort metrics and a discussion of how these metrics can be applied to spaces with dynamic façade components such as venetian blinds and switchable glazings.

Daylight availability metrics

Illuminance-based metrics are generally used to establish the 'daylight availability' in a space. They are mostly derived from a grid of upward-facing illuminance sensors at work plane height in order to describe the distribution of daylight within a space (see e.g. Figure 7.12).

Daylight factor

The oldest daylight availability metric is the daylight factor.[4] It is defined as the ratio of the internal illuminance at a point in a building to the unshaded, external horizontal illuminance under a CIE overcast sky (Figure 7.18) (Moon and Spencer 1942).

It is important to stress that the daylight factor is only defined under a CIE overcast sky and – since it is a ratio of inside to outside illuminances – it is independent of building site, time of day, and year. Given that the CIE overcast sky is rotationally invariant (Figure 7.9), the daylight factor is also independent of building orientation. It might seem odd to have a daylighting

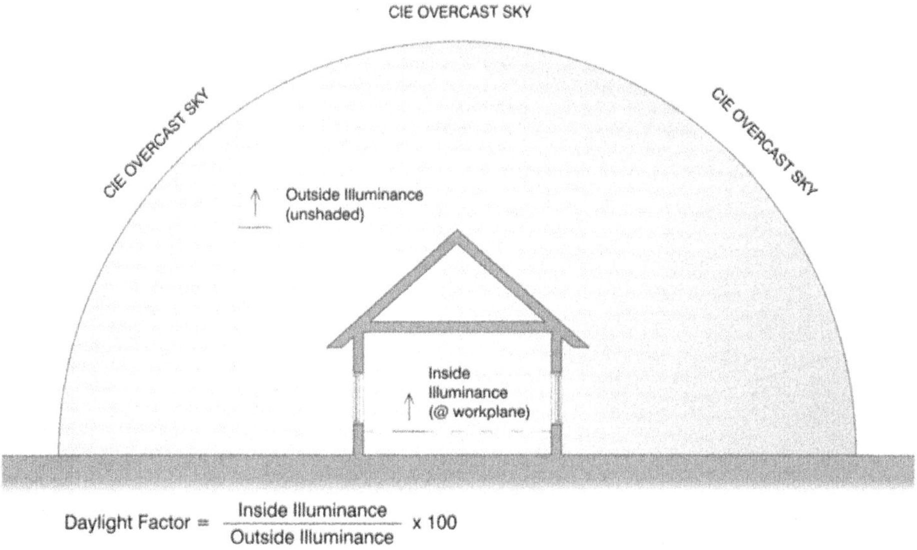

Figure 7.18 The daylight factor is defined as the ratio of an inside illuminance divided by the outside horizontal illuminance at an unshaded point under a CIE overcast sky

metric that does not include such key design parameters. However, this quality was the original motive for working with a ratio rather than with an absolute value as a way of dealing with the "frequent and often severe fluctuations in the intensity of daylight" (Waldram 1950). In fact, using the daylight factor as a preliminary design metric for spaces that primarily receive diffuse daylight or to verify whether a certain amount of daylight can be maintained under diffuse sky conditions makes perfect sense. The concept just becomes meaningless in the presence of direct sunlight. By itself, the daylight factor is also of limited use to ensure that a visually stimulating, healthful, and productive interior environment is accomplished.

One advantage of using the daylight factor for building design is that the designer can refer to various design guides and standards that recommend average daylight factor values for different space types. Most standards recommend a mean daylight factor of 2% for a 'primarily daylit' space (CIBSE 2014; IESNA 2011). Figure 7.19 shows the area with a daylight factor above 2% in the MIT reference office, a standardized sidelit space that is also shown in Figure 7.19 (Reinhart et al. 2013). Using a 2% daylight factor target level, the reference office is thus divided into a daylit and a non-daylit area with the former extending approximately 3.1 m or 1.2 times the window-head height away from the façade into the space.[5]

Due to its climate and orientation independence, the daylight factor for a south-facing office in Miami is the same as for a north-facing classroom in Anchorage – if both spaces have identical geometry and material properties. In order to make the daylight factor concept more climate-specific, Szokolay introduced the concept of the design 'sky illuminance' as the 15th percentile of outdoor illuminance during daylit hours (9am to 5pm) (Szokolay 2004). Once the design sky illuminance has been determined for a particular site, the product of the daylight factor and the design sky illuminance approximates the illuminance that should exceed 85% of the working hours. The design sky illuminance for Boston, MA, USA (42° N, 7° W) is about 5900 lux, which implies that the average illuminance in the sidelit space from Figure 7.19 is above 5900 lux × 4.6% = 271 lux for 85% of working hours between 9am and 5pm in Boston.

As a design metric, the daylight factor promotes several design measures that are known to enhance the daylight in a space, such as high window-head heights, high reflective ceiling

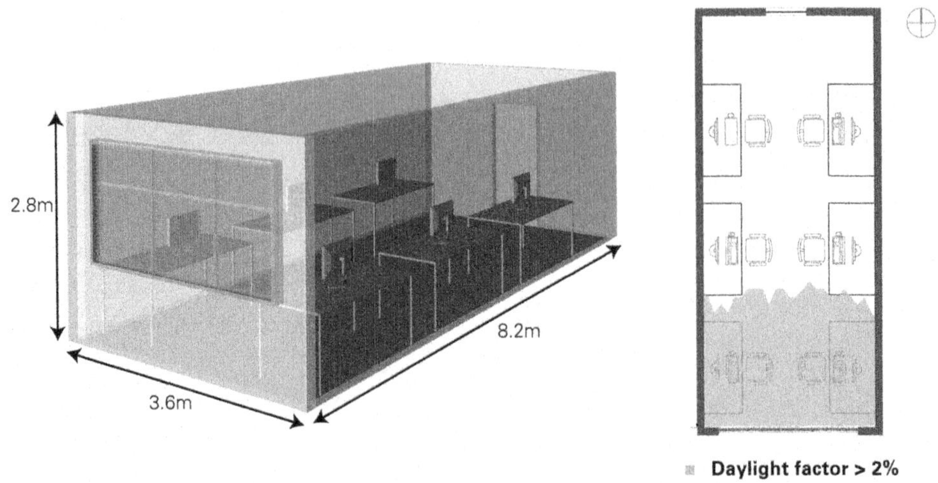

Figure 7.19 Visualization of the MIT reference office (left); daylight factor distribution in the MIT reference office (right)

and wall finishes, narrow floor plans, and large façade and skylight openings with high transmittance glazings. A weakness of the daylight factor, or any other daylight availability metric, is that it promotes one-dimensional, 'the more the better' approach to daylighting. Taking this to the extreme, *the daylight factor optimized building has a fully glazed building envelope*. Therefore, the use of other daylighting metrics is required to keep the daylight factor 'in check.'

CIE clear sky studies

A popular way of exploring how a space performs under clear sky conditions is to model a space at key moments in a year such as at 9am, noon, and 3pm on solstice and equinox days. Figure 7.20 shows the results for the reference office for Boston for ranges above 10 fc (107 lux) and 500 fc (582 lux). The two thresholds were introduced in LEED v3 and are meant to correspond to minimum and maximum levels to avoid glare and overheating (US Green Building Council 2009). Figure 7.20 shows that daylight penetrates deepest into the space during winter solstice but that the resulting levels might be too high near the window. During summer conditions, when the sun is highest in the sky, the first two rows of work stations seem adequately daylit. This analysis should be complemented with a thermal analysis of the space to establish whether the space overheats or requires substantial cool loads due to uncontrolled solar gains. The dates and times used in Figure 7.20 suggest that a space is used all year around as an office or classroom. For other space types – such as places of worship – the designer may instead concentrate on daylighting levels on key religious holidays.

The use of the daylight factor combined with CIE clear sky simulations leads to a more balanced daylighting design as far as solar gains and glare are concerned since the two metrics work 'against each other': Larger windows lead to higher daylight factors but also to more areas that are oversupplied under clear sky conditions. However, this approach has the drawback that specific climate data is not considered: What if it is virtually never sunny in December at the considered building site? The approach also requires a designer to conduct 10 individual simulations and to review the resulting figures. The next logical improvement is therefore to use a metric that explicitly considers annual climatic conditions.

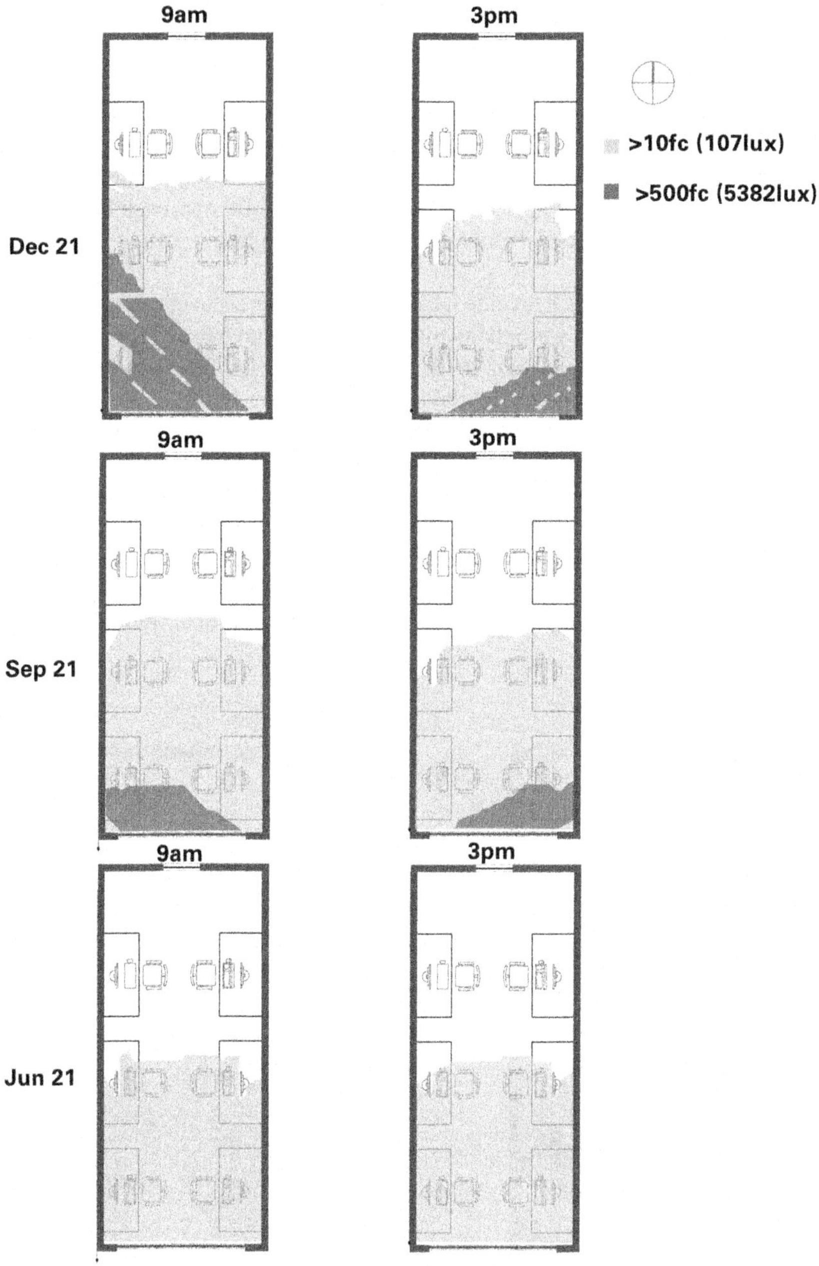

Figure 7.20 Indoor illuminance distribution in the sidelit space from Figure 7.19 under clear sky conditions on solar solstice and equinox days at 9am and 3pm

Climate-based metrics

Climate-based metrics are derived from the earlier discussed annual illuminance profiles, i.e. hourly or even sub-hourly time series of interior illuminances due to daylight. In order to become usable for design, this massive amount of data has to be converted into an intuitive metric.

The first step in this conversion process is to decide on which times of the year the analysis should be based. A common choice is to concentrate on the times when the investigated space will be occupied since daylight 'needs witnesses' to have an effect. This approach corresponds to that taken by energy simulation programs that also assume typical usage patterns. An alternative argument of why one might instead concentrate on the daylit hours of the year is that these hours will not change over the lifetime of a building, whereas usage patterns might change as warehouses are converted into apartments, etc. This 'loose fit' approach to building design can be especially attractive for developers who originally might not exactly know how a new building is going to be used.

The next step of the analysis is to decide what daylighting levels to consider 'adequate.' Over the years, several climate-based daylight availability metrics have been proposed (Mardaljevic et al. 2009; Reinhart et al. 2006). As of 2018, the two most established approaches were spatial daylight autonomy (sDA) and useful daylight illuminance (UDI). sDA is defined as "the percentage of the occupied hours of the year when a minimum illuminance threshold is met by daylit alone" (Reinhart and Walkenhorst 2001). The Illuminating Engineering Society of North American (IESNA), followed by the US Green Building Council (USGBC), have adopted sDA as a daylighting metric, assuming a standard occupancy schedule from 8am to 6pm all year around with daylight savings time lasting from March to November and a target illuminance level of 300 lux (IESNA 2012; US Green Building Council 2013). For a point to be considered 'daylit,' the sDA at the point has to be 50%, in short sDA_{300lux}[50%]. Figure 7.21 shows the

Figure 7.21 Daylit area distributions for the reference office in Boston according to sDA_{300lux}[50%] and UDDA and $UDI_{100lux–3000lux}$[80%]

$sDA_{300lux}[50\%]$ distribution for the reference office in Boston. All sensors in the gray, daylit area receive more than 300 lux of daylight for at least half of the occupied hours in the year.

UDI is an alternative daylighting metric to sDA that has lower and upper thresholds of 100 lux and 2000 lux, respectively (Nabil and Mardaljevic 2005). These two values divide the year into three bins. The upper bin ($UDI_{>2000lux}$) is meant to represent times when an oversupply of daylight might lead to visual and/or thermal discomfort, the lower bin ($UDI_{<100lux}$) represents times when there is 'too little' daylight, and the intermediate bin ($UDI_{100\text{–}2000lux}$) represents 'useful' daylight. Mardaljevic later modified the upper bound to 3000 lux. In 2014, the UK Education Funding Agency (EFA) established a UDI-based daylighting requirement for funding through its Priority Schools Building Programme. Being classroom-based, the EFA demands a space-averaged $UDI_{100lux\text{–}3000lux}$ value of 80% for a time basis of 8.30am to 4.00pm. Figure 7.21 shows the resulting $UDI_{100lux\text{–}3000lux}[80\%]$ distribution in the reference office (UK Education Funding Agency 2014).

The predictions of the two metrics in Figure 7.21 are quite different with sDA suggesting that the first two rows of workplaces near the façade are daylit, whereas UDI proposes that the useful daylit area only extends across for the second row because daylighting levels near the window regularly exceed 3000 lux. How should the reader interpret such results? sDA is a daylight availability metric that indicates how deep into the space sufficient daylight penetrates. A recent study compared $sDA_{300lux}[50\%]$ simulations in educational spaces in Brazil, Canada, Egypt, and the US against subjective student evaluations of these spaces and found that the two closely agree in many spaces, suggesting that designers can indeed use the metric to initially establish the 'daylit area' for a given design (Reinhart et al. 2014). A parallel exercise is then required to assess visual comfort conditions. As described below, high daylighting levels may lead to glare, which will in turn trigger occupants to deploy a shading device such as venetian blinds, if available. Once the status of a shading device throughout the year is known, sDA and UDI have to be recalculated, since venetian blinds will change the amount of daylight in a space. Daylight availability and visual comfort are thus closely intertwined and should further be complemented with a thermal analysis to account for overheating and/or space conditioning loads. Results can then be presented as separate metrics in a common dashboard (Reinhart and Wienold 2011).

A key question is how to evaluate visual comfort. Visual comfort has long been equaled to the 'absence of glare.' The IESNA recommends the use of annual solar exposure (ASE) to flag the occurrence of glare (IESNA 2012). ASE corresponds to the percentage of time that direct sunlight above 1000 lux is incident on a sensor. In contrast, UDI equals oversupply to more than 3000 lux total work plane illuminance. Since upward-facing illuminances are very different from the visual experience of a person talking to others or looking at a computer screen, metrics derived from luminance distributions in the field of view are likely to provide more reliable predictions of glare than either ASE or UDI. Some of these glare metrics are discussed in the "Glare metrics" section below.

Temporal maps

All climate-based illuminance metrics discussed earlier combine an annual time series of 8760 values into a single number for each sensor point to help a designer understand the spatial distribution of daylight. In order to react to a given outcome and to improve a design, it is instructive to also know the temporal distribution of daylight at a given sensor point. *Temporal maps* serve this purpose. A temporal map is a false color map with the day of the year mapped along the X axis and the time of day along the Y axis (Mardaljevic 2004; Kleindienst et al. 2008). Figure 7.22 shows temporal maps of the center point sensor of the first and second rows in the reference office. The first row sensor exhibits values above 3000 lux for most of the year between 10am and 4pm, suggesting the need for a shading device. The second row sensor is mostly well daylit

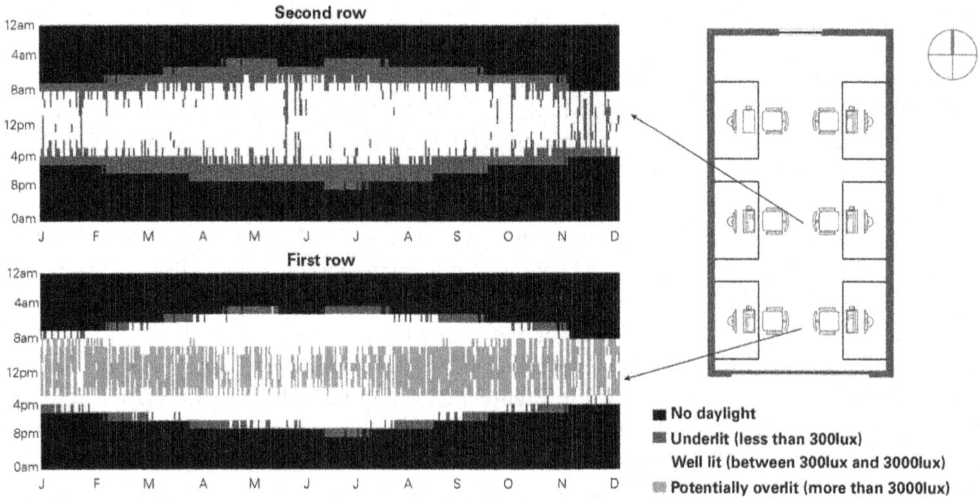

Figure 7.22 Temporal map of the first and second rows in the reference office

during midday with levels falling below 300 lux at the beginning and end of the workday. The temporal and spatial maps from Figures 7.21 and 7.22 should ideally be used in combination.

Annual light exposure

Before moving on to luminance-based metrics, the concept of annual light exposure is briefly described. Designing spaces for people is challenging because individuals have different requirements and tolerances regarding their visual and thermal environment and because of the uncertainty associated with the use of personal controls (movable shading devices and light switches). Designing spaces for artwork and/or plants instead is in that sense 'easier' because requirements and tolerances are better definable. Paintings and trees do not 'tinker' with their environment. As a consequence, the use of annual daylight simulations for the placement of artwork and indoor plants is more established (Franks 2005). Two frequently used design metrics for art galleries are annual light exposure, i.e. the cumulative amount of visible light incident on a point of interest over the course of a year, as well as a maximum illuminance level that may fall on the point at any given time. The International Commission on Illumination (CIE) Division 3 TC3.22 'Museum lighting and protection against radiation damage' recommends upper levels for both metrics for a variety of art forms (International Commission on Illumination (CIE) Division 3 TC-22, 2004). The earlier mentioned cumulative sky methods can be used for annual light exposure calculations, whereas annual illuminance profiles are required to verify that the second, maximum illuminance criteria is consistently met.

Glare metrics

This section presents a more refined approach to detect an 'oversupply' of daylight in a space. What constitutes an 'oversupply' of daylight? Too much daylight is associated with either unwanted solar gains or glare. Glare is a subjective human sensation that describes "light within the field of vision that is brighter than the brightness to which the eyes are adapted" (HarperCollins 2002). To establish whether solar gains are welcome at a particular point in time requires the use of a thermal balance analysis, ideally using computer simulations (Chapter 2). Glare is typically divided into *disability glare*, which is the inability of a person to see certain objects in a scene due to glare, *discomfort glare*, the premature tiring of the eyes due to glare, and *veiling reflections*, the ability to see the content of an electronic surface or glossy magazine due to too much reflected light off the surface. While disability glare in daylit interiors is relatively easy to identify, discomfort glare is a more subtle phenomenon that is closely linked to a person's overall indoor environmental satisfaction. Predicting the appearance of discomfort glare in daylit spaces has been studied for many years. As mentioned above, there is a widely shared notion that horizontal work plane illuminance – while easy to measure – is a poor predictor of discomfort glare, as the amount of light falling on a working area has little in common with a person's visual experience of a space. Instead, a glare analysis should be based on the luminance distribution in the field of view of an observer. A *glare index* is a numerical evaluation of high dynamic range images using a mathematical formula that has been derived from human subject studies. Example indices include the unified glare rating (UGR) and the daylight glare index (DGI). All of these equations were derived from experiments with artificial glare sources – none of them under real daylight conditions (Wienold and Christoffersen 2006). The reason for this is that until the mid-2000s, it had been next to impossible to collect high dynamic range images of daylit scenes under continuously changing lighting levels. Daylight glare probability (DGP) is an increasingly popular discomfort glare index that was derived by Wienold and Christoffersen from laboratory studies in daylit spaces using 72 test subjects in Denmark and Germany. In the experimental setup, two identical, side-by-side test rooms were used. In one of the rooms, a CCD camera-based luminance mapping technology was installed at the exact position and orientation as the head of the human subject in the other room (Wienold and Christoffersen 2006). DGP is defined by the following equation:

$$DGP = 5.87 \times 10^{-5} E_v + 9.18 \times 10^{-5} log\left(1 + \sum_i \frac{L_{s,i}^2 \omega_{s,i}}{E_v^{1.87} P_i^2}\right) + 0.16 \qquad (7.5)$$

where E_v is the vertical eye illuminance; $L_{s,i}$ is the luminance of source 'i'; $\omega_{s,i}$ is the solid angle of source i and P is the position index, a weighing function that varies with the distance of a glare source form the field of vision. The first two terms in the DGP equation correspond to brightness and contrast, respectively. The third term 'normalizes' the metric to 100% under severe glare conditions. The metric therefore weighs both causes of glare, making it superior to analyze glare in daylit spaces vis-à-vis conventional glare metrics that only consider contrast in the field of view (Jakubiec and Reinhart 2011b). As an example application, Figure 7.23 shows two visualizations and associated DGPs for the first row, east-facing workplace in the reference office under clear sky conditions on January 2 at 10am and July 4 at noon. The associated DGPs for the overcast and clear skies are "intolerable" (100%) and "imperceptible" (22%), respectively (Wienold 2009). As one might intuitively expect, the workspace experiences glare on a winter morning when the sun is in direct sight of the worker's viewpoint. However, during the summer months, the sun is high enough in the sky so that the resulting high daylighting levels near the window (shadow patches in Figure 7.23, right) do not bother the occupant.

While these points-in-time glare analysis figures are useful to visually understand the cause of glare in a particular situation, in order to evaluate the overall appearance of discomfort glare at a workspace it would be necessary to repeat the analysis from Figure 7.23 for all hours of the year using site-specific climate data. Wienold proposed a simplified method that uses daylight coefficient-based vertical eye illuminance and simplified images in order to calculate annual

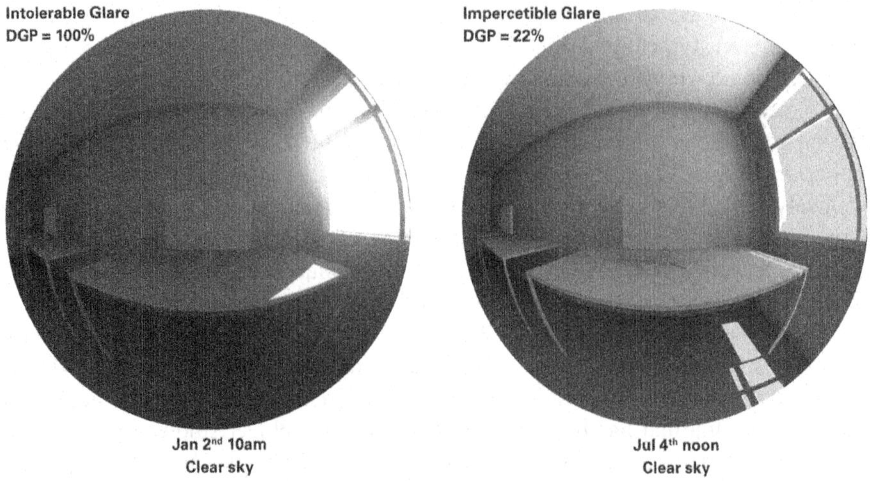

Figure 7.23 Radiance visualizations and associated DPGs for the first row east-facing workspace in the reference office (generated with Radiance and evalglare)

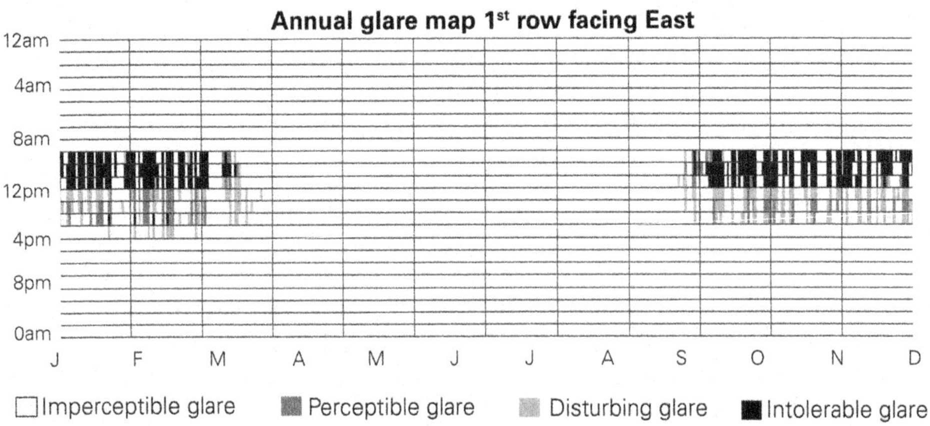

Figure 7.24 Annual glare map of a workspace in the reference office in Boston (generated with DAYSIM gen_dgp_profile and displayed using DIVA-for-Rhino v4)

DPG profiles more efficiently than running one full scene visualization at each time step (Wienold 2009).

Figure 7.24 shows an annual glare map for the workspace from Figure 7.23 using Wienold's *enhanced simplified DGP* method. Glare conditions are grouped into four categories, imperceptible, perceptible, disturbing, and intolerable. For the east-facing workspace, intolerable glare tends to occur on winter mornings between 9am and noon, followed by less severe conditions later in the day during clear sky conditions. Between mid-March and the end of September, the sun is too high in the sky to cause glare at the workspace. The figure provides nuanced, actionable information for designers as to when unacceptable glare conditions occur at the workspace. Running such an analysis iteratively allows gauging the benefit of different static façade elements, from overhangs to translucent glazing systems. From a simulation workflow standpoint, this type of analysis can be fully automated and requires minimal effort on behalf of the modeler, since the only required additional input are one or several representative viewpoints throughout a scene. However, as of 2017, annual glare simulations took about an hour to complete, which is quite disruptive to the design flow. The simulation times increase even further if dynamic façade elements are being tested (see next section).

Dynamic façades

It has by now been established in multiple ways that the reference office requires the addition of a glare protection device for occupants sitting in the vicinity of the window. Static external shading elements provide a valid option in cooling-dominated climates to block direct sunlight to avoid both glare and overheating. Since Olgay and Olgay's seminal work during the 1950s, a series of form-suggesting algorithms have been developed to support the design of such systems (Olgay and Olgay 1957; Marsh 2003; Sargent et al. 2011). However, in climates such as Boston with cold winters and warm summers, the benefits of a static shading system are intrinsically limited, since there will be different times in the year when the sun is exactly in the same spot in the sky but the building is once in heating and once in cooling mode. An example would be sunny days in mid-March (heating period) and mid-September (cooling period). In such situations, dynamic shading systems – from venetian blinds to electrochromics – offer a more climate responsive solution. In order to model the effect of dynamic façade systems, multiple annual illuminance and DGP profiles have to be carried out for at least two representative shading device settings, such as fully opened and fully lowered blinds.

Once indoor lighting levels for different shading device settings are known for all hours in the year, an algorithm is required to model the status of the shading device throughout the year. The task is straightforward for automated controls that are based on outside façade sensors, etc. However, understanding how building occupants manually control light switches or shading devices is more complicated and has been the subject of multiple field studies, starting with Arthur Rubin's investigation of venetian blinds in US office buildings and David Hunt's recordings of electric lighting use in offices and classrooms in the UK (Rubin et al. 1978; Hunt 1979). As mentioned above, ongoing efforts continue to generalize observations from such field studies into occupant behavior models that mimic how occupants interact with manually and automatically operated lighting controls and shading devices (Gunay et al. 2017). *Lightswitch 2002* is an early model that uses occupancy schedules, work plane illuminances, and the appearance of direct sunlight and/or glare at the workspace as input data (Reinhart 2004). A key assumption of the model is that occupant behavior can be described through two archetypical behavior patterns called 'active' and 'passive' users. For the active user, an expanded version of the original model assumes that the shading device is closed once the DGP at the workplace becomes 'disturbing' (> 40%) and opened after an extended lunch break and first thing in the morning upon arrival (Reinhart and Wienold 2011). Passive users keep the shading device closed all year and thus represent a 'worst case scenario.' Using the Lightswitch model, the annual glare map from Figure 7.24 can be used to predict the status of a shading system throughout the year. Since the space is occupied by up to six individuals, annual glare maps of all workspaces have to be determined to ensure that glare is avoided at all occupied workspaces. Figure 7.25 shows the resulting $sDA_{300lux}[50\%]$ distributions for the reference office for no shading device as well as venetian blinds operated by an active and a passive user. While mitigating glare, adding a

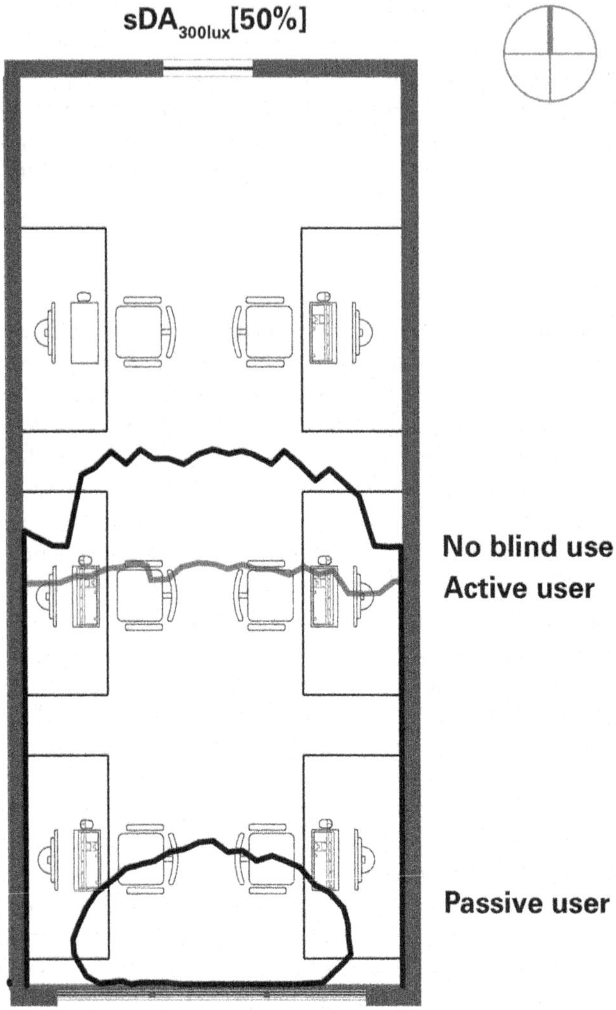

Figure 7.25 $sDA_{300lux}[50\%]$ distributions in the reference office in Boston for three different blind control scenarios according to the Lightswitch 2002 model

shading device to the reference office reduces its daylit area from 56% to 45% for an active and 9% for a passive user, respectively. There is a clear trade-off between visual comfort and daylight availability.

It is worth noting that in the presence of multiple users, distributed throughout a space at different distances to the windows, there exists an inherent conflict between those in the back who would like to get "more" daylight and those in the front who still get enough daylight even after closing the blinds to avoid glare. The resulting group dynamics between occupants is not well understood, yet. Another underexplored aspect of occupant behavior is the question of adaptability. If the occupant in Figure 7.23 is stoically looking forward all year around, he or she will frequently experience glare. However, just slightly rearranging the chair and monitor can mitigate glare under many sky conditions, meaning that the blinds need to be closed less often (Jakubiec and Reinhart 2011b). Given a severe scarcity of field data, there is little known about how occupants operate multiple sets of blinds in a space or how many occupants in a building a designer can expect to be either active or passive users. Despite these numerous caveats, the author believes that not considering the use of dynamic shading systems at all leads to misleading design interpretations: The reader might reconsider the UDI diagram from Figure 7.21. There seems to be plenty of usable daylight for the second row workspaces paired with an oversupply of daylight *near* the façade. However, once venetian blinds are added, the usable daylight is located *near* the façade, but there is limited daylight near the second row (Figure 7.25). The design conclusions are exactly opposite.

Discussion

This chapter presented a series of tips and insights into how to properly set up a daylight simulation and introduced metrics for daylight availability and glare. Additional metrics for a variety of other performance aspects are currently being developed, some of which are mentioned below. To further contextualize the use of daylight simulations in building design, the relationships between simulations and rules of thumb, physical scale models and electric lighting simulations are also discussed.

Evolving performance metrics

As daylight availability and glare metrics such as sDA, UDI, and DGP are making their way into standards and rating systems, daylighting researchers and practitioners are moving towards more 'occupant-centered' evaluations including view, visual interest, and non-visual effects (Amundadottir et al. 2017).

A view to the outside has long been acknowledged to be a key asset of a space, but limited research has gone into quantifying the quality of a view. LEED v4 includes a new view credit requirement that values if a point in a building can see windows facing in different directions and if the view through these windows includes nature and/or other objects of interest (US Green Building Council 2013). Another, more real estate–focused view evaluation approach considers the number of direct lines of sight between a point just outside of a façade and landmark objects such as other buildings or parks (Doraiswamy et al. 2015).

Yet another exciting research direction is concerned with the visual interest of a scene. Early work by John Flynn relied on human subject assessment of different electric lighting conditions in full-scale rooms to rate the scenes for brightness and interest (Flynn et al. 1979). In 2016, Shiobhan Rockcastle and her colleagues at EPFL linked contrast metrics derived from image processing techniques to human subject assessments of Radiance renderings of prominent daylit spaces. The authors then managed to correlate the RAMMG contrast algorithm to human subject perceptions of calm versus exciting (Rockcastle et al. 2016).

Finally, non-visual effects of light on humans refer to light absorbed by recently discovered photosensitive cells in the human eye, called ganglion cells, that are distinct from cones and rods that support our daytime and nighttime vision. Stimulated ganglion cells contribute to a variety of effects in humans, from suppressing melatonin to resetting our circadian (inner) clock and influencing alertness. Ganglion cells have a different spectral sensitivity curve that peaks at around 490 nm compared to 555 nm for daytime vision (Provencio et al. 2000). These differences in spectral sensitivity offer new opportunities for architectural (day)lighting design that we are only beginning to explore (Amundadottir et al. 2016). To be able to model these effects – once they are better understood – new spectral lighting tools have been developed that model lighting in multiple instead of the standard three (red, green, and blue) spectral bands (Solemma 2018).

As some of these metrics will become more established, they will likely serve as companion metrics to daylight availability and visual discomfort, which will lead to more complete evaluations of daylighting concepts during design.

Rules of thumb

As mentioned before, the use of daylight simulations has become increasingly common in architectural design, and this chapter aims to further this trend by providing novice users with information on how to get started. Many of the metrics introduced in the previous sections can only be calculated using computer simulations. However, while building design is getting increasingly digitized with electronic drawings becoming required construction documents in many jurisdictions, initial hand sketches and massing ideas are still informed by intuition, rules of thumb, and architectural precedence (Figure 7.1). It is therefore important to better link rules of thumb and simulations so that designers know when to 'naturally' switch from one analysis method to the other. One recent approach to bring rules of thumb and simulations closer together was to validate a series of interconnected rules of thumb based on simulations, making the predictions of both methods more complementary (Reinhart and LoVerso 2010).

Physical models

Physical or scale models of buildings can be analyzed *inside* using either a heliodome (artificial sun) or an artificial sky (emulated overcast sky) or *outside* under real sky conditions. Typical techniques for analyzing physical models are through direct observation, photography, and/or video animations and illuminance measurements. The use of physical models for daylighting design generally seems to be on a downward trend if findings from recent surveys can be trusted (Reinhart and Fitz 2006; Galasiu and Reinhart 2007). On the other hand, it is the author's experience that advanced daylighting consultants go back and forth between scale models and simulations, using the former when material imperfections and visual experience become important.[6] When one compares measurements under real sky conditions in a scale model and the corresponding full-size space, the errors introduced by the scale model tend to be comparable in magnitude to those introduced by a daylight simulation (Cannon-Brookes 1997). If one on the other hand uses a scale model under an artificial sky dome, a theoretical argument presented by Mardaljevic shows that "on the basis of a credible design goal for the sky simulator dome, high accuracy illuminance predictions (±10%) are practically unattainable" due to parallax errors (Mardaljevic 2002). Parallax errors are measurement inaccuracies that appear under artificial skies since the actual, physical light sources of the artificial sky structure are not infinitely far away from the physical model, which leads to within-model overshadowing of parts of the artificial sky.

The author believes that there is a very valid place for the use of physical models in daylighting design, especially in schools of architecture that anyhow tend to be producers of physical models for other purposes. For many designers, physical models are clearly more engaging than a two-dimensional model on a computer screen. At the same time, the author is cautioning the users of physical models not be lured into thinking that their measurements using these models are particularly accurate.

Electric lighting

Modeling electric light and daylight is largely the same process, and a number of mature programs are available to simulate most commercially available luminaires. The light emitted by an electric light source is called the light source's luminous intensity distribution. Several commonly available photometric file formats describe this distribution, with the IES format by the IESNA being the predominantly used file format (ANSI/IESNA; ANSI/IES_LM-63–02; ANSI/IESNA 2002). Most manufacturers provide IES files for all of their lighting products. The underlying data is mostly based on measurements, but several luminaire design software packages also write out IES files. Compared to daylight, electric lighting is somewhat 'easier' to model since lighting output is constant or scales linearly (for dimmed systems). Electric lighting and daylight can be easily combined in a simulation by defining both types of light sources in a scene. A remaining challenge is to accurately model the interaction between a photocell-controlled dimming system and the light in a space that is usually a mixture of daylight and electric light with varying spectral compositions. Most daylighting and energy simulation programs currently predict dimming levels based on work-plane illuminances essentially assuming a 'perfectly commissioned' dimming system that ideally dims the electric lighting so that a desired target illuminance is maintained on the work plane. A number of inaccuracies are associated with this idealization, since real photocells tend to be located on a wall or ceiling and made out of inexpensive, not cosine-corrected sensors that cannot accurately measure illuminances (Bierman and NLPIP 2007). Several researchers have presented methods of how to better model the true performance of a dimming system, including modeling the actual special sensitivity of a photo sensor (Ehrlich et al. 2002) and using simulations to identify sensor positions that yield an optimized correlation between the electronic signal at the photo sensor and the illuminance at the target level over the course of the year (Rogers et al. 2009).

Outlook

By combining metrics related to daylight availability, visual comfort (glare, view, etc.), and energy efficiency into an integrated framework to evaluate the daylighting concept for a building, architects and their consultants nowadays have an increasingly sophisticated framework at their disposition to design "well daylit" buildings (Reinhart and Wienold 2011). As new metrics are being added and occupancy behavior models for shading and lighting controls are being refined, this framework can be naturally extended over time. A remaining question is, How might daylight simulations be used and evolve going forward?

Following an optimistic trajectory, one could assume that buildings will become better and better because increasingly sophisticated diagnostics of different designs are becoming available. A more subdued interpretation of recent developments is that simulation results are increasingly challenging for a casual user to interpret and that – even though computers and thus simulations are faster than ever – the use of tools in design remains disruptive, since simulation times span between minutes and hours. Indeed, a recent field study by Nathaniel Jones and the author showed that users design and think more fluently, finding more optimized solutions, with a tool that provides feedback in less than a second vis-à-vis a similar simulation workflow that took 20 to 30 seconds to complete (Jones and Reinhart 2018). Such real-time simulations can be paired with virtual reality projections to further help designers to engage with environmental performance aspects (Rockcastle et al. 2017).

However tools end up evolving, concerns whether users actually understand what they are modeling remain, and there is a sustained need for effective training material, clear performance targets, and intuitive results visualizations.

Moving beyond individual buildings, the author foresees two developments in daylighting research and practice: an increased focus on individual users and their overall experience inside and outside of buildings as well as a push to create opportunities for daylighting buildings at the urban level. For the building performance simulation community, the former trend will require the development of agent-based occupancy modeling methods as well as links to smartphone apps that already follow our every step and action.

In order to expand the reach of daylight simulations to the urban level, Timur Dogan developed a method to effectively model floor-by-floor $sDA_{300lux}[50\%]$ distributions across whole neighborhoods (Dogan et al. 2012). Such methods can help urban planners to optimize daylight across larger, multi-building projects and municipalities to formulate zoning regulations that create long-term opportunities for daylighting across a city. As an example, Figure 7.26 shows an analysis of daylight availability versus urban density for a series of block typologies located in Manhattan, showing that certain massing typologies can support very high urban densities without unduly compromising access to daylight (Saratsis et al. 2017).

In the light of these developments, an emerging challenge for the field of daylight simulations will be to bridge the scales between the individual, the building, and the city to ensure that urban dwellers have access to sufficient and delightful daylight in a rapidly urbanizing world.

Synopsis

If carried out properly, validated daylight simulation programs are nowadays capable of correctly simulating annual daylighting levels in daylit spaces equipped with clear glazings, non-specular venetian blinds, and light shelves, as well as translucent panels. The resulting information can be (a) combined with occupant behavior models that mimic occupant use of movable shading devices and lighting controls and (b) used to predict the availability of daylight throughout the year, the likeliness of occupants experiencing discomfort glare, and the amount of energy savings due to automated controls. Further research is needed to adequately translate the results from daylight simulations into adequate design metrics and to increase the number of complex fenestration systems that can be reliably modeled. As the field evolves, researchers are increasingly developing metrics for other design aspects such as the quality of a view to the outside or the effect of (day)light on human health and alertness. Finally, tools have been developed to analyze the daylight availability at the urban level, facilitating the formulation of new zoning laws.

Notes

1 As a side note, while modeling a fully glazed façade explicitly as a 'mirror' material modifier in the Radiance raytracer may be required for an accurate glare analysis, this added realism comes at a price as far as the required simulation time and model handling are concerned and might not be required if work plane illuminances are calculated instead.

2 The weight assigned to a single ray usually depends on the optical properties of the material(s) from which it has been reflected before, as well as the number of reflections since the primary ray has been emitted from a sensor or viewpoint.

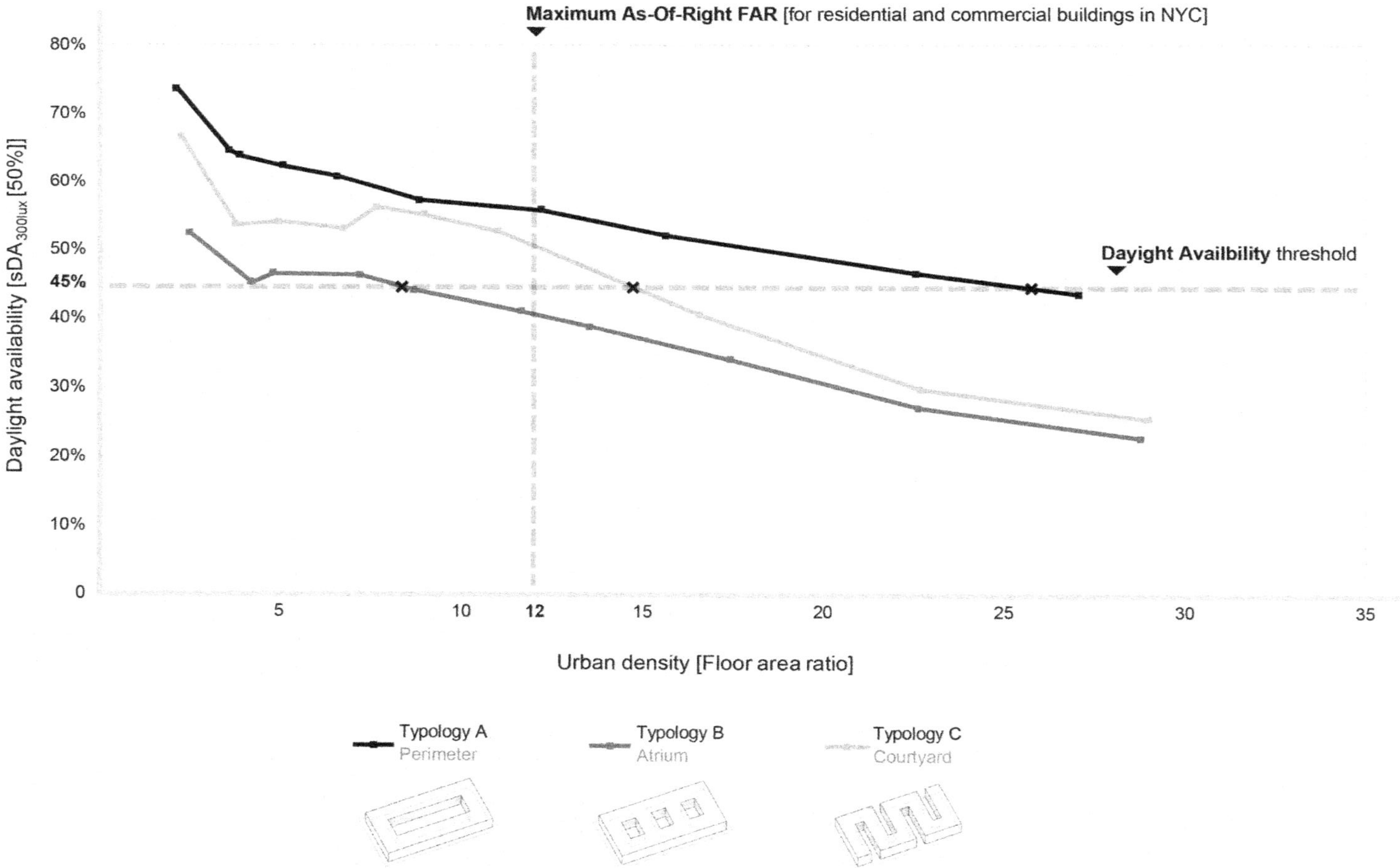

Figure 7.26 Daylight simulation study of urban density (floor area ratio) versus daylight availability (sDA300lux[50%]) for three block typologies in Manhattan

Source: Saratsis et al. 2017

3 A caveat of the study – as stated by the authors – is that invitations to participate in the survey (although they were disseminated as widely as possible) were somewhat skewed towards Radiance users due to the simple fact that the study authors are part of that community and therefore yielded a high feedback rate on the survey.
4 The 'glazing factor' promoted by the earlier mentioned LEED green building rating system calculates a quantity that is in essence identical to the mean daylight factor distribution in a space (Reinhart and LoVerso 2010).
5 Note that a target daylight factor level of 2% anywhere in the space differs from a mean daylight factor criteria of 2% for a whole space.
6 Private communication with C. Human (Loisos + Ubbelohde) and K. Yancey (Lam Partners).

References

Aizlewood, M.E. (1993) Innovative daylighting systems: An experimental evaluation. *Lighting Research & Technology* 25(4): 141–152.

Amercian Institute of Architects. (2012) An Architect's Guide to Integrating Energy Modeling in the Design Process. *Report by the American Institute of Architects (AIA)*. Washington, DC, pp. 1–86.

Amundadottir, M.L., Lockley, S.W. and Andersen, M. (2016) Unified framework to evaluate non-visual spectral effectiveness of light for human health. *Lighting Research & Technology* 49(6): 673–696. https://doi.org/10.1177/1477153516655844.

Amundadottir, M.L., Rockcastle, S., Khanie, M.S. and Andersen, M. (2017) A human centric approach to assess daylight in buildings for non-visual health potential, visual interest and gaze behavior. *Building and Environment* 113: 5–21.

Andersen, M. and deBoer, J. (2006) Goniophotometry and assessment of bidirectional photometric properties of complex fenestration systems. *Energy and Buildings* 38(7): 836–848.

ANSI/IESNA. (2002) *LM – 63–02 ANSI Approved Standard File Format for Electronic Transfer of Photometric Data and Related Information*. http://webstore.ansi.org.

ANSI/IESNA, ANSI/IES_LM-63–02, and ANSI/IESNA. (2002) LM – 63–02 ANSI approved standard file format for electronic transfer of photometric data and related information. In *Lighting Measurement Standard (LM)*. New York: Illuminating Engineering Society of North America. http://webstore.ansi.org.

ASHRAE. (2007) ANSI/ASHRAE Standard 140–2007 standard method of test for the evaluation of building energy analysis computer programs. American Society of Heating, Refrigerating and Air-Conditioning Engineers.

ASHRAE. (2017) *ASHRAE Handbook of Fundamentals*. Atlanta, GA: American Society of Heating, Refrigerating and Air-Conditioning Engineers.

Autodesk. (2017) Autodesk – Ecotect (Version 5.6). Autodesk Inc. h.

Bierman, A. and National Lighting Product Information Program (NLPIP) (2007) Photosensors – dimming and switching systems for daylight harvesting. *Specifier Reports*. Try, NY: National Lighting Product Information Program (NLPIP).

Bourgeois, D., MacDonald, I.A. and Reinhart, C.F. (2006) Adding advanced behavioral models in whole building energy simulation: A study on the total energy impact of manual and automated lighting control. *Energy and Buildings* 38(7): 814–823.

Bourgeois, D., Reinhart, C.F. and Ward, G. (2008) A standard daylight coefficient model for dynamic daylighting simulations. *Building Research & Information* 36(1): 68–82.

Cannon-Brookes, S.W.A. (1997) Simple scale models for daylighting design: Analysis of error illuminance prediction. *Lighting Research & Technology* 29(3): 135–142.

CIBSE. (2014) *Lighting Guide 10: Daylighting – A Guide for Designers: Lighting for the Built Environment (LG10/14)*. London, UK.

CIE. (1973) Standardization of luminance distribution on clear skies. *CIE Publication 22*. Vienna: International Commission on Illumination.

CIE. (2004) *ISO 15469/CIE S 011/E Spatial Distribution of Daylight – CIE Standard General Sky, 2nd edition*. Geneva, Switzerland: ISO/CIE.

CIE. (2006) *CIE 171 – Test Cases to Assess the Accuracy of Lighting Computer Programs*. Vienna, Austria: International Commission on Illumination (CIE).

Clarke, J. and Janak, M. (1998) Simulating the thermal effects of daylight-controlled lighting. *Building Performance (BEPAC), Issue 1*.

Cohen, M.F. and Wallace, J.R. (1993) *Radiosity and Realistic Image Synthesis*, 2nd edition. Boston, MA: Academic Press Professional Inc.

Compagnon, R. (2004) Solar and daylight availability in the urban fabric. *Energy and Buildings* 36: 321–328.

Covey, S.R. (1989) *The 7 Habits of Highly Effective People*. New York: Fireside.

Crawley, D.B., Hand, J.W. and Lawrie, L.K. (1999) Improving the weather information available to simulation programs. *Sixth International IBPSA Conference (BS '99) Kyoto, Japan* II, pp. 529–536.

Dogan, T., Reinhart, C.F. and Michalatos, P. (2012) Urban daylight simulation: Calculating the daylit area of urban designs. In *SimBuild 2012*. Madison, WI, August 1–3, pp. 1–8.

Doraiswamy, H., Ferreira, N., Lage, M., Vo, H.T., Wilson, L., Werner, H., Park, M. and Silva, C. (2015) Topology-based catalogue exploration framework for identifying view-enhanced tower designs. In *ACM Transactions on Graphics (SIGGRAPH Asia '15)* 230: 1–13.

Dufton, A.F. (1946) Protractors for the computation of daylight factors. *D.S.I.R. Building Research Technical Paper No. 28*. London: Her Majesty's Stationary Office.

Dufton, A.F. and Longmore, J. (1946) *BRS Daylight Protractors AP 68*. Building Research Establishment, pp. 1–25.

Ehrlich, C., Papamichael, K., Lai, J. and Revzan, K. (2002) A method for simulating the performance of photosensor-based lighting controls. *Energy and Buildings* 34: 883–889.

Figueiro, M.G., Brainard, G.C., Lockley, S.W., Revell, V.L. and White, R. (2008) Light and human health: An overview of the impact of optical radiation on visual, circadian, neuroendocrine, and neurobehavioral responses. *IES TM-18–08*.

Flynn, J., Hendrick, C., Spencer, T. and Martyniuk, O. (1979) A guide to methodology procedures for measuring subjective impressions in lighting. *Journal of Illuminating Engineering Society* 8: 95–110.

Franks, M. (2005) Daylighting in museums. In *Paper presented at the International Radiance Workshop*, Montreal. https://www.radiance-online.org/community/workshops/2005-montreal/PDF/Franks_ArupCaseStudies.pdf, last accessed 2018

Galasiu, A. and Reinhart, C.F. (2007) Current daylighting design practice: A survey. *Building Research and Information* 36(2): 159–174.

Guglielmetti, R., Macumber, D. and Long, N. (2011) Openstudio: An open source integrated analysis platform. In *Building Simulation 2011*. Sydney, Australia, pp. 442–449.

Gunay, H.B., O'Brien, W., Beausoleil-Morrison, I., Goldstein, R., S Breslav and Khan, A. (2014) Coupling stochastic occupant models to building performance simulation using the discrete event system specification formalism. *Journal of Building Performance Simulation* 7(6): 457–478.

Gunay, H.B., O'Brien, W., Beausoleil-Morrison, I. and Gilani, S. (2017) Development and implementation of an adaptive lighting and blinds control algorithm. *Building and Environment* 113 (Supplement C): 185–199. https://doi.org/https://doi.org/10.1016/j.buildenv.2016.08.027.

HarperCollins. (2002) *Collins Thesaurus of the English Language – Complete and Unabridged*, 2nd edition. New York City, NY: HarperCollins.

Hoes, P., Hensen, J.L.M., Loomans, M.G.L.C., deVries, B. and Bourgeois, D. (2009) User behavior in whole building simulation. *Energy and Buildings* 41: 295–302.

Hopkinson, R.G., Longmore J. and Petherbridge P. (1946) An empirical formula for the computation of the indirect component of daylight factor. *Transactions of the Illuminating Engineering Society* 19: 201. London.

Hunt, D.R.G. (1979) The use of artificial lighting in relation to daylight levels and occupancy. *Building and Environment* 14: 21–33.

Ibarra, D. and Reinhart, C.F. (2009) *Daylight Factor Simulations – How Close Do Simulation Beginners 'Really' Get? Building Simulation 2009*. Glasgow, Scotland: IBPSA.

Ibarra, D. and Reinhart, C.F. (2013) Teaching daylight simulations – improving modeling workflows for simulation novices. In *Proceedings of Building Simulation 2013*. Chambery, France.
IBPSA-USA. (2017) Building energy software tools directory. US Department of Energy. www.buildingenergysoftwaretools.com/.
IESNA. (2011) *IES Lighting Handbook, 10th edition*. New York: Illuminating Engineering Society of North America, ISBN 0-87995-150-8.
IESNA. (2012) IES LM – 83–12 IES Spatial Daylight Autonomy (SDA) and Annual Sunlight Exposure (ASE). New York City, NY.
Inanici, M. (2006) Evaluation of high dynamic range photography as a luminance data acquisition system. *Lighting Research and Technology* 38(2): 123–136.
Inanici, M. (2009) Applications of image based rendering in lighting simulation: Development and evaluation of image based sky models. *Building Simulation 2009* 1: 264–271.
Inanici, M., Brennan, M. and Clark, E. (2015) Spectral daylighting simulations: Computing circadian light. In *14th Conference of International Building Performance Simulation Association*. Hyderabad, India, December 7–9, pp. 1–8.
Jakubiec, J.A. (2016) Building a database of opaque materials for lighting simulation. In *Passive and Low Energy Architecture (PLEA)*, Los Angeles, July 11–13, pp. 1–8. Database URL: http://spectraldb.com/.
Jakubiec, J.A. and Reinhart, C.F. (2011a) Diva-for-rhino 2.0: Environmental parametric modeling in rhinoceros/grasshopper using radiance, daysim and energyPlus. *Building Simulation 2011*. Sydney, Australia: IBPSA.
Jakubiec, J.A. and Reinhart, C.F. (2011b) The 'Adaptive Zone' – A concept for assessing glare throughout daylit spaces. *Lighting Research and Technology* 44: 149–170.
Janak, M. (1997) Coupling building and lighting simulation. *5th International IBPSA Conference*, September 8–10, 1997, Prague II: 313–319.
Jarvis, D. and Donn, M. (1997) Comparison of computer and model simulations of a daylit Interior with reality. In *5th International Building Performance Simulation Association (IBPSA) Conference*.
Jones, N.L. and Reinhart, C.F. (2016) Real-time visual comfort feedback for architectural design. In *PLEA 2016–32nd International Conference on Passive and Low Energy Architecture*. Los Angeles.
Jones, N.L. and Reinhart, C.F. (2017) Experimental validation of ray tracing as a means of image-based visual discomfort prediction. *Building and Environment* 113: 131–150.
Jones, N.L. and Reinhart, C.F. (2018) Effects of teal-time simulation feedback on design for visual comfort. *Journal of Building Performance Simulation*. https://doi.org/10.1080/19401493.2018.1449889
Kittler, R., Perez, R. and Darula, S. (1997) A new generation of sky standards. In *Lux Europa*. Amsterdam: Netherlands Institution of Illuminating Engineering.
Kittler, R., Perez, R., Darula, S. and Bratislava, P. (1998) A set of standard skies characterizing daylight conditions for computer and energy conscious design. *US SK 92 052 Final Report, ICA SAS*. Bratislava.
Kleindienst, S., Bodart, M. and Andersen, M. (2008) Graphical representation of climate-based daylight performance to support architectural design. *LEUKOS* 5(1): 39–61.
Lafortune, E.P. and Willems, Y.D. (1993) Bi-directional path tracing. *Proceedings of CompuGraphics* 93: 145–153.
LBNL. (2017) International Glazing Database. http://windows.lbl.gov.
Leder, S.M., Pereira, F.O.R. and Moraes, L.N. (2007) Characterization of a medium reflection coefficient to urban vertical surfaces (in Portuguese). In *Encontro Nacional e Encontro Latino Americano de Conforto No Ambiente Construído*. Ouro Preto, Brazil, August 8–10.
Lynes, J. (1979) A sequence for daylighting design. *Lighting Research & Technology* 11(2): 102–106.
Mahdavi, A., Mohammadi, A., Kabir, E. and Lambeva, L. (2008) Shading and lighting operation in office buildings in Austria: A study of user control behavior. *Building Simulation* 1: 111–117.
Mardaljevic, J. (1995) Validation of a lighting simulation program under real sky conditions. *Lighting Research & Technology* 27(4): 181–188.

Mardaljevic, J. (2000a). Simulation of annual daylighting profiles for internal illuminance. *Lighting Research & Technology* 32(2): 111–118.

Mardaljevic, J. (2002b) Quantification of parallax errors in sky simulator domes for clear sky conditions. *Lighting Research & Technology* 34(4): 313–332.

Mardaljevic, J. (2004) Spatio-temporal dynamics of solar shading for a parametrically defined roof system. *Energy and Buildings* 36(8): 815–823.

Mardaljevic, J., Heschong, L. and Lee, E. (2009) Daylight metrics and energy savings. *Lighting Research and Technology* 41(3): 261–283.

Mardaljevic, J. and Rylatt, M. (2003) Irradiation mapping of complex urban environments: An image-based approach. *Energy and Buildings* 35(1): 27–35.

Marsh, A. (2003) Computer-optimised shading design. In Hensen, J. (ed.), *Building Simulation 2003*. Eindhoven, The Netherlands: International Building Performance Simulation Association (IBPSA).

McNeil, A. and Lee, E.S. (2013) A validation of the radiance three-phase simulation method for modelling annual daylight performance of optically complex fenestration systems. *Journal of Building Performance Simulation* 6(1): 1–14.

Mental-images. (2007) *Mental Ray(R) Functional Overview Version 1.5*. http://www.Mentalimages.com/Fileadmin/User_upload/PDF/Mental_ray_Functional_Overview.Pdf. Berlin.

Moon, P. and Spencer, D.E. (1942) Illumination from a non-uniform sky. *Illuminating Engineering (N.Y.)* 37(10): 707–726.

Nabil, A. and Mardaljevic, J. (2005) Useful daylight illuminance: A new paradigm to access daylight in buildings. *Lighting Research & Technology* 37(1): 41–59.

Newsham, G.R. (1994) Manual control of window blinds and electric lighting: Implications for comfort and energy consumption. *Indoor Environment* 3: 135–144.

Olgay, A. and Olgay, V. (1957) *Solar and Shading Devices Control*. Princeton: Princeton University Press.

Olseth, J.A. and Skartveit, A. (1989) Observed and modeled hourly luminous efficacies under arbitrary cloudiness. *Solar Energy* 42(3): 221–233.

Pechacek, C.S., Andersen, M. and Lockley, S.W. (2008) Preliminary method for prospective analysis of the circadian efficacy of (day)light with applications to healthcare architecture. *LEUKOS* 5(1): 1–26.

Perez, R., Ineichen, P., Seals, R., Michalsky, J. and Stewart, R. (1990) Modeling daylight availability and irradiance components from direct and global irradiance. *Solar Energy* 44(5): 271–289.

Perez, R., Seals, R. and Michalsky, J. (1993) All-weather model for sky luminance distribution – preliminary configuration and validation. *Solar Energy* 50(3): 235–245.

Preetham, A.J., Shirley, P. and Smits, B. (1999) A practical analytic model for daylight. *International Conference on Computer Graphics and Interactive Techniques Archive Proceedings of the 26th Annual Conference on Computer Graphics and Interactive Techniques*, pp. 91–100.

Provencio, I., Rodriguez, I.R., Jiang, G., Hayes, W.P., Moreira, E.F. and Rollag, M.D. (2000) A novel human opsin in the inner retina. *The Journal of Neuroscience* 20(2): 600 LP–605. www.jneurosci.org/content/20/2/600.abstract.

Reindl, D.T., Beckman, W.A. and Duffie, J.A. (1990) Diffuse fraction correlations. *Solar Energy* 45(1): 1–7.

Reinhart, C.F. (2014) *Daylighting Handbook I – Fundamentals + Designing with the Sun*. Cambridge, MA. www.daylightinghandbook.com.

Reinhart, C.F., Jakubiec, J.A. and Ibarra, D. (2013) Definition of a reference office for standardized evaluations of façade and lighting technologies. In *Proceedings of Building Simulation 2013*. Chambery, France.

Reinhart, C.F. (2004) Lightswitch 2002: A model for manual and automated control of electric lighting and blinds. *Solar Energy* 77(1): 15–28.

Reinhart, C.F. and Andersen, M. (2006) Development and validation of a radiance model for a translucent panel. *Energy and Buildings* 38(7): 890–904.

Reinhart, C.F. and Breton, P-F. (2009) Experimental validation of 3ds max® design 2009 and daysim 3.0. *LEUKOS*, no. July 2009.

Reinhart, C.F. and Fitz, A. (2006) Findings from a survey on the current use of daylight simulations in building design. *Energy and Buildings* 38(7): 824–835.

Reinhart, C.F. and Herkel, S. (2000) The simulation of annual daylight illuminance distributions- A state of the art comparison of six RADIANCE-based methods. *Energy and Buildings* 32(2): 167–187.

Reinhart, C.F. and LoVerso, V.M. (2010) A rules of thumb-based design sequence for diffuse daylight. *Lighting Research & Technology* 42: 7–32.

Reinhart, C.F., Mardaljevic, J. and Rogers, Z. (2006) Dynamic daylight performance metrics for sustainable building design. *LEUKOS* 3(1): 1–20.

Reinhart, C.F., Rakha, T. and Weissman, D. (2014) Predicting the daylit area – A comparison of students assessments and simulations at eleven schools of architecture. *LEUKOS* 10(4): 193–206.

Reinhart, C.F. and Walkenhorst, O. (2001) Dynamic radiance-based daylight simulations for a full-scale test office with outer venetian blinds. *Energy & Buildings* 33(7): 683–697.

Reinhart, C.F. and Wienold, J. (2011) The daylighting dashboard – A simulation-based design analysis for daylit spaces. *Building and Environment* 46(2): 386–396.

Robinson, D. and Stone, A. (2004) Irradiation modeling made simple: The cumulative sky approach and its applications. In *PLEA 2004 – The 21st Conference on Passive Low Energy Architecture*. Eindhoven, The Netherlands, September 19–22, 2004.

Rockcastle, S., Amundadottir, M.L. and Andersen, M. (2016) Contrast measures for predicting perceptual effects of daylight in architectural renderings. *Lighting Research & Technology* 49(7): 882–903. https://doi.org/10.1177/1477153516644292.

Rockcastle, S.F., Chamilothori, K. and Andersen, M. (2017) An experiment in virtual reality to measure daylight-driven interest in rendered architectural scenes. In *Building Simulation 2017*. San Francisco, CA, pp. 1–8.

Rogers, Z., Scheib, J. and Fitzpatrick, H. (2009) SPOT (Sensor Placement and Optimization Tool) Users Manual Ver. 4. www.archenergy.com/SPOT/support.html. Architectural Energy Corporation.

Rubin, A.I., Collins, B.L., Tibott, R.L., Collins, B.L. and Tibott, R.L. (1978) *Window Blinds as a Potential Energy Saver – A Case Study*. Washington, DC: NSB Building Science Series 112, National Bureau of Standards.

Sadeghipour Roudsari, M. and Pak, M. (2013) Ladybug: A parametric environmental plugin for grasshopper to help designers create an environmentally-conscious design. In *Building Simulation 2013*. Chamberry, France, pp. 3128–3135.

Saratsis, E., Dogan, T. and Reinhart, C.F. (2017) Simulation-based daylighting analysis procedure for developing urban zoning rules. *Building Research & Information* 45(5): 478–491. https://doi.org/10.1080/09613218.2016.1159850.

Sargent, J.A., Niemasz, J. and Reinhart, C.F. (2011) SHADERADE: Combining rhinoceros and energyplus for the design of static exterior shading devices. In *Building Simulation 2011*. Sydney, pp. 310–317.

Schregle, R. (2004) *Daylight Simulation with Photon Maps*. Saarbruecken: Universitaet des Saarlandes.

Solemma. (2016) *DIVA-for-Rhino V4*. Cambridge, MA: Solemma. http://solemma.com.

Solemma. (2018) *Adaptive Lighting for Alertness (ALFA) V1*. http://solemma.net/Alfa.php.

Szokolay, S.V. (2004) *Introduction to Architectural Science*. Oxford: Elsevier Ltd.

Tregenza, P.R. (1987) Subdivision of the sky hemisphere for luminance measurements. *Lighting Research & Technology* 19: 13–14.

Tregenza, P.R. and Waters, I.M. (1983) Daylight coefficients. *Lighting Research & Technology* 15(2): 65–71.

Tsangrassoulis, A., Santamouris, M. and Asimakopoulos, D. (1996) Theoretical and experimental analysis of daylighting performance for various shading systems. *Energy and Buildings* 24: 223–230.

UK Education Funding Agency. (2014) Priority School Program. www.gov.uk/government/collections/priority-school-building-programme-psbp.

US-DOE. (2017) *EnergyPlus*. https://energyplus.net/.

US Green Building Council. (2009) *LEED-NC* (Leadership in Energy and Environmental Design) V3. www.usgbc.org.
US Green Building Council. (2013) *LEED-NV* (Ledership in Energy and Environmental Design) V4. Washington, DC. www.usgbc.org.
US Green Building Council. (2017) *USGBC | LEED.* https://new.usgbc.org.
Waldram, P.J. (1909) A standard of daylight illumination of interiors. *Illuminating Engineering* 2: 469.
Waldram, P.J. (1950) *A Measuring Diagram for Daylight Illumination.* London: B.T. Batsford Ltd.
Walkenhorst, O., Luther, J., Reinhart, C.F. and Timmer, J. (2002) Dynamic annual daylight simulations based on one-hour and one-minute means of irradiance data. *Solar Energy* 72(5): 385–395.
Ward, G.J. (1994) The RADIANCE lighting simulation and rendering system (Computer Graphics). In *Proceedings of '94 SIGGRAPH conference*, pp. 459–472.
Ward, G. (2017) *Photosphere Software Version 1.8. Software Anyhere.* Albany, CA. https://doi.org/www.anyhere.com/.
Ward, G. and Shakespeare, R. (1998) *Rendering with Radiance. The Art and Science of Lighting Visualization.* San Francisco, CA: Morgan Kaufmann Publishers.
Wasilowski, H.A. and Reinhart, C.F. (2009) Modeling an existing building in designBuilder/energyPlus: Custom versus default inputs. *Building Simulation 2009.* Glasgow, Scotland: IBPSA.
Wienold, J. (2009) Dynamic daylight glare evaluation. *Building Simulation 2009.* Glasgow: IBPSA.
Wienold, J. and Christoffersen, J. (2006) Evaluation methods and development of a new glare prediction method for daylight environments with the use of CCD cameras. *Energy and Buildings* 38(7): 743–757.
Winkelmann, F.C. and Selkowitz, S. (1985) Daylight simulation in the DOE-2 building energy analysis program. *Energy and Buildings* 8: 271–286.

Recommended reading

Lam, W.M.C. (1986) *Sunlighting as Formgiver for Architecture.* New York: Van Nostrand Reinhold.
Reinhart, C.F. (2014) *Daylighting Handbook I – Fundamentals + Designing with the Sun.* Cambridge, MA: Building Technology Press. www.BuildingTechnologyPress.com.
Reinhart, C.F. (2018) *Daylighting Handbook II – Daylight Simulations + Dynamic Façades.* Cambridge, MA: Building Technology Press.www.BuildingTechnologyPress.com.
Ward, G. and Shakespeare, R. (1998) *Rendering with RADIANCE – The Art and Science of Lighting Visualization.* San Francisco, CA: Morgan Kaufmann Publishers.

Activities

Conducting daylight simulations without any background knowledge may lead to dubious interpretations. However, theoretical simulation knowledge without hands-on experience is also of limited use. The simulation exercise below is taken from http://buildingtechnologypress.com/Resources.html, where more advanced exercises are available as well.

Assignment: shading studies and sun path diagrams

Task 1: direct shading

The objective of this assignment is to validate the results of direct shading studies provided by Radiance and other daylight simulation environments. To do so, you are asked to take a photo of an object of your choice outside at least twice at different times of day under clear sky conditions. You are then asked to model the object in a CAD environment, such as Rhinoceros 3D, and generate two clear sky visualizations under the same sky conditions from about the same perspective as the photos. Present your findings. Tips: (a) Pay attention to daylight savings time. (b) Be sure to carefully align your physical model using, for example, a smartphone compass. An example study is shown in Figure 7.27.

Task 2: annual sun path diagram

Show an annual sun path diagram of your object and mark the two times in the year when the photos from Task 1 were taken. An example result is shown in Figure 7.28.

Figure 7.27 Comparison of photographs versus simulations of the shadows cast by an object on September 8 in Boston at noon and 3.45pm

Source: Photos and simulations by Jeff Geisinger

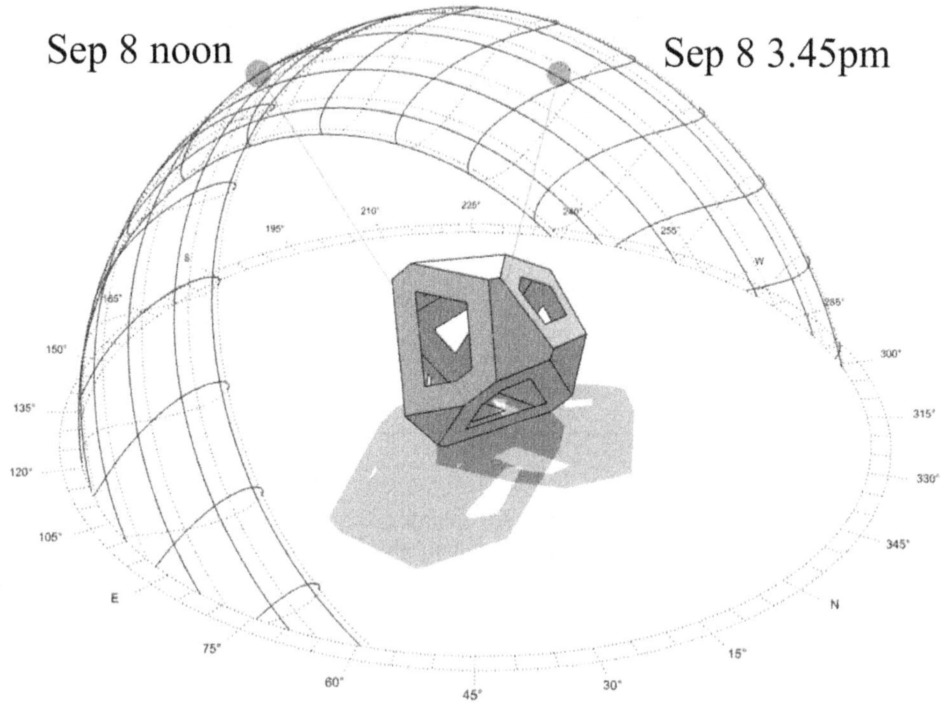

Figure 7.28 Annual sun path diagram in Boston (simulation: DIVA-for-Rhino)

8 Moisture modeling and durability assessment of building envelopes

Recent advances

Aytaç Kubilay, Xiaohai Zhou*, Dominique Derome and Jan Carmeliet*

Scope

In this chapter, recent advances in properly taking into account moisture phenomena in modeling the hygrothermal behavior of building envelopes are presented. These advances deal with the modeling and selection of climatic boundary conditions when modeling the interaction of the building envelope with the outdoor environment and with the modeling and assessment of the hygrothermal envelope conditions in terms of durability and performance.

Learning objectives

At the end of the chapter, the reader should have achieved the following objectives:

1 To understand that a main source of moisture on buildings is wind-driven rain;
2 To gain an overview of the methodology required for modeling wind-driven rain, especially using the Eulerian multiphase model;
3 To be familiarized with the different manifestations of wind-driven rain through studies on the influence of façade features and nearby buildings on rain deposition;
4 To understand the significance of properly selecting climatic conditions for hygrothermal simulation of the building envelope;
5 To gain an overview in the methodology to select a moisture reference year for hygrothermal simulations;
6 To gain an overview of coupled moisture and heat transport modeling in building components, including freezing and thawing;
7 To be familiarized with the different applications of hygrothermal simulation through studies of durability, e.g. biodegradation of wooden beams and freeze-thaw cycle of brick, of walls reinsulated from the interior.

Key words

Heat and moisture transfer/wind-driven rain/computational fluid dynamics/moisture damage

Introduction

Moisture plays an important role in building performance, especially the performance of building envelopes, and therefore moisture should be taken into account in the simulation of the behavior of buildings and building components. The deterioration of building structures due

* The two first authors contributed equally to this work.

to moisture often begins in regions where the building exposure to wind-driven rain (WDR) impingement is the highest. The chapter's first part, titled "Numerical Modeling of Wind-driven Rain," presents recent developments in WDR simulation. WDR is one of the most important moisture sources in the built environment (Blocken and Carmeliet 2004) and constitutes a significant part of the external moisture load, thus it should be considered as a boundary condition in heat and moisture transport models for the building envelopes. To properly capture the moisture-related effects of rain, different steps in the WDR phenomenon should be considered. The impinging WDR intensity is the total amount of rainwater that comes into contact with the building surface (Figure 8.1(a)). The interaction of wind-flow conditions and rain of various intensities may lead to very different rain deposition distributions within the urban environment. Rain deposition is not only influenced by the presence of large objects, such as buildings of different sizes and shapes and other elements of the surrounding environment, but also by small-scale features such as façade details.

After raindrops come into contact with the surface, phenomena such as splashing, bouncing, adhesion, runoff, evaporation, absorption, and the distribution of the moisture in the façade (rain penetration and wetting/drying) are observed (Figure 8.1(b)). The response of the wall to WDR is also determined by a diversity of parameters: the WDR intensity and raindrop-size distribution, the raindrop diameter, impact angle and impact speed, material-surface characteristics such as wall roughness, material characteristics such as moisture permeability and moisture retention, construction details, cracks and joints, etc.

The second part of the chapter, titled "Boundary Conditions for Hygrothermal Simulation of Building Envelope Assemblies," deals with the selection of the meteorological data to be used when simulating the hygrothermal performance of wall assemblies. It is good practice for building envelope hygrothermal simulations to subject the assembly to 2 or 3 years of meteorological data and then analyze the hygrothermal response of the last year to ensure that the response is independent from initial conditions. Also, it is common practice to select a year that has a certain level of moisture risk, commonly 10%. A recent methodology developed for the selection of this year is presented. This methodology is based on a comparison of evaporation and wetting potentials, as explained below.

The third part of the chapter, titled "Hygrothermal Performance of Building Envelope Assemblies," deals with the hygrothermal simulation and the evaluation of the risk of wood decay and of freeze-thaw damage in envelope assembly. Building materials in a wall assembly may be subjected to deterioration under the combined effect of temperature and moisture. Numerical hygrothermal models, simulating the coupled transport of heat and moisture over

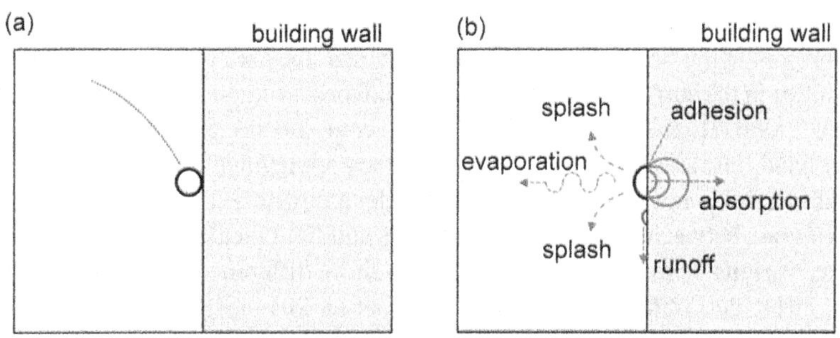

Figure. 8.1 Schematic representation of the two parts in wind-driven rain research: (a) assessment of the impinging wind-driven rain intensity (before rain impact) and (b) assessment of the response of the wall (at and after rain impact)

varying environmental conditions, are powerful tools to assess long-term heat and moisture performance of wall assemblies (Carmeliet et al. 2009) and predict the risk of moisture damage.

Finally, the chapter ends with some discussion and conclusions.

Numerical modeling of wind-driven rain

Introduction

Rain in the built environment is a complex phenomenon that has impact on, among others, the durability of the materials constituting the different urban surfaces such as building façades, roofs, vegetated areas, paved streets, and pedestrian ways. Raindrops that strike and are deposited on a porous surface can be absorbed or run off, both with potentially damaging consequences. As a result of absorption, moisture transfer can occur by capillary action, in the case of solid masonry (Hens 2010; Abuku et al. 2009), or by solar-induced thermal gradients, in the case of wood-framed walls (Carmeliet and Derome 2012a, 2012b). As a result of runoff, water can leak into the building façade through cracks or inadequate capillary breaks (Teasdale St-Hilaire and Derome 2007). Water in porous materials can lead to several undesired phenomena, such as freeze-thaw damage on exterior wall surfaces (Van Balen 1996), erosion of building materials (Tang et al. 2004), moisture-induced salt migration (Charola and Lazzarini 1986), discoloration by efflorescence (Franke et al. 1998), and surface soiling (Etyemezian et al. 2000), and may result in interior surface problems such as mold growth. Moreover, the conservation of historic buildings, the design of new buildings, and the development of new building materials rely on the correct prediction of moisture loads (Briggen et al. 2009). The correct prediction of WDR is also significant in the assessment of environmental risks related to the leaching of harmful biocides and nanoparticles from buildings (Burkhardt et al. 2012). These effects of WDR deposition do not occur only on buildings but also throughout urban and built environments.

Given the large relevance and significance of WDR in the built environment, in the past years, several reviews on the topic have been published. Blocken and Carmeliet (2004) provided an extensive review of the history of wind-driven rain research, with specific focus on assessment methods such as on-site experiments, wind-tunnel experiments, semi-empirical formulae, WDR maps, and numerical simulation based on computational fluid dynamics (CFD). A later review focused in detail on the evaluation of three calculation models for WDR (Blocken and Carmeliet 2010). More recent reviews can be found in Blocken et al. (2011), Moonen et al. (2012), and Blocken (2014).

Determining the distribution of WDR intensity can be performed with three general methods: experimental, semi-empirical, and numerical methods. Field measurements of WDR on building façades are difficult to set up, time consuming, and confined to the meteorological conditions present at the time of experiments. Moreover, they are prone to errors mainly due to the uncertainties in the surrounding environment and measuring instruments (Blocken and Carmeliet 2005, 2006b; Högberg et al. 1999; Kubilay et al. 2014a). Semi-empirical methods are, on the other hand, fast and easy to use. However, they are defined for specific building configurations and mostly do not take into account all relevant factors. Kubilay et al. (2014a) shows that the deviations between semi-empirical models and field measurements can be up to 88%. Furthermore, various semi-empirical models can show differences of up to 300% (Blocken et al. 2010, 2011; de Freitas et al. 2013; Kubilay et al. 2014a). Semi-empirical methods are generally suitable only for stand-alone buildings in simple configurations or for preliminary analysis. These methods will not give accurate results in cases of complex wind flow around multi-building configurations (Blocken et al. 2009). Besides, both measurements and semi-empirical methods are limited in terms of spatial resolution. CFD simulations can be used to obtain accurate and detailed spatial and temporal information on WDR. Although CFD simulations can be complex and time consuming, they can provide results for various types of rain events in much shorter times compared to experimental methods. The following subsections, first, describe how CFD simulations can be used to determine the distribution of WDR. A brief overview is given on two currently used methods, which are based on Lagrangian particle tracking (LPT) and Eulerian multiphase (EM) models. Second, the latter model is discussed in detail, and examples of time-averaged and time-resolved calculations of wind-driven rain are provided. Third, results from different applications are presented.

Scientific foundation: overview of numerical modeling approaches for wind-driven rain

WDR is the type of rain carried by wind and thus characterized by an oblique velocity vector as wind flow affects the droplet trajectories. Figure 8.2(a) shows the oblique rain intensity vector *R*. The horizontal component, R_{wdr}, of the rain vector is the WDR intensity (mm h^{-1}) that causes a rain flux through a vertical plane, whereas the vertical component, R_h, is the rainfall intensity (mm h^{-1}) that causes a rain flux through a horizontal plane.

Regardless of the WDR model used, several common steps in a WDR simulation are required to calculate the spatial and temporal distributions of catch ratio. These steps can be summarized as follows:

a The wind-flow pattern around the building is calculated. This step generally involves the calculation of steady-state wind-flow pattern based on Reynolds-averaged Navier-Stokes (RANS) turbulence models.

b The movement of raindrops is simulated using the calculated wind-flow field from step a. The calculation method developed by Choi (1993, 1997) and extended by Blocken and Carmeliet (2002, 2007b) is based on Lagrangian particle tracking (LPT), where trajectories for individual raindrops are obtained by solving their equation of motion. For this, raindrops are injected from numerous positions so that they cover the entire surface of a chosen façade. On the other hand, in the Eulerian multiphase model (Huang and Li 2010; Kubilay et al. 2013, 2014b), rain is regarded as a continuum, similar to the gas phase, i.e. air. Continuity and momentum equations are solved for different raindrop sizes, which give the distribution of rain fraction and rain velocity everywhere in the computational domain.

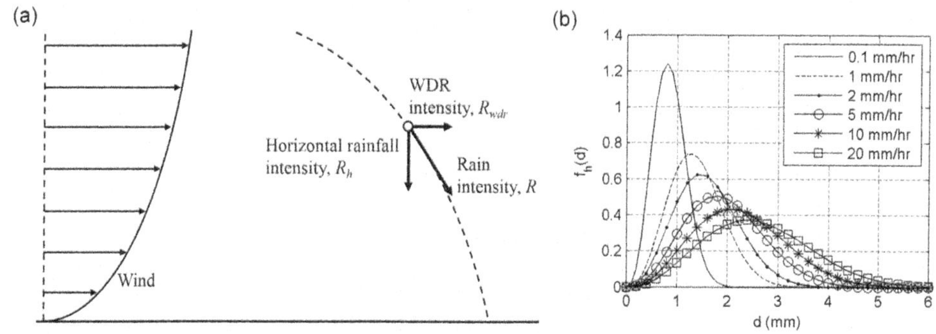

Figure 8.2 (a) Rain intensity vector and its components. (b) Raindrop size distribution through a horizontal plane with the rainfall intensity as a parameter – calculated from the raindrop size distribution in the air according to Best (1950).

c Once step b is completed, specific catch ratio, $\eta_d(d)$ (–), for each raindrop size, here the diameter, d (mm), can be determined. Specific catch ratio is obtained by normalizing the WDR intensity, R_{wdr}, on a surface due to a specific raindrop size with the unobstructed rainfall intensity away from buildings, R_h:

$$\eta_d(d) = \frac{R_{wdr}(d)}{R_h}. \tag{8.1}$$

The distribution of specific catch ratio differs for each raindrop size, as raindrops have different inertia and drag force depending on their size.

d The catch ratio, η (–), can be defined for the whole spectrum of raindrop sizes in the air as:

$$\eta = \int_d f_h(R_h, d)\eta_d(d)\mathrm{d}d \tag{8.2}$$

where $f_h(R_h, d)$ is the raindrop-size distribution through the horizontal plane (–), which is also a function of the rainfall intensity. An overview of various relationships describing raindrop-size distributions is given by Uijlenhoet and Stricker (1999). Figure 8.2(b) shows the raindrop size distribution through a horizontal plane based on the raindrop size distribution in the air by Best (1950). For lower rainfall intensities, the spread of the probability distribution curve is small and raindrops are of smaller size. As the rainfall intensity increases, larger raindrops tend to get more frequent and the spread of the curve gets larger. The distribution of catch ratios on building surfaces is influenced by building geometry, environment topography, position on the building façade, wind speed, wind direction, rainfall intensity, and raindrop-size distribution (Choi 1993; Blocken and Carmeliet 2002). This step creates a database of catch ratio distributions for reference values of wind direction, wind speed, and rainfall intensity for each location on the building façade.

e Finally, the catch ratio values are interpolated for each given position on the building using experimental data records for wind speed, wind direction, and horizontal rainfall intensity during a certain rain event. This approach is based on the assumption that the values calculated with steady-state airflow calculations for each experimental time step can capture the variations in the WDR intensity. The experimental time steps are mostly chosen to be 10 minutes, i.e. 10-minute averaged values for wind velocity and rainfall intensity, following the guidelines of van der Hoven (1957), Sumner (1981), and Blocken and Carmeliet (2007a, 2008) in order to minimize the errors related to the time resolution of meteorological datasets.

In order to model the complete moisture redistribution in the urban environment, a multiscale approach is necessary. At the droplet scale, the fate of rain droplets after impingement is studied based on the input coming from the WDR model. Droplet physics can be numerically modeled using methods such as the volume of fluid method (VOF) (Lee et al. 2016a). At the neighborhood scale, absorption in the built environment and the wetting and drying cycles of building materials, and eventually water runoff, are modeled. For this, heat and moisture transport in the built environment, i.e. building façades, ground, and vegetation, are coupled with WDR intensity using building-envelope heat-air-moisture (BE-HAM) (e.g. Janssen et al. 2007; Abuku et al. 2009). Because such analyses involve various modeling approaches at different scales, the efficiency of WDR models is particularly significant.

The calculation of WDR intensity using LPT models requires the surfaces to be fully covered for all sizes of raindrops. Therefore, for accurate WDR analyses, a very high number of raindrops is necessary, even in the case of stand-alone buildings with simplified façades (Blocken and Carmeliet 2006a, 2007b). This is a limitation as it becomes practically impossible to study the effects of small façade details. The EM model offers an option with less computational effort compared to LPT models with simpler boundary conditions (Kubilay et al. 2013). Furthermore, the EM model allows the estimation of the WDR intensity as well as the raindrop impact parameters, such as impact speed and impact angle, on all surfaces even in complex geometries. On the other hand, LPT models offer a more straightforward approach for droplet-wall interaction. This is because individual droplet trajectories are calculated in Lagrangian formulation, making it easier to implement empirical correlations related to phenomena at droplet impact. Besides, care must be taken when using the EM model against numerical diffusion especially in regions with large gradients of rain-phase fraction, e.g. in the presence of regions sheltered from rain. The following subsections focus mainly on the distribution of WDR intensity before raindrop impact on increasingly more complex cases, thus requiring the EM approach. Therefore, in the next subsection, the governing equations for the EM model are presented in detail.

Computational methods

In the Eulerian multiphase (EM) model, rain is considered to be composed of multiple phases, instead of a uniform phase, in addition to the air phase. Each class of raindrop size is treated as a different phase, as raindrops with similar size will interact with the wind-flow field in a similar way. For each separate rain phase defined in the EM model, the following continuity and momentum equations are solved (Kubilay et al. 2015):

$$\frac{\partial \alpha_d}{\partial t} + \frac{\partial \alpha_d \overline{u_{d,j}}}{\partial x_j} = 0 \tag{8.3}$$

$$\frac{\partial \alpha_d \overline{u_{d,i}}}{\partial t} + \frac{\partial \alpha_d \overline{u_{d,i}}\,\overline{u_{d,j}}}{\partial x_j} + \frac{\partial \alpha_d \overline{u'_{d,i} u'_{d,j}}}{\partial x_j} = \alpha_d g_i + \alpha_d \frac{3\mu_a}{\rho_w d^2} \frac{C_d \operatorname{Re}_R}{4} \left(\overline{u_{a,i}} - \overline{u_{d,i}}\right) \tag{8.4}$$

where α_d denotes the volume fraction of rain phase (–), $u_{d,i}$ the velocity component of rain phase in direction i (m s^{-1}), d the raindrop diameter (m, $u_{a,i}$ the velocity component of wind in direction i (m s^{-1}), ρ_w the density of the liquid water (kg m^{-3}), μ_a is the dynamic air viscosity (kg m^{-1}s^{-1}), g_i the gravitational acceleration (m s^{-2}), and C_d the drag coefficient (–). Overbar denotes Reynolds averaging and $u'_{d,i}$ denotes the fluctuating part of rain phase velocity (m s^{-1}). Re_R denotes the relative Reynolds number (–) and is calculated as follows:

$$\operatorname{Re}_R = \frac{\rho_a d}{\mu_a} \left\| \mathbf{u}_a - \mathbf{u}_d \right\| \tag{8.5}$$

where ρ_a is the density of air (kg m^{-3}), $\mathbf{u}_a$ the air phase velocity vector (m s^{-1}), and $\mathbf{u}_d$ the rain phase velocity vector (m s^{-1}). For each rain phase, the continuity and momentum equations are solved separately. The molecular mass flux is assumed negligible in comparison to the convective and turbulent mass fluxes (Gosman et al. 1992; Gousseau et al. 2011). The second term on the left-hand side in Equation 8.4 corresponds to the mean convective mass flux. The third term on the left-hand side is the turbulent mass flux, which is the transport of mass due to turbulent motions in the flow. Turbulent dispersion of droplets is due to these turbulent motions. Turbulent dispersion was neglected in many earlier WDR modeling efforts. The EM model provides a direct approach to include turbulent dispersion in WDR calculations. Turbulent flux in Equation 8.4 can be modeled according to the Boussinesq eddy viscosity approximation (Boussinesq 1887):

$$-\overline{u'_{d,i}u'_{d,j}} = \nu_{t,d}\left(\frac{\partial \overline{u_{d,i}}}{\partial x_j} + \frac{\partial \overline{u_{d,j}}}{\partial x_i} - \frac{2}{3}\frac{\partial \overline{u_{d,k}}}{\partial x_k}\delta_{ij}\right) - \frac{2}{3}k_d\delta_{ij} \tag{8.6}$$

where $\nu_{t,d}$ is the turbulent kinematic viscosity of the rain phase (m^2 s^{-1}), k_d is the turbulence kinetic energy of the rain phase (m^2 s^{-2}), and δ_{ij} is the Kronecker delta. Here, the Reynolds stresses are related to the mean shear rate by the turbulent viscosity, which should be modeled. Issa and Oliveira (1993), Shirolkar et al. (1996), Hill (1998), and Rusche (2002) provide a list of various approaches to relate the eddy viscosities of both phases to each other, including both solid-liquid and gas-liquid flows. The turbulent viscosity of the rain phase can be related to the one of the air phase as:

$$\nu_{t,d} = C_t^2\nu_{t,a} \quad C_t = \frac{u'_{d,RMS}}{u'_{a,RMS}}, \tag{8.7}$$

where $\nu_{t,d}$ is the turbulent kinematic viscosity of the air phase (m^2 s^{-1}) and C_t (–) is a scalar response coefficient defined as the ratio of the root mean square of the velocity fluctuations of the rain and air phases. Melville and Bray (1979) relate the turbulent viscosity of the dispersed phase and the continuous phase in a solid-liquid flow, which can be used to estimate the response coefficient as shown in Equation 8.8, where t_p denotes the particle relaxation time and t_{fl} denotes the Lagrangian fluid time scale. The particle relaxation time is defined as the rate of response of particle acceleration to the relative velocity between the particle and the carrier fluid. The Lagrangian fluid time scale can be viewed as the characteristic large eddy lifetime (Shirolkar et al. 1996), where k denotes the turbulence kinetic energy (m^2 s^{-2}) and ε the turbulence dissipation rate (m^2 s^{-3}) for the air phase.

$$\frac{\nu_{t,d}}{\nu_{t,a}} = \frac{1}{1+\dfrac{t_p}{t_{fl}}}, \quad t_p = \frac{4\rho_w d^2}{3\mu_a C_d \mathrm{Re}_R}, \quad t_{fl} \cong 0.2\frac{k}{\varepsilon} \tag{8.8}$$

In the limit of very small raindrops ($t_p/t_{fl} \rightarrow 0$), the droplets will fully follow the fluctuations in the wind flow and the response coefficient becomes $C_t \rightarrow 1$. On the other hand, larger raindrops are only weakly influenced by the fluctuations in the wind flow.

It is assumed that there is no mass transfer between air and rain phases, as well as between different rain phases, i.e. no evaporation, coalescence, or breakup is considered. Any change in the raindrop size distribution due to mass transfer between phases is assumed to be taken into account through the raindrop size distribution. Rain phase calculations are one-way coupled with the air phase, with the effect of raindrops on the wind flow thus ignored. This is a valid assumption, as the volumetric ratio of rain in air is below 1×10^{-3} for rainfall intensities up to 20 mm/h and below 1×10^{-2} for even the most severe cases according to the study by de Wolf (2001).

For the rain phases, inlet boundaries are defined as the boundaries from which rain enters the computational domain. The vertical component of rain phase velocity at the inlet is set equal to the terminal velocity for the raindrop. The horizontal component of rain phase velocity is set so that there is a local force equilibrium between the rain and wind phases. The volume fraction of rain phase at the inlet is calculated using the following relation: $\alpha_d = R_h f_h(R_h, d)/V_t(d)$, where $V_t(d)$ represents the terminal velocity of a raindrop with diameter d.

Once the governing equations of WDR are solved, the specific catch ratio for rain phase d on the surfaces of the computational domain can be calculated using the following relation (Huang and Li 2010):

$$\eta_d(d) = \frac{\alpha_d |V_n(d)|}{R_h f_h(R_h, d)} \tag{8.9}$$

where $|V_n(d)|$ denotes the velocity magnitude of the rain phase in the direction normal to the building surface. The rain phase volumetric fraction α_d and the normal velocity $|V_n(d)|$ are obtained by solving Equations 8.3 and 8.4. The probability of a raindrop size $f_h(R_h, d)$ is obtained based on the study of Best (1950).

Figure 8.3(a) shows the catch ratio on the surfaces of a cubic building, with height $H = 10$ m, and on the surrounding ground for a reference wind speed of 5 m/s and for a rainfall intensity of 1 mm/h. Significant wetting gradients are visible both on the windward façade and the rooftop of the cubic building, as well as on the ground surface. Figures 8.3(b) and (c) show the catch ratio distributions on the windward façade of a cubic building for rainfall intensities of 1 mm/h and 30 mm/h at a reference wind speed of 10 m/s. For both rainfall intensities, the highest values are found at the top edge and at the top corners of the façade. This is due to the higher wind speed values at higher altitudes along with the acceleration of wind flow around the edges of the building. As rainfall intensity increases, the wetting gradient gets smaller, so that the maximum catch ratio on the façade decreases and the minimum catch ratio increases. At higher rainfall intensity values, there are more of larger raindrops, as shown in Figure 8.2(b), which are less influenced by the local wind-flow features, e.g. acceleration around the edges of the building. The surface wetting on the ground also shows the effects of the local wind flow around the building. An area with high catch ratio is visible around the building with a shape resembling a horseshoe. This shape comes from the following: part of the raindrops at the lower parts of the windward façade is directed towards the ground surface due to the standing vortex. Similarly, part of the raindrops is directed away from the side edges of the building, showing the influence of the separated wind flow around the edges of the building.

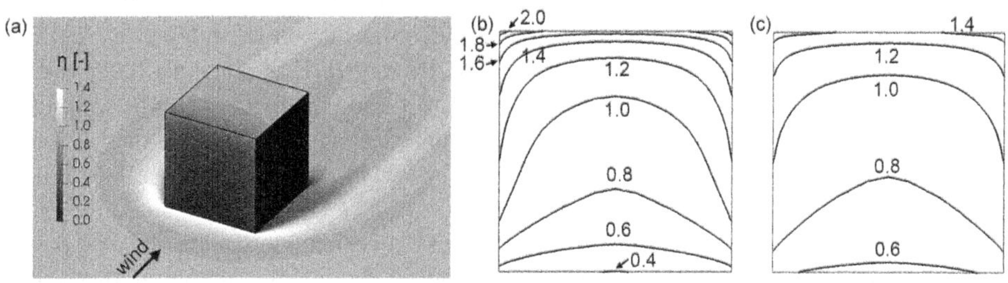

Figure 8.3 Distribution of catch ratio (a) on the surfaces of a cubic building, with height $H = 10$ m, and on the surrounding ground for wind speed at cube height $U_{10} = 5$ m/s and rainfall intensity $R_h = 1$ mm/h. Catch ratio on the windward façade for $U_{10} = 10$ m/s and rainfall intensity (b) $R_h = 1$ mm/h and (c) $R_h = 30$ mm/h.

Source: Derome et al. 2017

Time-averaged modeling of wind-driven rain

This section shows how time-averaged results of wind-driven rain for reference conditions are used to obtain catch ratio for a rain event. First, numerical and experimental catch ratio values are compared for a rain event on an array of low-rise cubic buildings. Second, numerical results are presented on a monumental tower building, namely the hunting lodge St. Hubertus in the Netherlands, and compared with available experimental data.

The array of low-rise buildings is composed of nine identical cubic building models of height of 2 m. Field measurements of WDR are conducted on the west façades of two cubes in the array by using 18 WDR gauges in total, nine gauges on each. The meteorological data for the rain event of September 16–19, 2013, are shown in Figure 8.4(a). The average wind speed at 5 m height is 1.6 m/s, whereas the average rainfall intensity is 1.8 mm/h. The total rainfall amount is 36 mm. The rain showers consist of high peaks in rainfall intensity for short periods. The wind direction is mostly between southwest (225°) and west (270°) except for some short periods when it switches to south-southeast.

For this study, first, steady-state WDR simulations for three reference wind directions (wind from northwest, west, and southwest) and five reference wind speeds are calculated. From these simulation results, the catch ratio database for relevant reference conditions is calculated, allowing the temporal study of any rain event. The validation studies are conducted by interpolating the results from the reference simulations using the 10-minute experimental data records. Figure 8.4(b) compares the measured and the numerical cumulative WDR amounts for the WDR gauges along the central axis of the façades during the rain event. The numerical values follow the experimental data successfully except for some discrepancies in the last part, where there are many short rain events with spikes of high rainfall intensity. Figure 8.5 compares the experimental and numerical catch ratio values by the WDR gauges on the west façades of the two cubes at the end of the rain event on September 16–19, 2013. The average discrepancy is 6.5% on the upstream and 12.0% on the downstream cube.

The second study is done for the St. Hubertus building consisting of a rectangular lower part and a 34.5 m high tower. The building is located in the Dutch National Park "De Hoge Veluwe." There is no other building in the immediate vicinity of the hunting lodge. The terrain southwest of the building is partly open, i.e. there is a grass field and a pond. The wind direction is mainly southwest with only limited fluctuations. The building is surrounded by a forest in the other directions. Most of the WDR gauges were located on the southwest façade of the tower, as this façade showed severe moisture-related deterioration due to the prevailing southwest wind direction during rain events. Therefore, in the numerical simulations, the wind direction is taken to be perpendicular to this façade of the tower.

Figure 8.6 shows numerical results of WDR on the tower façade facing southwest. The results obtained with the EM model including turbulent dispersion (Kubilay et al. 2015) are compared with existing numerical results using the LPT model (Briggen et al. 2009) and with available experimental data acquired on the windward façade of the tower of the rain events on September 17, 2007 (Briggen et al. 2009). The numerical simulations with the LPT model neglect the turbulent dispersion of raindrops, while the EM model includes the effect of turbulent dispersion.

At the higher positions of the façade, the differences between numerical results with the LPT model and measurements are relatively low. Considering the complex nature of WDR, it can be said that the numerical model is successful at capturing the WDR intensity at these parts of the building. However, at the lower positions, there is a relatively large difference between the LPT results and measurements. Given the relatively low wind speed during the rain event (lower than 2 m/s at 10 m height) and the higher wind-blocking effect of the tower compared to a low-rise building, the effect of turbulent dispersion becomes more important on the lower

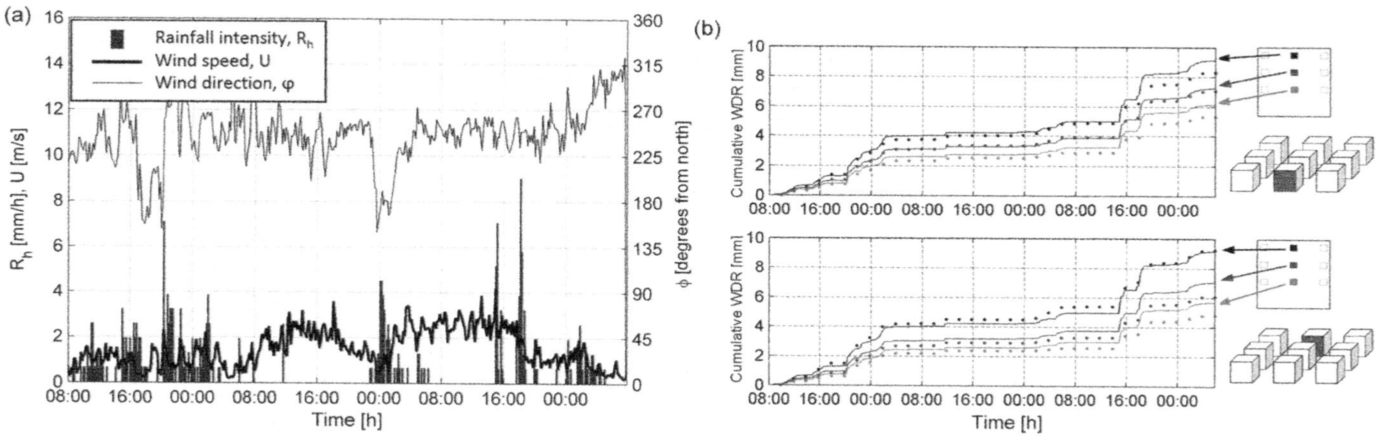

Figure 8.4 (a) 10-minute meteorological data for the rain event on September 16–19, 2013. (b) Comparison of cumulative WDR amounts on two cubes in the array during the rain event on September 16–19, 2013 (dots: measurements, solid lines: simulations).

Source: Kubilay et al. 2014b

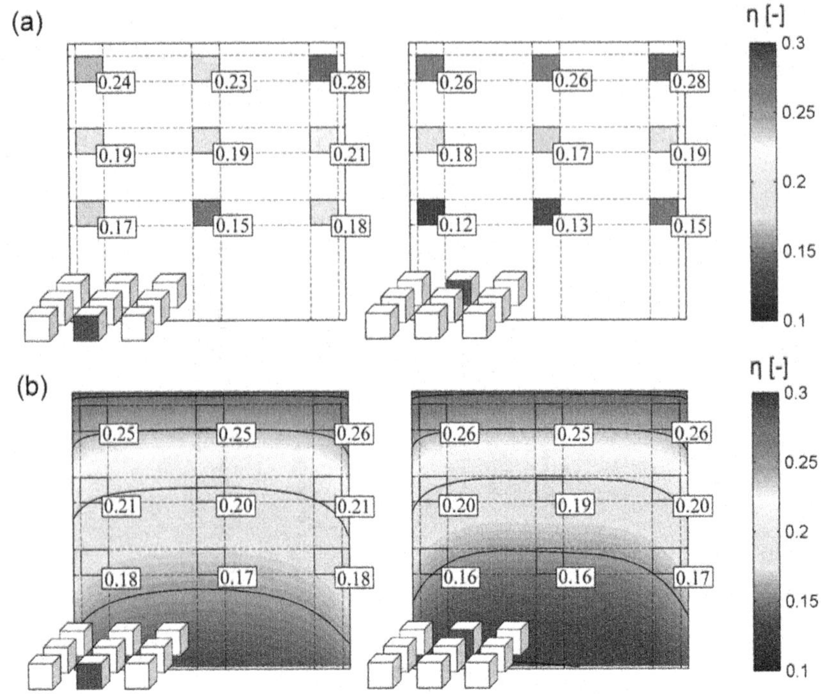

Figure 8.5 Comparison of (a) experimental and (b) numerical catch ratio data on the upstream and downstream cubes after the rain event on September 16–19, 2013

Source: Kubilay et al. 2014b

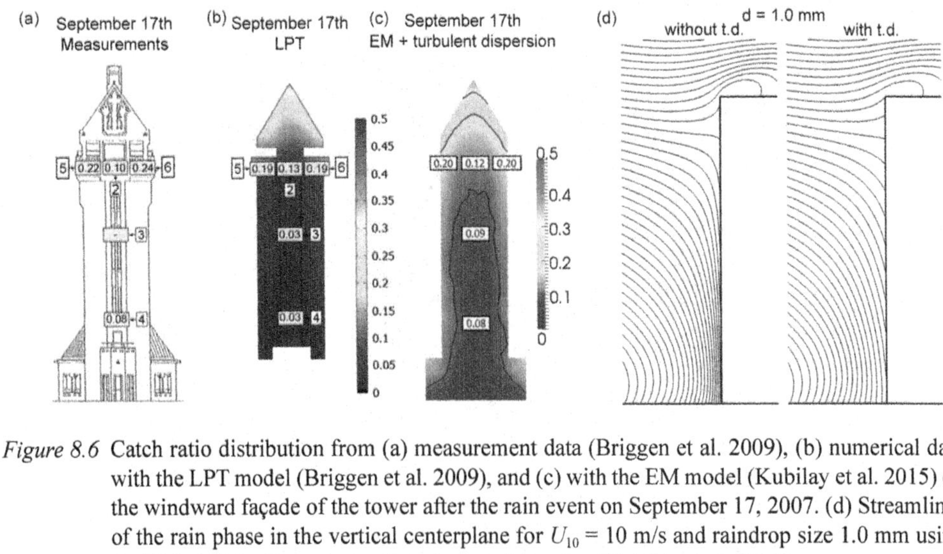

Figure 8.6 Catch ratio distribution from (a) measurement data (Briggen et al. 2009), (b) numerical data with the LPT model (Briggen et al. 2009), and (c) with the EM model (Kubilay et al. 2015) on the windward façade of the tower after the rain event on September 17, 2007. (d) Streamlines of the rain phase in the vertical centerplane for U_{10} = 10 m/s and raindrop size 1.0 mm using the EM model without and with turbulent dispersion (Kubilay et al. 2015).

parts of the building. Moreover, the horizontal rainfall intensity values in the rain events are relatively low (between 1mm/h and 4 mm/h), which means the raindrop size distribution shifts to smaller raindrops, increasing thus the influence of the local wind-flow field on the raindrop trajectories. Figure 8.6(c) shows the calculated catch ratio distribution with the turbulent dispersion modeling. The average deviation between simulations and measurements decreases from 18.0% to 11.4% by modeling turbulent dispersion with the EM model. The discrepancy decreases especially for the lower parts of the tower. For a detailed insight on the issue of turbulent dispersion, Figure 8.6(d) compares the streamlines of 1.0 mm raindrops for a wind speed of 10 m/s in the vertical centerplane of a high-rise building calculated with the EM model without and with turbulent dispersion. Without turbulent dispersion modeling, for the lower two-thirds of the building height, the raindrops travel almost parallel to the building façade. On the other hand, when turbulent dispersion is taken into account, the streamlines are less parallel to the building façade and the raindrops actually impinge on the façade in contrast to what is predicted without turbulent dispersion, thus increasing the predicted catch ratio. Therefore, the underestimation of WDR in Figure 8.6(b) at the bottom of the tower, where the streamlines of the rain phase are almost parallel to the façade, is mainly attributed to the omission of turbulent dispersion of raindrops.

Time-resolved modeling of wind-driven rain

A common aspect of the large number of WDR studies, using either LPT or EM models, is that the wind-flow field around buildings is calculated using steady RANS models. This subsection focuses on time-resolved WDR simulations using LES and EM models together on the geometry of low-rise cubic buildings. For simplicity, constant wind direction is considered. In this case, the EM model can provide the distribution of instantaneous catch ratios on all surfaces in the computational domain. Figure 8.7(a) shows instantaneous streamlines of 0.3 and 1.0 mm raindrops. The equidistant starting locations of the streamlines are at the inlet and top boundaries of the computational domain in the centerplane and away from the influence of buildings. Streamlines are tangential to the local raindrop velocity. The jaggy streamlines indicate that the 0.3 mm raindrops are influenced heavily by the local wind-flow field. In comparison, the instantaneous streamlines for the raindrop size of 1.0 mm in Figure 8.7(b) are more straight and parallel to each other. In general, it is found for larger droplets that the instantaneous streamlines do not show significant differences compared with the time-averaged streamlines, especially below the roof level. The deviations from the time-averaged streamlines are mostly due to the presence of larger turbulent scales above the roof level. On the other hand, small droplets can deviate strongly from the time-averaged streamlines both above the roof level, where the turbulent length scales are larger, and below the roof level, where the turbulent length scales are smaller.

Analyzing further these LES and EM model results, the resulting instantaneous specific catch ratio values as well as the running means over time from start ($t = 0$) are shown in Figure 8.7(b) for 0.3 and 1.0 mm raindrops. The specific catch ratio values are surface-averaged over the square-shaped areas A, B, and C, chosen for simplicity to be representing the location of WDR gauges used in the *in situ* measurements. The fluctuations in instantaneous specific catch ratio are significantly larger for the 0.3 mm droplets at position A. As mentioned above, this is because smaller raindrops are more influenced by the fluctuations in the wind-flow field than larger raindrops are. For both droplet sizes, the running mean of specific catch ratios are relatively stable after the initial 100–300 s. If one is interested in the time-averaged values (as mostly done in previous WDR studies), the findings confirm that the commonly used 10-minute averaging of meteorological data is valid. An exception may be rain events where the meteorological conditions, such as wind direction and wind speed, change a lot within

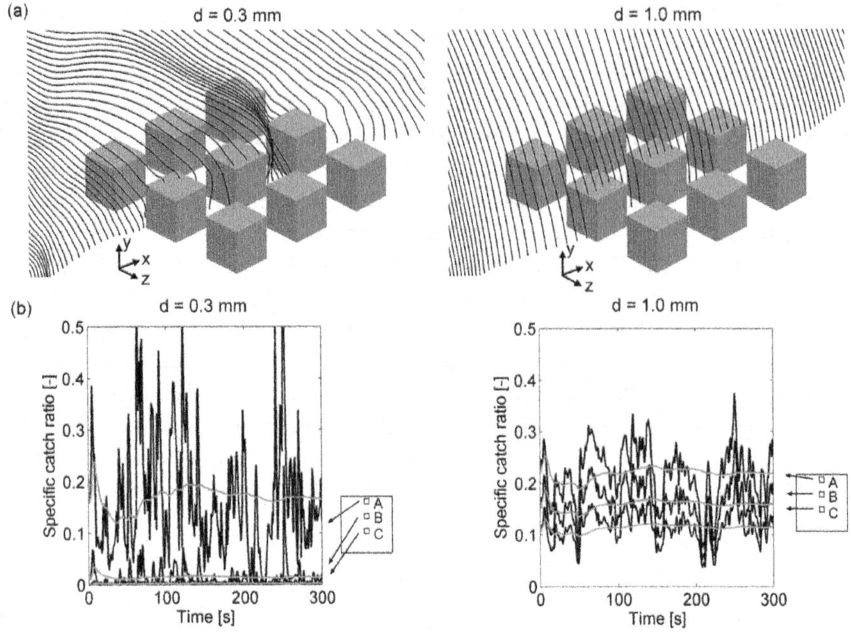

Figure 8.7 (a) Instantaneous streamlines for the raindrop sizes of 0.3 mm and 1.0 mm. (b) In dark-colored lines, instantaneous specific catch ratio values on the windward façade of the downstream cube, averaged over three WDR gauge positions (A, B, and C) for diameters of sizes of 0.3 mm and 1.0 mm (Kubilay et al. 2017a); in light-colored lines, running average of catch ratio values.

short time intervals. Such conditions may result in the actual catch ratios to be above or below the mean value, increasing the discrepancy between the numerical and experimental WDR intensities.

Application examples

We now present WDR studies using the EM model performed for different geometries and with different aims to illustrate the breadth of applications that can be explored with this tool, although many other studies could be contemplated. We also used these studies to familiarize further the reader with WDR and its different manifestations.

Rain deposition on buildings with different façade features

Numerical and semi-empirical models of rain exposure on building façades focus mainly on simplified flat building façades. Such studies take into account the large-scale influences of building size and shape and of the surrounding environment. On the other hand, small-scale façade details can also have a considerable impact on WDR intensity and moisture absorption into porous building materials. In this subsection, a numerical WDR analysis is performed on a mid-rise residential building with roof overhang, balconies, and window sills. The building has dimensions of height × length × width of 19.30 × 8.00 × 16.00 m^3. The roof of the building is flat and is bordered with a 1 m high parapet, as normally found in buildings with accessible roofs for safety reasons. Two separate case studies are performed on the building.

CASE STUDY 1: THE INFLUENCE OF ROOF-OVERHANG LENGTH

The first case study focuses on the influence of different roof-overhang lengths on WDR intensity. Case 1 is the reference case without roof overhang, whereas case 2 and case 3 have roof overhangs of 0.75 m and 1.50 m long, respectively. Figure 8.8(a) shows the distributions of catch ratio on the windward façades for a reference wind speed of U_{10} = 5 m/s at 10 m height and a reference rainfall intensity of R_h = 1 mm/h. The catch ratio distribution on the windward façade of case 1 shows the common characteristics of WDR surface wetting, i.e. top and side edges have larger catch ratio values while the maximum catch ratio values are at the top corners. As the roof-overhang length increases, the catch ratio values on the actual façade of the building decrease sharply at the upper part. Furthermore, the sheltered region grows towards the ground as the roof-overhang length increases.

Figure 8.8(b) presents the sheltering provided by different roof-overhang lengths. The sheltering ratio $\delta_{i\text{-}j}$ corresponds to the ratio between the decrease in catch ratio and the unsheltered catch ratio, which is defined as:

$$\delta_{i,j} = \frac{\eta_i - \eta_j}{\eta_i} \tag{8.10}$$

where η_i and η_j correspond to the catch ratio values obtained in cases *i* and *j*, respectively. The maximum sheltering due to the presence of roof overhangs is observed at the lowest reference wind speed at all floors. As wind speed increases, the raindrops gain a larger horizontal velocity and become more parallel to the ground, a change of direction that decreases the sheltering. A large difference in sheltering is observed at the fifth floor at all reference wind speeds and rainfall intensities indicating that sheltering increases with the roof-overhang length. For

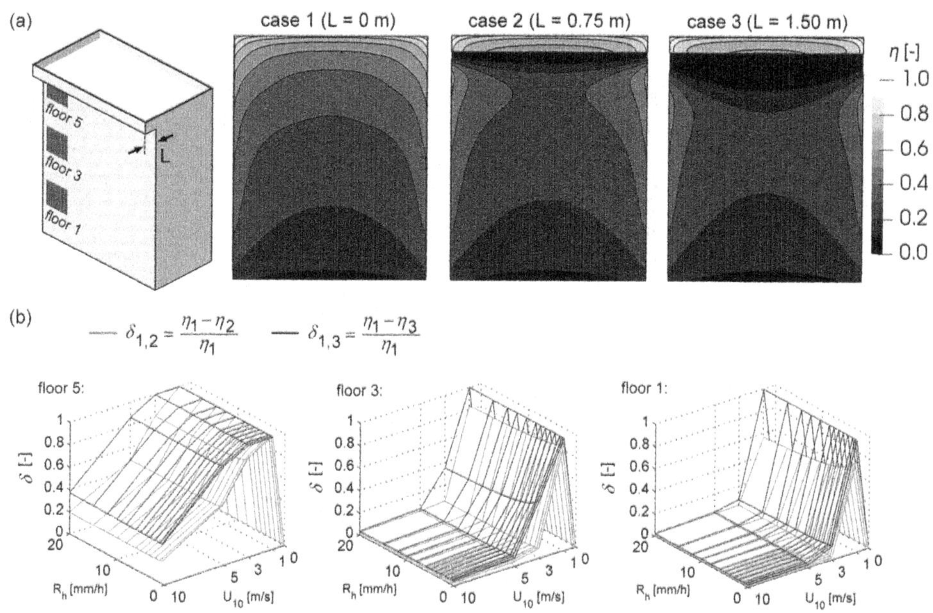

Figure 8.8 (a) Contours of catch ratio, η, for wind speed U_{10} = 5 m/s and rainfall intensity R_h = 1 mm/h on the windward façades of cases 1, 2, and 3. (b) Sheltering ratio charts surface-averaged over the indicated areas at floors 5, 3, and 1.

Source: Kubilay et al. 2017b

$U_{10} = 3$ m/s, case 3 still provides a sheltering between 35% and 51% at floor 3 and a sheltering between 8% and 12% at floor 1. Sheltering ratios are less sensitive to the changes in rainfall intensities than changes in wind speed. A small increase in sheltering ratios is observed as the rainfall intensity goes below 5 mm/h. This is because smaller droplets are pushed more towards the lower parts of the building, increasing the sheltered region below the roof overhang. For example, at the first floor, for reference wind speed of 1 m/s, case 2 provides about 74% sheltering for $R_h = 1$ mm/h and about 60% sheltering for $R_h = 10$ mm/h.

CASE STUDY 2: THE INFLUENCE OF FAÇADE DETAILS

The second case study analyzes the influence of façade details, namely balconies and window sills, on WDR intensity. This geometry is named case 4 and its roof-overhang dimensions are the same as in case 3 in Figure 8.8. The floor-to-floor distance is 3 m. The balconies are 1.3 m high, 1.5 m long, and 3 m wide and located on both sides of the building. The window sills are located on only one side of the building in order to see their impact on WDR intensity more clearly and to decrease the total number of simulations. The window sills are only 0.1 m high and projecting from the façade by 0.1 m. The catch ratio values on the façade of case 4 in Figure 8.9(a) are much lower than the ones on the façade of cases 1–3. However, the front walls of the balconies receive a higher amount of WDR compared to what the façade would receive, especially for the balconies at the top floor.

The sheltering ratio between the façades of cases 3 and 4 are given in Figure 8.9(b) with respect to reference rainfall intensity and reference wind speed values. The largest sheltering effect of the balconies on the façade is observed at the first and third floors. At these floors, parts of the building façade behind the balconies are largely sheltered by the balcony of the

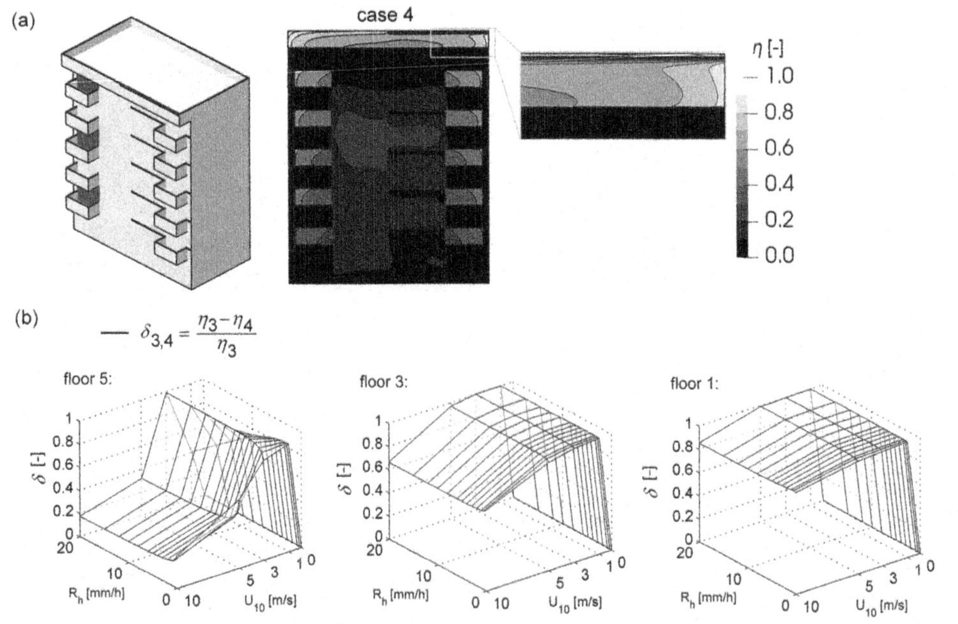

Figure 8.9 (a) Contours of catch ratio, η, for wind speed $U_{10} = 5$ m/s and rainfall intensity $R_h = 1$ mm/h on the windward façade of case 4. (b) Sheltering ratio charts, surface-averaged over the indicated areas at floors 5, 3. and 1.

Source: Kubilay et al. 2017b

upper floor, which acts as a local roof overhang, whereas, in case 3, there is no such sheltering for these floors. Sheltering is over 90% until up to a wind speed of 5 m/s. At higher speeds, sheltering decreases slightly, but still remains over 60% at U_{10} = 10 m/s. At the fifth floor, the balconies shelter the building façade further compared to case 3. Sheltering is over 85% at wind speed of 3 m/s for all rainfall intensities. The sheltering amount decreases to about between 30% and 40% at U_{10} = 5 m/s and is about 20% for U_{10} = 10 m/s.

Influence of nearby buildings

Figure 8.10 compares the rain phase streamlines for raindrop sizes d = 0.3 and 1.0 mm at reference wind speed U = 5 m/s in the centerplane for wind-facing cubic buildings arranged in a 3 × 3 array. The streamlines for the droplet size of 0.3 mm are largely influenced by the wind-flow patterns below the roof level, i.e. the recirculation regions. Furthermore, the size and location of the recirculation regions are also apparent by the way the streamlines move away from the windward façades. For example, the standing vortex windward of the upstream-most cube lies in the bottom half height of the cube and, hence, the 0.3 mm droplets start to move away from the façade roughly at that height. The air-recirculation regions between cubes are larger in size and, hence, the 0.3 mm droplets within these recirculation regions move away from the façade already at the roof level. In comparison, the 1.0 mm droplets are not so responsive to the wind-flow patterns below the roof levels. They are mostly uninfluenced by the standing vortex upstream of the row of the cubes and only slightly influenced by the recirculation regions between the cubes.

Figure 8.11 shows the catch ratio charts at two points, the top edge and in the middle, on the upstream and downstream central cubes in the array. The wind direction considered here is again perpendicular to the façades marked in the figure. The catch ratio charts give the catch ratio over the whole spectrum of droplet sizes at a certain position for different combinations of reference wind speed, reference wind direction, and reference rainfall intensity. The WDR intensity values at the top of the façades in Figure 8.11(a) are very close to each other for each couple of reference wind speed and reference rainfall intensity. On the other hand, in Figure 8.11(b), the difference in catch ratio due to the influence of the local wind-flow field is clear. The lower part of the downstream cube is exposed to less rain. The difference between the two cubes increases as the reference rainfall intensity decreases because, at low intensities, the spread of the drop size distribution curve is smaller and there are more raindrops of smaller size. Smaller raindrops have smaller inertia and, thus, they are more easily moved away from the lower parts of the façade by the wind-flow recirculation regions between the buildings. Smaller raindrops are also accelerated more easily towards the top part of the façade due to the accelerated wind flow over the roof. The difference in catch ratio values increases also with

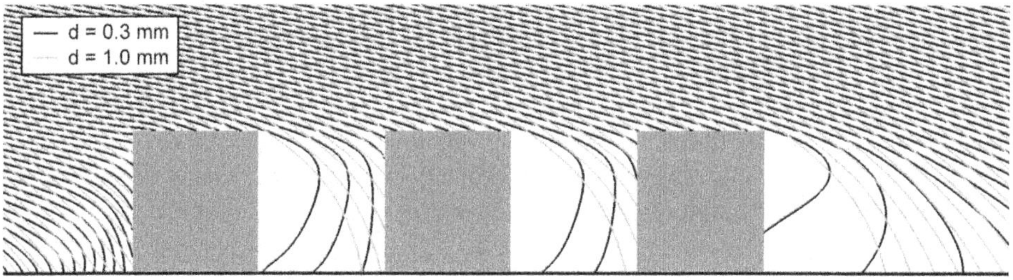

Figure 8.10 Streamlines of rain phases in the vertical centerplane of the domain for raindrop sizes d = 0.3 and 1.0 mm at reference wind speed U = 5 m/s. The domain counts nine building models of size 2 × 2 × 2 m^3 in a 3 × 3 array.

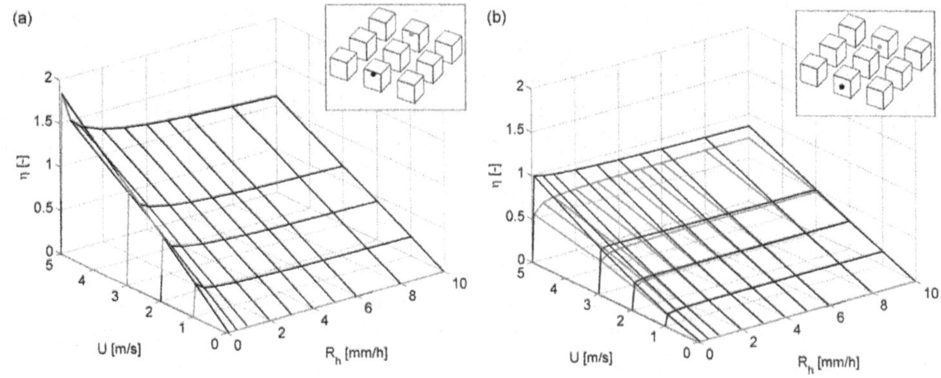

Figure 8.11 Comparison of catch ratio charts for positions (a) at the top edge and (b) at the center of the façades of the upstream and downstream cubes

Source: Kubilay et al. 2014b

increasing reference wind speed, as the recirculation region is getting stronger at higher reference wind speed values.

Droplet impact parameters

Figure 8.12 shows the raindrop impact speed on the façades of the upstream and downstream cubes in the array for a perpendicular wind direction for the reference wind speed $U = 3$ m/s. For the 0.3 mm droplets in Figure 8.12(a), the lowest impact speed on the upstream cube is found near the top of the façade, where the vertical component of the droplet velocity is slowed down by the acceleration of the wind flow over the roof. This effect is not as significant for the downstream cube. The highest impact speeds for both cubes are found at the lower parts of the façades. The different positions of highest impact speed can be related to the different positions of the recirculation regions in front of the cubes. The impact speed values for the droplet size of 1.0 mm in Figure 8.12(b) are higher than the ones for the droplet size of 0.3 mm, mainly due to the higher terminal velocity of the larger droplets. The distribution of impact speed is also more homogeneous than it is for the 0.3 mm droplets.

The impact angle of raindrops must be analyzed in complement to the impact speed in order to define the physics of the droplets that impinge on the façade. Thus, Figure 8.13 shows the impact angle between the raindrop trajectory and the surface normal, for the same wind conditions as in Figure 8.12. For the 0.3 mm droplets, the regions with higher impact speed in Figure 8.12, i.e. the lower part of the façade of the upstream cube and the center of the façade of downstream cube, are found to be with impact angles around 90°, which means the raindrops impinge with a direction almost parallel to the façade. Hence, these parts actually end up with low catch ratio values. The increase in impact speed is due to the increase in vertical velocity and not the velocity component normal to the façade, the latter being the factor affecting the catch ratio. For the 1.0 mm droplets in Figure 8.13(b), the impact angles are found to be lower than 70° for both cubes, meaning the 1.0 mm droplets impinge with smaller angles with the surface normal than the 0.3 mm droplets, with the exception of the top of the façades.

Discussion

This section discussed the physics behind the spatial and temporal distribution of WDR intensity in urban areas. The case studies presented here, both on single and multiple buildings,

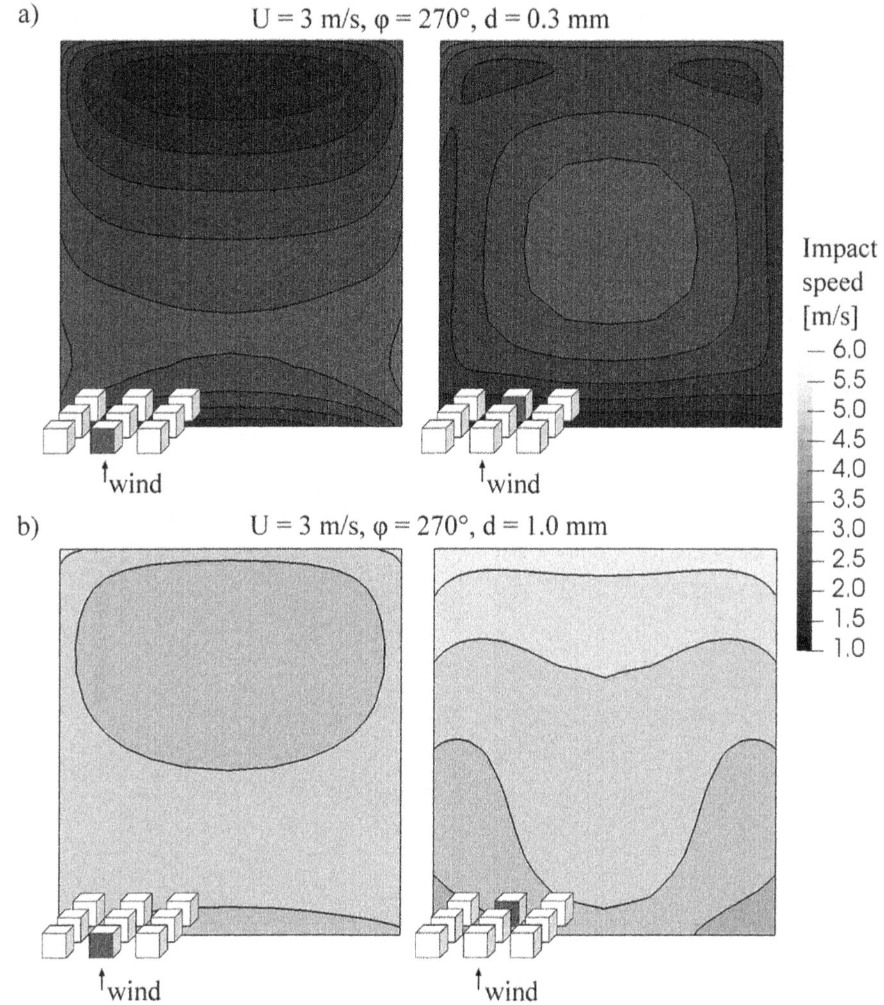

Figure 8.12 Raindrop impact speeds on the upstream and downstream cubes for the reference wind speed $U = 3$ m/s for the raindrop size (a) 0.3 mm and (b) 1.0 mm

show that significant variations of surface wetting due to WDR can occur depending on the atmospheric conditions and the arrangement of buildings. Furthermore, small façade features have an impact on which parts of the façade are sheltered and/or exposed. Of particular relevance, the WDR deposited on the façade is one of the inputs required for the assessment of the hygrothermal performance of building envelope assemblies.

CFD simulations can provide valuable detailed data where experimental data could have been significantly time consuming and expensive to obtain. We note that the accuracy of the CFD simulations depends on the errors due to simplifications in the physical modeling and due to numerical solution of the mathematical model. Physical modeling errors include factors such as the choice of turbulence model, geometrical complexity, and type and location of physical boundary conditions. Numerical errors can be due to factors such as computer round-off, convergence, and discretization. Therefore, care must be taken when the computational domain and grid are generated and discretization schemes are set for both wind and WDR calculations.

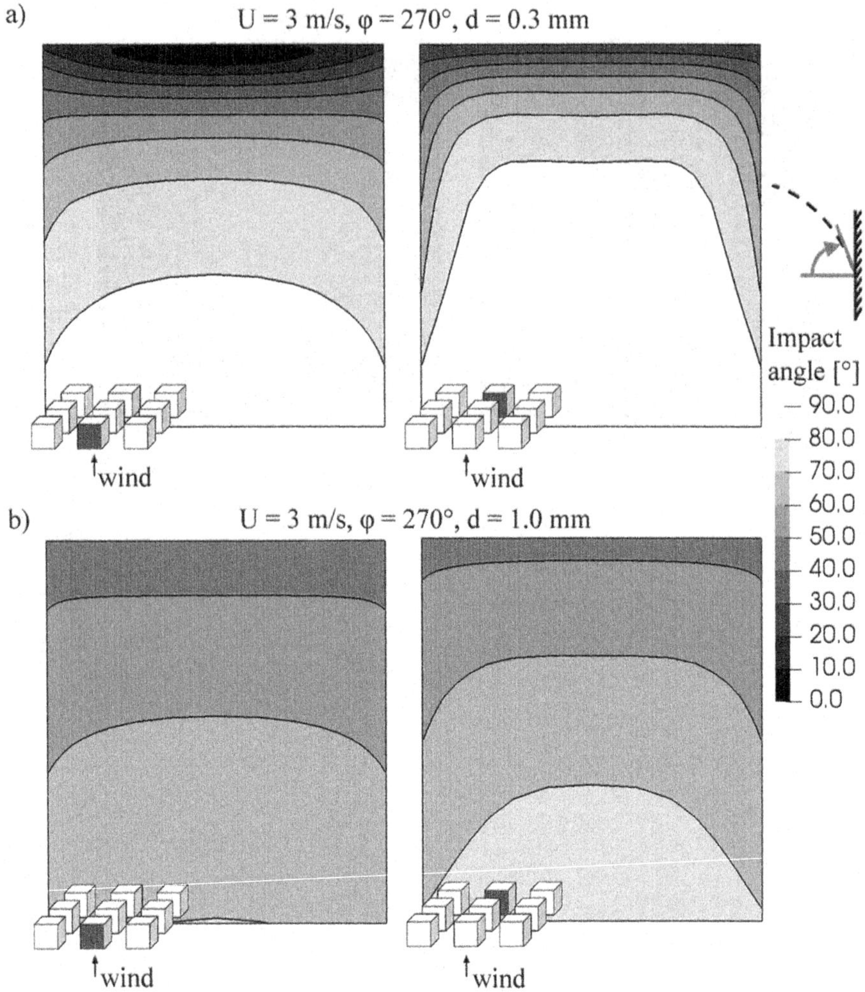

Figure 8.13 Raindrop impact angles on the upstream and downstream cubes for the reference wind speed $U = 3$ m/s for the raindrop size (a) 0.3 mm and (b) 1.0 mm. Impact angle is the angle between the raindrop trajectory and the surface normal.

While considering the spatial distribution of WDR, this section neglected the droplet physics before and after impact. The multiphase flow of air and rain was considered to be a dilute flow, and the splitting or coalescence of droplets was not directly modeled. This is a valid assumption, as also supported by comparison with the measurement data, for rainfall intensities of less than 100 mm/h, i.e. very heavy rainfall conditions. On the other hand, for porous building surfaces, film forming may occur after a sufficient amount of water is adsorbed in the material. On vertical or inclined surfaces, such thin water films may run off under the force of gravity. Runoff can change the distribution of surface wetting and can also cause water to leak into the building façade through cracks. Therefore, further work is needed to document and integrate the full spectrum of droplet physics after impact such as spreading, splashing, film forming, or runoff (Lee et al. 2016b, 2016c; Martin et al. 2015).

Boundary conditions for hygrothermal simulation of building envelope assemblies

Introduction

The heat and mass balances in hygrothermal models consider heat flows by conduction, convection, and radiation; moisture flows by vapor diffusion, convection, and liquid transport; and airflows are driven by natural, external, or mechanical forces. Hygrothermal models have been widely used to predict heat and moisture transport through multi-layer building assemblies. Delgado et al. (2013) presented a comprehensive review about hygrothermal models and their applications. An important difference between different hygrothermal models is on the retained degree of complexity of the considered physical processes. We present two examples below in the next subsection.

Under environmental loadings, excessive moisture accumulation can occur in building envelopes and may lead to structural damage, mold growth, decrease of thermal resistance of building materials and degradation of indoor air quality. For a specific problem, building designers need to use the specific moisture conditions of a climate location or region to assess correctly the durability of the building envelope. Numerical hygrothermal models are useful tools to assess long-term heat and moisture performance of wall assemblies and predict the risk of moisture damage. However, hygrothermal models need representative climatic data including rainfall, air temperature, relative humidity, solar radiation, longwave radiation, wind speed, and velocity as inputs. A common approach is to determine a reference year from available long-term climatic data. We propose a method for selecting a moisture reference year in the section "Application example: selecting a moisture reference year for hygrothermal simulations."

Scientific foundation: overview of numerical modeling approaches for coupled moisture and heat transport model in building envelope

Coupled moisture and heat transport model in building components

In traditional hygrothermal models, moisture is taken into account in liquid and vapor forms. In this hygrothermal study, only a solid masonry wall is considered. Therefore, air advection is not considered in this hygrothermal model. The governing equations for moisture and heat transfer are, respectively (Janssen et al. 2007),

$$\frac{\partial w_l}{\partial p_c}\frac{\partial p_c}{\partial t} + \nabla \cdot g_l + \nabla \cdot g_v = 0 \cdot \tag{8.11}$$

$$\begin{aligned}&(C_s \cdot \rho_s + C_l \cdot w_l) \cdot \frac{\partial T}{\partial t} + \nabla \cdot (C_l \cdot (\mathrm{T} - \mathrm{T}_{ref}) \cdot g_l \\ &+ (C_v \cdot (\mathrm{T} - \mathrm{T}_{ref}) + L_v) \cdot g_v) = -\nabla \cdot \lambda \nabla T\end{aligned} \tag{8.12}$$

where w_l is the liquid water content (kg m^{-3}), p_c is the capillary pressure (Pa), g_l is the liquid flow rate (kg m^{-2}s^{-1}), g_v is the vapor flow rate (kg m^{-2}s^{-1}), T is the temperature (K), C_s, C_l, and C_v are the specific heat capacity of solid-matrix, liquid water and vapor (J kg^{-1}K^{-1}), ρ_s is the bulk density of the solid-matrix (kg m^{-3}), T_{ref} is 273.15 K, L_v is the latent heat of vaporization (Jkg^{-1}), and λ is the thermal conductivity (W m^{-1}K^{-1}).

The liquid flow rate g_l is written as:

$$g_l = -K_l(p_c)\nabla p_c \tag{8.13}$$

where $K_l(p_c)$ is the liquid permeability (s), $p_c = p_a - p_l$ p_a is the atmospheric pressure (Pa), and p_l is the liquid water pressure (Pa).

The vapor flow rate g_v is written as:

$$g_v = -\delta_v \cdot \frac{p_v}{\rho_l \cdot R_v \cdot T} \cdot \nabla p_c - \delta_v \cdot \frac{p_v}{\rho_l \cdot R_v \cdot T^2}(\rho_l \cdot L_v) \cdot \nabla T \tag{8.14}$$

where δ_v is the water vapor permeability (s), p_v is the vapor pressure (Pa), ρ_l is the density of water (kg m^{-3}), and R_v is the gas constant of water (J kg^{-1}K^{-1}). The second term on the right-hand side in Equation 8.14 represents temperature-driven vapor diffusion.

The moisture and heat transfer equations are solved using the finite element program COMSOL. The time-dependent problem is solved by integration of the partial differential equations over time according to the implicit backwards differentiated formula (BDF) method. The COMSOL solver used is the direct solver MUMPS (MUltifrontal Massively Parallel Sparse direct Solver) with default solver options. Numerical time steps are automatically selected by the COMSOL solver, depending on a relative or absolute tolerance for the accuracy of the integration. The coupled moisture and heat transport model was validated in a HAMSTAD benchmark (Hagentoft et al. 2004). An example of application of the model is given in the section "Application example 1: evaluation of moisture problems in wooden beam embedded in masonry wall."

Coupled moisture and heat transport model with freezing and thawing processes in building components

When the problem of freeze-thaw damage of a building component is to be analyzed, the traditional coupled moisture and heat transport model may not be sufficient. As is well known, freezing-induced moisture migration significantly changes the moisture distribution in a freezing porous medium (Zhou et al. 2014). In addition, phase change between the liquid and solid states of water affects the temperature distribution in the porous medium. Therefore, a hygrothermal model considering the freezing and thawing processes in porous medium is necessary to analyze accurately the risk of freeze-thaw damage in the masonry wall.

The derived moisture and heat transport equations are based on the following hypotheses (Zhou et al. 2017): (1) the porous media are undeformable and no volume change occurs during freezing and thawing; (2) there is no salt present in the porous medium; (3) there is an analogy between capillary pressure curve (the relationship between capillary pressure and moisture content) and freezing curve (the relationship between subzero temperature and unfrozen liquid water content). If a solute is considered to be present in the porous medium, the melting point of water in porous medium is further depressed. The freezing point depression depends also on the concentration of the solutes. In reality, porous media are deformable and thus volume changes occur during freezing and thawing, leading to damage. Within the scope of this chapter, the influence of volume change on the freezing process is not considered.

$$\frac{\partial w_t}{\partial p_t}\frac{\partial p_t}{\partial t} + \nabla \cdot g_l + \nabla \cdot g_v = 0 \tag{8.15}$$

$$\left(C_s \cdot \rho_s + C_l \cdot w_l + C_i \cdot w_i\right) \cdot \frac{\partial T}{\partial t} - L_f \cdot \frac{\partial w_i}{\partial t}$$
$$+\nabla \cdot (C_l \cdot (\mathrm{T} - \mathrm{T}_{ref}) \cdot g_l + (C_v \cdot (\mathrm{T} - \mathrm{T}_{ref}) + L_v) \cdot g_v) = -\nabla \cdot \lambda \nabla T \tag{8.16}$$

where w_t is the total moisture content of liquid water and ice ($w_t = w_l + w_i$), p_t is the total water pressure corresponding to the total moisture content w_t, w_i is the ice content (kg m^{-3}), and L_f is the latent heat of freezing (J kg^{-1}).

The concept of freezing point depression is used for modeling the freezing of water and formation of ice in the porous medium (Williams and Smith 1989). The determination of whether a porous medium is at frozen condition is based on the following equation:

$$T^* = T_0 \cdot \exp\left(\frac{-p_t}{\rho_l \cdot L_f}\right) \tag{8.17}$$

where T^* is the freezing temperature (K) for a porous medium with a total water pressure of p_t and T_0 is the freezing reference temperature (273.15 K). When the temperature of the porous medium is larger than the freezing temperature ($T \geq T^*$), the porous medium is unfrozen and no ice exists. When temperature of porous medium is smaller than the freezing temperature ($T < T^*$), the porous medium is at frozen condition and ice exists.

The relation between capillary pressure and temperature in a frozen porous medium is described using the following equation:

$$p_c = -\rho_l \cdot L_f \cdot \ln\left(\frac{T}{T_0}\right) \approx -\rho_l \cdot L_f \cdot \frac{T - T_0}{T_0} = -S_f (T - T_0) \tag{8.18}$$

where $S_f = \dfrac{\rho_l \cdot L_f}{T_0}$.

The ice content at frozen condition is given as:

$$w_i = w_t - w_l = \Theta(p_t) - \Theta(p_c) = \Theta(p_t) - \Theta(-S_f (T - T_0)) \tag{8.19}$$

The moisture retention curve $\Theta(p)$ is used to characterize the relation between the liquid moisture content and capillary pressure and the relation between total moisture content and total water pressure.

$$w_l = \Theta(p_c) \tag{8.20}$$
$$w_t = \Theta(p_t) \tag{8.21}$$

where Θ is the moisture retention function, p_c is the liquid water pressure corresponding to liquid water content w_l, and p_t is the total water pressure corresponding to total moisture content w_t.

The capillary pressure for nonfrozen and frozen conditions is described using the following equation:

$$\begin{cases} p_c = p_t & \text{if } T \geq T^* \\ p_c = -S_f (T - T_0) & \text{if } T < T^* \end{cases} \tag{8.22}$$

where T is the temperature of the porous medium.

The coupled moisture and heat transport model with freezing and thawing processes was validated with experimental results from Zhou et al. 2014. An example of application of the model with freezing and thawing processes is given in the section "Application example 2: numerical evaluation of freeze-thaw damage risk of masonry walls after internal retrofitting."

Application example: selecting a moisture reference year for hygrothermal simulations

Wetting and drying processes determine the moisture conditions of building envelopes. A moisture reference year selection process should encompass both the wetting and drying loading. The moisture reference year should reflect the climatic variability subjected to the building envelope under consideration and provide the required level of safety with regard to moisture problems. Regarding the wetting component, wind-driven rain is the largest moisture source that impacts the durability of building envelopes. The orientation of the façade strongly affects the amount of wind-driven rain on the façade. On the other hand, the drying process deals with evaporation of water from the porous building materials to the surrounding environment. The potential evaporation is defined as a measure of the ability of the environment to remove water from a wet building envelope through the process of evaporation at its saturated surface. The selection is made among existing yearly weather data. Please refer to Chapter 6 for more information about weather data.

The Climatic Index approach

The Climatic Index can be considered as a measure of the balance between the wetting and drying components (Zhou et al. 2016). A simple way to describe the Climatic Index of a building envelope is to compare wetting load to drying potential, dividing the annual wind-driven rain load by the annual potential evaporation, as follows,

$$Climatic\ \ Index = \frac{annual\ wind\ driven\ rain\ load}{annual\ potential\ evaporation} \tag{8.23}$$

where the numerator represents the wetting component while the denominator represents the drying component. Therefore, a higher Climatic Index value indicates a higher moisture damage risk for the building envelope.

In the study here, we are interested in the calculation of WDR for a single location on the façade and, therefore, a semi-empirical method can be chosen. However, if one is interested in the variation of WDR at different positions on the façade or in a more accurate determination of the rain load, the CFD-based WDR approaches described in the first part of the chapter should be used to obtain an accurate spatial variation of rain load. The wind-driven rain load is calculated according to *ASHRAE Standard 160 – Criteria for Moisture Control Design Analysis in Buildings* (TenWolde 2008), where the WDR intensity (R_{WDR}) is empirically determined using the following equation:

$$R_{WDR} = R_h \times F_E \times F_D \times 0.2 \times V_{10} \times \cos\theta \tag{8.24}$$

where R_h is the horizontal rainfall amount (kg m^{-2}h^{-1}), F_E is the rain exposure factor, F_D is the rain deposition factor, V_{10} is the wind speed at 10 m above ground (ms^{-1}), and θ is the angle between the wind direction and the normal to the façade (rad). Factor F_E depends on the surrounding terrain and the height of the building, while Factor F_D describes the influence of the building itself. The annual WDR load (kg m^{-2}a^{-1}) is calculated as the sum of hourly R_{WDR}.

The Penman equation (Penman 1948) is a common method used in the field of agricultural and environmental physics for predicting potential evaporation from a surface. It considers both energy balance and convective vapor transport. The Penman equation for calculating the potential evaporation from a vertical surface is as follows (Zhou et al. 2016):

$$E = \frac{\Delta}{\Delta + \gamma} \frac{K + L}{I} + \frac{\gamma}{\Delta + \gamma} h_m (e_a - e) \tag{8.25}$$

where E is the potential evaporation flux (kg m^{-2}h^{-1}), Δ is the ratio between saturation vapor partial pressure and air temperature (PaK^{-1}), γ is the psychrometric constant (PaK^{-1}), K is the net short-wave radiation (Wm^{-2}), L is the net long-wave radiation (Wm^{-2}), I is the latent heat of vaporization (Jkg^{-1}), h_m is the convective vapor transfer coefficient (sm^{-1}), depending on wind speed, e_a is the saturated vapor partial pressure of the air (Pa), and e is the vapor partial pressure in the air (Pa).

The Penman equation consists of two terms: the radiation term ($\frac{\Delta}{\Delta + \gamma} \frac{K + L}{I}$) and the convective vapor transport term ($\frac{\gamma}{\Delta + \gamma} h_m (e_a - e)$). The radiation term is related to the radiative energy balance at the surface, while the convective term is affected by the atmospheric wind conditions. The annual potential evaporation (kg m^{-2}a^{-1}) is calculated by summing up the daily potential evaporation. More details on the derivation of potential evaporation can be found in Zhou et al. 2016.

Selecting moisture reference year

Moisture problems are often caused or exacerbated by weather conditions far from the average. A 10% level criterion is often suggested to select moisture reference years. Such 10% damage risk is justifiable, since this level refers to a return period of 10 years, which is appropriate for hygrothermal problems, as moisture accumulated during a bad year can dry out in the following years, preventing long-term deterioration (Sanders 1996). Four steps are used in the procedure for selecting a moisture reference year for hygrothermal simulations: (1) calculation of the Climatic Index for an extended set of years, e.g. 30 years; (2) selection of three years around the 10%-level largest Climatic Index from this set of years; (3) performance of hygrothermal simulations for the selected three years and determination of the corresponding relative humidity-temperature (RHT) values, and (4) selection of the year with the highest RHT value out of these three years.

As an example, three masonry wall assemblies are selected for hygrothermal modeling to consider the influence of wall structure (Figure 8.14), which are (1) a three-layer masonry wall without insulation, (2) an internally insulated masonry wall, and (3) an externally insulated masonry wall. The following materials are used in the masonry wall assemblies: lime plaster, brick, mortar, aerogel insulation plaster (except for wall 1), and interior plaster.

Four geographic locations in Switzerland, i.e. Zurich, Geneva, Lugano, and Davos, are selected to consider different climatic conditions. The mean annual temperatures in Zurich, Geneva, Lugano, and Davos are 8.8, 9.9, 11.9, and 3.9 °C, respectively. Weather data from 1981 to 2010 for each city from MeteoSwiss are used as climatic input data for the hygrothermal simulation. A meteorological year starting in October and ending in September is selected, as moisture accumulated within the building envelope during the winter will affect the moisture conditions during the summer.

A hygrothermal response indicator, called the RHT Index, is used in the analysis of hygrothermal simulation results for evaluating the risk of moisture problems (Kumaran et al. 2002).

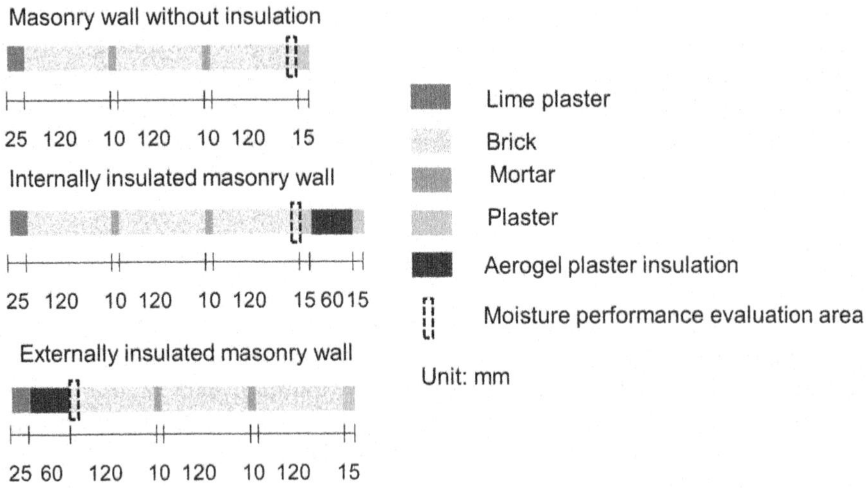

Figure 8.14 Schematic representation of three masonry wall assemblies

Source: Zhou et al. 2016

The RHT Index is calculated based on 10-day interval averaged relative humidities, RH (%), and temperatures, T (°C).

$$RHT = \sum (RH - RH_x) \times (T - T_x), \quad for\ RH > RH_x\ and\ T > T_x \tag{8.26}$$

where threshold values $RH_x = 80\%$ and $T_x = 5$ °C are used. The moisture degradation mechanism considered here is that of mold occurrence, possible at a relative humidity larger than 80% and a temperature larger than 5 °C. An evaluation region in the building envelope is selected for calculating the RHT Index. For the masonry wall without insulation and for the internally insulated masonry wall, the innermost 1 cm of the third brick is chosen as the evaluation region (Figure 8.14). Although this region is in contact with the interior finishing, it is still considerably affected by the external climate conditions. The selection of this region is based on practical experience, known as a critical zone for moisture problems for walls with inside insulation. For the externally insulated masonry wall, the outermost 1 cm of the first brick is selected as evaluation region (Figure 8.14), as this is the most critical region for the masonry wall with external insulation.

Figure 8.15 shows the Climatic Index for the year 1994 in the four cities. Zurich shows much larger Climatic Index values than do other cities, while Lugano shows the smallest Climatic Index values. The high Climatic Index values in Zurich are due to high WDR load, while the low Climatic Index values in Lugano are due to high potential evaporation. In order to select the orientation that gives the most severe moisture problems, the orientation with the largest Climatic Index is selected for hygrothermal simulation. Based on the criterion, the wall orientations selected for further analysis are 240, 240, 330, and 30 degrees from north, for Zurich, Geneva, Lugano, and Davos, respectively.

For the different years, we simulate the hygrothermal behavior using the coupled heat and moisture model and evaluate the hygrothermal performance using the RHT Index. The Climatic Index is compared with the RHT Index to evaluate how well the Climate Index predicts the risk for moisture damage. The RHT Index plotted in function of the Climatic Index from 1981 to 2010 is shown in Figures 8.16(a)–(c) for the three different masonry wall structures.

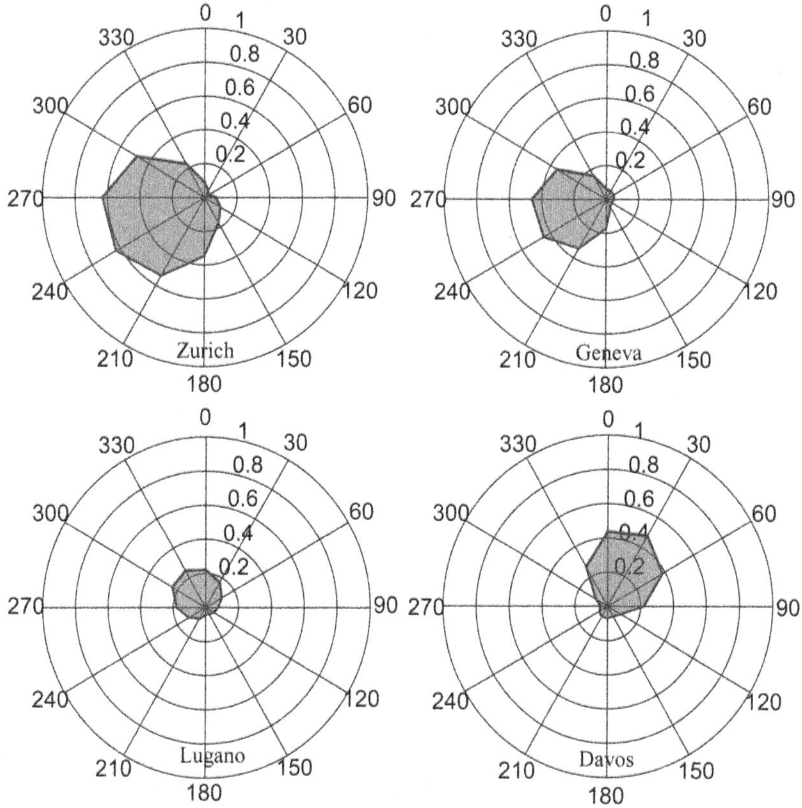

Figure 8.15 Climatic Index rose for Zurich, Geneva, Lugano, and Davos in 1994

In general, it is found that larger Climatic Indices are associated with larger RHT Indices while smaller Climatic Indices are associated with smaller RHT Indices.

Geneva shows the largest correlation coefficients between RHT and Climatic Indices compared to the other cities. For the four cities, the correlation coefficients between RHT and Climatic Indices are mostly above 0.65. The correlation coefficients between RHT and Climatic Indices for the internally insulated masonry wall are larger than are those for the other two types of masonry wall. In the case of masonry wall without insulation in Lugano, the correlation coefficient between RHT and Climatic Indices is only 0.62. The explanation for this lower correlation is that the RHT Index of many years in this city is equal to zero, while the Climatic Indices for these times have non-zero values. Note that when the relative humidity at the evaluation location is below a determined value, which is also the threshold relative humidity of 80% required for mold growth, the RHT Index becomes 0. Therefore, it can be concluded that, within some uncertainty, there is a clear correlation between the RHT and Climatic Indices.

As mentioned earlier, selecting a moisture reference year with a 10% failure level, or a 90% safety level, is a common procedure. For a dataset with 30 years, this 10% failure level year corresponds to the third wettest year. Using the Climate Index as a moisture damage risk indicator, the moisture reference years are the years with the third largest Climatic Index: 1993, 1983, 1993, and 1986 for Zurich, Geneva, Lugano, and Davos, respectively. The RHT Indices of these years are analyzed to evaluate whether these years have indeed 10%-level moisture risk. The RHT Indices for the selected years are indicated with a triangle in Figures 8.16(a)–(c).

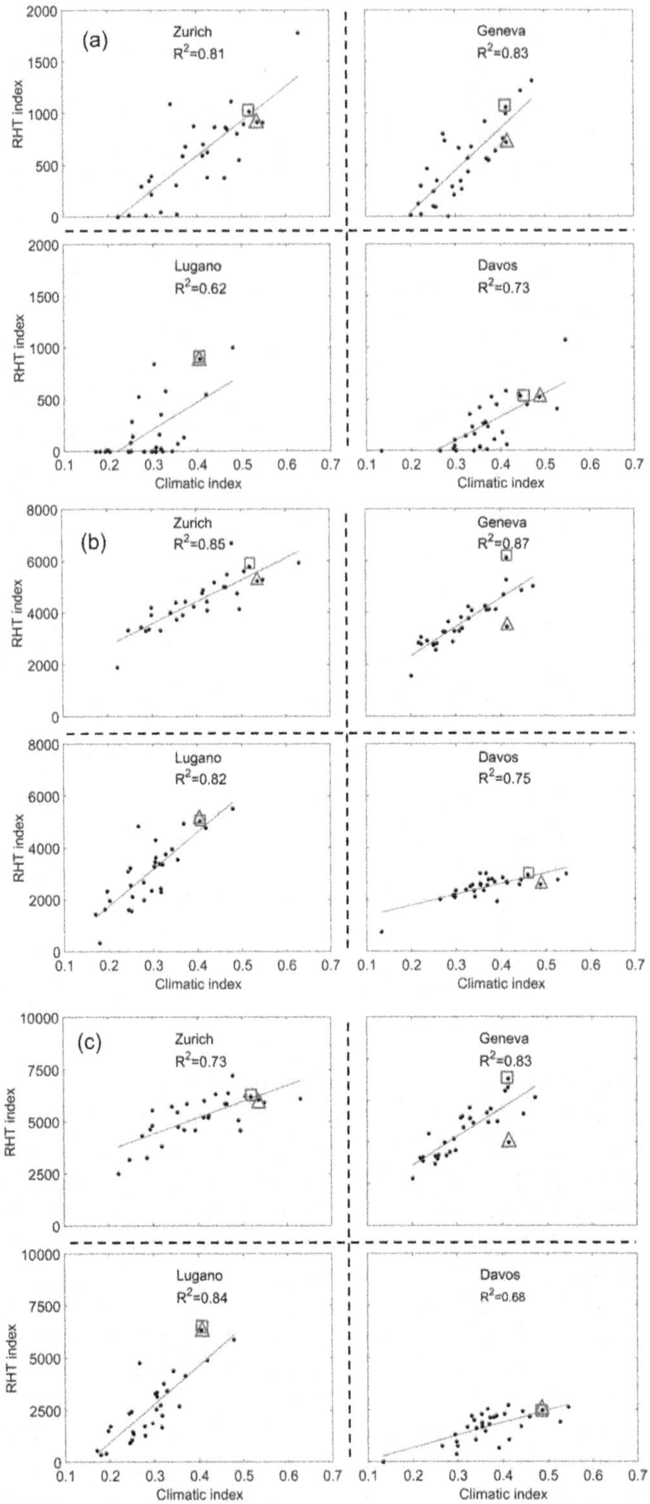

Figure 8.16 Relation between RHT Index and Climatic Index for the masonry wall without insulation (a), internally insulated masonry wall (b), and externally insulated masonry wall (c) from 1981 to 2010. The triangles indicate the years with the third largest Climatic Index and the squares indicate the chosen moisture reference years based on the new procedure.

Source: Zhou et al. 2016

It is observed that the RHT Indices of the selected years do not correspond necessarily to the third largest RHT values, meaning they are not the third wettest years. Especially for Geneva, the RHT Index for 1983 is much smaller than the third value (the triangle is actually situated below the correlation line). By comparison, the RHT values in Geneva for the years showing the fourth (2006) and fifth (2000) largest Climate Index are found to be higher for all wall types. Therefore, it can be concluded from the analysis that the selection of the moisture reference year based on the third largest Climate Index does not guarantee the selection of a proper moisture reference year.

A new procedure is needed due to the uncertainty in selecting the appropriate moisture reference year based on a 10%-level Climate Index criterion. The uncertainty in selecting the appropriate moisture reference year can be reduced by comparing a set of years. Therefore, the following procedure is proposed: first, select a set of years around the 10%-level Climatic Index; second, perform hygrothermal simulations for these years and determine the RHT values; third, select out of this set of years the year with the highest RHT value. Taking our case of a weather dataset of 30 years as example, we determine the RHT Indices for the years with the third, fourth, and fifth largest Climatic Indices, and then select the year with the largest RHT Index among these three years as moisture reference year. The result of this procedure is exemplified in Figure 8.16, where the squares indicate the RHT of the selected moisture reference years obtained following the new procedure. These years (squares) show for some wall envelopes and cities an RHT Index close to the value obtained from the 10% risk level based on Climatic Index (triangles), while for others the RHT Index is higher. For example, the year with the largest RHT Index is selected for internally and externally insulated masonry walls in Geneva. A larger RHT Index indicates higher risk of moisture problem. Therefore, the selected moisture reference year for internally and externally insulated masonry walls in Geneva has the highest moisture risk. Selecting the year with the highest moisture risk means the most conservative design criteria. However, the new procedure can reduce the uncertainty in selecting a moisture reference year with low moisture risk. Figure 8.17 shows the moisture reference years selected based on the new procedure and their corresponding RHT values.

A Climatic Index is proposed in this study to evaluate the moisture risk of building envelopes exposed to a typical climate region. The Climate Index evaluates the balance between wetting and drying potentials and is used for the selection of a moisture reference year. Three types of

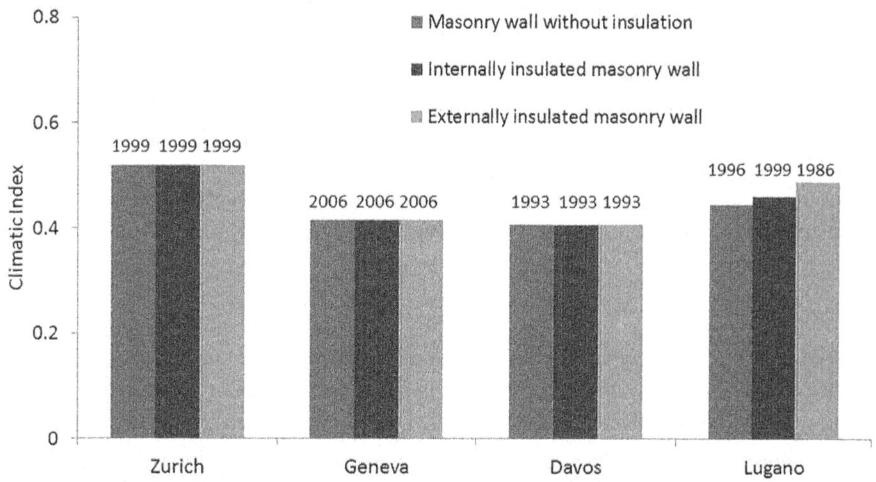

Figure 8.17 Selected moisture reference years based on the CI + RHT procedure

masonry wall assemblies for four cities in Switzerland were simulated with a coupled heat and mass transport model. The combination of Climatic Index and RHT Index allows for a more rational selection of moisture reference years in a two-step selection process. The possibility of selecting a year with low moisture risk is considerably reduced. This easy and reliable procedure can be applied for different wall orientations and wall compositions for different climate regions.

Hygrothermal performance of building envelope assemblies

Introduction

Excessive moisture in a building envelope can impact not only the durability of building materials, the indoor air quality, and the health of the inhabitants but also the energy efficiency of the building envelope system. However, moisture damage in building envelopes is not solely due to moisture conditions. Thermal conditions also have a large influence on moisture damage. For example, for wood decay to occur, the temperature must be above 0 °C. By comparison, for freeze-thaw damage of porous materials to occur, temperature must be at least below 0 °C. Therefore, it is essential to evaluate the combined moisture and temperature response of the building envelope.

Application example 1: evaluation of moisture problems in wooden beam embedded in masonry wall

Introduction

The occurrence of moisture problems in wooden beams in masonry walls is a research topic of high practical and scientific importance (Kehl 2013). High wind-driven rain loads may lead to high moisture content in load-bearing masonry walls and decay of wood beams embedded in such walls. Internal insulation of masonry walls may significantly increase the decay risk of wooden beams due to lower temperature and consequently lower drying potential in the existing wall (Harrestrup and Svendsen 2016; Morelli and Svendsen 2013). High moisture content will affect the performance and service life of wood. For instance, mold growth on a moist wooden surface will produce large numbers of spores, give off unpleasant odors, and impair air quality (Carll and Wiedenhoeft 2009). Furthermore, moisture is an indispensable factor for development of fungi rot in wood. The safe moisture content in wood regularly used in building regulations is 0.20 kg/kg to ensure the absence of growth of fungi.

Field investigation and numerical study both indicate that exterior render affects water uptake of external walls during driving rain events (Künzel et al. 2004). Exterior renders with high liquid water absorption coefficient tend to lead to high moisture content in the masonry (Künzel et al. 2004; Pazera and Bomberg 2010). Therefore, the response of masonry wall to internal insulation may be very different depending on the material properties of the rendering.

Model setup

The decay risk of a wooden beam embedded in an internally insulated masonry wall with different renders is studied. The masonry wall consists of render, brick, mortar, aerogel plaster, interior plaster, and wooden beam (Figure 8.18). There is an air gap at the end of the wooden beam, separating the beam from the masonry wall. The used exterior renders are lime mortar, lime plaster, cement lime plaster, lime cement plaster, and mineral plaster, respectively. The liquid permeability of the renders is shown in Figure 8.19. Lime mortar has the largest

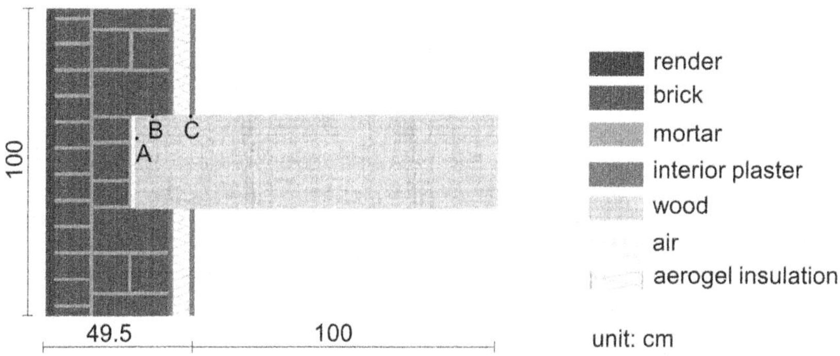

Figure 8.18 Schematic representation of the wood beam in an internally insulated masonry wall

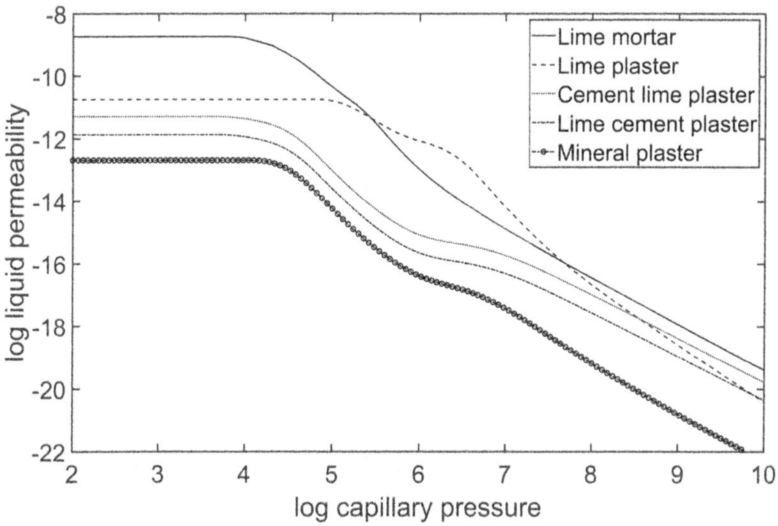

Figure 8.19 Liquid permeability of the five renders

liquid permeability while mineral plaster has the smallest one. A two-dimensional hygrothermal model with governing Equations 8.11 and 8.12 is applied to study the moisture problem in the wooden beam. The hygrothermal analysis is based on the moisture reference year of 1999 for Zurich.

Results

Figure 8.20 shows the simulated moisture content at three locations in the wooded beam for the different exterior renders. The moisture content at a given location is strongly affected by the type of exterior render. The render with larger liquid permeability leads to higher moisture content in the wood beam. For example, at location B, the highest moisture content is associated with lime mortar, which has the largest liquid permeability (Figure 8.19). By comparison, the lowest moisture content is associated with mineral plaster, which has the smallest liquid permeability.

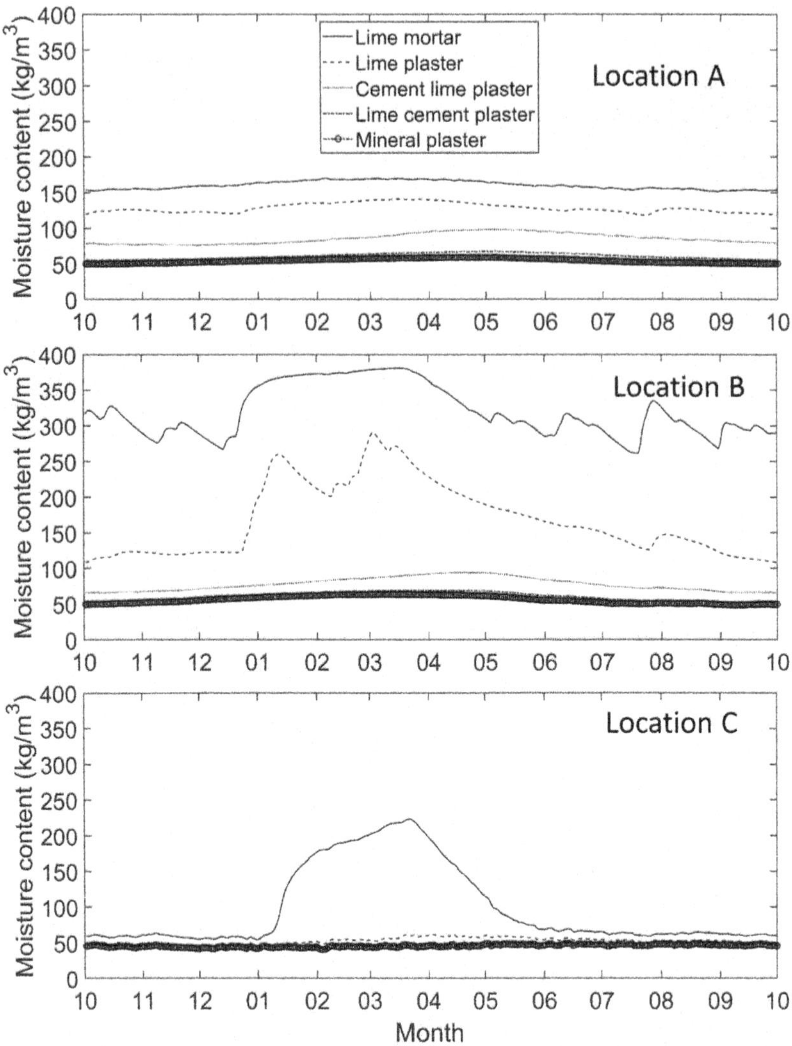

Figure 8.20 Simulated moisture content at three locations for different renders

There is a large difference of moisture content among the three locations. The moisture content is highest at location B, as the wood is in contact with brick at this location. At location A, the moisture content is lower compared to location B. The air gap next to location A works as a capillary break, which stops capillary-driven flow from outside. In addition, wood has very small liquid permeability. Moisture transports very slow in the wood. The high moisture content at location B hardly affects the moisture content at location A. At location C, the moisture content is low for all renders except lime mortar. As location C is close to an inside wall surface, it is mainly affected by the indoor environment. However, for the masonry wall with lime mortar as render, the high liquid permeability of lime mortar results in high moisture content in the masonry wall. The drying towards the indoors is insufficient to lower the moisture content at location C for this render. Among the three locations, location B is at a critical point in terms of a moisture problem.

In a next step, the model developed by Viitanen et al. (2010) for the assessment of biological deterioration of wood is used. In this model, the decay process is divided into two processes:

an activation process and a mass loss process. Both activation and mass loss occur when the temperature is above 0 °C and the RH is above 95%.

A parameter α is defined as a relative measure of the state of fungi growth with respect to its state at the initiation of the mass loss process. The initial value of α is 0.

$$\alpha = \sum_{0}^{t} \Delta\alpha$$

$$\text{where } \Delta\alpha = \begin{cases} \dfrac{\Delta t}{t_{crit}(\mathrm{RH},\mathrm{T})} & \text{when } T > 0°\mathrm{C} \text{ and } RH > 95\% \\ -\dfrac{\Delta t}{17520} & \text{otherwise} \end{cases} \tag{8.27}$$

$$t_{crit}(\mathrm{RH},\mathrm{T}) = \left[\frac{2.3T + 0.035RH - 0.024T \times RH}{-42.9 + 0.14T + 0.45RH}\right] \times 30 \times 24[\text{hours}]$$

The mass loss process starts when α reaches 1. The mass loss rate is dependent on RH and T at each time step according to the following relation:

$$ML(\mathrm{t}') = \sum_{t \text{ at } \alpha=1}^{t'} \frac{ML(\mathrm{RH},\mathrm{T})}{dt} dt \tag{8.28}$$

$$\frac{ML(\mathrm{RH},\mathrm{T})}{dt} = -5.96 \times 10^{-2} + 1.96 \times 10^{-4} T + 6.25 \times 10^{-4} RH[\% / \text{hour}] \tag{8.29}$$

The values of relative humidity and temperature determined at location B for different renders are used for the analysis of wood decay. The alpha value reaches 1 in mid-October for the lime mortar render and in the beginning of February for the lime plaster render (Figure 8.21). The alpha value for the wall envelopes with the other three renders remains 0. Mass decay is considerably large at location B for the walls rendered with lime mortar and lime plaster, with mass loss of 43.7% and 20.2% in one year, respectively. By comparison, the mass loss rate is null for the other renders. In this case study, the moisture conditions in the wooden beam are strongly affected by the type of exterior render. Moisture problems are likely in wall envelopes with capillary active renders such as lime mortar and lime plaster, which do not provide sufficient rain protection.

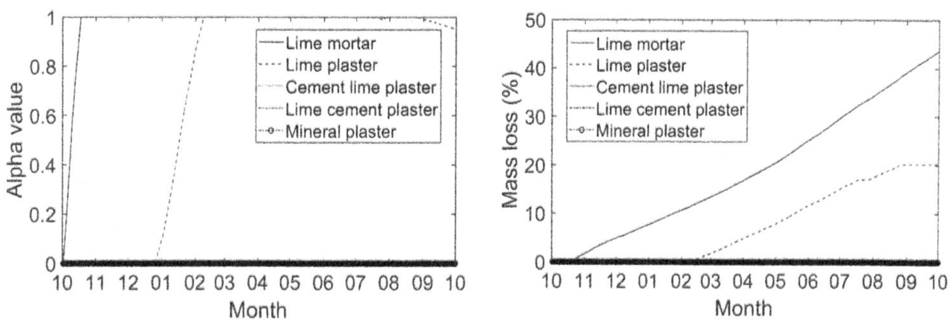

Figure 8.21 Alpha value (left) and mass decay (right) at location B for the wall envelopes with different renders

Application example 2: numerical evaluation of freeze-thaw damage risk of masonry walls after internal retrofitting

Introduction

Interior wall insulation is the solution to be considered to improve building energy efficiency for historical buildings with a façade that is worth preserving. However, the hygrothermal performance of the building envelope is significantly changed by applying internal thermal insulation. As the internal insulation reduces the heat flux from the interior into the exterior wall, the wall temperature and its drying ability decrease. Consequently, such intervention leads to increased moisture levels and lower temperature in the exterior masonry wall. This is the main reason of the occurrence of freeze-thaw damage in masonry walls after internal retrofitting. Therefore, it is essential to analyze the eventual risk of freeze-thaw damage in masonry walls subjected to internal retrofitting.

The main mechanism responsible for freeze-thaw damage to occur in porous building materials is phase change of water (Litvan 1988). In porous material, the volumetric expansion of water by 9% due to freezing exerts tensile stresses on the material matrix and causes damage when the stresses exceed the strength of the material matrix (Powers 1955; Everett 1961; Liso et al. 2007). One indicator, the number of freeze-thaw cycles, can be used to estimate the risk for freeze-thaw damage (Grossi et al. 2007). In this method, each freeze-thaw cycle is expected to add to the cumulative freeze-thaw damage similar to what occurs in mechanical fatigue with constant load cycles (Fagerlund 2000). Many studies often use, for simplicity, the number of zero crossings on a Celsius scale as the number of freeze-thaw cycles (Sedlbauer et al. 2000; Grossi et al. 2007; Klõšeiko et al. 2015; van Aarle et al. 2015). However, this approach fails to account for the fact that water in porous medium freezes at different temperatures below 0 °C, depending on the pore size.

Field observations of freeze-thaw damage in masonry structures suggest that the performance of bricks is affected not only by the material characteristics but also by the microenvironment to which they are exposed (Hansen and Kung 1988). Moisture content in a masonry wall is affected by environmental conditions such as building geographical location, wall orientation, and period of the year. The most likely meteorological condition for occurrence of freeze-thaw damage is a considerable temperature drop below 0 °C with a significant amount of rainfall in the previous days. Liso et al. (2007) developed a frost decay exposure index based on the number of freezing events and rainfall sums prior to freezing events to assess freeze-thaw damage risk.

Therefore, a hygrothermal model including freezing and thawing phenomena in porous medium is needed to study the changes in hygrothermal performance after adding an insulation layer on the internal side of masonry walls. A new method that uses the number of actual ice growth and melt cycles as an indicator of freeze-thaw cycles is proposed in this study, where one cycle of formation and disappearance of ice in a porous medium is defined as one freeze-thaw cycle. The results from such hygrothermal simulations could be used to analyze the risk of freeze-thaw damage in the masonry wall.

Model setup

For hygrothermal simulations, masonry wall assemblies with and without internal insulation are chosen to study the influence of adding internal insulation on freeze-thaw damage risk of a masonry wall (Figure 8.22). The masonry building envelope assemblies consist of the following materials: three-layer masonry, aerogel insulation plaster, and interior plaster. The hygrothermal modeling including freezing and thawing (Equations 8.15–8.16) is applied for

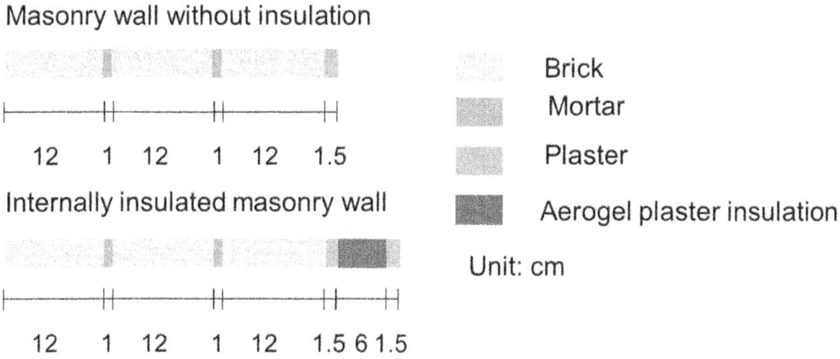

Figure 8.22 Selected masonry wall assemblies for hygrothermal simulation

Source: Zhou et al. 2017

two locations in Switzerland: Zurich and Davos. The climate in Zurich is characterized by mild winters and warm summers. By comparison, Davos, located at 1560 m above sea level, has a considerable cold climate. In winter, precipitation mainly occurs in the form of snow. To select a wall orientation subject to the most severe moisture conditions, the orientation with the largest WDR load is selected for hygrothermal simulation. The selected wall orientations for Zurich and Davos are 240 and 30 degrees, respectively. Meteorological data from 1981 to 2010 for the two cities from MeteoSwiss are used for hygrothermal modeling.

Results

COMPARISON BETWEEN TWO METHODS FOR CALCULATING FREEZE-THAW CYCLES

As mentioned, the traditional method for assessing freeze-thaw cycle is based on the number of temperature crossings below and above 0 °C. In this study, a new method that uses formation and disappearance of ice content in porous medium as freeze-thaw cycle is proposed. The two methods are compared for two cities for an uninsulated masonry wall for the period from 1981 to 2010. The meteorological year is defined from the beginning of October to the end of September.

Significant differences between the number of freeze-thaw cycles exist for the traditional and the new proposed methods (Figure 8.23). The traditional method based on temperature shows a much larger number of freeze-thaw cycles than the new method based on ice content. There is almost no correlation between the numbers of freeze-thaw cycles between the two methods. For Zurich, the correlation coefficient between the numbers of freeze-thaw cycles determined by the two methods is 0.079, and for Davos, the correlation coefficient is 0.069. The average number of freeze-thaw cycles based on temperature is around 44 for Zurich and around 100 for Davos. By comparison, the average number of freeze-thaw cycles based on ice content is below 10 for both cities. Taking the year 1981 for Zurich as example, the number of freeze-thaw cycles based on the traditional method is 41, while it is 20 according to the new method. The occurrence of diurnal temperature variation around 0 °C is the main reason for the large number of freeze-thaw cycles determined by the traditional method based on temperature (Figure 8.24). Temperature drops below 0 °C and returns above 0 °C in a short period in some periods of November, February, and March, while the ice content stays always at 0 kg/m^3 in these periods. No ice exists in the brick even though the temperature in the brick is very low. The reason for this absence of ice is that the total moisture content is very low in the brick. For example, the total moisture content is 51.0 kg/m^3 and the temperature is −7.2 °C on February 27. For the

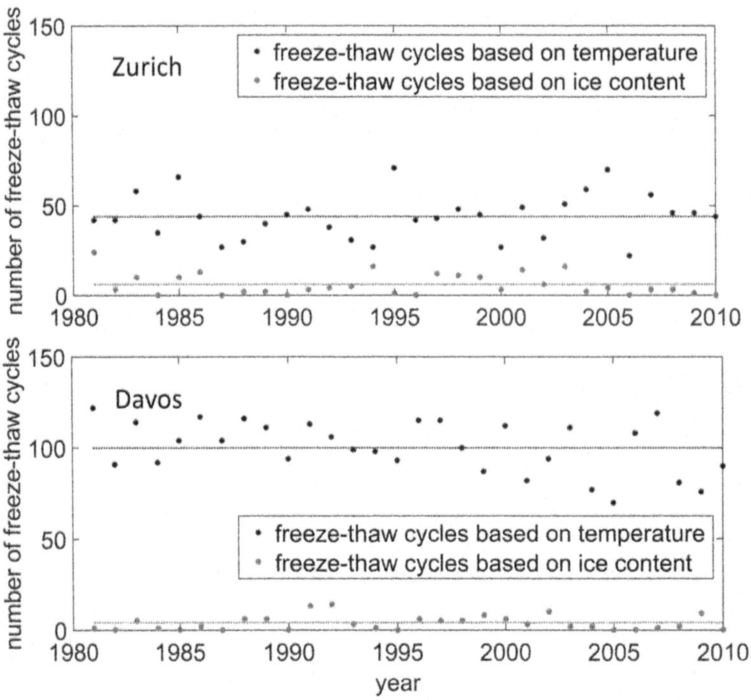

Figure 8.23 Comparison of freeze-thaw cycles based on two methods for the external part of the brick (0.5 cm from exterior surface) for an uninsulated masonry wall in Zurich and Davos. The line indicates the average number of freeze-thaw cycles.

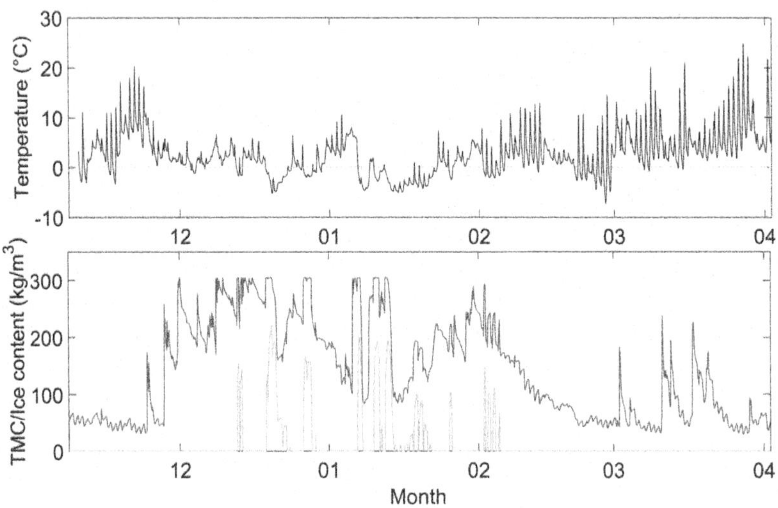

Figure 8.24 Temperature (above) and total moisture content (TMC) and ice content (below) variations for the external part of the brick (0.5 cm from exterior surface) in the cold period of 1981 for Zurich. The dashed line indicates temperature of 0 °C.

total moisture content of 51.0 kg/m³, the corresponding freezing temperature is −18.2 °C. As the freezing temperature is much lower than the temperature in the brick, no ice could exist at this time. By comparison, when the total moisture content is high, the corresponding freezing temperature also increases. For example, on February 3, the total moisture content is 254.8 kg/m³ and thus the corresponding freezing temperature is −0.63 °C. The temperature of the brick is −2.1 °C, which is lower than the freezing temperature. Therefore, ice exists at this condition: the ice content is 118.4 kg/m³ and the liquid water content is 136.4 kg/m³.

COMPARISON AMONG WALL ASSEMBLIES WITH AND WITHOUT INTERNAL INSULATION

A high number of freeze-thaw cycles does not necessarily mean high risk of freeze-thaw damage. However, a high number of freeze-thaw cycles together with high ice content tends to result in a high risk of freeze-thaw damage. Therefore, a year with a substantially high number of freeze-thaw cycles and high ice content is used for studying the influence of internal insulation on freeze-thaw damage risk. The selected years for Zurich and Davos are 1981 and 1989, respectively.

Figures 8.25 and 8.26 compare the saturation degree of ice content at four locations for wall assemblies with and without internal insulation for Zurich and Davos. The saturation degree of ice content, a normalized value from 0.0 to 1.0, allows comparison between results even for different materials. The internally insulated masonry wall shows in general higher ice contents. In addition, the internal insulation leads to longer occurrence of ice in the wall. The difference in ice content between the uninsulated masonry wall and the internally insulated wall becomes larger at a larger distance from the exterior surface. It is remarkable that internal insulation has a very different influence on freeze-thaw damage risk for the two cities. For the uninsulated masonry wall in Zurich, the maximum saturation degree of ice content at locations of 2, 8, 14, and 20 cm from the exterior surface is 0.72, 0.46, 0.37, and 0.0, respectively. By comparison,

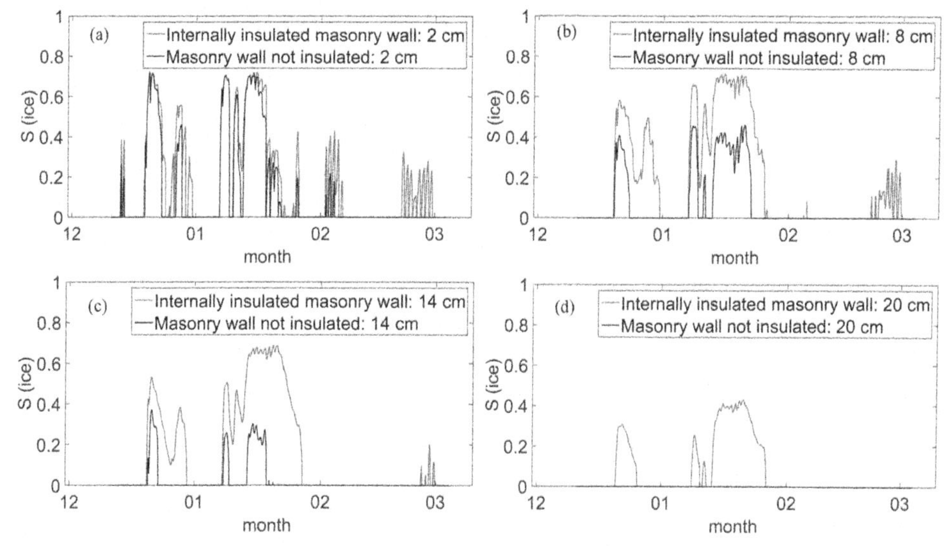

Figure 8.25 Comparison of saturation degree of ice content (S) at 2 (a), 8 (b), 14 (c), and 20 (d) cm from the exterior surface for the wall assemblies with and without internal insulation for Zurich in 1981

Source: Zhou et al. 2017

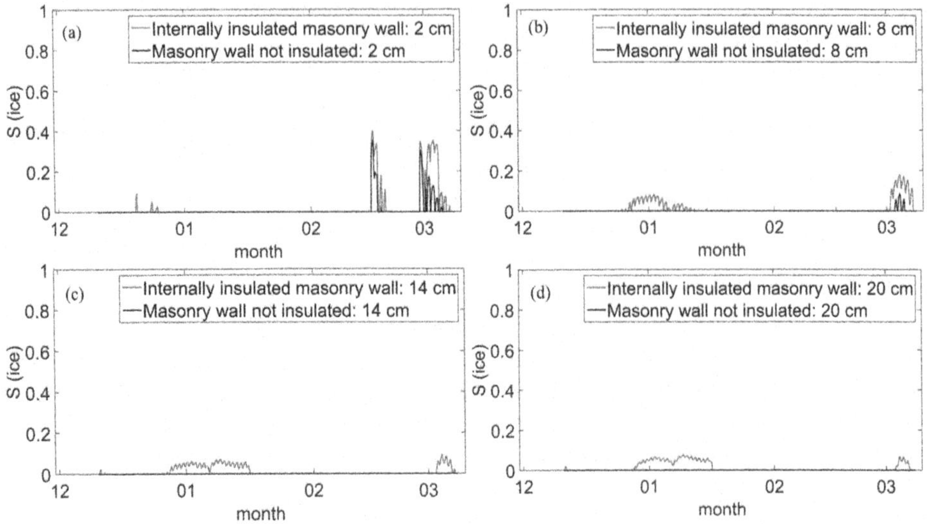

Figure 8.26 Comparison of saturation degree of ice content (S) at 2 (a), 8 (b), 14 (c), and 20 (d) cm from the exterior surface for the wall assemblies with and without internal insulation for Davos in 1989

Source: Zhou et al. 2017

for the internally insulated masonry wall, the same indicator at the same locations yields 0.72, 0.71, 0.69, and 0.42, respectively. The difference of saturation degree of ice content between the two wall assemblies is 0.0 at 2 cm from exterior surface, while it is up to 0.32 at 14 cm and 0.42 at 20 cm from exterior surface. The brick at 20 cm from the exterior surface suffers from freezing and thawing cycling after adding internal insulation while it shows no freezing and thawing in the uninsulated masonry wall. The internal insulation reduces the temperature and also the drying potential of the masonry wall, which is the reason for freezing of moisture in the inner wall in the cold period. Furthermore, the number of freeze-thaw cycles also increases after adding internal insulation. For Zurich, the number of freeze-thaw cycles at 2 cm from exterior surface is 14 for the uninsulated masonry wall, while it increases to 25 for the internally insulated masonry wall. However, the difference of the number of freeze-thaw cycles at deeper depths such as 8, 14, and 20 cm is quite small between the uninsulated masonry wall and the internally insulated masonry wall.

Compared to Zurich, the ice content and the number of freeze-thaw cycles are much lower in Davos. For Davos, the maximum saturation degree of ice content is 0.4 at 2 cm from the exterior surface for the internally insulated masonry wall. The ice content difference between the uninsulated masonry wall and the internally insulated wall is also much smaller in Davos. The maximum difference of degree of saturation of ice content between the two assemblies with and without internal insulation is only 0.09 at 8 cm from the exterior surface. No ice exists at 14 and 20 cm from the exterior surface for the uninsulated masonry wall. Even though ice exists at the same locations for the internally insulated masonry wall, the ice content is very low. The primary reason for the low ice content in Davos is the low air temperature, which leads to precipitation occurring in the form of snow in the cold period, and thus low wind-driven rain load to the masonry wall. As temperature in Davos stays mostly very low – below freezing in winter – the number of freeze-thaw cycles determined by ice melting and water freezing is also very low.

Discussion

Biodegradation risks in wooden beams embedded in masonry walls and freeze-thaw damage risk of masonry walls are analyzed in this study. The material properties of the exterior render have a large influence on the moisture damage risk of wooden beams. The risk of moisture problems is different among the different locations of a wooden beam. A hygrothermal model, accounting for moisture and heat transport in porous medium and tracking the occurrence of freezing and thawing processes, is developed. We propose a new method that uses the number of actual ice growth and melt cycles as an accurate indicator of the freeze-thaw loading. It is found that both the number of freeze-thaw cycles and ice content increase after adding internal insulation.

In this study, the analysis of damage risks is based on a combined assessment of the moisture and temperature conditions. In reality, the hygrothermal behavior of porous materials is also often coupled to its mechanical behavior, yielding for example swelling or deformation and eventually mechanical damage. For instance, the damage resulting from the shrinking and swelling of wood depends on the magnitude of the change in moisture content. The occurrence of freeze-thaw damage of porous materials is thus influenced by not only moisture content and temperature at the time of freezing but also its mechanical resistance of the porous materials. Therefore, it is essential to develop a numerical model coupling heat, moisture, deformation, and damage in porous materials (e.g. Moonen et al. 2010).

The assumption of perfect hydraulic contact is widely used in hygrothermal models. However, the capillary moisture transport is strongly affected by the interfaces between different building materials, although the influence of interface is still not well understood (Qiu et al. 2003; Derluyn et al. 2011), specifically in the cases presented in this chapter in terms of resistance due to pore size mismatch in masonry. Capillary moisture transport along the interface may play an important role in the water uptake in masonry and may affect the hygrothermal performance of the building envelope.

In addition to frost damage, salt crystallization is also a major cause of deterioration of porous building materials. The salt crystallization pressures may lead to spalling and cracking of porous building materials. The mechanism of salt crystallization in porous materials is still an active research topic (Flatt 2002; Scherer 2004; Derluyn et al. 2014; Koniorczyk et al. 2016). Future research is needed on the nucleation and growth kinetics of salt crystals. To further understand salt nucleation and growth, small-scale results from discrete models could be coupled to a macroscopic model of coupled heat, water, salt ion transport, and salt crystallization (Derluyn et al. 2014). Also, water transport in porous materials can lead to other types of material degradation, for example corrosion of embedded fasteners, where one approach couples heat and mass transport in wood to corrosion of steel fasteners (Zelinka et al. 2011, 2014).

Synopsis

The focus of this chapter has been mainly on the moisture-related effects of WDR on building materials and envelopes. Surface wetting due to WDR is also an integral part of studies related to evaluating mitigation solutions for local and urban heat islands in the built environment, energy consumption of buildings and urban systems, vitality of urban vegetation, management of storm water and comfort, and safety and health of the population. A more accurate modeling of moisture redistribution in urban environments can improve the evaluation of evaporative cooling and the resulting urban thermal comfort. Such consideration of moisture is particularly important because rain is the main moisture source in an urban environment (Moonen et al. 2012; Kubilay et al. 2016), especially on building envelopes. Transport of moisture within the building envelope must be considered conjointly with heat transport, and the risks associated with the presence of moisture can be assessed.

The first part of the chapter focused on the rain deposition distribution on building envelopes, while taking into account influences of nearby buildings and façade details at different meteorological conditions. The variation of droplet impact speed and angles was demonstrated, which would influence the physics once the droplets come into contact with the surface. The EM model with the turbulent dispersion of raindrops was used in time-averaged and time-resolved modeling of WDR. The model enables the determination of WDR intensity on more complicated computational domains with less computational effort and simpler boundary conditions compared to previous models.

The second part presented a developed methodology to select a moisture reference year to conduct hygrothermal simulations on building envelopes. The section introduced the Climatic Index, which is used to select the reference year based on annual wetting load due to WDR and annual potential evaporation.

The third part evaluated the risk of wood decay and of freeze-thaw damage in the envelope assembly. The simulations of a wood beam in an internally insulated masonry wall indicate that the capillary active renderings are likely to lead to moisture problems and wood decay. The second part of the section extended the hygrothermal modeling with freezing and thawing processes and revealed the increased risk of freeze-thaw damage after interior wall insulation.

References

Abuku, M., Janssen, H. and Roels, S. (2009) Impact of wind-driven rain on historic brick wall buildings in a moderately cold and humid climate: Numerical analyses of mould growth risk, indoor climate and energy consumption. *Energy Buildings* 41(1): 101–110.

Best, A.C. (1950) The size distribution of raindrops. *Quarterly Journal of the Royal Meteorological Society* 76(327): 16–36.

Blocken, B. (2014) 50 years of computational wind engineering: Past, present and future. *Journal of Wind Engineering and Industrial Aerodynamics* 129: 69–102.

Blocken, B. and Carmeliet, J. (2002) Spatial and temporal distribution of driving rain on a low-rise building. *Wind and Structures* 5(5): 441–462.

Blocken, B. and Carmeliet, J. (2004) A review of wind-driven rain research in building science. *Journal of Wind Engineering and Industrial Aerodynamics* 92(13): 1079–1130.

Blocken, B. and Carmeliet, J. (2005) High-resolution wind-driven rain measurements on a low-rise building – experimental data for model development and model validation. *Journal of Wind Engineering and Industrial Aerodynamics* 93(12): 905–928.

Blocken, B. and Carmeliet, J. (2006a) The influence of the wind-blocking effect by a building on its wind-driven rain exposure. *Journal of Wind Engineering and Industrial Aerodynamics* 94(2): 101–127.

Blocken, B. and Carmeliet, J. (2006b) On the accuracy of wind-driven rain measurements on buildings. *Building and Environments* 41(12): 1798–1810.

Blocken, B. and Carmeliet, J. (2007a) On the errors associated with the use of hourly data in wind-driven rain calculations on building facades. *Atmospheric Environment* 41(11): 2335–2343.

Blocken, B. and Carmeliet, J. (2007b) Validation of CFD simulations of wind-driven rain on a low-rise building facade. *Building and Environments* 42(7): 2530–2548.

Blocken, B. and Carmeliet, J. (2008) Guidelines for the required time resolution of meteorological input data for wind-driven rain calculations on buildings. *Journal of Wind Engineering and Industrial Aerodynamics* 96(5): 621–639.

Blocken, B. and Carmeliet, J. (2010) Overview of three state-of-the-art wind-driven rain assessment models and comparison based on model theory. *Building and Environments* 45(3): 691–703.

Blocken, B., Dezsö, G., van Beeck, J. and Carmeliet, J. (2009) The mutual influence of two buildings on their wind-driven rain exposure and comments on the obstruction factor. *Journal of Wind Engineering and Industrial Aerodynamics* 97(5–6): 180–196.

Blocken, B., Deszö, G., van Beeck, J. and Carmeliet, J. (2010) Comparison of calculation methods for wind-driven rain deposition on building facades. *Atmospheric Environment* 44(14): 1714–1725.

Blocken, B., Stathopoulos, T., Carmeliet, J. and Hensen, J.L.M. (2011) Application of computational fluid dynamics in building performance simulation for the outdoor environment: An overview. *Journal of Building Performance Simulation* 4(2): 157–184.

Boussinesq, J. (1887) Essai sur la théorie des eaux courantes. *Mémoires présentés par divers savants à l'Académie des Sciences* 23: 1–680.

Briggen, P.M., Blocken, B., Schellen, H.L. (2009) Wind-driven rain on the facade of a monumental tower: Numerical simulation, full-scale validation and sensitivity analysis. *Buildings and Environments* 44(8): 1675–1690.

Burkhardt, M., Zuleeg, S., Vonbank, R., Bester, K., Carmeliet, J., Boller, M. and Wangler, T. (2012) Leaching of biocides from Façades under natural weather conditions. *Environmental Science and Technology* 46(10): 5497–5503.

Carll, C. and Wiedenhoeft, A.C. (2009) Moisture-related properties of wood and the effects of moisture on wood and wood products. In Trechsel, H.R. and Bomberg, M. (eds), *Moisture Control in Buildings: The Key Factor in Mold Prevention*, MNL18–2nd, *2nd edition*. West Conshohocken, PA: ASTM International, pp. 54–79.

Carmeliet, J. and Derome, D. (2012a) Temperature driven inward vapor diffusion under constant and cyclic loading in small-scale wall assemblies: Part 1 experimental investigation. *Buildings and Environments* 48: 48–56.

Carmeliet, J. and Derome, D. (2012b). Temperature driven inward vapor diffusion under constant and cyclic loading in small-scale wall assemblies: Part 2 heat-moisture transport simulations. *Buildings and Environments* 47: 161–169.

Carmeliet, J., Roels, S. and Bomberg, M.T. (2009) Towards development of methods for assessment of moisture-originated damage. Chapter 28. In Trechsel, H.R. and Bomberg, M. (eds), *Moisture Control in Buildings: The Key Factor in Mold Prevention*, MNL18–2nd, *2nd edition*. West Conshohocken, PA: ASTM International, pp. 591–605.

Charola, A.E. and Lazzarini, L. (1986) Deterioration of brick masonry caused by acid rain. *ACS Symposium Series* 318: 250–258.

Choi, E.C.C. (1993) Simulation of wind-driven-rain around a building. *Journal of Wind Engineering and Industrial Aerodynamics* 46–47: 721–729.

Choi, E.C.C. (1997) Numerical modelling of gust effect on wind-driven rain. *Journal of Wind Engineering and Industrial Aerodynamics* 72(1): 107–116.

de Freitas, S.S., Barreira, E. and de Freitas, V.P. (2013) Quantification of wind-driven rain and evaluation of façade humidification. *Central European Symposium on Building Physics, September 9–11, Vienna, Austria.*

Delgado, J.M., Barreira, E., Ramos, N.M. and de Freitas, V.P. (2013) *Hygrothermal Numerical Simulation Tools Applied to Building Physics*. Springer Science and Business Media. Berlin, Germany.

Derome, D., Kubilay, A., Defraeye, T., Blocken, B. and Carmeliet, J. (2017) Ten questions concerning modeling of wind-driven rain in the built environment. *Buildings and Environments* 114: 495–506.

Derluyn, H., Janssen, H. and Carmeliet, J. (2011) Influence of the nature of interfaces on the capillary transport in layered materials. *Construction and Building Materials* 25(9): 3685–3693.

Derluyn, H., Moonen, P. and Carmeliet, J. (2014) Deformation and damage due to drying-induced salt crystallization in porous limestone. *Journal of the Mechanics and Physics of Solids* 63: 242–255.

de Wolf, D.A. (2001) On the Laws-Parsons distribution of raindrop sizes. *Radio Science* 36(4): 639–642.

Etyemezian, V., Davidson, C.I., Zufall, M., Dai, W., Finger, S. and Striegel, M. (2000) Impingement of rain drops on a tall building. *Atmospheric Environment* 34: 2399–2412.

Everett, D. (1961) The thermodynamics of frost damage to porous solids. *Transactions of the Faraday Society* 57: 1541–1551.

Fagerlund, G. (2000) *Fatigue Effects Associated with Freeze-thaw of Materials*. Division of Building Materials, LTH, Lund University.

Flatt, R.J. (2002) Salt damage in porous materials: How high supersaturations are generated. *Journal of Crystal growth* 242(3): 435–454.

Franke, L., Schumann, I., van Hees, R., van der Klugt, L., Naldini, S., Binda, L., Baronio, G., van Balen, K. and Mateus, J. (1998) Damage atlas: Classification and analyses of damage patterns found in brick masonry. *European Commission Research Report No 8, vol. 2, Fraunhofer IRB Verlag.*

Gosman, A.D., Lekakou, C., Politis, S., Issa, R.I. and Looney, M.K. (1992) Multidimensional modeling of turbulent two-phase flows in stirred vessels. *AIChE Journal* 38: 1946–1956.

Gousseau, P., Blocken, B. and van Heijst, G.J.F. (2011) CFD simulation of pollutant dispersion around isolated buildings: On the role of convective and turbulent mass fluxes in the prediction accuracy. *Journal of Hazardous Materials* 194: 422–434.

Grossi, C.M., Brimblecombe, P. and Harris, I. (2007) Predicting long term freeze-thaw risks on Europe built heritage and archaeological sites in a changing climate. *Science of the Total Environment* 377(2–3): 273–281.

Hagentoft, C., Kalagasidis, A., Adl-Zarrabi, B., Roels, S., Carmeliet, J., Hens, H., et al. (2004) Assessment method of numerical prediction models for combined heat, air and moisture transfer in building components: Benchmarks for one-dimensional cases. *The Journal of Thermal Envelope and Building Science*27: 327–352.

Hansen, W. and Kung, J. (1988) Pore structure and frost durability of clay bricks. *Materials and Structures* 21(6): 443–447.

Harrestrup, M. and Svendsen, S. (2016) Internal insulation applied in heritage multi-storey buildings with wooden beams embedded in solid masonry brick facades. *Building and Environment* 99: 59–72.

Hens, H. (2010) Wind-driven rain: From theory to reality. In *Thermal Performance of the Exterior Envelopes of Whole Buildings XI,* Clearwater, FL.

Hill, D.P. (1998) The computer simulation of dispersed two-phase flow. PhD Thesis, Imperial College of Science, Technology & Medicine, Department of Mechanical Engineering, London, UK.

Högberg, A., Kragh, M. and van Mook, F.J.R. (1999) A comparison of driving rain measurements with different gauges. In *5th Symposium of Building Physics in the Nordic Countries*. Göteborg, Sweden, August 24–26, 1999.

Huang, S. and Li, Q. (2010) Numerical simulations of wind-driven rain on building envelopes based on Eulerian multiphase model. *Journal of Wind Engineering and Industrial Aerodynamics* 98(12): 843–857.

Issa, R.I. and Oliveira, P.J. (1993) Modelling of turbulent dispersion in two phase flow jets. *Engineering Turbulence Modeling and Experiments* 2: 947–957.

Janssen, H., Blocken, B. and Carmeliet, J. (2007) Conservative modelling of the moisture and heat transfer in building components under atmospheric excitation. *International Journal of Heat and Mass Transfer* 50(5–6): 1128–1140.

Kehl, D., Ruisinger, U., Plagge, R. and Grunewald, J. (2013) Wooden beam ends in masonry with interior insulation-A literature review and simulation on causes and assessment of decay. In *Proceedings of 2nd Central European Symposium on Building Physics*. Vienna, Austria, September 9–11, pp. 299–304.

Klõšeiko, P., Arumägi, E. and Kalamees, T. (2015) Hygrothermal performance of internally insulated brick wall in cold climate: A case study in a historical school building. *Journal of Building Physics* 38(5): 444–464.

Koniorczyk, M., Gawin, D., Konca, P. and Bednarska, D. (2016) Modeling damage of building materials induced by sodium sulphate crystallization. *Bauphysik* 38(6): 366–371.

Kubilay, A., Derome, D., Blocken, B. and Carmeliet, J. (2013) CFD simulation and validation of wind-driven rain on a building facade with an Eulerian multiphase model. *Building and Environment* 61: 69–81.

Kubilay, A., Derome, D., Blocken, B. and Carmeliet, J. (2014a). High-resolution field measurements of wind-driven rain on an array of low-rise cubic buildings. *Building and Environment* 78: 1–13.

Kubilay, A., Derome, D., Blocken, B. and Carmeliet, J. (2014b) Numerical simulations of wind-driven rain on an array of low-rise cubic buildings and validation by field measurements. *Building and Environment* 81: 283–295.

Kubilay, A., Derome, D., Blocken, B. and Carmeliet, J. (2015) Numerical modeling of turbulent dispersion for wind-driven rain on building facades. *Environmental Fluid Mechanics* 15(1): 109–133.

Kubilay, A., Derome, D. and Carmeliet, J. (2016) Influences of wind-driven rain and radiation on urban microclimate. In *CESBP Central European Symposium on Building Physics*, September 14–16, Dresden, Germany.

Kubilay, A., Derome, D. and Carmeliet, J. (2017a) Analysis of time-resolved wind-driven rain on an array of low-rise cubic buildings using large eddy simulation and an Eulerian multiphase model. *Building and Environment* 114: 68–81.

Kubilay, A., Carmeliet, J. and Derome, D. (2017b) CFD simulations of wind-driven rain on a mid-rise residential building with various types of facade details. *Journal of Building Performance Simulation* 10(2): 125–143.

Kumaran, M., Lackey, J., Normandin, N., van Reenen, D. and Tariku, F. (2002) *Summary Report From task 3 of MEWS Project at the Institute for Research in Construction-Hygrothermal Properties of Several Building Materials*. Ottawa, Canada: National Research Council.

Künzel, H.M., Künzel, H. and Holm, A. (2004) Rain protection of stucco facades. In *Proceedings of the Thermal Performance of the Exterior Envelopes of Whole Buildings IX–International Conference*. Florida, USA.

Lee, J.B., Derome, D., Dolatabadi, A. and Carmeliet, J. (2016a) Energy budget of liquid drop impact at maximum spreading: Numerical simulations and experiments. *Langmuir* 32(5): 1279–1288.

Lee, J.B., Derome, D. and Carmeliet, J. (2016b) Drop impact on natural porous stones. *Journal of Colloid and Interface Science* 469: 147–156.

Lee, J.B., Derome, D. and Carmeliet, J. (2016c). Absorption of impinging water droplet in porous stones. *Journal of Colloid and Interface Science* 471: 59–70.

Liso, K.R., Kvande, T., Hygen, H.A., Thue, J.V. and Harstveit, K. (2007) A frost decay exposure index for porous, mineral building materials. *Building and Environment* 42(10): 3547–3555.

Litvan, G.G. (1988) The mechanism of frost action in concrete-theory and practical implications. In *Proceedings of Workshop on Low Temperature Effects on Concrete*. Sapporo, Hokkaido, Japan, pp. 115–134.

Martin, M., Defraeye, T., Derome, D. and Carmeliet, J. (2015) A film flow model for analysing gravity-driven, thin wavy fluid films. *International Journal of Multiphase Flow* 73: 207–216.

Melville, W.K. and Bray, K.N.C. (1979) A model of the two-phase turbulent jet. *International Journal of Heat and Mass Transfer* 22: 647–656.

Moonen, P., Defraeye, T., Dorer, V., Blocken, B. and Carmeliet, J. (2012) Urban physics: Effect of the microclimate on comfort, health and energy demand. *Frontiers of Architectural Research* 1(3): 197–228.

Moonen, P., Sluys, L.J. and Carmeliet, J. (2010) A continuous – discontinuous approach to simulate physical degradation processes in porous media. *International Journal for Numerical Methods in Engineering* 84(9): 1009–1037.

Morelli, M. and Svendsen, S. (2013) Investigation of interior post-insulated masonry walls with wooden beam ends. *Journal of Building Physics* 36(3): 265–293.

Pazera, M. and Bomberg, M. (2010) Applying lessons from clay – brick veneer to design a stucco mix. In *BEST Conference Building Enclosure Science and Technology*, Portland, OR.

Penman, H.L. (1948) Natural evaporation from open water, bare soil and grass. In *Proceedings of the Royal Society of London A: Mathematical, Physical and Engineering Sciences*.

Powers, T.C. (1955) Basic considerations pertaining to freezing-and-thawing tests. In *ASTM Proceedings*, pp. 1132–1155.

Qiu, X., Haghighat, F. and Kumaran, M.K. (2003) Moisture transport across interfaces between autoclaved aerated concrete and mortar. *Journal of Thermal Envelope and Building Science* 26(3): 213–236.

Rusche, H. (2002) Computational fluid dynamics of dispersed two-phase flows at high phase fractions. PhD Thesis, Imperial College of Science, Technology & Medicine, Department of Mechanical Engineering, London, UK.

Sanders, C. (1996) Heat, air and moisture transfer in insulated envelope parts: Environmental conditions. International Energy Agency, Annex 24 Final report, volume 2 Acco, Leuven.

Scherer, G.W. (2004) Stress from crystallization of salt. *Cement and Concrete Research* 34(9): 1613–1624.

Sedlbauer, K. and Kunzel, H. (2000) Frost damage of masonry walls: A hygrothermal analysis by computer simulations. *Journal of Thermal Envelope and Building Science* 23(3): 277–281.

Shirolkar, J.S., Coimbra, C.F.M. and McQuay, M.Q. (1996) Fundamental aspects of modeling turbulent particle dispersion in dilute flows. *Progress in Energy and Combustion Science* 22: 363–399.
Sumner, G.N. (1981) The nature and development of rainstorms in Coastal East Africa. *Journal of Climatology* 1(2): 131–152.
Tang, W., Davidson, C.I., Finger, S. and Vance, K. (2004) Erosion of limestone building surfaces caused by wind-driven rain: 1. Field measurements. *Atmospheric Environment* 38(33): 5589–5599.
Teasdale-St-Hilaire, A. and Derome, D. (2007) Comparison of experimental and numerical results of wood-frame wall assemblies wetted by simulated wind-driven rain infiltration. *Energy and Buildings* 39: 1131–1139.
TenWolde, A. (2008) ASHRAE standard 160 PcCriteria for moisture control design analysis in buildings. *Ashrae Transactions* 114(1): 167–171.
Uijlenhoet, R. and Stricker, J.N.M. (1999) A consistent rainfall parameterization based on the exponential raindrop size distribution. *Journal of Hydrology* 218(3–4): 101–127.
van Aarle, M., Schellen, H. and van Schijndel, J. (2015) Hygro thermal simulation to predict the risk of frost damage in masonry; Effects of climate change. In *Proceedings of 6th International Building Physics Conference*. Torino, Italy, pp. 2536–2541.
Van Balen, K. (1996) Expert system for evaluation of deterioration of ancient brick masonry structures. *Science of the Total Environment* 189: 247–254.
van der Hoven, I. (1957) Power spectrum of horizontal wind speed in the frequency range from 0.0007–900 cycles per hour. *Journal of Meteorology* 14(2): 160–164.
Viitanen, H., Toratti, T., Makkonen, L., Peuhkuri, R., Ojanen, T., Ruokolainen, L. and Raisanen, J. (2010) Towards modelling of decay risk of wooden materials. *European Journal of Wood and Wood Products* 68(3): 303–313.
Williams, P. and Smith, M. (1989) *The Frozen Earth: Fundamentals of Geocryology*. Cambridge: Cambridge University Press.
Zelinka, S., Derome, D. and Glass, S. (2011) Combining hygrothermal and corrosion models to predict corrosion of metal fasteners embedded in wood. *Building and Environment* 46: 2060–2068.
Zelinka, S.L., Glass, S.V. and Derome, D. (2014) The effect of moisture content on the corrosion of fasteners embedded in wood subjected to alkaline copper quaternary treatment. *Corrosion Science* 83: 67–74.
Zhou, X., Derome, D. and Carmeliet, J. (2016) Robust moisture reference year methodology for hygrothermal simulations. *Building and Environment* 110: 23–35.
Zhou, X., Derome, D. and Carmeliet, J. (2017) Hygrothermal modeling and evaluation of freeze-thaw damage risk of masonry walls retrofitted with internal insulation. *Building and Environment* 125: 285–298.
Zhou, X., Zhou, J., Kinzelbach, W. and Stauffer, F. (2014) Simultaneous measurement of unfrozen water content and ice content in frozen soil using Gamma ray attenuation and TDR. *Water Resources Research* 50: 9630–9655.

Recommended reading

ASHRAE. (2017) Building envelope. Chapters 25–27. In *Handbook – Fundamentals*. ASHRAE.
Hagentoft, C-E. (2001) *Introduction to Building Physics*. Lund, Sweden: Studentlitteratur AB.
Hens, H. (2017) *Building Physics – Heat, Air and Moisture: Fundamentals and Engineering Methods with Examples and Exercises*, 3rd edition. Chapters 2–3. Berlin, Germany: Ernst & Sohn.
Hutcheon, N.B. and Handegord, G.O. (1983) *NRC/CNRC Building Science for a Cold Climate*. Toronto, Canada: John Wiley & Sons.
Trechsel, H.R. and Bomberg, M. (2009) *MNL-18–2ND, Moisture Control in Buildings: The Key Factor in Mold Prevention*, 2nd edition. West Conshohocken, PA: ASTM International.

Activities

Exercise 1: Catch ratio charts are given below for two positions on the windward façade of a stand-alone cubic building with a height of 10 m. Pos. A is located at the top of the façade, while Pos. B is located at a height of 4 m.

Two curves are fitted for each position as indicated, one for a reference wind speed at 10 m height, U_{10} = 3 m/s and another for a reference rainfall intensity R_h = 1 mm/h:

a*exp(b*x) + c (for $R_h > 0$ and $U_{10} > 0$)

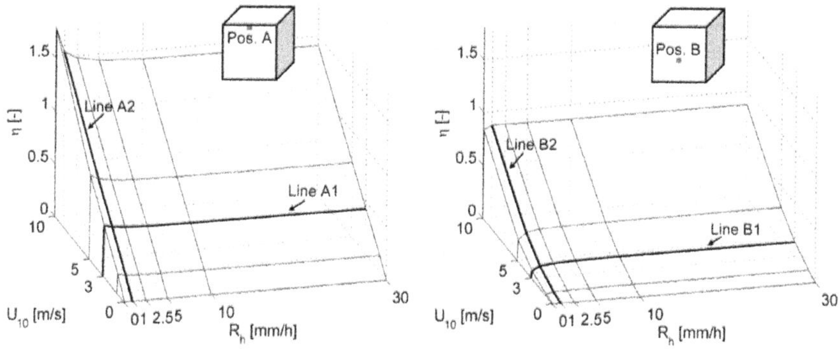

	A	*B*	*C*
Line A1	0.044	−0.402	0.435
Line A2	−156.8	−0.001	156.8
Line B1	−0.072	−0.781	0.137
Line B2	0.190	0.173	−0.198

Consider the two hypothetical rain events below. Wind direction is perpendicular to the façade during the rain events. For both rain events, the total rainfall amount is 17 mm. Assuming that no runoff occurs, (a) calculate the total wind-driven rain at Pos. A and Pos. B at the end of the rain events and (b) compare the catch ratio for the rain events.

Rain event 1:

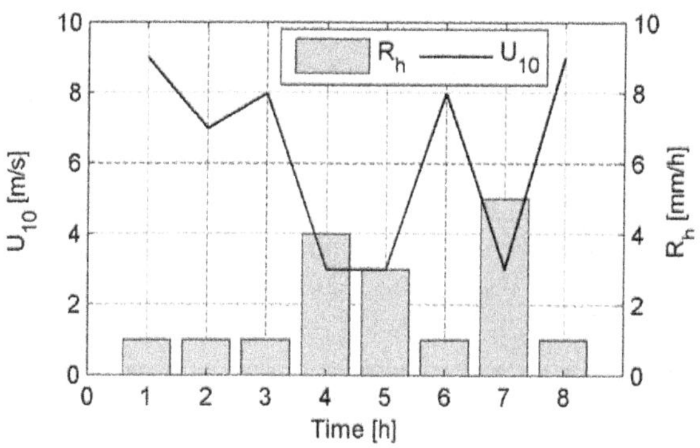

Time [h]	*R_h [mm/h]*	*U_{10} [m/s]*
1	1.0	9.0
2	1.0	7.0
3	1.0	8.0
4	4.0	3.0
5	3.0	3.0
6	1.0	8.0
7	5.0	3.0
8	1.0	9.0

Rain event 2:

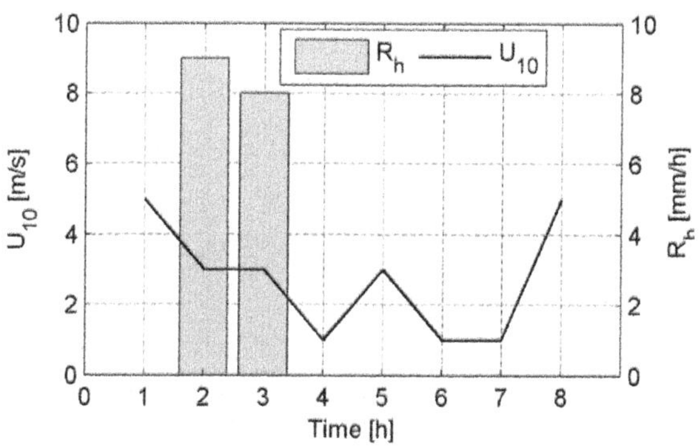

Time [h]	*R_h [mm/h]*	*U_{10} [m/s]*
1.0	0.0	5.0
2.0	9.0	3.0
3.0	8.0	3.0
4.0	0.0	1.0
5.0	0.0	3.0
6.0	0.0	1.0
7.0	0.0	1.0
8.0	0.0	5.0

Solution:

	Rain event 1		*Rain event 2*	
	Pos. A	*Pos. B*	*Pos. A*	*Pos. B*
Total WDR [mm]	11.73	4.57	7.42	2.33
Catch ratio [–]	0.69	0.27	0.44	0.14

Higher average wind speed during Rain event 1 results in higher catch ratio values. The average rainfall intensity is lower during Rain event 1, which leads to a larger difference between catch ratio at positions A and B than the difference for Rain event 2.

Exercise 2: Plot annual wind-driven rain load (WDR) and potential evaporation (PE) in Zurich and Lugano for different orientations with the following data and compare difference of WDR and PE between Zurich and Lugano. Calculate Climatic Index for each orientation and plot Climatic Index rose for Zurich and Lugano. Select the orientation with the largest Climatic Index.

Orientation (°)	*Zurich*		*Lugano*	
	WDR *($kgm^{-2}a^{-1}$)*	*PE* *($kgm^{-2}a^{-1}$)*	*WDR* *($kgm^{-2}a^{-1}$)*	*PE* *($kgm^{-2}a^{-1}$)*
0	96.2	856.1	310.8	1073.0
30	64.7	906.3	216.2	1122.5
60	38.0	990.5	116.2	1229.0
90	38.6	1075.3	65.1	1363.8
120	72.1	1136.7	62.6	1483.1
150	147.6	1162.3	58.5	1549.4
180	308.2	1154.1	55.1	1538.3
210	443.7	1138.1	62.6	1477.3
240	482.4	1093.9	105.8	1378.3
270	429.4	1022.7	200.8	1267.5
300	304.6	942.1	307.9	1167.0
330	159.4	877.9	347.7	1098.3

9 Computational modeling in architectural acoustics

Ardeshir Mahdavi

Scope

Building upon a brief discussion of the acoustics of the built environment, this chapter concentrates on the fundamentals of computational room acoustics.

Learning objectives

The chapter aims in the first instance to provide a general understanding of the building-related applications in environmental acoustics. More specifically, the chapter outlines the computational methods adopted in room acoustics simulation applications and explains the variety and utility of numeric performance indicators in room acoustics.

Key words

Room acoustics/sound propagation/absorption/ray tracing/image source method/auralization

Introduction

Acoustical applications in the built environment

Along with thermal and visual aspects, the quality of the built environment is substantially shaped by acoustical factors. To appreciate this, consider the following three illustrative instances related to outdoor sound propagation, building acoustics, and room acoustics:

1 Sound generated by vehicular traffic, trams, rail traffic and aircrafts, industry, and construction can negatively influence the well-being of people not only in the outdoor environment, but also – upon transmission via buildings' enclosure – within buildings.
2 Within buildings, air-borne and impact sound generated in one zone can travel to other zones and potentially cause annoyance. Moreover, depending on the room function, various levels of speech privacy might be necessary.
3 Speech intelligibility and the quality of musical performances in architectural spaces are decisively influenced by their properties vis-à-vis room acoustics.

Design and retrofit measures in areas exemplified through the above instances are complex and interrelated. For instance, the technical specification of buildings' envelope solutions must take the acoustical outdoor boundary conditions into consideration. Likewise, a good design in view of room acoustics may still be compromised due to undesired sound transmission from adjacent spaces. As with other areas of building performance, computational methods and tools can support complex design and retrofit measures related to the acoustical aspects of the built environment.

In this chapter, the focus lies on modeling approaches in room acoustics. Outdoor sound propagation (ISO 1996; Hornikx 2016) is not explicitly treated here. However, the underlying physical phenomena (involving mainly sound propagation in the air and the interaction of the acoustical wave front with obstacles) are arguably comparable, in principle, to those governing the indoor sound propagation as treated in room acoustics (Cremer and Müller 2016a, 2016b; Kuttruff 2017; Rossing 2007). Needless to say, as opposed to indoor sound propagation, consideration of meteorological circumstances (such as wind velocity) in outdoor sound propagation modeling is important.

Also not treated is the subject of noise (and vibration) control in buildings. As such, this subject represents a very important application area of applied acoustics in the built environment domain, typically referred to as building acoustics. This discipline is concerned with those situations when exposure to sound events is not desired, necessitating various noise control strategies. Typical instances of noise control applications in the built environment pertain to, for example, traffic noise exposure, air-borne and impact sound transmission between rooms, as well as noise and vibration due to buildings' mechanical installations and equipment (Beranek 1988; Crocker 2007; Fasold and Veres 2003; Hopkins 2007; Vigran 2008).

As alluded to before, proper solutions in building acoustics are a prerequisite for successful room acoustics. For example, musical performances in a concert hall require both adequate envelope sound insulation (in view of sound transmission from external environment and from adjacent rooms) and quietly operating building service systems (for heating, cooling, ventilation, and illumination). As such, a number of calculation methods and associated tools are available for assessments in building acoustics. However, they cannot be labeled as simulation tools in the strict sense of the word. Oftentimes, deployed physical models in this area rely on statistical energy analysis (SEA) (EN 2000a, 2000b, 2000c; Gerretsen 1986; Lyon 1994; Thaden 2005; Vorländer and Thaden 2000). Likewise, empirically calibrated gray-box-type calculation methods have been used in lieu of high-resolution numeric formalisms based on first principles (Mahdavi 1991; EN 2000a; 2000b, 2000c). Efforts of the latter kind are mostly situated in the specialized research domains and not yet pervasive in terms of practical computational design support tools (Clasen and Langer 2007).

Some basic observations concerning room acoustics

The term "room acoustics" evokes associations with concert halls, opera houses, theater spaces, auditoria, and churches (Ando 1998; Holden 2016). Prediction and evaluation of the acoustical quality of such spaces is indeed one of the main concerns of room acoustics as a science and as a source of knowledge for design support. However, good room acoustics is not important only regarding spaces for music and speech performances. More "ordinary" spaces, such as a living room, an open-plan office, a library reading room, or the waiting room of a clinic, may be experienced as possessing good or poor room acoustics, given requirements pertaining to speech intelligibility, acoustical privacy, and acoustical comfort.

A scientifically based, computationally supported evaluation of architectural designs in view of their room acoustics quality must meet two challenges: First, methods, tools, and reliable model input data are needed to correctly predict the properties of the sound field in a designed (as yet not existing) space. Second, predicted properties of the sound field in a space must be analyzed and interpreted in view of relevant acoustical performance criteria. Given the absence of quantitative methods up to the early 20th century, it was not possible to reliably predict the acoustical performance of architectural spaces in the design stage. There are indeed still unsolved problems concerning both the exact prediction of the sound field in a space and the relationship between objective parameters of the sound field and subjective perception of acoustical quality by inhabitants. Nonetheless, recent advances in room acoustics have reached a point so that satisfactory acoustics cannot only be defined rationally, but can be also predicted reliably. Such advances pertain to sophisticated mathematical methods, powerful computational implementations, and deeper insights concerning the relationship between objective (measureable) sound field parameters and subjective perception of room acoustics quality. In the following, a number of salient topics in the area of room acoustics and associated computational simulation approaches are discussed. However, the chapter does not address questions pertaining to electronic equipment and systems for sound reinforcement and control in rooms.

Chapter overview

The present chapter includes a brief introduction to modeling methods and tools in room acoustics. First, a compact description of fundamentals of room acoustics and the related performance indicators is provided, followed by a discussion of modeling approaches and techniques adapted in room acoustics simulation applications. Next, the issues of model uncertainty, validation, and calibration are briefly addressed as related to the application of simulation and auralization tools in the design and retrofit process. The chapter ends with summary and conclusions.

Tasks and indicators

Fundamental design considerations in room acoustics

In building retrofit scenarios, an actual detailed "base case" scenario for parametric simulations is already available in terms of an existing structure. However, in a typical building design process, the deployment of simulation tools can obviously start only after a preliminary design is generated. This is also the case in room acoustics simulation. To generate a preliminary design that represents a reasonable starting point for simulation-supported acoustical performance queries, a prior knowledge of the relevant processes and criteria is necessary, including at least a basic understanding of relevant topics in the physics, physiology, and psychology of sound, the dynamics of the sound field in architectural spaces, and acoustically relevant properties of building materials, elements, and assemblies. These topics cannot be covered here in depth, but related knowledge and insights can be gained, among other things, from a fairly extensive body of literature (Cremer and Müller 2016a, 2016b; Fasold and Veres 2003; Kuttruff 2017). Nonetheless, some basic abridged reflections about the functional requirements and creative design decisions and their implications for the acoustical quality of the resulting designs would be useful at this stage.

Acoustically relevant functional requirements predominantly pertain to

- a building's (more specifically, a room's) intended function (lecture room, theater, cinema, concert hall, multi-purpose space, etc.);
- the number of inhabitants the space is expected to accommodate.

Such basic information can provide a rough direction with regard to a design's geometry (size/volume) and materiality (finishing/furniture). However, with respect to this kind of preliminary functional requirements, possible detailed design solutions are entirely under-determined. For instance, volume and function do not fully constrain the space of a room's possible shapes, as a large number of very different designs can fulfill basic functional requirements. At this juncture, both logical and heuristic (experience-based) considerations can further guide the design development process. Consider, for instance, the following general guidelines concerning the distribution of sound energy in rooms such as lecture halls and performance spaces, in which acoustical information exchange plays an important role:

- Sound energy emitted from the source should be preferably distributed in the room in a uniform manner.
- With regard to sound energy distribution, both high focal concentrations and dead spots should be avoided.
- The time interval between the first arrival (direct source-receiver path) and distinct subsequent arrivals (reflections) of the sound energy at a listener's location should not exceed certain limits so as to avoid undesired effects (e.g. echoes).
- The direction from which the bulk of the reflected sound energy arrives at a listener's location (e.g. from the ceiling versus from the walls) may have ramifications for the subjective (qualitative) perception of the space's character (e.g. public versus intimate).

Naturally, such guidelines can help planners not only to refine candidate design solutions, but also to prepare those toward an iterative simulation-supported design generation and evaluation process. Thereby, advanced room acoustics simulation tools offer multiple utilities. For instance, they can graphically visualize the sound energy distribution. This capability, aside from providing a qualitative feel for the space's behavior, can support design diagnostics, i.e. support the detection of critical design problems – for instance identify the room enclosure component responsible for an unwanted reflection. Moreover, simulation provides the opportunity to efficiently compute, for fairly detailed design models, the values of a fairly comprehensive set of performance indicators. The following section provides an overview of such indicators.

Note that, for certain – narrowly defined – performance requirements, computer-aided simulation procedures can also support a kind of performance-to-design mapping (Mahdavi 1993a, 1993b; Mahdavi and Berberidou-Kallivoka 1993, 1994; Mahdavi et al. 1997a). For instance, the detailed geometry of a room's articulated ceiling (or the arrangement of acoustical reflectors) can be automatically derived from explicitly defined constraints (e.g. the maximum allowable arrival time difference between direct and reflected sound energy). As this option is not widely used in practice, it will not be discussed here in detail. However, it may be of interest for certain specialized design tasks.

Performance indicators

Once a sufficiently rich acoustical simulation model is generated, it can be used to support evaluative processes pertaining to room acoustics. A common approach to evaluate the acoustics of rooms involves the computation of the values of relevant room acoustical performance indicators and their comparison with recommended values (Fasold and Veres 2003). The definition and acceptance of such indicators typically result from long-term processes, whereby a consensus is reached that a certain proposed computable derivative of the sound field is relevant to and correlates with subjective (perceived) quality assessments by experts and participants of controlled tests. Table 9.1 provides examples of room acoustic indicators (adopted, in part, from Fasold and Veres 2003), divided into three broad groups, namely general, position-related, and space-related indicators. Included in this table are also recommended value ranges for each indicator. Note that the proper use and correct interpretation of numeric indicators and their computed values require both theoretical understanding and practical experience. For instance, most values in this table pertain to "default" room sizes in the relevant category and may need to be adjusted for different rooms of different volumes.

Table 9.1 Examples of room acoustics performance indicators together with corresponding recommended values

Type	*Room acoustics indicators*	*Recommended value for*	
		Speech	*Music*
General	Reverberation time T [s]	ca. 1 s	ca. 2 s
	Early decay time EDT [s]		ca. 2.2 s
	Bass ratio BR [–]	0.9–1.0	1.1–1.3
Position-related	Speech clarity C_{50} [dB]	≥ −2 dB	
	Articulation loss of consonants Al_{cons} [%]	< 15%	
	Clarity measure C_{80} [dB]		> −3 dB
	Direct sound measure C_7 [dB]		≥ −10 dB
	Speech transmission index STI [–]	> 0.5	
	Center time TS [ms]	60–80 ms	70–150 ms
	Echo criterion EK	< 1	2. . . 7
	Strength measure G [dB]	1–10 dB	1–10 dB
Space-related	Reverberance measure R [dB]		2–6 dB
	Lateral fraction LF [%]		25–40 %

The oldest and still broadly applied numeric indicator of room acoustic is the reverberation time T. It is defined as the time that it takes for the mean sound pressure level in a room to drop 60 dB after a stationary source of a sound in a room is turned off. Reverberation time in a room can be estimated, among others, via Equation 9.1 (Sabine 1927) and Equations 9.2 and 9.3 (Eyring 1930). Given the fact that absorption coefficients and equivalent absorption areas are frequency dependent, the resulting reverberation times are obviously frequency dependent as well. Sabine equation is applicable only to rooms with smaller values of mean absorption coefficient ($\alpha < 0.3$). Eyrings formula applies in principle also to rooms with higher sound absorption levels (Fasold and Veres 2003).

$$T = 0.163 \frac{V}{A_{tot}} [s] \tag{9.1}$$

$$T = 0.163 \frac{V}{-\ln(1-\bar{\alpha}) S_{tot}} [s] \tag{9.2}$$

$$\bar{\alpha} = A_{tot} / S_{tot} \qquad [-] \tag{9.3}$$

V Room volume [m³]
A_{tot} Total equivalent absorption area [m²]
$\bar{\alpha}$ Mean absorption coefficient of the room [–]
S_{tot} Total room surface area [m²]

The early decay time EDT relates likewise to the sound pressure level decay in a room. However, it is based on the initial 10 dB sound pressure level drop (Jordan 1980; ISO 2009). EDT is suggested to better correlate with the subjective perception of the duration of reverberation.

A further indicator, namely the bass ratio BR, was meant to characterize the frequency aspect of reverberation time as a single-number indicator (Beranek 1962). It is calculated using

Equation 9.4, based on the reverberation time T values at four (octave-band) mid-frequencies (125, 250, 500, and 1000 Hz).

$$BR = \frac{T_{125} + T_{250}}{T_{500} + T_{1000}} \quad (9.4)$$

Speech clarity C_{50} (Januska 1968; Lochner and Burger 1961; Niese 1965; Thiele 1953; ISO 2009) provides a measure of speech intelligibility (see Equation 9.5).

$$C_{50} = 10\log\left(\frac{E_{0...50ms}}{E_{50ms...\infty}}\right)[dB] \quad (9.5)$$

$E_{0...50ms}$ Sound energy reaching the listener in the interval from 0 ms (direct sound arrival) to 50 ms

$E_{50ms...\infty}$ Rest sound energy (reaching the listener after 50 ms)

Articulation loss of consonants Al_{cons} is an alternative to C_{50} toward objective assessment of speech intelligibility in rooms or with sound reinforcement systems (Peutz 1971). Based on previous work by Peutz (1974) and Peutz and Kok (1984), Bistafa and Bradley (2000) introduced the following formulation of AL_{cons} (see Equations 9.6 and 9.7) (see also Long 2006):

$$AL_{cons} = 9T_{60}\left[\frac{1}{1+(r_l / r)^2}\right] + a\ [\%] \quad (9.6)$$

$$r_l = 0.21\sqrt{\frac{V}{T_{60}}}\ [m] \quad (9.7)$$

r_1 Distance at which the direct field level is 10 dB below the reverberant field level for a directivity of one [m]

r Distance between sound source and listener [m]

a Correction factor: 1.5% for a "good" listener up to 12.5% for a "bad" listener

The clarity measure C_{80} (see Equation 9.8) is relevant for temporal and register clarity of music (particularly musical passages of high tempo) (Abdel Alim 1973).

$$C_{80} = 10\log\left(\frac{E_{0...80ms}}{E_{80ms...\infty}}\right)\ [dB] \quad (9.8)$$

$E_{0...80ms}$ Sound energy reaching the listener in the interval from 0 ms (direct sound arrival) to 80 ms

$E_{80ms...\infty}$ Rest sound energy (reaching the listener after 80 ms)

Direct sound measure C_7 (see Equation 9.9) is assumed to constitute an objective correlate to the subjective perception of "nearness" and "directness" of the sound source (Ahnert and Schmidt 2007).

$$C_7 = 10\log\left(\frac{E_7}{E_\infty - E_7}\right)\ [dB] \quad (9.9)$$

E_7	Sound energy reaching the listener in the interval from 0 ms up to 7 ms (direct sound)
$E_\infty - E_7$	Sound energy reaching the listener after the direct sound (reverberations)

Speech transmission index *STI* (Steeneken and Houtgast 1980; IEC 2011) is computed based on the reduction of the signal modulation from sound source location to receiver with octave center frequencies of 125 Hz to 8000 Hz (see Equations 9.10 to 9.12). To ascertain this influence, the modulation transfer function (MTF) is employed. Thereby, the available useful signal *S* is related to the prevailing interfering noise *N*. The modulation reduction factor *m*(*F*) characterizes the interference with speech intelligibility: Modulation frequencies from 0.63 Hz to 12.5 Hz in third octaves are used for the calculation. In addition, the modulation transfer function is subjected to a frequency weighting (WMTF – weighted modulation transfer function). Thereby, the modulation transfer function is divided into seven frequency bands that are each modulated with the modulation frequency. This results in a matrix of 98 (7 × 14) modulation reduction factors m_i. Table 9.2 provides a key to the interpretation of STI values in view of the implied speech intelligibility.

$$STI = \frac{\bar{X}_i + 15dB}{30dB} \; [-] \tag{9.10}$$

$$X_i = 10\log\left(\frac{m_i}{1 - m_i}\right) \; [dB] \tag{9.11}$$

$$m(F) = \frac{1}{\sqrt{1 + (2\pi F \cdot T / 13{,}8)^2}} \cdot \frac{1}{1 + 10^{-\frac{S/N}{10dB}}} \tag{9.12}$$

F	Modulation frequency [Hz]
m(*F*)	Modulation reduction factor
T	Reverberation time [s]
S/*N*	Signal/noise ratio [dB]

The center time *TS* (Kürer 1971; ISO 2009) is a quality criterion relevant for both speech intelligibility and musical clarity (see Equation 9.13).

$$TS = \frac{\int_0^\infty t \cdot p^2(t)dt}{\int_0^\infty p^2(t)dt} \; [ms] \tag{9.13}$$

Table 9.2 Perceived speech intelligibility for various ranges of STI values

Syllable intelligibility	*STI*
Poor	0 to 0.3
Satisfactory	0.3 to 0.45
Good	0.45 to 0.6
Very good	0.6 to 0.75
Excellent	0.75 to 1.0

Source: After Ahnert and Schmidt 2007; IEC 2011

t Time [ms]
$p(t)$ Instantaneous sound pressure [Pa]

Echo criterion (*EK*, see Equations 9.14 and 9.15) for speech EK_{Speech} and music EK_{Music} (Dietsch and Kraak 1986) is calculated from the room impulse response and allows for the examination of the occurrence of high-energy late reflections (echoes). One requirement of good room acoustics implies that at all seats, the reflection sequence is uniform and tight and that no high-energy late reflections occur. Strong reflections arriving later than 50 ms after the direct sound (not preceded by any or only few weaker reflections) can be subjectively perceived as distinct from the direct sound (Ahnert and Schmidt 2007).

$$EK = \frac{\Delta t_S(\tau)}{\Delta t_E} \tag{9.14}$$

$$t_S(\tau) = \frac{\int_0^{\tau} t \cdot |p(t)|^n \, dt}{\int_0^{\tau} |p(t)|^n \, dt} \tag{9.15}$$

For calculation purposes related to speech, Δt_E and n are suggested to have the values 9 ms and 2/3 respectively. Corresponding values for music are suggested to be 14 ms and 1. Thereby, Δt_E denotes the critical time period, in which the $t_s(\tau)$ increases to echo threshold level. The limit for EK_{speech} is assumed to be around 1. In case of EK_{music}, a limit around 2 is suggested to cover "critical" musical motives (Dietsch and Kraak 1986). Depending on the genre and style of music, the limit can be larger (Ahnert und Schmidt 2007).

Strength measure *G* (Jamagushi 1972; Lehmann 1976; ISO 2009) is a normalized measure of overall sound energy at a listener's position as related to the energy resulting from a power-equivalent omnidirectional sound source in a distance of 10 m from the source (see Equation 9.16).

$$G = 10 \log \frac{E_{tot}}{E_{tot10}} \ [dB] \tag{9.16}$$

E_{tot} Total sound energy at the listener's location
E_{tot10} Total sound energy of a reference sound source (in free field) at 10 m distance from the source

Reverberance measure *R* (see Equation 9.17) is an indicator of the acoustic liveliness or "reverberance" of the acoustic impression in case of musical performances (Beranek and Schultz 1965).

$$R = 10 \log \frac{E_{\infty} - E_{0...50ms}}{E_{0...50ms}} \ [dB] \tag{9.17}$$

Lateral energy fraction *LF* (see Equation 9.18) (Barron 1993) denotes the ratio of the laterally arriving sound energy components and those arriving from all sides (within an interval of 80 ms). It serves to gauge the subjective appraisal of the extension of a (musical) sound source. The higher the lateral efficiency, the acoustically broader the sound source appears.

$$LF = \frac{E_{5...80ms(lateral)}}{E_{0...80ms}} \, [\%] \tag{9.18}$$

Advanced room acoustic simulation applications allow for the computation of the values of most of the above-listed indicators. Moreover, they offer multiple possibilities and formats for the structured illustration and visualization of the simulated data. However, as mentioned earlier, the judicious selection of such indicators for any given design or retrofit task as well as the correct interpretation of the results require domain knowledge and experience. Measurements of such parameters in existing spaces, combined with regular auditory experiences of the sound field in actual rooms, can effectively enrich the proficiency of the users of room acoustic simulation applications toward proper construction of computational room models and competent interpretation of simulation results.

Computational methods

Introduction

Sound propagation in rooms may be modeled in different ways. Given a specific inquiry, criteria such as room size, features, and critical frequency range, computational capacity, and necessary (or appropriate) level of precision are among the factors that may give one method an advantage over another. Generally speaking, acoustical phenomena could be described either as waves or as particles. As such, numeric solutions can be applied to solve the wave equation. However, while wave-based models can provide accurate results for single frequencies, their application for large and complex rooms and higher frequencies are still challenging from the computational standpoint, given – among others – the problems in obtaining eigenvalues and associated normal modes. However, ongoing developments (particularly with regard to computational capacities) may expand the practical feasibility of wave-based methods in the future (Hornikx 2015; Hornikx et al. 2015; Sheaffer and Fazenda 2014). An alternative computational approach for describing the characteristics of sound fields in architectural spaces (suitable also for large complex rooms and for high frequencies) involves the notion of sound particles whose propagation paths in the space are abstracted in terms of sound rays. This idea goes back to the discipline of geometrical room acoustics, in which the concept of waves is replaced by that of sound rays. Analogous to light rays in geometrical optics, a sound ray denotes a small portion of a spherical wave with vanishing aperture that originates from a specific point and follows a well-defined direction of propagation (Kuttruff 2017).

Despite the fact that geometrical acoustics can only provide an incomplete description of the inherently complex dynamics of the sound field in a room, it is considered to be highly useful from the practical standpoint. It also represents the underlying theoretical basis of many computational simulation applications in room acoustics. Specifically, deployment of simulation methods based on geometrical room acoustics is justified when a room's (and its boundary's) dimensions are considerably larger than the wavelength of the sound.

Different geometrically based methods have been applied to simulate sound propagation in large rooms. The rapidly evolving research and development in this area makes the comprehensive and up-to-date coverage of such methods a challenging task. In the following, two methods (the "ray tracing method" and the "image source method") as well as combinations of those (hybrid solutions) will be briefly discussed. Note that these methods, which imply high order reflections, do not explicitly embody the wave property of sound and thus do not reflect the fundamentally complex behavior of the sound field in a space. As a consequence, the assumption of purely geometrical behavior should be restricted to low order reflections, necessitating alternative approaches for computational description of higher order reflections.

The ray tracing method

The ray tracing method (Krokstad et al. 1968; Kulowski 1985; Lewers 1993; Mahdavi et al. 1997b; Vian and van Maercke 1986) involves the emission of sound particle rays from a source in various directions (see Figure 9.1). The collision of a particle with a surface results in:

- an energy loss whose magnitude is a function of the surface's absorption coefficient α (i.e. the ratio of non-reflected to incident sound intensity);
- the change of the direction of propagation, to be determined – in case of a specular reflection – according to Snell's law.

Certain conditions must be met in order to ensure that, in the course of modeling, neither false reflections are generated nor valid reflections are ignored. The probability that a ray will find a surface of area A after a traveling time of t is quite high, if the wave front area per ray is smaller than $A/2$. From this fact, the required minimum number of rays N may be estimated using Equation 9.17 (Rindel 2000).

$$N \geq 8\pi c^2 t^2 A^{-1} \quad (9.17)$$

Here, c denotes the speed of sound propagation in air.

Obviously, a room's enclosure elements and objects in the sound field are not completely smooth. Details of acoustical wave front's interactions with irregularities can be rather complex (Choi et al. 2015; Cox and D´Antonio 2004; Kuttruff 2017). In most currently available room acoustics simulation applications that utilize ray tracing, the respective sound reflection processes are captured in simplified ways. Thereby, to characterize the reflective behavior of a surface in a simulation environment, commonly the notion of a scattering coefficient (s) is applied (Hodgson 1991; Lam 1996). It denotes the ratio between reflected sound intensity in non-specular directions and the total reflected sound intensity. Thus, $s = 0$ denotes fully specular reflection, while $s = 1$ suggests complete diffusivity.

To simulate the direction of a diffuse reflection, stochastic methods can be used, involving random numbers and a probability function based on Lambert's cosine law (Stephenson 1985). The scattering coefficient enables the allocation of weights to the diffuse and specular components of a reflection. There is currently a lack of comprehensive empirical

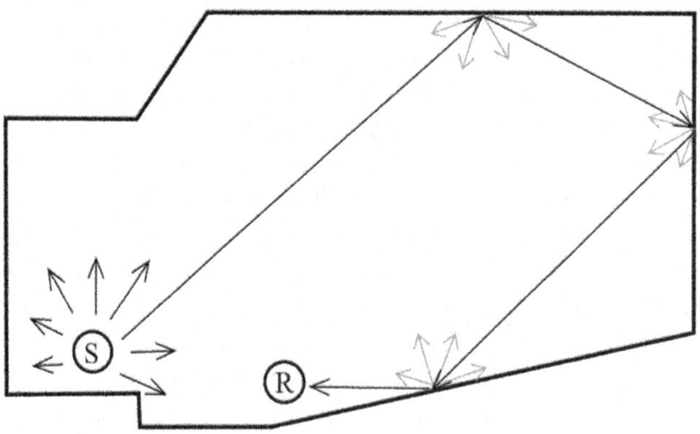

Figure 9.1 Schematic illustration of the ray tracing method

Source: Adapted from Lövstad 2003

data on the frequency-dependent scattering coefficients of various architecturally relevant surfaces. Generally speaking, higher values of (mid-frequency) scattering coefficients are assigned to audience areas (0.6–0.7) and rough building structures (0.4–0.5), whereas smooth surfaces are assumed to have lower scattering coefficients (0.01–0.05). Ongoing efforts to assemble comprehensive acoustical material properties in general and scattering coefficient information in particular are expected to improve the practical utility of room acoustical simulation tools.

In a typical ray tracing scenario, the paths of particles are followed over multiple reflections, whereby the initial energy loss due to air attenuation and absorption losses are accounted for. Using specified particle counters, the energy and arrival time (possibly together with arrival direction) are registered and processed in terms of histograms depicting energy impulse responses. According to Kuttruff (2017), a total of 10^5 to 10^6 sound particles can offer reasonably accurate results given a histogram with a 3 ms timestep resolution.

As such, the ray tracing method allows for the inclusion of multiple (and different) non-omnidirectional sound sources as well as a variety of reflection patterns (from fully specular to ideally diffuse). Moreover, to increase efficiency, ray bundles can be applied instead of single sound rays. Such bundles can display a circular or polygonal cross-section (Kuttruff 2017). The respective procedures are typically referred to as cone tracing or pyramid tracing (Stephenson 1996).

The image source method

Particularly in those cases where enclosure boundaries of a space are composed of planar surfaces, the concept of image sources facilitates the efficient construction and pursuit of sound rays about the space. Given a point sound source in front of a plane surface element, each reflection from this surface can be imagined to originate from a virtual source located behind this surface and at the same distance from it as the real source. The concept of image sources can be extended to include image sources of higher order. To illustrate this, Figure 9.2 shows, using the example of a sound source S in a simple room, the image sources of first order S' and second order S".

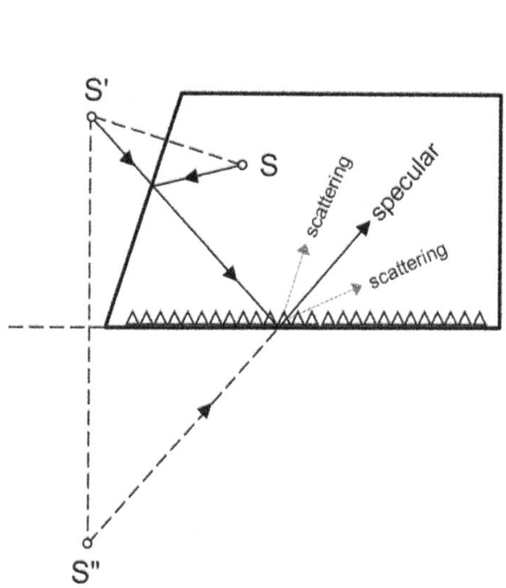

Figure 9.2 Schematic illustration of first and second order image sources

Source: Modified from Kuttruff 2017

The number of image sources within a radius of *ct* (i.e. sound path length from source to receiver) in a room (with volume *V*) can be estimated using Equation 9.18 (Rindel 2000). It provides an approximate value for the number of reflections that arrive at the location of a receiver up to the time *t* after the emission of the sound signal.

$$N_{refl} = 4\pi c^3 t^3 (3V)^{-1} \tag{9.18}$$

Assuming a room with *N* (plane) surfaces and a corresponding number of first-order source images (mirrored by each surface), $N(N-1)$ new second-order images emerge (Kuttruff 2017). This procedure can be repeated so that the total number $v(i_0)$ of images of order up to i_0 is obtained as follows (see Equation 9.19):

$$v(i_0) = N((N-1)^{i_0} - 1)(N-2) - 1 \tag{9.19}$$

The resulting number of possible image sources can thus rapidly increase, making image source methods practically applicable only in cases where it is sufficient to only consider reflections of low order and in case of simple rooms with regular geometry.

Hybrid methods

Hybrid methods attempt to combine the positive features of both ray tracing and image source methods (Naylor 1993; Vorländer 1989; Välimäki et al. 2012). Thereby, efficient solutions are worked out to determine if image sources are valid or not. Toward this end, rays are traced from the source until they hit a surface. The resulting reflections are then examined ("visibility test") to see if they are relevant for (energetically contribute to) the receiver location. This operation involves tracking back from the receiver toward the image source. Assuming an image source has been consequently deemed to be valid, the corresponding reflection's energy can be computed as the product of source energy (in the direction of radiation) and the reflection coefficient of the surfaces involved. Typically, an "image tree" is constructed in the process to keep track of the emerging early images, avoiding thus that each valid image is accounted for more than once. Otherwise, the ensuing reflectogram could include redundant reflections and become erroneous as a result (Rindel 2000).

The temporal length of a reflectogram itself represents a basic computational challenge: There is a limit to the number of rays that can be used in a computation. Thus, the sequence of reflections in a reflectogram can only be accurately predicted until a certain time limit, constrained by the finite number of the rays considered. To construct the "reverberation tail" in the reflectogram, other approaches are needed (Heinz 1993; Naylor 1993; Mahdavi et al. 1997b). For example, the "secondary source method" (Naylor 1993) deploys different methods to treat early and late reflections. While early reflections are obtained via image source generation, late reflections are modeled in terms of secondary sources that emerge when a ray hits a surface. Secondary sources act as elemental area radiators whose energy is calculated based on the product of the energy of the primary source (per ray) and the reflection coefficient of the surfaces involved in the reflection sequence. The energy arrived at a receiver from a secondary source is proportional to the cosine of the angle between the surface normal at the location of the secondary source and the vector from the secondary source to the receiver point. Moreover, the energy level attenuates from the source to the receiver in accordance with the inverse square law. Given the total path length of a reflection (from the primary source through intermediate reflections to the secondary source and thereafter to the receiving point), the reflection arrival time can be determined.

According to Rindel (2000), a very high transition order (from early to late reflection) does not necessarily lead to more accurate simulation results: The likelihood that images would be visible from the receiver point decreases at higher reflection orders. Moreover, as compared to real circumstances, geometrical acoustics overestimates the strength of reflections from smaller surfaces (relevant for high-order reflections). Thus, it is suggested that missing such reflections due to the application of secondary source method does not necessarily introduce significant errors in predictions. Ray numbers of up to 1000 and transition orders of two or three have been claimed to be sufficient to provide reliable simulation results for most architectural spaces.

The problem of diffraction

In their basic varieties, ray tracing and image source methods ignore diffraction phenomena (see Figure 9.3). This is a considerable deficiency, given the fact that diffraction is a common feature of acoustical wave front behavior in spaces (Pierce 1974). As such, obstacles in the path of a wave front do not create perfect "shadows." Diffraction can be thus envisioned as the process by which sound energy distribution deviates from the straight path (i.e. the virtual sound ray "bends") at the edges of objects and openings. The magnitude of this deviation (bending angle) is primarily a function of the sound frequency.

There have been a number of theoretical suggestions and corresponding implementations to address the problem of diffraction in room acoustics simulation (Pohl and Stephenson 2011; Rindel 1986; Rindel et al. 2009; Stephenson 1996). Thereby, a typical approach involves the application of a simplified diffraction pattern to every sound ray passing across an obstacle at a certain distance. As Kuttruff (2017) suggests, individual diffraction processes create multiple secondary rays in different directions, resulting in additional complexity and computational cost. This drastically increases the complexity of the process and also the processing time. Whereas there have been a number of proposals to improve the computational efficiency of diffraction modeling (see, for example, Stephenson 1996), the general problem of computational

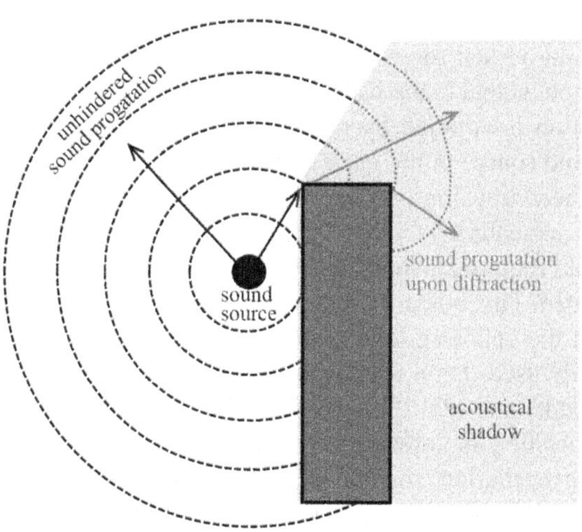

Figure 9.3 Sound diffraction

Source: Modified from DLR 2017

representation of diffraction in room acoustics simulation is still not fully resolved. Nonetheless, solutions implemented in commercially available simulation tools can – for a large segment of typical simulation tasks – provide reasonable modeling opportunities. Thereby, appropriate configuration of the associated simulation settings is crucial (e.g. number of rays, number of reflections, impulse response length, treatment of scattering). Further information regarding the state of diffraction treatment in room acoustics can be found in Savioja and Svensson 2015.

Auralization

The conventional use scenario of room acoustics simulation is, highly simplified, as follows: A proposed design is modeled using a simulation application. Simulation results are expressed in terms of numeric values of a number of the previously discussed acoustical performance indicators. Designers and consultants typically compare such numeric results with applicable requirements (recommendations, standards, etc.) to decide whether a particular design meets mandated performance criteria. If it does not, modifications are made in the design and the performance is iteratively tested until satisfactory results have been obtained. Recently, the possibility has emerged that this process could be enriched by approaches that rely on auralization, enabling the users to "directly" experience and evaluate the acoustical conditions in the built environment (Vorländer 2008; Vorländer et al. 2015; Yang 2006; Pätynen and Lokki 2010).

Analogous to visualization, auralization (Kleiner et al. 1993) allows, in principle, to go beyond the utility of quantitative (numeric) performance indicators, via facilitation of "direct" perception of relevant qualitative (in this case, acoustical) features of the environment already in the design stage. The idea is to subject a "dry" signal to the same transformation that the sound propagation in a real room would do, whereby the simulated (virtual) room model would act as a surrogate for the real room. The transformation itself would be captured in terms of the aforementioned room's computationally generated impulse response. This impulse response model can be used toward modeling the head-related transfer function of the listener. The binaural room impulse response (BRIR) can be computed as a pair of impulse responses for a listener's ears whose head is located at the receiver position. Thereby, the head-related transfer function (HRTF) is used as obtained from an artificial head representing an "average" human head (Rindel 2000). Note that it is also possible to measure individualized HRTFs (ÖAW 2017). Thus, all kinds of (anechoically recorded) signals such as music and speech passages could be implanted in the virtual room by combining the signal with the computed BRIR. In order to present the output signal to the listener, high-quality headphones can be used. As such, this mode of presentation is afflicted by the "in-head-localization" effect, which means the listener locates the sound source in the back of his/her head. However, individualized HRTFs together with optimal head tracking may facilitate outside head localization (Hendrickx et al. 2011). A more realistic presentation mode – involving a better presentation of the directionality of the exposure – would require an anechoic room, in which (typically two) loudspeakers are stereophonically operated. This mode, however, faces the problem that the ears would receive cross-talk signals from the opposing loudspeakers. To avoid this, the "crosstalk cancellation" (CTC) technique can be used, i.e. a filter that adds cancellation signals to the input signals (Neu et al. 1992; Schroeder and Atal 1963; Urbach et al. 1992). Furthermore, to account for the circumstance that the position and orientation of listeners' heads are not typically fully steady, "dynamical cross-talk cancellation" method may be deployed involving four loudspeakers that can be arranged into six different CTC configurations. Thereby, the choice of a configuration at any given time is guided by the information provided by a head-tracking device.

Ongoing work in the area of advanced auralization is inspired in part by the promise of the combination of real-time auralization and visualization toward virtual reality applications (see, for example, Gardner 1995; Garcia 2002; Schröder 2011; Schröder and Vorländer 2007; Schröder et al. 2010; Vorländer et al. 2015; Wefers and Schröder 2009; Wefers and Vorländer 2012). It is conceivable that related systems, once they will have reached a sufficient level of efficiency and practical usability, could provide exciting design support functionalities. As such, high-quality virtual reality applications could offer planners and users of future built environments the opportunity to realistically experience the acoustical and visual consequences of alternative design solutions.

Uncertainty, validation, calibration

Sources of error

As with simulation models in other domains, room acoustical models are subject to reliability and validity questions (Vorländer 2013). There are multiple sources of uncertainty to be considered:

- Formalisms pertaining to the description of underlying physical phenomena in room acoustical simulation applications that are based on ray tracing and image sources are frequently incomplete. Many aspects of such phenomena (mode of waves, diffraction and interference processes, resonance effects) are either not considered at all or merely simplistically represented.
- Simulation models can be of insufficient or inappropriate level of resolution in terms of geometry and associated details. Likewise, assumptions pertaining to material properties are prone to uncertainties due to the missing or imprecise information.
- Advanced simulation tools typically include a large number of adjustable settings and options such as number of rays, number of reflections, impulse response length (duration of ray tracing), and transition order (number of reflections with image source method). Users' decisions regarding the configuration of such settings can significantly influence the simulation results. To avoid associated pitfalls, comprehensive and coherent documentations of simulation tools' algorithms and input data requirements are essential. However, they cannot substitute for sufficient levels of understanding and experience on the side of the users.

In the following, a number of topics are briefly addressed that are relevant for the evaluation and proper use of acoustical simulation applications. First, the round robin test methodology for simulation model validation is discussed, followed by a brief report on the experiences with the comparison of simulation and measurement results in case of existing buildings. Next, the congruence level between subjective impressions from real spaces and those gained from simulation-based auralizations are addressed. Finally, the potential of simulation model calibration in case of acoustical retrofit projects is discussed.

Round robin tests

A common way to explore the validity of a simulation tool is to conduct a round robin test, whereby multiple developers simulate the same (real) object independently. Subsequently, all results are compared with the measurement results from the real object. Regarding room acoustical simulation programs, one round robin test (Vorländer 1995) involved an 1800 m^3 auditorium for which eight acoustical parameters (ISO 2008) were computed by 16 participants. Simulations were compared with measurement results (average of multiple measurements as performed by seven participants). Three programs were found to produce accurate results: The deviations of these tools' predictions from the measurements were not larger than the difference between multiple measurements.

In the course of a subsequent round robin test (Bork 2005), a number of room acoustical parameters (T_{30}, EDT, D_{50}, C_{80}, TS, G) were simulated for a 78 m² recording studio ($V = 400$ m³). The room acoustical parameters were also measured in this room using two source positions and three different receiver positions. Simulations were performed using six different commercially available room acoustical simulation programs. Thereby, two different room models were considered. The first involved a simplified representation of the geometry of diffusers and absorbers as plane surfaces, whose absorption and scattering data were derived based on measurements. The second representation involved the detailed modeling of diffuser and wooden absorber geometries. The difference between the simulation results for these two representations was not found to be significant. The deviation of the simulated parameter values from the respective measurements were derived for all programs in terms of mean relative errors at 1000 Hz (averaged over all six positions) with the dimension of "limen" (i.e. the smallest audible sensation) for each parameter. These are 0.05 s for T_{30} and EDT, 5% for D_{50}, 1 dB for C_{80} and G, and 10 ms for TS. The relative errors, as averaged over all frequencies, revealed similar magnitudes. The relative errors in prediction of T_{30} and EDT ranged from 2% to 5%. Simulation errors regarding the other parameters, as calculated with four of the six programs, were found to be below the limit of auditory detectability.

Certain errors (for example deviation of simulated reverberation times at 1000 Hz) were suggested to be the consequence of input assumptions concerning absorption coefficients. Laboratory measurements of absorption coefficients are typically performed under conditions that favor diffuse reflection. The materials' reflective behavior in common architectural situations can be to some extent directional, leading to underestimation of the reverberation time in simulation. Moreover, locally varying parameter ranges at lower frequencies cannot be detected with these applications, as they do not consider the sound wave phases. Nonetheless, the results of this round robin test suggest that a number of commercially available simulation applications can, under controlled test conditions and given reliable material information, provide results that show a good agreement with measurements. Additional insights in this area are expected from more recent round robin efforts (Round Robin 2017).

Realistic simulation deployment scenarios

Application of simulation tools in the context of actual design processes differs from the controlled test scenario described earlier, as detailed design specifications (final geometry configuration, definitive data on material properties) may not be available. A comparable situation is faced when acoustical simulations are performed for existing spaces, for which detailed documentations of acoustic properties (specifically of room surfaces) are not available. An examination of the reliability of room acoustic simulation in such circumstances may thus provide clues as to the reliability of simulation-based predictions of acoustical performance indicator values in the architectural design phase. To exemplify this point, data from recent acoustical measurements and simulations in 14 spaces (Mahdavi et al. 2008a, 2008b) are considered in the following discussion. These spaces include a lecture room and a ceremonial hall in a university, a small recording studio, five atria in different building complexes (museums, office buildings, etc.), as well as six bath buildings of different scales.

For these objects, frequency-dependent reverberation times and sound pressure levels (for various source and receiver positions) were measured and simulated. Simulations were performed using a commercially available room acoustical simulation and auralization tool (see Christensen et al. 2016 for a more recent release of this application). To compute the reverberation times and sound pressure levels, virtual loudspeakers were placed in the model in the same location as the real loudspeakers in the course of measurements. Likewise, the location of virtual microphone positions in the model matched those of the real microphones during the measurements.

The input data assumptions concerning the absorption coefficient data for surface finishes were based mostly on general literature and the simulation tool's material database. Material properties for simulation input were defined independent of (and unaffected by) the measurement results. Figure 9.4 shows the mean relative error and respective standard deviations (in percent) of simulated frequency-dependent reverberation times. Figure 9.5 shows the correlation between measured and simulated reverberation times (all objects, all frequencies). Figure 9.6 shows the mean difference (and respective standard deviations in dB) between measured and simulated frequency-dependent sound pressure levels for all objects. Figure 9.7 shows the correlation between measured and simulated sound pressure levels (all objects, all frequencies).

These results imply, for the objects studied, a generally good agreement between measurements and simulations of reverberation times and sound pressure levels. The frequency-dependent mean relative error of reverberation time simulations over all objects did not exceed 20%. Individual relative errors (for different objects and frequencies) of reverberation time simulations were only in about 10% of the cases more than 20%. The mean relative error of the mean reverberation times (over 125, 500, and 2000 Hz), as calculated for all objects, is 3% (with a standard deviation of 15%). The comparison of measured and simulated sound distribution levels displays likewise a reasonable agreement. The mean difference between the measured and simulated sound pressure levels (over 125, 500, and 2000 Hz) aggregated over all objects is −0.9 dB (with a standard deviation of ±2.3 dB).

It is likely that the simulation errors are primarily the result of inaccuracies in modeling the material properties. In this study, similar to most application instances of simulation in the early stages of the design process, absorption coefficients were primarily obtained from general information in literature. Acoustical properties of building elements are indeed seldom specified in the pertinent technical design documents. Nonetheless, despite difficulties in accurately determining the absorption coefficients of surfaces, it was possible to reproduce measured reverberation times and sound distribution patterns with reasonable accuracy.

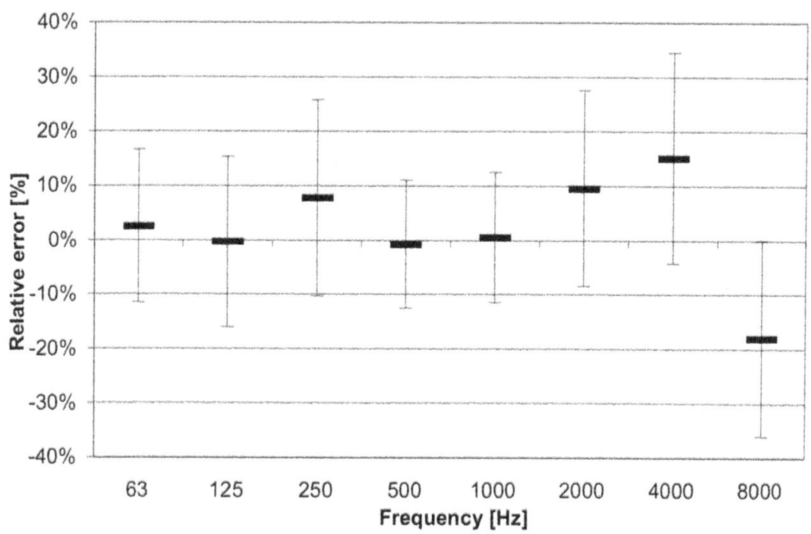

Figure 9.4 Mean relative error and respective standard deviations of simulated frequency-dependent reverberation times

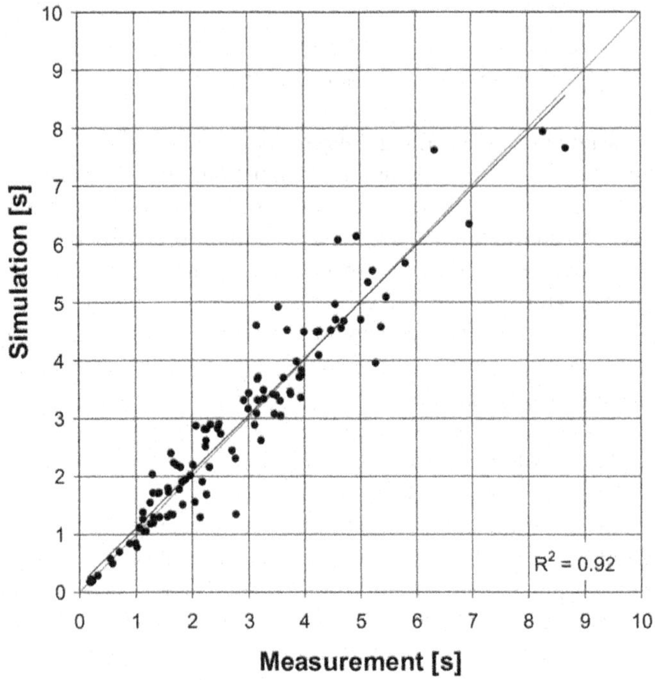

Figure 9.5 Measured versus simulated reverberation times (all objects, all frequencies)

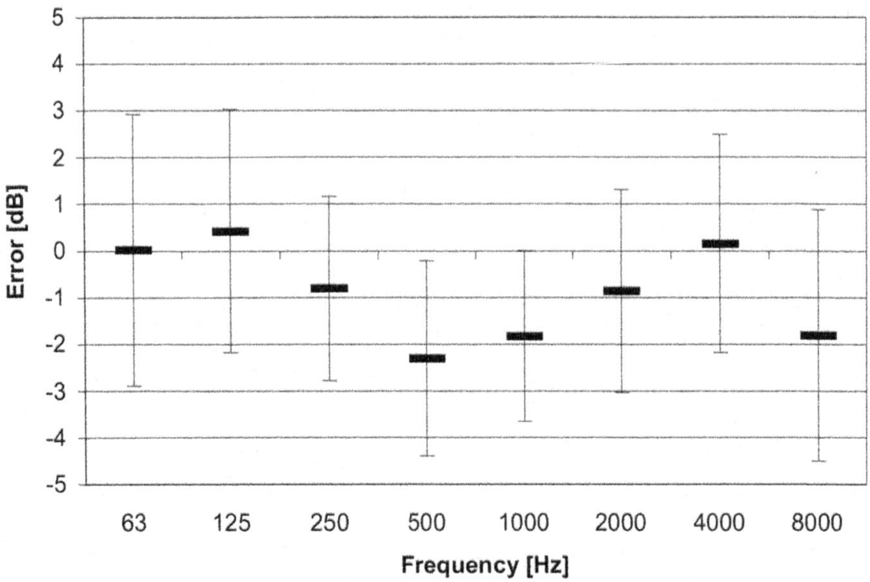

Figure 9.6 Mean differences (together with respective standard deviations) between measured and frequency-dependent sound pressure levels (all objects)

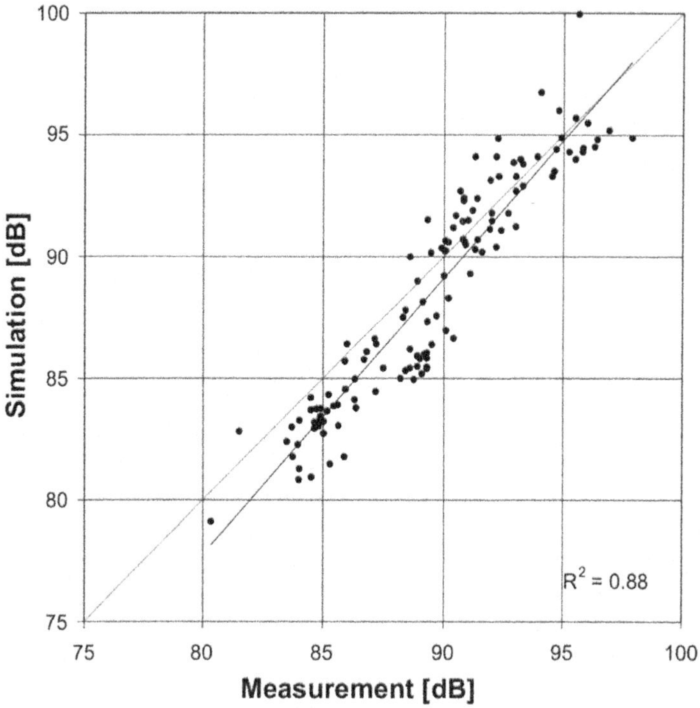

Figure 9.7 Measured versus simulated sound pressure levels (all objects, all frequencies)

Given carefully constructed and adequately detailed models of room geometry and availability of relevant information on acoustical surface properties, room acoustical simulation can yield reasonable performance assessments of alternative design and retrofit options. Toward this end, precise definition of performance requirements in design briefs and accurate documentation of design attributes (including detailed information on the acoustical properties of materials and components applied) are critical. Finally, more comprehensive empirically obtained databases of acoustical properties of architectural elements need to be incorporated in acoustical simulation applications in order to expedite the simulation and analysis process.

Real and auralized rooms

Given increased availability of auralization tools, a relevant question arises: To which extent can subjective assessments of the acoustical properties of a real space be reproduced by assessments that are based on auralization? In one related study (Mahdavi et al. 2005), previous research methods and results (Fasold and Winkler 1976; Mahdavi and Eissa 2002) were used to select a metric that captures certain subjective dimensions of the acoustical experience of spaces (see Table 9.3). Three rooms with differing acoustical attributes were selected for the study, including a ceremonial hall (aula) in a university, a mid-size lecture room, and a small recording studio. Reverberation time, sound level distribution, and ambient sound level were measured in all three spaces. A digital recording of a sample of acoustical events was prepared including speech, music, and various other sounds. The recording also included a reading of a set of logatomes (Fasold and Winkler 1976) for speech intelligibility test purposes. The

Table 9.3 Evaluation axes and corresponding bi-polar scales with associated points (from −2 to +2) for the evaluation of the acoustical quality of space

Evaluation axis	*Bi-polar terms*
Reverberation	Reverberant (+2) – dry (−2)
Loudness	Loud (+2) – quiet (−2)
Clarity	Clear (+2) – blurry (−2)
Sharpness	Shrill (+2) – dull (−2)
Warmth	Warm (+2) – cool (−2)
Brightness	Bright (+2) – dark (−2)
General impression	Pleasant (+2) – unpleasant (−2)
Speech intelligibility	Good (+2) – poor (−2)

acoustical properties of these rooms were evaluated by a group of 30 test participants using the above-mentioned metric, as the pre-recorded sequence of acoustical events was being played in the rooms. Additionally, a speech intelligibility test was performed using the logatomes. Moreover, the participants were asked to specify the location of the sound source by specifying the prevailing direction from which they believed the sound was coming.

Auralizations of the spaces were generated using a room acoustics simulation tool (see Christensen et al. 2016) and the aforementioned pre-recorded sequence of acoustical events. The underlying simulation models were tested and found to correctly reproduce measured sound distribution pattern and reverberation times. The auralizations of the three spaces were presented (via headphones attached to computers) to and evaluated by a second demographically similar group of 75 test participants, who again made use of aforementioned acoustical quality metric and also located the sound source. As with the real spaces, a speech intelligibility test was performed based on a logatomes list as included in the sound sample recording.

The evaluation results of room acoustics in the real spaces were compared with those based on the virtual spaces. Figure 9.8 shows, for all three spaces, the correlation between the evaluation means of the real spaces versus the evaluation means of the auralizations for the eight categories considered. The results show a general agreement between acoustical evaluations based on real and auralized rooms, particularly for the ceremonial hall and the recording studio. However, the degree of correlation is different from space to space. A stronger congruence between the two evaluation methods was found in acoustically more "pronounced" cases. The ceremonial hall is rather large and reverberant, whereas the recording studio is very small and acoustically "dry." The lecture room, however, is moderate in both size and acoustical properties. This plausibly suggests that the computational reproduction of acoustical properties of real spaces is in view of evaluative utility (as expressed in terms of a quality metric) more effective in rooms with pronounced acoustical attributes. A closer look at the individual scales of the quality metric reveals different levels of "expressiveness": whereas the numeric values of some scales change noticeably from space to space, others change only marginally (see Figure 9.8). In this sense, the scales for reverberation, clarity, speech intelligibility, and general impression may be said to be more expressive. Sharpness, warmth, and brightness scales, however, appear to be less expressive.

Auralization displayed in this case a reasonable reproduction of the speech intelligibility circumstances in the spaces. Note that failure in reproduction of the perceived direction of the incoming sound is, in this case, likely due to the use of headphones for the presentation of the auralized signal.

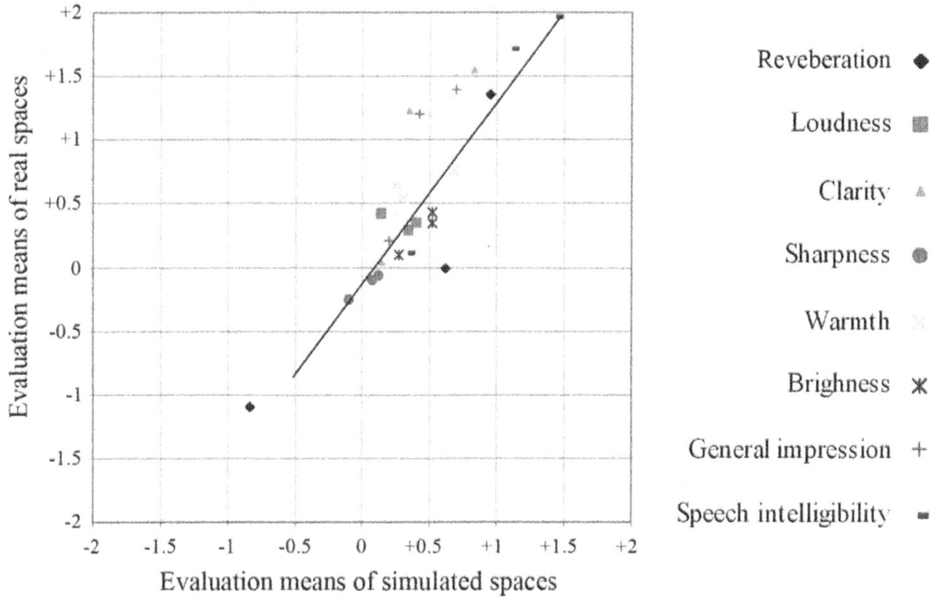

Figure 9.8 Evaluation means based on real and auralized spaces (regression analysis across all spaces)

A note on model calibration

Ideally, simulation studies should be carried in a transparent and reproducible fashion with regard to model attributes and input assumptions. However, in specific use cases pertaining to the retrofit of existing buildings, the reliability of simulation-based analyses can be improved to some extent via model calibration (see, for example, Taheri and Mahdavi 2014; Penna et al. 2015). The idea is to generate a simulation model corresponding to the building's existing condition and compare the model's predictions with measurements. Thereby, certain systematic errors can be detected. Such errors could be the results of the resolution issues of the model geometry, assumptions regarding material properties, or selected simulation settings. In these kinds of cases, appropriate model adjustments could be beneficial. Due to the partially opaque nature of such adjustment processes, the improved predictive performance on the part of the calibrated model of the existing building cannot be expected to translate one-to-one to models of retrofit scenarios. Retrofit scenarios introduce new features into the simulation models, which are not necessarily accounted for in the calibration process. Nonetheless, the base case calibration process does seem to increase the odds for better general predictive performance due, in part, to fine tuning of factors such as geometry representation and simulation settings (Tugrul et al. 2013; Jakic et al. 2017).

Conclusion

Acoustically relevant planning interventions in the built environment in general and the design and retrofit of architectural spaces with good acoustical attributes in particular have become, thanks in part to advances in computational methods and techniques, an accountable process. Using advanced simulation applications, it is now possible to predict the behavior of sound fields in built environments in a reliable manner. Thus, numeric values of a comprehensive set of performance indicators can be computed and compared to desired (recommended) ranges. Moreover, sound propagation processes can be visualized not only in terms of sound level distribution maps, but also in terms of rays that, emanating from the sound source, travel throughout the room and bounce off the surfaces. Using such representations and associated reflectograms, critical reflections and arrival times can be appraised and necessary adjustments in designs can be initiated. At a minimum, the simulation-supported design process can ensure that essential performance requirements are fulfilled and design failures are avoided. Furthermore, the advent of powerful auralization tools opens new and exciting perspectives: Design and retrofit proposals can henceforth not only be evaluated based on predicted values of numerically expressible performance indicators, but also be made the direct objects of acoustical experience. This development can support the refinement of designs in view of numerically intangible yet subjectively evident (experiential) space qualities.

Needless to say, simulation applications must be further improved in the future. More elaborate algorithmic solutions will have to be implemented to better capture the wave nature of sound phenomena. Likewise, more detailed descriptions of the interactions between sound waves and room enclosure are needed, addressing, among others, the coincidence and resonance frequencies. Moreover, comprehensive and reliable databases of acoustical material properties must be implemented as the integral feature of any effective simulation tool. Ultimately, the utility of simulation-based design support process is quintessentially dependent on the resolution, consistency, and fidelity of the input information, i.e. the design model itself.

Acknowledgment

The author would like to acknowledge the support of Josef Lechleitner toward the collection of material for the preparation of the present chapter.

References

Abdel Alim, O. (1973) Abhängigkeit der Zeit- und Registerdurchsichtigkeit von raumakustischen Parametern bei Musikdarbietungen (Dependence of time and register definition of room acoustical parameters with music performances). PhD Thesis, TU Dresden, Germany.

Ahnert, W. and Schmidt, W. (2007) Fundamentals to perform acoustical measurements. *Appendix to EASERA (version 1.1) Tutorial* (www.easera.de).

Ando, Y. (1998) *Architectural Acoustics*. New York: Springer Verlag, ISBN 0-387-98333-3.

Barron, M. (1993) *Auditorium Acoustics and Architectural Design*. London: E & FN SPON.

Beranek, L.L., ed. (1988) *Noise and Vibration Control*. Washington, DC: Institute of Noise Control Engineering, ISBN 0-9622072-0-9.

Beranek, L.L. (1962) *Music, Acoustics and Architecture*. New York/London: John Wiley and Sons.

Beranek, L.L. and Schultz, T.J. (1965) Some recent experiences in the design and testing of concert halls with suspended panel arrays. *Acustica* 15: 307.

Bistafa, S.R. and Bradley, J.S. (2000) Revisiting algorihms for predicing the articulation loss of consonants alcons. *Journal of the Audio Engineering Society* 48(6): 539, Fig. 2, p. 537, and Figure 4.

Bork, I. (2005) Report on the 3rd Round Robin on room acoustical computer simulation – Part II: Calculations. *Acta Acustica United with Acustica* 91: 753–763.

Choi, Y.J., Bradley, J.S. and Jeong, D.U. (2015) Experimental investigation of chair type, row spacing occupants and carpet on theatre chair absorption. *The Journal of the Acoustical Society of America* 137: 105.

Christensen, C.L., Koutsouris, G. and Gil, J. (2016) *ODEON Room Acoustics Software*. Version 13. User's manual. www.odeon.dk.

Clasen, D. and Langer, S. (2007) Finite element approach for flanking transmission in building acoustics, *Building Acoustics* 14(1).

Cox, T.J. and D'Antonio, P. (2004) *Acoustic Absorbers and Diffusers*. London: Spon Press.

Crocker, M.J., ed. (2007) *Handbook of Noise and Vibration Control*. Hoboken, NJ: John Wiley and Sons, ISBN 978-0-471-39599-7.

Cremer, L. and Müller, H.A. (2016a) *Principles and Applications of Room Acoustics*, Vol. 1, Newport Beach, CA: Peninsula Publishing. ISBN 0932146295.

Cremer, L. and Müller, H.A. (2016b) *Principles and Applications of Room Acoustics*, Vol. 2, Newport Beach, CA: Peninsula Publishing. ISBN 0853341141.

Dietsch, L. and Kraak, W. (1986) Ein objektives Kriterium zur Erfassung von Echostörungen bei Musik- und Sprachdarbietungen (An objective criterion for capturing echo disturbances with music and speech performances). *Acustica* 60: 205.

DLR (2017) www.pa.op.dlr.de/acoustics/essay/beugung.html.

EN (2000a) *Acoustic Properties of Building Elements and of Buildings. Building Acoustics – Estimation of Acoustic Performance of Buildings from the Performance of Elements – Part 1: Airborne Sound Insulation between Rooms*. Standard Reference EN 12354–1:2000.

EN (2000b) *Acoustic Properties of Building Elements and of Buildings. Building Acoustics – Estimation of Acoustic Performance of Buildings from the Performance of Elements – Part 2: Impact Sound Insulation between Rooms*. Standard Reference EN 12354–2:2000.

EN (2000c) *Acoustic Properties of Building Elements and of Buildings. Building Acoustics – Estimation of Acoustic Performance of Buildings from the Performance of Elements – Part 3: Airborne Sound Insulation against Outdoor Sound*. Standard Reference EN 12354–3:2000.

Eyring, C.F. (1930) Reverberation time in dead rooms. *Journal of the Acoustical Society of America* 1(2A): 217–241.

Fasold, W. and Veres, E. (2003) *Schallschutz und Raumakustik in der Praxis*. Berlin: HUSS-MEDIEN, Verlag Bauwesen.

Fasold, W. and Winkler, H. (1976) *Bauphysikalische Entwurfslehre: Band 5 Raumakustik*, 1. Auflage. Berlin: VEB Verlag für Bauwesen.

Garcia, G. (2002) Optimal filter partition for efficient convolution with short input/output delay. *Audio Engineering Society*, Convention Paper 5660, Los Angeles, CA.

Gardner, W.G. (1995) Efficient convolution without input-output delay. *Journal of the Audio Engineering Society* 43(3): 127–136.

Gerretsen, E. (1986) Calculation of airborne and impact sound insulation between dwellings. *Applied Acoustics* 19(4): 245–264.

Heinz, R. (1993) Binaural room simulation based on the image source model with addition of statistical methods to include the diffuse sound scattering of walls and to predict the reverberant tail. *Applied Acoustics* 38: 145.

Hendrickx, E., Stitt, P., Messonnier, J.C., Lyzwa, J.M., Katz, B. and Boishéraud, C. (2011) Influence of head tracking on the externalization of speech stimuli for non individualized binaural synthesis. *The Journal of the Acoustical Society of America* 141.

Hodgson, M. (1991) Evidence of diffuse surface reflections in rooms. *The Journal of the Acoustical Society of America* 89: 765.

Holden, M. (2016) *Acoustics of Multi-Use Performing Arts Centers*. New York: CRC Press, ISBN 798-0-415-51719-5.

Hopkins, C. (2007) *Sound Insulation*. Oxford, UK: Elsevier Ltd. ISBN 9780750665261.

Hornikx, M. (2015) Acoustic modelling for indoor and outdoor spaces. *Journal of Building Performance Simulation* 8(1): 1–2. doi: 10.1080/19401493.2015.1001616.

Hornikx, M. (2016) Ten questions concerning computational urban acoustics. *Building and Environment* 106(2016): 409–421.

Hornikx, M., Hak, C. and Wenmaekers, R. (2015) Acoustic modelling of sport halls, two case studies. *Journal of Building Performance Simulation* 8(1): 26–38, doi: 10.1080/19401493.2014.959057.

IEC. (2011) *Sound System Equipment – Part 16: Objective Rating of Speech Intelligibility by Speech Transmission Index*. IEC 60268–16 Ed. 4.0.

ISO. (1996) *Acoustics – Attenuation of Sound during Propagation Outdoors General Method of Calculation*. ISO 9613–2.

ISO. (2008) *Measurement of the Reverberation Time of Rooms with Reference to Other Acoustical Parameters*. ISO 3382.

ISO. (2009) *Acoustics – Measurement of room Acoustic Parameters, Part 1: Performance Spaces*. ISO-3382–1.

Jakic, N., Kiesel, K., Lechleitner, J. and Mahdavi, A. (2017) Exploring the potential of simulation model calibration: An acoustical retrofit case study. In *Proceedings of Envibuild 2017*. Vienna, Austria.

Jamagushi, K. (1972) Multivariate analysis of subjective and physical measures of hall acoustics. *Journal of the Acoustical Society of America* 52(5A): 1271–1279.

Januska, I. (1968) Experimentally stated correlation between objective echogram evaluation and speech intelligibility. *Archivum Akustiki* 140.

Jordan, V.L. (1980) *Acoustical Design of Concert Halls and Theaters*. London: Applied Science Publishers Ltd.

Kleiner, M., Dalenbäck, B.I. and Svensson, P. (1993) Auralization – An overview. *Journal of the Audio Engineering Society* 41(11): 861–875.

Krokstad, A., Stroem, S. and Soersdal, S. (1968) Calculating the acoustical room response by the use of a ray tracing technique. *Journal of Sound and Vibration* 8(1): 118–125.

Kulowski, A. (1985) Algorithmic representation of the ray tracing technique. *Applied Acoustics* 18: 449–469.

Kürer, R. (1971) Einfaches Messverfahren zur Bestimmung der Schwerpunktzeit raumakustischer Impulsantworten. (A simple measuring procedure for determining the "center time" of room acoustical impulse responses). In *Proceedings of the 7th Internal Congress on Acoustics*. Budapest.

Kuttruff, H. (2017) *Room Acoustics*. New York: CRC Press, Taylor & Francis Group.

Lam, Y.W. (1996) A comparison of three diffuse reflection modeling methods used in room acoustics. *The Journal of the Acoustical Society of America* 100: 2181.

Lehmann, P. (1976) Über die Ermittlung raumakustischer Kriterien und deren Zusammenhang mit subjektiven Beurteilungen der Hörsamkeit. PhD Thesis, Technische Universität Berlin, Germany.

Lewers, T. (1993) A combined beam tracing and radiant exchange computer model of room acoustics. *Applied Acoustics* 38(2–4): 161–178.

Lochner, J.P.A. and Burger, J.F. (1961) The intelligibility of speech under reverberant conditions. *Acustica* 11: 195.

Long, M. (2006) *Architectural Acoustics*. London: Elsevier Academic Press, CA.

Lövstad, A. (2003) Evaluation of objective echo criteria. Diploma Thesis. Norwegian University of Science and Technology, Department of Telecommunications.

Lyon, R. (1994) Theory and application of statistical energy analysis. *The Journal of the Acoustical Society of America* 98(6): 3021.

Mahdavi, A. (1991) Sound transmission between rooms: A comparative analysis of calculation methods. *Journal of Acoustical Society of America* 90(2): 927–936.

Mahdavi, A. (1993a) Open simulation environments: A preference-based approach. In *Proceedings of CAAD Futures '93*. Pittsburgh, PA: CMU, pp. 195–214.

Mahdavi, A. (1993b) "Simulation environments for architectural acoustics based on the "Two-Way" feedback approach. In *The International Conference on Noise and Vibration Control*, St. Petersburg, Russia.

Mahdavi, A. and Berberidou-Kallivoka, L. (1993) A two-way inference approach to daylighting simulation. *Journal of the Illuminating Engineering Society* 22(1): 66–74.

Mahdavi, A. and Berberidou-Kallivoka, L. (1994) GESTALT: A prototypical Realization of an open daylighting simulation environment. *Journal of the Illuminating Engineering Society* 23(2): 62–71.

Mahdavi, A. and Eissa, H. (2002) Subjective evaluation of architectural lighting via computationally rendered images. *Journal of Illuminating Engineering Society* 32(2): 11–20.

Mahdavi, A., Kainrath, B., Orehounig, K. and Lechleitner, J. (2008b) Measurement and simulation of room acoustics parameters in traditional and modern bath buildings. *Building Simulation* 1(3).

Mahdavi, A., Lechleitner, J. and Pak, J. (2008a) Measurements and predictions of room acoustics in atria. *Journal of Building Performance Simulation* 1(2): 67–74.

Mahdavi, A., Liu, G. and Ilal, M.E. (1997b) CASCADE: A novel computational design system for architectural acoustics. In *Proceedings of IBPSA (International Building Performance Simulation Association) Conference*. Prague, Vol. II, pp. 173–180.

Mahdavi, A., Mathew, P., Kumar, S. and Wong, N.H. (1997a) Bi-directional computational design support in the SEMPER environment. *Automation in Construction* 6: 353–373.

Mahdavi, A.M., Becker, G. and Lechleitner, J. (2005) An empirical examination of the evaluative utility of design auralization. In Beausoleil-Morrison, I. and Bernier, M. (eds), *Building Simulation 2005*, Ninth International IBPSA Conference, August 15–18. Montreal, Canada, pp. 677–684.

Naylor, G.M. (1993) ODEON – Another hybrid room acoustical model. *Applied Acoustics* 38: 131–143.
Neu, G., Mommertz, E. and Schmitz, A. (1992) Investigation on the true directional sound reproduction by playing head-related recordings over two loudspeakers, part I. *Acoustica* 76: 183.
Niese, H. (1965) Vorschlag für die Definition und Messung der Deutlichkeit nach subjektiven Grundlagen. *Hochfrequenztechnik und Elektroakustik* 65–64.
ÖAW. (2017) *Austrian Academy of Sciences, Acoustics Research Institute, "ARI HRTF Database"*. www.kfs.oeaw.ac.at/index.php?option=com_content&view=article&id=608:ari-hrtf-database&catid=158&Itemid=857&lang=en.
Pätynen, J. and Lokki, P. (2010) Evaluation of concert hall auralization with virtual symphony orchestra. In *Proceedings of the International Symposium on Room Acoustics*. ISRA 2010, August 29–31, 2010, Melbourne, Australia.
Penna, P., Cappelletti, F., Gasparella, A., Tahmasebi, F. and Mahdavi, A. (2015) Multi-stage calibration of the simulation model of a school building through short-term monitoring. *Journal of Information Technology in Construction* 20(Special Issue): 132–145.
Peutz, V.M.A. (1971) Articulation loss of consonants as a criterion for speech transmission in a room. *Journal of the Audio Engineering Society* 19(11): 915–919.
Peutz, V.M.A. (1974) Designing sound systems for speech intelligibility. *Journal of the Audio Engineering Society, (Abstracts)* 22: preprint 967.
Peutz, V.M.A. and Kok, B.H.M. (1984) Speech intelligibility. *Journal of the Audio Engineering Society, (Abstracts)* 32: preprint 2089.
Pierce, A.D. (1974) Diffraction of sound around corners and over wide barriers. *The Journal of the Acoustical Society of America* 55: 941–955.
Pohl, A. and Stephenson, U.M. (2011) Introduction of diffraction into geometrical acoustics comparison of different sound energy reunification techniques. *Proceedings of Forum Acusticum; European Conference of the EAA*. Aalborg, Denmark.
Rindel, J.H. (1986) Attenuation of sound reflections due to diffraction. *Proceedings of NAM-86*. Aalborg, Denmark, pp. 257–260.
Rindel, J.H. (2000) The use of computer modeling in room acoustics. *Journal of Vibroengineering* 3(4): 41–72; *Proceedings of the International Conference BALTIC-ACOUSTIC 2000*, pp. 219–224.
Rindel, J.H., Nielsen, G.B. and Christensen, C.L. (2009) Diffraction around corners and over wide barriers in room acoustic simulations. In *Proceedings of the 16th International Congress of Sound and Vibration*, July 5–9. Kraków, Poland.
Rossing, T.D., ed. (2007) *Handbook of Acoustics*. Dordrecht, Heidelberg, London and New York: Springer, ISBN 9780387304465.
Round Robin. (2017) International round-robin on auralisation. *Technische Universität Berlin Audio Communication Group*. Einsteinufer 17c, 10587 Berlin. http://rr.auralisation.net/.
Sabine, W.C. (1927) *Collected Papers on Acoustics*. Cambridge, MA: Harvard University Press.
Savioja, L. and Svensson, P. (2015) Overview of geometrical room acoustic modeling techniques. *The Journal of the Acoustical Society of America* 138(2): 708–730.
Schröder, D. (2011) Physically based real-time auralization of interactive virtual environments. PhD Thesis, RWTH Aachen University, Aachen, Germany.
Schröder, D. and Vorländer, M. (2007) Hybrid method for room acoustic simulation in real-time. In *Proceedings of the 20th International Congress on Acoustics (ICA)*, Madrid, Spain.
Schröder, D., Wefers, F., Pelzer, S., Rausch, D., Vorländer, M. and Kuhlen, T. (2010) Virtual reality system at RWTH Aachen. *Room Acoustics (ISRA)*, Melbourne, Australia.
Schroeder, M.R. and Atal, B.S. (1963) Computer simulation of sound transmission in rooms. *IEEE International Convention Record* 7.
Sheaffer, J. and Fazenda, B. (2014) WaveCloud: An open source room acoustics simulator using the finite difference time domain method. *Forum Acousticum*, Krakow.
Steeneken, H.J.M. and Houtgast, T. (1980) A physical method for measuring speech-transmission quality. *The Journal of the Acoustical Society of America* 67(1): 318–326.
Stephenson, U. (1985) Eine Schallteilchen-computer-simulation zur Berechnung der für die Hörsamkeit in Konzertsälen massgebenden Parameter. *Acustica* 59(1): 1–20.
Stephenson, U. (1996) Quantized beam tracing: A new algorithm for room acoustics and noise immission prognosis. *Acustica/Acta Acustica* 82: 517.

Taheri, M. and Mahdavi, A. (2014) Optimization-based monitoring-supported calibration of a thermal performance simulation model. *International Journal of Optimization in Civil Engineering* 4(2): 261–271.
Thaden, R. (2005) Auralisation in building acoustics. PhD Thesis, RWTH Aachen University, Aachen, Germany.
Thiele, R. (1953) Richtungsverteilung und Zeitfolge der Schallrückwürfe in Räumen. *Acustica* 3: 291.
Tugrul, A., Lechleitner, J., Schuss, M. and Mahdavi, A. (2013) Simulation-supported room acoustics retrofit of office spaces. BS2013–*13th International Conference of the International Building Performance Simulation Association*, ISBN: 978-2-7466-6294-0. pp. 339–344.
Urbach, G., Mommertz, E. and Schmitz, A. (1992) Investigation on the true directional sound reproduction by playing head-related recordings over two loudspeakers, part II. *Acoustica 1992* 77: 153.
Välimäki, V., Parker, J.D., Savioja, L., Smith, J.O. and Abel, J.S. (2012) Fifty years of artificial reverberation. *IEEE Transactions and Language Procession* 20(5).
Vian, J.P. and van Maercke, D. (1986) Calculation of the room impulse response using a ray-tracing method. In *Proceedings of ICA Symposium on Acoustics and Theatre Planning for the Performing Arts*. Vancouver, Canada, pp. 74–76.
Vigran, T.E. (2008) *Building Acoustics*. London and New York: Taylor & Francis, ISBN: 9780415428538.
Vorländer, M. (1989) Simulation of the transient and steady-state sound propagation in rooms using a new combined ray-tracing/image-source algorithm. *The Journal of the Acoustical Society of America* 86(1): 172–178.
Vorländer, M. (1995) International Round Robin on room acoustical computer simulations. In *Proceedings of the 15th International Congress on Acoustics*. Norway: Trondheim, Vol. II, pp. 689–692.
Vorländer, M. (2008) *Auralization; Fundamentals of acoustics, modelling, simulation, algorithms and acoustic virtual reality*. Berlin Heidelberg: Springer-Verlag, e-ISBN 978-3-540-48830-9.
Vorländer, M. (2013) Computer simulations in room acoustics: Concepts and uncertainties. *The Journal of the Acoustical Society of America 133*. doi: 10.1121/1.4788978.
Vorländer, M., Schröder, D., Pelzner, S. and Wefers, F. (2015) Virtual reality for architectural acoustics. *Journal of Building Performance Simulation* 8(1): 15–25.
Vorländer, M. and Thaden, R. (2000) Auralisation of airborne sound insulation in buildings. *Acustica United with Acta Acustica* 86(1): 70–76.
Wefers, F. and Schröder, D. (2009) Real-time auralization of coupled rooms. In *Proceedings of the EAA Auralization*. Symposium, Espoo, Finland.
Wefers, F. and Vorländer, M. (2012) Optimal filter partitions for non-uniformly partitioned convolution. In *Audio Engineering Society Conference: 45th International Conference: Applications of Time-Frequency Processing in Audio*. Helsinki, Finland: Audio Engineering Society.
Yang, W. and Hodgson, M. (2006) Auralization study of optimum reverberation times for speech intelligibility for normal and hearing-impaired listeners in classrooms with diffuse sound fields. *The Journal of the Acoustical Society of America* 120: 801. doi: 10.1121/1.2216768.

Recommended reading

Given the paucity of comprehensive textbooks dedicated solely to computational room acoustics, the reader should consult the technical references above. As to the introductory texts to room acoustics in general, many sources exist. (Examples include Barron 1993; Beranek 1962; Fasold and Veres 2003; Jordan 1980; Kuttruff 2017.)

Activities

A simulation expert in the area of the room acoustics must command the scientific foundations of room acoustics, understand the logic and limitations of implemented algorithms in computational tools, and be familiar with empirical measurement techniques in architectural acoustics. A recurrent cross-checking between simulation and measurement results is essential for the development of expertise and a sense of professional judgment in this area. Moreover, there exist, for a number of performance indicators in the area of room acoustics (such as reverberation time), simple calculation formulas. The comparison of simulation results with those obtained via such simple formulas provides a number of useful learning opportunities.

Part II

Design

10 The role of simulation in performance-based building

Godfried Augenbroe

Scope

This chapter explores how simulation supports performance-based building through different lifecycle stages, with a focus on the design stage. As design evolves, a building project invokes many intermittent and recurring dialogues between stakeholders that engage in the definition and subsequent negotiation of performance requirements and their fulfillment. This requires, among others, an agreement about how to quantify performance, i.e. how to choose an adequate measurement method. Measurement methods not only quantify performance criteria but also de facto define them. The aim of this chapter is to show how any objective specification of performance criteria and their measurements can be derived while avoiding subjective bias in their interpretation. In any real life design case, a set of criteria needs to be introduced to manage the many performance aspects of design situations. As these need to be dealt with simultaneously, design decision making rests on the careful balancing of different, and sometimes conflicting, performance targets. It is shown that a rigorous, system-theoretic definition of performance measures is necessary to prepare a rational decision process. The treatment in this chapter is an attempt to lay the basis for this and at the same time explore the role of building simulation in the quantification of the performance criteria.

Learning objectives

Simulation does not exist in a vacuum. It has many purposes, and one particular purpose, the support of design evolution, is discussed in this chapter. If simulation is used in performance-based building, it is essential to position it in a decision making framework. In doing so, this chapter has the following learning objectives:

- Understand the essence of building simulation as a "virtual experiment" and recognize the importance of the role of the experiment in studying and quantifying behavioral outcomes of competing design options.
- Understand the need to define quantifiable expressions of performance to objectively compare design options; "raw" output data of simulations is typically not adequate for this purpose.
- Understand the basics of a system framework that identifies Aspect Systems, which will be defined as conceptual data models of physical building systems that achieve a certain function, and which are hence translated into a computational model that is subjected to simulation or any other experiment in order to study building behavior in relation to the stated requirement.
- Recognize that every real life design decision requires a multi-criterion approach where some criteria may need simulation for their quantification, while others rely on other methods, and some may lack an accepted objective quantification.

- Be able to analyze a design problem, determine the important criteria and prepare multi-criterion decision making.
- Recognize the potential role of uncertainties in the outcome of building simulation and have general knowledge about how it may influence decision making.

Key words

Performance criteria/performance indicators/decision making/design evolution/system framework

Introduction

This chapter provides an engineering perspective on performance-based building design. Statements of performance play a major role in expectations expressed by owners and future occupants. Traditionally, design solutions were driven by prescriptive terms, i.e. specifying properties of the solution rather than the expected performance of the solution. Building codes and regulations have long contributed to this by adopting prescriptive specifications. This is no longer the case, as many countries are moving parts of their regulations and supporting standards to the performance domain (Foliente 2000), while a research networking effort at the beginning of the century produced a roadmap towards their broad adoption in the industry (Pebbu 2005). The key ingredient of performance-based building is the up-front formulation of statements of requirements and the subsequent management of a process that guarantees to come as close as possible to their fulfillment. The fulfillment process takes place through dialogues between owners, developers, designers, engineers and building operators. The focus in this chapter is on the designers that seek to fulfill the client's performance requirements. We posit that only a proper framework that anchors the definition of measures and measurement methods can avoid the many disconnects in the dialogues between demand (client and operator) and supply side (designers and engineers). In current practice, these disconnects are rampant throughout the whole building delivery process, especially where the expression of expectations and the expressions of fulfillment lack a formal basis. Systematic procedures for matching fulfillments with expectations is therefore considered an important target of the building industry to become more client-driven and provide better customer satisfaction and better value overall. The need to move the industry in this direction has fostered the introduction of performance-based building methods (Foliente et al. 1998). Building teams are also adopting integrated project delivery, an approach which involves a greater number of stakeholders making design decisions earlier in the building procurement process (Barrett 2008). At this stage, we need a workable definition of performance provided in the first axiom.

> **Axiom 1**: *Performance is an expression of how well an artifact fulfills a function or meets a need.*

A performance-based approach can become the key enabler of rational decision making across many stakeholders. Building simulation is a key tool to quantify performance criteria that inform decisions. Building simulation models observe the building's behavior under a specific usage scenario (Augenbroe 2004) as depicted in Figure 10.1. A piece of reality is translated into a model, which is then studied in a variety of experiments (simulation runs) that are conducted in an "experiment box" (the simulation tool). The experiment is set up to generate observable states that reveal behavior that determines in a quantifiable way the performance under study. It should be noted that the design of the experiment is far from trivial. It must be such that the observable states generated by the experiment reveal indeed the meaningful behavior that

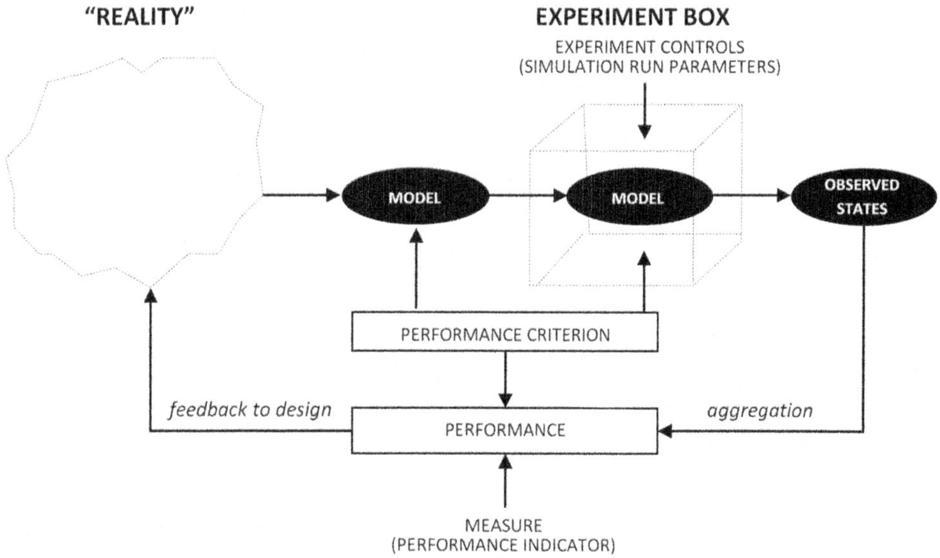

Figure 10.1 Performance testing requires a (virtual) experiment

aggregates into a quantification of the performance criterion. It is obvious that this requires a deep understanding of the physical domain, the performance measure and the experiment box. This leads to the insight that performance is not a lazy person's post-processing of simulation results or loose framing of the outcome of a simulation.

> **Axiom 2**: *Performance thinking requires formalization of building behavior informed by deep domain knowledge culminating in a mathematical model that captures how the building responds to external and internal "forces".*

As Figure 10.1 shows, the performance under study is at the heart of the experimental set-up: it demands a specific model, controls the set-up of the experiment and determines the choice of quantification of the experiment outcomes. When looked at this way, simulation is no longer the art of replicating reality in the machine, but rather the art of performing the right type of virtual experiment with the right computational model and software. In fact, building performance simulation is no longer confined to the computational tool but covers a process that starts much earlier and ends later. The main steps in this process are (1) agreement about relevant performance criteria, (2) agreement about ways to express and elicit quantified expectations at the outset of the project, (3) agreement about ways to measure and express them as levels of performance of proposed solutions and (4) supporting rational design decisions driven by the outcomes of the previous steps. The last step involves all stakeholders to contemplate trade-offs between potentially conflicting or hard to meet performance targets within the imposed time and budget limits. There are no well-established step-by-step methods to guide client and design team through this process, although some progress has been made in supporting the second stage of the process, see for instance Becker (2008). The author works with intuitive performance concepts falling short of defining them rigorously and placing them in an expectation-fulfillment framework. This is in fact a deficiency of most, if not all, publications in the field of building performance. Some authors take another perspective and specifically deal with the first stage of the process. Huovila et al. (2005) reports the development of EcoProp as a tool for client involvement in the early stage, whereas Verheij (2005) focuses on dedicated dialogue support between stakeholders in the discovery and agreement finding on performance criteria and related issues and conflicts to be considered. He asserts that the asymmetry in knowledge between client and design team members requires a rich platform that can structure the dialogue in a goal-driven way, reaching consensus about criteria and measurement. Ozsariyildiz (2006) advocates tools that remove the barrier between client and design team altogether, thus blurring the distinction between demand and supply side, but in many situations this will not be practical or feasible. Things are negatively affected by the lack of transparent demand-supply roles in the building process, as explained in Beard et. al. (2001) which shines light on many of the ineffective dialogues between clients and design teams in current practice. We argue that this is in part due to the lack of a framework for the categorization of functionality and its mapping into sets of performance criteria with mutually agreed measures. The third stage of the process has received a lot of attention lately as multiple stakeholder, multi-criteria decision making has become a fashionable research topic. Tools like analytical hierarchy process (AHP), as in Saaty (2013), and utility theory (Quiggin 1997) are now routinely used to arrive at an optimal decision. Hazelrigg (2003) offers an interesting view on the adequacy of these approaches for rational decision making. For a comprehensive and definitive overview of the performance literature through the last 50 years and its role in different performance-based settings, I refer the reader to de Wilde (2018).

Scientific foundation

Dialogues about performance between stakeholders require the systematic introduction of functional requirements, their linkage and further refinement into performance aspects, and their translation into individual performance criteria. Each individual criterion must be accompanied by a measure and measurement method. The actual measurement outcome (i.e. its value) is often presented as a "performance indicator", or PI, but here we will use "performance measure" as the more generic naming convention. There is no established theory or formal classification that underpins the definition of performance criteria at the whole building scale although several general schemes have been developed, mostly linked to research networking activities initiated by the International Council for Building (CIB) published originally as a report (CIB 1982) and then further explained in Gibson (1982). Other relevant foundations can be found in the outcomes of the earlier mentioned thematic research networking effort (Pebbu 2005). Some other attempts at a fundamental approach linking functions and requirements are worth mentioning. Inspired by design theories of Herbert Simon (1996) and his followers, Wim Gielingh in his General AEC Reference Model (Gielingh 1988) hypothesizes a design process in which the design problem is reduced to smaller manageable functional units. For each of these units, a designer looks for a technical solution that satisfies the functional requirements of the unit. Another closely aligned framework is the function-behavior-structure ontology developed by the same community (Gero and Kannegiesser 2014). This chapter borrows from these frameworks, looks at one particular way of classifying building systems and discusses its use in a performance-based design framework. For this, however, we need to explain our particular worldview on design in order to make the intrinsic limitations of our approach more explicit.

> **Axiom 3**: *Performance-based design implies a worldview adoption of design as discretely incremental.*

From the body of literature on design theory going back more than 40 years, it emerges that there is no agreement of how design can be defined or how a formal theory for it can be constructed. One of the most cited is Simon's assertion, "Everyone designs who devises courses of action aimed at changing existing situations into preferred ones" (Simon 1996). But it is now commonly understood as a mere starting point based on the realization that the preparation and planning as well as carrying out the actual design (changes) of an artifact need deep inspection of the cognitive processes in the brain as well as the human-to-human interaction with all its dynamic sociological contexts, as elucidated by Galle (2001) and Love (2000). A practical approach to performance-based design removes itself from these theoretical considerations by postulating design as a sequence of discrete events where performance can be measured and compared to either previous measurements for comparative analysis or to an *a priori* stated requirement. It is left to the design protocol and management process to decide at which of these discrete events a formal decision is invoked with stakeholder involvement. The selection and then enforcement of these discrete decision moments is a major goal of the management of a design project. In the scope of this chapter, we will not deal with the process management aspect. This consequently isolates our treatment from actual execution in a given design process. The goal is to articulate the formal definition of performance criteria and the role of simulation in their measurement. It should be well understood that the formalization of design decisions around performance targets is only a first step of their realization in practice. The presumed dynamic management throughout the process of all design partners around discrete decision events informed by performance measures is as yet unproven as practical and efficient. It is to be expected that in real life, the execution will be driven mostly by heuristics and only at a small number of predefined decision moments by formal decision making informed by performance measures.

Axiom 4: *Practical execution of performance-based design demands dedicated process management.*

Some practitioners claim that a form of this mixed mode process management is already mainstream in industry. Based on my experience with decision making in practice, I argue that there is no validity to this claim. In fact, even when the macro steps of design evolution are managed as consecutive discrete "gates" (as defined in the in-house workflow template that many design firms use), the decision making is not formalized and in most cases informed only by loosely defined "fly-by" performance measures. Most importantly, the verification of the fulfillment of client expectations is mostly clouded in biased conceptualizations and not based on agreed-upon verifiable measures.

Building system classification

Figure 10.2 shows a functional decomposition that is derived from the work cited above. The main functions of a building (such as "provide shelter" and "facilitate organizational processes") are decomposed top-down into lower level functions such as "provide safety", "habitability", and "sustainability". During this decomposition, one will arrive at functional needs that can be associated with a criterion that is used to express how well that need is met. The measurement attached to the criterion quantifies the value of an achieved performance or states the required value for a performance that is to be achieved.

The lower half of the figure shows the aggregation of technical systems from basic elements (window, door, chiller, finishing, etc.) until we reach the level of the standard "technical building systems" such as lighting, sprinkler, HVAC. The classification of technical subsystems in Figure 10.2 corresponds to the established system and trade classifications like Masterformat (CSI 2017) and OmniClass (Omniclass 2017). An overview of different classification systems is given by Afsari and Eastman (2016). Elements principally belong to only one technical subsystem, but many connections between subsystems are in fact elements themselves and therefore belong to more than one subsystem. There exists a multitude of constraint relationships

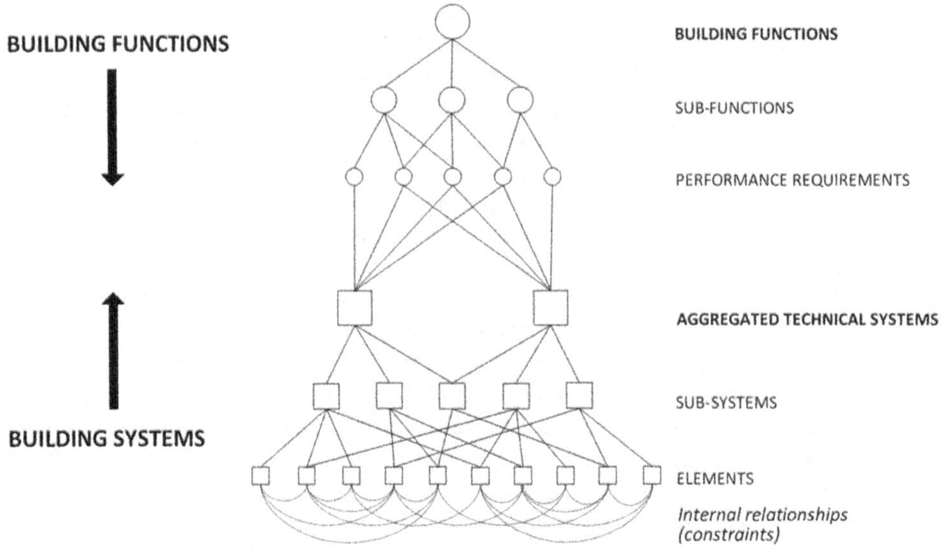

Figure 10.2 Top-down functional decomposition and bottom-up assembly of building systems

across elements of a system and between elements of different subsystems. These constraints are related to composition, compatibility and assembly rules. The complexity and dependencies of the constraint relationships are the primary reason that design is a complex and knowledge intensive activity, requiring deep domain expertise. As the figure indicates, in the middle, where top-down functional decomposition and bottom-up technical system aggregation meet, a natural connection between functions and technical systems exists. Indeed, technical systems contribute to many building functions, and building functions are supported by many technical systems. Referring to the previous work on function-behavior-structure, the association with Figure 10.2 becomes clear when one realizes that structure (bottom) generates behavior which satisfies function (top). This forces us to make the representation of behavior more explicit.

The mapping of a design program into functional requirements at increasing levels of detail is a complex job, not supported by an established methodology that can be broadly applied to any type of situation. The closest methodologies come from the systems engineering world, where requirements engineering has been developed at an advanced level. Gilb (2005), for example, presents one of the most insightful contributions to the field, although it must be stated that the application of his thinking requires re-invention and re-discovery in every project based solely on experience and domain expertise. It should be noted that building requirements engineering, and fundamental thinking about criteria and their management, is not taught, at least not at the fundamental level, in building design schools. For the bottom half of the picture, the situation is quite different. Building design students are trained to develop integral design solutions from existing part solutions. Most of the attention goes into recognizing the complexity of system integration and knowing when to develop new solutions, i.e. when it is necessary to depart from the existing systems. Again, there are few authors that have tried to establish a systems integration theory that can be applied to practical design problems. The systems integration handbook (Rush 1986) and studies by Habraken (1998) are good guiding rods for practical and theoretical work in systems integration that over time could support the lower half of the figure better. On the other hand, it should be said that the lower half of the figure is not essential to the application of performance-based design, even if the systems generation is done by informal design generation.

In spite of the lack of a workable framework for both halves of Figure 10.2, it is still useful to understand and discuss the process of matching performance requirements with design proposals, sometimes boiling down to selecting that design among proposed alternatives that best meets the requirements. In fact, the center of the figure depicts where demand and supply traditionally interface in building projects. It is therefore relevant to use this figure as a template to explore the role that simulation plays in this interface.

Axiom 5: *Performance is only measurable at the interface of function and system.*

First, it is necessary to understand performance requirements, i.e. what can and should be formulated as performance statements? The goal of specifying required performance, rather than prescribing properties, is to guarantee that a building system is capable of fulfilling the functional needs at a defined level of performance. There are seven main concepts in the performance approach (Foliente et al. 1998; Vanier et al. 1996), as identified in Figure 10.3.

1 Goal (occupancy, or use) is usually a qualitative statement that addresses the needs of a stakeholder (in our case: owner, occupant, developer, society) or consumer and determines the required level of performance.
2 A subset of functional requirements are mandatory requirements that must be fulfilled to ensure that the facility meets stakeholder's essential needs.
3 Performance requirements are the user requirements expressed in terms of the performance of a product. Performance requirements are translated into criteria that are "measured" by the same PIs that we use for achieved performance.
4 PIs are quantified measures that indicate a particular performance. A PI, by definition, is an agreed-upon measure that is quantified by an agreed-upon (and sometimes standardized) measurement method.
5 Verification methods are used to evaluate whether performance requirements have been met. Verification methods can be experiments (tests), calculations, or a combination of both. Each criterion has its own measurement method. A number of different measurement methods may exist for the same performance criterion, i.e. one criterion may have alternative PIs attached to it. Indeed, many different ways to measure a certain performance can be introduced, each with its own merit and each requiring a different experimental set-up,

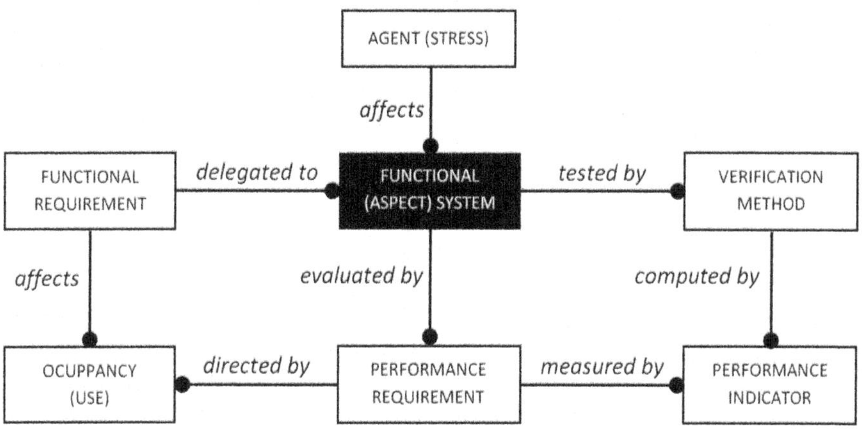

Figure 10.3 Relationship among concepts in the performance approach

Source: Adapted from Vanier et al. 1996

model or aggregation method. An elaboration of this discussion can be found in Augenbroe et. al. (2004).

6 Functional (or "Aspect") Systems are parts of a building that each fulfill a given function. Elements of Aspect Systems can be any building element.

7 Agents alter the ability of functional elements to meet the functional needs. One major example of agent is the external weather conditions, or more precisely the micro climate in which the building is going to operate. For instance, the probability of hurricane wind conditions will alter the ability of the structural system to maintain stability.

Figure 10.3 shows how the seven concepts are related to each other. The small filled circles indicate the cardinality of the relationship, where the cardinality could be specified as one-to-one or one-to-many. The latter detail is not shown in the figure, but most cardinalities can be assumed to be one-to-many. For example, *one* performance requirement can be measured by *many* performance indicators, whereas every performance indicator is related to exactly *one* verification method (i.e. performance measure) for its quantification.

In order to determine whether a particular design satisfies a performance requirement, a certain measure is used to quantify the required or expected performance as well as the achieved performance. Our focus is primarily on the latter, as this is where simulation plays a cardinal role. As explained earlier, simulation executes a virtual experiment on a model of the designed artifact. This notion can now be refined for a predefined performance measure, because we don't need to run the experiment for the full system, i.e. for all its objects and relationships. Instead, we can limit the simulation to that part of the system that plays a role in the performance measure that we want to quantify. This implicitly defines a reduced part of the system that we will refer to as "Aspect System". To perform a simulation on this Aspect System, its objects and relationships need to be translated into a computational model on which the simulation is performed in the manner as depicted in Figure 10.1. It should be noted that it requires a substantial infusion of domain expert knowledge to develop the computational model for the objects (and their properties) and relationships in the Aspect System. An obvious reason is that the Aspect System is incomplete. Indeed, current conceptual building representations are poor in the representation of deep domain knowledge. An excellent foray into how more expressive building representations can remove the limitation can be found in a recent PhD dissertation (Cavieres 2018).

It is a sobering thought to realize that very few, if any, software applications currently used in building design have been developed through the formal derivation steps indicated above, i.e. through definition of an Aspect System and consequent translation into a computational model. In fact, most computational models operate on a part of the full system that is not explicitly stated. It is only through reverse engineering of the actual code that one can confidently identify the Aspect System. The original developers of the model have made assumptions about which objects can be ignored, for instance because their impact on the core physical relationships treated in the model is deemed negligible. This means that the computational model implies a reduced Aspect System based on expert knowledge based considerations. Reversely, it should be noted that any performance measure, quantified by a particular computational model, defines exactly one Aspect System. This will be further elaborated in the next section.

Although the framework of Figure 10.3 provides a useful conceptualization of a high level classification of performance concepts, it does not offer a classification of building functions. With the initially infinite variability that characterizes the current style of building design, it is doubtful whether such a functional classification can ever be made with sufficient generality, and moreover one should ask whether this would be of much practical use. In current practice, performance dialogues are typically generated in a "middle-out" fashion. Indeed, performance requirements are defined when certain design issues come into focus, i.e. when pertinent issues need to be resolved and selections between discrete design variants need to be made. At that point, one is typically dealing with a small subset of functions and associated criteria and a limited (but not small) subset of systems. At the interface of functions and systems, the stakeholder dialogues with respect to expected and fulfilled performances take place. It is important to support these dialogues with criteria and their quantified expressions of performance and to identify the role of simulation in adequate quantification methods.

The role of simulation in the system framework

A closer look at Figure 10.2 reveals that the role of simulation is still not defined. In fact, it is not clear at all *how* at a certain level of aggregation, a technical system, relates to a function. The many-to-many relationship at the interface between function and technical system is too vague to identify an appropriate virtual experiment that reveals how well a system fulfills a function. The example of visual comfort will clarify this. Imagine that Figure 10.4 shows a particular decomposition until the level is reached where the visual comfort criterion is addressed. Understanding the physics of the problem (i.e. the building environmental and human aspects of visual comfort), the following technical systems can be identified as having a role in producing a certain level of visual comfort. In parenthesis, we state the most important elements or parameters of these systems impacting on the realization of visual comfort:

- external enclosure system (orientation, size and location of windows, shading devices, shading control)
- neighborhood system (nearby buildings, vegetation, weather, especially cloud cover)
- electric lighting system (type of fixtures, location, control)
- internal enclosure (surface finishes (color, reflectivity))
- organization system (workplace location and orientation, internal partitions height and finishes)

Axiom 6: *Building performance can only be observed through a real or virtual experiment.*

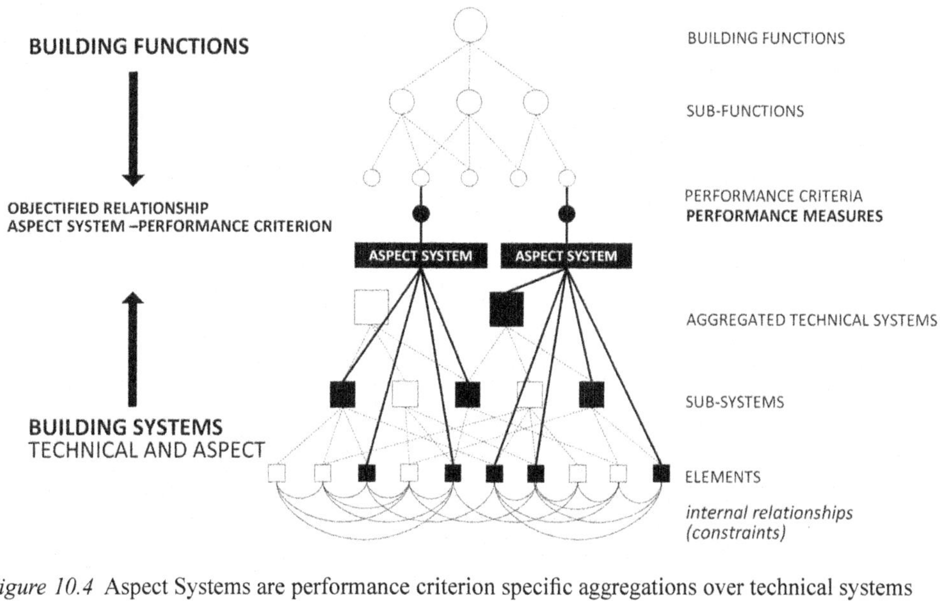

Figure 10.4 Aspect Systems are performance criterion specific aggregations over technical systems

The implication is that a collection of elements, and more explicitly element parameters, from diverse technical systems delivers the level of visual comfort. Together they form a functional system (Figure 10.4), which is typically composed of elements from diverse technical (sub)systems. The choice of elements is based on the rule of whether they contribute to building behavior that is relevant for a certain measure. As introduced above, we denote this functional system as "Aspect System". The implication is that a quantification method (in most cases a simulation) of a performance measure operates on an Aspect System that we compose in the way indicated earlier. Operating on an Aspect System has been defined more precisely as performing a simulation with a computational model that represents all objects and obeys all relationships between them as defined in the Aspect System. It should be noted that any performance can only be studied at the whole building (or more precise Aspect System) level. No component or element of a building has a performance, unless that component or part can be subjected to a test in which it operates in total isolation from the building. In that case, it is a mono-functional system.

> **Axiom 7**: *Building performance does not aggregate. The overall performance of a building cannot be derived from the separate performance of its components and parts.*

In principle, there are as many types of simulations (each operating on their corresponding Aspect Systems) as there are performance criteria and their corresponding measures. Figure 10.4 shows that each function can be linked to a criterion and thus to one Aspect System. There can be multiple criteria associated with the same function (not shown in the figure) in which case each criterion has in principle its own Aspect System. The relationship between a function and the Aspect System is thus "objectified" into the performance measure. The quantification of a performance measure is delegated to a virtual experiment as explained above, i.e. involving a simulation, normative calculation, or any type of quantification method. The way the criterion and its (normalized) measure are defined determine what type of simulation needs to be performed, i.e. what external conditions, what control, what time period, etc. To continue the example: assume that among the criteria linked to visual comfort is daylight autonomy, which can be regarded as one of many criteria related to visual comfort. As long as there is no standard that defines the measurement method for daylight autonomy, each project will have to negotiate an accepted measurement method between involved stakeholders. Let's assume that all parties have agreed on the following measure: "the average percentage of office hours in the month of March that all workers (at their workstations) in a given space get sufficient daylight to perform required eye tasks with certain ease (to be defined) and without the need for artificial lighting". This implies that the unit of the measure is percent. The quantification can be done with a variety of lighting simulation tools, none of which offer our custom daylight autonomy measure as one of its standard outcomes. But that is not a problem, as most retail lighting tools allow such configuration by the modeler that its outcome can be processed to quantify our measure. The more important issue is whether the agreed measure leaves room for interpretation of how the measure is defined and how its quantification should be conducted. Even if we are satisfied that the above definition is unambiguous, how the quantification should be conducted requires more detail. This usually takes the form of a specification of the deemed experiment, for example as follows:

> One should select a lighting simulation tool that takes the effect of all elements of the Aspect System into account, run hourly calculations for local average solar load and sky conditions, and take shading configurations, static blinds control (e.g. closed only when direct solar strikes on window) and shading and reflection from surrounding buildings, vegetation and ground into account. This will deliver the hourly incident solar shortwave

radiation on the fenestration. The next step is to calculate luminance, on desks and/or computer screens, given the workstation arrangements and worker tasks expressed in the office system as well as the reflectance of internal surfaces, and use accepted thresholds for eye comfort, given the eye tasks for each worker. Every office-hour that thresholds are not exceeded for any worker is counted as a daylight autonomy hour.

The accumulated outcome divided by the total number of office hours leads to the intended quantification of the performance measure. It should be noted that this virtual experiment is advanced in nature and will require an advanced "experimentalist" (in this case an experienced light simulationist).

Meanwhile, we have bypassed another major hurdle, which is to ascertain that the performance measure thus defined is in fact an adequate way to measure the daylight autonomy criterion. This issue most often forms the major bottleneck in practical design projects. It typically boils down to a lack of understanding of the deeper significance of "criterion". It is easiest to explain with an everyday example. A criterion for obesity is BMI (body mass index). Another criterion is waist circumference. Both can be measured in simple experiments. With the interest in the effect of overweight on health, medical science has introduced these measures as a quick proxy to the health effect of extra kilos. There are obviously more elaborate methods to define weight-related health issues, but research has shown that the BMI measure has a very strong correlation with health effects and because of its simplicity is now commonly used. Let's apply the same to our daylight example. We could have defined the calculation method differently, i.e. based on whole year or even multi-year climate data, reflecting vacation periods, daily periods that workers actually spend at their desk (stochastic), actual dynamic blinds control (stochastic), etc. Quantifying this measure would surely require more work but also be more precise. The latter is intuitively obvious but not necessarily the better choice if one can prove that the outcome of the measure as defined (i.e. based on a set of normative assumptions) is strongly correlated with the outcome of the more elaborate experiment in which one replicates the actual office scenarios without resorting to normative assumptions. I argue that the true attraction of performance measures is the introduction and verification of quantification methods that are supported by simplified calculations enabled by normative assumptions. An example is given by Kim et al. (2013), in which the LEED rating of the Energy and Atmospheric category (EAc) is done with two different experiments. The first one is based on ASHRAE 90.1, which requires dynamic simulation. The second method is based on a monthly quasi-stationary calculation with normative assumptions. Both can be used as measure of energy performance. Figure 10.5 shows the results of both measures for a large set of diverse buildings. Each dot represents the two measures as x and y values. As the LEED score is essentially based on a comparative ranking, the authors produce a ranking of the set of buildings for both measures separately. Applying the Wilcoxon test on the two rankings confirms a very strong correlation which identifies both rankings as equal. They conclude that therefore the simplified calculation is fully adequate for the given LEED rating purpose. Unfortunately, there is very little similar work to report in other building performance domains.

In the daylight performance case, an important role of the lighting simulationist is to communicate with the project's overall "criterion manager". Such a manager does not exist in current projects, or at least not by this job designation. The criterion manager steers and manages the stakeholder dialogues, suggesting the infusion of quantified measures in the dialogue where this is needed. This dialogue management involves:

- the systematic discovery of performance criteria and related quantification
- identifying the Aspect System for each criterion and identifying the relevant parameters of the objects

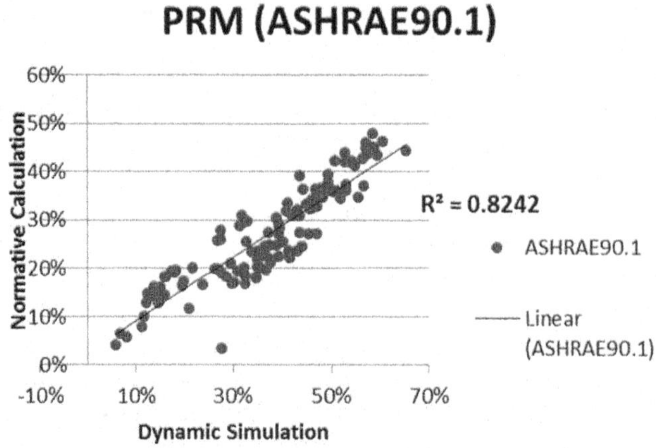

Figure 10.5 Normative performance rating method (PRM) versus ASHRAE 90.1 based EnergyPlus simulation

- setting up the virtual experiments that generate behavior that is relevant input to the criterion
- the quantification and feedback to stakeholders

As we will explain soon, it is becoming increasingly more important to do the latter with the disclosure of uncertainties and assumptions in the quantification process. This leads to certain demands on the set-up of the experiment.

Illustration of the ad hoc development of a normative calculator

This section will show the ad hoc development of a normative calculator of the urban heat island (UHI) effect. It typically proves to be a difficult task for students to perform. The following shows the rough outline of the calculator that was developed in the form of an Excel spreadsheet to be included in a sustainability rating method. Every new building adds to the UHI of an existing urban setting. A performance criterion is that we want this addition to be as small as possible, or even negative. For that reason, we introduce the UHI performance criterion which is measured by calculating the marginal addition of the building to the UHI effect. The effect is understood as the result of heat generated from urban structures, as they absorb and reflect solar radiation and re-radiate long-wave radiation, and from the anthropogenic heat sources as depicted in Figure 10.6.

We assume that the marginal contribution to the heat island effect of any new building is predominantly caused by the additional amount of trapped solar irradiation (mostly during the day) and inhibited long-wave losses to the sky (mostly during the night). This is an assumption that has to be verified, either from literature sources or by using a high fidelity model for many different urban settings in the application region of the rating system. In this particular case, the criterion is part of the GSAS rating system (GSAS 2017). After it was established that for these settings the trapped solar energy and reduced long-wave night radiation were indeed the leading drivers, it was clear that the normative calculation should focus primarily on these effects. Fortunately, a tool that calculates the net energy balance of absorbed solar and long-wave radiation to the sky is fairly simple. It only requires information about the location and height of surrounding buildings, and absorptance and emissivity as well as sky view factor and *U*-value

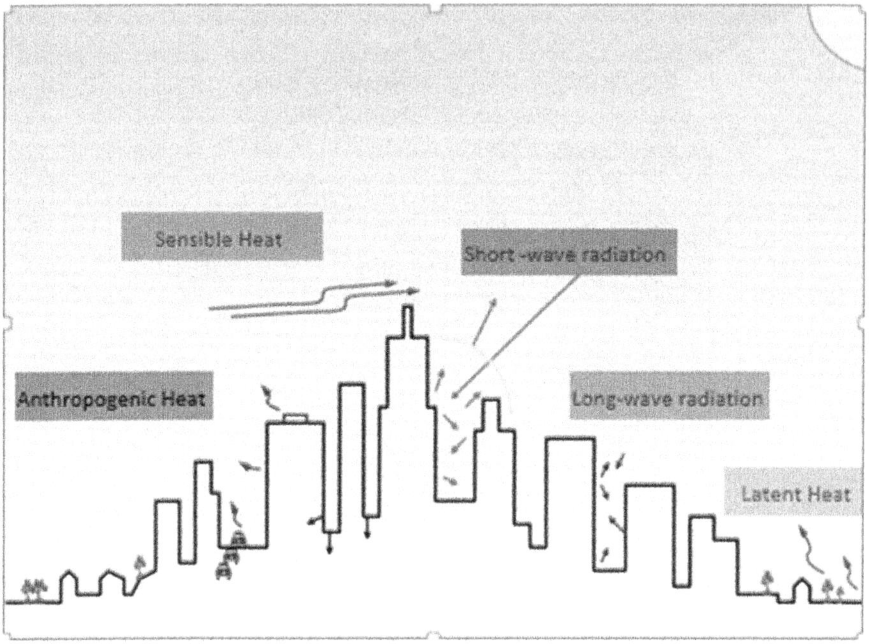

Figure 10.6 Typical urban heat island modeling elements

of all external surfaces of the proposed building. This information defines the elements, their (geometric) relationships and their parameters of the Aspect System for the UHI criterion. The resulting calculator compares the effect of the pre-development site and the post-development site with the proposed building. It calculates surface temperatures for the building façades based on climate data, view factors and a simple heat balance. It then determines heat gain from the sun and heat loss to the sky and to the air. All adjacent buildings within the 250 m radius are taken into account in the calculation. Figure 10.7 shows the geometry definition for the calculation.

Based on simple geometric and solar angle calculations, the marginal contribution to UHI is calculated as the difference between the pre-development apparent albedo of the project site and the post-construction apparent albedo. The difference between the two outcomes quantifies the UHI measure. The smaller the increase of the albedo, the higher the performance. In the best case, the albedo decreases, which would constitute the highest performance. The developed calculator proved to be simple enough to be implementable in a spreadsheet.

The above serves to underline the need for a revival of simplified calculation tools, not to limit the computational expense or to reduce the work of the modeler, but mostly to conduct normative calculations that have a proven strong correlation with high fidelity outcomes for the relevant case set. The major advantage is that they calculate performance measures with a fairly small effort and a small Aspect System. Such calculators are highly promising to be incorporated into computer-aided design environments, because they are transparent and avoid the use of over-engineered models that require more data as input than is typically available at the opportune moment that the performance criterion enters into the design decision making.

Axiom 8: *Performance measures need normative calculators to become efficient and mainstream.*

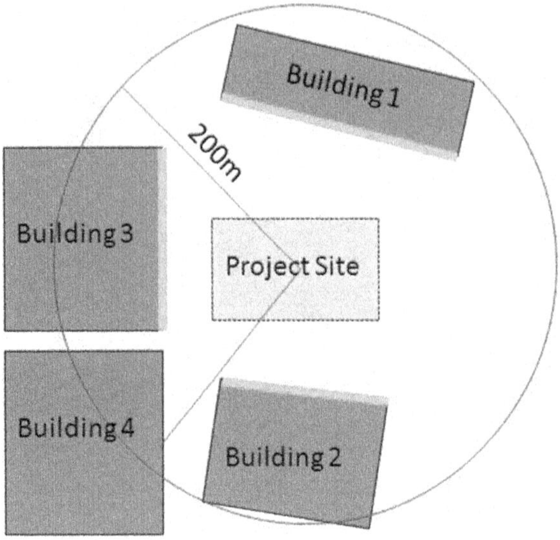

(a)

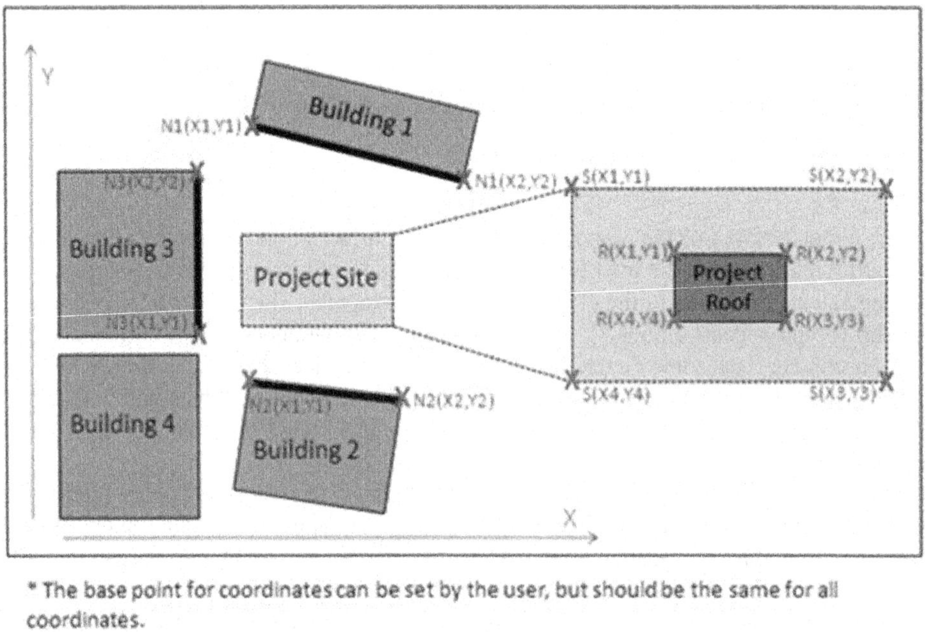

(b)

Figure 10.7(a–b) Urban geometry used by the normative UHI calculator

Computational methods

The art of modeling and simulation is leaving things out that don't affect the answer (Sokolowski, et al. 2009). Translating this to performance simulation in light of the previous section leads to a number of observations that are explored in depth below.

Observation 1: design decisions never deal with one single performance aspect; design is always inherently about satisfying many performance requirements at the same time. This implies that multiple stakeholders are engaged in the dialogue and that performance verification deals with many performance criteria simultaneously. Informing the decision making while recognizing trade-offs between different criteria is the primary objective. Delivering the most thorough and most detailed simulation in itself is not. Attempting the "best" simulation may in fact be counterproductive, as it will often result in over-engineered models and time-consuming simulation runs. Three issues should determine the ambition level of the performance analysis: the extent/scope of the Aspect System, the complexity of the measure and the needed fidelity of the computational model. The elaboration of the daylight autonomy example can serve as explanation:

- One could reduce the extent/scope of the Aspect System, for example, by ignoring surrounding buildings or by ignoring the office system parameters. A reduction of the Aspect System usually leads to *conditional statements of performance compliance*, e.g. of the type "PI (x |empty room AND no surrounding buildings) meets $PI_{required}$" which should be read as "the performance for given design x, under the condition of empty room and no surrounding buildings, meets the performance requirement". In reality, one will usually condition the outcome on a worst case approach, i.e. "PI (x |fully furnished room AND high urban density of surrounding buildings) meets $PI_{required}$". This approach represents a safety factor as we usually apply to limit-state performance requirements, for example in structural safety requirements conditioned on worst case load scenarios.
- One could attempt a simpler performance measure, such as the office time-averaged luminance level on a standard work plane in the center of the space; this would represent a particular case of a *simplified calculation* based on the consideration that such simplification does not alter the outcome more than one deems acceptable.
- One could use a coarser lighting model that does not deal explicitly with the effect of internal reflections; for instance, one could calculate the hourly radiation coming into the room (Q_{tot}) and use an empirical formula producing the average luminance value as a function of Q_{tot}, depth of the office space, average reflectance. This approach represents a *normative approach* similar to what was introduced and illustrated in the previous section. It is normative in the sense that the calculated outcome may not necessarily represent a physical (experimental) outcome but is rather used as a surrogate for a true outcome or a useful proxy measure for the criterion. This approach is defendable when one can ascertain that the value of the surrogate outcome has a strong correlation with the true outcome of the performance measure. Standardization bodies typically choose this route because normative calculations are transparent and carry no bias related to the individual expert or hidden assumptions in the experiment simulation tool. The development of a normative performance toolkit is described by Augenbroe and Park (2005). The example in the previous section serves as an insightful example of how to construct a normative approach.

Any one of these choices leaves something out and thereby reduces the complexity of simulation or calculation that is used for the quantification. This can only be justified if there is proof that outcomes based on these reductions still inform design decisions adequately without a bias in certain situations. In many cases, simulation only needs to be adequate for a comparative analysis of design variants. In such cases there is typically no high demand for absolute accuracy. On the other hand, the outcome of the method should reflect changes in design parameters adequately and reflect all major causalities between design parameters and the measure. The reduced experiment relieves the demand on the sophistication of the simulation tool and the needed expertise of the user.

Observation 2: performance quantification and comparisons take place in various stages of design, i.e. from conceptual design when the elements of the Aspect System are not well developed yet, to late design when most design choices have been made and hence most of the elements and parameters of the Aspect System are fully known. Research has generated a variety of tools that claim to support early design decisions, but very few, if any, are used in practice. Practice usually regards them as naïve, or not integrated, or too simplistic. It could be argued from a theoretical perspective that these tools make little sense anyway. To elucidate this, it should be understood that performance quantification in the conceptual design stage is by definition a conditional statement as explained above. Indeed, by definition the calculated performance is conditioned on the fact that certain assumptions about as yet unknown system parameters remain valid throughout the design process. None of the current generation of early design tools reflect this design uncertainty properly as probability distributions. Instead, the calculations are performed with static defaults which makes one wonder how much confidence one can have in the outcomes. Sanguinetti et al. (2009) show in a façade design study that explicit statements about conditional performance can be formulated if one uses conditional probabilities of ensuing design variations, assuming one can make a reasonable guess about the uncertainties of these as yet unknowns. The paper shows that such statements express the true nature of uncertainty in early design decisions. In a recent thesis, this theme is further developed in the context of sample early design decision scenarios (Rezaee 2016). The author's research starts from the recognition that early decisions have to incorporate the uncertainty in future, yet to be chosen design parameters. She shows that this uncertainty more often than not overwhelms the separation in performance between competing design options in such a way that the results of the early design simulation carry so much uncertainty that they lose the ability to rank one option over another with a reasonable degree of confidence.

Observation 3: irrespective of the design stage, every time a performance verification uses simulation to approximate (and forward-project) the real behavior of a building, one has to confront the fact that no building ever behaves as its model (*which is at best an idealization of the eventual real artifact*) predicts. Many uncertainties play a role when constructing the model. Unfortunately, the state of the art of current simulation is that uncertainties are rarely dealt with explicitly, hence parameter values are based on best guesses or, if some statistical data are available, their expected mean values at best. The reality is that best guesses are just that and are fraught with assumptions and expert bias and in some cases informed by bad or incomplete science. In the actual realization and operation of a building, many things will not perform as expected. Compounding the effect of model uncertainties, unexpected operation, physical parameter variations, unpredictable deteriorations of systems, etc., any outcome generated by a deterministic simulation has to be regarded as an arbitrary sample from a large statistical distribution. Figure 10.8(a) shows the example of a deterministic result for design options A and B. Figure 10.8(b) shows the result of the same simulation model with uncertainty analysis of the two design options. It needs little explanation that the deterministic preference for A over B has only limited validity when the full range of possible outcomes under uncertainty is known. Indeed, as the figure shows, there is a considerable chance that A underperforms B. A corollary of this treatment is that the original performance requirement may have no meaning. After all, requiring that a certain measure (PI value) for a proposed design with many possible outcomes for the population x, should surpass a minimum value PI_{min} makes less sense when $PI(x)$ is a stochastic variable.

The natural extension is therefore to formulate risk-oriented performance requirements, for example in the following form, $P\,(PI(x) < PI_{min}) < 5\%$, which expresses that the probability that the design performance does not meet the minimum required value shall be less than 5%. It is fairly straightforward to propagate the uncertainty ranges of parameters through a simulation model using a Monte Carlo approach as early studies have shown (Wit 2001; Wit and Augenbroe 2002; Moon and Augenbroe 2007; Hopfe 2009; Hu and Augenbroe 2009; MacDonald and Strachan 2001). The bottleneck in these studies was the lack of good *a priori* estimates of uncertainties. More recent work has made a fundamental inquiry into the uncertainty ranges of building model and scenario parameters (Sun 2014; Wang 2016). Several

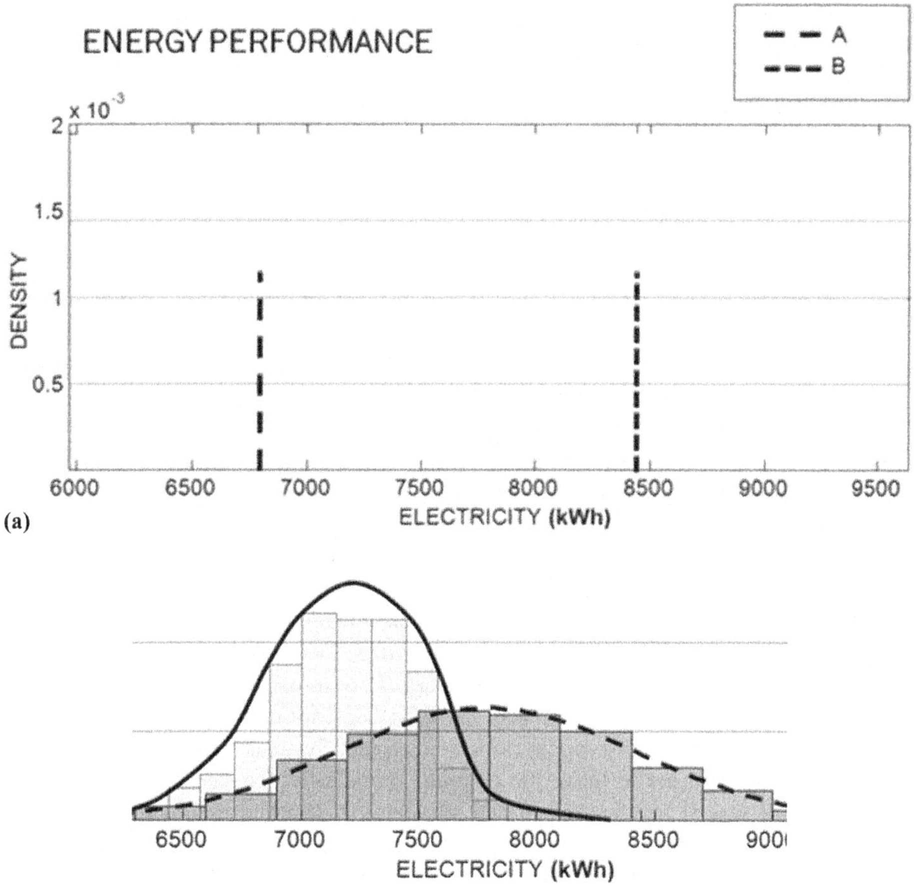

Figure 10.8(a–b) Deterministic versus probabilistic performance outcomes

papers report comprehensive research into the uncertainty and sensitivity analysis of building models, offer a generic repository of parameter and scenario uncertainty as well as quantify the model discrepancy of 11 EnergyPlus modules (Wang et al. 2016a; Sun et al. 2015; Lee et al. 2013). The full uncertainty analysis of performance should contain the screening of dominant parameters (Saltelli 2008; Lee et al. 2013). An iterative improvement of the *a priori* estimates is necessary in those cases where uncertainty ranges in certain parameters are shown to have a dominant effect on the resulting performance risk. Recent advances have also enabled a full risk analysis of critical performances, as for example shown in (Wang et al. 2016b) for the inherent risk in energy saving performance contracts. In spite of these advances, one can hardly expect practice to embrace uncertainty analysis any time soon in their routine projects. For the time being they seem more preoccupied by a rush to higher fidelity tools and tailored parameter exploration frameworks.

Axiom 9: *Performance uncertainty analysis will not become mainstream unless proven to reduce risks.*

Observation 4: the Aspect System view can be captured in a conceptual data model with all elements and their properties that contribute to a certain performance measure. This model is de facto a formal representation of the input data of the associated performance quantification method. If this quantification requires simulation, the Aspect System is in fact the formal representation (a conceptual data model) of the input data supplied to the simulation tool used for the quantification. This insight has a number of consequences:

- Testing the adequacy of simulation tools: the current generation of simulation tools was not developed with the aim to quantify performance criteria. Instead, most tools were developed to carry out virtual experiments in a generic sense, i.e. with emphasis on the experiment box rather than on the experiment. The consequence is that simulation tools have defined their own representations of buildings, although in most cases not explicitly available. Future generations of building simulation tools should be developed with the explicit goal to support specific experiments and hence support explicit Aspect Systems. Current simulation tools can be added to this set if a formal representation of their input model is tested against consolidated Aspect Systems. This was the fundamental approach of the Design Analysis Integration project (Augenbroe and de Wilde, 2003) which conceptually defined simulation tools that could respond to specific performance quantification requests. The key to this was the pre-definition of the Aspect System and an automatic mapping from an overall building model to local Aspect System.
- Many recent efforts map building data from a building information model (BIM) (www.iai-tech.org/) to the input models for simulation. In doing so, they are mapping to the Aspect Systems that we have defined in Figure 10.3. From the very nature of the Aspect System, it follows that these mappings are hard if not impossible to develop in a generic way. The main reason is that currently the BIM does not explicitly define performance criteria and hence does not support associated Aspect Systems (or in BIM terms: Aspect System views). In order to make the mapping work, BIM interfaces are developed that "roam" the technical building systems in the current BIM and pick and choose the elements and their properties that are needed for a particular performance assessment, i.e. quantification method. The "invisible" Aspect System is constructed on the fly by such a mapping. Much of this work will become obsolete if a future generation of the BIM offers explicit support for Aspect Systems. Given the vast amount of resources that would have to be spent on this, the more likely scenario is that future mappings will be driven by explicit specification of Aspect Systems as target schema for the mapping. The expert directs the mapping through a so-called constructive interface that allows user supply of missing data in the BIM source model or user intervention to resolve ambiguities.

Axiom 10: *Current BIM contains insufficient information to automatically support performance analysis.*

In summary, we can conclude that the systematic framework introduced in this chapter enables us to identify the role of simulation in design performance assessment and, moreover, identify the crucial choices that one faces in defining different measures and the appropriate simulation method for their quantification.

Application examples

This section gives an account of applications of the performance dialogue process on real-world projects. The focus is to show how design problems are governed by performance criteria in multiple domains. Typically, only a subset of those can be quantified with existing simulation tools. Other criteria require the introduction of specific customized measures for which often no established measurement method exists. The purpose of this section is to show that the role of simulation in performance design is still limited. The examples serve as an agenda for enlarging the scope of building simulation and become more universal in the support of performance dialogues in the design phase. Every example deals with the choice between two discrete design options. The treatment aims at a comparison of the two options in such a way that an ensuing decision process with different stakeholders with different value systems could be effectively informed. All examples are the result of student work in a graduate course in the High Performance Building program of the School of Architecture at Georgia Tech.

Case 1: choice between granite stone and aluminum cladding

Case 1 background: this case is part of a range of design choices for an office building. The location of the project is in a city with high pollution and in a climate zone with severe winters and hot summers. The client wanted the building to be energy efficient and to have a building cladding material that would be durable and require little maintenance. The client's needs focused on the durability of the material, i.e. the wall cladding should be able to withstand pollution and should not change in color, texture and tensile properties for at least a period of 50 years.

Table 10.1 shows the criteria, measures (PI) and quantification methods chosen in this case. The selected performance criteria are a reduction of a longer list offered initially by the client. The reduction took place in a dialogue with the designer to select those that were deemed most relevant in the choice between the two competing materials. It was left to the architectural engineer to come up with adequate PIs for the selected criteria and select the appropriate quantification methods. It is worth noting that no simulation was used in this case. The energy performance calculation was done using a normative calculation tool (Augenbroe et al. 2004), which was adequate for the comparative analysis. All other PI calculations were based on normative approaches as well (not detailed here).

Figure 10.9 shows the Aspect Systems for each of the performance requirements. The black boxes show the macro elements of the two competing technical systems. Three types of lines

Table 10.1 Case 1 criteria and performance indicators (subset shown)

Performance criteria	*PI #*	*PI name/description*	*Quantification method*
Durability	1	Defacement because of weather and pollution	Assessment based on industry and manufacturer's average (based in part on durability experiments)
	2	Product's service life	Assessment based on industry and manufacturer's data; service life prediction
Energy performance	3	Energy required for heating and cooling	Building performance normative toolkit; produces accurate comparative energy consumption results
Constructability	4	Recycling capacity	Based on Eco-Impact calculator from ATHENA institute
	5	Waste production	Assessment based on industry data (normative)
Maintenance	6	Time taken to lay out stone	Calculation of time and people required to set up X m^2 of wall cladding (based on contractor survey)
	7	Replacement	Use of Markov chain maintenance intervention matrix

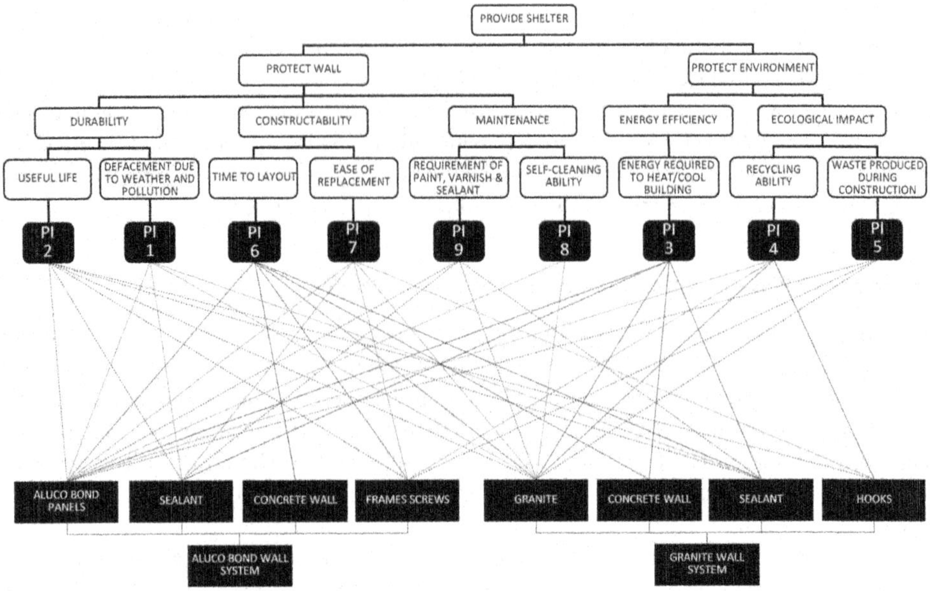

Figure 10.9 Aspect Systems

are used to show which element is part of which Aspect System. Figure 10.9 is used in the quantification as the schedule for the data (element properties) that need to be available to support the performance quantification. In practice, this would constitute a significant help in servicing the different consultants with the information needed to perform their assessment. As debated earlier, the explicit specification of the Aspect Models for the quantification method could also be used to drive constructive BIM interfaces, i.e. generating the input data from available BIM data.

After normalization of the PI scales, the radar chart in Figure 10.10 shows the outcomes.

Granite had been the first choice, not based on performance but merely on availability and cost. In the performance assessment, both products fare well on some performance indicators, but not all the criteria are met at the expected or required level. Both products are viable in terms of cost and availability and the skills required for installing the cladding systems. Both products have a service life of at least 50 years and can last longer with routine cleaning and maintenance. In the assessment, the granite wall system causes more air pollution during its life cycle than the Aluco Bond aluminum system, but it also saves more energy. The aluminum system is cheaper and quicker to install. It turns out that, based on a weighted average of the indicators in Figure 10.10, the aluminum cladding option offers the better choice in this case.

Case 2: choice between two car-park layouts

Case 2 background: the main design intention in this case was to offer a variety of apartment layouts within a single residential building. Because of the code regulations in the city where the project is located, the buildings' capacity (number of units) became constrained by the number of parking spaces provided. Two design options offering the highest number of car park units were evaluated, as point of departure, to develop the rest of the building, including the number of possible apartment types and façade variations of an apartment type.

Table 10.2 shows the criteria and their quantification method. For each performance criterion, multiple measures (PIs) were defined representing alternative methods to quantify the

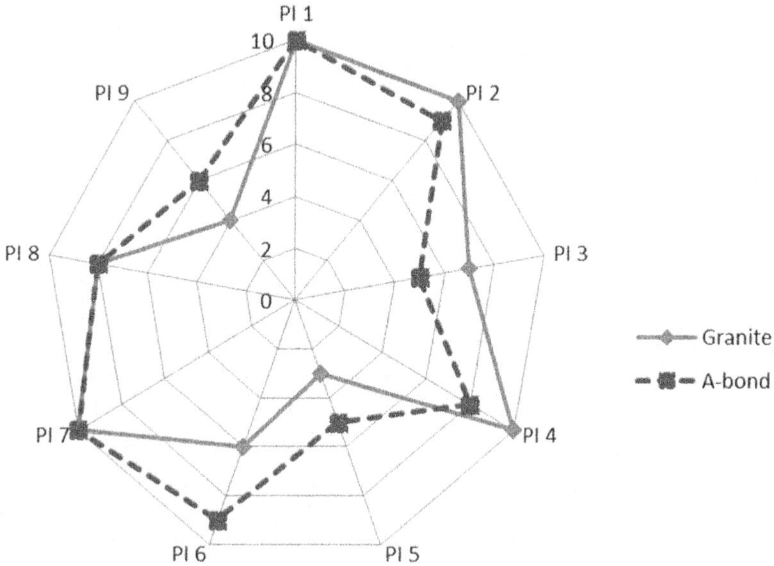

Figure 10.10 Normalized results

Table 10.2 Case 2 criteria and performance indicators

Performance criteria	*PI #*	*PI name/description*	*Quantitative method*
Area efficiency	1–5	Utility of design in terms of area usage	Measure of geometric properties of shapes
Façade constructability	6–9	Time needed to construct the façade design	Measure provided by contractors and suppliers
Capacity	10–12	Number of tenants that can be accommodated in the building	Count of car park units, shops and structural bays
Morphology	13–20	Flexibility of the design to allow for variation	Measure based on building matrices of design options
Fire safety	21–23	Ability of design to meet fire safety regulation in the city	Egress calculation based on simple shape properties
Ventilation	24	Pedestrian wind hinder	CFD analysis (simulation software package)
Real estate	25	Lease/sale revenue	Real estate consultants (expert judgment)
Energy performance	26	Energy consumption	Energy analysis (normative calculation)

criterion. Most of the resulting PIs within one criterion differed in their way of calculation, e.g. simulation or normative, low resolution or high resolution, first order approximation or high fidelity simulation. The last column shows the PI calculation method eventually used to inform the decision making. The column indicates three PIs that potentially need advanced simulation, i.e. fire safety/egress, ventilation/wind hinder and energy analysis. In the course of the project, it was decided that egress and energy analysis could be based on normative calculations and only ventilation/wind hinder needed full-blown simulation. These decisions were reached based on time, convenience and the trust that normative methods were good enough for the comparative analysis. Figure 10.11 shows the sets of elements that compose the various Aspect Systems, whereas Figure 10.12 shows the normalized results.

As is obvious from Figure 10.12, there is no clear winner. It will largely depend on the preferences of the stakeholders to arrive at a consensus decision.

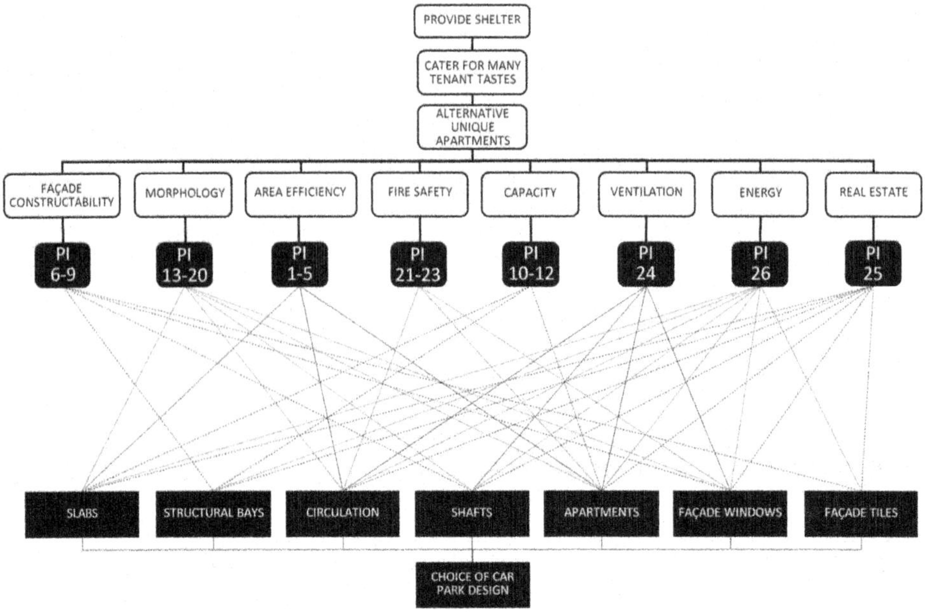

Figure 10.11 Aspect Systems

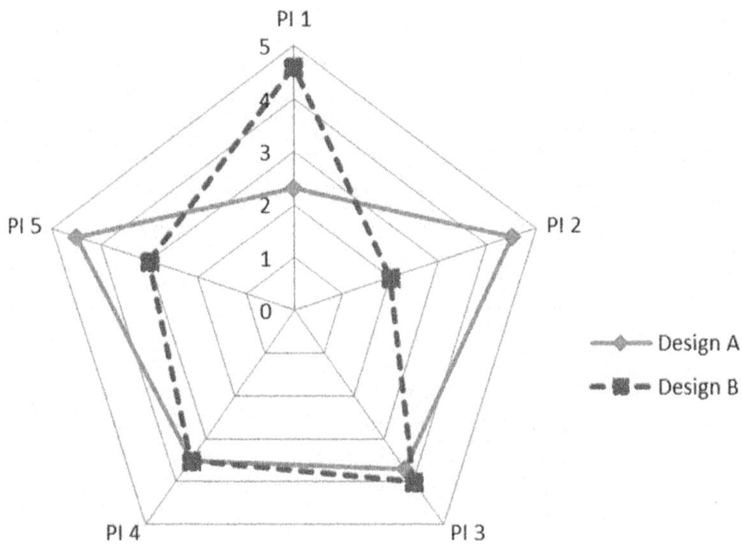

Figure 10.12 Normalized results

Case 3: choice between off-grid PV panel system and green roof

Case 3 background: in this case, the owner provided a list of design goals and gave his preference for achieving a goal, as given in Table 10.3. In this particular instance, a SWOT analysis of the two options was used to select five main criteria and define one PI for each criterion as shown in Table 10.4. In the early analysis, it was found that the PV panel solution performed well in terms of low environmental impacts, low load to structure and low maintenance, but the green roof was better from an energy conservation and acoustics perspective.

Figure 10.13 shows the macro Aspect Systems. Involved simulation domains in this case are lighting, energy and acoustics. The LCA and maintenance domains in principle require

Table 10.3 Ranking of owner's needs

No.	*Goals*	*Weights*
1	Strength of concept	7
2	Simplicity of design	6
3	Creativity of design	7
4	Challenging solution	4
5	Connection of interior and exterior	4
6	Material use	6
7	Quality of indoor space	6
8	Natural lighting	8
9	Energy saving	8
10	Sustainability	8
11	Initial cost	4
12	Maintenance and operation cost	8
13	Short schedule	6
14	Efficient use of space on site	5
15	Prefabrication	7
16	Reusable contents	8
	TOTAL	**100**

Table 10.4 Project E criteria and performance indicators

Performance criterion	*PI #*	*PI name/description*	*Quantitative method*
Energy conservation	1	Annual energy saving	Lighting simulation in Ecotect Energy simulation (normative calculation) Green power estimation in PowerWatt
Acoustics	2	Acoustical comfort	Calculate sound reflection and transmission with simple acoustics simulation
Environmental impacts	3	Lifecycle greenhouse gas emission	LCA (simplified to GHG impact category)
Impact to structure	4	Extra structure load effect on cost	Calculate total weight of options (used as surrogate outcome for cost)
Maintenance	5	Maintenance costs	Estimate annual average maintenance and operation cost based on published data

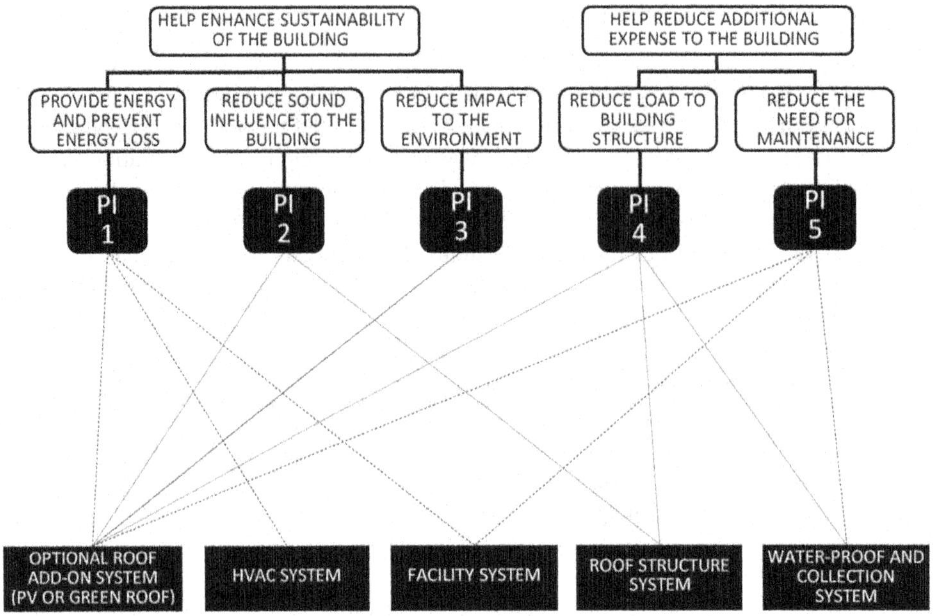

Figure 10.13 Aspect Systems

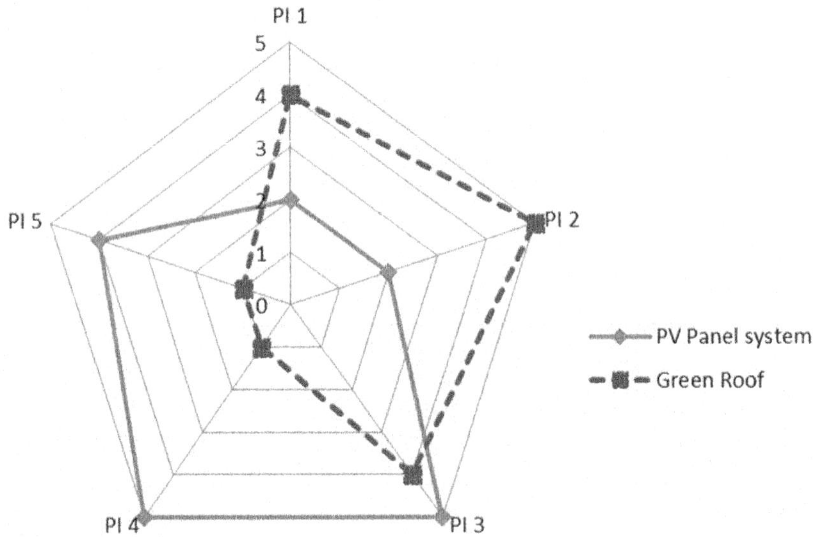

Figure 10.14 Normalized results

an extensive long-term experiment for accurate quantification of performance, and simulation would hence be appropriate. However, for obvious reasons, standardized calculation procedures were used to quantify lifecycle impact (limited in this case to greenhouse gas) and maintenance cost.

Figure 10.14 shows the mixed outcome which was used to inform the stakeholder's decision making.

The role of simulation

The three cases show clearly that simulation is a dispersed activity throughout the performance-based design process. Performance simulation expertise is necessary from the start of the analysis of the options, through criteria definition and choice of quantification method. The actual execution of simulation is mostly defined by the earlier steps in the process. This defines the role of simulation in performance-based design as an integral part of the planning and preparation of decisions. The other remarkable fact is that the need for sophisticated simulation seems to be less than is often assumed. There are several reasons: (1) the multitude and integrality of the combined criteria become the core of attention rather than any single criterion on its own, (2) normative comparative analysis and parametric fast turnaround tools respond better to the dynamics of multi stakeholder decision making and (3) the systematic framing of the quantification in terms of Aspect Systems explains the role of simulation to the client and across different domains of expertise in the design team.

It is clear that the radar chart with normalized performance indicator values needs to be followed up by a method that helps stakeholders to reach a "best" decision, based on their individual ranking of criterion importance and individual weighting of the PI outcomes on each criterion. A brief overview is given in the next section.

Multi-criterion decision making

Since Keeney and Raiffa (1976) proposed the multi-criterion decision analysis (MCDA) approach in their seminal work, decision theory has steadily evolved. The analysis of the decision problem introduces and processes a formal, yet possibly subjective judgment during the alternatives' evaluation (Keeney 1982). Current MCDA procedures structure the decision problem in a sequence of steps that provide clarity for those who will make the decision. The MCDA starts when its aims and parties are identified. The parties are composed by three main players: the Decision Maker, who makes the decision and assumes the consequences, the Decision Analyst, who facilitates the process, and the Stakeholder, who will be affected by the consequences of the decision (Huovila et al. 2005). The next step is to develop the decision analysis system based on the exchange of information provided by the parties. Then, objectives and criteria, which express the means by which the alternatives create value, are identified (Dodgson et al. 2000). Once the parties agree on the decision problem organization, the subsequent step is to weight the criteria in relation to each alternative and to rank their importance in relation to the decision problem. Therefore, this process means weighting both the differences between options and "how much that difference matters" (Dodgson et al. 2000). Hence, a consistent weighting scale should be used in order to compare different criteria subjects. After the results are calculated, questions such as "what if . . ." are common, and an analysis of them should be made in order to provide an understanding of the consequences of the decision. This can be made through a sensitivity analysis that aims to evaluate how the model is sensitive to changes and clarify conflicts of interests between stakeholders.

There are different decision techniques when involving multiple criteria, and this will depend on the type of decision, how the technique deals with the data provided, the time available for the analysis, the nature of the data available and the analytical skills of the stakeholders. All techniques that approach the MCDA present criteria and alternatives in its structure and require subjective judgment to deal with the problem complexity. This judgment is typically done as a numerical analysis by scoring the most preferred options higher on the scale. The cases presented in this chapter have three main characteristics: there are a limited number of alternatives, uncertainty and risk are not considered and the criteria are formally independent of each other. Therefore, techniques based in the linear additive model are suitable to this approach. Linear additive models are the basis of the MCDA; they combine individual values into one overall output. This is done with multiplications of each option's score on each criterion by the weights of the criteria, and then summing these weighted scores. The well-known analytical hierarchy process also uses a linear additive model, but the weights and the scores obtained by alternatives are based, respectively, on paired comparisons of criteria and alternatives (Saaty 2013). AHP is one of the most frequently applied multi-criterion decision techniques (Dodgson et al. 2000). There are several software tools on the market that support AHP, which requires a several hour session with the stakeholders in the room. Before the session, the criteria and participants are preloaded into the software, while the outcome of the performance assessments is distributed and explained to all stakeholders. One then proceeds using the stakeholders' ability to rank the criteria in relation to each alternative and to rank their importance in relation to the decision problem through a paired comparison. The tool then ranks the two options based on all inputs and allows a sensitivity analysis of the outcome. An interesting example of the application can be found in Lima (2007). As stated before, AHP foundation is justifiably criticized especially by those that prefer utility theory, with its strong theoretical foundation (Hazelrigg 2003).

After stipulating previously that uncertainty in the performance outcomes is unavoidable and necessary to consider in decision making, it should be observed that the treatment in this section has been purely based on deterministic results. As discussed earlier, our tools are increasingly expanded to deal with decision making under uncertainty, but more work needs to be done in conveying uncertainties or rather the risks resulting from them to the stakeholders in such a way that the decision analysis can transparently consider them.

Discussion

In this chapter, we have shown that contemplating the role of simulation in performance-based design forces us to look beyond offering traditional simulation skills to the project. We argue that the simulationist must in fact be an integral partner in the performance criteria management and ensuing dialogue concerning the definition of measures for the individual criteria and choice of appropriate quantification methods. In some cases, the measures depend on advanced simulation; in other instances, a simple calculation or expert judgment will be sufficient. The role of these different methods is in principle interchangeable and only determined by the special need and context of the decisions that the project team is facing.

A system theoretic approach has been shown to lead the way in defining the interface between physical building elements and performance criteria. A closer look at this interface reveals that the Aspect System plays an important conceptual role. It is introduced as a composition of elements and parameter subsets over (in principle) all technical building systems. Each Aspect System is uniquely associated with a particular performance measure. As such it is a representation of the physical model on which the experiment (simulation) is performed in order to study behavior through outcomes and quantify a particular measure. One could imagine that this Aspect System is also used in conformance testing the simulation software. If the simulation (input) model is a superset of the Aspect System, it is suitable for the quantification of the given performance measure. The explicit management of all Aspect Systems for which a performance requirement has been specified would greatly enhance integral data management across a design team. Future versions of BIM could explicitly support the Aspect Systems as "views", and BIM-driven interfaces would replace the current generation which haphazardly harvests elements from different technical systems to be processed in the mapping software. For the latter there is little common ground across different interface developments, leading to much unwarranted duplication of work.

Perhaps the biggest challenge of simulation in performance-based design is to provide a variety of normative calculations when an advanced high fidelity simulation is not necessary to provide the answer, either because of the presence of uncertainties or the lack of information, or because the context of the decision simply does not demand it. The latter statement is strongly influenced by the author's 40-year exposure to overly complicated, wasteful, inefficient and over-engineered building simulation models.

Final Axiom: "*Everything is going to be fine in the end. If it's not fine it's not the end*".
– Oscar Wilde

Synopsis

The chapter has placed building simulation in a design context. It has shown that comparison of different design variants requires carefully crafted performance criteria with their measures. It also emphasized that a system theoretic approach is needed to define objective performance measures as the "outcome" of simulation. For this reason, the concept of Aspect System was introduced. It was shown that every simulation is in fact an operation on an Aspect System, the outcome of which can be expressed as a mathematical formula over the properties of elements in the Aspect System. It was shown that the approach leads to unambiguous definitions of performance measures that are used to prepare multi-criterion decision making. If all parties are satisfied with the definition of the measures and measurement methods, they will use them to express performance targets and to check whether design proposals meet these targets. It was shown that quantification, or any measurement method for that matter, can be supported by different types of building simulation, calculations, expert judgment, or other. Diverse examples have been presented to show that a well-structured set of performance criteria can deliver the needed support for rational design decision making.

In addition, the explorations in this chapter have touched on issues of design evolution and decision making, the anachronism between the importance of early design decisions and the deficiencies of current simulation tools to support them, the aggregation of simulation results into meaningful expressions of performance, the role of uncertainty and the need to distinguish performance evaluation from software execution.

Acknowledgments

Thanks to P. Sanguinetti for her valuable contributions to the paper as a whole, and to T. Chatterjee, M. El-Khaldi and F. Zhao for their contribution of the case studies used in this chapter.

References

Afsari, K. and Eastman, C.M. (2016) A comparison of construction classification systems used for classifying building product models. In *52nd ASC Annual International Conference Proceedings.*

Augenbroe, G. (2004) Chapter 1 Introduction. In Ali Malkawi and Godfried Augenbroe (eds), *Advanced Building Simulation*. London: Taylor & Francis.

Augenbroe, G., Malkawi, A. and de Wilde, P. (2004) A performance based language for design analysis dialogues. *Journal of Architectural Planning and Research* 21(4): 321–330.

Augenbroe, G. and de Wilde, P. (2003) *Design analysis interface (DAI)*. Final Report. Atlanta, GA: Georgia Institute of Technology. http.dcom.arch.gatech.edu/dai/.

Augenbroe, G. and Park, C.S. (2005) Quantification methods of technical building performance. *Building Research and Information* 33(2): 159–172.

Barrett, P. (2008) *Revaluing Construction.* Oxford, UK: Blackwell Publishing.

Becker, R. (2008) Fundamentals of performance-based building design. *Building Research and Information* 1: 356–371.

Beard, J.L., Wundram, E.C. and Loulakis, M.C. (2001) *Design-Build: Planning Through Development.* New York: McGraw-Hill.

Cavieres, A. (2018) A functional modeling framework for interdisciplinary building design. PhD dissertation, Georgia Institute of Technology, School of Architecture.

CIB. (1982) *Working with the Performance Approach in Building.* Rotterdam, The Netherlands: CIB Publication No. 64.

CSI. (2017) Masterformat. www.csiresources.org/practice/standards/masterformat.

Dodgson, J., Spackman, M., Pearman, A. and Phillips, L. (2000) *Multi-criteria Analysis Manual.* London.

Foliente, G. (2000) Developments in performance-based building codes and standards. *Forest and Products Journal* 50(7/8).

Foliente, G.C., Leicester, R.H. and Pham, L. (1998) *Development of the CIB Proactive Program on Performance Based Building Codes and Standards, BCE Doc 98/232.* Australia: International Council for Research and Innovation in Building and Construction (CIB).

Galle, P. (2001) Viewpoint: Metatheory in practice. *Design Studies* 22(1): 101–102.

Gibson, E.J. (1982) Working with the performance approach in building. CIB Report Publication 64. Rotterdam, The Netherlands.

Gielingh, W. (1988) General AEC reference model. ISO TC184/SC4/WG1 document 3.2.2.1 (also TNO report BI-88–150, October 1988).

Gilb, T. (2005) *Competitive Engineering.* Hoboken, NJ: Elsevier.

Gero, J.S. and Kannengiesser, U. (2014) The function-behaviour-structure ontology of design. In Chakrabarti, A. and Blessing, L.T.M. (eds), *An Anthology of Theories and Models of Design.* Springer, pp. 263–283.

GSAS. (2017) www.gord.qa/trust-gsas-resource-center-overview.

Habraken, N.J. (1998) *Structure of the Ordinary; Form and Control in the Built Environment*, ed. Jonathan Teicher. Cambridge, MA: MIT Press.

Hazelrigg, G. (2003) Validation of engineering design alternative selection methods. *Journal of Engineering Optimization* 35(2): 103–120.

Hopfe, C. (2009) Uncertainty and sensitivity analysis in building performance simulation for decision support and design optimization. Doctoral dissertation TU Eindhoven, The Netherlands.

Hu, H. and Augenbroe, G. (2009) Right sizing an off-grid solar house. In *Proceedings of Building Simulation 09*, July 2009. Glasgow.

Huovila, P., Porkka, J., Gray, C. and Al Bizri, S. (2005) Conclusions and recommendations on decision support tools for performance based building: Performance based building thematic network. Collaborative effort of VTT building and transport (Finland) and University of Reading (U.K.).

Keeney, R.L. (1982) Decision analysis: An overview. *Operations Research* 30(5): 36.

Keeney, R.L. and Raiffa, H. (1976) *Decisions with Multiple Objectives: Preferences and Value Tradeoffs.* New York: John Wiley and Sons.

Kim, J.H., Augenbroe, G. and Suh, H.-S. (2013) Comparative study of the LEED and ISO-CEN building energy performance rating methods. In *BS2013 IBPSA Conference*. Chambery.

Lee, B.D., Sun, Y., Augenbroe, G. and Paredis, C.J.J. (2013) Towards better predication of building performance: A workbench to analyze uncertainty in building simulation. In *Proceedings of the 13th International Building Performance Simulation Association Conference*. Chambery, France.

Lima, C. (2007) The use of formal methods for decision making in the planning phase of healthcare facilities. MS Thesis Georgia Institute of Technology. https://smartech.gatech.edu/handle/1853/14587.

Love, T. (2000) Philosophy of design: A meta-theoretical structure for design theory. *Design Studies* 21(3): 293–313.

Macdonald, I. and Strachan, P. (2001) Practical application of uncertainty analysis. *Energy and Buildings* 33(3): 219–227.

Moon, H.J. and Augenbroe, G. (2007) Application of the probabilistic simulation and Bayesian decision theory in the selection of mold remediation actions. In *Proceedings Building Simulation 07*, Beijing, China.

Omniclass. (2017) www.omniclass.org/.
Ozsariyildiz, S. (2006) Inception support for large-scale construction projects. A knowledge modeling approach. Doctoral Thesis TU Delft, The Netherlands.
PeBBu. (2005) Pebbu thematic network, conceptual framework. www.irbnet.de/daten/iconda/CIB22199.pdf.
Preiser, W. and Jacqueline V. (2005) *Assessing Building Performance*. New York: Routledge.
Quiggin, J. (1997) *Generalized Expected Utility Theory: The Rank-Dependent Model*. Dordrecht, The Netherlands: Kluwer Academic Publishers.
Rezaee, R. (2016) Application of inverse modeling to performance-based architectural design in the early stage. PhD dissertation Georgia Institute of Technology.
Rush, R.D. (ed.) (1986) *The Building Systems Integration handbook*. Heinemann: AIA and Butterworth.
Saaty, T.L. (2013) *Fundamentals of Decision Making and Priority Theory with the Analytic Hierarchy Process*. Analytic Hierarchy Process Series, Vol. 6, 1st edition. Pittsburgh, PA: RWS Publications, 1994.
Saltelli, A., Ratto, S., Andres, T., Campolongo, F., Cariboni, J., Gatelli, D., Saisana, M. and Tarantola, S. (2008) *Global Sensitivity Analysis*. Hoboken, NJ: Wiley & Sons.
Sanguinetti, P., Eastman, C. and Augenbroe, G. (2009) Courthouse energy Evaluation: BIM and simulation model interoperability in concept design. In *Proceedings Building Simulation BS09*, Glasgow, July 2009.
Simon, H.A. (1996) *The Sciences of the Artificial, 3rd edition*. Cambridge, MA: MIT Press.
Sokolowski, J.A. and Banks, C.M. (2009) *Principles of Modeling and Simulation. A Multi Disciplinary Approach*. Hoboken, NJ: John Wiley & Sons.
Sun, Y. (2014) Closing the building energy performance gap by improving our predictions. PhD dissertation Georgia Institute of Technology.
Sun, Y., Su, H., Wu, C.F.J. and Augenbroe, G. (2015) Quantification of model form uncertainty in the calculation of solar diffuse irradiation on inclined surfaces for building energy simulation. *Journal of Building Performance Simulation* 8(4), 253–265.
Vanier, D.J., Lacasse, M.A. and Parsons, A. (1996) Modeling of user requirements using product modeling. In Becker, R. and Paciuk, M. (eds), *3rd International Symposium – Applications of the Performance Concept in Building*, 2: 6–73 to 6–82. Tel-Aviv, Israel: The National Building Research Institute.
Verheij, J.M. (2005) Collaborative planning of AEC projects and partnerships. PhD Thesis, Georgia Institute of Technology.
Wang, Q. (2016a) Accuracy, validity and relevance of probabilistic building energy models. PhD Thesis, Georgia Institute of Technology.
Wang, Q., Augenbroe, G., Kim, J.H. and Gu, L. (2016) Meta-modeling of occupancy variables and analysis of their impact on energy outcomes of office buildings. *Applied Energy* 174: 166–180.
Wang, Q., Lee, B.D., Augenbroe, G. and Paredis, C.J.J. (2016) An application of normative decision theory to the valuation of energy efficiency investments under uncertainty. *Automation in Construction* 73: 78–87.
Wilde, P. de. (2018) *Building Performance Analysis*. Wiley (in press).
Wit, M.S. de (2001). Uncertainty predictions of thermal comfort in buildings. PhD dissertation, TU Delft.
Wit, S. de and Augenbroe, G. (2002) Analysis of uncertainty in building design evaluations and its implications. *Energy and Buildings* 34(9): 951–958.

Recommended reading

First and foremost, one should read the brand new textbook by de Wilde (2018), as it brings all relevant literature and many valuable insights together in one place.

For further reading on performance thinking in general, the author recommends the cited papers by Foliente and the 2005 paper by Augenbroe. A good overview from a broader perspective is given by Preiser and Vischer (2005). For follow-up reading on specific issues in performance simulation, decision making and standardization, one should consult the special issue of *Building Research and Information* on performance-based building, Vol. 33, March–April 2005.

Assignment

A good exercise to train oneself in the use of the performance framework is to revisit and rationalize a design decision problem that one was part of. This should be a real problem where a real decision was made and enacted, preferably concerning a discrete decision case during building design, construction or operation. The following assignment challenges the reader to rationalize the original design decision problem.

Starting point: take a design, construction or operation context that you have been personally immersed in. You should single out a discrete decision problem that concerns a choice between design option A and design option B.

Now perform the following steps and answer the stated questions in each step:

Step 1: context

- Describe the context of the case and the options.
- Describe the original decision: why was it taken? On what grounds?
- Describe the two competing options that you are going to consider.
- Who (which stakeholders) is involved in the decision making?
- Was risk analysis used in the original decision making?

Step 2: rationalization

- What functional requirements drive the decision? Use decomposition of functions if needed until you arrive at the granularity that is adequate for your analysis.
- Match each function with a performance aspect and (more specifically) a criterion.
- Choose and explain the performance measures that you are going to use to measure each criterion.

Step 3: methodology

- Describe the decision problem in the context of matching each decomposed function with an Aspect System (show this in a system diagram).
- Describe every Aspect System in terms of aggregations over components and properties of technical subsystems (show in system diagram).
- Explain the connection between Aspect Systems and functions and locate your performance measures at those connections.

Step 4: method

- Develop the performance measures and their quantification methods (at least five) and do a thorough quantification of at least three of them using some type of building simulation tool.

Step 5: informing the decision making process

- Produce a radar chart plotting options A and B.

Step 6: evaluation

- How will your approach affect the decision making process? Is it likely that another outcome will result compared to the original decision?
- How would the recognition of uncertainty affect your analysis?

Answer

The chapter treats three case studies that show the narrative as requested by the assignment. Read and study the case studies in the section titled "Application examples" and apply to your own case.

11 BIM and BPS

A case study of integration cost metrics and design options

Timothy Hemsath, Matthew Goldsberry and Joel Yow

Scope

This chapter highlights building information modelling (BIM) relating to building performance simulation (BPS), reviewing several ways that BPS and BIM interact in design and operations. Highlighted is an integrated project delivery case study project leveraging a custom build real-time BPS/BIM modelling tool.

Learning objectives

- What is BIM?
- Relationship of BIM to BPS
- Role of interoperability
- Integrated project delivery case study

Key words

Building information modelling/cost modelling/integrated project delivery

Introduction

Building information modelling (BIM) is a broad concept; it's not software, not something you download, but a way of creating and accessing information about a building. The idea of BIM originated in the early 1970s by authors Chuck Eastman, Nicolas Negroponte and others who talked about how we could structure an information environment around buildings and what one could do with that information. Today we see how that vision is becoming a reality in BIM, "building information at large is a digital simulation of the building itself" (Bernstein and Jezyk 2011).

BIM imbues a model with information about building systems for simulation, whether accessing the geometry, the materiality, the operations, the equipment, the construction processes, the schedule, the project's location, the envelope composition or any of many other attributes of a building. As a result, the simulation and analysis becomes part of a building information model. Today there are many ways that internal or external BIM BPS can leverage this information to inform design processes and understand operational impacts. Using simulation and analysis provides insight to improve and coordinate building design, operations and construction.

The authors of the *BIM Handbook* establish that building information modelling is "not a thing or a type of software" (Eastman et al. 2008). This distinction is important, since a model, being a thing, can then be leveraged for BPS. Analysis and simulation completed as part of BPS could and should then be part of the BIM effort.

Eastman et al. (2008) additionally identifies that BIM allows building architects, engineers, operators and owners to create "an accurate virtual model of a building . . . the computer-generated model contains precise geometry and relevant data needed to support the construction, fabrication, and procurement activities". The model can represent many different design and operational objectives and goals.

Building performance simulation is the computational modelling of a building system simulated to provide approximations of the modelled system's behavior. The results help us understand more about what is simulated, such as "heat and mass transfer in the building fabric, airflow in and through the building, daylighting, and a vast array of system types and components" (Hensen and Lamberts 2011). Hensen and Lamberts (2011) posits the benefits of BPS in practice, recognizing that predicting and analyzing building behavior is more efficient and economical than resolving these issues once the building is built.

Offering a perspective from the practice of building design and engineering, BPS is leveraged in the many ways for design and validation throughout the many different stages of project delivery. A few examples are:

- Operational energy use
- Daylighting and illumination
- Airflow
- Structural evaluation
- Acoustic performance
- Envelope hygrothermal performance
- Building lifecycle analysis
- Material and product quantities
- Construction cost estimating
- Construction processes, 4-D, 5-D modelling
- Building operations

These items illustrate the utilization of information in BIM, for the purpose of BPS. Outcomes from the examples can aid in understanding building performance for design and operations.

Current BIM

Understanding BIM as integral to BPS for building design and operation necessitates an understanding about the flow of information. To begin, this section discusses how data is generated by BIM and how that same data is exchanged through various platforms, a process typically referred to as interoperability. Next, this data is used by researchers and professionals in BPS to highlight the variety of ways that BPS relates to BIM. Data truly becomes information once it is analyzed through BPS. This transformation is crucial to gaining insight from a BIM model.

Data is input into a building model representing the physical and functional characteristics of a building. Commonly this begins with standard or custom object-based definitions of building components (walls, doors, windows) and project templates based on industry standards, professional agreement and company policy. Built geometry relates to the programmatic space needs and the necessities of enclosure and use.

In BIM objects, spaces and information are based on parameters and rules as a method to integrate them into a comprehensive model. Associations between various objects and parametric rules define the geometric and non-geometric aspects of a building. Since the objects are parametric, they can change automatically based on modifications to the building or associated objects. Users embed the parametric objects with properties and attributes that become the information in BIM. These attributes allow interfacing with other applications for BPS.

Once information is embedded in the model, this information can be shared. Interoperability is the flow of information from one model to another, from BIM to BPS. Different types of data are more interoperable than are others. The National BIM Standard (NIBS 2018) and the International Organization for Standardization (ISO) level established the Industry Foundation Class (IFC) facilitating interoperability as "the ability of two or more systems or components to exchange information and to use the information that has been exchanged". The IFC is a neutral exchange format for interoperability, therefore allowing the exchange of information from one software to another. Important to BPS is the way that the BIM defines "structural and energy analyses, cost estimation and schedule the construction, fabrication issues for each subsystem, plus much more. Interoperability identifies the need to pass data between applications, and for multiple applications to jointly contribute to the work at hand" (Eastman et al. 2008). Therefore, an interoperable BIM eliminates the need to replicate data input to facilitate smooth workflows and automation with BPS.

Ideally then the challenges of creating information and sharing it between BPS and BIM is effortless. However, this is not always the case as often there is a disaggregation of BPS from BIM, "as different applications and model formats across different domains co-exist in a building's lifecycle" (Yan 2011). The information exchange into BPS or into BIM is not clean. BPS is often used to aid in solving for multiple criteria or variables associated with building design, construction or operations. In addition, BPS interrogates aspects of physics, such as lighting, structure and thermodynamics, in relationship to the building. Depending on the methodology chosen to achieve these outcomes, a specific software or analysis approach is needed. Therefore, engaging BPS requires assessing the interoperability of the BIM.

Constructing a BIM ecosystem also depends upon the integrative approach connecting software, information and partners leveraging BIM on a wider spectrum (Grilo and Jardim-Goncalves 2010). Consequently, from the authors' own experience, we propose five interoperable methods linking BIM and BPS that architecture, engineering, construction, owners (AECO) and academic researchers utilize.

- Intrinsic – within BIM software
- Add-on – integrated into BIM with plugins
- Hybrid – direct from BIM to analysis tools and returned
- Adscititious – extract and analyze information from BIM
- Exogenous – completely outside of a BIM

Each represents different aspects within a BIM ecosystem. Intrinsic BIM software is often limited and may only able to integrate certain aspects of BPS, for example structural load calculations or quantity take-offs. Because of these limits, others create add-on items to enhance or utilize information in the software for design, analysis and operations. Teams utilize these customized plugins to leverage information within BIM for their BPS purpose. Tally (Timberlake 2018), for example, leverages intrinsic BIM quantity take-off information for lifecycle analysis (LCA). Another approach is the hybrid, where the software is linked to an online source for information, simulation or analysis, which is how energy simulation platforms Trimble's Sefaira and Revit's Insight 360 work. The information is often one-directional or binary, in which it is sent out and flows back and forth between BIM and an online repository. Expanding further, the adscititious analysis builds on extrinsic information from the BIM for BPS, often then communicating externally with the BIM. An example here is dRofus – an external database of information that is flowing in and out of the BIM live, allowing multiple stakeholders access to information simultaneously. Finally, the exogenous approach lies outside of BIM, not interoperability. Here there is no connection, and engaging BPS is wholly a separate endeavour from BIM, such as energy modelling in eQuest, where the entire building representation is recreated to achieve BPS results. These five methods describe ways software may operate and integrate BPS and BIM.

BIM in building lifecycle

Phases of a building's design and operation encompass programming, conceptual design, schematic design, design development, detailed design, construction documentation, construction administration, construction, commissioning and operation phase. Ideally, BIM occurs throughout this lifecycle, offering advantages for BPS through collaboration, parametric tools and construction-level information.

Collaboration occurs between project participants such as the architects, engineers and constructors to reduce errors and field changes, leading to an efficient and reliable building delivery process that reduces project time and cost (Eastman et al. 2008). Owners and managers of buildings benefit from this collaboration improving the value of a project. Traditional project delivery can suffer from lags and losses of information between the various lifecycle phases. Requiring BIM usage and early team engagement can reduce if not eliminate this loss. Once a project is built, owners continue to benefit from being able to use the BIM for operations, facility improvements and future renovations.

Parametric design tools aid in early and late phases of design and documentation. As part of BIM, the parametric features allow definition and measurement of the physical or real building traits. Analysis of functional aspects of building program, construction and operating cost constraints along with environmental considerations allows planning and coordination of these various systems. Design teams benefit from high-level collaboration; Eastman et al. (2008) explains that parametric tools aid in strengthening the design decisions by providing quick feedback on complex problems.

Construction-level information embedded during the BIM design phases speeds up the documentation and delivery process. The model itself represents details of construction integrating design, fabrication, document review and installation. During construction, the BIM continues benefiting the collaborators by integrating their efforts during design into construction and throughout operation.

As an example of collaboration using parametric tools and construction-level information, the following part of this chapter describes a case study, computational methods, examples of the application, discussion and a synopsis of the relation/interaction between BIM and BPS. The concluding synopsis highlights BPS and BIM workflow and demonstrates their inter-relationship, expanding on how simulation and analysis support design and construction activities.

Scientific foundation (case study)

The University of Pennsylvania is building a $1.5 billion new hospital on Penn Medicine's West Philadelphia Campus. The Pavilion (Figure 11.1) is being designed and constructed by HDR, Foster+Partners, BR+A, L.F. Driscoll and Balfour Beatty using the integrated project delivery model. The building will house 500 private patient rooms, 47 operating rooms and an emergency department in a 1.5 million square foot facility. Built on the site that was previously occupied by Penn Tower and its parking garage, this prominent urban space on the Hospital of the University of Pennsylvania (HUP) campus will link HUP to Perelman Center for Advanced Medicine. This will be the largest capital project in Penn Medicine's history and Philadelphia's most sophisticated and ambitious health care building project, expected to be completed in 2021.

Figure 11.1 Rendering of The Pavilion

Source: Courtesy of PennFIRST

Computational methods

Currently, traditional cost estimating is time consuming and difficult on large projects, especially during conceptual design when decisions are fluid and few things are considered "final". Several weeks of manually calculating square footages and making assumptions on design intent, sequencing and market costs yields a gross number with a generous margin of error. The cost estimating team uses developed plans, sections and elevations, mechanical/electrical/plumbing/structural (MEPS) system selections, material choices and unique project conditions, e.g. tunnels, bridges and underground parking, to develop an overall cost estimate. During the later phases of traditional delivery, project teams may even stop work during the cost estimating phases to avoid moving ahead at risk.

Therefore, a future state reliant on integrated project delivery (IPD) methods and the associated metrics have a profound impact on the design process. Projects designed and built using IPD share both risk (uncertainty) and reward (profit). Cost certainty, a dynamic metric, is closely tied to overall team performance, requiring design teams to gather detailed feedback on multiple design options much earlier in the design process. Since dynamic metrics are inherently twitchy, even with a project of this scale and complexity, teams must either create detailed two- or three-dimensional representations of conceptual design intent, typically a very labor-intensive task, or associate data to simplified representations of design intent. Decisions made fall into the same categories listed above, with the team using target value design (Macomber and Barberio 2018) principles to refine iterations and expedite stakeholder decision-making. Rapid, accurate cost feedback during the conceptual design phases is paramount, since integrated project teams can be penalized by schedule if they slow down or pause between design iterations or even project phases.

As outlined in the previous section, expediting cost estimates during the conceptual design phases requires dedicating significant resources to the digital representation of elements often reserved for later phase BIM, e.g. explicitly modelling lifts in a patient room compared to a yes/no toggle for a patient lift in lieu of modelling. From the data perspective, both approaches yield the same level of detail and result in the same level of accuracy for the cost estimate.

The same third-party cost information is associated to either geometry or a data field, the key difference being the process of association. The opportunities and constraints of each approach become more apparent when teams simultaneously carry multiple design options through a single phase. The traditional approach of representing objects or items in BIM may move faster on day one as items are modelled and seen on screen or paper immediately, but it quickly loses effectiveness when wholesale changes need to be made or previous iterations are scrapped entirely. Data-driven design necessitates a deliberate approach, with a focus on front-loading information and reducing the time between critical feedback sessions. Rather than immediately jumping into what it takes to represent a data point, the team must take a critical look at the key decisions needed at each step in the project delivery process. In the case of the project team, using BIM and BPS inspired new tools to fill the gap between detailed digital representations and data-rich design options.

Application examples

For The Pavilion, HDR's Computational Design team developed a solution that utilized McNeel's Rhinoceros (Rhino) software and David Rutten's Grasshopper visual scripting interface to assign data keys and values to geometric objects, rendering the resulting data on screen in a live heads-up display (HUD). This customized process allowed the project team to quickly create data-rich schematic models without the need for extraneous modelling, interoperability workflows or extensive coding experience. Much like the metadata associated with a digital photograph, e.g. date, time, device type and resolution, geometric items could by modified and adapted without the loss of important data fields. By maintaining data integrity and structure within the model, the impact of change was shown live on-screen, and screenshots of the model and its associated HUD were used in project reports.

The IPD team included several representatives with construction scheduling, sequencing and cost estimating expertise. By teaming with these subject matter experts to create a seamless data connection between their knowledge and the design model, the project team was able to reliably utilize cost information and mitigate risk.

Process

The successful adoption of diverse construction metrics into design models necessitates a change in design methodology and process. Project teams can realize higher value and minimize uncertainty by using real-time data assignment, queries and visualizations. By leveraging these concepts, the project team was able to move the project forward with more cost and schedule certainty, without sacrificing design excellence.

A – Assigning Data. By incorporating object data into design models, project teams leverage computational processes and data structure adopted by many other industries. As design concepts are optimized to create a solution that addresses multiple constraints, the level of detail determined by the user should be consistent with the level of data used to drive decisions. The project dashboard, shown in Figure 11.2, breaks down design options by program areas, unit

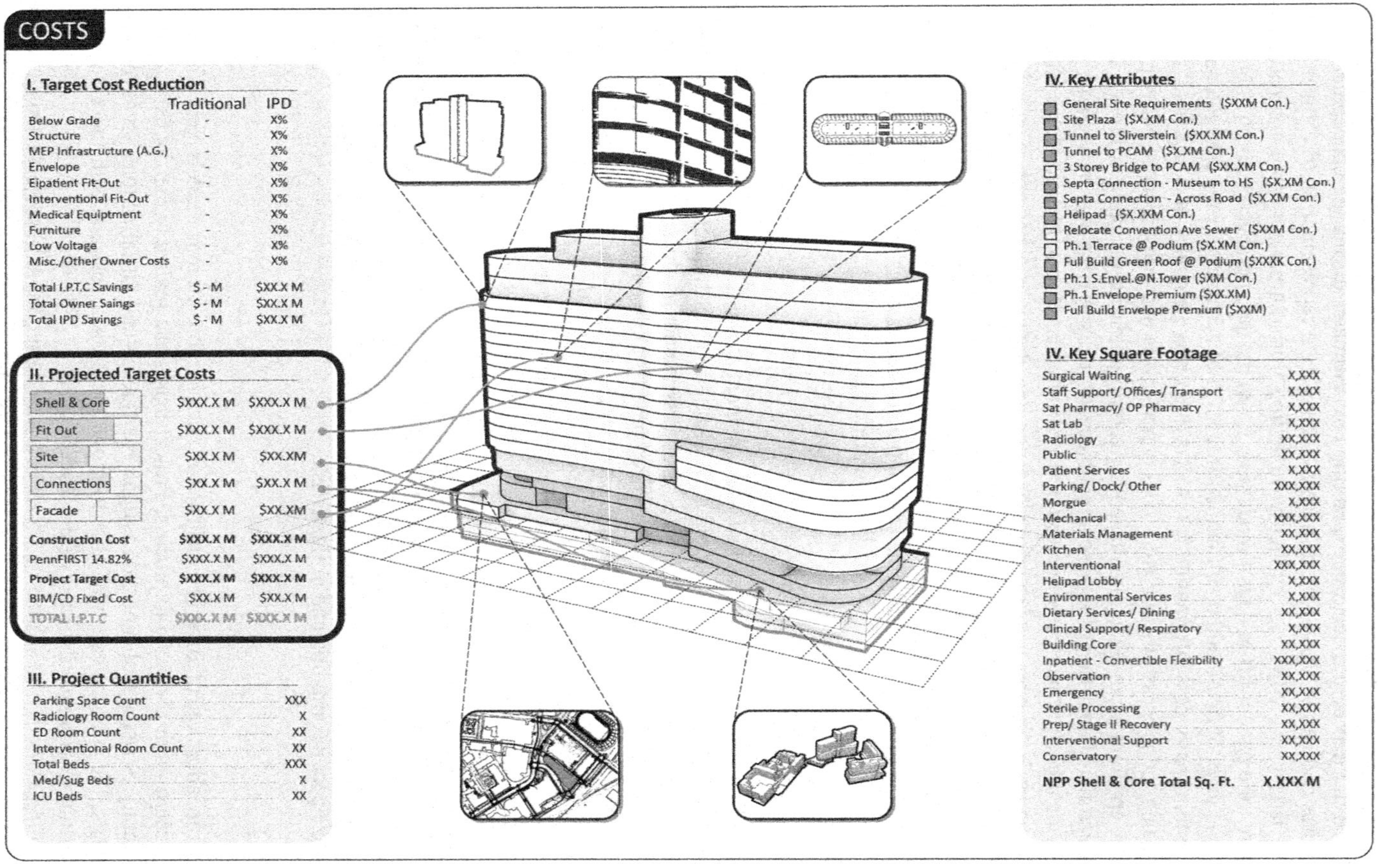

Figure 11.2 Dashboard for cost visualization

Source: Courtesy of ©HDR, Inc.

costs, delivery savings, façade types and unique conditions – each metric selected by the IPD team to simulate cost performance throughout the lifecycle of the project.

B – Querying Data. Object data is useful only when the data can be standardized, structured and quickly queried. Figure 11.2 shows the many ways design teams can visualize object data once it is logically structured. This real-time visualization of object data queries becomes the heads-up display.

Recording object data using strictly manual take-off methods is tedious and resource-intensive, with a higher probability of error. Assigning cost data to objects is not a holistic replacement of traditional cost estimating, but rather a computationally aided enhancement that can be leveraged to enable stronger overall team performance.

C – Reviewing Data. By querying object data and reviewing the results in an interactive fashion, manual "fact-checking" can occur at a much higher rate. Representatives from each specialty within the integrated team work in an open setting where design iterations are developed, discussed and modified in real time and supported by instantaneous data feedback and expert review (Figure 11.3). This creates an environment of transparency for decisions to be made with all key stakeholder feedback included.

The object-enriched design model and HUD are used as the primary source of data review to maintain schedule certainty. When further detail or investigation is necessary, all design options and the associated data contained are exported for breakdown and analysis by the cost estimating team. As more detail is created within the model, the cost estimating team can provide additional information that can be imported into the model.

Assigning, querying and reviewing design data in quicker feedback loops enables integrated project teams to increase the value of design services by moving quickly with a high degree of cost certainty. Each unique aspect of the design process is refined using both art and science, encouraging creative solutions that solve multiple issues and enabling teams to reconcile potential conflicts through communication, transparency and iteration.

Figure 11.3 Design team reviewing data

Source: ©HDR, Inc.

Discussion

Because there is not a singular approach to successful integrated project delivery, companies like HDR innovate. These innovators, according to Bohlen and Beal (1957), fall on Rogers' bell curve for innovation adoption lifecycle as an early majority adopter. Traditionally, large architecture/engineering/construction (A/E/C) companies are significantly more risk-averse than companies of similar size in other industries, with a much lower investment into research and development. This deliberation can lead to short-term stability but threatens the long-term value proposition of those services if adjustments to process and investments into technology are not encouraged and explored.

In the context of the industry, disruptions may take a significant amount of time to gain traction and realize results, but once new technology and methods are learned, changes become permanent and what was once considered a radical concept becomes standard practice. This concept is illustrated by using the S-curve as the idealized general form of all learning curves. Slowly accumulating smaller steps at first are followed by larger steps and then successively smaller ones later as the learning activity reaches its limit and saturation occurs within an industry (Ritter and Schooler 2002).

Parallels between software development and alternative delivery models can be drawn when models like IPD are viewed as a product. Typically, software goes through an extensive internal development and testing cycle before public release. In an IPD setting, project teams are challenged to set the correct targets and goals for the project that will determine performance. In each scenario, design teams are required to frontload information in BIM and test with BPS, refining the proposed solutions based on subject matter expert feedback and focus group testing. Each team member must buy into the change in design process to maximize the reward potential and minimize risk. As adoption drives value, value drives innovation and innovation begins the adoption lifecycle once more.

Synopsis: the relation/interaction between BIM and BPS

The Pavilion is a case study connecting BIM and BPS for the purposes of analysis, development and adoption of new tools and processes to support design solutions, build consensus among stakeholders and exceed team performance expectations. Modelling with data in a BIM ecosystem enables integrated project teams to be more nimble and responsive to key stakeholders much earlier in the design process, but also requires the development of new BPS tools and the adoption of a BIM design methodology. As the lines between the traditional design phases dissolve and deliverables become more fluid, it becomes paramount to critically analyze the needs of each project and frontload data inputs, establishing a BIM ecosystem. Collaborators, parametric design and construction information align key decisions with methods for building visualization and reporting expanding the value of AEC services.

The demand for target value design, integrated project delivery and performance-based contracts is being driven by the client side of the industry and will continue to evolve over time, as metrics for success are developed and tested within the market. To realize the most potential within this market, project teams will need to integrate specialties, data and design methodologies within BIM for BPS. While it may be counter-intuitive to delay design iterations when delivering projects faster is a priority, the avoidance of re-work and uninformed decisions using BIM increases the probability of a higher quality product being delivered on time and under budget. From the technical requirements of computation to the business strategy of each client, BIM and BPS are intertwined throughout any truly integrated project, ultimately creating a fundamental shift in how projects are designed, delivered, built and occupied.

References

Bernstein, P. and Jezyk, M. (2011) Models and measurement: Changing design value with simulation, analysis, and outcomes. In Noble, D. and Kensek, K. (eds), *BIM in Current and Future Practice*. Hoboken, NJ: John Wiley & Sons, p. 89.

Bohlen, J.M. and Beal, G.M. (May 1957) The diffusion process (PDF). Special Report No. 18. Agriculture Extension Service, Iowa State College 1: 56–77.

Eastman, C., Teicholz, P., Sacks, R. and Liston, K. (2008) *BIM Handbook: A Guide to Building Information Modeling for Owners, Managers, Designers, Engineers, and Contractors*. Hoboken, NJ: John Wiley & Sons, Inc.

Grilo, A. and Jardim-Goncalves, R. (2010) Value proposition on interoperability of BIM and collaborative working environments. *Automation in Construction* 19(5): 522–530.

Hensen, J. and Lamberts, R. (2011) Introduction. In Hensen, J. and Lamberts, R. (eds), *Building Performance Simulation for Design and Operation*. London and New York: Spon Press.

Kieren Timberlake. (2018) Tally® LCA App for Autodesk® Revit®. www.kierantimberlake.com/pages/view/95/tally/parent:4.

Macomber, H. and Barberio, J. (2018) Target-Value Design: Nine Foundational Practices for Delivering Surprising Client Value. www.lcicanada.ca/wp-content/uploads/2015/11/3-Target-Value-Design-LPC.pdf.

National BIM Standard – United States® Version 3 ©2018 National Institute of Building Sciences buildingSMART alliance™.

Ritter, F.E. and Schooler, L.J. (2002) The learning curve. In *International Encyclopedia of the Social and Behavioral Sciences*. Amsterdam: Pergamon, pp. 8602–8605.

Yan, W. (2011) Parametric BIM SIM: Integrating parametric modeling, BIM, and simulation for architectural design. In Noble, D. and Kensek, K. (eds), *Building Information Modeling*. Hoboken, NJ: John Wiley & Sons, Inc.

Recommended reading

Nobel, D. and Kensek, K. (eds) (2011) *Building Information Modeling: BIM in Current and Future Practice*. Hoboken, NJ: John Wiley & Sons, Inc.

Hemsath, T.L. and Banhosseini, K.A. (2018) *Energy Modeling in Architectural Design*. New York and London: Routledge.

Activity

Each recommended reading provides examples of information relevant to buildings. Diagram below or on a piece of paper using an example from the reading where this data becomes part of a BIM.

12 Building simulation for policy support

Drury B. Crawley

Scope

This chapter discusses the use of simulation in supporting building policy decision-making. Examples of building simulation for policy setting in this chapter include development and evaluation of building energy efficiency standards, setting building performance levels for utility incentives, evaluating potential impacts and direction for energy efficiency building retrofit programs, and sector-wide technology performance studies. The critical role of building simulation in policy making is emphasized.

Learning objectives

The reader is introduced to use of building simulation to embody the performance of a portion of the building stock to evaluate options and directions for building policy. The reader should understand the key components of multi-building model policy studies including energy, environmental, and financial decision-making.

Key words

Policy analyses/building stock/building energy standards/R&D planning/building technology assessments/policy analysis methodologies/sample design

Introduction

Building performance simulation is regularly used to support decision-making in the design or retrofit of individual buildings. Yet, one of its most powerful uses lies beyond the performance of individual buildings in supporting building policy setting and decision-making: to evaluate performance levels for minimum standard regulations, assess the value of improved building performance for utilities or governments, or support high-level, public decision-making. Examples of policy targets include:

- defining a cost-effective performance level for various aspects of a new building energy efficiency standard,
- evaluating the performance of an existing or proposed building energy standard,
- establishing financial incentives for improved building energy performance, i.e. the value to the utility or political entity,
- evaluating the potential impact of and direction for voluntary programs encouraging energy-efficient new building design or existing building retrofit,
- evaluating the applicability of specific technologies or systems for the new building design or existing building retrofit markets,

- evaluating the potential for introduction of renewable energy technologies at the building, community, or regional level,
- evaluating the potential impact of changes in community, regional, national, or international policy,
- evaluating requirements for existing or new energy supply at a community, regional, or national level, and
- evaluating the potential impact on building performance of changes relating to environment.

Coupling simulation with building models, designed to represent a range of building types and locations, can embody a portion of a building stock (existing or new, domestic, public, commercial, industrial, large, medium, or small) or the entire stock, which allows discrete modelling of building policy objectives and trends. Simulation provides policy decision makers a means of assessing what-if scenarios across a spectrum of buildings, ensuring that regulations and policy are set at the most financially and environmentally beneficial levels for individuals and the public.

In the context of building policy, performance simulation has been in wide use for more than 40 years in North America. These include evaluating the performance of new or proposed building energy standards, evaluating the value and design of energy savings programs for a utility or government entity, and evaluating the potential impact on the built environment of actions relating to climate change. Swan and Ugursal (2009) provide a comprehensive review of the various modelling techniques used for modelling residential sector energy consumption, and they note,

> Two distinct approaches are identified: top-down and bottom-up. The top-down approach treats the residential sector as an energy sink and is not concerned with individual end-uses. It utilizes historic aggregate energy values and regresses the energy consumption of the housing stock as a function of top-level variables such as macroeconomic indicators (e.g. gross domestic product, unemployment, and inflation), energy price, and general climate. The bottom-up approach extrapolates the estimated energy consumption of a representative set of individual houses to regional and national levels, and consists of two distinct methodologies: the statistical method and the engineering method. Each technique relies on different levels of input information, different calculation or simulation techniques, and provides results with different applicability.

Swan and Ugursal say that the most appropriate use of these three approaches in modelling residential buildings corresponds to their strongest attribute:

- top-down approaches for supply analysis based on projections of energy demand,
- bottom-up statistical techniques for determining energy end-uses including behaviour based on energy bills and simple surveys, and
- bottom-up engineering techniques for explicitly calculating energy end-uses based on detailed characteristics, enabling impact of new (or alternative) technologies.

This chapter reviews related commercial building research in support of building policy where comparison of alternative technologies, systems, or climatic response is required – a bottom-up engineering approach – building energy standards, building stock models, utility incentives, and climate modelling.

Evaluation of building energy standards

Building energy standards are a set of definitions of minimum performance of building components and equipment. They can be as simple as a table of minimum efficiency levels for a packaged air-conditioning unit, minimum insulation levels for walls and roofs, or maximum lighting power density. Building energy standards also can be performance-based, requiring that a new or existing building has energy consumption no higher than a prescribed value. But where does building performance simulation fit into building energy standards? Primarily, performance simulation has been to evaluate the energy savings associated with a proposed standard (as compared to typical building practice or existing standards).

National building energy standards date back to the 1970s. For example, ASHRAE developed Standard 90–75 (ASHRAE 1975) to improve the efficiency of buildings in direct response to the oil embargo of the early 1970s. In the late 1970s, the US DOE proposed Building Energy Performance Standards (BEPS) (US DOE 1979) that set maximum energy performance levels for both residential and commercial buildings. A key feature of both the development and the proposed standards was the innovative use, at that time, of building performance simulation. Characteristics of typical buildings were collected throughout the United States, and baseline simulation models were constructed for more than 20 building types. Simulations were then performed and target levels established. Designers could then use building simulation models to demonstrate that their building performed at an energy level less than the maximum for that building type. Unfortunately, BEPS was too far ahead of the market and was buried under a landslide of adverse public comments.

Subsequent research by Pacific Northwest Laboratory (PNL 1983) used building simulation to propose aggressive updates to the then current Standard 90A–1980 in ASHRAE Special Project 41 (SP41) (Pacific Northwest Laboratory 1983). The SP41 proposals were based on life-cycle cost evaluation of the improved energy performance, evaluating ten buildings in six locations. This research was used as the starting point for the development of the next major update of Standard 90.1 in 1989 (ASHRAE 1989). Many of the public review drafts were tested by Pacific Northwest Laboratory for US DOE from 1985 through 1989, using five building prototypes and five locations (weather data) to evaluate energy savings levels. Interim energy simulation results were reported by Crawley and Briggs (1985a, 1985b), US DOE (1987), and Crawley and Boulin (1989), among others. A regional evaluation of state building energy codes in comparison with Standard 90.1–1989 was reported by Johnson et al. (1988). The five building prototypes were developed to be typical of commercial buildings rather than representative of the entire stock of buildings. With only five building prototypes and five weather locations, the results from these studies could not be extrapolated to represent the value of potential savings for the new standard across the entire commercial sector.

Carlo et al. (2003) reported on an analysis of 12 prototype buildings to establish minimum requirements for a building energy-efficiency code in Brazil. While the code was based on Standard 90.1–1999, simulation allowed the researchers to customize the requirements to the climate and economic situation in Brazil. The goal was to create a multi-variable regression equation from the simulation results for calculating the annual energy consumption of a building in that location. The regression equation was tuned to that location with 10 coefficients. With 12 widely varying prototype buildings (in floor area and number of stories), this required 1,616 combinations of envelope characteristics to be simulated. Carlo proposed to replicate this analysis for other Brazilian locations – requiring a similar number of simulations for each. Subsequently, Carlo and Lamberts (2008) reported on the impacts on electricity consumption of various envelope characteristics using six prototype buildings. Carlo developed a series of equations from simulation results which define the envelope efficiency level through a comparison of the proposed building envelope with a low and high efficiency level. These equations are the basis of new building energy efficiency regulations in Brazil.

Recently, the European Union (EU) has established a Directive on the Energy Performance of Buildings (CEC 2001), which requires member states to create and deploy calculation methods for rating energy performance for buildings larger than 1,000 m^2. This rating has a scale of

A to H, where A is the lowest-energy and H the highest-energy buildings. Throughout Europe, this directive is requiring the use of building performance modelling of many more buildings than in the past. In the United Kingdom, the Netherlands, and a few other countries, the building performance modelling required to rate non-domestic buildings has been implemented in a simple set of equations for predicting energy use, known as Simplified Building Energy Model (SBEM) (BRE 2008a, 2008b). Rather than a model of energy use of a specific building, the SBEM is a comparative model for rating the relative carbon emissions of the proposed building. The SBEM uses monthly utilization factor calculations and is limited in the technologies that it can evaluate.

Building models, building stock, and utility incentives

In a policy framework relating to building energy performance, the targets are individual buildings, but there is a need to establish how the proposed policy or standard will perform in a broader context – of new or existing buildings within a region or country. This requires an understanding of the building stock:

- the number and size of buildings being built and their operating and thermo-physical characteristics;
- the number and size of existing buildings and their operation schedules, energy consumption, energy costs, and thermo-physical characteristics.

Internationally, detailed data relating new or existing commercial building stocks to actual building energy performance are relatively scarce in comparison to what Swan and Ugursal found for the residential sector. Macmillan and Köhler (2004) found that national residential and commercial sector-level energy use data were generally available throughout the world. But they also found that detailed data on the building stock and its energy use was very limited – citing studies in North America and Europe. Barrett (2009) supports this in a review of recent building stock research – saying that papers "concentrate on domestic buildings in detail with less about the non-domestic buildings." Barrett goes on to say that the non-domestic sector accounts for a large and growing proportion of energy use (as evidenced in Figure 12.1), but the literature he reviewed provides few details of this growth. He asserts that this is due to the more heterogeneous nature of the commercial building stock and that empirical data is sparse.

In the United States, the Energy Information Administration has conducted quadrennial energy consumption surveys since 1979 for commercial, industrial, and residential buildings. These surveys provide a wealth of information about the numbers and consumption of the entire building stock based on a statistical representation of the sector and the representativeness of several thousand buildings for each sector. The Commercial Building Energy Consumption Survey (CBECS) comprises survey data on more than 6,700 US buildings (EIA 2015). But these data are not complete thermo-physical models of each building. For example, there is little detail on the energy-consuming systems or the thermal characteristics of the building envelope. Further, EIA masks certain data such as number of workers, number of floors, and floor area in large buildings (more than 20 stories and/or 45,000 m^2) so that individual buildings cannot be identified. Griffith et al. (2008) discusses the limitations of using the survey data in the context of a bottom-up research study – requiring supplementing the data with assumptions, defaults, data from other literature, and probabilistic assignments.

Recently, Natural Resources Canada has published a similar commercial building survey for Canada (NRCan 2006), the Commercial and Institutional Consumption of Energy Survey (CICES). CICES includes information on 7,349 buildings, but it focuses on the floor area,

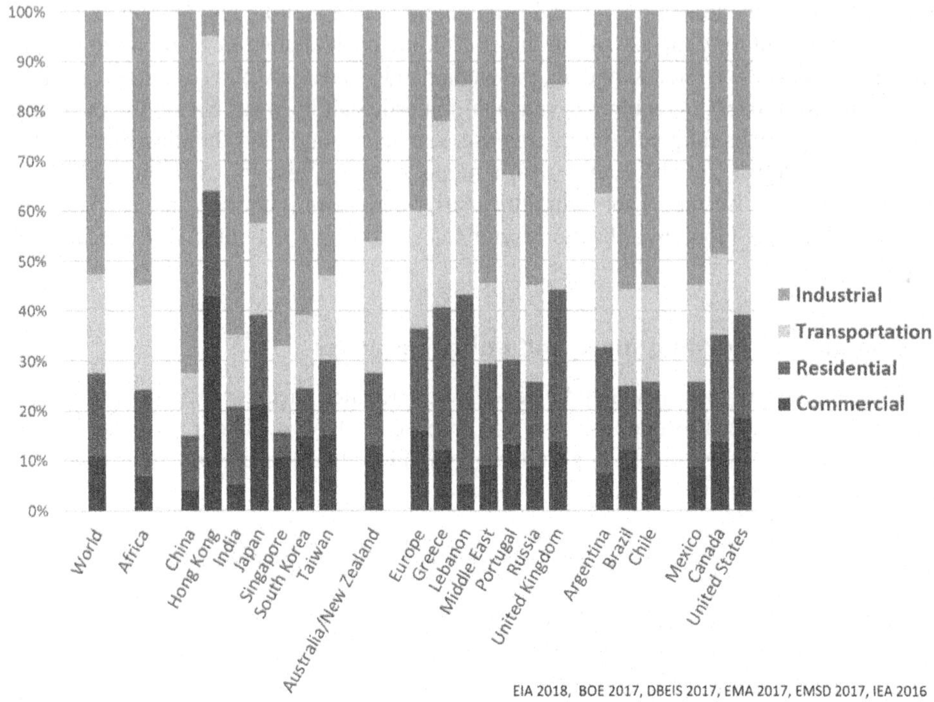

Figure 12.1 Percentage energy use by sector for the world and selected countries

Sources: EIA 2018, BOE 2017, DBEIS 2017, EMA 2017, EMSD 2017 and IEA 2016

building type, and the forms and amounts of energy used in the survey year. Key information such as the number of floors or any information about the building envelope or heating and cooling systems is not collected. This limits the use of the CICES as a source of input for building simulation.

California has created a similar survey with more details on the surveyed buildings (PG&E 1999). This survey was recently updated and extended state-wide as reported by Kinney and Piette (2002). The updated survey included the collection of sufficient building characteristics so that calibrated simulation models of each of the more than 2,800 commercial buildings could be created (Ramirez et al. 2005).

The Building Owners and Managers Association (BOMA) publishes an annual report on US office markets (BOMA 2018). BOMA publishes information obtained from more than 3,000 office buildings including income, expenses, energy consumption, rent, and occupancy rate. But like the surveys noted above, the minimum geometric and other key information required for simulating the performance of those buildings is not collected by BOMA.

In Scotland, the Energy Systems Research Unit (ESRU) Domestic Energy Model (EDEM) (Clarke et al. 2003, Clarke et al. 2008, ESRU 2008) supports energy policy formulation for the residential sector. It provides representations of the entire domestic sector for Scotland, which allows policy makers to quickly apply a wide range of improvements analytically and evaluate which will provide the best energy performance or reduction in carbon. Analyses performed include national housing stock fabric upgrade strategies, local community upgrade strategy and carbon roadmap formulation, carbon and energy performance, and energy labelling in compliance with the EU Energy Performance of Buildings Directive. But the EDEM is limited to the domestic sector. An equivalent model for the non-domestic sector is not available.

The International Energy Agency (IEA) also publishes energy use data for buildings in member countries (IEA 2017), but this is a statistically disaggregated estimate, a top-down inference based on energy production and supply data. Little information is available on the average energy performance of buildings by type or climate zone.

While survey data are extremely useful for a building stock snapshot, they usually are limited by the information collected. Only rarely are there enough data to create a building model in energy simulation software. Several projects have worked to use existing survey data to create prototypes that represent large portions or building types and regions. Briggs et al. (1987, 1992) created 20 existing office building prototypes, and Crawley and Schliesing (1992) created 10 new office building prototypes for use by the Gas Research Institute in research and market assessment. The existing building models were based on the 1979 Nonresidential Buildings Energy Consumption Survey (NBECS), the first survey in the CBECS. The new building models were based on Standard 90.1. In both cases, substantially more input data were required for the models, and sources are documented in their reports. Huang et al. (1991) extended the prototypes to the entire commercial sector for 20 urban markets in the United States. Huang and Franconi (1999) updated the prototypes to evaluate the contributing components of commercial building loads. In all cases, these prototypes were limited by the assumptions that the authors had to make to create complete simulation models. Often these assumptions were not well documented.

Rather than producing prototypes to represent multiple buildings, Griffith et al. (2007, 2008) created building simulation models for each of the more than 4,000 buildings in the 2003 CBECS. These building models were developed first as a baseline with energy features and performance consistent with Standard 90.1–2004 (ASHRAE 2004). Then low-energy technologies and renewable energy systems anticipated to be available by 2025 were applied to the models. The result was an assessment of the technical potential for achieving zero-energy buildings throughout the commercial building sector in the United States. The conclusion was that the zero-energy building goal could be, on average, achieved in the commercial building sector. Building types which can reach the zero-energy target easily are offices, warehouses, schools, and retail. Restaurants, hospitals, and other energy-intensive building types will be the hardest to bring to net zero energy. While not creating prototypes improved the breadth and representativeness of the analysis, it also created a logistic and quality assurance issue – how to deal with multiple thousands of building models. Griffith used an XML schema associated with a database of inputs to define each model. Since the inputs were derived directly from the CBECS data or related assumptions, the inputs could be automatically verified before a series of simulations were started.

Deru et al. (2011) published a set of 16 benchmarks for new commercial buildings based on the 2003 CBECS and ASHRAE Standard 90.1–2004. The benchmark buildings' descriptions include a scorecard comprising detailed documentation of all inputs and assumptions as well as information and data sources. Each benchmark building also has a corresponding EnergyPlus input file (US DOE 2018). These are regularly updated for analysis of the impacts of new versions of Standard 90.1 (PNNL 2017).

For utilities, the economic benefit of incentives paid for improved energy efficiency is often obscured by their regulated environment and the complex valuation of equitable sharing of net benefits, cost capitalization, and risks with their stakeholders. Rather than evaluating individual incentives, they aggregate the energy efficiency measures into a portfolio, essentially a top-down approach to the utility sector. For example, Cappers et al. (2009) present an analysis of energy efficiency incentives in the context of a prototypical investor-owned utility. This analysis adapts a spreadsheet-based financial model (known as the Benefits Calculator) developed in support of the National Action Plan for Energy Efficiency (Jensen 2007).

The literature is full of similar portfolio-centric approaches to energy efficiency with a focus on program costs, costs and benefits to ratepayers, and similar utility rate case calculations.

Many consultants offer services to develop and evaluate incentives for utilities but few publish in the peer-reviewed literature, which may indicate that they see their utility incentive calculation methodologies and procedures as a business advantage. Instead, utilities focus on supporting their customers and administering the incentive programs. Ter-Martirosyan (2003) says that this is due to changes which began in the late 1980s, with public utility commissions and other regulators focusing on regulation of the incentives themselves rather than the rate of return the utility could earn. For utilities, this has a two-fold result – decreased quality and cuts in service (as described by Ter-Martirosyan) and less interest in detailed engineering approaches that are overwhelmed by their heavily regulated environment.

Another critical issue for simulation for building policy is which climate data to use. Climate data issues and resources are discussed in Chapter 6.

Data issues in multi-building studies

The largest challenge of research involving large numbers of simulation models is dealing with the immense amount of data that building performance simulation programs can create. For example, a single EnergyPlus simulation with 10-minute time step output can easily produce more than 600 megabytes of data. Handling the data from hundreds or thousands of simulations can quickly overwhelm a researcher.

In their studies of Standard 90.1 in the 1980s, Crawley and Briggs (1985a, 1985b) and Crawley and Boulin (1989) created five building type DOE-2.1E models for the commercial sector. They were based on the earlier models created for ASHRAE Special Project 41 and included variants tied to five climate regions throughout the United States. Three variations on the base model were created – Standard 90–75, Standard 90A–1980 (ASHRAE 1980), and the proposed draft standards, which became Standard 90.1–1989. This large number of simulations (5 buildings × 5 climates × 3 standards) required that the structure of the analysis be carefully designed, including a systematic file-name convention that included abbreviations for building type, climate zone, and variant of the standard. The researchers also created batch scripts to automate running the simulations and extracting summary energy results. When the ASHRAE committee developing the standard created three drafts in three years, this structure and automation proved invaluable. When Johnson et al. (1988) studied the potential for upgrading building energy standards in the Pacific Northwest, they quickly adapted the 90.1 scripts and input files and completed the analysis in less than two months.

In the study for the Gas Research Institute described above, Briggs et al. (1992) chose to create a spreadsheet that included the key characteristics' definitions for the building prototypes and all the parts of the simulation input files. Invoking a macro in the spreadsheet automatically generated the input files for the existing office building prototypes. This provided the added advantage of having all the input in tabular form, ready for insertion into the research report. A further advantage was that the prototypes could be modified easily and the data checked carefully.

When Huang and Franconi (1999) created new prototypes, they constructed these as snippets of simulation input files. The user simply invoked a batch script that assembled the correct parts into a complete input file.

Griffith et al. (2007, 2008) took a more robust approach in studying the technical potential for achieving zero-energy commercial buildings. Because this study involved creating a building simulation model for each of the more than 5,000 buildings in the 2003 CBECS, it necessitated the development of a method for automatically creating the input files. Griffith created a set of high-level XML definitions for all the key inputs, such as floor area, weather data, floor-to-floor height, lighting power intensity, and HVAC system. He created a specific part of the input file for each of these XML definitions. When the XML scripting tool was passed an XML key, a specific input file could be created. Further scripting tools were created to automatically submit simulations on multi-core computer clusters and extract summary results from the completed simulations.

Creating a structured approach to studies involving multi-building simulations is also vital to quality assurance. Checking the values of individual inputs is not easily accomplished if several hundred or thousand individual input files are involved. Summary tables of the values of key inputs that can be checked easily and then automatically transferred to input files are critical.

Donn (2007) studied the role of quality assurance (QA) in environment design decision support tools for architecture, including a variety of building performance simulation tools for daylighting, thermal design, and acoustics. He found that not only was it critical to have QA procedures as a standard part of any building analysis, it was critical to have QA measures that are codified and incorporated into the simulation tools themselves. These QA measures would be reality tests to examine whether outcomes from the design decision tools behave in a believable manner – like a real building. Donn also proposed the establishment of a shared database on QA performance data available to all design decision tool users.

Synopsis

This chapter has presented an overview of simulation in building policy studies in several areas:

- development and evaluation of building energy standards,
- buildings models and building stock,
- utility incentives, and
- QA and data volume issues in multi-building studies.

For the development and evaluation of building energy standards, having a few prototype buildings that could represent the wider building stock was important. Several of the studies (US DOE 1987, Johnson et al. 1988) had few buildings, and the results could not be extrapolated to the wider commercial building sector. Similarly, using multiple buildings raises the level of complexity, as seen in the work by Carlo et al. (2003), which has implications for data management and QA. On the other extreme, a simple model such as the SBEM can inhibit the adoption of new technologies. It is important to be able to adequately cover the range of characteristics expected for the buildings to be simulated.

In setting policy for commercial buildings, the ultimate targets are individual buildings. But how the proposed policy will perform in a broader context of a region or country or a building type or entire building sector is important as well. Policy makers need to be able to relate individual decisions to their impacts. In that context, this requires an understanding of the target building stock including the size and number of buildings as well as their operating and physical characteristics. Few sources of detailed data below sector-level energy use are available. A few building characteristics surveys were available in North America and Europe, but they all had limitations in the depth or quality of data available for building modelling. The exception was in California, where the commercial building survey was coupled with calibrated building simulation models. Further review found work that filled in the data missing from the surveys with assumptions, rules of thumb, and other data. Often these supplementary data are not described in the reports or other documentation, making it difficult for potential users to fully understand the models. More recent benchmark buildings (Deru et al. 2011) are working to overcome these limitations by fully documenting all input assumptions and sources of data in the building models.

Interestingly, there were very few published studies on setting the level and technical attributes for utility incentives in the peer-reviewed literature. Most utility incentives literature focused instead on the delivery and administration of incentive programs for energy efficiency in commercial buildings.

Several studies noted that despite a substantial increase in the number of locations for which climate data are available over the last few years, limited availability and temporal resolution in many areas throughout the world is still an important issue.

Finally, studies that dealt with a large number of building simulations found several means of automating the creation of building models. Some used spreadsheets, others used scripts, but in all cases, it was critical to having a structured approach for the validation and error checking of individual inputs. This structure forms a critical part of the framework necessary for building research supporting policy makers. Crawley (2008) proposed a generalized framework for policy analysis using building performance simulation. This is shown in Table 12.1.

Table 12.1 Generalized framework for policy analysis using building performance simulation

Parameters			*Policy study*			
			Minimum building energy standard development	*Utility incentives beyond minimum standards*	*Upgrading existing buildings*	*Sector-wide technology or performance study*
Research and policy focus						
Research focus	Existing building	Single building			•	
		Multiple buildings or stock			•	•
	New building	Single building	•	•		
		Multiple buildings or stock				•
Research parameters	Building envelope (walls, roofs, fenestration)		•	•	•	•
	Lighting		•	•	•	•
	Internal loads		•		•	•
	HVAC		•	•	•	•
	Renewable technologies			•	•	•
Building model						
Climate data	Climatic design conditions		•	•	•	•
	Typical meteorological data		•	•	•	•
	Observed hourly data				•	•
Baseline model	Simple prototype thermo-physical model		•	•		
	Existing building prototypes based on building stock				•	•
	New building prototypes based on minimum standards and building stock					•
	Zero- and low-energy prototypes including renewable technologies					•

Source: Crawley 2008

References

ASHRAE (1975) Energy conservation in new building design. *ASHRAE/IES Standard 90–75*, New York: ASHRAE, Inc.

ASHRAE (1980) Energy efficiency design of new buildings. *ANSI/ASHRAE/IES Standard 90A–1980*. Atlanta, GA: ASHRAE, Inc.

ASHRAE (1989) Energy efficient design of new buildings except low-rise residential buildings. *ANSI/ASHRAE/IES Standard 90.1–1989*. Atlanta, GA: ASHRAE, Inc.

ASHRAE (2004) Energy efficient design of new buildings except low-rise residential buildings. *ANSI/ASHRAE/IESNA Standard 90.1–2004*. Atlanta, GA: ASHRAE, Inc.

Barrett, M. (2009) Policy challenges for building stocks. *Building Research and Information* 32(2): 201–205.

Briggs, R.S., Crawley, D.B. and Belzer, D.B. (1987) *Analyses and Categorization of the Office Building Stock*, GRI-87/0244. Chicago: Gas Research Institute.

Briggs, R.S., Crawley, D.B. and Schliesing, J.S. (1992) *Energy Requirements for Office Buildings, Volume 1, Existing Buildings*, GRI-90/0236.1. Chicago: Gas Research Institute.

British Research Establishment (2008a) *National Calculation Method*. www.ncm.bre.co.uk/

British Research Establishment (2008b) *Simplified Building Energy Model Technical Manual*. Garston, UK: British Research Establishment.

Building Owners and Managers Association International (2018) *2018 Office Experience Exchange Report*. Washington, DC: BOMA.

Bureau of Energy (2017) *Energy Statistics Handbook 2016*. Taipei, Taiwan: BEA. https://web3.moeaboe.gov.tw/ECW/english/content/SubMenu.aspx?menu_id=1537.

Cappers, P., Goldman, C., Chait, M. Edgar, G., Schlegel, J. and Shirley, W. (2009) *Financial Analysis of Incentive Mechanisms to Promote Energy Efficiency: Case Study of a Prototypical Southwest Utility*. Berkeley, CA: Lawrence Berkeley National Laboratory.

Carlo, J., Ghisi, E. and Lamberts, R. (2003) The use of Computer Simulation to Establish Energy Efficiency Parameters for a Building Code of a City in Brazil. In *Proceedings of Building Simulation 2003*. Eindhoven, The Netherlands, pp. 131–138.

Carlo, J. and Lamberts, R. (2008) Development of envelope efficiency labels for commercial buildings: Effect of different variables on electricity consumption. *Energy and Buildings* 40(11): 2002–2008.

Clarke, J.A., Ghauri, S., Johnstone, C.M., Kim, J.M. and Tuohy, P.G. (2008) The EDEM methodology for housing upgrade analysis, carbon and energy labelling and national policy development. In *Proceedings of the eSim 2008 Conference*. Quebec City, Canada, pp. 135–142.

Clarke, J.A., Johnstone, C.M., Lever, M., McElroy, L.B., McKenzie, F., Peart, G., Prazeres, L. and Strachan, P.A. (2003) Simulation support for the formulation of domestic sector upgrading strategies. In *Proceedings of Building Simulation 2003*, Eindhoven, The Netherlands, pp. 219–226.

Commission of the European Communities (2001) *Directive on the Energy Performance of Buildings*. Brussels, Belgium: Commission of the European Communities.

Crawley, D.B. (2008) Building performance simulation: A tool for policymaking. PhD Thesis, University of Strathclyde, Glasgow, Scotland.

Crawley, D.B. and Briggs, R.S. (1985a) Standard 90: The value. *ASHRAE Journal* 27(11): 18–23.

Crawley, D.B. and Briggs, R.S. (1985b) Envelope impacts of ASHRAE standard 90.1P: A case study view. In *Proceedings of Thermal Performance of the Exterior Envelopes of Buildings III*. Clearwater Beach, FL: ASHRAE/DOE/BTECC, pp. 1128–1140.

Crawley, D.B. and Boulin, J.J. (1989) Standard 90.1's ENVSTD: Both a compliance program and an envelope design tool. In *Proceedings of Thermal Performance of the Exterior Envelopes of Buildings IV*. Orlando, FL: ASHRAE/DOE/BTECC/CIBSE, pp. 332–333.

Crawley, D.B. and Schliesing, J.S. (1992) *Energy Requirements for Office Buildings, Volume 2, Recent and Future Buildings*, GRI-90/0236.2. Chicago: Gas Research Institute.

Department for Business Energy and Industrial Strategy (2017) *Energy Consumption in the UK*. London, UK: BEIS. www.gov.uk/government/statistics/energy-consumption-in-the-uk.

Deru, Michael, Field, K., Studer, D., Benne, K., Griffith, B., Torcellini, P., Liu, B., Halverson, M., Winiarski, D., Rosenberg, M., Yazdanian, M., Huang, J. and Crawley, D. (2011) *U.S. Department of Energy Commercial Reference Building Models of the National Building Stock*. Technical Report NREL/TP-5500–46861, February 2011. Golden, CO: NREL.

Donn, M.R. (2007) Simulation of imagined realities: Environmental design decision support tools in architecture. PhD Thesis. Victoria University of Wellington., New Zealand.

Electrical and Mechanical Services Department (2017) *Hong Kong Energy End-Use Data 2017*. Hong Kong SAR, China: EMSD. www.emsd.gov.hk/en/energy_efficiency/energy_end_use_data_and_consumption_indicators/hong_kong_energy_end_use_data/data/index.html.

Energy Information Administration (2015) *A Look at the U.S. Commercial Building Stock: Results from EIA's 2012 Commercial Buildings Energy Consumption Survey (CBECS)*. Washington, DC: Energy Information Administration. www.eia.gov/consumption/commercial/reports/2012/buildstock/

Energy Information Administration (2018) *International Energy Outlook 2017*, DOE/EIA-0484 (2017). Washington, DC: Energy Information Administration. www.eia.gov/outlooks/ieo/pdf/0484(2017).pdf and www.eia.gov/outlooks/ieo/ieo_tables.php.

Energy Markets Authority (2017) *2017 Singapore Energy Statistics*. Singapore: EMA. www.ema.gov.sg/Singapore_Energy_Statistics.aspx.

Energy Systems Research Unit (2008) *ESRU Domestic Energy Model*. Glasgow: Energy Systems Research Unit, University of Strathclyde. www.esru.strath.ac.uk/Programs/EEff/index.htm.

Griffith, B., Long, N., Torcellini, P., Judkoff, R., Crawley, D. and Ryan, J. (2007) *Assessment of the Technical Potential for Achieving Net Zero-Energy Buildings in the Commercial Sector*, NREL/TP-550–41957. Golden, CO: National Renewable Energy Laboratory.

Griffith, B., Long, N., Torcellini, P., Judkoff, R., Crawley, D. and Ryan, J. (2008) *Methodology for Modeling Building Energy Performance Across the Commercial Sector*, NREL/TP-550–41956. Golden, CO: National Renewable Energy Laboratory.

Hensen, J. (1999) Simulation of building energy and indoor environmental quality – Some weather data issues. In *Proceedings of the International Workshop on Climate Data and Their Applications in Engineering*. Prague: Czech Hydrometeorological Institute.

Huang, Y.J., Akbari, H., Rainer, L. and Ritschard, R. (1991) *481 Prototypical Commercial Buildings for Twenty Urban Market Areas*, GRI-90/0326. Chicago, IL: Gas Research Institute.

Huang, Y.J. and Franconi, E.M. (1999) *Commercial Heating and Cooling Loads Component Analysis*, LBNL-37208. Berkeley, CA: Lawrence Berkeley National Laboratory.

International Energy Agency (2016) *Country Energy Balances by Sector*. Paris: IEA. www.iea.org/statistics/statisticssearch/

International Energy Agency (2017) *Key World Energy Statistics 2017*. Paris: International Energy Agency. www.iea.org/publications/freepublications/publication/KeyWorld2017.pdf.

Jensen, V. (2007) *Aligning Utility Incentives with Investment in Energy Efficiency: A Product of the National Action Plan for Energy Efficiency*. Fairfax, VA: ICF International, Inc.

Johnson, B.K., Crawley, D.B. and Perez, R.C. (1988) Comparison of energy conservation standards for new commercial buildings in the Pacific Northwest. In *Proceedings of the Pacific Northwest Regional Economic Conference*. Boise, Idaho.

Kinney, S. and Piette, M.A. (2002) *Development of a California Commercial Building Energy Benchmarking Database*, LBNL-50676. Berkeley, CA: Lawrence Berkeley National Laboratory.

Macmillan, S. and Kohler, J. (2004) *Modelling Energy Use in the Global Building Stock: A Pilot Survey to Identify Available Data Sources*. Technical Report 6. Norwich: Tyndall Centre for Climate Change Research.

Natural Resources Canada (2006) *2004 Commercial and Institution Consumption of Energy Survey*. Ottawa, ON: Natural Resources Canada. http://oee.nrcan.gc.ca/Publications/statistics/cices05/pdf/cices05.pdf.

Pacific Gas and Electric Company (1999) *Commercial Building Survey Report*. San Francisco, CA.

Pacific Northwest Laboratory (1983) *Recommendations for Energy Conservation Standards and Guidelines for New Commercial Buildings*. PNL-4870. October 1983. Richland, WA: Pacific Northwest Laboratory.

Pacific Northwest National Laboratory (2017) *Energy Savings Analysis: ANSI/ASHRAE/IES Standard 90.1–2016*, DOE/EE-1614, October 2017. Richland, WA: Pacific Northwest National Laboratory. www.energycodes.gov/sites/default/files/documents/02202018_Standard_90.1-2016_Determination_TSD.pdf.

Prototype Models (2018) www.energycodes.gov/development/commercial/prototype_models.

Ramirez, R., Sebold, F., Mayer, F.T., Ciminelli, M. and Abrishamim, M. (2005) A building simulation Palooza: The California CEUS project and DrCEUS. In *Proceedings of Building Simulation 2005*, Montreal, Canada, pp. 1003–1010.

Swan, L.G. and Ugursal, V.I. (2009) Modeling of end-use energy consumption in the residential sector: A review of modeling techniques. *Renewable and Sustainable Energy Reviews*, in press.

Ter-Martirosyan, A. (2003) The effects of incentive regulation on quality of service in electricity markets. Working Paper. Washington, DC: George Washington University.

US DOE (1979) *Notice of Proposed Rulemaking for Building Energy Performance Standards*. Washington, DC: US DOE.

US DOE (1987) *Notice of Proposed Rulemaking, Interim Standard, Voluntary Performance Standards for New Commercial and Multi-Family High Rise Residential Buildings*. Washington, DC: US DOE.

US DOE (2018) *EnergyPlus Version 8.9*. www.energyplus.net/.

Activities

1 Find out how the performance requirements for the applicable building energy standards in your area were determined using the categories listed in Table 12.1.

2 Learn what representations of building stock energy performance are available for your region from surveys or prototype buildings.

3 Determine the energy use of various sectors of your country such as the data shown in Figure 12.1.

Part III

Operational optimization

13 Building simulation for practical operational optimization

David E. Claridge and Mitchell T. Paulus

Scope

Whole-building simulation is increasingly used in the design of building heating, ventilating, and cooling (HVAC) systems but has seldom been used subsequently to optimize the operation of the building. However, there is now an increased number of cases where simulation has/is being used in some significant way to optimize the operation of buildings. This chapter provides an overview of these applications with specific examples. These uses of simulation typically require that the inputs of the simulation program be adjusted so the simulated values of heating and cooling approximately match the measured consumption values, a process typically called "calibrating" the simulation to the measured performance.

When calibrated simulations are used for practical operational optimization of building systems, this normally occurs as part of the existing building commissioning (EBCx) process. This process will be new to many readers, so it is described in some detail. A systematic and relatively simple procedure for rapidly calibrating simulations is presented and illustrated with detailed examples. The chapter concludes with a description of eight different applications of the use of calibrated simulation in the EBCx process for practical operational optimization of buildings.

Learning objectives

To systematically and quickly calibrate a whole-building simulation to measured heating and cooling consumption data. How a calibrated simulation may be used to optimize building operation and to maintain optimized operation of a building.

Key words

Operational optimization/calibrated simulation/existing building commissioning

Introduction

The continuous use of simulation throughout the life cycle of a building has been envisioned for more than two decades (e.g. Selkowitz et al. 1992), but the authors are unaware of instances where simulation was used in the building design and subsequently incorporated to maintain and optimize the routine on-going operation of the building.

There are a significant number of reported cases where simulation has/is being used in some way after the design of a building. Most of these post-design uses of simulation occur as a part of the many forms of building commissioning.

This chapter begins with a description of several applications of building simulation for operational optimization that have found application primarily in a research setting. Most of the practical applications of building simulation for operational optimization have occurred using what are termed "calibrated" simulations as part of the existing building commissioning process, so the chapter continues by introducing a process for calibrating simulations. An overview of the existing building commissioning process is given to provide context for the description of the different applications of calibrated simulation within the process that concludes the chapter.

Post-design applications of simulation

Post-design applications of simulation may be considered in five different categories.

These categories include the use of:

- Design simulation in post-construction commissioning of new buildings
- Calibrated simulation for existing building commissioning (EBCx)
- Simulation to evaluate new control code
- Calibrated simulation for fault detection and diagnostics
- Calibrated simulation as a baseline for savings determination

The first three applications of simulation listed above are typically related to operational optimization in some manner and will be discussed in significant detail. The use of simulation for diagnostic purposes may be a part of optimization independent of the EBCx process or may be utilized within the EBCx process; it will be discussed in the section titled "Using calibrated simulation for practical optimization of operational performance," where the discussion is within the context of the EBCx process. The use of calibrated simulation as an essential element of a savings determination process (e.g. Option D of the *International Performance Measurement and Verification Protocol* (EVO 2012) and ASHRAE Guideline 14 (ASHRAE 2014)) may be used in any savings determination process, but will be discussed under EBCx since it can readily be used in this application.

Important factors in whole-building simulation for operational optimization

Use of whole-building simulation for operational optimization is conceptually most practical in cases where a simulation has already been performed as part of the design process or as part of a retrofit evaluation. The input values for the existing simulation may then be used to predict building performance and explore opportunities for optimization; likewise, if the operation has been optimized, deviations will indicate the need for an investigation to bring the building back to optimum operation. However, the specific comparisons that can be performed will be dictated by the capabilities of the simulation model and by the performance data available for comparison with the simulation. If energy management control systems (EMCS) data are used for comparison with the simulation, appropriate energy consumption sensors must be available on the EMCS.

The simulation will generally be used to evaluate what may be termed "passive testing" or "active testing" (Claridge 2004). The term "passive testing" refers to the use of data collected during normal operation of the building, without any intervention to extend the range of operating variables implemented during any particular time interval. In contrast, active testing entails the use of specified control sequences to determine response to an extended range of operating variables or a particular dynamic sequence of operating variables.

Passive testing

Passive testing for operational optimization has generally involved examining simulated data and measured consumption data. These comparisons are sometimes performed with time series data, but often with plots of heating/cooling consumption plotted as functions of outside air temperature. If the simulation is an idealized simulation, then significant differences between the simulated and measured consumption are taken as indications that the system is not performing optimally or as designed.

An illustrative sequence of passive tests is described below. This sequence is not exhaustive. Many more tests are used, and many others will undoubtedly be developed and used.

a *Check room temperatures and humidity levels.* Trend logs of the temperature and humidity in individual zones can be tracked over one or more days. If these values stay within the setpoints (and any control undershoot, overshoot, or throttling range), temperature and humidity control are deemed acceptable. If not, diagnostics (that may or may not involve simulation) are needed to diagnose reasons for the excursions observed. If this test is passed,
b *compare energy use with predictions* over a period of at least a few days. If measured consumption is within an acceptable range of predicted consumption, this test is passed. However, it is non-trivial to develop practical "passing" criteria for this type of test. If "pass" criteria are narrowly missed, this may indicate a need to change inputs; and
c *extrapolate performance from limited trend data to design conditions.* It is certainly necessary to determine whether equipment capacity is adequate.

Depending on the capability of the simulation and the sensors available on the EMCS, it is desirable to verify a wide range of operating parameters, such as airflows and supply temperatures to individual zones, water-side system parameters, and primary system performance.

Active testing

Active testing involves specific active tests implemented for diagnostic purposes; these may be triggered in response to the failure of one or more passive tests. Active testing may also include functional tests devised to explore the comfort control capabilities of the system and its dynamic response over a wide range of operating conditions. Active tests normally provide some empirically determined input variables to be used in the simulation program being used.

Active testing will normally control an input variable with a significant impact on building comfort and energy performance through a specified sequence of values. It may include the cycling of a known lighting load or another major load to test the response of the HVAC system and test system performance. Likewise, it may involve varying the space temperature setpoints in a way that will test system capacity and control.

Multiple tests are needed to test system response to a range of loads in the spaces and to determine the efficiency with which the primary and secondary systems are working with the control system to meet the space loads.

Different applications of building-level simulation for operational optimization

Four different applications of simulation for operational optimization are described in this section.

1 *Use of design simulation in post-construction commissioning of new buildings.* A design simulation of the building may be used to predict heating and cooling performance and

the predictions may be compared with measured use – significant deviations then serve as clues to identify problems in the building. The design simulation should have the occupancy schedules changed if necessary to reflect the actual occupancy of the building. Simulation may also be used at this stage to refine and optimize the control strategy. Relatively complex simulations are utilized for this purpose. This will be discussed later in this section.

2 *Use of a design simulation or calibrated simulation for ongoing commissioning.* The same simulation developed in the design process may then be run at specified intervals, e.g. weekly, monthly, etc., and the model predictions compared with the measured energy consumption. This simulation must be modified to reflect any differences between the design assumptions and actual use of the building. Deviations then serve to trigger an alarm when building performance degrades. The nature of the deviations observed may also be used to help diagnose reasons for performance degradation. Alternatively, if no design simulation was performed, subsequent to existing building commissioning to ensure that the building is operating optimally, a simulation may be calibrated to the operation of the building and used to trigger an alarm when building performance degrades. These simulations may be run on-line or off-line. The response time of maintenance staff to issues affecting only energy use and not comfort is normally measured in days or weeks, so daily or weekly use of such simulations is appropriate. The use of calibrated simulations within the on-going commissioning phase of the EBCx process will be discussed in the "Using calibrated simulation for practical optimization of operational performance" section.

3 *Use of calibrated simulation for existing building commissioning.* A rapidly calibrated simulation may be used in multiple steps in the EBCx process. These applications include screening buildings to identify good candidates for further examination, predicting the savings that will be achieved from implementing proposed EBCx measures, creating baselines for savings determination, and detecting and diagnosing faults after the process has been implemented. This application of calibrated simulation will be extensively discussed in the "Using calibrated simulation for practical optimization of operational performance" section.

4 *Use of simulation to evaluate new control code.* Either the design simulation or a calibrated simulation may be used to test the energy impact of proposed changes in control code before implementation. This will generally be done off-line and is discussed later in this section.

Examples using building-level simulation for operational optimization

Each application of building-level simulation for operational optimization that does not require a calibrated simulation is illustrated in this section in the context of one or more specific applications. Applications using calibrated simulation are discussed in in the "Using calibrated simulation for practical optimization of operational performance" section.

Use of design simulation in post-construction commissioning of new buildings

Keranen and Kalema (2004a, 2004b) have utilized simulations of the 9500 m^2 IT-Dynamo Building in the commissioning process. This building, which houses the Department of Information Technology at Jyvaskyla Polytechnic in Jyvaskyla, Finland, was completed in May 2003 and placed in service in August 2003. The building was simulated using the programs IDA-ICE and RIUSKA based on DOE-2. The heating and electricity use of the building are monitored, and the electricity used for HVAC is separately monitored. Indoor conditions, including numerous temperatures and CO_2 levels, are monitored and stored for two days on the building automation system. During commissioning, these variables were transferred to long-term storage.

Comparison of the simulated and measured heating consumption showed measured heating consumption almost twice the simulated consumption during the first three months of operation. Investigation revealed that heating and cooling thermostat dead bands were too narrow in about 20% of the building area, resulting in continuous operation of either the heating or the cooling at maximum values in these areas. Correction of this problem reduced the difference between measured and simulated heating consumption to less than 10% in each of the next three months.

Carling et al. (2004a, 2004b) have tested the use of whole-building simulation in the commissioning of an office building in Katsan, Sweden. This building utilizes multiple innovative HVAC systems. They used the IDA simulation environment to assess these innovative systems during design and for component sizing. During the initial commissioning, they evaluated the performance using extensive measurements from the building energy management system. Five-minute values for about 200 signals were collected during a full year following the initial occupancy of the tenants. A detailed whole building simulation model was calibrated with adjusted internal loads based on measured electrical power and measured weather inputs. The results of the calibrated model were used to determine whether the HVAC systems performed as intended by comparing the measurements of the whole-building energy use as well as with some important temperatures and control signals.

Comparisons between the extensive measurements and the calibrated model were made and led to the detection and correction of five problems with control setpoints and operation. Some of the problems were the supply air temperature setpoint was not correctly implemented, the outdoor temperature sensor was biased, and cooling started at a lower temperature than expected in one zone.

On-line simulation as a commissioning follow-up tool

A design simulation or a calibrated simulation may be embedded in the EMCS. It can serve as an alarm when consumption deviates beyond an alarm limit. It may also be used to evaluate the impact of any control changes implemented – comparison of measured performance with simulation results would show whether performance has improved or degraded because of the changes.

Liu et al. (2002a) present the results of a study of the potential for using simulation programs for on-line fault detection, problem diagnosis, and operational schedule optimization for large commercial buildings with built-up HVAC systems. They specifically examine the potential for the use of calibrated whole-building simulation for existing building commissioning and on-going commissioning.

This study reviewed over a dozen simulation programs and determined that AirModel[1] (now WinAM) and EnergyPlus (LBL 2001) were most suitable for initial use in the on-line simulation applications that were the focus of the study. These programs cover both ends of a spectrum from a relatively simple program that can be used quickly and embedded in an EMCS for on-line simulation (WinAM) to one of the more detailed and flexible simulation programs available (EnergyPlus).

Others performing investigations of on-line simulation include Claridge and Liu (2004) and Tsubota and Kawashima (2004).

Use of simulation to evaluate new control code

Simulation can test new control code before actually putting it into the system and activating the new or modified mode of control. Baumann (2003) used this approach to optimize control of the water supply temperature for heating and cooling in a new school. Wang and Xu (2003) have developed a new simulation tool particularly intended to permit rapid evaluation of the dynamic performance of a building at short time steps so it is suitable for evaluation of control code.

Baumann (2003) used a TRNSYS simulation in conjunction with the GenOpt program to develop an optimal control strategy for the heating and cooling water supply temperatures in a 10,000 m^2 vocational school in Biberach, Germany. This school was completed in the summer of 2004 and incorporates an embedded hydronic heating and cooling (EHHC) system consisting of flexible tubes embedded in the massive concrete ceilings. Water heated by a heat pump or cool ground water is supplied to the EHHC system, depending on whether heating or cooling is required in the building.

The massive ceiling has a very long time constant, so controlling the temperature of the water supplied to the system is critical to minimize the number of occupied hours when the space temperatures are either too hot or too cold. The control scheme adopted uses the median temperature values for the last three days (with the most recent day double-weighted) to define the "median" outdoor temperature that determines the temperature of the supply water. The supply water temperature varies between a maximum value of 28 °C and a neutral value of 21 °C as the "median" outdoor temperature varies between two specified values. The supply temperature is maintained at 21 °C until the "median" outside temperature increases to an initial cooling temperature, at which the supply temperature starts to decrease linearly toward 18 °C at a higher specified temperature. The supply temperature is held constant at 18 °C for all higher ambient temperatures and is held constant at 28 °C for temperatures below the lower specified heating temperature.

The simulation was used to optimize the values of the specified temperature ranges for adjusting the supply water temperature to minimize heating and cooling energy subject to the constraints that the room temperature never is below 21 °C or goes above 26 °C for a small number of hours. It was found that the optimum heating consumption was only half the maximum value that produced equivalent comfort.

Wang and Xu (2003) developed a hybrid model suitable for rapid short time-step simulation of the dynamic performance of buildings – hence suitable for control simulation and optimization. This hybrid model uses a 3-resistor, 2-capacitor (3R2C) model of the building envelope, with the sum of the three resistances constrained by the total resistance of the envelope and the total capacitance constrained by the total thermal capacitance of the envelope. Wang and Xu use a genetic algorithm to choose optimum values for the resistances and capacitances so the frequency response and phase lag of the 3R2C model closely approximates that of the theoretical model for the walls at all but the highest frequencies. The same technique is used to develop optimal parameters for a 2R2C model of the interior mass in the building by searching in the time domain.

This model was developed in part to determine optimal control strategies for night ventilation and off-peak precooling to take advantage of a time-of-use rate that was implemented in Hong Kong in 2001.

The following applications of building-level simulation for operational optimization have been described and illustrated:

- Design simulation in post-construction commissioning of new buildings
- Design simulation for on-going commissioning
- Simulation to evaluate new control code

Examples of each of these applications are provided. The only application that has found use outside research projects to date is the use of calibrated simulation for existing building commissioning. This application is discussed in the "Using calibrated simulation for practical optimization of operational performance" section. The other examples have been applied in a research setting, and costs must be lowered for routine application.

A simple process for calibrating simulations

Building energy simulation has become increasingly important both in new building design and in evaluating potential energy efficiency measures in existing buildings. Simulations can be valuable tools in many applications, such as calculating energy savings from proposed or implemented retrofits, existing building and new building commissioning, fault detection and diagnosis, and program evaluation. Today many simulation tools exist, ranging from complex, whole-building modeling programs, such as DOE-2 and BLAST, to simplified modeling programs like WinAM. Each of these programs requires various input parameters from the user to model the building, or specific systems within the building. The accuracy in using these simulation programs depends on the ability of the user to input parameters that result in a good model of actual building energy use. The initial parameters used are generally obtained from building design data, which can result in output errors of 50% or more when compared with actual building performance. An example of simulated energy use versus actual use is shown below in Figures 13.1 and 13.2. Figure 13.1 shows the daily simulated and measured heating usage in Harrington Tower on the Texas A&M University campus for the year 2003, while Figure 13.2 is a plot of the same data as a function of average daily dry bulb temperature.

As evidenced in these figures, a simulation may not accurately reflect the performance of the building; in this case, the simulated consumption reasonably approximates the building consumption at temperatures above 20 °C, but is about 2/3 the actual consumption at 0 °C. Any conclusions made about heating conservation measures and their savings based on data from this simulation have a high probability of significant error. The simulated data must closely

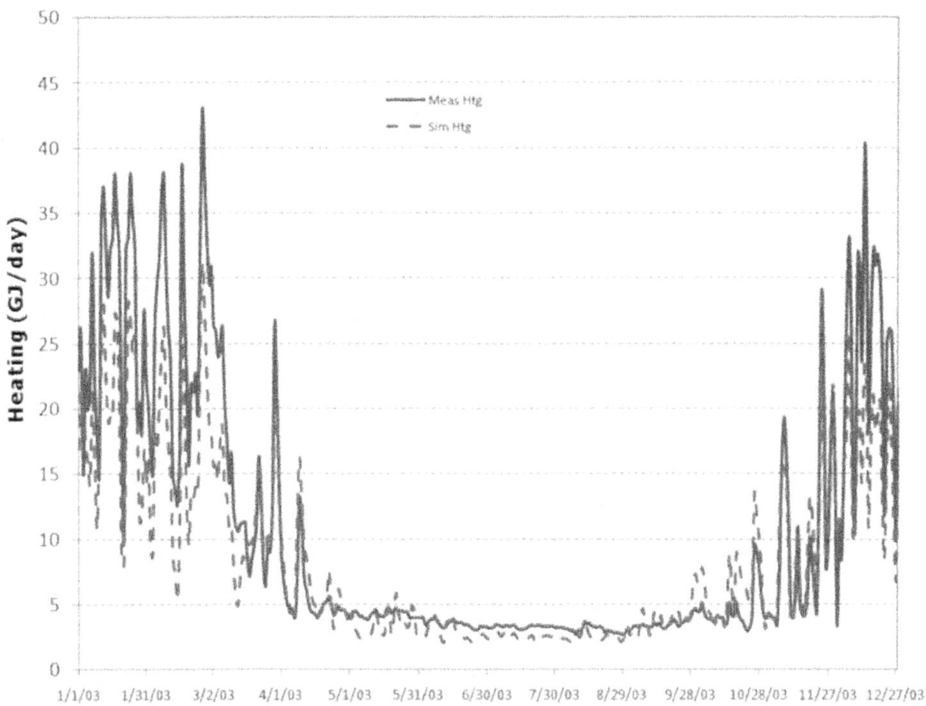

Figure 13.1 Simulated and measured heating usage in Harrington Tower for 2003

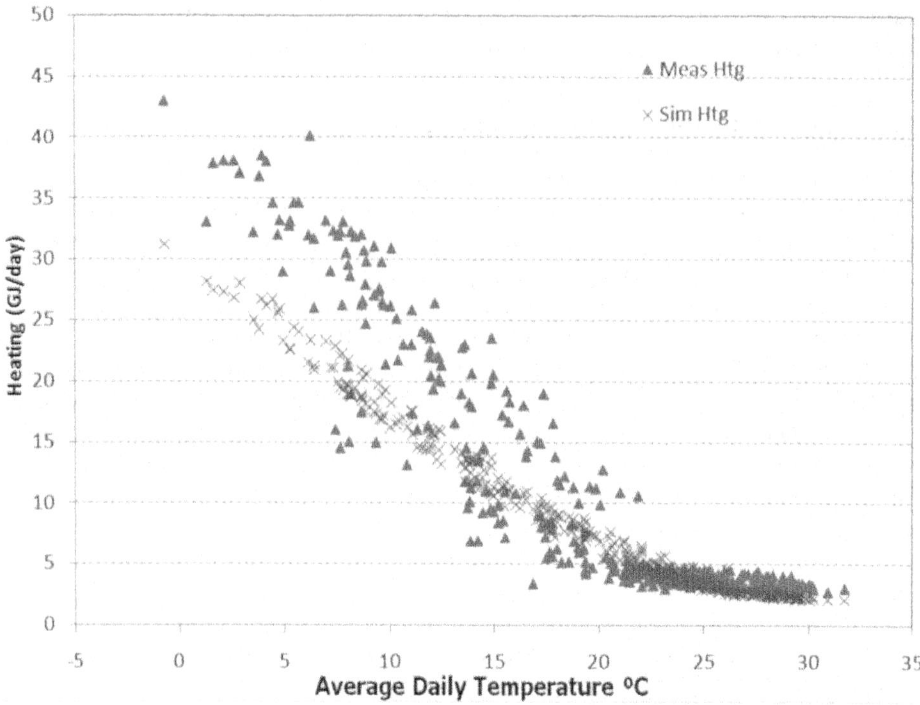

Figure 13.2 Simulated and measured daily heating usage for Harrington Tower versus daily average dry bulb temperature

match the measured data to allow useful analyses to be made, and the closer the data match, the better the expected results of analyses. Unfortunately, in many cases the process of calibrating a simulation can be extremely tedious and time consuming, especially with simulation programs that require hundreds of input parameters. Because changing two or more different parameters in the input could result in similar changes to the output, it is difficult to know which parameters to change to achieve the most representative model with the desired results, particularly for inexperienced users. Without a straightforward method of calibrating, the process can be too expensive and time consuming to be deemed worthwhile.

A method has been developed by Wei et al. (1998) as modified by Claridge et al. (2003) to calibrate simulations with relative ease based on a concept called "calibration signatures." Experienced users of this method have shown the ability to calibrate a two-zone simulation in a few hours, significantly less time than previously needed with a "trial and error" approach.

Definitions

The term "calibration signature" is defined as follows:

$$Calibration\,signature = \frac{Residual}{Maximum\,measured\,energy} \times 100\% \tag{13.1}$$

$$Residual = Measured\,consumption - Simulated\,consumption \tag{13.2}$$

The "maximum measured energy" is the maximum heating or cooling energy use recorded in the data set examined. The calibration signature is a normalized plot of the difference between measured energy use and simulated energy use over a specified temperature range. For each temperature, a measured energy use value and a simulated energy value exist. The difference in these values for each point is divided by the maximum measured energy use and multiplied by 100%. These values are then plotted versus temperature. An example of a pair of calibration signatures for heating and cooling use is shown in Figure 13.3.

Due to the scatter that exists, it is often helpful to draw a line representing the average value of each signature at each temperature, as will be illustrated later. This technique allows the trend of the signature to be evaluated. If a simulation were perfectly calibrated, the calibration signatures would be horizontal lines at 0%. The two signatures may also be plotted in separate charts.

It is now useful to define another term, "characteristic signature." A characteristic signature is identical to a calibration signature, except that instead of comparing simulated and measured values, it compares two simulations. One simulation is taken to be the baseline or "measured" value. Then, by varying parameters one by one, signatures can be plotted and compared. Characteristic signatures are defined as:

$$Characteristic\ signature = \frac{Change\ in\ energy\ consumption}{Maximum\ energy\ consumption} \times 100\% \tag{13.3}$$

where the change in energy consumption is the changed simulation value minus the baseline simulation value. As mentioned, the baseline model is treated as the "measured" case, and maximum energy consumption comes from this model. Characteristic signatures are generated for a range of parameters for the building being calibrated. An example of a pair of characteristic signatures is shown in Figure 13.4. This signature was generated for a dual-duct variable-air-volume (DDVAV) system in Harrington Tower by changing the simulation minimum flow rate from the baseline value of 1.75 L/(s·m^2) to 1.25 L/(s·m^2).

Notice that for a change in a parameter, characteristic signatures should be generated for both heating and cooling energy use. Ideally, characteristic signatures should be generated for the

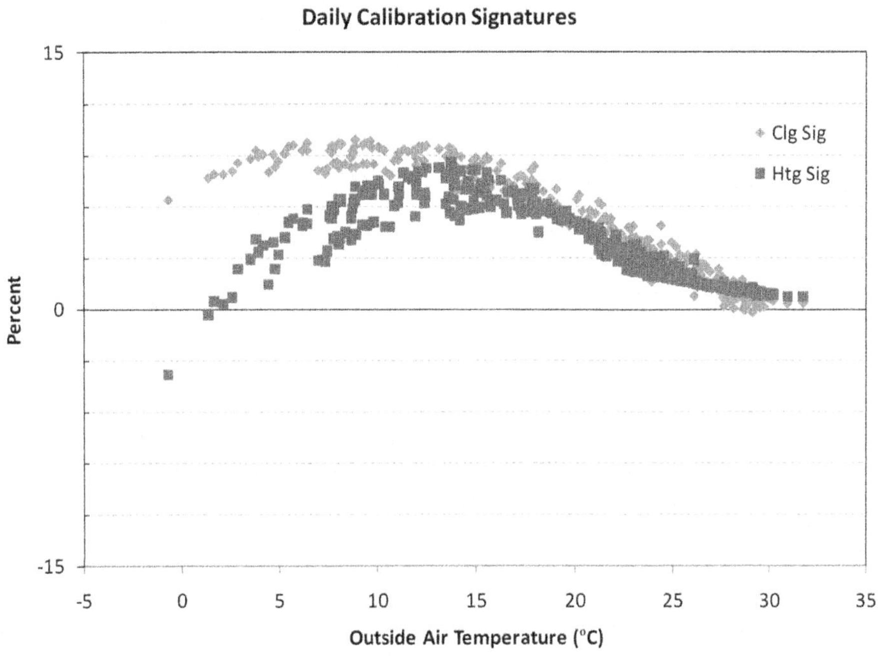

Figure 13.3 Example of calibration signatures

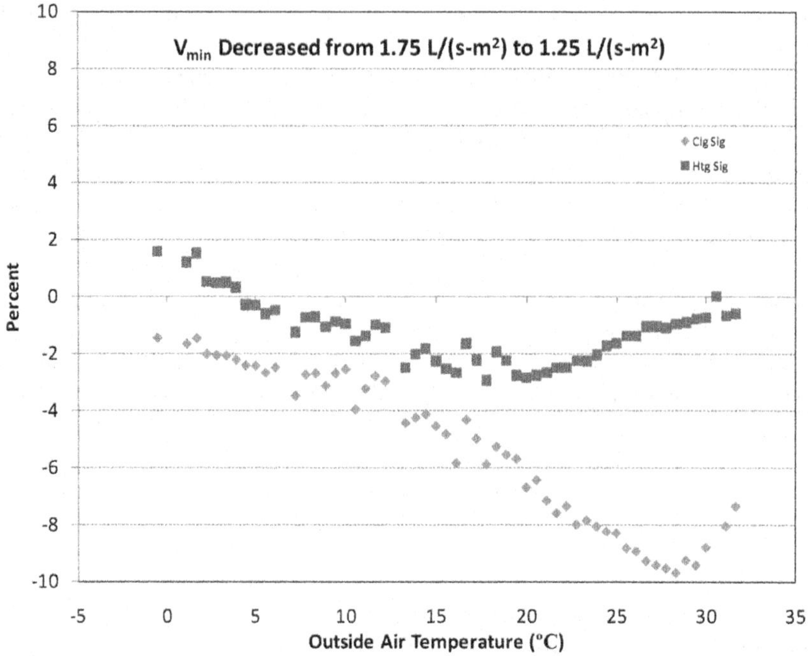

Figure 13.4 Characteristic signatures for a dual duct VAV system in Harrington Tower with a change in minimum flow rate

building to which the simulation is being calibrated. However, it is still helpful to use characteristic signatures for a similar building in a similar climate if it is not practical to generate characteristic signatures for the building to be calibrated. Characteristic signatures for the dual-duct variable-air-volume system in Harrington Tower are given in the Appendix. The parameters of major importance for which characteristic signatures should be generated include cold deck temperature, hot deck temperature (dual-duct systems), supply airflow rate (constant volume systems), minimum airflow rate (variable-air-volume systems), floor area, preheat temperature, internal gains, outside airflow rate, room temperature, envelope U-value, and economizer settings. Characteristic signatures from a climate with different correlations between relative humidity and dry bulb temperature are unlikely to work well.

Two indices used for evaluating the accuracy of a simulation are the root mean square error and the mean bias error. The root mean square error (RMSE) is defined as:

$$RMSE = \sqrt{\frac{\sum_{i=1}^{n} (Residual_i)^2}{n}} \tag{13.4}$$

where n is the total number of data points. The RMSE is a good measure of the overall magnitude of the errors, but does not give any reflection of bias, since no indication is made as to whether the errors are positive or negative. A good simulation will minimize the RMSE, though it is generally difficult to reduce this below 10% of the mean of the larger of the heating and cooling consumption. The mean bias error (MBE) is defined as:

$$MBE = \frac{\sum_{i=1}^{n} Residual_i}{n} \tag{13.5}$$

where n is the number of data points. The MBE is an overall measure of the prediction bias, since positive and negative errors cancel each other. The relative importance of MBE and RMSE varies depending on the purpose of the calibrated model. Both should be considered if the model will be used to evaluate energy efficiency measures, while MBE becomes more important if the model will be used as the baseline for evaluating energy savings after efficiency measures have been implemented.

The approach used in the remainder of this chapter will minimize a combined error measure defined as:

$$ERROR_{TOT} = \left[RMSE_{TOT}^2 + MBE_{TOT}^2\right]^{1/2} = \left[\left(RMSE_{CLG}^2 + RMSE_{HTG}^2\right) + \left(MBE_{CLG} + MBE_{HTG}\right)^2\right]^{1/2} \tag{13.6}$$

This error measure arbitrarily combines the root mean square heating and cooling errors in quadrature and the heating and cooling mean bias errors algebraically before combining both in quadrature. The algebraic combination of mean bias errors is a combination that reduces the mean bias in the total cost of heating and cooling energy.

Implementing the calibration process

Calibration signatures combined with characteristic signatures are used to calibrate a simulation quickly. The calibration signatures for heating and cooling generated for the simulation are compared with the characteristic signatures from the corresponding system and climate type, to see which change of parameter or parameters most closely resemble the calibration signature. Normally one parameter is changed at a time in the correct direction and according to the magnitude needed. For example, if the calibration signature is in the range of 20% at low temperatures and a similar characteristic signature shows the same trend, but is in the range of only 5%, the parameter adjustment would need to be significantly greater than what was done to get the characteristic signature to increase the magnitude. The adjustment is, of course, limited by reasonable values – a cold deck setpoint would not be 3 °C, for example. Once the parameter to be varied has been chosen, it is changed and the simulation is run again. The $ERROR_{TOT}$ is calculated again, and calibration signatures are again generated and compared with the characteristic signatures. This process is repeated until an acceptable low value of $ERROR_{TOT}$ is achieved and the calibration signatures approximate a normal distribution around zero. At this point, the simulation can be considered calibrated to the measured data. As noted above, a good target for a calibrated simulation is RMSE values of the heating and cooling data that are about 10% of the average value of the larger of the heating or cooling consumption. It is not always possible to reach this target, as illustrated in the second calibration example.

This process of calibration is presented more logically in Figure 13.5 in the form of a flow chart.

Calibration example #1

The Harrington Education Tower shown in Figure 13.6 is an eight-floor, 9,430 m² building housing the College of Education at Texas A&M University. It consists primarily of offices, but there are some classrooms and computer labs. There are 1,770 m² of glazing, with 50% of it being on the first, second, and eighth floors, as they are predominantly glass sided. Harrington Tower receives hot water and chilled water from the Texas A&M Physical Plant for its HVAC

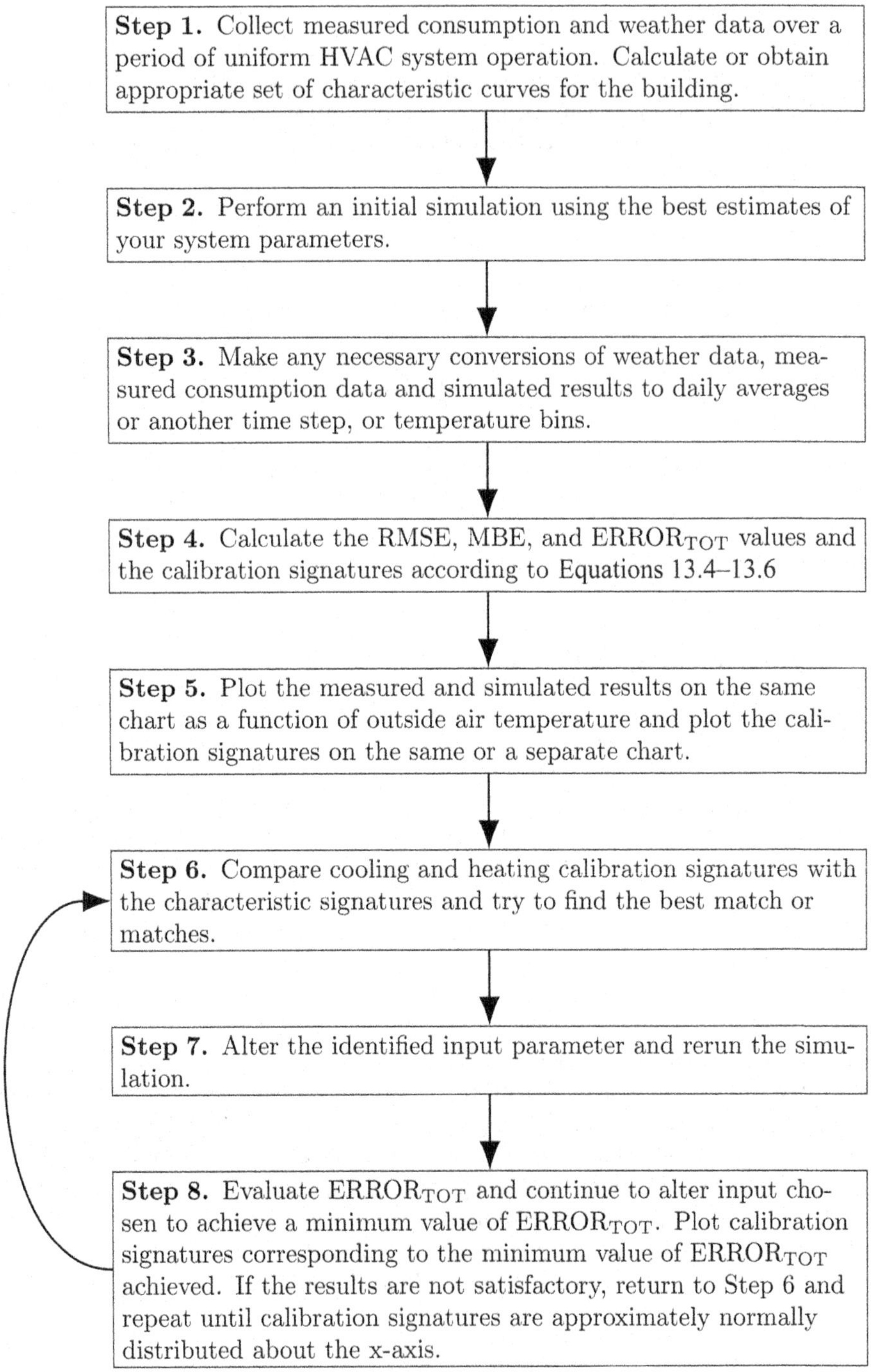

Figure 13.5 Flow chart detailing the overall process of calibration

systems and provides heating and cooling to about 90% of the building with a single dual-duct variable-air-volume air handler powered by a 150 kW fan providing up to 65,000 L/s of air. To illustrate the calibration procedure, the building is simulated using WinAM. To provide a cleaner example of the procedure, one or more inputs to the simulation were changed, and the output of that simulation is used as the "measured" consumption data for this example.

Figure 13.6 Harrington Tower on the Texas A&M University campus

Source: Photo courtesy of Texas A&M University (https://brandguide.tamu.edu/visual-style/photo-video.html)

Example #2 will show the calibration of this building to measured data instead of simulated data. Table 13.1 shows initial simulation inputs for Example #1. The entry "Linearize" means that the hot deck temperature varies according to $T_{hd} = 43.3 - 1.33(T_{oa} - 4.4)$ when $4.4 \leq T_{oa} \leq 21.1$ °C where T_{oa} is the outdoor air dry bulb temperature. Figure 13.7 compares the "measured" and initially simulated heating and cooling consumption as functions of daily average temperature, while Figure 13.8 shows the initial set of calibration signatures for this example.

Figure 13.7 shows that the simulated values of both heating and cooling consumption appear to be systematically lower than measurements at outdoor air temperatures below about 20 °C. The calibration signatures in Figure 13.8 show that both heating and cooling consumption are 5%–10% (of the maximum measured value) lower than the measured values over much of this range. $ERROR_{TOT}$ is 3.72 GJ/day for this initial simulation with initial mean bias error values of −1.76 GJ/day for cooling and −1.05 GJ/day for heating corresponding to −8.8% and −14.8% of the average measured values, respectively.

The calibration signatures should now be compared with the characteristic signatures for the DDVAV system in Harrington Tower shown in the Appendix. When comparing the signatures, a similarity in the shape of calibration signatures and the characteristic signatures for decreasing the cold deck temperature from 12.8 °C to 10 °C can be seen. The characteristic signatures are shown in Figure 13.8(b). Note that decreasing slopes exist for both the chilled water and the hot water.

Figure 13.8 shows the initial calibration signatures together with the three most relevant characteristic signatures. Note that Figure 13.8(c) is relevant as a mirror image, i.e. it implies that an increase may be needed in V_{min} rather than a decrease. Examining the characteristics signatures, decreasing T_{cl} seems to show the most similarity. We decrease T_{cl} and get a minimum value of $ERROR_{TOT}$ for $T_{cl} \sim 7$ °C. 7 °C is likely not a realistic value since we know the chilled water is supplied to the building at ~6 °C.

Table 13.1 Initial values of key simulation inputs for Example #1

Input		*Value*		*Unit*
Floor area	A_{floor}	9,430		m^2
Wall area	A_{walls}	3,457		m^2
Window area	$A_{windows}$	1,767		m^2
Roof area	A_{roof}	1,520		m^2
Outside air percentage	X_{oa}	15		%
Minimum supply airflow rate	V_{min}	1.78		$L/(s \cdot m^2)$
Maximum supply airflow rate	V_{max}	5.6		$L/(s \cdot m^2)$
Interior zone percentage	$X_{IntZone}$	58		%
Wall overall heat transfer coefficient	U_{walls}	0.85		$W/(m^2 \cdot K)$
Window overall heat transfer coefficient	$U_{windows}$	6.25		$W/(m^2 \cdot K)$
Roof overall heat transfer coefficient	U_{roof}	0.28		$W/(m^2 \cdot K)$
Internal heat gain	Q_{int}	18.2		W/m^2
Room temperature setpoint	T_{room}	22.2		°C
Cooling coil leaving temperature setpoint	T_{cl}	12.8		°C
Preheat coil temperature setpoint	T_{ph}	1.7		°C
Hot deck temperature setpoint	T_{hd}	43.3	$T_{oa} < 4.4$	°C
		Linearize	$4.4 \leq T_{oa} \leq 21.1$	
		23.9	$T_{oa} > 21.1$	

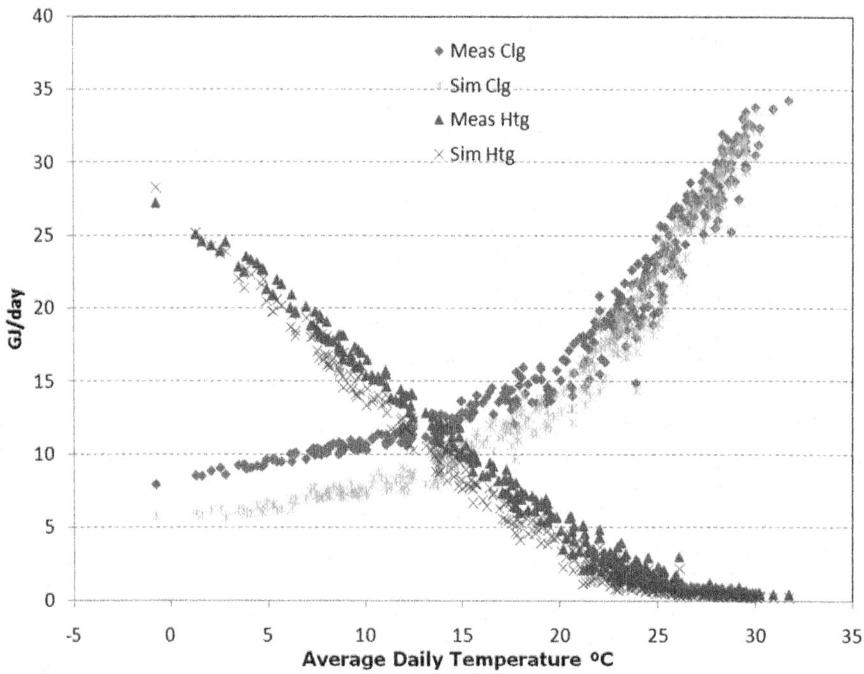

Figure 13.7 Comparison of initial simulated and "measured" heating and cooling consumption values as functions of daily average temperature for Example #1

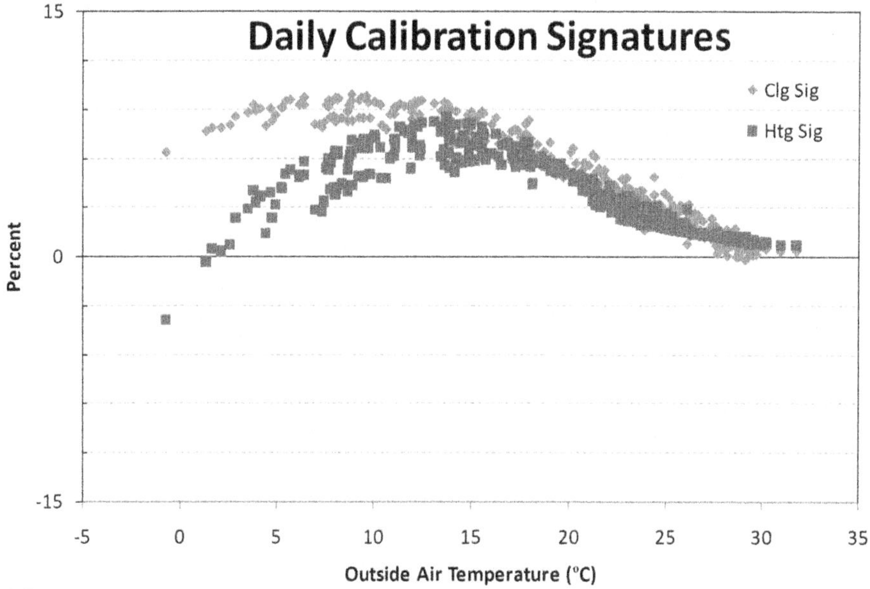

(a)

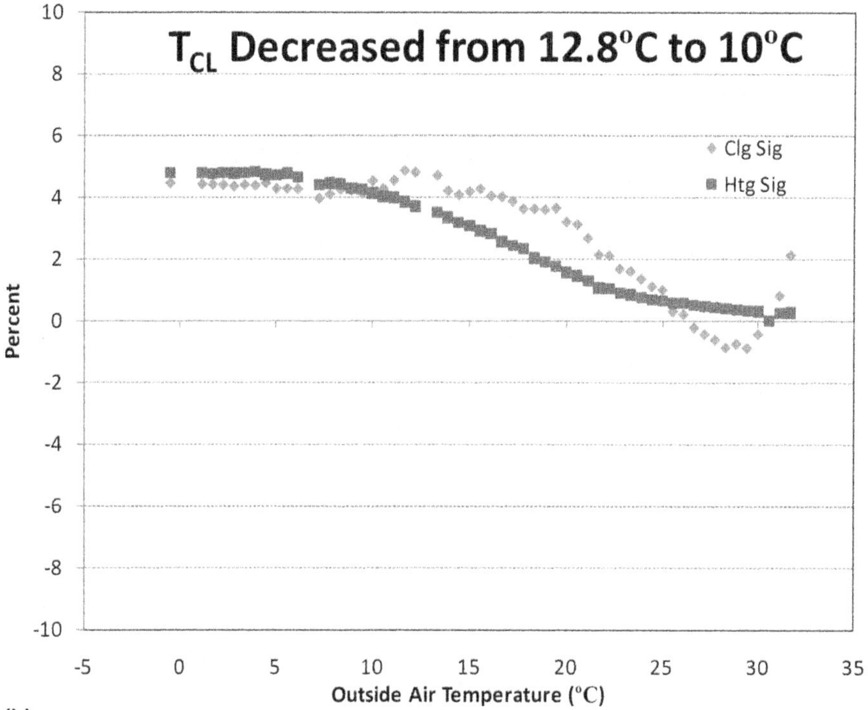

(b)

Figure 13.8 Initial calibration signatures (a) and relevant characteristic signatures (b–d)

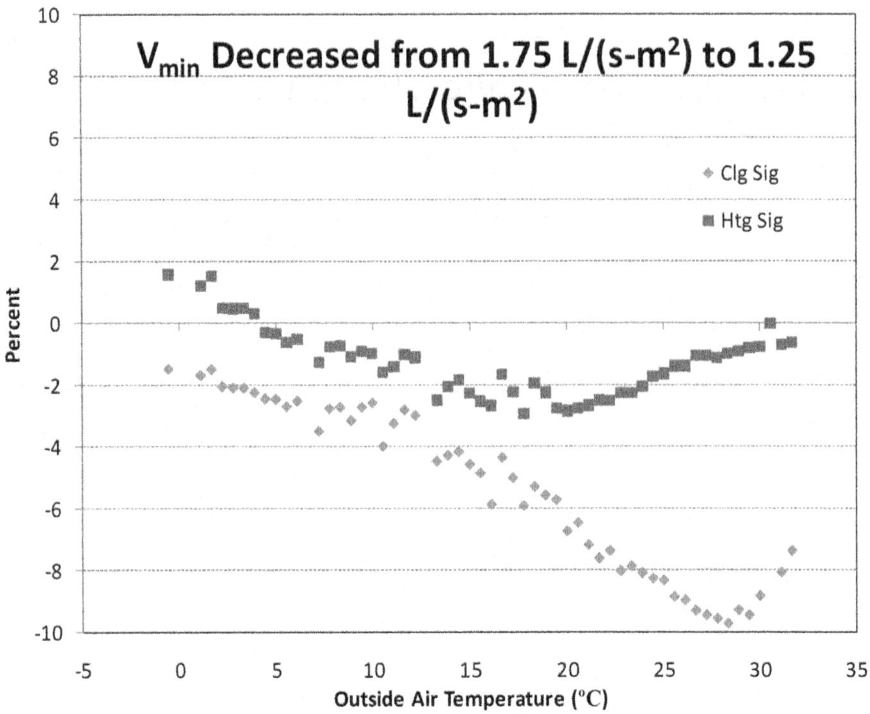

(c)

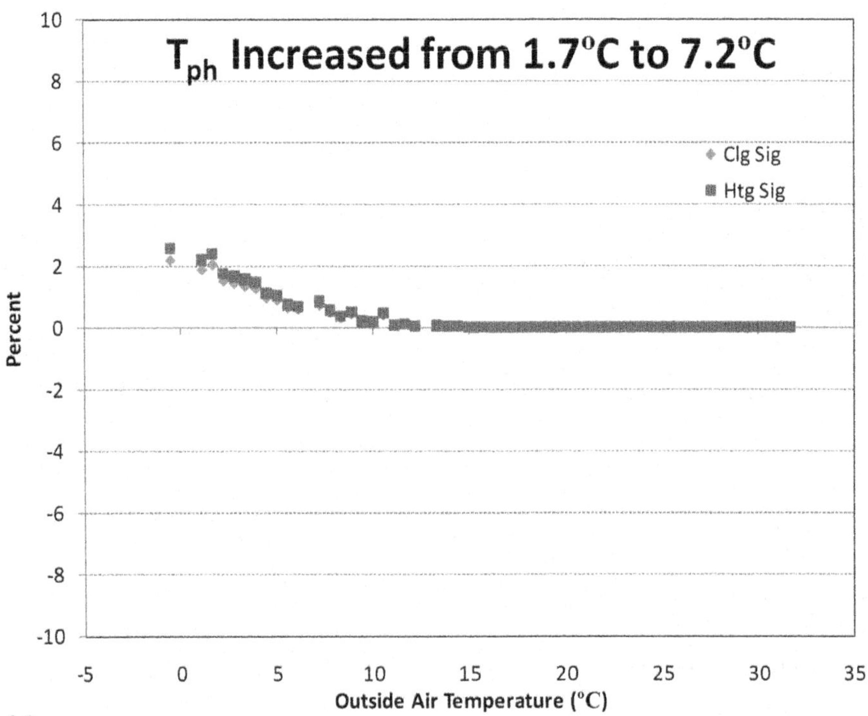

(d)

Figure 13.8 (Continued)

The hot water characteristic signature for decreasing V_{min} goes from approximately zero at high ambient temperatures to a minimum value and then becomes positive at low ambient temperatures, approximately the mirror image of the hot water calibration signature. The preheat change required to have any chance of zeroing the calibration signatures appears to be large so we then try increasing V_{min}. The value of $ERROR_{TOT}$ reaches a minimum value of 1.13 GJ/day when V_{min} is increased to 2.34 L/(s·m²). The calibration signatures for V_{min} = 2.34 L/(s·m²) are shown in Figure 13.9(a).

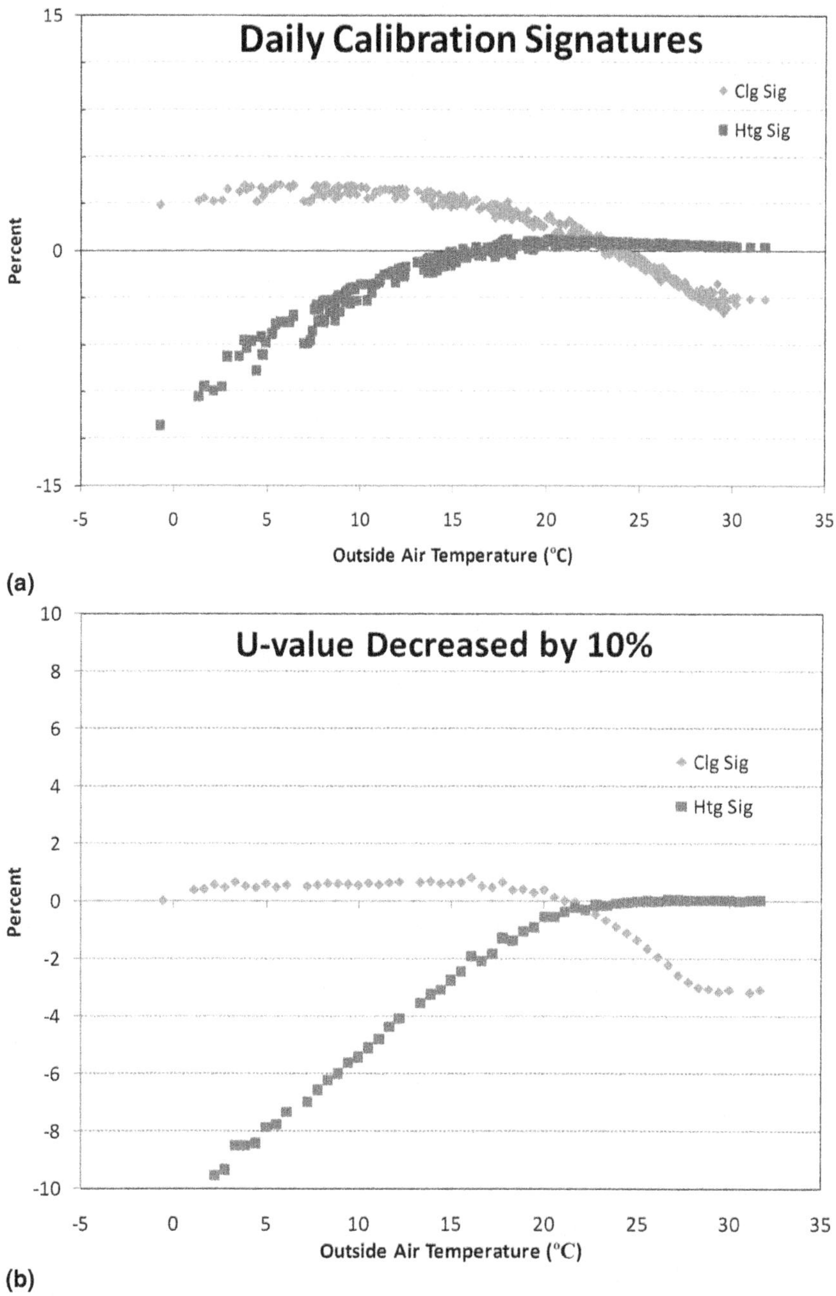

Figure 13.9 (a) Calibration signatures with V_{min} = 2.34 L/(s·m²) and (b) characteristic signatures for decreasing the *U*-value

Reducing the *U*-values by 8% brought the $ERROR_{TOT}$ to 0.84 GJ/day and the calibration signatures suggested further increasing V_{min}. This iteration lowered $ERROR_{TOT}$ to 0.61 GJ/day and the calibration signatures suggested further decreasing the *U*-values. Three more iterations reduced $ERROR_{TOT}$ to 0.14 GJ/day and resulted in the calibration signatures shown in Figure 13.10. This figure resulted from V_{min} = 2.46 L/(s·m²) with *U*-values reduced by 16%. These values may be compared with the values V_{min} = 2.49 L/(s·m²) and *U*-values reduced by 14% used to generate the "measured" data for this example. This figure shows much less scatter than can be expected from measured data as will be shown in Example #2.

Calibration example #2

The same building simulated in Example #1 is simulated again, but this time, it will be calibrated to the actual measured heating and cooling data. As noted earlier, one large DDVAV air handler serves about 90% of the building. The remaining 10% is served by a small single-duct constant volume air handling unit (AHU) with terminal reheat. Table 13.2 shows the key initial inputs used for AHU1, the DDVAV AHU. The major changes from Example #1 are the inclusion of the measured non-HVAC electric power for Q_{int} and a reduction in various areas to account for the area served by AHU2. Table 13.3 shows the inputs used for AHU2, the single-duct constant volume (SDCV) AHU.

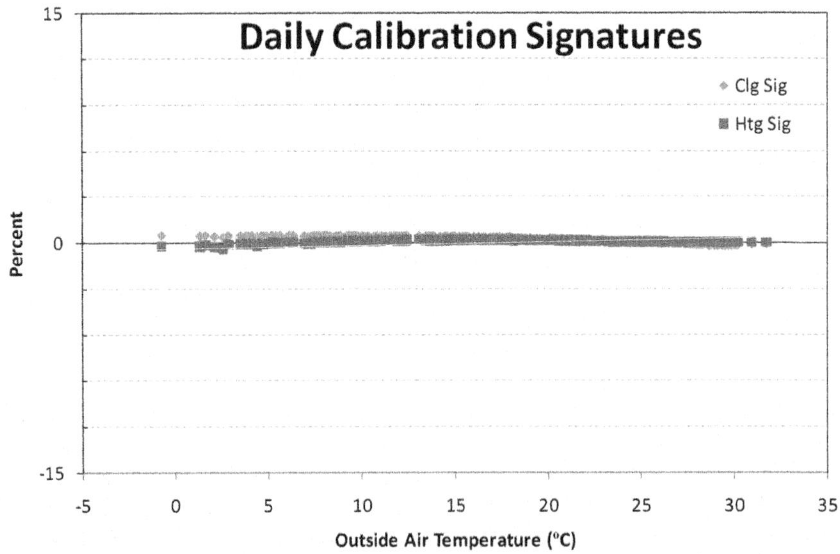

Figure 13.10 Calibration signatures for Example #1 after calibration

Table 13.2 Initial inputs for the DDVAV AHU

Input		*Value*	*Unit*
Floor area	A_{floor}	8,780	m²
Wall area	A_{walls}	3,320	m²
Window area	$A_{windows}$	1,360	m²
Roof area	A_{roof}	1,520	m²
Outdoor air volumetric flow rate	V_{oa}	4,950	L/s
Minimum supply airflow rate	V_{min}	1.78	L/(s·m²)
Maximum supply airflow rate	V_{max}	5.6	L/(s·m²)

Input		*Value*		*Unit*
Interior zone percentage	$X_{IntZone}$	58		%
Wall overall heat transfer coefficient	U_{walls}	0.85		W/(m²·K)
Window overall heat transfer coefficient	$U_{windows}$	6.25		W/(m²·K)
Roof overall heat transfer coefficient	U_{roof}	0.28		W/(m²·K)
Internal heat gain	Q_{int}	26.9		W/m²
Room temperature setpoint	T_{room}	22.2		°C
Cooling coil leaving temperature setpoint	T_{cl}	12.8		°C
Preheat coil temperature setpoint	T_{ph}	1.7		°C
Hot deck temperature setpoint	T_{hd}	43.3	$T_{oa} < 4.4$	°C
		Linearize	$4.4 \le T_{oa} \le 21.1$	
		23.9	$T_{oa} > 21.1$	

Table 13.3 Inputs used for the SDCV AHU

Input		*Value*	*Unit*
Floor area	A_{floor}	650	m²
Wall area	A_{walls}	50	m²
Window area	$A_{windows}$	400	m²
Roof area	A_{roof}	0	m²
Outdoor air volumetric flow rate	V_{oa}	660	L/s
Maximum supply airflow rate	V_{max}	5.6	L/(s·m²)
Interior zone percentage	$X_{IntZone}$	50	%
Wall overall heat transfer coefficient	U_{walls}	0.85	W/(m²·K)
Window overall heat transfer coefficient	$U_{windows}$	6.25	W/(m²·K)
Roof overall heat transfer coefficient	U_{roof}	0.28	W/(m²·K)
Internal heat gain	Q_{int}	26.9	W/m²
Room temperature setpoint	T_{room}	22.2	°C
Cooling coil leaving temperature setpoint	T_{cl}	12.8	°C
Preheat coil temperature setpoint	T_{ph}	1.7	°C

Harrington Tower was simulated using this set of inputs, and Figure 13.11 plots the simulated and measured heating and cooling consumption values as functions of ambient temperature. The calibration signatures for this simulation are shown in Figure 13.12, and the error measures for the initial simulation vs. the measured data are given in Table 13.4.

Examining the initial calibration signatures for Example #2 and comparing with the characteristic signatures in the Appendix, the heating calibration signature matches the characteristic signature for going from no economizer to an economizer quite well. The cooling signatures do not match as closely as the heating signatures, but after implementing an economizer, $ERROR_{TOT}$ is reduced somewhat to 5.31 GJ/day. It is known that the building has an economizer, but it has frequently been turned off since additional heating often offsets cooling savings. The calibration signatures (Figure 13.13) show some resemblance to the characteristic signatures for increasing V_{oa}.

V_{oa} is increased from 1.78 to 2.08 L/(s·m²), reducing $ERROR_{TOT}$ slightly to 5.13 GJ/day. Subsequently, increasing interior zone area to 73% and increasing the room temperature to 22.5 °C reduces $ERROR_{TOT}$ to 4.94 GJ/day as shown in Table 13.5.

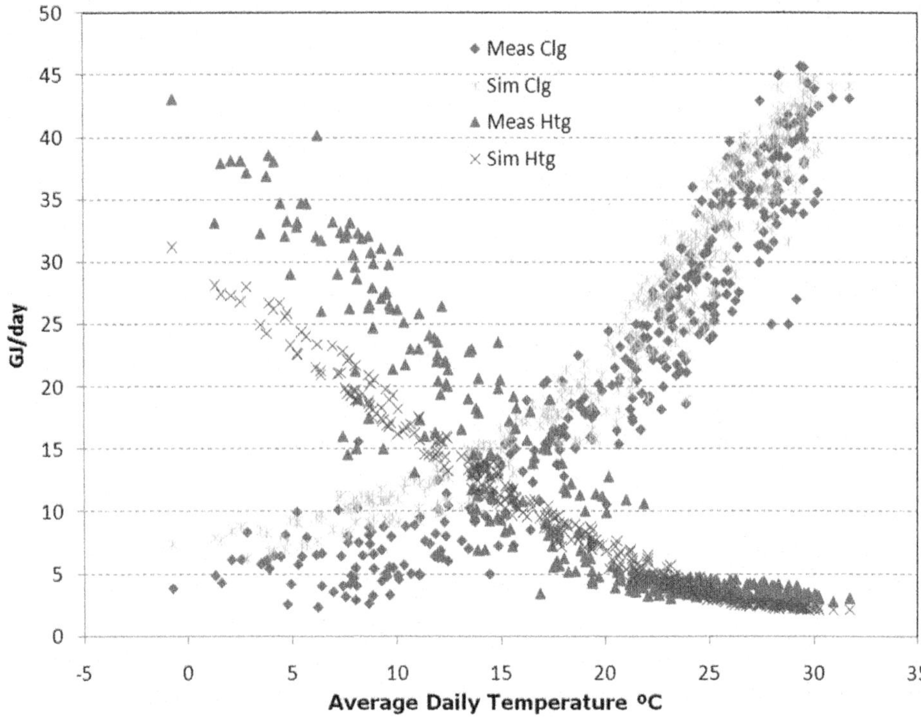

Figure 13.11 Measured and initial simulated heating and cooling consumption as functions of average daily temperature for Example #2

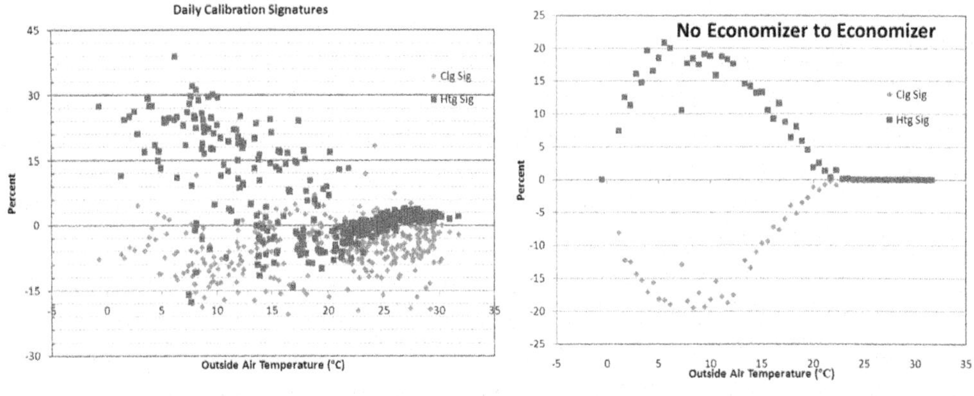

Figure 13.12 Initial calibration signatures for Example #2 and characteristic signatures for going from no economizer to an economizer

Table 13.4 Initial error measures for Example #2

	GJ/day		GJ/day
$RMSE_{Htg}$	5.02	$RMSE_{TOT}$	6.14
$RMSE_{Clg}$	3.54	MBE_{TOT}	0.18
MBE_{Htg}	−2.21	$ERROR_{TOT}$	6.14
MBE_{Clg}	2.40		

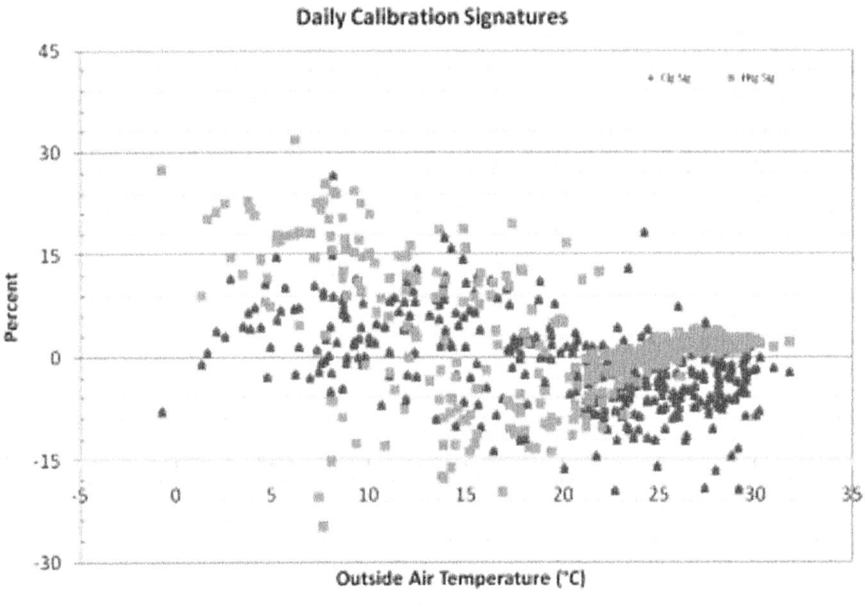

Figure 13.13 Calibration signatures after economizer was added

Table 13.5 Error measures for the calibrated simulation

	GJ/day		GJ/day
$RMSE_{Htg}$	3.81	$RMSE_{TOT}$	4.90
$RMSE_{Clg}$	3.08	MBE_{TOT}	0.61
MBE_{Htg}	0.49	$ERROR_{TOT}$	4.94
MBE_{Clg}	0.11		

The final calibration signatures are shown in Figure 13.14 and the simulated heating and cooling values for the calibrated simulation are compared with the measured values in Figure 13.15. The average cooling consumption for this building was 23.25 GJ/day. So, the heating RMSE is a relatively high 16.3% of the average cooling consumption. Since it is known that the economizer is not always operated in this building, it is possible that the relatively high scatter in the final calibrated model is due to this irregular operation of the building. The measured heating points in Figure 13.15 for temperatures from approximately 10 °C to 20 °C appear to fall into two groups that may correspond to the economizer operating or not operating.

Summary – a simple process for calibrating simulations

Calibrated simulations can be powerful tools for analyzing building performance and predicting future behavior. Using the process of calibration signatures and characteristic signatures that has been explained, the calibration of a simulation can be greatly simplified. The best way to be successful with this approach is to practice it until familiarity with the procedures and most important parameter variations is achieved. It is also helpful to generate characteristic signatures for the specific building to be calibrated, to better match the climate and system type of the building. In all simulations, the more that is known about actual building conditions being simulated, the more accurately the building can be simulated. But even without extensive knowledge of the building, this process can provide acceptable results in a reasonable amount of time when done correctly. It is therefore useful to become acquainted with the procedures and practiced in the application of the process.

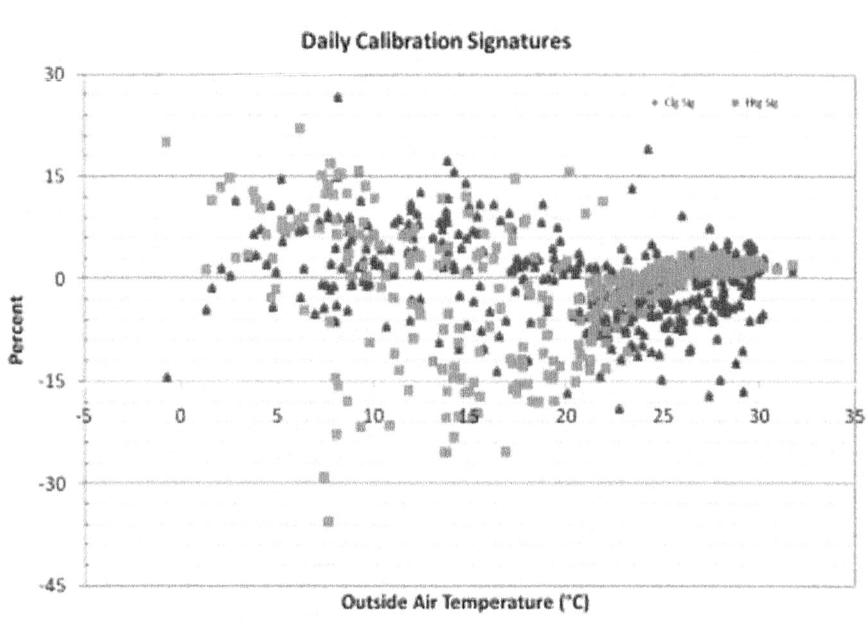

Figure 13.14 Final calibration signatures for Example #2

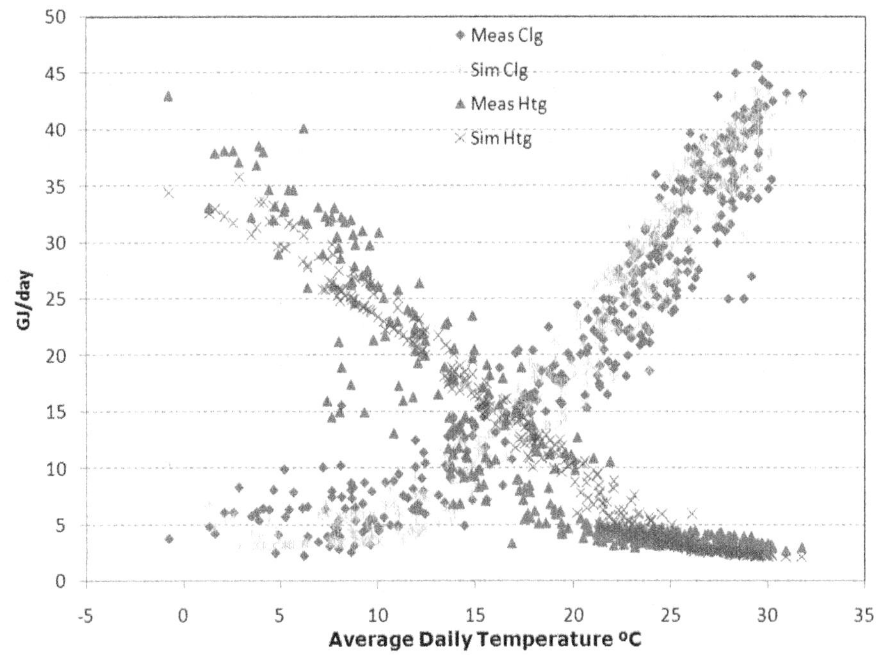

Figure 13.15 Measured and calibrated simulated heating and cooling consumption as functions of average daily temperature

Using calibrated simulation for practical optimization of operational performance

Optimizing operational performance of a building can be defined in many ways. The fundamental approach for optimization is to first minimize the loads in the building, minimize waste in the secondary system, and then optimize plant performance. Operational measures in the residential sector tend to emphasize moving interior temperatures to the edges of the comfort zone and turning off lights and appliances when not needed. Retrofits tend to emphasize improving envelope performance by adding insulation and replacing inefficient lighting, appliances, and heating and cooling equipment with more efficient equipment. The same approach has historically been taken throughout the commercial sector as well; the primary difference has been greater emphasis on improved lighting and HVAC equipment and less emphasis on envelope retrofits, since the performance of these buildings tends to be dominated by internal heat gains rather than envelope performance.

Within this chapter, we consider only internal gain dominated buildings. In many such buildings, energy waste that can be eliminated, without changing the equipment in the building, primarily occurs in the secondary systems and the plant.

Optimization in the strict sense requires the optimization of an objective function. Energy use constrained by the requirement that comfort is maintained is the objective function to be optimized. The millions of different variations in building systems make it impractical (at least today) to use simulation to truly optimize the performance of an individual building. However, insights afforded by simulation can be used to substantially improve the performance of an individual building – hence our use of the term "practical optimization."

The simplest detection of faulty operation might be a comparison of a calibrated simulation with a design simulation. This comparison might immediately show significant differences between the design system settings and apparent operation. After verifying the differences with a field visit, the settings may be changed and operation improved.

It is critical to note that we have said "apparent operation." A calibrated simulation should never be taken as truth without verifying differences in the field. A simulation model never models every facet of a building's operation, so there are always deviations between predicted behavior and actual behavior, even with the best calibrated simulation. There are often multiple sets of input variables that will give an equally close approximation to a building's performance – sometimes with a large difference in one or more parameters.

The use of calibrated simulation for operational optimization of buildings will be discussed in this chapter within the existing building commissioning process, so we next provide an overview of the EBCx process to give context for the presentation of different uses of calibrated simulation in the process.

Building commissioning and calibrated simulation

The term "commissioning" has been used for generations to refer to the process of making a navy ship completely ready for active duty. Over the last few decades, the term has come to refer to the process that makes a building or some of its systems completely ready for use. In the context of this chapter, commissioning will refer to making the HVAC systems of a building fully functional.

It is reasonable to expect that any new building will operate properly when it is finished, but the manager of a large medical center once said, "I don't want any more new buildings. I want three-year-old buildings where the problems have been fixed." The objective of the HVAC commissioning process that is now increasingly being used in new buildings is to ensure that when a building is delivered to the owner, the problems have been fixed and the systems operate as they were designed to operate. When the term "new building commissioning" is used in this chapter, it will refer to this process.

There is also an increasing recognition that most existing buildings do not operate as well or as efficiently as they could and should. A study at Texas A&M University found that simply applying a "commissioning" process to over 60 existing buildings had produced average reductions in energy use of 14% (Bynum et al. 2008). These savings were not achieved by installing more efficient equipment in the buildings; they were achieved simply by fixing or replacing broken components and changing control settings to optimize the operation of the HVAC systems for the way the building was being used while maintaining, and in many cases improving, the comfort conditions in the buildings. The process of identifying HVAC system problems, diagnosing the problem solutions, determining optimum control algorithms, and then implementing these changes in existing buildings is more complex than it may appear and often becomes an iterative process. For example, many constant volume air distribution systems operate with more airflow than required, using more fan power and more heating and cooling energy than necessary. Reducing this airflow can make one zone not cool properly due to an imbalance in the air distribution system. These competing objectives necessitate going through the complete problem identification-diagnosis-optimization-implementation sequence again.

This commissioning process goes by multiple names, including retro-commissioning, re-commissioning, Continuous Commissioning®,[2], and existing building commissioning.

ASHRAE Guideline 0.2 (ASHRAE 2015), *Commissioning Process for Existing Systems and Assemblies*, has adopted the term "existing building commissioning," so this term will be used throughout this chapter.

The version of EBCx practiced by the authors is Continuous Commissioning® (CC®). CC is an on-going process developed at Texas A&M University to resolve operating problems, improve comfort, optimize energy use, and identify retrofits for existing commercial and institutional buildings and central plant facilities. CC focuses on improving overall system control and operations for the building, as it is currently utilized, and on meeting existing facility needs. It is not intended to ensure that a building's systems function as originally designed, but it ensures that the building and its systems operate optimally to meet the current uses of the building. Near-optimal operational parameters and schedules are developed based on actual building conditions and current occupancy requirements. Once the operation has been optimized, optimization is maintained by a continuous on-going process that monitors building performance and corrects problems that arise.

Once a building is operating properly, various HVAC failures, control changes, and building usage changes that do not cause comfort problems will generally reduce the operating efficiency over time unless there is an active program that identifies and responds to these changes. For example, in a volleyball arena on the Texas A&M campus, control failures in two of the 12 air supply units in the arena caused them to go into a continuous heating mode. This unneeded heating caused no comfort problems since the other 10 units readily compensated by providing additional cooling – but the operating cost increased by US $75,000/year (Claridge et al. 2004). Different procedures are used to identify and diagnose such problems – they will be called "on-going commissioning" in this chapter.

Overview of the existing building commissioning (EBCx) process

Commissioning of existing buildings has become a key energy management activity over the last decade, often resulting in energy savings of 10%, 20%, or sometimes 30% without significant capital investment. It generally provides an energy payback of fewer than three years. In addition, building comfort is improved, systems operate better, and maintenance cost is reduced. Commissioning measures typically require no capital investment, though the process often identifies maintenance that is required before the commissioning can be completed. Potential capital upgrades or retrofits are often identified during the commissioning activities, and knowledge gained during the process permits more accurate quantification of benefits than is possible with a typical audit. Involvement of facilities personnel in the process can also lead to improved staff technical skills.

Existing building commissioning of HVAC systems is typically defined as the commissioning of HVAC systems in an existing building, and we will only consider the EBCx process for HVAC systems in this chapter. The focus of EBCx is usually on energy-using equipment such as mechanical equipment and related controls. It may or may not bring the building back to its original design intent, since the occupancy may have changed or the original design documentation may no longer exist. The major phases of the EBCx process according to the proposed ASHRAE Guideline 1.2, *Technical Requirements for the Commissioning Process for Existing HVAC & R Systems and Assemblies* (ASHRAE 2018) are:

1 Multi-facility planning
2 Assessment
3 Investigation

4 Implementation
5 Hand-off
6 On-going commissioning

A calibrated simulation may be utilized in each of the six steps of the EBCx process. Screening buildings for the overall level of potential savings can be most effectively performed utilizing calibrated simulation. The assessment of the building operation identifies major potential EBCx measures, the overall potential savings from implementing a package of measures, and the cost of implementation. The potential savings are most readily evaluated using a rapidly calibrated simulation, though it is still often done using spreadsheet methods. This early calibrated simulation can be very helpful in developing the project plan, and a calibrated simulation completed after the detailed investigation can form an excellent baseline for savings determination. Planning the detailed investigation can be aided by using this early calibrated simulation as well, while a refined calibrated simulation based on findings of the detailed investigation provides valuable guidance on measure selection and implementation order. Calibrated simulation is an approved method for creating baselines to determine energy savings and has been found to be an effective way to track continuing performance and detect and diagnose faults that degrade building performance during the on-going commissioning period.

These applications of calibrated simulation within the EBCx process are summarized in Table 13.6, which shows how various tasks in each phase of the EBCx process are performed without and with the use of calibrated simulation. Use of calibrated simulation in each of these phases is discussed in detail in this section. Case study examples of the use of calibrated simulation in the EBCx process are provided for several phases.

Table 13.6 Comparison of the EBCx process without and with calibrated simulation

Phase	*W/o calibrated sim.*	*With calibrated sim.*
Multi-facility planning	Prioritize using energy use intensities (EUIs)	Prioritize using estimated range of potential savings
Assessment	Measures investigated selected using engineering experience; estimate savings with spreadsheet or uncalibrated simulation; baseline regression models created	Calibrated simulation inputs indicate some improper operation and guide selection of measures investigated; improved savings estimate with calibrated simulation; calibrated simulation used for baseline models
Investigation	Measures selected based on experience of engineer	Investigation priority determined by savings potential from calibrated simulation; savings updated with calibrated simulation
Implementation	Plan based on experience	Calibrated simulation helps prioritize effort in implementation
Hand-off	Document benefits provided with regression model	Document benefits provided with calibrated simulation
On-going commissioning	Periodically visit site and examine system operation	Visit site and correct problems when significant degradation occurs as indicated by calibrated simulation

Multi-facility planning: screen buildings for potential level of opportunity before a site visit

When multiple buildings are potential candidates for the optimization using the EBCx process, it is common to prioritize buildings for a site visit to conduct an assessment based on a simple performance index like energy usage in GJ/(m^2·yr). However, simple calibrated simulations can provide a much more accurate screening of buildings for energy savings opportunities before visiting the building in person. Opportunity Assessor is a building assessment tool developed by the Energy Systems Laboratory (ESL) at Texas A&M Engineering Experiment Station to identify the potential impact of commissioning an existing building. It can quickly determine if a building has a large or small potential savings from implementation of the CC process with only a fraction of the inputs normally used for building simulations.

Opportunity Assessor uses a statistical approach with limited inputs including utility bills, local weather, and information about the building to quickly estimate the energy savings opportunity using the CC process.

Once a building has been identified as an opportunity, a detailed assessment will investigate the building's operation, identify measures, and estimate savings from those measures.

By relying on a calibrated simulation, many potential pitfalls when using metrics such as EUIs for ranking target buildings for commissioning are eliminated. EUIs without context typically do not normalize well for climate, space usage, and occupancy concerns.

The questionnaire for Opportunity Assessor is simple enough to be completed using limited system information and conversations with building operators. The types of questions include those shown in Table 13.7. Utility bills for the different energy sources for heating and cooling, which are commonly available, are the only other inputs from the user. Weather data is collected automatically using the supplied zip code and weather data provided by the National Oceanic and Atmospheric Administration (NOAA).

Opportunity Assessor uses the answers to the questions to build a simplified steady-state simulation model using WinAM. It creates up to 50 different semi-random plausible simulation models and calibrates each one separately to the utility bill data provided.

With the 50 calibrated models, a series of common EBCx measures are applied to the simulations. The difference in the annual energy consumption between the models is the energy savings, and the average cost savings of the 50 models are presented to the user. This information aids the user in the binary decision of whether to proceed with the project, collecting more

Table 13.7 Questions to answer for using calibrated simulation in Opportunity Assessor

Question	*Possible answers*
ZIP code of building	
Total conditioned floor area	
Number of floors	
Type of building	Office, school, hospital, laboratory, etc.
% of external wall with windows	
Window type	Single pane, double pane
# of hours per day the AHUs are typically on	
How does the building get its cooling?	Chilled water, DX A/C units
How does the building get its heating?	Electric resistance heating, heat pump, hot water, steam, gas furnace
Is the HVAC system primarily single or dual duct?	Single duct, dual duct
Constant air volume or variable air volume?	CAV, VAV

information and improving the calibrated simulation, or move on to investigate other potential buildings. Figure 13.16 shows a diagram of this process.

As an example, the preliminary phases of the EBCx process were applied to three different buildings in the Bryan Independent School District (BISD). The buildings included Kemp Elementary School, Davila Middle School, and Rudder High School (shown in Figure 13.17), located in Bryan, Texas.

The answers to the input questionnaire for Rudder High School are provided in Table 13.8, along with the utility bill data for electricity and natural gas in Tables 13.9 and 13.10. The final savings estimates provided by Opportunity Assessor are given in Table 13.11.

Even with limited information, using calibrated simulations in the preliminary assessment phase has indicated that Rudder High School (16% savings) has potential as an EBCx project, while Kemp Elementary School (5% savings) may not be investigated further.

Calibrated simulation in the assessment

The assessment phase of the EBCx process "develops a plan for the heating, ventilating, air conditioning, and refrigeration (HVAC & R) EBCx Process that includes the estimated cost and impacts of the process" (ASHRAE 2018). This typically involves two to five days spent on

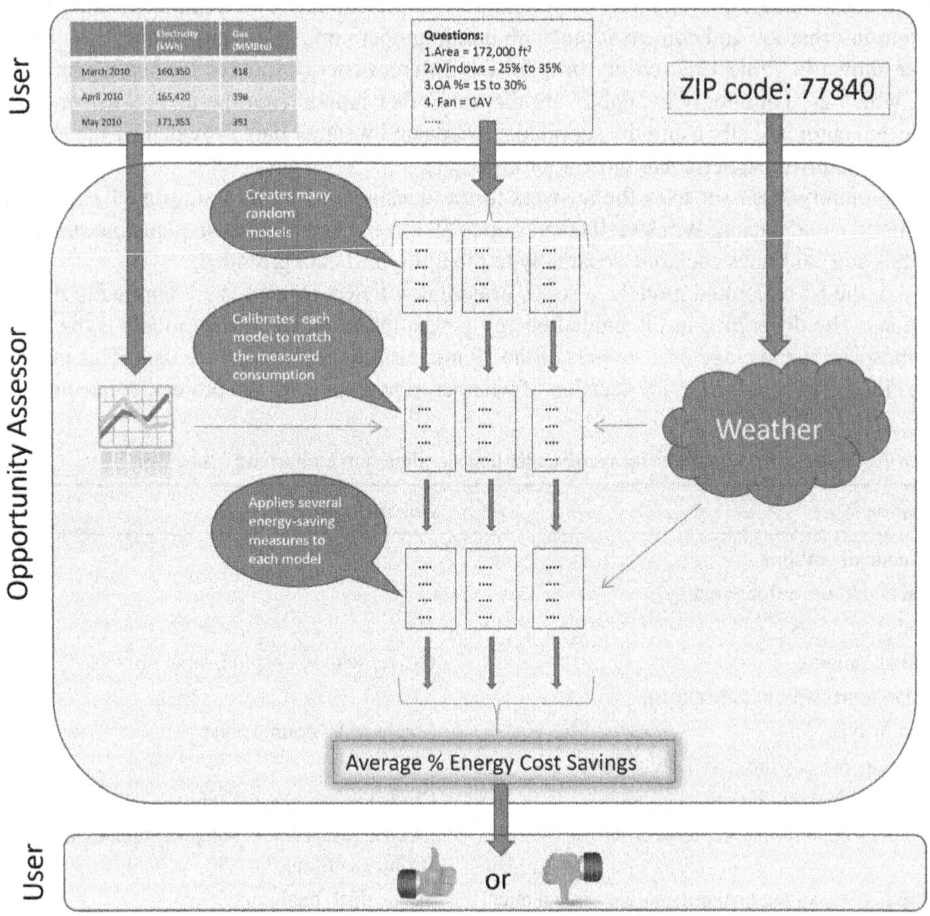

Figure 13.16 Diagram of how Opportunity Assessor uses calibrated simulation to estimate energy and cost savings potential

Figure 13.17 Rudder High School

Source: Photo courtesy of Bryan ISD. Used by permission

Table 13.8 Answers to Opportunity Assessor questionnaire for Rudder High School

Question	*Answer*
Cooling type	Chilled water
On-site chiller or district CHW source	On-site chillers
Is CHW provided to any other buildings?	Not sure
Heating type	Hot water
On-site boiler or district HW source	On-site boiler
Is CHW provided to any other buildings?	Not sure
Is there outside lighting?	Yes
Conditioned area (m^2)	27,380
Number of floors	1
Type of building	High school – grades 9–12
% window area	Average (15%–40%)
Window type	Not sure
HVAC system type	Single duct
Fan control type	CAV

Table 13.9 Rudder High School electricity utility bills

Start date	*End date*	*Consumption [kWh]*	*Bill cost*
1/21/2016	2/18/2016	270,048	$27,593
2/18/2016	3/21/2016	320,160	$30,912
3/21/2016	4/20/2016	340,704	$34,719

(*Continued*)

Table 13.9 (Continued)

Start date	*End date*	*Consumption [kWh]*	*Bill cost*
4/20/2016	5/19/2016	371,424	$36,998
5/19/2016	6/21/2016	410,688	$38,239
6/21/2016	7/20/2016	342,720	$31,837
7/20/2016	8/19/2016	371,232	$34,581
8/19/2016	9/20/2016	422,688	$38,168
9/20/2016	10/20/2016	387,936	$36,739
10/20/2016	11/18/2016	323,328	$30,996
11/18/2016	12/20/2016	282,336	$27,965
12/20/2016	1/19/2017	262,080	$25,972

Table 13.10 Rudder High School natural gas utility bills

Start date	*End date*	*Consumption [GJ]*	*Bill cost*
1/23/2016	2/22/2016	1,133	$5,205
2/22/2016	3/22/2016	1,023	$4,569
3/22/2016	4/22/2016	994	$4,238
4/22/2016	5/23/2016	681	$3,169
5/23/2016	6/22/2016	192	$970
6/22/2016	7/22/2016	111	$742
7/22/2016	8/22/2016	115	$819
8/22/2016	9/23/2016	173	$1,184
9/23/2016	10/24/2016	177	$1,268
10/24/2016	11/21/2016	603	$4,314
11/21/2016	12/21/2016	1,139	$7,743
12/21/2016	1/24/2017	1,369	$9,997

Table 13.11 Estimated total dollar savings for three schools in BISD

School	*Estimated annual potential dollar savings*
Rudder High School	16%
Davila Middle School	10%
Kemp Elementary School	5%

the site interviewing building operators, examining building control system screens, and making measurements in the building mechanical rooms. The measures investigated are typically selected by the team leader based on their engineering judgment applied to the information collected including the HVAC system types, the information obtained from the building operators, and the past utility bills. After the various measures are investigated during the site visit, the potential savings from the process or "impacts of the process" is normally carried out using some form of uncalibrated simulation – many providers use simple spreadsheets. The creation of the baseline to be used for measuring the savings from the EBCx process may be created during the assessment phase or during the investigation phase. This chapter will discuss it under the assessment phase. The normal process for creating the baseline for savings determination is to develop a piece-wise linear regression model of the past energy use, typically using outside air temperature as the independent variable.

The following sections discuss the use of calibrated simulation during the assessment phase in three tasks: (a) guide selection of the EBCx measures investigated, (b) estimate energy savings and comfort improvements possible from the EBCx process, and (c) develop performance baselines for savings measurement.

Guide selection of EBCx measures investigated during assessment

While the measures investigated during the assessment are normally determined primarily by the experience and judgment of the lead engineer, the HVAC system parameters determined as the inputs needed to obtain a well-calibrated simulation may immediately identify areas where the building is not operating as expected. The calibrated simulation used for this purpose may be selected from the Opportunity Assessor (OA) simulation runs, or an individually developed calibrated simulation may be used for this purpose.

This application is illustrated with a case study. When the on-site assessment of the case study building was conducted, OA had not been set up to make the input files of its runs available to the EBCx engineer. In this case, the engineer leading the assessment developed a calibrated WinAM assessment prior to arriving on site from the utility bills and other information that had been provided. This information included an earlier energy audit of the building that had identified some leaking preheat coils in the building, and it was considered likely by the site personnel that these preheat coils had not been repaired.

The case study building was a 13,800 m^2 office building in the eastern United States. It had a plant with gas boilers and electric chillers and used single-duct variable-air-volume (SDVAV) air handlers. Additional equipment information included:

- 14 AHUs with 2 MW tons total cooling capacity
- 2 screw chillers with 1 MW capacity each
- 2 MW cooling tower
- 2 boilers @ 1 MW output
- 2 HW pumps @ 41 m^3/hr, 120 kPa head
- 2 boiler pumps @ 41 m^3/hr, 60 kPa head
- Water-side economizer

The available electric and gas consumption data are plotted in Figures 13.18 and 13.19.

It is evident that the summer gas consumption is very high relative to the peak winter consumption. There was no information available regarding the shape of the building or the window and wall properties, so these were estimated from a picture of the building that was located on the web. WinAM did not have water-side economizer modeling capabilities, so an air-side economizer was assumed instead.

There were two very striking characteristics of the simulation inputs for this building once calibrated. The minimum flow setting of the VAV boxes based on the calibrated simulation was 4.6 L/(s-m^2) and the maximum economizer flow was 1.0 L/(s-m^2) of outside air. These inputs provided two major items to investigate during the assessment.

The investigation found that the water-side economizer was being used – the chillers were not on when the team arrived at the site. They did come on later in the day, and it was found that a flow restriction in the water-side economizer was severely limiting its cooling capacity – consistent with the calibrated simulation result showing maximum outside airflow of 1.0 L/(s-m^2).

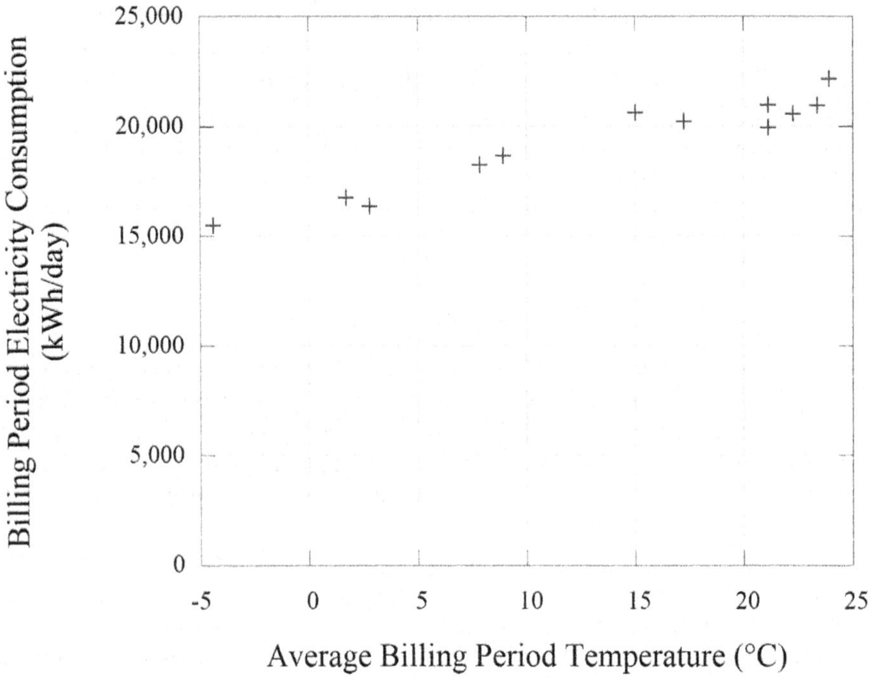

Figure 13.18 Average daily electricity consumption as a function of average billing period temperature

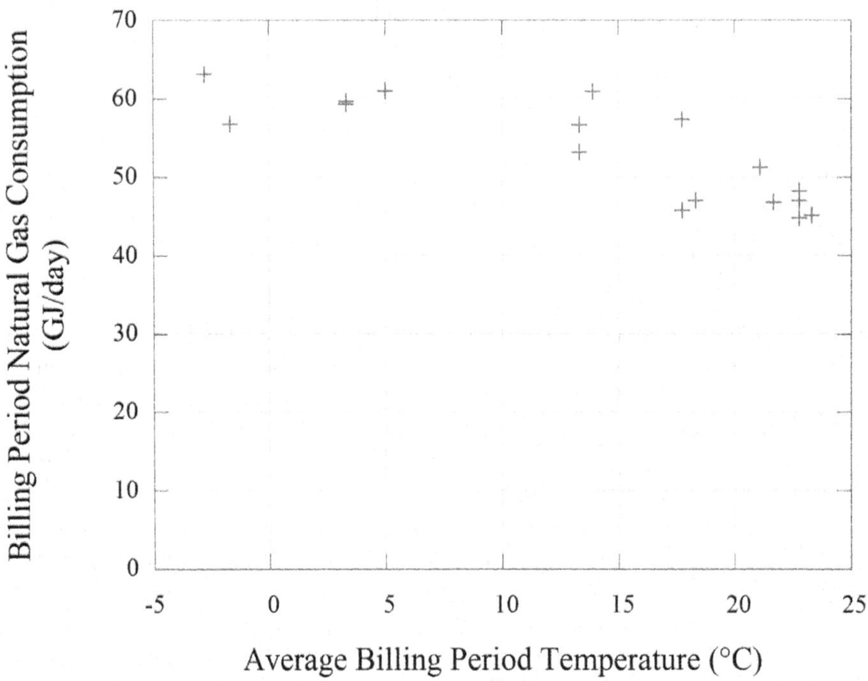

Figure 13.19 Average daily natural gas consumption as a function of average billing period temperature

Although it was a relatively cool day, the VFDs were all running from 65% to 100% of full speed, with only one below 75%. Half were above 90%. The AHU leaving temperatures were all ~13 °C, but the air supplied to the spaces ranged from 16 °C to 22 °C, consistent with the high minimum flow settings found in the calibrated simulation and with the very high summer heating consumption.

Estimate the energy savings possible from the EBCx process

The normal approach used to estimate the energy savings from the EBCx process during the assessment is to use some form of simple simulation or spreadsheets that simulate individual portions of the HVAC system.

Calibrated simulation has been shown to be beneficial in the initial assessment and in planning the preliminary on-site assessment. With the additional information gathered during the on-site assessment, a more accurate and detailed assessment of the savings potential using calibrated simulation is possible.

In the initial assessment using Opportunity Assessor, systems are typically aggregated together into a simplified model. However, for more accurate prediction of savings, the different systems in the building can be specified in greater detail. For example, at the Rudder High School, the initial model assumed that the building used a single-duct variable-air-volume system. In reality, Rudder High School has three constant volume air handling units in addition to variable-air-volume units as shown in Table 13.12.

During the on-site visit, the total number of air handling units, the type of air handling units, and the number of each type of air handling units used in the building was obtained, along with the floor area served. Accurate information regarding the number of occupants and the schedule for each AHU and the events schedule for Rudder High School was acquired. The setpoints for the space temperature, supply temperature, and preheat and precool temperature were set according to gathered trend data.

Once the model was calibrated, the effect of different energy conservation measures can be estimated by changing the various system parameters in the simulation. For Rudder High School, the following commissioning measures were applied:

1 The design supply airflow for the constant air volume air handling units was reduced from 1.2 CFM/ft² to 1 CFM/ft² for CAV 1 and 1.1 CFM/ft² for CAV 2.
2 The minimum occupied outside airflow was reduced from 0.55 CFM/ft² to 0.26 CFM/ft².

Table 13.12 Modeled air handling units at Rudder High School

Name	*Floor area (m²)*	*Interior zone percentage*	*Window + wall area (m²)*	*Window percentage*
CAV 1	400	60	200	25
CAV 2 fieldhouse	300	60	100	25
CAV 3	2,200	60	1,100	25
Dance studio	200	0	100	40
VAV 1	1,000	70	500	20
VAV 2	900	60	400	40
VAV 3 fieldhouse	2,300	60	1,100	40
VAV 3 fieldhouse OAU	1,000	60	500	40

3 The preheat coil setpoint was reduced from 18.3 °C to 7.2 °C.
4 The entering condenser water temperature setpoint was changed from a constant 28 °C to an offset of 2.8 °C from the outdoor wet bulb temperature (a 2.8 °C approach).

The total savings amounted to approximately $64,000, which corresponds to 15% savings from the baseline energy cost total. The details are shown in Table 13.13. The estimated potential using OA using the limited information before the on-site visit was 16%±2%, which matches the 15% estimation from the more detailed model. The breakdown of the energy end uses providing the cost savings is shown in Figure 13.20.

Table 13.13 Estimated savings from commissioning measures at Rudder High School

Scope	*Parameter name*	*Change*	*Yearly savings*
Central utility plant	Condenser water schedule type	Constant ⇨ Wet bulb plus offset	$2,780
Central utility plant	Constant condenser water temperature	28 °C ⇨ Not used	
Central utility plant	Condenser water control approach	Not used ⇨ 2.8 °C	
Central utility plant	Condenser water minimum temperature	Not used ⇨ 15.6 °C	
CAV 1	Design supply airflow	6 ⇨ 5 L/(s·m²)	$761
CAV 1	Min occupied outside airflow	2.83 ⇨ 1.32 L/(s·m²)	$551
CAV 1	Constant preheat coil setpoint	12.8 ⇨ 7.2 °C	$49
CAV 2 fieldhouse	Design supply airflow	6 ⇨ 5.5 L/(s·m²)	$398
CAV 2 fieldhouse	Min occupied outside airflow	2.83 ⇨ 1.32 L/(s·m²)	$704
CAV 2 fieldhouse	Constant preheat coil setpoint	18.3 ⇨ 7.2 °C	$171
CAV 3	Min occupied outside airflow	2.83 ⇨ 1.32 L/(s·m²)	$3,153
CAV 3	Constant preheat coil setpoint	18.3 ⇨ 7.2 °C	$869
Dance studio	Design supply airflow	6 ⇨ 3.8 L/(s·m²)	$1,328
Dance studio	Min occupied outside airflow	2.33 ⇨ 1.32 L/(s·m²)	$242
Dance studio	Constant preheat coil setpoint	18.3 ⇨ 7.2 °C	$2,538
VAV 1	Minimum occupied supply airflow	2.8 ⇨ 1.5 L/(s·m²)	$839
VAV 1	Min occupied outside airflow	2.33 ⇨ 1.32 L/(s·m²)	$875
VAV 1	Constant preheat coil setpoint	18.3 ⇨ 7.2 °C	$354
VAV 2	Minimum occupied supply airflow	2.8 ⇨ 1.5 L/(s·m²)	$1,147
VAV 2	Min occupied outside airflow	2.33 ⇨ 1.32 L/(s·m²)	$766
VAV 2	Constant preheat coil setpoint	18.3 ⇨ 7.2 °C	$254
VAV 3 fieldhouse	Minimum occupied supply airflow	2.8 ⇨ 1.5 L/(s·m²)	$3,509
VAV 3 fieldhouse	Min occupied outside airflow	2.28 ⇨ 1.32 L/(s·m²)	$2,679
VAV 3 fieldhouse OAU	Minimum occupied supply airflow	2.8 ⇨ 1.5 L/(s·m²)	$1,604
VAV 3 fieldhouse OAU	Min occupied outside airflow	2.28 ⇨ 1.32 L/(s·m²)	$1,295
VAV 3 fieldhouse OAU	Constant preheat coil setpoint	18.3 ⇨ 7.2 °C	$604
VAV 4 cafeteria	Minimum occupied supply airflow	2.8 ⇨ 1.5 L/(s·m²)	$542
VAV 4 cafeteria	Min occupied outside airflow	2.23 ⇨ 1.32 L/(s·m²)	$206
VAV 4 cafeteria	Constant preheat coil setpoint	18.3 ⇨ 7.2 °C	$584
VAV 5	Minimum occupied supply airflow	2.8 ⇨ 1.5 L/(s·m²)	$20,619
VAV 5	Min occupied outside airflow	2.3 ⇨ 1.32 L/(s·m²)	$14,865
TOTAL			$64,290

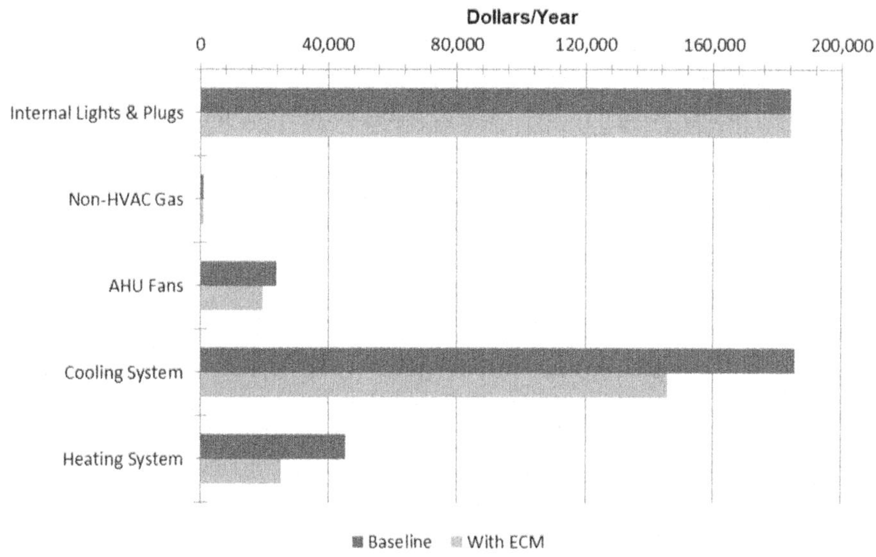

Figure 13.20 Energy cost breakdown of end-energy use

Use a calibrated simulation as the baseline for measurement of energy savings

The two major documents used as the basis for savings determination when a building has undergone the EBCx process are ASHRAE Guideline 14 *Measurement of Energy, Demand and Water Savings* (ASHRAE 2014) and the *International Performance Measurement and Verification Protocol* (EVO 2012). Option C within these guidelines is normally used to measure energy savings for EBCx projects, since it creates baseline models from historic energy consumption data using simple regression models. However, both have included the use of calibrated simulation to create the baseline for savings measurement for over a decade. Guideline 14 notes, "This technique is particularly suited to accounting for multiple energy end uses, especially where interactions occur between measures. Additionally, this technique is useful for situations where baseline shifts may be encountered and where future energy impacts may need to be assessed."

The EBCx process routinely has "interactions occur between measures," so the use of calibrated simulation to create baselines for savings determination is particularly appropriate when the process is applied to a building, provided the cost of baseline creation is acceptable. While the authors are aware of the occasional use of calibrated simulation for baseline creation, the automation or near automation of the process as described in this chapter is expected to substantially increase the use of calibrated simulation in this application in the future.

Assist in planning and conducting the detailed investigation

The investigation phase is often conducted without the use of any simulation. This sometimes leads to significant time being spent investigating measures that may represent non-ideal behavior with minimal impact on energy use or comfort. This relatively unproductive investigation can be minimized by continuing to evaluate the savings of a measure before spending considerable time investigating a measure with minimal impact.

After the on-site assessment has been conducted, the calibrated simulation used to estimate the savings potential during the assessment provides an ideal starting point for planning the detailed investigation. This simulation will normally include the data collected on site to inform its input, and it will have been used to estimate the total project savings potential. The simulation also provides useful initial estimates of the savings potential from individual measures. Hence when combined with knowledge of typical effort to implement measures, a plan can be created for the investigation that allocates appropriate amounts of time to investigate and confirm the savings potential of individual measures, as well as the ability to implement these measures.

The case study in the "Case study #1: off-line simulation in commissioning" section illustrates how measures that were identified using a calibrated simulation were investigated and subsequently implemented.

Assist in implementation

Simulation is not normally used during measure implementation, though it can potentially be beneficial. The building staff are often unwilling to have some measures implemented. Calibrated simulation estimates of the savings expected from a measure sometimes help convince the staff to permit implementation. If a measure is not implemented, this will often change the savings that will be realized by other measures. The results of an updated calibrated simulation can provide savings estimates that may be used to reprioritize the time and effort to be expended in the implementation process.

Use of calibrated simulation in hand-off

The normal use of calibrated simulation in the hand-off phase is only for the baselines used to document the savings achieved. It is important to convincingly demonstrate to the staff the improvements in energy efficiency and comfort that have already been achieved.

Use of calibrated simulation in on-going commissioning

Use of calibrated simulation for on-going commissioning

A simulation calibrated to a commissioned building can be compared to the measured consumption on an on-going basis. The comparison may be performed at any desired interval, but once a week will often be adequate. Significant consumption increases can serve as an alarm to indicate when additional commissioning follow-up is justified. Building operators generally are not interested in following up on an alarm that does not directly impact comfort or create occupant complaints unless it has a rather substantial cost impact. Hence, there is little need for this information on an hourly or even a daily basis when tracking whole-building consumption. This type of tracking may also be performed with simple regression models of consumption, as well as with more detailed physical models. If the simulation is coupled to a diagnostic system that can indicate probable causes of deviations, the value of the simulation is further increased.

The need for this type of tracking has been explored by Turner et al. (2001) and Claridge et al. (2004), and findings in two specific buildings have been investigated and reported by Chen et al. (2002) and Liu et al. (2002b). Lin and Claridge have investigated statistical criteria for fault detection in these applications (2012, 2013, 2015) and have also extended its use to diagnostics (2017).

The overall objective of the on-going commissioning (OCx) process is to ensure that the benefits obtained from EBCx process implementation are sustained and improved if possible with the passage of time. This is often conducted by having an expert engineer periodically visit the building, examine control sequences, and follow up on any indications of inefficient operation.

However, studies have shown that degradation in system performance after commissioning, specifically the decay of energy savings achieved, is highly variable from building to building (Claridge, Turner, Liu, et al. 2004). Hence if the energy performance of a building can be monitored over time, there is generally no real benefit to putting engineering effort into visiting and examining the performance of the HVAC systems unless there is evidence that performance is or has degraded. The OCx process may be viewed as the process of detecting, diagnosing, and curing faults that occur after the EBCx process has been initially implemented in a building. The OCx process may also seek to improve the performance of the building further as well.

However, this section will emphasize the application of calibrated simulation to detect and provide a partial diagnosis of faults that occur in a building after commissioning.

The first example of this application of calibrated simulation known to the authors occurred at Texas A&M University (Curtin et al. 2007, 2010, Bynum et al. 2010, 2012). Sbisa Dining Hall (Figure 13.21) is a 7,600 m^2 single-story building with a partial basement on the Texas A&M University campus. Its primary function is as a dining facility, although a small fraction of the building is dedicated to office space and a convenience store. A total of 19 AHUs, 12 on the main floor and 7 in the basement, supply the heating and cooling needs of the building. Four of these units are strictly for makeup air in the kitchen and are interlocked with individual fume hoods, two are SDVAV and the rest are SDCV with terminal reheat units. Three constant volume dedicated outdoor air handling units provide pretreated makeup air for the majority of the AHUs. Thermal energy is supplied to the building in the form of hot and chilled water from the central utility plant, both of which are metered. The electricity consumption of the building is also metered and recorded hourly.

Case-study calibration

The simulation was calibrated to the baseline consumption period of 02/02/2004–12/31/2004, the results of which are presented in Figure 13.22 and Table 13.14. As a reminder, RMSE is the root mean squared error (Equation 13.4), MBE is the mean bias error (Equation 13.5), and the

Figure 13.21 Sbisa Dining Hall

Source: Photo courtesy of Texas A&M University (https://brandguide.tamu.edu/visual-style/photo-video.html)

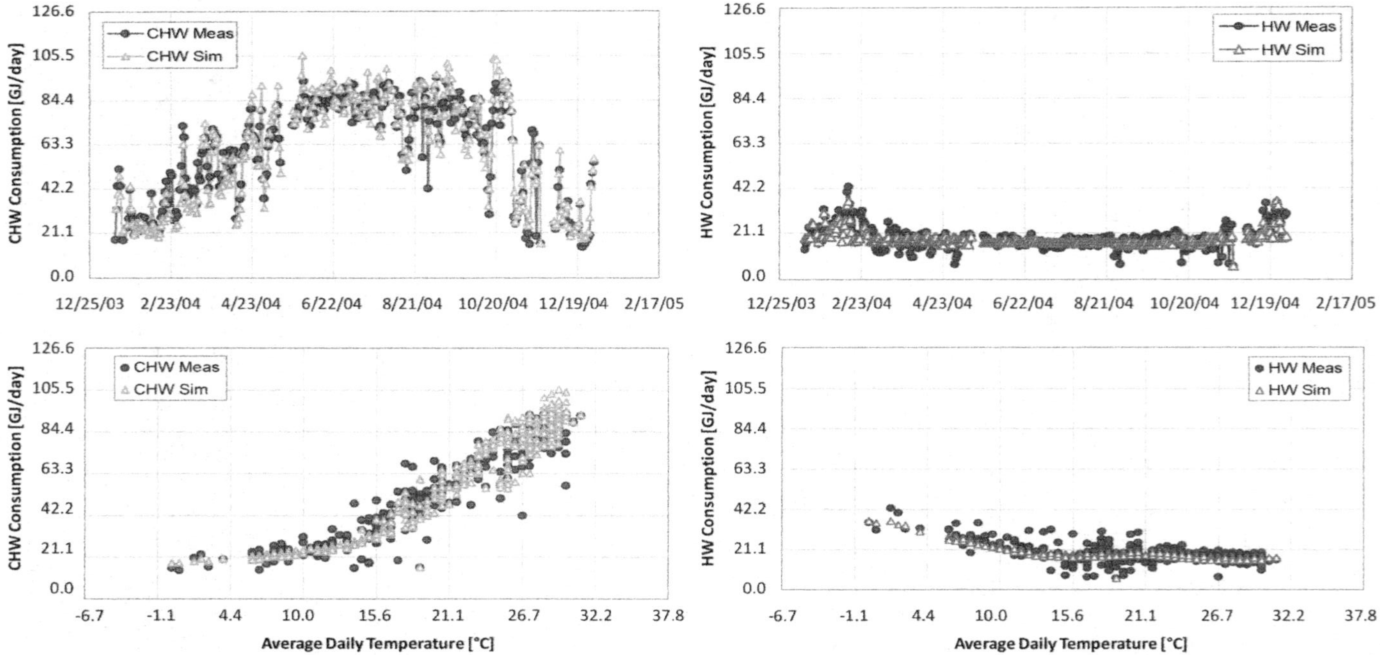

Figure 13.22 Time series and OA temperature-dependent plots of the measured and simulated cooling and heating consumption for the period used to calibrate the Sbisa Dining Hall simulation

Table 13.14 Calibration statistics of the Sbisa Dining Hall

	RMSE	*MBE*	*Max*	*Average*	*CV-RMSE*	
CHW:	7.34	0.63	95.46	59.00	12.45%	GJ/day
HW:	3.54	−0.30	43.04	18.74	18.92%	GJ/day

CV-RMSE is the coefficient of variation with respect to the RMSE, or the RMSE divided by the average value of the measured dependent variable data.

Initial findings

During the early spring of 2006, an unexpected increase in cooling energy consumption was detected, at an additional cost to the campus of approximately $950/week. The fault was determined to be the result of excessively low discharge air temperatures in two of the three dedicated OAHUs in the building. Several figures are presented below to explain how, with the calibrated simulation, the conclusion was drawn that this fault existed. Also described is how the simulation, along with the addition of some key trended control system data, assisted in narrowing down the likely cause of the fault. Note that all the plotted data are presented as daily averages.

Fault detection

Figure 13.23 displays the measured and simulated chilled water energy consumption for the period of 2/20/2006–6/4/2006, which spans the period during which this cooling energy consumption fault appeared.

The significance of the daily deviation seen in Figure 13.23 may be difficult to gauge unless it is also accumulated and compared with past performance. The cumulative effect of this increased consumption can be seen in Figure 13.24, where the daily difference between the measured and simulated consumption is multiplied by an estimated cooling energy cost of $12.74/GJ and added to the previous day's total. This figure shows that costs had increased by nearly $9,500 over those expected for the 10-week period (3/27/2006–6/4/2006).

Figure 13.25 is a plot of the difference (Meas−Sim), as a percent of the maximum consumption during the baseline period (2004), which was 95.5 GJ/day. This means that a 10% difference in Figure 13.25 corresponds to a measured cooling energy consumption of 10% of 95.5 or 9.55 GJ/day greater than that which was simulated. This error plot shows that the seven-day moving average of the percent difference had been within the ±10% range for nearly the entire period of the plot (more than two years) except for the period starting around April 25, 2006, and continuing through the beginning of June.

Diagnosis

To assist in identifying the cause of the fault, Figure 13.26 compares the WB_{Elec} consumption for this period of increased cooling energy consumption, alongside the WB_{Elec} from the same dates in 2005. The similarity of the two series in Figure 13.26 indicates that the cause of this increase in cooling energy is not likely related to an increased electric load in the building. Additionally, referring back to Figure 13.24, it can be seen that the measured HW heating energy consumption did not vary from the simulation nearly as much as the CHW consumption, and actually was consistently slightly less than the simulated consumption since the beginning of 2006. This

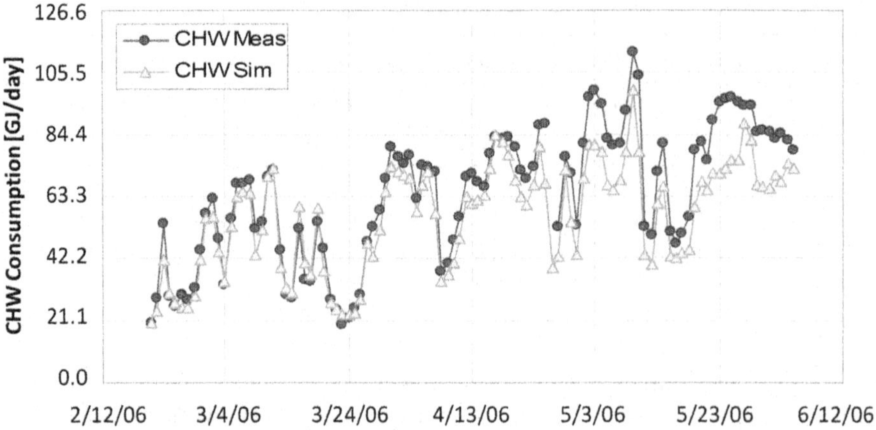

Figure 13.23 Measured and simulated chilled water energy consumption 2/20/2006–6/4/2006

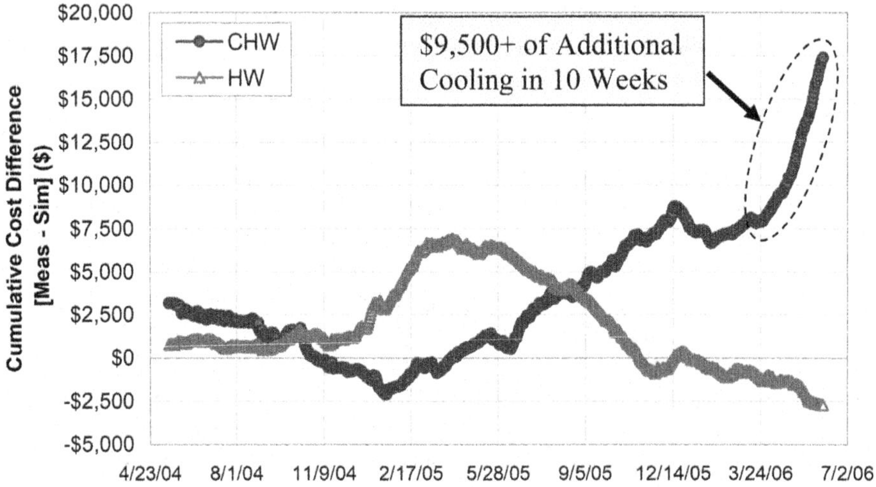

Figure 13.24 Cumulative cost difference (assuming $12.74 and $16.73/GJ for CHW and HW respectively)

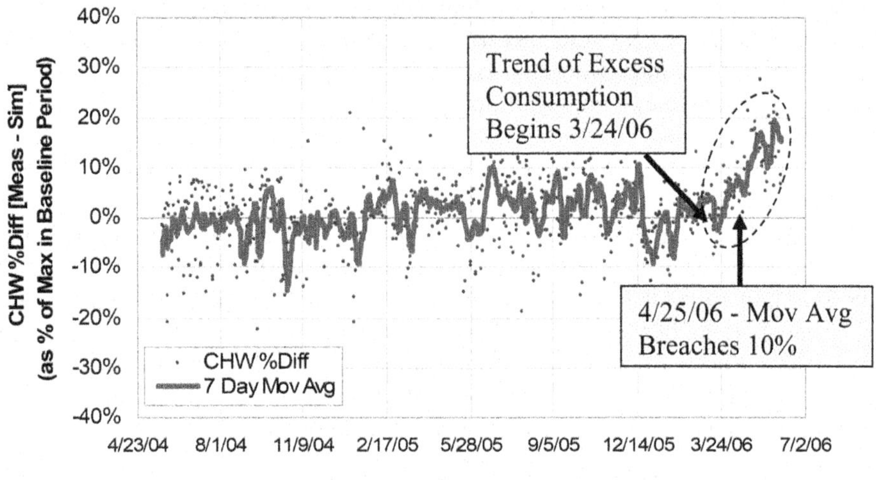

Figure 13.25 CHW % difference plot (10% error → (Meas−Sim)) of 9.55 GJ/day

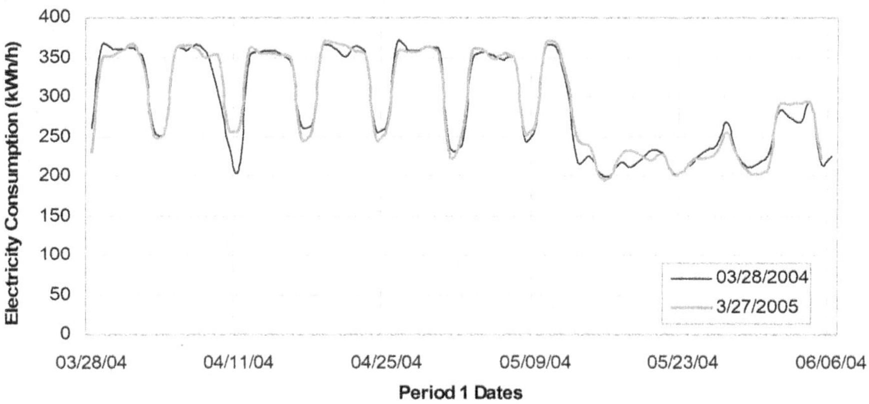

Figure 13.26 Electricity consumption for two 10-week periods one year apart; start dates of periods are 3/28/2005 and 3/27/2006

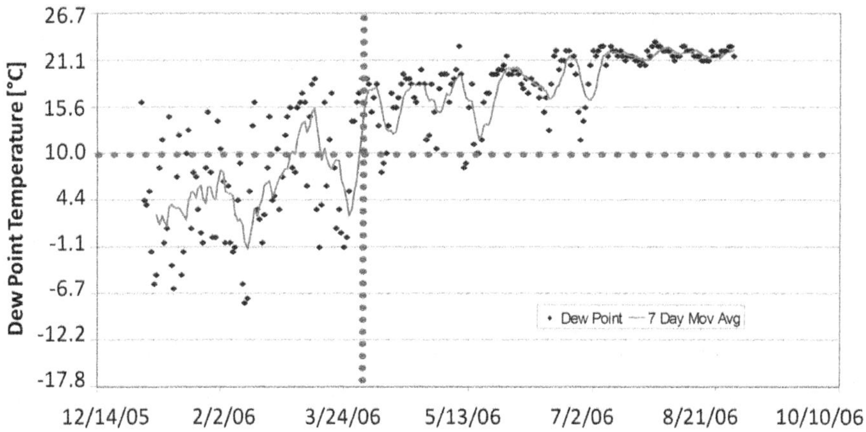

Figure 13.27 Jan–Aug 2006 average daily dew point temperature in College Station, TX

revealed that the increased cooling energy was not caused by an increase in heating energy consumption.

Further investigation into trended control data points led to the discovery of exceptionally low discharge air temperatures (extended periods at < 10 °C) in two of the three OAHUs (OAHUB1 and OAHU2) in the building. From analyses of the trend data, it was determined that these same operating conditions had been in place for an extended period (at least since November 2005), and therefore the change in discharge air temperature of these units did not coincide directly with the identified period of increased cooling consumption. But this does not mean that the two were unrelated. The average ambient dew point temperature in College Station does not tend to consistently rise above 10 °C until late March/early April (Figure 13.27). This means that the low discharge air temperatures for the outside air handling units did not result in a significant increase in latent cooling loads until spring arrived.

Case study conclusion

This investigation surmised that the OAHU's discharge air temperature control setpoints were lowered in late September/early October 2005 during a time when the outside dew point had already fallen well below its summer highs. Although the trended data is not available from that period to prove this conjecture, there are other reasons to believe this was a latent cooling issue. First, if it was an increase in sensible cooling, either hot water reheat energy would have increased to compensate for the increased cooling, or a noticeable drop in space temperatures would have occurred. It was noted previously that heating energy use in the building was actually slightly lower than expected for this period. Also, since space temperatures were trended and investigated during this period and no significant changes were identified, an increase of sensible cooling as the determining factor in the excess consumption was ruled out. Therefore, it was recommended that the discharge air temperature schedules for OAHUB1 and OAHU2 be reset to higher levels that maintain adequate humidity control in the building, without adding a penalty from excessive latent cooling.

Using calibrated simulation for fault detection and diagnosis

Fault detection and diagnosis (FDD) plays a major role in energy management and can effectively reduce operating costs and related problems. Modern advances in building automation systems (BAS) and the improvements in data handling, including collection, storage, and analysis, make detailed, on-line fault detection using calibrated simulations a cost-effective measure to maintain savings after commissioning.

The field of FDD with regards to buildings is vast. Kim and Katipamula (2017) have done a detailed literature review that presents an analysis of all the different classifications of FDD, which can include model-based systems, rules and alarms, and "black box" techniques. Well-calibrated models of a system can offer significant advantages over other approaches concerning FDD. When calibrated, they provide the best estimation of building performance under any operation. Because they are derived from fundamental engineering formulas, they can model both "normal" and "faulty" operation and discern what specific components may be the cause.

Case study #1: off-line simulation in commissioning

The following case study illustrates the use of off-line simulation in the process of commissioning an existing building. However, it also serves to illustrate the diagnostic capability of simulation, whether used off-line or on-line. It is taken from Claridge and Liu (2004).

The Basic Research Building (BRB) at the M. D. Anderson (MDA) Cancer Center is a seven-story building with 11,400 m^2 gross floor area, which includes 8,600 m^2 for the laboratory and office section, 1,900 m^2 for a library, and 900 m^2 for mechanical rooms and other purposes. The HVAC systems operate 24 hours per day.

Four single-duct constant volume air handling units (AHUs) provide cooling and heating to the laboratory and office section. The design airflow rate is 71,000 L/s with 100% outside air. The preheat deck setpoint is 12.8 °C. If the outside air temperature is below 12.8 °C, the preheat coil warms the air temperature to 12.8 °C. If the outside air temperature is higher than 12.8 °C, the preheat valve is closed. The cold deck temperature is set at 12.8 °C. The room temperature is controlled using reheat. If the room temperature is below the setpoint, which varies from 22.2 °C to 23.9 °C from room to room, the reheat coil is turned on to maintain the room temperature.

In addition to the single-duct system that serves most of the building, there is one dual-duct constant volume air handling unit, which provides cooling and heating to the library section. The design airflow is 12,700 L/s with 50% outside air intake. The cold deck setpoint is 12.8 °C. The hot deck setpoint varies from 29.4 °C to 43.3 °C as the outside air temperature decreases from 29.4 °C to 4.4 °C.

Three single-duct air handling units provide heating and cooling to mechanical rooms and other spaces. The design airflow is 6,600 L/s with 100% return air. The cold deck setpoint is 12.8 °C. If the room temperature is satisfied, the AHUs will be turned off.

Simulation development

WinAM was used to simulate the building heating and cooling energy consumption using simplified building and system models. The building was divided into two parts: the laboratory section, which uses 100% outside air, and the library section, which uses 50% outside air. Each part was simplified to two zones: interior and exterior. The operational design schedules were used in the simulation.

Figures 13.28(a) and 13.28(b) compare the measured heating and cooling with projected heating and cooling energy consumption. The projected energy consumption was

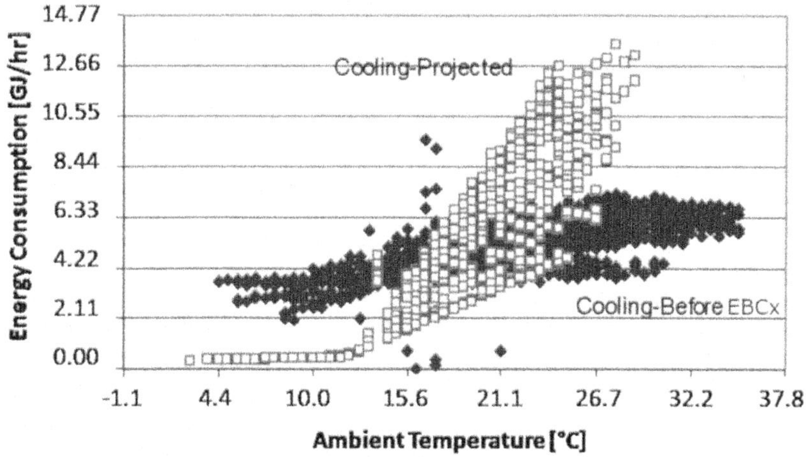

Figure 13.28(a) Comparison of projected and measured cooling energy consumption

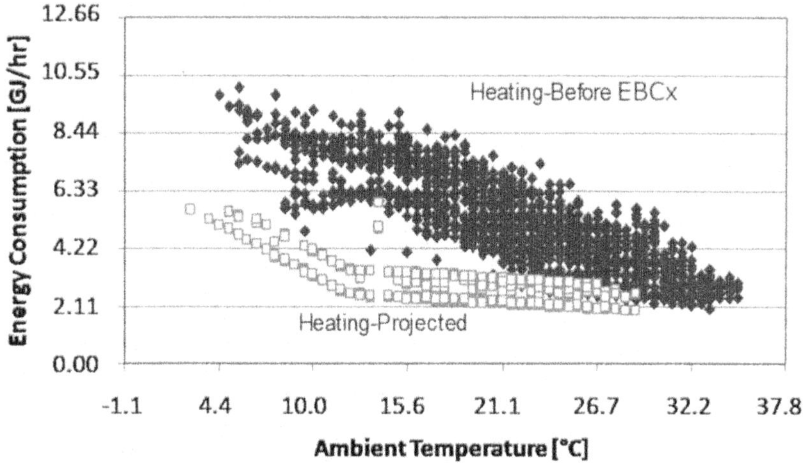

Figure 13.28(b) Comparison of projected and measured heating energy consumption

simulated using actual Houston weather but not the weather data corresponding to the measured energy consumption, since the measured dew point temperature was missing for the measured energy consumption period. The projected heating is significantly less than the measured heating, while the projected cooling is significantly higher than the measured cooling.

Fault detection and diagnosis

The projected cooling is approximately twice as high as the measured values during the peak summer period, while the projected heating is slightly lower than the measured values during winter. The cooling difference indicates that a fault may exist in the cooling energy metering system. Since the measured value is approximately 50% of the projected value, it was hypothesized that the scaling factor or the engineering conversion was set incorrectly. The measured cooling energy consumption was then multiplied by a factor of 2. Figure 13.29 presents the corrected heating and cooling energy consumption.

Later, field inspection found that the by-pass line for the utility meter was fully open. Consequently, the utility meter only measured half of the chilled water flow.

Analysis of the difference between the measured and simulated cooling energy consumption after the metering error was corrected indicated the probable presence of one or more leaking chilled water valves, so inspection of the control valves was recommended.

Field inspection

A field inspection was conducted and found that (1) all control valves were less than three months old; (2) existing pneumatic lines were used when the valves were replaced; (3) all chilled water valves are normally open with a range of 20 to 55 kPa gauge; and (4) the maximum control pressure to the valves was 35 kPa gauge due to old, leaking pneumatic lines. As a result, it was not possible to completely close the valves.

This confirmed that the leaking chilled water control valves were the primary cause of the poor performance. Fixing the leaking pneumatic lines was expected to reduce the heating and cooling energy consumption to the projected level.

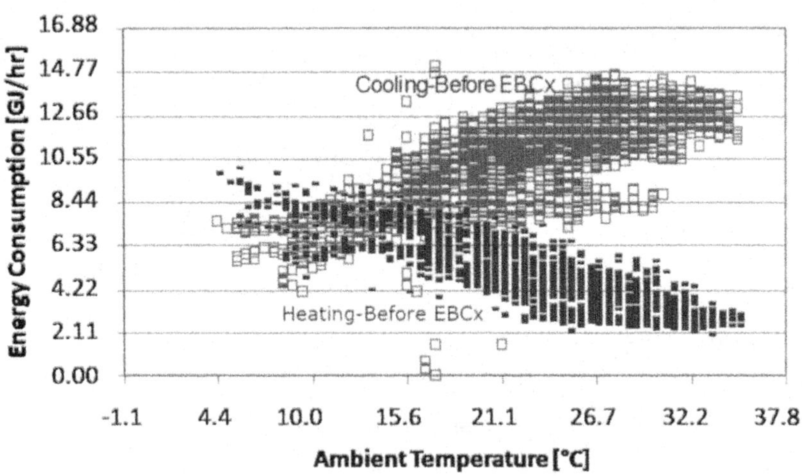

Figure 13.29 Measured heating and cooling energy consumption after meter correction

Operational optimization

It was recommended to reset the supply air temperature from 13.9 °C to 15 °C as the outside air temperature decreases 38 °C to 15 °C. This was shown through simulation to decrease simultaneous heating and cooling significantly with a moderate room humidity level increase.

Implementation

The implementation included replacing the pneumatic lines and programming the reset schedule into the building automation system. These changes were made at the same time.

Figure 13.30 compares the measured chilled water consumption before fault detection and diagnosis, the chilled water consumption after fixing the pneumatic lines and implementing the optimized schedule, and the simulated optimal consumption. Figure 13.31 provides the same comparisons for heating water.

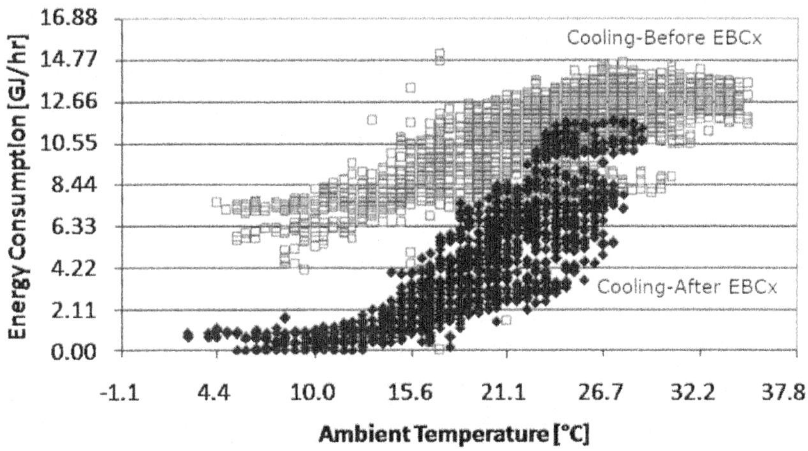

Figure 13.30 Comparison of measured cooling energy consumption before and after EBCx or repair of leaky pneumatic lines and implementation of optimal reset schedule

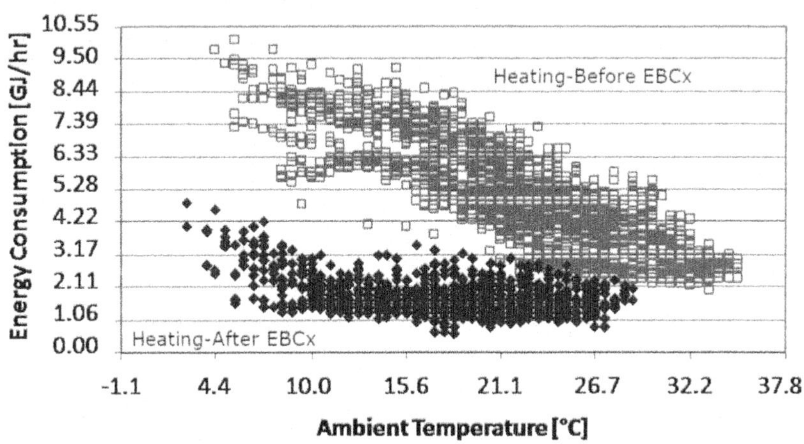

Figure 13.31 Comparison of measured heating energy consumption before and after repair of leaky pneumatic lines and implementation of optimal reset schedule

The measured annual cooling energy savings are 30,500 GJ/yr, and heating energy savings are 17,050 GJ/yr. The total annual cost savings are $369,000/yr, which includes heating savings of $129,000 and cooling savings of $240,000.

When the ambient temperature is lower than 10 °C, the measured energy consumption agrees with the simulated energy consumption. When the ambient temperature is higher than 10 °C, the measured energy consumption is somewhat higher than the simulated energy consumption. It appears that the building has other problems such as leaking reheat valves and excessive airflow.

Off-line simulation case study summary

The simulation effectively identified the metering and valve leakage problems successfully in this case. Reheat valve leakage problems and excessive airflow problems were identified after fixing the leaking chilled water valve. This suggests that on-line fault detection should be an on-going process.

The simulation effectively identified HVAC component problems and was used to develop optimized HVAC operation and control schedules in another available case study. It further indicated that building thermal energy consumption would be reduced by 23% or $191,200/yr by using the optimized operating schedules in this building. The measured energy savings after implementing the optimized schedules were consistent with the simulated savings.

These results, coupled with similar experience in other buildings, strongly support the value of simulation in the commissioning process, both for off-line and on-line applications as a diagnostic and optimization tool for building operation.

Case study #2: DET-Toa method

An example of a calibrated simulation model being applied to complete FDD is the Days Exceeding Threshold-Toa (DET-Toa) method (Lin and Claridge 2015). Experience shows that detailed dynamic envelope simulation is not needed for the operational optimization implemented in the EBCx process (Liu and Claridge 1998). Hence, the calibrated simulation used is based on the Simplified Energy Analysis Procedure as described earlier (Knebel 1983), which led to software developed at Texas A&M named WinAM, the same type of model used in the first case study.

The model is calibrated using cooling and heating data from a baseline period in which the building is assumed to be operating properly or optimal. This period would normally be after some existing building commissioning has occurred. The baseline period also normally must be long enough to have captured the entire range of expected outdoor air temperatures and weather conditions.

The fault detection criterion of the DET-Toa method goes as follows:

> An abnormal energy consumption fault is identified if the deviation between the measured and simulated consumption is greater than one standard deviation of the residuals between measured and simulated consumption in the baseline period and persists for at least 20 days which are consecutive when ordered according to increasing or decreasing outside air temperature within the data being analyzed.

By ordering the data by outdoor air temperature, faults that are temperature dependent can be more easily separated than when analyzing them sequentially. HVAC operational changes affect the energy use in different ways depending on the environmental conditions. This is to say that each control change has its own *characteristic signature*, as described in the "A simple process for calibrating simulations" section.

The method was tested on synthetic data related to two buildings located on the Texas A&M campus. The first was the Eller Oceanography and Meteorology (EOM) building and the second was the Veterinary Research (VR) building.

The EOM building has 14 floors with a basement and 16,723 m² of conditioned space. It contains offices, classrooms, and laboratories. The commissioning work on this building was finished in March 1997. The simulation model was calibrated to the baseline consumption period from March 19 to August 31, 1997. The daily average temperature during the baseline period is between 46 °F and 88 °F, which covers nearly 75% of the temperature variation during the following year (32 °F–90 °F). The standard deviations of the residuals in the baseline period were 1465 kWh/day and 967 kWh/day respectively for CHW and HW energy consumption. They are 7.7% and 5.1% of the average daily combined CHW and HW energy consumption during the baseline calibration period.

The VR building is a five-story building with 10,684 m² of conditioned space. The building is comprised primarily of laboratories but also contains classrooms and offices. The building is served by five SDVAV AHUs with terminal reheat boxes. The commissioning was completed in November 1997. The simulation model was calibrated to the baseline period from January 1 to July 20, 1998. The daily average temperature during this period is between 41 °F and 90 °F. The standard deviations of the residuals in the baseline period were 1524 kWh/day and 1084 kWh/day respectively for CHW and HW energy consumption. They are 7.9% and 5.7% of the average daily combined CHW and HW energy consumption during the baseline calibration period.

Table 13.15 shows the summary of the changes that were made to the buildings for the sake of testing. The annual energy impact (AEI) of each change was defined to be

$$AEI = \frac{\textit{Total energy consumption difference in one year caused by the change}}{\textit{Annual energy consumption when there is no fault}}$$

or essentially the energy change due a control change, normalized to the annual total energy use. These changes included the supply temperature setpoints, amount of ventilation air, minimum airflow setpoints, and room temperature setpoints.

Tables 13.16 and 13.17 show the FDD results for the EOM and VR buildings. The DET-Toa method was able to identify a fault for 19 of the 20 faults introduced.

This example of fault detection was accomplished using a calibrated simulation to predict consumption, and when a substantial difference appeared between the cumulative measured

Table 13.15 Synthetic changes to the buildings

EOM building			*VR building*		
ID	*Synthetic control changes*	*AEI (%)*	*ID*	*Synthetic control changes*	*AEI (%)*
1	Outside airflow ratio increase of 3.1%	2.1	1	Outside airflow ratio increase of 6.5%	3.4
2	Outside airflow ratio decrease of 3.1%	−2.1	2	Outside airflow ratio decrease of 6.5%	−3.3
3	Cold deck leaving temperature increase of 2.2 °C	−1.4	3	Cooling coil leaving temperature increase of 0.9 °C	−7.0
4	Cold deck leaving temperature decrease of 2.5 °C	4.1	4	Cooling coil leaving temperature decrease of 0.8 °C	7.6
5	Hot deck leaving temperature increase of 5.6 °C	9.5	5	Preheat temperature increase of 1.8 °C	1.8

(*Continued*)

Table 13.15 (Continued)

EOM building			*VR building*		
ID	*Synthetic control changes*	*AEI (%)*	*ID*	*Synthetic control changes*	*AEI (%)*
6	Hot deck leaving temperature decrease of 6.7 °C	−4.9	6	Preheat temperature decrease of 2.2 °C	−1.5
7	Minimum airflow ratio increase of 16%	6.8	7	Minimum airflow ratio increase of 4.9%	8.5
8	Minimum airflow ratio decrease of 20%	−5.0	8	Minimum airflow ratio decrease of 6.5%	−8.2
9	Room cooling set-point temperature increase of 1 °C	−4.0	9	Room cooling set-point temperature increase of 1.6 °C	−2.3
10	Room cooling set-point temperature decrease of 0.9 °C	4.9	10	Room cooling set-point temperature decrease of 0.9 °C	2.3

Note: The denominator of outside airflow ratio and minimum airflow ratio is the maximum design airflow volume in the system.

Table 13.16 Summary of fault detection results for the EOM building

ID	*Synthetic control changes*	*Consumption*	*Detection*		*Fault identification day*	
			DET-Date	*DET-Toa*	*DET-Date*	*DET-Toa*
1	Outside airflow ratio increase of 3.1%	CHW Increase	Y	Y	6/27/1998	6/24/1998
2	Outside airflow ratio decrease of 3.1%	CHW Decrease	Y	Y	6/27/1998	6/24/1998
3	Cold deck leaving temperature increase of 2.2 °C	CHW Decrease HW Increase	N	Y		11/23/1997 (CHW fault) 12/1/1997 (HW fault)
4	Cold deck leaving temperature decrease of 2.5 C	CHW Increase	Y	Y	2/21/1998	12/6/1997
5	Hot deck leaving temperature increase of 5.6 C		N	N		
6	Hot deck leaving temperature decrease of 6.7 °C	HW Decrease	N	Y		12/20/1997
7	Minimum airflow ratio increase of 16%	HW Increase	N	Y		12/15/1997
8	Minimum airflow ratio decrease of 20%	HW Decrease	Y	Y	12/18/1997	11/12/1997
9	Room cooling setpoint temperature increase of 1 °C	CHW Decrease	Y	Y	6/27/1998	6/21/1998
10	Room cooling setpoint temperature decrease of 0.9 C	CHW Increase	Y	Y	6/27/1998	6/21/1998

Note: Consumption "Increase/Decrease" means the measured consumption is higher/lower than the fault-free simulated consumption.

Table 13.17 Summary of fault detection results for the VR building

ID	*Synthetic control changes*	*Consumption*	*Detection*		*Fault identification day*	
			DET-Date	*DET-Toa*	*DET-Date*	*DET-Toa*
1	Outside airflow ratio increase of 6.5%	CHW Increase	Y	Y	6/29/2000	6/4/2000
2	Outside airflow ratio decrease of 6.5%	CHW Decrease	Y	Y	6/29/2000	6/4/2000
3	Cooling coil leaving temperature increase of 0.9 °C	CHW Decrease	N	Y		4/2/2000
4	Cooling coil leaving temperature decrease of 0.8 °C	CHW Increase	N	Y		3/15/2000
5	Preheat temperature increase of 1.8 °C	HW Increase	N	Y		11/9/2000
6	Preheat temperature decrease of 2.2 °C	HW Decrease	N	Y		12/4/2000
7	Minimum airflow ratio increase of 4.9%	HW Increase	N	Y		12/28/2000
8	Minimum airflow ratio decrease of 6.5%	HW Decrease	Y	Y	12/29/2000	12/27/2000
9	Room cooling setpoint temperature increase of 1.6 °C	CHW Decrease	Y	Y	7/15/2000 (CHW fault)	7/14/2000 (CHW fault)
		HW Increase			10/30/2000 (HW fault)	2/14/2000 (HW fault)
10	Room cooling setpoint temperature decrease of 0.9 °C	CHW Increase	Y	Y	6/28/2000 (CHW fault)	5/30/2000 (CHW fault)
		HW Decrease				3/18/2000 (HW fault)

Note: Consumption "Increase/Decrease" means the measured consumption is higher/lower than the fault-free simulated consumption.

consumption and the simulated consumption, it was concluded that a fault occurred. Careful engineering examination of the pattern of the excess consumption was used to diagnose the probable cause of the fault remotely.

Subsequently, Lin and Claridge have developed statistical procedures and criteria for detecting the presence of a fault (2012, 2013, 2015). They presented two Days Exceeding Threshold (DET) methods – DET-Date and DET-Toa, respectively, as objective metrics to detect abnormal building cooling or heating energy consumption faults. The DET methods can detect persisting small increases or decreases in the normal energy consumption. They quickly detected the fault in the Sbisa Dining Hall that was described above and have been shown to reliably detect numerous "faults" created by changing one or more inputs in a calibrated simulation to degrade the performance of a building by a small amount.

For fault diagnosis, Lin and Claridge (2017) investigated the use of two similarity measure methods – namely cosine similarity and Euclidean distance similarity to isolate the cause(s) of identified abnormal energy consumption. These methods require only building energy consumption and weather data as inputs. Cosine similarity measures the similarity between two vectors based on the cosine value of the angle between them. Euclidean distance similarity evaluates the similarity between two vectors based on the distance between them. Both can quantify the level of similarity between the observed fault pattern and different reference fault patterns. The most similar fault pattern in the reference database is identified as the cause of the observed fault. When applied to the Sbisa data, cosine similarity identified "outside airflow temperature precooling decrease" as the most probable of 12 possible causes, while Euclidean distance similarity indicated that "outside airflow temperature precooling decrease" was slightly less probable as the cause than "outside airflow ratio increase." When applied to data from another campus building where a preheat valve had been detected to be leaking, both methods correctly identified the leaking preheat valve as the most probable cause of the fault.

Hence, we conclude that calibrated simulation can provide valuable input to the on-going commissioning process. A client has requested the use of these techniques in a commercial application to evaluate the on-going performance of several existing buildings in the eastern United States that were commissioned in 2017.

Synopsis

This chapter introduces the concept of whole-building simulation for operational optimization and describes four different post-design uses of simulation for operational optimization:

- Design simulation in post-construction commissioning of new buildings
- Design simulation or calibrated simulation for on-going commissioning
- Calibrated simulation for existing building commissioning
- Simulation to evaluate new control code

The majority of the chapter is devoted to the use of calibrated simulation in the process of commissioning existing buildings and the use of calibrated simulation in the follow-up on-going commissioning process. The existing building commissioning (EBCx) process is introduced and a detailed description of a process for calibrating simulations that is suitable for use in the EBCx process follows. The calibration process is further clarified with two case study examples.

The use of calibrated simulation to screen buildings for potential opportunities before a site visit is described and illustrated with an example. Its use in the on-site assessment and the detailed investigation and implementation within the EBCx process is described with multiple examples. Then the application of calibrated simulation to create baselines for savings determination and in the on-going commissioning activity is described with examples.

The chapter includes a comprehensive case study in which off-line simulation was used in the EBCx process to aid in the diagnosis of operating problems and to predict the savings potential of these problems. It also includes a detailed case study in which on-line calibrated simulation was used to track the performance of a building and detect the appearance of an operational problem and diagnose a change in the outside air pre-treatment temperature as the cause of the problem.

The use of calibrated simulation in the EBCx process is increasing rapidly and is likely to become nearly universal in the years ahead.

Acknowledgments

Portions of this chapter are based on the work of Subtask D2 of the International Energy Agency ECBCS Annex 40 "Commissioning of Buildings and HVAC Systems for Improved Energy Performance." Subtask D2 is titled "Use of Whole Building Simulation in Commissioning." Regular participants of the subtask meetings included Oliver Baumann (Germany), David Claridge, Chair (USA), Jorgen Eriksson (Sweden), Nestor Fonseca (Belgium), Johnny Holst (Norway),

Per Isaakson (Sweden), Timo Kalema (Finland), Hannu Keranen (Finland), Jean Lebrun (Belgium), Noriyasu Sagara (Japan), Makato Tsubaki (Japan), and Sheng Wei Wang (Hong Kong). Eighteen others attended at least one meeting. While this chapter has been written by the authors, it reflects contributions from all of these individuals as well as the contributions specifically referenced.

Notes

1 AirModel is a DOS implementation of the ASHRAE Simplified Energy Analysis Procedure (Knebel 1983) developed at Texas A&M University. It has been superseded by a modern Windows version known as WinAM, and the term WinAM will be used henceforth regardless of which program was actually used.

2 Continuous Commissioning and CC are registered trademarks of the Texas A&M Engineering Experiment Station.

References

ASHRAE. (2014) *Guideline 14–2014 Measurement of Energy, Demand, and Water Savings*. Atlanta, GA: ASHRAE, Inc.

ASHRAE. (2015) *Guideline 0.2–2015, Commissioning Process for Existing Systems and Assemblies*. Atlanta, GA: ASHRAE, Inc.

ASHRAE. (2018) Public Review Draft, Proposed *Guideline 1.2, Technical Requirements for the Commissioning Process for Existing HVAC & R Systems and Assemblies*. Atlanta, GA: ASHRAE, Inc.

Baumann, O. (2003) Design and optimization of control strategies and parameters by building and system simulation. In *Proceedings of the International Conference for Enhanced Building Operations*. Berkeley, CA.

Bynum, J., Claridge, D.E., Turner, W.D., Deng, S. and Wei, G. (2008) The cost-effectiveness of continuous commissioning over the past ten years. In *Proceeding of 8th International Conference for Enhanced Building Operation*. Berlin, Germany, October 22–24, Session 6.3, CD.

Bynum, J.D., Claridge, D.E. and Curtin, J.M. (2010) Development and testing of an Automated Building Commissioning Analysis Tool (ABCAT). *Proceeding of ASME 2010 4th International Conference on Energy Sustainability ES2010*. Phoenix, AZ, May 17–22, Paper ES2010–90389, CD.

Bynum, J.D., Claridge, D.E. and Curtin, J.M. (2012) Development and testing of an Automated Building Commissioning Analysis Tool (ABCAT). *Energy and Buildings* 55, 607–617.

Carling, P., Isakson, P., Blomberg, P. and Eriksson, J. (2004a) Simulation-aided commissioning of the Katsan Building. IEA Annex 40 Working Paper A40-D-M5-SWE-AF/KTH.

Carling, P., Isakson, P., Blomberg, P. and Eriksson, J. (2004b) Experiences from evaluation of the HVAC performance in the Katsan building based on calibrated whole-building simulation and extensive trending. IEA Annex 40 Working Paper A40-D-M6-SWE-AF/KTH-2.

Chen, H., Deng, S., Bruner, H.L., Claridge, D.E. and Turner, W.D. (2002) Continuous Commissioning results verification and follow-up for an institutional building – A case study. In *Proceedings of the 13th Symposium on Improving Building Systems in Hot and Humid Climates*. Houston, TX, pp. 87–95.

Claridge, D.E. (ed.) (2004) Using models in commissioning – using simulation models at the building level. In Visier, J.C. (ed.), *Commissioning Tools for Improved Energy Performance. Final Report of IEA ECBCS Annex 40*. International Energy Agency.

Claridge, D.E., Bensouda, N., Lee, S.U., Wei, G., Heinemeier, K. and Liu, M. (2003) *Manual of Procedures for Calibrating Simulations of Building Systems*, submitted to the California Energy Commission Public Interest Energy Research Program, Report HPCBS#E5P23T2b.

Claridge, D.E. and Liu, M. (2004) Using whole building simulation models in commissioning. IEA Annex 40 Working Paper A40-D2-M4-US-TAMU-02. https://oaktrust.library.tamu.edu/bitstream/handle/1969.1/5057/ESL-IC-04-10-31.pdf.

Claridge, D.E., Turner, W.D., Liu, M., Deng, S., Wei, G., Culp, C., Chen, H. and Cho, S.Y. (2004) Is commissioning once enough? *Energy Engineering* 101(4): 7–19.

Curtin, J., Claridge, D.E. and Bynum, J. (translated and submitted by H. Peitsman). (June, 2010). Een analyse-tool voor commissioning van gebouwen. *TVVL Magazine Commissioning (Dutch)*, pp. 2–6.

Curtin, J.M., Claridge, D.E., Painter, F., Verdict, M., Wang, G. and Liu, M. (2007) *Development and Testing of an Automated Building Commissioning Analysis Tool*. California Energy Commission, PIER Energy-Related Environmental Research Program. CEC-500-01-044-2007.

EVO. (2012) *International Performance Measurement and Verification Protocol: Concepts and Options for Determining Energy and Water Savings*, Vol. 1, Efficiency Valuation Organization (EVO) 10000–1.2012. San Francisco, CA. 94121–0518.

Keranen, H. and Kalema, T. (2004a) Commissioning of the IT-dynamo building. IEA Annex 40 Working Paper A40-D2-M6-FI-TUT-01.

Keranen, H. and Kalema, T. (2004b) Commissioning educational building IT-dynamo. IEA Annex 40 Working Paper A40-D2-M7-FI-TUT-01.

Kim, W. and Katipamula, S. (2017) A review of fault detection and diagnostics methods for building systems. *Science and Technology for the Built Environment* 24(1): 3–21.

Knebel, D. (1983) *Simplified Energy Analysis Using the Modified Bin Method*. Atlanta, GA: American Society of Heating, Refrigerating and Air-Conditioning Engineers.

LBL. (2001) *The Reference to EnergyPlus Calculations*. Energy Plus Engineering Document, Lawrence Berkeley National Laboratory.

Lin, G. and Claridge, D.E. (2012) Comparison of two statistical approaches to detect abnormal building energy consumption with simulation test. In *Proceeding 12th International Conference for Enhanced Building Operation*, October 24–25. Manchester, UK, CD.

Lin, G. and Claridge, D.E. (2013) A statistical approach to detect abnormal building energy consumption with ABCAT. *ASHRAE Transactions* 119, Pt.1, Paper D-DA-13–005, 9 pp.

Lin, G. and Claridge, D.E. (2015) A temperature-based approach to detect abnormal building energy consumption. *Energy and Buildings* 93, 110–118.

Lin, G. and Claridge, D.E. (2017) Whole building fault diagnosis methods for HVAC systems. to be published.

Liu, M. and Claridge, D.E. (1998) Use of calibrated HVAC system models to optimize system operation. *Journal of Solar Energy Engineering* 120(2): 131–138.

Liu, M., Song, L. and Claridge, D.E. (2002a) Potential of on-line simulation for fault detection and diagnosis in large commercial buildings with built-Up HVAC systems. IEA Annex 40 Working Paper A40-D2-M4-US-TAMU-01.

Liu, C., Turner, W.D., Claridge, D., Deng, S. and Bruner, H.L. (2002b) Results of CC follow-up in the G. Rollie White Building. In *Proceedings of the 13th Symposium on Improving Building Systems in Hot and Humid Climates*. Houston, TX, pp. 96–102.

Selkowitz, S.E., Haberl, J.S. and Claridge, D.E. (1992) Future directions: Building technologies and design tools. In *Proceedings of the ACEEE 1992 Summer Study on Energy Efficiency in Buildings*, Volume 3. Washington, DC, pp. 269–290.

Tsubota, Y. and Kawashima, M. (2004) Online building energy evaluation system. IEA Annex 40 Working Paper A40-D-M5-J-TEPCO-2-rev1.

Turner, W.D., Claridge, D.E., Deng, S., Cho, S., Liu, M., Hagge, T., Darnell, C. and Bruner, H. (2001) Persistence of savings obtained from continuous commissioning. In *Proceedings of 9th National Conference on Building Commissioning*. Cherry Hill, NJ.

Wang, S. and Xu, X. (2003) Hybrid model for building performance diagnosis and optimal control. In *Proceedings of the International Conference for Enhanced Building Operations*. Berkeley, CA.

Wei, G., Liu, M. and Claridge, D.E. (1998) Signatures of heating and cooling energy consumption for typical AHUs. In *Proceedings of the Eleventh Symposium on Improving Building Systems in Hot and Humid Climates*. Fort Worth, Texas, pp. 387–402.

Activity

Consider the four-floor building shown in Figure 13.32, with the SW wall in the left portion and the SE wall in the right portion of the picture. The NW and SE faces are 108 m long and consist of four floors, each 4 m high, while the northeast- and southwest-facing walls are 69 m long. The ground floor is level with the sidewalk in the foreground.

The building is supplied with hot water and chilled water from a central plant and distributes heating and cooling throughout the building using 12 dual-duct variable-air-volume air

Figure 13.32 Sample building for calibration

Source: Photo courtesy of Texas A&M University (https://brandguide.tamu.edu/visual-style/photo-video.html)

handlers with design flows of 19,000 L/s powered by 30 kW fans on variable speed drives. *Calibrate a simulation to the measured daily heating, cooling, and electric consumption data available in the file at https://cc-compass.tamu.edu/Content/buildingdata.xlsx.*

This file also contains measured daily average temperature and humidity data corresponding to the measured energy data. In addition, the measured daily pump and fan power for the building is provided, along with a week of typical hourly electrical consumption data.

For your initial simulation, you may use the following assumptions:

- The building is located at 30° N latitude
- 14% of the wall area is composed of clear single-glazed windows approximately 2 m wide, recessed by 1.25 m, and fitted with venetian blinds
- The wall U-value is 0.018 W/(m²·K)
- The roof U-value is 0.009 W/(m²·K)
- Minimum flow setting is 9,000 L/s for each AHU
- Fixed cooling coil leaving air temperature is 13 °C
- Hot deck setpoint is 50 °C
- Exterior zone is 6 m wide around the exterior of the building
- No economizer
- Mixed air is preheated to 7 °C if it falls below 7 °C at the preheat coils
- 15% outside air
- Interior space temperature setting is 23 °C

Use the characteristic signatures for the Harrington Building to assist in your calibration.

Answer

Your calibrated simulation should have an $ERROR_{TOT}$ value of 17 GJ/day or less.

Appendix

Characteristic signatures for DDVAV system in the Harrington Building in College Station, Texas, USA

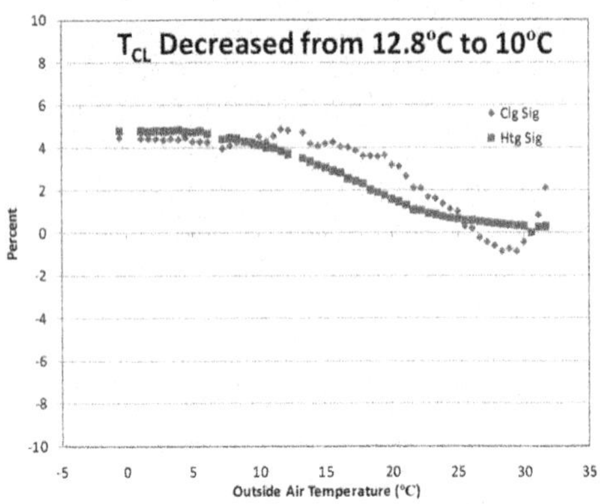

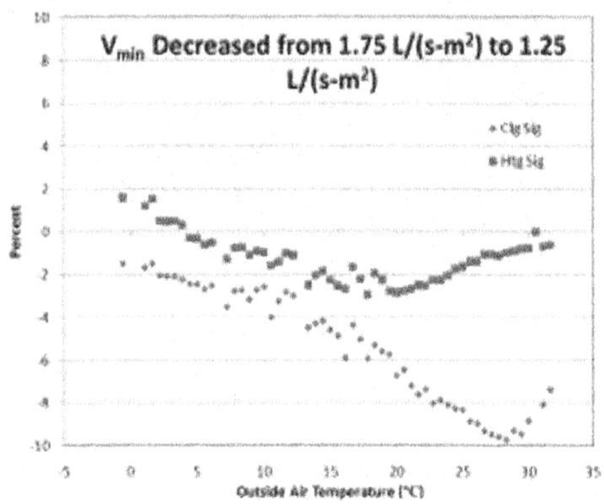

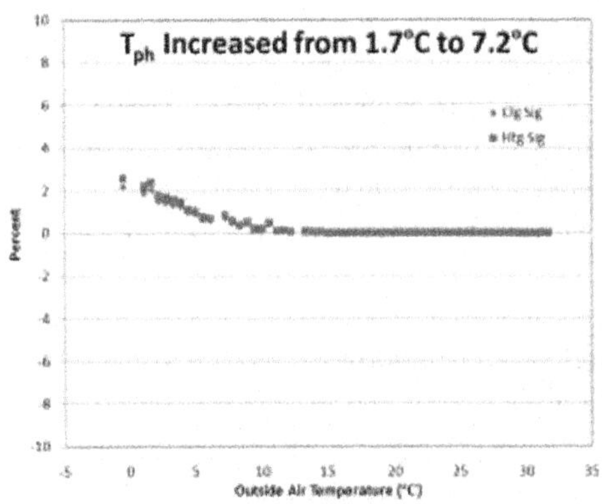

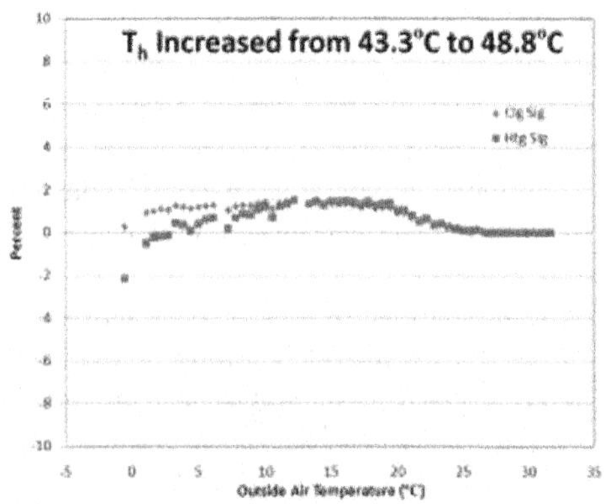

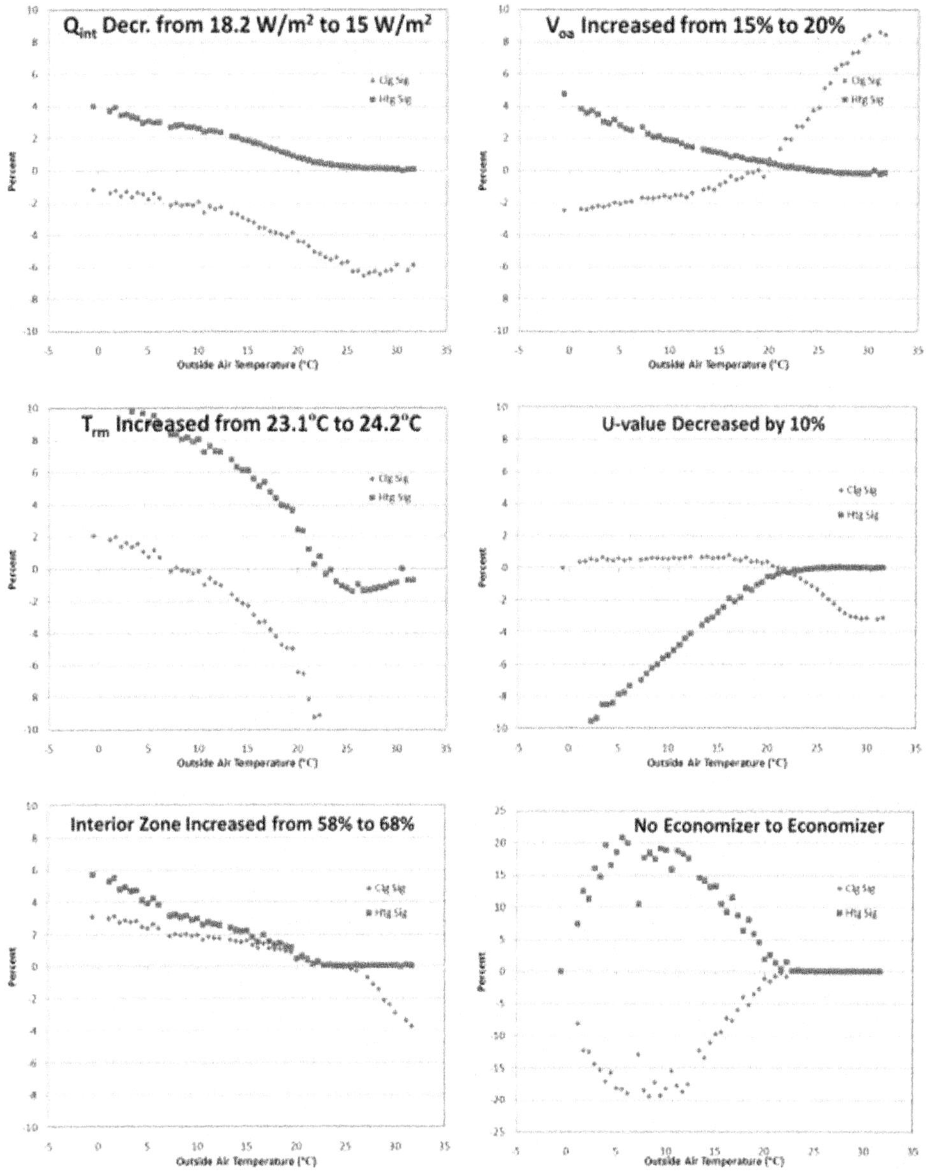
Q_int Decr. from 18.2 W/m² to 15 W/m²
V_oa Increased from 15% to 20%
T_rm Increased from 23.1°C to 24.2°C
U-value Decreased by 10%
Interior Zone Increased from 58% to 68%
No Economizer to Economizer
Clg Sig
Htg Sig
Percent
Outside Air Temperature (°C)

14 Modeling and simulation in building automation systems

Gregor P. Henze

Scope

This chapter offers a treatment of the role of building automation and control systems in the context of building performance modeling. In particular, we motivate the use of models for the advanced control of building energy systems by offering application examples employing dynamic models for online and offline predictive optimal control tasks. The chapter introduces model categories that reflect the degree to which physical first principles have been employed and compares procedures for calibrating such models using least squares and probabilistic approaches, both extensively illustrated by an example. Moreover, the chapter describes several current building-related examples of offline and online model predictive control as well as the application of rule extraction to the control of commercial buildings employing mixed mode ventilation.

Learning objectives

At the end of this chapter, the reader will be able to:

a Differentiate between white box, gray box and black box models
b Assess the relative model characteristics for any application scenario and select the proper approach
c Apply the model development and calibration process to any linear time-invariant building energy problem using lumped parameter analysis and equivalent RC network representation
d Extrapolate to other model predictive control opportunities based on the examples provided

Key words

Building automation and control/online and offline model predictive control/rule extraction from optimal control

Introduction

Commercial building performance simulation requires not only a description of building geometry, construction materials, energy systems, and equipment but also a characterization of the building utilization through occupancy, lighting, and equipment schedules. In addition, a detailed specification of the building operation is necessary, commonly through a definition of setpoint schedules (such as for temperature and/or relative humidity), reset schedules that relate control set points with load conditions such as approximated by outside air temperature, HVAC system availability, and sequencing of multiple pieces of HVAC equipment (e.g. boiler and chiller staging). Achieving even greater fidelity in modeling building controls through control sequences of several devices for a single control loop (e.g. outside air damper, followed by heating coil, and then by cooling coil for supply air temperature control), control cascades, and PID control loop behavior may be only warranted (time and effort) and possible (simulation program capability) in very select cases.

While high-end building performance simulation programs such as EnergyPlus (with the Energy Management System), Modelica, and TRNSYS allow for an approximate representation of common control approaches found in commercial buildings today, modeling sophisticated energy management strategies as found in well-run modern facilities (e.g. occupancy-based override of minimum airflow settings on terminal boxes, demand controlled ventilation, or complex reset schedules that adjust control set points in response to current building states) may be difficult to impossible to model.

The emphasis of this chapter is model-based control, i.e. building control problems where a model of the process under control is required and utilized to guide the operation of either an entire building or selected subsystems. The intriguing facet is that such a control-oriented model would be used *in* the existing building automation system, but relevant control aspects *of* the existing BAS must be embedded in the control-oriented model itself. Thus, the title of this chapter could have been either "Modeling and simulation *in* building automation systems" or "Modeling and simulation *of* building automation systems."

Regarding the control of an entire building, model-based control implies a model to be used by the BAS for real-time control as well as a representation of the BAS in that model. Simply consider how the performance of a building under model-based control could be simulated. The modeled building at high fidelity would include a BAS, which for its control would use another (lower complexity controls oriented) model of the same building! But don't worry, we will not take matters to this extreme; instead, we will focus on relevant applications of model-based control for commercial buildings. Yet, modeling and simulation of a controlled system with a BAS will typically require two models of the same system: one control-oriented model for control purposes of the BAS (prediction model) and one model for representation of the controlled system (simulation model).

Applications for modeling in building automation systems

The strongest justification for modeling in general and model-based control in particular is that we can exploit such models to guide us toward a performance optimum, which could be lowest energy use, lowest energy cost, carbon emissions, or other performance metrics, while observing constraints such as maintaining acceptable indoor environmental conditions and HVAC system cycling limitations. Forty percent of the source energy and 75% of the electricity are consumed in buildings in the US (22% for residences and 18% for commercial buildings),[1] and similar figures can be found for European buildings. Hence, the existing building stock is a treasure chest for improved operation through optimal control, at least in terms of the available potential (realizing this potential yet is much more difficult).

Moreover, with new construction becoming more and more complex, model-based control in real time using a detailed building energy model holds the promise of serving as a guidepost for maintaining the building operation close to the design intent. Consider that modern buildings are increasingly equipped with sophisticated HVAC systems, controllable building envelope components (e.g. dynamic shading, façade-integrated ventilation), integration of energy storage (thermal and electrical), on-site renewable energy sources, and more. Such complex systems require coordinated and optimized operation at a system-level in order to achieve the desired level of comfort and energy efficiency. This is the main motivation for the application of advanced, i.e. model-based, building automation systems.

Examples of model-based building control include (a) instantaneous optimal building control *without* storage effects such as the global setpoint optimization of a combination of chiller, cooling tower, and associated air handling units (Braun et al. 1989, Brandemuehl et al. 1998), operation of building power plants (Van Schijndel 2002), or central chilled water plant control (Ma and Wang 2009), as well as (b) predictive optimal building control *with* storage effects including building thermal mass control (Braun 2003), ice and chilled water storage control (Henze 2003, Kintner-Meyer and Emery 1995, Massie 2002), and mixed-mode ventilation building control (Spindler 2004, Spindler and Norford 2009b, May-Ostendorp et al. 2011, Tanner and Henze 2014). Moreover, model predictive control has been demonstrated in a university campus in California, where a central plant with thermal energy storage tanks was controlled along with multiple campus buildings (Ma et al. 2012). The absence of storage allows for the use of time-independent equipment models such as polynomial curve fits or performance maps, while the presence of storage requires the use of time-dependent dynamic models that account for the memory associated with the massive building construction and its thermal response. A comprehensive review of supervisory and optimal control applications has been provided by Wang and Ma (2008) and is recommended for further case studies.

For an optimal control application to be possible, we require a building energy system over which we have control and whose operational faults have been removed previously. Through optimization, the energy efficiency of the system and the indoor climate is improved given the current boundary conditions such as weather, comfort requirements, and the presence of occupants. Thus, optimization can be described as the targeted improvements of the building operation subject to the currently imposed boundary conditions and constraints. Typical approaches for optimizing building operation include the adjustment of HVAC system operating time periods, modes of equipment operation, set points, and energy sources.

Such model-based control can be accomplished either as:

- Offline optimization: In this case, the building operation is optimized externally, i.e. the building is simulated offline, where appropriate occupant profiles and a suitable weather data set are assumed. The optimized control parameters are then passed back (possibly manually or through control rules) and used for the building operation. The aim here is to determine general control strategies for the systems in the building to minimize annual energy consumption or costs.
- Online optimization: Model-based predictive and learning controllers aim to optimize the system by accounting for current and future building operational conditions, e.g. in terms of weather, indoor climate, or presence of occupants. From this information, the actual operation can be optimized, conceivably every hour or 15 minutes. Yet, this type of optimization requires significant effort and a close coupling with the building automation system.

In the scientific community, numerous approaches have been developed and tested for optimizing the operation of HVAC systems (Wetter 2004). In practice, however, the system operation is often optimized manually. Even during manual optimization, computer models are sometimes called upon for assistance in a trial and error approach (Claridge 2004).

A possible taxonomy for model-based building control is provided in Figure 14.1.

Fault detection and diagnosis (FDD) is a necessary prerequisite for achieving a fault-free operation and is therefore a prerequisite to optimization, although it will need to be conducted on a continuous basis. As a result, a two-step process should be adopted, which is schematically shown in Figure 14.2.

The process begins with data collection, which can be acquired through the building automation system in the ideal case or with short-term metering equipment. Information about the building in the form of measurement data, set points, and status information as well as stock data are gathered with the help of an extended building energy audit.

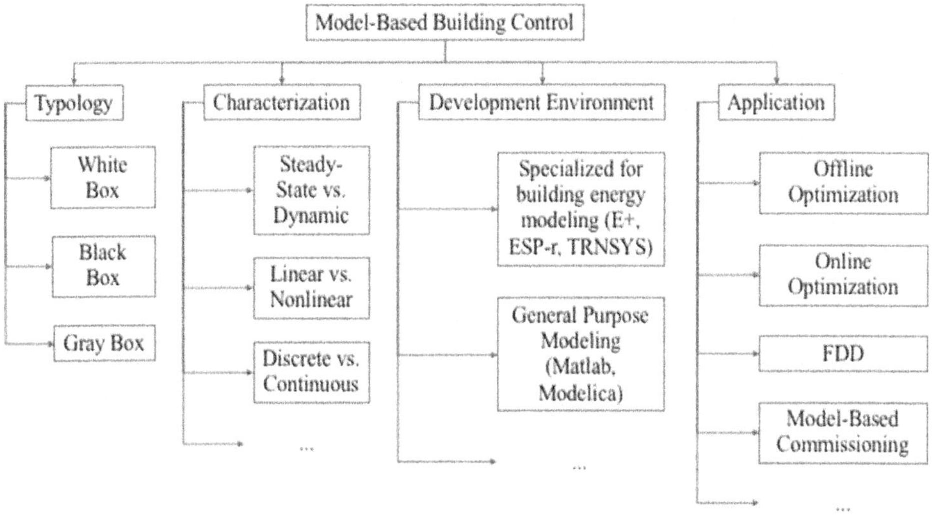

Figure 14.1 Taxonomy of model-based building control

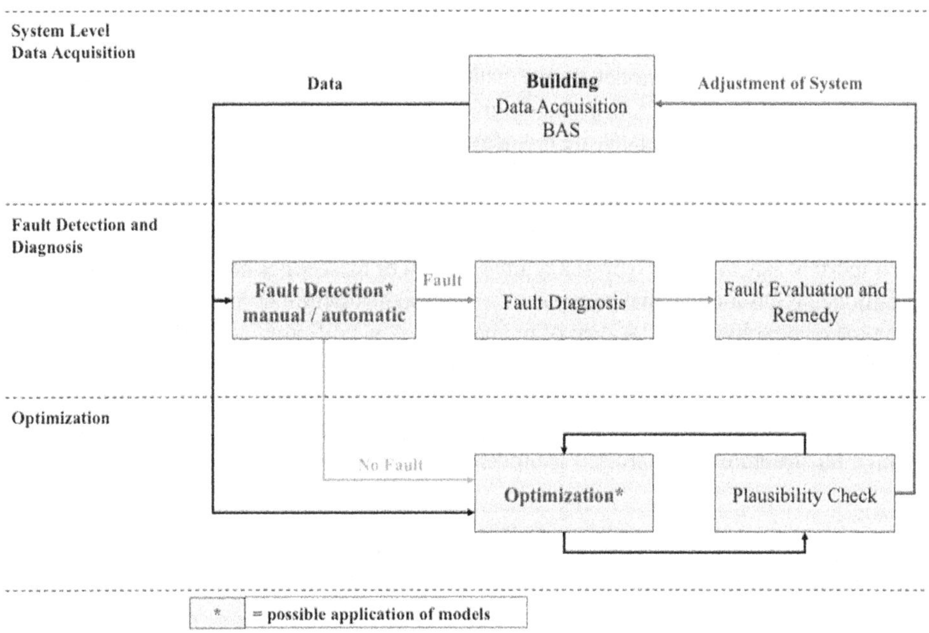

Figure 14.2 Process for fault detection and optimization in building operation

Fault detection and diagnosis is part of the first step. The fault detection can take place either manually (e.g. by the operating personnel) or automatically. When a fault is detected during operation, a fault diagnosis is carried out whereby the cause for the fault is investigated. Whether a fault is judged to be critical depends on excess energy consumption and costs as well as thermal comfort impacts. These factors are weighted according to their importance from the viewpoint of the building operator or owner.

Once all critical faults are addressed and remedied, the optimization can commence to generate specifications for the set points of the system operation. A plausibility test should be carried out before these optimized set points are fed into the building automation system.

To ensure the energy efficient operation over the entire building lifetime, the FDD process should be performed at regular intervals in order to be able to react promptly to any changes. These changes can be intentional (such as building renovation, reconstruction of part of the building) as well as unintentional (such as degradation of an energy system due to age or a fault.)

Model types, development tools, and their implementation

Models are representations of systems, which when properly developed allow conclusions to be drawn about the system performance under specified boundary conditions. A suitable model attempts to reduce the complexity as much as possible without losing those details and features, which are important for sufficient system representation. Model complexity should be reduced to simplify the model development and its validation as well as to achieve fast calculation time, which is crucial for control purposes. The actual systems are described in terms of mathematical formulas as physical or conceptual models, as shown in Figure 14.3. The behavior of real systems is described using physical principles, which are abstracted in the development of a conceptual model, then articulated mathematically and implemented numerically for computer simulation. To validate the computer model, the physical concepts employed in the conceptual model need to first be validated, followed by a verification of the numerical and numerical computation (often using other, previously developed computer models), followed by an empirical validation of the computer model predictions against observations.

The central elements of a model are the mathematical expressions, which represent relationships between inputs and outputs based on given parameters. The representation of the inputs and outputs and the parameters at the boundary of the model are shown in Figure 14.4. Input variables and parameters are specified by analysts or taken from other models or processes. In the same way, the output data can be used for other models as inputs or for model evaluation by analysts. The generic structure, on the other hand, remains unchanged.

An important aspect of this representation is that all of these types of models can also be used as sub-models. Sub-models make up parts of the model structure in larger models. Depending on the type or the physical relevance of the parameter, we differentiate between the following types of models:

- White box model: physical model with exclusively physically meaningful parameters (internal structure of the process is modeled),

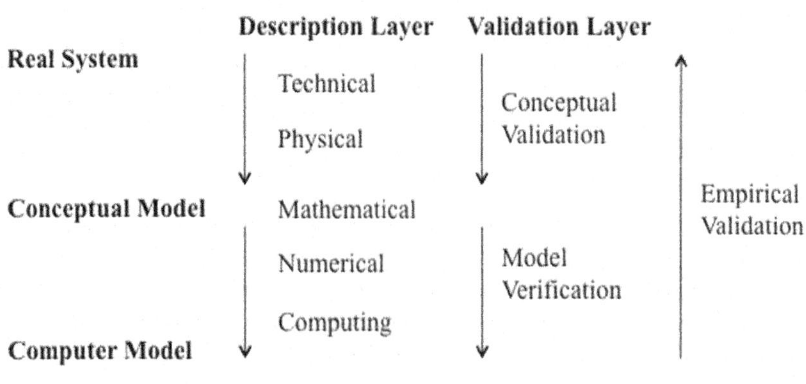

Figure 14.3 Model and validation levels

Source: Palomo del Barrio and Guyon 2003

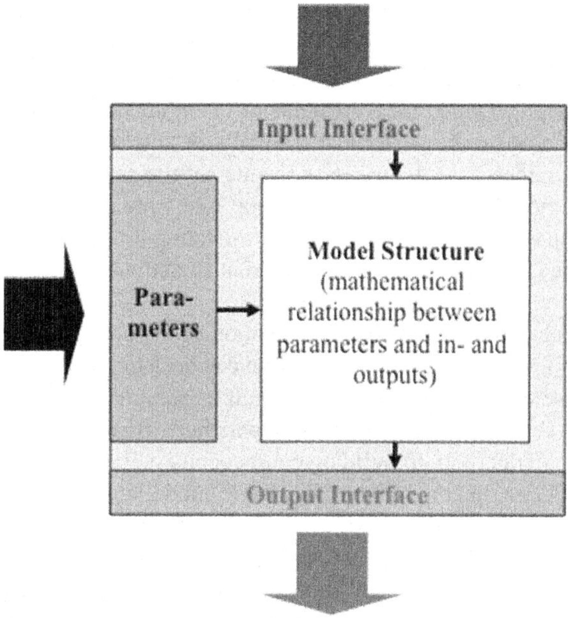

Figure 14.4 General structure of a model

- Black box model: non-physical model (e.g. statistical/stochastic) without physically relevant parameters (internal structure is not modeled), and
- Gray box model: model form combining both the white and black box models. Commonly, fundamental dynamic characteristics are approximated using physically motivated (white) low-order (first and second) dynamics, while steady-state performance characteristics are represented by performance maps (black), forming together a blended, gray, representation.

Before the individual types of models are described in more detail, the calibration of models is presented first in the following section.

Model calibration

Good building energy models can provide parameter values for the expected or optimized operation of an overall building or a specific building energy system; these calculated values are used for comparison with the currently implemented ones and are thus of special interest for fault detection and optimization.

In general, it is necessary that the parameters of the model be fitted to the actual physical system. This process is called parameter estimation or model calibration. In creating the model, it can be differentiated between two different types with respect to the parameter specification (ASHRAE 2013):

- *Forward modeling*: In forward modeling, it is assumed that all of the necessary model parameters are adequately known so that a realistic model of the actual system can be created. For most of the models, this assumption is valid for only some of the parameters. Therefore, this type of modeling is limited, for the most part, for use in the design phase of new systems. The term "forward" here shall imply that there is no coupling or correspondence to measurements. Most of the models that fall under the forward modeling category are based on fundamental principles and therefore require parameters like material characteristics, which tend to be known with relative confidence.

- *Inverse modeling*: In contrast to forward modeling, some or all of the parameters in inverse modeling are estimated using measurements of the input and output quantities of the system. These estimates can be performed manually or automatically. Here the aim is to fit the calculated parameters to the measured values of the system in the best possible way, a process called model calibration. As a general rule, only simple models are suitable for determining the parameters using the inverse modeling procedure. This is especially true for models where the parameters are estimated in an automated procedure.

One often finds models calibrated with a combination of the above-named parameter estimation methods. Many models contain parameters that can be determined with sufficient accuracy based on the expertise of analysts (forward) as well as parameters that are determined using measurement data (inverse). Typical examples for parameters that are suitable for forward modeling are the thermal conductivity of building components such as the external wall or the roof. An example for a parameter that is suitable for inverse modeling is the effective building mass.

In order to make reliable prognoses about the behavior of real systems, it is imperative to adequately calibrate models that are to be used for control or FDD. As a result, inverse modeling is always chosen (at least for parts of the modeling) for models that are used in the area of fault detection and optimization. Certainly, the degree of the automation in the estimation of the parameters can vary greatly depending on the model and the available measurement data.

Figure 14.5 shows a schematic of the calibration of a black box model. For the calibration, a training data set is used, which contains the measured input and output quantities of the physical system over a given time period. The model parameters must now be adjusted so that the model can replicate the relationship between the input and output quantities of the training data set with sufficient accuracy.

As a rule, calibration is an iterative process by which the output variables of the system can be calculated with the assistance of the model structure, the measured inputs, and the known, or

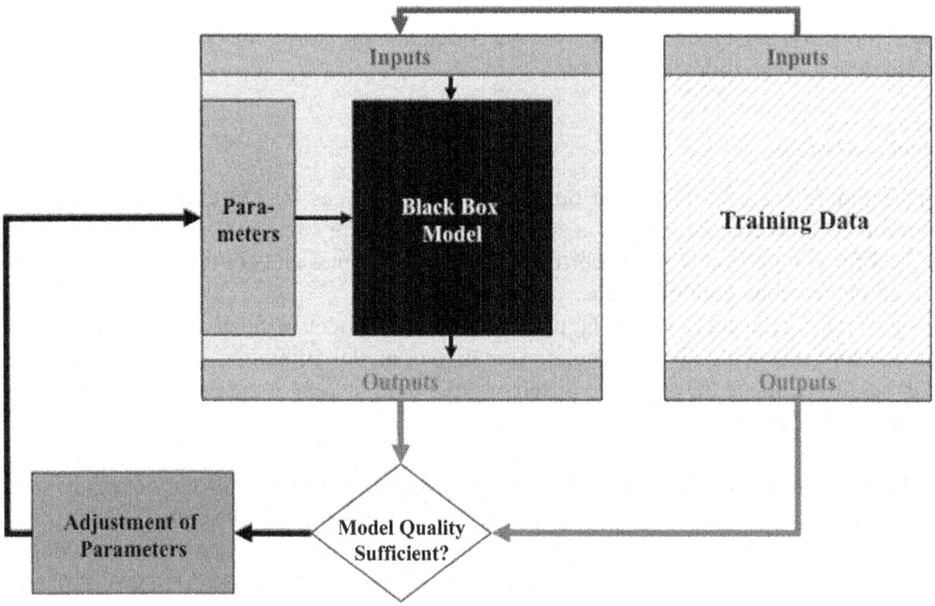

Figure 14.5 Schematic of model calibration

estimated, parameters. The calculated outputs of the model are compared to those quantities in the training data set. Common metrics are normalized mean bias error, coefficient-of-variation, or a combined goodness-of-fit as proposed by ASHRAE Guideline 14. If significant variations occur, the parameters must be fitted appropriately and the system output then newly calculated. If the discrepancy is small enough and thus the model output acceptable, then this process can be aborted. This process is commonly automatically conducted by nonlinear least squares minimization. In some cases, the model structure may have to be adapted, a step which must be performed manually in most cases. Often an additional validation data set is used to see how the calibrated performs on previously unseen data. At the end of this process, the calibrated model is available for use in model-based control.

Operational building energy modeling, i.e. the use of models for building control, is still in its infancy, and applications are not currently scalable. In harmony with Occam's razor (Jaynes 2003), it makes sense to consider the calibration of lower dimensional models before moving onto the high dimensions associated with typical building simulation tools. In the operational context, an emphasis is placed on higher resolution rather than solely capturing bulk dynamics. A desired application of operational models is model predictive control, embedded in the controller hardware rather than in a separate optimization infrastructure, where speed, accuracy, and robustness are crucial, e.g. performing thermal zone identification using a rooftop unit's (RTU) controller for superior scheduling and fault diagnostics. Computationally efficient models that capture critical system dynamics are known as gray box models and have been successfully used to forecast cooling loads and energy consumption for optimal control strategy evaluation (Braun and Chaturvedi 2002; Braun et al. 2001; Zhou et al. 2008). They allow blending of physical (or semi-physical) building knowledge and performance measurements, and the experimental study in the "Least squares and Bayesian calibration of gray box models" section utilized the CEN-ISO standard thermal network model (ISO 2007) to avoid a biased model structure.

Pavlak et al. (2013) asked whether the computational expense of probabilistic (Bayesian) techniques is warranted, specifically in the context of lower-order dynamic models. They extend previous research and development of a Bayesian calibration technique applied to building energy models (Neumann et al. 2011) and compare performance to a traditional, nonlinear least squares parameter estimation method of calibration. More recent research on Bayesian calibration of building energy models, also using the CEN-ISO representation Heo et al. (2012), used a popular framework and code (Kennedy and O'Hagan 2001) for retrofit analysis. Others using the same framework (Higdon et al. 2008) have successfully shown how computational challenges of high-dimensional data problems can be significantly reduced. Although Bayesian approaches are tremendously successful for high-dimensional problems, the question posed is whether such techniques are computational excess if all one desires is a calibrated, low-dimensional model for operational concerns.

White box models

White box models are based on the physical description of the system; their parameters have a physical meaning. Such models are also known as "first principle models." In developing these models, often the conservation laws (e.g. energy and mass balance) of classical engineering are applied.

The amount of detail contained in white box models can vary quite a bit, in spite of the use of the physical laws. If all phenomena typical for a building are modeled, the number of required parameters for an adequate description can quickly increase to several hundred. With increasing resolution in the models, for example starting from the simple one-zone model up to models that simulate indoor airflow and daylight distribution in individual rooms, the models can contain from several equations up to several thousand equations. In the same way, the effort for modeling and solving the equations as well as the application and usefulness spans a wide range.

White box models can be categorized according to the equations on which they are based as summarized in Table 14.1:

- Steady-state vs. dynamic models,
- Linear vs. nonlinear models, and
- Differentiable, continuous, and discontinuous models.

In steady-state models, the output does not depend on the dynamic course of events over time. On the other hand, the output of dynamic models depends on time since storage effects need to be considered, which in turn requires the use of differential equations. Depending on the type of differential equation, a further differentiation can be made: ordinary, partial, linear, and nonlinear.

The system of equations that is used to describe the dynamic building performance can usually not be solved with analytical methods. Furthermore, it must be noted that the choice of the solution method and the numerical parameters (e.g. tolerance, step size, maximum number of iterations, accuracy) impact the results, e.g. for instable or chaotic systems. For such models, the dependence on solver features can dominate the results.

Since evaluating the target function in calculations for building optimization requires, in part, great computational effort, it can be shown that convergence can be achieved faster when the accuracy is variable (Wetter 2004). For this purpose, the optimization algorithm must control the accuracy of the entire model, and it is very important that the target function is continuous

Table 14.1 Examples for different models or types of equations

Type	*Example*	*Application*
Steady-state linear equations	$\dot{q} = \frac{T_2 - T_1}{R} U(T_2 - T_1)$ Transmission	Steady-state heat loss
Steady-state nonlinear equations	$\dot{q} = \varepsilon K(T_2^4 - T_1^4)$ Radiation	Building simulation (radiative heat flow between walls)
Dynamic linear equations Ordinary differential equation ODE	$C\frac{dT}{dt} = UT - Ta$ Heat storage	Passive/active heat storage
Dynamic linear equations Partial differential equation PDE	$\frac{\partial}{\partial t}u(x,t) = a\frac{\partial^2}{\partial x^2}u(x,t)$ Heat diffusion equation	Thermal bridge calculations
Dynamic nonlinear equations Partial nonlinear PDE	$\rho\frac{\partial}{\partial t}u + \rho(u \cdot \nabla)u$ $= -\nabla p + \eta\Delta u + (\lambda + \eta)\nabla(\nabla \cdot u) + f$ Navier-Stokes equation	Computational fluid dynamics (CFD)
Discontinuous relations	Switches/if-clauses	Building control system simulation

and differentiable, which avoids chaotic behavior. Many building simulation programs do not fulfill these requirements (controllable accuracy, continuous, differentiable target function); therefore, the building simulation program Modelica was developed that meets these specifications. Certain control features (such as an on and off switching, switching functions with hysteresis) prevent these requirements (continuous differentiable target function) from being met.

A greater amount of effort is not necessarily needed for modeling white box models as opposed to other model types. Above all, the effort depends on the amount of detail in the model. Very simple white box models can be developed that, in spite of their simplicity, accurately model the behavior of the actual system. Yet, a prerequisite for a white box model is that the underlying mathematical descriptions are known and that a numerical solution is attainable.

For white box models, it is known *a priori* what the boundaries are for extrapolating results. Even classical methods for uncertainty analysis can be applied to white box models in order to determine the accuracy of the results. For a calibrated white box model, therefore, the domain of validity and thus the extrapolation behavior within the desired range of accuracy is known.

The values of white box model parameters are available over a wide range of confidence. In part, exact measurement values are available (e.g. thermal conductivity of certain materials) and, in part, only rough estimates are available for the values, since exact measurement values require a great deal of effort (e.g. window openings and air exchange rates due to natural ventilation). For the calibration of white box models, it makes sense to fit the parameters that are known less confidently. However, plausibility checks for these parameters must be carried out (correct order-of-magnitude). If the model cannot be calibrated with this method, then the module structure must be reconsidered. Moreover, the literature points out that too many parameters represent a potential source of error (Lebrun 2001; Déqué, Ollivier et al. 2000) as parameter uncertainties may compound significantly to give erroneous results.

Black box models

Black box models are characterized by the fact that they have no physically significant parameters. The model structure generally does not reflect the structure of the actual system and only serves to arrive at a suitable model, which is able to reproduce the mapping between the output and input variables of the actual system correctly, i.e. with sufficient accuracy. The parameters are usually automatically estimated for this purpose (Ljung 1999).

This automatic parameter estimation represents a significant advantage over white box models. The setup cost and required computational effort is small, by comparison, for black box models. The downside, however, is the fact that the models allow little if any insight into the inner structure of the actual system. As a result, the analytical capability of black box models is limited.

Black box models have been used for fault detection as well as optimization; however, the ability to extrapolate or generalize beyond the area of training data is often limited. The advantage of black box models lies in their fast and automated identification of patterns in the input-output relationship.

Regarding the structure of black box models, similar to the white box counterparts, a distinction can be made between steady-state and dynamic as well as between linear and nonlinear models. Depending on the model structure, various methods are available for estimating the parameter values. Table 14.2 provides a summary of the various approaches.

Steady-state, linear models

Models of steady-state processes are time independent, i.e. they describe the mapping between input and output for steady-state or quasi steady-state conditions. Linear black box models for steady-state processes therefore represent simple linear equations:

Table 14.2 Summary of black box models

Model type	*Model structure*	*Parameter estimation*	*Example*
Steady-state			
Linear	Linear function	Linear regression (least squares method)	Energy signatures
Nonlinear	Polynomial	Nonlinear regression (least squares method)	Pump and fan curves
	Arbitrary nonlinear functions	Iterative methods, e.g. Levenberg-Marquardt	
Dynamic			
Linear	Transfer function model (ARMA, ARMAX, etc.)	Linear regression (least squares method), iterative methods	Unsteady heat flow through plane wall
Nonlinear	Polynomials (e.g. Wiener/ Hammerstein model, Volterra model)	Linear regression (least squares method)	Linear system with nonlinearity in input or output (e.g. control element with saturation)
Nonlinear	Neural networks, wavelet, radial basis networks, support vector machines	Damped Gauss-Newton, back-propagation	Arbitrary nonlinear systems

$$y = a_0 + a_1 u_1 + a_2 u_2 + \ldots + a_n u_n \tag{14.1}$$

where:

y outputs;
u_1–u_n inputs;
a_0–a_n parameter.

The parameters for these models can be calculated easily and quickly with the aid of linear multiple regression (ordinary least squares [OLS] method).

An example of this type of model is the energy signature of a building. This represents a linear connection between the energy consumption and various determining factors such as outside temperature, radiation, and operating times (see e.g. Kissock et al. 2003). Linear models that exhibit clear breaks in their relationship as function of an independent variable are known as change point models and can be used to determine the balance point temperatures for heating and cooling of a building.

Steady-state, nonlinear models

Nonlinear, steady-state black box models produce nonlinear mappings between inputs and outputs. Polynomials are often used as the model structure. A simple example with only one input and output is, for example,

$$y = a_0 + a_1 u + a_2 u^2 + \ldots + a_n u^n \tag{14.2}$$

Not only this particular model but also many other nonlinear models are linear in their parameters or can be transformed correspondingly (e.g. by logarithmic transformation). As with linear models, therefore, the parameter calculation can be carried out in this case using OLS methods.

If the model cannot be transformed so that it is linear in its parameters, iterative processes such as the Gauss-Newton or Levenberg-Marquart procedures can be used for calculating the parameters. These are, however, numerically more complex and, in some circumstances, convergence is not guaranteed.

Examples of a nonlinear, steady-state model are performance maps such as pump and fan curves.

If the parameters of steady-state models are calculated using data from actual systems, it is necessary to ensure that the data represents steady-state conditions. For this reason, prior to the actual parameter calculation, a steady-state filter is often used in order to extract steady-state conditions from the data.

Dynamic, linear models

It is characteristic of dynamic systems that the mapping from inputs to outputs is time dependent. A plethora of black box models exists for dynamic linear systems, which can be described as transfer function models and for which there is an all-embracing theory (Ljung 1999).

A common structure of these models implies that the current output variable is calculated using linear functions of previous input and output variables.

The easiest model of this type is the finite impulse response model (FIR), which produces the mapping between the current output and the current and previous inputs:

$$\begin{aligned} y(t) &= B(q)\, u(t) + e(t) \\ &= b_1 u(t-1) + \ldots + b_n u(t-n) + e(t) \end{aligned} \tag{14.3}$$

where:

$u(t)$ measured inputs;
$y(t)$ measured outputs;
$e(t)$ disturbance variable (normal distribution);
b_1–b_n parameters.

The black box models of linear, dynamic systems used in practice are all modifications of this model. For the general case, the following connection arises:

$$A(q)\,y(t) = \frac{B(q)}{F(q)}u(t) + \frac{C(q)}{D(q)}e(t) \tag{14.4}$$

The five matrices or polynomials *A*, *B*, *C*, *D*, and *F* represent the transfer functions. In practical applications, one or more of these functions is unity, which gives rise to various model structures. The most common are listed in Table 14.3. Figure 14.6 shows the structure of an ARX model.

The term "autoregressive" describes the inclusion of previously measured data in the model for the outputs, while "moving average" relates to the inclusion of external disturbance variables.

Generally, simple and tested methods are used for calculating parameters for models of this type (ordinary least squares), and it is assumed that the disturbance variable is a normally distributed variable with zero mean.

These models are, however, limited to linear, time-invariant systems. Since actual systems – at least in the building sector – are commonly time-variant and nonlinear, the usefulness of

Table 14.3 Typical linear black box models for dynamic systems

Polynomials in use	*Short form of the model structure*	*Remarks*
B	FIR	Finite impulse response
AB	ARX	Autoregressive with external input
ABC	ARMAX	Autoregressive moving average and external input
AC	ARMA	Autoregressive moving average
BF	OE	Output error
BFCD	BJ	Box-Jenkins

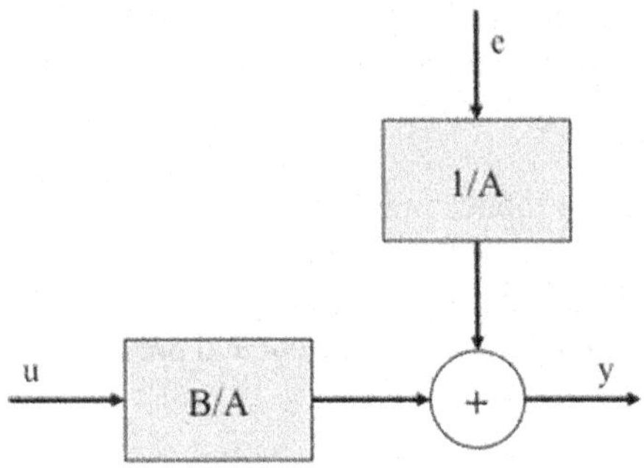

Figure 14.6 Structure of an ARX model

these models is limited. One typical application is for modeling the time-dependent transfer of heat through flat walls.

To a certain extent, temporal variance (e.g. with recursive models or weighting functions for "older" inputs) or nonlinearity (e.g. inclusion of steady-state nonlinearities of the inputs or outputs in otherwise linear models, Wiener or Hammerstein models) can be handled by reducing the problem back to a linear model (Ljung 1999).

Dynamic, nonlinear models

Nonlinear models clearly offer more freedom with regard to the structure of the model, which on the one hand makes them suitable for describing real systems, but on the other hand is numerically expensive. Apart from the choice of variables (measured data) and regressors for the model, the most important step is the selection of a model structure that can reflect any desired nonlinear feature of the variable space.

Classical methods for identifying nonlinear dynamic systems are largely based on polynomial models. These have the advantage of having linear parameters, which means that the estimation of the parameters can simply be carried out by means of linear regression.

Nevertheless, other forms of nonlinear models, such as neural nets (in particular the nonlinear autoregressive type with exogenous input NARX) and support vector machines, are now being employed extensively. Judging by the sheer number of publications dealing with the creation and application of neural networks, support vector machines, and wavelet models, there is a strong interest apparent in these approaches.

The advantages of nonlinear black box models such as neural nets are obvious: In principle, they can be used to represent any real nonlinear, dynamic system with a manageable numerical effort.

A few distinct disadvantages must be mentioned, though, which are closely associated with the black box character, i.e. these models require large amounts of data to estimate the parameters, while at the same time an extrapolation beyond the domain of training data is potentially hazardous. As the model parameters are not physical, they are not suitable for optimization of building parameters (Katipamula and Brambley 2005a).

Gray box models

Gray box models are a hybrid form of white box and black box models and they are distinguished by the fact that they have physically significant as well as physically insignificant parameters (Lebrun 2001, Braun and Chaturvedi 2002).

Gray box models can be developed for individual components or they are derived from the combination of white box and black box models for a large complete system (Figure 14.7). In addition, black box models can complement white box models, in order to incorporate higher-order features that are not captured by the physical model (Vilim et al. 2001) (Figure 14.8).

In physics and engineering, black box elements in gray box models are known as empirical models and have a long tradition in these fields. In the past, statistical methods were used for this, which have been enhanced by newer developments such as neural nets.

In the literature, there are also examples of gray box models that are derived from the calibration of black box models with the aid of white box models (Déqué et al. 2000). In these cases, linear, polynomial, or harmonic functions are matched to the simulation results. In this way, not only the number of parameters, but also the calculation requirements can be substantially reduced. The parameters of the matched functions no longer have physical significance but are derived from physical models (see Figure 14.9). Depending on the problem set and the selection of the basic functions, the robust extrapolation characteristics of white models can be retained.

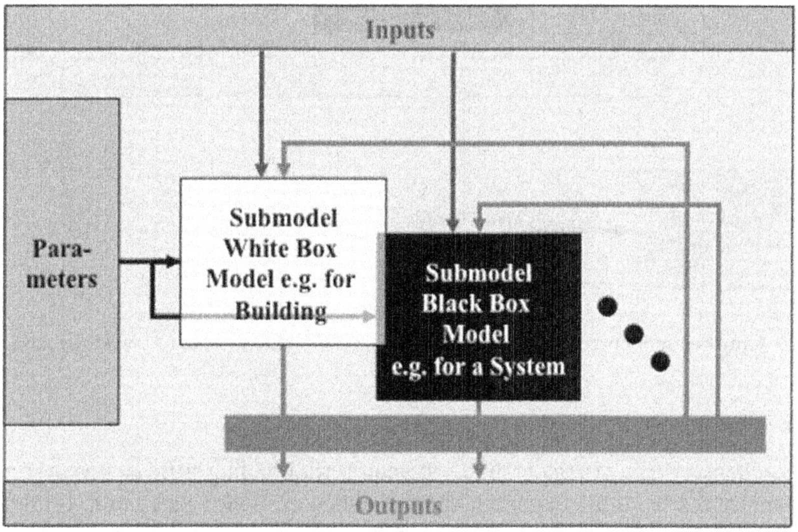

Figure 14.7 Gray box models that subsume models with mixed characteristics are combined into one overall model of different parts

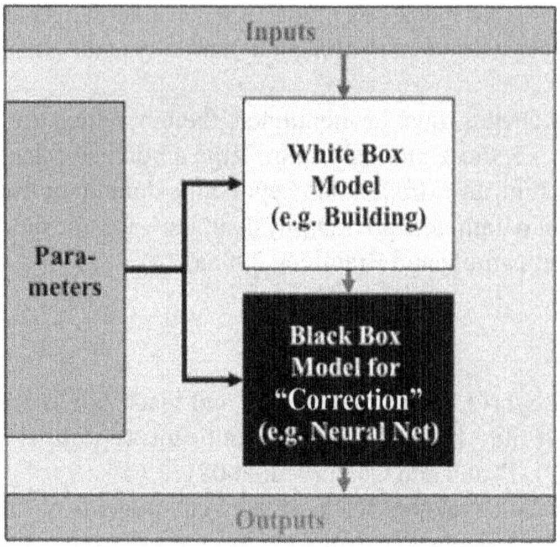

Figure 14.8 Gray box model that enhances a white box model using a black model for higher-order error correction

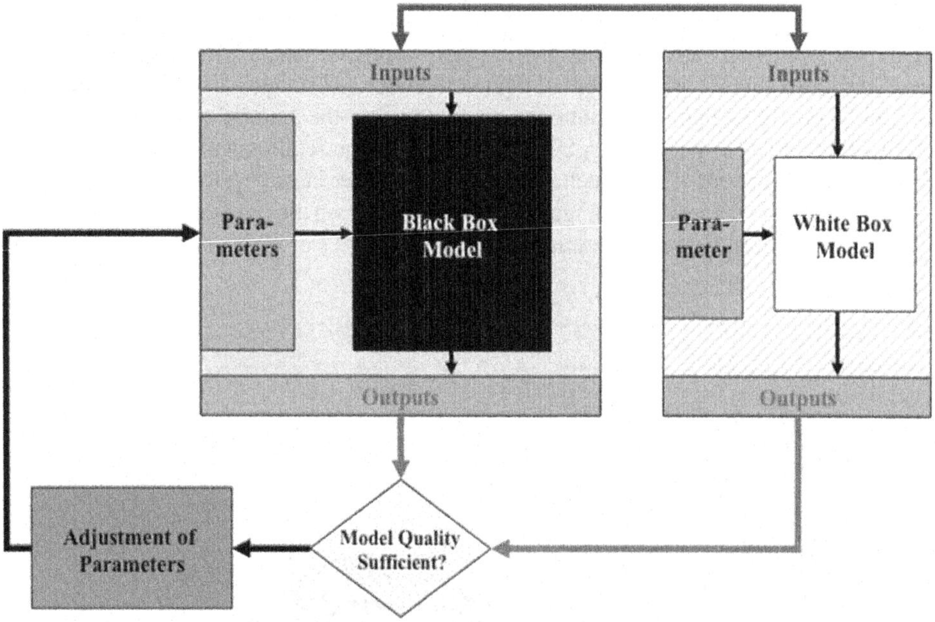

Figure 14.9 Gray box model created by using a white box model to produce a black box model

Summary of model characteristics

In the preceding sections, a range of model structures for building systems was described. This section summarizes the most important characteristics of model generation, calibration, and application for fault detection and optimization. See Table 14.4 for an overview.

Essentially, white and black box models are opposites: While white box models have very good analytical capabilities, they are often costly to produce and calibrate; the reverse is true

Table 14.4 Summary of the advantages (+) and disadvantages (–) of various model types; "o" implies an intermediate satisfaction of the feature

Feature	*White box*	*Gray box*	*Black box*
Insight into physical processes	+	o	–
Number of parameters → error source	–	o	+
Effort in model generation	+ –	+ o	+
Calculation speed	+ o	+ o	+
Need for training data	+	o –	–
Complexity of calibration	+ –	+ –	+
Generalization or extrapolation	+	o	–
Portability to other systems	+	+ –	– *
Usable for FDD	+	+	+
Usable for optimization and control	+	o	o

* Only with new training data.

for black box models. It is not necessarily true, however, that white box models (when of simple structure and resultant reduced prediction accuracy) are more difficult to produce than black box models.

To gain an understanding of the system, white box models are initially a good choice. It is hence advisable to search for a way to allow a white box model to be generated with simplified assumptions, which can be enhanced later, either with greater detail or with a matching black box model (Lebrun 2001).

On the other hand, black box models are most suitable for rapid and easy identification of patterns (e.g. usage profiles for buildings) and the recognition of outliers from the expected pattern. The patterns identified can additionally be used as input values for other, more detailed models. In particular, for complex processes of which the internal structure is unknown or not known in detail, black box models can prove to be helpful.

Development environments for whole building models

Whole building models are assemblies of sub-models each of which may be categorized as black, white, or gray box. Specialized building energy modeling programs, such as TRNSYS or OpenStudio, can be viewed as toolboxes (similar to the general computational environment Matlab/Simulink) including libraries of all sub-models for individual building components and associated numerical solvers. From this perspective, a whole building model is basically the co-simulation of numerous white, gray and (rarely) black box models. Whole building models for the purpose of model predictive control can thus also be based on specialized simulation programs. A comprehensive overview of building energy simulation tools is provided by IBPSA-USA,[2] where hundreds of software tools are documented, ranging from simulation environments that are specifically designed for building simulation to general simulation tools.

Typical examples of these traditional building simulation programs are OpenStudio,[3] EnergyPlus,[4] eQuest,[5] ESP-r,[6] or TAS.[7] Many of these simulation programs operate with models having a medium to high level of detail, i.e. they are multiple zone models with algorithms for heat transfer in multilayer building components, ventilation, solar and internal gain, building energy systems in buildings and control. It must be appreciated, however, that building energy system models often do not have the same level of detail and scope as building models.

Due to the increased numerical performance of computers, the costs of the calculation requirements of many projects/cases is now much lower than the cost of generating simulation models and integrating new models. As a result, a high level of modeling flexibility or a particularly clear and comprehensible implementation of the physical principles is of much higher importance. A modeling approach for this is the equation-based approach, where the physical equations are also included in the simulation program as a component of the simulation model. This enables them to be matched to the current problem in terms of the level of detail and in relation to the interconnections (Sahlin 2000). The disadvantage of equation-based structures is the high calculation expense.

At scientific conferences related to buildings (building physics, IAQ, HVAC, air conditioning, building simulation), a large number of new mathematical/physical models (in the form of equations) have been presented (Sahlin 2000; Sahlin et al. 2004, Wetter et al. 2014). The advantage of equation-based simulation structures is that such new models can be directly integrated into simulations. This means that the community that can participate in the further development of the simulation tools can be substantially expanded. For the integration of new physical models, no special numerical knowledge is necessary.

With TRNSYS,[8] developed in 1973 (Klein 1973) for the simulation of solar installations, certain elements of equation-based models were put into practice. It allows the arbitrary interconnection of subordinate models and the integration of algebraic equations. Modeling with equations was not seen as the primary interest here, but merely as an enhancement opportunity. The only example of an equation-based simulation tool produced specifically for buildings and installations simulation is IDA-ICE.[9]

In principle, *general* simulation tools allow any system to be modeled. Typical examples of this group are Matlab/Simulink,[10] Dymola/Modelica,[11] Wolfram System Modeler,[12] IDA-SE, or EES.[13] Most of these tools belong to the equation-based type and were used successfully for buildings and in particular in the field of building control (André et al. 2008). While Matlab/Simulink is a de facto standard (in particular in the domain of controls analysis), Dymola/Modelica has gained significant prominence in the last few years.

Implementation of model-based control

We differentiate between two types of model-based control implementations: (1) *Offline optimization* simulates and optimizes the control parameters independently of the current building operation (Figure 14.10); by contrast, (2) *online optimization* is a continuous process coupling both building and controllers. Current building state and possibly predictions (e.g. daily power or climate predictions) are considered for this optimization approach (Figure 14.11).

Simple feedforward control only works adequately when the system can be accurately represented with an instantaneous algebraic equation. However, when an accurate dynamic model is to be used for control, model-based predictive control (MPC) is the approach of choice. With MPC, the dynamic model will be used to find a control strategy over a sliding planning horizon that minimizes a performance criterion of choice, such as the cumulative root mean square (RMS) error of the outlet water temperature deviation from the setpoint. Once the optimal strategy is found, the first action of the strategy at the current time is selected as the current control action. After the next time increment, the process is repeated.

As its name suggests, model predictive control automates a target system (the "plant") by combining a prediction and a control strategy. An approximate, linear plant model, as described earlier, provides the prediction. The control strategy compares predicted plant states to a set of objectives, and then adjusts available actuators to achieve the objectives while respecting the plant's constraints. Such constraints can include the actuators' physical limits, boundaries of safe operation, and lower limits for indoor environmental quality. This constraint tolerance

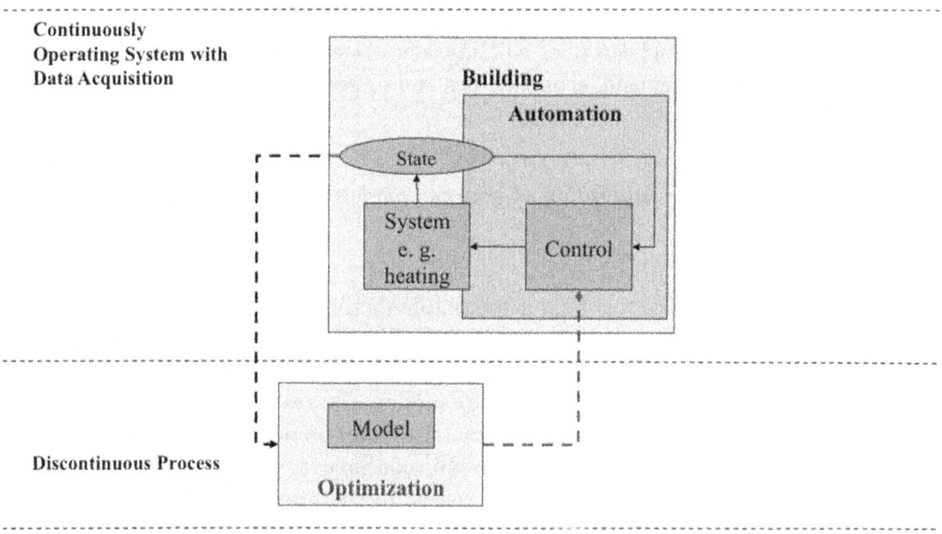

Figure 14.10 Process structure for offline building optimization

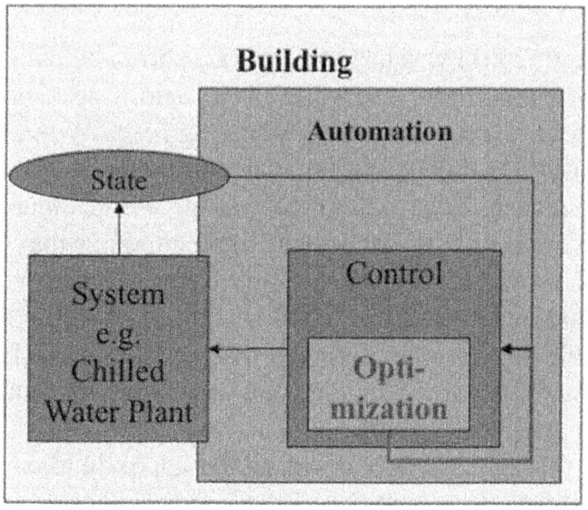

Figure 14.11 Process structure for online optimization

of model predictive control differentiates it from other optimal control strategies, such as the linear quadratic Gaussian (LQG) approach.

Case studies of model-based control applications

This section introduces a set of illustrative examples of least squares versus Bayesian model calibration as well as online and offline optimization approaches applied to commercial and residential buildings using model predictive control. MPC is a control methodology which seeks strategies through time that minimize an objective or cost function, based on the predictions of an energy model. In the context of building systems, MPC allows for discovery of near-optimal operation strategies that minimize, for example, the energy use, carbon dioxide emissions, or dollar cost of a facility. The first examples describes global temperature setpoint optimization for a large office building, the second example illustrates the development of optimal and near-optimal window opening strategies for a small mixed mode ventilated office building, and the third example describes building control strategies in consideration of energy use, energy expense, peak demand, economic demand response revenue, and frequency regulation revenue.

Least squares and Bayesian calibration of gray box models

Approach

To examine model calibration techniques for simple thermal models that can be embedded on RTU or terminal unit control hardware, traditional least squares and Bayesian inference are used to calibrate a simple thermal resistance-capacitance (RC) network to time series data generated from a detailed building energy model. The results of the two methods are compared directly by plotting least squares point estimates on posterior distribution contour slices to contrast and discuss the difference in information available from each method. The robustness of the methods to perform under a variety of conditions is also evaluated through repeating the calibration using training data sets of varying length, noise magnitude, and noise color.

Gray box model of retail building

EnergyPlus (Crawley et al. 2001) was used to generate surrogate data from a five-zone retail building consisting of a single floor area of 2,300 m^2 and a peak occupancy, lighting, and appliance power equal to 7.11 m^2/person, 32.3 W/m^2, and 5.23 W/m^2, respectively. The retail building model was chosen for its relative simplicity, allowing straightforward abstraction to a single-zone approximation reduced-order model as shown in Figure 14.12. The packaged DX systems typically associated with stand-alone retail buildings may be good candidates for embedded model-based control algorithms as well. Surrogate data was preferred here over real measurements so that latent variables could be controlled in the experimental study and calibration methods could be directly compared and contrasted. Only hourly sensible zone loads (estimated by the calibration techniques) and corresponding temperatures were utilized as surrogate data for the gray box modeling.

The gray box model presented here is based on the approximation of heat transfer mechanisms by an analogous electrical lumped resistance-capacitance network. Gray box models

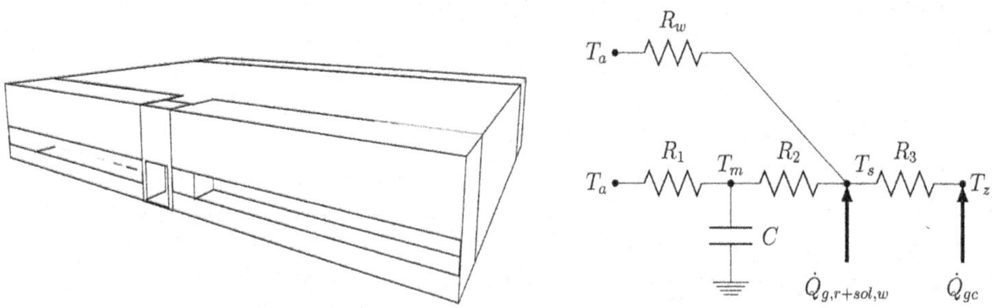

Figure 14.12 Retail building simulation models: a detailed EnergyPlus model (used to produce surrogate data) and corresponding gray box model

can lump multi physics phenomena together and be considered purely physical, or can lump variables/parameters together such that a semi-physical model is obtained. In any case, a simplified model is sought containing only the dynamics of interest. The only requirements of the parameter estimations, i.e. the Bayesian and nonlinear least square approaches considered, is that values are bounded in accordance with feasibility. A five-parameter model based on the RC network used in the CEN-ISO 13790 "Simple Hourly Method" ISO (2007) of Figure 14.12 was used to forecast winter heating loads for the retail building.

The five-parameter RC network heat transfer and storage characteristics of the opaque building envelope materials are represented by R_1, R_2, and C. These elements link the ambient temperature node (T_a) to a pseudo interior surface temperature node (T_s), accounting for potential heat storage of the mass materials. Glazing heat transfer is represented by a single resistance (R_w) connecting the ambient temperature node to the surface temperature node, as thermal storage of glazing is typically neglected. R_3 represents a lumped convection/radiation coefficient between the surface temperature node and zone air temperature node (T_z). The convective portion of internal gains (lighting, occupants, and equipment) are applied as a direct heat source to the zone temperature node, shown as $\dot{Q}_{gc}$, and the radiant fraction along with glazing transmitted solar gains ($\dot{Q}_{g,r+sol,w}$) are applied to the surface node.

Traditional calibration

A number of researchers have considered the calibration of thermal network models using nonlinear least squares approaches, or similar techniques, to parameter estimation (Dewson et al. 1993; Braun and Chaturvedi 2002; Fraisse et al. 2002; Armstrong et al. 2006; Zhou et al. 2008). It has been demonstrated that both time and frequency domain least squares minimization via traditional (e.g. Gauss-Newton) or metaheuristic (e.g. genetic) algorithms are beneficial. However, these methods result in point estimates of the parameters without consideration of uncertainties and can be sensitive to initial optimizer seeds or starting conditions. From an operational modeling perspective, extended Kalman filters (EKF) have been incorporated to improve real-time load estimates using building automation system data (O'Neill et al. 2010).

Inverse gray box model parameters were identified using least squares algorithms that minimize the root-mean-squared error (RMSE) between the reduced order model predictions and surrogate time series data. A two-stage optimization was implemented that performs a direct search over the parameter space and executes a nonlinear least squares algorithm starting from the best location identified through direct search. The direct search is performed over uniformly random points located within a bounded parameter space. Matlab optimizer lsqnonlin, based on trust-region Newton methods (Matlab 2012), was used to solve the constrained nonlinear least squares problem. The parameter space was constrained to consider only physically plausible values (e.g. no negative resistances), while allowing room for estimates to incorporate geometry and construction uncertainty as well (e.g. lower resistances could be chosen to compensate for an underestimation of actual external surface area). To evaluate the reliability of this method, the least squares estimation was repeated 2,500 times with each iteration starting from a new set of 500 randomly generated direct search points.

Bayesian calibration

High-dimensional integrals associated with problems in computational physics lead to the development of Markov Chain Monte Carlo (MCMC) algorithms, which are capable of efficiently sampling from probability distributions by exploiting the Markov property (Jaynes 2003). This lead to the explosion of Bayesian techniques, with the Metropolis algorithm, the breakthrough approach, being named as one of the top 10 algorithms of the 20th century (Cipra 2000).

The probabilistic perspective not only provides insight into the relationship between sets of model parameters, revealing trade-offs and compensating interactions, but also lends itself to continuous model uncertainty quantification and tuning where the posterior distribution of an initial parameter estimation can be used as the prior for a subsequent parameter estimation update once new building performance data has been collected. Dodier used Bayesian (belief) networks for whole building energy diagnostics (Dodier et al. 1998; Dodier and Kreider 1999). Lauret et al. (2006) demonstrated improvements over traditional parameter estimation methods through the application of Bayes's theorem to determine better estimates of convection coefficients for a radiant barrier roof system model. More recently, Booth et al. (2012) makes use of London housing stock models for hierarchical modeling, with considerations of internal heating set points, fraction of space heating, air leakage, heating system COP, window *U*-value, and window-to-wall ratio.

Here the Bayesian calibration of the gray box model relied on the extension of a previously developed technique (Neumann et al. 2011). It has benefits over traditional methods because prior knowledge of the system is directly incorporated into the estimation task. The prior probability distribution is updated with any measured data to form the posterior probability distribution, which represents the state of knowledge in any inference task. The inference can essentially be thought of as fitting a joint probability distribution to a measured data set. Specifically, conditional probabilities are related through the product rule to derive Bayes's theorem and allow consideration of "before data" and "after data."

$$p(A \mid B) = \frac{p(A)\,p(B \mid A)}{p(B)} \tag{14.5}$$

From a calibration perspective, the probability of parameter set Θ given measured data D and a knowledge of the system K can be written as posterior probability $p\ (\Theta \mid D\ K)$. Bayes's theorem then allows the conditional probability $p\ (\Theta \mid D\ K)$ to be computed from $p\ (\Theta \mid K)$, $p\ (D \mid \Theta K)$ and $p\ (D \mid K)$ as in

$$p(\Theta \mid DK)\frac{p(\Theta \mid K)\,p(D \mid \Theta K)}{p(D \mid K)} \tag{14.6}$$

Where $p\ (\Theta \mid K)$ represents prior knowledge about parameter values, $p\ (D \mid \Theta\ K)$ represents the likelihood of observing the measured data set D given a particular parameter set Θ and knowledge of the system K, and $p\ (D \mid K)$ is the probability of observing the data set. To express uninformed belief that the best parameters lie within a particular range, uniform priors $p\ (\Theta \mid K)$ were placed on each resistance and capacitance parameter with the same upper and lower bounds used in the least squares approach (Table 14.5).

Results

To compare results between least squares parameter estimation and the Bayesian methods, the 2,500 solutions from the nonlinear least squares (NLSQ) results were plotted on top of the 2-D contour slices of the posterior distribution. This allows for direct comparison of the probability of a least squares solution from a Bayesian perspective. It is noted that the figures presented herein have been cropped to show least squares solutions and the most likely Bayesian estimation (p_{max}); the entire parameter space is not represented and additional areas of lower probability may exist outside the plotted region.

Figure 14.13 (top left) shows that several least squares runs were in agreement with the Bayesian inference for R_1, although the median value is slightly lower. R_2 values were in close agreement as well, although slightly higher for the Bayesian results. Since R_1 and R_2 represent the envelope material resistances, choosing higher values for both parameters is not directly explainable from that figure. However, looking ahead, the Bayesian inference chose a lower

Table 14.5 Reduced order model training parameter bounds

	R_1	R_2	R_3	R_w	C
Min	0	0	0.0303	0.05	78
Max	4.99	4.99	33.33	3	536659
Units	m²K/W	m²K/W	m²K/W	m²K/W	J/m²K

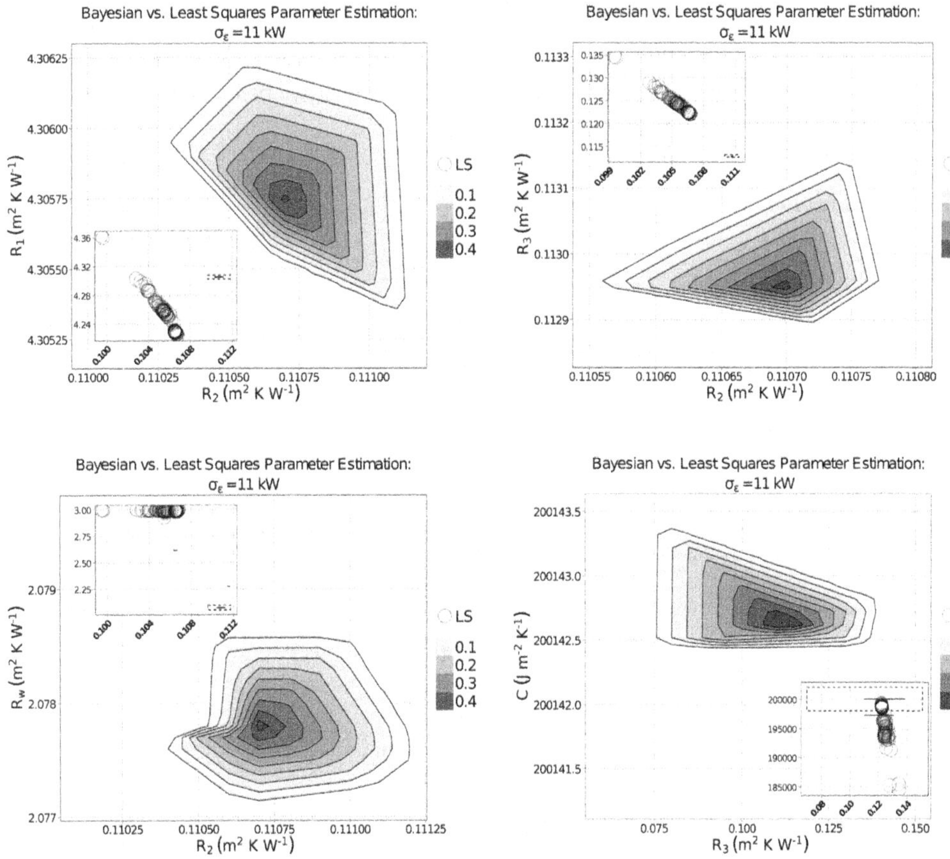

Figure 14.13 Posterior contour with least squares solutions: R_1 vs. R_2 (top left), R_3 vs. R_2 (top right), R_w vs. R_2 (bottom left), C vs. R_3 (bottom right)

glazing resistance R_w, which may result in similar overall envelope performance. Figure 14.13 (top right) again shows slightly higher values for the Bayesian estimate of R_2 and lower values for the internal convection resistance R_3. This inverse relationship between envelope resistance and internal convection resistance may result in similar heat transfer to the zone air node. Figure 14.13 (bottom left) directly highlights an inverse relationship between R_w and R_2 with the Bayesian inference choosing reduced glazing resistance in favor of increased external wall resistance. Figure 14.13 (bottom right) shows the close agreement between thermal capacitance values (C). Although no least squares points appear directly on the posterior, the difference between chosen C values is less than 1%.

Least squares and Bayesian calibration methods performed similarly in terms of mean bias error and root mean square error, regardless of noise, when given an adequate length of data. Bayesian methods did show the potential to outperform on short and noisy data sets when utilizing informative priors. Bayesian calibration was also able to provide further insight into parameter interactions and trade-offs, as well as parameter uncertainty. However, the additional information does not come without added computational cost. The Bayesian calibration required approximately 100 times more CPU time than did the traditional calibration. Additionally, storage and memory requirements would be higher as well since all information is stored in the Bayesian analysis.

Given the intended application of embedded model-based control algorithms and the comparable performance of both methods in a variety of trials, traditional calibration methods appear sufficient for the desired thermal model identification. Future work should consider Bayesian methods for more complex buildings with more parameters and greater uncertainty, where the simple thermal model used in this work may be inadequate. The Bayesian analysis can also be extended to include uncertainty in model structure, which may prove beneficial when considering more complex scenarios and diverse applications.

Control optimization of a large office building

Background

In Corbin et al. (2013), a model predictive control environment is developed. The environment integrates Matlab and EnergyPlus with a modified particle swarm optimizer to predict optimal

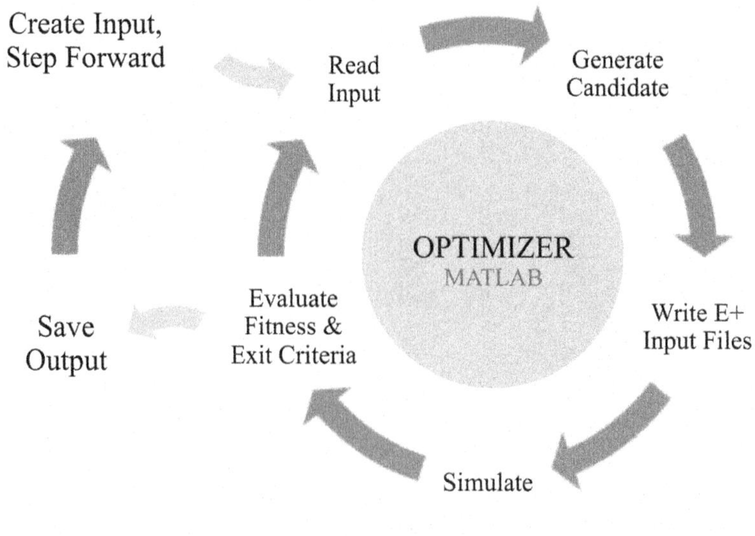

Figure 14.14 Offline optimization process using Matlab and EnergyPlus

building control strategies, continuously, over timeframes ranging from hours to months. A supporting framework is described that couples the environment to a building automation system, allowing real-time optimization considering operator overrides and updated weather forecasts. Application of the environment to optimal HVAC control is demonstrated. First, the environment is used to determine hourly cooling set points, incorporating weather forecasts and day-ahead electricity prices, to minimize daily energy cost for the reference large office building supplied with EnergyPlus.

MPC optimization environment

The model predictive control environment described in Corbin et al. (2013) and illustrated in Figure 14.14 consists of an optimization algorithm, a black box function evaluator and a master program that orchestrates their interaction. The latter is a Matlab program that manipulates EnergyPlus input (IDF) files, initiates simulations, and parses the resulting output files. EnergyPlus is used as the black box function evaluator; the optimization algorithm is a variant of particle swarm algorithm (PSO) described by Kennedy and Eberhart (1995).

Optimization process

Optimization follows a relatively simple iterative process, as shown below. The process is referred to as offline as it is not controlling a physical building. The program first loads building model template and weather template files. Historical weather information and forecasts provided in the program input file are substituted into the weather template file, which is then saved to disk. The weather template is an EnergyPlus Weather (EPW) formatted file containing TMY2 or TMY3 weather data for the building's location.

The building energy model template, an EnergyPlus IDF file, contains the building description with selected values replaced by unique identifiers. Typically, these identifiers represent the decision variables chosen by the optimization algorithm, but are also used for writing other dynamic information to the IDF such as simulation timeframe and output variable objects.

The program then uses an optimization algorithm to generate a candidate decision vector and replaces each of the tokens in the template IDF with the decision vector values. The candidate building model is then saved as a separate case file, and EnergyPlus is executed using the case file and weather file as inputs. Upon completion of the case file simulation, the program then opens and parses the resulting output, extracting the values required to calculate the objective function.

The results of the objective function calculation are then used by the optimization algorithm to generate a new candidate decision vector. This cycle repeats until the optimization algorithm reaches one of several user configurable exit criteria. When an exit criterion is satisfied, the process then steps forward one execution horizon and the process of optimization starts again for the new planning horizon. A description of the various optimization horizons is presented below.

The program has been written in a way that allows different optimizers and building energy simulation programs to be substituted with minimal effort. Although the program is still under development, we have successfully demonstrated its application to several optimization problems, two of which are presented as case studies later. Current research efforts using this program include the optimal control of mixed-mode buildings and associated natural and mechanical ventilation systems, thermally activated building systems, and passive thermal building mass.

Online optimization framework

Coincident to the creation of the MPC optimization program, a server-client system was developed that allows optimized control strategies to be exchanged between the optimization program and a building automation system. When optimized in this way, this process is referred to as online optimization. The system is composed of three components: a building sited client computer, a remote database server written in Java with a SQL database, and a remote cluster of simulation computers. The system has been successfully tested in several commercial buildings since 2009. The system will follows the basic process described below, which is described graphically in Figure 14.15.

At the beginning of the hour, the server creates an XML input file containing weather history, control history, weather forecasts from NOAA, and day-ahead energy prices provided by an energy supplier or regional transmission organization. This input file is then sent over the internet to the cluster of simulation computers where a compiled version of the MPC optimization program and EnergyPlus are installed. One of these computers then uses the input file and the building and weather templates already present to predict a control strategy that minimizes energy cost. The process of optimization then follows that described previously. In an online optimization, the optimization program is limited to a fixed length of optimization time, e.g. one hour, but it may converge and terminate before that time limit applies. The optimal control strategy is saved as an XML file and sent back to the database where it is stored.

At the end of the hour, the optimized control strategy is sent to the client computer installed in the building where it communicates through a proprietary interface with the BAS. The BAS then implements the strategy and the process repeats each hour of the day. Building operator overrides of previously suggested control set points are communicated back to the database server and the next input file is updated to reflect the change in operation.

Case description

The purpose of this study is to determine the electricity cost savings associated with optimal pre-cooling of a building's thermal mass using cooling set points as decision variables. By

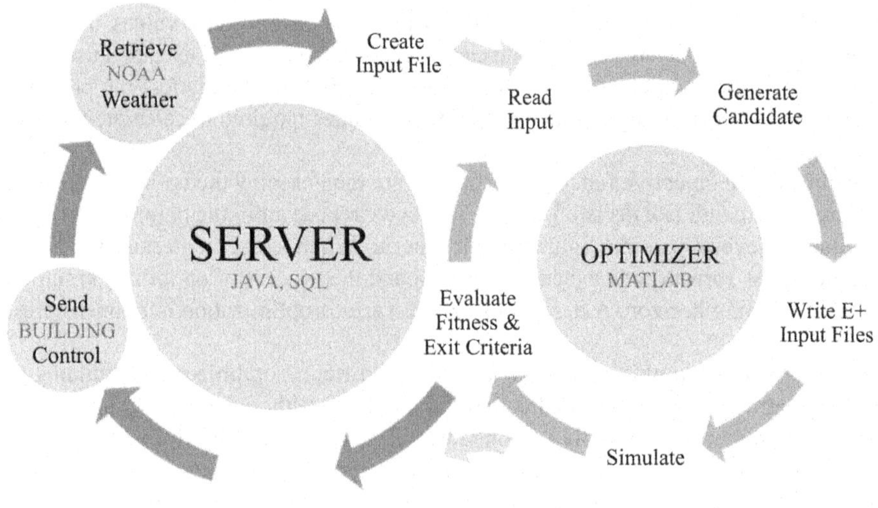

Figure 14.15 Online optimization process using Matlab and EnergyPlus

pre-cooling during early morning hours, we hypothesize that HVAC electricity costs may be minimized when price differentials exist. An optimization timeframe spanning the first week in August during the peak cooling season is selected for this study.

The building chosen for optimization is the reference large office building included with the EnergyPlus installation. The building is hard-sized according to Chicago design day data. Setpoint depressions below the free-float temperature of any zone will force HVAC equipment on and thus pre-cool the building during unoccupied hours. The building is occupied between 8:00am and 6:00pm on weekdays, and between 9:00am and noon on Saturday. All other model parameters remain unchanged. Historical Chicago weather data obtained from NOAA (2011) and real-time electricity prices obtained from PJM (2010) from the summer of 2007 are used.

The optimizer is configured to use a planning and execution horizon of 24 hours beginning at midnight, with two neighborhoods of 15 particles each. The pre-conditioning horizon is set to seven days, but no termination horizon is included. One neighborhood is seeded with a setpoint schedule corresponding to default building operation; the other is seeded with a heuristic pre-cooling strategy. Decision variables corresponding to cooling schedule set points are distributed across the planning horizon in the following way: one decision variable per hour is allocated for the first 12 hours of the day, one decision variable is allocated for the remainder of the occupied period, and one decision variable is allocated for the unoccupied period at the end of the day.

Decision variables are bounded between upper and lower temperature limits that depend on occupancy. These boundaries are 32 °C and 16 °C during unoccupied times, and 22 °C and 24 °C during occupied times, respectively. Each day of optimization is limited to two hours of simulation time. Following optimization, a single base case representing default building operation was simulated for comparison. The cooling setpoint schedule used for this base case corresponds to the upper cooling setpoint boundary.

Results from the optimization are shown in the four charts below. The uppermost chart (Figure 14.16) shows temperature boundaries, average zone temperature, average operative temperature, and optimized cooling setpoint schedule for each hour of the week. Below this is a chart (Figure 14.17) showing hourly chiller and fan electricity consumption, followed by a third chart (Figure 14.18) showing the real-time electricity price, and finally a fourth chart (Figure 14.19) showing cumulative HVAC electricity cost savings calculated with respect to the base case. Note that August 1 is a Wednesday.

Electricity price follows a somewhat regular periodic pattern with the lowest prices occurring in the early morning hours between 2:00am and 4:00am. The highest prices occur later in the afternoon, with the peak at approximately 4:00pm. Prices on Saturday and Sunday, August 4 and 5 respectively, are slightly lower than weekday prices. The hourly difference in price exposes a cost incentive to the optimization, which it exploits by depressing cooling set points during the early morning hours and forcing HVAC equipment to run prior to occupancy.

The resulting setpoint depressions and HVAC operation can be clearly seen in the two uppermost charts. In the three to four hours prior to occupancy on weekdays, the cooling setpoint is lowered, forcing the HVAC equipment to turn on when it would otherwise be off in the base case. The effect on average zone air and operative temperatures can be seen in the early morning hours as the building is cooled in the absence of the large cooling loads normally present in the afternoon.

Interestingly, the HVAC equipment is not forced on earlier in the morning when electricity prices are at their lowest. This suggests that there are trade-offs present between the cost of

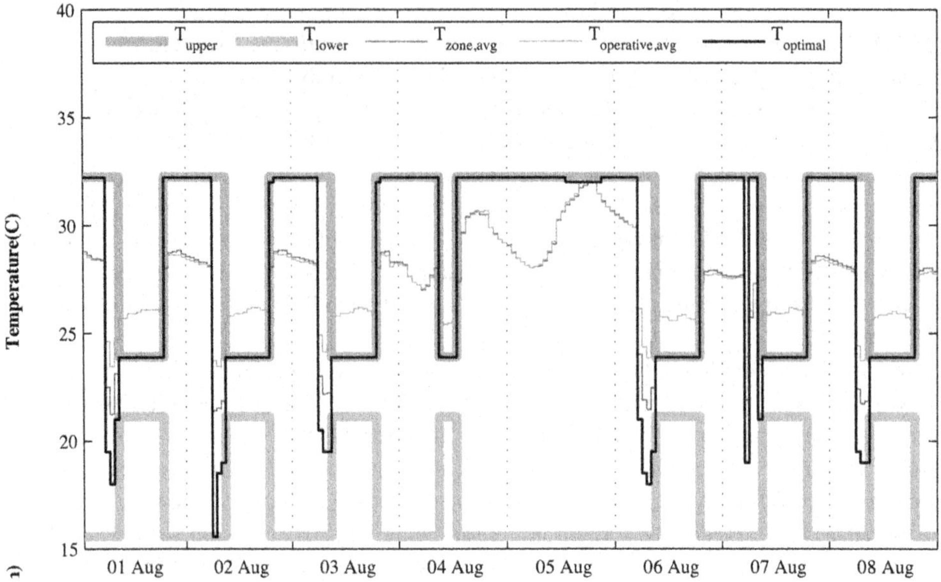

Figure 14.16 Optimal set points, temperatures, and temperature boundaries: August 1–8, 2007

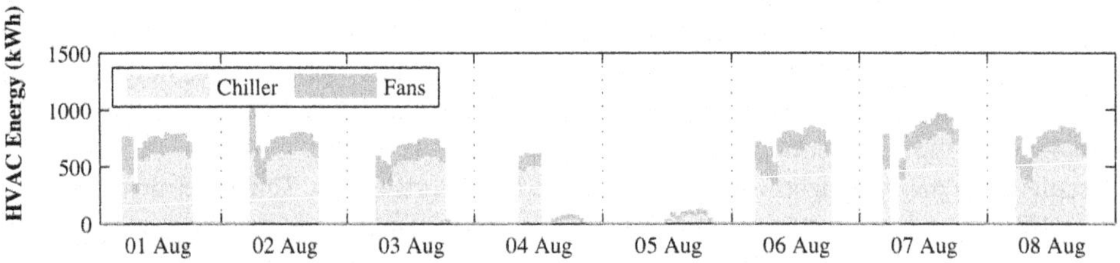

Figure 14.17 Optimal fan and chiller energy consumption: August 1–8, 2007

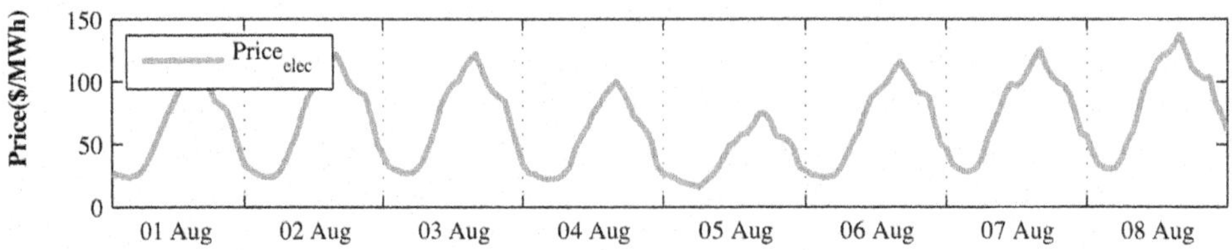

Figure 14.18 Prices: August 1–8, 2007

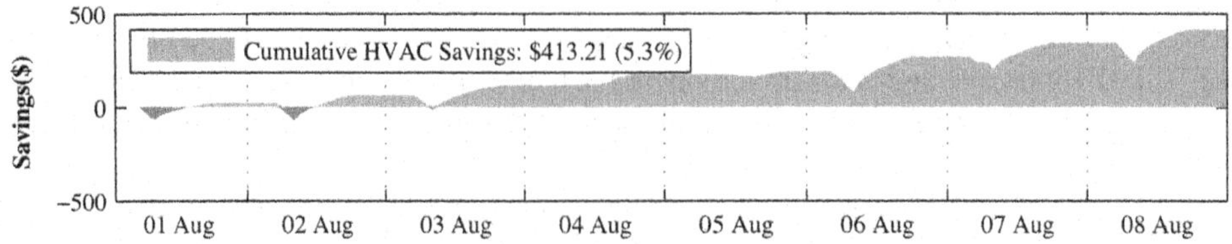

Figure 14.19 Cumulative savings compared to base case operation: August 1–8, 2007

pre-cooling and the 'efficiency' of the building mass as thermal storage. No pre-cooling is observed on Saturday or Sunday because of reduced cooling loads and a weaker cost incentive.

A plot of cumulative HVAC savings reveals an interesting pattern associated with pre-cooling. In early morning hours, cumulative savings drop as the optimized set points force HVAC equipment to run. During the same hours, HVAC equipment remains off in the base case. However, over the course of the day, cumulative savings rise as the pre-cooled building absorbs thermal gains, resulting in decreased loads on the HVAC equipment. When prices are highest, avoided HVAC electricity costs outweigh the cost of pre-cooling, resulting in net positive cost savings on each day. Over the course of the week, approximately 5.3% of HVAC-related electricity costs have been avoided.

MPC of mixed mode buildings with rule extraction

Background

Mixed mode (MM) buildings represent a hybrid approach to space conditioning, employing a combination of natural ventilation and mechanical systems alongside each other and intelligently switching between systems to minimize energy use, while preserving the comfort and well-being of occupants. MM buildings have demonstrated reductions in cooling- and ventilation-related energy use from 20% to 50% over code buildings (Torcellini et al. 2004, Heiselberg 2006) and consistently outperform conventional buildings on thermal comfort and occupant satisfaction (Brager and Baker 2009). However, the performance gains and promise of MM buildings hinge to a large degree on their controls; this is what distinguishes MM buildings from a mere conventional building with operable windows. The effectiveness of the MM control strategy directly determines the extent to which natural ventilation is able to displace mechanical cooling and ventilation, systems that on average account for a quarter of commercial building energy use in the United States today.

Control schemes for mixed mode buildings

MM building controls have generally been classified into three topologies. Under *zoned* control, natural ventilation and mechanical conditioning are allowed to occur simultaneously, but in different zones of the building. For example, perimeter offices may be naturally ventilated and core zones mechanically conditioned. In *concurrent* operation, natural ventilation and mechanical conditioning may operate in the same space at the same time. Finally, *changeover* control allows natural ventilation and mechanical conditioning in the same space, but never at the same time. Most MM buildings will not fall cleanly into one of these categories, mainly because at least some amount of zoning is require to provide dedicated mechanical cooling to certain high load spaces like server rooms.

The building industry's experience with MM buildings is limited, as this style of high-performance building has only emerged in the US in the past two decades. Some very helpful design guidelines have been published in Chartered Institute of Building Services Engineers (CIBSE) application manuals (CIBSE 2005, CIBSE 2000) and in the US trade press (McConahey 2008), but there is no consensus on best practice regarding controls. As such, control sequences must currently be developed on a project-by-project basis. Algorithms usually involve a series of simple heuristics and if/then statements usually developed by an HVAC designer for the building's sequence of operations. For example, "if the outdoor temperature drops below 68 °F, open all automated windows and turn off mechanical cooling." It should be noted that most MM buildings are not fully automated. Occupants retain a great deal of control in MM buildings, particularly in the control of window positions for private offices.

This *adaptive* approach can reduce the complexity of the control system and has been shown to improve occupant thermal comfort by affording them greater latitude to adapt to thermal disturbances (Brager and Baker 2009).

To provide a benchmark against which the performance of building control strategies can be measured and as a means toward developing more energy efficient control strategies for MM buildings, model predictive control has been applied in recent years to the study of MM buildings. Although MPC has been applied extensively in the HVAC engineering field in the past years (Braun 2009, Henze et al. 2004, Henze et al. 2007), it has only recently been applied to MM buildings by Spindler and Norford through the optimization of neural network models specifically trained on two unique buildings (Spindler and Norfored 2009a; Spindler 2004).

The work conducted by May-Ostendorp et al. (2011) expands on the work of Spindler and Norford in two ways. First, it serves as a proof of concept for MPC conducted on physical/white box building models rather than the inverse/gray box approach taken in previous research. This allows the use of freely available and validated building energy simulation tools like EnergyPlus to develop control strategies. Second, it seeks to generalize results for use in typical MM buildings, hence the building models have evolved out of the reference commercial buildings developed by Deru et al. (2006).

Rule extraction

In developing generalized control guidelines for MM buildings, May-Ostendorp et al. (2011) sought to use MPC results to extract simplified control rules that are implementable in buildings today. To achieve truly near-optimal results in real buildings, one could of course couple MPC to a building automation system to direct the optimal control actions of the building in real time. However, this approach would require a set of specialized technologies to communicate MPC decisions to a BAS and, therefore, may not be practical in all facilities.

As an alternative, they used a novel data mining process to extract generalized control rules and decision models from the MPC results. These rules can then serve as the basis for control logic in actual MM buildings, achieving nearly the same energy performance as the actual MPC, but at a fraction of the computational cost. Generalized linear models (GLM) for decision modeling purposes were employed, which has been attempted in various forms in the water management field, but never in the context of MPC in buildings (Bessler et al. 2003, Wei and Hsu 2009).

MPC implementation

The goal of MPC in the context of MM buildings is to minimize the energy use, energy cost, or CO_2 emissions of a building (while preserving thermal comfort) by manipulating operable window positions. Naturally, one can simultaneously investigate optimal control of the building's mechanical systems as well, but for demonstration purposes, the decision space was limited to window openings. The problem can be mathematically formulated as the minimally constrained, integer optimization problem:

$$\begin{aligned} &\min Z\left(\vec{x}_t\right) = E + P \\ &\text{subject to } \vec{x}_t \in \{0,1\} \end{aligned} \tag{14.7}$$

Where $\vec{x}_t$ is a vector of binary decisions (in time) regarding window positions, E is either the energy use or cost over a planning horizon (determined through building energy simulation), and P is a general penalty term applied to discourage certain undesirable characteristics in the solutions, such as thermal discomfort or excessive switching between open and closed window states.

The cost function that results does not lend itself well to traditional gradient or pattern search techniques, because it can contain many local minima. As a result, a meta-heuristic search technique, particle swarm optimization, has been adopted to quickly and robustly search the decision space for near-optimal solutions. PSO combines simple rules with randomized weighting factors to generate complex search behavior in a population of "particles" evaluating the search space. The action is akin to the flocking behavior of birds and schooling behavior of fish.

As with these organisms, information shared between individuals in the swarm affects the decisions of others, all of whom eventually converge on the best solution found by the group. The algorithm is non-deterministic and therefore the search pattern of any swarm is impossible to determine *a priori*. This characteristic of the algorithm decreases the likelihood that it becomes stuck in local minima, at the expense of guaranteed convergence upon the true global minimum. The particular implementation of the PSO algorithm used in this study is a variant of the algorithm presented in the foundational work conducted by Eberhart and Kennedy (1995).

In Figure 14.20, a block diagram schematic of the overall optimization environment is presented to demonstrate the general solution approach. Building models are read in and modified by the PSO algorithm in Matlab by manipulating schedules that control window openings. The resulting models are evaluated using the EnergyPlus simulation engine. Results are read back into Matlab, where the cost function is computed, and the PSO algorithm decides how to proceed to the next decision vector. The algorithm recurses until a predetermined exit criterion is reached.

Figure 14.21 illustrates some representative results for a small office building located in Boulder, Colorado, USA, from a "swing season" week in June. A base case and reference case were run over the same period for comparison. The base case is the standard DOE reference building, without any natural ventilation, whereas the reference case is the corresponding MM building, but with occupant-driven window control per the modified Humphrey's algorithm (Rijal et al. 2008).

The upper graph shows ambient temperatures over the week. The middle graph shows the optimal solution alongside the mean window opening behavior in the reference model. The base case building is, by definition, sealed, so no window positions are shown. The hashed portions of the optimizer solution represent concurrent natural ventilation and mechanical cooling. Finally, the bottom graph shows cumulative energy savings for the optimized case compared to both the base and reference cases.

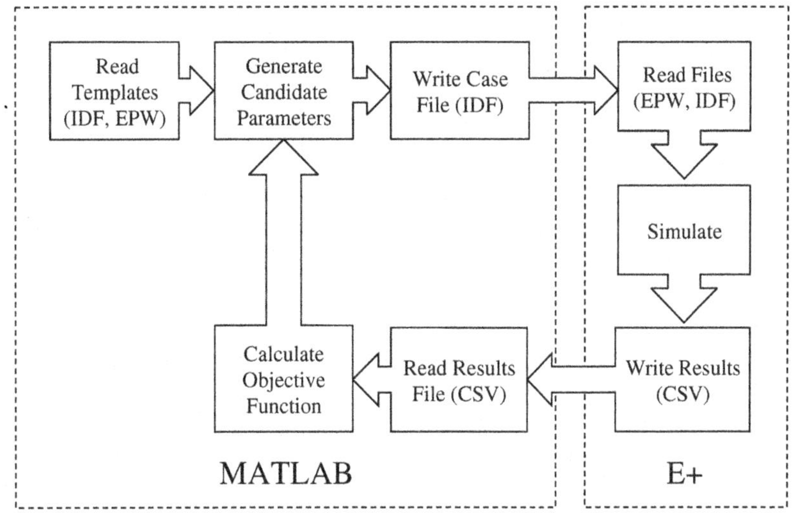

Figure 14.20 The optimization environment coupling Matlab and EnergyPlus

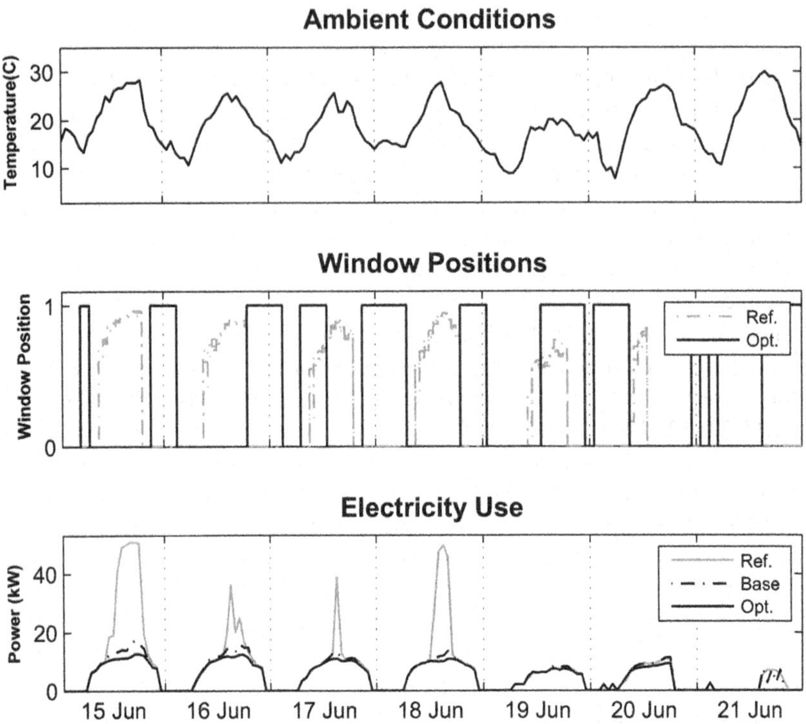

Figure 14.21 Optimal solution and energy use profile for a typical week in mid-June. The optimal solution outperforms the base and reference cases through a night ventilation strategy. In the reference case, occupant window openings during peak solar gains and daytime temperatures result in a dramatic spike in fan and direct expansion cooling energy use.

The optimal solution found during the early summer manifests itself as a night ventilation strategy, with the optimizer opening windows during cooler nighttime periods. This form of passive thermal energy storage utilization allows the building to ride out some of the daytime cooling loads without the need for mechanical cooling. The energy savings accrue most noticeably on the cooling peak in the afternoon for all days except June 19, which is a somewhat more mild day with lower cooling loads to begin with. The optimizer arrived at these solutions with minimal constraints and absolutely no expert knowledge about the decision space, yet the solution mimics a heuristic approach in use in many mixed-mode buildings.

It is interesting to note that the simulated occupants in the reference case building frequently make use of the operable windows during relatively warm periods of the day that also coincide with peak solar gains, resulting in increased cooling loads on the mechanical system. More importantly, the occupants do not have the ability to recreate the night cooling strategy, except perhaps by leaving windows open at the end of the occupied period, so as a result of these two conditions, the reference building actually results in *increased* energy use compared to both the optimal solution and the base case.

Rule extraction implementation

Four models were formulated to compare different parameter sets and pruning approaches. For models 1 through 3, a stepwise regression approach was used to find the parameter set that minimized the model cost/complexity metric AIC (Akaike's information criterion). With each successive model, a larger number of lagged predictor variables were included to attempt to capture process memory. Model 1 utilized only current time step predictors (x_t), whereas model 2 included 1-h lagged predictors as well (x_{t-1}). Model 3 included the previous hour's optimal window state (y_{t-1}) as a predictor in addition. The fourth model was manually pruned based on the results of models 1 through 3 in an attempt to eliminate some of the redundancy present in the zone temperature variables. The large number of parameters for models 1 through 3 results from the inclusion of multiple zone temperatures in the best model predictor set.

The resulting model predictions and the original optimizer sequence are presented in Figure 14.22 for the week of June 15 through 21, with probabilistic predictions as a dashed line. All models perform remarkably well when compared to the original MPC data, tracking periods of opening and closing accurately, even very brief periods, such as the 2 h opening occurring on the first night of the week. However, models 1 and 2 also miss several long periods

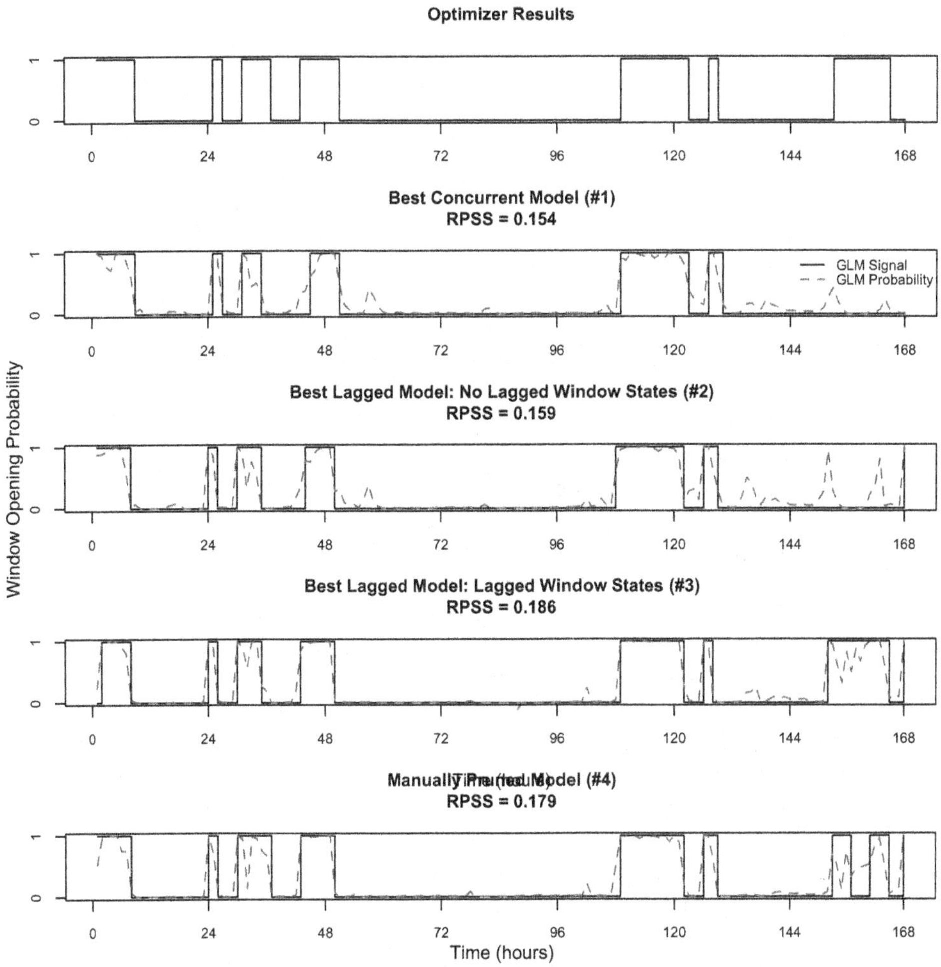

Figure 14.22 Predictions of window position for models 1 through 4 are presented above alongside offline MPC optimal results for a typical week in mid-June. Models 3 and 4 demonstrate the overall best performance. Both models contain the autoregressive window state term.

during which the optimizer windows were open, namely the beginning of the sixth and seventh nights. In the case of model 2, these openings are missed as a result of the hysteresis placed on the output signal. The RPSS (rank probability skill score) shown in the figure compares the accuracy of model predictions against chance, but rather than simply compare model against a 50:50 chance of a window opening, it compares against the probability of window openings seen in the optimizer results, a rather weighted coin. This provides for a more rigorous test of model performance. The RPSS is negative if model results are worse than chance, 0 if model results reproduce chance events, and positive if model results are closer to the original observations than chance.

The GLM rule extraction technique presented by May-Ostendorp et al. (2011) worked best when some time-lagged information is present as a predictor variable to ensure that some of the process memory is preserved. Even with a simplified model that does not incorporate this autoregressive component, one can very closely mimic the general characteristics of the optimizer results, achieving 70%–90% of optimizer energy savings, but at a small fraction of the computational expense. Given the simple mathematical formulation of the GLM, it would be possible to implement this sort of decision model into modern direct digital control systems to control MM buildings in a near-optimal manner in real time.

Commercial building control for energy and ancillary service markets

Background

Operating a reliable and effective large-scale electric power system requires the procurement and scheduling of resources over several time scales. Long-term planning secures the availability of adequate generation capacity to meet changing consumer needs, while short-term scheduling and dispatch activities ensure real-time stability through continuously balancing electricity supply and demand. In addition to the scheduling of generation capacity, various ancillary services, such as frequency regulation, spinning reserves, voltage control, and black start, are required to maintain power system reliability (Heffner 2008).

Frequency regulation ancillary service is responsible for correcting small deviations between supply and demand that occur on minute or sub-minute time scales, while spinning and non-spinning reserve services provide a safeguard against generator failures and contingency events. Providing ancillary services through flexible load response has the potential to increase electric grid reliability and efficiency while offering loads a revenue-generating opportunity. The large power draw of commercial buildings, along with thermal mass characteristics, has sparked interest in providing ancillary services through intelligent operation of building mechanical systems.

As a precursor to participating in ancillary service markets, the quantity of service available must be estimated. Pavlak et al. (2014) presented a model-based approach for estimating commercial building frequency regulation capability. For commercial buildings to participate in ancillary service markets it is necessary to estimate the quantity of service that can be provided. It is also necessary to develop operational strategies that optimally balance revenue-generating opportunities with cost- (or energy-) reducing objectives (e.g. following a frequency regulation signal may increase peak demand if not planned carefully).

A model predictive control framework was developed by Pavlak et al. (2014) to determine optimal operating strategies in consideration of energy use, energy expense, peak demand, economic demand response revenue, and frequency regulation revenue. The methodology was demonstrated through simulation for medium office and large office building applications, highlighting its ability to merge revenue-generating opportunities with traditional demand- and cost-reducing objectives.

Modeling

Thermal zone and HVAC models were developed for a 14,000 m^2 six-story (medium) office building with packaged direct expansion (DX) variable-air-volume cooling system. Thermal zone dynamics were modeled using inverse gray box resistance-capacitance networks (Braun and Chaturvedi 2002) and were identified using output from detailed, multi-zone, EnergyPlus (Crawley et al. 2001) representations. The 11-parameter RC structure used for the medium office thermal model is shown in Figure 14.23. Model structures were chosen based on sensible load and temperature prediction performance after considering a variety of forms for each building.

The packaged DX variable-air-volume system of the medium office building, outlined in Figure 14.24, shows a VAV air handling unit with DX cooling coil, hot water heating coil, and dedicated outdoor air system (DOAS). Equipment models were based on the quasi steady-state physical formulations used by several mainstream whole building simulation programs (Brandemuehl et al. 1993). Equipment models are simulated in sequence so the impact of each individual component affects the others downstream.

The overall simulation scheme can be described by the following "predictor-corrector" approach: (1) perform an ideal load calculation to determine zone sensible load that meets the zone temperature setpoint, (2) simulate the HVAC system using the ideal load as a requested system load, and (3) update zone temperature and load based on actual system performance and capabilities. This approach is repeated for each simulation time step allowing HVAC performance, control strategies, and capacity limits to affect zone conditions. The simulation also includes a simplified moisture balance to compute zone humidity levels as well.

Frequency regulation capability estimation

To participate in frequency regulation (FR) markets, it is necessary to quantify the potential regulating capacity of a building. At a high level, the problem is similar to that posed in the

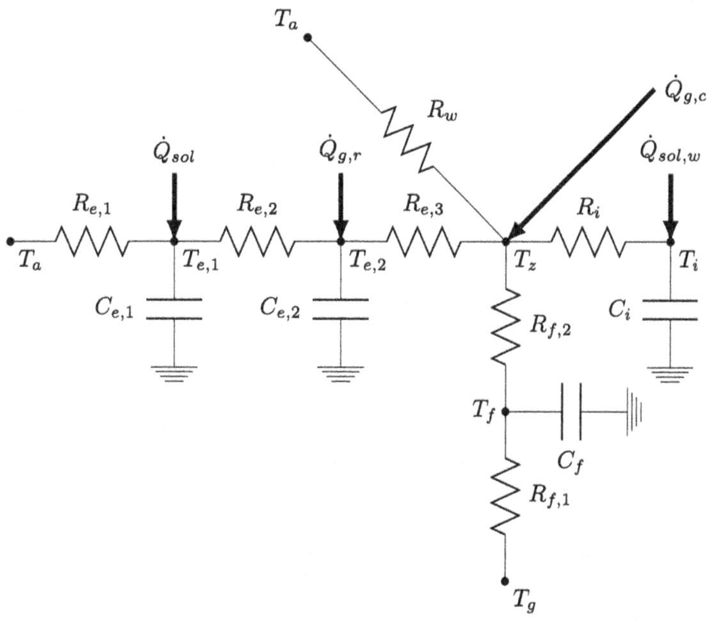

Figure 14.23 Eleven-parameter RC network used for thermal modeling of the medium office building

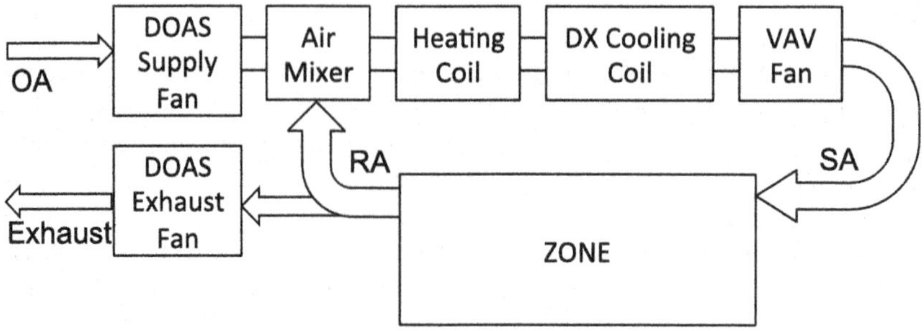

Figure 14.24 Packaged variable-air-volume HVAC system

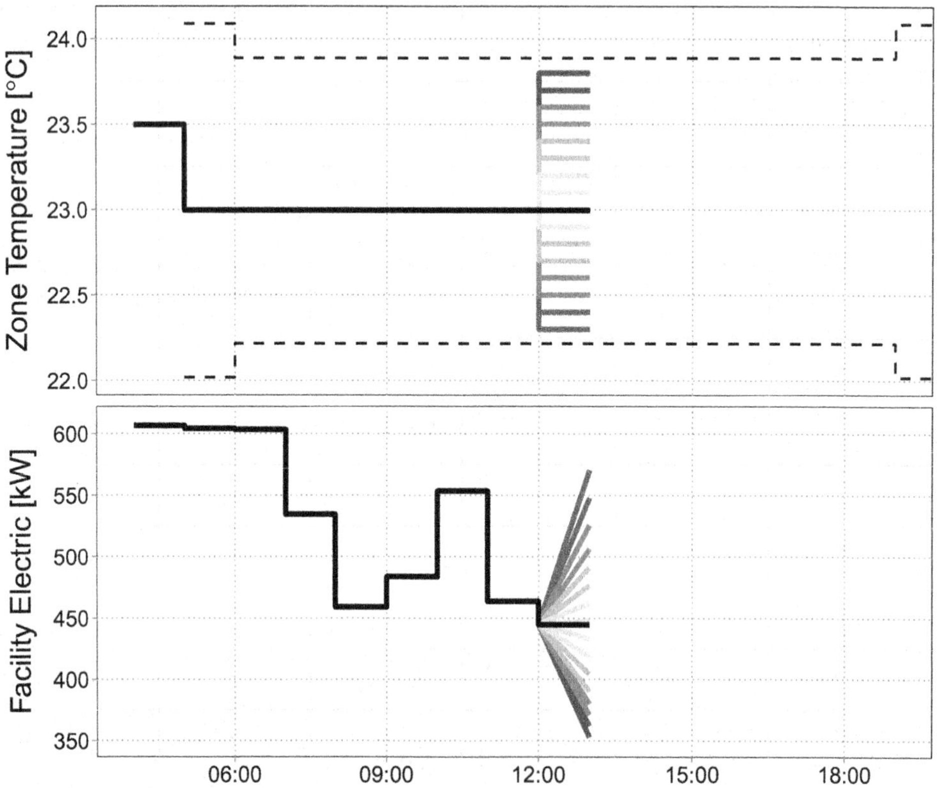

Figure 14.25 Zone setpoint perturbation example and the observed power response

electric vehicle literature in that it is desired to determine the potential of a building to increase or decrease its power draw with respect to a particular baseline (i.e. preferred operating point). However, due to the complexity and challenges, a neat linear formulation is precluded. Consequently, a model perturbation approach is used to determine the relationship between building power response and changes in control input.

This concept is illustrated for FR via zone setpoint modulation in Figure 14.25. In the example, the model is assumed to be tracking a baseline temperature setpoint of 23 °C, and it is desired to determine the regulating capability for hour ending 1 PM. The model is used to evaluate the impact of zone setpoint changes from the baseline by simulating 0.1 K increments between a lower temperature limit of 22.2 °C and an upper limit of 23.9 °C as shown in the top window of Figure 14.25. Simulation results are compared with the original baseline demand profile to determine potential changes in whole building power. Increasing the temperature set point results in a reduction in zone sensible load and HVAC power, providing regulation up. Decreasing the zone temperature set point increases HVAC power resulting in regulation down. The bottom window of Figure 14.25 shows the potential for approximately ±100 kW of power change by modulating the set point between the upper and lower limit.

An advantage of using more complex models is that the power response can be grouped based on other simulation output variables. Depending on the HVAC system in use, it may be undesirable to allow the frequency regulation signal to cycle equipment on and off. In Figure 14.26, the power response is grouped based on the number of active cooling stages and shows that limiting the modulation to a range of 22.6–23.4 °C would prevent turning stages on or off. Limiting the setpoint changes to this smaller range also reduces the regulating potential to approximately ±50 kW.

The model perturbation analysis is repeated for each hour it is desired to determine the regulating capability. Figure 14.27 provides an example of performing the FR estimation for a period from hour ending 9 AM to 4 PM, assuming a constant baseline of 23 °C. The example highlights the fact that regulating potential may vary each hour depending on changes in HVAC equipment loading and efficiency.

Results for medium office building

The multi-market optimization was applied to a medium office building for low target demand (TDL) and high TDL cases. The low TDL case creates the scenario of a potentially peak setting day, while the high TDL case simulates a scenario in which peak demand is not critical. The building was assumed to be available for regulation from 9am to 5pm, using the zone temperature setpoint as the modulation variable. Temperature bands were limited to a range that produces symmetric power response (i.e. equal regulation up and down) to satisfy ISO requirements. Figure 14.28 shows the results for a low TDL of 325 kW (i.e. demand limiting).

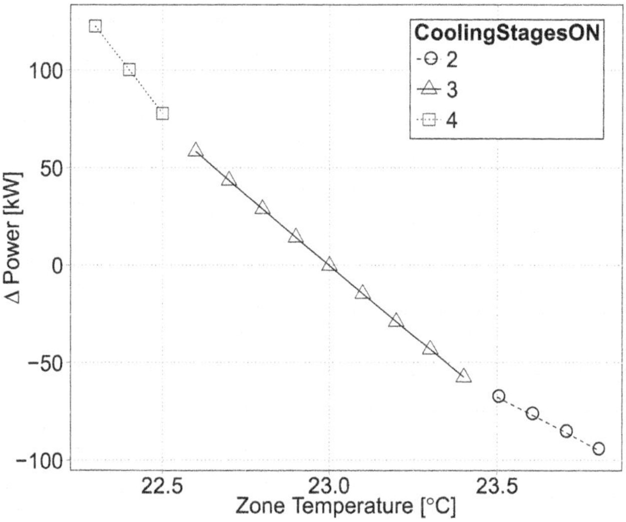

Figure 14.26 Setpoint perturbation power response grouped by active cooling stages

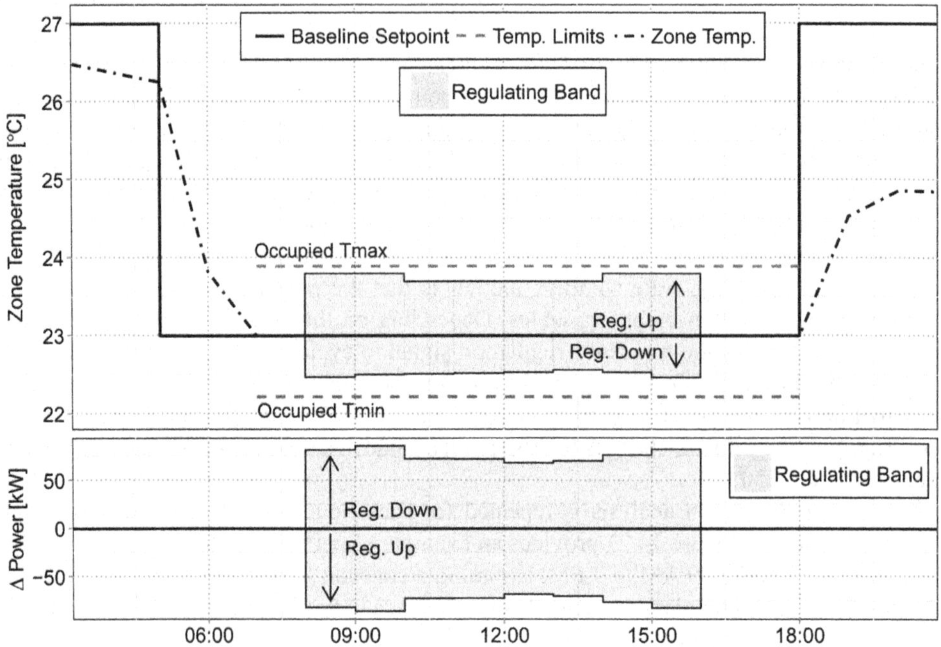

Figure 14.27 Setpoint perturbation performed for each hour from 9am to 4pm

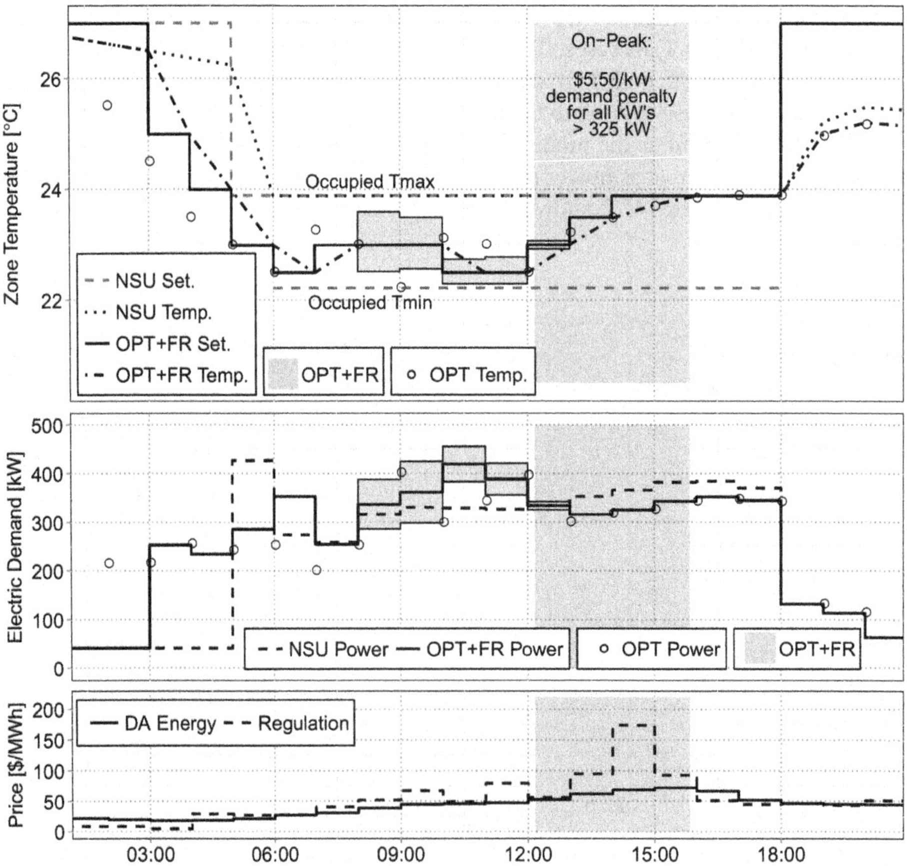

Figure 14.28 Medium office zone setpoint optimization with FR and low TDL

The nighttime setup (NSU) case represents typical building operation without optimization, while the OPT+FR case represents the result of the multi-market optimization process. In the OPT+FR case, the optimizer chose to start the HVAC system at hour ending 3am, pre-cooling the zone to a temperature of 22.5 °C at hour ending 7am. The temperature is held near the middle and bottom of the occupied temperature range preceding the on-peak period to maintain stored thermal energy, and then stepped to the upper bound (occupied T_{max}) during the on-peak period to release thermal energy back into the zone. With the OPT+FR baseline zone setpoint strategy, regulation was available for five of nine potential hours ranging from ±8 kW to ±$60 kW.

Overall, the primary objective of this scenario was to reduce on-peak demand since a significant $5.50 penalty is applied for each kW over the TDL. The minimal impact of regulation in this scenario is also visible, contributing only $12 of revenue. The temperature and demand profiles of the OPT case (i.e. zone setpoint optimization without regulation) are also plotted for reference and show a similar strategy when regulation opportunities are excluded.

The results for the high TDL scenario are shown in Figure 14.29. Looking at the OPT temperature and demand profiles shows an earlier start-up with slight pre-cooling before 6am, followed by near NSU operation for the remaining hours. Expending an extra 82 kWh earlier in the day results in slight overall energy cost savings ($5.69) by reducing consumption during high-price hours. No regulation would be possible in the OPT case since the temperatures remain near the upper bound during hours available for regulation. In the

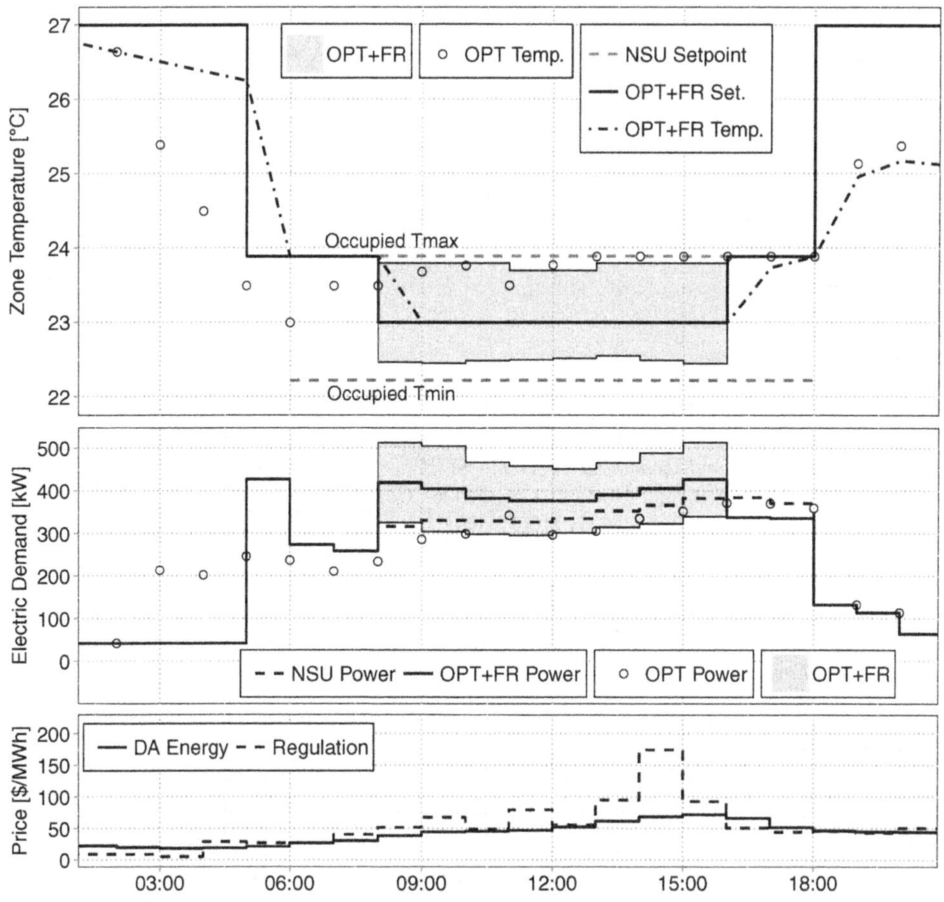

Figure 14.29 Medium office zone setpoint optimization with FR and high TDL

OPT+FR case, the optimizer chose to follow the NSU strategy until hour ending 8am, after which the building is available to provide regulation. During regulation hours, the baseline zone set point was kept in the middle of the acceptable temperature range in order to create room for providing regulation. Keeping the zone set point 1 K lower during regulating hours uses 360 kWhs more than the NSU case, however the regulation revenue more than offsets the additional energy cost resulting in an overall savings of 15.7%. On average, an estimated ±85 kW of regulation were available, with a maximum of ±100 kW available during hour ending 10am.

Overall, the methodology developed by Pavlak et al. (2014) and presented here allows for the estimation of commercial building regulating potential since detailed interactions and constraints are captured through the use of whole-building energy models. The MPC framework, considering both cost and revenue, appropriately determined opportunities when buildings could participate in FR. The optimizer was able to increase the potential regulation capability by moving buildings away from setpoint and system capacity limits to a more flexible state.

Notes

1 www.eia.doe.gov
2 www.buildingenergysoftwaretools.com
3 www.openstudio.net
4 https://energyplus.net
5 http://doe2.com
6 www.esru.strath.ac.uk/Programs/ESP-r.htm
7 www.edsl.net/main/Software/Designer.aspx
8 http://sel.me.wisc.edu/trnsys
9 IDA-ICE 4.7, 2017, www.equa.se
10 www.mathworks.com
11 www.3ds.com/products-services/catia/products/dymola and www.modelica.org
12 www.wolfram.com/system-modeler/
13 www.fchart.com/ees/

References

André, P., Hannay, C., Hannay, J., Lebrun, J., Lemort, V. and Teodorese, I.V. (2008) A contribution to the audit of an air-conditioning system: Modelling, simulation and benchmarking. *Building Services Engineering Research and Technology* 29(1): 85–98.

Armstrong, P.R., Leeb, S.B. and Norford, L.K. (2006) Control with building mass-Part I: Thermal response model. *ASHRAE Transactions* 112(1): 449.

ASHRAE. (2013) *ASHRAE Handbook Fundamentals 2013*. ASHRAE, Inc.

Bessler, F.T., Savic, D.A. and Walters, G.A. (2003) Water reservoir control with data mining. *Journal of Water Resources Planning and Management* 129(1): 26–34.

Booth, A.T., Choudhary, R. and Spiegelhalter, D.J. (2012) Handling uncertainty in housing stock models. *Building and Environment* 48: 35–47.

Brager, G. and Baker, L. (2009) Occupant satisfaction in mixed-mode buildings. *Building Research & Information* 37(4): 369–380.

Brandemuehl, M.J., Bradford, J.D. and Seaton, W.W. (1998) *Implementation of On-line Optimal Supervisory Control of Cooling Plants Without Storage*. Colorado: University of Colorado.

Brandemuehl, M.J., Gabel, S. and Andresen, I. (1993) *HVAC2 Toolkit, Algorithms and Subroutines for Secondary HVAC System Energy Calculations*. Atlanta, GA: ASHRAE, Inc.

Braun, J.E. (2003) Load control using building thermal mass. *Journal of Solar Energy Engineering* 125(3): 292–301.

Braun, J.E. (2009) Smart HVAC&R equipment-coming to a building near you? *HVAC&R Research* 15(1): 1.

Braun, J.E. and Chaturvedi, N. (2002) An inverse gray-box model for transient building load prediction. *HVAC&R Research* 8(1): 73–99.

Braun, J.E., Klein, S.A., Mitcel, J.W. and Beckman, W.A. (1989) Applications of optimal control to chilled water systems without storage. *ASHRAE Transactions* 95: 663–675.

Braun, J.E., Montgomery, K.W. and Chaturvedi, N. (2001) Evaluating the performance of building thermal mass control strategies. *HVAC&R Research* 7(4): 403–428.

Chartered Institution of Building Services Engineers. (2000) *CIBSE AM13–2000: Mixed Mode Ventilation*. Page Bros.

Chartered Institution of Building Services Engineers. (2005) *CIBSE AM10: Natural Ventilation in Non-domestic Buildings*. Page Bros.

Cipra, B.A. (2000) The best of the 20th century: Editors name top 10 algorithms. *SIAM News* 33(4): 1–2.

Claridge, D.E. (2004) Using simulation models for building commissioning. In *International Conference on Enhanced Building Operation*, Energy Systems Laboratory, Texas A&M University.

Corbin, C.D., Henze, G.P. and May-Ostendorp, P. (2013) A model predictive control optimization environment for real-time commercial building application. *Journal of Building Performance Simulation* 6(3): 159–174.

Crawley, D.B., Lawrie, L.K., Winkelmann, F.C., Buhl, W.F., Huang, Y.J., Pedersen, C.O., . . . and Glazer, J. (2001) EnergyPlus: Creating a new-generation building energy simulation program. *Energy and Buildings* 33(4): 319–331.

Del Barrio, E.P. and Guyon, G. (2003) Theoretical basis for empirical model validation using parameters space analysis tools. *Energy and Buildings* 35(10): 985–996.

Déqué, F., Ollivier, F. and Poblador, A. (2000) Grey boxes used to represent buildings with a minimum number of geometric and thermal parameters. *Energy and Buildings* 31(1): 29–35.

Deru, M., Griffith, B. and Torcellini, P. (2006) Establishing benchmarks for DOE commercial building R&D and program evaluation. ACEEE Summer Study, Pacific Grove, California.

Dewson, T., Day, B. and Irving, A.D. (1993) Least squares parameter estimation of a reduced order thermal model of an experimental building. *Building and Environment* 28(2): 127–137.

Dodier, R.H., Curtiss, P.S. and Kreider, J.F. (1998) Small-scale on-line diagnostics for an HVAC system. *ASHRAE Transactions* 104: 530.

Dodier, R.H. and Kreider, J.F. (1999) Whole building energy diagnostics. *ASHRAE Transactions* 105(1): 579–589.

Eberhart, R.C. and Kennedy, J. (1995) Particle swarm optimization. In *Proceeding of IEEE International Conference on Neural Network*. Perth, Australia, 1942–1948.

Fraisse, G., Viardot, C., Lafabrie, O., and Achard, G. (2002). "Development of a simplified and accurate building model based on electrical analogy." Energy Build., 34(10), 1017–1031.

Gouda, M.M., Danaher, S. and Underwood, C.P. (2002) Building thermal model reduction using nonlinear constrained optimization. *Building and Environment* 37(12): 1255–1265.

Heffner, G. (2008) *Loads Providing Ancillary Services: Review of International Experience*. Lawrence Berkeley National Laboratory.

Heiselberg, P. (2006) Hybrid ventilation in non-residential buildings. *Building Ventilation: The State of the Art* 191.

Henze, G.P. (2003) An overview of optimal control for central cooling plants with ice thermal energy storage. *Journal of Solar Energy Engineering* 125(3): 302–309.

Henze, G.P., Felsmann, C. and Knabe, G. (2004) Evaluation of optimal control for active and passive building thermal storage. *International Journal of Thermal Sciences* 43(2): 173–183.

Henze, G.P., Le, T.H., Florita, A.R. and Felsmann, C. (2007) Sensitivity analysis of optimal building thermal mass control. *Journal of Solar Energy Engineering* 129(4): 473–485.

Heo, Y., Choudhary, R. and Augenbroe, G.A. (2012) Calibration of building energy models for retrofit analysis under uncertainty. *Energy and Buildings* 47; 550–560.

Higdon, D., Gattiker, J., Williams, B. and Rightley, M. (2008) Computer model calibration using high-dimensional output. *Journal of the American Statistical Association* 103(482): 570–583.

ISO. (2007) 13790: 2007. Energy performance of buildings-Calculation of energy use for space heating and cooling.

Jaynes, E.T. (2003) *Probability Theory: The Logic of Science*. Cambridge: Cambridge University Press.

Katipamula, S. and Brambley, M.R. (2005a) Review article: Methods for fault detection, diagnostics, and prognostics for building systems – a review, part I. *HVAC&R Research* 11(1): 3–25.

Katipamula, S. and Brambley, M.R. (2005b) Review article: Methods for fault detection, diagnostics, and prognostics for building systems – a review, part II. *HVAC&R Research* 11(2): 169–187.

Kennedy, M.C. and O'Hagan, A. (2001) Bayesian calibration of computer models. *Journal of the Royal Statistical Society: Series B (Statistical Methodology)* 63(3): 425–464.

Kintner-Meyer, M. and Emery, A.F. (1995) Optimal control of an HVAC system using cold storage and building thermal capacitance. *Energy and Buildings* 23(1): 19–31.

Kissock, J.K., Haberl, J.S. and Claridge, D.E. (2003) Inverse modeling toolkit: Numerical algorithms (RP-1050). *ASHRAE Transactions* 109(2): 425–434.

Klein, S.A. (1973) TRNSYS – A transient simulation program. Solar Energy Laboratory, University of Wisconsin-Madison.

Lauret, P., Miranville, F., Boyer, H., Garde, F. and Adelard, L. (2006) Bayesian parameter estimation of convective heat transfer coefficients of a roof-mounted radiant barrier system. *Journal of Solar Energy Engineering* 128(2): 213–225.

Lebrun, J. (2001) Simulation of a HVAC System with the Help of an Engineering Equation Solver. Building Simulation, Rio de Janeiro, Brazil, August 13–15.

Ljung, L. (1999) *System Identification*. New York: John Wiley & Sons, Inc.

Ma, Y., Borrelli, F., Hencey, B., Coffey, B., Bengea, S. and Haves, P. (2012) Model predictive control for the operation of building cooling systems. *IEEE Transactions on Control Systems Technology* 20(3): 796–803.

Ma, Z. and Wang, S. (2009) An optimal control strategy for complex building central chilled water systems for practical and real-time applications. *Building and Environment* 44(6): 1188–1198.

Massie, D.D. (2002) Optimization of a building's cooling plant for operating cost and energy use. *International Journal of Thermal Sciences* 41(12): 1121–1129.

MATLAB. (2012) version 8.0.0.783 (R2012b). The MathWorks Inc., Massachusetts.

May-Ostendorp, P., Henze, G.P., Corbin, C.D., Rajagopalan, B. and Felsmann, C. (2011) Model-predictive control of mixed-mode buildings with rule extraction. *Building and Environment* 46(2): 428–437.

McConahey, E. (2008) Mixed mode ventilation: Finding the right mix. *ASHRAE Journal* 50(9): 36–48.

Neumann, C., Jacob, D., Burhenne, S., Florita, A., Burger, E. and Schmidt, F. (2011) Model-based methods for fault detection and optimization in building operations. Report, Fraunhofer ISE.

NOAA. (2011) NOAA's National Weather Service. www.nws.noaa.gov.

O'Neill, Z., Narayanan, S. and Brahme, R. (2010) Model-based thermal load estimation in buildings. *IBPSA-USA Journal* 4(1): 474–481.

Pavlak, G.S., Florita, A.R., Henze, G.P. and Rajagopalan, B. (2013) Comparison of traditional and Bayesian calibration techniques for gray-box modeling. *Journal of Architectural Engineering* 20(2): 04013011.

Pavlak, G.S., Henze, G.P. and Cushing, V.J. (2014) Optimizing commercial building participation in energy and ancillary service markets. *Energy and Buildings* 81: 115–126.

PJM. (2010) *PJM Markets and Operations*. www.pjm.com.

Rijal, H.B., Humphreys, M.A. and Nicol, J.F. (2008) How do the occupants control the temperature in mixed-mode buildings? Predicting the use of passive and active controls. In *Proceedings of Conference: Air Conditioning and the Low Carbon Cooling Challenge*. Windsor, UK, pp. 27–29.

Sahlin, P.P. (2000) The methods of 2020 for building envelope and HVAC systems simulation – Will the present tools survive? In *CIBSE/ASHRAE Conference*. Dublin.

Sahlin, P., Eriksson, L., Grozman, P., Johnsson, H., Shapovalov, A. and Vuolle, M. (2004) Whole-building simulation with symbolic DAE equations and general purpose solvers. *Building and Environment* 39(8): 949–958.

Spindler, H. (2004) System identification and optimal control for mixed-mode cooling. PhD dissertation, Massachusetts Institute of Technology.

Spindler, H.C. and Norford, L.K. (2009a) Naturally ventilated and mixed-mode buildings – Part I: Thermal modeling. *Building and Environment* 44(4): 736–749.

Spindler, H.C. and Norford, L.K. (2009b) Naturally ventilated and mixed-mode buildings – Part II: Optimal control. *Building and Environment* 44(4): 750–761.
Tanner, R.A. and Henze, G.P. (2014) Stochastic control optimization for a mixed mode building considering occupant window opening behaviour. *Journal of Building Performance Simulation* 7(6): 427–444.
Torcellini, P.A., Deru, M., Griffith, B., Long, N., Pless, S., Judkoff, R. and Crawley, D.B. (2004) Lessons learned from field evaluation of six high-performance buildings. In *Proceedings of the ACEEE Summer Study on Energy Efficiency in Buildings*. Pacific Grove, CA, pp. 1–13.
Van Schijndel, A.W.M. (2002) Optimal operation of a hospital power plant. *Energy and Buildings* 34(10): 1055–1065.
Vilim, R.B., Garcia, H.E. and Chen, F.W. (2001) An identification scheme combining first principle knowledge, neural networks, and the likelihood function. *IEEE Transactions on Control Systems Technology* 9(1): 186–199.
Wang, S. and Ma, Z. (2008) Supervisory and optimal control of building HVAC systems: A review. *HVAC&R Research* 14(1): 3–32.
Wei, C.C. and Hsu, N.S. (2009) Optimal tree-based release rules for real-time flood control operations on a multipurpose multireservoir system. *Journal of Hydrology* 365(3); 213–224.
Wetter, M. (2004) *Simulation-Based Building Energy Optimization. Engineering; Mechanical Engineering*. Berkeley, CA: University of California PhD.
Wetter, M., Zuo, W., Nouidui, T.S. and Pang, X. (2014) Modelica buildings library. *Journal of Building Performance Simulation* 7(4): 253–270.
Zhou, Q., Wang, S., Xu, X. and Xiao, F. (2008) A grey-box model of next-day building thermal load prediction for energy-efficient control. *International Journal of Energy Research* 32(15): 1418–1431.

Activity

Modeling of a multilayer wall

Problem statement

First, using lumped parameter analysis, you are to develop a state space model of a multilayer opaque construction element, something that experts refer to as a wall. Assume that the overall transmission heat transfer resistance R_T from an interior temperature T_i to an exterior temperature T_o can be broken into 10% interior convection resistance, 40% conduction resistance, and 50% exterior convection resistance. Further, assume that the total wall capacitance C_T can be broken into 15% capacitance of the internal mass node (inside surface) and 85% on the exterior mass node (outside surface). The result will be lumped parameter model with three resistors and two capacitors. To simplify the solar radiation treatment, assume that a radiative flux q_r acts on the inside surface only. Begin with an energy balance on both inside and outside surface temperature nodes and develop a state space model for use in a modeling environment such as Python, Matlab, or Simulink.

Next, develop a model to compare the thermal response of a "high thermal capacity" element and a "low thermal capacity" element using the information given below. Excite the three input variables (internal air temperature; radiant flux; external air temperature) in turn and compare the responses of the two construction types using graphical outputs of the time responses with both results shown on the same graph. Comment on the practical significance of this by giving examples of applications in which the two construction types would be most suited.

"High" thermal capacity:

20 mm light plaster
100 mm heavy concrete block

70 mm mineral wool insulation
25 mm air gap
100 mm outer brick

"Low" thermal capacity:

12 mm slag
10 mm air gap
100 mm light concrete block
70 mm mineral wool insulation
25 mm air gap
2 mm metal clad

Find the parameters (R_T, C_T) for the above using the standard data listed in Table 14.11 on chapter 28 page 19 of the 1997 *ASHRAE Handbook – Fundamentals* (SI units).

Answer

The solution to this activity borrows from an article by Gouda et al. (2002), and their contribution is gratefully acknowledged. We visualize the problem of transient heat transfer through an opaque multilayer construction element as an equivalent electric *RC* circuit as illustrated in Figure 14.30:

This can represent a wall, floor, or roof of n layers using two capacitors, C_i, C_o and three resistors R_i, R_m, R_o from the problem statement:

$$R_i = 0.1R_T; R_m = 0.4R_T;\ R_o = 0.5R_T \quad \text{and} \quad C_i = 0.15C_T; C_o = 0.85C_T \tag{14.8}$$

(in which R_T, C_T are the overall resistance and capacitance for the construction element). Note that q_r is a radiant source term (e.g. solar radiation).

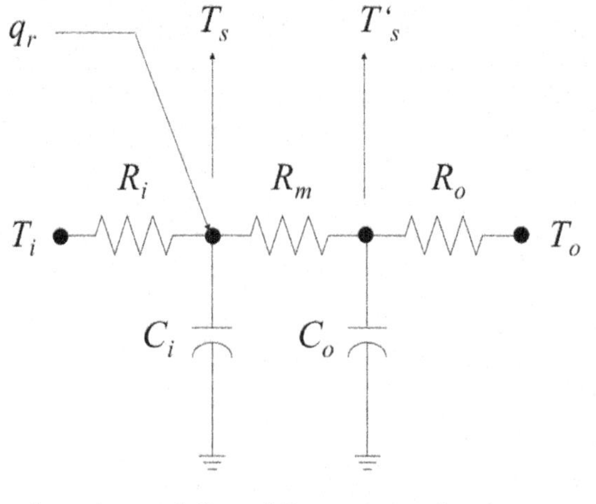

Figure 14.30 Equivalent thermal network for multilayer construction element

With $\dot{T} = dT / dt$, the circuit can be described by the following energy balances about the internal temperature nodes when substitutions for C_i, C_o and R_i, R_m, R_o are made:

$$0.15C_T\dot{T}_s = q_r + \frac{T_i - T_s}{0.1R_T} - \frac{T_s - T_s^{'}}{0.4R_T}$$
$$0.85C_T\dot{T}_s^{'} = \frac{T_s - T_s^{'}}{0.4R_T} - \frac{T_s^{'} - T_o}{0.5R_T} \tag{14.9}$$

Rearranging:

$$\dot{T}_s = \frac{q_r}{0.15C_T} + \frac{T_i}{0.015R_TC_T} - \frac{T_s}{0.06R_TC_T} + \frac{T_s^{'}}{0.06R_TC_T} \tag{14.10}$$

Table 14.6 Material properties of individual layers and composite wall

High thermal mass							
Code number	*Description*	*L*	*cp*	*R*	*m*	*C*	
E0	Inside surface resistance	0	0.00	0.121	0.00	0.00	
E1	20 mm plaster	20	0.84	0.026	30.74	25.82	
C3	100 mm high density concrete block	100	0.84	0.125	99.06	83.21	
B24	70 mm insulation	70	0.84	1.76	6.34	5.33	
B1	Air space resistance	0	0.00	0.16	0.00	0.00	
A2	100 mm face brick	100	0.92	0.076	203.50	187.22	
A0	Outside surface resistance	0	0.00	0.059	0.00	0.00	
Total				**2.327**	**m²K/W**	**301.58**	**kJ/m²K**
						3.02E+05	**J/m²K**

Low thermal mass							
Code number	*Description*	*L*	*cp*	*R*	*m*	*C*	
E0	Inside surface resistance	0	0.00	0.121	0.00	0.00	
E2	12 mm slag	20	0.84	0.026	30.74	25.82	
B1	Air space resistance	0	0.00	0.16	0.00	0.00	
C2	100 mm low density concrete block	100	0.84	0.266	61.98	52.06	
B24	70 mm insulation	70	0.84	1.76	6.34	5.33	
B1	Air space resistance	0	0.00	0.16	0.00	0.00	
A3	2 mm steel siding	2	0.42	0	11.71	4.92	
A0	Outside surface resistance	0	0.00	0.059	0.00	0.00	
Total				**2.552**	**m²K/W**	**88.13**	**kJ/m²K**
						8.81E+04	**J/m²K**

$$\dot{T}_s^{'} = \frac{T_s}{0.34R_T C_T} - \frac{T_s^{'}}{0.1445R_T C_T} + \frac{T_o}{0.425R_T C_T} \tag{14.11}$$

(note that R_T, C_T are the overall thermal resistance and capacitance of the construction element *per unit surface area*, respectively). The outputs, $T_s, T_s^{'}$, are the state variables; $\dot{T}_s, \dot{T}_s^{'}$ are state variable derivatives; T_i, q_r, T_O are input variables, and R_T, C_T are parameters.

We can express Equations 14.10 and 14.11 in standard state space notation:

$$\dot{x} = Ax + Bu$$
$$y = Cx + Du$$

in which $\dot{x}$ is a vector of state derivatives; y is a vector of output variables and x and u are vectors of state and input variables; A, B, C, D are matrices of parameters.

As a result, we obtain:

$$\begin{bmatrix} \dot{T}_s \\ \dot{T}_s^{'} \end{bmatrix} = \begin{bmatrix} \dfrac{-0.075}{0.0009R_T C_T} & \dfrac{1}{0.06R_T C_T} \\ \dfrac{1}{0.34R_T C_T} & \dfrac{-0.765}{0.1445R_T C_T} \end{bmatrix} \begin{bmatrix} T_s \\ T_s^{'} \end{bmatrix} + \begin{bmatrix} \dfrac{1}{0.015R_T C_T} & \dfrac{1}{0.15C_T} & 0 \\ 0 & 0 & \dfrac{1}{0.425R_T C_T} \end{bmatrix} \begin{bmatrix} T_i \\ q_r \\ T_o \end{bmatrix} \tag{14.12}$$

$$\begin{bmatrix} y_1 \\ y_2 \end{bmatrix} = \begin{bmatrix} 1 & 0 \\ 0 & 1 \end{bmatrix} \begin{bmatrix} T_s \\ T_s^{'} \end{bmatrix} + \begin{bmatrix} 0 & 0 & 0 \\ 0 & 0 & 0 \end{bmatrix} \begin{bmatrix} T_i \\ q_r \\ T_o \end{bmatrix} \tag{14.13}$$

Using the information suggested in the problem statement and as detailed in Table 14.16, the total resistance and total capacitance for the two construction types can be found to be $R_{T,hi} = 2.327$ m²K/W, $C_{T,hi} = 3.02 \times 10^5$ J/m²K, and $R_{T,lo} = 2.552$ m²K/W, $C_{T,lo} = 8.81 \times 10^4$ J/m²K.

Results shown here have been generated with the help of Matlab/Simulink. The state space model of the multilayer construction element is shown in Figure 14.31 with the model parameters being shown in the dialog form in Figure 14.32. Both construction elements can be simultaneously investigated using a block model shown in Figure 14.33. Here, only the inside surface temperature T_s is of interest and the outside surface temperature $T_s{}'$ is ignored. Figure 14.34 shows sample results for a diurnal outside air temperature variation and a diurnal step change in inside air temperature.

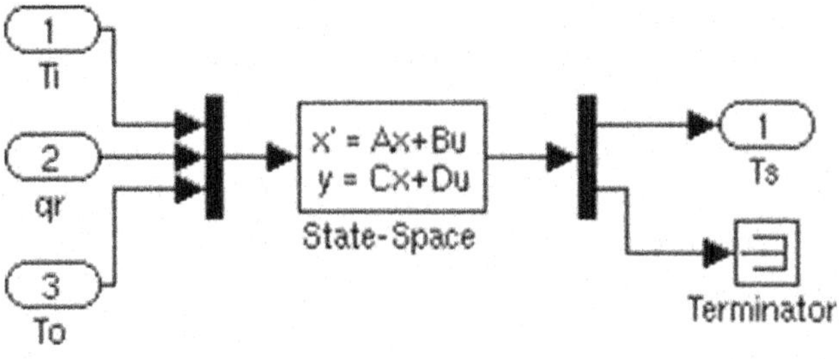

Figure 14.31 State space model simulation block

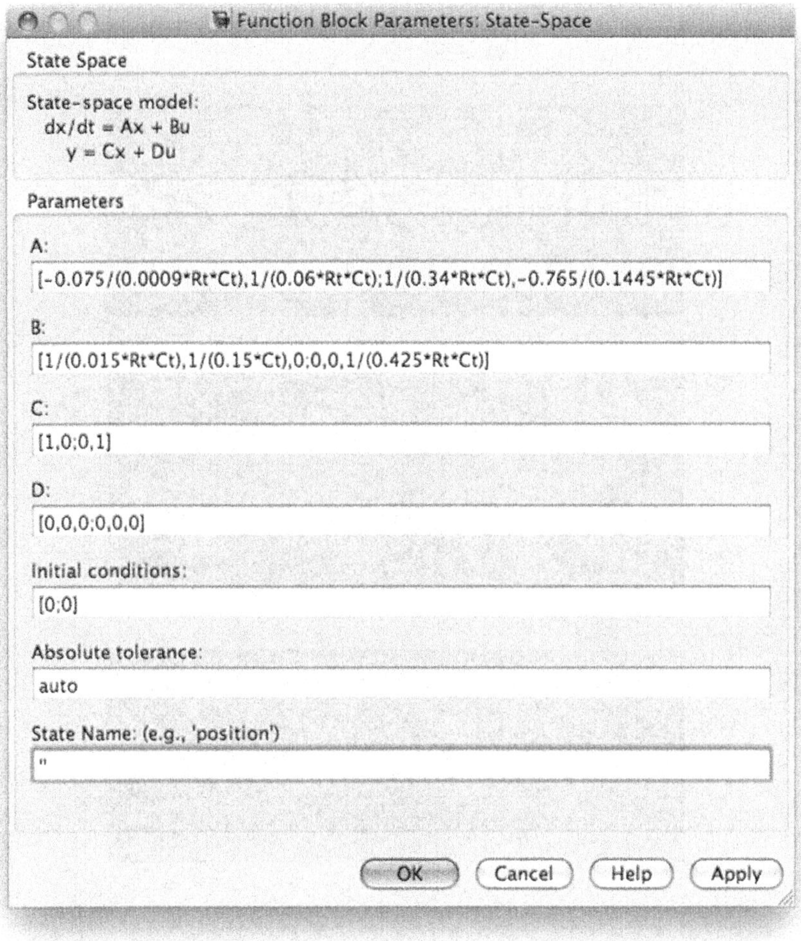

Figure 14.32 State space model parameter dialog form

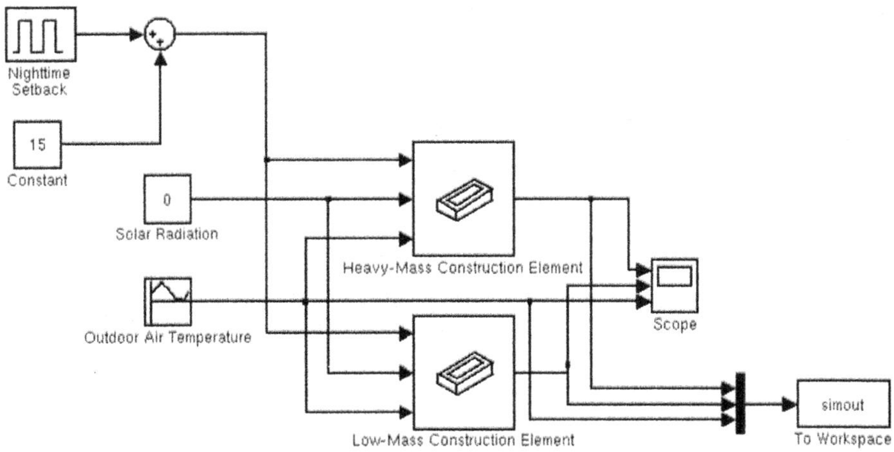

Figure 14.33 Simulation program to evaluate the thermal response of high-mass and low-mass construction elements subject to inside and outside air temperature variations

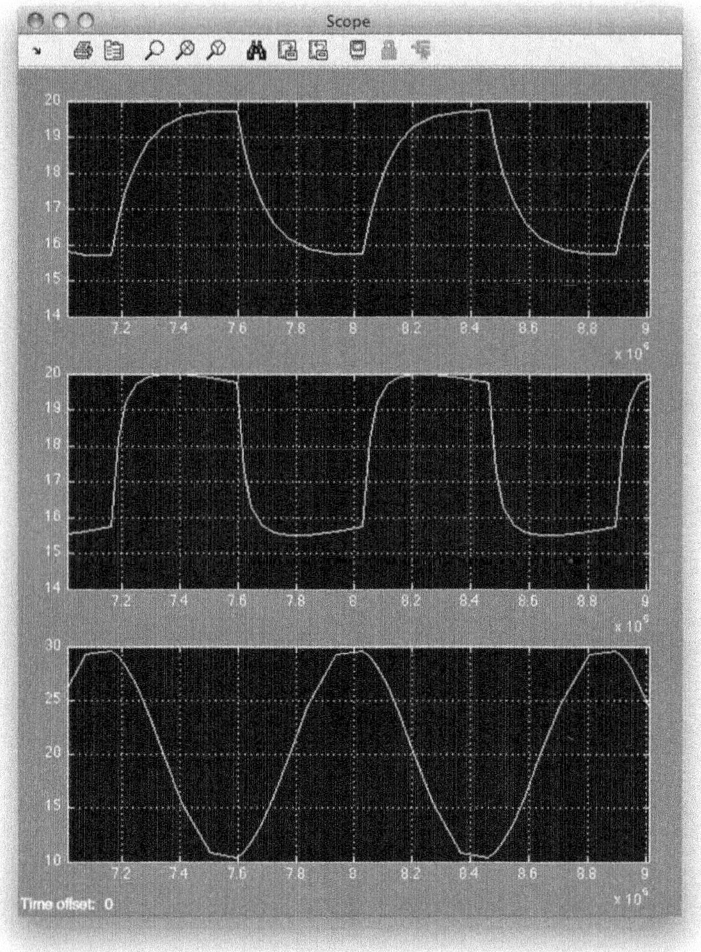

Figure 14.34 Sample thermal response of high-mass (top) and low-mass (middle) construction subject to diurnal outside air temperature variation (bottom) and diurnal inside temperature setback (not graphed)

Part IV

Systems

15 HVAC systems performance prediction

Jonathan Wright

Scope

This chapter provides an introduction to the principles of heating, ventilating, and air-conditioning (HVAC) system performance modeling, including a description of the alternative forms of steady-state model, the computational methods for the solution of the model equations, and the potential sources of uncertainty in performance prediction. The application of HVAC performance models to HVAC system fault detection and optimization is also described.

Learning objectives

To develop an understanding of the principles of HVAC system modeling, in terms of the level of detail required for a particular application and the alternative modeling approaches available.

Key words

HVAC system performance/steady-state component models/system modeling/solution of simultaneous equations/sources of uncertainty

Introduction

HVAC system performance prediction has a number of applications in evaluating the impact of a particular HVAC system configuration on building energy use and the extent to which the system is able to maintain the desired indoor-air conditions, in HVAC system condition monitoring, and in the study of supervisory and local-loop control strategies.

Where the impact of a particular system type on building energy use is to be investigated, the HVAC system performance simulation is integral to the simulation of the building thermal performance. However, for some applications, such as HVAC system condition monitoring, the performance of the system is simulated independently of the building thermal performance. The integrated simulation of building and HVAC system performance is not described here; this chapter is focused instead on the development of HVAC component and system models.

There are a number of alternative types of HVAC system available, but each of these is formed from a similar set of HVAC components. This observation led to the development of two approaches to HVAC system modeling. In some simulation environments, system models are selected from a list of pre-configured system types, while in other environments, the system is defined component by component. Pre-configured system models have the advantage that the user is spared the effort and difficulty of defining the system model component by component. However, restricting the user to selecting from a choice of pre-configured system types has the major disadvantage that often the system to be modeled is not available in the list of pre-configured systems. The probability of this increases with the development of innovative design solutions, and so for this reason, contemporary simulation approaches are based on the definition of the HVAC system by linking the input-output variables of individual system components to form a model of the complete HVAC system. Therefore, this chapter focuses on the development of HVAC component models and the connection of their input-output variables in forming an HVAC system model. Although the approaches described in this chapter have been adopted in a number of existing simulation programs, the concepts described here also act as an introduction to contemporary research into equation-based simulation languages (Wetter et al. 2016).

The particular application of an HVAC system model can influence the extent to which the non-linear and dynamic characteristic of the system are modeled. Where the aim of the simulation is to evaluate the building energy use, it is often assumed that the HVAC systems and their controls have an idealized linear characteristic and operate under perfect control. In this case, a simple steady-state linear model can be used to represent the relationship between the system input and output variables, although a non-linear efficiency relationship may also be used to model the impact of part-load operation on the systems energy use. Conversely, the application of HVAC system models to condition monitoring and the study of control system behavior demand the use of models that encapsulate the full non-linear performance characteristics of the system; the system dynamics must also be modeled in the analysis of HVAC control system performance, particularly when local-loop control strategies are being investigated. This chapter describes the formulation of models that include the non-linear static characteristic of HVAC systems.

HVAC component models consist of the variables to be simulated at the input and output of the component, one or more generically applicable equations defining the relationship between the input-output variables, and a set of equation parameters (constants) that define a specific HVAC component (size). The component model equations are formulated in one of two ways, in a way that either fixes the input and output variables or provides some flexibility in the choice of input-output variables. An "input-output free" model has the flexibility of being used in a variety of applications. For example, an input-output free pump model could be used to simulate the pump output pressure during system design and, say, the pump speed during commissioning. The model equations can be derived directly from first principles, a simple curve-fit of the measured performance of a specific product, or a combination of both. This chapter describes the different types of model equation and their use in formulating both fixed input-output and input-output free models. The use of polynomial curve-fitting of an HVAC component manufacturer's data is also described.

The HVAC system component models are linked through their input-output variables to form a model of the HVAC system (or sub-system). The choice of linking variables must be done in such a way that the set of system equations resulting from the linking of the components can be solved for the selected set of simulation variables. This is particularly the case for the linking of input-output free models, where there is a greater choice of potential linking variables than fixed input-output variable models. Although greater care must be taken in selecting the linking variables when using input-output free models, the solution of the input-output free model equations is potentially more robust than the solution of the fixed input-output model equations (Sowell and Haves 2001; Wetter et al. 2016). The linking of the components to form a system, methods for checking that the resulting set of system equations can be solved for the simulation variables, and the methods for the solutions of the system equations are described in this chapter.

Building performance simulation is an inherently uncertain process, with the uncertainty in the performance modeling reducing confidence in the application of the predicted performance.

As a result, uncertainty analysis in building performance prediction has been the focus of wide-ranging research (Tian et al. 2018). This chapter introduces sources of uncertainty associated with HVAC system performance prediction.

Scientific foundation

The time constant of an HVAC system is generally in the order of minutes,[1] whereas the time constant of the building fabric is in the order of hours and as such, in comparison to the thermal response of the building fabric, the HVAC system can be considered to be in a quasi-steady-state condition for the majority of its operation (the exception being during the system start and stop operation). Since the weather and other simulation boundary conditions are normally specified at hourly intervals, steady-state models are often used to represent the HVAC system performance in building thermal simulations (particularly where the focus of the simulation is on the building energy use). However, where the focus of the simulation is on the control system performance, dynamic models of the HVAC systems should be used (with the boundary conditions being either specified or interpreted to be in sub-hourly intervals). This chapter describes the principles of developing steady-state HVAC component models, together with the formation of HVAC system models and the solution of the model equations.

HVAC component models are defined in terms of the simulation variables, the model equations, and the model parameters (Figure 15.1). The simulation variables in HVAC component models are normally those that represent the thermo-fluid properties of mass flow-rate $(\dot{m})$, temperature (T), humidity ratio (w), and pressure (P); variables related to the control of the component, such as a control signal (u) or motor speed (n), are also included in the set of variables for some component models. Note that separate sets of input-output variables are specified for each fluid stream entering (i, j), or leaving (o, k), the component. For a given type of component, the model equations $(F = \{f_1(\cdot), f_2(\cdot), \ldots, f_l(\cdot)\})$, represent the physical relationship between the simulation variables at the input and output of the component and are normally of a generically applicable form. The equations are of an algebraic form in steady-state models, but include differential terms in dynamic models. The model parameters $(A = \{\alpha_1, \alpha_2, \ldots \alpha_m\})$, define a specific product or "size" of component. The form of model parameters depends on the form of model equations; in some instances, they can have a physical meaning (such as a heat transfer coefficient), while in other models, they can be a dimensionless coefficients of, for example, a polynomial curve-fit. In addition to the model equations that define the relationship

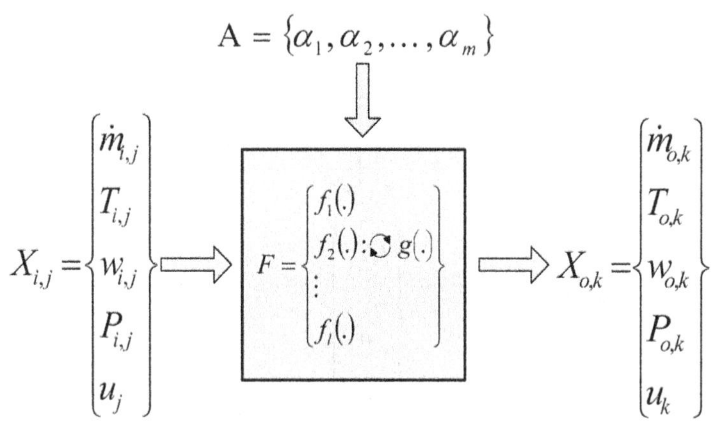

Figure 15.1 HVAC component model structure

between the variables at input and output, it is likely that the model will require the calculation of thermodynamic property relationships ($g(\cdot)$), and other intermediate calculations which can be iterative (↻). This section focuses on defining the main categories of model equation (and associated model parameters).

The steady-state model equations can be categorized as being derived from first principles, from empirical data, or from a mix of first principles and empirical data.

First-principle models

First-principle models are derived from fundamental physical relationships and do not have any empirically derived element. An example is a model of an air-mixing section (Figure 15.2), where three model equations can be derived from a mass and energy balance on the fluid streams (Equations 15.1 to 15.3).

$$\dot{m}_o = \dot{m}_{i,1} + \dot{m}_{i,2} \tag{15.1}$$

$$\dot{m}_o w_o = \dot{m}_{i,1} w_{i,1} + \dot{m}_{i,2} w_{i,2} \tag{15.2}$$

$$\dot{m}_o h_o \left(T_o, w_o\right) = \dot{m}_{i,1} h_{i,1} \left(T_{i,1}, w_{i,1}\right) + \dot{m}_{i,2} h_{i,2} \left(T_{i,2}, w_{i,2}\right) \tag{15.3}$$

where $h_i(\cdot)$ is enthalpy, this being a function of the fluid temperature and humidity ratio. Note that for clarity, the model equations are given in the form of the mass and energy balance. These equations would normally be transformed to be either in a form that allowed the direct calculation of the component output variables (generally taken to be $\dot{m}_o$, T_o, and w_o) or in an input-output free form. Note also that there are no model parameters in this case.

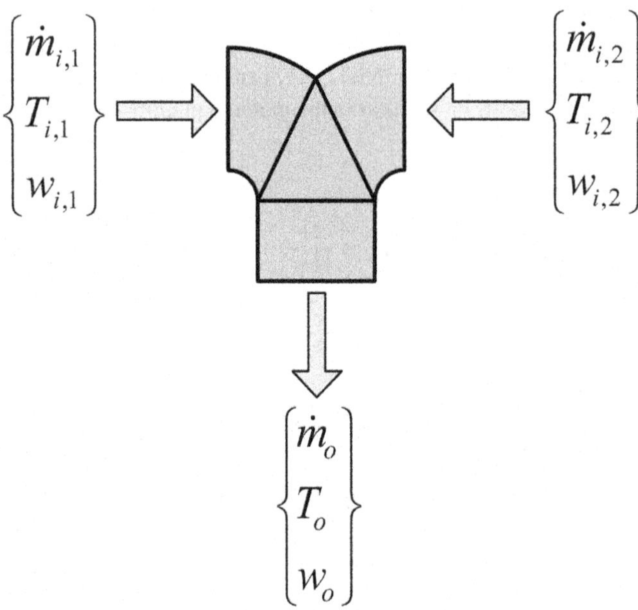

Figure 15.2 Air-mixing duct section

Unfortunately, there are very few instances where a model can be developed entirely from first principles. In fact, a complete model of a ducted air-mixing section would include the air pressure loss across the duct section, the pressure loss coefficient being derived from experimental (empirical) measurement.

Empirical models

Empirical models are used where the HVAC components' internal processes are not well understood or where the effort required to generate a detailed first-principle model of the component is not justified. As such, this type of model is frequently used to model the performance of turbo machines such as compressors, pumps, and fans. Although it is possible to develop a detailed model of a pump or fan that is based on a geometric specification of the component's impeller, such detail is not readily available and is not of importance in building design. The approach can also been used to model the part-load performance of refrigeration sub-systems, where a detailed simulation of the refrigeration components and cycle is not warranted (Braun 1988).

Many empirical models take the form of a polynomial curve-fit of the component manufacturer's test data, with one or more polynomial expressions forming the model equations, and the model parameters being the polynomial coefficients (Wright 1986). Although not described here, it is also possible to use other "black box" models such as those based on fuzzy logic (Angelov et al. 2000).

In the case of pumps and particularly fans, two elements of the component's performance are modeled, fluid pressure rise and fluid temperature rise, both as a function of flow rate. Both fan and pump performance data can be provided for a range of impeller speeds and several impeller diameters. Figure 15.3 illustrates the pressure-flow performance curves for three impeller

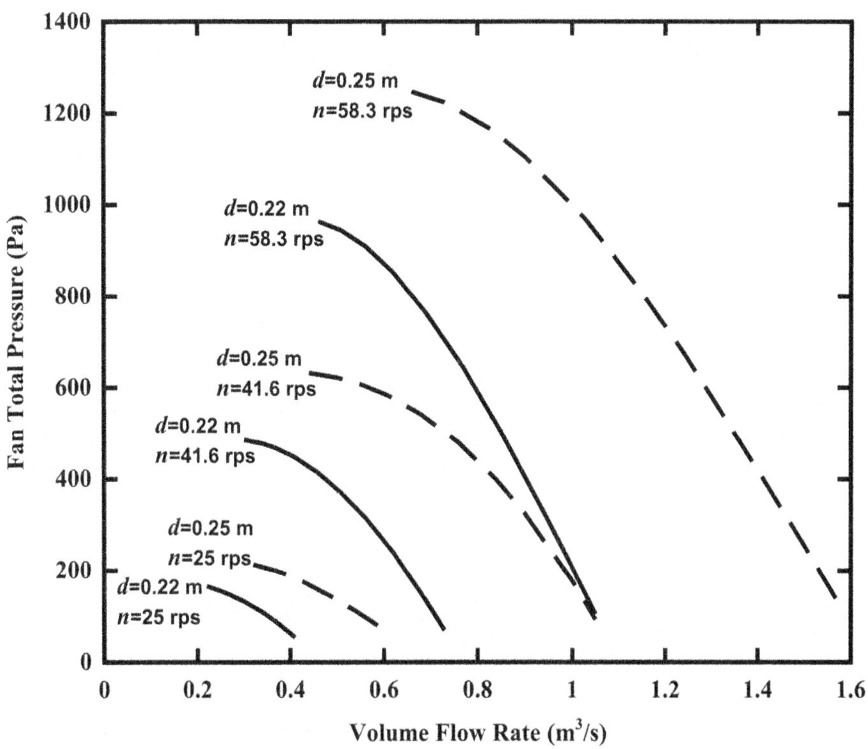

Figure 15.3 Centrifugal fan performance curves

speeds (n) and two diameters (d) of a centrifugal fan (a similar set of performance characteristics is normally given for the fan power or efficiency as a function of air flow rate).

Rather than generate a separate curve-fit for each impeller diameter and speed, the number of curve-fits can be reduced in one of two ways: by generating a curve-fit for one impeller speed and diameter and predicting the performance at other diameters and speeds by using fan and pump scaling rules, or by normalizing the performance data, the component performance for a number of impeller speeds and diameters often then being represented by a single normalized performance curve. Both the scaling rules and expressions for normalization have been derived from dimensional analysis. Normalizing the data generally results in the complete range of speeds and diameters being represented by a single non-dimensional curve (the exception being for axial-flow fans where the performance data is given as a function of the impeller blade angle as well as the impeller diameter and speed; this requires the use of a two-dimensional curve-fit, as it is not possible to scale or normalize for the blade angle (Wright 1991)).

The applicability of using scaling rules depends on the complete pump or fan characteristic being available at a given speed; this may not be available when the component performance data is given in a tabular form (since such data would normally be presented with the rows and columns of the table being formed from the air-flow rate and pressure rise, with the speed data being fragmented as a result). Since the complete characteristic is not required in order to generate a normalized performance curve, performance modeling using data normalization is the more widely applicable approach. The non-dimensional volume flow rate (ϕ), total pressure rise (ψ), and absorbed power (λ), are given by:

$$\phi = \frac{\dot{m}}{\rho_i \pi^2 d^3 \, n/4} \tag{15.4}$$

$$\psi = \frac{2P}{\rho_i \left(\pi d n\right)^2} \tag{15.5}$$

$$\lambda = \frac{8W}{\rho_i \pi^4 d^5 n^3} \tag{15.6}$$

where $\dot{m}$ is the fluid mass-flow rate through the component, P is total pressure rise of the fluid, W is the absorbed power, ρ_i is the fluid density at the inlet to the component, d is the impeller diameter, and n is the impeller speed.[2]

Figure 15.4 shows the non-dimensional fan flow-pressure rise curve for the fan performance data illustrated in Figure 15.3. The six performance curves have been reduced to one curve which can be modeled, in this case, using a second order polynomial having only three parameters, $\{\alpha_1, \alpha_2, \alpha_3\}$; $\psi = \alpha_1 \phi^2 + \alpha_2 \phi + \alpha_3$.

The scatter in the data about the curves is a result of the digitizing processes used to extract numerical values for flow rate and pressure rise from the fan performance curves. The scatter contributes to the uncertainty in the performance prediction. Uncertainty (scatter) can also occur due to a lack of geometric similarity among the different component diameters (Wright 1991) or the effects of the working fluid being compressible (although this effect is limited for the range of fluid pressure rise occurring in most HVAC systems). Uncertainty in the model prediction can also occur due to the poor or unreliable performance measurement by the component manufacturer. Normalization of the data has the benefit that it makes such errors and uncertainty apparent and tends to average their effect.

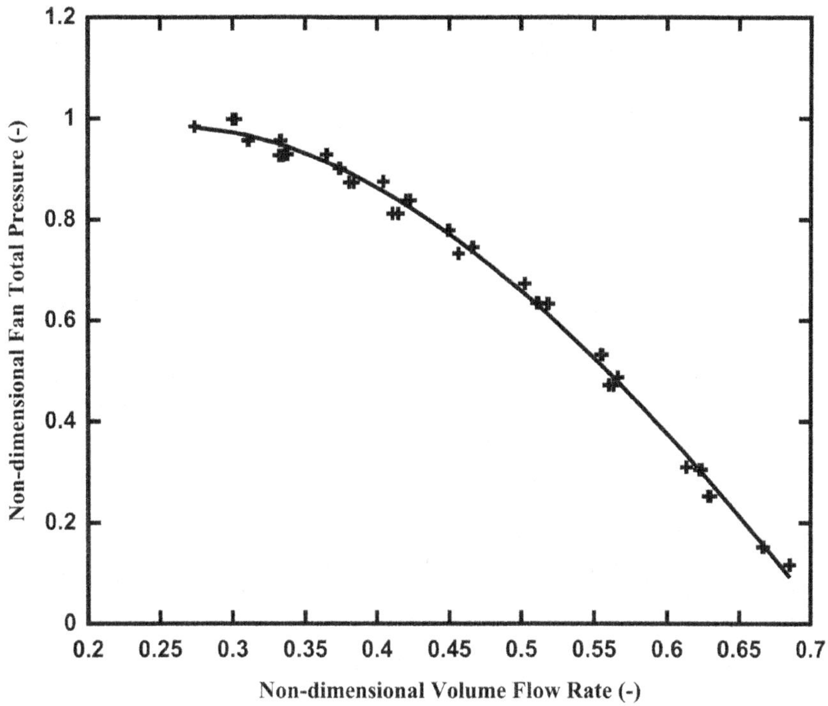

Figure 15.4 Centrifugal fan non-dimensional performance curves

Empirical models are only valid provided that the simulated performance lies within the range of the measured performance. Wright (1991) suggested that when the empirical model is used as part of an optimization study, then constraints could be developed and applied to ensure that the model performance lies within the range of measured performance data. Alternatively, Wetter (2013) described a fan and pump model that increases the reliability of the performance simulation by ensuring that the model provides a unique solution in each region of operation.

Quasi-first-principle models

The majority of HVAC component models can be classified as being quasi-first-principle models, particularly models of components that promote heat exchange (heating and cooling coils, evaporators, condensers, and boilers). The equations of quasi-first-principle models are derived from fundamental principles but require some empirical input (generally in relation to the model parameters). They are distinguished from purely empirical models in that the equations describing the performance characteristic of the components are derived from fundamental principles, whereas in a purely empirical model, the performance characteristic is a result of a curve-fit of measured performance data. Since the model equations are derived from fundamental principles, this class of model is also sometimes referred to as a "first-principle" model (for instance, by Haves et al. 1996).

A model of duct friction loss can be classified as a quasi-first-principle model, it being developed from a combination of theoretical relationships and empirically determined performance data. Duct friction loss can be calculated using the Darcy equation:

$$\Delta P = \frac{fL}{D_h} \times \frac{\rho v^2}{2} \tag{15.7}$$

where ΔP is the pressure loss (Pa), f is the friction factor, L is the length of duct, D_h is the hydraulic diameter (m), ρ is the air density (kg/m^3), and v is the air velocity (m/s). The friction factor f can be obtained from the Colebrook equation:

$$\frac{1}{\sqrt{f}} = -2\log\left(\frac{\varepsilon}{3.7Dh} + \frac{2.51}{Re\sqrt{f}}\right) \tag{15.8}$$

where Re is the Reynolds number and ε is the empirically derived absolute roughness factor (m).

Figure 15.5 illustrates an input-output model for friction loss in rectangular ducts (where W, H, L are the width, height, and length of duct (m) and μ is the dynamic viscosity of the air (Pa.s). As well as the model equation for the duct outlet pressure, the model requires a number of preliminary calculations to determine the air velocity (v), hydraulic diameter (D_h), Reynolds number (Re), and friction factor (f), the friction factor in Equation 15.8 being transcendental and solved by an iteration performed within the model calculation routine. Note that the equation for outlet pressure is an implicit function of the mass-flow rate as well as being an explicit function of the inlet pressure. Note also that since duct leakage is not included in the model, the air mass-flow rate is treated as a "through" variable ($\dot{m}$), its value being the same at the duct outlet as at the inlet; "through" variables are often illustrated as being a model input variable, as in the case of the air mass-flow rate in Figure 15.5.

Uncertainty in HVAC component models

Uncertainty is defined as being either epistemic or aleatoric. Aleatoric uncertainty is due to the natural and random variation in a parameter value and in general is unavoidable. Epistemic uncertainty is associated with subjective choice and as such has the potential to be changed and reduced. Both forms of uncertainty are evident in HVAC system performance modeling.

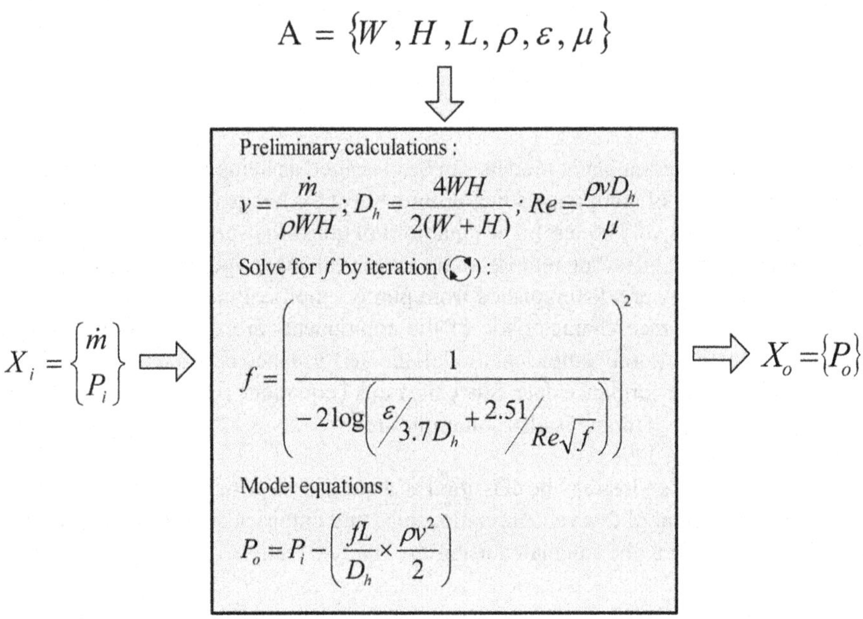

Figure 15.5 Rectangular duct friction loss model

Figure 15.1 illustrates the elements and structure of a HVAC component model. The uncertainty in the value of the output variables is dependent on the uncertainty in the value of the input variables and model parameters, as well as the extent to which the model equations represent the physical processes associated with the component.

Where an *input variable* is linked to the output variable of another component, the *input uncertainty* is a result of the distribution of the value of the linked output variable. In effect, the uncertainty propagates from one component to the next. For instance, the uncertainty in water flow rate leaving a three-port valve would form the input uncertainty in flow rate entering a heating coil (as in the system illustrated in Figures 15.6 and 15.11). It should be noted that in many simulations the measurement sensors are not explicitly modeled, the sensed variables simply being treated as linking variables. For instance, in the heating coil sub-system example illustrated in Figure 15.6, the temperature sensor has not been modeled, with the temperature of the air leaving the coil being linked directly to the input variable of the controller (Figure 15.11). However, temperature, pressure, and flow sensors are prone to a range of possible measurement errors that contribute to the uncertainty in HVAC system performance prediction. These sensors have a finite accuracy, can be biased in the measurement, and have a limit on their sensitivity (this being the minimum change in measured variable that results in a change in the sensor output). Sensor accuracy is a form of aleatoric uncertainty, but in many cases, the bias is a form of epistemic uncertainty as it relates to the installation position of the sensor in the system. The characteristics of HVAC sensor measurements and their contribution to the uncertainty of HVAC performance prediction is described further by Buswell (2001).

Some components in a HVAC system will have input variables that map to the boundary conditions on the system. Where the boundary condition is the temperature, humidity, and pressure of the ambient (outdoor) air, then the uncertainty in the variable values may include the probability of the condition occurring as well as the uncertainty in respective sensor measurements. For instance, the weather conditions that map to the system input variables are often chosen to represent the weather conditions that result in "typical" energy use; the extent to which these weather conditions are "typical" is uncertain and so should be treated as such.

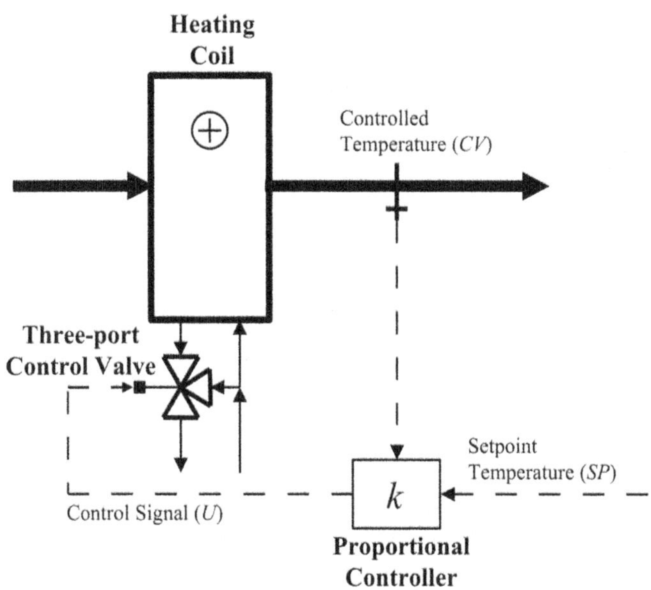

Figure 15.6 Heating coil sub-system

Uncertainty also exists in the extent to which the *model equations* represent the performance of the component. For empirical models, uncertainty in the equations relates to the extent to which the selected form of curve-fit matches the data, this uncertainty being compounded by the uncertainty in the measured data used in the empirical model. Although first-principle and quasi-first-principle models are formed from the laws of physics, the complexity and interaction of the physical processes (heat transfer, mass transfer, and fluid flow) dictate that the component performance could be represented by more than one set of equations (the uncertainty in the choice of equation set being a form of epistemic uncertainty). Heat exchangers are mechanically simple components, yet the complexities of the heat and mass transfer processes lead to uncertainty in the extent to which the selected equations model the heat exchanger performance. An analysis of the uncertainty in the equations for a fin-tube, air-to-water heat exchanger is given by Buswell (2001). An insight into the differences in performance prediction of these heat exchangers can also be found in the work of Zhou and Braun (2007), which compares the prediction of a detailed and a simplified heat exchanger model. Uncertainty also arises due to the tolerances applied in the numerical methods used to solve the model equations, the equations not being solved to meet an absolute level of accuracy but to lie within a tolerance band.

The form of *uncertainty* associated with the *model parameters* depends on the form of model and choice of equations (and whether the parameters are, for example, the coefficients of a curve-fit or the geometric dimensions of the component). Where the parameter relates specifically to performance (for instance, the leakage or "let-by" of a three-port valve), then there is uncertainty in how the parameter value will change with time (as, for instance, valve leakage increases with use and wear). The change in HVAC component parameters with time has been researched as part of fault-detection and diagnosis studies, this work including an approach that attempts to automatically track the parameter values with time (Norford et al. 2002). For some components, there is also uncertainty in the published (catalogue) component performance, as this performance is commonly measured under standard test conditions and uses a standardized system configuration. It is highly probable that the system in which the component is finally installed will have a different configuration to the standardized test system. The difference in test and final system configurations can result in the installed component performance being different to the test performance (in design, the model parameter having been identified from the test performance). For instance, it is acknowledged that the performance of a fan is influenced by the duct configuration up- and downstream from the fan, and although some standards attempt to address this by testing the fan performance for a range of system configurations (ISO 2017), there is still uncertainty in which test configuration best matches the installed system.

Computational methods

The computational tools required for use in HVAC system performance prediction are concerned with curve-fitting empirical data and the solution of the model equations.

Polynomial curve-fitting

Methods for identifying the coefficients of a polynomial curve-fit are widely available and are not described here. However, for model developers who are faced with polynomial curve-fitting for a large number of different components, it may be useful to develop a dedicated curve-fit software package that is able to search for an optimum order of polynomial, this being particularly the case when a curve-fit is required for a function of more than one variable (Wright 1986).

Model developers who employ a polynomial curve-fit should also be aware that a curve-fit is only reliable within the range of the curve-fit data; extrapolation beyond the range of data can result in the prediction of component operation that is infeasible in practice. This is particularly the case when higher order polynomials are used to curve-fit the data, or where the curve-fit is a function of more than one variable (Wright 1986). When this is the case, it is wise either to add extra data points to the curve-fit in an attempt to ensure the extrapolated component performance curves represent practicable operation (Wright 1986) or to form functions that can be used to check that the simulated performance lies within the original measured performance (Wright 1991).

Formation of system models and solution of model equations

HVAC component models consist of a set of algebraic and differential equations; in many instances, the dynamic performance of the component is not modeled, so only the algebraic equations exist. For a given set of input variables (X_i) and model parameters (A), the model algebraic equations (F) are formulated in one of two ways: either to calculate the value of the output variables directly (F: $(X_i, A) \rightarrow X_o$) or, given the current estimate of the output variable values (X_o), to calculate the residual error (R) in solving the equations (F: $(X_i, A) \rightarrow R$), this being equal to zero on solution of the equation. For example, the equation for duct friction loss illustrated in Figure 15.5 is written in input-output form (F: $(X_i, A) \rightarrow X_o$):

$$P_o = P_i - \left(\frac{fL}{D_h} \times \frac{\rho v^2}{2} \right) \tag{15.9}$$

The residual form (F: $(X_i, X_o, A) \rightarrow R$), of the equation is:

$$R = (P_i - P_o) - \left(\frac{fL}{D_h} \times \frac{\rho v^2}{2} \right) \tag{15.10}$$

Note that since the residual form of equations are "input-output free," the equations can be solved for any of the simulation variables; in the case of Equation 15.10, it can be solved for the inlet pressure or the outlet pressure (whereas in Equation 15.9, it can only be solved for the outlet pressure).

Regardless of the form of equation, the equations for individual components are linked through the input-output variables to form a set of simultaneous algebraic equations. The type of algorithm used to solve the equations depends on the form of equation, directed input-output equations generally being solved using approaches based on successive substitution, and input-output free equations using methods based on a Newton-Raphson iteration.

The linking of the component models to form a set of simultaneous equations, the choice of system boundary conditions, and the form of solution algorithm are illustrated here using the components of a heating coil sub-system.

Example heating coil sub-system component models

The example heating coil sub-system model (Figure 15.6) is formed here from three component models: a model of the heating coil, a model of the proportional controller, and a model of the three-port valve. Although the potential bias error and measurement accuracy will contribute to the uncertainty in performance prediction, the temperature sensor has not been modeled, it being assumed that the sensor bias error is zero and measurement perfect. Similarly, the valve actuator has also been omitted from the model on the assumption that the actuator hysteresis is zero, but in practice, if unknown, the potential hysteresis will also contribute to the prediction uncertainty. The model equations are written in input-output form (F: $(X_i, A) \rightarrow X_o$), with the input-output variables relating directly to the physical processes and the conditions at inlet and outlet of the components.

The *heating coil model* is illustrated in Figure 15.7. The model is based on the "NTU-effectiveness" method, this form of heat exchanger model having been widely used in HVAC system performance modeling (Holmes 1982; Buswell 2001; Zhou and Braun 2007). The model input variables are the air and water mass-flow rates ($\dot{m}_a, \dot{m}_w$), although these are strictly "through" variables as the fluid-flow also appears at the component outlet, and the air and water inlet temperatures (T_{ai}, T_{wi}). The model output variables are the air and water outlet temperatures (T_{ao}, T_{wo}). The model parameters are the width (W) and height (H) of the coil that are exposed to air-flow; the number of water-tube rows (N_r), the number of water-tube circuits (N_c),

$$A=\{W,H,N_r,N_c,D,r_a,r_m,r_w\}$$

$$X_i=\begin{Bmatrix}\dot{m}_a\\ T_{ai}\\ \dot{m}_w\\ T_{wi}\end{Bmatrix} \Rightarrow$$

Preliminary calculations:

$$v_a=\frac{\dot{m}_a}{\rho_a \times W \times H};\ v_w=\frac{\dot{m}_w}{\rho_w N_c \frac{\pi D^2}{4}}$$

$$UA=\frac{N_r \times W \times H}{r_a v_a^{-0.8}+r_m+r_w v_w^{-0.8}}$$

$$C_a=\dot{m}_a C_{pa};\ C_w=\dot{m}_w C_{pw}$$

$$z=\frac{\min(C_a,C_w)}{\max(C_a,C_w)}$$

$$NTU=\frac{UA}{\min(C_a,C_w)}$$

$$\varepsilon=\begin{cases}0, & \text{if } (NTU)=0\\ \dfrac{NTU}{1+NTU}, & \text{else if } (1-z)=0\\ 1-e^{-NTU}, & \text{else if } (z)=0\\ \dfrac{1-e^{-NTU(1-z)}}{1-ze^{-NTU(1-z)}}, & \text{else}\end{cases}$$

Model equations:

$$T_{ao}=T_{ai}+\frac{\varepsilon \min(C_a,C_w)(T_{wi}-T_{ai})}{C_a} \quad \text{(HC.1)}$$

$$T_{wo}=T_{wi}-\frac{C_a(T_{ao}-T_{ai})}{C_w} \quad \text{(HC.2)}$$

$$\Rightarrow X_o=\begin{Bmatrix}T_{ao}\\ T_{wo}\end{Bmatrix}$$

Figure 15.7 Heating coil model

and the water-tube internal diameter (D); and the heat transfer resistance coefficients relating to the air-side thermal resistance (r_a), the thermal resistance of the water-tubes and fins (r_m), and the water-side thermal resistance (r_w) (typical values for the thermal resistances are given in Table 15.1, Holmes 1982).

The overall heat transfer conductance ($U A$) is a function of the coil dimensions and fluid velocities (v_a, v_w); note that the fluid densities (ρ_a, ρ_w) are internally computed variables, as are the fluid specific heat capacities (C_{pa}, C_{pw}).

The model (Figure 15.7) has two final equations; the first (HC.1) is derived from an energy balance between the conditioned air and the rate of heat transfer through the metal surfaces of the coil (this being a function of the coil's effectiveness, ε). The second equation (HC.2) is derived from an energy balance between the air and water. Both equations are written in input-output form (F: (X_i, A) $\rightarrow X_o$).

The *three-port valve model* (Figure 15.8) has one input variable, the valve stem position (S, having a value in the range 0–1), and one output variable, the water mass-flow rate through the control port ($\dot{m}_w$). The model parameters are the control port curvature (β), with a value of 0.0 being equivalent to a "linear valve"; the valve authority (γ); the fractional leakage through the valve when closed (l); and the flow rate through the valve when the control port is fully open ($\dot{m}_{max}$).

Table 15.1 Heating coil model thermal resistance parameters

Coil type	*Fin density*	*Water turbulator*	r_a $\left(\dfrac{k.s^{-0.8}}{kg^{-0.8}m^2kW.row}\right)$	r_m $\left(\dfrac{k}{m^2kW.row}\right)$	r_w $\left(\dfrac{k.s^{-0.8}}{kg^{-0.8}m^2kW.row}\right)$
Low "efficiency"	Low	×	1.32	0.68	1.1
Nominal coil	Low	✓	0.44	0.2	0.2
High "efficiency"	High	✓	0.49	0.38	0.38

Source: Holmes 1982

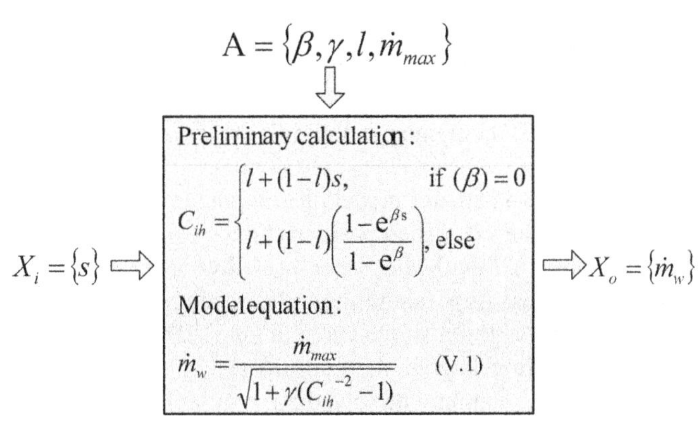

Figure 15.8 Three-port valve model

Source: Salsbury 1996

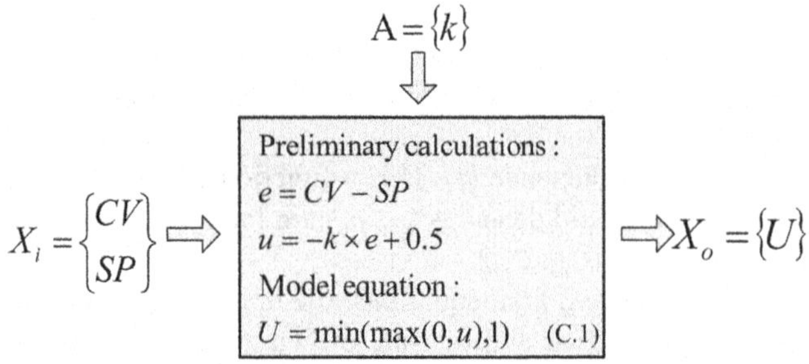

Figure 15.9 Proportional controller model

The internally computed parameter C_{ih} relates to the inherent characteristic of the valve, whereas the final model equation (V.1), relates to the installed characteristic (Salsbury 1996).

The *proportional controller model* is illustrated in Figure 15.9. It has two input variables, the controlled variable (*CV*) and the controlled variable setpoint (*SP*); the setpoint has been defined as a variable rather than a parameter as it is normally set by a supervisory controller, which could, at some point, be included in the sub-system model. The model output variable is the control signal (*U*, having a value in the range 0–1), and the model parameter is the proportional gain (*k*). The model has one final equation (C.1), which in this case ensures that the output signal is in the range of 0 to 1.

A detailed description of the *modeling uncertainty* associated with the coil and valve models is described by Buswell (2001), the coil model in this case including cooling and wet heat transfer as well as dry heat transfer. The discussion includes a description of the propagation of the uncertainty from the model input variables and model parameters, and through the model equations to the predicted output.

Formation of system models

There are two tasks in forming a system model: the first is to identify the variables used to link the component models (and in effect, the model equations); the second task is to identify the set of boundary variables (or alternatively, the simulation variables).

The components are linked to form a system model through the input-output variables, the matching of pairs of variables mirroring the physical processes. With reference to the sub-system schematic diagram (Figure 15.6) and the component model descriptions (Figures 15.7 to 15.9), the air temperature at the outlet of the heating coil is the controlled variable; hence, in the absence of a sensor model, the air temperature at the outlet of the heating coil model (T_{ao}) is linked to the controlled variable (*CV*) at the "inlet" to the controller model (Figure 15.10). The linking results in T_{ao} and *CV* being mapped to a single variable ($\{T_{ao}, CV\} \rightarrow T_{ao}$), in this case having the notation T_{ao}.

Similarly, in the absence of an actuator model, the output signal (*U*), from the proportional controller is linked directly to the valve stem position (*S*), the two variables being mapped ($\{U, S\} \rightarrow U$) to a single variable (*U*). Finally, the water mass-flow rate leaving the three-port valve ($\dot{m}_w$) is linked to the water mass-flow rate through the heating coil ($\dot{m}_w$), the two flow rates being mapped ($\{\dot{m}_w, \dot{m}_w \rightarrow \dot{m}_w$) to a single variable ($\dot{m}_w$). The connected system model is illustrated by the information flow diagram in Figure 15.11.

A set of *n* algebraic equations can only be solved for *n* variables. Figure 15.11 indicates that there are eight unique variables ($\dot{m}_a$, T_{ai}, T_{wi}, T_{wo}, T_{ao}, SP, U, $\dot{m}_w$) and four algebraic equations (HC.1, HC.2, V.1, C.1), and therefore, four of the variables must be defined to be boundary variables. A poor choice of boundary variables can result in an insoluble set of equations.

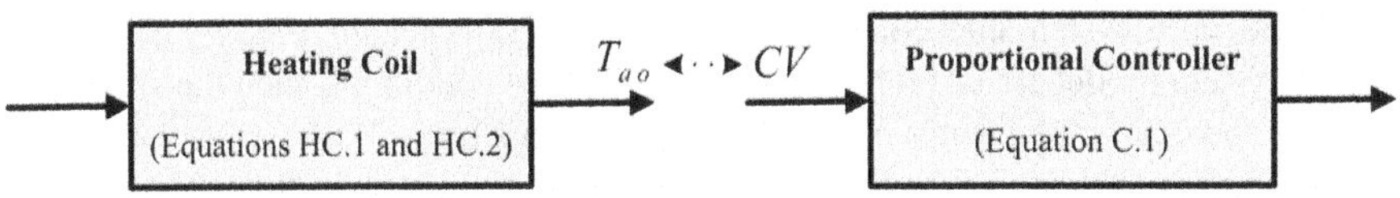

Figure 15.10 Heating coil and controller linking variables

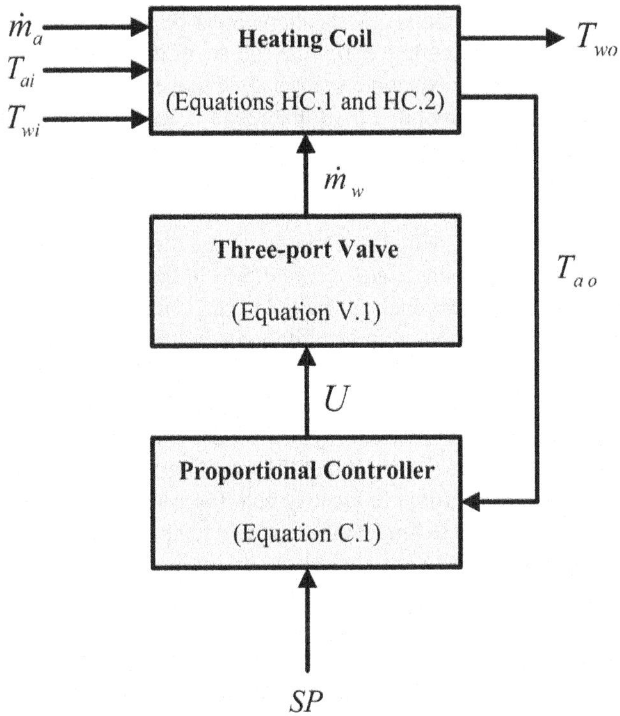

Figure 15.11 Heating coil sub-system information flow diagram

Table 15.2 System adjacency matrix (well-formed model)

Equation	*Simulation variable*			
	T_{wo}	T_{ao}	U	$\dot{m}_w$
HC.1	0	**1**	0	**1**
HC.2	**1**	**1**	0	**1**
V.1	0	0	**1**	**1**
C.1	0	**1**	**1**	0

Although, in most simulation studies, the system inlet conditions naturally form the boundary variables and result in a solvable set of equations, the viability of the simulation should be checked. This can be achieved through examination of an adjacency matrix (Table 15.2), which describes the relationship between the simulated variables and the model equations. The matrix indicates the presence of a variable in an equation by a non-zero entry in the matrix (a zero entry indicating that there is no relationship between the variable and equation).

For example, if the system inlet conditions and controller setpoint ($\dot{m}_a$, T_{ai}, T_{wi}, SP) of the heating coil sub-system model are taken as the system boundary conditions, then the simulated variables become the coil air and water outlet temperatures, together with the control signal and the water mass-flow rate (T_{wo}, T_{ao}, U, $\dot{m}_w$). The resulting adjacency matrix is illustrated in Table 15.2. The key indication that the simulation model and the choice of boundary conditions have led to a well-formed simulation model is that every row of the adjacency matrix has at least one non-zero entry (this being the case in Table 15.2). The fact that every equation has more than one non-zero entry also suggests that the set of equations must be solved simultaneously, although graph-theoretic methods can be used to identify the sub-set of the equations that must be solved simultaneously, with the remaining equations being solved sequentially (Sowell and Haves 2001).

To illustrate a choice of boundary conditions that result in an insoluble set of equations, consider the case, perhaps during system commissioning, where the control setpoint and control signal are known together with measurements of the air temperature at inlet and outlet of the coil, but the air and water flow rates and water temperatures are unknown. If the known conditions (T_{ai}, T_{ao}, U, SP) are set as the boundary conditions, and the unknown ($\dot{m}_a$, $\dot{m}_w$, T_{wi}, T_{wo}) as the simulated variables, the adjacency matrix (Table 15.3) indicates that the equations are insoluble, as none of the simulation variables are linked to the controller equation (C.1). In effect, this reduces the number of equations to three, with the result that the equation set can only be solved for three unknowns rather than four. Inspection of the information flow diagram (Figure 15.11) indicates that the assertion that the controller equation is redundant is correct, as all of the controller's input and output variables (T_{ao}, U, SP) are known and have been set as boundary conditions. Intuition would also suggest that the resulting set of simultaneous equations is insoluble, as it is physically impossible to identify both the water inlet and outlet temperatures (the best that could be achieved is to model the change in temperature across the heating coil).

The linking of the component models to form a system model is generally independent of the form of model equations, as the linking variables are normally those related to the physical processes (outlet temperature links to inlet temperature; outlet humidity ratio links to inlet humidity ratio; and so on). However, the form of algorithm that can be used to solve the set of simultaneous algebraic equations is dependent on the form of equations.

Solution of directed input-output algebraic equations

System models consisting of directed input-output equations (of the form F: $(X_i, A) \rightarrow X_o$) are generally solved using methods based on the successive substitution algorithm. The solution procedure begins by identifying one or more iteration variables and an associated information flow diagram. For instance, consider the example heating coil sub-system (Figure 15.6) and associated component models (Figures 15.7 to 15.9). If the boundary conditions are taken to be the air mass-flow rate ($\dot{m}_a$), the air and water inlet temperatures (T_{ai}, T_{wi}), and the controller setpoint (SP), and the water mass-flow rate ($\dot{m}_w$) is selected as the iteration variable, then the resulting information flow diagram is illustrated in Figure 15.12. Figure 15.12 makes clear that the coil outlet air

Table 15.3 System adjacency matrix (insoluble model)

Equation	*Simulation variable*			
	T_{wo}	T_{wi}	$\dot{m}_a$	$\dot{m}_w$
HC.1	0	1	1	1
HC.2	1	1	1	1
V.1	0	0	0	1
C.1	0	0	0	0

temperature (T_{ao}) and the control signal (U) are also candidates for the iteration variable. However, the coil water outlet temperature (T_{wo}) is not a viable iteration variable, as it is not used to link any components (and therefore equations). Inspection of the adjacency matrix for the system (Table 15.2) indicates that the water outlet temperature is only linked to heat exchanger equation HC.2, and as such can be calculated directly from HC.2 provided that values for the water mass-flow rate ($\dot{m}_w$) and outlet air temperature (T_{ao}) are known. This makes it clear that only three equations need to be solved simultaneously using successive substitution (HC.1, V.1, and C.1).

An implementation of the successive substitution algorithm for the heating coil sub-system is given in Figure 15.13. The algorithm begins by specifying an initial guess for the iteration

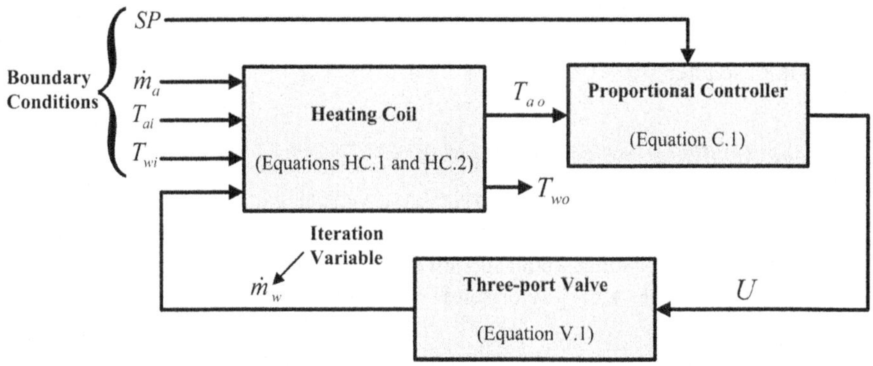

Figure 15.12 Heating coil sub-system information flow diagram for successive substitution

```
//Set boundary conditions for current time period,t
set SP^t, ṁ_a^t, T_ai^t, T_wi^t;

//Initialize iteration
guess ṁ_w;                  //Iteration variable
set ζ;                      //Convergence tolerance
set ṁ_w^p = ṁ_w/(1 - ζ);   //Previous value of iteration variable

//Iteration

while ( |ṁ_w^p - ṁ_w| / ṁ_w^p > ζ )

    set ṁ_w^p = ṁ_w;  //Reset previous value of iteration variable

    calculate T_ao;  //From the coil model, Equation HC.1

    calculate U;  //From the controller model, Equation C.1

    calculate ṁ_w;  //From the valve model, Equation V.1

endwhile

calculate T_wo;//From the coil model, Equation HC.2

//Solution = T_ao, U, ṁ_w, T_wo;
```

Figure 15.13 Heating coil sub-system successive substitution pseudo-code variable, together with a convergence tolerance on the iteration, given here as a fractional change (ζ), in the iteration variable. The algorithm continues by sequentially calculating the model output variables from the model equations, the sequential iteration continuing until the change in iteration variable is less than or equal to the convergence tolerance.[3] In the case of the heating coil sub-system, the water outlet temperature is calculated on completion of the iteration.

The successive substitution algorithm is seemingly simple to implement, although it can fail to converge due to divergence or oscillation of the iteration variable value. The robustness of the algorithm is dependent, in part, on the choice of iteration variable and the equation iteration sequence, the choice of a robust iteration sequence increasing in difficulty with the size and complexity of the system being modeled. The difficulty in forming a robust iteration sequence has led to an increase in the use of residual model equations and more robust solvers based on the Newton-Raphson iteration.

Solution of input-output free algebraic equations

Input-output free models (F: $(X_i, X_o, A) \rightarrow R$) can be solved using methods based on the Newton-Raphson algorithm. The Newton-Raphson algorithm is derived from a first-order Taylor-series expansion of the model equations about the solution. For the case of four residual equations (and four simulation variables), the Taylor-series expansion would result in Equation set 15.11a:

$$\begin{bmatrix} \frac{\partial R_1}{\partial x_1} & \frac{\partial R_1}{\partial x_2} & \frac{\partial R_1}{\partial x_3} & \frac{\partial R_1}{\partial x_4} \\ \frac{\partial R_2}{\partial x_1} & \frac{\partial R_2}{\partial x_2} & \frac{\partial R_2}{\partial x_3} & \frac{\partial R_1}{\partial x_4} \\ \frac{\partial R_1}{\partial x_1} & \frac{\partial R_1}{\partial x_2} & \frac{\partial R_1}{\partial x_3} & \frac{\partial R_1}{\partial x_4} \\ \frac{\partial R_1}{\partial x_1} & \frac{\partial R_1}{\partial x_2} & \frac{\partial R_1}{\partial x_3} & \frac{\partial R_1}{\partial x_4} \end{bmatrix} \begin{bmatrix} x_1 - x_1^s \\ x_2 - x_2^s \\ x_3 - x_3^s \\ x_4 - x_4^s \end{bmatrix} = \begin{bmatrix} R_1 \\ R_2 \\ R_3 \\ R_4 \end{bmatrix} \tag{15.11a}$$

where $x_1^s, x_2^s, x_3^s, x_4^s$ represent the values of the simulation variables at the solution (these being unknown); R_1, R_2, R_3, R_4 are the residual equation values which are computed for the current estimate of the solution (x_1, x_2, x_3, x_4); similarly, the gradients of the residual equations $\left({\partial R_i}/{\partial x_j}\right)$, with each simulation variable are calculated about the current estimate of the solution.

Equation 15.11b gives Equation 15.11a in matrix form, where J is the Jacobian matrix of partial derivatives; D the vector of differences between the current estimate of the solution and the value at the solution; and $\vec{R}$ the vector of residual equation values.

$$J \times D = \vec{R} \tag{15.11b}$$

Having computed the partial derivatives and residual equation values, Equation 15.11b can be solved for the vector of differences D, between the variable's current values and the values at the solution. The differences can then be used to correct the current value of the simulation variables:

$$(X - D) \rightarrow X \tag{15.12}$$

where $X = \begin{bmatrix} x_1 \\ x_2 \\ x_3 \\ x_4 \end{bmatrix}$.

Figure 15.14 illustrates the Newton-Raphson procedure, which begins by assigning values to the model parameters (A), the convergence tolerance (ζ), and the differencing intervals (∂X); in some algorithms, the differencing interval is also adapted within the do-while iteration loop. The algorithm continues with the boundary conditions (X_t^b) being set for the current time-step (t). An initial guess of the value of the simulation variables (X) must also be given for each time-step; the closer the guess to the solution, the more likely it is that a solution will be found. Once a solution has been found for the initial time-step ($t = 0$), the variable values found in the previous time-step ($t - 1$) can be used to initialize the guess of the solution in the current time-step (t), the rationale being that, in many instances, the system operation does not change significantly from one time-step to the other.

The iteration sequence begins by calculating the value of the residual equations ($\vec{R}$) for the current estimate of the solution (X), the value of the residuals also being a function of the model parameters (A) and the boundary variables (X_t^b). The iteration continues with the calculation of the Jacobian matrix of partial differentials (J), this also being a function of the model parameters, boundary variables, current estimate of the solution, as well as the differencing intervals (∂X). The set of simultaneous equations (Equation 15.11b) can then be solved for the estimated

//Begin

set A; //Vector of model parameters

set ζ; //Convergence tolerance

set $\partial X = \{\partial x_1, \partial x_2, \cdots, \partial x_n\}$; //Variable differencing intervals

//Initialize iteration for current time period, t

set X_t^b; //Vector of boundary variables

guess $X = \{x_1, x_2, \cdots, x_n\}$; //Simulation variables

//Iteration

do

  calculate $\vec{R}^T = [R_1(A, X_t^b, X), R_2(A, X_t^b, X), \cdots R_n(A, X_t^b, X)]$; //Residual equations

  calculate $J = f(A, X_t^b, X, \partial X)$; //Jacobian matrix

  solve $J \times D = \vec{R}$; //Solve for D

  set $X = X - D$; //Update simulation variable values

while $\left(\sum_{i=1}^{n} |R_i| > \zeta \right)$

//Solution = $X = \{x_1, x_2, \cdots, x_n\}$

Figure 15.14 Newton-Raphson pseudo-code

distances of the variable values from the solution (*D*), these then being used to update the current estimate of the solution. The iteration continues until the absolute sum of the residual equation values lies within the convergence tolerance (ζ).

The residual equations for the heating coil sub-system (Figure 15.11) can be formed by subtracting the left-hand side of the model equations from the right-hand side of the equations:

$$R_1 = T_{ai} + \frac{\varepsilon \min(w_a, w_w)(T_{wi} - T_{ai})}{w_a} - T_{ao} \tag{HC.1R}$$

$$R_2 = T_{wi} - \frac{w_a(T_{ao} - T_{ai})}{w_w} - T_{wo} \tag{HC.2R}$$

$$R_3 = \frac{\dot{m}_{max}}{\sqrt{1 + \gamma(C_{ih}^{-2} - 1)}} - \dot{m}_w \tag{V.1R}$$

$$R_4 = \min(\max(0, u), 1) - U \tag{C.1R}$$

If the components are connected in the same manner as illustrated in Figure 15.11, and we choose the boundary conditions to be equal to the inlet conditions and the control setpoint ($\dot{m}_a$, $\dot{m}_w$, T_{ai}, T_{wi}, *SP*), the adjacency matrix for the sub-system (Table 15.4) is equivalent to that for the solution of the directed input-output equations (Table 15.2).

Since the residual equations are not a function of all the simulation variables, the Jacobian matrix will be sparse, as it can contain many zero entries (the zero entries in the Jacobian matrix corresponding to the zero entries in adjacency matrix). The sparse nature of the Jacobian indicates that simultaneous equations (Equation 15.11b) are best solved using sparse matrix methods (Sowell and Haves 2001).

Application examples

The use of HVAC component models is illustrated with two applications, HVAC system condition monitoring[4] and HVAC system optimization. The factors having most impact on the success of the modeling differ between the two applications.

HVAC system condition monitoring often requires models that are easily calibrated to a correctly operating system and as such they can have some model parameters that have no real physical meaning. Conversely, models used in HVAC system optimization must have parameters that represent the manufactured components.

Table 15.4 System adjacency matrix

Residual equation	*Simulation variable*			
	T_{wo}	T_{ao}	U	$\dot{m}_w$
HC.1R	0	1	0	1
HC.2R	1	1	0	1
V.1R	0	0	1	1
C.1R	0	1	1	0

Since the control signals are often recorded during condition monitoring, there is no need to model the local-loop controls. However, since the controller outputs are not available as boundary conditions during the design stage, the local-loop controllers must be modeled in system optimization studies. The availability of the control signals during condition monitoring can allow the system equations to be solved sequentially rather than simultaneously, but simultaneous solution is required when the control loops are modeled in optimization studies.

Since the modeled outputs are used to assess the state of a real system in condition monitoring schemes, the uncertainty arising from measurements and the modeling procedure are critical to the success of the scheme (Buswell 2001; Buswell and Wright 2004). As well as accounting for the uncertainty during the system optimization process, it is also critical to develop optimization constraints that ensure the optimization search remains within the component performance envelopes (the performance envelopes being defined by the manufacturer's measured data and/or the valid region of the component model).

Reference models for HVAC system commissioning and fault detection

HVAC component models have been used in the commissioning of HVAC systems, as well as the detection of HVAC system faults. In both cases, a model is used as a reference for correct operation of the system (Figure 15.15).

Figure 15.15 illustrates the approach, in which, given a set of measured input conditions[5] (X_i^m) and a set of model parameters (A), the model is used to predict the value of a single quantity at the "outlet" of the component or system (x_o). This is compared to the measured value of the same quantity (x_o^m), the difference in the measured and modeled values (e) being used to infer the existence of a fault (or in the case of system commissioning, that the system is poorly commissioned or that incorrect equipment has been installed).

The form of reference model and associated parameters depends on the application. In the case of system commissioning, the model must encapsulate the design intent, and therefore it is important for the model to have parameters that relate to the design specification of the installed system components (for instance, the diameter of a fan or the number of tube rows in a heating coil). However, following the completion of the system commissioning, models for use in fault detection are often calibrated to a correctly operating system to produce near-zero prediction error (e), the rationale being that an increase in the model error following calibration would indicate the presence of a new fault condition. In this case, the model equations and parameters may be simplified and should be of a form that enables their calibration to the system; for instance, a heating coil model may include a heat transfer correction parameter that can be tuned to account for the difference in predicted and installed performance.

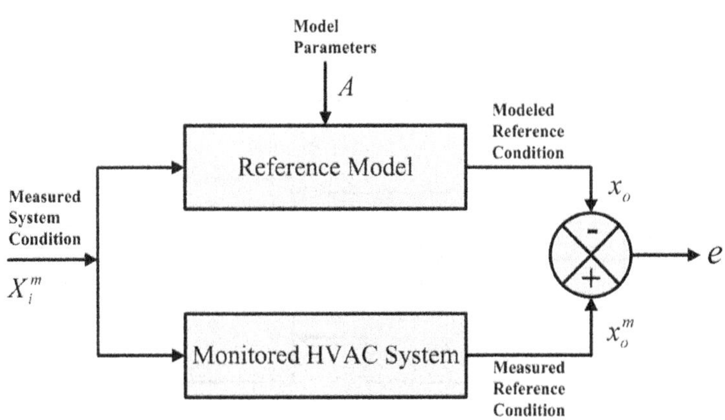

Figure 15.15 HVAC system condition monitoring scheme

The form of model equations also depends on which system conditions are being measured, these often being restricted to measurements used in the direct control of the system. This can result in the simplification of the model equations or the required model inputs being estimated from other measurements; for instance, a flow temperature or flow rate being estimated through an energy balance on different flow streams.

Given that the measured conditions can include the control signals (U) at the output of the local-loop controllers, the model equations can often be solved sequentially. For example, consider the heating coil sub-system in Figure 15.6 and the associated information flow diagram in Figure 15.11; if the outlet air temperature (T_{ao}) is used as a modeled reference condition, and the control signal (U) is measured together with the air mass-flow rate ($\dot{m}_a$) and the air and water inlet temperatures (T_{ai}, T_{wi}), then the iterative loop associated with the outlet air temperature is broken and the proportional controller model is redundant. This enables the equations for the three-port valve and heat exchanger to be solved sequentially (Figure 15.16).

Sources of uncertainty have a significant impact on the effectiveness of HVAC system condition monitoring schemes. Uncertainty exists not only in the measured values, but also in the model parameters and the form of model equations. Uncertainty in the model parameters can occur in the manufacturer's measurement of component performance or through the procedure used to calibrate the model parameters to a set of measured data.

The uncertainty in the model equations is a result of the assumptions made during the model development. For example, the equations for heat exchanger effectiveness (ε) used in the heating coil model (Figure 15.7) are for a pure counter-flow heat exchanger, whereas the flow regime in an installed heating coil is more likely to be a combination of cross-flow and counter-flow. Where the model equations are solved simultaneously, "uncertainty" also arises due to the convergence tolerance used by the solution algorithm (the equations rarely being solved exactly using numerical solution procedures).

If the condition monitoring scheme is based on the use of steady-state models, then there is also some uncertainty about the extent to which the system was in steady-state when the performance measurements were taken. The combined impact of several sources of uncertainty can mask faults that have a magnitude equivalent to 14% of the peak capacity of the coil in a cooling coil sub-system (Buswell and Wright 2004).

Component models for HVAC system optimization

The design and operation of HVAC systems can be optimized by using an optimization algorithm to search for the value of the model parameters and/or model input variables that minimize a specified criterion such as energy use, capital cost, or life-cycle cost (Wright 1996). The optimization algorithm successively generates new trial solutions in a manner that minimizes the objective function ($f(X_o, X_i, \mathrm{A})$), with each new trail solution being evaluated using the HVAC system model (Figure 15.17). Where the optimization is concerned with optimizing the size of the HVAC components, the optimization variables (X^{opt}) are mapped onto the

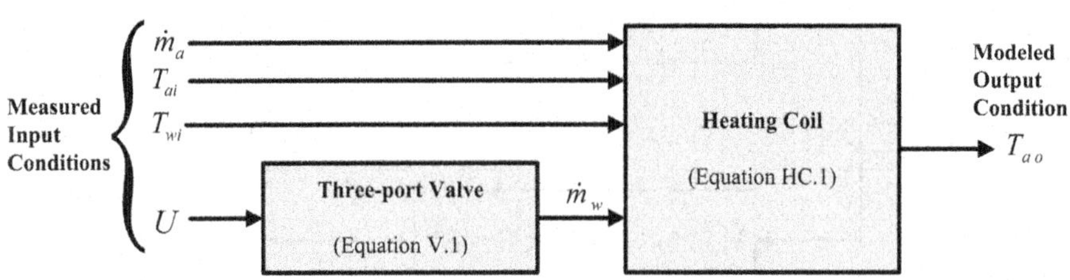

Figure 15.16 Information flow diagram for sequential equation solution

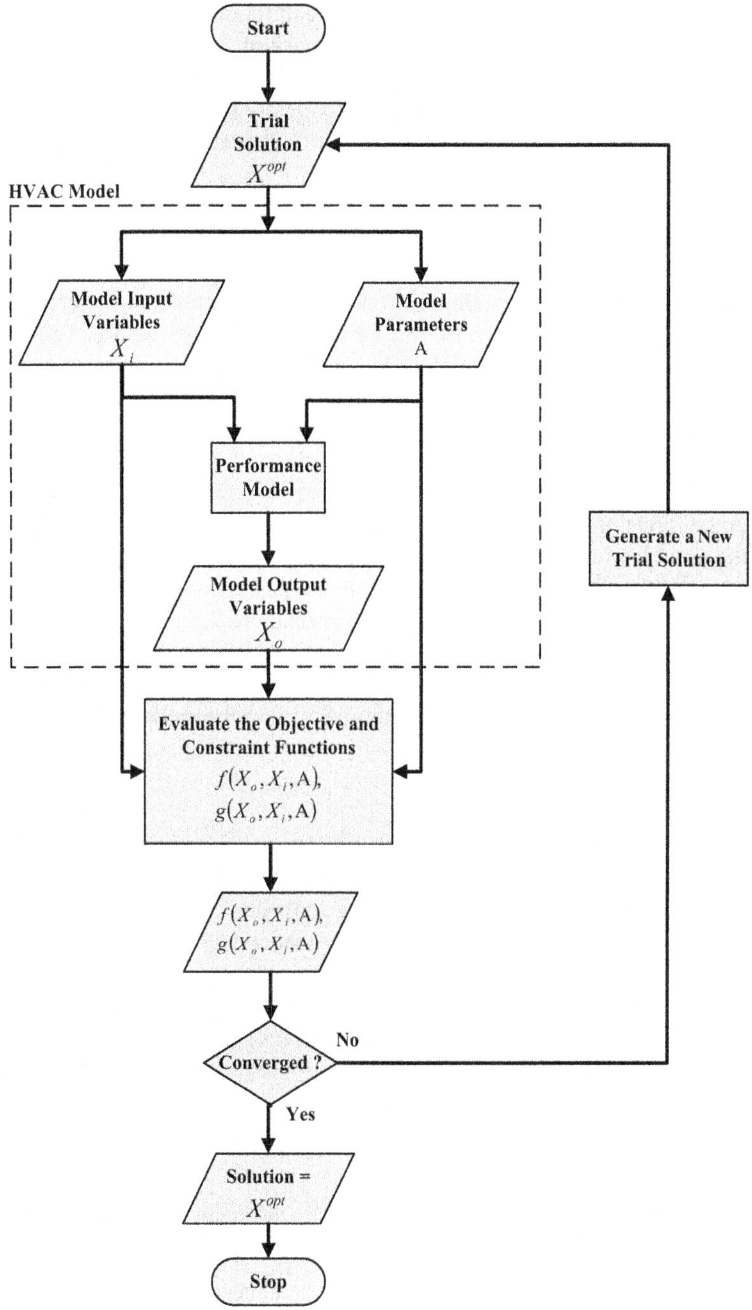

Figure 15.17 Optimization process

model parameters (since the physical size of the components are normally defined by the model parameters). Similarly, since the control setpoints are normally specified as a sub-set of the model input variables, optimization of the system's supervisory control strategy is achieved by mapping the optimization variables onto the model input variables. In some instances, the size and operation of the HVAC systems are optimized simultaneously with the form of building fabric construction (Wright et al. 2002).

Since the optimization is often concerned with sizing the components, the component models must be formulated to represent a range of possible sizes of component. Quasi-first-principle models (such as of coils and other heat exchangers) tend to be generic and therefore valid for range of components. However, empirical models (typically used for pumps and fans) require some form of data normalization or use of scaling rules in order to represent a range of components (Wright 1991).

Optimization of HVAC system size and operation is normally performed subject to one or more constraints (g (X_o, X_i, A)) that ensure the design solutions do not exceed practicable performance limits (such as those set on fluid velocities) and are within the measured and/or valid performance envelope of the component. Constraint functions that represent the limits of the measured component performance envelope are particularly important for fully empirical models (Wright 1991), as extrapolation beyond the measured range of performance is likely to be impracticable (as the optimized size of a component may not be manufactured and available in practice). It may also be necessary to place constraints on the solutions obtained from quasi-first-principle models. For example, the Colebrook equation for duct friction factor (Equation 15.8) is only valid for fully turbulent flow, and in this respect, a duct size optimization based on the model in Figure 15.5 should include a constraint on the Reynolds number to ensure that the optimization solution is for fully turbulent flow.

In summary, HVAC component models for use in HVAC system optimization must be able to model the performance of a range of component sizes (and products from different manufacturers). Separate post-processed models must also be developed for calculation of derived quantities used in formulation of the optimization objective and constraint functions, these being a function of the model input variables and model parameters, as well as the model output variables.

Discussion

This chapter describes the fundamental forms of HVAC component models, the connection of the component models to form a HVAC system model, and two approaches to solving the system algebraic equations. These principles can be applied to the modeling of a wide range of HVAC components and systems. However, growing concerns over the difference in predicted (modeled) building performance and the performance measured during operation means that these models should now be extended to include the uncertainty in performance prediction. Further, it is important to consider the sources of available component and system performance data before developing a component model, as the form of available data will inevitably dictate the form of the component model parameters; an understanding of the relationship between the data and the form of modeling will contribute to the quantification of the epistemic uncertainty associated with the choice of model equations.

Uncertainty in HVAC system performance prediction

To date, very little research has been conducted into the uncertainty quantification and analysis of HVAC system performance. The most significant work is in the field of fault detection and diagnosis (Buswell 2001; Buswell and Wright 2004), this work considering the uncertainty in performance measurement and in heat exchanger modeling. The work illustrates the difficulty in quantifying the uncertainty associated with the model equations and addresses the problem of comparing the prediction from steady-state models with the measured performance of the system, this being in a dynamic and not steady-state. Further research is required to quantify the uncertainty associated with a wider range of HVAC components, including the effect of installation on the standardized performance and the degradation of performance with time. The uncertainty quantification for dynamic performance models is more challenging than for steady-state models, but where control system behavior is to be analyzed, it is likely that dynamic models will be used; the poor tuning of system controls can have a significant effect on system performance (the control parameter tuning being a form of epistemic uncertainty).

Performance data and model development

Given that there are very few HVAC components that can be modeled using purely first principles, the majority of models rely on some measured performance data. The development of HVAC component models therefore depends on the availability and form of measured performance data. The test procedure used to obtain the data can also have an impact on the development of the model. For instance, one goal of a performance test of a fin-tube heating or cooling coil is to identify the air-side heat transfer coefficient. However, the air-side heat transfer coefficient could be defined in several ways. For use in design practice, the air-side heat transfer coefficient might be given as a function of the coil face-area and associated air-velocity (with different functions being given for each fin-density). However, a more detailed performance evaluation might give the air-side heat transfer coefficient as a function of the air velocity in the passage between the fins. In both cases, it is also necessary to make assumptions about the form of heat transfer coefficient inside the water-tubes and for the fin efficiencies. The form of the coil model will therefore be restricted by the form of the air-side heat transfer coefficients and must also incorporate the same assumptions made about the water-side heat transfer coefficient and fin efficiency (as the measured heat transfer coefficients are an implicit function of these assumptions). Similar observations can be made regarding the performance testing of fans as the measured performance is a function of the particular test duct, the implication being that the model should attempt to include the "installation effect." To date, the format of HVAC performance data has varied from one manufacturer to another, making it difficult to develop a generic model structure. This deficiency is being addressed through initiatives in "building information modeling" (BIM), and in particular, the development of HVAC industry foundation classes (Bazjanac 2004).

Synopsis

HVAC systems are constructed from components. In the same manner, HVAC system models can be constructed from models of the individual system components. This chapter describes the principles used in the development of steady-state component models together with the procedures for linking components to form a system model. Two forms of model equation (directed input-output equations and input-output free residual equations) are described together with the common algorithms used to solve the set of simultaneous system-model equations. The principles described in the chapter can be applied to a wide range of HVAC equipment and systems, but require extension to include the systems dynamic response when the performance of the control systems is to be evaluated. Increasing concern over the difference in predicted and measured performance indicates that contemporary HVAC component and system models should be extended to include the aleatoric and epistemic uncertainty in the modeled outputs. The relationship between the format of the HVAC equipment manufacturers' measured performance data and the development of HVAC component models is also described. Two applications of HVAC system modeling are described in the chapter: the use of HVAC system models in system condition monitoring and the application of model-based optimization methods in the optimum sizing of HVAC systems.

Notes

1 Some HVAC systems are strongly coupled to thermal storage elements, which significantly increases the time constant of this part of the system.
2 Note that an impeller speed is in revolutions per second.
3 The check for convergence can also be extended to identify divergence or oscillation of the iteration.

4 The calibration of simulation models and their use in building performance monitoring and performance optimization is also discussed in Chapter 13, "Building Simulation for Practical Operational Optimization."
5 A "condition" is taken to be a "measured" quantity, including temperature, relative humidity, flow rate, pressure, and control signal.

References

Angelov, P.P., Hanby, V.I., Buswell, R.A. and Wright, J.A. (2000) A methodology for modelling HVAC components using evolving fuzzy rules. In *IEEE International Conference on Industrial Engineering, Control and Instrumentation*, IECON-2000, Nagoya, Japan, pp. 247–252.

Bazjanac, V. (2004) Building energy performance simulation as part of interoperable software environments. *Building and Environment* 39: 879–883.

Bourdouxhe, J.-P., Grodent, M. and Lebrun, J. (1998) *Reference Guide for Dynamic Models of HVAC Equipment*. Atlanta, GA: ASHRAE, ISBN: 1883413605.

Braun, J.E. (1988) Methodologies for the design and control of central cooling plants. PhD Thesis, University of Wisconsin-Madison, WI.

Brandemeuhl, M. (1993) *HVAC2 Toolkit: Algorithms and Subroutines for Secondary HVAC Systems Energy Calculations*. Atlanta, GA: ASHRAE, Inc.

Buswell, R.A. (2001) Uncertainty in the first principle model based condition monitoring of HVAC systems. PhD Thesis, Loughborough University. https://dspace.lboro.ac.uk/2134/7559.

Buswell, R.A. and Wright, J.A. (2004) Uncertainty in model based condition monitoring. *Building Services Engineering Research and Technology* 25(1): 65–75.

Haves, P., Salsbury, T.I. and Wright, J.A. (1996) Condition monitoring in HVAC subsystems using first principles models. *ASHRAE Transactions* 102(1): 519–527, ISSN: 0001–2505.

Holmes, M.J. (1982) The simulation of heating and cooling coils for performance analysis. In *Proceedings of the 1st International Conference on System Simulation in Buildings*. Liege, Belgium, pp. 245–280.

ISO. (2017) Fans – Performance testing using standardized airways. ISO 5801:2017(en), International Organization for Standardization.

Lebrun, J., Bourdouxhe, J.-P., Grodent, M. (1999) *HVAC 1 Toolkit: A Toolkit for Primary HVAC System Energy Calculation*. Atlanta, GA: ASHRAE, ISBN 1-883413-73-7.

Norford, L.K., Wright, J.A., Buswell, R.A., Luo, D. Klaassen, C.J., Suby, A. (2002) Demonstration of fault detection and diagnosis methods for air-handling units. *HVAC&R Research* 8(1): 41–71. https://doi.org/10.1080/10789669.2002.10391289.

Salsbury, T.I. (1996) Fault detection and diagnosis in HVAC systems using analytical models. PhD Thesis, Loughborough University, Loughborough, Leicestershire, LE11 3TU, UK.

Sowell, E.F. and Haves, P. (2001) Efficient solution strategies for building energy system simulation. *Energy and Buildings* 33: 309–317.

Tian, W., Heo, Y., de Wilde, P., Li, Z. Yan, D., Park, C.S., Feng, X. and Augenbroe, G. (2018) A review of uncertainty analysis in building energy assessment. *Renewable and Sustainable Energy Reviews* 93: 285–301. https://doi.org/10.1016/j.rser.2018.05.029.

Underwood, C.P. (1999) *HVAC Control Systems: Modelling, Analysis and Design*. London: E&FN Spon.

Wetter, M. (2013) Fan and pump model that has a unique solution for any pressure boundary condition and control signal. In *Proceedings of BS2013: 13th Conference of International Building Performance Simulation Association*. Chambery, France, August 26–28, pp. 3505–3512.

Wetter, M., Bonvini, M. and Nouidui, T. (2016) Equation-based languages – A new paradigm for building energy modeling, simulation and optimization. *Energy and Buildings* 117: 290–300. http:/dx.doi.org/10.1016/j.enbuild.2015.10.017.

Wright, J.A. (1986) The application of the least squares curve fitting method to the modelling of HVAC component performance. In *Proceedings of the Second International Conference on System Simulation in Buildings*. University of Liege, Liege, pp. 27–45.

Wright, J.A. (1991) HVAC optimisation studies: Steady-state fan model. *Building Services Engineering Research and Technology* 12(4): 129–135.

Wright, J.A. (1993) HVAC simulation studies: Solution by successive approximation. *Building Services Engineering Research and Technology* 14(4): 179–182, ISSN: 0143–6244.

Wright, J.A. (1996) HVAC Optimisation Studies: Sizing by Genetic Algorithm. *Building Services Engineering Research and Technology* 7–14, 17(1), ISSN: 0143–6244.

Wright, J.A., Loosemore, H.A. and Farmani, R. (2002) Optimization of building thermal design and control by multi-criterion genetic algorithm. *Energy and Buildings* 34(9): 959–972, ISSN: 0378–7788.
Zhou, X. and Braun, J.E. (2007) A simplified dynamic model for chilled-water cooling and dehumidifying coils-part 1: Development. *HVAC&R Research* 13(5): 785–824.

Recommended reading

Numerous academic journal papers can be found that describe the development and application of HVAC component performance models. However, in particular, descriptions of several steady-state HVAC system component models can found in two reference "toolkits" developed by the American Society of Heating, Refrigerating and Air-Conditioning Engineers (ASHRAE); there being separate toolkits for primary systems (Lebrun et al. 1999) and secondary systems (Brandemeuhl 1993). Both toolkits give the theoretical basis for each component model as well as each model coded as a FORTRAN routine. The models described in these toolkits have been adopted for use in several established building performance simulation programs.

The principles of developing dynamic HVAC component models are described by Bourdouxhe et al. (1998). The models described in this reference are a result of an extensive survey of dynamic model development, with models being described for ducts and pipes, heat exchangers, boilers, heat pumps, and chillers. This publication also gives models for control system components such as sensors and actuators; control system component models are also described in the textbook by Underwood (1999).

Mathematical methods for the solution of algebraic equations can be found in any mathematics textbook. The extent to which the computational performance of these methods can be improved through the application of graph-theoretical methods in reducing the set of simultaneous equations is described by Sowell and Haves (2001).

Activities

1 Using the centrifugal fan performance data given in Table 15.5, develop an input-output model based on a curve-fit of the non-dimensional fan performance. Use the model to predict the total pressure and air temperature at the fan outlet for a 0.25 m diameter fan

Table 15.5 Centrifugal fan performance data

D (*m*)	*n* (*rpm*)	$\dot{m}$ (*kg/s*)	*P* (*Pa*)	*W* (*W*)
0.22	1500	0.26	166	52
0.22	1500	0.32	145	51
0.22	2500	0.55	417	266
0.22	2500	0.36	490	216
0.22	3500	0.83	759	728
0.22	3500	0.55	974	649
0.25	1500	0.72	71	95
0.25	1500	0.65	109	102
0.25	2500	1.28	97	326
0.25	2500	1.09	313	483
0.25	3500	1.85	146	752
0.25	3500	1.5	670	1352

running at a speed of 3000 rpm when the air mass-flow rate is 0.9 kg/s, the inlet air temperature is 20.0 °C, and the inlet air pressure is 0.0 Pa. Assume a fixed inlet air density of 1.2 kg/m^3 and an air specific heat capacity of 1.005 kJ/kg K.

2 Implement the rectangular duct friction loss model illustrated in Figure 15.5 and use it to calculate the outlet pressure of a 0.5 m by 0.5 m square duct when the inlet mass-flow rate is 0.9 kg/s and the inlet pressure is 100 Pa. Take the air density and dynamic viscosity to be 1.2 kg/m^3 and 1.85 × 10^5 Pa.s respectively. Assume that the duct is constructed from galvanized sheet steel having an absolute roughness of 0.09 mm.

Answers to activities

1 Figure 15.18 illustrates an input-output centrifugal fan model based on a polynomial curve-fit of the non-dimensional performance data. The model parameters are the fan diameter, polynomial coefficients, inlet air density, and specific heat capacity. The fan speed has also been treated as a parameter, although it could equally be a model input variable, which would allow the model to be used in variable-volume system simulations. The air density and specific heat capacity would also normally be calculated within the model as functions of the inlet air temperature. Figure 15.19 illustrates the normalized fan performance curves; a '+' indicates a normalized data point from Table 15.5, and a 'o' the normalized data point for the example operating conditions. The normalized fan curves can be modeled using a second-order polynomial for the relationship between normalized flow and pressure rise, and a third-order fit for the relationship between normalized flow and power. The model predicts that a 0.25 m diameter fan running at 3000 rpm produces a fan total pressure of 802 Pa and a power output of 816 W; this translates to an outlet temperature of 20.9 °C and pressure of 802 Pa.

2 The air outlet pressure is 99.38 Pa, resulting from an air friction loss of 0.62 Pa (with a friction factor *f* of 0.019; an air velocity v of 3.0 m/s; a hydraulic diameter D_h of 0.5 m; and a Reynolds number *Re* of 9.73 × 10^4). These values were obtained using the calculation sequence illustrated in Figure 15.5. Full implementation of the model requires the

$$A=\{d,n,\alpha_1,\alpha_2,\alpha_3,\alpha_4,\alpha_5,\alpha_6,\alpha_7,\rho_i,C_p\}$$

$$X_i=\begin{Bmatrix}\dot{m}\\T_i\\P_i\end{Bmatrix} \Rightarrow \boxed{\begin{gathered}\phi=\frac{\dot{m}}{\rho_i\pi^2d^3\,n/(60\times4)}\\ \psi=\alpha_1\phi^2+\alpha_2\phi+\alpha_3\\ \lambda=\alpha_4\phi^3+\alpha_5\phi^2+\alpha_6\phi+\alpha_7\\ P_o=P_i+\frac{\psi\rho_i\left(\pi d\,n/60\right)^2}{2}\\ T_o=T_i+\frac{\lambda\rho_i\pi^4d^5\left(n/60\right)^3}{8\dot{m}C_p}\end{gathered}} \Rightarrow X_o=\begin{Bmatrix}T_o\\P_o\end{Bmatrix}$$

Figure 15.18 Input-output centrifugal fan model

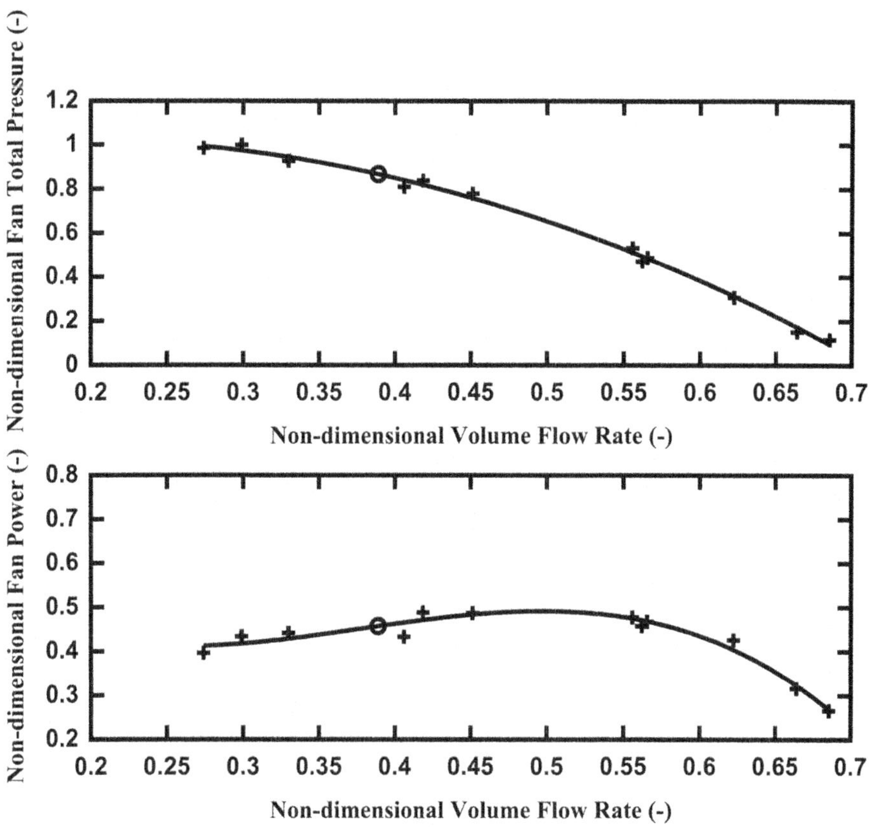

Figure 15.19 Normalized fan performance curves

development of an iteration procedure to solve the transcendental friction factor equation (Equation 15.8). Solution of this equation was achieved here using a using a successive approximation algorithm (Wright 1993). The algorithm is robust and easily implemented, and it can be used to solve a range of iteration problems, including those associated with calculating the psychrometric properties of air. However, the method relies on an understanding of the relationship between the iteration variable (termed the "estimate" in Figure 15.20) and the value calculated from the equation being solved. Figure 15.21 illustrates that for this problem, when the estimated value is lower than the solution, the calculated friction factor is higher than the estimate. Given this relationship, the algorithm has been initialized with an initial guess of the solution for the friction factor being equal to the maximum likely friction factor (Figure 15.20). The algorithm proceeds by continuously subtracting a specified increment (Δf) from the iteration variable (f), until the iteration variable has a value less than that calculated from Equation 15.8 ($f_t > f$). At this point, Figure 15.21 indicates that too much has been subtracted from the iteration variable, so it is reset to its previous value, so that a smaller increment can be subtracted from the iteration variable (the increment (Δf, being halved in each iteration). The iteration continues until either the estimated or the calculated values of the friction factor are equal, or when the increment falls below the convergence tolerance ($\Delta f \leq \zeta$). Figure 15.22 illustrates that for the example problem, the friction factor is found in eight iterations of the algorithm.

```
//Initialize iteration parameters
Δf = 0.05;       //Increment size
f  = 0.1;        //Maximum friction factor
ζ = 0.00025; //Convergence tolerance
//Begin iteration
while (Δf > ζ)
   f_o = f;      //Store current estimate
   f = f-Δf;  //Reduce estimate by current increment
   //Calculate friction factor
```

$$f_t = \left(\frac{1}{-2\log\left(\varepsilon / 3.7D_h + 2.51 / Re\sqrt{f} \right)} \right)^2$$

```
   //Compare estimated and calculated values
   if (f_t ==f) break;   //Solution found, stop
   elseif (f_t >f) f=f_o; //Estimate too low, reset
   endif
   //Halve increment size
   Δf = Δf/2.0;
endwhile
//Solution = f
```

Figure 15.20 Pseudo-code for the friction factor successive approximation

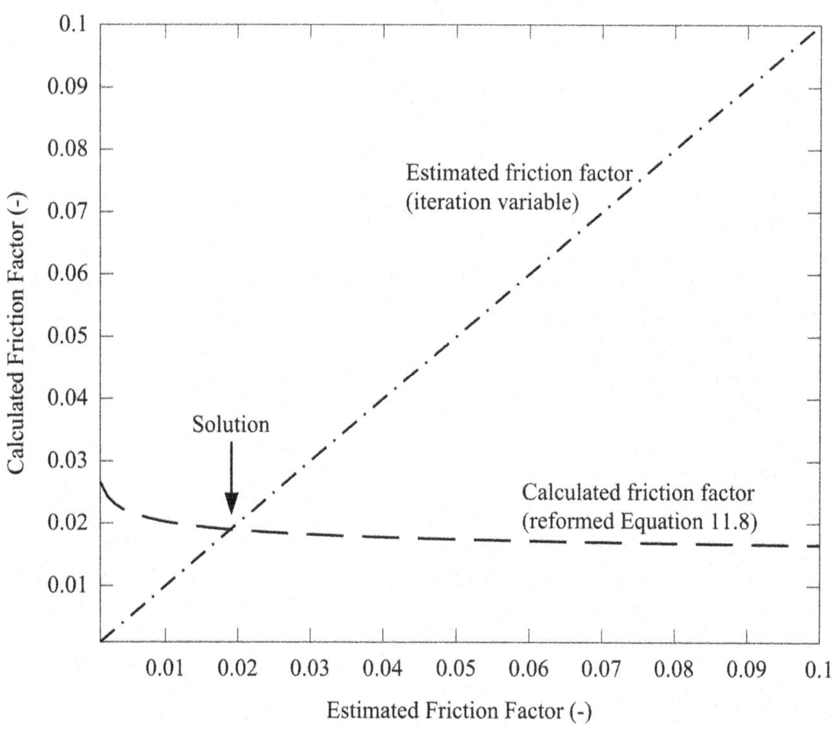

Figure 15.21 Comparison of estimated (iterated) and calculated friction factor

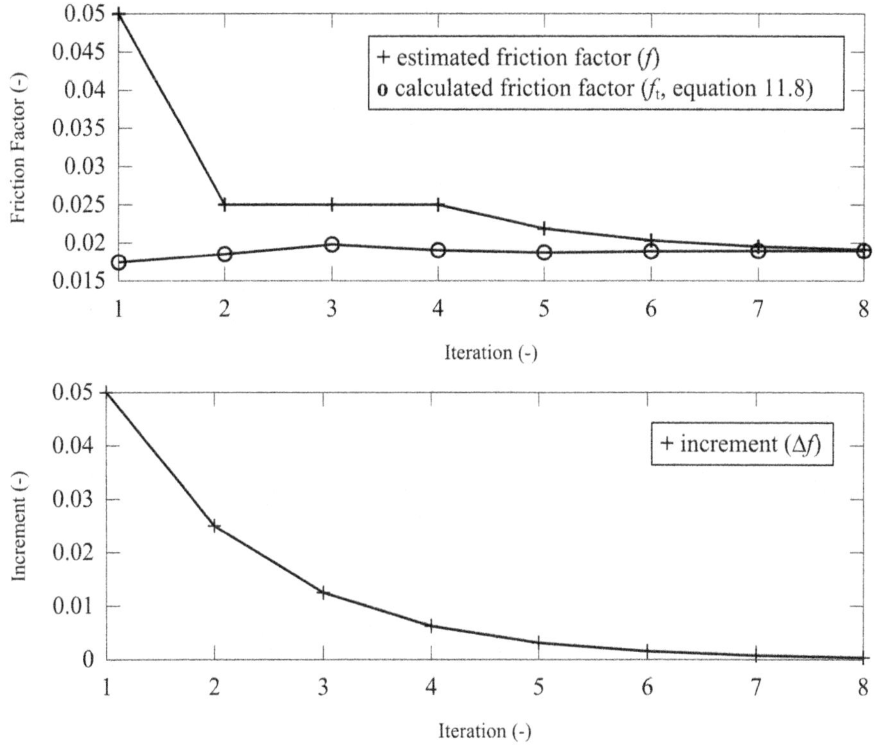

Figure 15.22 Successive approximation convergence

16 Micro-cogeneration system performance prediction

Ian Beausoleil-Morrison

Scope

This chapter discusses fuel cell and combustion-based cogeneration devices with electrical outputs under 15 kW.

Learning objectives

Develop an understanding of small-scale cogeneration (micro-cogeneration) technologies and their potential applications for heating, cooling, and powering buildings. Develop an understanding of modelling the performance of these devices concurrently with building thermal and electrical systems.

Key words

Combined heat and power/cogeneration/micro-cogeneration/fuel cell/Stirling engine/internal combustion engine

Introduction

Micro-cogeneration, residential cogeneration, and small-scale combined heat and power are synonyms for technologies that concurrently produce electricity and heat from a single fuel source. By convention, these terms apply to systems with electrical outputs less than 15 kW. These emerging technologies offer a number of potential benefits: reduced primary energy consumption, reduced greenhouse gas emissions, reduced electrical transmission and distribution losses, and alleviation of electrical grid congestion during peak periods. Numerous companies are actively developing fuel cell (FC), internal combustion engine (ICE), and Stirling cycle engine (SE) micro-cogeneration devices and introducing these to the market (Knight and Ugursal 2005; Onovwiona and Ugursal 2006). Although these technologies are technically immature and their potential benefits remain unproven, promising indications of their performance have led to financial incentives and favourable electricity tariff structures in many countries to encourage their adoption (Hawkes et al. 2014).

Micro-cogeneration devices have only modest fuel-to-electrical conversion efficiencies. The existing prototypes have been found to have efficiencies in the range of 5% to 30% in terms of the net alternating current (AC) electrical output relative to the source fuel's lower heating value, or LHV (Beausoleil-Morrison 2008; Carbon Trust 2007; Entchev et al. 2004). These electrical efficiencies are relatively low compared to state-of-the-art natural-gas-fired central power generation: combined-cycle central power plants can achieve efficiencies in the order of 55% (Colpier and Cornland 2002; DeMoss 1996). As such, it is critical that the thermal portion of the micro-cogeneration device's output be well utilized, otherwise micro-cogeneration will compare poorly to the best-available central power generation technologies (Adam et al. 2015; Gandiglio et al. 2014; Napoli et al. 2015). Potential uses of this thermal output include space heating, domestic hot water (DHW) heating, and space cooling. The latter could be accomplished by using the micro-cogeneration device's thermal output to activate a thermal refrigeration cycle (the candidate cycles are outlined in Henning 2007, for example).

Residential buildings have a strong diversity in both electrical and thermal loads, and there is no inherent coincidence between these loads. This fact is illustrated in Figure 16.1, which plots space heating and electrical demands for a house located in Ottawa (Canada) over a typical spring day. (The electrical demand was measured while the space heating demand was simulated.) The space heating demand is modest on this day due to relatively warm weather. As can be seen, this thermal demand is greatest from late evening to early morning and low during the day when passive solar gains are available. In contrast, the electrical demand peaks during the morning and before and after the dinner hour, this due to the operation of appliances for cooking and cleaning. Each combination of house, occupants, and climate will produce different patterns, and these will vary day to day. However, what will be consistent is the diversity of these loads and their non-coincidence.

This reality necessitates some kind of thermal (and perhaps electrical) storage in any practical micro-cogeneration system. The volume and thermal characteristics of the thermal store, occupant electrical usage patterns, the house's thermal characteristics, prevailing weather, and dispatch strategies all influence whether the micro-cogeneration device's thermal output will be exploited or wasted. Accurately assessing this highly integrated system can only be achieved within the context of whole-building performance simulation and only when the micro-cogeneration device is simulated concurrently with the building's thermal and electrical demands and with supporting HVAC components.

Scientific foundation

Stirling engines and internal combustion engines are heat engines. They first convert the internal energy associated with the bonds between atoms in the reacting fuel to thermal energy through chemical reactions (combustion). The heat released by the combustion is then converted to mechanical energy in a thermodynamic cycle. The well-known Otto and Diesel cycles are employed in internal combustion engines, whereas Stirling engines are external combustion devices operating with closed cycles. The interested reader is referred to Finkelstein and Organ (2001) for a detailed treatment of the Stirling cycle and the design and operation of Stirling engines. The mechanical energy produced by both ICE and SE devices is then converted to electrical energy in a generator. Thermal energy can be recovered from exhaust gases and cooling water circuits to provide the device's useful thermal output. Both ICE and SE devices can be fueled from a variety of sources, such as gasoline, petrodiesel, natural gas, bioethanol, and biodiesel. SE devices can also be fueled from biomass since they employ external combustion.

In contrast to the heat engines, fuel cells are energy conversion devices that directly convert the internal energy associated with the bonds between atoms in the reacting fuel into electrical energy. This is accomplished through the electrochemical oxidation of a fuel and the electrochemical reduction of oxygen. These electrochemical reactions occur at electrodes which are continuously fed with fuel and oxygen and which are separated by an electrolyte layer. A number of fuel cell varieties exist, and these are typically classified according to the electrolyte material employed. Two types are pertinent for micro-cogeneration: solid oxide fuel cells (SOFC) and proton exchange membrane fuel cells (PEMFC), also referred to as polymer electrolyte membrane fuel cells.

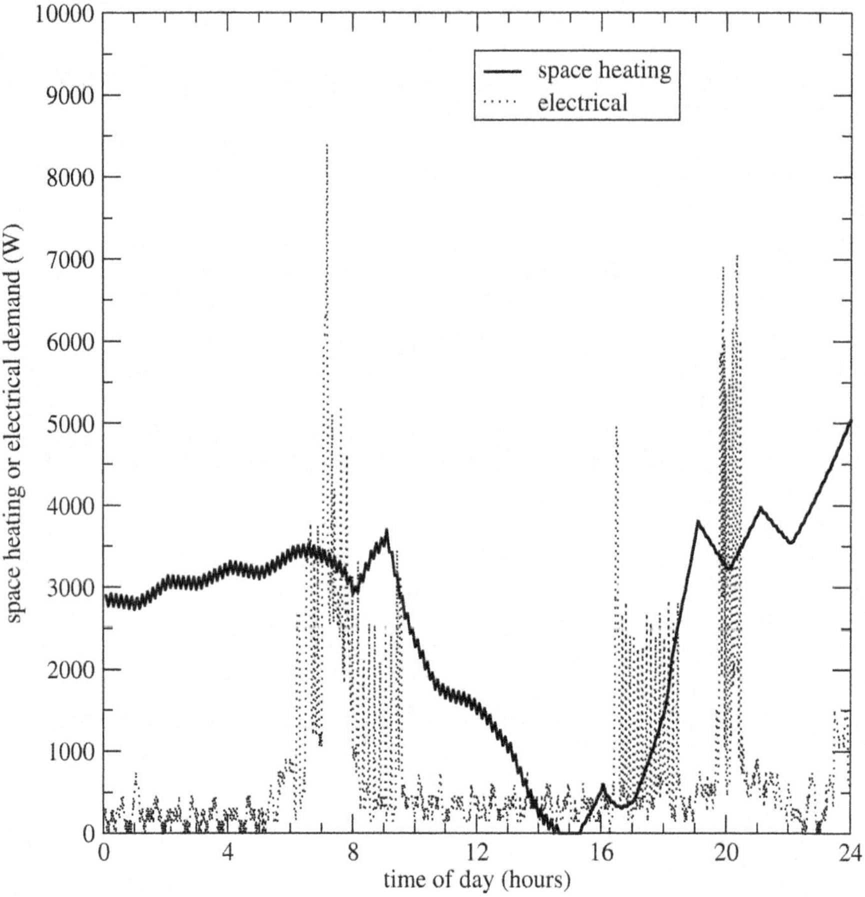

Figure 16.1 Space heating and electrical demands of a house over a typical spring day

SOFCs use a solid metal oxide as the electrolyte. Due to their high operating temperature (800 to 1000 °C), they are able to operate directly on natural gas and other light hydrocarbon fuels. These temperatures are sufficient to internally reform the gas's constituent hydrocarbon molecules (methane, ethane, propane, etc.) to hydrogen and carbon monoxide, which are then supplied to the anode where they are partially oxidized with the oxygen ions crossing the electrolyte.

In contrast, PEMFCs employ a polymer as the electrolyte to conduct hydrogen ions. These are low-temperature fuel cells that operate at temperatures below 100 °C. Because of these low operating temperatures, PEMFCs must be fueled by hydrogen. This hydrogen can either be delivered and/or stored at the building site or it can be produced from a hydrocarbon fuel (e.g. natural gas) by an external reformer.

The electrochemical reactions occurring in both types of fuel cells produce a direct current (DC), which must then be inverted to AC in a power conditioning unit to produce the micro-cogeneration device's electrical output. Due to thermodynamic irreversibilities, some of the source fuel's chemical energy is converted to thermal energy. Additionally, with SOFCs, fuel that is unreacted at the fuel cell electrodes is often combusted in an after-burner resulting in more heat production. SOFC micro-cogeneration devices are typically equipped with heat exchangers to extract much of this thermal energy from the fuel cell's product gases prior to their exhaust. In the case of PEMFCs, the stack is typically cooled with an external water circuit to remove the thermal energy resulting from the irreversibilities, and this warmed water stream can form the device's useful thermal output.

The interested reader is referred to Mench (2008) and Larminie and Dicks (2003) for a thorough review of fuel cell electrochemistry and fuel processing.

It is important to note that for SOFC and PEMFC systems, the fuel cell stack itself is only a single component within a micro-cogeneration device. Similarly, the thermodynamic cycle (Otto, Diesel, Stirling) is only one component of combustion-based systems. This reality is illustrated in Figure 16.2, which shows one possible system configuration for a SOFC micro-cogeneration device. Note that many other system configurations are possible, which will vary from manufacturer to manufacturer. The interested reader is referred to Larminie and Dicks (2003) for detailed explanations of these components and their arrangements.

Besides the fuel cell stack (shown in grey), the system might include an after-burner to combust unreacted fuel; an air filter and pre-heater; a fuel desulfurizer, pre-heater, pre-reformer,

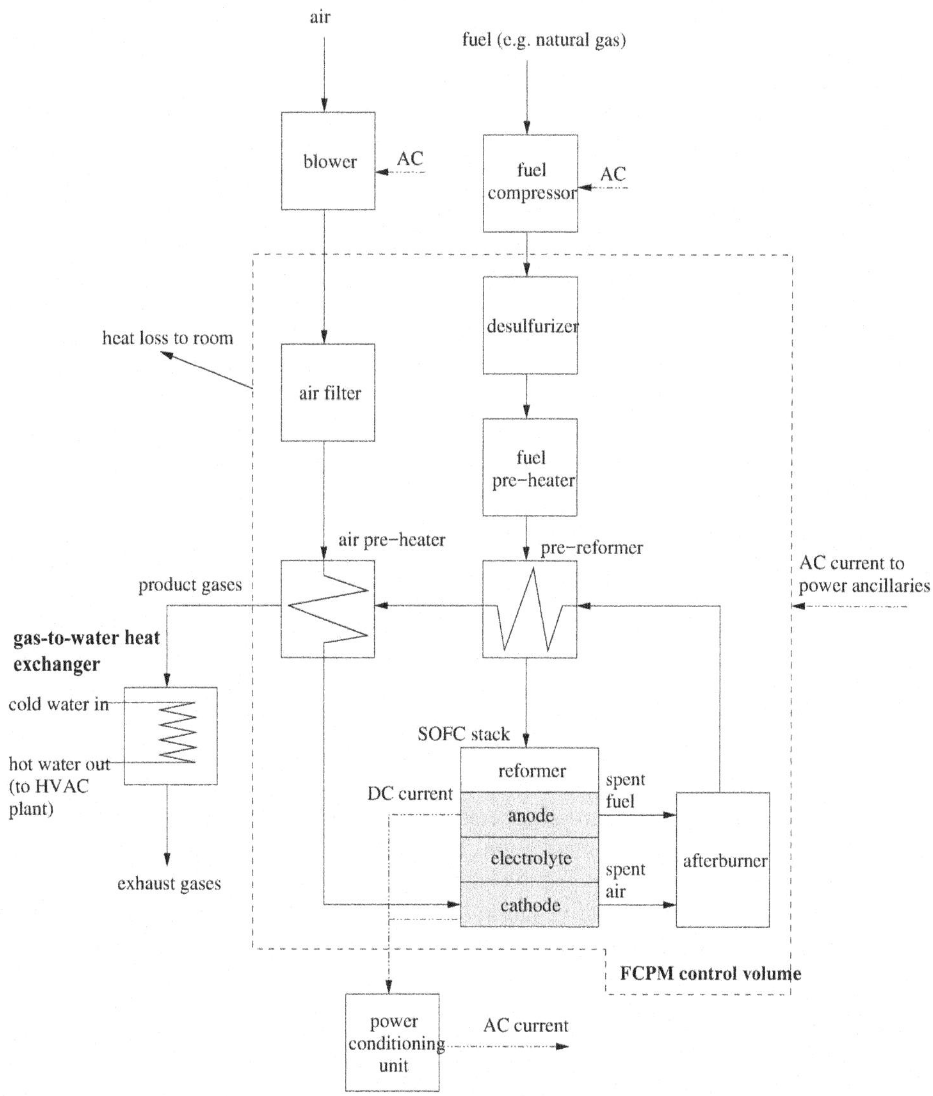

Figure 16.2 One possible system configuration of a SOFC micro-cogeneration device

and reformer; and water preparation. A compressor may be required to supply pressurized fuel while a blower will likely be present to supply air to provide oxygen to support the electrochemical and combustion reactions. A pump may also be required to supply liquid water for steam reformation purposes. A battery could be used for buffering the fuel cell stack's DC electrical production and for meeting load transients, and the system will include a power conditioning unit to convert the electrical output to AC. All SOFC microcogeneration systems will include a heat recovery device that transfers the heat of the hot product gases to the building's HVAC system. Some systems may include an integrated auxiliary burner that is activated when the fuel cell cannot satisfy the building's thermal loads. The complexities of these systems must be considered in the modelling of micro-cogeneration devices.

The computational methods that are recommended for treating micro-cogeneration devices within the context of building-performance simulation are elaborated in the next section.

Computational methods

The pragmatic modelling of micro-cogeneration devices within the context of building-performance simulation requires models at an appropriate level of resolution.

In the past it has been common to model the performance of micro-cogeneration devices using performance-map methods, wherein the device's electrical and thermal efficiencies are treated as constant or are a parametric function of the device's loading (e.g. Alanne et al. 2006; Boait et al. 2006; Cockroft and Kelly 2006; De Paepe and Merten 2007; Dorer et al. 2005; Hawkes et al. 2007; Hawkes and Leach 2005, 2007; Maalla and Kunsch 2008; Peacock and Newborough 2006; Pearce et al. 2001; Possidente et al. 2006; Sibilio et al. 2007; Sicre et al. 2005). This approach essentially decouples the modelling of the micro-cogeneration device from that of the house and other aspects of the thermodynamic system and as such precludes an accurate treatment of the thermal coupling to the building and its HVAC system, as demonstrated by Ribberink et al. (2009).

A research project (known as Annex 42) undertaken within the International Energy Agency's Energy Conservation in Buildings and Community Systems Programme (IEA/ECBCS) addressed this shortcoming by developing two new models for micro-cogeneration devices. One of these models is suitable for the combustion-based ICE and SE micro-cogeneration devices while the other treats SOFC and PEMFC devices. These are recommended for use in simulating the performance of micro-cogeneration devices while concurrently modelling building thermal and electrical demands within the context of whole-building simulation. Both models are available in the publicly released versions of ESP-r and EnergyPlus, while TRNSYS TYPE is based upon the ESP-r source code as well as an IDA-ICE version of the fuel cell model, and can be acquired upon request.

The following subsections describe the methodologies employed by the IEA/ECBCS Annex 42 models by focusing on a few key aspects. Additionally, the methods recommended for calibrating these models to represent actual micro-cogeneration devices are treated. The interested reader is referred to IEA/ECBCS Annex 42 final reports and associated journal publications for a more complete description of these models, their validation, and their calibration (Beausoleil-Morrison 2007, 2008; Beausoleil-Morrison and Ferguson 2007; Beausoleil-Morrison and Lombardi 2009; Beausoleil-Morrison et al. 2006; Ferguson et al. 2009; Kelly and Beausoleil-Morrison 2007).

Modelling fuel-cell-based devices

Topology of fuel cell micro-cogeneration model

The model topology of the IEA/ECBCS Annex 42 FC micro-cogeneration model is illustrated in Figure 16.3.

The model discretizes the FC micro-cogeneration system into groupings of components that comprise major sub-systems, such as those that produce electrical power, supply air, provide auxiliary heat, capture heat from hot product gases, convert DC to AC, etc. In this manner, once the model is calibrated for a specific FC micro-cogeneration device, analyses can be conducted to explore the benefits of improving the performance of individual sub-systems. For example, the impact of improving the DC-AC power conditioner upon overall system performance can be simulated without recalibrating the portions of the model that represent the fuel cell power module, gas-to-water heat exchanger, and other sub-systems. Additionally, such a structure facilitates the future development of more detailed modelling methods for specific sub-systems.

The model includes the 12 control volumes shown in Figure 16.3, a subset of which will be required to represent any given FC micro-cogeneration device. For example, the treatment of an internally reforming SOFC micro-cogeneration system that operates on standard natural gas delivery pressures would not require the inclusion of the control volumes that represent the fuel compressor and pump. Likewise, as an SOFC does not require stack cooling, the control volumes representing the pump, air cooler, and water-water heat exchanger on the stack cooler would be neglected. The control volumes representing the auxiliary burner, dilution air system, and battery may or may not be required depending upon the system configuration.

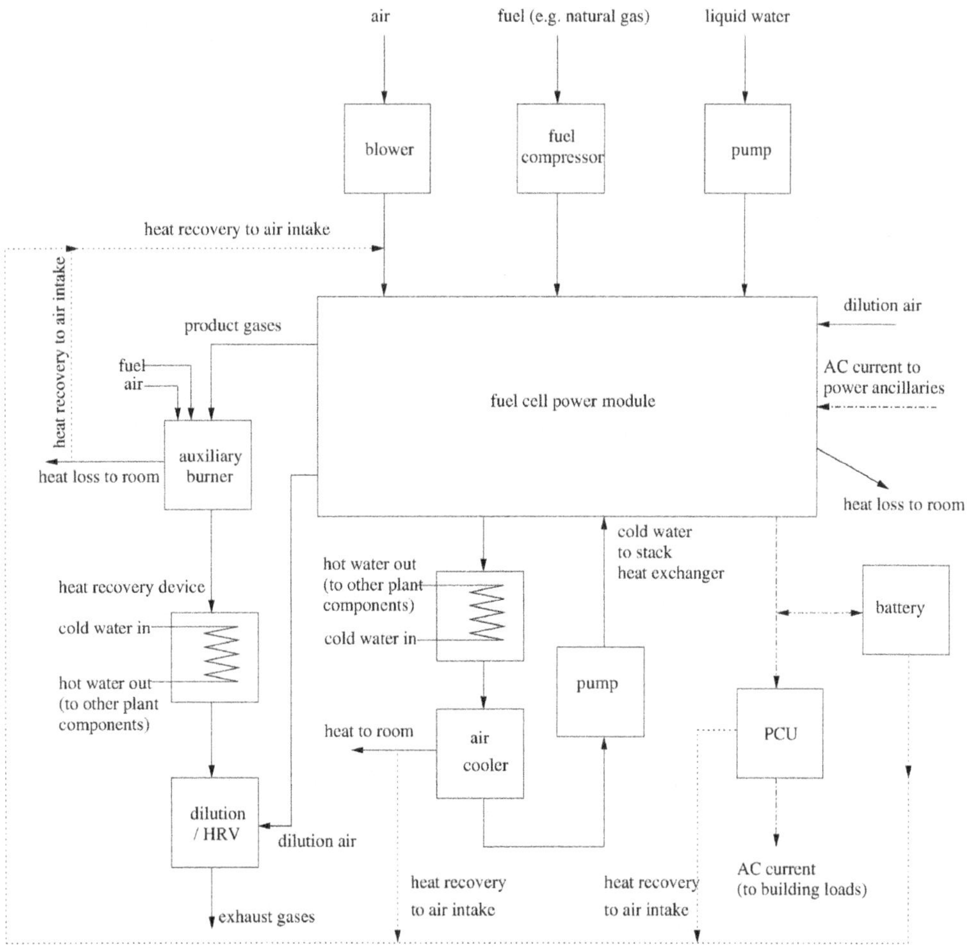

Figure 16.3 Topology of FC micro-cogeneration model

Each of the 12 control volumes is modelled in as rigorous a fashion as possible given the constraints of computational efficiency and the need to calibrate model inputs based upon the testing of coherent systems. It is worth noting that the equations described in this section will be recalculated hundreds of thousands of times to perform a single annual simulation at two minute time-steps. (Such short time-steps are typically required for explicitly simulating the performance of plant systems.)

The following subsections illustrate the modelling methods employed by focusing on two control volumes. One represents the exhaust-gas-to-water heat exchanger that produces the useful thermal output. The other control volume represents the fuel cell power module (FCPM) which includes the stack, after-burner, and the other components enclosed by the dashed line in Figure 16.2.

Exhaust-gas-to-water heat exchanger

The gas-to-water heat exchanger that transfers energy from the hot gases exiting the FCPM and/or the auxiliary burner is shown on the left side of Figure 16.3. A schematic representation of the control volume encapsulating this heat exchanger is shown in Figure 16.4. The state point labels shown in the figure are used in the development that follows.

The sensible component of the heat transfer from the hot gases to the water is characterized with the log mean temperature difference (LMTD) method for counterflow heat exchangers:

$$q_{sensible} = (UA)_{eff} \cdot \frac{(T_{g-in} - T_{w-out}) - (T_{g-out} - T_{w-in})}{ln \dfrac{T_{g-in} - T_{w-out}}{T_{g-out} - T_{w-in}}} \tag{16.1}$$

where $T_{g\text{-}in}$ and $T_{g\text{-}out}$ are the temperatures of the gases entering and exiting the heat exchanger and $T_{w\text{-}in}$ and $T_{w\text{-}out}$ are the entering and exiting water temperatures. $(UA)_{eff}$ is the effective product of the heat exchanger's heat transfer coefficient and area (W/K).

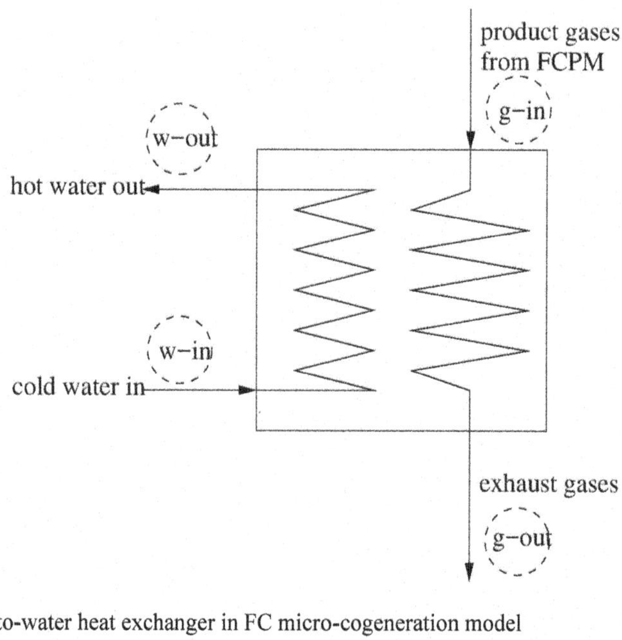

Figure 16.4 Gas-to-water heat exchanger in FC micro-cogeneration model

If it is assumed that heat loss from the heat exchanger to the ambient is negligible and that the heat capacity of each fluid stream remains constant through the heat exchanger, then the following energy balance can be written for the heat transfer between the fluid streams:

$$q_{sensible} = (\dot{N}\hat{c}_P)_g \cdot [T_{g-in} - T_{g-out}] = (\dot{N}\hat{c}_P)_w \cdot [T_{w-out} - T_{w-in}] \tag{16.2}$$

where $\hat{c}_p$ is the fluid's molar heat capacity (J/kmolK) and is its molar flow rate.

The rearrangement of Equations (16.1) and (16.2) allows the solution of the temperature of the gas exiting the heat exchanger based upon the entering gas and water temperatures:

$$T_{g-out} = \left\{ \frac{1 - \dfrac{(\dot{N}\hat{c}_P)_g}{(\dot{N}\hat{c}_P)_w}}{exp\left[(UA)_{eff} \cdot \left(\dfrac{1}{(\dot{N}\hat{c}_P)_g} - \dfrac{1}{(\dot{N}\hat{c}_P)_w}\right)\right] - \dfrac{(\dot{N}\hat{c}_P)_g}{(\dot{N}\hat{c}_P)_w}} \right\} \cdot T_{g-in} + \left\{ \frac{exp\left[(UA)_{eff} \cdot \left(\dfrac{1}{(\dot{N}\hat{c}_P)_g} - \dfrac{1}{(\dot{N}\hat{c}_P)_w}\right)\right] - 1}{exp\left[(UA)_{eff} \cdot \left(\dfrac{1}{(\dot{N}\hat{c}_P)_g} - \dfrac{1}{(\dot{N}\hat{c}_P)_w}\right)\right] - \dfrac{(\dot{N}\hat{c}_P)_g}{(\dot{N}\hat{c}_P)_w}} \right\} \cdot T_{w-in} \tag{16.3}$$

Once T_{g-out} is solved with Equation 16.3, the sensible heat transfer from the gas to the water can be determined with Equation 16.2.

The water flow rate through the heat exchanger can vary in time, depending upon the arrangement and control of HVAC components that are coupled to the FC micro-cogeneration device. The flow rate and composition of product gases through the heat exchanger will vary in response to the flow rate of fuel and air supplied to the FC stack as well as the operation of the auxiliary burner (refer to Figure 16.3). These changing conditions required that the $(UA)_{eff}$ value required by Equation 16.3 be evaluated at each time-step of the simulation. An approach is employed which casts it as a parametric relation of the water and product gas flow rates:

$$(UA)_{eff} = hx_{s,0} + hx_{s,1} \cdot \dot{N}w + hx_{s,2} \cdot \dot{N}_w^2 + hx_{s,3} \cdot \dot{N}_g + hx_{s,4} \cdot \dot{N}_g^2 \tag{16.4}$$

where $hx_{s,i}$ are empirical coefficients characterizing a specific heat exchanger. The form of Equation 16.4 facilitates the determination of the $hx_{s,i}$ coefficients from experimental data, as will be shown in the section "Model calibration."

The interested reader is referred to Beausoleil-Morrison et al. (2006) and Kelly and Beausoleil-Morrison (2007) for a description of the methods employed to treat latent heat transfer as well as alternate methods to the one outlined here.

Fuel cell power module

The energy balance for the FCPM (shown in the centre of Figure 16.3) can be written in the following form:

$$\begin{aligned} &\dot{H}_{fuel} + \dot{H}_{air} + \dot{H}_{liq-water} + P_{el-ancillaries-AC} \\ &= P_{el} + \dot{H}_{FCPM-gas-out} + q_{stack-cooling} + q_{skin-loss} + q_{FCPM-to-dilution} \end{aligned} \tag{16.5}$$

where $\dot{H}_{fuel}$ and $\dot{H}_{air}$ represent the enthalpy carried into the control volume by the fuel and air required to support electrochemical and combustion reactions (including excess air). Similarly, $\dot{H}_{liq-water}$ represents the enthalpy of the liquid water that might be supplied for fuel reformation purposes. The last term on the left side of the equation represents the power draw of ancillaries that are included within the control volume and that are powered by AC electricity that is supplied to the FC micro-cogeneration device. All terms in Equation 16.5 are in terms of Watts.

P_{el} is the net DC power production, that is the stack power less ohmic losses in cabling and the power draw of DC-powered ancillaries included within the control volume. $\dot{H}_{FCPM-gas-out}$ represents the enthalpy carried out of the control volume by the exiting gas stream. These are the hot products of the electrochemical and combustion reactions as well as the excess air and the inert constituents of the fuel.

$q_{stack-cooling}$ represents the heat transfer from the FCPM to the stack cooler, as would be present in a PEMFC. The final two terms in Equation 16.5 represent thermal losses: $q_{skin-loss}$ is the radiant and convective heat transfer from the skin of the FC micro-cogeneration device to the containing room while $q_{FCPM-to-dilution}$ represents the heat transfer from the FCPM to the air stream, which might be drawn through the micro-cogeneration device's cabinet to limit skin temperatures and for gas venting purposes.

The enthalpy flow rates in Equation 16.5 represent summations of the enthalpies of constituent gases, e.g.:

$$\dot{H}_{air} = \sum_i (\dot{N}_i \hat{h}_i)_{air} \tag{16.6}$$

where $\hat{h}_i$ is the molar enthalpy (J/kmol) and $\dot{N}_i$ is the molar flow rate (kmol/s) of gas constituent i (e.g. N_2, O_2).

Standardized enthalpies are used since chemical reactions are occurring within the FCPM. This ensures that the enthalpy of each reactant or product is properly related to the enthalpies of other elements and compounds by using a standard reference state. By convention, the standard state is taken to be 25 °C and 1 atmosphere pressure and the enthalpies of all elemental substances (e.g. O_2, H_2) are taken to be zero at the standard state.

The enthalpy of each reactant or product gas can be expressed as a sum of its enthalpy at the standard state (i.e. the standard enthalpy of formation) and the deviation between its enthalpy and that at the standard state. Furthermore, the definition of the specific heat can be introduced to provide the following relationship,

$$\begin{aligned} \hat{h}_i &= \Delta_f \hat{h}_i^o + \left[\hat{h}_i - \Delta_f \hat{h}_i^o \right] \\ &= \Delta_f \hat{h}_i^o + \int_{T^o}^{T} \hat{c}_p dT \end{aligned} \tag{16.7}$$

where $\Delta_f \hat{h}_i^o$ is the i standard molar enthalpy of formation (J/kmol) and $\hat{c}_p$ is the molar specific heat (J/kmolK) for gas i.

The fuel's lower heating value (LHV) can be conveniently introduced into the energy balance represented by Equation 16.5 if it is assumed that the reactions of the fuel constituents are complete, as given by the following stoichiometric reaction:

$$C_x H_y + \left(x + \frac{y}{4} \right) \cdot O_2 \rightarrow x \cdot CO_2 + \frac{y}{2} \cdot H_2O \tag{16.8}$$

The LHV (J/kmol) of a fuel is expressed using the standard enthalpies of formation of the reactants and products:

$$LHV_{fuel} = \frac{\Delta_f H^o_{fuel} + \Delta_f H^o_{O_2} - \Delta_f H^o_{CO_2} - \Delta_f H^o_{H_2O}}{\dot{N}_{fuel}} = \frac{\Delta_f H^o_{fuel} - \Delta_f H^o_{CO_2} - \Delta_f H^o_{H_2O}}{\dot{N}_{fuel}} \quad (16.9)$$

where $\Delta_f H^o_{fuel}$ is the total flow rate of the standard enthalpy of formation of the fuel entering the FCPM control volume and $\Delta_f H^o_{CO_2}$ and $\Delta_f H^o_{H_2O}$ are the total flow rates of the enthalpies of formation of the product gases created by the complete reaction of the fuel (all in W). $\dot{N}_{fuel}$ is the molar flow rate of the fuel (kmol/s).

The FCPM's efficiency at converting the energy released through the chemical reaction of the fuel to electrical energy is expressed as follows:

$$\varepsilon_{el} = \frac{P_{el}}{\dot{N}_{fuel} \cdot LHV_{fuel}} \quad (16.10)$$

where ε_{el} is the electrical conversion efficiency (-).

The electrical conversion efficiency can be introduced into the energy balance by substituting Equations 16.6, 16.7, 16.9, and 16.10 into Equation 16.5:

$$\begin{aligned} &\sum_i \left(\dot{N}_i \textstyle\int_{T^o}^{T} \hat{c}_p dT\right)_{fuel} + \sum_i \left(\dot{N}_i \textstyle\int_{T^o}^{T} \hat{c}_p dT\right)_{air} + \sum_i \left(\dot{N} \textstyle\int_{T^o}^{T} \hat{c}_p dT\right)_{liq-water} \\ &- \dot{N}_{liq-water} \cdot \hat{h}^o_{fg} + P_{el-ancillaries-AC} + (1 - \varepsilon_{el}) \cdot \dot{N}_{fuel} \cdot LHV_{fuel} \\ &= \sum_i \left(\dot{N}_i \textstyle\int_{T^o}^{T} \hat{c}_p dT\right)_{FCPM-gas-out} + q_{stack-cooling} + q_{skin-loss} + q_{FCPM-to-dilution} \end{aligned} \quad (16.11)$$

where $\hat{h}^o_{fg}$ is the latent heat of vapourization of water at the standard state (J/kmol).

This term represents the energy required to vapourize the liquid water which is introduced for fuel reformation.

Once the flow rates and temperatures of the fuel, air, and liquid water streams that are introduced into the control volume are defined the first three terms of Equation 16.11 can be calculated. These temperatures are established through the solution of energy balances that represent the fan, fuel compressor, and water pump that are represented in Figure 16.3.

Empirical relations are used to establish a number of terms in Equation 16.11. For example, the air supply rate is cast as a parametric function of the electrical production and the entering air temperature:

$$\dot{N}_{air} = (a_0 + a_1 \cdot P_{el} + a_2 \cdot P_{el}^2) \cdot (1 + a_3 \cdot T_{air}) \quad (16.12)$$

where T_{air} is the temperature of the air (°C) supplied to the FCPM.

The form of Equation 16.12 facilitates the determination of the a_i coefficients from experimental data, as will be shown in the section "Model calibration."

Another example of an empirical relation that is used to establish the terms of Equation (16.11) is that for the electrical conversion efficiency:

$$\varepsilon_{el} = \left[\varepsilon_0 + \varepsilon_1 \cdot P_{el} + \varepsilon_2 \cdot P_{el}^2\right] \cdot \left[1 - N_{stops} \cdot D\right] \cdot \left[1 - L \cdot MAX\left|\textstyle\int_0 dt - t_{threshold}, 0\right|\right] \tag{16.13}$$

The $[1 - N_{stops} \cdot D]$ term in Equation (16.13) represents the degradation of the efficiency as a result of start-stop cycling while the $[1 - L \cdot MAX\,|\int_0 dt - t_{threshold}, 0|]$ term represents operational degradation.

The ε_{el} terms of Equation (16.13) along with D and L are empirical constants that characterize the performance of a given FC micro-cogeneration device. The methods that can be used to establish these values are treated in section "Model calibration."

Techniques similar to those elaborated above for the air supply and electrical conversion efficiency are employed to establish the individual terms in Equation 16.11 each time-step of the simulation. This equation is then rearranged and solved each time-step of the simulation to predict $T_{FCPM-gas-out}$, the temperature of the hot gases exiting the FCPM and thus entering the subsequent control volume (see Figure 16.3). Similar methods are used to treat the other control volumes represented in the figure.

Modelling combustion-based devices

The FC micro-cogeneration model described in the section titled "Modelling fuel-cell-based devices" employed a topology wherein the system was discretized into sub-systems (see Figure 16.3). As indicated in the sections "Exhaust-gas-to-water heat exchanger" and "Fuel cell power module," with this approach state points at the couplings between the subsystems are solved. As will be shown in section "Model calibration," calibrating this type of model necessitates data that can only be gathered through invasive measurements.

A more idealized representation is recommended for combustion-based micro-cogeneration devices for the pragmatic reason that invasive measurements are more difficult to effect in these technologies. Additionally, unlike their fuel cell counterparts, combustion-based micro-cogeneration devices tend to be configured to thermal load follow. In this mode of operation, the devices will spend significant amounts of time in warming up and cooling down modes. Consequently, a representation that considers the thermal mass of the device is required in order to accurately treat these transient situations.

The IEA/ECBCS Annex 42 combustion micro-cogeneration model is based upon three abstract control volumes. One control volume represents the conversion of internal energy to electrical and thermal energy, while the other two control volumes represent the thermal mass of the engine and its cooling water.

The micro-cogeneration device's electrical conversion efficiency is given by:

$$\varepsilon_{el} = \frac{P_{el-net-AC}}{\dot{N}_{fuel} \cdot LHV_{fuel}} \tag{16.14}$$

where $P_{el-net-AC}$ is the net AC electrical production from the micro-cogeneration device (W). As with the FC micro-cogeneration model, the denominator represents the product of the fuel flow rate (kmol/s) and the fuel's LHV (J/kmol).

When in steady-state operation (i.e. following completion of the warm-up cycle), the electrical conversion efficiency is determined using a parametric relation of the following form (higher-order terms are not shown here for the sake of clarity):

$$\begin{aligned}\varepsilon_{el} &= \varepsilon_0 + \varepsilon_1 \cdot (\dot{N}_{fuel} \cdot LHV_{fuel}) + \varepsilon_2 \cdot (\dot{N}_{fuel} \cdot LHV_{fuel})^2 \\ &+ \varepsilon_3 \cdot (\dot{m}_w) + \varepsilon_4 \cdot (\dot{m}_w)^2 + \varepsilon_5 \cdot (T_{win-in}) + \varepsilon_6 \cdot (T_{w-in})^2\end{aligned} \tag{16.15}$$

where $\dot{m}_w$ is the mass flow rate (kg/s) of the water stream transporting the device's useful thermal output and T_{w-in} is the temperature of that water at the inlet to the device.

An adjustment to the efficiency calculated with Equation 16.15 is made during the warm-up period to account for the variation of fuel flow and power production as the engine temperature rises to its steady-state value. During the cool-down period, it is assumed that there is no electrical production but rather that electricity is consumed for controls and auxiliary systems.

The useful thermal output is calculated as follows:

$$\begin{aligned}q_{thermal} &= (\dot{m}c_P)_w \cdot (T_{w-out} - T_{w-in}) \\ &= \eta_{th} \cdot \dot{N}_{fuel} \cdot LHV_{fuel} - (UA)_{losses} \cdot (T_{engine} - T_{room}) \\ &- (Mc_P)_{engine} \frac{dT_{engine}}{dt} - (Mc_P)_{HX} \frac{dT_{w-out}}{dt}\end{aligned} \tag{16.16}$$

where $q_{thermal}$ is the useful thermal output from the device (W). $(\dot{m}c_P)_w$ represents the product of the flow rate (kg/s) and heat capacity (J/kgK) of the water stream transporting the device's thermal output and T_{w-in} and T_{w-out} represent the temperatures of this fluid at the inlet and exit of the device.

The first term in the second line of Equation 16.16 represents the gross conversion of internal energy to thermal energy in the device. As with the electrical conversion efficiency, η_{th} is expressed as a parametric equation (again, the higher-order terms are not shown here for the sake of clarity):

$$\begin{aligned}\eta_{th} &= \eta_0 + \eta_1 \cdot (\dot{N}_{fuel} \cdot LHV_{fuel}) + \eta_2 \cdot (\dot{N}_{fuel} \cdot LHV_{fuel})^2 \\ &+ \eta_3 \cdot (\dot{m}_w) + \eta_4 \cdot (\dot{m}_w)^2 + \eta_5 \cdot (T_{w-in}) + \eta_6 \cdot (T_{w-in})^2\end{aligned} \tag{16.17}$$

Some of this gross thermal energy liberation is lost due to radiant and convective heat transfer from the skin of the micro-cogeneration device to the containing room. This is represented by the second term in the second line of Equation 16.16.

The final two terms of Equation 16.16 represent the transient energy storage of the micro-cogeneration device. It was found that two separate thermal storage terms were required to correctly represent the thermal storage of both the engine and the cooling water contained within the engine's heat exchanger.

The form of Equation 16.16 is such that it can calculate the thermal output during steady-state operation as well as during the warm-up and cool-down phases.

The following section will treat the methods that can be employed to calibrate the model by establishing the values of the coefficients in Equations 16.15, 16.16, and 16.17. The interested reader is referred to Ferguson et al. (2009) and Beausoleil-Morrison and Ferguson (2007) for a complete description of the IEA/ECBCS Annex 42 combustion micro-cogeneration model.

Model calibration

The sections "Modelling fuel-cell-based devices" and "Modelling combustion-based devices" described the recommended methodologies for simulating the performance of fuel cell and combustion-based micro-cogeneration devices within the context of building-performance simulation. As was described, these models rely upon empirical coefficients that are unique to specific micro-cogeneration devices (see Equations 16.4, 16.12, 16.13, 16.15, and 16.17). The process of establishing these coefficients (i.e. inputs) for a particular device is known as model calibration.

Ideally, a series of experiments are conducted on a particular micro-cogeneration device under controlled conditions in order to gather the necessary data to calibrate the models. Beausoleil-Morrison and Lombardi (2009) describe the experimental program that was designed and conducted to calibrate the FC model to represent the performance of a prototype 2.8 kW_{AC} SOFC micro-cogeneration device. In these experiments, for example, the values of $\dot{N}_w$ and $\dot{N}_g$ in Equation 16.4 were varied and measurements were made on the performance of the gas-to-water heat exchanger. The measured temperatures and flow rates were used to assemble a set of derived $(UA)_{eff}$ values using Equations 16.1 and 16.2. A non-linear regression was then performed to determine the values of the $hx_{s,i}$ coefficients that produced the best fit between the controlled $\dot{N}_w$ and $\dot{N}_g$ and the derived $(UA)_{eff}$ values.

Figure 16.5 compares the $(UA)_{eff}$ values determined with Equation 16.4 and the calibrated coefficients with the $(UA)_{eff}$ values derived from the measurements. The error bars represent the uncertainty of the values derived from the measurements at the 95% confidence level. The left side of the figure contrasts the calibration with variations in $\dot{N}_w$ while the right side contrasts the calibration with variations in $\dot{N}_g$.

Similar experiments must be conducted to gather the necessary data to calibrate each aspect of the FC model, such as Equations 16.12 and 16.13. Although powerful and accurate, this calibration approach necessitates invasive measurements. For example, the above-mentioned calibration of the gas-to-water heat exchanger required measuring the temperatures and flow rates of the water and gas streams at the heat exchanger inlets and outlets.

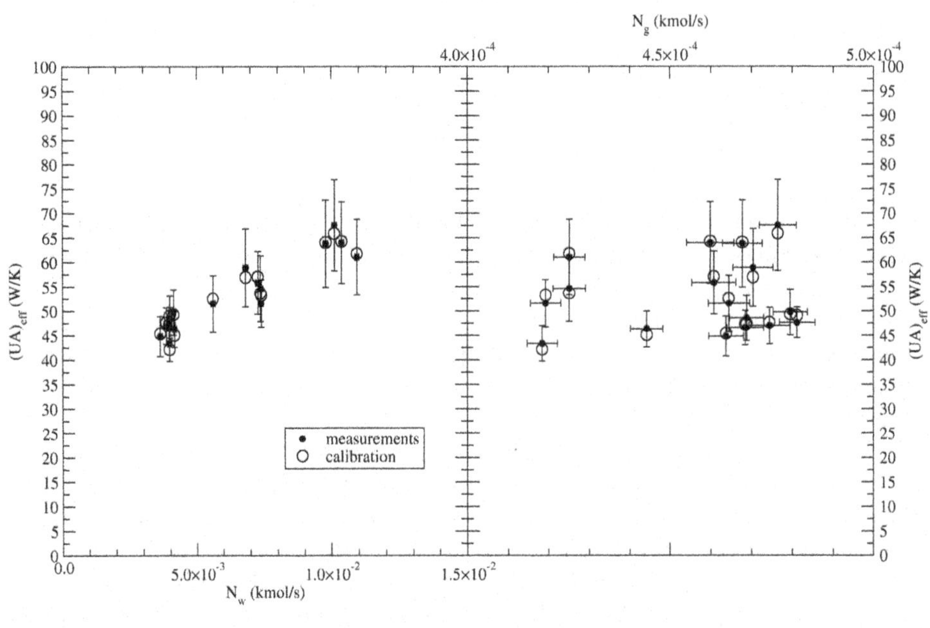

Figure 16.5 Calibration of Equation 16.4 using experimental data

In contrast, the design of the combustion-based model averts the need for invasive measurements. Rather, measurements taken external to the micro-cogeneration device – the electrical output, fuel supply rate, and the temperature and flow rate of water entering and exiting the heat exchanger – are sufficient to calibrate Equations 16.15 and 16.17. Ideally, a series of experiments are conducted on a particular micro-cogeneration device under controlled conditions in order to gather the necessary calibration data. However, as demonstrated by Ferguson et al. (2009), it is possible to calibrate the model using less complete data with a dynamic parameter estimation method.

The interested reader is referred to Beausoleil-Morrison (2007), which provides the full set of calibration parameters for prototype SOFC, SE, and ICE micro-cogeneration devices. Johnson et al. (2013) also provides calibration parameters for a 1 kW_{AC} PEMFC micro-cogeneration device, while various other calibration efforts are documented in Darcovich et al. (2014).

Application example

The introduction to this chapter provided the motivation for modelling the behaviour of micro-cogeneration devices within the context of building-performance simulation. This section demonstrates how the models described in this chapter can be applied to concurrently simulate the performance of micro-cogeneration devices, supporting HVAC components and the building's thermal and electrical demands. This application focuses upon a representative detached house served by a hypothetical SOFC micro-cogeneration system. ESP-r is used for the purposes of this demonstration, but as explained earlier the same modelling capabilities are available in other simulation programs.

Figure 16.6 illustrates how the hypothetical SOFC micro-cogeneration device is coupled to the house's convective heating system in this example. The thermal output of the SOFC micro-cogeneration device is transferred to a water storage tank. The pump that circulates water between the tank and the SOFC runs continuously and draws 20 W of electrical power. When

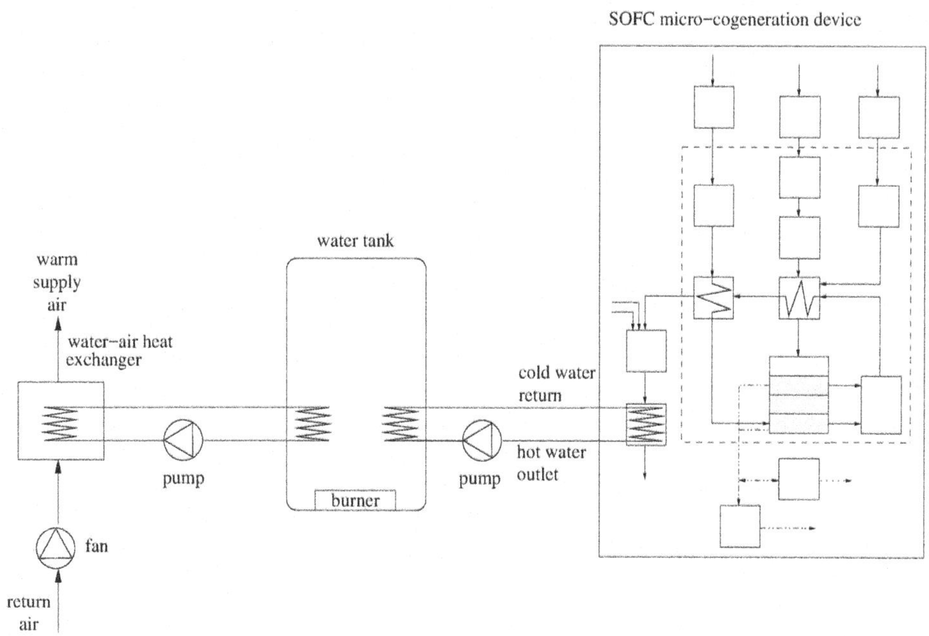

Figure 16.6 Configuration of SOFC micro-cogeneration heating plant

the house's temperature drops below 25.0 °C, the second pump cycles on to circulate hot water from the tank through a water-to-air heat exchanger. Likewise, the fan cycles on to circulate return air through the heat exchanger and to deliver this warmed air to heat the house. When operating, this pump and fan draw 20 W and 50 W of electrical power, respectively.

The tank's make-up burner is assumed to have an efficiency of 95% (higher heating value [HHV] basis, which corresponds to the best-available technology condensing system) and cycles on when then tank temperature drops below 50 °C. The tank has a volume of 1000 L and is well insulated. To protect the tank from excessive thermal input, energy is extracted from the tank and rejected to the environment when its temperature rises to 92 °C. This fictitious representation of a heat dump accurately accounts for the need to periodically reject excessive thermal production to the environment but neglects the electrical draw of ancillary devices that would be required to provide this functionality.

Rather than conducting this simulation using the calibrated parameters for the prototype 2.8 kW_{AC} SOFC micro-cogeneration device treated in Beausoleil-Morrison and Lombardi (2009), for this example the model was configured to represent a hypothetical device. This hypothetical device has the same electrical conversion efficiency as the tested prototype. Similarly, the prototype's measured data were used to calibrate many of the model's terms that were previously described, such as the air supply rate, heat exchanger, and transient response. However, unlike the tested prototype, this hypothetical device is given the capability of modulating its output between 500 W and 2.5 kW (net DC production). As well, it is assumed to have a constant DC-AC power conditioning efficiency of 95% and to have only 100 W of convective and radiative heat losses.

Initially, the SOFC micro-cogeneration device is assumed to operate continuously throughout the year producing 1 kW of AC power. It is assumed that the house can export power to the central electrical grid when there is a surplus of power and import from the grid when there is a deficit. A simulation was conducted for a typical house using Ottawa (Canada) weather data. The non-HVAC electrical demands imposed by the occupants were assumed to follow one of the demand patterns recommended by Armstrong et al. (2009).

The monthly integrated electrical balance for this initial simulation is illustrated in Figure 16.7. Over the year, the SOFC micro-cogeneration device generates 31.5 GJ of electrical energy (net AC) while the occupants and HVAC equipment consume 30.5 GJ. Considerable grid interaction is required due to temporal mismatches between production and demand: over the year, 11.7 GJ of electrical energy are exported to the grid, while 10.6 GJ are imported. As shown in the figure, there is little seasonal variation in these observations.

The monthly integrated thermal energy balance on the water storage tank (see Figure 16.6) for this initial simulation is illustrated in Figure 16.8. As can be seen, during the winter months the SOFC micro-cogeneration device's thermal output is insufficient to meet the house's space heating demand. Consequently, extensive use of the tank's make-up burner is required. The make-up burner is also required in the spring and fall. The impact of temporal mismatches between the SOFC micro-cogeneration device's thermal production and the building's space heating demand can be seen in April and October. During each of these two months, the space heating demand is greater than the SOFC micro-cogeneration device's thermal production. Notwithstanding, some of this thermal production must be dumped due to overheating of the tank during some periods. And during the summer months, the vast majority of the micro-cogeneration device's thermal output is wasted through the heat-dump facility.

This initial simulation predicts that, over the year, 65.8% of the fuel's LHV is converted to useful electrical and thermal output, as defined by:

$$\eta_{overall} = \frac{E_{netAC} + E_{gross\ thermal} - E_{tank\ skin\ losses} - E_{dump}}{N_{fuel} \cdot LHV} \tag{16.18}$$

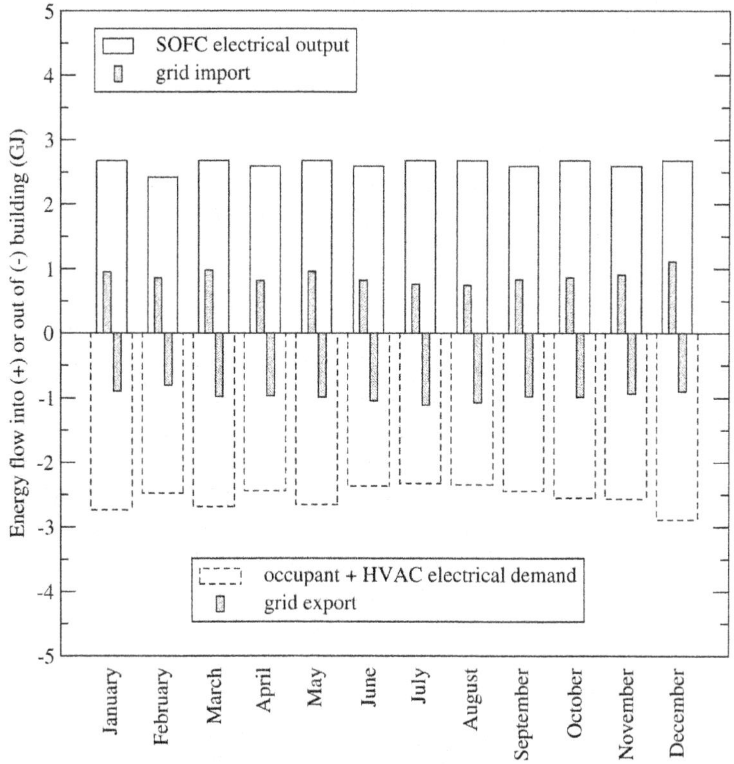

Figure 16.7 Monthly integrated electrical balance when SOFC produces a constant 1 kW_{AC} output

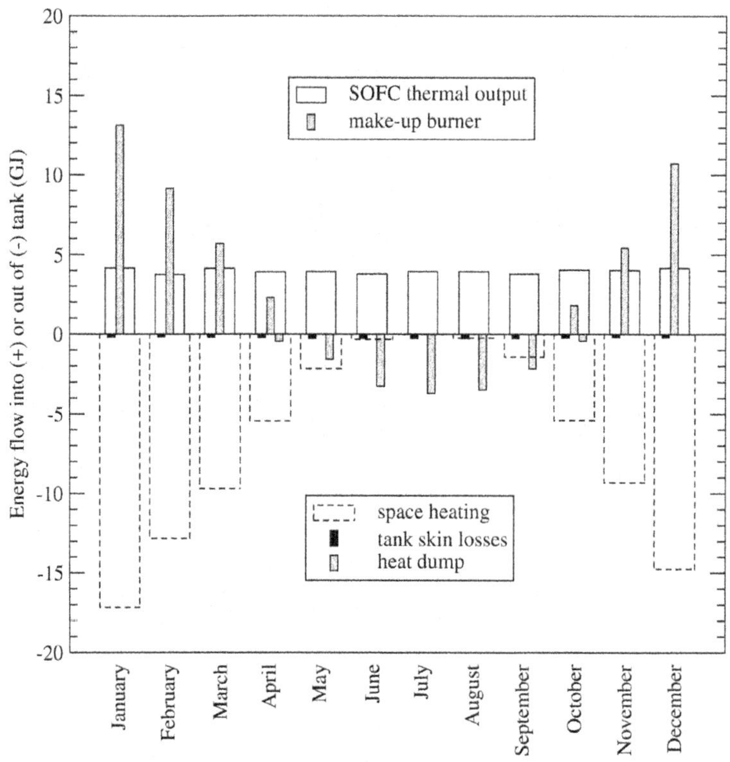

Figure 16.8 Monthly integrated thermal balance on tank when SOFC produces a constant 1 kW_{AC} output

where $\eta_{overall}$ is the overall efficiency (LHV basis), N_{fuel} is the fuel consumption over the year (kmol), and LHV is the fuel's lower heating value. E_{netAC} is the SOFC micro-cogeneration device's net AC electrical production and $E_{gross\ thermal}$ is its gross thermal output, both in GJ. $E_{tank\ skin\ losses}$ represents the thermal losses from the tank to the containing room, while E_{dump} is the amount of energy that must be dumped from the tank due to overheating, both in GJ.

To demonstrate how the modelling methods outlined in this chapter can be used to study the impact of dispatch strategy upon system operation, two additional simulations were performed with this example. In one case, rather than producing a constant 1 kW of electrical output, the SOFC micro-cogeneration device was made to follow the electrical demand of the occupants and HVAC equipment. However, its load following capabilities were constrained to the hypothetical device's modulating range of 500 W to 2.5 kW (net DC). As well, measured data from the tested prototype were used to impose limits on how quickly the device could modulate its electrical output in response to load variations. In the second case, the SOFC micro-cogeneration device was made to follow the house's demand for space heating.

Figure 16.9 contrasts the annually integrated performance of these three dispatch strategies. In this figure, the useful thermal output is taken to be the gross output less tank skin losses and the amount of energy that must be dumped due to tank overheating:

$$E_{useful\ thermal\ output} = E_{gross\ thermal} - E_{tank\ skin\ losses} - E_{dump} \tag{16.19}$$

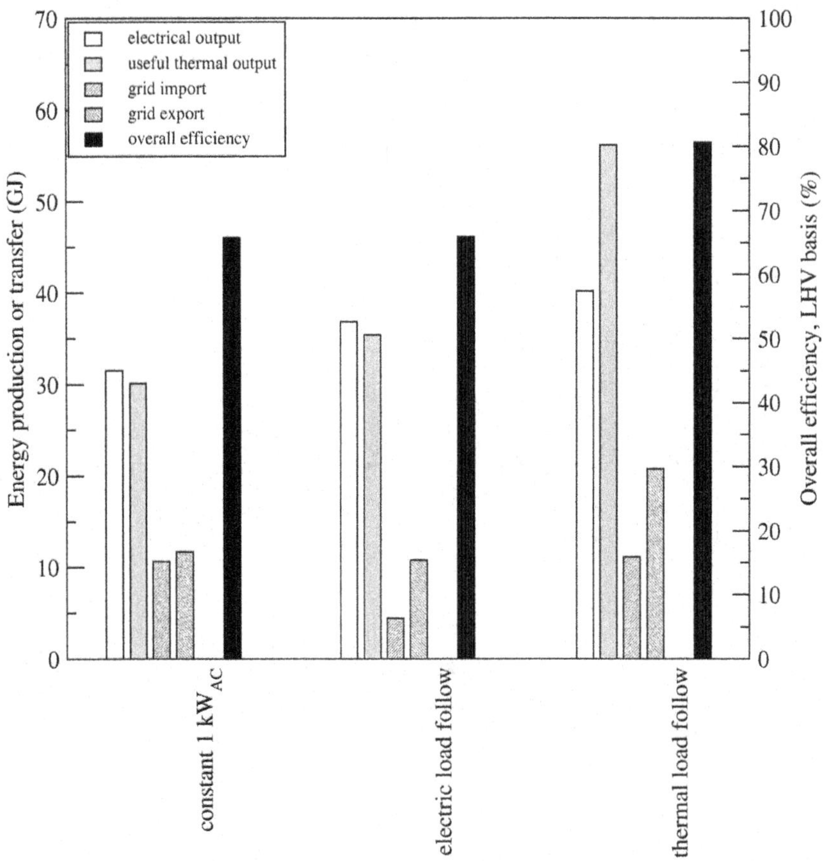

Figure 16.9 Comparison of annual results for three dispatch strategies

As can be seen, both alternate dispatch strategies result in greater electrical and useful thermal production from the SOFC micro-cogeneration device. The thermal load following scenario requires the greatest reliance upon the electrical grid, whereas grid interaction is minimized with electrical load following, as expected. The thermal load following strategy results is a substantially higher overall system efficiency (see Equation 16.18): 80% versus 65%–66%.

This section demonstrates only a single application of the micro-cogeneration device modelling capabilities treated in this chapter. These methods can be used to examine various system configurations and dispatch strategies and to assess the performance of specific or hypothetical devices in various climates and buildings and with various occupancy scenarios.

Discussion

The accurate treatment of micro-cogeneration devices requires that the user configure building simulation programs to treat building thermal and electrical demands concurrently with electrical and thermal energy conversion systems. This demands the user to operate across multiple modelling domains and that the interactions between the modelling domains be considered.

The greatest challenge in modelling the electrical and thermal performance of these technologies is in acquiring sufficient data to calibrate the recommended models. These technologies are only emerging onto the market so few detailed performance data exist. Notwithstanding, this chapter has enumerated the existing data sets and has treated the methods that are recommended for model calibration.

The assignments that follow allow the reader to develop the necessary knowledge and skills to treat these nascent technologies within building-performance assessments.

Synopsis

This chapter has provided an introduction to micro-cogeneration technologies based upon fuel cells, internal combustion engines, and Stirling engines for the combined production of heat and electricity within buildings. These emerging technologies have the potential to deliver energy services more efficiently while reducing environmental impacts. However, due to the complex performance characteristics, these technologies can only be accurately assessed by building performance simulation programs that treat building thermal performance concurrently with electrical and HVAC systems.

This chapter has outlined the recommended methods for modelling micro-cogeneration devices as well as techniques for calibrating these models to represent the performance of specific devices using measured data.

References

Adam, A., Fraga, E.S. and Brett, D.J. (2015) Options for residential building services design using fuel cell based micro-CHP and the potential for heat integration. *Applied Energy* 138: 685–694.

Alanne, K., Saari, A., Ugursal, V. and Good, J. (2006) The financial viability of an SOFC cogeneration system in single-family dwellings. *Journal of Power Sources* 158: 403–416.

Armstrong, M., Swinton, M., Ribberink, H., Beausoleil-Morrison, I. and Millette, J. (2009) Synthetically derived profiles for representing occupant-driven electric loads in Canadian housing. *Journal of Building Performance Simulation* 2(1): 15–30.

Beausoleil-Morrison, I. (ed) (2007) *Experimental Investigations of Residential Cogeneration Devices and Model Calibration*. IEA/ECBCS Annex 42 Report. ISBN No. 978-0-662-47523-1.

Beausoleil-Morrison, I. (ed) (2008) *An Experimental and Simulation-Based Investigation of the Performance of Small-Scale Fuel Cell and Combustion-Based Cogeneration Devices Serving Residential Buildings*. IEA/ECBCS Annex 42 Report. ISBN No. 978-0-662-47923-9.

Beausoleil-Morrison, I. and Ferguson, A. eds) (2007) *Inter-Model Comparative Testing and Empirical Validation of Annex 42 Models for Residential Cogeneration Devices*. IEA/ECBCS Annex 42 Report. ISBN No. 978-0-662-47562-0.

Beausoleil-Morrison, I. and Lombardi, K. (2009) The calibration of a model for simulating the thermal and electrical performance of a 2.8 kw_{AC} solid-oxide fuel-cell micro-cogeneration device. *Journal of Power Sources* 186(1): 67–79.

Beausoleil-Morrison, I., Schatz, A. and Maréchal, F. (2006) A model for simulating the thermal and electrical production of small-scale solid-oxide fuel cell co-generation systems within building simulation programs. *Journal of HVAC&R Research Special Issue* 12(3a): 641–667.

Boait, P., Rylatt, R. and Stokes, M. (2006) Optimisation of consumer benefits from micro-combined heat and power. *Energy and Buildings* 38: 981–987.

Carbon Trust. (2007) *Micro-CHP Accelerator: Interim Report*. Technical report, Carbon Trust, UK. CTC726.

Cockroft, J. and Kelly, N. (2006) A comparative assessment of future heat and power sources for the UK domestic sector. *Energy Conversion and Management* 47: 2349–2360.

Colpier, U.C. and Cornland, D. (2002) The economics of the combined cycle gas turbine – an experience curve analysis. *Energy Policy* 30(4): 309–316.

Darcovich, K., Entchev, E. and Tzscheutschler, P. eds) (2014) *Current Updates on the Development and Implementation of Micro-Cogeneration System Models for Building Simulation Programs*. IEA/ECBCS Annex 54 Report.

De Paepe, M. and Mertens, D. (2007) Combined heat and power in a liberalised energy market. *Energy Conversion and Management* 48: 2542–2555.

DeMoss, T. (1996) They're he-e-re (almost): The 60% efficient combined cycle. *Power Engineering* 100(7): 17–21.

Dorer, V., Weber, R. and Weber, A. (2005) Performance assessment of fuel cell micro-cogeneration systems for residential buildings. *Energy and Buildings* 37: 1132–1146.

Entchev, E., Gusdorf, J., Swinton, M., Bell, M., Szadkowski, F., Kalbfleisch, W. and Marchand, R. (2004) Micro-generation technology assessment for housing technology. *Energy and Buildings* 36(9): 925–931.

Ferguson, A., Kelly, N., Weber, A. and Griffith, B. (2009) Modelling residential-scale combustion-based cogeneration in building simulation. *Journal of Building Performance Simulation* 2(1): 1–14.

Finkelstein, T. and Organ, A. (2001) *Air Engines*. New York: ASME Press. ISBN No. 0-7918-0171-3.

Gandiglio, M., Lanzini, A., Santarelli, M. and Leone, P. (2014) Design and optimization of a proton exchange membrane fuel cell CHP system for residential use. *Energy and Buildings* 69: 381–393.

Hawkes, A., Aguiar, P., Croxford, B., Leach, M., Adjiman, C. and Brandon, N. (2007) Solid oxide fuel cell micro combined heat and power system operating strategy: Options for provision of residential space and water heating. *Journal of Power Sources* 164: 260–271.

Hawkes, A., Entchev, E. and Tzscheutschler, P. (eds) (2014) *A Comparative Review of Microgeneration Policy Instruments in OECD Countries*. IEA/ECBCS Annex 54 Report.

Hawkes, A. and Leach, M. (2005) Solid oxide fuel cell systems for residential micro-combined heat and power in the UK: Key economic drivers. *Journal of Power Sources* 149: 72–83.

Hawkes, A. and Leach, M. (2007) Cost-effective operating strategy for residential micro-combined heat and power. *Energy* 32: 711–723.

Henning, H.-M. (2007) Solar assisted air conditioning of buildings – an overview. *Applied Thermal Engineering* 27(10): 1734–1749.

Johnson, G., Beausoleil-Morrison, I., Strathearn, B., Thorsteinson, E. and Mackintosh, T. (2013) The calibration and validation of a model for simulating the thermal and electrical performance of a 1 kw_{AC} proton-exchange membrane fuel-cell micro-cogeneration device. *Journal of Power Sources* 221: 435–446.

Kelly, N. and Beausoleil-Morrison, I. eds) (2007) *Specifications for Modelling Fuel Cell and Combustion-Based Residential Cogeneration Devices within Whole-Building Simulation Programs*. IEA/ECBCS Annex 42 Report. ISBN No. 978-0-662-47116-5.

Knight, I. and Ugursal, V. (eds) (2005) *Residential Cogeneration Systems: A Review of the Current Technologies*. IEA/ECBCS Annex 42 Report. ISBN No. 0-662-40482-3.

Larminie, J. and Dicks, A. (2003) *Fuel Cells Explained*, 2nd edition. New York: John Wiley & Sons. ISBN No. 978-0-470-84857-9.

Maalla, E.M.B. and Kunsch, P.L. (2008) Simulation of micro-CHP diffusion by means of system dynamics. *Energy Policy* 36(7): 2308–2319.

Mench, M. (2008) *Fuel Cell Engines*. New York: John Wiley & Sons. ISBN No. 978-0-471-68958-4.

Napoli, R., Gandiglio, M., Lanzini, A. and Santarelli, M. (2015) Techno-economic analysis of PEMFC and SOFC micro-CHP fuel cell systems for the residential sector. *Energy and Buildings* 103: 131–146.

Onovwiona, H. and Ugursal, V. (2006) Residential cogeneration systems: Review of the current technology. *Renewable and Sustainable Energy Reviews* 10(5): 389–431.

Peacock, A. and Newborough, M. (2006) Impact of micro-combined heat-and- power systems on energy flows in the UK electricity supply industry. *Energy* 31: 1804–1818.

Pearce, J., Zahawi, B.A. and Shuttleworth, R. (2001) Electricity generation in the home: Modelling of single-house domestic combined heat and power. In *Proceeding of Science, Measurement and Technology*, volume 148. Institute of Engineering and Technology, pp. 197–203. ISSN 1350–2344.

Possidente, R., Roselli, C., Sasso, M. and Sibilio, S. (2006) Experimental analysis of micro-cogeneration units based on reciprocating internal combustion engine. *Energy and Buildings* 38: 1417–1422.

Ribberink, H., Bourgeois, D. and Beausoleil-Morrison, I. (2009) A plausible forecast of the energy and emissions performance of mature-technology Stirling engine residential cogeneration systems in Canada. *Journal of Building Performance Simulation* 2(1): 47–61.

Sibilio, S., Sasso, M., Possidente, R. and Roselli, C. (2007) Assessment of micro-cogeneration potential for domestic trigeneration. *International Journal of Environmental Technology and Management* 7(1–2): 147–164.

Sicre, B., Bühring, A. and Platzer, B. (2005) Energy and financial performance of micro-CHP in connection with high-performance buildings. In *Proceeding of Building Simulation 2005*. Montréal, Canada, pp. 1131–1138.

Recommended reading

Beausoleil-Morrison, I. (ed.) (2008) An experimental and simulation-based investigation of the performance of small-scale fuel cell and combustion-based cogeneration devices serving residential buildings. IEA/ECBCS Annex 42 Report. ISBN No. 978-0-662-47923-9.

Assignments

Assignment 1

In this assignment you will develop familiarity with modelling micro-cogeneration devices by incorporating an FC micro-cogeneration device into a building and HVAC system model. A simple idealized arrangement has been selected to allow efforts to be focused upon adding the micro-cogeneration component to a model and configuring the model using supplied data.

Use any of the simulation programs that have implemented the IEA/ECBCS Annex 42 FC micro-cogeneration model (see the section "Computational methods"). In this assignment, the building is inconsequential and a simple HVAC system is utilized in which a pump circulates a fixed flow rate of constant-temperature water through the micro-cogeneration device. The model's stack heat exchanger and gas-to-water heat exchanger should be configured in a cascading arrangement, as illustrated in Figure 16.10. Configure the micro-cogeneration model as follows:

- The device is able to modulate its electrical output at the rate of 100 W/s such that the FCPM can produce from 500 W to 2.5 kW of DC power. However, in this simulation a constant 1 kW of net AC power is produced by the micro-cogeneration device.
- The electrical conversion efficiency (see Equations 16.10 and 16.13) is a constant 35%.

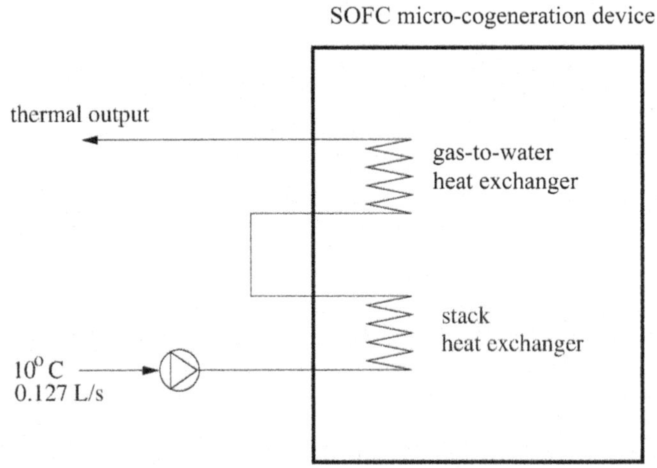

Figure 16.10 HVAC configuration for Assignment 1

- There is no battery and the efficiency of the DC to AC power conditioner is 100%.
- The blower, fuel compressor, and water pump do not draw any power.
- The fuel is 100% methane and the air is 79% nitrogen and 21% oxygen.
- Air is supplied to the FCPM at a constant rate of $1 \cdot 10^{-4}$ kmol/s (see Equation 16.12).
- Air and fuel are supplied to the micro-cogeneration device from the building, which is maintained at a constant 10 °C.
- The gas-to-water heat exchanger has a constant $(UA)_{eff}$ of 9 W/K.
- There are no skin losses from the FCPM and no liquid water supply, ancillary devices consume no electrical power, and there is no dilution air system or stack cooling system (refer to Equation 16.5).

Conduct a simulation for a single day using a time-step between 1 and 10 minutes. Examine the results file generated by the simulation program and plot the net electrical output and thermal output from the micro-cogeneration device as a function of time. Determine the fuel consumption rate, the rate of heat transfer from the micro-cogeneration device to the HVAC system's water stream, and the fraction of the fuel's energy (LHV basis) that is converted to useful electrical and thermal output using Equation 16.18.

Assignment 2

In this assignment you will examine how the micro-cogeneration model can be calibrated to represent the performance of a specific (hypothetical) device. This assignment is built upon Assignment 1 but drops the assumption of a constant electrical conversion efficiency. Rather, the data given in Table 16.1 are to be used to calibrate this model input using the methods outlined in the section "Model calibration." Use these data to calibrate the SOFC's electrical conversion efficiency using Equation 16.13. For the purposes of this assignment, ignore the degradation terms in this equation.

Configure the micro-cogeneration model using these calibrated parameters. All other inputs should be as in Assignment 1. Then conduct a simulation for a single day using a time-step between 1 and 10 minutes. Determine the fuel consumption rate and the rate of heat transfer from the micro-cogeneration device to the HVAC system's water stream.

Table 16.1 Electrical efficiency data for Assignment 2

P_{el} (W)	ε_{el} (–)
500	0.3025
700	0.3150
900	0.3443
1100	0.3641
1300	0.3797
1500	0.3900
1700	0.3832
1900	0.3629
2100	0.3440
2300	0.3212
2500	0.2974

Assignment 3

In the previous assignments, the micro-cogeneration device operated with a constant electrical and thermal output. In this assignment you will explore how controls can be employed in your chosen simulation platform to allow the micro-cogeneration device to respond to building loads.

Configure the micro-cogeneration model as in Assignment 2. However, rather than producing a constant 1 kW net AC output, have the micro-cogeneration device follow the building's electrical demand profile as illustrated in Figure 16.11.

All other inputs should be as in Assignment 2. Conduct a simulation for a single day using a time-step between 1 and 10 minutes. Plot the FCPM's electrical efficiency and the rate of heat transfer from the micro-cogeneration device to the HVAC system's water stream as a function of time over the day. Calculate the total net AC electrical production and thermal production over the day.

Assignment 4

The previous assignments have considered hypothetical micro-cogeneration devices. In this assignment you will configure the model to represent the performance of an actual prototype SOFC micro-cogeneration device.

Using the calibration parameters provided in Beausoleil-Morrison and Lombardi (2009), reconfigure the model of Assignment 2 to represent the performance of the tested 2.8 kW_{AC} SOFC micro-cogeneration device. Configure the model so that the micro-cogeneration device produces a constant 2.8 kW of net AC power. In the previous assignments, the temperature of the water supplied to the micro-cogeneration device's heat exchanger was 10 °C. The temperature of this water is raised to 40 °C in this assignment. All other inputs should be as in Assignment 2.

Conduct a simulation for a single day using a time-step between 1 and 10 minutes. Calculate the total fuel consumption and the thermal production and skin losses over the day.

Assignment 5

Configure a model to represent the application example given in the section "Application example" and conduct simulations to represent the three control scenarios demonstrated.

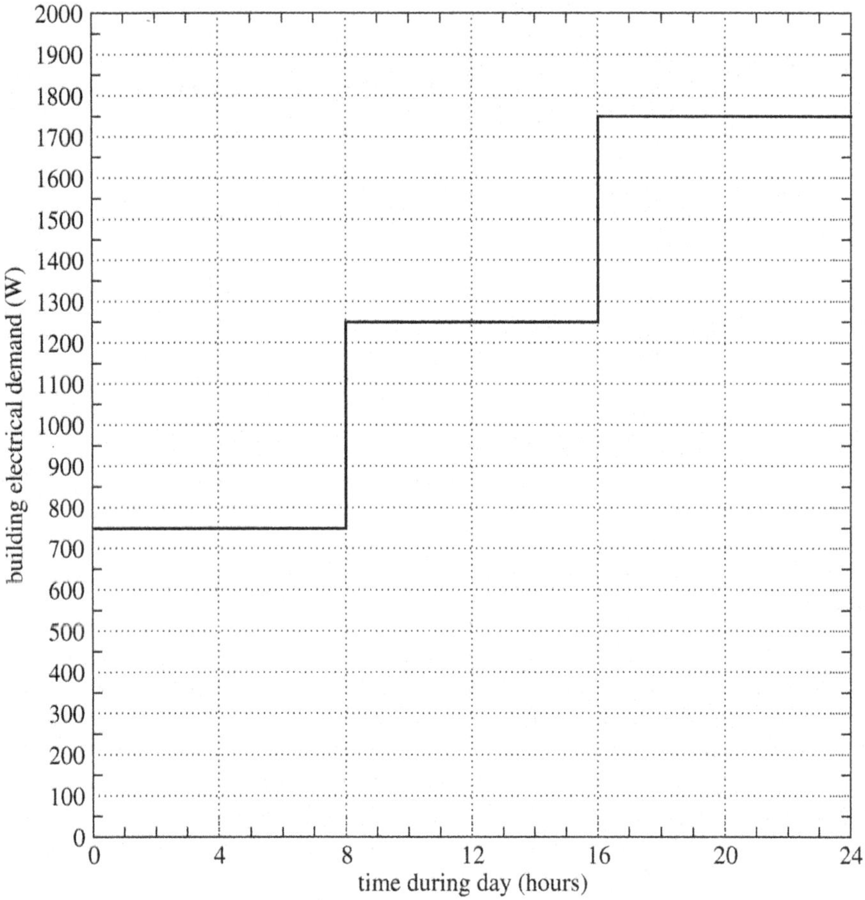

Figure 16.11 Building electrical demand for Assignment 3

Assignment 6

The previous assignments have focused upon FC micro-cogeneration systems. In this assignment you will develop familiarity with modelling a SE micro-cogeneration device.

Replace the FC in Assignment 1 with an SE. Ferguson et al. (2009) provide the calibrated model inputs for a natural-gas-fired SE micro-cogeneration that produces 699 W of net AC output under steady conditions. Use the calibrated parameters provided in this chapter to configure the model to represent this device. Furthermore, configure the device to operate at a constant electrical output of 699 W. All other inputs should be as in Assignment 1.

Conduct a simulation for a single day using a time-step between 1 and 10 minutes. Calculate the rate of useful thermal output and the rate of heat loss.

Answers

Assignment 1

- $\dot{N}_{fuel} = 3.56 \cdot 10^{-6}$ kmol/s
- $q_{sensible} = 1\ 822$ W
- $\dfrac{P_{net-AC} + q_{sensible}}{\dot{N}_{fuel} \cdot LHV_{fuel}} = 98.8\%$

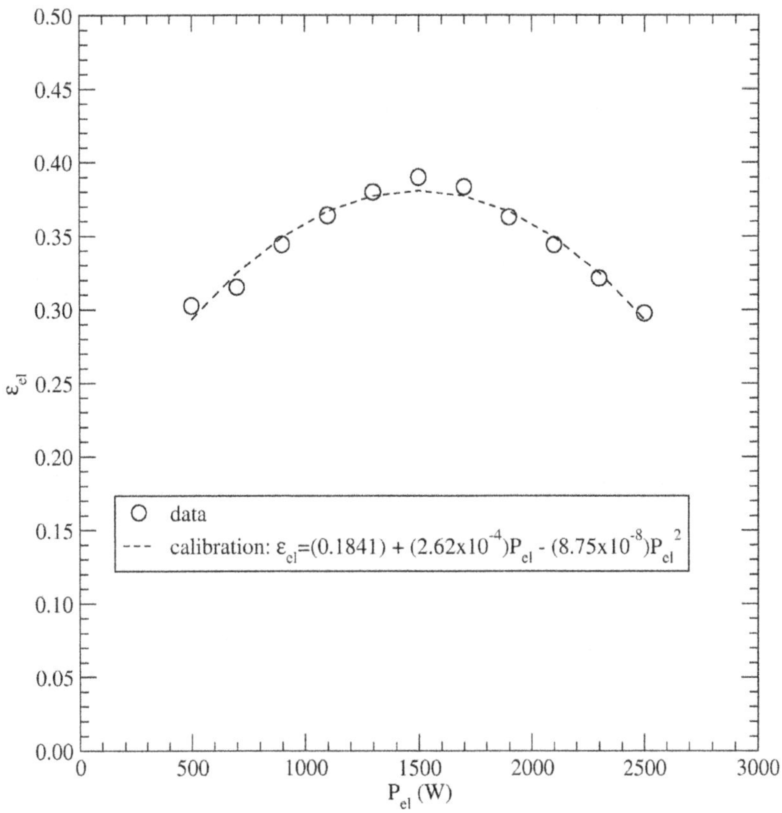

Figure 16.12 ε_{el} calibration for Assignment 2

Assignment 2

The calibration of the FCPM's electrical efficiency is illustrated in Figure 16.12. The figure's legend gives the calibration parameters for Equation 16.13.

- $\dot{N}_{fuel} = 3.48 \cdot 10^{-6}$ kmol/s
- $q_{sensible} = 1\ 754$ W

Assignment 3

Figure 16.13 plots the FCPM's electrical efficiency and the rate of heat transfer from the micro-cogeneration device to the HVAC system's water stream as a function of time over the day.

- $E_{netAC} = 108.0$ MJ
- $E_{gross\ thermal} = 184.3$ MJ

Assignment 4

- $\int_{day} \dot{N}_{fuel} = 1.14$ kmol
- $\int_{day} q_{sensible} = 296$ MJ
- $\int_{day} q_{skin-loss} = 63$ MJ

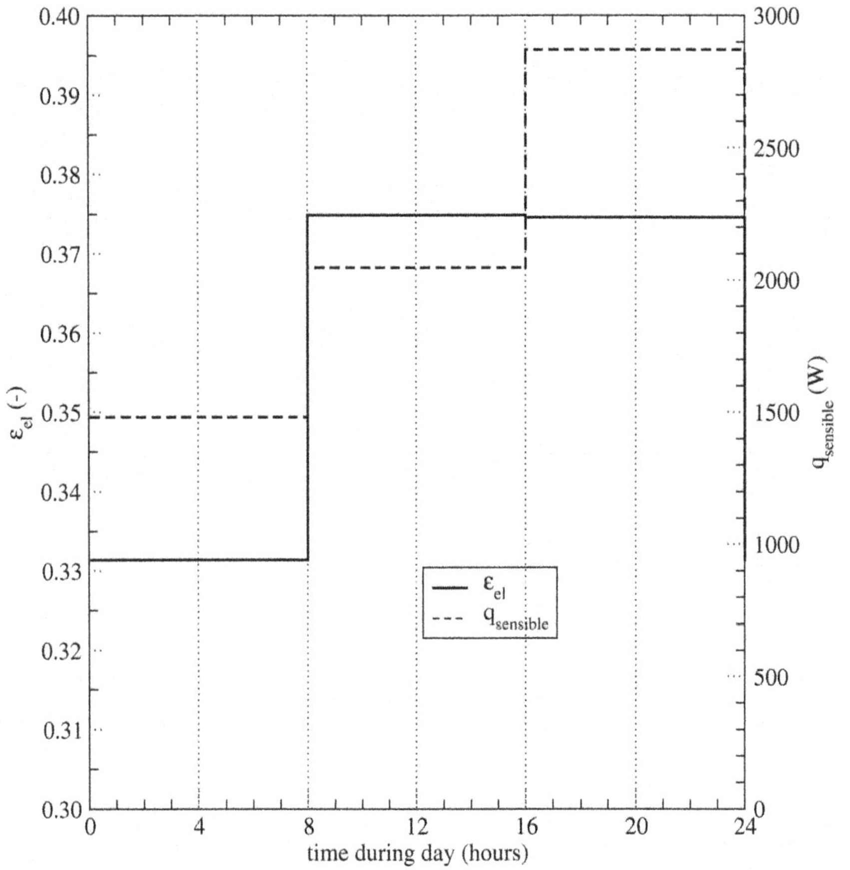

Figure 16.13 Electrical conversion efficiency and thermal output results for Assignment 3

Assignment 5

Refer to the results presented in the section "Application example."

Assignment 6

- $q_{thermal} = 6\ 321$ W
- $(UA)_{losses} \cdot (T_{engine} - T_{room}) = 977$ W

17 Modeling in building-to-grid integration

Sen Huang, Thomas Sevilla and Wangda Zuo

Scope

This chapter introduces system modeling in the context of building-to-grid integration. Relevant concepts with a focus on requirements for system modeling are provided as background information. Then, methods to address those requirements are elaborated and demonstrated with two case studies.

Learning objectives

This chapter helps readers to:

- comprehend concepts and requirements for system modeling in the building-to-grid integration;
- understand application of system modeling in the building-to-grid integration activities at present; and
- appreciate future directions in system modeling to support the building-to-grid integration.

Key words

Building-to-grid integration/distributed energy resources/demand and supply balance

Introduction

Conventional approaches to design and operate the power grid face many challenges, including negative environmental impacts (Zhu et al. 2003), susceptibility to severe weather events (American Geophysical Union 1997), and vulnerability to physical or cyber-attacks (Liang et al. 2017). Adding distributed energy resources (DERs), such as renewable energy generation and energy storage, to the power grid helps overcome the above challenges (U.S. Department of Energy 2015; North American Electric Reliability Corporation 2017). DERs not only reduce the power grid's dependence on fossil fuels but also improve the power grid's reliability and resilience (Driesen and Katiraei 2008; Johnson 2017; Maharjan et al. 2015). However, DERs also pose new challenges to the design and operation of the power grid. In the design phase, DERs increase the need for the integrated design, yet seldom considered in practice. For instance, the economic storage capacity of a behind-the-meter battery may vary with different profiles of power usage (power consumed by end-uses) (Wu et al. 2017). Therefore, it is necessary to optimize the storage capacity together with the power usage profiles instead of treating the latter as an input in the optimization on the storage capacity.

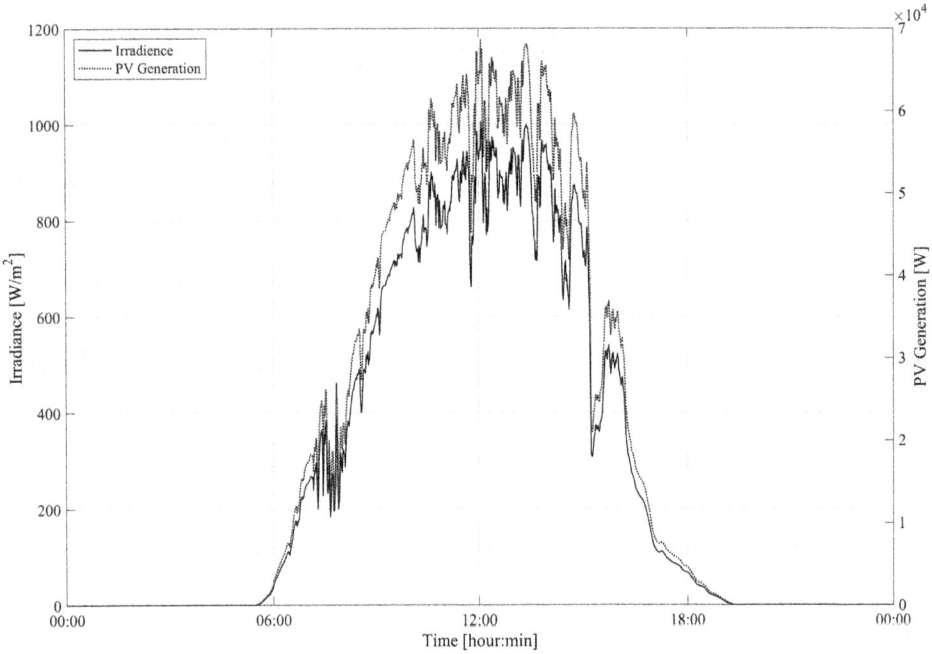

Figure 17.1 Received solar irradiance and power generation of a PV panel installed in Anna Maria Island, Florida, U.S. on June 1, 2017

In the operation phase of the power grid, DERs may mitigate some existing problems but exacerbate others. For example, driven by the unstable demand (Ji et al. 2014), balancing the power supply and demand is a complex problem even without DERs. Expanding the power grid to include DERs, especially renewable generation, increases uncertainties on the supply side. Unlike traditional power generation facilities, the renewable generation has less controllability and can highly depend on the weather conditions. As shown in Figure 17.1, the power generation from a photovoltaic (PV) panel is dominated by the received solar irradiance. Likewise, only limited ways are available to actively adjust the power output for such a DER. For instance, we may modulate the power output by changing the tilt angle of PV panels. However, the extent to which the power output can be modulated is subject to many factors, such as the time and the location of the PV panels. In addition, we notice some intermittencies in the power output in Figure 17.1. Those intermittencies are caused by random sky variations and contribute significantly to the uncertainty in the power supply from the DERs.

Building-to-grid integration

To support the large-scale adoption of DERs, one approach is to integrate the demand side into the design and operation of the power grid. Among end users in the demand side, buildings are promising candidates for being integrated in the power grid for three reasons (U.S. Department of Energy 2016): (1) Buildings are major consumers of electricity. For example, nearly 75% of the electricity use in the U.S. is attributed to buildings (U.S. Department of Energy 2016). (2) Buildings are "power-flexible." Nearly one-fourth of the U.S. building-related electricity usage can be adjusted to balance the supply and demand for the power grid (Dodrill 2011). (3) Buildings can be readily manipulated via the building energy management systems. These systems provide not only the required measurement but also implementation platforms for integrating the buildings to the power grid.

As a result, researches were conducted to integrate buildings to power grids, or building-to-grid integration. Lu et al. (2015) optimized the building design parameters (e.g. building orientation) and parameters for PV panels (e.g. the rated power output) to minimize the total operational cost considering the dynamic electricity price provided by the power grid. Hao et al. (2014) showed that it was feasible to adjust the power of the supply fan in a commercial building HVAC system to absorb the uncertainty from the renewable energy generation. In our view, building-to-grid integration employs buildings as already deployed, cost-effective solutions for passive energy storage. Specifically, during the design phase, buildings can be considered as a low-cost energy storage solution for an economic design. During the operation phase, buildings can be employed to filter uncertainties from the supply side to balance the power grid.

Nevertheless, there are many technical challenges to be addressed before large-scale applications of the building-to-grid integration. One of them is the realization of the building-to-grid integration. The building-to-grid integration can be realized by different methods but primarily via control. The realization of building-to-grid integration should be built based on a deep understanding of buildings and grids, since both buildings and grids are complicated, non-linear, dynamic systems. In addition, many constraints need to be considered simultaneously in the decision-making process. To support the realization of building-to-grid integration, system modeling can deepen the understanding on those complex systems by simulating their behaviors and interaction. The remainder of the chapter elaborates on these details.

Scientific foundation

This section elaborates the fundamental problems in the *design*, *evaluation*, and *deployment* of building control strategies for the building-to-grid integration. Those problems ultimately determine the requirement for system modeling in the building-to-grid integration.

Problems in design

There are two fundamental problems in designing building control strategies for the building-to-grid integration. The first problem is how to estimate "power flexibility" of studied objects. The power flexibility defines the extent to which the power usage of the studied objects is allowed to vary. The studied object can be a device, a building, or even a group of buildings and devices. Using one building as an example, its power flexibility can be described by

$$\Delta P_{hi} = P_{max} - P_{base}, \tag{17.1}$$

$$\Delta P_{lo} = P_{min} - P_{base}, \tag{17.2}$$

where ΔP_{hi} and ΔP_{lo} are the high and low bounds of the building's power flexibility, respectively. The P_{base} is the building's power usage under normal operation. The P_{max} and P_{min} are the maximum and minimum allowable power usages of the building, obtained by solving:

$$J = \underset{\overrightarrow{cu}}{argmax}\left(P = f\left(\overrightarrow{cu},\vec{u},\vec{x}\right)\right), \tag{17.3}$$

$$s.t.\, g\left(\overrightarrow{cu},\vec{u},\vec{x}\right) \in \Theta, \tag{17.4}$$

$$J = \underset{\overrightarrow{cu}}{argmin}\left(P = f\left(\overrightarrow{cu}, \vec{u}, \vec{x}\right)\right), \tag{17.5}$$

$$s.t.\, g\left(\overrightarrow{cu}, \vec{u}, \vec{x}\right) \in \Theta, \tag{17.6}$$

where the building's power usage, P, is determined by controllable factors $\overrightarrow{cu}$ (e.g. room temperature set points), uncontrollable factors $\vec{u}$ (e.g. weather condition), and states $\vec{x}$ (e.g. current room temperatures). As a constraint, g is a function determining the quality of the service provided by the building (e.g. maintaining the room temperature at a desired value), and Θ is a set containing all acceptable values g.

The second problem is how to achieve the desired building power usage under different constraints (e.g. comfort and cost). Feedback controls are commonly used to maintain desired values for certain variables. However, they may not guarantee the optimal values for the $\overrightarrow{cu}$ that not only meet the desired power usage, but also minimize the negative impacts, such as the extent to which the thermal comfort is sacrificed. This problem can be reformulated to minimize the negative impact ω under constraints of building service quality and desired power usage P_{ds}:

$$J = \underset{\overrightarrow{cu}}{argmin}\left(\omega\left(\overrightarrow{cu}, \vec{u}, \vec{x}\right)\right), \tag{17.7}$$

$$s.t.\, g\left(\overrightarrow{cu}, \vec{u}, \vec{x}\right) \in \Theta, \tag{17.8}$$

$$f\left(\overrightarrow{cu}, \vec{u}, \vec{x}\right) = P_{ds}. \tag{17.9}$$

Solving the first problem determines the capacity of a building to enable the building-to-grid integration beforehand. Based on this capacity, the power grid operator can set an appropriate goal for the building-to-grid integration. Solving the second problem helps decide the potential impacts of the building-to-grid integration on buildings and their occupants. The challenges in solving these two problems are the same: (1) the studied system has multi-inputs and multi-outputs; (2) there exist complex nonlinear relationships between the inputs and outputs.

System modeling can play an important role in solving the above two problems. In this context, we consider system modeling as a mathematical way of describing the relationship between inputs and outputs, based on the theoretical knowledge and/or empirical insights. When developing system models for the design of the building-to-grid integration control strategies, we should consider two aspects:

1. *System models should be able to run quickly with limited computational resources.* This is necessary for enabling real-time control via various computing platforms (from a single board computer to a high-performance workstation computer) used by building control systems.
2. *It is preferable if system models can provide derivatives of the objective functions.*

The key of the building-to-grid integration is to solve optimization problems, as defined in Equations 17.3 to 17.9. We can use derivative-based optimization algorithms if system models can provide derivatives of the objective functions. Compared to derivative-free algorithms, the derivative-based algorithms often have better performance, in terms of computational cost and finding a global optimal solution.

Problems in evaluation

Evaluation is crucial for the building-to-grid integration control strategies to avoid any adverse effects on both the buildings and the grid. A rigorous evaluation under realistic operating conditions can reduce those adverse effects. A fundamental problem herein is *how to realistically represent the operating conditions of the building system.*

On one hand, field tests are commonly used evaluation methods, mainly because they provide the operating conditions close to the real ones. For example, Su and Norford (2015) employed a real chiller plant to evaluate a controller that modulated the chiller power for the power grid. However, some limitations exist in the field test method, and the primary one is the high cost. In Su and Norford's (2015) test, the costs included not only the operational cost for the chiller plant but also the time and labor cost for setting up and performing the test. Another limitation is that the field test often doesn't provide fully arbitrary operating conditions needed for performing a complete evaluation. For example, operating conditions subject to seasonal weather conditions cannot be evaluated if the field test is conducted only for a short period of time.

On the other hand, system modeling provides a virtual testbed for the control strategies. The cost of performing simulations is often lower compared to the field test. In addition, we can set arbitrary operating conditions in models to conduct a comprehensive evaluation. However, there are two aspects to be considered when developing system models for the control strategies evaluation:

1 *System models should have sufficient fidelity according to the evaluation objective.* A faithful representation of the required operating conditions is crucial for the evaluation. Thus, we should define the model fidelity to provide necessary details and accuracy in the corresponding operating conditions. Requiring more details of system operating conditions often leads to higher fidelity models.
2 *It is preferable to have building models that can be coupled with grid models.* Since control strategies impact both the buildings and the grid, having both of them in the system modeling allows studying their dynamic interaction. To couple the building and grid models, some new contents should be considered in the building models. For example, the voltage and current should be treated as inputs when evaluating the building's response to the grid, which is not the case usually.

Problems in deployment

Buildings generally respond slowly to control actions due to their thermal and mechanical inertias. In other words, there may exist a significant delay between the time when a control action is taken and the time when a corresponding response is observed (Schachter and Mancarella 2014). For the real-world deployment, we may need to utilize control strategies which can generate control actions in advance. Based on Equations 17.3 to 17.9, in order to make control actions ahead of time, predictions of future operating conditions are indispensable. As a result, a fundamental problem is raised for the deployment of control strategies: how can future building operating conditions be predicted?

There are two typical methods to address the above problem: time series method and system modeling. The time series method indexes characteristics of a series of data points in time order, such as trends. The indexed characteristics are then recognized by methods such as regression. Afterwards, the recognized characteristics can be used for predicting the data points in the future. For example, Reikard (2009) designed a logarithmic regression with time-varying coefficients to predict the hourly solar radiation for the next hour based on the solar radiation at the current hour and that at 24 hours ago. Those time series methods assume that there are no intermittent changes in the driving forces of the operating conditions. These assumptions are acceptable for some applications, such as the prediction of the outdoor dry bulb temperature (Ren and Wright 2002). However, it may lead to significant inaccuracy in other applications, such as the short-term (minute level) prediction of solar radiation (Reikard 2009).

Compared to the time series method, system modeling usually doesn't rely on the above assumption. For example, Lurwan et al. (2014) developed a system model which associates the solar radiation on an hourly basis with day type and geography of the location. However, three aspects should be considered when developing the system models:

1 *The temporal resolution of the prediction should be determined based on the goal of the control strategies.* Take the prediction of the solar radiation as an example. If the deployed strategies aim to limit the studied system's power demand on the grid over a 15-minute interval, a minute-level prediction may be sufficient. However, a finer prediction will be needed if the deployed strategies are to balance the instantaneous power supply and demand for the grid.
2 *Stochastic models are more preferable to deterministic models.* There are always uncertainties in the driving force of building thermal load. For example, it is well known that the occupant behavior contributes significantly to the uncertainty of building operating conditions (Yan et al. 2015). This indicates that a deterministic prediction may be unrealistic. To reduce the deviation in the prediction, caused by the stochastic nature, it is preferred to use stochastic models, such as probability-based models (Huang et al. 2017).

System modeling for building-to-grid integration

This section provides a general introduction to system modeling of buildings under the building-to-grid integration context. We first summarize methods to establish the system models. We then elaborate on applications of those methods in the building-to-grid integration to cater to various needs.

Methods to establish the system models

As we discussed in the previous section, system modeling is essentially finding mathematical formulas to quantitatively describe the relationship between inputs and outputs. A key step in doing so is to determine the model structure, which is a set of candidate formulas. There are three types of model structures: black box, gray box, and white box. They are built based on different levels of understandings of the studied system, as shown in Table 17.1.

When using the black box structure, one common practice is to select the formulas that can approximate any relationship arbitrarily provided appropriate parameters. Examples of these formulas include polynomial formulas, Taylor series, and so on. When selecting the approximation formulas, the ones easy to solve are often preferred. As a result, the models with the black box model structure are computationally efficient.

Both gray box and black box structures need to determine their model parameters, usually by observing values of inputs and outputs. The process to obtain the values from the observation is called "training" while the observed values are called "training set." A common practice to conduct the training for linear model structures is to use the least square regression. For nonlinear

Table 17.1 Requirements of understandings on the studied system for different model structures

Model structure	*Levels of understandings on the studied system*	*Example for the level of understandings*
Black box	Not sufficient to derive any formulas	Inputs and outputs are known
Gray box	Only sufficient to derive formulas with unknown parameters	The change rates of outputs are proportional to the value of inputs
White box	Sufficient to derive formulas	There exist known offsets between inputs and outputs

model structures, such as artificial neural networks, more sophisticated training methods, such as the backward propagation method, are needed. Readers can find more information about the model training in the literature (Schalkoff 1997).

To sum up, there are three remarks on the above model structures:

1 The prerequisite for using the white box model structure is to have sufficient information for the studied system. As a comparison, the prerequisite for using the other two model structures is a sufficient training dataset.
2 The formulas in the black box model structure may only make sense numerically as they contain no physical meaning. Therefore, they are not able to provide the insights for some studies, e.g. obtaining the derivative information that is required for certain optimization algorithms. This limitation doesn't apply for the white box model structure. Whether this limitation exists in the gray box model structure depends on the amount of information involved in the model construction.
3 Using the white box model structure ensures that the models have high fidelity. For the other two structures, high fidelity is not guaranteed.

Under the building-to-grid integration context, different structures may be preferable depending on application needs. For instance, the white box model structure is preferable in the evaluation of the control strategies where high fidelity is required. The black box model structure is more feasible in the control-oriented model since the computational speed is critical in this case.

In addition, based on its inclusion of inherent randomness, system modeling can also be divided into two groups: deterministic and stochastic modeling. The deterministic modeling assumes that all the variables affecting the outputs are considered in the inputs. Thus, the outputs are fully determined by the inputs. In the stochastic modeling, some factors for the outputs are considered as noises with stochastic characteristics. In this case, we can only expect the probability distributions for the outputs.

Applications of system modeling

As discussed before, system modeling can be used to address the problems at different phases of building-to-grid integration, including estimating the flexibility of changing the power usage in the design, representing the operating conditions during the evaluation, and predicting the operating conditions during the deployment. Based on their application scope, existing models can be divided into two groups: (1) device-to-building modeling if the studied subjects are within the boundary of a single building; (2) building-to-community modeling if they are beyond the boundary of a single building.

Current device-to-building models are mainly built via deterministic modeling but with different model structures. One example is a model-based performance evaluation of different demand response strategies to reduce the building peak power usage (Ma et al. 2011). One of the strategies is to reset the zone temperature set point based on a predefined schedule as:

$$T_{set} = \begin{cases} 32, & t_h \geq 18 \mid t_h < 7 \\ 25, & t_h \geq 7 \;\&\; t_h < 12 \\ \dfrac{2(t_h - 12)}{3} + 21, & t_h \geq 12 \;\&\; t_h < 18,\; t_h \in [0, 23], \end{cases} \tag{17.10}$$

where T_{set} is the zone temperature set point for the studied building and t_h is the hour index for each day. Since the information shown in Equation 17.10 is sufficient to quantitatively describe the relationship between t_h and T_{set}, the authors modeled this strategy with the white box model structure in EnergyPlus (Crawley et al. 2001).

The second example utilizes a group of air conditioning units, which serve residential buildings, to provide service to the grid (Hao et al. 2017). For the studied air conditioning units, the only information considered by the authors was the change of thermal energy in the system should follow the first thermodynamic law. Based on such limited information, they developed a gray box model structure as:

$$C\frac{dT_z}{dt} = \frac{(T_a - T_z)}{R} + m\dot{q} + w, \tag{17.11}$$

where C is the thermal capacity, R is the thermal resistance, T_z is the average zone temperature, T_a is the outdoor air temperature, $\dot{q}$ is the rated cooling rate for the air conditioning unit, w is the process noise, and m is a state variable which indicates the on/off status of air conditioning unit. The parameters to be determined are C, R, and w for each air conditioning unit.

The third example is to develop a chiller power demand response strategy (Xue et al. 2015). In this case, the authors selected a black box model structure, which estimates the chiller power, P, using:

$$\frac{Q_{sup}}{P} = \sum_{i=1}^{6} d_i \left(\frac{Q_{sup}}{Q_{rat}}\right)^{i-1}, \tag{17.12}$$

where Q_{sup} is the cooling energy supplied by the chiller, Q_{rat} is the rated cooling capacity of the chiller, and d_i are performance coefficients.

On the other hand, building-to-community models mainly deal with a group of buildings. Although a group of buildings can be modeled with the device-to-building models, simulating a large number of device-to-building models can often take a substantial amount of time and/or computing resources, which may be unacceptable for the control-oriented modeling (Kalsi et al. 2012). Currently, most building-to-community models are designed to handle the thermostatically controlled loads (TCLs), such as air conditioning units and water heaters. This is due to the following reasons: (1) TCLs are major consumers of the electricity in residential buildings; (2) TCLs are relatively simple since their dynamics can be well approximated with linear models, such as Equation 17.11; (3) the power of the TCLs can be assumed to be constant.

However, there are still challenges when modeling the TCLs. For example, in real applications, TCLs tend to be heterologous, meaning the *C*, *R*, and *w* in Equation 17.11 may be different for each TCL. The diversity in the parameters makes it difficult to represent the aggregated dynamics without considering the dynamics of individual TCL. As shown in Figure 17.2(a), the simulated time evaluations (with device-to-building modeling) of two heterologous electric water heaters are different although the initial temperatures are the same. For homogeneous TCLs, it is still difficult to capture their responses since their initial temperatures are likely different (Figure 17.2(b)). We will first discuss how to address the homogeneous cases, and then elaborate how the heterologous cases can be handled.

In general, there are two methods to deal with the issue related to the initial state problem in the homogeneous case. Using the water heater as an example, the first method is to estimate the probability that the temperatures of the TCLs are at certain values with stochastic modeling while initial temperatures are treated as noises. Based on the probability of the temperatures,

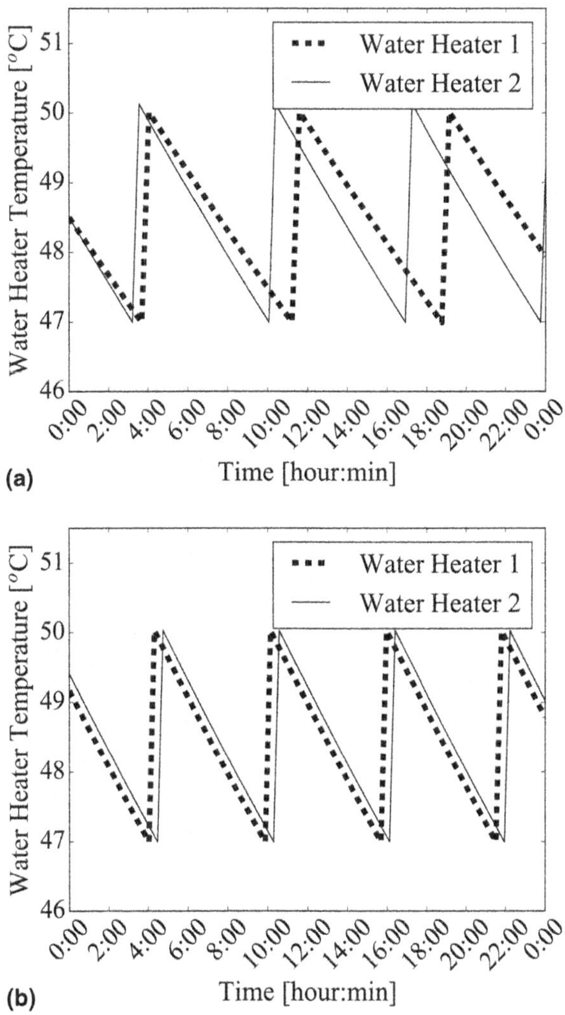

Figure 17.2 Simulated temperature evolution of two electric water heaters: (a) heterogeneous device with same initial temperatures and (b) homogeneous device with different starting temperatures

we can estimate the probability of TCLs' on/off statuses. To implement this method, researchers often adopt white box structures which use Fokker-Planck equations (Risken 1996). The Fokker-Planck equation describes the time evaluation of the probability density function of the studied variable under the influence of random forces, as shown in Figure 17.3. For example, Callaway (2009) derived a Fokker-Planck equation to describe the time evolution of the probability distribution of the operating statuses of 10,000 TCLs.

The second method is to develop a linear state-space model which describes the evolution of the state change for the studied population with deterministic modeling (Widén et al. 2012). Depending on the mechanisms of the state-space model generation, it can be further categorized as two groups. The first group aims to derive the state-space model with a white box structure and the second group uses black box structures to obtain the state change.

An example of the state-space model with a white box structure is the research of Kalsi et al. (2012). While fixing the ambient condition, they modeled the steady states of 1,000 TCL (air conditioners) in three steps: (1) they derived an analytical description regarding the changing rate of TCL's temperature; (2) based on the analytical description obtained from the previous step, they obtained a ratio of time when a TCL is on during a duty cycle, φ; and (3) they created a state-space representation of TCLs operation as

$$\frac{d\vec{n}}{dt} = \psi\vec{n}, \tag{17.13}$$

$$\vec{n} = \begin{bmatrix} n_{off}^{us} \\ n_{off}^{s} \\ n_{on}^{us} \\ n_{on}^{s} \end{bmatrix}, \tag{17.14}$$

$$\psi = \begin{bmatrix} -\varphi & \varphi & 0 & 0 \\ 0 & -\varphi & 0 & 1-\varphi \\ \varphi & 0 & \varphi-1 & 0 \\ 0 & 0 & 1-\varphi & \varphi-1 \end{bmatrix}, \tag{17.15}$$

where n is the number of TCLs under different states, s means that the temperature requirement of the TCLs is satisfied, while us means unsatisfied. The *off* and *on* are the states of TCLs. The

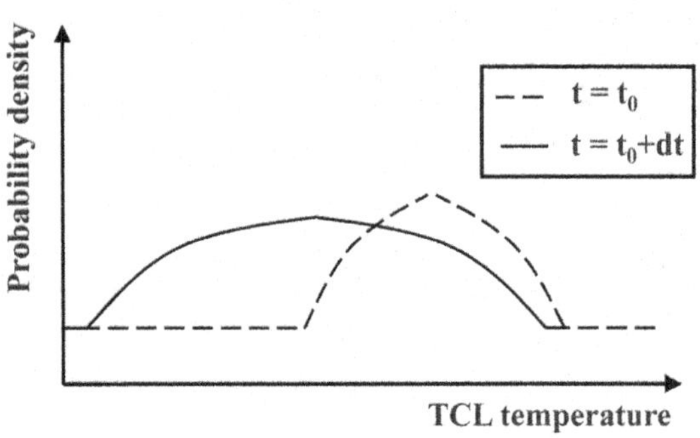

Figure 17.3 Time evaluation of the probability density function of the temperature of TCLs under the influence of random forces

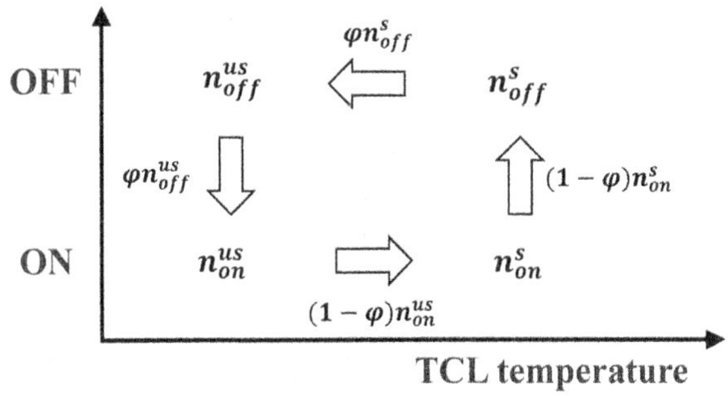

Figure 17.4 The state transition diagram of TCL operation

above formulas can be visualized in Figure 17.4. The numbers of TCLs under different states are treated as state variables, e.g. n^{us}_{off} and n^{us}_{on}. The change of state variables is determined by the current states as well as the duty cycle. Take the n^{us}_{on} as an example, there are always $(1-\varphi)n^{us}_{on}$ TCLs leaving the state while φn^{us}_{off} TCLs are entering the state. Based on the above information, we can estimate the number of running TCLs at any given time.

Using n^{us}_{on} and n^{s}_{on}, the aggregated power was calculated as

$$P_{aaa} = \left(n^{us}_{on} + n^{s}_{on}\right)P_0. \tag{17.16}$$

An example of the black box model structure is Mathieu et al.'s (2012) study. They proposed

$$P_{aaa}(t) = P_0\vec{x}(t), \tag{17.17}$$

$$\vec{x}(t+1) = A\vec{x}(t) + B\vec{u}(t), \tag{17.18}$$

where $\vec{x}(t)$ is the state vector of the TCLs, $\vec{u}$ contains the operating conditions of TCLs, and A and B are the Markov transition matrices. To determine A and B, they generated the training set by simulating each TCL with Equation 17.11.

Regarding the heterologous TCL population, a common practice is to divide the heterologous TCLs into several clusters using the k-mean algorithm (Lattin et al. 2003), then represent each cluster as a homogeneous TCL population. As a result, we convert a heterologous TCL population into a set of homogenous populations and use the methods above to handle each homogenous population.

Case studies

In this section, we demonstrate how to apply the system modeling to the building-to-grid integration with two case studies. The first case study (Hao et al. 2016) evaluates a transactive control strategy that realizes demand response with a commercial building. Transactive control is a distributed control strategy to provide demand response by engaging various self-interested DERs. The second case study explores the possibilities of mitigating a duck curve problem with batteries in a residential neighborhood. The duck curve problem is caused by the imbalance between the peak demand and renewable energy generation (Miller et al. 2014). Both case studies will answer these questions: (1) what the application problems are; (2) how to solve the problems using system modeling; (3) how system modeling is implemented.

Case study 1: transitive control of an HVAC system

Problem description

The studied HVAC system serves an office building, located in Washington, U.S. It consists of one chiller, one air-handling unit (AHU), ducts, and 17 variable air volume (VAV) terminals that provide conditioned air to offices and open spaces. The AHU is composed of one supply fan and one cooling coil. The ducts are equipped with two dampers to adjust the amount of exhaust air and fresh air, respectively.

The objective is to reduce the peak electric power demand of the HVAC system and maximize the overall thermal comfort level of the zones during the cooling season. Assuming the power of the pump that serves the chiller is constant during operation, the objective can then be achieved by solving the following optimization problem,

$$J = \underset{\vec{\dot{q}}}{argmax}\left(\sum_{i=1}^{17} U^i\right), \tag{17.19}$$

$$\text{s.t. } P_{ch} + P_{fan} \leq P_{lim}, \tag{17.20}$$

where $\vec{\dot{q}}$ is a set of cooling energy, $\dot{q}^i$, provided to the ith VAV terminal, and U^i expresses the satisfaction of the occupants with the $\dot{q}^i$. The P_{ch} and P_{fan} are the power of the chiller and fan, respectively. The P_{lim} is the total power limit for the chiller and fan.

Methodology

The above optimization problem can be solved in a distributed manner with a transactive control strategy (Hao et al. 2016). The transactive control uses market mechanisms to engage self-interested responsive devices to achieve power balance in the power grid. As shown in Figure 17.5, the transactive control strategy creates two virtual markets: An *Air* market with

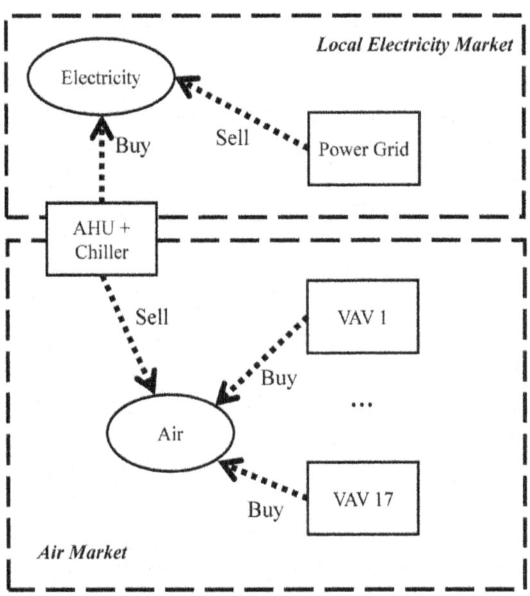

Figure 17.5 The transactive control strategy

the commodity being the conditioned air and a *Local Electricity* market with the commodity being the electricity. In the *Air* market, each VAV terminal is the buyer while the AHU is the seller. In the *Local Electricity* market, the buyer is the chiller/AHU while the seller is the power suppliers, e.g. the power grid. Since there is only one AHU and one chiller in this study, they are treated as a whole (*AHU+Chiller*) to simplify the problem. The participants of the two markets, such as VAV terminal and *AHU+Chiller*, are named as *agents*.

In each market, agents are self-interested, i.e. they attempt to maximize their revenue from selling certain amount of resources at a high price or minimize the cost for buying resources at a low price. As shown in Figure 17.6, each buyer agent expresses the amount of resource it is willing to purchase under different prices with a curve, named "demand curve." Similarly, the seller agents create "supply curves" to show the amount of resources they will sell for a given price. Then the market determines an equilibrium price, named "clear price," which balances the supply and demand. After that, based on the clear price, each agent determines the amount of resource it will buy or sell such that a Pareto optimality can be achieved. At this stage, the demand and supply in the market will be balanced and the market is considered to be cleared.

The transactive control strategy can be implemented in a five-step procedure.

Step 1: create the demand curve for buyers in the Air market

As shown in Figure 17.7, the demand curve is designed as a piecewise curve and characterized by the four variables: p^-, p^+, q^+, and q^-. The slope between p^- and p^+ determines how sensitive the buyer agents are to the change of the price, i.e. the steeper the slope, the buyer agent is

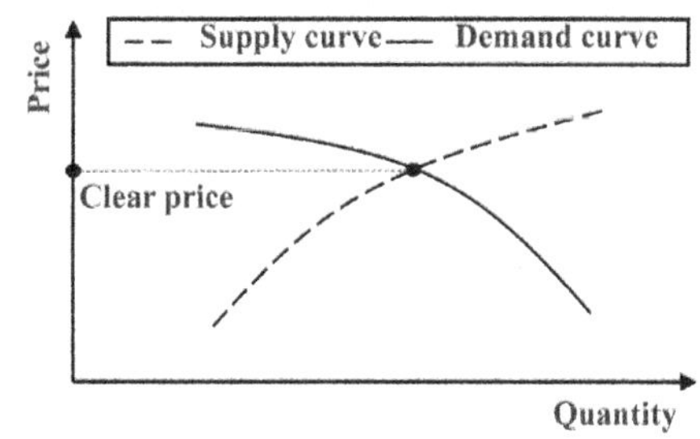

Figure 17.6 The supply and demand curves of the markets

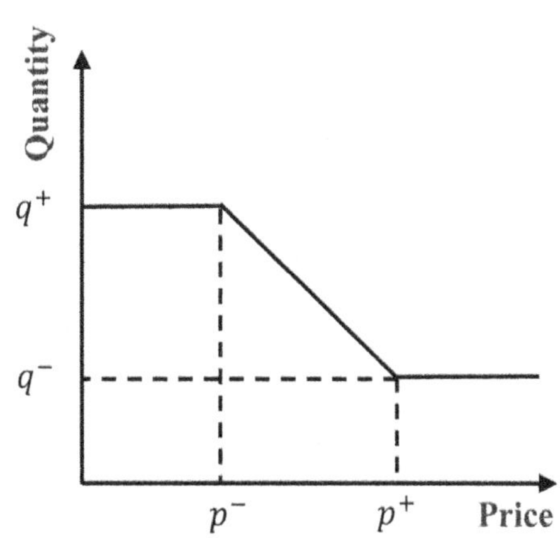

Figure 17.7 The demand curve for the Air market

more willing to adjust its demand. To quantify this sensitivity, we introduce a parameter, k, to calculate the p^- and p^+ for each buyer based on its preference:

$$p^- = p_0 - k\Delta p, \tag{17.21}$$

$$p^+ = p_0 + k\Delta p, \tag{17.22}$$

where p_o and Δp are predefined parameters. The value of k is within the range from 0 to 1 and the smaller k is, the more willing the buyer agent will be to change the demand. The q^+ and q^- represent the demand flexibility of the buyer, subject to thermal constraints and physical limits. In this case study, the buyer is the VAV terminal. The constraint is that the conditioned air should be sufficient to maintain the zone temperature within the control dead band. The air flow rate should be also bounded by the minimum and maximum air flow rates provided by each VAV terminal.

Step 2: aggregate the demand curves for the Air market

The *Air* market then summarizes the individual demand curves submitted by all buyers to an aggregated curve. As shown in Figure 17.8, the demands under the same price in all curves are summed to form an aggregated demand curve of the *Local Electricity* market. In this case, the aggregated demand curve indicates the flexibility of the total demand for the cooling energy at the AHU level. It also expresses the sensitivities of the demand flexibility to the price change.

Step 3: create a demand curve for the Local Electricity market

In this step, the *AHU+Chiller* agent converts the aggregated demand curve for conditioned air in the *Air* market to the demand curve for electricity in the *Local Electricity* market. The *AHU+Chiller* agent converts the price information of the aggregated demand curve from the *Air* market to the price of demand curve in the *Local Electricity* market:

$$p_{ele} = \eta p_{air}, \tag{17.23}$$

where p_{ele} and p_{air} are the prices for the electricity and the conditioned air, respectively. The η is the conversion factor. Then, the *AHU+Chiller* agent converts the quantity from the *Air* market to *Local Electricity* market by calculating the required electricity for generating a given amount of cooling energy.

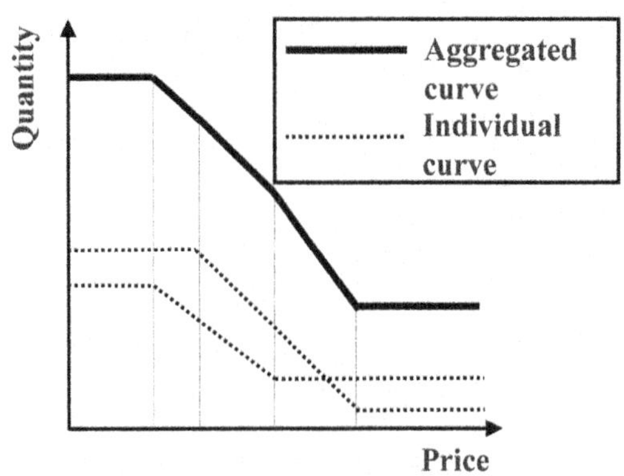

Figure 17.8 The aggregation of the demand curves for the Air market

Step 4: clear the Local Electricity market

In this step, the *Local Electricity* market is cleared based on the received electricity demand curve from the *AHU+Chiller* agent and a predefined supply curve, which represents the power limit. For simplicity, the supply curve is defined as a vertical line in Figure 17.9, although the price may vary with the power limit in other cases. The intersection of the supply curve and the demand curve shows that we are able to adjust the demand to meet the desired demand by the supply side at a cleared price for electricity. However, it is possible that there is no intersection, when the building demand is too low or too high. For both cases, the studied system doesn't have the sufficient flexibility to meet the desired demand by the supply side. As a result, the studied system will use default control set points and there is no demand change caused by the building-to-grid integration.

Step 5: clear the Air market

After clearing the *Local Electricity* market in Step 4, the last step is to clear the *Air* market. The clear price for the conditioned air is defined as

$$p_{air,clear} = \frac{1}{\eta} p_{ele,clear}. \tag{17.24}$$

Meanwhile, the clear quantity is determined based on the curve defined in Figure 17.7. The temperature set point for the controlled thermal zone to realize the clear quantity can then be calculated correspondingly.

Involved building models

To implement the above transactive control strategy, three questions need be answered with the help of building models:

1 In Step 1, how are q^{+} and q^{-} determined, subject to the thermal constraints?
2 In Step 3, how is the amount of required electricity for generating a given amount of conditioned air calculated?
3 In Step 5, what are the corresponding temperature set points to realize the clear amount of conditioned air?

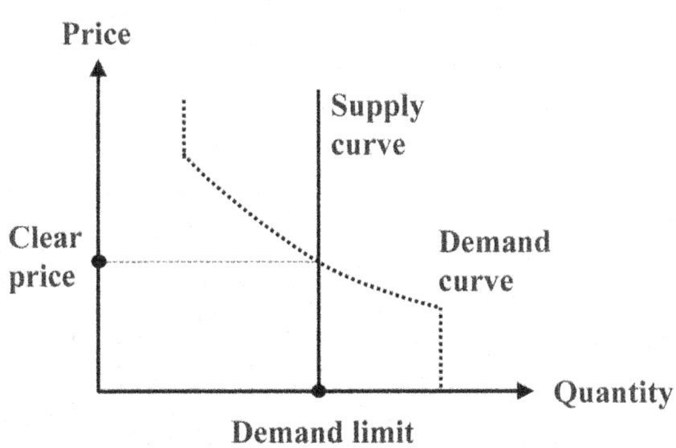

Figure 17.9 The clear of the Local Electricity market

To answer the first question, the black box model structure was adopted to create a model as:

$$f_1(T) = q(t+dt) = a_0 + a_1T_z(t) + a_2T_{out}(t) + a_3T_z(t+dt) + a_4q(t), \quad (17.25)$$

where t is the time, dt is the time interval, q is the supplied cooling energy, and T_z and T_{out} are the zone temperature and outdoor air temperature, respectively. The parameters a_0 to a_4 can be determined by training data provided by measurements or high-fidelity models.

Equation 17.25 allows us to map the temperature set point for the next time step with the corresponding amount of conditioned air. As a result, if only the cooling mode is considered, the first question can be answered by:

$$q_{max} = f_1(T^-), \quad (17.26)$$

$$q_{min} = f_1(T^+), \quad (17.27)$$

where T^- and T^+ are the high and low bounds of the zone temperature control dead band.

Likewise, the second question can be answered by:

$$f_2(q) = P_{ch} + P_{fan}, \quad (17.28)$$

$$P_{ch} = \frac{q}{COP\varepsilon}, \quad (17.29)$$

$$P_{fan} = b_1\frac{q}{C(T_{dis} - T_{mix})} + b_2\left(\frac{q}{C(T_{dis} - T_{mix})}\right)^2 + b_3\left(\frac{q}{C(T_{dis} - T_{mix})}\right)^3, \quad (17.30)$$

where b_0 to b_3 are parameters and P_{ch} and P_{fan} are the power for the chiller and fan, respectively. The COP is the coefficient of performance and ε is the sensible heat ratio of the chiller, respectively. The T_{mix} is the mixed air temperature and T_{dis} is the discharge air temperature. Additionally, the last question is answered by employing the following model:

$$f_3(T) = T(t+dt) = c_0 + c_1T_z(t) + c_2T_{out}(t) + c_3q(t+dt) + c_4q(t), \quad (17.31)$$

where $c_{0...}$, c_4 are parameters.

Model-based evaluation

In this study, the researchers establish a system model using EnergyPlus for the studied system and use that model as a virtual testbed to evaluate the proposed control strategies. In the evaluation, the TMY3 file for Pasco is adopted. The test period is from 7:00 to 17:00 on August 20. The control horizon length is 5 min and k for each VAV terminal is 1. In addition, η, COP, and ε are 1, 6.16, and 1, respectively. The parameters in Equations 17.25 to 17.31 are determined based on the regression analysis of the simulation results from the EnergyPlus. Figure 17.10 shows the validation results of f_3 () for one VAV agent.

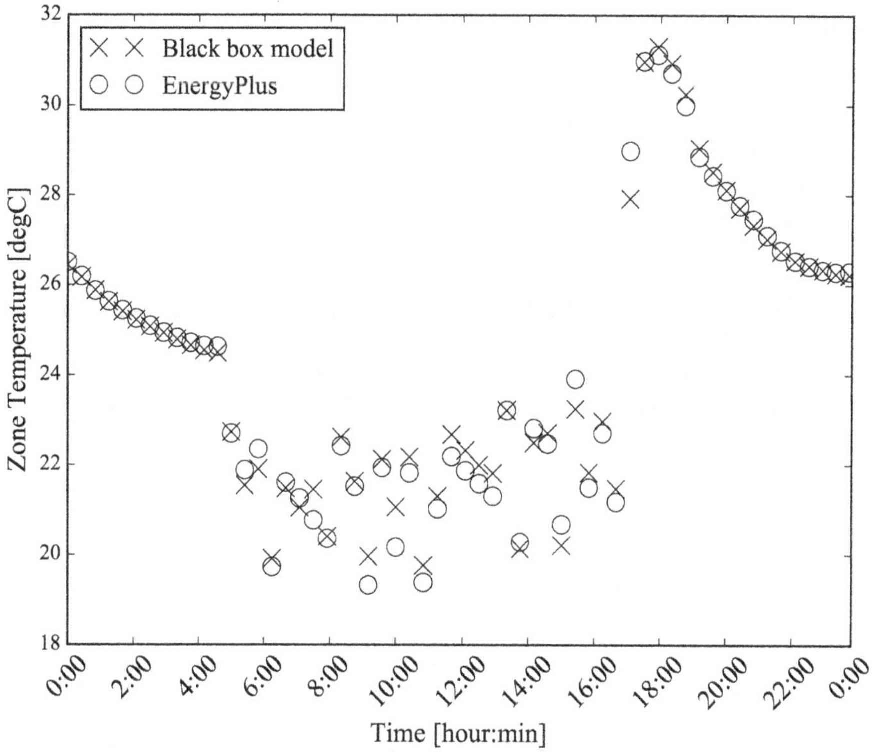

Figure 17.10 Comparison of zone temperature prediction by different models

Based on the above procedure and models, the HVAC system with the transactive control strategy is evaluated and its performance is compared with the same system without the transactive control, which adopts a constant temperature set point of 21 °C for all zones. Figure 17.11 shows that without the transactive control, the peak demand is 6.5 kW around 14:00. With the transactive control, the peak demand is restrained to 6.0 kW and shifts to around 10:00. As a result, the transactive control continues changing set points for VAV terminals instead of the fixed set points adopted by the one without transactive control.

To conclude this case study, we can see that:

1 system modeling can provide the estimation of the building response which is critical to realize the building-to-grid integration;
2 system modeling can also provide the testbed for evaluating the performance of the building-to-grid integration.

Case study 2: model-based evaluation on operations of homes with DERs

As mentioned in the introduction section, adding DERs may introduce new challenges in the power grid operation. One challenge is the duck curve problem (Miller et al. 2014). This case study first elaborates how the duck curve occurs, and its consequential issues are elaborated. Final, a potential solution is proposed and evaluated with system modeling.

Duck curve

The duck curve describes the power demand on the power grid (named *power draw*) under a high PV power penetration. As shown in Figure 17.12, the power draw begins to drop slowly

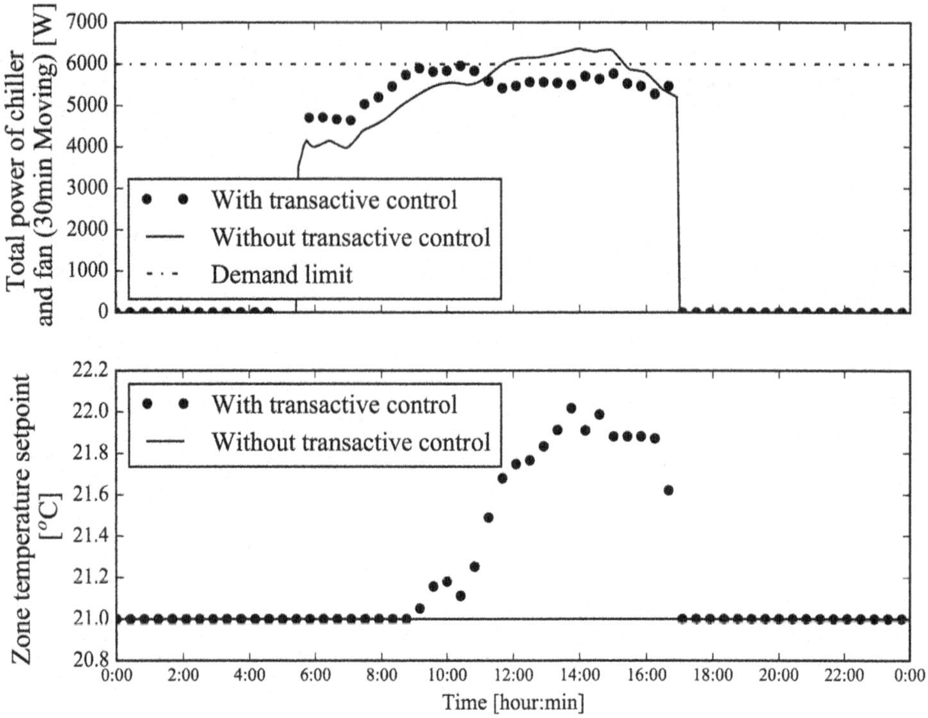

Figure 17.11 Comparison of the system behaviors with and without the transactive control

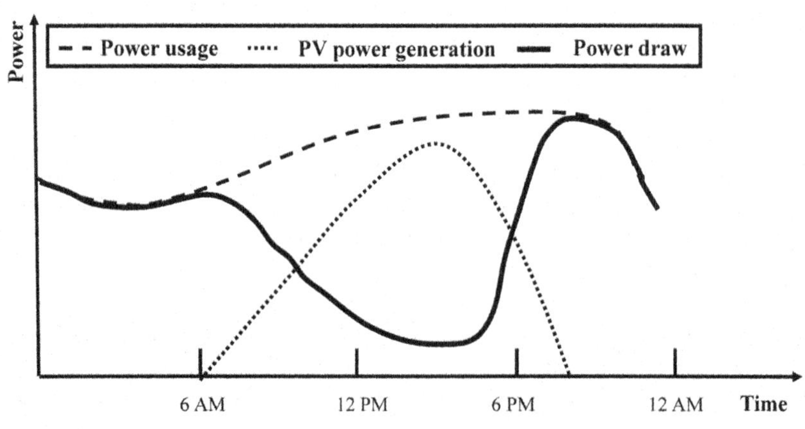

Figure 17.12 Duck curve

in the morning when the power generation from the PV becomes available. It then reaches its valley in the afternoon as the PV power generation reaches its peak. After that, the power draw starts to increase since the PV power generation decreases. In the early evening, the power draw ramps up quickly when the power usage peaks and the contribution from PV power falls due to the sunset. Since the power draw curve resembles a duck, it is named duck curve.

The duck curve affects the balance between the supply and demand in the power grid. First, it requires a quick increase of energy production in the evening to meet the high-power usage. Such rapid increase of the energy production may be dangerous to the power grid infrastructure and not easily achievable with current infrastructure. Second, the PV power generation may be more than the power usage. In that case, the PV generation may have to be curtailed, which reduces economic and environmental benefits from using the PV.

The duck curve problem can be resolved by using storage technologies to store the excessive power generated by PV panels and use it later on. As shown in Figure 17.13, we can store a part of the PV energy generated during the day and discharge the stored energy in the evening. As a result, the resulting power draw profile is smoother than the one without any storages.

Model-based evaluation on homes with DERs

In this study, we conducted the evaluation with a simulated residential neighborhood consisting of six homes as shown in Figure 17.14. Based on their power usages, the homes can be divided into two groups and each group contains three types of homes:

- The first type (A1/B1) purely relies on the power grid to meet its power usage;
- The second type (A2/B2) is equipped with a PV system. They use the PV system as their primary power supply and the power grid as their secondary supply. In addition, the excessive energy generated by the PV will flow back to the power grid;

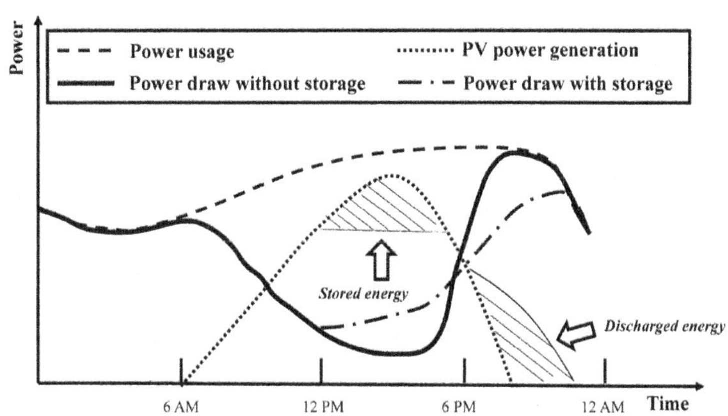

Figure 17.13 Resolving duck curve problem with storage

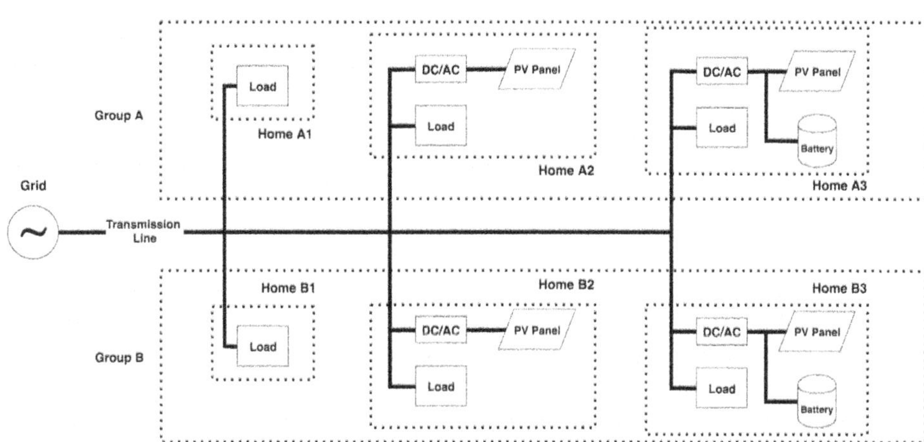

Figure 17.14 Studied residential neighborhood

- The third type (A3/B3) contains both a PV system and a battery. This type of home will use the PV power generation as the primary power supply, then the battery as their secondary supply and the power grid as the tertiary supply. When there is excessive power from PV generation, they will charge the battery until it is full. After that, they will send the energy back to the grid.

This studied neighborhood is modeled using the Modelica Buildings Library (Wetter et al. 2014). Figure 17.15 shows the diagram of Modelica models for the neighborhood. Each icon represents a component model. Taking the model for Home A3 as an example, it contains sub-component models, such as the battery and the PV system, which are shown in Figure 17.16. The inputs for the home model are the weather data and the power usage profile while the major output is the power draw by the home.

In the simulation, two power usage profiles for homes are adopted. Power usage profile A is obtained from literature (Hill 1978) which measured power usages for homes in California, U.S. This profile is used as an input for homes in group A and is the same as the power draw of A1 in Figure 17.17. Power usage profile B is generated by adding randomized deviation (up to 15%) to the power usage profile A. It is used as inputs for homes in group B and is plotted as the power draw of Home B1 in Figure 17.17. The PV system is sized to be 3.5 kW. In addition, the battery is sized as 13.5 kWh and can discharge until it is completely empty with the maximum discharge rate and maximum charge rates both set as 5 kW.

The simulation was run for a whole day (June 1) with the TMY3 file of San Francisco, California, U.S., and the results are shown in Figure 17.17. With a large PV capacity but no energy storage, the power draw profiles for the second homes (A2 and B2) are duck curves. Their power draws increase by around 6.8 kW from 12:18 to 19:09. In addition, the power draw is negative from 6:50 to 17:06, which means that the power grid needs to absorb the over-generated energy. Thanks to the battery, the power draws for the third homes (A3 and B3) are almost a flat line starting from 6:50. This means that batteries are able to absorb the excessive energy generation from the PV and discharge the energy during the evening to smoothen the ramping power draw.

Figure 17.18 illustrates the detailed operation status of the Home A3. We use a state of charge (SOC) to indicate the amount of stored energy in the battery. In the early morning, SOC is zero because all of the energy stored in the battery was used in the previous night. Then the battery starts to charge when there is excessive energy from the PV and discharge around 17:00.

The above results indicate that adding a battery smoothens the duck curve as expected. However, as shown in Figure 17.19, we can still observe the duck curve at the aggregated power draw by the entire neighborhood since only a few homes are equipped with energy storage and control. To solve this problem, we may consider two options: (1) enabling the interaction between homes so that the energy storage can be shared by all homes; or (2) utilizing the home as "virtual batteries." The first option requires a mechanism to coordinate the allocation of the stored energy for all homes to achieve a desired power draw at the neighborhood level. Depending on the diversity of the power usage profiles, finding the optimal allocation that minimizes the negative impacts on the operation (e.g. voltage stability) can be challenging. Potentially, game theory (Ross and Mehdi 1996) can be used to resolve this problem. On the other hand, the second option uses the power flexibility of each home by charging/discharging it as a virtual battery. To do so, however, the power flexibility of each home needs to be quantified so that we can accurately control the amount of stored energy. Building models serve that purpose, as discussed in case study 1.

This evaluation focuses on whether the proposed battery and its control can resolve the duck curve. To simplify the system modeling, we ignore several side effects from using the battery

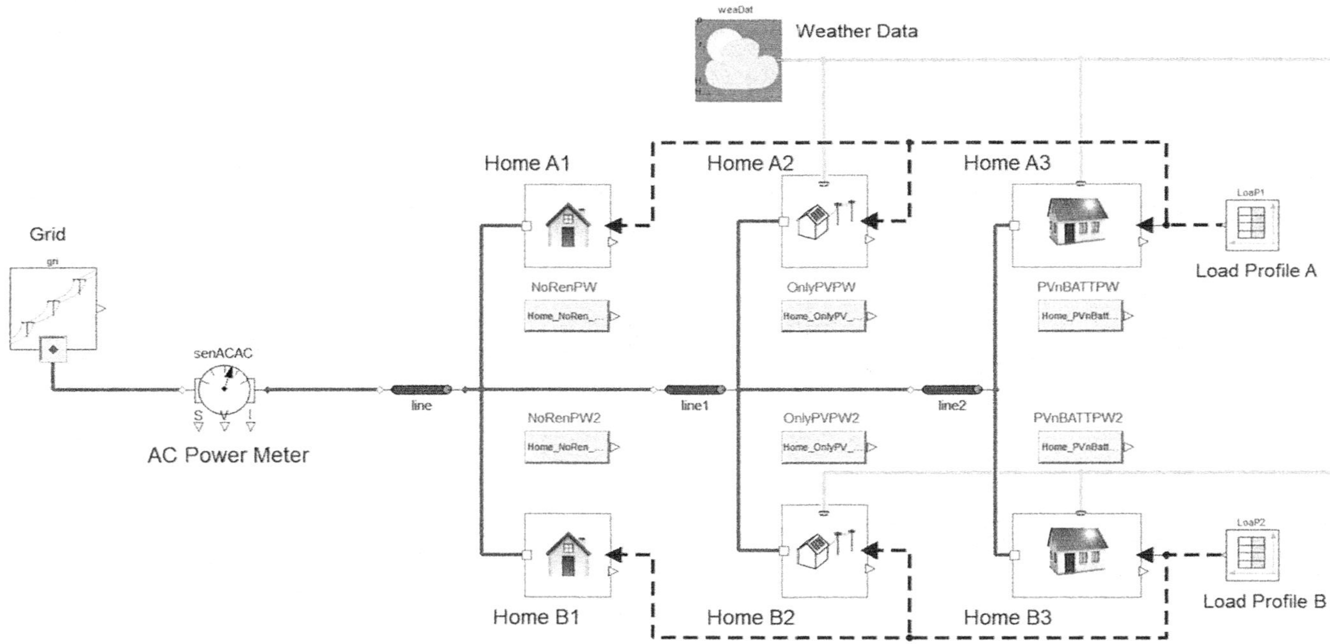

Figure 17.15 Diagram of Modelica model for the neighborhood

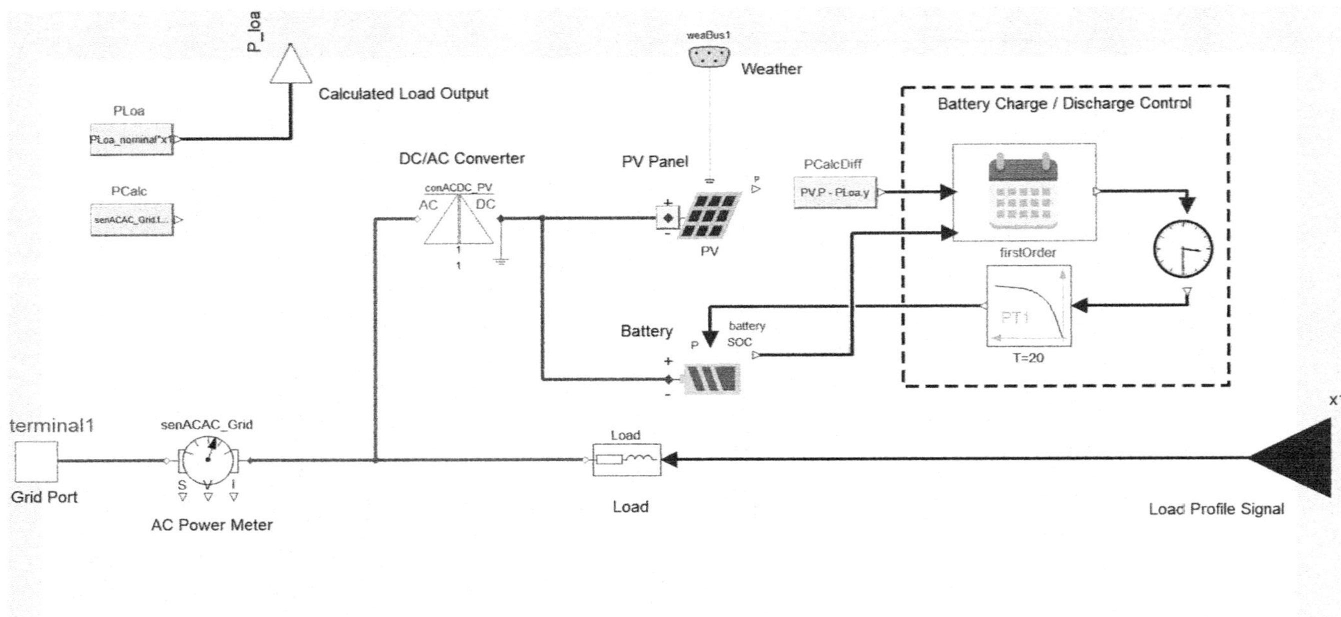

Figure 17.16 Diagram of Modelica model for Home A3

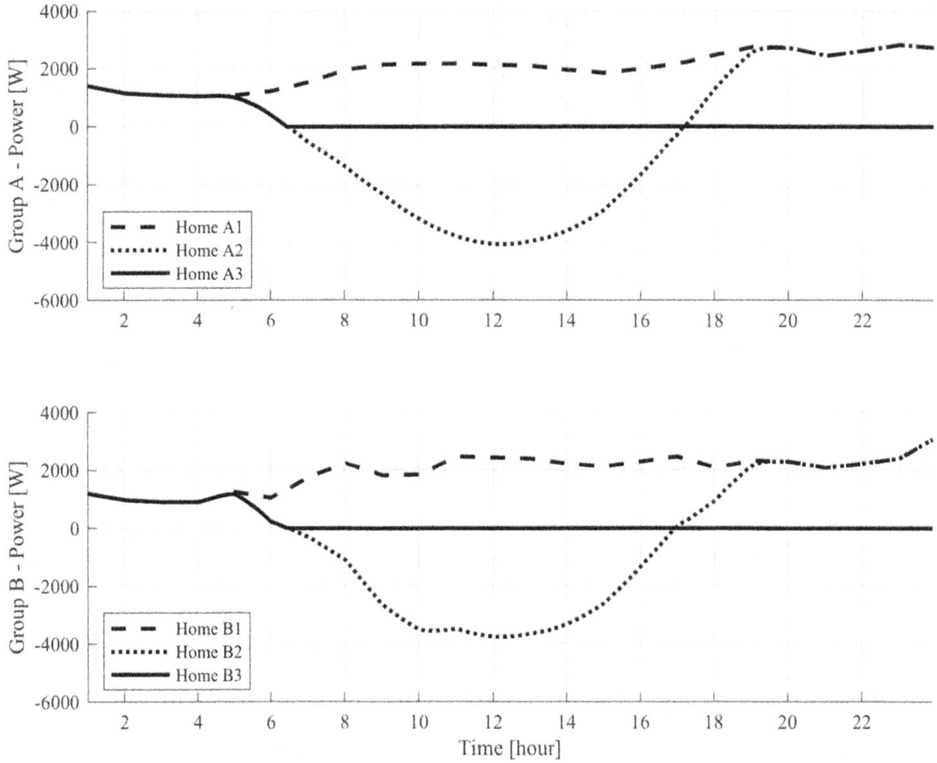

Figure 17.17 Simulated power draw of the six homes

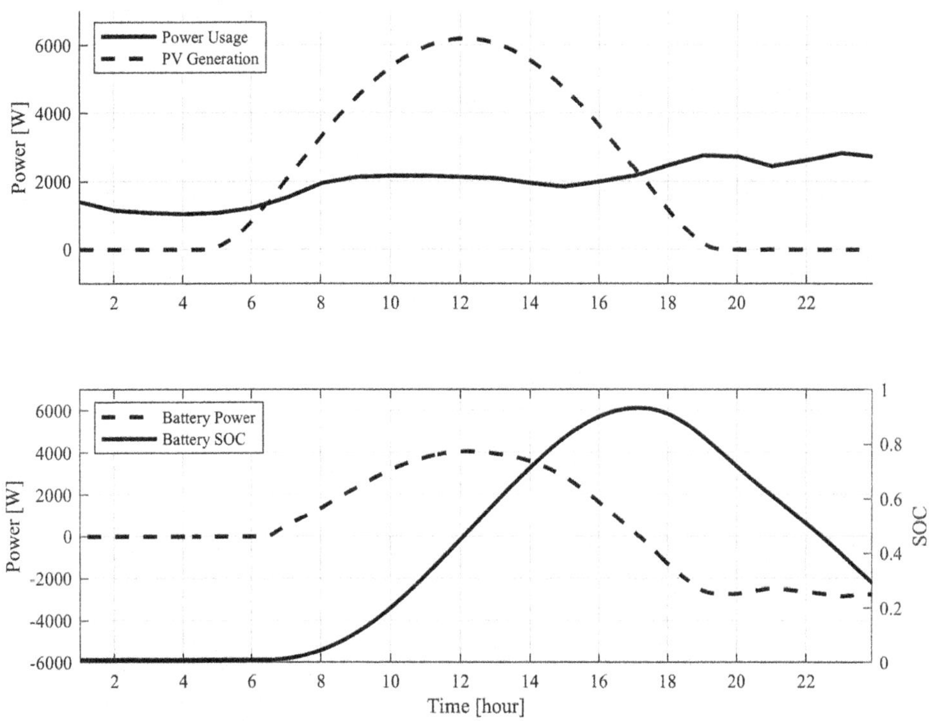

Figure 17.18 Results for the battery operation for Home A3

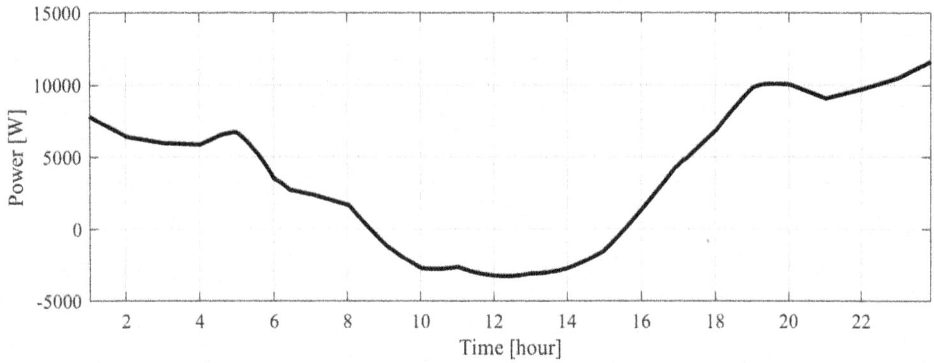

Figure 17.19 Aggregated power draw by the studied neighborhood

control, e.g. the voltage stability of the micro-grid and the power grid. Further improvements on the model fidelity would be necessary if a comprehensive evaluation on the proposed control or other advanced controls is preferred.

Discussion

In the context of building-to-grid integration, system modeling can be employed in the design, evaluation, and deployment of relevant control strategies. However, modeling technologies must be carefully selected for individual applications based on a trade-off between the accuracy and cost associated with modeling and simulation. Therefore, sufficient domain knowledge is necessary for deciding when and how to apply different models. In addition, advances in technology and increased availability of complex algorithmic science have created a trend to employ technologies such as machine learning in building modeling to address challenges raised by building-to-grid integration (Tsanasa and Xifarab 2012).

While this chapter focuses on building control strategies, there are many more challenges to be considered in the building-to-grid integration. For example, the building control strategies discussed in this chapter are to address the short-term imbalance between the demand and supply. However, we still need to address the long-term imbalance as the energy demand is outpacing the supply of the current infrastructure. Due to the physical limits, such as aging infrastructure, space limitations, and environmental constraints, however, this long-term imbalance can't be resolved by simply adding more capacities. Thus, the future infrastructure needs to be designed with considerations on both supply and demand, where the building system modeling can play an important role. Owing to the page limitation, this chapter doesn't elaborate on them in detail. For more information, readers are suggested to read Chen et al. (2016) and Shawhan et al. (2015).

Synopsis

This chapter discusses the needs for system models in the building-to-grid integration applications. It also elaborates different application requirements by corresponding modeling technologies. After that, it uses two case studies to demonstrate how building models should be selected, set, and applied for specific applications. As an emerging research topic, building-to-grid integration is still evolving with new concepts and applications. Although the authors have tried their best to summarize the current state-of-art building modeling in building-to-grid integration, more related information are available in the recommended readings.

References

American Geophysical Union. (1997) Geomagnetic storms can threaten electric power grid. *Earth in Space* 9(7): 9–11.

Callaway, D.S. (2009) Tapping the energy storage potential in electric loads to deliver load following and regulation, with application to wind energy. *Energy Conversion and Management* 50(5): 1389–1400.

Chen, Y., Hong, J. and Liu, C.C. (2016) Modeling of intrusion and defense for assessment of cyber security at power substations. *IEEE Transactions on Smart Grid* 9(4): 2541–2552.

Crawley, D.B., Lawrie, L.K., Winkelmann, F.C., Buhl, W.F., Huang, Y.J., Pedersen, C.O., Strand, R.K. et al. (2001) EnergyPlus: Creating a new-generation building energy simulation program. *Energy and Buildings* 33(4): 319–331.

Dodrill, K. (2011) *Demand Dispatch – Intelligent Demand for a More Efficient Grid*. National Energy Technology Laboratory.

Driesen, J. and Katiraei, F. (2008) Design for distributed energy resources. *IEEE Power and Energy Magazine* 6(3): 30–39.

Eto, J.H. (1996) *The Past, Present, and Future of U.S. Utility Demand-side Management Programs*. Lawrence Berkeley National Laboratory [LBNL-39931].

Hagerman, J. (2001) *Buildings-to-Grid Technical Opportunities: Introduction and Vision*. Building Technologies Office, U.S. Department of Energy.

Hammerstrom, D.J., McDermott, T.E., Huang, Q., Sivaraman, D., Kalsi, K., Tang, Y., Lian, J. Veeramany, A., Makhmalbaf, A. and Woodward, J.T. (2017) *Transactive Systems Simulation and Valuation Platform Trial Analysis*. Pacific Northwest National Laboratory [PNNL-26409].

Hao, H., Corbin, C.D., Kalsi, K. and Pratt, R.G. (2016) Transactive control of commercial buildings for demand response. *IEEE Transactions on Power Systems* 32(1): 774–783.

Hao, H., Lin, Y., Kowli, A.S., Barooah, P. and Meyn, S. (2014) Ancillary service to the grid through control of fans in commercial building HVAC systems. *IEEE Transactions on Smart Grid* 5(4): 2066–2074.

Hao, H., Wu, D., Lian, J. and Yang, T. (2017) Optimal coordination of building loads and energy storage for power grid and end user services. *IEEE Transactions on Smart Grid* 8(5): 4335–4345.

Hill, D.H. (1978) Home production and the residential electric load curve. *Resources and Energy* 1(4): 339–358.

Huang, S., Zuo, W. and Sohn, M. (2017) A Bayesian network model for predicting cooling load of commercial buildings. *Building Simulation* 11(1): 87–101.

Ji, L., Niu, D.X. and Huang, G.H. (2014) An inexact two-stage stochastic robust programming for residential micro-grid management-based on random demand. *Energy* 67(2014): 186–199.

Johnson, M. (2017) Controlling and optimizing resilient distributed energy resources and microgrids with a demand-side operation platform. *The Electricity Journal* 30(2017): 12–15.

Kalsi, K., Elizondo, M., Fuller, J., Lu, S. and Chassin, D. (2012) Development and validation of aggregated models for thermostatic controlled loads with demand response. In *Paper Presented at the 45th Hawaii International Conference on System Sciences*. Maui, Hawaii, U.S.

Lattin, J.M., Douglas, C. and Green, P.E. (2003) *Analyzing Multivariate Data*. Pacific Grove, CA: Thomson Brooks Cole.

Liang, G., Weller, S.R., Zhao, J., Luo, F. and Dong, Z. (2017) The 2015 Ukraine blackout: Implications for false data injection attacks. *IEEE Transactions on Power Systems* 32(4): 3317–3318.

Lu, Y., Wang, S., Zhao, Y. and Yan, C. (2015) Renewable energy system optimization of low/zero energy buildings using single-objective and multi-objective optimization methods. *Energy and Buildings* 89(2015): 61–75.

Lurwan, S.M., Mariun, N., Hizam, H., Radzi, M.A.M. and Zakaria, A. (2014) Predicting power output of photovoltaic systems with solar radiation model. In *Paper Presented at the 2014 IEEE International Conference Power & Energy*. Kuching, Malaysia.

Ma, J., Qin, S.J., Li, B. and Salsbury, T. (2011) Economic model predictive control for building energy systems. In *Paper Presented at the 2011 IEEE Power & Energy Society General Meeting*, Anaheim, CA.

Maharjan, S., Zhang, Y., Gjessing, S., Ulleberg, Ø. and Eliassen, F. (2015) Providing microgrid resilience during emergencies using distributed energy resources. In *Paper Presented at the 2015 IEEE Globecom Workshops*. San Diego, CA.

Mathieu, J.L. and Callaway, D.S. (2012) State estimation and control of heterogeneous thermostatically controlled loads for load following. In *Paper Presented at the 45th Hawaii International Conference on System Science Maui*. Hawaii, HI.

Miller, N.W., Shao, M., Pajic, S. and D'Aquila, R. (2014) Western wind and solar integration study phase 3 – Frequency response and transient stability. National Renewable Energy Laboratory [NREL/SR-5D00–62906].

Moreno-Munoz, A. (2017) *Large Scale Grid Integration of Renewable Energy Sources*. Institution of Engineering and Technology.

North American Electric Reliability Corporation. (2017) *Distributed Energy Resources Connection Modeling and Reliability Considerations*. North American Electric Reliability Corporation.

Reikard, G. (2009) Predicting solar radiation at high resolutions: A comparison of time series forecasts. *Solar Energy* 83(2009): 342–349.

Ren, M.J. and Wright, J.A. (2002) Adaptive diurnal prediction of ambient dry-bulb temperature and solar radiation. *HVAC&R Research* 8(4): 383–402.

Risken, H. (1996) Fokker-Planck equation. In *The Fokker-Planck Equation*. Berlin, Heidelberg: Springer.

Ross, S.M. and Mehdi, J. (1996) *Stochastic Processes*. Wiley.

Schachter, J. and Mancarella, P. (2014) A short-term load forecasting model for demand response applications. *Paper presented at the 11th International Conference on the European Energy Market*. Krakow, Poland.

Schalkoff, R.L. (1997) *Artificial Neural Networks*. New York: McGraw-Hill Higher Education.

Shawhan, D., Taber, J. and Zimmerman, R. (2015) A detailed power system planning model: Estimating the long-run impact of carbon-reducing policies. In *Paper Presented at the 48th Hawaii International Conference on System Sciences*. Kauai, Hawaii, U.S.

Su, L. and Norford, L.K. (2015) Demonstration of HVAC chiller control for power grid frequency regulation – Part 1: Controller development and experimental results. *Science and Technology for the Built Environment* 21(8): 1134–1142.

Taylor, Z.T., Gowri, K. and Katipamula, S. (2008) GridLab-D technical support document: Residential end-use module version 1.0. Pacific Northwest National Laboratory [PNNL-17694].

Tsanasa, A. and Xifarab, A. (2012) Accurate quantitative estimation of energy performance of residential buildings using statistical machine learning tools. *Energy and Buildings* 49(2012): 560–567.

U.S. Department of Energy (2016) *Quadrennial Energy Review: Energy Transmission, Storage, and Distribution Infrastructure*. U.S. Department of Energy.

Wetter, M., Zuo, W., Nouidui, T.S. and Pang, X. (2014) Modelica buildings library. *Journal of Building Performance Simulation* 7(4): 253–270.

Widén, J., Molin, A. and Ellegård, K. (2012) Models of domestic occupancy, activities and energy use based on time-use data: Deterministic and stochastic approaches with application to various building-related simulations. *Journal of Building Performance Simulation* 5(1): 27–44.

Wu, D., Kintner-Meyer, M., Yang, T. and Balducci, P. (2017) Analytical sizing methods for behind-the-meter battery storage. *Journal of Energy Storage* 12(2017): 297–304. doi: 10.1016/j.est.2017.04.009.

Xue, X., Wang, S., Yan, C. and Cui, B. (2015) A fast chiller power demand response control strategy for buildings connected to smart grid. *Applied Energy* 137(2015): 77–87.

Yan, D., O'Brien, W., Hong, T., Feng, X., Gunay, H.B., Tahmasebi, F. and Mahdavi, A. (2015) Occupant behavior modeling for building performance simulation: Current state and future challenges. *Energy and Buildings* 107(2015): 264–278.

Zhu, F., Zheng, Y., Guo, X. and Wang, S. (2003) Environmental impacts and benefits of regional power grid interconnections for China. *Energy Policy* 33(14): 1797–1805.

Recommended reading

This chapter focuses on the applications of system modeling in the building-to-grid integration. To better understand those applications, a comprehensive view for the building-to-grid integration is critical. Therefore, readers are suggested to read the following literature:

1 "The past, present, and future of U.S. utility demand side management" (Eto 1996). This document elaborates the evolution of demand side management and its applications.

2 "Buildings-to-grid technical opportunities" (Hagerman 2001). This document discusses opportunities to mitigate certain "building-to-grid" issues.
3 *Large Scale Grid Integration of Renewable Energy Sources* (Moreno-Munoz 2017). This book introduces a variety of approaches to overcome the interconnection challenges.
4 "GridLab-D Technical Support Document: Residential End-Use Module Version 1.0, 2008" (Taylor et al. 2008). This document details the implementation of typical residential building models for power system simulation.

Activities

To practice the use of the system modeling in the building-to-grid context, readers are suggested to run the Smart Grid Interoperability Panel using Case 1 provided by the transactive energy simulation platform (Hammerstrom et al. 2017). This case study includes a transmission and generation system, a large commercial building with responsive HVAC thermostat, and single-family residences that have responsive HVAC thermostats. The control strategy is tested using a double-auction market to achieve the balance between power supply and demand.

18 Modelling HVAC and renewable energy plant and control

Chris Underwood and Simon Rees

Scope

The chapter firstly presents an introduction to modern approaches to both steady-state and dynamic modelling of HVAC and related plant components and systems. Approaches to representations of components such as water-air heat exchangers are dealt with and their representation in systems that include control elements are described. Common approaches to solving the resulting equations are also presented. A series of simulation case studies are provided to illustrate the application of the methods. The chapter also presents examples of approaches to modelling the key components of a number of renewable energy systems that can be found integrated in buildings.

Learning objectives

To be able to describe approaches to the modelling of HVAC, plant and renewable energy systems and their key components and how they can be represented mathematically and possible solution approaches. To be able to describe the data requirements of such modelling approaches and to execute calculations to find the state of example component models.

Key words

HVAC/plant/control/renewable energy/photovoltaic/geothermal/solar thermal/wind power

Introduction

In this chapter, 'plant' is considered to be equipment included in a building or a group of buildings for the purpose of generating and distributing energy, with the ultimate aim of maintaining comfort as well as the supply of essential services such as power, water supply and so on. A distinction is usually drawn to distinguish between energy generating plant or 'primary plant' and energy distributing plant or 'secondary plant'. Examples of the former include boilers, chillers, combined heat and power (CHP) modules, heat pumps, primary air handling units, solar collector arrays and photovoltaic arrays. Examples of secondary plant include room equipment such as radiator-convectors (or baseboard heaters), fan coil units and secondary air handling units, chilled ceilings, chilled beams and, conceivably, luminaires. Control systems designed to match the capacity to demand of these systems are, of course, to be found at both primary and secondary levels. The distinction can be important because whereas the secondary plant both influences and is influenced by the building rooms and zones within which the plant is located, the primary plant merely responds to the demand placed on it by the secondary systems. The primary plant does not directly interact with the building, which means that it can often be modelled in isolation provided the load demands to be met are known. Developments in equation-based methods (in which the fundamental objects are mathematical equations) can therefore play a role in certain types of plant modelling to get around an otherwise rigid and inflexible approach when using standalone energy simulation programs. For further discussions on this, see Sahlin et al. (2004), Chow et al. (1997), Trčka and Hensen (2010) and Bertagnolio and Lebrun (2008). For examples of equation-based methods, see F_Chart (2011), Modelica (2010) and SPARK (2010).

When choosing a modelling method for plant and control modelling, complexity and rigour are to be traded with computational cost (i.e. computer processing times). Since this affects time and money in terms of both data preparation and turnaround times due to computer processing, busy practitioners require to exercise judgement when making a choice of modelling method, which usually amounts to selecting the least complex (and rigorous) approach that will deliver the level of output detail, precision and accuracy needed to meet a given modelling objective. In most modelling applications, it is the level of rigour associated with plant and control modelling that will dominate computation cost rather than the modelling of the building envelope itself.

Rigour in relation to modelling plant and control has a strong association with the simulation time horizon over which the modelling is to be conducted – indeed the two considerations relate to one another inversely. This, in turn, has a clear association with the problem-solving application. Some examples can be found in Table 18.1.

Strategies for modelling plant and control

Not surprisingly, three strategic approaches are available for modelling plant and control that map onto the three sets of applications in Table 18.1. In practice, many commercially available and freeware computer codes that enable plant modelling involve a degree of hybridization between at least two of the methods. Table 18.2 summarizes the three possibilities with comments on very typical computational cost based on today's personal computer hardware. Each of the methods will be explored in detail in the subsequent sections of this chapter.

Table 18.1 Some typical applications for plant and control modelling

Time horizon	*Typical applications*	*Complexity*
Long (seasons to years)	• Seasonal energy use • Seasonal carbon emissions • Control strategy • Building zoning • Summer overheating and air conditioning feasibility • Limited optimization studies	Low
Moderate (months to seasons)	• Plant selection (strategic) • Plant sizing • Selection of working fluids • Adaptive comfort • Primary plant performance • Weather data impacts • Need for humidity control • Comparative plant performance • Optimization studies	Moderate to high

(*Continued*)

Table 18.1 (Continued)

Time horizon	*Typical applications*	*Complexity*
Low (seconds to days)	• Control design and stability • 'Smart' controller design • Decision-making algorithms • Optimal design and response • Detailed plant response analysis • Thermal storage system design • Critical selection of working fluids • Real-time analysis and system status • System fault detection and diagnosis • Artificial intelligence applications	High to very high

Table 18.2 Some typical applications for plant and control modelling

Method	*Typical features*	*Typical computational cost*
Balanced energy	1 Participating building is modelled rigorously/ dynamically. 2 Un-modelled plant. 3 Perfect control set point tracking imposed. 4 Constant plant performance metrics imposed.	Seconds to a few minutes (*)
Quasi-steady-state modelling	1 Participating building is modelled rigorously/ dynamically. 2 Plant modelled using algebraic energy and mass balances. 3 Plant capacity proportional to control system demands. 4 Makers 'equation-fit' plant models or look-up tables for plant performance metrics.	Minutes up to one or two hours (*)
Fully dynamic modelling	1 Any participating building may, in some instances, use a simple steady-state model. 2 Plant modelled using differential equations. 3 Plant capacity adjusted by control algorithm. 4 Plant performance metrics determined by the model.	Minutes up to several hours or days (*)

*Broadly, the computer elapse time needed to solve a moderately complex problem. For balanced energy and quasi-steady-state methods, this will mainly depend on the number of participating building zones. Fully dynamic modelling is usually either at plant level only or involving coupling with a simplified single-zone building.

Plant modelling using energy balance method

The plant is not modelled in this approach, though the building is rigorously modelled using dynamic methods as described elsewhere in this book. With target heating and cooling (as relevant) control set points defined for each zone of the building, it is simply assumed that the plant always tries to maintain the set point at all times when the plant is made available up to the limit of the pre-defined plant capacity. When the plant is first switched on, the maximum capacity of the secondary plant is imposed on the building rooms and zones and propagated back to the primary plant. The zone temperatures will rise (winter) and usually fall if air-conditioned (summer) until it reaches set point and then be maintained there. Thus, the plant capacity to meet the set point in these conditions is determined by room/zone energy balances.

The room/zone energy balance is given by Equation 18.1.

$$V\rho_a c_a \frac{dT_{ai}}{dt} = \sum_{i=1}^{N} A_i h_i \left(T_{s,i} - T_{ai}\right) + m_{infil} c_a \left(T_{ao} - T_{ai}\right) + q_{cas} + q_{plant} \tag{18.1}$$

where:

- V is the room or zone volume (m^3);
- ρ_a is the air density (kgm^{-3});
- c_a is the specific heat of air ($Jkg^{-1}K^{-1}$);
- A_i is the area of room surface i (m^2);
- h_i is the convection coefficient of surface i ($Wm^{-2}K^{-1}$);
- m_{infil} is the infiltration mass flow rate (kgs^{-1});
- $T_{s,i}$ is the surface temperature of surface i (°C);
- T_{ai} is the room air temperature (°C);
- T_{ao} is the external air temperature (°C);
- q_{cas} is the heat added due to casual (uncontrolled) sources or heat gains (W);
- N is the total number of surfaces making up the room enclosure;
- q_{plant} is the heat added to the room by the plant (W).

Thus, at all times when the plant is made available and the room/zone temperature, T_{ai}, is above (cooling) or below (heating) the pre-defined set point value, the maximum (rated) capacity of the plant, q_{plant}, is added (heating) or deducted (cooling) and T_{ai} free-floats towards its target value. When it gets there, it is held at its set point value and Equation 18.1 is now used to calculate a balancing value of q_{plant} (less than or equal to its rated value) to maintain T_{ai}. These calculations are repeated at each time interval appropriate to the building simulation and, over time, q_{plant} is integrated to give energy use in Wh or kWh. When the plant is not offered, its value is set at zero and Equation 18.1 is then used to calculate a free-float value of T_{ai}. Of course, by this method, it is possible for q_{plant} to be held at its rated value indefinitely so long as T_{ai} has not reached target, and this situation will provide evidence of an under-sized plant. So, some useful results can be determined by a modelling approach of this kind, though its utility is limited as far as the analysis of plant behaviour is concerned.

Example 18.1 Energy balance method

A room measures 5 m × 5 m × 3 m (high). All room surfaces are at a steady 18 °C and have a convection coefficient of 8 $Wm^{-2}K^{-1}$. There is a constant infiltration air change rate of 0.5 h^{-1} and a constant casual heat gain to the room of 10 Wm^{-2} of floor area. The density and specific heat capacity of the room air are 1.2 kgm^{-3} and 1005 $Jkg^{-1}K^{-1}$, respectively.

a What will be the plant load to maintain the room at 20 °C when the external air temperature is 0 °C?

b What will the room temperature settle at when the plant (and casual heat gains) are off, the room surface temperature is at a steady 14 °C and all other conditions are unchanged?

c In summer the temperature of all room surfaces on a particular day is at a steady 28 °C. What will be the plant load to maintain the room at 24 °C when the external air temperature is 20 °C and all other conditions remain the same?

Solution:

At a maintained (i.e. steady-state) room temperature, the LHS of Equation 18.1 will be zero.

Total area of room surfaces: $4 \times 5 \times 3 + 2 \times 5 \times 5 = 110$ m²
Room volume: $5 \times 5 \times 3 = 75$ m²
$\therefore m_{infil} = 1.2 \times 75 \times 0.5 / 3600 = 0.0125$ kgs^{-1}
Casual heat gain: $q_{cas} = 10 \times 5 \times 5 = 250$ W

Part (a) – the plant load can be calculated as follows:

$0 = 8 \times 110 \times (18{-}20) + 1005 \times 0.0125 \times (0{-}20) + 250 + q_{plant}$
from which $q_{plant} = 1761$ W

Part (b) – a settled room temperature implies a steady-state so the LHS of Equation 18.1 is again zero. q_{plant} (and q_{cas}) are now zero so Equation 18.1 can now be used to find T_{ai}:

$0 = 8 \times 110 \times (14 - T_{ai}) + 1005 \times 0.0125 \times (0 - T_{ai}) + 0 + 0$
from which $T_{ai} = 13.8$ °C

Part (c) – the plant load is obtained for these conditions:

$0 = 8 \times 110 \times (28{-}24) + 1005 \times 0.0125 \times (20{-}24) + 250 + q_{plant}$
from which $q_{plant} = -3470$ W (i.e. negative sign implies a cooling plant load).

End of solution.

The above is a somewhat trivial example because all of the boundary variables will tend to vary all the time, but it does illustrate the simplicity of this approach to plant modelling.

Steady-state modelling

Greater rigour and modelling utility than is possible using balanced energy modelling can be achieved when each plant component is given a functional relationship in which the output variables of interest are made a function of its input variables and certain physical parameters. It is appealing to note that, in many instances of plant modelling, the rates of change of plant components are much faster than are those of the building envelope. This means that where the building envelope can be adequately modelled using time intervals that are larger than the response time of the plant, then the plant can be treated as if it were a steady-state problem. Such an approach is sometimes called quasi-steady-state modelling. Many familiar and reliable methods of modelling can then be used.

Control functions

The simplest approach is the use of control functions, which assumes that the plant will respond to control signals in a perfectly linear manner. As an illustration, consider the control of heating and cooling functions in a room or zone of a building. A control function diagram is shown in Figure 18.1, which is based on a two-channel control governing heating and cooling. There is a heating control set point, $T_{H\text{-}SP}$, and a cooling set point, $T_{C\text{-}SP}$, and there are corresponding proportional bands, *HCB*, *CCB*, which define the range of the control variable for which the output of the system is directly proportional. If the control system is perfectly linearized, then the signal forming the vertical axis will be directly proportional to plant output. With this method, then, the control variable (e.g. room or zone temperature) is not maintained constant but can float on either side of a target set point value. This gives a more realistic prediction of plant response than is afforded by the balanced energy approach.

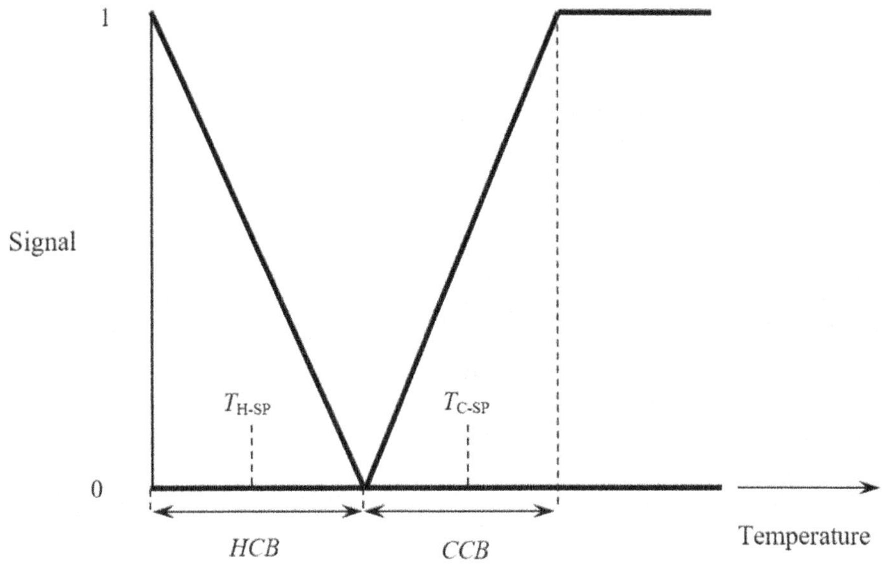

Figure 18.1 Typical control function diagram for room heating and cooling

It is possible to define a formal expression for the heating and cooling plant outputs to meet the conditions defined in the control function diagram (Equations 18.2 and 18.3).

$$q_{\text{H}}(t) = \min\left(q_{\text{H-design}}, \max\left(0, q_{\text{H-design}} \times \left(\frac{T_{\text{H-SP}} + HCB/2 - T_{\text{ai}}(t)}{HCB}\right)\right)\right) \tag{18.2}$$

$$q_{\text{C}}(t) = \min\left(q_{\text{C-design}}, \max\left(0, q_{\text{C-design}} \times \left(\frac{T_{\text{ai}}(t) - T_{\text{C-SP}} + CCB/2}{CCB}\right)\right)\right) \tag{18.3}$$

where $q_{\text{H}}(t)$, $q_{\text{C}}(t)$, $T_{\text{ai}}(t)$ are respectively the heating load, cooling load and room air temperature at some point in time, t, and $q_{\text{H-design}}$, $q_{\text{C-design}}$ are design, heating and cooling capacities of the plant, respectively.

Example 18.2 Control functions

Suppose that the control function diagram of Figure 18.1 is used to control heating and air conditioning terminals in a room with design capacities matching the values obtained in the previous example (i.e. 1761 W heating and 3470 W cooling). The heating set point is 20 °C and the cooling set point is 24 °C. Both proportional bands are 2 K. What will be the plant status when prevailing room temperatures are (a) 19.3 °C and (b) 23.8 °C?

Solution:

a The heating control range is 20 °C ± 2/2 = 19–21 °C. So a prevailing temperature of 19.3 °C is within the heating control band and the cooling will therefore be off. The heat output of the plant will be:

$$q_{\text{H}}(t) = \min\left(1761, \max\left(0, 1761 \times \left(\frac{20 + 2/2 - 19.3}{2}\right)\right)\right) = 1497\text{W}$$

b The cooling control range is 24 °C ± 2/2 = 23–25 °C. A prevailing temperature of 23.8 °C is within the cooling control band so the heating will now be off. The cooling load on the plant will, therefore, be:

$$q_C(t) = \min\left(3470, \max\left(0, 3470 \times \left(\frac{23.8 - 24 + 2/2}{2}\right)\right)\right) = 1388\text{W}$$

End of solution.

Steady-state theory

For greater modelling precision and detail, the plant can be described using established steady-state theory provided that the central assumption of quasi-steady-state modelling applies (i.e. plant disturbance response times within one building fabric modelling sampling interval). An example of this for heat transfer processes is the use of the effectiveness method (Kays and London 1964), in which the effectiveness (E) is defined by Equations 18.4–18.8.

$$E(t) = \begin{cases} \dfrac{T_{\text{co}}(t) - T_{\text{ci}}(t)}{T_{\text{hi}}(t) - T_{\text{ci}}(t)} & (m_{\text{h}}(t)c_{\text{h}} > m_{\text{c}}(t)c_{\text{c}}) \\ \dfrac{T_{\text{hi}}(t) - T_{\text{ho}}(t)}{T_{\text{hi}}(t) - T_{\text{ci}}(t)} & (m_{\text{h}}(t)c_{\text{h}} < m_{\text{c}}(t)c_{\text{c}}) \end{cases} \tag{18.4}$$

where T, m and c refer to temperature (°C), mass flow rate (kgs^{-1}) and specific heat capacity (Jkg^{-1}K^{-1}), respectively. Subscripts ci and co refer to the cold fluid inlet and cold fluid outlet connections to the heat exchanger and subscripts c and h refer to the cold fluid and hot fluid, respectively. In Equation 18.4 the heat exchange effectiveness at some point in time, t, for processes undergoing sensible counter-flow heat transfer, for example, is given by Equation 18.5.

$$E(t) = \frac{1 - \exp[-NTU(t)(1 - C(t))]}{1 - C(t)\exp[-NTU(t)(1 - C(t))]} \tag{18.5}$$

Alternatively, where one fluid is undergoing isothermal boiling or condensing heat transfer, Equation 18.6 may be used.

$$E(t) = 1 - \exp(-NTU(t)) \tag{18.6}$$

In the above, NTU is the number of (heat) transfer units (Equation 18.7) and C is the heat exchanger capacity ratio (Equation 18.8). For a given heat exchanger with heat exchange area A and overall heat transfer coefficient, U (Wm^{-2}K^{-1}), NTU and C are given as follows:

$$NTU(t) = \frac{AU}{\min(m_{\text{c}}(t)c_{\text{c}}, m_{\text{h}}(t)c_{\text{h}})} \tag{18.7}$$

$$C(t) = \frac{\min(m_{\text{c}}(t)c_{\text{c}}, m_{\text{h}}(t)c_{\text{h}})}{\max(m_{\text{c}}(t)c_{\text{c}}, m_{\text{h}}(t)c_{\text{h}})} \tag{18.8}$$

Example 18.3 Steady-state analysis

The condenser of a heat pump has an overall *AU*-value of 35 kWK^{-1} and serves a hot water space heating system with a circulating mass flow rate of 5 kgs^{-1}. The mass flow rate remains constant at all part-load conditions and the heating water temperature range at design conditions is 10 K. Heating control is achieved by varying the flow temperature to the heating system using a schedule which has a heating water temperature range of 30 to 55 °C (design) corresponding to an external air temperature range of −3 °C (design) to 15 °C. On a particular day, the external air temperature is 5 °C at the start of the heating cycle and rises to 10 °C by the end of the heating cycle. Find the range of heat pump condensing temperatures. (Take the specific heat capacity of water to be 4.2 $kJkg^{-1}K^{-1}$.)

Solution:

A schedule allows the controlled variable set point to be adjusted with respect to some other reference variable. Here, the heating water set point at the heat pump condenser outlet is scheduled with respect to the external air temperature as shown in Figure 18.2. This can be interpreted as a linear equation with a negative gradient of 15 K / 18 K and an intersection at 0 °C external temperature of 52.5 °C heating water temperature.

When the heating first switches on, the external air temperature is 5 °C and so the schedule will fix the required heating flow water temperature at 52.5−15 × 5 / 18 = 48.3 °C.

Because the 'hot' side of the condenser involves a vapour condensing nominally isothermally, the specific heat of this vapour, $dh/dt \rightarrow \infty$ so $m_c \times c_c = 5 \times 4.2 = 21$ kJK^{-1} is the smaller of the two heat capacities which results in $NTU = 35 / 21 = 1.667$.

The relevant calculation for the effectiveness will be Equation 18.6 since this heat exchanger involves condensing heat transfer.

$$\therefore\ E\left(5\ ^\circ\mathrm{C}\right) = 1 - \exp\left(-1.667\right) = 0.811$$

The water temperature difference across the condenser at these conditions can be calculated by assuming that it varies in proportion to the scheduled load since the heating water mass flow rate is constant. Thus, the present water temperature range will be 10 × (15 − 5)/(15 − −3) = 5.56 K. So, the present heating water return temperature will be 48.3 − 5.56 = 42.74 °C.

Applying Equation 18.4 in which $m_h \times c_h > m_c \times c_c$: $0.811 = 5.56 / (T_{hi}(t) - 42.74)$, from which the condensing temperature (i.e. $T_{hi}(t)$) at these conditions is found to be 49.6 °C.

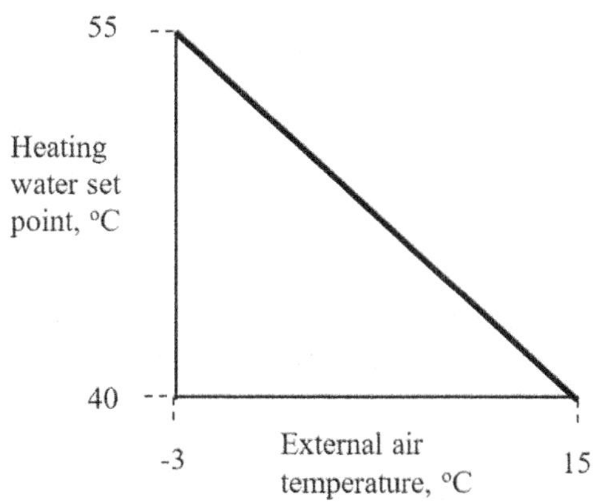

Figure 18.2 Heating schedule (used in Example 18.3)

Re-assessing for the conditions later in the day, in which the external air temperature has become 10 °C, the following are obtained:

Updated heating water flow temperature: 44.17 °C.
NTU and, hence, *E* are unchanged.
Updated heating water inlet temperature: 41.39 °C.
Revised condensing temperature: 44.8 °C.

Greater detail (and precision) in modelling information is evident in this method. The prediction of heat pump condensing temperatures means that the heat pump condensing pressure can be obtained. If the evaporating temperature (and, hence, compressor suction pressure) can be obtained by a similar method to that illustrated in this example, then the pressure ratio can be used to predict part-load performance of the compressor itself leading to an accurate prediction of heat pump performance and power consumption. (In a complete building simulation, the load on the heating system would be known from other variables, so the rather crude assumption of heating demand variation used in the above illustration would not be necessary.)

End of solution.

Though the above method works well for inflow heat exchangers, room-based heat exchangers such as radiator-convectors (i.e. 'baseboard' heaters) with mixed mode radiant and convective heat transfer on the room side of the heat exchanger require a different approach. A simple but adequate prediction of room emitter output can be obtained following the methods in BS EN 442–3 (BSI 2008a) and the Hydronics Institute (1990) as summarized here in Equation 18.9.

$$q(t) = K\left(T_{\mathrm{m}}(t) - T_{\mathrm{r}}(t)\right)^{n} \tag{18.9}$$

where:

- q is the heat emitter output (W);
- K is the overall heat transfer coefficient (WK^{-1});
- T_{m} is the meant heat emitter surface temperature (°C);
- T_{r} is the room air temperature (°C);
- n is an exponent dependent on the type of emitter.

Catalogue-fit modelling

The availability of comprehensive manufacturers' catalogue data describing the performance of certain types of plant can offer a simple and convenient route to robust plant modelling – so-called 'catalogue-fit' models. The more comprehensive examples of data can be found for certain types of primary plant – in particular heat pumps and chillers. Because these plants can be both complex and difficult to fully parameterize for the purpose of modelling from first principles, catalogue-fitted models have considerable appeal. The comprehensiveness of the data arises from the international standards that manufacturers are required to report their product performances to – examples for heat pumps and air conditioners include BS EN 14825 in Europe (BSI 2008b) and the ARI standards in the USA (ARI 1998).

The common structure is a polynomial relating an output variable such as a performance metric (e.g. coefficient of performance, *CoP*) to a single input such as source or sink temperature as depicted in Equation 18.10.

$$CoP(t) = a_0 + a_1 T(t) + a_2 T(t)^2 + ... + a_m T(t)^m \tag{18.10}$$

in which $a_0 \ldots a_{\mathrm{m}}$ are regression fitting constants.

For more detailed models, the bi-quadratic structure given in Equation 18.11 may be used. Here, an output variable, $z\,(x,y)$, is fitted to two or more independent variables using a multiple regression fitting method ($a_0 \ldots a_2$ and $b_0 \ldots b_2$ are fitting constants).

$$z(x,y) = \left(a_0 + a_1x + a_2x^2\right) \times \left(b_0 + b_1y + b_2y^2\right) \tag{18.11}$$

As an example, Yik et al. (2012) offer the following relationship for chiller power demand, W, as a function of air conditioning load, q_c, and condenser air or water inlet temperature, T_i, for oil-free centrifugal chillers (Equation 18.12). The fitting constants for the equipment in their study can be found in Table 18.3.

$$\begin{aligned} W(t) &= a_0 + a_1q_c(t) + a_2q_c(t)^2 + a_3T_i(t) + a_4T_i(t)^2 + a_5q_c(t)T_i(t) \\ &+ \ldots\, a_6q_c(t)^2\,T_i(t) + a_7q_c(t)T_i(t)^2 + a_8q_c(t)^2\,T_i(t)^2 \end{aligned} \tag{18.12}$$

Another common application for catalogue-fit type models is for use in modelling fans and pumps. Third-order polynomial fitting usually gives a very good accuracy and representation. For example, fan/pump pressure (P) can be correlated with inlet volume flow rate (V) using the form shown in Equation 18.13, and fan/pump power consumption (W) can be correlated with inlet volume flow rate V as indicated in Equation 18.14. Again, $a_0 \ldots a_3$ and $b_0 \ldots b_3$ are regression fitting constants and P_0, V_0 and W_0 are reference values of fan/pump pressure, inlet volume flow rate and power respectively.

$$P(t) = a_0 + a_1\left(\tfrac{P_0}{V_0}\right)V(t) + a_2\left(\tfrac{P_0}{V_0^2}\right)V(t)^2 + a_3\left(\tfrac{P_0}{V_0^3}\right)V(t)^3 \tag{18.13}$$

$$W(t) = b_0 + b_1\left(\tfrac{W_0}{V_0}\right)V(t) + b_2\left(\tfrac{W_0}{V_0^2}\right)V(t)^2 + b_3\left(\tfrac{W_0}{V_0^3}\right)V(t)^3 \tag{18.14}$$

In common with examples such as Equation 18.13, these equations merely need to be solved at building modelling time intervals to provide an updated value of plant status and output. One criticism of this approach is that the manufacturers' data upon which these component models are based are themselves usually based on a predictive model rather than comprehensive test data. In effect, a model is being fitted to another model. However, the manufacturers' data have

Table 18.3 Fitting constants for the catalogue-fit example model in Equation 18.12

Constant	*Air-cooled*	*Water-cooled*
a_0	0.007000	0.048346
a_1	−0.008249	0.184133
a_2	0.461093	0.201874
a_3	−0.0005403	−0.002417
a_4	0.00002641	0.00006864
a_5	0.018328	0.002683
a_6	−0.019843	−0.001912
a_7	−0.0001393	0.000432
a_8	0.0006012	0.0001234

Source: Yik et al. 2012

to be accurate and robust in order to comply with the standards to which they relate (not to mention the risk of litigation). A further criticism is that these models are only suitable strictly within the ranges of the data to which they are fitted. This is a matter for the user/modeller to take into consideration.

Dynamic-state modelling

For a more rigorous and detailed treatment of modelling building plant, recognize that variables change in reality with both time and spatial position. In practice, the rates of change of variables describing plant behaviour with respect to time tend to be much faster than similar variables describing behaviour in the fabric of buildings. This results in a wide range of time rates-of-change when all of the equations for both building fabric (as described elsewhere in this book) and plant are put together – a so-called 'stiff' system of equations. Stiff models are often tricky to solve and require high computational power (often alleviated through variable-step solvers), so ideas for simplifying the problem can be helpful. For example, relaxing the plant modelling technique to quasi-steady-state as described in the previous section. Alternatively, when solving problems over short time horizons, the building model can be relaxed to a fixed boundary or steady-state problem allowing an efficient focus to be placed on the plant. As far as spatial rates of change are concerned, this will tend to be of greatest interest with larger distributed systems, for example large complex heat exchangers, underfloor heating, chilled ceilings, etc. For a rigorous approach in which both time and spatial position play roles as independent variables, problem descriptions require resorting to partial differential equations (PDEs).

Linear systems of equations

In many cases, it will be possible to restrict the spatial domain to one dimension only – usually that along the flow path, thus assuming perfect lateral mixing. For example, a room-based heating operation using hot water as a dynamic spatially distributed problem can be described by Equation 18.15.

$$\frac{C_{\mathrm{hm}}}{x}\frac{\partial T}{\partial t} = m_{\mathrm{h}} c_{\mathrm{h}} \frac{\partial T}{\partial x} - \frac{AU}{x}\left(T - T_{\mathrm{r}}\right) \tag{18.15}$$

where:

- C_{hm} is the heat capacity of the fluid and its bounding heat exchange surface ($\mathrm{WK^{-1}}$);
- m_{h} is the mass flow rate of hot water ($\mathrm{kgs^{-1}}$);
- c_{h} is the specific heat capacity of the hot water ($\mathrm{Jkg^{-1}K^{-1}}$);
- A is heat transfer area ($\mathrm{m^2}$);
- U is an overall heat transfer coefficient ($\mathrm{Wm^{-2}K^{-1}}$);
- T is the fluid temperature (°C);
- T_{r} is the room temperature (°C);
- x is the length of the heat exchanger along the flow path (m).

For linear time-invariant problems in which model parameters remain constant (i.e. C_{hm}, m_{h}, c_{h}, AU are constant), a convenient approach is to discretize the problem spatially which results in a set of ordinary differential equations in which time is the only independent variable. The advantage of this approach is that a wide variety of numerical integration algorithms can be used to solve the resulting equation set. By dividing the spatial domain, x, into N equally sized zones each of width, Δx, a set of N ordinary differential equations will arise. The partially discretized form of Equation 18.15 for the nth zone can be expressed as Equation 18.16:

$$\frac{C_{\mathrm{hm}}}{N\Delta x}\frac{\mathrm{d}T_n}{\mathrm{d}t} = \frac{m_{\mathrm{h}}c_{\mathrm{h}}}{\Delta x}\left(T_{n-1} - T_n\right) - \frac{AU}{N\Delta x}\left(T_n - T_{\mathrm{r}}\right) \tag{18.16}$$

which reduces to Equation 18.17:

$$\frac{\mathrm{d}T_n}{\mathrm{d}t} = \alpha\left(T_{n-1} - T_n\right) - \beta\left(T_n - T_{\mathrm{r}}\right) \tag{18.17}$$

in which $\alpha = \frac{Nm_{\mathrm{h}}c_{\mathrm{h}}}{C_{\mathrm{hm}}}$ and $\beta = \frac{AU}{C_{\mathrm{hm}}}$, where C_{hm} is the overall thermal capacity of the water occupying the emitter and the material forming the heat transfer surface of the emitter. At the inlet zone of the emitter, T_{n-1} is replaced with the inlet water temperature. As a set of N ordinary differential equations will arise, a matrix differential equation can be written (Equation 18.18):

$$\dot{\mathbf{T}} = \mathbf{Ax} + \mathbf{Bu} \tag{18.18}$$

where $\dot{\mathbf{T}}$ is an $N \times 1$ vector of derivatives, **A** is an $N \times N$ matrix of coefficients, **x** is an $N \times 1$ vector of input variables, **B** is an $N \times 2$ matrix of boundary variable coefficients. (There are two boundary variables – the emitter inlet water temperature and the room air temperature.) **u** is a 2×1 vector of boundary variable values.

An expansion of Equation 18.18 is given in Equation 18.19. Note that the **A** matrix of coefficients is diagonally dominant.

$$\begin{bmatrix} \dot{T}_n \\ \dot{T}_{n+1} \\ \cdot \\ \cdot \\ \cdot \\ \dot{T}_N \end{bmatrix} = \begin{bmatrix} -(\alpha+\beta) & 0 & \cdot & \cdot & \cdot & 0 \\ \alpha & -(\alpha+\beta) & 0 & \cdot & \cdot & 0 \\ 0 & \alpha & -(\alpha+\beta) & 0 & \cdot & 0 \\ 0 & \cdot & \cdot & \cdot & \cdot & 0 \\ 0 & \cdot & \cdot & \cdot & \cdot & 0 \\ 0 & \cdot & \cdot & 0 & \alpha & -(\alpha+\beta) \end{bmatrix} \times \ldots$$

$$\ldots \begin{bmatrix} T_n \\ T_{n+1} \\ \cdot \\ \cdot \\ \cdot \\ T_N \end{bmatrix} + \begin{bmatrix} \alpha & \beta \\ 0 & \beta \\ \cdot & \cdot \\ \cdot & \cdot \\ \cdot & \cdot \\ 0 & \beta \end{bmatrix} \times \begin{bmatrix} T_{\mathrm{inlet}} \\ T_{\mathrm{r}} \end{bmatrix} \tag{18.19}$$

A key issue is the choice of the number of modelling zones, N, which will tend to vary from application to application. Further consideration will be given to this in an application case study in the next section and for a detailed treatment of this applied to room heat emitters, see Fong et al. (2015).

Nonlinear systems of equations

An inspection of Equation 18.16 shows two obvious sources of nonlinearity. First, if the heating system mass flow rate, m_{h}, is not constant, then a product of two variables arises (mass flow rate and temperature). In fact, the variation of heating water mass flow rate is a common method of plant capacity control, so this will arise in many practical modelling applications. Second, a sophisticated plant model is likely to incorporate a function for the calculation of the overall heat transfer coefficient, AU, which is likely to vary independent of time as a function of other variables in the model. These nonlinearities mean that a more flexible approach will be needed for the treatment of Equation 18.16. This is afforded by fully discretizing in both time and space and solving the resulting algebraic equation set iteratively. Re-casting Equation 18.16 with the time term discretized using a backward-in-time method and simplifying results in Equation 18.20.

$$T_{n,t} = \frac{C_{\mathrm{hm}} T_{n,t-\Delta t} + \Delta t N m_{\mathrm{h}} c_{\mathrm{h}} T_{n-1,t} + \Delta t AU T_{\mathrm{r},t}}{C_{\mathrm{hm}} + \Delta t N m_{\mathrm{h}} c_{\mathrm{h}} + \Delta t AU} \tag{18.20}$$

An iterative loop must be visited at each time instant to solve an equation such as 18.20. This is because, though the temperatures along the heat exchanger in each zone are known at the previous time instant, they are not known at the current time instant. A glance at Equation 18.20 shows temperatures at the current time instant on both sides of the equation ($T_{n,t}$ and $T_{n-1,t}$) and so an iterative approach is used. A common method is to use a Gauss-Seidel scheme (also known as successive displacement or successive substitution). This method works as follows with reference to Equation 18.20.

Step 1 – at the current time instant, t, $T_{n\,...\,N,t-\Delta t}$ will be known and $T_{\mathrm{r},t}$ will be supplied from other model equations. (If m_{h} and/or AU are time-variant; $m_{\mathrm{h},t}$, AU_t, they too will be updated from other model equations.)

Step 2 – visit each zone, n, and update $T_{n,t}$ using Equation 18.20 holding $T_{n\,...\,N,t-\Delta t}$ constant. This will results in an updated set of values $T_{n\,...\,N,t}$. Store the results.

Step 3 – check the updated values $T_{n\,...\,N,t}$ with the previously stored values. If all values are within an acceptable tolerance of the previously stored values at each zone location, replace $T_{n\,...\,N,t-\Delta t}$ with $T_{n\,...\,N,t}$, advance time by Δt and continue. If not, go to Step 2.

Modelling case studies in plant and control

In this section, consideration will be given to several contrasting plant modelling case studies making use of some of the theories described in the foregoing sections. The first of these will deal with the question of spatial discretization in plant modelling. The second will consider a primary plant application involving a heat pump coupled to a thermal store. The third case study will consider the modelling of a simple feedback control loop applied to air heating and cooling plant.

Case study 1: heat emitter model discretization analysis

In this case study, the influence of the number of modelling zones for a room emitter model of the type described in the previous section will be considered. Consider a hot water heat emitter with a design-rated capacity of 3 kW when receiving hot water at 50 °C (inlet) and 40 °C outlet. The heat emitter is in a room space to be maintained at a nominal design temperature or 19 °C when the external air temperature is −3 °C. Compare the results of an arbitrary step change in heating water inlet temperature from 50 °C to 60 °C when the heat emitter is divided into 3 equally sized modelling zones and 10 equally sized modelling zones. Assume, for this exercise, that the room heat transfer can be described using a steady-state heat loss model.

Model used: Equation 18.17 interpreted as Equation 18.19.

Constants: α and β evaluated to be $4.524 \times 10^{-3}\ s^{-1}$ and $5.75 \times 10^{-4}\ s^{-1}$ respectively (3rd-order model) and α will be $1.508 \times 10^{-2}\ s^{-1}$ for the 10th-order model (β unchanged).

Implementation: Use a Matlab-Simulink state-space block to implement Equation 18.19 with a Rosenbrock variable-step solver with the maximum step size fixed at 1 s.

The model is given in Figure 18.3 based on the step change in inlet hot water temperature occurring after 300 s of initial steady-state plant response. Results are shown in Figure 18.4 giving a comparison of the two room air temperature predictions (°C) and the two heat emitter output predictions (kW).

Conclusion: The room temperature settles approximately 1200 s after the step change in inlet heating water temperature at 22.165 °C for the 3rd-order model and 22.568 °C using the 10th-order model. The final heat transfer values from the emitter are 3.43 kW (3rd-order) and 3.49 kW (10th-order). Though the final heat transfer values are very similar, the higher order model is noted to take longer to settle. Though the difference between final temperatures appears relatively small (0.4 K), this may be significant if the modelling application is used for control system simulation. This is because relatively small deviation in controlled variable

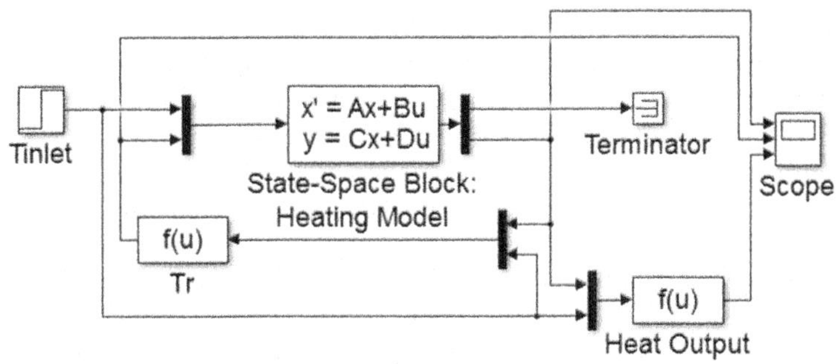

Figure 18.3 Information flow diagram for modelling case study 1

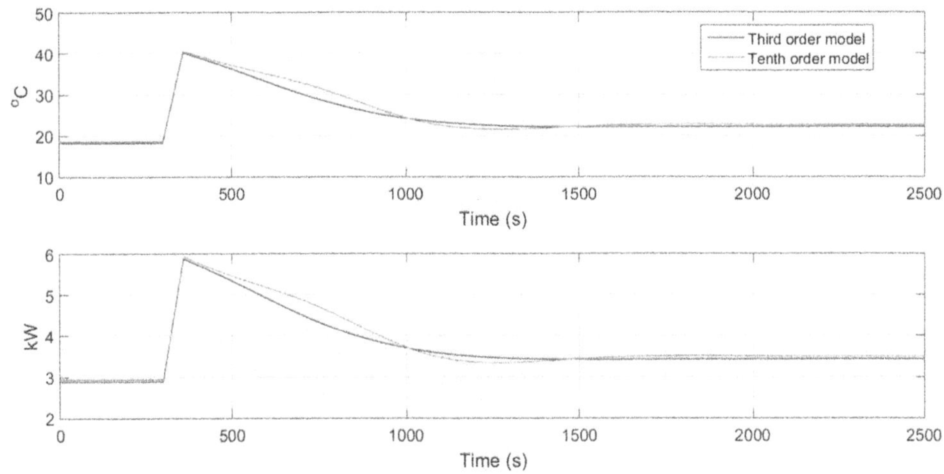

Figure 18.4 Results for case study 1

value (i.e. likely to be room temperature in this case) can have a relatively high effect in control signal value. For a more detailed treatment of this subject, refer to Fong et al. (2015).

Case study 2: heat pump coupled to a hybrid thermal store

This case study considers the charge and discharge transient of a novel hybrid thermal store coupled to an air-source heat pump compared with a conventional water tank of the same volume. The thermal store consists of a salt-hydrate phase change material encapsulated in a set of stainless steel coils mounted in a water tank. The tank is designed so that it will fit into a standard kitchen under-bench module in a residential application. Full details of the modelling methods used and parameter and input data sets can be found in Underwood et al. (2018).

The model information flow diagram in this case is shown in Figure 18.5. The input vector for the heat pump defines the external (source) air temperature, primary heating mass flow rate and the primary heating inlet water temperature to the heat pump. For the hybrid thermal store, the input vector comprises the primary heating inlet temperature and mass flow rate (from the heat pump), secondary heating system temperature and mass flow rate, ambient surrounding temperature and heating status switch (charging or discharging).

Two simulations are compared here – the hybrid thermal store and an alternative thermal store consisting of a water tank of the same volume. The simulations give the charge phase transient tank outlet temperature and the corresponding discharge phase temperature over one complete cycle. Results are plotted in Figure 18.6 – the top graph showing the tank temperature during charge and discharge based on the two alternative tank types and the bottom graph showing the heat transfer discharged to the heating system (kW).

Results: Thermostat switching is evident during charge in the case of the hybrid thermal store. This is because the phase change material takes a finite time to fully melt, which prompts several heat pump switching events. During discharge, it can be seen that the hybrid tank is able to hold a load of approximately 4 kW for at least 2 h, whereas the conventional tank has fully discharged in this time. This is significant because a modern well-insulated house of medium size will have a space heating design load of, typically, 4–5 kW in the UK (as an example). This means that the hybrid tank would be able to satisfy heating demand during the crucial electrical grid-stress phase in later afternoon/early evening (Underwood et al. 2018). The heating spike at the start of discharge is due to cold heating system water initially displacing warm tank water

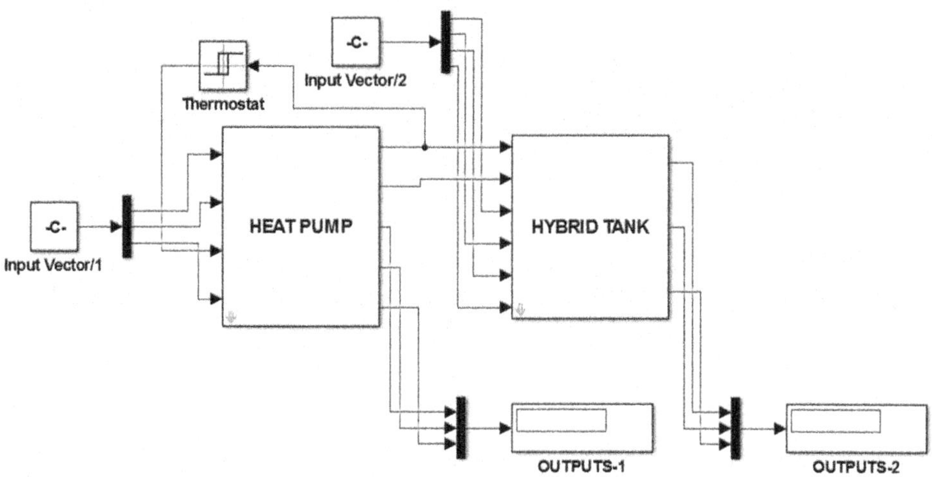

Figure 18.5 Information flow diagram for case study 2

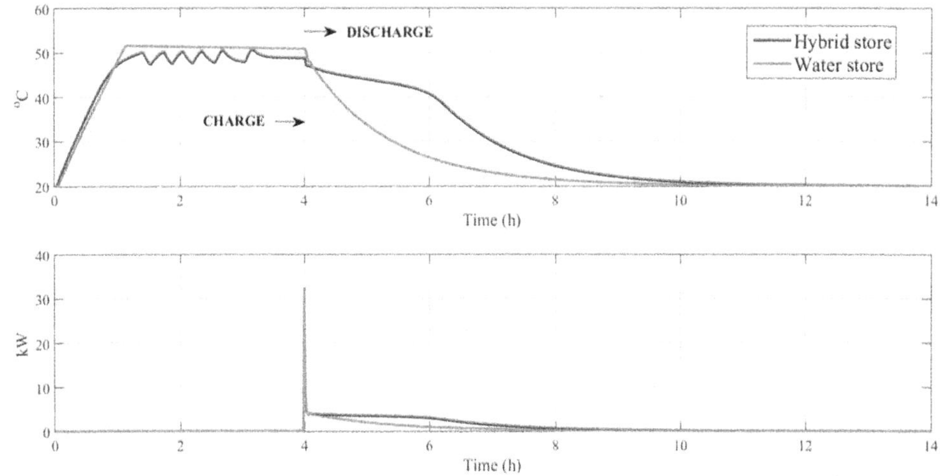

Figure 18.6 Results for case study 2

and once the heating system water catches up, this spike quickly settles down to the nominal heating system output value.

Case study 3: feedback control of an air cooling coil

In this final case study, consideration is given to a simplified treatment of the modelling of modulating feedback control of a common plant operation – that of air cooling using a chilled water cooling coil. The coil model is a 3rd-order model adopting a relatively simple contact factor approach (essentially, similar to the effectiveness concept described in Equation 18.6 but with the coil sub-divided into three equally sized zones for dynamic modelling). The coil is modelled using the heat transfer correlations presented in Gomaa et al. (2002).

The coil is parameterized with a 1 m^2 frontal area and an air face velocity of 2 ms^{-1}. Chilled water at a constant 7 °C is admitted at a design maximum flow rate of 3 kgs^{-1}. The flow rate is varied in response to the error between the coil supply air outlet temperature and its set point value. The control signal is generated using a fixed gain value (which can be adjusted) – essentially proportional control. A saturation block is used to ensure that the chilled water mass flow rate cannot stray beyond its limits of 0–3 kgs^{-1}. The complete model is shown in the information flow diagram of Figure 18.7.

The plant was simulated over a 1 h period using a uniform random number sequence for the disturbed inlet air temperature (varying in the range 18–28 °C). The coil air outlet set point temperature was set to 12 °C and, at 30 minutes, the set point was increased at 14 °C in order to further explore the control tracking behaviour. Results are shown in Figure 18.8.

Results: Using an arbitrary controller gain value of −1 (the negative sign to correct the control action for negative feedback), the control tracking is reasonably good especially when it is considered that the disturbance sequence is somewhat arbitrary for this illustration. This can be seen in graph ‘A’ of Figure 18.8 (lower curve). It is also possible to see the increase in outlet air temperature starting at 30 minutes due to the increase in control set point. The increase in set point now means that the plant can operate more comfortably within its range (at the lower set point it was operating at close to its design capacity), and this is evidenced by a slightly lower variance in the controlled variable value as well as in chilled water flow rate (graph ‘B’ in Figure 18.8) during which the flow rate spends less time at its upper limit value. Of course, this is far from optimal control and more sophisticated control methods could be

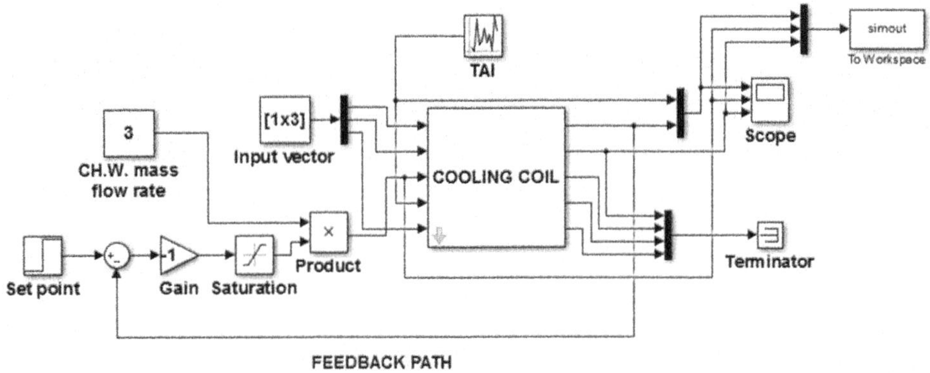

Figure 18.7 Information flow for case study 3

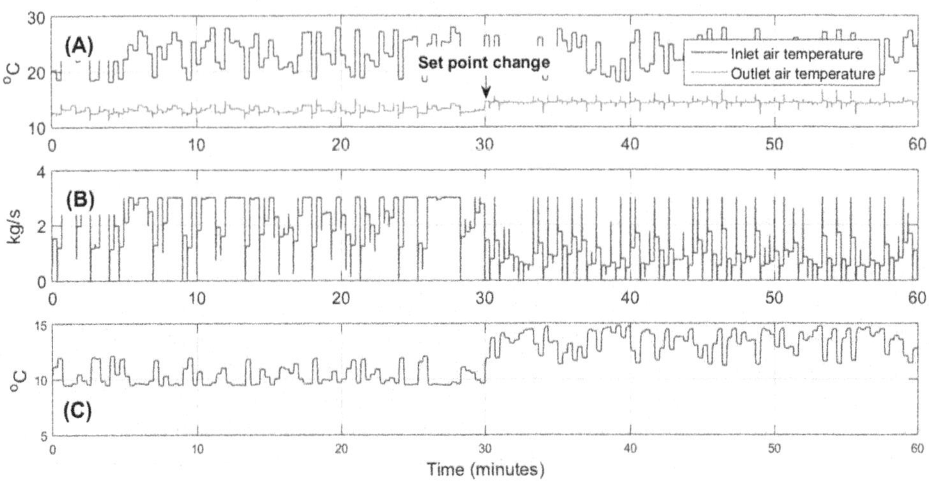

Figure 18.8 Results for case study 3. (A) inlet and outlet air temperature; (B) chilled water mass flow rate; (C) chilled water outlet temperature.

considered, but this case study illustrates the potential for plant modelling at a detailed level of enquiry.

For a more detailed treatment of the distributed modelling of cooling coils, refer to Underwood and Yik, chapter 5 (2004), and for a more in-depth look at control systems and control system modelling for these types of application, refer to Underwood (1998).

Solar thermal collectors

A heat balance approach to modelling solar collectors needs to consider the incoming solar radiation normal to the solar collector and the losses to surroundings from the collector surfaces. In sunny but cold climates, the influence of the latter is particularly important. The solar radiation striking the surface-normal of the array will be reduced according to the transmissivity of the collector glass cover plate(s) and the absorptivity of the collector absorber surface before it reaches the heat transfer fluid. The losses from the collector on the other hand will be dependent of whether the absorber space is evacuated, which reduces collector losses to, predominantly, skywards long-wave radiation. The energy balances about the collector fluid can be expressed with a modified form of Equation 18.17 as given in Equation 18.21.

$$\frac{dT_n}{dt} = \alpha(T_{n-1} - T_n) - \beta(T_n - T_a) + A_{abs}S \qquad (18.21)$$

in which A_{abs} is the absorber area and S is the solar radiation absorbed by the collector absorber plate (kWm^{-2} or Wm^{-2}). α and β can be defined in a similar manner as for Equation 18.17 with the exception that the AU component will represent the overall collector loss coefficient with its surroundings (kWK^{-1} or WK^{-1}). Note also that the fluid specific heat capacity in this case (i.e. in α) will not necessarily be for water but will depend on any antifreeze additive used.

For a simplified steady-state analysis of collector behaviour, Equation 18.21 can be reduced to Equation 18.22.

$$q_{utilised} = A_{abs}S - A_{col}U_L(\bar{T}_{cf} - T_a) \qquad (18.22)$$

in which:

$\bar{T}_{cf} = (T_{cf,o} + T_{cf,i}) / 2;$

$T_{cf,o}$ is the collector fluid temperature at outlet (°C);
$T_{cf,o}$ is the collector fluid temperature at inlet (°C);
T_a is the ambient temperature (°C);
A_{col} is the collector area (m^2 – note that A_{abs} is usually slightly smaller than A_{col});
U_L is the collector heat loss coefficient to surroundings (Wm^{-2}K^{-1}).

The overall heat utilization from the collector can be obtained from Equation 18.23.

$$q_{system} = \max\left(0, m_{cf}c_{cf}\left(T_{cf,o} - T_{cf,i}\right)\right) \qquad (18.23)$$

Related to this, an alternative modelling approach is to compare the actual heat delivered to the hot water system with the heat utilized by the collector had it been uniformly at a temperature equivalent to the inlet water temperature, $T_{cf,i}$. This defines a heat removal factor for the collector, F_R (Equation 18.24).

$$F_R = \frac{m_{cf}c_{cf}\left(T_{cf,o} - T_{cf,i}\right)}{A_{abs}S - A_{col}U_L(T_{cf,i} - T_a)} \qquad (18.24)$$

It is now possible to define a collector thermal effectiveness, η_{col} (Equation 18.25):

$$\eta_{col} = \frac{q_{system}}{A_{col}\Sigma G} \qquad (18.25)$$

in which $\Sigma G = G_{dir,col\_} + G_{dif,col} + G_{gr,col}$ (WK^{-1} or kWK^{-1}) where $G_{dir,col}$ is the direct solar radiation corrected to the normal of the solar collector, $G_{dif,col}$ is the diffuse solar radiation corrected to the normal of the solar collector and $G_{gr,col}$ is the surroundings-reflected solar radiation corrected to the normal of the solar collector. For the solar absorbed by the collector absorber (S in Equations 18.21 and 18.22), the direct, diffuse and ground-reflected components require to be further corrected to account for the proportion of radiation at the collector surface-normal that is transmitted through the glass cover plate(s) of the collector and for the proportion of the transmitted radiation that is absorbed by the absorber plate – the so-called $\tau\alpha$ product. These corrections are given in Equation 18.26.

$$S = G_{\text{dir,col}}\tau\alpha_{\text{dir}} + G_{\text{dif,col}}\tau\alpha_{\text{dif}} + G_{\text{gr,col}}\tau\alpha_{\text{gr}} \tag{18.26}$$

The transmissivity of the essentially clear glass used for cover plates varies to some extent with the angle of incidence of the beam radiation and with the number of cover plates used. Two or more glass cover plates will reduce the transmitted solar radiation to the collector but will also reduce the top-loss component of the heat loss coefficient so a trade-off will be sought. Typically, the transmissivity will vary in the range 0.7–0.9 corresponding to four cover plates and one cover plate respectively. A larger number of cover plates is most likely to make sense in situations with long periods of excellent bright sunshine but with low ambient temperatures (i.e. certain high-altitude locations). Further details on the variation of $\tau\alpha$ can be found in the *ASHRAE Handbook* (ASHRAE 2016). The transmissivity will remain more or less constant for angles of solar incidence down to about 50° but will fall sharply at higher values of incidence angle. Corrections for angles of incidence off-normal, θ, can be obtained from Equation 18.27, where K is a non-dimensional correction factor applied to $\tau\alpha$. The empirical coefficient b_o in this equation is approximately −0.1 for single glass covers (Klein 1975).

$$K_{\tau\alpha} = 1 + b_o\left(\frac{1}{\cos\theta} - 1\right) \tag{18.27}$$

As to the absorber plate absorptivity, typical values of around 0.92 can be expected for plates with high-absorptivity coatings.

The overall thermal transmittance value, U_L, in Equation 18.22 and Equation 18.24 will consist of a loss term at the top of the collector and, where relevant, the back and sides. The top loss will be dominant because it will also exhibit a significant contribution of skyward radiation loss besides convection. Wind flow will also play a strong role in the latter. For flat plate collectors, insulation at the back and edges of the collector will help to minimize these losses. Equations 18.28 and 18.29 can be used to account for these losses.

$$U_{\text{edge}} = \frac{A_{\text{edge}}}{A_{\text{col}}} \times \left(R_{\text{so}} + R_{\text{si}} + \tfrac{x_{\text{edge}}}{k_{\text{edge}}}\right)^{-1} \tag{18.28}$$

$$U_{\text{back}} = \left(R_{\text{so}} + \tfrac{x_{\text{back}}}{k_{\text{back}}}\right)^{-1} \tag{18.29}$$

where:

R_{so}, R_{si}	are the external and internal surface resistances respectively (m^2KW^{-1});
k	is thermal conductivity ($Wm^{-1}K^{-1}$);
x	is length (m);
A	is area (m^2).

As to the dominant top-loss U-value, probably the most enduring loss model for flat plate collectors is due to Klein (1975) as given in Equation 18.30 and comprising a convective term and a longwave loss term.

$$U_{top} = \left[\frac{N}{\frac{C}{\bar{T}_{cf}} \times \left(\frac{\bar{T}_{cf} - T_o}{N + f} \right)^e} + \frac{1}{h_o} \right]^{-1} + \frac{\sigma \times (\bar{T}_{cf} + T_a) \times (\bar{T}_{cf}^2 + T_a^2)}{(\varepsilon_P + 0.00591 N h_o)^{-1} + \left(\frac{2N + f - 1 + 0.133\varepsilon_P}{\varepsilon_g} \right) - N} \qquad (18.30)$$

in which:

- ε is surface emissivity (subscript 'g' refers to the glass cover plate and 'p' to the absorber plate);
- σ is the Stefan-Boltzmann constant;
- N is the number of glass cover plates;
- h_o is the external surface convection coefficient ($Wm^{-2}K^{-1}$).

An overall loss coefficient for flat plate collectors is then obtained by summing the three terms (Equation 18.31).

$$U_L = U_{top} + U_{edge} + U_{back} \qquad (18.31)$$

For evacuated tube collectors, convective losses are largely eliminated (other than surface convection from the outer glass surface depending on the amount of solar radiation absorbed in the glass but likely to be relatively minor) and so allowance for longwave loss only is generally sufficient. The complex geometries of these types of collectors means that it is usually preferable to obtain loss factors from test data, which will usually be available from manufacturers.

Simulation case study 4: solar domestic water heating

This treatment of solar collectors is concluded with a case study consisting of an example simulation of a flat plate collector in two contrasting locations. In common with the other case studies considered in this chapter, it should be noted that these examples are presented here as illustrations of the potential uses and applications of the plant modelling methods described in this chapter rather than as robust and definitive solutions to the particular problems presented.

The model used is based on Equations 18.21–18.31. Annual simulations are carried out with increasing collector array area in two contrasting climate locations – the hot city of Havana in Cuba (latitude 23° N) and the cooler city of Newcastle upon Tyne in the UK (latitude 54° N). The objective is to find how much of the hot water demand for a small family could be met by the array and at what overall annual efficiency. A new term is defined to support this – the demand removal factor, D_R, which is defined as the proportion of the hot water demand (including storage tank losses) that is met by the solar collector array. The information flow diagram for this system model is shown in Figure 18.9. A break heat exchanger is included because, for the cold climate application, the collector array fluid would most likely contain an antifreeze

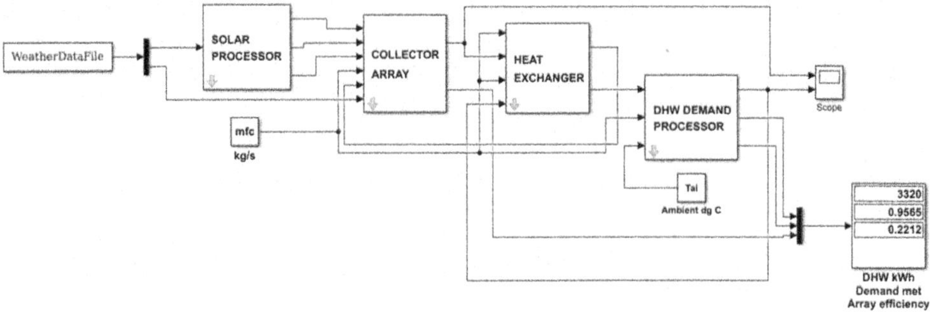

Figure 18.9 Information flow for solar water heating case study

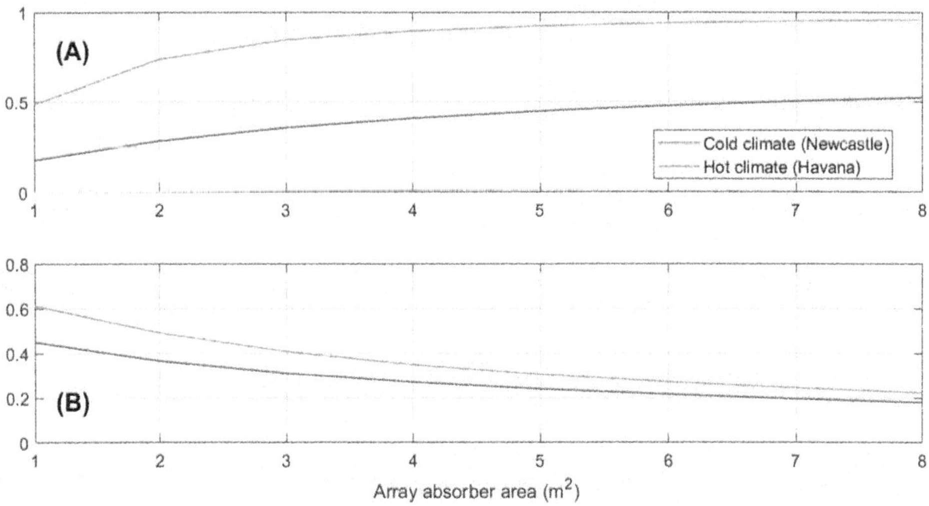

Figure 18.10 Results for solar water heating case study. (A) demand removal factor; (B) collector efficiency.

solution for winter protection. Though this would not be strictly necessary in Havana, it was also included in these simulations for a consistent comparison (though it is unlikely to have had much impact on results). The demand processing block in Figure 18.9 simulates (using a 10th-order model) a hot water storage tank with a cold water inlet and hot water outlet together with primary heat input coming from the break heat exchanger (i.e. solar collector array). The hot water demand is linked to the cold water and hot water inlet and outlet rates which are determined using one of a number of standard daily 'tapping cycles' which are assumed to repeat each day. The tapping cycle used in this case study is defined in BS EN 13203–2 (BSI 2015). The 'medium' family demand set was used from this standard and fed from a 150-litre tank. Annual weather data appropriate for the two city locations was used.

Results: The results are presented in Figure 18.10. The top graph ('A') shows an increasing proportion of demand removal with increasing collector area for both cities which is to be expected. The demand removal is almost fully achieved with the larger collector array size in the hot city of Havana though the rate of improvement slows considerably for array areas greater than 2–3 m^2. For the cooler city of Newcastle, only 50% of the demand can be met with an array size of 6 m^2 and there seems little point in further increases in size. Though the efficiency of the Havana collector array is higher than for Newcastle (where losses will be higher), both locations show reducing efficiency as the collector array size increases due to losses increasing at a higher rate than improvements in demand removal.

Solar photovoltaic systems

Solar photovoltaic (PV) systems are becoming increasingly important elements in building systems and impactful on building performance. Consequently, modelling of such systems has become increasingly important in predicting building performance and in designing both residential and non-residential buildings. Here we provide an introduction to modelling such systems and present a model that has been used in a number of applications to represent realistic system behaviour.

The approach taken to modelling the performance of PV systems depends on the objectives of the modelling exercises and the level of detail of design information (parameter values) available. Whichever model is chosen, we can express the electrical power delivered to the buildings (AC) system as a function of module area, A_{mod}, total solar irradiation on the module (panel) $G_{T,i}$ and functions of conversion efficiency as follows:

$$p_i = A_{mod} G_{T,i} \eta_{mod}(I_i, V_i, T_{c,i}, \ldots) \eta_e(I_i, V_i, \ldots) \tag{18.32}$$

where η_{mod} is the efficiency of conversion of solar to DC electrical power of the module, and η_e accounts for the efficiency of conversion of DC to AC power in the combination of electrical system components often referred to as the 'balance of system' (principally the inverter efficiency). Both these efficiencies are functions of a number of other variables such as the instantaneous currents (I) and voltages (V) but also the thermal state of the panel and other environmental factors. The main difference between the available models is the degree of sophistication with which these functions are represented and also the length of integration of the power and solar fluxes. Possible calculation objectives and corresponding approaches include:

- assessment of PV contributions to the overall annual energy demands using synoptic climate data;
- simulation of the PV system assuming generic parameter values and fixed efficiencies;
- simulation of the renewable system using specific component data and hourly climate data;
- detailed simulation of the PV and electrical system components using short time-step data and explicit representation of the electrical/electronic sub-systems.

The ready availability of hourly climate data means that hourly simulation of PV systems that capture the primary climate-related factors affecting performance are feasible and implemented in many whole-building energy simulation tools. In such tools, prediction of incident solar fluxes and shading is straightforward (i.e. calculated by other models) and so the primary factors of interest – sensitivity to incident solar flux, coincident external temperatures and coincident demands – can be considered. Calculations of variation in performance on an hourly basis also allow tariff variations and import/export balances to be considered in the analysis.

At the early stages of building design – before specific PV products have been specified – it can be useful to make performance calculations with generic performance parameters and simple models. In this case the functions η_{mod} and η_e can be simply fixed values. An increased level of realism can be achieved by calculating efficiencies according to module current and voltage in particular solar, environmental and load conditions. Before describing the details of such approaches, it is necessary to define some characteristics of PV cells and modules.

In general, the current-voltage (*I-V*) characteristics of PV modules are very nonlinear. At a given level of irradiation, the current is nearly constant over much of the operating voltage range and near the upper limit of the voltage range, the current drops dramatically. The power output ($p = IV$) correspondingly has a distinct peak towards the upper end of the voltage range. This is illustrated in Figure 18.11. Such characteristic curves can be defined by three pairs of I and V values at conditions of open circuit (zero current), short circuit (zero voltage) and maximum power: ($I_{oc} = 0$, V_{oc}), (I_{sc}, $V_{sc} = 0$), (I_{mp}, V_{mp}) as also shown in Figure 18.11. The relationship between V_{oc} is very close to linear in most cases as illustrated in Figure 18.12 (left). Data defining these three *I-V* points are published for standard conditions by module manufactures and are typically of the form shown in Table 18.4.

Some characteristics are sensitive to temperature and so power conversion efficiency is also ultimately sensitive to operating temperature. Open circuit voltage and short circuit current are often taken to vary linearly with temperature about test conditions so that coefficients of variation (μ_{Voc} and μ_{Isc} respectively) with temperature are defined and published by manufacturers (for example, as in Table 18.4). Similarly a coefficient of variation of maximum power efficiency is given ($\mu_{p,mp}$) along with the value ($\eta_{mp,ref}$) at the reference conditions. These data are commonly used to derive other model parameters for simulation and design purposes.

Several models suitable for simulation and that use parameters derived primarily from manufacturers' data, like that given in Table 18.4, have been derived based on conceptions of the module cells as equivalent diode circuits. These are typically based on a single-diode representation with either one or two resistances or two-diode representations as illustrated in Figure 18.13. In these representations, the stimulated photocurrent (I_{ph}) is offset by current through the diode (I_d) and leakage current (I_{sh}) through a parallel shunt resistance (R_{sh}) so that the net current (I) is given by

$$I = I_{ph} - I_d - I_{sh} \tag{18.32}$$

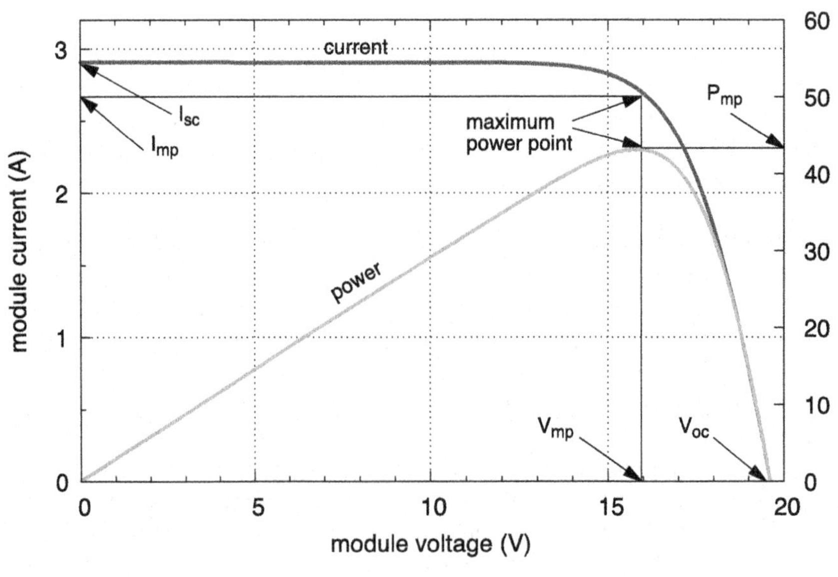

Figure 18.11 Typical characteristics of a photovoltaic module at standard test conditions

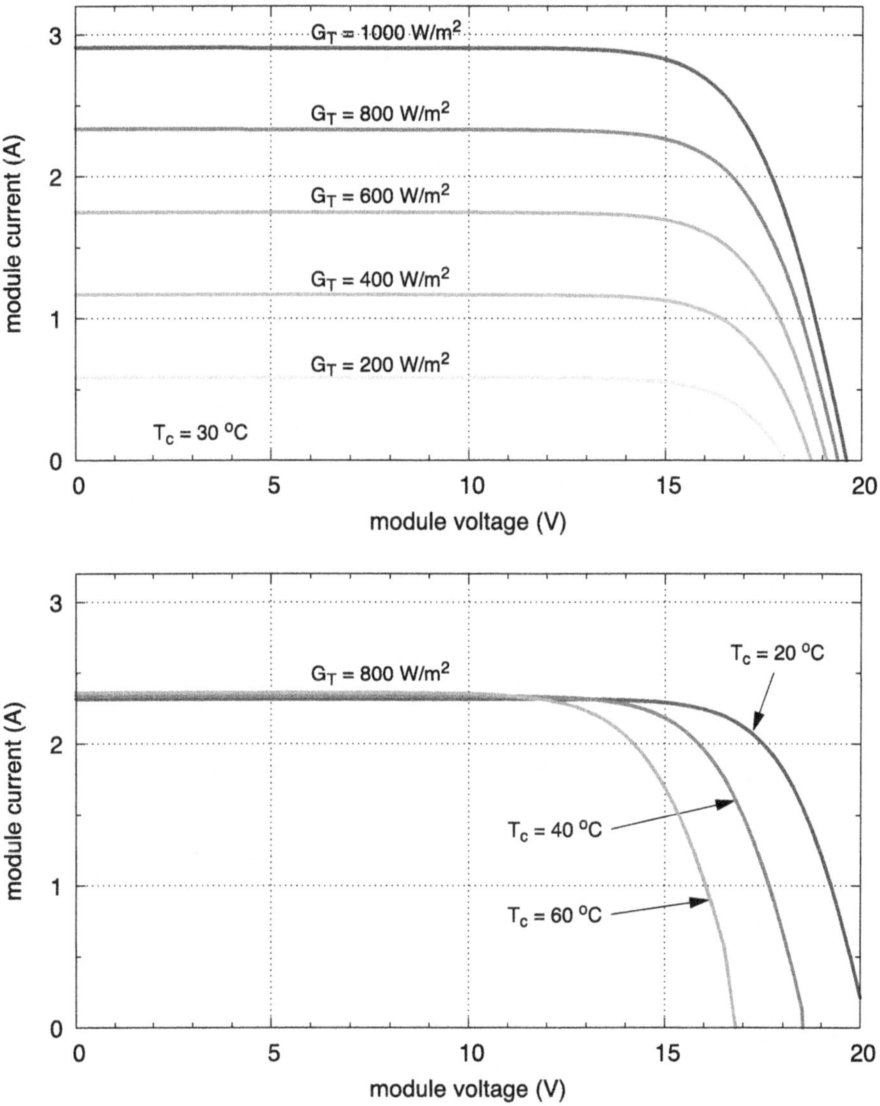

Figure 18.12 Typical variation in photovoltaic module characteristics according to solar irradiation (top) and cell temperature (bottom)

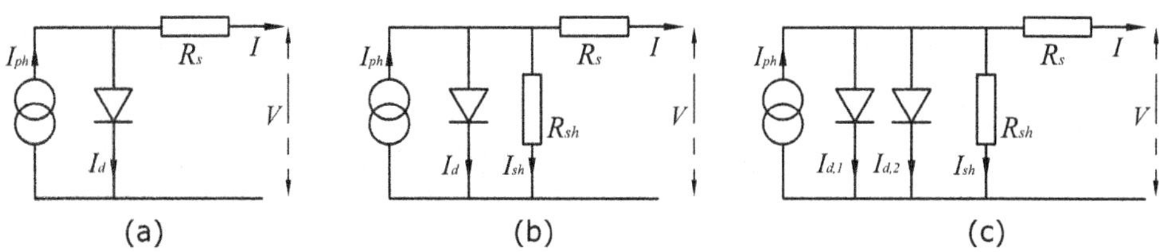

Figure 18.13 Representations of photovoltaic cells based on single diode circuits without (a) and with (b) shunt resistances and two diodes (c)

The relationships between voltage and current for each of these terms can be substituted to give an overall voltage-current expression for the panel. For the diode, the relationship is

$$I_d = I_o \left[\exp\left(\frac{V + IR_s}{nV_T} \right) \right] \tag{18.33}$$

For the shunt resistance current,

$$I_{sh} = \frac{V + R_s I}{R_{sh}} \tag{18.34}$$

with:

$$V_T = \frac{kT_c}{q} \tag{18.35}$$

The overall *I-V* expression becomes:

$$I = I_{ph} - I_o \left[\exp\left(\frac{V + IR_s}{nV_T} \right) - 1 \right] - \frac{V + R_s I}{R_{sh}} \tag{18.36}$$

where:

I_o = diode reverse saturation current (A);
n = the diode ideality factor (between 1 and 2 for real diodes);
k = the Boltzman constant (1.381×10^{-23} J/K);
q = the electron charge (1.602×10^{-19} Coulomb);
T_c = cell operating temperature (K).

As n is effectively an empirical factor that has to be estimated from the known data at the three measurement points, nV_T is sometimes regarded as a combined empirical factor, a (e.g. Duffie and Beckman 1991). The extension of this formulation to a two-diode representation is straightforward but results in a further model parameter that has to be estimated (Masmoudi et al. 2016).

It is possible to make some simplifications of the one diode representation of a photovoltaic module. For example, for many modules the shunt resistance is very high and assuming infinite resistance results in small errors: this eliminates the right-hand term in Equation 18.36. Similarly, the 1 in the second term can be ignored with little error. This is the approach taken in the 2nd edition of Duffie and Beckman (1991). This results in a model with four parameters that have to be estimated and models following this approach have been implemented in popular simulation tools (Klein et al. 1997; Crawley et al. 2001). If the shunt resistance is retained (e.g. as in Duffie and Beckman 2013), then a fifth parameter has to be found.

The three empirically established reference *I-V* points – open circuit, short circuit and maximum power – can be substituted into Equation 18.36 to give relations for the model parameters: I_o, R_s, R_{sh}, n (or a) etc. However, other relationships need to be established to form a set of equations defining all four or five parameters. Details of different approaches as to how this may be done is explained elsewhere (e.g. Schroder 1998). It is necessary to find these using numerical methods.

As the *I-V* characteristic is dependent also on cell operating temperature, it is further necessary to incorporate a method for calculating this in any model of a PV module or cell. A number of approaches are possible by defining a heat balance on the module and relating this to environmental variables. The heat balance can be formulated and a suitable heat transfer parameter found by considering conditions at which the Nominal Operating Cell Temperature (NOCT) is measured (Table 18.4). Absorbed solar irradiation can be balanced against the portion converted into electrical power and heat lost from the module as follows:

$$\tau\alpha G_T = \eta_{mp} G_T + U_L (T_c - T_a) \tag{18.37}$$

where $\tau\alpha$ is the proportion of incident solar irradiation G_T (W/m^2) that is transmitted and absorbed (as in the discussion of thermal collectors above), U_L is an overall heat transfer coefficient (W/m^{2}K) and T_a is the air temperature. The NOCT reference conditions are $T_a = 20$ °C; $G_T = 800$ W/m^2; wind speed 1 m/s. As there is no current drawn from the module in test conditions ($\eta_{mp} = 0$) and substituting the test values, the empirical parameters can be found from

$$\tau\alpha / U_L = (T_{c,NOCT} - T_{a,NOCT}) / G_{T,NOCT} \tag{18.38}$$

For the usual high transmittance low iron glass covering, it is reasonable to take a value of 0.9 for $\tau\alpha$ (as Duffie and Beckman (1991) suggest) so that we can use Equation 18.38 to find the overall module heat transfer coefficient (U_L). The cell temperature can then be calculated from the current air temperature and solar irradiation using

$$T_C = T_a + \frac{G_T \tau\alpha}{U_L}\left(1 - \frac{\eta_{mp}}{\tau\alpha}\right) \tag{18.39}$$

Table 18.4 Performance electrical characteristics of a typical 0.5 m^2 PV panel

PV panel parameter	*Value and units*
Maximum power (p_{max})**	65 W
Warranted minimum power	60 W
Voltage at p_{max}	17.6 V
Current at p_{max}	3.69 A
Short circuit current	3.99 A
Open circuit voltage	22.1 V
Temperature coefficient of short circuit current (μ_{Isc})	(0.065±0.015)%/K
Temperature coefficient of open circuit voltage (μ_{Voc})	−(80±10) mV/K
Maximum power temperature coefficient ($\mu_{p,mp}$)	−(0.5±0.05)%/K
Nominal operating cell temperature (NOCT)*	47±2 °C
Module area: 36 cells 114 mm square	0.468 m^2
Efficiency at standard test conditions (STC)** power ($\eta_{mp.ref}$)	0.139 (13.9%)

*Irradiance of 800 W/m^2, wind speed 1 m/s and air temperature of 20 °C.

**Irradiance of 1000 W/m^2 at an AM1.5G solar spectrum and air temperature of 25 °C.

Source: Data by BP Solar BP365U

Other formulations of heat balance involving more parameters (e.g. wind speed) have been suggested and are available in some software. Other forms of heat balance may furthermore be more appropriate for modules closely integrated with the fabric rather than those mounted on frames.

It should be noted that the equivalent diode representations of PV modules, as outlined earlier, result in an implicit equation (Equation 18.36) that has to be solved numerically along with that for cell temperature at each time step to find the operating point on the *I-V* curve. This requires some representation of the characteristics of the load, e.g. pump or battery, if the system for an 'off-grid' application. Otherwise, assuming the maximum power point is being tracked, the maximum power point has to be found by searching on the *I-V* curve to find the maximum power at each step.

In design calculations (e.g. using monthly climate data), it may be sufficient to use a model of intermediate complexity between one assuming fixed conversion efficiencies and one with a detailed circuit representation. Such a modelling approach is described later following that of Duffie and Beckman (1991). This takes account of the effect of environmental temperature and irradiation on cell temperature – and therefore conversion efficiency – but does not calculate current and voltage explicitly. This could be thought of as a model capturing the first-order effects of variations in cell temperature on conversion efficiency. This modelling approach necessarily assumes that the module is always operating at its maximum power point and is therefore only suitable for building integrated systems using maximum power point tracking (MPPT) equipment. Although first presented as a design method using synoptic climate data, it can be used for hourly analysis where such data are available and is readily implemented in a spreadsheet.

The model is formulated firstly so that the variation in cell conversion efficiency is calculated according to cell temperature. The difference between the efficiency in test conditions and that at a particular cell operating temperature depends can be related to the published maximum power temperature coefficient ($\mu_{p,mp}$) and the corresponding temperature difference as follows:

$$\eta_{mp} = \eta_{mp,ref} + \mu_{p,mp}\left(T_c - T_{ref}\right) \tag{18.40}$$

where η_{mp} is the maximum power efficiency at the current time step, $\eta_{mp,ref}$ is the maximum power efficiency at standard test conditions (STC) and T_{ref} is the STC reference temperature (25 °C). Although some manufacturers give a value for the coefficient ($\mu_{p,mp}$), it is most accurately calculated as $\mu_{p,mp} = \eta_{mp,ref}\mu_{Vocp,mp} / V_{mp}$.

In order to arrive at a formulation that requires no iteration, the expression for cell temperature, Equation 18.39 can be substituted to give

$$\eta_{mp} = \eta_{mp,ref} + \left[1 + \frac{\mu_{p,mp}}{\eta_{mp,ref}}\left(T_a - T_{ref}\right) + \frac{\mu_{p,mp}G_T}{\eta_{mp,ref}}\frac{\tau\alpha}{U_L}\left(1 - \frac{\eta_{mp}}{\tau\alpha}\right)\right] \tag{18.41}$$

The last term $\eta_{mp}/\tau\alpha$ can be approximated by $\eta_{mp,ref}$ with little error. This results in the final equation for module power conversion efficiency:

$$\eta_{mp} = \eta_{mp,ref} + \left[1 + \frac{\mu_{p,mp}}{\eta_{mp,ref}}\left(T_a - T_{ref}\right) + \frac{\mu_{p,mp}G_T}{\eta_{mp,ref}}\frac{\tau\alpha}{U_L}\left(1 - \eta_{mp}\right)\right] \tag{18.42}$$

This equation has parameters that can be entirely derived from manufacturers' data. Efficiency is expressed as a linear function of dry-bulb temperature and incident solar irradiation. As the coefficient $\mu_{p,mp}$ is always negative, this formulation suggests a fall in efficiency on warm bright days as expected. In winter conditions, cold atmospheric conditions will be shown to be advantageous for a given level of solar irradiation.

Example 18.4 Solar photovoltaics

Using the data in Table 18.4, it is possible to calculate the overall heat transfer coefficient and the cell temperature when the ambient temperature is 3 °C and the incident solar irradiation is 150 W/m². The result of this calculation can be compared with that representing a summer condition: 28 °C and 650 W/m².

Solution:

Firstly, it is necessary to calculate the overall heat transfer coefficient using Equation 18.38 and the data from Table 18.4 for the normal operating cell temperature (800 W/m² irradiation and 20 °C). A value of 0.9 for $\tau\alpha$ is assumed as noted earlier. Hence:

$$0.9 / U_L = (47.0 - 20.0) / 800 = 0.03375$$

so that $U_L = 26.67$ W/m².K.

The cell temperature can then be estimated from Equation 18.39 using the nominal efficiency (13.9%) as follows:

$$T_c = 3.0 + 150 \times 0.03375\ (1 - 0.139/0.9) = 7.28\ °C$$

Similarly, for the summer condition when the air temperature is 28 °C and the incident irradiation is 650 W/m²:

$$T_c = 28.0 + 650 \times 0.03375\ (1 - 0.139/0.9) = 46.55\ °C$$

The significant difference in temperatures between winter and summer can be seen to be strongly driven by the solar irradiation. Note that these calculations assumed the nominal efficiency. A better solution could be obtained using Equation 18.42 for the overall efficiency and then calculating the cell temperature, as suggested below in the chapter activity.

End of solution.

In order to make some assessment of PV system output to the building AC power system, it is furthermore necessary to consider other factors detrimental to overall efficiency arising in the 'balance of system' and represented cumulatively by the term η_e in Equation 18.32. A number of potential causes of loss can be identified. These may amount to very modest reductions in efficiency when considered alone, but cumulatively represent a noticeable effect that should be considered. A number of causes have been identified in the literature and have been noted in Table 18.5. These losses include those relating to the limitations in power point tracking, cable losses as well as those usually acknowledged relating to limited inverter efficiency. Some allowance for degradation from test conditions due to dust and dirt (soiling) may also be considered. Taken together these may give values for η_e in the range 0.85–0.95.

Although it is difficult to deal with some forms of inefficiency other than by estimated constant parameters, one important element in the overall efficiency of the balance of system is the behaviour of the inverter, which is known to be rather nonlinear. Over much of the working range of an inverter, the efficiency of conversion of DC to AC power can be expected to be approaching a constant high value (greater than 95% in many instances). However, conversion

Table 18.5 Possible PV system operational losses

Type of losses	*Description*	*Report*
Production variations and mismatch	PV module efficiency deviates from STC specification due to manufacturing process variations (+/– 5% over a sample of modules).	2, 3, 4, 6
DC wiring losses	DC cable transmission losses – power transmitted at low voltage and high current results in noticeable losses.	3, 4, 6
Inverter maximum power point tracking (MPPT)	MPPT may be imperfect where loads or irradiation vary rapidly.	1, 2, 3, 6
Inverter DC-AC conversion	Loss in conversion of DC power to AC power.	1, 3, 4, 6
Inverter saturation	Occurs when the power input to the inverter exceeds the inverter rating (poor sizing).	5
Dirt and dust	The panel glass cover transmittance is reduced over time by dirt and dust.	6

1 – Oozeki et al. (2003); 2 – Stettler et al. (2005); 3 Reinders et al. (1999); 4 – Kiefer et. al. (1995); 5 – Ransom and Funtan (2005); 6 – CEC (2011)

Source: After Firth 2006

efficiencies tend to drop off to much lower values approaching the bottom of the working range. Example efficiency characteristics of inverter products are shown in Figure 18.14.

In view of the importance of inverter performance and selection, a good deal of test data is available and models are available that make use of such data to define empirical model parameters. The model developed at Sandia National Laboratory (King et al. 2007) is well supported by manufacturers' part-load performance data and is utilized in popular design software. The model represents behaviour in terms of a second-order polynomial to define the AC power output according to DC input at a given system voltage as follows:

$$P_{ac} = (P_{dc} - B)\left[\left(P_{aco} / (A - B)\right) - C(A - B)\right] + C(P_{dc} - b)^2 \quad (18.43)$$

where:

$$A = P_{dco}\left[1 + C_1\left(V_{dc} - V_{dco}\right)\right]$$

$$B = P_{so}\left[1 + C_2\left(V_{dc} - V_{dco}\right)\right]$$

$$C = C_o\left[1 + C_3\left(V_{dc} - V_{dco}\right)\right]$$

and:

p_{ac} = AC-power output from the inverter system;
p_{dc} = DC-power input;
V_{dc} = DC voltage input (a design value that may be different from the reference value);
P_{aco} = maximum AC power rating;
P_{dco} = DC power at which the maximum AC power rating is achieved;

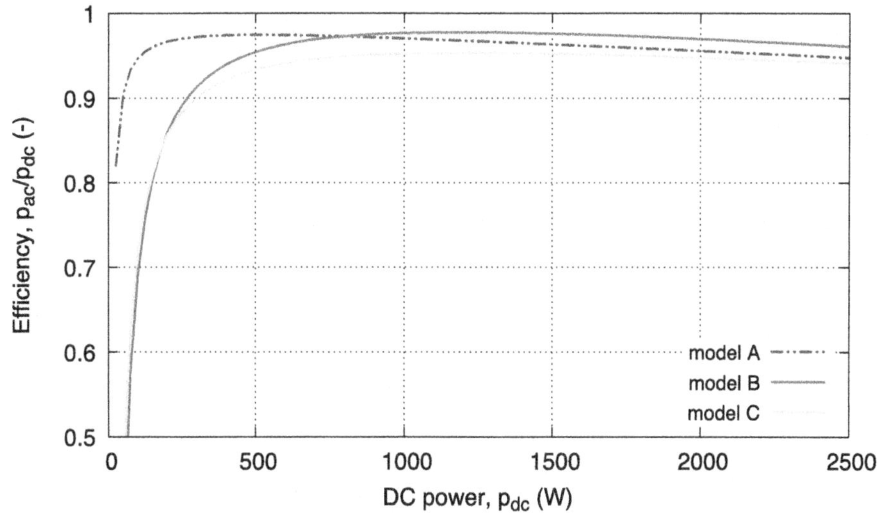

Figure 18.14 Example DC-AC inverter system efficiency characteristics (nominal capacity 2.5 kW)

V_{dco} = DC voltage at the reference condition;
P_{so} = DC power required to start the inversion process;
P_{so} = DC power required to start the inversion process;
C_0 = empirical parameter defining the parabolic curvature about the reference power condition;
C_1 = empirical parameter describing variation in p_{dco};
C_2 = empirical parameter describing variation in p_{so};
C_3 = empirical parameter describing variation in C_0 with DC voltage input.

This model, along with a sample of manufactures' data, have been used to generate the characteristics shown in Figure 18.14. This shows some noticeable variation in performance at the bottom of the operating range. The nature of the characteristic also highlights the need for inverter capacity to be carefully matched with expected DC output. When this model is combined with the PV module model of conversion efficiency described by Equation 18.42, the combined effect of variations in environmental conditions and variations in inverter efficiency can be considered. The combined effects are well illustrated by plotting the ratio of the model output to that predicted by assuming a constant conversion efficiency.

Example system performance ratio data are shown in Figure 18.15. These data are shown firstly as DC output from the modules (before inverter). Here the variations – that is the fall in efficiency with increased solar irradiation – are shown by the negative slope to the data and the scatter is due to variations in coincident dry-bulb temperature around a given level of solar irradiation. The further sets of data show the impact of variations in inverter efficiency and illustrate the variation between different manufacturers' models with the same nominal capacity.

It can be seen that different levels of sophistication of the representation of photovoltaic panel systems are possible. Some models are amenable to spreadsheet calculation outside of building simulation tools and others may be available integrated into such tools or equation solving environments. Probably the level of sophistication chosen will depend on the modeller's objectives at a given stage in a project and whether specific modules and inverters are selected and consequently if parameter data can be readily obtained.

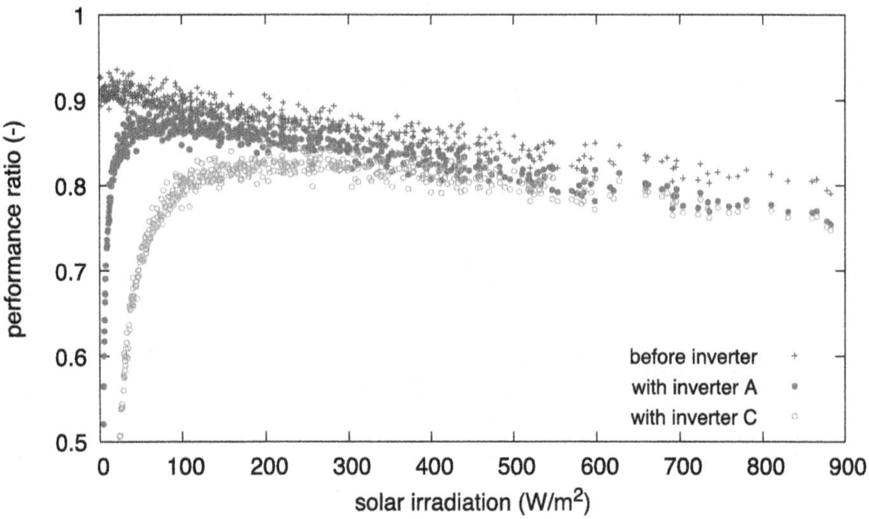

Figure 18.15 PV system performance ratio variation. Each point represents one hour in a summer month. Variations assuming different inverter models are shown along with modelled DC output (before inverter).

Building wind power systems

Micro-scale wind turbines (say less than 50 kW for current purposes) for off-grid applications are a relatively well understood technology and as reduction in carbon emissions has been of increasing concern, so to, attempts to integrate such wind turbines into building projects have grown. Consequently, in assessing demands and emissions from buildings, it is desirable to make assessments of wind power generation potential in building simulation studies and so some of the key modelling issues are highlighted here. Details of particular turbine technologies are not dealt with, as our focus is on overall power generation and relatively simple approaches to representing particular devices are all that is required.

In an ideal project, one would want to establish the variations in wind speeds at a particular site using anemometer devices and long-term data logging. This approach is rarely economic or practical and often some reliance must be made on other sources of wind data – at least at the early stages of project analysis. Using modelling methods to simulate operation of wind energy systems can be thought of as a five-step process:

1 selection of a wind data source;
2 calculation of wind speed at the wind turbine position;
3 calculation of power generation variations;
4 correction for other power conversion losses;
5 integration and reporting of time-varying outputs.

It is important to recognize that there are uncertainties associated with each step in this process. In particular, the difficulties with estimating wind conditions in certain complex urban environments can mean that the uncertainty in any modelling result is high – possibly unacceptably so (Phillips et al. 2007). These uncertainties should be acknowledged in any analysis.

It is possible to state, in very general terms, that the power available from a wind turbine is a function of its swept area (A), air density (ρ) and effective wind speed (U_e), which can be expressed as

$$P_{WT} = \eta_e f\,(A, \rho, U_e) \tag{18.44}$$

The term η_e is included to indicate the useful power output is reduced by other inefficiencies in the power conversion and distribution processes – in a similar way to the parameter defined for PV power production. The difference between modelling approaches depends on how this function is represented and, in particular, what is assumed about the effective wind speed. As the power available from the wind is a function of its kinetic energy, the function is often expressed as a fraction of the theoretically available energy modified by a power coefficient C_P so that power output at a particular hub wind speed can be represented as

$$P_{WT} = \eta_e \frac{1}{2} C_P A_s U_e^3 \tag{18.45}$$

There is a theoretical upper limit to the value of C_P for horizontal axis wind turbines (HAWT) known as the Betz limit: 0.59. In reality, and for micro-scale devices in particular, the value of C_P changes with wind speed. This theoretical relationship nevertheless highlights the sensitivity of any prediction of power output to wind speed. This sensitively is compounded by the steep variation in wind speed with height near the ground.

Recorded wind data are measured at a particular height at the weather station (often 10 m) in open and flat locations (e.g. airports). Consequently, recorded wind speeds are most likely to be higher than those experienced locally in urban locations. It is accordingly necessary to consider the difference in local speed between the values recorded in standard weather records and that at the height of the wind turbine hub. In principle this can be done by applying mathematical functions that describe boundary layer velocity profiles.

In open landscapes the wind speed gradually increases from virtually zero at ground level and the boundary layer develops according to a recognizable logarithmic profile. In areas where there are trees, hedges and buildings, the airflow patterns near the ground become disrupted as the wake of the airflow around such objects generates turbulence and the regular wind speed profile does not develop at ground level but at some distance closer to the top of the obstructions. Such land features are said to increase effective 'surface roughness' and this can be expressed according to 'roughness length' (z_0) as illustrated in Table 18.6.

To find the wind speed at the height of the wind turbine hub, it is necessary to be able to relate the local velocity U_z, at height z, to the velocity at the point of measurement U_{ref}. Both power-law and log-law relationships are used in wind engineering to describe this relationship. A common logarithmic profile definition is

$$U_z = \frac{\ln(z / z_0)}{\ln(z_{ref} / z_0)} U_{ref} \tag{18.46}$$

The effects of turbulence near the ground, and the corresponding modification of the wind boundary layer, is most significant in the urban environment where building, road and vegetation variations constitute a very 'rough' surface. This is particularly so in the centre of large cities where tower blocks or skyscrapers make for very large variations in surface level. This can be considered in the calculations by identifying a 'displacement' height (d) at which the effective wind velocity is taken to be zero. Panofsky and Dutton (1984) and Mertens (2003) suggest that this displacement can be taken to be 75% of the average building height ($\bar{h}_b$) so that

$$d = 0.75\bar{h}_b \tag{18.47}$$

Table 18.6 Variations in surface roughness length (z_0) according to terrain type

Roughness class	*Roughness length (m)*	*Terrain type*
0	0.0002	Water surface
0.5	0.0024	Completely open terrain with a smooth surface, e.g. concrete runways in airports, mowed grass, etc.
1	0.03	Open agricultural area without fences and hedgerows and very scattered buildings. Only softly rounded hills
1.5	0.055	Agricultural land with some houses and 8 m tall sheltering hedgerows with a distance of approx. 1250 m
2	0.1	Agricultural land with some houses and 8 m tall sheltering hedgerows with a distance of approx. 500 m
2.5	0.2	Agricultural land with many houses, shrubs and plants, or 8 m tall sheltering hedgerows with a distance of approx. 250 m
3	0.4	Villages, small towns, agricultural land with many or tall sheltering hedgerows, forests and very rough and uneven terrain
3.5	0.8	Larger cities with tall buildings
4	1.6	Very large cities with tall buildings and skyscrapers

After taking the displacement into account, the logarithmic relationship between height and reference velocity can be written as

$$U_z = \frac{\ln\left((z-d)/z_0\right)}{\ln\left(zref/z_0\right)}U_{ref} \tag{18.48}$$

The formula is not valid at heights below the displacement height. Mertens (2003) furthermore suggested prediction of wind speed using this approach (or similar analytical profiles) is not accurate below a height defined according to roughness length:

$$z \geq 20z_0 + d \tag{18.49}$$

In urban terrain, this suggests the wind profile cannot be reliably predicted using functions like this unless the height of interest (e.g. the turbine hub height) is some distance above the average roof line. Profiles calculated according to these guidelines are illustrated in Figure 18.16. In this figure, the profile is shown dotted below the height given by Equation 18.49. This suggests some caution should be taken in calculating wind velocities at turbine hub height where the hub is relatively close to the roof line.

In order to estimate the available wind resources at a particular building location, it is important to consider the relationship with adjacent objects and conditions immediately surrounding the building. In collections of buildings of similar height, a large part of the flow passes over the roofs, and lower parts of the buildings experience swirling flow and some degree of shelter. Where building heights are more variable, tall buildings can induce wakes that significantly affect adjacent buildings depending on wind direction. Consequently, in view of the uncertainty in calculating flows very near and directly above the roofs of buildings, applications of calculation methods such as those noted above should be applied cautiously.

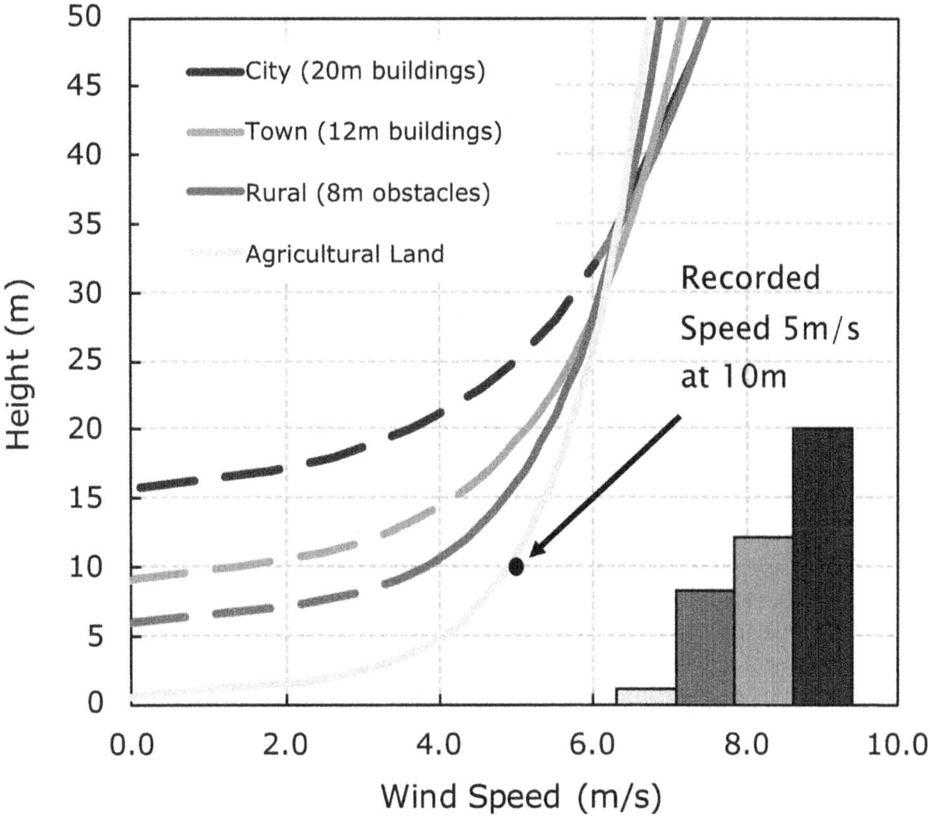

Figure 18.16 Logarithmic wind profiles calculated for different environments and corresponding roughness lengths. The profiles are calculated for a reference velocity of 5 m/s at a standard reference height of 10 m. The rectangles indicate the respective mean building heights. Dotted lines indicate uncertainty in application of the profile as suggested by Equation 18.49 and Mertens (2003).

Example 18.5 Wind power

Calculate the wind speed at a height of 16 m in a small town environment with average building height 10 m and where the local weather station recorded wind speed is 5 m/s at 10 m height.

Solution:

Firstly, the displacement height is estimated from Equation 18.47 as $d = 0.75 \times 10 = 7.5$ m.

For a small town environment, the roughness length from Table 18.4 is 0.4 so that Equation 18.47 gives

$$U_z = \frac{\ln\left((18-7.5)/0.4\right)}{\ln\left(10/0.4\right)}5.0 = \frac{3.268}{3.219}5.0 = 5.08\,\text{m/s}$$

Referring to Equation 18.49, $z \geq 20z_O + d$ suggests the minimum height we should consider this reliable would be $20 \times 0.4 + 7.5 = 15.5$ m. Note that this is somewhat above average roof level.

End of solution.

Having noted that there is likely to be considerable uncertainty in predicting fluctuations in wind speeds in complex urban environments, it is nevertheless possible to adjust recorded wind speeds and arrive at time series data that can be used in power conversion calculations at a particular building site and design hub height. The question, then, is how this can be related to power output?

Returning to Equation 18.45, it was noted that C_p tends to be variable in general. Fortunately, many micro-turbine manufactures provide data that allow power to be expressed as a function of hub wind speed – much as indicated in Equation 18.46. This data can be used in modelling the wind turbine device where the particular type of device is known. In this approach, power output can be interpolated from the power curve data at each step using the wind data from a weather file that has been corrected for the turbine location and height. An example power curve is plotted in Figure 18.17. This shows comparison with the theoretical limit (Equation 18.45 with $C_P = 0.59$) and illustrates that in practical turbines the wind speed has to rise above a cut-in value before it rotates. The profile is deliberately designed to flatten at higher wind speeds to ensure that the device is not damaged by severe wind conditions.

The power curve data will have been measured at standard atmospheric conditions (288.15 K and 101325 Pa). In hourly calculations, it is possible to make corrections to the power data to account for the fact that the air density will be different at the site depending on climatic conditions. Coincident temperature and air pressure data are usually available from the same weather data source as recorded wind data. The power output at a particular step can be corrected according to the prevailing temperature and pressure as follows:

$$P_{WT,i} = \eta_e \cdot P_{u,i}\left(U_{z,i}\right)\left(\frac{P_i}{P_0}\right)\left(\frac{T_0}{T_i}\right) \tag{18.50}$$

where, at time step i, $p_{WT,i}$ is the uncorrected power at local wind speed $U_{z,i}$, P_i and T_i are the atmospheric pressure and absolute temperature and P_0 and T_0 are the pressure and absolute temperature at standard measurement conditions (Gipe 2004). This approach to calculation is

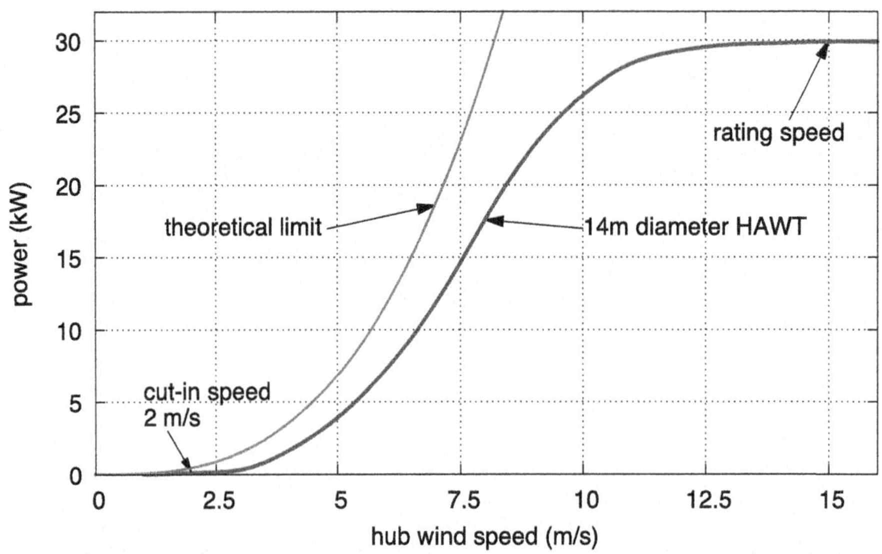

Figure 18.17 Example wind turbine power curve data for a horizontal axis wind turbine with nominal 30 kW capacity at 15 m/s

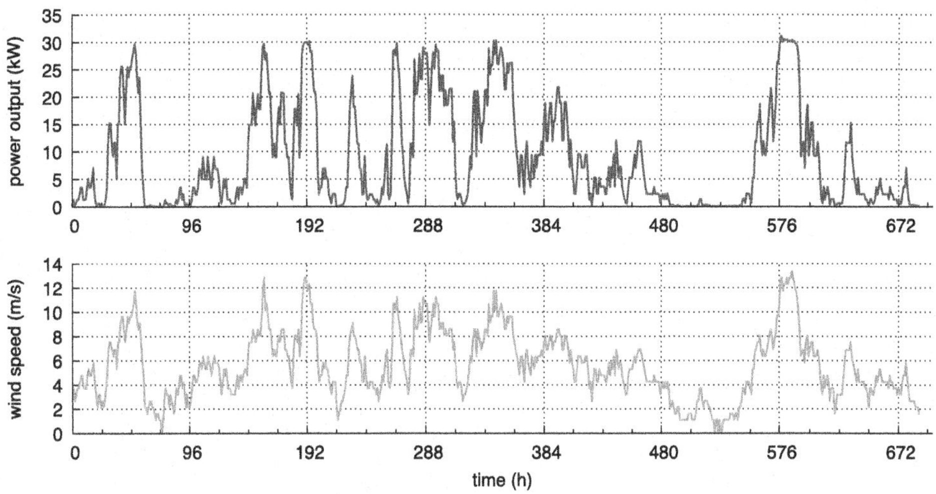

Figure 18.18 An example of wind power conversion calculated from manufacturers wind power curve data (Figure 18.17) and hourly wind speeds (London, December data) calculated at hub height of 7 m

not difficult to implement in spreadsheets or hourly building simulation tools. Data from an example calculation using the power curve in Figure 18.17 is shown in Figure 18.18. These data illustrate how wind power output falls to zero when the wind speed falls below the turbine cut-in speed of 2 m/s.

Geothermal heating and cooling

The interaction between the ground and the adjacent atmospheric environment, along with the superior thermal capacity of the ground, are such that fluctuations in ground temperatures are always reduced in amplitude compared to the fluctuations in dry-bulb temperature. Making use of the ground's relatively high thermal capacity, compared to either the air or typical building fabrics, is correspondingly thermodynamically advantageous when it comes to possible sources or sinks for heat pump systems. This is one of the main reasons for the abiding interest in geothermal heating and cooling using heat pumps (ground source heat pumps) as potentially one of the most efficient means of heating and/or cooling buildings.

Modelling and simulation of air-source heat pumps can be said to be little more complex than other air-source refrigeration systems, as it is straightforward to use the specified external air temperatures as boundary conditions when calculating heat transfer conditions on the source-side. Defining the fluid temperature conditions in ground-coupled heat pumps systems[1] is rather more complex, and further models of ground heat transfer have to be employed in order to capture realistic variations in temperatures.

Transferring heat between a heat pump and the ground by embedding pipes in either shallow horizontal configurations (e.g. slinky loops), in much deeper vertical boreholes (e.g. using U-tubes) or in concrete elements of the building foundation system (e.g. concrete piles) allows coupling with a thermal capacity that is orders of magnitude greater than that of the building itself. This means that the behaviour of the system also has to be considered on much longer timescales than the building fabric. In fact, to model a ground-coupled heat pump and its ground heat exchanger, both short timescales have to be considered to capture heat transfer in the fluid and the region of ground close to the pipes (i.e. where ground temperature gradients are the greatest) and very long timescales in order to consider the total heat exchange with the ground environment around the heat exchanger. It is fairly well understood that, over seasonal timescales, the ground heat exchanger will shift the temperatures in the surrounding region of the ground (on the scale of tens of metres) depending on whether the demands on the system amount to net heat rejection or heat extraction. For example, a building that rejects heat, when considered over the whole season, may take as long as 10 years before the average ground heat exchanger temperature reaches a steady-periodic pattern. This means that it is not feasible to design ground heat exchanger systems based on peak load conditions as might be done for other system components, and some form of long-term simulation is inevitable for both design (e.g. deciding on borehole depth and spacing) and performance prediction purposes (e.g. estimating system annual energy demands).

A number of models of borehole ground heat exchange systems have been developed. These can be broadly categorized as:

1 analytical models;
2 two-dimensional numerical models (e.g. Wetter and Huber 1997);
3 response factor models (e.g. Fisher et al. 2006);
4 three-dimensional numerical models (e.g. Diersch et al. 2011).

An example of the temperature field calculated after prolonged (50 year) application of a heat flux and calculated with a fully three-dimensional heat conduction solver is shown in Figure 18.19. This illustrates a fully developed temperature field after net heat rejection in an array of 4 × 4 borehole heat exchangers. The variability of heat transfer between neighbouring boreholes can be seen in the horizontal cross-section such that the temperature in the central heat exchangers is greater than those at the perimeter. In the vertical cross-section, it can be seen that at this timescale, the horizontal heat transfer is not two-dimensional but shows heat flow below the array and from the array towards the ground surface far from the centre of the array.

The most common approaches applied in design tools and system modelling (hourly or sub-hourly) are two-dimensional numerical models or forms of response factor methods. A number of variations of two-dimensional numerical models represent the flow of fluid in a single U-tube as a series of vertical nodes that are connected horizontally with resistances representing conduction between the pipes, the borehole wall and the surrounding ground. Heat capacity can be represented at some or all of the nodes and often heat transfer in the vertical direction in the surrounding ground is ignored, i.e. the model is quasi two-dimensional. Such models are applicable to single boreholes, as they often do not represent interaction between neighbouring boreholes.

In a number of design tools and widely used simulation tools, response factor approaches are used. These approaches have the advantage of being relatively fast to execute and capable of representing the important three-dimensional effects in arrays of borehole heat exchangers. In some of these approaches, and in design calculations, it is common to divide the timescales under consideration and to regard heat transfer within the borehole as steady-state, such that it can be defined by a thermal resistance which when averaged over the depth of the borehole can be defined as R_b (K.W/m):

$$R_b = \frac{\left(\overline{T}_f - T_b\right)}{q_b} \tag{18.51}$$

where the mean borehole fluid temperature, T_b, is the borehole wall temperature and q_b (W/m) is the heat flux per unit depth at the borehole wall. This resistance is dependent on a number

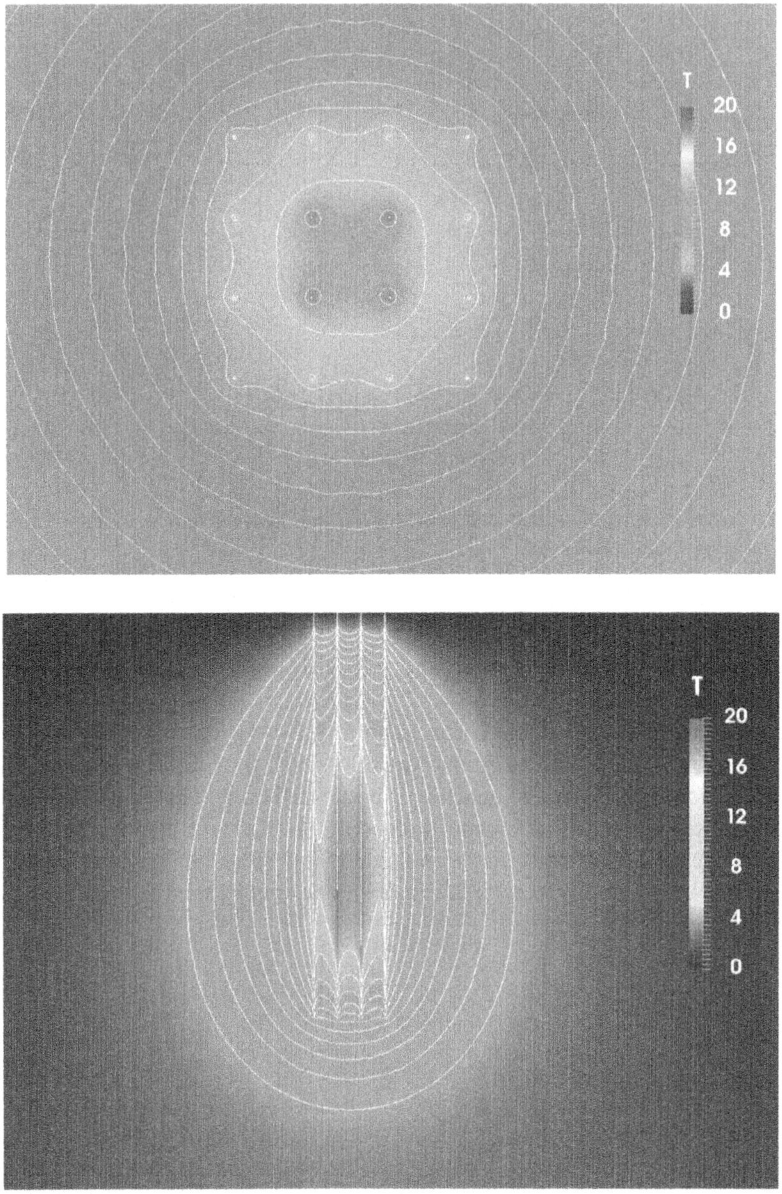

Figure 18.19 The long-term temperature field calculated for heat rejection in a 4 × 4 borehole heat exchanger array. A horizontal contour plane (top) and vertical plane (bottom). The contour lines show degree intervals.

of factors in a typical U-tube borehole heat exchanger. Besides the pipe and grout thermal conductivities, it is also dependent on the spacing between the pipes and their relative position in the borehole as well as borehole size and pipe dimensions. It should also be noted that this resistance includes the convective resistance between the fluid and the pipe walls and so is dependent on the flow conditions and fluid properties – these being noticeably variable with the low flow rates and high glycol content found in many installations. A typical borehole with a single U-tube is illustrated in Figure 18.20 along with the resistance network representation.

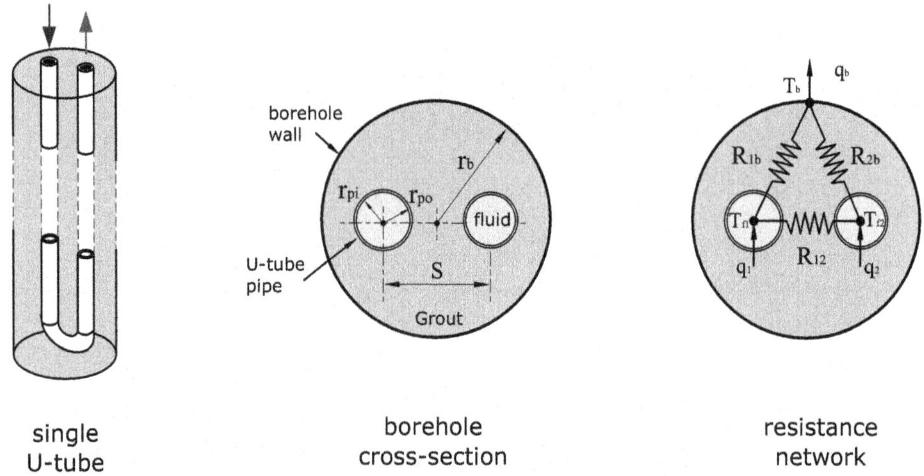

Figure 18.20 A typical single U-tube borehole heat exchanger and the representation of thermal resistances inside the borehole

Javed and Spitler (2016) have given a systematic overview of the methods that can be used to calculate borehole resistance. These range from simplified correlations to highly accurate analytical methods.

Heat transfer around borehole heat exchanger arrays is commonly taken to be by conduction only (homogeneous material and no groundwater flow) and is represented mathematically by a linear diffusion equation: Fourier's equation. This allows the solution of time series to be found using super-position principles (Duhamel's theorem). This can be applied both spatially and temporally. It can further be taken that responses to input steps can be normalized by thermal conductivity and timescales by thermal diffusivity. For example, a normalized timescale, t_s, can be defined as

$$t_s = \frac{H^2}{9\alpha} \tag{18.52}$$

where H is the depth of the borehole (m) and α is the thermal diffusivity of the ground (m^2/s). It is firstly useful to apply super-position in the spatial sense. This allows the response (temperature at the end of an applied heat step) of a complete borehole array to be calculated by superimposing the response to a single borehole. The calculation of the response of a single borehole to a step in heat flux can be found by making a numerical calculation in cylindrical coordinates (i.e. 2-D about the borehole axis) and the effective 3-D effect found by super-position (Hellström 1991). Corresponding analytical approaches have also been developed (Javed and Spitler 2016). The responses consequently reflect the proper long-term interaction between borehole heat exchangers. Such response factors are termed 'g-functions' and examples are shown with respect to the log of t/t_s in Figure 18.21. One noticeable feature of these data is that at short timescales (a number of weeks in practice), the responses all converge to the response of a single heat exchanger. This is to be expected, as at short timescales heat has not penetrated to the point of having a noticeable effect on neighbouring boreholes.

Applying the 'g-function' response factor consequently allows the borehole temperature, T_b, to be found relative to the undisturbed (initial) ground temperature, T_g, from application of a single heat step as follows:

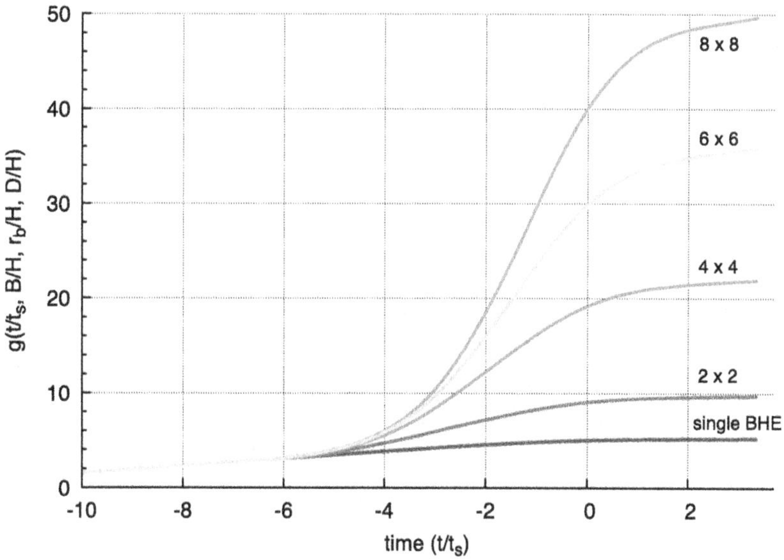

Figure 18.21 Examples of borehole heat exchanger array 'g-function' data

$$T_b = T_g + \frac{q}{2\pi k*} \cdot g(t/t_s) \tag{18.53}$$

where q is the heat flux of the pulse per unit length and $k*$ is the effective thermal conductivity of the ground and the function g is for a particular borehole array geometry. In a calculation with time-varying boundary conditions, as in a system simulation or design calculation, the time-varying temperature can be found by super-imposing the response to individual steps in heat flux as follows:

$$T_b = T_g + \sum_{i=1}^{n} \frac{(q_i - q_i - 1)}{2\pi k*} .g(t / t_s) \tag{18.54}$$

In general, g-functions are not simply functions of time but also borehole geometric configuration. They can then be normalized for given ratio of borehole spacing to depth, B/H, and the ratio of borehole radius to depth, r_b/H. Furthermore, in general the top of the heat exchanger does not rise to the surface but is some depth D below the surface. Hence, the functions should also be differentiated by the ratio D/H. Accordingly, in design and simulation tools, it is convenient to refer to a library of such functions depending on key design parameters and possibly interpolate between some of the data.

At short timescales (of the order of hours), most of the changes in thermal conditions occur within or close to the borehole. In a system simulation that works with small time steps (hourly or sub-hourly), a constant borehole resistance is not a good assumption. This is not an issue in design tool calculations where only long timescales are applicable. For system simulations, the response functions must be adapted to account for short timescale effects. Yavuzturk and Spitler (1999) showed how g-functions can be extended to short times by calculating responses numerically with a model that included the details of the components inside the borehole, i.e. the pipes and grout material. This is less generalizable (i.e. the calculated response is specific to a particular size/arrangement of pipes) but can be blended with the respective long timescale g-function data to allow both short and long timescales to be dealt with.

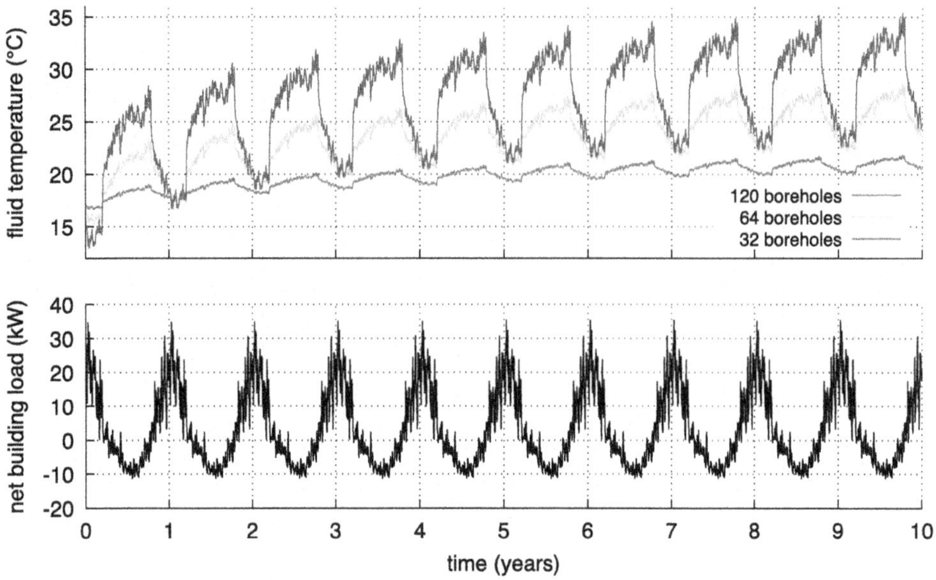

Figure 18.22 Borehole heat exchanger system fluid temperatures predicted over a 10-year period with different array sizes. The corresponding building loads are shown in the lower part of the figure.

Source: After Fisher et al. 2006

An example of a long-term calculation made with such a model (Fisher et al. 2006) is shown in Figure 18.22. Results using three different sizes of borehole array are shown for a building with predominant cooling loads. The average fluid temperature can be seen to follow the annual building loads and also shows year-on-year increases until a steady-periodic pattern is approached after approximately 10 years. This highlights the need to carry out multi-year simulations if the sustainability of the system is to be predicted realistically.

Simulation of borehole heat exchanger response is necessary in a simplified way in selecting or sizing ground heat exchangers. A common approach would be to calculate long-term response using large time steps, e.g. derived from monthly energy exchange and peak heat transfer rates imposed over a few hours. If the objective is to design so as to maintain the minimum or maximum temperatures within certain limits (perhaps defined by heat pump limiting temperatures or a target heat pump efficiency), the response can be simulated and the borehole length adjusted until a design size is arrived at (Spitler and Bernier 2016).

In whole system simulations, models are formulated to deterministically predict heat transfer rates and outlet temperatures given specified inlet flow rates and temperature. As simulation progresses, responses are calculated by keeping track of heat exchange at previous time steps and 'looking up' the response of each according to the appropriate g-function. After some long period, assumptions can be made about the sensitivity to past responses and historic fluxes aggregated together (Yavuzturk and Spitler 1999). The types of parameters that are typically required in such models is shown in Table 18.7 along with possible data sources.

As models such as those using the g-function approach rely on the relationships between mean borehole fluid temperature and heat exchange, some assumptions are necessary about the relationship between this temperature and both the inlet and outlet temperature (Rees 2015). A further question in modelling ground heat exchange systems is the impact of the transport of fluid along the long pipe lengths usually involved – both in terms of the large heat capacity of the fluid and also the fact that it may take fluid a few minutes or more to traverse the loop. In the type of quasi-3-D nodal models described earlier, fluid transport up and down the borehole is

Table 18.7 Parameters typically defined for borehole heat exchanger models and possible data sources

Model parameter	*Possible data source*
Borehole length, H	Design tool output or specialist contractor proposal
Borehole radius, R_b	Design tool output or specialist contractor proposal
Borehole spacing, B	Design tool output or specialist contractor proposal
Borehole sub-surface depth, D	Design assumption or specialist contractor proposal
Array configuration, (A × B) or total number of boreholes	Design tool data
Borehole thermal resistance	Design tool output
Ground thermal conductivity	Thermal response test or geological survey
Grout/backfill thermal conductivity	Grout manufacturers' data
Ground and grout specific heat capacity	Grout manufacturers' data
Ground and grout density	Response test data or geological survey
U-tube pipe size	Hydraulic system design data
U-tube pipe spacing	Design tool output or contractor proposal
Pipe thermal properties	Pipe manufacturers' or design guide data
Fluid properties	Antifreeze manufacturers' or design guide data

accounted for explicitly and the physical effects are captured to some degree. Otherwise some representation of the effects can be made by coupling the heat exchanger model with discretized pipe models (Rees 2015).

Note

1 Only closed-loop systems are considered here.

References

ARI. (1998) *ARI Standard 550/590–98 – Water Chilling Packages using the Vapour Compression Cycle*. Arlington, VA: Air Conditioning and Refrigeration Institute.

ASHRAE. (2016) *Handbook of Systems and Equipment*. Solar Energy and Equipment, p. 37.9. Atlanta, GA: American Society of Heating Refrigeration and Air-conditioning Engineers.

Bertagnolio, B. and Lebrun, J. (2008) Simulation of a building and its HVAC system with an equation solver: Application to benchmarking. *Building Simulation* 1: 234–250.

BSI. (2008a) *BS EN 442–3: Specification for Radiators and Convectors*. London: The British Standards Institution.

BSI. (2008b) *BS EN 14825: 2008. Air Conditioners, Liquid Chilling Packages and Heat Pumps with Electrically-driven Compressors for Space Heating and Cooling: Testing and Rating at Part Load Conditions*. London: The British Standards Institution.

BSI. (2015) *BS EN 13203–2:2015. Gas-Fired Domestic Appliances Producing Hot Water. Part 2: Assessment of Energy Consumption*. London: The British Standards Institution.

CEC. (2011) *Emerging Renewables Program Guidebook*, 12 edition. California Energy Commission, 52 p.

Chow, T.T., Clarke, J.A. and Dunn, A. (1997) Primitive parts: An approach to air-conditioning component modelling. *Energy and Buildings* 26: 165–173.

Crawley, D.B., Lawrie, L.K., Pedersen, C.O., Strand, R.K., Liesen, R.J. and Winkelmann, F.C. (2001) EnergyPlus: Creating a new-generation building energy simulation program. *Energy and Buildings* 33(4): 319–331.

Diersch, H.-J.G., Bauer, D., Heidemann, W., Ruhaak, W. and Schatzl, P. (2011) Finite element modeling of borehole heat exchanger systems Part 1. *Fundamentals. Computers & Geosciences* 37: 1122–1135.

Duffie, J.A. and Beckman, W.A. (1991) *Solar Engineering of Thermal Processes*, 2nd edition. New York: Wiley, 919 p.

Duffie, J.A. and Beckman, W.A.(2013) *Solar Engineering of Thermal Processes*, 4th edition. Hoboken, NJ: Wiley, 910 p.

F_Chart Software. (2011) *EES – Engineering Equation Solver*. www.fchart.com/ees/.

Firth, S.K. (2006) Raising efficiency in photovoltaic systems: High resolution monitoring and performance analysis. PhD Thesis, De Montfort University.

Fisher, D.E., Rees, S.J., Padhmanabhan, S.K. and Murugappan, A. (2006) Implementation and validation of ground-source heat pump system models in an integrated building and system simulation environment. *International Journal of HVAC&R Research* 12(3a): 693–710.

Fong, J., Edge, J.S., Underwood, C.P., Tindale, A. and Potter, S.E. (2015) Performance of a dynamic distributed element heat emitter model embedded into a third-order lumped parameter building model. *Applied Thermal Engineering* 80: 279–287.

Gipe, P. (2004) *Wind Power: Renewable Energy for Home, Farm and Business*, 2nd edition. White River Junction, VT: Chelsea Green Publishing, 512 p.

Gomaa, A.G., Underwood, C.P., Bond, T.J. and Penlington, R. (2002) Experimental investigation of sensible heat transfer in fin-and-tube cooling coils. *Building Services Engineering Research and Technology* 23(3): 197–204.

Hellström, G. (1991) Ground heat storage: Thermal analysis of Duct storage systems. PhD Thesis, University of Lund.

Hydronics Institute. (1990) *Testing and Rating Standard for Baseboard Radiation*. Berkeley Heights, NJ: Hydronics Institute.

Javed, S. and Spitler, J.D. (2016) Calculation of borehole thermal resistance. In Rees, S.J. (ed.), *Advances in Ground-Source Heat Pump Systems* (Chapter 3). Oxford: Woodhead Publishing, pp. 63–95.

Kays, W.M. and London, A.L. (1964) *Compact Heat Exchangers*. 2nd edition. New York: McGraw-Hill.

Kiefer, K., Körkel, T., Reinders, A., Rössler, E. and Wiemken, E. (1995) PV roofs in Germany – Operating results from intensified monitoring and analysis through numerical modeling. In *13th European Photovoltaic Solar Energy Conference*. Nice, pp. 575–579.

King, D.L., Gonzalez, S., Galbraith, G.M. and Boyson, W.E. (2007) Performance model for grid-connected photovoltaic inverters. Sandia Technical Report SAND2007–5036. Albuquerque.

Klein, S.A. (1975) Calculation of flat plate collector loss coefficient. *Solar Energy* 17, 79–80.

Klein, S.A., Beckman, W.A. and Mitchell, J.W. (1997) TRNSY – A transient system simulation program. Madison, WI.

Masmoudi, F., Ben Salem, F. and Derbel, N. (2016) Single and double diode models for conventional mono-crystalline solar cell with extraction of internal parameters. In *13th International Multi-Conference on Systems, Signals and Devices*, SSD 2016, pp. 720–728.

Mertens, S. (2003) The energy yield of roof mounted wind turbines. *Wind Engineering* 27(6): 507–516.

Modelica. (2010) A unified object-oriented language for physical systems modeling. www.modelica.org/documents/ModelicaSpec32.pdf.

Oozeki, T., Izawa, T., Otani, K. and Kurokawa, K. (2003) An evaluation method of PV systems. *Solar Energy Materials and Solar Cells* 75(3–4): 687–695.

Panofsky, H.A. and Dutton, J.A. (1984) *Atmospheric Turbulence, Models and Methods for Engineering*. New York: John Wiley & Sons.

Phillips, R., Blackmore, P., Anderson, J., Clift, M., Aguilo-Rullan, A. and Pester, S. (2007) *Micro-Wind Turbines in Urban Environments: An Assessment*. BRE Trust Report FB17, ISBN 978-1-84806-021-0.

Ransome, S. and Funtan, P. (2005) Why hourly averaged measurement data is insufficient to model PV system performance accurately. In *20th European Photovoltaic Solar Energy Conference*. Barcelona.

Rees, S.J. (2015) An extended two-dimensional borehole heat exchanger model for simulation of short and medium timescale thermal response. *Renewable Energy* 83: 518–526.

Rees, S.J. (ed.) (2016) *Advances in Ground-Source Heat Pump Systems*. Oxford: Woodhead Publishing, p. 460. ISBN 978-0-08-100311-4.

Reinders, A., Van Dijk, V.A.P., Wiemken, E. and Turkenburg, W.C. (1999) Technical and economic analysis of grid-connected PV systems by means of simulation. *Progress in Photovoltaics: Research and Applications* 7(1): 71–82.

Sahlin, P., Eriksson, L., Grozman, P., Johnsson, H., Shapovalov, A. and Vuolle, M. (2004) Whole-building simulation with symbolic DAE equations and general purpose solvers. *Building and Environment* 39: 949–958.

Schroder, D.K. (1998) *Semiconductor Material and Device Characterization*, 3rd edition. New York: Wiley, p. 522.

SPARK. (2010) *Simulation Problem Analysis and Research Kernel (spark)*. http://eta.lbl.gov/resources/software/spark.

Spitler, J.D. and Bernier, M. (2016) Vertical borehole ground heat exchanger design methods. In Rees, S.J. (ed.), *Advances in Ground-Source Heat Pump Systems*. Oxford: Woodhead Publishing. pp. 29–61.

Stettler, S., Toggweiler, P., Wiemken, E., Heydenreich, W., de Keizer, A.C. and van Sark, W. (2005) Failure detection routine for grid-connected PV systems as part of the PVSAT-2 project. In *Proceedings of the 20th European Photovoltaic Solar Energy Conference & Exhibition*. Barcelona, Spain, pp. 2490–2493.

Trčka, M. and Hensen, J.L.M. (2010) Overview of HVAC system simulation. *Automation in Construction* 19: 93–99.

Underwood, C.P. (1998) *HVAC Control: Modelling, Analysis and Design*. London: Taylor & Francis.

Underwood, C.P., Shepherd, T., Bull, S.J. and Joyce, S. (2018) Hybrid thermal storage using coil-encapsulated phase change materials. *Energy and Buildings* 159: 357–369.

Underwood, C.P. and Yik, F.W.H. (2004) *Modelling Methods for Energy in Buildings*. Chapter 5. Oxford: Wiley-Blackwell.

Wetter, M. and Huber, A. (1997) *TRNSYS Type 451: Vertical Borehole Heat Exchanger EWS Model, Version 3.1 – Model Description and Implementing into TRNSYS*. Stuttgart, Germany: Transsolar GmbH.

Yavuzturk, C. and Spitler, J.D. (1999) A short time step response factor model for vertical ground loop heat exchangers. *ASHRAE Transactions* 105(2): 475–485.

Yik, F.W.H., Lai, J.H.K., Fong, N.K., Leung, P.H.M. and Yuen, P.L. (2012) A case study on the application of air- and water-cooled oil-free chillers to hospitals in Hong Kong. *Building Services Engineering Research and Technology* 33(3): 263–279.

Recommended reading

Readers seeking further details of the material presented here on HVAC modelling and control are referred to the book by Underwood and Yik (2004). The book *Solar Engineering of Thermal Processes* (2013) has been through a number of editions and remains a very useful source of reference material regarding solar thermal and photovoltaic technologies. For readers seeking further details of both wind power technology and models, the book by Paul Gipe (2004) serves as a good introduction. A multi-author reference on geothermal systems, *Advances in Ground Source Heat Pump Systems* (Rees et al., 2016), provides much further detail on borehole heat exchanger modelling, calculation of borehole resistance and a range of geothermal technologies.

Activities

This suggested activity can be used to gain some insight into how PV output can be calculated from basic simulation data and how it can be expected to vary. This can be carried out in a spreadsheet tool. Firstly, choose a solar panel product and extract the relevant performance coefficients such as those given in Table 18.4. Either from a simulation weather file or simulation tool output, select a time series of solar irradiation and coincident dry-bulb temperatures and insert them into two columns of the spreadsheet. From the performance data, calculate the heat transfer coefficient data as shown in the example. You can then use Equation 18.42 to calculate the efficiency at each step. It is convenient to express Equation 18.42 as follows when using a spreadsheet:

$$\eta_{mp} = \eta_{mp,ref}\left[A + B \times T_a + C \times I_T\right] \qquad (18.55)$$

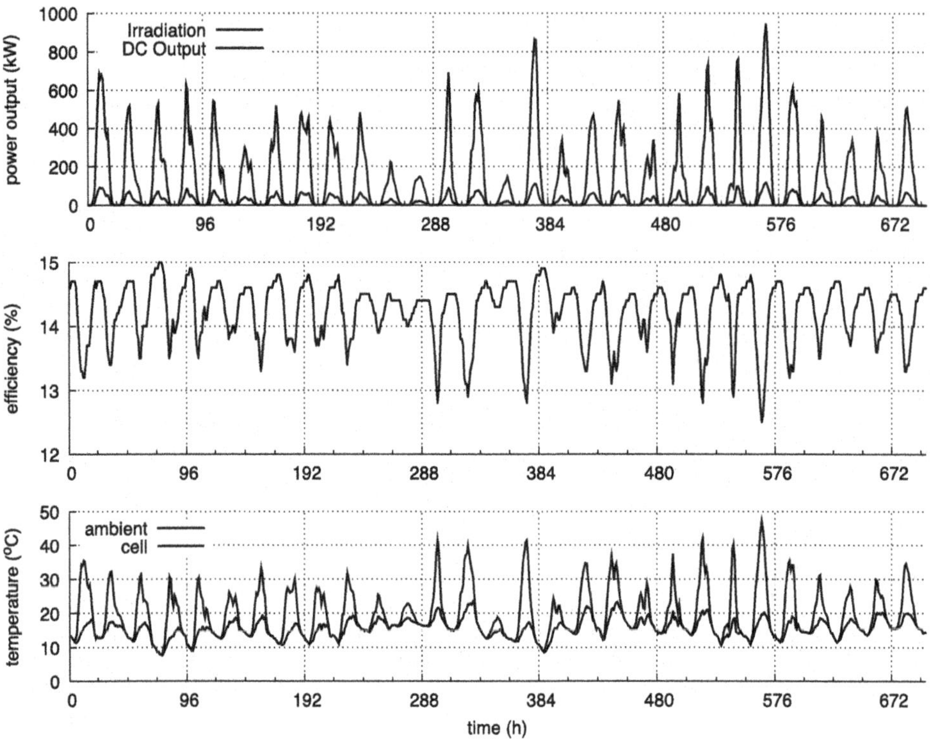

Figure 18.23 Example output from the chapter activity

The constants *A*, *B* and *C* are then

$$A = 1 - \frac{\mu_{P,mp}}{\eta_{mp,ref}} T_{ref} \quad B = \frac{\mu_{P,mp}}{\eta_{mp,ref}}, \quad C = \frac{\mu_{P,mp}}{\eta_{mp,ref}} \frac{\tau\alpha}{U_L} \left(1 - \eta_{mp,ref}\right) \tag{18.56}$$

These constants are calculated from the manufacturers' data (Table 18.4) and need be computed only once (e.g. at the start of the spreadsheet). You can then generate a time series for the efficiency. This can be supplemented by a calculation of cell temperature using Equation 18.39 and overall output using Equation 18.32. For additional interest, readers can add an inverter model to calculate AC output. Example output that could be expected from this activity is shown in Figure 18.23.

19 A view on future building system modeling and simulation

Michael Wetter

Scope

This chapter presents what a future environment for building system modeling and simulation may look like. As buildings continue to require increased performance, more flexible load shifting, better integration of renewable energy and waste heat sources, and better comfort, their energy and control systems are becoming more integrated and complex. We therefore focus in this chapter on the modeling, simulation and analysis of building energy and control systems. Such systems can be classified as heterogeneous systems because they involve multiple domains, such as thermodynamics, fluid dynamics, heat and mass transfer, electrical systems, control systems and communication systems. Also, they typically involve multiple temporal and spatial scales, and their evolution can be described by coupled differential equations, discrete equations and events. Modeling, simulation and optimization of such systems requires a higher level of abstraction and modularization to manage the increased complexity compared to what is used in today's building simulation programs. Therefore, the trend towards more integrated building systems, and the need to use models in design, delivery and operation, is likely to be a driving force for changing the status quo of today's building simulation programs.

A range of topics that would require many additional pages of discussion has been omitted or is mentioned only briefly in relation to new software developments. Examples include computational fluid dynamics for air and particle flow in and around buildings, people movement, daylight simulation, uncertainty propagation and optimization methods for building design and controls. For different discussions and perspectives on the future of building modeling and simulation, we refer to Sahlin (2000), Augenbroe (2001), Malkawi and Augenbroe (2004) and Clarke (2015).

Learning objectives

The learning objective of this chapter is to better understand how building simulation programs can advance to better support modeling for design and operation of integrated building energy and control systems, with a perspective on how recent advances can overcome current limitations. In this chapter, the reader will be exposed to (i) motivating factors that may change the way we model and simulate building energy and control systems, (ii) technologies that enable such a change and (iii) a path for realizing a next generation modeling and simulation environment. While an in-depth discussion of the various enabling technologies is not possible in a single chapter, the reader will be exposed to the main concepts, and references to further literature are provided.

Key words

Requirements for building system simulation/modeling versus simulation/modular environments/object-oriented equation-based modeling/agent-based modeling/multi-domain modeling/management of complexity/building information model

Introduction

Imagine a possible future workflow for building design and operation. To start, an integrated design team sketches a building envelope in an object-oriented CAD system. Generative design algorithms turn customer requirements and design constraints into designs, by generating different building and energy system architectures and optimizing these competing design variants. These optimizations include sizing of components while taking into account the need for electrical load shifting, the temporal availability of solar thermal energy, the building's active and passive storage and the impact of active façade systems. A dashboard provides immediate feedback about the environmental impact and life-cycle costs of these different design variants. The dashboard also allows the user to insert design variants into the family of design candidates and to adjust design preferences to further guide the optimization. Once the team selected a design, it saves its representation in a building information model (BIM). The BIM will be sent to code authorities for performance-based code compliance checking within minutes. The BIM will also be sent to builders, mechanical system providers and controls providers to bid on the project. Once equipment providers are selected, the simplified component models will be replaced with equipment models as published by manufacturers in their electronic catalogues. During construction, building elements are automatically prefabricated based on the digital model. Control sequences and functional verification tests are automatically translated from the design specification to the building automation system. During building operation, the BIM data will be used to generate models for use in model predictive control (MPC), fault detection and diagnostics (FDD) and preventive maintenance and to monitor actual versus expected energy use. The generation of these models, in some cases involving model simplification to make them better suited for the application at hand, will be initiated from the digital twin of the building, e.g. a digital representation that evolves with the building.

What building simulation program will support such a life-cycle approach to simulation from product R&D to building operation? This is a focus of this chapter. Central to our discussion is the belief that the next generation tools that interface with the workflow of simulators and that become part of building automation systems are not building simulation programs in the conventional sense. To explain this, we use the illustration from Cellier and Kofman (2006), shown in Figure 19.1, which distinguishes between the *mathematical model* and the *simulation program*. Cellier and Kofman state that the whole purpose of a mathematical model is to provide the human user of the modeling and simulation environment with a means to present knowledge about the physical system in a way that is as convenient to the user as possible. This may be for a building envelope an object-oriented CAD system, for an HVAC system a graphical schematic editor and for control systems an editor for block diagrams and finite state machines. Thus, the mathematical model represents the user interface. The mathematical model does not depend on how it is going to be used by the simulation program. The primary purpose of the simulation program is to compute the system response, typically in the form of time trajectories. The simulation program is automatically generated from the mathematical model. It typically provides a run-time interface with which the user can control the simulation process.

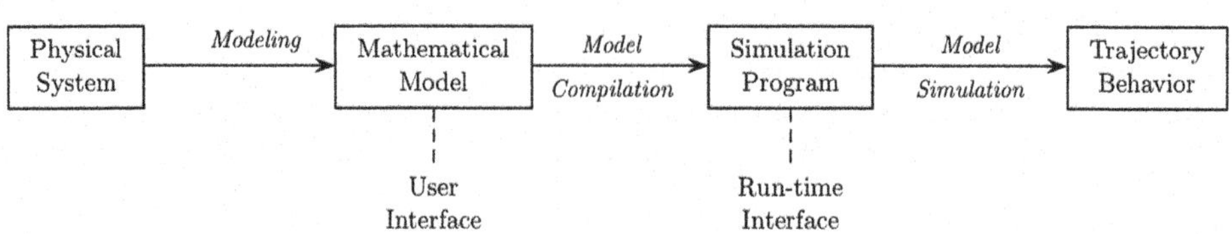

Figure 19.1 Modeling vs. simulation

Source: Adapted from Cellier and Kofman 2006, with kind permission of Springer Science+Business Media

Such a separation between mathematical model and simulation program is a critical underlying principle for a future modeling and simulation environment for building systems, as we will see in this chapter. By following this principle, it is possible to structure the problem more naturally in the way a human thinks, and not how one computes a solution.

Motivating factors for a future modeling and simulation environment

To increase performance and improve comfort, building systems are becoming increasingly integrated. Integration is achieved physically by using systems that allow better energy recovery, energy storage and increased use of ambient energy sources and sinks for heating, cooling, ventilation and lighting. Integration is also achieved operationally by means of integrating controls for façade, lighting, HVAC, combined heat and power, solar systems, electrical grid and security. The characteristics of these systems are that they involve multiple functional domains (heat and mass transfer; fluid dynamics; daylighting; electricity generation and conversion; controls; communication across networks), and their temporal evolution may be described in the continuous time domain, in the discrete time domain and in a discrete event domain. Typical time-scales range from the order of subseconds (communication networks) to years (geothermal heat sources). This heterogeneity poses new challenges for modeling and simulation, which are used to reduce time-to-market during research and development, and to improve performance and cost-effectiveness during the design and operation of innovative building energy systems and associated algorithms for controls, fault detection and diagnostics.

Given the different topologies in which building systems can be configured, such as using a double façade as a ventilation air exchange path or using a thermally activated ceiling slab for thermal storage, a future modeling environment needs to allow a more intuitive representation of such complex systems that allows a user to quickly add new component models and reconfigure components to form new systems. The current separation between model developer and model user is not conducive to stimulate such innovation.

It is generally recognized that properly designed and implemented control sequences can reduce energy consumption by around 20% to 30% (Fernandez et al. 2017). However, in particular for large buildings that have built-up HVAC systems with custom control sequences, control sequences are often implemented incorrectly. For such systems, it is customary for the mechanical engineer to write the control sequence in English language. This typically involves copying and adapting a part of the control sequence from similar projects. These sequence specifications are generally ambiguous, leave considerable room for interpretation and may miss part of the sequence. The control provider uses this English language description for cost estimation, and later for implementation. The control provider typically implements the control sequence by combining parts of sequences from previous projects that appear to have a similar controls intent. Consequently, the controls intent is often not realized. Therefore, to realize the energy savings and load shifting capabilities of control sequences, tools and a process should be put in place that allow a mechanical engineer to select and adapt a control sequence from a library of sequences, test its performance using a simulation model of the HVAC and building, and export the control sequence in a control-vendor neutral format. The exported control sequence should then be used by control providers to conduct cost estimates and ultimately implement the sequence on their control platform through machine-to-machine translation. Figure 19.2 shows such a process, in which the actual control implementation is then verified against the formal specification that is exported from the design specification. To enable such a process, the control representation in building simulation programs needs to resemble how actual control algorithms are implemented, and the numerical solution methods need to be capable of correctly simulating feedback control, e.g. they need to be able to adapt their time step to maintain accuracy during fast transients, and they need to handle time events and state events.

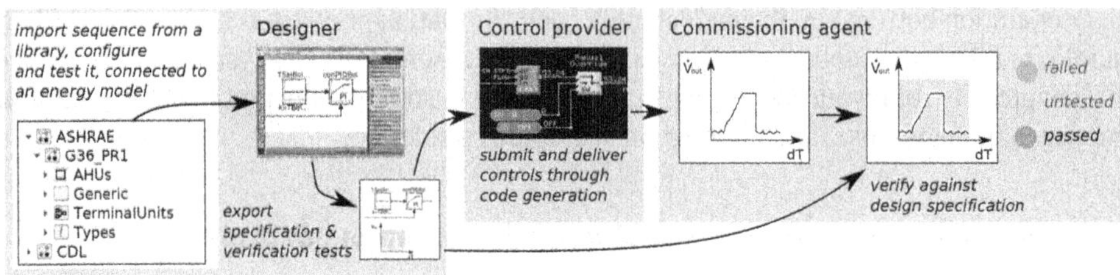

Figure 19.2 Overview of process for control sequence design, export of a specification, implementation on a control platform and verification against the specification

Source: Wetter et al. 2018a

Given the increased integration of different domain disciplines, a future environment for modeling and simulation also needs to enable the concurrent collaboration of multi-disciplinary teams, as opposed to a single person, the sharing of models to integrate different domains, and linking these models to executable requirements that verify whether the design intent is met. This may become increasingly important because much of the innovation in building science is likely to happen at the interface between different disciplines; for example, building physics for passive energy storage, mechanical systems for efficient energy conversion, controls systems for reducing inefficiencies caused by fluid flow friction and heat transfer, control methods that shift loads to periods of low electricity rates, communication systems for propagation of sensor and actuator signals and mathematics to extract better quality information from the vast amount of data measured by building automation systems. Addressing the need for different discipline experts to collaborate more effectively will require designs of future tools that are different from today's tools.

Because embedded computing, access to real-time sensor data, mathematical optimization and machine learning is becoming cheaper and more powerful, and because the performance of low-energy buildings is sensitive to controls, occupant interaction and equipment degradation, it is likely that computational models combined with algorithms for mathematical optimization and machine learning will be used for commissioning, fault detection, diagnostics, model-based controls and building analytics. In this context, the use of models can range from an individual piece of equipment, such as for a chiller fault detection, to whole communities that may be connected by energy grids for electricity, heating and cooling. The models may be data-driven, physics-based or a combination thereof. Since sustainable buildings will have a lifespan that exceeds multiple generations of programming languages, operating systems and building simulation programs, models used during operation should be expressed at a high enough level of abstraction that ensures maintainability and portability and facilitates upgrading to a newer technology as buildings are retrofitted. Moreover, as some of these mathematical methods are considerably more efficient and robust if they can analyze and manipulate the model's equation, expressing models in the form of equations rather than as executable code is favorable (Wetter et al. 2016).

With respect to using building simulation for training of service technicians, commissioning agents and building operators, today's building simulation programs fall short in teaching how mechanical systems and their control systems operate. At present, models of these systems are often so simplified that they do not capture the dynamic behavior and part-load operation of the mechanical system or the response of feedback control systems. Also, in many building simulation programs, the governing physical equations cannot be readily inspected by the user, which renders them as a black box. Immersive simulation has the potential to improve the knowledge about the operational behavior of buildings by presenting results of building simulation in a form that is closer to the human than a graph on a computer screen. Work to date has been primarily focused on computation and visualization of spatial distribution of temperature, flow and comfort in rooms (Malkawi 2004). A versatile environment that would allow a realistic modeling and simulation of the mechanical and control system, and that could be reconfigured by the user to examine a variety of systems, would extend the promise of immersive simulation to training professionals in the operation of building systems.

The transition towards integrated systems introduces new challenges. Effective modeling strategies need to manage the increased complexity, the increased fragmentation of the supply chain and the need for working in multi-disciplinary design teams using shared models. In the electronics design automation community, which faces challenges that are similar to the ones of the building industry, there are calls for a new design science that addresses system-level design challenges (Sangiovanni-Vincentelli 2007). The basic tenet of the advocated design flow is the identification of design as a meet-in-the-middle process, where successive refinements of specifications meet with abstractions of potential implementations. The refinement and abstraction process takes place at precisely defined layers. The embedded systems community also called for a new system science that is jointly physical and computational. To address challenges posed by networking physical devices that contain embedded control systems, the computing and networking abstractions need to be rebuilt to embrace physical dynamics and computations in a unified way (Lee 2006, 2011; Lee and Sangiovanni-Vincentelli 2011). In this community, the model semantic is expressed at a high enough level of abstraction to allow formal analysis of system properties and to generate code that can be used for design, synthesis and operation. It would be a surprise if advances in these related disciplines would not also cause the field of building simulation to embrace a higher level of abstraction that better addresses challenges in research, design and operation of integrated building systems.

A closer look at the software architecture of today's building simulation programs reveals that they are ill-suited to address the above needs. Many building simulation programs do not exploit advances in computer science that allow systems and their evolution to be expressed at a higher level of abstraction than that of imperative Fortran or C/C++ function calls. Most of today's building simulation programs are large, monolithic programs that intertwine physics equations, statements that control the program flow logic, numerical solution methods and data input and output routines. This makes extending these programs for applications that were not envisioned by their developers difficult. Lack of modularity has forced developers of building simulation programs to try to catch up by adding modeling capabilities that allow simulating new technologies. However, it may be more beneficial to develop a modular platform that can rapidly be extended and customized by users and equipment providers to drive the innovation of new systems. Such modular platforms would also facilitate extracting subsystem models for use during building operation, for example, for FDD or model predictive control, possibly after conducting some semi-automated model simplifications. In view of the gap between current building simulation programs and modern modeling and simulation environments, we will now take a step back to describe some basic needs that motivate a new approach to building system modeling and simulation.

Needs for more natural system representations

"Object programs produced by Fortran will be nearly as efficient as those written by good programmers." This statement from the Fortran manual (IBM 1956), released in 1956, shows that as early as 50 years ago, there was a trend towards languages that were closer to the problem at hand. Even then, experts questioned whether the "high-level" programming language would be adequately efficient. Today, most building simulation programs are still written using imperative programming languages such as Fortran or C/C++ to encapsulate causal assignments in programming procedures. In these programs, procedures call other procedures, thereby transferring the locus of control from one procedure to another. Inside these procedures are imperative statements for algebraic equations, differential equations, difference equations and numerical solution algorithms. This syntax for formulating physical phenomena was motivated by the programming languages that were available decades ago. However, the syntax is far removed from how a physicist would describe the laws that relate physical quantities to each other or how an electrical engineer would describe communication in a network.

It turns out that physical systems are more conveniently modeled by (i) defining an object, (ii) exposing its boundary conditions and (iii) encapsulating inside the object mathematical constraints between the boundary conditions, the state variables and their derivatives. Consider for example a model for a heat conducting rod that stores no energy. It is not possible to determine whether heat flows through the thermal conductor because its ends are at different temperatures, or whether the temperatures at its ends are different because of the heat flow. So why should a user have to pick one or the other view when writing a model for a thermal conductor? It is more natural to create an object that exposes heat flow rate and temperature, i.e. (Q_1, T_1) and (Q_2, T_2), and encapsulates the constraints $Q_1 + Q_2 = 0$ and $Q_1 = K(T_1 - T_2)$, where K is the heat conductance (see Figure 19.3). How these equations are solved should be left entirely to an algorithm that generates a simulation program.

Such an equation-based, object-oriented model formulation is convenient to model physical systems that can be described by differential algebraic systems of equations. But is it also convenient to model *cyber-physical* systems? Cyber-physical systems are systems in which embedded computers and networks monitor and control physical devices with feedback loops, and the physical processes affect computations and vice versa (Lee 2006). Cyber-physical systems are becoming increasingly prevalent in buildings, and their size, in terms of nodes that send and receive signals, is growing rapidly. Components in cyber-physical systems exchange a sequence of discrete messages with each other through wired or wireless networks, which may lose or delay some messages. The sequence of discrete messages affects the evolution of each component's state variables. Such communication networks can be conveniently modeled by *finite state machines*. Finite state machines are objects that, if given the current state and an input, return the new state and an output. For computer modeling, it is convenient to represent such systems with actor-oriented component models.

Figure 19.3 illustrates the disconnect between how building simulation programs are written and how the evolution of physical systems and communication systems can be described. In

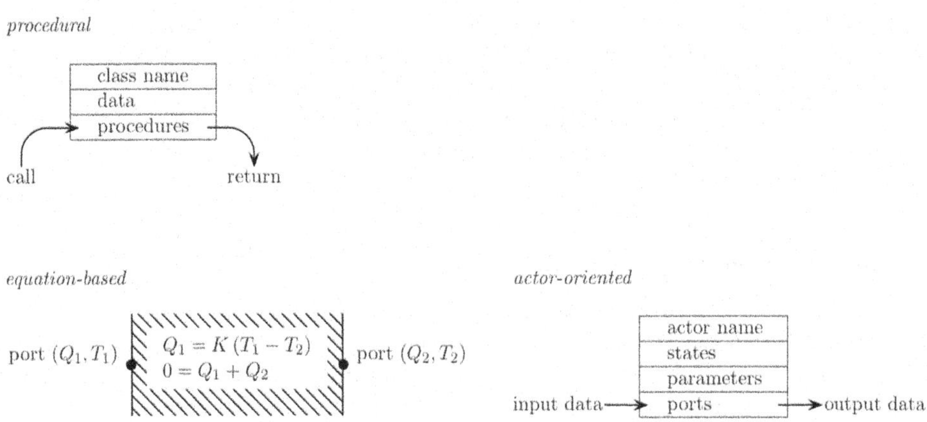

Figure 19.3 Comparison between procedural, equation-based and actor-oriented component models

Source: Procedural and actor-oriented illustration adapted with permission from Lee 2003 procedural programs, a procedure that belongs to a class is called. This procedure manipulates data, may call other procedures and then returns to transfer the locus of control back to the calling procedure. This is how typical building simulation programs are written. In equation-based programs, classes are typically encapsulated based on thermodynamic system boundaries. Ports are used to expose potential and flow variables (e.g. temperature and heat flow rate) without specifying what is input and output, as this is not needed to describe physics. Acausal relations define constraints between the port data and between the thermodynamic states of the system. In actor-oriented programs, actors contain ports that receive and send tokens. A *model of computation* (Lee 2003) specifies operational rules that determine when actors perform internal computations, update their states and perform external communications, which can, for example, be at fixed time intervals or whenever new tokens arrive. The ports are connected to other components' ports to communicate with each other.

Needs to reduce software cost and development time

Cellier (1996) states the main factors that reduce cost and development time of software as follows:

Reusability: A software design methodology that ensures optimal reusability of software components is the most essential factor in keeping the software development and maintenance cost down.

Abstraction: Higher abstraction levels at the user interface help reduce the time of software development as well as debugging. The conceptual distance between the user interface and the final production code needs to be enlarged. Software translators can perform considerably more tasks than they traditionally did.

The reusability of code can be significantly increased through the use of equation-based object-oriented models that separate acausal relations between model variables from their solution algorithms. For example, Figure 19.4 shows models of a heat conducting wall in which the heat input is regulated by a feedback control loop. Only half of the wall depth has been modeled because the wall is assumed to be symmetric and exposed to the same boundary conditions. In Figure 19.4(a), heat is added from the object called "source" to the wall's axis of symmetry, i.e. to the object called "mass1" which is at the center of the wall. In Figure 19.4(b), heat is added to the surface of the wall. The top row shows the implementation using equation-based object-oriented models, where each icon encapsulates a model for an individual physical phenomenon (such as heat convection, heat conduction and heat storage). With this representation, the system being modeled is easily recognized. In contrast, the bottom row shows the same physical models, but they have been implemented using causal models in the form of block diagrams. Figure 19.4(a) and Figure 19.4(c) describe the same system, as do Figure 19.4(b) and Figure 19.4(d). Note that the top row also contains causal models, but only for the control system since this is most naturally represented using input/output blocks. The equation-based object-oriented system models are easier to read because the schematic model representations only define the connectivity between the component models, but there is no notion of causality needed. In contrast, the block diagrams in the bottom row impose causality for each component and for the information flow within the system models. This limits model readability and reuse. It also led to more work for creating the models.

When developing the equation-based object-oriented model shown in the top row of Figure 19.4, each individual component model, such as for the heat conductor or the heat storage, could be tested individually in isolation of the overall system. Such unit tests help detect possible programming errors earlier when it is cheaper to fix them. Assembling the equation-based

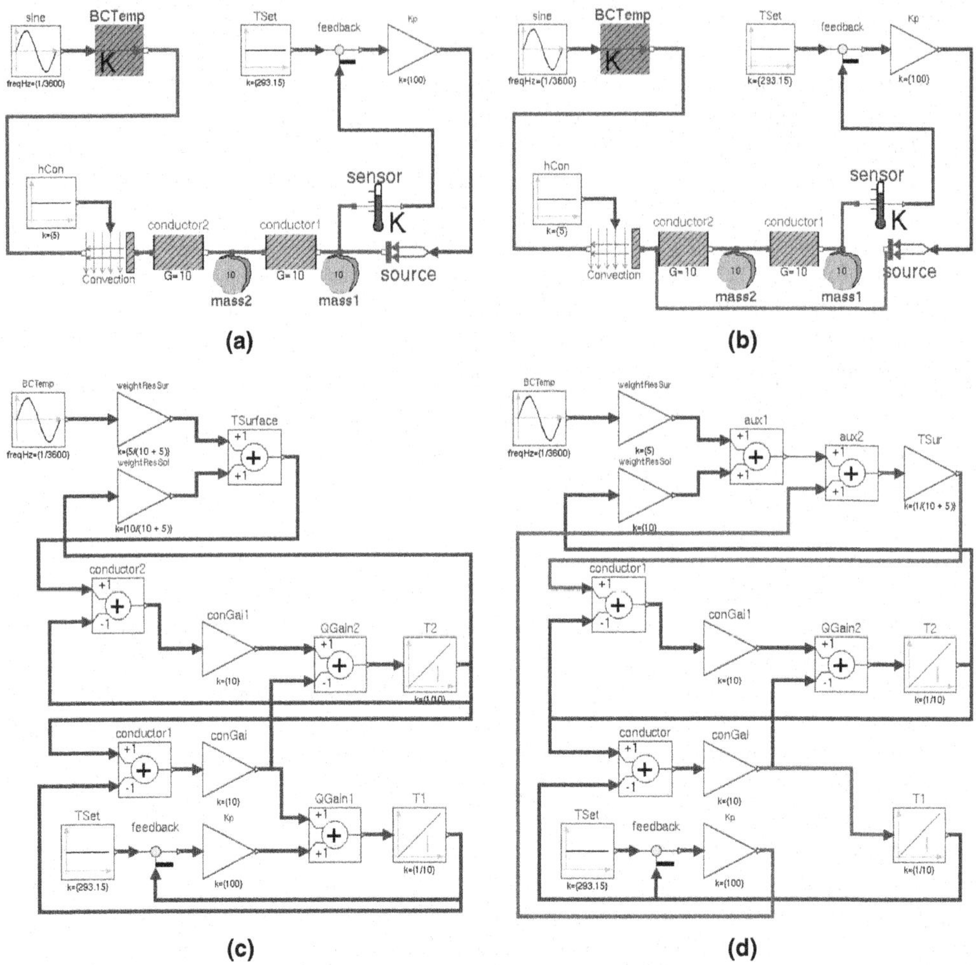

Figure 19.4 Representation of heat flow models with feedback control

object-oriented system model by connecting lines between the components then poses little risk for introducing errors. To assemble the causal system model, however, gains must be computed and heat flows must be added and connected to the corresponding integrator, which is time consuming and presents opportunities to make mistakes. The left and right columns of the figure differ only in the location where heat is added to the construction. For the equation-based object-oriented model, all a modeler must do is reconnect one line; for the causal model the modeler must add and delete blocks and recompute gains, thus reusing a significantly smaller part of the model and possibly introducing mistakes. The example illustrates that the equation-based object-oriented system representation not only reduces development time but also allows a more natural system representation as discussed in the previous section.

Needs to integrate tools in a coherent workflow

Researchers and practitioners represent different user groups who use modeling and simulation within different workflows and therefore have different requirements that are best addressed by a modular environment. With a modular approach, the same underlying core of the environment can be used for both user groups while customized interfaces can address the individual requirements. An example of a modular software environment is Linux, which can be customized to run in embedded systems such as cell phones, in graphical desktop environments, in server farms and in super computers. Modularization also makes it significantly easier for users to participate in program development; an example is the modular design of TRNSYS's simulation components that allows a large user community to develop simulation components that extend the scope of the program.

Researchers require an environment that integrates well with computational-based rapid prototyping and system design processes that reduce cost and product time-to-market. The environment must also provide formal ways for managing the increased system complexity and integration challenges in heterogeneous design and supply chains (Sangiovanni-Vincentelli 2007; Barnard 2005). For product development, it is common that simulation models are progressively replaced with more detailed models and with hardware as development progresses. Such processes typically include automatic code generation that translates a control algorithm developed in a modeling and simulation environment into code that can be downloaded to control hardware. New environments should allow users to add novel component and system models rapidly and to combine models in new system topologies to analyze new configurations for integrated building systems. Meeting these requirements mandates an environment that consists of modular models with standardized interfaces that can be individually tested and replaced with more detailed models as more information becomes available during the design, or supplanted by hardware that is linked with the simulation model.

Practitioners require that a new environment for modeling and simulation integrates well with the building delivery process. Since BIMs are becoming increasingly used, mandated by large building owners, supported for energy code compliance checking and profitable for design teams to use, it is important that the building simulation program can be generated using BIM inputs. Integrating BIMs into the building life cycle will require that mathematical models are instantiated and parameterized using data specified in or referenced by a BIM. As controls become increasingly important in reducing energy consumption, peak electrical power and downsizing or even eliminating mechanical equipment (such as eliminating a chiller if night ventilation of a building suffices to meet the comfort requirements), control-relevant information also needs to be specified by the BIM. To avoid data inconsistency between design and operation, and to avoid duplicate data entry, the environment should also be able to use a BIM to instantiate and parameterize models for model-based controls, fault detection and diagnostics. In this context, BIM will be used in a broader sense than it is today and will include formal specifications of control sequences and functional verification tests that the HVAC system has to meet. The specifications will have a semantics that is expressive enough so they can be used during design to simulate the system performance, during bidding for estimating construction costs, during commissioning to verify the proper operation and during operation for controlling the building system. Model instantiation and parameterization is easiest to achieve if the components specified in the BIM and in the simulation model have the same modularization and if the connectivity between components is only constrained by how physical laws permit real components to be connected. To meet these requirements, a more general approach for component connectivity rules is needed than what can be found in many building simulation programs.

Enabling technologies

The previous section showed that building systems are becoming increasingly integrated across different functional domains. This increases design and operational complexity. A future modeling and simulation environment for building systems will need to address this challenge by using technologies that enable an intuitive representation of the system topology, the intrinsic behavior of individual components and the interaction among the components once they become part of a system. This section discusses these technologies.

We will make a distinction between *modeling* and *simulation* as introduced by Cellier and Kofman (2006) and illustrated in Figure 19.1. Since building energy analysis requires large amounts of input data, it is convenient for our discussion to further separate mathematical modeling into behavioral modeling and data modeling. By *behavioral modeling* we mean specifying the mathematical equations that model a phenomenon, which Polderman and Willems (2007) call the behavioral equations. Behavioral models may be based on physical laws, performance curves or functions that update states and outputs of a finite state machine. By *data modeling*, we mean representing the static data of a system without specifying the mathematical model that describes how the system responds to inputs and evolves in time.

The purpose of modeling is to create a mathematical model from a physical system. From the mathematical model, a simulation program can be generated by using symbolic manipulations to sort equations, invert equations and replace derivative operators with numerical discretization schemes. Generating a simulation program can be done manually by a programmer, as is the current practice in most building simulation programs such as in TRNSYS or EnergyPlus. Alternatively, it can be automated, as done in equation-based modeling environments such as in Dymola, OpenModelica, JModelica, IDA or EES. The manual approach requires a model developer to write a simulation program (or subroutine), interface it with numerical solution methods, integrate it into the program kernel and implement methods to retrieve parameters and time-varying inputs and to write outputs. Manually generated programs are error prone and take considerably more time to write than specifying the actual physical relations that govern the object to be modeled. Equation-based modeling, in contrast, allows a model builder to concentrate the effort on describing the physical constraints and state transitions that define the evolution of the system. Generating efficient code for numerical simulation is left to a code generator.

Behavioral models

In the building simulation community, the need for more flexible means of constructing simulation models was expressed three decades ago by Sowell et al. (1986). A draft research plan was assembled by an international team working at the Lawrence Berkeley National Laboratory in 1985 for the collaborative development of a next generation building simulation program called the Energy Kernel System (EKS) for building simulation. A prototype software system called SPANK (Simulation Problem Analysis Kernel) that eventually became SPARK was created that contained some of the features envisioned for EKS (Sowell et al. 1986, 1989). Another development that embraced equation-based model formulation for building simulation is the Neutral Model Format (NMF) (Sahlin and Sowell 1989; Sahlin 1996). NMF is the primary modeling language used in IDA, an equation-based modeling and simulation environment that is used by researchers and practitioners and has recently been used to translate and simulate models written in the Modelica language described later (Sahlin and Grozman 2003). The Engineering Equation Solver (EES) is another equation-based modeling and simulation environment with an interface to thermophysical property functions, such as for steam and refrigerants (Klein and Alvarado 1992).

In 1996, a significantly larger industry-driven effort has been started that includes a wider range of industrial sectors and academic partners. A consortium formed to develop Modelica, a freely available, equation-based object-oriented modeling language that is designed for component-oriented, multi-domain modeling of dynamic systems. The goal of the consortium is to combine the benefits of existing modeling languages to define a new uniform language for model representation (Mattsson and Elmqvist 1997; Fritzson and Engelson 1998). Over the past decade, the Modelica language has gained significant adoption in various industrial sectors and has been used in demanding industrial applications. It is well positioned to become the de facto standard for modeling complex physical or communication systems. Modelica libraries for multi-domain physics include models for control, thermal, electrical and mechanical systems, as well as for fluid systems and different media (Elmqvist et al. 2003; Casella et al. 2006).

The rise of Modelica provided an opportunity to revive the efforts of the International Energy Agency (IEA) Task 22 to create a model library for building simulation, which was in Task 22 implemented in NMF. The intent of Task 22 was to develop a library of component models that can be shared among different modeling and simulation environments (Vuolle et al. 1999). From 2012 to 2017, the IEA conducted, under its Energy in Buildings and Communities Program, the project Annex 60, which developed new generation computing tools for building and community energy systems based on the Modelica and Functional Mockup Interface standards. This led to an open-source Modelica library that is now used as the core of four Modelica libraries for building and district energy systems (Wetter et al. 2015a). In 2017, this Annex 60 library was renamed to IBPSA library, and will be further developed within the IBPSA Project 1.

Modelica models can contain differential equations, algebraic equations and discrete equations. Using standardized interfaces, a model's mathematical relations among its interface variables are encapsulated, and the model can be represented graphically by an icon. The encapsulation facilitates model reuse and exchange and allows connecting component models to system models using a graphical or textual editor. The tenets of Modelica models are that each component represents a physical device with physical interface ports, such as heat flow rate and temperature for solid ports and pressure, species flow and enthalpy for fluid ports. Defining a device- and physics-oriented system architecture enables intuitive model construction. The Modelica approach is realized by following the *object-oriented modeling paradigm*, a term that was coined by Elmqvist (1978) and summarized by Cellier et al. (1995) as follows:

Encapsulation of knowledge: The modeler must be able to encode all knowledge related to a particular object in a compact fashion in one place with well-defined interface points to the outside.

Topological interconnection capability: The modeler should be able to interconnect objects in a topological fashion, plugging together component models in the same way as an experimenter would plug together real equipment in a laboratory. This requirement entails that the equations describing the models must be declarative in nature, i.e. they must be acausal.

Hierarchical modeling: The modeler should be able to declare interconnected models as new objects, making them indistinguishable from the outside from the basic equation models. Models can then be built up in a hierarchical fashion.

Object instantiation: The modeler should have the possibility to describe generic object classes and instantiate actual objects from these class definitions by a mechanism of model invocation.

Class inheritance: A useful feature is class inheritance, since it allows encapsulation of knowledge even below the level of a physical object. The so encapsulated knowledge can then be distributed through the model by an inheritance mechanism, which ensures that the same knowledge will not have to be encoded several times in different places of the model separately.

Generalized networking capability: A useful feature of a modeling environment is the capability to interconnect models through *nodes*. Nodes are different from regular models (objects) in that they offer a variable number of connections to them. This feature mandates the availability of *across* and *through variables*, so that power continuity across the nodes can be guaranteed.

Figure 19.5 illustrates how a multizone building model was created by graphically connecting physical ports for heat flow rate, airflow rate and control signals of thermal zone models, airflow models and HVAC system models. Each icon encapsulates a model that may encapsulate other models, thus allowing a model builder to manage the complexity of large system models through hierarchical modeling. Hierarchical model building also facilitates model reuse and testing of submodels before they are assembled as a large system model that may be difficult to debug. Object-oriented model construction with clearly defined encapsulations accelerates model development by allowing concurrent construction of domain-specific models by their respective domain experts. For example, the HVAC and controls engineer can construct their models simultaneously and combine them later to a system model. The concurrent workflow with subsequent model combination is illustrated by Figure 19.5(a). On the left side of the figure, the three separate models represent the HVAC systems, the interzonal airflow and the building heat transfer (Wetter 2006a, 2006b). The different domain models are integrated with each other by drawing lines between the model ports. The lines automatically generate equality constraints for across variables (temperature, pressure) and conservation equations for through variables (heat flow rate, mass flow rate).

Libraries of component models that allow assembling such systems are created using objects with standardized interfaces that define all independent variables necessary to describe the desired effect. Using these standardized interfaces makes models compatible with each other without having to add converters that convert one set of interface data into another one. For example, in Modelica's thermal library there is a connector called `HeatPort` that defines an interface for elements that transfer or store heat by introducing the tuple temperature and heat flow rate as

```
partial connector
Modelica.Thermal.HeatTransfer.Interfaces.HeatPort
"Thermal port for 1-D heat transfer";
SI.Temperature T "Port temperature";
flow SI.HeatFlowRate Q_flow
"Heat flow rate (positive if flowing into the component)";
end HeatPort;
```

The keyword `flow` declares that all variables connected to `Q_flow` need to sum to zero. For example, if two `HeatPorts` that are called `port1` and `port2` are connected, a Modelica translator will generate the equality constraint `port1.T = port2.T` and the conservation equation `port1.Q_flow + port2.Q_flow = 0`. Similar connectors are defined for other physical interfaces including fluid ports, electrical signals and translational and rotational

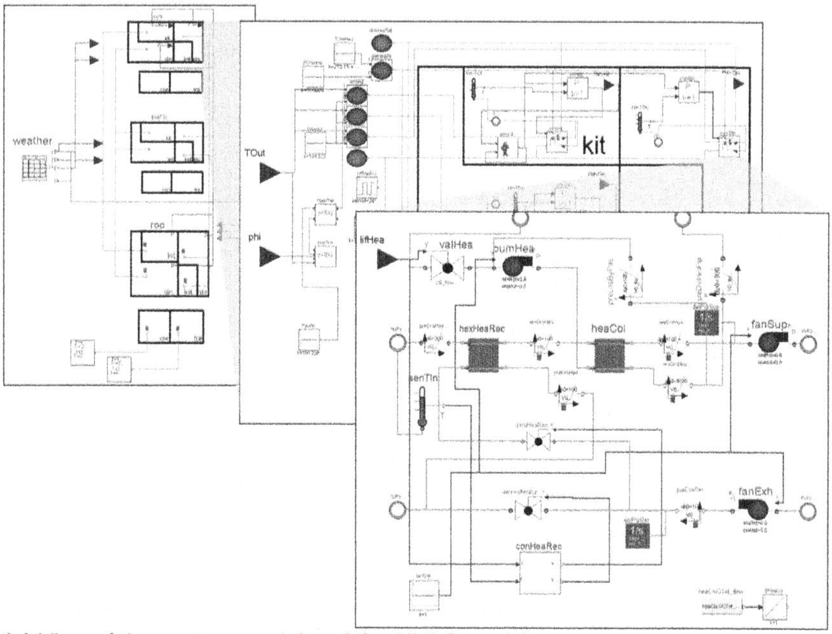

(a) View of the system model and the HVAC model

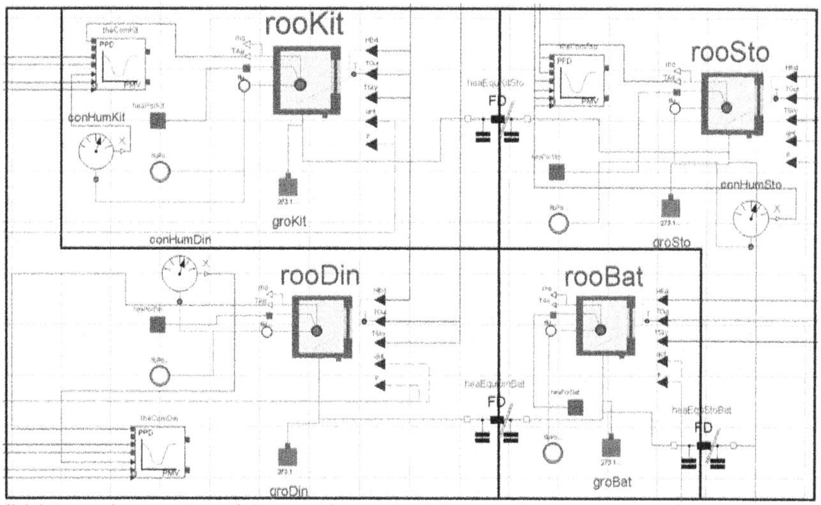

(b) View of a section of the multizone building model

Figure 19.5 Schematic view of a building system model in Modelica that was constructed using different hierarchical layers to manage the model complexity, and that decomposes different domains. In the top figure (a), on the left, there are models for the HVAC system (top), for the interzonal airflow (middle) and for the building envelope heat transfer (bottom), which are connected to each other by fluid loops and control signals to form a coupled system of equations. The top right shows the inside of the HVAC system definition for each thermal zone, with an icon for the HVAC secondary loop, for the controller and for the user that may adjust the set points to maintain comfort. The bottom right shows the actual implementation of the HVAC secondary loop with the heat exchangers, the fans and pumps. The bottom figure (b) shows a section of the multizone building model with models for the rooms that are connected to each other with finite difference heat transfer models. Also shown are models for the thermal comfort.

bodies. The above heat port connector is used to define an interface for one-dimensional heat transfer elements with no energy storage in the following partial model:

```
partial model Element1D
"Partial heat transfer element that does not store energy"
SI.HeatFlowRate Q_flow "Heat flow rate";
SI.TemperatureDifference dT "port_a.T-port_b.T";
HeatPort port_a;
HeatPort port_b;
equation
dT = port_a.T - port_b.T;
port_a.Q_flow = Q_flow;
port_b.Q_flow = -Q_flow;
end Element1D;
```

This model is incomplete because it does not specify how temperatures and heat flow rate depend on each other. This incompleteness is denoted by the keyword `partial`, which has two purposes: It shows to a user that the model cannot be instantiated, and it tells a Modelica translator that the model is allowed to have more variables than equations, which is used for automatic error checking. To define a complete model, the above partial model is used to define a thermal conductor as

```
1 model ThermalConductor
2 "Thermal element transporting heat without storage"
3 extends Interfaces.Element1D;
4 parameter SI.ThermalConductance G "Thermal conductance";
5 equation
6 Q_flow = G*dT;
7 end ThermalConductor;
```

The thermal conductor model is graphically represented by an icon that is used, for example, in the system model shown in Figure 19.4. Note that by replacing the parameter declaration on line 4 and the equation on line 6, the semantics can be changed to represent other one-dimensional heat transfer elements, such as a model for convective heat transfer.

Equation-based modeling languages have been shown to significantly reduce model development time compared to conventional programming languages such as C/C++ or Fortran, as shown in Figure 19.6. A comparison of the development time for the BuildOpt multizone thermal building model which is implemented in C/C++ and a comparable implementation in Modelica showed a development time five to ten times faster when using Modelica (Wetter 2005, 2006a; Wetter and Haugstetter 2006). The Modelica development is faster primarily because Modelica is an acausal, equation-based, object-oriented language. This automated the work for writing routines for data input and output, sorting equations and procedure calls, implementing numerical solution methods and managing the large amount of data that is involved in building simulation. The shorter development time also manifests itself in a code size four times smaller (6,000 lines instead of 24,000 lines of code, not counting freely available libraries). In another experiment, the author asked three engineers familiar with Fortran and Modelica to implement the system shown in Figure 19.4(a) in Fortran and in the Modelica modeling and simulation

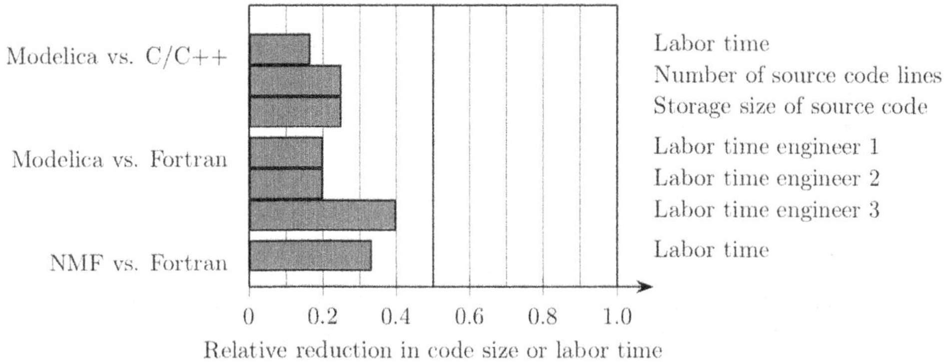

Figure 19.6 Comparison of relative reduction in labor time and code size for equation-based vs. procedural programming languages

environment Dymola. Even for such a simple problem where data input, output and storage are easy to implement, the model in Modelica was built 2.5 to 5 times faster than in Fortran. Similar increase in productivity has also been reported by Sahlin (2000), who was three times faster in implementing a thermal building simulator using the equation-based language NMF. These benchmarks indicate significant time savings for modeling when using equation-based modeling languages.

Pollok, Klöckner and Zimmer conducted a study about how users understand different model formulations. Preliminary analysis indicates that models represented as graphical, equation-based models with acausal ports are understood fastest, followed by equation-based textual representations, algorithm-based textual representations and graphical block-diagram-based representations. This holds for both model-recognition and -prediction tasks, as well as rated modeler confidence (Pollok 2017).

Data models

A BIM is an instance of a data model that describes a specific building unambiguously. Several data models are used by the construction industry. The only open, object-oriented data model that covers the whole building life cycle are the Industry Foundation Classes (IFC), developed by the International Alliance for Interoperability and formalized by the International Organization for Standardization (ISO) as the ISO 16739-1:2018 standard. Economic motivations for using BIMs include an increase in productivity through data interoperability among the participants of the building delivery process, as well as data integrity during project planning, cost analysis, energy analysis and automatic code compliance checking. This section focuses upon how a future modeling and simulation environment for building systems can make BIM use throughout the building life cycle more seamless. Therefore, our discussion is tool-oriented to show how building performance assessment tools can best address requirements that originate from the use of BIM. We refer the reader to Augenbroe et al. (2004) and Lam et al. (2004) for a process-oriented discussion of simulation models within the building process, to Bazjanac (2008) for a discussion of semi-automated data transformation between BIM authoring tools and building simulation tools and to Wimmer et al. (2017) for a discussion on model view definition for energy simulation.

This discussion assumes that future BIMs will contain specifications for building controls that will be used during design for building simulation, as well as during operation for model-based algorithms for controls, fault detection and diagnostics. Since controls are an essential tool for increasing building system efficiency, they would be a natural extension to a BIM, and the IFC2x platform specification indeed mentions building automation as one discipline that it aims to support. To enable a seamless combination of data models represented by a BIM and behavioral models represented in an equation-based object-oriented language that can generate a building simulation program, the component modularization and the variables that connect components should be identical in the data model and the behavioral model. Furthermore, data and behavioral models need to allow connecting components in the same way as one would connect actual components. Otherwise, the set of possible HVAC systems is larger than the set of HVAC systems that can be simulated, which makes seamless BIM use difficult if not impossible for some HVAC systems.

Many building simulation programs are based on software principles that do not match how buildings and energy systems are actually built and controlled. Also, their modeling language is not expressive enough to allow computing the evolution of heterogeneous systems in a way that properly accounts for concurrency of event-driven systems. This can make the use of models for simulation-based controls and commissioning difficult. In addition, the model modularization and model connectivity rule is imposed by the program developers. They decide what should be lumped in a monolithic subsystem model, what causality should be imposed on the equations to translate them into procedural code and what connections should be required for an object. These decisions are not motivated by how a BIM represents an HVAC system, and they do not necessarily reflect real HVAC systems. Rather, the selection of component modularization, component causality and connection variables is motivated by accommodating the data flow and solution algorithms imposed by the program architecture. For example, the room model of EnergyPlus computes a heating or cooling load in units of power. This load is then sent to zone equipment units, such as a fan coil, that try to meet the load. Thus, the control input of the fan coil is a load in watts, as opposed to a temperature signal from a thermostat. This type of component connectivity is different from the connectivity of actual components. Furthermore, in EnergyPlus, a valve that adjusts a cooling coil's water flow rate is implicitly assumed to be part of the coil, which represents a component modularization that does not correspond with that of actual systems. Further difficulties in generating a simulation model from a BIM can arise in non-conventional systems that do not follow the fluid loop structure of some building simulation programs. For example, because EnergyPlus does not allow equipment between thermal zones, the double façade system shown on the left side in Figure 19.7 had to be implemented using two air loops, and the exhaust fan had to be replaced by a supply fan to satisfy the EnergyPlus system structure. Automatically mapping the actual topology on the left from a BIM into the topology on the right seems to be an impossible task. The BIM translator would first need to understand in what system context the components are used, and then map one system topology, representing the actual building expressed in the BIM, into another system topology that can be simulated by the particular building simulation program. Because of this difficulty, early implementations of BIM to building simulation program translators are based on restricting the BIM instances to system topologies that can be simulated by the downstream simulation program (Bazjanac and Maile 2004). In view of these limitations, more fundamental research should be done to understand the computability of mappings between BIM and BIM-processing applications, as also suggested by Augenbroe et al. (2004).

To generate a mathematical model from a BIM, a more generic component modularization and connectivity rule than what is employed in today's building simulation programs needs to be used. It is most natural to base the modularization rule on how components are modularized in reality; typically, modular units are defined by how they are packaged when shipped to the construction site. Furthermore, the connectivity rules of models should follow the same rules in which components can be connected in reality. This is easiest to realize by using acausal models that follow the object-oriented modeling paradigm and whose interface variables are

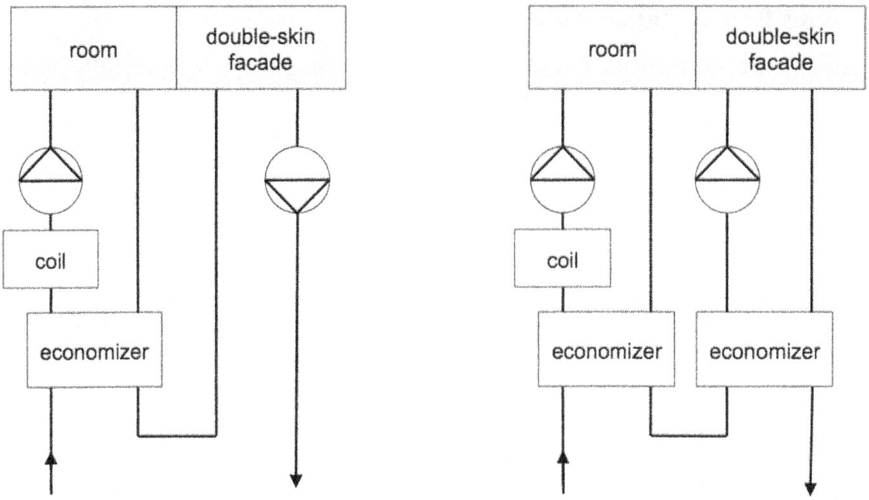

Figure 19.7 Schematic diagram of an HVAC system that uses a double-skin façade for the exhaust air (left side) and implementation in EnergyPlus that required two air loops (right side)

Source: Courtesy of Tobias Maile, Lawrence Berkeley National Laboratory

sufficient to describe the interaction of connected objects. Models with these traits can be arbitrarily assembled to build systems as long as their connections have the same physical properties, such as temperature and heat flow rate for a solid-to-solid interface. Such a connection framework is indeed realized in the Modelica language.

To enable the workflow shown in Figure 19.2 that integrates energy modeling with control deployment and verification, Wetter et al. (2018a) developed a Control Description Language (CDL). As control consists of not only communication but also logic, the intention of CDL is to standardize how to express the control logic through a vendor-independent language, thereby augmenting BACnet that already standardizes communication through a vendor-independent protocol. The language has been designed to simplify translation to actual control product lines. Wetter et al. (2018b) demonstrated that CDL suffices to express control sequences in a vendor-independent format that can be used within an annual building simulation. Moreover, the authors showed that two different control sequences, both published by ASHRAE, yield a 30% difference in annual HVAC energy consumption under comparable comfort. This 30% difference raises the question of whether the idealization of control sequences that is common practice in today's building energy simulation programs leads to trustworthy energy calculations.

A further difficulty arises when modeling cyber-physical systems. Cyber-physical systems will become more common as it becomes cheaper to densely instrument buildings with networks of sensor and actuators that control the building system. In cyber-physical systems, physical processes can affect when computations are performed and messages are sent across networks. For some systems, event detection, event ordering and other techniques for modeling and simulation of heterogeneous systems that properly model concurrency are needed. These needs may be easiest to meet by generating an actor-oriented system model for the communication network in which a model of computation controls the interactions between the different actors. The model may then be coupled with equation-based models for the physical components to form a hybrid system, which then can be used for designing and commissioning such networks.

Algorithms for translation and simulation

The translation of mathematical models into a simulation program can be fully automated. When translating a model, symbolic processing is important to reduce computing time since many building system simulation problems lead to large, sparse differential algebraic equation systems (DAE systems). Symbolic processing is typically used to reduce the index of the DAE system and to exploit sparsity. Pantelides (1988) presents an algorithm that reduces the index of DAE systems so that the equations can be solved using ordinary differential equation solvers. This algorithm is, for example, used by Dymola (Mattsson et al. 2000). It is current practice in modern simulators to exploit sparsity during model translation or during run-time. For example, Modelica simulation environments use symbolic processors during model translation to reduce the system of equations to block lower-triangular form. This process is called *partitioning* and is discussed in detail by Duff et al. (1989). Partitioning algorithms can guarantee that it is not possible to further reduce the dimensionality of the system of equations by permuting variables and equations. After partitioning, *tearing* can be used to break the dependency graph of equations and variables to further reduce the dimensionality of the system of equations. The resulting systems of equations are typically small but dense. Tearing can be done automatically or it can be guided by using physical insight with language constructs that can be embedded in a model library (Elmqvist and Otter 1994). A further reduction in computation time can be obtained by symbolically inserting the discretization formulae that represent the numerical integration algorithm into the differential algebraic equation model, a process called *inline integration* that was introduced by Elmqvist et al. (1995). The IDA solver, in contrast, employs algorithms for sparse Jacobian factorization during simulation. It relies less on global symbolic manipulations during model translation, but it also employs tearing and automatic generation of Jacobians (Sahlin and Grozman 2003). Fast model compilation is an advantage of IDA's approach that is particularly attractive for whole building simulations.

There are several technologies in place that decrease the computation time and lead to increased numerical robustness. But how do they compare to conventional building simulation programs? In general, comparing the run-time efficiency of different simulators is a challenging endeavor because of the different temporal and physical resolution of the tools, and because solvers can be tuned to be more efficient for certain classes of problems at the expense of robustness. Nevertheless, the following comparison shows that commercial general purpose simulators can be as efficient in performing a building simulation as commercial special tailored building simulation programs. Sahlin et al. (2004) compared the computation time of IDA ICE with EnergyPlus. In their numerical experiments, IDA ICE required approximately half the computation time that was required by EnergyPlus for a three-zone building with natural ventilation and twice the computation time in initial experiments for a 53-zone building without natural ventilation. Sowell and Haves (2001) compared the computation time of SPARK and HVACSIM+ for a variable air volume flow system that serves six thermal zones. They report that SPARK computes about 15 to 20 times faster than HVACSIM+, and they attribute the decrease in computation times to SPARK's symbolic processing. Another variable air volume flow network model with 26 zones shows that the computation time of SPARK and Dymola are similar for this class of problems (Wetter et al. 2008). Wetter and Haugstetter (2006) compared the simulation time of a multizone building model written in Modelica with a building model using TRNSYS's TYPE 56 that gives comparable energy use and transient response. The Modelica model was simulated using the solver of Dymola and Simulink. They report that Dymola computed four times slower and Simulink three times slower than TRNSYS. In these experiments, TRNSYS computed the heat conduction with conduction transfer functions, which are considered to be computationally faster than the finite difference scheme used in the Modelica implementation. Wetter et al. (2015b) compared a novel time integration method called quantized state systems (Kofman and Junco 2001; Cellier 2006) that discretizes the state variables rather than time. This converts the differential equations into a discrete event simulation, and thereby naturally allows combining the continuous time dynamics of the HVAC system with the discrete time and discrete event dynamics of the control system. Compared to EnergyPlus, their prototype implementation simulated 30% to 50% faster. Faster simulations are expected once their implementation better exploits sparsity, as done by Fernandez and Kofman (2014).

One should keep in mind that these benchmarks have been simulated using a single processor. The recent switch to computers with parallel processors requires, however, that programs be written differently to take advantage of the new hardware architecture. It is reasonable to expect that models formulated in equation-based, high-level programming languages, as opposed to existing simulation programs, are in a better position to take advantage of the shift towards parallel hardware. One reason for this expectation is that equation-based simulation environments naturally separate the algebraic and numerical solvers from the model equations. This facilitates replacing algebraic and numerical solvers, as well as model translators, with newer versions that exploit new hardware architectures.

Standard for run-time data exchange among tools

The aforementioned modeling technologies have been developed for initial-value ordinary differential equations in time, coupled to algebraic and discrete equations. Many problems, however, require coupling of such models to models for physical processes that require solution of equations within spatial domains. Examples include computational fluid dynamics for airflow in or around buildings, ray tracing for daylighting and shading calculations for complex façades. Such applications often exploit the problem structure to develop domain-specific numerical methods, code parallelizations and compilation that take advantage of particular hardware (Zuo and Chen 2010; Jones and Reinhart 2016; Jones et al. 2012). In addition, there is also the need to couple domain-specific legacy code with more modern implementations. These couplings often require co-simulation, in which each domain is represented by its own simulator, and boundary conditions that couple the simulators are synchronized as time progresses in the respective simulators. Various attempts for co-simulation, using custom interfaces that are added to the simulators on a case-specific manner, have been done by the building simulation community (Trcka et al. 2006, 2009; Beausoleil-Morrison et al. 2012; Dols et al. 2016; Wetter 2011). Unfortunately, this led to duplicative efforts and fragmentation in simulator interoperability. Over the past years, however, with the Functional Mockup Interface (FMI) standard, a standard has been developed for co-simulation and for exchanging models among simulators (Blochwitz et al. 2011). The FMI standard defines an application programming interface (API) in the C language to interact with the simulator or model, an XML scheme that allows for discovery of the interface variables and the capabilities of the simulator, and a structure for how the simulator or model is to be shared in a zip file. This zip file is called a Functional Mockup Unit (FMU). Tools that support the FMI standard can interact with the FMU to conduct an orchestrated simulation that can involve multiple FMUs, and do so in a manner that is agnostic to the tool that generated the FMU. As of 2019, the FMI standard is supported by more than 100 tools.

Moreover, as FMI standardizes how a simulator can be packaged, distributed and interacted with, it can be used in web services to make simulation models accessible to an audience that has no expertise in simulation. FMUs have also been used to encapsulate models for use in applications for energy monitoring, fault detection and diagnostics and in MPC.

Collaborative development

Over the past decade, Web 2.0 applications that enable distributed, concurrent, collaborative development of code and documentation have become a commodity. A characteristic of many successful open-source projects is that they employ a modular design that consists of small units with standardized interfaces. These units are good at doing one particular task, thereby providing a core of program functionalities around which further features can be built by a community of developers. Structuring the code into small units allows units to be developed and tested individually and then integrated into a larger system. It also reduces barriers to participation since only a small part of the program architecture needs to be understood by most contributors. This modularization facilitates reusing code, distributing development efforts and using experts for the tasks that they are best at, such as developing numerical solvers, mathematical models or user interfaces. Involving different experts with a variety of backgrounds brings diverse points of view for solving a problem, which can further increase the quality of the code.

Many building simulation programs are developed by teams that do not always involve experts from other disciplines such as computer science, scientific computation and applied mathematics. Therefore, many advances in these fields are transferred to the building simulation community at a rather slow rate. To facilitate a better transfer of science and technologies, future developers of an environment for modeling and simulation should integrate, and where necessary, further develop existing technologies with other communities that are active in system modeling and simulation. This should enable sharing of models, solvers and interfaces to accelerate the development of more advanced tools. In such a model, significantly bigger investments for tool and library development can be leveraged.

A path towards a next generation environment for building simulation and computational engineering

Equation-based object-oriented modeling and simulation environments have been applied successfully in numerous demanding industrial applications. In the building simulation community at the time of this writing, they are primarily used in academia and for product development, but not yet by industry for whole building energy simulation. The primary reason for this is that more research and development is needed for high-performance model translators and improved numerical methods for time integration of whole building energy simulations. Furthermore, model libraries need to include more component models in order for this technology to be competitive with conventional building simulation programs that have been developed and optimized over several decades, often involving hundreds of person years. Work that addresses these items has started (Wetter et al. 2015a, 2015b). In the meantime, Modelica and FMI have been used successfully for applications in buildings such as model predictive control (De Coninck and Helsen 2016; Wetter et al. 2016), implementation of sequences of operations (Schneider et al. 2017; Wetter et al. 2018a, 2018b), FDD (Bonvini et al. 2014; Sterling et al. 2014), controls design (Wetter et al. 2014), design of non-conventional HVAC systems (Maccarini et al. 2017) and analysis of district energy systems (Fuchs et al. 2016).

These are applications in which new use cases pose new requirements on tools, generate new flows of data and give rise to new mechanical and energy system-level topologies. This poses challenges for legacy tools, as they have not been developed for such applications. It therefore makes the modular approach of equation-based tools and of co-simulation technologies attractive and competitive, and it is to be expected that through such applications the technologies will be further matured. In parallel, the performance is expected to increase through advances on tools and numerical methods that often occur outside of the building simulation community but are applicable to our problems (Fernandez et al. 2014; Schuchart et al. 2015; Bergero et al. 2015). Once such performance gains have materialized, it is expected that these technologies will make inroads for more conventional applications in building simulation.

Specific applications in which a new modular environment for building modeling and simulation is competitive and already well suited include the following:

- *Analysis and design of new HVAC system architectures.* New HVAC system architectures at the scale of a building or a district energy system often require creating models of HVAC and control that require custom assembly of new or existing components. Examples of such new systems include adaptive façade systems that are integrated with the HVAC system (Loonen et al. 2017), two-pipe systems for simultaneous heating and cooling (Maccarini et al. 2017), new generation district energy systems with bi-directional water flow (Kräuchi et al. 2015) and novel control algorithms (Bünning et al. 2017).
- *Model predictive control.* When solving the optimal control problem in a model predictive control algorithm, all state variables of the simulation model need to be reset to initial values prior to evaluating the performance of different control actions. However, in many traditional building simulation programs, resetting state variables is difficult if not impossible without significant code changes. Furthermore, for computational efficiency and to prove that the optimization algorithm finds the optimal control sequence, the differential equations need to be smooth in the control input, which is typically not the case in building simulation programs (Wetter and Wright 2004). Also, some optimization methods are considerably more efficient if they can access and manipulate the underlying equations (Åkesson et al. 2010).
- *Fault detection and diagnostics.* For fault detection and diagnostics of components and subsystems, typically only a subset of the building system needs to be simulated. Realistic modeling of the component dynamics can be important to discern between faults and transient effects. Furthermore, for dynamic systems, FDD algorithm typically require estimation of the state, and adjusting the state variables in the model based on measurements and state estimation algorithms (Bonvini et al. 2014). This can be hard to accomplish in large monolithic building simulation programs.

These applications have in common that they require a flexible modeling environment that allows a user to rapidly add new models, symbolically invert models and extract a subsystem of a model for further analysis. By targeting these applications, modular modeling technologies can be further developed by users who typically have technical skills associated with the particular application. This leads to maturation of modeling technologies in terms of model scope, validation and formulation.

Synopsis

In a future environment for modeling and simulation, models will be expressed at a higher level of abstraction, which will allow a modeler to formulate the system closer to how the actual system behaves. Mathematical models and simulation programs will be distinct system representations; the first accommodates a human's way of expressing and structuring systems and allows computer-algebra tools to analyze and manipulate the model, while the second enables computation of numerical approximations to the solution of the system of equations. Separating mathematical models from their solution processes will also facilitate assembling system models that combine different domains. Combining such models will be realized by formulating equality and conservation equations that link shared interface variables on an equation-level, as opposed to integrating different programs, each with its individual solver, data management, error handling and input/output routines. Combining models on an equation-level is easier from the point of view of software engineering, mathematics (to ensure a consistent solution) and usability. Following the object-oriented modeling paradigm enables managing the complexity of large systems through hierarchical modeling. It also enables model assembly the same way as an experimenter would connect real components. This, in turn, allows for the creation of a modeling environment where equation-based object-oriented models have the same modularity and connectivity rules as a modular BIM, facilitating the use of BIM to create simulation programs at different stages of the building life cycle. In addition, the higher level of abstraction allows for joint development of not only models, but also modeling and simulation environments and their associated technologies with other industrial sectors. This includes numerical and symbolic solvers and standards for modeling and simulation.

Acknowledgments

This research was supported by the Assistant Secretary for Energy Efficiency and Renewable Energy, Office of Building Technologies of the U.S. Department of Energy, under Contract No. DE-AC02–05CH11231.

References

Åkesson, J., Årzen, K.-E., Gäfvert, M., Bergdahl, T. and Tummescheit, H. (2010) Modeling and optimization with Optimica and JModelica.org – languages and tools for solving large-scale dynamic optimization problem. *Computers and Chemical Engineering* 34(11): 1737–1749.

Augenbroe, G. (2001) Building simulation trends going into the new millennium. In Lamberts, R., Negrão, C.O.R. and Hensen, J. (eds), *Proceeding of the 7th IBPSA Conference*, vol. I. Rio de Janeiro, Brazil, pp. 15–27.

Augenbroe, G., de Wilde, P., Moon, H.J. and Malkawi, A. (2004) An interoperability workbench for design analysis integration. *Energy and Buildings* 36(8): 737–748.

Barnard, P.A. (2005) Software development principles applied to graphical model development. In *Modeling and Simulation Technologies Conference and Exhibit*. San Francisco, CA: American Institute of Aeronautics and Astronautics.

Bazjanac, V. (2008) IFC BIM-based methodology for semi-automated building energy performance simulation. In *Proceeding of CIB-W78, 25th International Conference on Information Technology in Construction*. Santiago de Chile, Chile.

Bazjanac, V. and Maile, T. (2004) IFC HVAC interface to EnergyPlus – a case of expanded interoperability for energy simulation. In *Proceeding of SimBuild*. Boulder, CO.

Beausoleil-Morrison, I., Kummert, M., Macdonald, F., Jost, R., McDowell, T. and Ferguso, A. (2012) Demonstration of the new ESP-r and TRNSYS co-simulator for modelling solar buildings. *Energy Procedia* 30: 505–514.

Bergero, F., Botta, M., Campostrini, E. and Kofman, E. (2015) Efficient compilation of large scale dynamical systems. In Fritzson, P. and Elmqvist, H. (eds), *11th International Modelica Conference*. Modelica Association: Paris, France, pp. 449–458.

Blochwitz, T., Otter, M., Arnold, M., Bausch, C.,. Clauß, C., Elmqvist, H., Junghanns, A., Mauss, J., Monteiro, M., Neidhold, T., Neumerkel, D., Olsson, H., Peetz, J.-V. and Wolf, S. (2011) The functional mockup interface for tool independent exchange of simulation models. In *Proceeding of the 8th International Modelica Conference*. Dresden, Germany: Modelica Association.

Bonvini, M., Sohn, M.D., Granderson, J., Wetter, M. and Piette, M.A. (2014) Robust on-line fault detection diagnosis for HVAC components based on nonlinear state estimation techniques. *Applied Energy* 124(0): 156–166.

Bünning, F., Sangi, R. and Müller, D. (2017) A modelica library for the agent-based control of building energy systems. *Applied Energy* 193: 52–59.

Casella, F., Otter, M., Proelss, K., Richter, C. and Tummescheit, H. (2006) The Modelica Fluid and Media library for modeling of incompressible and compressible thermo-fluid pipe networks. In Kral, C. and

Haumer, A. (eds), *Proceeding of the 5-th International Modelica Conference*, vol. 2, Vienna, Austria, pp. 631–640.

Cellier, F.E. (1996) Object-oriented modeling: Means for dealing with system complexity. In *Proceeding 15th Benelux Systems and Control Conference*. Mierlo, The Netherlands, pp. 53–64.

Cellier, F.E., Elmqvist, H. and Otter, M. (1995) Modeling from physical principles. In Levine, W.S. (ed.), *The Control Handbook*, ISBN:0849385709. Boca Raton, FL: CRC Press.

Cellier, F.E. and Kofman, E. (2006) *Continuous System Simulation*. New York: Springer.

Clarke, J.A. (2001) *Energy Simulation in Building Design*, 2nd edition. Oxford, UK: Butterworth-Heinemann.

Clarke, J. (2015) A vision for building performance simulation: A position paper prepared on behalf of the IBPSA Board. *Journal of Building Performance Simulation* 8(2): 39–43.

De Coninck, R. and Helsen, L. (2016) Practical implementation and evaluation of model predictive control for an office building in Brussels. *Energy and Buildings* 111: 290–298.

Dols, W.S., Emmerich, S.J. and Polidoro, B.J. (2016) Using coupled energy, airflow and indoor air quality software (TRNSYS/CONTAM) to evaluate building ventilation strategies. *Building Services Engineering Research and Technology* 37(2): 163–175.

Duff, I.S., Erisman, A.M. and Reid, J.K. (1989) *Direct Methods for Sparse Matrices*. Monographs on Numerical Analysis, Oxford: Oxford Science Publications.

Eastman, C., Liston, K., Sacks, R. and Teicholz, P. (2008) *BIM Handbook: A Guide to Building Information Modeling for Owners, Managers, Designers, Engineers and Contractors*. New York: John Wiley & Sons.

Elmqvist, H. (1978) A structured model language for large continuous systems. PhD thesis, Lund Institute of Technology, Lund, Sweden.

Elmqvist, H. and Otter, M. (1994) Methods for tearing systems of equations in object-oriented modeling. In *European Simulation Multiconference*. Barcelona, Spain, pp. 326–332.

Elmqvist, H., Otter, M. and Cellier, F. (1995) Inline integration: A new mixed symbolic/numeric approach for solving differential-algebraic equation systems. In *Proceeding of European Simulation Multiconference*. Prague, Czech Republic, pp. xxiii–xxxiv.

Elmqvist, H., Tummescheit, H. and Otter, M. (2003) Object-oriented modeling of thermo-fluid systems. In Fritzson, P. (ed.), *Proceeding of the 3rd Modelica Conference*. Linköping, Sweden, 269–286.

Fernandez, J. and Kofman, E. (2014) A stand-alone quantized state system solver for continuous system simulation. *Simulation* 90(7): 782–799.

Fernandez, N., Nicholas, E.P., Katipamula, S., Wang, W., Xie, Y., Zhao, M. and Corbin, C. (2017) Impacts of commercial building controls on energy savings and peak load reduction. PNNL Tech. Report 25985, doi:10.2172/1400347, Richland, WA.

Fritzson, P. (2004) *Principles of Object-Oriented Modeling and Simulation with Modelica 2.1*. Piscataway, NJ: John Wiley & Sons.

Fritzson, P. and Engelson, V. (1998) Modelica – A unified object-oriented language for system modeling and simulation. *Lecture Notes in Computer Science* 1445: 67–90.

Fuchs, M., Teichmann, J., Lauster, M., Remmen, P., Streblow, R. and Mueller, D. (2016) Workflow automation for combined modeling of buildings and district energy systems. *Energy* 117(2): 478–484.

IBM Corporation. (1956) *The Fortran Automatic Coding System for the IBM 704 EDPM*. New York, NY: IBM Applied Science Division and Programming Research Department.

Jones, N.L., Greenberg, D.P. and Pratt, K.B. (2012) Fast computer graphics techniques for calculating direct solar radiation on complex building surfaces. *Journal of Building Performance Simulation* 5(5): 300–312.

Jones, N.L. and Reinhart, C.F. (2016) Parallel multiple-bounce irradiance caching. *Computer Graphics Forum* 35(4): 57–66.

Klein, S.A. and Alvarado, F.L. (1992) Engineering equation solver (EES). *F-Chart Software*, Madison, WI.

Kofman, E. and Junco, S. (2001) Quantized-state systems: A DEVS approach for continuous system simulation. *Transactions of The Society for Modeling and Simulation International* 18(1): 2–8.

Kräuchi, P., Schluck, T. and Sulzer, M. (2015) Modelling of low temperature heating networks with IDA-ICE. In *Proceeding of International Conference CISBAT*. Lausanne, Switzerland, September, pp. 827–832.

Lam, K.P., Wong, N.H., Mahdavi, A., Chan, K.K., Kang, Z. and Gupta, S. (2004) SEMPER-II: An internet-based multi-domain building performance simulation environment for early design support. *Automation in Construction* 13(5): 651–663.

Lee, E.A. (2003) Model-driven development – from object-oriented design to actor-oriented design. In *Workshop on Software Engineering for Embedded Systems: From Requirements to Implementation* (a.k.a. The Monterey Workshop), Monterey, CA.

Lee, E.A. (2006) Cyber-physical systems – are computing foundations adequate? In *Position Paper for NSF Workshop on Cyber-Physical Systems: Research Motivation, Techniques and Roadmap*, Berkeley, CA.

Lee, E.A. (2011) Cyber physical systems: Design challenges. In *11th IEEE International Symposium on Object Oriented Real-Time Distributed Computing (ISORC)*, 2008, pp. 363–369. https://doi.org/10.1109/ISORC.2008.25.

Lee, E.A. and Sangiovanni-Vincentelli, A.L. (2011) Component-based design for the future. In *Design Automation & Test in Europe Conference & Exhibition*, pp. 1–5, ISSN 1530–1591.

Loonen, R.C.G.M., Favoino, F., Hensen, J.L.M. and Overend, M. (2017) Review of current status, requirements and opportunities for building performance simulation of adaptive facades. *Journal of Building Performance Simulation* 10(2): 205–223.

Maccarini, A., Wetter, M., Afshari, A., Hultmark, G., Bergsøe, N.C. and Vorre, A. (2017) Energy saving potential of a two-pipe system for simultaneous heating and cooling of office buildings. *Energy and Buildings* 134: 234–247.

Malkawi, A.M. (2004) Immersive building simulation. In Malkawi, A.M. and Augenbroe, G. (eds), *Advanced Building Simulation*, 1st edition, New York, NY: Spon Press.

Malkawi, A.M. and Augenbroe, G. (2004) *Advanced Building Simulation*, 1st edition. New York, NY: Spon Press.

Mattsson, S.E. and Elmqvist, H. (1997) Modelica – An international effort to design the next generation modeling language. In Boullart, L., Loccufier, M. and Mattsson, S.E. (eds), *7th IFAC Symposium on Computer Aided Control Systems Design*. Gent, Belgium.

Mattsson, S.E., Olsson, H. and Elmqvist, H. (2000) Dynamics selection of states in Dymola. *Modelica Workshop*, Lund, Sweden, pp. 61–67.

Pantelides, C.C. (1988) The consistent initialization of differential-algebraic systems. *SIAM Journal on Scientific and Statistical Computing* 9(2): 213–231.

Polderman, J.W. and Willems, J.C. (2007) *Introduction to Mathematical System Theory*. Texts in Applied Mathematics, vol. 26, New York, NY: Springer.

Pollok, A. (2017) Institute of System Dynamics and Control German Aerospace Center, German Aerospace Center. Private communication, April 18.

Ptolemaeus, C. (ed.) (2014) *System Design, Modeling, and Simulation Using Ptolemy II*. Ptolemy.org.

Sahlin, P. (1996) Modeling and simulation methods for modular continuous systems in Buildings. PhD thesis, KTH, Stockholm, Sweden.

Sahlin, P. (2000) The methods of 2020 for building envelope and HVAC systems simulation – will the present tools survive? In *Dublin Conference*, The Chartered Institution of Building Services Engineers, Dublin, Ireland.

Sahlin, P., Eriksson, L., Grozman, P., Johnsson, H., Shapovalov, A. and Vuolle, M. (2004) Whole-building simulation with symbolic DAE equations and general purpose solvers. *Building and Environment* 39(8): 949–958.

Sahlin, P. and Grozman, P. (2003) IDA simulation environment – a tool for Modelica based end-user application development. In Fritzson, P. (ed.), *Proceeding of the 3rd Modelica Conference*. Linköping, Sweden, pp. 105–114.

Sahlin, P. and Sowell, E.F. (1989) A neutral format for building simulation models. In *Proceeding of the Second International IBPSA Conference*. Vancouver, BC, Canada, pp. 147–154.

Sangiovanni-Vincentelli, A.L. (2007) Quo vadis, SLD? Reasoning about the trends and challenges of system level design. *Proceeding of the IEEE* 95(3): 467–506.

Schneider, G.F., Peßler, G. and Steiger, S. (2017) Modelling and simulation of standardised control functions from building automation. In *Proceedings of 12th International Modelica Conference*. Prague, Czech Republic, May 15th–17th. DOI: 10.3384/ecp17132209.

Schuchart, J., Waurich, V., Flehmig, M., Walther, M., Nagel, W.E. and Gubsch, I. (2015) Exploiting repeated structures and vectorization in Modelica. In Fritzson, P. and Elmqvist, H. (eds.) *Proceeding of the 11th International Modelica Conference*. Paris, France, pp. 265–272.

Sowell, E.F., Buhl, W.F., Erdem, A.E. and Winkelmann, F.C. (1986) A prototype object-based system for HVAC simulation. Technical Report LBL-22106, Lawrence Berkeley National Laboratory.

Sowell, E.F., Buhl, W.F. and Nataf, J-.M (1989) Object-oriented programming, equation-based submodels, and system reduction in SPANK. In *Proceeding of the Second International IBPSA Conference*. Vancouver, BC, Canada, pp. 141–146.

Sowell, E.F. and Haves, P. (2001) Efficient solution strategies for building energy system simulation. *Energy and Buildings* 33(4): 309–317.

Sterling, R., Provan, G., Febres, J., O Sullivan, D., Struss, P. and Keane, M. (2014) Model-based fault detection and diagnosis of air handling units: A comparison of methodologies. *Energy Procedia* 62(0): 686–693.

Tiller, M.M. (2001) *Introduction to Physical Modeling with Modelica*. Norwell, MA: Kluwer Academic Publisher.

Trcka, M., Hensen, J.L.M. and Wetter, M. (2009) Co-simulation of innovative integrated HVAC systems in buildings. *Journal of Building Performance Simulation* 2(3): 209–230.

Trcka, M., Hensen, J.L.M. and Wijsman, A.J.T.M. (2006) Distributed building performance simulation – a novel approach to overcome legacy code limitations. *ASHRAE HVAC&R* 12(3a): 621–640.

Vuolle, M., Bring, A. and Sahlin, P. (1999) An NMF based model library for building thermal simulation. In Nakahara, N., Yoshida, H., Udagawa, M. and Hensen, J. (eds), *Proceeding of the 6th IBPSA Conference*. Kyoto, Japan.

Wang, H. and Zhai, Z. (2016) Advances in building simulation and computational techniques: A review between 1987 and 2014. *Energy and Buildings* 128: 319–335.

Wetter, M. (2005) BuildOpt – a new building energy simulation program that is built on smooth models. *Building and Environment* 40(8): 1085–1092.

Wetter, M. (2006a) Multizone building model for thermal building simulation in Modelica. In Kral, C. and Haumer, A. (eds), *Proceeding of the 5th International Modelica Conference*, vol. 2. Vienna, Austria, pp. 517–526.

Wetter, M. (2006b) Multizone airflow model in Modelica. In Kral, C. and Haumer, A. (eds), *Proceeding of the 5th International Modelica Conference*, vol. 2. Vienna, Austria, pp. 431–440.

Wetter, M. (2011) Co-simulation of building energy and control systems with the building controls virtual test bed. *Journal of Building Performance Simulation* 4(3): 185–203.

Wetter, M., Bonvini, M. and Nouidui, T.S. (2016) Equation-based languages – a new paradigm for building energy modeling, simulation and optimization. *Energy and Buildings* 117(1): 290–300.

Wetter, M., Fuchs, M., Grozman, P., Helsen, L., Jorissen, F., Lauster, M., Müller, D., Nytsch-Geusen, C., Picard, D., Sahlin, P. and Thorade, M. (2015a) IEA EBC Annex 60 Modelica Library – An international collaboration to develop a free open-source model library for buildings and community energy systems. In *Proceeding of the 14th IBPSA Conference*. Hyderabad, India, pp. 395–402.

Wetter, M., Grahovac, M. and Hu, J. (2018a) Control Description Language. In *1st American Modelica Conference*. Cambridge, MA.

Wetter, M. and Haugstetter, C. (2006) Modelica versus TRNSYS – A comparison between an equation-based and a procedural modeling language for building energy simulation. In *Proceeding of SimBuild*. Cambridge, MA.

Wetter, M., Haves, P., Moshier, M.A. and Sowell, E.F. (2008) Using SPARK as a solver for Modelica. In *Proceeding of SimBuild*. Berkeley, CA.

Wetter, M., Hu, J., Grahovac, M., Eubank, B. and Haves, P. (September 2018b) OpenBuildingControl: Modeling feedback control as a step towards formal design, specification, deployment and verification of building control sequences. In *Building Performance Modeling Conference and SimBuild*. Chicago, IL, pp. 775–-782.

Wetter, M., Nouidui, T.S., Lorenzetti, D., Lee, E.A. and Roth, A. (2015b) Prototyping the next generation EnergyPlus simulation engine. In *Proceeding of the 14th IBPSA Conference*. Hyderabad, India, pp. 403–410.

Wetter, M. and Wright, J. (2004) A comparison of deterministic and probabilistic optimization algorithms for nonsmooth simulation-based optimization. *Building and Environment* 39(8): 989–999.

Wetter, M., Zuo, W., Nouidui, T.S. and Pang, X. (2014) Modelica Buildings Library. *Journal of Building Performance Simulation* 7(4): 253–270.

Wimmer, R., Pinheiro, S., O'Donnell, J., Maile, T., Muhic, S., Frisch, J., Bazjanac, V. and van Treeck, C. (2017) Realizing openBIM – development of a BIM model view definition for advanced building energy performance simulation. *GI – Gebäudetechnik in Wissenschaft & Praxis* 138(04): 276–291.
Zuo, W. and Chen, Q. (2010) Simulations of air distributions in buildings by FFD on GPU. *HVAC&R Research* 785–798.

Recommended reading

For a computational approach to heterogeneous systems, the reader is referred to the textbook of Ptolemaeus (2014). For a discussion about trends and challenges of system-level design, we refer to Sangiovanni-Vincentelli (2007). The prerequisites for generating efficient simulation code from a general purpose solver are discussed by Elmqvist et al. (1995). Cellier (1996) discusses common pitfalls and misconceptions about the modeling of physical systems. Tiller (2001) and Fritzson (2004) discuss different equation-based modeling approaches for dynamical systems. Cellier and Kofman (2006) present in a rigorous way methods for transforming mathematical models into computational code. The textbook has been written for an engineering audience. Mattsson et al. (2000) describe how symbolic and numerical methods are combined in Dymola to solve reliably and efficiently high-index DAE systems. For a detailed discussion of BIM, see for example Eastman et al. (2008). Clarke (2001) gives a detailed exposition of models that are typically found in building simulation programs. Malkawi and Augenbroe (2004) present recent trends in building simulation with a focus on combined air and heat flow, uncertainty propagation and integration of simulation into the design process. Clarke (2015) presents a vision for future building performance simulation. Wang and Zhai (2016) review advances in building simulation between 1987 and 2014.

Assignments

1 Implement two versions of the system model shown in Figure 19.4(a). For the first version, use causal block diagrams, and for the second version, use acausal models for the physical components. As parameters, use for each heat capacitor *10 J/K*, for each heat conductor *10 W/K*, for the convective heat transfer *5 W/K*, for the setpoint and the initial temperatures *20 °C* and for the control gain *100 W/K*. For the boundary condition, use $T(t) = 20 + 5 \sin(2 \pi t/3600)$, where t denotes time in seconds and the temperature is in degree Celsius. Compute the maximum control error. Next, reconfigure the system to the topology shown in Figure 19.4(b). Compute again the maximum control error. Discuss the time it took to develop and reconfigure the two model representations.
Hint: You may use Modelica or MATLAB/Simulink with the SimScape™ extension to create acausal and causal models.
2 What do you think a next generation building energy analysis tool will look like?
3 How do you increase the value of a model during the building life cycle? What are its applications?

Answers

1 For configuration (a), the maximum control error is *0.12 K* and for configuration (b), it is *0.24 K*.

Part V

Urban

20 Integrated resource flow modelling of the urban built environment

Darren Robinson

Scope

Around three quarters of the energy and materials to sustain human life are consumed in urban settlements, and this has profound environmental consequences. With the urban population forecasted to double by 2050, there is an urgent need to develop techniques to better understand this urban metabolism and ways of improving its efficiency. After briefly introducing the key factors influencing urbanisation and its consequences, the nature of the integrated urban modelling challenge is discussed. Techniques for modelling the urban climate, land use and the transport of people are then described, together with challenges in developing a platform for the social simulation of inhabitants' responses to policies and strategies destined to accelerate the penetration of measures to reduce urban resource throughputs. After discussing promising model integration technologies and associated data standards, alternative urban building energy modelling strategies are described in detail and examples of their application are presented. The chapter closes by discussing where efforts should be prioritised to develop a comprehensive urban modelling capability, on the one hand, and to improve the rigour with which energy use in buildings is modelled on the other. The International Building Performance Simulation Association (IBPSA) is well placed to lead this interdisciplinary challenge.

Learning objectives

Although very much in its infancy, the modelling of resource flows at the urban scale is a vast and complex topic. Of greatest international concern is the use of energy and the emissions arising from it, with the building, transport and industrial sectors playing the dominant roles. Relative to the other key sectors, building simulation has a long tradition of supporting predictions of energy use and emissions; but endeavours to increase the scale of application from individual buildings to urban stocks of buildings are relatively recent. Nevertheless, this has already led to advances in simulating energetic couplings between buildings and with district energy centres as well as of the stochastic behaviours of occupants that influence energy use. Other innovations relate to workflows to support the efficient calibration and execution of complex urban building energy models.

The aims of this chapter are to give an overview of the achievements that have thus far been made in this new modelling domain as well as the many challenges that await us. The objectives here are both to provide an insight into the importance of considering buildings in their urban context and to stimulate further research in this domain, and in the development of comprehensive urban models to better support the testing of policies and strategies to improve the sustainability of urban settlements and thus of *Homo urbanus*.

Key words

Urban modelling/climate/land use/transport/energy/social simulation/integration technologies/data exchange/urban building energy modelling

Urbanisation and its consequences

It is thought that the first substantial human settlements started to form around 10,000 years ago, as humans cooperated in the sharing of agricultural practices and the trading of goods. This cooperation increased steadily, so that by 1800 the urban fraction of the 0.5 B strong global population stood at around 10% (Sachs 2008). The Industrial Revolution transformed how societies were organised. As new industries grew, they required and attracted large, concentrated pools of labour and associated urban infrastructure. The urban centres that developed provided dense new markets for new commodities and became hubs of trade and commerce and the drivers of economic growth (Coutard et al. 2014). At the same time, mechanisation of agriculture significantly increased the area of land that could be sown and harvested and reduced the numbers of workers needed to work the land, fertilisers replenished soil nutrients and pesticides reduced crop losses, increasing productivity. This in turn enabled a greater proportion of the population to live in urban settlements and to engage in other (non-agricultural) productive activities. Consequently, the urban fraction of the global population increased from around 13% in 1900 until parity was reached with the rural population in 2007, when the global population was estimated to be 6.7 B.

By 2050 it is projected (assuming medium fertility) that the population will increase to 9.5 B and its urban fraction to three quarters (UN 2012): the urban population is set to more than double. The majority of this urban growth is expected to take place in developing countries, which are expected to accommodate more than 80% of the global urban population by 2050, as rural-urban migrants search out a better quality of life for themselves and their families. Under business as usual conditions, this larger, more economically active urban population is expected to exert greater stress on the natural environment; during a period in which the 196 signatories of the Paris Agreement agreed to reduce their carbon output, to strive to keep global warming to well below 2 °C.

Other concerns are socio-economic. Both the size of cities and the distribution of wealth within them are thought to follow a Pareto (power law) distribution in capitalist society, so that the majority of wealth is owned by a small number of people and most are relatively poor. As wealth is increasingly concentrated, income inequality and levels of poverty are rising, leading to social polarisation and exclusion. Sections of society are being denied full participation in everyday activities and marginalised with respect to resources such as housing, work, social services and the political sphere (Robinson et al. 2015). This is often manifest in social stratification and fragmentation, segregation to certain parts of the city and alienation and it undermines social cohesion, leading in the worst of cases to social unrest, protest and riots.

Since the majority of our future population will reside in towns and cities, it stands to reason that these settlements will play a pivotal role in the quest for future more sustainable living. New ways of accommodating more and wealthier people must be invented, having increased standards of living in towns and cities, with reduced environmental impacts, whilst simultaneously improving upon resilience to social, economic and environmental disruptions.

Achieving the transition towards smarter more sustainable, liveable and resilient urban futures requires that we make considerable progress on (at least) two fronts. We need:

- A comprehensive and rigorous theoretical framework with which to characterise the sustainability and resilience of urban settlements (their points of origin). In this, we also need to understand how the social, economic and environmental pillars interrelate and how conflicts between them can be managed, as well as the means to establish ambitious yet achievable targets to transition towards (their destinations).
- To understand the context-specific policies, strategies and business models that should be put in place to drive the transition process; what these new more sustainable urban futures may look like; how resilient they may be to disturbances.

For this, we need parsimonious systemic urban models; indeed, a framework for the integration and co-simulation of a suite of parsimoniously formulated urban system models. This is an extraordinarily complicated and ambitious task. This is mainly because cities are complex non-linear non-equilibrium systems whose emergent (social, economic, physical) structures depend upon the bottom-up interactions of the actors (institutions, firms and individuals) that inhabit them. These features make the task of predicting future city configurations practically impossible, though we may be able to predict likely configurations. But it is also because (i) cities are spatially and functionally diverse, undermining the feasibility of a framework that can suitably transcend spatial scales spanning increasingly close to three orders of magnitude and also accommodate the range of social, economic and physical structures and contexts, (ii) there is a dichotomy between the existence and availability of data with which to calibrate and validate urban system models, and the spatiotemporal granularity with which that data is available and (iii) there is a lack of knowledge regarding the behaviours of urban actors and their responses to policies, strategies and business models that may be conceived to positively influence their investment and operational (daily, seasonal, etc.) behaviours.

But, if we can develop suitably complex models of the range of phenomena influencing these future states and their associated resource throughputs and consequences, accounting for the (regulatory, financial, educational, social and technological) stimuli that influence them and at an appropriate spatiotemporal resolution, we will have an extremely powerful resource at our fingertips. This resource would enable us to answer city-specific questions such as:

- How should the city be spatially structured, in terms of the distribution of density and of diversity of land and building uses and their forms to minimise resource demands and promote social cohesion?
- How should transport, energy and water networks be configured and which technologies should be employed with them?
- Which industrial process and corresponding technologies should be accommodated; how can they best be coupled with one another and with urban networks?
- From which materials should urban infrastructure, networks, buildings and technologies be comprised?
- How should buildings be designed and controlled to reduce net resource demands and to what extent can behaviour reduce demands?
- Which combinations of policy measures, strategies and business models are likely to stimulate development along these lines and how should they be phased?
- What would be the likely health and wellbeing impacts of these interventions?
- How robust is our city system functioning to exogenous or endogenous disruptions, etc.?

Developing and maintaining an urban systems modelling framework that enables us to satisfactorily answer these (and other) urban policy, planning, design and management questions is a long-term inter- and trans-disciplinary endeavour, one for which the scale and scope of research investment should match the scale and scope of the challenge. IBPSA is well positioned to lead the way here.

The following section unpicks this urban modelling challenge, identifying the systems that need to be integrated within a suitably comprehensive urban systems modelling framework, strategies for modelling these systems and actors' responses to policy measures conceived to influence them and the data and computational modelling techniques that need to be called upon to facilitate this. The penultimate section focusses on the subject that is at the heart of this book: building energy simulation – albeit now at the urban scale. This describes alternative modelling strategies, distinguishing these by their level of spatial aggregation, and sets out a number of recommendations to augment their usefulness. As elsewhere in this book, the chapter closes with a brief synopsis.

The urban modelling challenge

Figure 20.1 identifies the principal systems and some of the primary interconnections between them that influence urban resource flows and should thus be modelled and integrated within a suitably comprehensive urban systems modelling framework.

Let us begin with (1) building and activities, which refers to the physical composition of the building and its HVAC and lighting systems, the activities of the occupants accommodated within it and their corresponding ownership and use of electrical appliances in support of their activities. Clearly, these buildings are located on (2) parcels of land, which are themselves interconnected and serviced by transport and utility infrastructure. Whether developed upon or not, these parcels of land have their own surface properties: they may be partially or fully vegetated, or conversely partially or fully impervious, with corresponding implications for evaporative mass and latent heat fluxes and for the conduction and storage of heat within the ground. They may accommodate buildings of varying horizontal and vertical massing, with corresponding implications for albedo and thus for shortwave and longwave radiation exchange. These surface fluxes influence (3) the urban climate, as does the rejection of heat from buildings' HVAC systems, which is in turn influenced by the urban climate. Goods and people (4) are also transported, using our transport infrastructure, between the parcels of land on which are buildings are located, with corresponding heat gains influencing the urban climate. As implied above, whether they are homes, workplaces or retail outlets or accommodate health, education, leisure or service facilities, these buildings temporarily accommodate people; as do the vehicles used to transport people between these buildings. For sound ethical and practical reasons, it is neither possible nor desirable to build a detailed representation of the socio-demographic characteristics of a city's real population. But it is both necessary and appropriate to build a

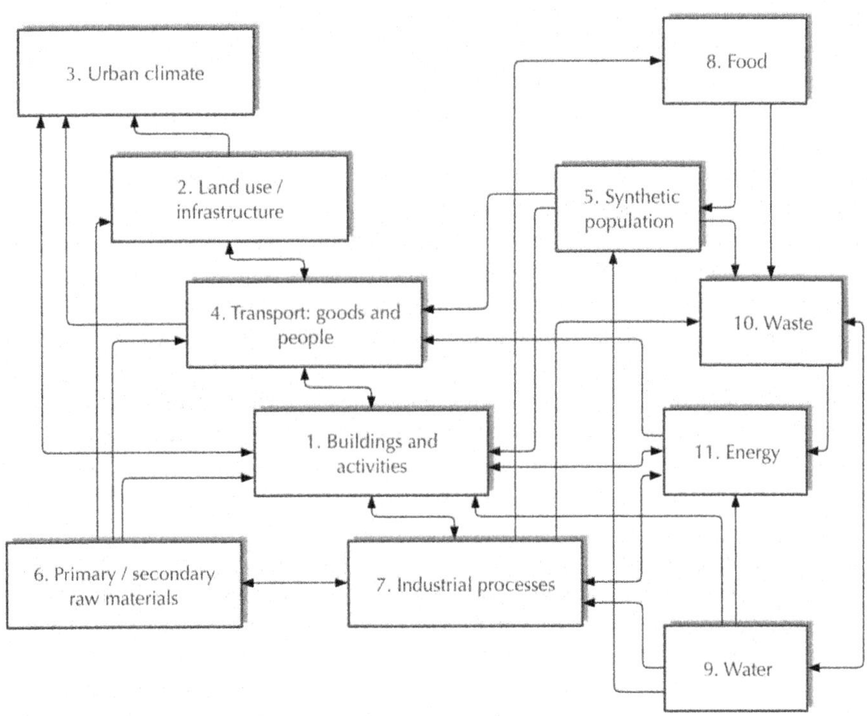

Figure 20.1 Urban systems diagram: a system of (the principal) urban systems and interconnections between them that influence urban resource flows

(5) synthetic representation of the population that replicates the aggregate characteristics of the real population without enabling individual citizens' identities and attributes to be identified. Our infrastructure, buildings and their fixtures and fittings and transport vehicles are constructed from (6) materials, components and products that entail unique life-cycle impacts and that have properties impacting upon their performance; some of which may be fabricated using (7) industrial processes accommodated within buildings within our city's boundaries. Foods (8) may also be produced within similar industrial premises, some of which will be consumed by members of our synthetic population, which also consume (9) water, as do some of our building activities. Waste (10) from this food and water consumption, as well as from industrial and construction processes, requires treatment, and these treatment processes may lead to usable waste heat (thermal energy). Finally, our buildings and associated activities, industrial processes and transport vehicles use (11) thermal and electrical energy. This may be centrally generated (or recovered) and distributed to our city, or locally recovered and/or converted and distributed.

This summary of the systems and their interconnections that influence urban resource flows is far from exhaustive, but it does serve to illustrate the complexity of the challenge in terms of the range of paradigms to be modelled and integrated. And to this we need in principle to add a further layer of complexity, for physical models of these systems need to be accompanied by social simulations of suitable abstractions of the decision-making processes of the actors that influence them, and how these processes can in turn be influenced by appropriately conceived policy measures.

In preparing this summary, we have supposed that our spatial scale of analysis is the entire city and that our time horizon is the medium term of, say, 30–50 years. But this, of course, will not normally be the case; indeed, an integrated city-scale urban model does not yet exist, for urban modelling studies typically focus on the smaller scales of an urban block, neighbourhood or district and on short time horizons. In these cases, we might assume that the city structure is stationary so that land use changes may be neglected and transport and other (e.g. utility) network infrastructure topologies will be essentially unchanged (at least beyond our constrained system boundaries), likewise with respect to transport modes and technologies and food supply chains. Thus, it may be reasonable to neglect systems 2, 4 and 8, whilst others may only be addressed in a highly localised and/or abstracted sense.

Nevertheless, given the focus of this book on building simulation combined with the ambition of this chapter to stimulate research at higher spatial scales and longer time horizons, in what follows we describe in more detail the physical phenomena, together with candidate strategies for modelling them, that influence building and activities (in particular on systems 2–4 and 11), which is itself discussed in detail in the next main section ("Urban building energy modelling"). We also briefly introduce the challenges to be tackled by a social modelling and simulation platform that would address actors' decision-making responses to policies and strategies conceived to improve upon the efficiency of city functioning. We then discuss some of the data and computational modelling techniques that need to be called upon to facilitate integrated urban modelling at higher spatiotemporal scales.

Modelling tasks and strategies for the principle urban sub-systems

Urban climate

Urban terrain, in which built and natural forms are extruded from the ground, is often highly geometrically complex, having greater rugosity than is found in rural settings where the climatic measurements that form the basis of standard meteorological databases tend to be made. Due to inter-reflections between the surfaces constituting this terrain, its albedo tends to be lower, so that shortwave radiation is more efficiently absorbed relative to rural terrain. Furthermore, obstructions between surfaces diminish views to the sky, so that heat is less efficiently transferred at night by longwave exchange to the relatively cold sky. In consequence, more heat is absorbed and retained in urban than in rural terrain. Added to this is the heat gain due to anthropogenic sources, for example due to transport systems and heat sources in buildings, although this may be partially offset by evapotranspiration from vegetation. In consequence, the mean temperature tends to be higher in urban than in rural terrain, with the intensity of the urban heat island (UHI) effect tending towards a maximum at the centre of a city, as the air entrained into the buoyant plume (which may be both deflected and diluted by wind) is progressively heated. This type of behaviour has been observed in numerous experimental field studies (Santamouris 1988; Palmer et al. 2000; Graves et al. 2001). However, urban surfaces also tend to have greater thermal inertia than do rural surfaces. Indeed, Swaid and Hoffman (1998) estimate the difference in time constant to be six to eight hours. The urban temperature profile is thus shifted in phase with respect to the rural profile, so that the UHI profile (the difference between the two, $T_{urban}(t) - T_{rural}(t)$, tends to reach a minimum in the late afternoon/early evening (to the extent that this instantaneous UHI may be negative) and a maximum during the night. Due to the above mechanisms, the measured intensity of the London UHI varies from −2 °C to +6 °C with corresponding implications for buildings' energy performance.

However, the current practice in building simulation is to directly use the velocity and temperature from a climate file, which is normally based on rural measurements, so that the influence of the UHI on the building energy balance is ignored.

But modelling the urban climate to resolve this shortfall is complicated by the diverse time and length scales involved in atmospheric flow. The length scale for example can vary from a few metres (buildings) to a few kilometres (mountains). It is practically impossible to resolve all the scales in a single model with the present computational power available. A way to overcome this problem is to couple different models capable of resolving different scales.

A range of global numerical weather forecasting models having a resolution of a few hundred kilometres have been developed, such as the Global Forecast System "GFS" (www.emc.ncep.noaa.gov/modelinfo) or the Global Environmental Multiscale model "GEM" (http://collaboration.cmc.ec.gc.ca/science/rpn/gef_html_public/index.html). These global models may provide boundary conditions to a mesoscale model having a resolution of a few hundred metres, so that an entire city and its local topographic context may be represented. Examples include the open-source models MM5 (www.mmm.ucar.edu/mm5/) and WRF (www.mmm.ucar.edu/weather-research-and-forecasting-model). The exchange of energy, mass and momentum with the urban texture may then be modelled by embedding an urban canopy model (Brown and Williams 1998; Masson 2000; Kusaka et al. 2001; Martilli et al. 2002; Rasheed 2009) within the mesoscale model. Such a nested (global to urban mesoscale) model can be used to provide the necessary climate input data to both building and urban resource flow simulation software (Rasheed et al. 2009) to improve the accuracy of their energy predictions.

As a related example, Rasheed (2009) and Rasheed et al. (2009, 2011) simulated the mesoclimate of Basel in Switzerland for a 24-hour period, for two cases: the first representing the region of Basel, but assuming it to be fully rural; the second representing the city both in terms of land cover and in the impact on energy, mass and momentum transfer of the urban canopy. From these results (Figure 20.2), it was possible to estimate the relative cooling demands within the city, suggesting an increase of up to 40%, due to Basel's UHI.

The utility of predictions from nested multiscale numerical modelling workflows such as that of Rasheed et al. (2011) are contingent upon the accuracy with which the urban structure is represented within the urban canopy parameterisation scheme. These schemes abstract real complex urban textures into repeated idealisations, either in the form of symmetric homogenous canyons or as cuboids that may or may not be symmetric. The complication arises from the difficulty in identifying parameters for these idealised representations that mimic the real

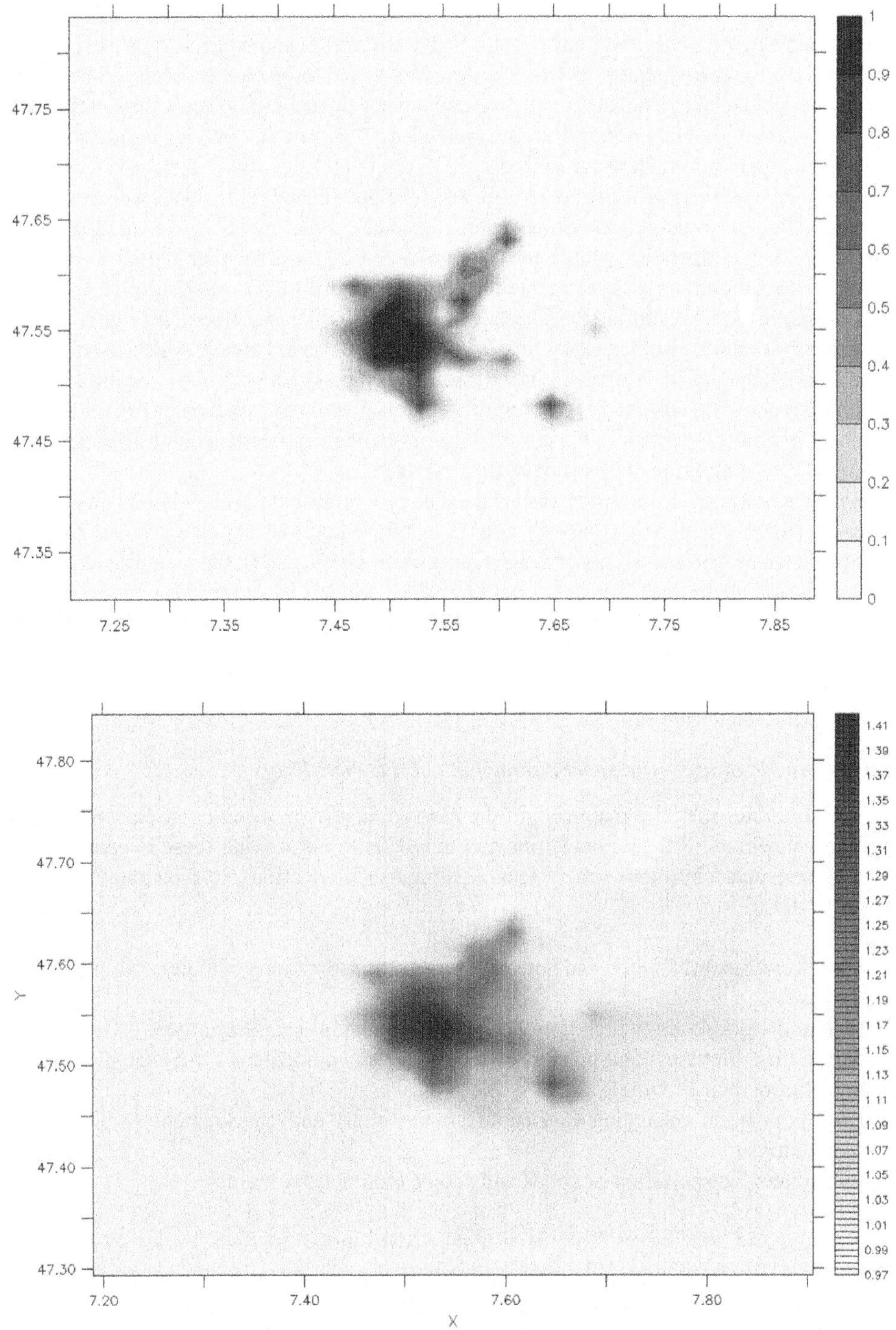

Figure 20.2 Fraction of land use classed as urban (above) and cooling energy demands (below) for the city of Basel and its bounding region; these latter relative to its local rural context (quotient of urban and rural cooling degree-hours): both spatially represented by longitude (x) and latitude (y)

Source: Rasheed et al. 2009, 2011

complex geometry of the urban texture being modelled, that the idealisation exhibits similar radiative and aeraulic properties (e.g. net shortwave absorption and drag). Rasheed et al. (2011) tackle this *equivalent geometry problem* by trial and error: for each mesoscale grid cell, they compare annual incident irradiation and drag and modify the simplified equivalent until the differences are approximately minimised. It would clearly be more suitable to employ an optimisation algorithm to minimise these differences. To our knowledge, this has yet to be achieved.

An alternative to this multiscale workflow, and its modelling of equivalent urban geometries, is the modelling of a relatively simplified energy balance. To this end, Bueno et al. (2013) effectively translate a temperature profile measured at a nearby rural weather station to the urban context. Called the urban weather generator, this is based on the modelling of vertical heat diffusion above the rural site and the corresponding contribution to temperature within an urban boundary layer that is also heated by flux from within the urban canopy, which itself depends on the urban boundary layer temperature. Restricted comparisons with observed urban canopy temperatures show favourable results, but this approach addresses exclusively local microclimate thermal processes and does not consider the entrainment of heated/cooled air from beyond the immediate canyon: the mesoscale urban heat island.

Whether employing a simplified energy balance or a multiscale urban climate model, there is a need, when considering relatively long-term future horizons, to also account for future climate change scenarios. To this end, Kikumoto et al. (2014, 2016) have employed dynamic downscaling of predictions from a global circulation model, accounting for future emission scenarios, to provide initial and boundary conditions to a mesoclimate model. This strategy, however, assumes that land use and anthropogenic heat gains remain constant during the simulation period.

Land use and land use – transport interaction (LUTI) modelling

To simulate urban spatial dynamics and the associated environmental consequences arising from the behaviours of the actors (firms and individuals) influencing these dynamics, a city may be represented by eight sub-systems, distinguished according to their rates of change (Wegener 1994):

- Slow: land use distribution and networks (utility, transport and communication) to support these uses.
- Medium: workplaces (which may accommodate several changes in type of employment during their lifetimes) and housing (which may accommodate several changes in household during their lifetime).
- Fast: population (changing household composition) and employment (changing firm composition).
- Immediate: transportation of goods and people (and other resources).

A broad range of approaches have been employed to model these processes (Waddell and Ulfarsson 2004). The more popular early approaches were based on *spatial interaction* or *spatial input-output* models. Spatial interaction models use a gravitational analogy to model travel between origin and destination zones, which increases with the amount of activity within them and decreases with the square of the distance between them. Putman (1983) has applied this to model choices of trips, residential relocation and employment location, but this type of model lacks spatial resolution and many of the behavioural factors influencing the choices made. Spatial input-output models define regional zones which are treated as sub-economies that engage in production, consumption, import and export within and between them. In this way, real estate and labour markets as well as travel demand are modelled (de la Barra 1989; Echenique et al. 1990), but the spatial aggregation level is high relative to that used in conventional transport modelling.

More recent models are based on much lower levels of spatial aggregation, for example using *cellular automata* (CA) or *multi-agent systems* (MAS). Cellular automata have emerged from Conway's Game of Life (Gardner 1970), in which simple behavioural rules are applied to cells within a grid (or cellular space), whereby the future state of a given cell depends upon its current state and that of its neighbours. In this way, CA have been used to study a broad range of complex emergent systems (Wolfram 1994, 2002) ranging from the formation of star clusters to the behaviour of bacteria and insect populations.

In the present context, they have also been applied widely to the study of land cover or land use change (White and Engelen 1993, 2000; Batty 1997; Torrens and O'Sullivan 2001). But as Benensen et al. (2005) point out, these models are limited by the immobility of the cells, whereas housing and transport dynamics necessitate mobile or non-fixed entities. Representing the urban milieu as a regular grid of cells is also constraining; features that are larger than a single cell cannot be directly represented and features that are smaller need to be aggregated. MAS have emerged as a more consistent response to these limitations, as individual agents may be associated with arbitrary shape and location and respond to more sophisticated interactions between agents (such as between individual, firms, workplaces, housing and modes of transport). On this basis, Miller et al. (2004) have developed a highly detailed MAS-based land use and transport interaction model, resolving real estate market demand-supply interactions as well as travel demand for a range of activities. As an alternative to full MAS, Batty (2005) suggests that the characteristics of both CA and MAS may be combined, with cells representing the physical and spatial structure and agents the human and social units that function within the cellular space, so resolving Benensen's key criticism.

Transport modelling

Originally developed to study the impacts of large-scale transport infrastructure projects, the zonal (or four-step) model is by far the popular. Common to all models of this type is the identification of a boundary within which the effects of changes to the systems are assumed to be contained. Inside this boundary, the geographic area is split into zones. A region of interest is identified within which finer spatial scale zones are used. This region is where changes to the system are predominately applied and the largest effects are expected to be felt. Zones larger in spatial scale are placed between the region of interest and the boundary of the system. Outside the boundary, the external system is represented by one or more external zones. These external zones are the method by which the system within the boundary can interact with the outside world (McNally 2007).

The objective of the four-step model is to produce a matrix of trips which displays the number of trips between each zone over a defined period of time:

$$\begin{pmatrix} T_{11} & \cdots & T_{1N} \\ \vdots & \ddots & \vdots \\ T_{N1} & \cdots & T_{NN} \end{pmatrix}$$

and a mapping of these trips to a representation of the transport network. Here T_{ij} is the total number of trips from zone i to zone j over the defined period. The matrix may be segmented into additional matrices by mode type, trip purpose and person type (e.g. car ownership status and sometimes by socio-economic status).

One of many reasons that the four-step strategy has traditionally prevailed over other modelling strategies is due to the underpinning data collection methods employed. Relatively high-quality demographic data and to a degree socio-economic and land use data is available at the zonal level and can be obtained relatively inexpensively. Along with trip frequency data and system constraints, this data allows trip generation to be estimated: the total number of trips originating from each *i*th zone (O_i) and those destined for each *j*th zone (D_j).

To determine how these trips are joined together, expressed as the number of people travelling between two zones, T_{ij}, the trips must be joined together in a process known as trip distribution. This can either be achieved by using costly survey data (Anon 2014) or be modelled (preferably supported by some survey data for calibration purposes). The generalised gravity model is a common approach to performing trip distribution (Wilson 1970). In this approach, people living in each *i*th zone will be attracted to all other zones. The strength of attraction to these zones depends on the employment opportunities and services available and the difficulty and expense of travelling there. The ability of people in zone *i* to secure employment opportunities and services in zone *j* will depend on the relative competition from people in other zones for those opportunities and services. The difficulty and expense of travelling between two zones, *i* to *j*, can be encapsulated as a generalised cost C_{ij}. This generalised cost is unlikely to be the same for all travellers between the two zones, because different people will have differing mobility options (for example, households who do and don't own cars). For car trips, the generalised cost will include the journey time (dependent on the trip distance, velocity restrictions and congestion), the cost of fuel and perhaps maintenance costs. Public transport generally has fixed routes and hence the cost attributed to journey times will additionally depend on the public transport routes, interchanges and the distance of the stops to the desired (within zone) destination. The financial cost of public transport is simply the price of the ticket. However, if socio-economic factors are also considered, variations to the weights used in determining the generalised costs may be needed. The mathematical form of the generalised gravity models is dependent on the degrees of freedom used in the model. The doubly constrained model has the form:

$$T_{ij} = A_i O_i B_j f(C_{ij})$$

With A_i and B_j acting as weighting factors for the relative competition between zones. There are a number of possible forms for the function $f(C_{ij})$. Popular amongst these is a form derived from the principles of entropy maximisation: $f(C_{ij}) = \exp(-\beta C_{ij})$, where β represents travellers' sensitivity to changes in the generalised cost, which needs to be calibrated (Wilson 1970). The general form can be separated by transport mode, trip type and person type.

In the original four-step models, trip distribution typically follows the generation of trips. The split by mode follows, leading to the first three steps of the four steps of the chain: (i) trip generation, (ii) trip distribution, (iii) mode choice and (iv) assignment.

The result at this point is a series of origin-destination (O-D) matrices, indicating the number of trips between different zones. As noted earlier, separate matrices can be prepared for each mode choice and trip purpose. Assignment is the final step in the four-step modelling process, where the trips defined by the O-D matrices compete for space on the transport network. This transport network is usually represented by a node link graph model, the nodes generally representing intersections and the links the roads or rail lines between them. Trips may now be split by the routes taken through the network, with final route assignment being achieved when an equilibrium state is achieved, whereby traffic arranges itself on a network such that the cost of travel on all routes used between each O-D pair is equal to the minimum cost of travel and all unused routes have equal or greater cost (Anon 2014).

Once assignment has been specified, the generalised cost may be re-estimated (the initial run uses an estimate, often based on the shortest path between two zones and using the free speed of the network) and the process repeated until the system has converged.

The energy demands arising from these trips are typically determined in a post-process, given mean values of energy use or emissions per km travelled, as a function of mode-dependent vehicle type and the estimated distances travelled between zones. This may be refined using network link speeds in conjunction with speed-dependent estimates of energy use/emission rates for each vehicle archetype. Estimates of emissions within zones can, in conjunction with estimates of emissions arising from activities and processes accommodated within buildings, be input to urban climate models, which themselves incorporate models of pollutant dispersion to estimate pollutant concentrations and associated health hazards.

With the purpose of supporting the rigorous formulation of Nottingham City Council's Sustainable Energy Action Plan, this zonal strategy was employed to good effect by Long and Robinson (2019). In this, they modelled the energy use and emissions arising from the transport and building (of which more in the "Urban building energy modelling" section) sectors and their dependency on the energy sector (within and without the city boundaries), for a range of scenarios covering the period 2014 to 2030.

The zonal transport model includes processes for trip generation from land use information, choice of mode between car and public transport and routing choices for both modes. It calculates emissions and energy use, split by trips between zones and by vehicle type (Figure 20.3), providing a detailed breakdown of results. Other outputs include emissions of CO_2, hydrocarbons and small particles (PM10).

Results from the modelled scenarios provided quantitative inputs to a multi-criteria decision-making process that were complemented by qualitative considerations, with stakeholders ranking the scenarios according to their relative merits through pairwise comparisons.

Zonal models, then, are well suited to support relatively large spatial scale and long time horizon planning decisions, for they are agile, flexible and efficient in terms of both data and computational overheads. In contrast, they lack in spatiotemporal resolution as well as the resolution with which travellers' motives are represented; likewise, travellers' propensity to undertake multi-modal trips as well as multi-purpose chains of trips during any given day. For this, an activity-based model is needed.

It is beyond the scope of this chapter to describe this type of model in detail (for this we refer the reader to Castiglione et al. (2015) and Horni et al. (2016)), but in brief they are designed to model and execute daily activity plans, consisting of (i) the order in which activities (e.g. working, shopping, leisure) are undertaken, (ii) the desired location of the activities, (iii) the preferred arrival time and duration of the activities and (iv) the transport mode(s) and route for travelling between these activities. The success of these plans is determined by a utility function representing the time spent travelling to and subsequently spent at travellers' activities, with a disutility accounting for delays. Plans (sequences, timings, modes and routes) are iteratively adjusted until this utility is maximised (a cumulative-utility based equilibrium is reached), with transport along the links of the network being explicitly modelled using a queue-based method. The speed at which the queue moves is dependent on the link velocity, which in turn depends on network conditions (e.g. congestion).

Although activity-based models are considerably more onerous than their zonal counterparts, they are also considerably more powerful (in particularly for detailed network design decision-making), lending themselves well, for example, to the future modelling of autonomous electric vehicles and the associated location of charging points.

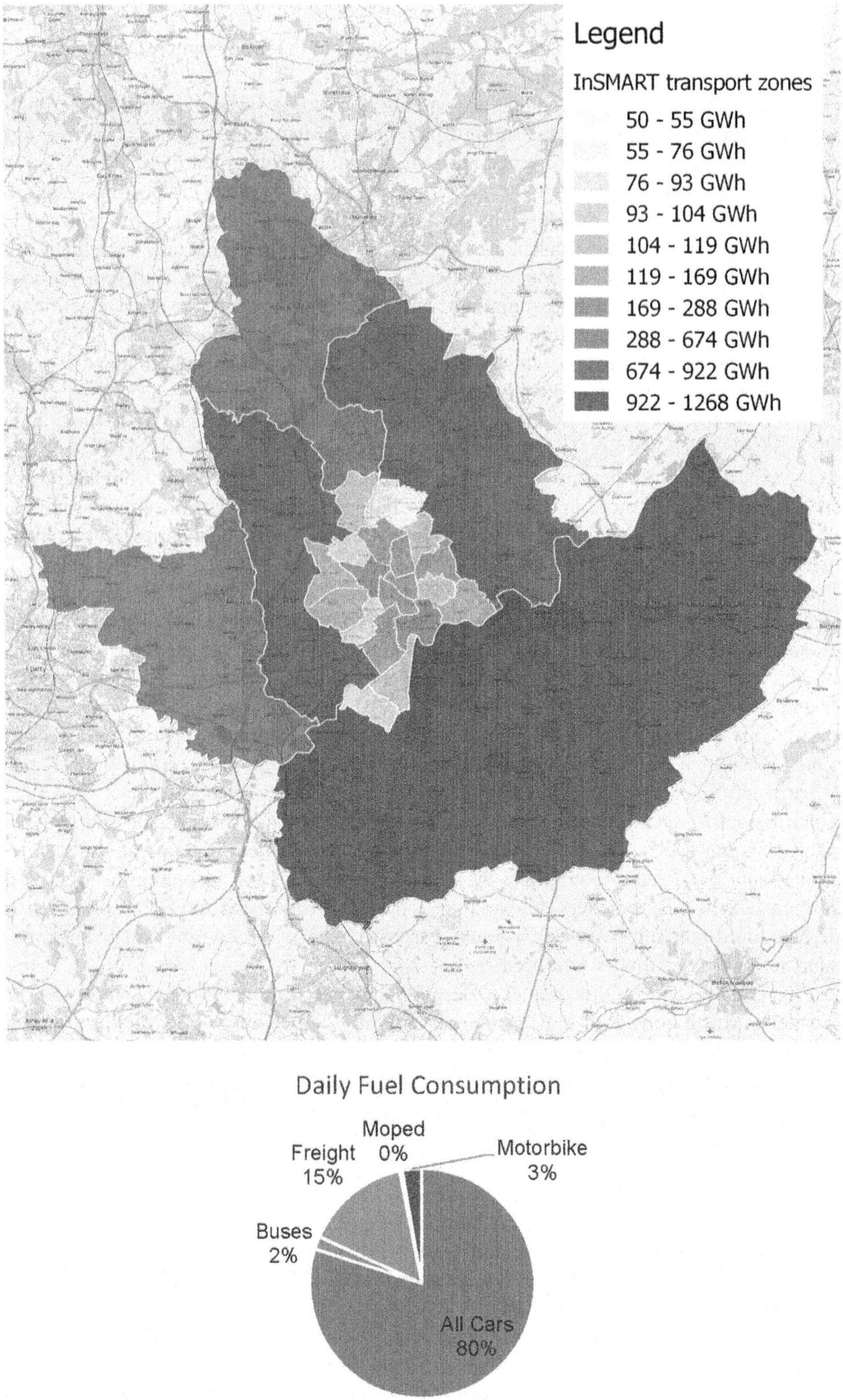

Figure 20.3 (Top) Total transport energy (GWh) per origin zone for the base year (2014) in the Nottingham travel to work area. (Bottom) Percentage split of average daily fuel use by vehicle type in the base year.

Social modelling and simulation

As noted earlier, cities (indeed any urban settlement) are complex non-equilibrium systems whose emergent (social, economic, physical) structure is determined from the bottom-up actions of and interactions between the institutions, firms and individuals that inhabit them. Thus, if we are to model cities' emergent structures, we must represent the decision-making processes of their actors; decisions that impact not only on their day-to-day operational behaviours, but also, indeed more importantly, those that impact on their investments; ranging from homeowners replacing façade or HVAC system components to municipalities investing in large infrastructure projects, such as the installation of new energy systems, roads or light rail lines. These investments may be motivated by policy measures influencing regulatory interventions, financial instruments such as levies or subsidies, educational campaigns, social (peer) pressure or technological innovation.

A coherent strategy for modelling these investment and, indeed, operational decisions is through multi-agent simulation, with agents acting independently (financial), in consultation with peers (social) and in response to top-down campaigns (educational) or statutes (regulatory); weighing up the costs and benefits, both social (e.g. warmer or cooler buildings following from improved energy conservation) and financial (e.g. increased disposable income arising from lower energy bills).

This in turn can be taken to imply the existence of a conceptual framework for the characterisation of social, economic and environmental sustainability, of the principle factors influencing these three pillars and the interrelationships between them, and of the potential responses of our archetypes of interventions destined to moderate these sustainability impacts. Thus, the modelling of agents' behaviours should form part of a broader strategy to model both perceptions and behavioural responses to them. This is a highly complex and long-term endeavour, which may be informed by the recent progress that has been made in modelling building occupants' behaviours (of which more later).

Integration: data and computational modelling techniques

Even when restricting our scope to the urban sub-systems described above, it will be evident that our modelling landscape is heterogeneous, addressing phenomena of diverse spatiotemporal resolution. Models of land use and transport interactions (LUTI) may discretise space using either regular (e.g. grid cells) or irregular (e.g. an administrative unit such as a super output area) lattices, predicting annual or multiannual transitions in land use. Subsequent modelling of the transport of goods and people may use a similar spatial discretisation, if based on a zonal modelling strategy, but a far finer temporal resolution. Urban climate models may be nested with global circulation models that represent future climate change (emission) scenarios. These too may be multiannual and will require downscaling to a regular mesoscale lattice of typically 1 km^2, with the energy, mass and momentum transfers within the urban canopy represented by a parameterisation that is roughly equivalent to (and thus should be derived from) the real urban geometry. This urban geometry will then be required as input to an urban scale model of building energy use, either explicitly in the case of a microsimulation strategy or implicitly if based on the modelling of archetypal abstractions. Both the urban mesoclimate and building energy models will have an annual timescale and (sub-)hourly resolution. For relatively small spatial scales, these building energy models may also be coupled with urban thermal and hydraulic and/or electrical network models, the latter having a temporal resolution of minutes or even seconds if being used to study network integrity.

Thus, we need to prepare a semantically attributed geometric representation of our urban scene at diverse resolutions, ideally contained within a common spatiotemporal database and using a common standard for data exchange between this and our diverse models. We also need to be able to orchestrate calls to these models and to share data, aggregated if necessary, between them during their execution. Thus (unless employing a classical approach of hard-coded software integration), we need to employ model integration technologies and workflows to prepare descriptions of our urban scene that will be imported in a suitably structured format to our integrated models from a suitably structured database. We introduce these concepts and candidate technologies in what follows.

Integration technologies

An integrated approach enables dynamic linking of heterogeneous simulation tools during their execution, even of multiple instances of the same tool, each employed to simulate different parts of an urban scene. Integration technologies may be classified as employing co-simulation (meaning co-operative simulation) or distributed simulation strategies.

Co-simulation (Sicklinger et al. 2014) refers to the orchestration of interactions between two or more multi-physical simulators, to ensure data is exchanged between them as required. When using a co-simulation approach, a number of variables are defined that need to be shared between different simulation tools, accessed via the co-simulation platform. With its roots in the automotive industry, the Functional Mock-up Interface (FMI) (Blochwitz et al. 2011) is a co-simulation standard that has gained significant traction within the building simulation community. FMI is an open-source standard by which two or more Functional Mock-up Units may be coupled, one of which acts as the Master Algorithm: controlling the execution of each FMU and the exchange of data between them. Each FMU consists of a. zip archive containing an XML file describing the variable definitions used by the FMU and a compiled executable of the source code containing the equations being solved.

Distributed simulation (Chandy and Misra 1979), on the other hand, is an approach where a number of dissimilar simulations running inside the same environment are split to run over two or more nodes that are connected through a network. Originally developed by the U.S. Department of Defense, the High Level Architecture (HLA) is one such approach (IEEE 2010). In HLA terminology, a federate is the component that is coupled following a specified Federate Object Model (FOM), while a federation refers to a set of federates that follow a common FOM to achieve a defined objective. The HLA standard is comprised of rules to be followed by all HLA-compliant federates: the modelling schema describing information common to a federation and the Run-Time Infrastructure (RTI) that represents the software infrastructure required by federates, to exchange information in a controlled way. The RTI (the HLA equivalent of the FMI Master Algorithm) plays an important role in providing services to federates as a distributed operating system specific to co-simulation.

Awais et al. (2013) show how the virtues of both HLA and FMI can be combined, using HLA as the master to coordinate between distributed FMUs running in the cloud. This approach appears to have much to commend it for urban simulation.

A complement to distributed simulation is hardware acceleration, whereby the simulation task can be broken down into discrete units that can be simulated in parallel, using either multicore processors or graphics processing units. These latter lend themselves well to massive parallelisation of rendering tasks, as may be required for example by radiosity-based radiation models in urban scale building energy models.

Data standards for the built environment

A standardised digital representation is critical to the urban modelling task. Building information modelling (BIM) and CityGML are two examples, both of which are widely adopted in academia and industry.

BIM has its roots in the construction community. A BIM represents both the physical and functional characteristics of a facility (or asset) and thus aims to provide a repository of its 2-D, 3-D and non-spatial information that can be shared among stakeholders throughout the facility's life-cycle (NBIM 2007). To ease model sharing and interoperability, efforts to standardise BIM have resulted in Industry Foundation Class (IFC) software schemas (BuildingSMART 2013).

CityGML in contrast is an Open Geospatial Consortium data model and exchange standard that is focussed on modelling urban areas (Groger et al. 2012). The standard supports descriptions of both built and natural features, ranging from the city scale down to internal structures in buildings (walls and floors). A hierarchy of levels of detail defines the fidelity of a CityGML model, from 2-D footprint polygons to complex 3-D representations of external and internal structure (Figure 20.4).

BIM may be thought of as taking an introverted view of the world, describing a building in considerable detail from the external surface of the envelope inwards, whereas CityGML takes a more extroverted view, focussing more on the external envelope outwards. BIM models typically contain substantially more information than is required for urban scale models, but there is interest in combining the virtues of both (Deng et al. 2016), where detailed building models need to be complemented, for example, with representations of urban infrastructure.

On its own, CityGML is a format for generic urban modelling, analysis and visualisation tasks, but it may also be extended by defining Application Domain Extension (ADE) classes. One such example, the EnergyADE, defines buildings' energy-related attributes, such as the construction materials used, their properties, occupancy details and so forth. Deriving a large-scale spatiotemporal database of the urban area using these standards has much to commend it for integrated urban modelling applications.

As an XML schema, the CityGML specification supports the representation and exchange of complete urban scenes using a single XML document. However, use of an appropriate database management system (DBMS) allows for efficient query and storage of the scenes without the need to read and parse the entire document into memory when it is employed for analysis, simulation or visualisation. The adaptation of the CityGML schema to an entity-relationship model suitable for database implementation has been undertaken as an open-source project. This is released as 3DCityDB (www.3dcitydb.org/), currently in v3.3.1 and jointly developed by TU München and two software companies, virtualcitySYSTEMS and MOSS Computer Grafik Systeme. 3DCityDB also offers tools to help with access and dissemination of CityGML scenes, including browser-based WebGL visualisation and OGC Web Feature Service.

The following section focusses on urban building energy modelling strategies and their applications. There, we also describe a candidate workflow to populate an integrated urban modelling database, with a specific focus on building energy microsimulation.

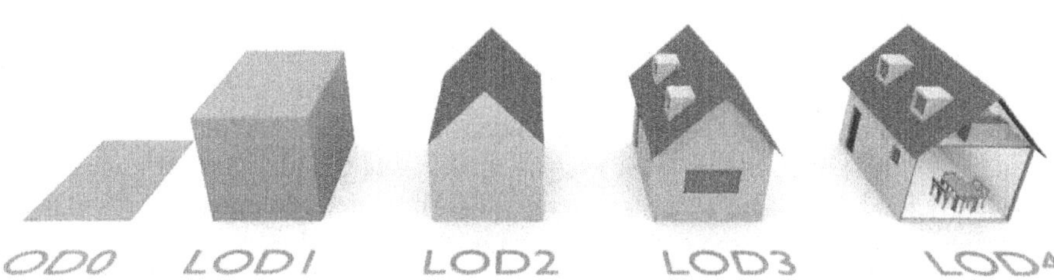

Figure 20.4 Levels of detail of the CityGML standard

Source: Biljecki et al. 2014

Urban building energy modelling

Strategies to simulate the energy flows arising from urban stocks of buildings can be broadly categorised as (a) statistical and/or system dynamics representations of the aggregate stock, performance improvements to it and, in the latter case, inflows to [new construction] and outflows from [demolition] this stock, (b) simulation of archetypes and extrapolation to the remainder of the stock and (c) explicit microsimulation of the entire stock in its urban context.

Of these, the detailed stock modelling strategies (b and c) are by far the most powerful, since they are based on detailed representations of the geometric forms of which the stock is comprised and the semantic attribution of these forms. Thus, the effectiveness of detailed scenarios for the improvement or replacement of existing or the installation of new façade elements, HVAC and renewable energy systems and small power appliances can be tested.

Such detailed stock modelling workflows involve a number of steps. Archetypal modelling techniques require (i) a sampling strategy to identify archetypes, (ii) surveys to define the representative 3-D geometry and (distributions of) semantic attributes of these 3-D models (construction, systems, appliances and activities), (iii) the creation of a synthetic stock that suitably represents the characteristics (based on step ii) of the set of archetypes, and finally (iv) the simulation of this synthetic stock, extrapolated to the full stock.

Microsimulation techniques also require steps ii to iv of the above, but now rather than choosing archetypes and constructing representative 3-D models of them, spatial (e.g. cadastral and LiDAR or photogrammetry) data are used to develop an explicit 3-D representation of the entire stock at an appropriate level of detail, with each element semantically attributed, employing a process similar to step iii above.

Of these techniques, archetypal sampling is the most commonly employed, whether based on relatively simple energy balance models or on complex single-building simulation programs. Nevertheless, approaches to both archetypal modelling and microsimulation are described in the following sections, alongside examples of their application.

The sector that has been most studied to date, and for which simulation techniques have been best established, is the relatively homogeneous residential sector. This is largely due to the more widespread availability of survey data describing the composition of this stock. The highly inhomogeneous non-residential sector is comparatively poorly understood – with modelling efforts typically focussing on the use of benchmark data of questionable validity or on the relatively detailed archetypal simulation of a specific sub-sector (normally offices). The following sections are therefore focussed on the modelling of housing stocks, though the implications for modelling non-residential stock are also considered.

Archetypal sampling

Urban housing stocks have conventionally been modelled in much the same way as national housing stocks (for a review of national housing stock energy models, we refer the interested reader to Sousa et al. 2017), employing simplified energy balance models to calculate annual energy use and emissions of archetypes, extrapolated to the stock of buildings represented by them. Such is the case of Jones et al.'s (1998, 1999, 2007) Energy and Environmental Prediction (EEP) model and the subsequent Solar Energy Planning (SEP) tool (Rylatt et al. 2001; Gadsden et al. 2003), though this latter assigns survey data to individual buildings which are then explicitly modelled using a similar energy balance model. Shimoda et al. (2003) employ a similar approach, based on the definition and modelling of archetypes (460 combinations of building type, building size and household composition) to model the housing stock of Osaka, but now using a dynamic simulation program and a representation of occupants' presence and use of appliances. In what follows, we describe a methodology for simulating city housing stocks that takes its inspiration from the work of Shimoda et al. (and the subsequent work of Taniguchi et al. 2007).

The InSmart workflow

As noted earlier when discussing zonal transport modelling, a team at the University of Nottingham was charged with the development and application of an integrated urban energy modelling workflow with which to test alternative city decarbonisation scenarios, to support the reformulation of Nottingham's Sustainable Energy Action Plan. An important part of this work, which was undertaken as part of the European Commission funded project InSmart (Gargiulo et al. 2017), involved the preparation and deployment of a workflow (Figure 20.5) to estimate reference case city housing stock energy use and emissions and to test strategies for reducing them (Long et al. 2018). The steps involved in this workflow are as follows:

1 Define building archetypes representing the diversity of housing built form (apartment, terraced/row house, detached or semi-detached house) and construction period (pre-1920: Victorian, 1920–1945: inter-war, 1946–1964: post-war, 1965–1979: sixties/seventies, post-1979: modern). These 20 combinations were reduced to 16, based on their relative prevalence. A combination of spatial, building use and age classification data enabled all instances of these archetypes to be spatially identified.
2 An average of around 35 surveys were conducted for each of the archetypes, recording (using a dedicated tablet app) their location, tenure, construction characteristics, façade glazing ratio, heating systems and setpoints and an inventory of lighting and electrical appliances. Dimensioned plans and photographs were also recorded, together with observations of the use of attic spaces.
3 The archetypes were enriched by defining candidate subtypes using the survey data, to account for variations including the presence of heated attics, conservatories, whether terraces were mid- or end and a non-standard number of storeys.
4 Using the data from step 1, candidate archetypal subtypes were retained if they were sufficiently prevalent.

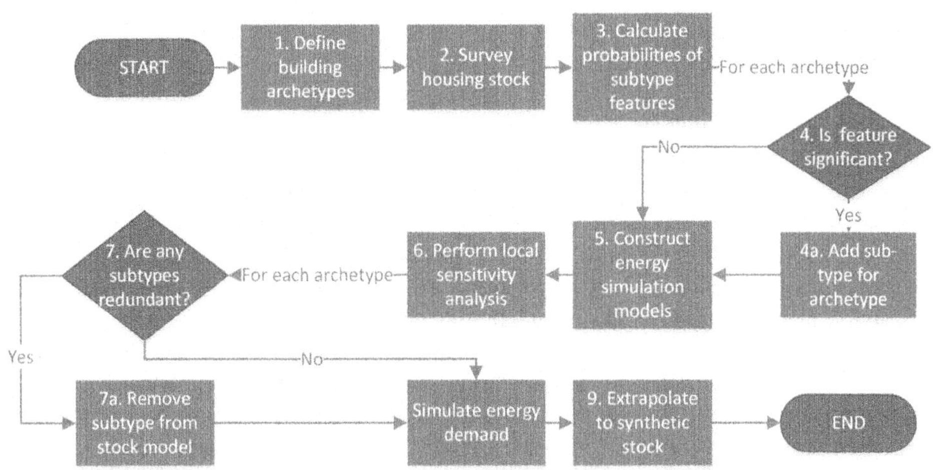

Figure 20.5 Flowchart describing the InSmart housing stock energy modelling strategy

Source: After Long et al. 2018

5 Around 70 unique models, in the form of EnergyPlus input data files (.idf), were manually created for each retained sub-archetype, attributed using median parameter values from the survey data.
6 Local sensitivity analysis was then conducted to identify the parameters to which simulated annual energy use was most sensitive. For this, scripts were prepared using the statistical computing package *R* to automate the modification of. idf parameters, the execution of models and the subsequent extraction and analysis of results.
7 If simulated energy use and the outcomes from sensitivity analyses for one subtype closely matched those of another, then this subtype was removed.
8 A synthetic stock of archetypal subtypes was then created, with each element of these stocks having parameters assigned using a Latin Hypercube Sampling (LHS) strategy, sampling from cumulative distribution functions of the parameters to which performance was found in step 6 to be sufficiently sensitive. The size of the stock *S* was determined when a t-test was able to reject the null hypothesis that successive (comparing *S-1* and *S*) mean annual energy use estimations were statistically significantly different. This process led to the creation of a little over 3,000 unique energy models.
9 Finally, the data from step 1 (now distinguishing between archetypal subtypes) was employed to associate a normalised annual energy use, using Monte Carlo sampling of results from the relevant sub-stock from step 8, to each archetypal subtype in the real housing stock. This relationship, between the subtype represented in the spatial map and the element of the synthetic stock of subtypes, was preserved, so that subsequent variants in the subtype properties arising from decarbonisation scenarios and the associated impacts on energy use and emissions could be efficiently represented.

InSmart workflow: results and discussion

The total housing thermal energy demand simulated using this workflow (Figure 20.6) is within 3% of the corresponding estimate using historical data (a combination of measured and estimated metre readings) provided to the UK Department for Business Energy and Industrial Strategy (BEIS) by energy utility companies.

The reductions in energy use of Nottingham's housing stock arising from the insulation of both solid (e.g. Victorian) and cavity walls, the addition or augmentation of loft insulation and improvements in draught proofing, were estimated to be 18%, 3% and 9% respectively. Assuming these savings to combine additively and ignoring the potential sharing of co-benefits arising from these measures, between reduced energy use and associated bills and improved comfort (a higher mean indoor temperature) then the combined reduction is about 30%. This implies that significant further improvements, in particular concerning heating systems, lights and appliances, are required to meet future decarbonisation targets.

The InSmart workflow has proven to be both accurate and relatively efficient, in terms of both human effort (given the level of process automation) and computational effort. However, it does rely on surveys that are costly to perform initially and to quality assure subsequently. One solution that has general and immediate applicability is to use a new dynamic housing stock energy modelling hub *EnHub* (Sousa et al. 2018). Originally developed for the dynamic modelling of national housing stocks and using national survey datasets (in particular the English Housing Survey), this can be readily deployed to model city housing stocks and to test corresponding decarbonisation scenarios. Indeed, work is underway to enrich *EnHub* by coupling it in the first instance to the multi-agent stochastic simulation platform No-MASS to model occupants' behaviours. Work is likewise underway to develop a new prototype social simulation platform to evaluate households' likely uptake of candidate decarbonisation investments in response to policy measures designed to stimulate them. This has particular promise,

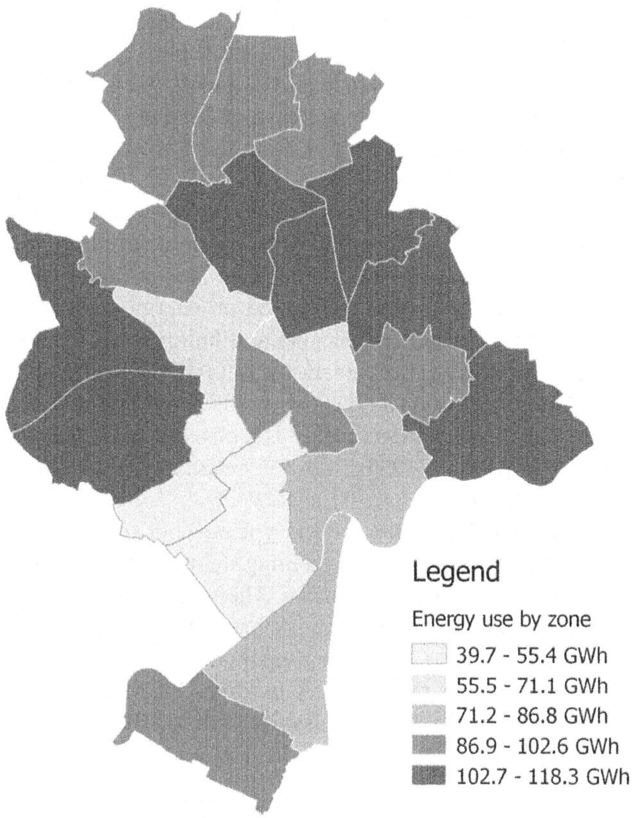

Figure 20.6 Total simulated residential energy demand by city ward

Source: After Long et al. 2018

but caution will be needed in deploying *EnHub* to model city housing stocks where local construction practices deviate from those represented in national survey data. In such cases, hybrid calibration datasets may be needed.

But whether using urban-specific workflows of the type developed in the InSmart project or adapting national housing stock energy modelling workflows for urban applications, the fact remains that archetypal sampling and simulation methods fail to account for interactions between buildings and urban climate impacts. Such methods also lack the versatility to support future co-simulation of urban energy and transport systems (indeed of the range of systems identified in Figure 20.1). This is important given the future trajectory in which buildings, energy and transport systems will be increasingly well integrated, as nations seek to decarbonise their transport and power sectors.

The case of non-residential stocks

As noted earlier, non-residential building stocks are highly heterogeneous. Examples include retail outlets (e.g. hardware, white goods, clothing, department store, supermarket, pharmacy), cultural facilities (e.g. cinemas, pubs and clubs, sport venues, theatres, galleries and museums), medical facilities (e.g. nursing homes, medical centres, hospitals), educational buildings (e.g. schools, further and higher educational institutions, libraries), public buildings (e.g.

governmental offices, assemblies and embassies, police stations, courts, prisons) and industrial buildings (e.g. light and heavy manufacturing, water processing, power plants, breweries and wineries). There is also considerable heterogeneity amongst individual subtypes of this non-exhaustive list of categories, particularly in the case of hospitals and universities (which are themselves frequently organised into diverse campuses), and many buildings incorporate mixes of use, particularly in city centres.

The chief complication concerns the lack of availability of data, both for model calibration and validation, whether based on top-down statistical methods or more bottom-up archetypal sampling methods. In the former case, there is simply a lack of high-quality benchmark data describing energy use (preferably by end use) based on energy audit or energy survey data for representative samples of the range of categories of building use (Bruhns et al. 2011). For example, in the UK, CIBSE's energy benchmark dataset (CIBSE 2008), which is the statutory benchmark and widely used, addresses a relatively broad set of building use categories but is, in many instances, based on very small samples. Also, many of the benchmarks used are based on building surveys carried out in the 1990s (Mortimer et al. 2000) and thus may not accurately reflect current energy use patterns.

Top-down models using benchmark data of this type have, of course, the limitation that they cannot be used to reliably test hypotheses concerning improvements to building performance arising from specific decarbonisation interventions. They can, however, usefully serve to evaluate the ability of bottom-up endeavours to do so, based on the simulation of representative typologies (as in the residential case). But bottom-up modelling strategies require the availability of comprehensive survey data, which is sorely lacking. One future solution to this dilemma would be the hosting by an independent organisation such as IBPSA of a database of (ideally survey-based) model attribution data, categorised for example by building location, use and age.

But there is one further complication (at least in the UK), and this relates to the lack of spatial and semantic data to enable existing non-residential building uses to be identified, on the one hand, and the extent of mixed-use buildings that are occupied by these uses on the other; likewise, where in the building these uses are located. A promising emerging solution to the latter is the 3DStock model proposed by Evans et al. (2017). This combines cadastral data describing building footprints with valuation office data describing commercial and public floor space, to allocate internal space in building footprints to explicitly represent the uses on different building floors.

It is because of these severe historic data limitations that efforts to model non-residential buildings have thus far focussed on the relatively prevalent category of commercial office buildings. For example, Yamaguchi et al. (2003, 2007a, 2007b) have progressively increased the scope of their workflow for modelling office buildings in Osaka based on building-specific survey data: originally restricted to two office buildings, then increasing in scale to a district of 55 buildings connected to district co- and tri-generation systems and finally to the entire city based on extrapolation from 612 archetypes.

Microsimulation

Urban energy microsimulation is complex for it implies that, no matter the size of the building stock being simulated, each building will be explicitly simulated accounting for its energetic couplings with its urban context (which are largely represented in Figure 20.1). At the simplest level, adjacent buildings interact with one another through surface radiation exchanges: shortwave, visible and longwave. But they also influence and are in turn influenced by the urban climate as well as local heat and power networks, which may in their turn be influenced by local industrial activities and energy from waste plants, providing heat (and potentially power) to these networks. Buildings and local energy networks are also increasingly being integrated with transport networks and the vehicles flowing through them.

The first attempt to handle these complex building-urban energy interactions was the sustainable urban neighbourhood modelling tool, SUNtool (Robinson et al. 2003, 2007). The purpose of SUNtool was to support urban designers to minimise net resource (particularly energy) use at the neighbourhood scale, a scale that is normally too small to influence transport network topologies. It comprises both a graphical user interface (GUI) and an integrated resource flow solver.

The SUNtool GUI supports the user in localising and describing the three-dimensional structure of the urban scene to be simulated, using a dedicated simplified 3-D modeller. These objects are then semantically attributed, associating each building with the relevant iDefaults archetype; each archetypal dataset corresponding to default occupancy, constructional, equipment etc. attributes. These attributes may be overridden at building or surface hierarchical levels, whereas at higher levels new or modified archetypes may be defined. Buildings may also be associated with district energy centres, and (e.g. solar) energy conversion systems can be directly embedded into building surfaces. Complete description are then parsed in. xml format to the solver.

The SUNtool solver has at its core a reduced dynamic thermal model that is coupled with a simplified radiosity algorithm for handling surface radiation exchanges, a family of energy conversion (e.g. boiler, solar thermal, photovoltaic, wind turbine, heat pump) system models and a psychrometric approach to modelling changes in the enthalpy of air, wherever air is used as the medium for delivering zone heating/cooling needs. The solver also integrates simplified stochastic models accounting for occupants' presence and their use of blinds, lights and windows, though these interactions could also be represented using schedules defined by a schedule generator in the GUI. Following an unconvincing attempt to adapt Swaid and Hoffman (1990) Cluster Thermal Time Constant model, the urban thermal climate was not represented in SUNtool, nor was the local wind climate.

SUNtool's successor CitySim (described in detail below) follows much the same philosophy as SUNtool, explicitly micro-simulating buildings in their urban context that are subject to stochastic perturbations arising from occupants' interactions.

In something of a departure from these approaches, two alternative microsimulation platforms have recently emerged: SimStadt (Nouvel et al. 2015a, 2015b) and UMI (Reinhart et al. 2013; Davila et al. 2016). In both cases, workflows have been developed to develop 2.5-D/3-D geometric models and to semantically attribute these models using default parameters relating to standard archetypes. SimStadt provides the option of employing either a statistical top-down method to estimate the annual thermal energy demands of these buildings or a more detailed bottom-up approach; the latter is based on a simplified monthly energy balance model. UMI on the other hand prepares EnergyPlus. idf files for the building to be simulated, as well as other nearby buildings within a 300 m radius that shade this building, calling the solver and reading in results to display them within the Rhino environment in which UMI is a plugin. This is more thermally rigorous than the SimStadt approach, but it relies on EnergyPlus's simplistic radiation model (occlusions to the sky are handled in a crude way, and they only contribute first-order diffuse reflections) and is based on an approximate criterion for determining potential radiative interactions. Neither SimStadt nor UMI explicitly represent occupants' stochastic behaviours.

CitySim

At the time of writing, CitySim (Robinson et al. 2009; Robinson 2011) remains the most energetically complete and rigorous platform for the explicit microsimulation of buildings in their urban context that are subject to stochastic perturbations from occupants. In common with its predecessor SUNtool, the *CitySim Pro* GUI can be used to import or prepare detailed 3-D models describing the scene to be simulated and to attribute these buildings, to write the scene description in CitySim's native. xml file format, to call the CitySim solver and to read and interrogate the simulation results. Alternatively, a GIS-based workflow (Perez et al. 2011; Perez and Robinson 2012; Perez 2014) can be employed for the modelling of relatively larger scenes, sanitising and enriching existing datasets to populate a PostgreSQL database, from which a CitySim. xml file is exported and the corresponding results are imported and then visualised using the Open Source GIS software QuantumGIS.

Reflecting the growing popularity and maturity of the CityGML data exchange standard and the associated application domain extensions, the CitySim solver (Figure 20.7) has recently been modified to import CityGML files, created using the workflow described later (see the section "Data processing workflow for urban energy microsimulation").

Upon populating CitySim's internal data model describing the scene to be simulated and the models to be employed in doing so, a radiation pre-process is first called. This computes view factors (sun to surface, sky patch to surface and surface to surface) and from these creates matrices describing sky to surface and surface to surface radiative couplings. The radiation model then proceeds to calculate shortwave irradiances incident on each surface from the sun, sky and occluding surfaces. The same view information is also used to calculate internal illumination at two points within a hypothetical room that is extruded inward from the external

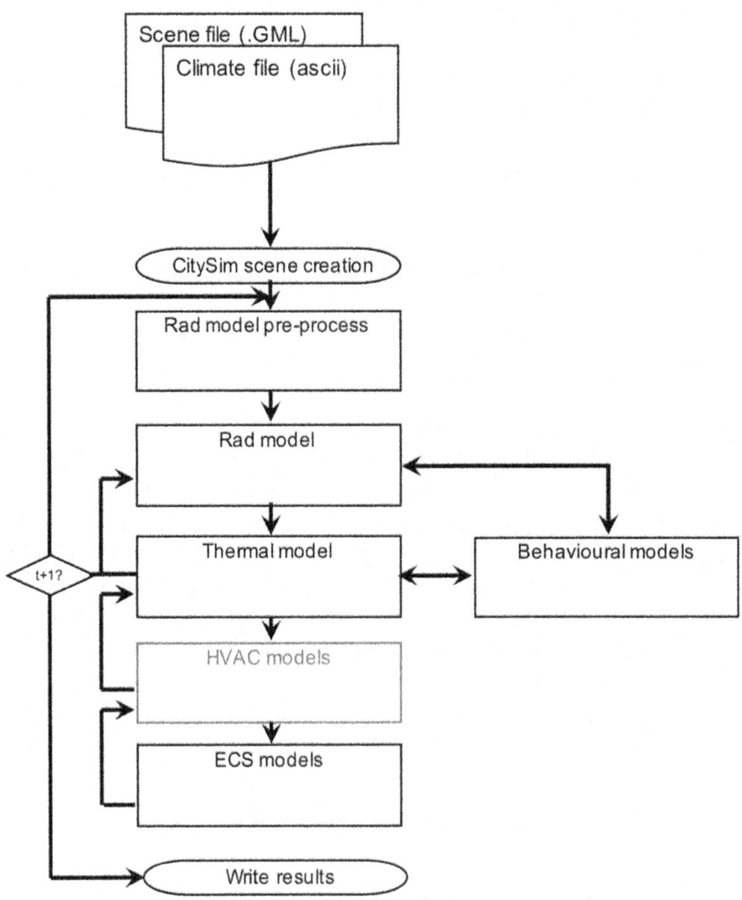

Figure 20.7 CitySim solver structure

surface centroid, provided that this surface contains glass. A simplified dynamic thermal model is then called to calculate zonal (the building by default) space heating and cooling energy demands. The radiation and thermal models each provide inputs to stochastic models of occupants' presence and dependent behaviours (use of lights, blinds and windows), which in turn influence transmitted radiation and the space conditioning demands, as does an estimation of surface longwave exchange. For the latter, the surface view information in conjunction with the external surface temperature from the preceding simulation timestep are used. If air is utilised as the medium for delivering space heating and cooling, then psychrometric models are called to calculate the associated enthalpy changes and the corresponding loads to be met by HVAC equipment. The associated energy conversion system (ECS) models are then called and, if their capacity is insufficient, the supply state is adjusted and the predicted room thermal state (temperature) is corrected (using the thermal model). This sequence is employed for all buildings within a timestep and for all timesteps throughout the period (normally a year) being simulated.

We briefly describe below the principles of each category of model in CitySim's solver. We then discuss data processing workflows for urban building energy modelling (UBEM), with a specific focus on microsimulation using CitySim for the city of Nottingham, and present initial results from its application.

Thermal model

The thermal model is based on analogy with an electrical circuit (Lefebvre et al. 1987), in which the heat flow between a wall and the outside air can be represented by an electric current through a resistor linking the two corresponding nodes (wall and outside air), and the wall's inertia can be represented by a capacitance linked at that node.

In our model (Kämpf and Robinson 2007), which is a refinement of that due to Nielsen (2005), an external air temperature node T_{ext} is connected with an outside surface temperature node T_{os} via an external film conductance K_e, which varies according to wind speed and direction (Figure 20.8). T_{os}, which also experiences heat fluxes due to shortwave and longwave exchange, is connected to a wall node T_w of capacitance C_w via a conductance defined by the external part of the wall. In fact, this node resembles a mirror plane, so that we have similar connections to an internal air node T_a of capacitance C_i via an internal surface node T_{is}. T_{is} may also experience shortwave flux due to transmitted solar radiation and longwave flux due to radiant heat gains from internal sources (people, lights and appliances), and T_a may experience convective gains due to absorbed shortwave radiation, internal casual gains and heating/cooling systems. Finally, our internal air node may be connected with our external air temperature node via a variable resistance due to infiltration and ventilation.

For a whole building with many subspaces, the air nodes of each zone are linked via the separating wall conductance; interzonal airflow can also be handled through this conductance. To account for the corresponding inertia, the capacitance of the separating wall is subdivided and allocated to the neighbouring zones' capacitances (Figure 20.9).

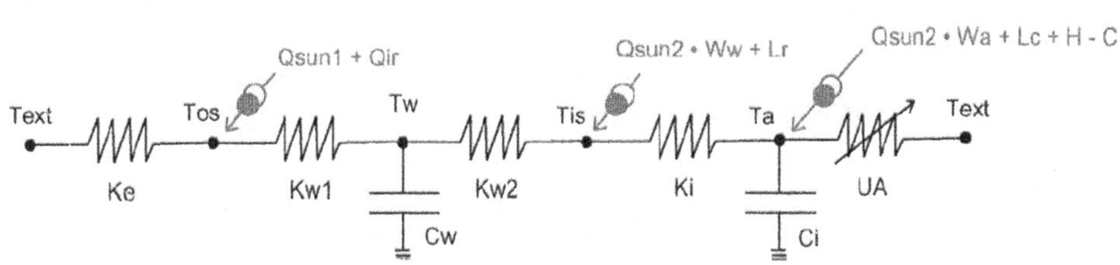

Figure 20.8 Monozone form of the CitySim thermal model

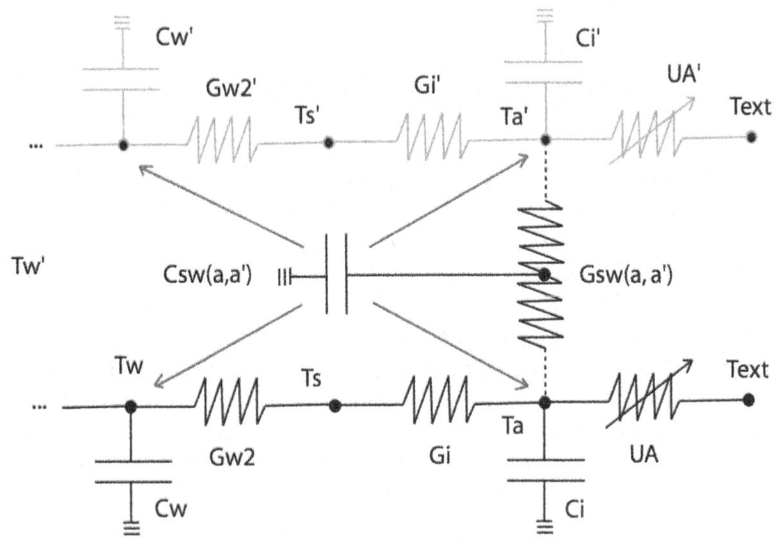

Figure 20.9 Interzonal connection between zones represented by the two-node thermal model

Predictions from this simplified model compare well with those of ESP-r (Clarke 2001) for a range of monozone and multizone scenarios.

Radiation models

As with SUNtool, the simplified radiosity algorithm (SRA) of Robinson and Stone (2004) is used to solve for the shortwave irradiance incident on the surfaces defining our urban scene. For some set of p sky patches, each of which subtends a solid angle Φ (Sr) and has radiance R ($Wm^{-2}Sr^{-1}$) then, given the mean angle of incidence ξ (radians) between the patch and our receiving plane of slope β, together with the proportion of the patch that can be seen σ ($0 \leq \sigma \leq 1$), the direct sky irradiance $(Wm^{-2})\, I_{d\beta} = \sum_{i=1}^{p} (R\Phi\sigma\cos\xi)_i$.

For this, the well-known discretisation scheme due to Tregenza and Sharples (1993) is used to divide the sky vault into 145 patches of similar solid angle, and the Perez all-weather model (Perez et al. 1993) is used to calculate the radiance at the centroid of each of these patches. The direct beam irradiance $I_{b\beta}$ is calculated from the beam normal irradiance I_{bn} which is incident at an angle ξ to our surface, of which some fraction ψ is visible from the sun, so that $I_{b\beta} = I_{bn}\psi\cos\xi$.

These direct sky and beam irradiances influence each receiving surface's radiance R, which in turn influences the irradiance incident at other surfaces that are visible to it and vice versa. To solve for this, a similar equation to that used for the sky contribution gives the reflected diffuse irradiance. In this case, two discretised vaults are used, one above and one below the horizontal plane, so that $I_{\rho\beta} = \sum_{i=1}^{2p} (R^*\Phi\omega\cos\xi)_i$, where ω is the proportion of the patch that is obstructed by urban (reflecting) surfaces and R^* is the radiance of the surface which dominates the obstruction to this patch (in other words, that which contributes the most to ω).[1] Of course, R^* itself depends on reflected diffuse irradiance as well as on the direct sky and beam irradiances. For this a set of simultaneous equations relating the beam and diffuse sky components to each surface's irradiance, which itself effects the reflected irradiance incident at other surfaces, may be formulated as a matrix and solved either by inversion or iteratively until a convergence criterion is passed (Robinson and Stone 2004, 2005a).

The principal complication in the above algorithm lies in determining the necessary view factors. For obstruction view factors, views encapsulating the hemisphere are rendered from each surface centroid, with every surface having a unique colour. Each pixel is then translated into angular coordinates to identify the corresponding patch as well as the angle of incidence. For sky view factors, $\Phi\sigma\cos\xi$ is treated as a single quantity obtained by numerical integration of $\cos\xi \cdot d\Phi$ across each sky patch. Likewise, for $\Phi\omega\cos\xi$, for which the dominant occluding surface is identified as that which provides the greatest contribution. A similar process is repeated for solar visibility fractions for each surface, for which a constant size scene is rendered from the sun position. See Robinson and Stone (2005a) for further details.

As noted earlier, this view information may also be used to calculate longwave irradiance, given the corresponding surface and sky temperatures (Robinson and Stone 2005a). Furthermore, additional renderings may be computed to determine the view information necessary to determine the direct and reflected contributions to internal illuminance (at known points) as well as the incoming luminous flux for internal reflection calculations (Robinson and Stone 2005a, 2006).

As Robinson and Stone (2005b) demonstrate, this simplified radiosity algorithm may be readily integrated into standard single-building simulation programs to improve the accuracy of their treatment of radiation exchange in the urban context.

Behavioural models

Key types of behaviour and their main impacts on urban resource flows are as follows:

- Presence: metabolic heat gains and pollutants.
- Lights: heat gains and electrical power demand.
- Windows: ventilation rates.
- Blinds: illuminances (and associated impacts on lighting control) and transmitted irradiance.
- Appliances (water/electrical): heat gains, electrical power demand, water use (and potential conversion of black water into biogas).
- Waste: combustible solids.

CitySim has hardcoded within it models of occupants' presence and their presence-dependent use of windows, blinds and lights.

Based on the hypotheses that all occupants act independently and that their actions at time $t+1$ depend only upon the immediate past (t), short-term chains of presence and absence are modelled as a first-order time-inhomogeneous Markov chain (Page et al. 2007). This model takes as input a time-varying profile of the probability with which occupants will be present and a time-invariant parameter that represents occupants' mobility.

Interactions with lights are based on the same models as those integrated with Lightswitch-2002 (Reinhart 2004). In particular, the probability of switching on lights at arrival is expressed (Hunt 1980) as a logit function of minimum internal workplane illuminance ($E_{i,min}$). Switching on lights at intermediate times is also predicted as a function of $E_{i,min}$ (Reinhart and Voss 2003), whereas switching off at departure is simply a function of the expected duration of absence (Pigg et al. 1996): if we expect to be absent for a long time, we will tend to switch lights off.

In common with the light switching model, the window opening model (Haldi and Robinson 2009) also differentiates between actions at arrival, at intermediate times and upon departure. Actions at arrival (or rather, 5 minutes after arrival) take place in two stages. Transitions in window opening status are predicted using logit models for which data is filtered to the initial state (open or closed) and the occupancy state (arrival, intermediate or departure). If a transition to open is predicted, the duration for which the window remains open is predicted using a Weibull distribution; unless this occurs at departure, in which case the window remains open until the following day. This simply reduces the number of redundant calculations (e.g. of no change of state).

The model for blind usage (Haldi and Robinson 2010) also makes the distinction between actions at arrival and at intermediate times.[2] The procedure is common for each of these occupancy types; only the probabilities differ. At a given moment, the probabilities of both lowering and raising a blind from its current position are predicted. It is then determined whether this will be a full or partial raising action. If partial, the change in unshaded fraction is predicted, again using a Weibull distribution.

For a more detailed description of these models and quantifications of their impacts on building energy performance, we refer the interested reader to Haldi and Robinson (2011).

Although this family of behavioural models is based on rigorous foundations and each model has performed well in validation tests, it is far from complete. Rather than continue to hardcode models within CitySim, thus restricting somewhat their usefulness, a new platform has been developed that can be coupled with any building or urban energy simulation program: the Nottingham Multi-Agent Stochastic Simulation platform (No-MASS).

No-MASS combines data-driven stochastic models, for the modelling of relatively commonplace behavioural phenomena for which data is abundant (Chapman et al. 2018a), with a range of strategies for the modelling of phenomena for which data is less abundant (Chapman et al.

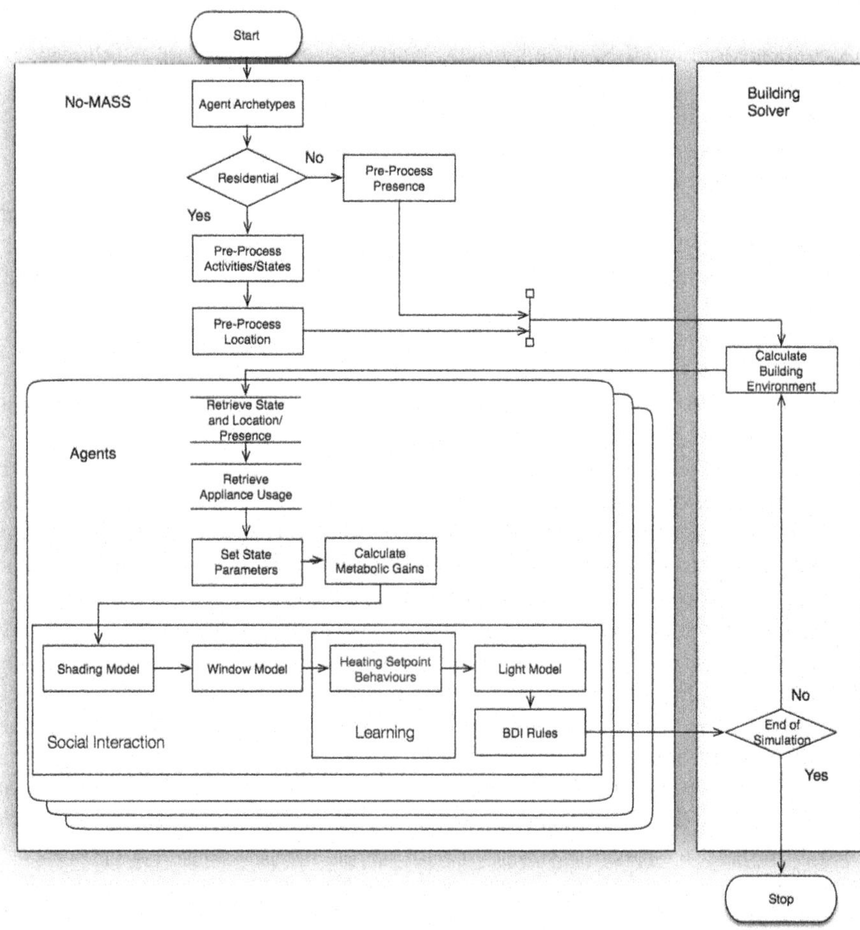

Figure 20.10 No-MASS solver structure

2018b): Figure 20.10. These include the use of belief-desire-intention (BDI) strategies for modelling straightforward behaviours (e.g. opening bedroom windows upon awakening or following bathing), agent learning for more complex behaviours (e.g. spatiotemporal home heating choices: timing and temperature preferences) and a system for the processing of votes cast by multiple agents (each vote representing the outcome from a data-driven model) to emulate negotiation mechanisms amongst the members of social groups (e.g. colleagues, friends or family members). In this way, No-MASS comprehensively simulates the behaviours of occupants of both residential and non-residential building types, and it is readily extensible. This is important, for there remain gaps in No-MASS's functionality, including the representation of domestic hot water use, the modelling of long absences and the relatively habitual behaviours that arise in residential settings (such as the positioning of blinds before departing for a workday in summer and the ventilation of bedrooms upon awakening: behaviours that can be readily accommodated using BDI rules).

Although No-MASS was originally coupled with EnergyPlus using FMI, this same technology has been used to couple No-MASS with CitySim (Wang et al. 2017).

HVAC and ECS models

In common with SUNtool (Robinson 2007), the HVAC model simply computes the psychrometric state (temperature, moisture content and hence enthalpy) of the air at each stage in its supply (e.g. outside, heat recovered, cooled, re-heated, supplied). Given the required mass flow rate (which may be defined by the energy to be delivered or by the room fresh air requirement), the total delivered sensible and latent heating and cooling loads can then be easily calculated.

Also similar to SUNtool (Robinson et al. 2007), the family of ECS models in CitySim comprise a range of technologies that provide/store heat and/or electricity to buildings. These include a thermal storage tank model for hot/cold fluids, boilers, heat pumps, cogeneration systems, combined cogeneration and heat pump systems, solar thermal collectors, photovoltaic cells and finally wind turbines. These ECS models are in general based on performance curve regression equations, whereas a simplified thermal model simulates the change of state of the sensible/latent heat store (so that phase change materials are also accounted for).

For the interested reader, the structure and theoretical basis of CitySim is described in considerably more detail in Robinson (2011).

Data processing workflow for urban energy microsimulation

Producing a data model for urban energy microsimulation requires the collation and integration of multiple datasets: spatial data describing the scene complemented with data describing the energy-related attributes of buildings, their systems and appliances as well as their occupants. For large urban models, semi-automated workflows need to be developed to process the source data into a sufficiently detailed urban data model; the resolution depends upon the specific modelling strategies employed.

A spatial model of the urban scene can be prepared by combining 2-D mapping of building footprints with digital elevation models (DEMs). Figure 20.11 shows an example using UK Ordnance Survey (OS) datasets (Mastermap, its associated Building Height Attribute (BHA) dataset and the OS Terrain 5 Digital Terrain Model (DTM)) for the city of Nottingham. This 2.5/3-D model directly feeds into a CityGML model. It is also a prerequisite for the accurate calculation of the total floor area of buildings, which is essential in any UBEM.

Figure 20.11 Integrating DEM (left) with 2-D building footprints (middle) to produce a 2.5-D model of urban scene (right)

Detailed spatial data on buildings and their elevation may not be available in some countries. In this instance, remote sensing data from satellites and other open-source mapping may be used to produce a coarser 2.5-D model that may be sufficient for lower resolution modelling strategies.

With the volumetric model defined (the steps shown in the upper left part of the workflow shown in Figure 20.12), the non-spatial building features of relevance (e.g. envelope multilayer constructions and elemental thermophysical properties) can be attributed from statistical models. Such models are typically based on survey data. For example, the English Housing Survey provides detailed construction, system, appliance, household and tenure data for a representative biannual sample of about 14,000 UK dwellings.

To attribute buildings according to these energy-related features, buildings need first to be categorised into archetypes matching the available data. The initial step is to identify the building use (e.g. housing, commercial, industrial) by integrating the 2.5-D model with a land use model. Built form (e.g. detached, semi-detached or terraced in the case of housing) can be identified using spatial analysis techniques such as those described by Rylatt et al. (2003), Hecht Meinel and Buchroithner (2015) and Beck et al. (2018). Identifying construction periods is a more difficult task. In many countries, including the UK, spatial datasets of building age are not readily available. Time-consuming manual digitisation and analysis of historic maps can be undertaken, but the data across all the relevant construction periods may not be available. To address this, approaches based on analysing the scene geometry have shown promise. Typically, these take the approach of learning predictive models of the building geometry and wider urban structure in order to infer age or construction period (Lannon et al. 2007; Tooke et al. 2014; Rosser et al. 2018).

With the urban scene categorised by building archetypes, the attribution of building-specific features can be performed using survey data. Buildings can be assigned values stochastically, using Latin Hypercube Sampling (LHS) from cumulative distribution functions of the relevant survey parameters (akin to the InSmart workflow described earlier) – preferably following a sensitivity analysis to isolate the parameters to which predicted performance quantities of interest are most sensitive – derived from survey data, or deterministically using median values for the relevant parameter. A stochastic approach will show greater diversity in energy use that better reflects the distribution of energy use amongst the real stock of archetypes, but a deterministic model will likely be sufficient for producing aggregated energy use estimates.

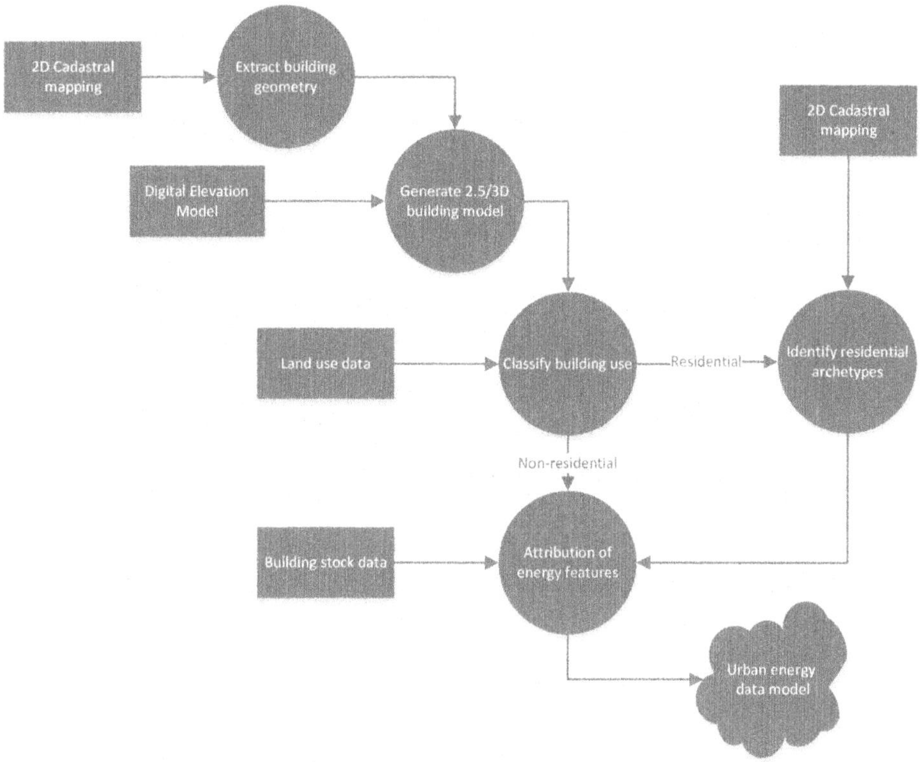

Figure 20.12 Data processing workflow for urban energy microsimulation

Example application: a district of Nottingham

Experiments are underway to deploy the microsimulation workflow introduced above (Figure 20.12) in Nottingham, UK. Using a common geospatial database enables simulation results to be visualised in 3-D and in the urban context. In this example, a CityGML model of a residential part of Nottingham has been created from Ordnance Survey and building heights map data. The CityGML model has then been enriched with EnergyADE features that are used as input to CitySim, in this case to simulate the shortwave irradiation incident on the building envelope. The results are colour-mapped and attached to the building geometry (Figure 20.13, top). Shown right in Figure 20.13 is an irradiation histogram for the simulated surfaces. This is broadly a bimodal distribution, with a lower peak representing east-/west-facing vertical surfaces and a higher peak representing the roof surfaces.

Outstanding UBEM and more general urban modelling challenges

Arguably the most complex and demanding task in developing new models and workflows to support their deployment is the identification of the most parsimonious strategies to employ. Indeed, no systematic study has been undertaken to this effect, such that there remain a number of quite fundamental open questions. Examples include: (i) What is the appropriate level of geometric detail? (ii) What is the appropriate field survey sample size to support the simulation of residential and non-residential building stocks? Which parameters should be surveyed

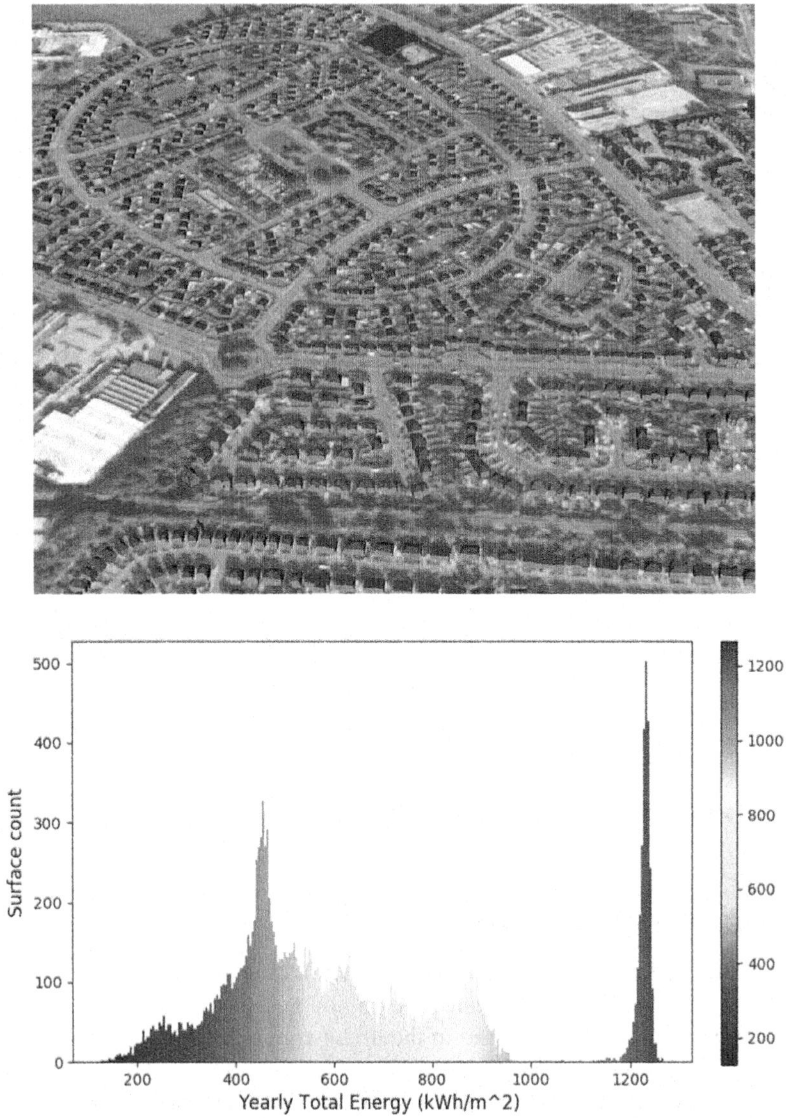

Figure 20.13 3-D visualisation of building irradiation (kWh/m^2) by surface (above) and distribution of annual total surface irradiation (below)

and thus defined for each building or archetype, depending upon prediction sensitivity? (iii) How can uncertainties in both deterministic and stochastic phenomena be best quantified and decomposed? (iv) What is the appropriate strategy for representing occupants' behaviours as a function of the stock size? (v) What are the appropriate strategies for representing the urban climatic context: radiative interactions, wind pressures (and associated modelling of surface convection, infiltration and ventilation), temperatures? (vi) What are the most promising technologies to support integrated urban modelling? Is it viable to have an integration platform and associated data processing workflows that can be sufficiently adaptable to transcend spatial scales, ranging from neighbourhoods to megacities? Is this even desirable? (vii) How can the factors influencing the temporal evolution of an urban scene (e.g. physical characteristic of buildings, the demographic characteristics of their populations) best be modelled? (viii) Is it feasible to define and agree upon international standards to achieve consistency in the provision of open access to data for urban modelling applications? Which data and at which level of spatiotemporal aggregation?

Clearly, the answers to many of these questions depend themselves on the questions to which the models and their workflows are to be employed to address. And these may vary depending upon context, the spatiotemporal scales in question and the degree of ambition of those posing the questions. Nevertheless, there is a need to chart this landscape, to determine the question-dependent strategies to be employed and to embark upon making them generally available. IBPSA has a potentially important role to play here.

Synopsis

The vast majority of buildings are constructed in urban environments. And this urban context has a direct impact on buildings' resource flows. Urban temperatures are on average warmer, the air is more contaminated and its velocity is lower. Adjacent buildings also reduce the availability of daylight and solar radiation. So, whether we are concerned with designing just one or many buildings, this urban context needs to be considered. The urban context also offers great opportunities for reducing urban resource consumption and the associated adverse environmental consequences. For example, complementary building activities can be located adjacent to one another to reduce transport needs and/or to promote synergetic resource exchanges. The siting of buildings can also be arranged to maximise natural resource availability, subject to local density constraints. Energy generation can be locally centralised, for example, to benefit from the more stable aggregate resource demand profile (to improve the quality of fit between demand and supply) as well as economies of scale compared with the building-embedded case. In order to maximise the utilisation of these opportunities, there is a compelling case for the use of computer modelling, both for detailed masterplanning and to support the formulation of sustainable urban planning policy and guidance. In this chapter, we have attempted to give a flavour of the progress that has been made in this respect.

Particularly good progress has been made in urban scale building energy modelling, and this has led to innovations in modelling strategies, in particular relating to radiative couplings between buildings, the stochastic behaviours of their occupants and the energetic couplings between buildings and district energy centres. Progress has also been made in the preparation of workflows to improve the rapidity and quality with which the data describing urban scenes is processed, represented and exchanged, likewise in the technologies that can be employed to integrate and orchestrate calls to and the data exchange between complementary models.

Of course, significant further work is required to develop a comprehensive platform for the modelling and optimisation of urban resource flows, even for the spatially stationary case; that is, for a particular annual snapshot in time (i.e. land use does not change). But, of course, cities are not stationary. They are complex systems whose macroscopic emergent structures depend on the actions of the firms and individuals accommodated within them. To understand whether particular policy interventions impacting on social, regulatory, financial and educational stimuli as well as on technological innovations and urban planning interventions are likely to have a positive impact on the ways in which a city develops, our actors' decisions need to be simulated; likewise, the processes accommodated within buildings and the transport of people and goods between them. This is clearly a long-term interdisciplinary endeavour, requiring the collaboration between sociologists, climatologists, transport planners, geographers, engineers, computer scientists, mathematicians and physicists. This work will keep us busy for quite some time to come!

Acknowledgements

Contributions from Dr Andrew Allen, Philipp Heyken, Dr Gavin Long, Dr Julian Rosser, Dr Yong Mao and Dr Sameh Zakhary on land use and transport modelling and on data and computational modelling techniques are gratefully acknowledged.

Notes

1 Note that $\omega = 1 - \sigma - \sigma_{self}$, where σ_{self} represents self-obstructions (the proportion of a patch that is not visible because it is behind the receiving surface).
2 Behaviour at departure is not special in this case, so it is simply treated as an intermediate timestep.

References

Anon. (2014) *TAG Unit M1–1, Principles of Modelling and Forecasting*. Department for Transport, U.K. Government. www.gov.uk/guidance/transport-analysis-guidance-webtag

Awais, M., Palensky, P., Elsheikh, A., Widl, E. and Matthias, S. (2013) The high level architecture RTI as a master to the functional mock-up interface components. In *Proceeding International Conference on Computing, Networking and Communications (ICNC)*, pp. 315–320.

Batty, M. (1997) Cellular automata and urban form: A primer. *Journal of the American Planning Association* 63(2): 266–274.

Batty, M. (2005) *Cities and Complexity: Understanding Cities with Cellular Automata, Agent-Based Models, and Fractals*. Cambridge, MA: The MIT Press.

Beck, A., Long, G., Boyd, D.S., Rosser, J.F., Morley, J., Duffield, R., Sanderson, M. and Robinson, D. (March 2018) Automated Classification Metrics for Energy Modelling of Residential Buildings in the UK with Open Algorithms. *Environment and Planning B: Urban Analytics and City Science*. doi:10.1177/2399808318762436.

Benensen, I., Aronovich, S. and Noam, S. (2005) Let's talk objects: Generic methodology for urban high-resolution simulation. *Computers, Environment and Urban Systems* 29(4): 425–453.

Biljecki, F., Ledoux, H., Stoter, J., Zhao, J.M. (2014) Formalisation of the level of detail in 3D city modelling. *Computers, Environment and Urban Systems* 48: 1–15.

Blochwitz, T., Otter, M., Arnold, M., Bausch, C., Elmqvist, H., Junghanns, A., Mauß, J., Monteiro, M., Neidhold, T., Neumerkel, D. et al. (2011) The functional mockup interface for tool independent exchange of simulation models. In *Proceeding 8th International Modelica Conference Dresden*. Linköping University Electronic Press, pp. 105–114.

Brown, M. and Williams, M. (1998) An urban canopy parameterisation for mesoscale meteorological models. In *Proceedings of the 2nd AMS Urban Environment Symposium*. Albuquerque, NM.

Bruhns, H., Jones, P. and Cohen, R. (2011) *CIBSE Review of Energy Benchmarks For Display Energy Certificates – Analysis of DEC Results to Date*. London: Energy Institute, University College London.

Bueno, B., Norford, L., Hidalgo, J. and Pigeon, G. (2013) The urban weather generator. *Journal of Building Performance Simulation* 6(4): 269–281.

BuildingSMART. (2013) Industry Foundation Classes Release 4, 2013. www.buildingsmart-tech.org/ifc/IFC4/final/html/index.htm.

Castiglione, J., Bradley, M. and Gliebe, J. (2015) *Activity-Based Travel Demand Models: A Primer*. Washington, DC: Transportation Research Board.

Chandy, K.M. and Misra, J. (1979) Distributed simulation: A case study in design and verification of distributed programs. *IEEE Transactions Software Engineering SE-5*(5): 440–452.

Chapman, J., Siebers, P-O. and Robinson, D. (2018a) Multi-agent stochastic simulation of occupants in buildings. *Journal of Building Performance Simulation*. 11(5): 604–621.

Chapman, J., Siebers, P-O. and Robinson, D. (2018b) Data-scare behavioural modelling and the representation of social interactions in building simulation. *Applied Energy* (under review).

CIBSE. (2008) *TM46 Energy Benchmarks – CIBSE*. CIBSE Publications.

Clarke, J.A. (2001) *Energy Simulation in Building Design*. Oxford: Butterworth-Heinemann.

Coutard, O., Finnvedan, G., Kabisch, S., Kichin, R., Matos, R., Nijkampo, P., Pronello, C., Robinson, D. (2014) *Urban Megatrends: Towards a European Research Agenda*. JPI Urban Europe: http://kth.diva-portal.org/smash/record.jsf?pid=diva2%3A739117&dswid=2029.

Davila, C.C., Reinhart, C.F., Bemis, J.L. (2016) Modeling Boston: A workflow for the efficient generation and maintenance of urban building energy models from existing geospatial datasets. *Energy* 117(2016): 237–250.

de la Barra, T. (1989) *Integrated Land Use and Transport Modelling*. Cambridge: Cambridge University Press.

Deng, Y., Cheng, J.C.P. and Anumba, C.J. (2016) Mapping between BIM and 3D GIS in different levels of detail using schema mediation and instance comparison. *Automation in Construction* 67: 1–21.

Echenique, M.H., Anthony, D.J.F., Hunt, D.J., Mayo, T.R., Skidmore, I.J. and Simmonds, D.C. (1990) The MEPLAN models of Bilbao, Leeds and Dortmund. *Transport Reviews* 10(4): 309–322.

Evans, S., Liddiard, R. and Steadman, P. (2017) 3DStock: A new kind of three-dimensional model of the building stock of England and Wales, for use in energy analysis. *Environment and Planning B* 44(2): 227–255.

Gadsden, S., Rylatt, M., Lomas, K. and Robinson, D. (2003) Predicting the urban solar fraction: A methodology for energy advisers and planners based on GIS. *Energy and Buildings* 35(1): 37–48.

Gardner, M. (1970) Mathematical Games – The fantastic combinations of John Conway's new solitaire game "life". *Scientific American* 223: 120–123.

Graves, H., Watkins, R., Westbury, P. and Littlefair, P. (2001) Cooling building in London: Overcoming the heat island. BRE Report BR 431, London: CRC Ltd.

Gröger, G., Kolbe, T.H., Nagel, C. and Häfele, K.-H. (2012) OpenGIS City Geography Markup Language (CityGML) Encoding Standard. Version 2.0.0, Open Geospatial Consortium, OGC Doc. No. 12–019.

Haldi, F. and Robinson, D. (2009) Interactions with window openings by office occupants. *Building and Environment* 44(12): 2378–2395.

Haldi, F. and Robinson, D. (2010) Modelling actions on shading devices. *Building and Environment* (in press).

Haldi, F. and Robinson, D. (2011) The impact of occupants' behaviour on building energy demand. *Journal of Building Performance Simulation* 4(4): 323–338.

Hecht, R., Meinel, G. and Buchroithner, M. (2015) Automatic identification of building types based on topographic databases – a comparison of different data sources. *International Journal of Cartography* 1(1): 18–31.

Horni, A., Nagel, K. and Axhausen, K.W. (eds) (2016) *The Multi-Agent Transport Simulation MATSim*. London: Ubiquity Press.

Hunt, D.R.G. (1980) Predicting artificial lighting use – a method based upon observed patterns of behaviour. *Lighting Research and Technology* 12(1): 7–14.

IEEE. (2010) IEEE standard for modeling and simulation (M&S) high level architecture (HLA) – framework and rules. *IEEE Std 1516–2010* 1–38.

Jones, P., Patterson, J., Lannon, S. and Prasad, D. (1999) An energy and environmental prediction tool for planning sustainable cities. In *Proceedings of the 16th International Conference on Passive and Low Energy Architecture* (PLEA). Brisbane, Australia, pp. 789–794.

Jones, P., Patterson, J. and Lannon, S. (2007) Modelling the built environment at an urban scale – energy and health impacts in relation to housing. *Landscape and Urban Planning* 83(1): 39–49.

Jones, P., Williams, J. and Lannon, S. (1998) An energy and environmental prediction tool for planning sustainability in cities. In *Proceedings of the 2nd European Conference REBUILD*. Florence, Italy.

Kämpf, J. and Robinson, D. (2007) A simplified thermal model to support analysis of urban resource flows. *Energy and Buildings* 39(4): 445–453.

Kikumoto, H., Ooka, R., Arima, Y. and Yamanaka, T. (2014) Study on the future weather data considering the global and local climate change for building energy simulation. *Sustainable Cities and Society* 14(2014): 404–413.

Kikumoto, H., Ooka, R. and Arima, Y. (2016) A study of urban thermal environment in Tokyo in summer of the 2030s under influence of global warming. *Energy and Buildings* 114(2016): 54–61.

Kusaka, H., Kondo, H., Kikegawa, Y. and Kimura, F. (2001) A simple single-layer urban canopy model for atmospheric models: Comparison with multi-layer and SLAB models. *Boundary Layer Meteorology* 101(3): 329–358.

Lannon, S.C., Alexander, D.K. and Jones, P.J. (2007) Housing stock surveys in GIS systems using pattern recognition. In *Proceeding of Geographical Information Science Research UK Conference*. NUI Maynooth, pp. 430–434.

Lefebvre, G., Bransier, J. and Neveu, A. (1987) Simulation of the thermal behaviour of a room by reduced order numerical methods [Simulation du comportement thermique d'un local par des methodes numeriques d'ordre reduit]. *Revue Generale de Thermique* 26(302): 106–114.

Long, G., Alalwany, M. and Robinson, D. (2018) Decarbonisation scenario modelling of city housing stocks by typological sampling. *Energy and Buildings* (in press).

Long, G. and Robinson, D. (2019) A comprehensive workflow for the evaluation of urban decarbonisation scenarios: Modelling, optimisation and multi-criteria decision analysis. *Energy Policy* (in press).

Gargiulo, M., Chiodi, A., De Miglio, R., Simoes, S., Long, G., Pollard, M., Gouveia, J, P. and Giannakidis, G. (2017) An integrated planning framework for the development of sustainable and resilient cities – The case of the InSMART Project. *Procedia Engineering* 198: 444–453, ISSN 1877–7058,

Martilli, A., Clappier, A. and Rotach, M.W. (2002) An urban surfaces exchange parameterisation for mesoscale models. *Boundary Layer Meteorology* 104(2): 261–304.

Masson, V. (2000) A physically-based scheme for the urban energy budget in atmospheric models. *Boundary Layer Meteorology* 94(3): 357–397.

McNally, M.G. (2007) The four-step model. In Hensher, D.A. and Button, K.J. (eds), *Handbook of Transport Modelling* (Handbooks in Transport, Volume 1), pp. 35–53. London: Elsevier.

Miller, E.J., Hunt, J.D., Abraham, J.E. and Salvini, P.A. (2004) Microsimulating urban systems. *Computers, Environment and Urban Systems* 28(1–2): 9–44.

Mortimer, N.D., Elsayed, M.A. and Grant, J.F. (2000) Patterns of energy use in nondomestic buildings. *Environment and Planning B: Planning and Design* 27: 709–720.

NBIM. (2007) *National Building Information Modeling Standard (NBIMS): Overview, Principles and Methodologies, Version 1.0 – Part 1*, Washington, DC: National Institute of Building Sciences.

Nielsen, T.R. (2005) Simple tool to evaluate energy demand and indoor environment in the early stages of building design. *Solar Energy* 78(1): 73–83.

Nouvel, R., Brassel, K.-H., Bruse, M., Duminil, E., Coors, V., Eicker, U. and Robinson, D., (2015a) Simstadt, a new workflow-driven urban energy simulation platform for CITYGML city models. In *Proceeding of CISBAT 2015*. Lausanne: EPFL 2015, pp. 889–894.

Nouvel, R., Mastruccib, A., Leopold, U., Baume, O., Coors, V. and Eicker, U. (20115b) Combining GIS-based statistical and engineering urban heat consumption models: Towards a new framework for multiscale policy support. *Energy and Buildings* 107(2015): 204–212.

Page, J., Robinson, D., Morel, N. and Scartezzini, J.-L. (2007) A generalised stochastic model for the prediction of occupant presence. *Energy and Buildings* 40(2): 83–98.

Palmer, J., Littlefair, P., Watkins, R. and Kolokotroni, M. (2000) Urban heat islands. *Building Services Journal* 2000(5): 55–56.

Perez, D. (2014) A framework to model and simulate the disaggregated energy flows supplying buildings in urban areas. *Thèse EPFL* No. 6102.

Perez, D., Kämpf, J.H., Wilke, U., Papadopoulou, M. and Robinson, D. (2011)*CitySim simulation The Case Study of Alt-Wiedikon, a Neighbourhood of Zürich City*. Lausanne: CISBAT.

Perez, D. and Robinson, D. (2012) Urban energy flow modelling: A data-aware approach. In Muller-Arizona et al. (eds), *Digital Urban Modelling and Simulation*. London: Springer (ISBN 978-3-642-29757-1).

Perez, R., Seals, R. and Michalsky, J. (1993) All-weather model for sky luminance distribution – preliminary configuration and validation. *Solar Energy* 50(3): 235–243.

Pigg, S., Eilers, M. and Reed, J. (1996) Behavioral aspects of lighting and occupancy sensors in private offices: A case study of a university office building. In *Proceedings of the 1996 ACEEE Summer Study on Energy Efficiency in Buildings*, pp. 8161–8171. American Council for an Energy-Efficient Economy.

Putman, S.H. (1983) *Integrated Urban Models*. London: Pion.

Rasheed, A. (2009) Multiscale modelling of urban climate. Unpublished PhD Thesis. EPFL, Lausanne, Switzerland.

Rasheed, A. and Robinson, D. (2009) Multiscale modelling of urban climate. *Proceeding of Building Simulation 2009.*

Rasheed, A., Robinson, D., Clappier, A., Chidambran, N. and Lakehal, D. (2011) Representing complex urban geometries in mesoscale modelling. *International Journal of Climatology* 31(2): 289–301.

Reinhart, C. (2004) Lightswitch-2002: A model for manual and automated control of electric lighting and blinds. *Solar Energy* 77(1): 15–28.

Reinhart, C.F. and Voss, K. (2003) Monitoring manual control of electric lighting and blinds. *Lighting Research and Technology* 35(3): 243–260.

Reinhart, C., Dogan, T., Jakubiec, A., Rakha, T. and Sang, A. (2013) UMI – an urban simulation environment for building energy use, daylighting and walkability. In *Proceeding of Building Simulation.*

Robinson, D. (2011) Computer Modelling for Sustainable Urban Design: Physical Principles, Methods and their Applications. London: Taylor & Francis (ISBN: 9781844076796).

Robinson, D., Bylund, J., Coutard, O., Finnveden, G., Hooimeijer, P., Kabisch, S., Matos, R., Noll, M., Ponds, R., Pronello, C., Riegler, J. (2015) The strategic research and innovation agenda of JPI Urban Europe. *JPI Urban Europe*: https://jpi-urbaneurope.eu/app/uploads/2016/05/JPI-Urban-Europe-SRIA-Strategic-Research-and-Innovation-Agenda.pdf

Robinson, D., Campbell, N., Gaiser, W., Kabel, K., Le-Mouele, A., Morel, N., Page, J., Stankovic, S. and Stone, A. (2007) SUNtool – a new modelling paradigm for simulating and optimising urban sustainability. *Solar Energy* 81(9): 1196–1211.

Robinson, D., Haldi, F., Kämpf, J., Leroux, P., Perez, D., Rasheed, A. and Wilke, U. (2009) CitySim – comprehensive micro-simulation of resource flows for sustainable urban planning. In *Proceedings of Building Simulation 2009.* Glasgow, Scotland, pp. 1083–1090.

Robinson, D., Stankovic, S., Morel, N., Deque, F., Rylatt, M., Kabele, K., Manolakaki, E. and Nieminen, J. (2003) Integrated resource flow modelling of urban neighbourhoods: Project SUNtool. In *Proceedings of Building Simulation 2003.* Eindhoven, The Netherlands, pp. 1117–1122.

Robinson, D. and Stone, A. (2004) Solar radiation modelling in the urban context. *Solar Energy* 3(77): 295–309.

Robinson, D. and Stone, A. (2005a) A simplified radiosity algorithm for general urban radiation exchange. *Building Services Engineering Research and Technology* 26(4): 271–284.

Robinson, D. and Stone, A. (2005b) Holistic radiation modelling with a fast simplified radiosity algorithm. In *Proceedings of Building Simulation 2005.* Montreal, Canada.

Robinson, D. and Stone, A. (2006) Internal illumination prediction based on a simplified radiosity algorithm. *Solar Energy* 80(3): 260–267.

Rosser, J.F, Long, G., Boyd, D., Zakhary, S., Mao, Y. and Robinson, D. (2018) Predicting the construction period of residential buildings using map data. *Computers Environment and Urban Systems* 73(2019): 56–67. https://doi.org/10.1016/j.compenvurbsys.2018.08.004.

Rylatt, M., Gadsden, S. and Lomas, K. (2001) GIS-based decision support for solar energy planning in urban environments. *Computers, Environment and Urban Systems* 25(6): 579–603.

Rylatt, R.M., Gadsden, S.J. and Lomas, K.J. (2003) Methods of predicting urban domestic energy demand with reduced datasets: A review and a new GIS-based approach. *Building Services Engineering Research and Technology* 24(2): 93–110.

Sachs, J.D. (2008) *Common Wealth: Economics for a Crowded Planet.* New York: Penguin.

Santamouris, M. (1998) The Athens urban climate experiment. In *Proceedings of the 15th International Conference on Passive and Low Energy Architecture (PLEA).* Lisbon, Portugal.

Shimoda, Y., Fujii, T., Morikawa, T. and Mizuno, M. (2003) Development of residential energy end-use simulation model at city scale In *Proceedings of. Building Simulation 2003.* Eindhoven, The Netherlands. pp. 1201–1208.

Sicklinger, S., Belsky, V., Engelmann, B., Elmqvist, H., Olsson, H., Wüchner, R. and Bletzinger, K.-U. Interface Jacobian-based co-simulation. *International Journal for Numerical Methods in Engineering* 98(6): 418–444.

Sousa, G., Mirzaei, P., Jones, B. and Robinson, D. (2017) A review and critique of UK housing stock energy models, modelling approaches, and data sources. *Energy and Buildings* 151: 66–80.

Sousa, G., Mirzaei, P., Jones, B. and Robinson, D. (2018) An open-source simulation platform to support the decarbonisation of the UK housing stock. *Energy and Buildings* (in press).

Swaid, H. and Hoffman, M.E. (1998) The prediction of impervious ground surface temperature by the Surface Thermal Time Constant (STTC) model. *Energy and Buildings* 13(2): 149–157.

Swaid, H. and Hoffman, M.E. (1990) Prediction of urban air temperature variations using the analytical CTTC model. *Energy and Buildings* 14: 313–324.

Taniguchi, A., Shimoda, Y., Asahi, T., Yamaguchi, Y. and Mizuno, M. (2007) Effectiveness of energy conservation measures in residential sector of Japanese buildings. In *Proceedings of Building Simulation 2007*. Beijing, China, pp. 1645–1652.

Tooke, T.R., Coops, N.C. and Webster, J. (2014) Predicting building ages from LiDAR data with random forests for building energy modeling. *Energy and Buildings* 68(PARTA), 603–610. https://doi.org/10.1016/j.enbuild.2013.10.004.

Torrens, P.M. and O'Sullivan, D. (2001) Cellular automata and urban simulation: Where do we go from here?, *Environment and Planning B* 28(2): 163–168.

Tregenza, P. and Sharples, S. (1993) Daylighting algorithms. ETSU S 1350–1993, UK.

United Nations. (2012) *World Urbanization Prospects, the 2011 Revision: Highlights*. Department of Economic and Social affairs, Population Division. New York: United Nations.

Waddell, P. and Ulfarsson, G.F. (2004) Introduction to urban simulation: Design and development of operational models. In Haynes, K., Stopher, P., Button, K. and Hensher, D. (eds), *Handbook 5: Transport Geography and Spatial Systems*. Oxford, UK: Pergamon Press.

Wang, K., Siebers, P.-O. and Robinson, D. (2017) Towards generalized co-simulation of urban energy systems. *Procedia Engineering* 198(2017): 366–374.

Wegener, M. (1994) Operational urban models. *Journal of the American Planning Association* 60(1): 17–29.

White, R. and Engelen, G. (1993) Cellular automata and fractal urban form: A cellular modelling approach to the evolution of land use patterns. *Environment and Planning A* 25(8): 1175–1199.

White, R. and Engelen, G. (2000) High-resolution integrated modelling of the spatial dynamics of urban and regional systems. *Computers, Environment and Urban Systems* 24(5): 383–400.

Wilson, A.G. (1970) *Entropy in Urban and Regional Modelling*. London: Pion.

Wolfram, S. (1994) Cellular Automata as models of complexity. *Nature* (311): 419–424.

Wolfram, S. (2002) *A New Kind of Science*. Wolfram Media.

Yamaguchi, Y., Shimoda, Y. and Mizuno, M. (2003) Development of district energy simulation model based on detailed energy demand model. In *Proceedings of Building Simulation 2003*. Eindhoven, The Netherlands, pp. 1443–1450.

Yamaguchi, Y., Shimoda, Y. and Mizuno, M. (2007a) Transition to a sustainable urban energy system from a long-term perspective: Case study in a Japanese business district. *Energy and Buildings* 39(1): 1–12.

Yamaguchi, Y., Shimoda, Y. and Mizuno, M. (2007b) Proposal of a modelling approach considering urban form for evaluation of city level energy management. *Energy and Buildings* 39(5): 580–592.

Recommended reading

Muller-Arizona et al. (eds) (July 2012) *Digital Urban Modelling and Simulation*. London: Springer (ISBN 978-3-642-29757-1).

Reinhart, C.F. and Davila, C.C. (2016) Urban building energy modeling – A review of a nascent field. *Building and Environment* 97: 196–202.

Robinson, D. (2011) *Computer Modelling for Sustainable Urban Design: Physical Principles, Methods and their Applications*. London: Taylor & Francis (ISBN: 9781844076796).

Activities and solution strategy

The most challenging task in urban building energy modelling lies with the design and implementation of workflows for (i) acquiring and sanitising data describing the three-dimensional structure of buildings and their semantic attributes, (ii) the preparation of these models in the format required by the corresponding simulation software, (iii) the calling of this software and the subsequent export, interpretation and visualisation of results and (iv) repeating the above steps to test hypotheses for the improvement of the buildings' performance. To more fully appreciate what is involved, we suggest that you complete the following activities:

- Choose a residential building on a street in the town or city that you live in.
- Review the availability of existing geospatial (e.g. cadastral and/or digital elevation model) datasets. Utilising this data, or your own measured survey data if this does not exist or is unavailable to you, prepare a suitably abstracted three-dimensional model of your chosen building, in your preferred building simulation software.
- Review the availability of existing datasets describing the semantic attributes required of your building simulation software (envelope construction properties; leakage/infiltration characteristics; HVAC systems; occupancy, lights and appliances). From these datasets and/or utilising your own survey observations, assign the necessary attributes to the three-dimensional model that you previously prepared in your software.
- Acquire the weather data needed by your software, enter the simulation parameters required for an annual simulation at say 5-minute timesteps and execute the simulation.
- Interrogate the simulation results to determine the total annual energy use for heating, hot water, lighting, appliances, ventilating and cooling. Informed by appropriate benchmark statistics for the type of building, set an annual energy use target that you would like to reach. Test strategies for reducing the building's annual energy use until you have reached this target.
- Prepare a flowchart describing a workflow that would enable you to scale up your analysis to model the entire housing stock of your town or say a district of your city (and to verify the plausibility of the corresponding predictions). Identify the datasets that would be required or that would need to be augmented to support this workflow, as well as the computational technologies that would be required, bearing in mind that you would like this workflow to be as automated as possible. Likewise, consider the computational hardware that would be required for the simulations to be executed in a reasonable amount of time.
- Roughly how many resources and which skillsets would be required to fully implement your workflow? Is it viable? Can your workflow be simplified to improve its viability?

21 Urban building energy modeling

Christoph Reinhart and Carlos Cerezo Davila

Scope

Over the past decades, detailed individual building energy models (BEM) on the one side and regional and country-level building stock models on the other side have become established modes of analysis for building designers and energy policymakers, respectively. More recently, these two toolsets have begun to merge into hybrid methods that are meant to analyze the energy performance of neighborhoods, i.e. several dozens to thousands of buildings. This chapter reviews emerging simulation methods and implementation workflows for such bottom-up urban building energy models (UBEM). Simulation input organization, thermal model generation and execution, as well as result validation, are discussed successively, followed by select application case studies and an outlook for future developments.

Learning objectives

This chapter is concerned with physics-based simulation methods to predict the thermal performance of an ensemble of buildings in an urban context. The reader will learn about emerging approaches, how to thermally characterize multiple buildings based on new and existing datasets as well as how to efficiently run a simulation and interpret results.

Key words

Urban building energy modeling/archetype definitions/sustainable cities

Introduction

The World Bank estimates that in 2016 the number of city-dwellers worldwide grew at a net rate of 2.039%, or about 2.9 million per week (The World Bank 2017). If this unprecedented urban growth continues to be largely ad hoc via informal settlements, sprawl and haphazard densification, global and local consequences for the environment, the economy and the quality of life of billions could be severe. In 2015, the United Nations formulated 17 interrelated sustainable development goals for their *Agenda 2030* to "transform our world" (United Nations General Assembly 2015). The agenda was adopted by all 193 countries of the UN General Assembly. Three of these goals provide specific actions related to buildings and their energy support infrastructure, namely to enhance energy efficiency and to increase the share of renewables in the energy supply mix,[1] to provide access to safe and affordable housing[2] and to reduce carbon emissions.[3] Such policy measures at the international and national levels provide a larger framework for individuals and institutions to operate in. However, at those higher levels, these goals are more aspirational than practical, as the actual implementation of building-related infrastructure upgrades happens locally at the municipal, neighborhood or individual building levels.

Fortunately, many city governments and institutions worldwide have responded to the aforementioned call for action by presenting ambitious greenhouse gas (GHG) emission reduction targets such as 60% by 2025 (London), 36% by 2030 (Singapore) and 80% by 2050 (New York City and Boston) (Blasio 2012; Mayor of London 2011; Walsh 2014; Singapore 2015). Other goals have already been realized by a series of universities and corporations from Harvard University (30% reduction in 2016) to Google (100% renewable in 2017) (Harvard University 2016; Hölzle 2017). GHG emission reduction targets nowadays command substantial political willpower, especially in regions that have already suffered from increased hurricane activity, drought and/or prolonged summer heat waves. While the significance of GHG emissions from transportation and industrial activities varies among cities, building-related emissions are always a key contributor. Fortunately, a plethora of measures exist to significantly reduce the energy use of virtually any building from energy audits and retrofits to the accelerated deployment of renewable energy, district heating and cooling, as well as combined heat and power systems. To prioritize, implement or incentivize such a wide range of measures not only requires political consensus and innovative financing mechanisms, but also an intimate technical understanding of how a city's energy ecosystem may change over time as a result of these interventions. For example, when the use of solar panels, home battery storage and electric cars rises throughout a city, electricity load curves will change along with the economics of auxiliary power plants, creating financial uncertainty for utilities operating those systems. Also, even with all of these technology measures to consider, some jurisdictions will likely experience an increase in energy use if more residents will be able afford air-conditioning systems and rely on them to stay healthy and comfortable during prolonged heat waves (Davis and Gertler 2015).

To manage these diverse and interrelated phenomena, city governments, utilities and owners of large real estate require planning tools that provide spatially and temporally defined energy demand data for their building portfolio and that facilitate the evaluation of "what if" scenarios. To address this need, a new genre of bottom-up building performance simulation-based methods has been developed in recent years that aims to predict energy use, load curves and indoor environmental conditions for several dozen to many thousand existing or new buildings. Given that these methods use – at their core – the same heat and mass flow models as the individual building energy models (BEM), discussed in Chapter 2, they have been coined *urban building energy models* (UBEM) (Reinhart and Cerezo Davila 2016). UBEMs are "bottom-up" engineering models that are distinctly different from both "top-down" building stock and bottom-up statistical models (Figure 21.1). The former link the building energy demand for a region with macroeconomic variables, such as population trends and economic activity, and increasingly subdivide the existing stock into smaller subsections (Swan and Ugursal 2009; Kavgic et al. 2010). Top-down models can provide estimates of what would happen if more buildings of a certain type were to be built or converted into another type (Howard et al. 2012). However, top-down models necessarily extrapolate from the status quo and are hence less suitable when more integrated energy supply-demand scenarios are being investigated or when an analysis focuses on a specific neighborhood.

At this meta-scale, bottom-up statistical or bottom-up physics-based (UBEM) models can be employed. Both model types represent each building individually. Statistical methods link building-level descriptors such as vintage, usage type and occupancy to measured building energy use via regression models (J Z Kolter and Ferreira 2011). They are "robust" since they are based on measured building energy data and thus able to accurately incorporate occupant behavior, which is notoriously difficult to mimic for UBEMs (Wilke et al. 2013). On the other

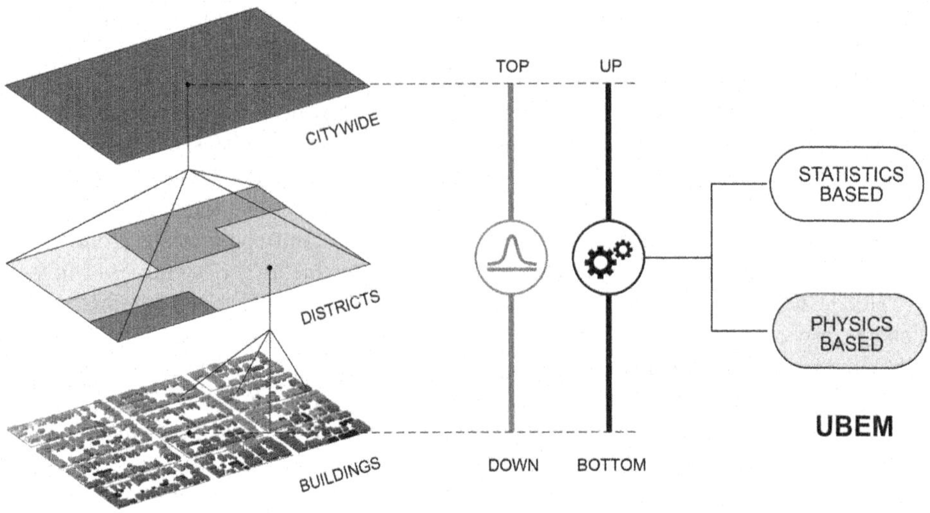

Figure 21.1 Top-down versus bottom-up statistical and physics-based (UBEM) urban simulation approaches

hand, statistical models that are based on annual metered data are unable to predict energy use in monthly or hourly time steps or to simulate the combined impact of several energy efficiency measures in buildings. UBEMs are therefore expected to establish themselves as key planning tools to answer specific questions such as how much energy could be saved if the oldest buildings in a neighborhood were to be weatherized or how electricity load curves would change if centrally controlled, programmable thermostats were distributed throughout a neighborhood. For new developments and/or urban densification projects, UBEMs further offer sophisticated urban design advice from street grid layout to load balancing between different building types. Setting up a UBEM requires considerable effort, because the same amount of simulation input as for an individual BEM is – in principal – needed for each building. Since UBEMs may contain thousands of buildings, conventional BEM approaches become unacceptably resource intensive. A reconceptualization and automation of conventional BEM workflows is required instead.

A critical reader, who is aware of the substantial time commitment and expert knowledge required to calibrate a BEM for an individual building – from adjusting schedules to fine-tuning construction standards, may be skeptical whether repeating the same process for many buildings is feasible or whether the resulting models are doomed to be unfit to base design decisions on. As will be shown throughout this chapter, those concerns are valid and – as is always the case when using a model – an UBEM user should be well aware of the capabilities and limitations of a model to answer specific questions. However, this chapter also presents an argument that in the age of large public urban datasets, new powerful simulation workflows can be created to answer some of the questions laid out above. Indeed, the authors would argue that there is a pressing societal need to expand the reach of building performance simulation methods to the urban realm: Translating a net worldwide growth of 2.9 million city dwellers per week to the Boston context would require the design, modeling and construction of some 400,000 buildings per week. To put this number in perspective, in 2016 there were close to 80,000 LEED-certified buildings worldwide (Shutters 2016). To become globally relevant and affordable, BEM needs to expand its scope to the urban realm.

Another reader with a different outlook may suggest that in order to model a city, it suffices to "simply" build a few representative BEMs for the main building archetypes in that city, including residential, commercial and retail, and then estimate citywide energy use by multiplying the energy use intensity (EUI)[4] of each building type with the total floor area of that programmatic type across the city. Such a most basic UBEM relies on the fact that building program and size are usually the two main factors determining the energy use of a building. Indeed, as will be shown below, if the desired use of an UBEM is to evaluate the impact of upgrades on all buildings of a given programmatic use, this approach can be justifiable. However, if an UBEM is to be used to make building-by-building-level decisions, differentiators beyond program have to be considered as well, including architectural massing, window sizes, microclimatic effects, building age, upkeep and occupant behavior. The impact of all of these effects will be addressed below.

The chapter is organized as follows. The "Computational models" section reviews the main elements required to build and execute an UBEM. This process requires the reconceptualization and automation of certain building energy modeling workflows as well as the validation of UBEM predictions against measured energy use. The "Applications" section offers a series of UBEM use cases. While the chapter focuses on operational building energy use of neighborhoods based on building-by-building simulations, the same approach has also been applied to other urban performance criteria such as embodied energy use (Cerezo Davila and Reinhart 2013), daylight availability (Dogan et al. 2012), walkability (Rakha and Reinhart 2013) and even cash flow modeling (Turan et al. 2017). Perez and Robinson called this larger field of exploration, which UBEM is a part of, "urban micro-simulation" (Perez and Robinson 2012; Robinson 2012). While the UBEM methods discussed in this chapter focus on synthesizing building load profiles, an example of a complementary model to design matching building energy supply systems is provided in the "Applications" section.

At this point, it is worthwhile noting that UBEM is a young but rapidly evolving field and that this chapter outlines the basics steps and principals used by most UBEMs without claiming to exhaustively cover all ongoing projects and initiatives. The example UBEMs shown throughout the chapter were generated using the urban modeling interface (UMI) plugin for Rhinoceros 3D (Reinhart et al. 2013).

Computational models

The task of creating an UBEM of a new or existing neighborhood can be broken down into the following subtasks: simulation input organization (data input), thermal model generation and execution (thermal modeling) as well as result verification (validation).

Data input

As shown in Figure 21.2, an UBEM requires the combination of several datasets – some identical and some abstractions of those for a BEM – including urban microclimate data, simplified building massings as well as "non-geometric" building/occupant parameters. All three input groups are described in the following.

To give the reader a sense of an UBEM project for new construction, this chapter presents a series of input datasets and simulation results for a proposal for a mixed-use development near Boston's Innovation District that was developed as a class project at MIT (Brown et al. 2017) (Figure 21.3). The project occupies a site area of 168,213 m^2 and includes program for 226,65 m^2 residential, 60,540 m^2 office and 59,2130 m^2 retail spaces, which add up to a floor area ratio (FAR) of 3.1 excluding roads.

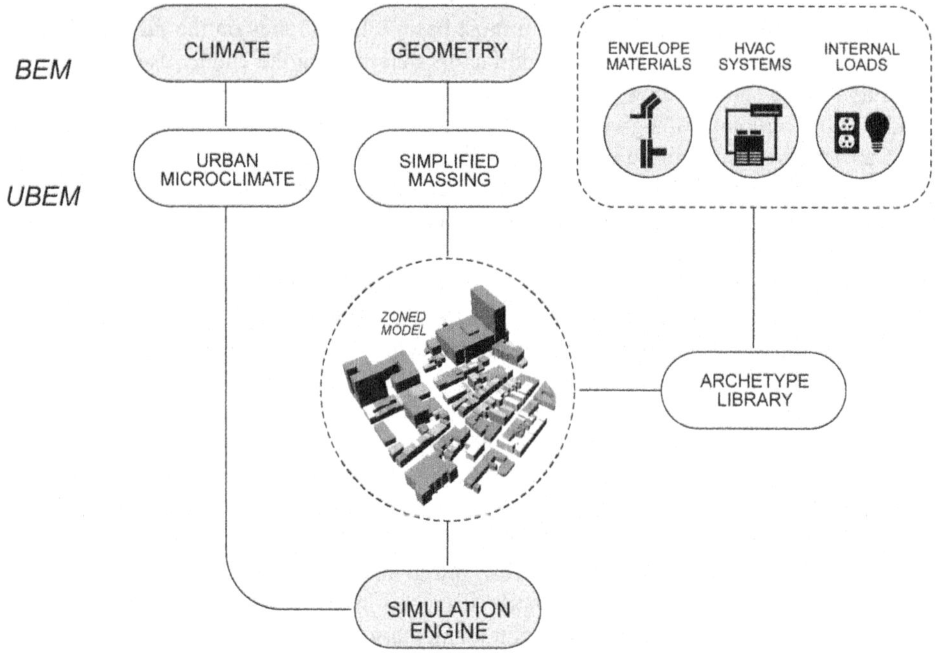

Figure 21.2 Data input required for a UBEM versus a BEM

Figure 21.3 Neighborhood proposal in Boston

Source: Brown et al. 2017

Climate data

As discussed in Chapter 6, climate datasets for building performance simulation have been available for a number of years following the initial establishment of a viable data format, the typical meteorological year (TMY) (Hall et al. 1978; Crawley et al. 1999), and the subsequent provision of data available in this format for multiple regions worldwide (US-DOE 2017). Apart from improving the worldwide coverage of these datasets, researchers have recently been exploring methods of how to model local microclimatic phenomena within cities, such as the urban heat island effect (Bueno et al. 2012). For the City of London, Mavrogianni et al. coupled locally measured temperature profiles with an UBEM in order to resolve the effect of the urban heat island effect on building energy use and resident health (Mavrogianni et al. 2009). Predicting local wind patterns (Liu et al. 2015) and linking Intergovernmental Panel on Climate Change (IPCC) climate change predictions to current day TMYs (Jentsch et al. 2013) are equally active areas of research with direct implications for UBEM. Figure 21.4 shows the impact of those effects on simulated annual energy use for the neighborhood from Figure 21.3 organized by building type. Three microclimatic conditions are considered, an unperturbed *Rural site* near Boston, a central site in *Urban Boston* that experiences urban heat island effects and the same site in 2080 (*Boston 2080*) according to the IPCC's HadCM3 scenario (Jentsch et al. 2013). Due to warmer temperatures at nighttime, heating loads are reduced by 9% as one moves from rural to urban while cooling loads increase by 2%. The trend continues as the climate heats up, leading to additional reductions in heating and increases in cooling of 43% and 67%, respectively, by 2080. Figure 21.4 clearly shows the dramatic changes in building load curves that utilities may expect going forward.

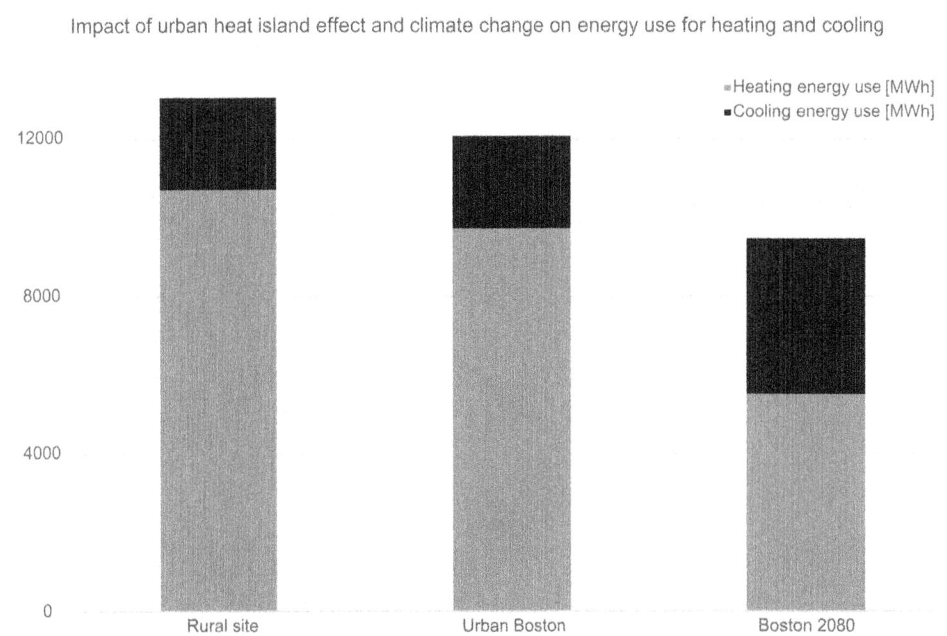

Figure 21.4 Effect of urban heat island effect and climate change on heating and cooling energy use for the neighborhood from Figure 21.3

Geometry

The minimum geometric input data required by an UBEM consists of three-dimensional building envelope models and window opening ratios, as well as terrain data. In case a building houses multiple programmatic uses, separate massing blocks are preferable. A common way of representing "water-tight" enclosures is through boundary representations or "B-reps." Depending on whether a new or existing neighborhood is the subject of investigation, geometric information can either be extracted from existing datasets or generated from scratch. For example, Figure 21.5 shows the urban massing model for the neighborhood from Figure 21.3 with overlaid programmatic uses. This type of massing model is used by urban planning teams from the earliest planning stages onwards and serves for communicating design ideas to clients or within a team. Massing models may also be used to evaluate water runoff, shading studies and access to daylight. Using design environments such as Grasshopper for Rhinoceros 3D (McNeel 2010b, 2010a), massing models can be set up parametrically, which in turn means that derived UBEMs can also be explored dynamically to – for example – determine ideal street grid width and orientation.

For existing neighborhoods, a variety of urban datasets is available to create massing models of buildings and landscape. Over the past decades, citywide geographic information systems (GIS) databases have not only become commonplace in many regions of the world but are also increasingly accessible to the general public, especially in the US. Larger cities such as Boston typically employ a dedicated GIS department that centrally maintains a digital dataset of all city assets. Because of this institutional presence, city employees are familiar with GIS-related

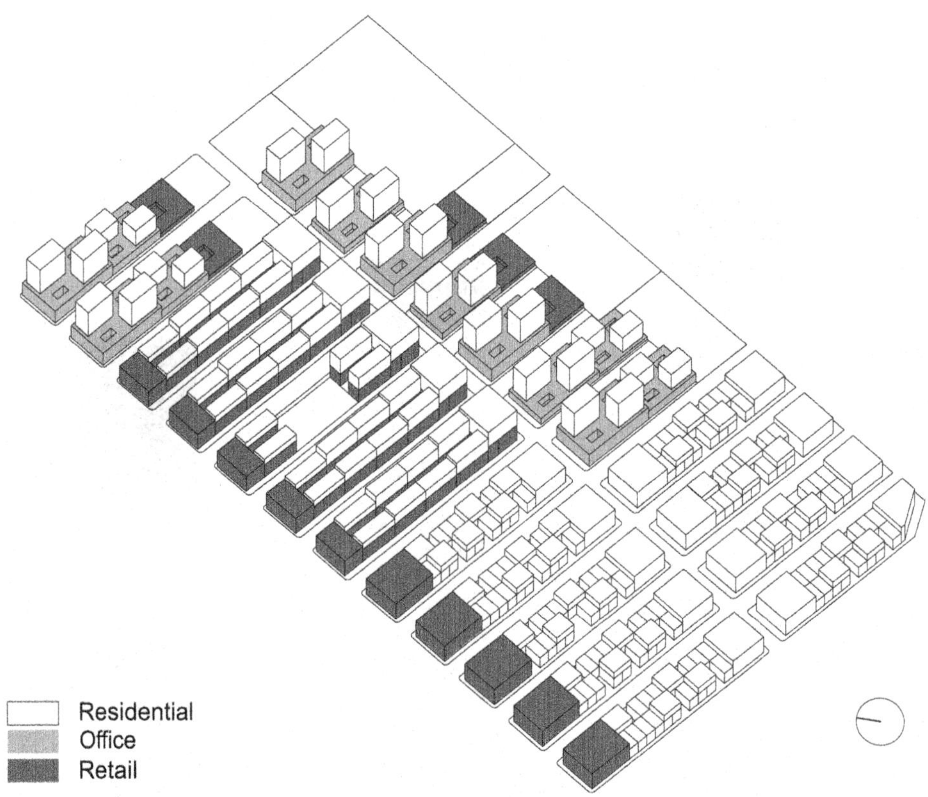

Figure 21.5 Urban massing model with program of the Boston neighborhood from Figure 21.3

tools, and new plans for urban development tend to be designed in compatible environments. Typical GIS entry fields are parcel and building footprints, as well as building height, age and programmatic use. Footprints and heights can be combined into so-called 2.5-D massing models.

As shown in Figure 21.6, this process is less straightforward than it might initially sound. A key quality measure for an urban massing model used for UBEM is that each building accurately represents the real structure's volume, i.e. the amount of conditioned space. It is equally important that adjacencies between neighboring structures are accurately modeled, as the size of the external envelope area directly scales with the amount of heat exchange with the outside. It is therefore critical to know whether two structures touch in a townhouse-type fashion or are separated by a narrow aisle. Since GIS footprint polygons are usually automatically extracted from satellite or flyover imagery, they tend to include a large variety of kinks, curves and details. Simplifying these complex shapes before extruding them poses a challenge to many thermal simulation engines, which is why polygon simplification techniques need to be applied to GIS floorplan shape files. These techniques reduce the number of points or segments in the original footprint without changing the built area or perturbing adjacencies between neighboring structures (Figure 21.6) (Tian et al. 2015). This type of geometric manipulation can be equally labor intensive and tedious.

It should further be noted that variables such as building heights are oftentimes not consistently defined even within the same GIS model and may range from the top of a gable to the mean building height or the base of a roof (Figure 21.6). Fortunately, in order to avoid gross errors, extruded massing models can often be crosschecked against complementary urban data. In the US, these are "tax assessment" databases that include building floor area, number of bedrooms, assessment value and the dates of construction and major renovations. Given that they determine local property taxes, these datasets tend to be up to date and may further help to determine non-geometric properties of a building. Another practical challenge when dealing with city-specific datasets is that many entry fields are non-standardized and require a detailed understanding of their meaning along with some creative thinking about how to treat missing entries (Figure 21.6). Fortunately, this mapping is a onetime task, so cities can rely on an expert consultant to perform this step. Example mappings for Boston, London and San Francisco are described in Cerezo Davila et al. (2016), Chen et al. (2017) and Coffey et al. (2015).

Yet another geometric urban data source may come from LiDAR data. LiDAR stands for "light detection and ranging" and is a remote sensing technique in which a grid of points with height information is determined via laser from flyovers. LiDAR datasets are available for

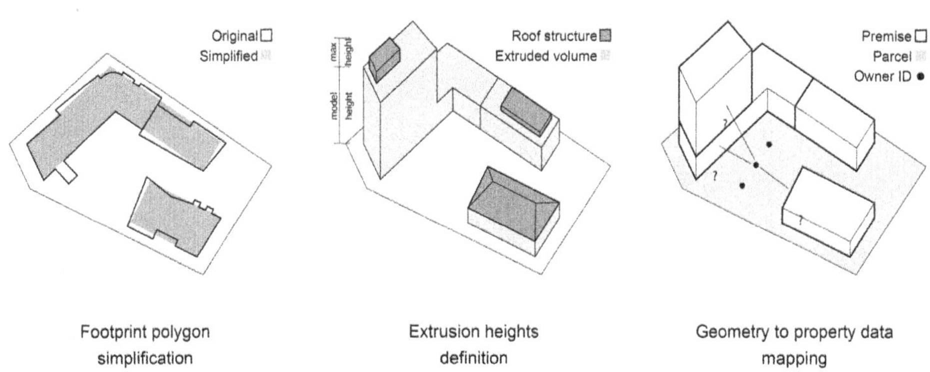

Figure 21.6 Steps required to convert building footprints and heights from a GIS file into a 2.5-D UBEM massing model

many parts of the world both free of charge and commercially. While LiDAR data provide the highest geometric fidelity with a median grid point spacing of 0.77 m and height uncertainties of less than 0.1 m,[5] the resulting geometry has to be further processed to separate buildings from trees and/or to link building massings to the earlier mentioned other urban datasets.

Apart from building form, window areas are also a key geometric UBEM input. While this information is not routinely collected by municipalities, creating digital representations of façades based on urban datasets is a well-established field (Brenner 2005). Some algorithms are based on images such as photographs taken at street level or via satellite, whereas others use point clouds collected either by terrestrial laser scanning (TLS) or from an airplane (LiDAR). Cao introduced an automated procedure that converts low resolution aerial urban photographs into geometry models for BEM (Figure 21.7) (Cao et al. 2017). At the urban level, such approaches can be combined with the earlier discussed massing models to yield more complete individual building representations.

In order to accommodate the range of geometric detail that can be stored for an individual building, from its exterior to interior walls and windows, open semantic formats such as CityGML have been developed and widely adopted in different parts of the world, especially Europe (Open Geospatial Consortium 2012; Eicker et al. 2014; Kaden and Kolbe 2013). Figure 21.8 shows

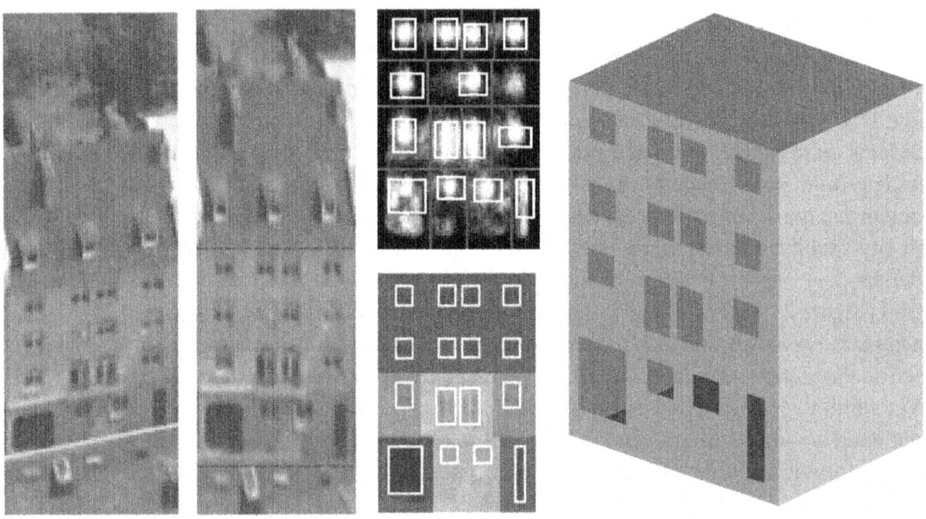

Figure 21.7 Image sequence that shows the segmentation a façade image and conversion into a BEM model

Source: Cao et al. 2017

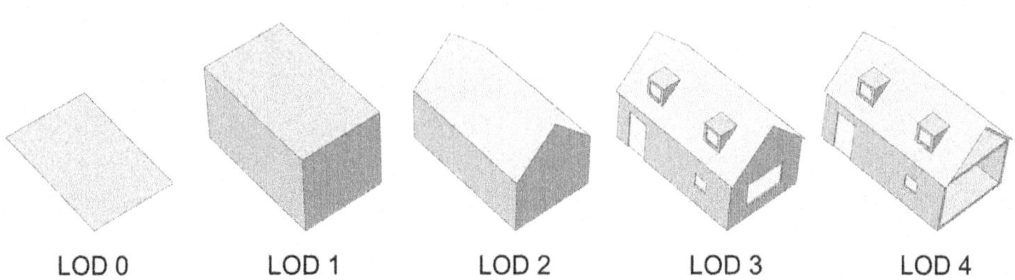

Figure 21.8 Building models in CityGML based on varying levels of detail (LOD)

varying levels of detail (LOD) in which building geometry can be stored in CityGML 2.0 from extruded 2.5-D models (LOD1) to detailed building envelopes and wall/roof material specifications (LOD2), doors and window positions (LOD3) and even interior features (LOD4) (Biljecki et al. 2016). As of 2018, most CityGML datasets were at LOD1 or LOD2.

Non-geometric building information

As shown in Figure 21.2, in addition to the outer shell and climate, non-geometric building properties have to be defined as well, including construction assemblies, programmatic use and HVAC systems. At the individual building level, this step routinely takes about a third of the modeling effort (Cerezo Davila et al. 2014) and constitutes one of the main sources of discrepancies between simulated and measured energy use due to uncertainty regarding infiltration rates, equipment loads and occupant behavior (ASHRAE 2014). As mentioned above, while these quantities can be measured for a small group of existing buildings, such detailed data collection efforts become impractical for larger urban areas. It is therefore customary to abstract a given building stock into "building archetypes," i.e. groups of buildings with similar properties. The archetype approach has also been used in the context of national or regional bottom-up building stock models to understand the aggregated impact of energy efficiency policies (Firth and Lomas 2009) or new technologies (Dall'o' et al. 2012). The generation of archetypes requires two steps called *segmentation* and *characterization*. In segmentation, the investigated building stock is divided into archetype groups (Figure 21.9). During characterization, complete sets of thermal properties have to be defined for each archetype. Details on both steps are provided in the following.

While the actual division of a building stock into archetypes is obviously of paramount importance for the reliability of the resulting UBEM, the process typically remains ad hoc, relying on somewhat plausible (and available) building properties such as use type, year built, construction typc and HVAC system (Ballarini et al. 2014; Mavrogianni et al. 2009; Mastrucci et al. 2014; Dascalaki et al. 2011; Filogamo et al. 2014; Fracastoro and Serraino 2011). An ideal way to conduct this step would be through regression analysis plus clustering using measured energy use of a large number of buildings plus a range of building properties. Unfortunately, access to individual building energy is usually restricted due to privacy laws so that such an analysis cannot be conducted. In cases where such datasets have been (at least partially) available, interesting patterns emerged. For example, Aksoezen et al. showed that older buildings

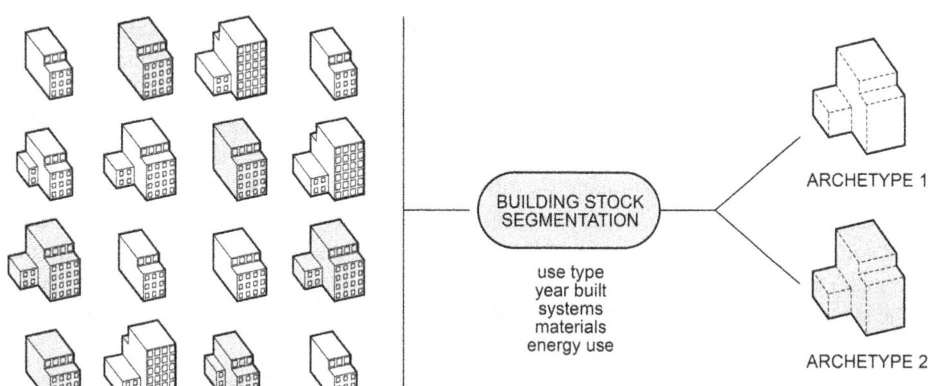

Figure 21.9 Segmentation of an ensemble of buildings into two groups represented by archetype buildings 1 and 2

can use less energy than one might expect due to their lower quality thermal properties because of modified, more energy-conscious occupant behavior (Aksoezen et al. 2015). Sokol et al. showed that the year of last renovation was a better predictor of building energy use for residential buildings in Cambridge, Massachusetts, than year of construction, even though the city did not know whether a renovation was energy-related (Sokol et al. 2017). Before, Kolter and Ferreira had built a regression model based on the same Cambridge dataset that explained 75% of the variance in measured monthly energy use for 6,500 buildings (J. Zico; Kolter and Ferreira 2011). Famuyibo et al. reduced the number of required archetypes from 81 to 13 by clustering actual individual building energy use (Famuyibo et al. 2012). It should be noted that being able to meaningfully reduce the number of archetypes is extremely valuable, since the characterization step is usually the most time-consuming and ambiguous part of an UBEM setup.

The reader may wonder how to conduct the segmentation step if individual building energy use is not available for a representative sample of a building stock under investigation. Since building constructions and systems change locally and over time – and are typically not individually documented – their definition in an archetype library requires intimate knowledge of the local building stock and its evolution. In a best case scenario, a country's building stock has already been broken into archetypes and characterized through some national research effort, a prominent example being the European TABULA project for which residential building typologies were developed for 13 European countries (Loga et al. 2012). In the absence of such building stock studies, UBEM archetypes should be classified and compiled in consultation with local experts, building codes and other relevant datasets. This remains a somewhat open-ended process as – in addition to constructions and systems – an archetype library also requires data inputs regarding occupant behavior, which has long been known to have a decisive impact on building energy use (Robinson et al. 2009). In the case of new construction, defining building archetypes is less ambiguous as the design team has control over the desired construction standard. In that case a modeler may start with building energy standards such as ASHRAE 90.1 and define potential energy upgrades in additional to those standards (ASHRAE 2013). Future resident behavior may be even more challenging to model than for existing buildings, as demographics can only be anticipated.

How many buildings should typically be represented in an archetype group? A review of previous urban modeling studies yielded that most studies used from several dozen up to around 150 archetypes to model between hundreds to tens of thousands of buildings (Reinhart and Cerezo Davila 2016). Given the effort level required to characterize each archetype, the latter number seems to constitute somewhat of an upper effort bound. For national building stock models, an individual archetype can thus represent up to 500,000 buildings. For neighborhood-level studies with several thousand buildings, the number of archetypes should depend on the specific questions investigated, i.e. if occupant behavior is key, the modeler should define several versions of the same building type but with a variety of occupant behavior patterns. If potential envelope upgrades are being explored for a given neighborhood, a finer distinction of envelope properties between different buildings can be helpful. In the latter case, the modeler may further rely on energy audit data to estimate quantities such as infiltration rates for buildings of different age and type, etc. An archetype building can then either be characterized as a *virtual building* or a *sample building*. In the former case, statistical data from several buildings along with expert input is used to set the archetpe's thermal properties (Ballarini et al. 2014). For the sample building approach, the modeler selects a representative, actual building which can then be characterized via a building audit.

A fundamental conflict between the concept of building archetypes – separated by programmatic use and building age – as well as actual urban land use is that cities assign programmatic use by parcel whereas many buildings are often mixed use, accommodating retail at the bottom as well as offices and/or apartments on higher floors. In such cases, many cities only keep the main use of a building in the GIS database. For UBEM purposes, it would be beneficial if cities started referencing spatial units with individual IDs that go beyond parcels and include buildings and floors as well as organizational groupings such as "premises" that may be part of a building or extend across several buildings. Evans et al. developed an example model for the British building stock what was compiled from existing national datasets (Evans et al. 2016).

Once a set of archetypes has somehow been determined, all non-geometric building properties required for a thermal model can be stored in an archetype "template." This can be achieved by expanding established building information modeling (BIM) formats such as gbXML (gbXML 2018), by adapting a GIS database format, or by using a custom data format (Cerezo Davila et al. 2014; Robinson et al. 2009).

Thermal modeling

Once climate data, building massing models and archetype templates are available, they need to be combined into a thermal model, which then needs to be executed and the results communicated back to the user in an intelligible format. Previously published UBEM workflows differ in the type and detail of thermal models used as well as whether the effect of surrounding buildings is taken into account. A number of these workflows are described below, going from low to increasingly higher complexity.

In the simplest case, an UBEM consists of single-zone, steady-state heat balance models of each building archetype. Simulation results for the archetypes are scaled up to the ensemble level by multiplying them with either the number of buildings per archetype (Firth and Lomas 2009) or a floor area-weighted function of that number (Dall'o' et al. 2012). As mentioned in the introduction, this modeling approach ignores that the urban context of a building can significantly affect its performance, e.g. through shading, local wind patterns, etc. To consider shading as well as building compactness, tools such as SIMSTADT/INSEL apply a single-zone steady-state model to each building separately, considering its actual urban surrounding (Strzalka et al. 2011). While steady-state methods are generally known to reliably predict heating loads, dynamic thermal simulation engines such as EnergyPlus, DOE2, TRNSYS and IDA-ICE are preferable for locations with notable cooling needs. Mata and Caputo accordingly used contextless single-zone dynamic models to analyze archetypes in France, Germany, Italy, Spain and the UK (Mata et al. 2014; Caputo et al. 2013). For investigations of detailed urban design choices including mixed-use buildings, multi-zone dynamic thermal models may become necessary. In practice, this requires converting a massing model into a network of volumetric thermal zones as shown in Figure 21.5. Same as for single-zone models, multi-zone models can either be generated for archetype buildings only (Heiple and Sailor 2008; Ascione et al. 2013; Mastrucci et al. 2014) or for each building individually so that solar shading can be considered as well. While building a multi-zone model for select archetypes is still feasible manually, this process has to be automated if applied to all buildings. For simple, rectangular buildings, this is a relatively straightforward geometric operation. For arbitrary building forms, Dogan et al. developed an "Autozoner" algorithm which automatically generates ASHRAE 90.1 Appendix G compliant multi-zone energy models from B-reps (Dogan et al. 2016).

While simple steady-state simulation models for several thousand buildings can be executed in a matter of an hour on a standard laptop, the simulation time for thousands of dynamic multi-zone models may take days. Fortunately, the process can be fully parallelized and thus sped up using cloud computing. Less resource-intensive approaches were recently proposed, such as envelope simplifications and model order reduction in Modelica (Kim et al. 2014), aggregating internal zones into thermal mass elements (Koene et al. 2014), clustering programmatic spaces in a neighborhood into shoebox models (Dogan and Reinhart 2017) and others (Lauster et al. 2014). To provide the reader with a general sense of the involved computational effort of different thermal modeling approaches, Figure 21.10 compares aggregate annual energy use for the neighborhood from Figure 21.3 based on several methods. As explained above, the *Autozoner* approach fully zones each building into ASHRAE 90.1 compliant perimeter and core zones and serves as the "reference case" for the other methods. The *Shoeboxer* abstracts all buildings as a linear supposition of sidelit two-zone perimeter core models. The *5 zone block* model creates for each building type a square BEM building with four perimeter and a core zone and then scales simulation results with the size of their program floor area. It thus ignores building geometry. *1 zone per floor* creates individual zones for each floor and building and thus takes as long to compute as the *Autozoner* variant, since solar shading masks have to be calculated by EnergyPlus for each zone separately. All four methods have identical lighting, equipment and domestic hot water (DHW) users since these only depend on floor area (no dimming controls). Aggregate heating and cooling loads are very close for the three approaches that include core zones. Only the *1 zone per floor* method has appreciably lower condition loads since solar gains, conduction losses and equipment gains average out across floor areas. The *5 zone block* model has similar results as the *Autozoner* for aggregate results but cannot reliably predict energy use for individual buildings, especially those that are envelope dominated.

Finally, in dense urban settings, being able to model local wind speeds and longwave radiation exchange between buildings in addition to direct shading can become relevant to quantify the impact of urban microclimate on building energy use and/or occupant health. This is further discussed in Chapter 22.

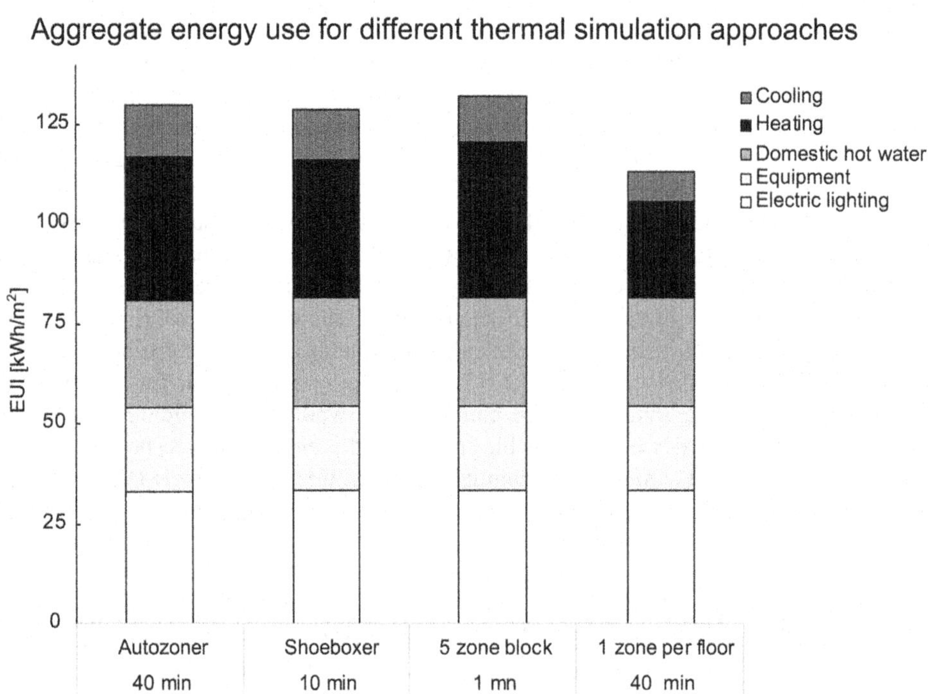

Figure 21.10 UBEM simulation results for the neighborhood from Figure 21.3 for different thermal modeling approaches (simulations based on UMI)

Validation

The ability of an UBEM to support different design or policy decisions naturally depends on how reliable the simulation results are. At this point, the reader should remember that the ultimate purpose of an UBEM is not just to model the status quo but to predict what would happen if energy upgrades were implemented in select or all buildings in a neighborhood. Given that even individual BEM predictions may significantly differ from measured results due to uncertainties such as infiltration rates and occupant behavior, it may initially seem unlikely that an UBEM will be capable of faithfully predicting the energy use of many buildings. However, it turns out that when comparing aggregated measured versus simulated annual energy use of multiple buildings, simulation errors at the individual building level tend to average each other out, leading to UBEM errors of 7% to 21% for heating loads and 1% to 19% for EUI predictions (Figure 21.11) (Reinhart and Cerezo Davila 2016). The authors found that for a residential neighborhood in Kuwait – when the same type of envelope and equipment/lighting upgrades were applied to all buildings – similar averaging effects took place in both the original as well as the upgraded UBEM (Cerezo Davila et al. 2017). In other words, predicted aggregate energy savings across a neighborhood are likely to lie in a similar range as aggregate energy use predictions of current conditions. Understanding the overall saving potential of a technology is

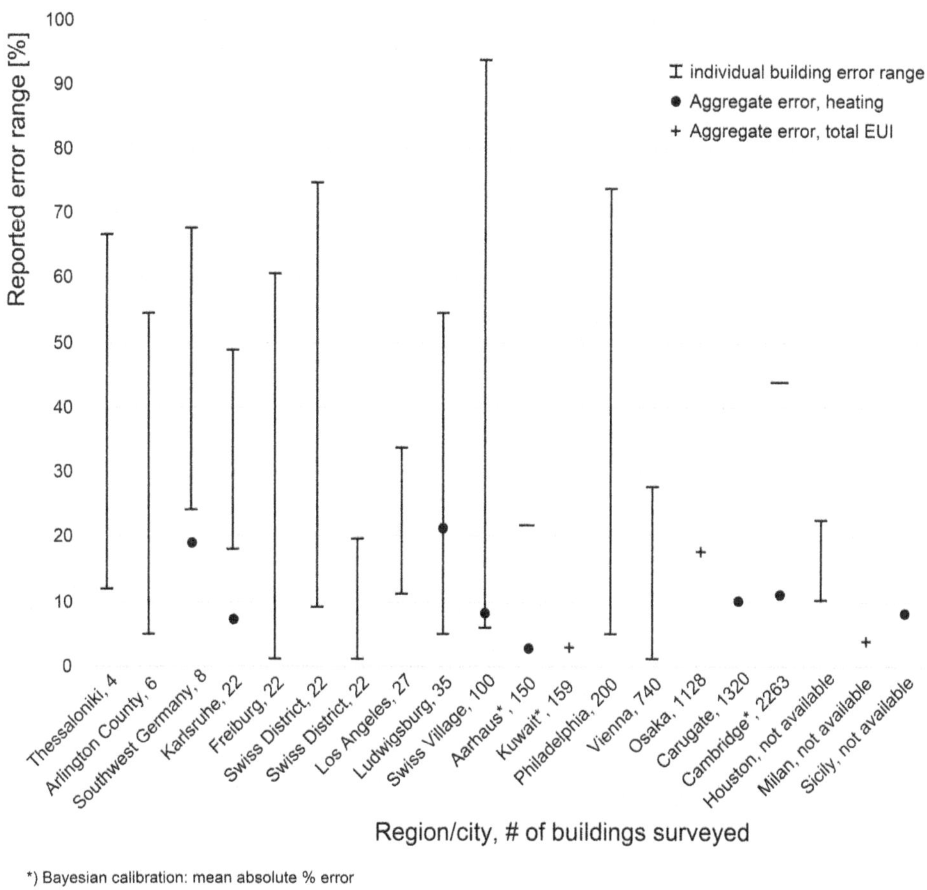

Figure 21.11 Reported aggregate and individual building simulation error ranges from a collection of UBEM studies

valuable for municipalities or campus operators who are exploring different upgrade options for their portfolio.

However, low simulation errors at the aggregate level do not translate into low errors at the building level. Indeed, multiple studies have reported that the simulation accuracy of an UBEM significantly decreases if annual energy uses are compared for individual buildings with errors ranging from 5% to as high as 99% (Figure 21.11). An owner of an individual building is naturally only interested in the energy savings from a given technology for his or her building. To help an owner to better understand the investment risk, an UBEM needs to report a range of expected savings for each building. To accomplish this, the UBEM should first be able to reproduce the full spread of energy use found in buildings throughout a neighborhood. This range can be surprisingly large. For example, Figure 21.12 (top left) shows the measured EUI distribution of a residential neighborhood in Kuwait City that consists of 164 single-family homes. While the mean EUI across the neighborhood is 209 kWh/m^2, individual EUIs range from 80 to 460 kWh/m^2.

While building owners are interested in *annual* savings, utilities need to also understand potential load curve changes at the substation level at hourly or even sub-hourly time steps. Given how difficult it is to estimate hourly equipment use, it is not surprising that Heiple and Sailor (2008) reported differences between measured and simulated load curves of up to 40%. In summary, depending on the questions asked, UBEMs need to be calibrated to varying degrees, although the ultimate goal is to find calibrating techniques that lead to

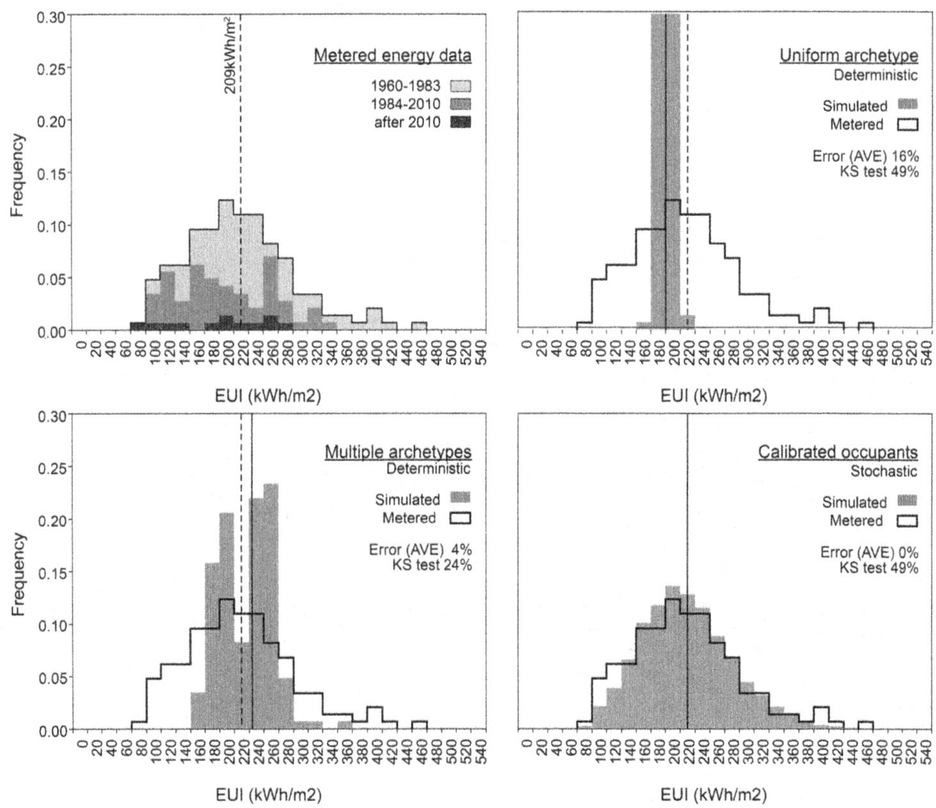

Figure 21.12 Comparison between metered and simulated EUI distributions for a residential neighborhood in Kuwait City

reliable predictions of the range of energy use across a neighborhood at hourly simulation time steps.

To work towards the first goal, reliable annual EUI distributions, the authors proposed a two-step calibration process that consists of initially classifying the buildings by construction type followed by a Bayesian calibration to mimic resident behavior (Cerezo et al. 2017). To illustrate the process, Figure 21.12 (top right and bottom row) compares UBEM results for the aforementioned Kuwaiti neighborhood for varying levels of calibration. For the *Uniform archetype* variant, a single construction type and fixed occupant behavior are assumed, i.e. all residents exhibit the same behavior at all times and live in equally constructed buildings of varying shape. The simulated mean EUI is 16% lower than the measured value and the Kolmogorov–Smirnov (KS) test yields that the two EUI distributions differ by 49%. The *Multiple archetypes* variant uses four different construction types as well as actual window-to-wall-ratios (WWRs), but occupant behavior remains deterministically defined. For this case, the mean EUI error is reduced to 4% and two peaks for older and newer buildings emerge (KS test = 24%). Finally, the *Calibrated occupants* variant considers construction types and varying occupant behavior based on a Bayesian calibration.

Bayesian calibration requires a modeler to have access to measured data for a subset of sample buildings either in the UBEM or of similar buildings close by. The modeler then needs to select simulation parameters, which are difficult to estimate and have a high impact on simulation results. This selection is ideally conducted through a sensitivity analysis of archetype models for a given building stock. Uncertainty ranges for each parameter then have to be defined. For the Kuwaiti example, the selected parameters were cooling set point temperature (STP), peak installed lighting power density (LPD), peak average occupant density (OCC) and an equipment load multiplier (MLT) (Cerezo et al. 2017). The next step is to train the model by running multiple simulations of the sample buildings for different combinations of input parameters and to "keep" combinations that yield sufficiently small simulation errors. This process leads to probability distributions for these input parameters (Figure 21.13, left). Using these probability distributions, each building in an UBEM can be modeled multiple

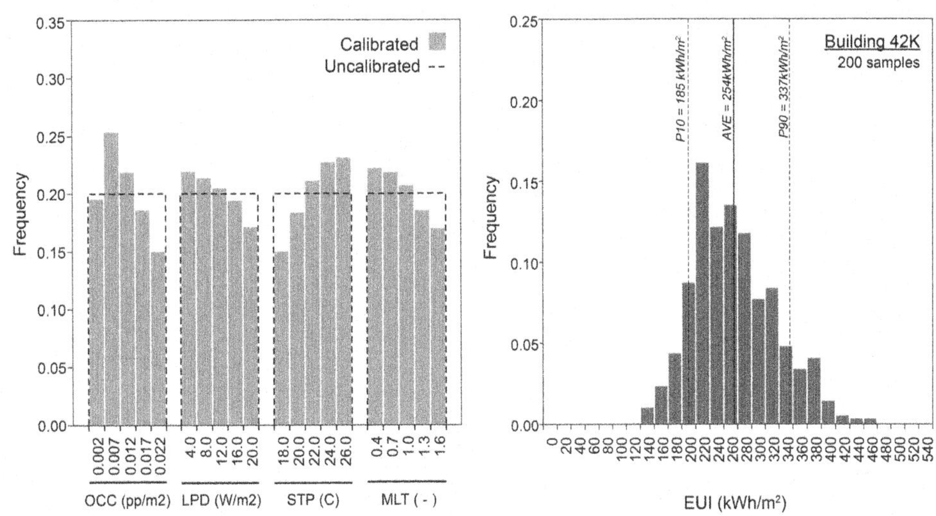

Figure 21.13 (Left) Probability distributions for occupancy (OCC), peak lighting power density (LPD), set point temperature (STP) and equipment multiplication factors (MLT) for the neighborhood from Figure 21.12. (Right) EUI distribution for building 42K based on 200 sample simulations.

times by stochastically drawing parameters from these distributions. Figure 21.13 (right) shows the resulting EUI distribution for one of the buildings (42K) from Figure 12.12. The predicted energy use of the building ranges from 185 to 337 kWh/m². Overlaying all of these individual building EUI distributions leads to the *Calibrated occupants* variant in Figure 21.12, which has the same mean EUI as the measured data with a KS test of 4%. Visual inspection confirms that the two EUI distributions are very similar.

The above laid out two-step Bayesian calibration approach can also be applied to non-occupant-related uncertainty parameters. A study of 2,263 residential buildings in Cambridge, Massachusetts, led to comparable results (Sokol et al. 2017) and further showed that – as one would expect – the time interval of the data to which an UBEM is calibrated corresponds to the time interval at which the UBEM will yield reliable results. In other words, if one calibrates an UBEM using annual energy data, one should not expect to be able to correctly model monthly energy data. On the flipside, the same Bayesian approach can also be used to calibrate to monthly data, if available.

Unfortunately, the above laid out Bayesian calibration process is not directly transferable to hourly load curve predictions, since hourly schedules have too many degrees of freedom. To avoid unrealistically strong peak loads resulting from identical hourly schedules, He et al. proposed to use stochastic behavioral models to model aggregate load curves more realistically (He et al. 2015). Figure 21.14 accordingly compares heating load curves for 125 residential buildings in the UK using fixed heating schedules on weekdays and weekends versus occupancy-based schedules using a Markov model derived from time of use surveys. The stochastic model exhibits significantly fewer peaks than the deterministic model.

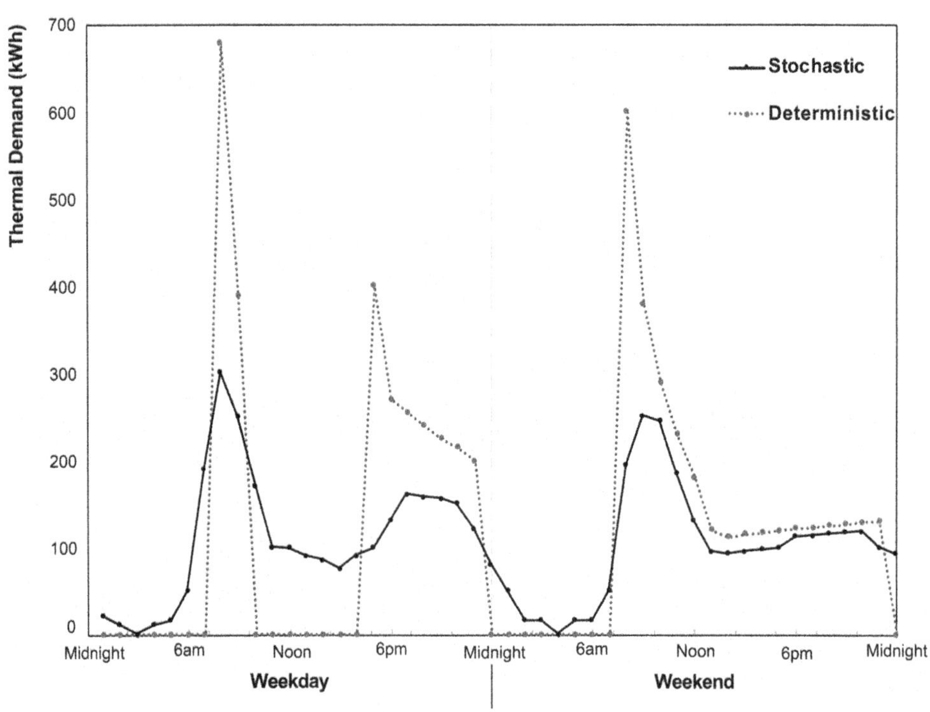

Figure 21.14 Comparison of simulated heat load curves for 125 residential buildings using deterministic schedules and a Markov model

Source: He et al. 2015

Apart from the earlier described archetype calibration methods, several research groups have focused on automatic BEM calibration techniques, which mostly rely on an optimization function to reduce the difference between measured and simulated data by iteratively adjusting the values of unknown parameters. Parameter selection routines range from brute-force sampling to machine learning and graphical pattern recognition (Robertson et al. 2013; Sanyal et al. 2014; Sun et al. 2016). The aim of this work is to replace a labor-intensive trial-and-error approach, which requires experienced auditors and modelers, with calculation power. While the computational expense per building is currently quite large, more optimized routines combined with smart meter data may over time lead to calibrated citywide UBEMs that are based on individual BEMs.

Applications

This section provides a series of case study applications for UBEM ranging from load management and integrated supply/demand side analysis to energy policy evaluations and multi-faceted urban planning. As will be shown, these different scenarios require a variety of ways to visualize and communicate massive amounts of energy data to stakeholders – including non-experts – as actionable information.

Load management

According to its Greenovate Climate Action Plan, the City of Boston aims to reduce its GHG emission by 25% by 2020 and 80% by 2050 (Walsh 2014). With Boston's commercial and residential sectors accounting for 75% of the city's total emissions, it is evident that the city needs to focus on buildings and their energy supply infrastructure to achieve these goals. In 2015, the authors collaborated with the Boston Redevelopment Authority to develop a citywide UBEM comprised of over 80,000 buildings divided into 52 archetypes (Cerezo Davila et al. 2016). Model output was hourly load data for electricity and gas for all buildings. In absence of measured individual building energy data for Boston, EUI distributions for each archetype could only be compared to measured energy use data in the 2003 Commercial Building Energy Consumption Survey (CBECS) (Michaels 2003). To demonstrate the potential value of the UBEM, the authors concentrated their analysis on three neighborhoods around the Back Bay area in West Boston. Figure 21.15 (left) shows that commercial high-rises dominate the energy use of the neighborhood. Figure 21.15 (right, dark gray area) shows predicted electricity peak loads on the hottest day of the year in the Boston weather file (July 7). Demand is highest in the late afternoon when residential and commercial buildings rely heavily on air conditioning. The light gray area shows how the load curve would change if half of the roof area was covered with photovoltaic modules (152 MW_{peak}). While the overall daily peak would be reduced by around 15%, the new load curve would have a midday dip followed by a steep load increase after 2pm. This shape is called a "duck curve" and results from a mismatch between the times when PV supply and building demand peak. Duck curves lead to well-documented technical challenges for the grid operator (CAISO 2016). Figure 21.15 (white area) shows that the effect could be alleviated if all thermostats in the commercial buildings were centrally controlled and temperature set points temporarily relaxed by up to 4 K during the hours from 5pm to 7pm. This demand response strategy effectively uses the thermal mass of commercial buildings as heat storage and would reduce the daily electricity peak by another 15% compared to current conditions. This type of analysis can only be conducted using an UBEM. However, it should be noted that this analysis relies on the UBEM to provide reliable hourly predictions which – as discussed in the previous section – is not necessarily true.

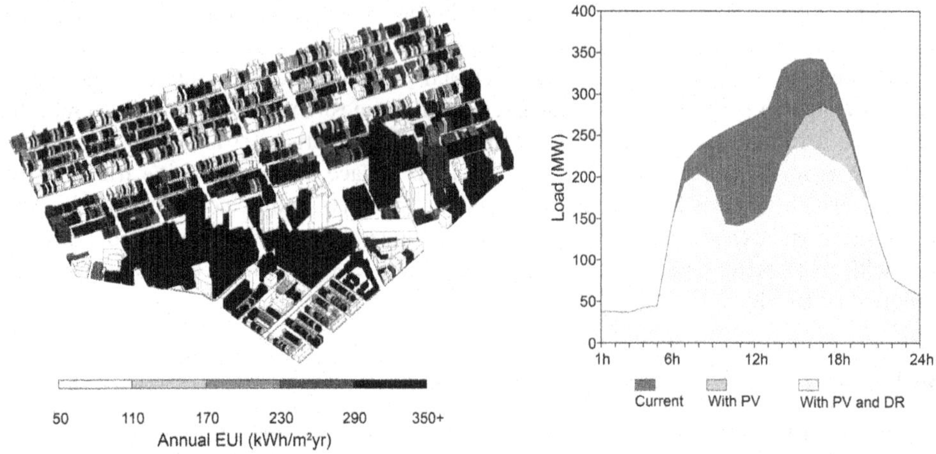

Figure 21.15 UBEM of Back Bay Boston along with aggregated electricity load curves on July 7, including widespread PV deployment (PV) combined with a thermostat-based demand reduction strategy (PV + DR)

Integrated supply/demand side management

The previous example showed how an UBEM could be used to estimate peak load reductions in existing neighborhoods. In the case of new developments, where buildings and their energy supply systems still have to be designed and constructed, simulated building load curves can be combined with energy supply models ranging from building-level systems to micro grids. Different approaches to model micro-cogeneration systems are reviewed in Chapter 16. To demonstrate the approach, Figure 21.17 compares carbon emissions and operational costs for three energy supply scenarios for the neighborhood proposal from Figure 21.3 based on a district energy system plugin for UMI (Letellier-Duchesne et al. 2018). The *Baseline* scenario assumes gas-fired building-by-building furnaces for heating and domestic hot water (DHW) and grid-supplied electricity for everything else. *All Electric* assumes that heating and DHW needs are met with heat pumps powered by the grid. Finally, *All Gas* assumes that a centralized combined heat and power (CHP) plant with a gas-fired backup boiler are in place to meet all heating and electricity demands for the site. For the example analysis shown in Figure 21.16, *All Gas* has the lowest operational energy costs with assumed gas and electricity prices of 0.04 US$/kWh and 0.19 US$/kWh, respectively. In contrast, annual carbon emissions are lowest for *All Electric* because in 2018 close to 50% of electricity was being generated by renewable energy sources and nuclear power plants, which translated into associated GHG emissions of 0.257 kgCO_2e/kWh for electricity and 0.180 kgCO_2e/kWh for natural gas (US-EPA 2016). Within a carbon pricing scheme, the cost of switching from *All Gas* to *All Electric* would lead to an equivalent carbon price of 58 US$/ton CO_2e. In comparison, the California Cap and Trade Program price for carbon in January 2018 was listed as 15 US$/ton$CO_2$e (Climate Policy Initiative 2018). This type of integrated cost/carbon analysis can be complemented with sophisticated financing schemas, allowing developers to understand the long-term financial and environmental viability of a given project.

Energy efficiency policies for building retrofits

Energy efficiency incentive programs have been around since the 1970s (Geller et al. 2006). They have the goal of providing focused financial support for building owners to retrofit a

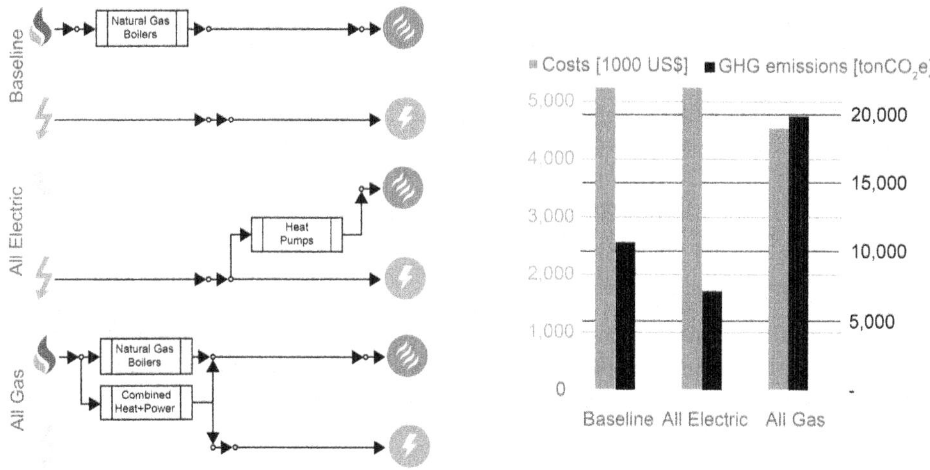

Figure 21.16 Comparison of different energy supply scenarios for the neighborhood proposal from Figure 21.3

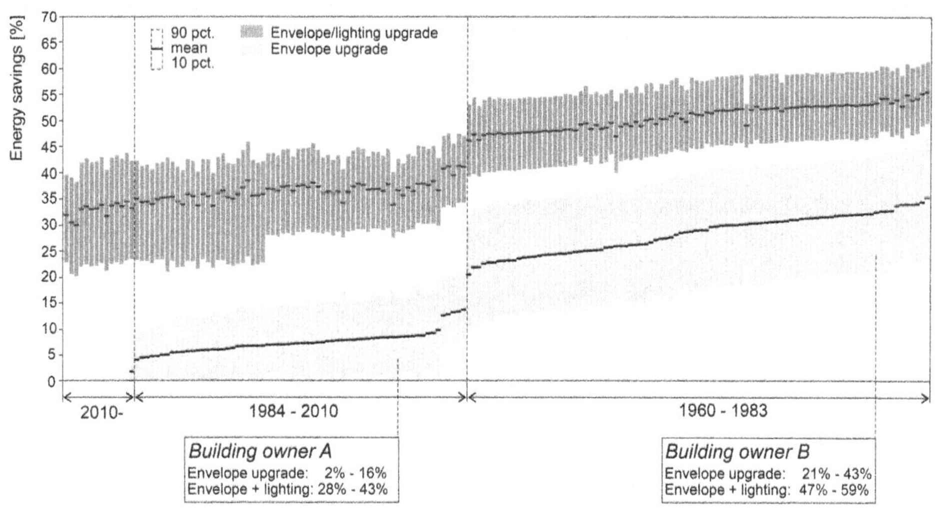

Figure 21.17 Uncertainty analysis of energy savings for envelope and lighting upgrades for all buildings from Figure 21.12 as well as a two select buildings

building in order to reduce regional GHG emissions and operational costs. Unfortunately, the uptake of these programs has been historically low for a number of reasons, from uninformed owners to high upfront costs as well as uncertainty regarding expected savings for different upgrade options. UBEMs have the potential to reduce this uncertainty. As an example, Figure 21.17 shows the uncertainty ranges for energy savings due to different retrofitting strategies for individual buildings in the Kuwaiti neighborhood from Figure 21.12 (Cerezo Davila et al. 2017). Buildings are ordered by construction period. By examining the overall graph for all buildings, a municipality or local utility can develop impactful incentive programs. In the example neighborhood, envelope upgrades should only be incentivized for buildings constructed before 1984. For newer buildings, the risk of not capturing any savings is too large. In contrast, upgrades to lighting (difference between the two incentive curves) lead to reliable savings for all building types.

Figure 21.17 also shows these results from the perspective of two owners, A and B. Owner A is likely to only consider lighting upgrades, whereas owner B might consider envelope plus lighting improvements with expected overall savings lying between 47% and 59%.

Multi-dimensional urban analysis

As mentioned in the introduction, urban massing models can also be combined with other environmental performance evaluations, allowing decision makers to develop a multi-dimensional view of a neighborhood proposal that goes beyond resource efficiency and also considers occupant comfort and well-being. As an initial placeholder metric for the latter, one may use the daylit area of all buildings in a neighborhood according to spatial daylight autonomy (Chapter 7). Figure 21.18 shows a plot of individual building EUI versus daylit area for the neighborhood proposal from Figure 21.3. The residential buildings, being generally located on top of retail spaces and offices, exhibit a large range in daylight availability from fully daylit to none. Due to the consistent construction standard assumed throughout the neighborhood (*Uniform archetype* approach), all residential EUIs only vary from 110 kWh/m² to 120 kWh/m². Deep plan retail spaces tend to have worse daylight and higher energy use due to higher lighting and equipment loads. A handful of retail spaces with a central atrium have the highest EUIs due to increased heating loads. Figure 21.18 offers insight into the tradeoffs urban planners make when designing mixed-use neighborhoods.

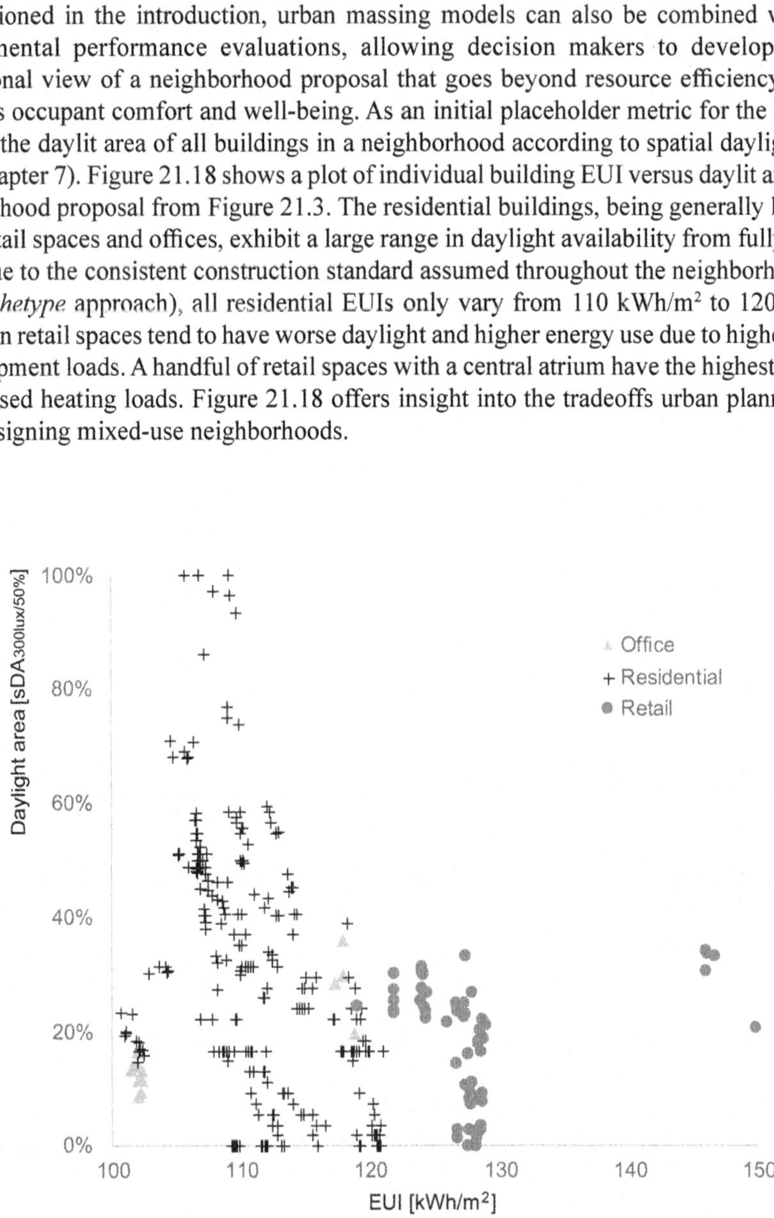

Figure 21.18 Integrated daylighting and energy analysis of the neighborhood from Figure 21.3

Outlook

The preceding sections have shown that significant progress has been made towards the development of simulation workflows to estimate overall operational building energy use of neighborhoods. There is mounting evidence that – once properly calibrated – the current generation of UBEMs is capable of truthfully modeling the spread of energy use across different buildings in a neighborhood as well as predicting potential savings from implementing energy upgrades to these buildings. Ongoing efforts aim to further predict energy use at hourly time steps. Given the valuable insight that urban planners, utilities and municipalities can already gain from such simulations, the required effort level to set up, calibrate and run such models seems justifiable. However, for UBEM to distinguish itself as a reliable, widely used urban analysis method, several challenges remain related to data access, generating high-resolution building load profiles and educating urban stakeholders.

Compiling high-quality template libraries for building archetypes is costly, is time consuming and requires local building expertise, as well as access to detailed audits and/or metered energy use. The question is, who should compile such libraries and how should those individuals be granted access to the underlying datasets? Cities will oftentimes lack the capacity to independently develop local template libraries, so it seems incumbent on national and international organizations to commission the development of such libraries and to orchestrate their maintenance and dissemination using a common data format. In the US, this could, for example, be an extension of the US-DOE Commercial Reference Buildings (Deru et al. 2011). While data privacy of individual building energy use remains a widespread concern, the earlier laid out Bayesian calibration process culminates in anonymized probability distributions that can be freely disseminated (Figure 21.13). More locally, initiatives such as the Building Energy Reporting and Disclosure Ordinance (BERDO) in Boston, which requires large- and medium-sized buildings to publicly report their annual energy and water use, are complementary steps that improve access to building energy data and should lead to more accountable facility management practices (The City of Boston 2018). Another type of information that is surprisingly absent in most city databases is related to building construction as well as the overall upkeep of individual buildings. While a wealth of related data is routinely collected during building audits or property sales, the authors are not aware of any city that systematically captures and links this information to existing databases. Another untapped opportunity for municipalities to document their building stock is through building and construction permits. In European Union member countries, such a process already happens through Energy Efficiency Certificates that tend to be nationally collected and can be geo-referenced (EU Directorate-General for Energy 2016).

As mentioned above, in order to provide reliable information for utilities, UBEMs should provide hourly load predictions down to the substation level. To get to this point, different data-harvesting technologies need to be developed and tested for a variety of urban typologies. One possible approach is to cluster hourly smart meter profiles for a large number of households and abstract the data through a series of "Eigen users" that make up the demographic fabric of a neighborhood (Xu et al. 2017). Another, exciting opportunity for calibrating UBEMs and employing them for continuously commissioning is to link them to urban sensor networks from weather station measurements to occupancy patterns from Wi-Fi data. In the age of the internet of things and smart sensing, such efforts could make UBEM an integral component of digital city or institutional campus models that municipalities and owners maintain to optimize the use of their assets.

Finally, in order for UBEMs to have the larger societal impact that was proposed in the introduction, stronger intellectual engagement between planners, policymakers, utility representatives and the building modeling community is necessary. For an UBEM user, it is important to realize that – while the underlying physical concepts of BEM and UBEM are very similar – the contexts in which these methods are applied are very different. Cities are complicated organizations, and consequential urban changes (should) undergo a lengthy consultation process that involves many stakeholders. To be effective, the information coming from an UBEM therefore needs to be constantly adapted to the knowledge levels of the audience. For this to happen, it will require training a new generation of individuals with adequate domain knowledge in all of the above areas.

Synopsis

The work presented in this chapter suggests that the effort level to set up and maintain citywide UBEMs can be streamlined to the point at which a city can maintain and query these models going forward. Remaining technical barriers include a need to further automate data conversion and mapping procedures, as well as to establish initiatives for collecting and sharing documentation of the building stock and its performance.

Acknowledgments

We thank Paige Xiomara Alvarez and Shreshth Nagpal for help with the literature review, figures and some of the simulations in the manuscript. The writing of this chapter has been supported by the Center for Complex Engineering Systems at MIT and KACST as well as Exelon Corporation.

Notes

1 Goal 7, Affordable and clean energy.
2 Goal 11, Sustainable cities and communities.
3 Goal 13, Climate action.
4 Site energy use divided by building floor area.
5 Based on ~1,800 topographic LiDAR surveys in the 2018 NOAA interagency elevation inventory.

References

Aksoezen, M., Daniel, M., Hassler, U. and Kohler, N. (2015) Building age as an indicator for energy consumption. *Energy and Buildings* 87: 74–86. https://doi.org/10.1016/j.enbuild.2014.10.074.

Ascione, F., De Masi, R.F. de Rossi, F., Fistola, R., Sasso, M. and Vanoli, G.P. (2013) Analysis and diagnosis of the energy performance of buildings and districts: Methodology, validation and development of urban energy maps. *Cities* 35: 270–283. https://doi.org/10.1016/j.cities.2013.04.012.

ASHRAE. (2013) *Standard 90.1–2013 Energy Standard for Buildings except Low-Rise Residential Buildings*. Atlanta, GA: American Society of Heating, Refrigerating and Air-Conditioning Engineers.

ASHRAE. (2014) *Standard 189.1–2014 Standard for the Design of High-Performance Green Buildings*. Atlanta, GA: ASHRAE, Inc.

Ballarini, I., Corgnati, S.P. and Corrado, V. (2014) Use of reference buildings to assess the energy saving potentials of the residential building stock: The experience of TABULA project. *Energy Policy* 68: 273–284. https://doi.org/10.1016/j.enpol.2014.01.027.

Biljecki, F., Ledoux, H. and Stoter, J. (2016) An improved LOD specification for 3D building models. *Computers, Environment and Urban Systems* 59: 25–37. https://doi.org/10.1016/j.compenvurbsys.2016.04.005.

Blasio, Mayor Bill De. (2012) *One City New York City's Built to Last*.

Brenner, C. (2005) Building reconstruction from images and laser scanning. *International Journal of Applied Earth Observation and Geoinformation* 6(3): 187–198. https://doi.org/https://doi.org/10.1016/j.jag.2004.10.006.

Brown, N., Giampieri, M., Jungmin Han, E. and Namju Lee, N.J. (2017) Boston seaport district. Class Project for MIT 4.433 Modeling Urban Energy Flows – Towards More Sustainable Cities and Neighborhoods. http://web.mit.edu/sustainabledesignlab/projects/UMIverse/UMIverse.html.

Bueno, B., Norford, L., Pigeon, G. and Britter, R. (2012) A resistance-capacitance network model for the analysis of the interactions between the energy performance of buildings and the urban climate. *Building and Environment* 54: 116–125. https://doi.org/10.1016/j.buildenv.2012.01.023.

CAISO. (2016) *What the Duck Curve Tells Us about Managing a Green Grid.* Folsom, CA. www.caiso.com.

Cao, J., Metzmacher, H., O'Donnell, J., Frisch, J., Bazjanac, V., Kobbelt, L. and van Treeck, C. (2017) Facade geometry generation from low-resolution aerial photographs for building energy modeling. *Building and Environment* 123: 601–624. https://doi.org/https://doi.org/10.1016/j.buildenv.2017.07.018.

Caputo, P., Costa, G. and Ferrari, S. (2013) A supporting method for defining energy strategies in the building sector at urban scale. *Energy Policy* 55: 261–270. https://doi.org/10.1016/j.enpol.2012.12.006.

Cerezo, C., Sokol, J., AlKhaled, S., Reinhart, C., Al-Mumin, A. and Hajiah, A. (2017) Comparison of four building archetype characterization methods in Urban Building Energy Modeling (UBEM): A residential case study in Kuwait City. *Energy and Buildings* 154: 321–334. https://doi.org/10.1016/j.enbuild.2017.08.029.

Cerezo Davila, C., Dogan, T. and Reinhart, C.F. (2014) Towards standardized building properties template files for early design energy model generation. In *2014 ASHRAE/IBPSA-USA Building Simulation Conference*. Atlanta, GA, September 10–12.

Cerezo Davila, C., Jones, N., Al-Mumin, A., Hajiah, A. and Reinhart, C.F. (2017) Implementation of a calibrated Urban Building Energy Model (UBEM) for the evaluation of energy efficiency scenarios in a kuwaiti residential neighborhood. In *15th International Conference of the International Building Performance Simulation Association (BS2017)*. San Francisco, CA: IBPSA.

Cerezo Davila, C. and Reinhart, C.F. (2013) Urban energy lifecycle: An analytical framework to evaluate the embodied energy use of urban developments. In *13th International Conference of the International Building Performance Simulation Association (BS2013)*. August 25–28. Chambery, France: IBPSA.

Cerezo Davila, C., Reinhart C.F. and Bemis, J. (2016) Modeling Boston: A workflow for the efficient generation and maintenance of urban building energy demand models from existing geospatial datasets. *Energy* 117: 237–250.

Chen, Y., Hong, T. and Piette, M.A. (2017) Automatic generation and simulation of urban building energy models based on city datasets for city-scale building retrofit analysis. *Applied Energy* 205: 323–335. https://doi.org/https://doi.org/10.1016/j.apenergy.2017.07.128.

The City of Boston. (2018) Building Energy Reporting and Disclosure Ordinance. *City Ordiance*. www.boston.gov/environment-and-energy/building-energy-reporting-and-disclosure-ordinance.

Climate Policy Initiative. (2018) *California Carbon Dashboard.* 2018.

Coffey, B., Stone, A., Ruyssevelt, P. and Haves, P. (2015) An idemiological Approach to Simulation – Based Analysis of Large Building Stocks. In *14th International Conference of the International Building Performance Simulation Association (BS2015)*. December 2015, 7–9. Hyderabad, India: IBPSA.

Crawley, D.B., Hand, J.W. and Lawrie, L.K. (1999) Improving the weather information available to simulation programs. In *Sixth International IBPSA Conference (BS '99) Kyoto, Japan* II, pp. 529–536.

Dall'o', G., Galante, A. and Torri, M. (2012) A methodology for the energy performance classification of residential building stock on an urban scale. *Energy and Buildings* 48: 211–219. https://doi.org/10.1016/j.enbuild.2012.01.034.

Dascalaki, E.G., Droutsa, K.G., Balaras, C.A. and Kontoyiannidis, S. (2011) Building typologies as a tool for assessing the energy performance of residential buildings – a case study for the hellenic building stock. *Energy and Buildings* 43(12): 3400–3409. https://doi.org/10.1016/j.enbuild.2011.09.002.

Davis, L.W. and Gertler, P.J. (2015) Contribution of air conditioning adoption to future energy use under global warming. *Proceedings of the National Academy of Sciences* 112(19): 5962–5967. https://doi.org/10.1073/PNAS.1423558112.

Deru, A., Field, K., Studer, D., Benne, K., Griffith, B., Torcellini, P., Liu, B. et al. (2011) *U.S. Department of Energy Commercial Reference Building Models of the National Building Stock*. Golden, CO. https://digitalscholarship.unlv.edu/renew_pubs/44.

Dogan, T. and Reinhart, C. (2017) Shoeboxer: An algorithm for abstracted rapid multi-zone urban building energy model generation and simulation. *Energy and Buildings* 140: 140–153. https://doi.org/10.1016/j.enbuild.2017.01.030.

Dogan, T., Reinhart, C.F. and Michalatos, P. (2012) Urban daylight simulation calculating the daylit area of urban designs. In *Fifth National Conference of IBPSA-USA*, 613–620.

Dogan, T., Reinhart, C. and Michalatos, P. (2016) Autozoner: An algorithm for automatic thermal zoning of buildings with unknown interior space definitions. *Journal of Building Performance Simulation* 9(2): 176–189. https://doi.org/10.1080/19401493.2015.1006527.

Eicker, U., Nouvel, R., Duminil, E. and Coors, V. (2014) Assessing passive and active solar energy resources in cities using 3D city models. In *Energy Procedia* 57: 896–905. https://doi.org/10.1016/j.egypro.2014.10.299.

EU Directorate-General for Energy. (2016) *Energy Certificates and Inspections*. https://ec.europa.eu/energy/en/topics/energy-efficiency/%0Abuildings/certificates-and-inspections.

Evans, S., Liddiard, R. and Steadman, P. (2016) 3DStock: A new kind of three-dimensional model of the building stock of england and wales, for use in energy analysis. *Environment and Planning B: Urban Analytics and City Science* 44(2): 227–255. https://doi.org/10.1177/0265813516652898.

Famuyibo, A.A., Duffy, A. and Strachan, P. (2012) Developing archetypes for domestic dwellings – An Irish case study. *Energy and Buildings* 50: 150–157. https://doi.org/10.1016/j.enbuild.2012.03.033.

Filogamo, L., Peri, G., Rizzo, G. and Giaccone, A. (2014) On the classification of large residential buildings stocks by sample typologies for energy planning purposes. *Applied Energy* 135: 825–835. https://doi.org/10.1016/j.apenergy.2014.04.002.

Firth, S.K. and Lomas, K.J. (2009) Investigating CO2 emission reductions in existing urban housing using a community domestic energy model. *Building Simulation* 2098–2105.

Fracastoro, G.V. and Serraino, M. (2011) A methodology for assessing the energy performance of large scale building stocks and possible applications. *Energy and Buildings* 43(4): 844–852. https://doi.org/10.1016/j.enbuild.2010.12.004.

gbXML. (2018) GbXML. (2018) www.gbxml.org/.

Geller, H., Harrington, P., Rosenfeld, A.H., Tanishima, S. and Unander, F. (2006) Polices for increasing energy efficiency: Thirty years of experience in OECD countries. *Energy Policy* 34(5): 556–573. https://doi.org/10.1016/j.enpol.2005.11.010.

Hall, I., Prairie, R., Anderson, H., Boes, E. and Sandia National Laboratories. (1978) Generation of typical meteorological years for 26 SOLMET stations, SAND78–1601. Albuquerque, NM.

Harvard University. (2016) *2006–2016 Harvard Climate Goal Report*. Cambridge, MA. https://issuu.com/greenharvard/docs/2006_2016_harvard_climate_goal_repo.

He, M., Lee, T., Taylor, S., Firth, S.K. and Lomas, K.J. (2015) Coupling a stochastic occupancy model to energyplus to predict hourly thermal demand of a neighbourhood. In *14th International Conference of the International Building Performance Simulation Association (BS2015)*. December 7–9, 2015, Hyderabad, India: IBPSA. www.ibpsa.org/proceedings/BS2015/p2655.pdf.

Heiple, S. and Sailor, D.J. (2008) Using building energy simulation and geospatial modeling techniques to determine high resolution building sector energy consumption profiles. *Energy and Buildings* 40(8): 1426–1436. https://doi.org/10.1016/j.enbuild.2008.01.005.

Hölzle, U. (2017) *Google Environmental Report – 2017 Progress Update*. Mountain View, CA. https://static.googleusercontent.com/media/environment.google/en//pdf/google-2017-environmental-report.pdf.

Howard, B., Parshall, L., Thompson, J., Hammer, S., Dickinson, J. and Modi, V. (2012) Spatial distribution of urban building energy consumption by end use. *Energy and Buildings* 45: 141–151. https://doi.org/10.1016/j.enbuild.2011.10.061.

Jentsch, M.F., James, P.A.B., Bourikas, L. and Bahaj, A.B.S. (2013) Transforming existing weather data for worldwide locations to enable energy and building performance simulation under future climates. *Renewable Energy* 55: 514–524. https://doi.org/10.1016/j.renene.2012.12.049.

Kaden, R. and Kolbe, T.H. (2013) City-wide total energy demand estimation of buildings using semantic 3D city models and statistical data. In *ISPRS 8th Conference*. Istambul, Turkey: ISPRS.

Kavgic, M., Mavrogianni, A., Mumovic, D., Summerfield, A., Stevanovic, Z. and Djurovic-Petrovic, M. (2010) A review of bottom-up building stock models for energy consumption in the residential sector. *Building and Environment* 45(7): 1683–1697. https://doi.org/10.1016/j.buildenv.2010.01.021.

Kim, E.J., Plessis, G., Hubert, J.L. and Roux, J.J. (2014) Urban energy simulation: Simplification and reduction of building envelope models. *Energy and Buildings* 84: 193–202. https://doi.org/10.1016/j.enbuild.2014.07.066.

Koene, F.G., Bakker, L.G., Lanceta, D. and Narmsara, S. (2014) Simplified building models of districts. In *BauSIM 2014*. Aachen, Germany: IBPSA.

Kolter, J.Z. and Ferreira, J. (2011) A large-scale study on predicting and contextualizing building energy usage. *Proceedings of the Conference on Artificial Intelligence (AAAI), Special Track on Computational Sustainability and AI, 2011*, 8.

Lauster, M., Teichmann, J., Fuchs, M., Streblow, R. and Mueller, D. (2014) Low order thermal network models for dynamic simulations of buildings on city district scale. *Building and Environment* 73: 223–231. https://doi.org/10.1016/j.buildenv.2013.12.016.

Letellier-Duchesne, S., Nagpal, S., Kummert, M. and Reinhart, C.F. (2018) Balancing demand and supply: Linking neighborhood-level building load calculations with detailed district energy network analysis models. *Energy* 150: 913–925.

Liu, J., Heidarinejad, M., Gracik, S. and Srebric, J. (2015) The impact of exterior surface convective heat transfer coefficients on the building energy consumption in urban neighborhoods with different plan area densities. *Energy and Buildings* 86: 449–463. https://doi.org/10.1016/j.enbuild.2014.10.062.

Loga, T., Diefenbach, N., Stein, B., Dascalaki, E., Balaras, C.A., Droutsa, K., Kontoyiannidis, S., Villatoro, O. and Wittchen, K.B. (2012) Typology Approach for Building Stock Energy Assessment. Main Results of the TABULA Project. Final Project Report : Appendix Volume. No. June 2009: 172. www.building-typology.eu/downloads/public/docs/report/TABULA_FinalReport_AppendixVolume.pdf.

Mastrucci, A., Baume, O., Stazi, F. and Leopold, U. (2014) Estimating energy savings for the residential building stock of an entire city: A gis-based statistical downscaling approach applied to Rotterdam. *Energy and Buildings* 75: 358–367. https://doi.org/10.1016/j.enbuild.2014.02.032.

Mastrucci, A., Baume, O., Stazi, F., Salvucci, S. and Leopold, U. (2014) A GIS-based approach to estimate energy savings and indoor thermal comfort for urban housing stock retrofitting. In *BauSIM2014 Conference Proceedings and Keynote Lecture Presentations*. RWTH Aachen University, Germany, September 22–24, 2014, p. 8. www.ibpsa.org/proceedings/bausimPapers/2014/p1136_final.pdf.

Mata, É., Sasic Kalagasidis, A. and Johnsson, F. (2014) Building-stock aggregation through archetype buildings: France, Germany, Spain and the UK. *Building and Environment* 81: 270–282. https://doi.org/10.1016/j.buildenv.2014.06.013.

Mavrogianni, A., Davies, M., Kolokotroni, M. and Hamilton, I. (2009) A GIS-based bottom-up space heating demand model of the london domestic stock. *International Building Performance Simulation Association (IBPSA) Building Simulation* 1061–1067. http://discovery.ucl.ac.uk/1302452/.

Mayor of London (2011) *Delivering London's Energy Future: The Mayor's Climate Change Mitigation and Energy Strategy*. London: The Greater London Authority, pp. 1–266.McNeel, R. (2010a) *Grasshopper – Generative Moeling with Rhino*. Seattle, WA: McNeel North America. www.grasshopper3d.com/.

McNeel, R. (2010b) Rhinoceros – NURBS Modeling for Windows (Version 4.0). Seattle, WA: McNeel North America. www.rhino3d.com/.

Michaels, J. (2003) Commercial Building Energy Consumption Survey (CBECS). Edited by US-Energy Information Administration. Washington DC: US Department of Energy.

Open Geospatial Consortium. (2012) *OGC City Geography Markup Language (CityGML) Encoding Standard*. OGC. www.opengis.net/spec/citygml/2.0.

Perez, D. and Robinson, D. (2012) Urban energy flow modelling: A data-aware approach. *Digital Urban Modeling and Simulation* 200–220. https://doi.org/10.1007/978-3-642-29758-8-11.

Rakha, T. and Reinhart, C. (2013) A carbon impact simulation-based framework for land use planning and non-motorized travel behavior interactions. *Proceedings of Building Simulation*. www.tarekrakha.com/s/CarbonImpact_BS2013_TR_CR.pdf.

Reinhart, C.F. and Cerezo Davila, C. (2016) Urban building energy modeling – a review of a nascent field. *Building and Environment* 97: 196–202. https://doi.org/10.1016/j.buildenv.2015.12.001.

Reinhart, C.F., Dogan, T., Jakubiec, J.A., Rakha, T. and Sang, A. (2013) Umi – an urban simulation environment for building energy use, daylighting and walkability. In *Proceedings of BS2013: 13th Conference of International Building Performance Simulation Association*, pp. 476–483.

Robertson, J., Polly, B. and Collis, J. (2013) *Evaluation of Automated Model Calibration Techniques for Residential Building Energy Simulation*. Golden, CO. www.nrel.gov/research/publications.html.

Robinson, D. (2012) *Computer Modeling for Sustainable Urban Design*. London: Earthscan.

Robinson, D., Haldi, F., Kampf, J., Leroux, P., Perez, D., Rasheed, A. and Wilke, U. (2009) CitySim – Comprehensive micro simulation of resource flows for sustainable urban planning. In *11th International Conference of the International Building Performance Simulation Association (BS2009)*. July 27–30, Glasgow, Scotland: IBPSA.

Sanyal, J., New, J., Edwards, R.E and Parker, L. (2014) Calibrating building energy models using super-computer trained machine learning agents. *Concurrency and Computation: Practice and Experience* 26(13): 2122–2133. https://doi.org/10.1002/cpe.3267.

Shutters, C. (2016) *LEED by the Numbers: 16 Years of Steady Growth*. Washington, DC. www.usgbc.org/articles/leed-numbers-16-years-steady-growth.

Singapore, Government of. (2015) Singapore's intended nationally determined contribution (indc) and accompanying information. No. 2005: 7.

Sokol, J., Davila, C.C. and Reinhart, C.F. (2017) Validation of a Bayesian-based method for defining residential archetypes in urban building energy models. *Energy and Buildings*. https://doi.org/10.1016/j.enbuild.2016.10.050.

Strzalka, A., Bogdahn, J., Coors, V. and Eicker, U. (2011) 3D City modeling for urban scale heating energy demand forecasting. *HVAC&R Research* 17(4): 526–539. https://doi.org/10.1080/10789669.2011.582920.

Sun, K., Hong, T., Taylor-Lange, S.C and Piette, M.A. (2016) A pattern-based automated approach to building energy model calibration. *Applied Energy* 165: 214–224. https://doi.org/https://doi.org/10.1016/j.apenergy.2015.12.026.

Swan, L.G. and Ismet Ugursal, V. (2009) Modeling of end-use energy consumption in the residential sector: A review of modeling techniques. *Renewable and Sustainable Energy Reviews*. https://doi.org/10.1016/j.rser.2008.09.033.

Tian, W., Rysanek, A., Choudhary, R. and Heo, Y. (2015) High resolution energy simulations at city scale. In *14th International Conference of the International Building Performance Simulation Association*, pp. 239–246.

Turan, I.,Chegut, A. and Reinhart, C.F. (2017) Connecting environmental performance analysis to cash flow modeling for financial valuation of buildings in early design. In *Building Simulation 2017*. San Francisco, CA, pp. 327–336.

United Nations General Assembly. (2015) Transforming our world: The 2030 agenda for sustainable development. https://Sustainabledevelopment.Un.Org/Content/Documents/7891Transforming%20Our%20World. Pdf, no. 1: 1–5. https://doi.org/10.1007/s13398-014-0173-7.2.

US-DOE. (2017) *EnergyPlus Weather Data*. Washington, DC: US Department of Energy. https://energyplus.net/.

US-EPA. (2016) *Inventory of U.S. Greenhouse Gas Emissions and Sinks: 1990–2014*. Washington DC.

Wilke, U., Haldi, F., Scartezzini, J.L. and Robinson, D. (2013) A bottom-up stochastic model to predict building occupants' time-dependent activities. *Building and Environment* 60: 254–264. https://doi.org/10.1016/j.buildenv.2012.10.021.

Walsh, M.J. (Mayor). (2014) Housing a changing city: Boston 2030. In *Report by the City of Boston*. Boston, pp. 1–139.

The World Bank. (2017) Urban population growth (Annual %). The World Bank|Data. https://data.worldbank.org/indicator/SP.URB.GROW.

Xu, S., Barbour, E. and González, M.C. (2017) Household segmentation by load shape and daily consumption. In *ACM SigKDD 2017 Conference*. August 13–17, Halifax, NS. https://doi.org/10.475/123 4.

22 Urban physics simulation for climate change adaptation of buildings and urban areas

Bert Blocken, Twan van Hooff, Yasin Toparlar, Christof Gromke, Hamid Montazeri and Wendy Janssen

Scope

This chapter discusses the simulation of urban thermal microclimate with a focus on heat waves in urban areas, the simulation of overheating of buildings and the effects of adaptation measures to limit temperatures in buildings and urban areas during heat waves. The spatial scales are the meteorological microscale (neighborhood scale) and the building scale; the methods are computational fluid dynamics and building energy simulation. Adaptation measures investigated at the neighborhood scale are avenue trees, green facades and green roofs; adaptation measures at the building scale are increased thermal resistance, increased thermal mass, increased short-wave reflectivity of facades and roofs, peak ventilation, vegetated roofs and exterior solar shading.

Learning objectives

Learning objectives are:

- What factors contribute to amplification of heat waves in urban areas?
- What factors contribute to overheating in buildings?
- Can we correctly simulate urban thermal microclimate, and, if so, what are the most important guidelines for doing so?
- How effective are adaptation measures like avenue trees, green facades and green roofs to locally reduce air temperatures in urban areas?
- What are the most effective adaptation measures to avoid overheating in buildings?

Key words

Urban physics/climate change/heat waves/urban heat island effect/climate change adaptation/vegetation/computational fluid dynamics/building energy simulation

Introduction

Climate change, mitigation and adaptation

Climate observations and analyses by the Intergovernmental Panel on Climate Change (IPCC) and national meteorological organizations show that climate change is occurring. The global temperature has increased by 0.56–0.92 °C in the last century and is expected to increase by 1.1 to 6.4 °C until the year 2100, depending on the greenhouse gas emission scenario (van den Hurk et al. 2006; Pachauri and Reisinger 2007). It is therefore important to apply both climate change mitigation and climate change adaptation.

Climate change mitigation is defined by the IPCC as "an anthropogenic intervention to reduce the sources or enhance the sinks of greenhouse gases." For the built environment, this includes reducing the amount of energy necessary to heat, cool and illuminate buildings and urban areas. In addition, renewable energy sources such as wind and solar energy can help to reduce the emission of greenhouse gases. Climate change adaptation is defined as "adjustments in natural or human systems in response to actual or expected climatic stimuli or their effects, which moderates harm or exploits beneficial opportunities" (IPCC 2001).

People spend around 90% of their time indoors (Awbi 2003), while most residential buildings in many western, central, eastern and northern European countries often do not have air-conditioning systems or any other active cooling systems to reduce the indoor air temperature during warm periods (Entranze 2010; van Kempen 2000). However, the number and the intensity of heat waves are expected to increase due to climate change (van den Hurk et al. 2006). It is undesirable to install this type of active cooling system on a large scale, as these will only increase energy consumption and will also increase climate change if fueled by non-renewable resources (Li et al. 2012). Climate change adaptation measures for buildings but also for urban areas are therefore important in order to minimize the exposure of people to excessively high air temperatures and hence to limit the negative impact of climate change on health, comfort, productivity and energy consumption.

Heat waves and the urban heat island

A heat wave is defined as a prolonged period of exceptionally hot weather. The increased frequency of heat waves can be considered an indicator of climate change. In the period between 1901 and 2009, 38 heat waves have been recorded in the Netherlands, seven of which took place in the last decade of this period (1999–2009) (Salcedo Rahola et al. 2009). The frequency and the intensity of heat waves are expected to increase throughout the current century as a result of climate change (van den Hurk et al. 2006).

Inside urban areas, the higher air temperatures due to climate change can be aggravated by a series of processes that give rise to the urban heat island (UHI) effect. These processes were described by Oke (1982) as follows (Figure 22.1): (1) Urban areas have amplified short-wave radiation gain due to the multiple reflections in the urban environment in combination with the relatively high short-wave absorptivity of urban surfaces. (2) Urban areas are characterized by amplified long-wave radiation gain mostly due to increased air pollution in and above urban areas. (3) The urban geometry often blocks the outgoing long-wave radiation due to small view factors to the sky, keeping part of this long-wave emission trapped inside street canyons, resulting in higher temperatures. (4) Urban areas also exhibit increased anthropogenic heat sources, including transport, industry and building heating, cooling and air-conditioning. (5) The construction materials in the urban environment (pavement, asphalt, concrete, stone, etc.) usually have a high heat capacity and high thermal conductivity, resulting in increased heat storage in urban areas, which is mainly released during nighttime. (6) Urban areas are characterized by limited evapotranspiration due to the relatively low number or even absence of water bodies and vegetation. (7) The urban geometry decreases convective heat transfer due to lower wind velocities in street canyons and courtyards and the resulting reduced ventilation rates of urban areas. The globally rising temperature, the increased frequency and intensity of heat waves and the amplification of heat waves by the above-mentioned seven processes all contribute to higher air temperatures in urban areas and demonstrate the importance of climate change adaptation of buildings and urban areas in the context of heat waves.

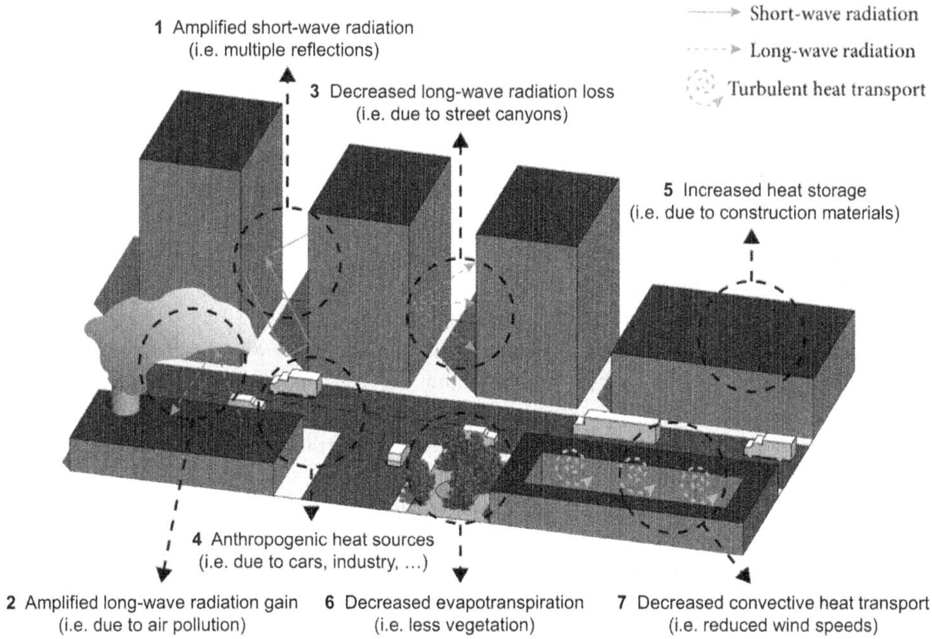

Figure 22.1 Seven physical processes leading to the urban heat island effect and the amplification of heat waves in urban areas

Climate change adaptation of buildings

Introduction

Due to the changing climate with rising temperatures, buildings will need to adapt in order to provide a healthy and comfortable indoor environment with a minimum or no increase in energy consumption. Several studies have been performed on the climate change adaptation of buildings with different simulation tools. Porritt et al. (2011, 2012) employed the building energy simulation programs IES VE and EnergyPlus to investigate a wide range of passive climate change adaptation measures for terraced houses. Both programs have been extensively validated (Henninger and Witte 2011; Zhai et al. 2011) and have an airflow network to investigate natural ventilation. EnergyPlus also allows for a large number of parametric simulations to be run in batch configuration and showed a very good agreement with analytical solutions and other airflow network models (Gu and Crawley 2009). Porritt et al. (2011, 2012) found that one or multiple adaptation measures can reduce the number of overheating hours by 32%–99%. In addition, a study by Coley et al. (2012) showed that behavioral adaptation measures (opening/closing windows, shifting work hours) can be just as efficient as (costly) structural adaptation measures (increasing thermal mass, adding shading). By painting the roof surface of a building in California white, Akbari et al. (1997) reported that the short-wave reflectivity increased from 0.18 to 0.73 resulting in cooling savings of 69% in the following period. Most of these studies were performed for only one building type and construction period. Therefore, a new study was set up by van Hooff et al. (2014) to study the effectiveness of six passive climate change adaptation measures applied to three generic residential buildings with characteristics of two different construction periods. This study is reported in the next subsections.

Building construction types and adaptation measures

Van Hooff et al. (2014) studied the effectiveness of six passive climate change adaptation measures applied to three generic residential buildings: a detached house, a terraced house and an apartment (Figures 22.2–22.4).

For these buildings, two reference case construction periods and types were considered: a construction typical of the 1970s, which is characterized by a low thermal resistance of the building envelope, and a construction according to the 2012 building regulations, with much higher insulation levels. The construction characteristics of both reference cases are listed in Tables 22.1 and 22.2.

The following adaptation measures were contemplated (Table 22.3): (1) Increasing the thermal resistance of the building envelope to R_C = 5.0 (RC50) and 6.5 m^2K/W (RC65). (2) Lowering the thermal capacity by replacing the reference case limestone inner leaf and concrete materials with wooden constructions (TM_low). (3) Increasing the short-wave reflectivity (albedo value) of the external surfaces from 0.3 to 0.6 (SWR06) and 0.8 (SWR08). (4) Increasing natural peak ventilation by opening the windows when the indoor air temperature is both above 24 °C and higher than the outdoor temperature. This is done for either the entire day (NV_all) or for the situation where the windows can only be opened during the day between 8:00 and 20:00 (NV_day). (5) Adding vertical exterior solar shading devices with a short-wave reflectivity of 0.9 and where the solar shading is engaged starting from 150 W/m^2 radiation on the window. (6) Employing an extensive vegetated roof (VR) with a leaf area index (LAI) of 5.

Figure 22.2 Facades, floor plans and dimensions of the detached house (modified from (AgentschapNL 2013)). Triangles in windows and doors indicate operable windows/doors for the additional ventilation measure. Dimensions in mm.

Source: van Hooff et al. 2014, ©Elsevier, reproduced with permission

Figure 22.3 Facades, floor plans and dimensions of the terraced house (modified from (AgentschapNL 2013)). Triangles in windows and doors indicate operable windows/doors for the additional ventilation measure. Dimensions in mm.

Source: van Hooff et al. 2014, ©Elsevier, reproduced with permission

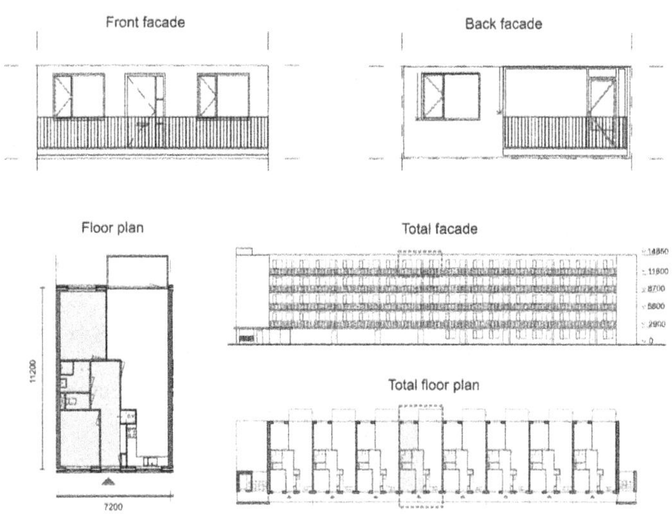

Figure 22.4 Facades, floor plans and dimensions of the apartment (modified from (AgentschapNL 2013)). The light gray area in the floor plan indicates the zone with the bedrooms. The dashed boxes in the figures of the total facade and total floor plan indicate the apartment under study. Triangles in windows and doors indicate operable windows/doors for the additional ventilation measure. Dimensions in mm.

Source: van Hooff et al. 2014, ©Elsevier, reproduced with permission

Table 22.1 Overview of construction characteristics for the reference case building from the 1970s

Element	*Details*	*R_C value (m^2K/W)*
External walls	Cavity walls with (inside to outside): limestone inner leaf, air cavity, brick outer leaf	0.4
Internal wall	Limestone wall	–
Roof (pitched)	Inside to outside: wooden sheeting, insulation layer, air cavity, roof tiles	0.8
Roof (flat)	Inside to outside: concrete, insulation layer, roofing material	0.8
External floor	Concrete	0.17
Internal floor	Concrete	–
Windows	Single pane glazing. Solar transmittance coefficient 0.7.	*U*-value: 5.2 W/m^2K

Table 22.2 Overview of construction characteristics for the reference case building from 2012

Element	*Details*	*R_C value (m^2K/W)*
External walls	Cavity walls with (inside to outside): limestone inner leaf, insulation, air cavity, brick outer leaf	3.5
Internal wall	Limestone wall	–
Roof (pitched)	Inside to outside: wooden sheeting, insulation layer, air cavity, roof tiles	4
Roof (flat)	Inside to outside: concrete, insulation layer, roofing material	4
External floor	Inside to outside: concrete, insulation	3.5
Internal floor	Concrete	–
Windows	Double pane glazing. Solar transmittance coefficient 0.7	*U*-value: 1.65 W/m^2K

Table 22.3 Overview of adaptation measures

Adaptation measure	*Description*	*Abbreviation*
Increased thermal resistance	The thermal resistance of all external building surfaces is increased to $R_C = 5.0$ m^2K/W and $R_C = 6.5$ m^2K/W, for cases RC50 and RC65, respectively. This measure is implemented by increasing the thickness of the insulation layers.	RC50, RC65
Changed thermal capacity	The thermal capacity is lowered, since the reference case is a heavy building. The thermal capacity is changed by replacing the limestone inner leaf by an inner leaf of wooden sheeting. In addition, concrete ceilings are replaced by wooden constructions.	TM_low
Increased short-wave reflectivity (albedo)	The short-wave reflectivity of the external surfaces is increased from the default value of 0.3 to 0.6 and 0.8, for cases SWR06 and SWR08, respectively.	SWR06, SWR08

Adaptation measure	*Description*	*Abbreviation*
Additional natural ventilation	Additional natural ventilation is provided by opening (parts of) the windows. The windows will be opened when the indoor air temperature is above 24 °C, but only when the indoor air temperature is higher than the outdoor air temperature. In one case (NV_all), the windows can be opened the entire day (24 hours); in the other case (NV_day), the windows can only be opened between 08:00 and 20:00.	NV_all, NV_day
Vegetated roof	The default roof constructions are extended to incorporate a vegetated roof with a leaf area index of 5.	VR
Solar shading	Exterior solar shading is applied for all windows on the east, south and west side of the facades. The solar shading is automatically lowered when the solar radiation on the window is equal to or above 150 W/m^2.	SH

Building energy simulations

Dynamic thermal simulations were performed with the software EnergyPlus. The simulations were performed with a 10-minute time step, except for the vegetated roof where a time step of 1 minute was needed for accurate simulations. The simulations were conducted for four different building orientations, with the front facade facing the cardinal wind directions. The buildings all have an exhaust ventilation flow rate of 0.7 dm^3/sm^2 and an infiltration flow rate of 0.2 ACH according to the Dutch building regulations. The KNMI weather data from De Bilt, the Netherlands, during July 2006 were used. This year was an exceptionally hot year with multiple heat waves (see Figure 22.5 for the air temperature during the month of July) and can represent a shape of things to come in future years. The adjusted adaptive temperature limit for residential buildings was used as a thermal comfort indicator (Peeters et al. 2009). Two different temperature limits were defined, for the living room (occupied between 06:00 and 23:00) and for the bedrooms (occupied between 23:00 and 6:00), respectively. The resulting number of overheating hours and degree hours (level of overheating) were used to compare the results for the reference case and the cases with the adaptation measures.

Results: detached house

Figure 22.6 shows the results of the adaptation measures for the detached house with the 1970s construction characteristics. By increasing the thermal resistance (RC50/RC60), the number of overheating hours and degree hours increase significantly compared to the reference case. The reason is that the short-wave solar radiation passes through the glazing and is then absorbed by the indoor surfaces and emitted as long-wave radiation, which cannot pass through the glazing. The related heat is therefore trapped to some extent, cannot be removed by radiation and needs to be removed by convection and conduction, i.e. the so-called greenhouse effect. Heat removal by conduction is a very slow process, particularly when insulation levels are high. As a result, increasing thermal insulation will increase the indoor temperature. When the thermal mass is reduced (TW_low), the number of overheating and degree hours increase for the ground floor and decrease for the first floor. This can be attributed to

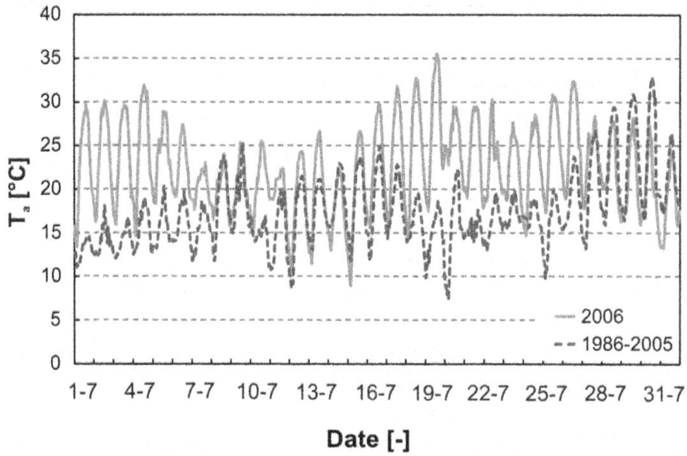

Figure 22.5 Measured hourly air temperatures in De Bilt, the Netherlands, in July 2006 (solid line) compared to averaged values over the period 1986–2005 (dotted line) (KNMI)

Source: van Hooff et al. 2014, ©Elsevier, reproduced with permission

the larger temperature fluctuations during the day, which is characteristic for a low thermal mass building. Therefore, compared to the reference case, there is a higher temperature during the day (when occupants are present on the ground floor) and a lower temperature during the night (when occupants are on the first floor). Increasing the short-wave reflectivity (SWR06 and SWR08) decreases the number of overheating hours, as it leads to lower exterior surface temperatures. Opening the windows during the day (NV_all) and applying exterior solar shading (SH) are the most effective adaptation measures for the detached house from the 1970s. The addition of a vegetated roof only provides a relatively small reduction (up to 17%) in the number of overheating hours. The vegetated roof increases the thermal resistance, which was already shown to have a negative effect as an adaptation measure. This effect appears to counteract to a large extent the benefit obtained by the lower surface temperature of the vegetated roof.

Figure 22.7 shows the results of the adaptation measures for the detached house built according to the 2012 building regulations. Due to the higher thermal resistance and the related amplified greenhouse effect, the number of overheating and degree hours for the reference case is significantly higher than that from the 1970s. This shows that newly built well-insulated buildings will have larger overheating problems if they are not properly equipped with measures to reduce this overheating. This increases the need for effective climate change adaptation measures for these buildings. By applying exterior solar shading or opening the windows for natural ventilation, the average number of overheating hours can be reduced to a very large extent. For the TM_low case, there is a significant increase in degree hours for the ground floor, while the number of overheating hours is only slightly higher. The lower thermal mass causes the indoor air temperature to reach very high temperatures, sometimes even above 40 °C. The results also show that increasing the short-wave reflectivity has less effect for buildings with a higher thermal resistance. This is because the heat flow through the building envelope is much lower due to this higher thermal resistance, regardless of the short-wave reflectivity value. The effects of the other measures effects are similar as for the detached house from the 1970s.

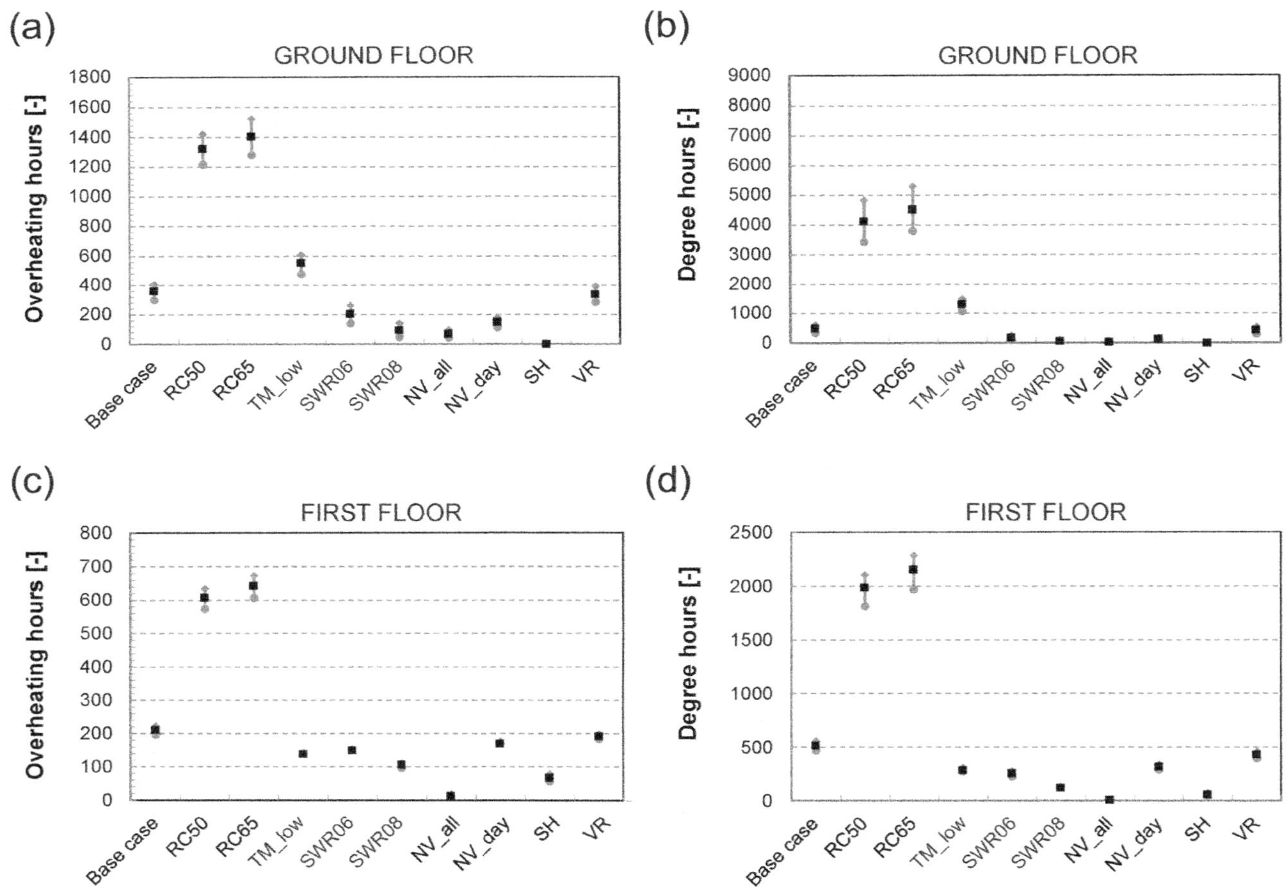

Figure 22.6 Number of overheating hours (a, c) and degree hours (b, d) for different cases of the detached house with construction characteristics from the 1970s. (a, b) Ground floor. (c, d) First floor.

Legend: ■ = the average of the four orientations, ● = minimum value, ♦ = maximum value

Source: van Hooff et al. 2014, ©Elsevier, reproduced with permission

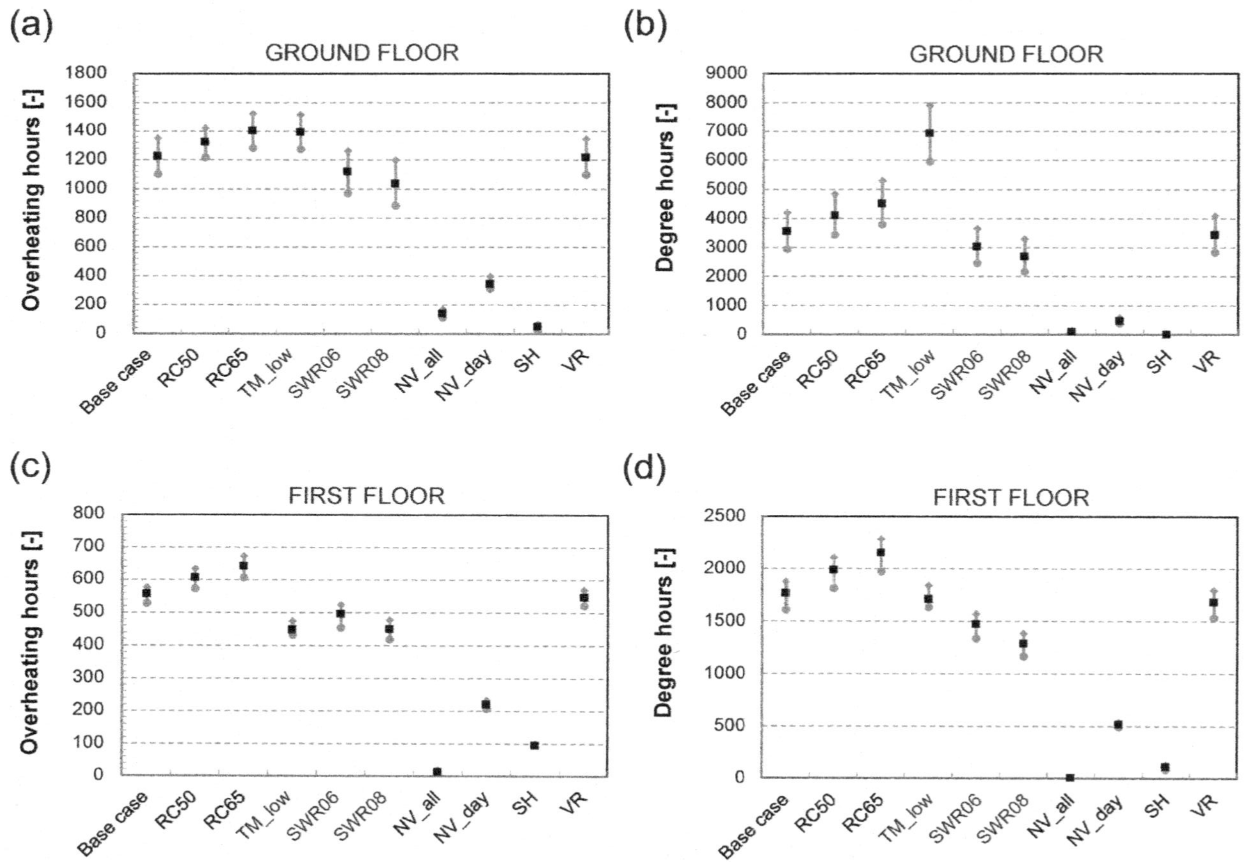

Figure 22.7 Number of overheating hours (a, c) and degree hours (b, d) for different cases of the detached house with construction characteristics from 2012. (a, b) Ground floor. (c, d) First floor.

Legend: ■ = the average of the four orientations, ● = minimum value, ♦ = maximum value

Source: van Hooff et al. 2014, ©Elsevier, reproduced with permission

Results: terraced house

Figures 22.8 and 22.9 show the results for the terraced house with construction characteristics from the 1970s and 2012, respectively. The largest difference compared to the detached house is the large spread between the minimum and maximum number of overheating and degree hours. This difference is attributed to the fact that the terraced house only has two building sides exposed to the solar radiation. The orientation of the front facade therefore plays a larger role in the number of overheating hours. A west or an east orientation has almost two times the overheating hours for a south- or north-orientated front facade. Another difference compared to the detached house is an increase of the number of degree hours on the first floor of the terraced house from 2012, when a lower thermal mass is used. Also the positive effect of increasing the short-wave reflectivity values is less pronounced due to the smaller surface area that is exposed to the outdoor environment and thus to solar radiation. As for the detached house, the two most effective measures are peak ventilation and exterior solar shading.

Results: apartment building

Figures 22.10 and 22.11 show the results of the adaptation measures for the apartment building with construction characteristics from the 1970s and 2012, respectively. Overall, the number of overheating and degree hours is higher for the living room and bedrooms than in the other building types. This is mainly attributed to the location of the living room and bedrooms directly below the roof construction. The effect of the other adaptation measures is similar to the other building types. Again, the two most effective measures are peak ventilation and exterior solar shading.

Overall, using peak natural ventilation and applying exterior solar shading are the most effective adaptation measures. Increasing the thermal resistance of the building has a negative effect on the overheating hours. Reducing the thermal mass in a building could have a positive effect for the indoor temperatures during the night, but only when the heat inside the building can be released either through the building envelope or by ventilation.

Climate change adaptation of urban areas: simulation and validation

Introduction

Several studies have been performed in the past decades on measures to reduce the air temperature in urban areas, especially during heat waves (e.g. Mochida and Lun 2008, Liu et al. 2012, Tominaga et al. 2015, Yang et al. 2017, Toparlar et al. 2018). One option is providing water facilities, which decrease the air temperature by evaporative cooling. Different possibilities include water ponds, fountains, waterfalls and mist spray. Nishimura et al. (1998) studied the effect of a water pond in a park and showed that the air temperature downwind of the water pond was reduced by 1–2 °C. The addition of waterfalls and fountains reduced the air temperature further (up to 4–5 °C) as the cooling effect by moving water is often greater due to a larger portion of the water's surface area being exposed to air. However, in areas with water scarcity, this type of measure will often be difficult to implement. Other measures are vegetation (trees, green roofs and facades), increased short-wave reflectivity of urban surfaces, etc. However, before the effectiveness of such measures can be computationally assessed, it is important to assess the simulation approach itself by validation actions.

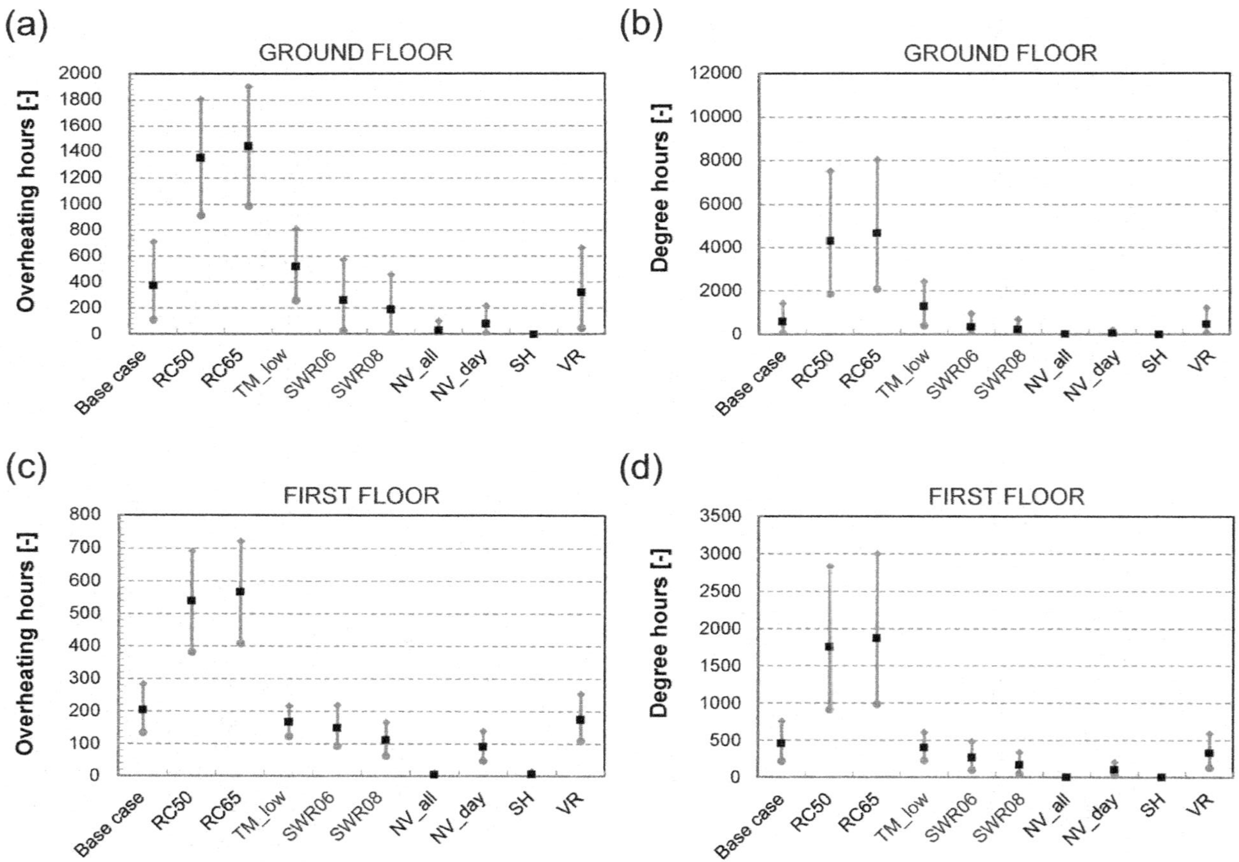

Figure 22.8 Number of overheating hours (a, c) and degree hours (b, d) for different cases of the terraced house with construction characteristics from the 1970s. (a, b) Ground floor. (c, d) First floor.

Legend: ■ = the average of the four orientations, ● = minimum value, ♦ = maximum value

Source: van Hooff et al. 2014, ©Elsevier, reproduced with permission

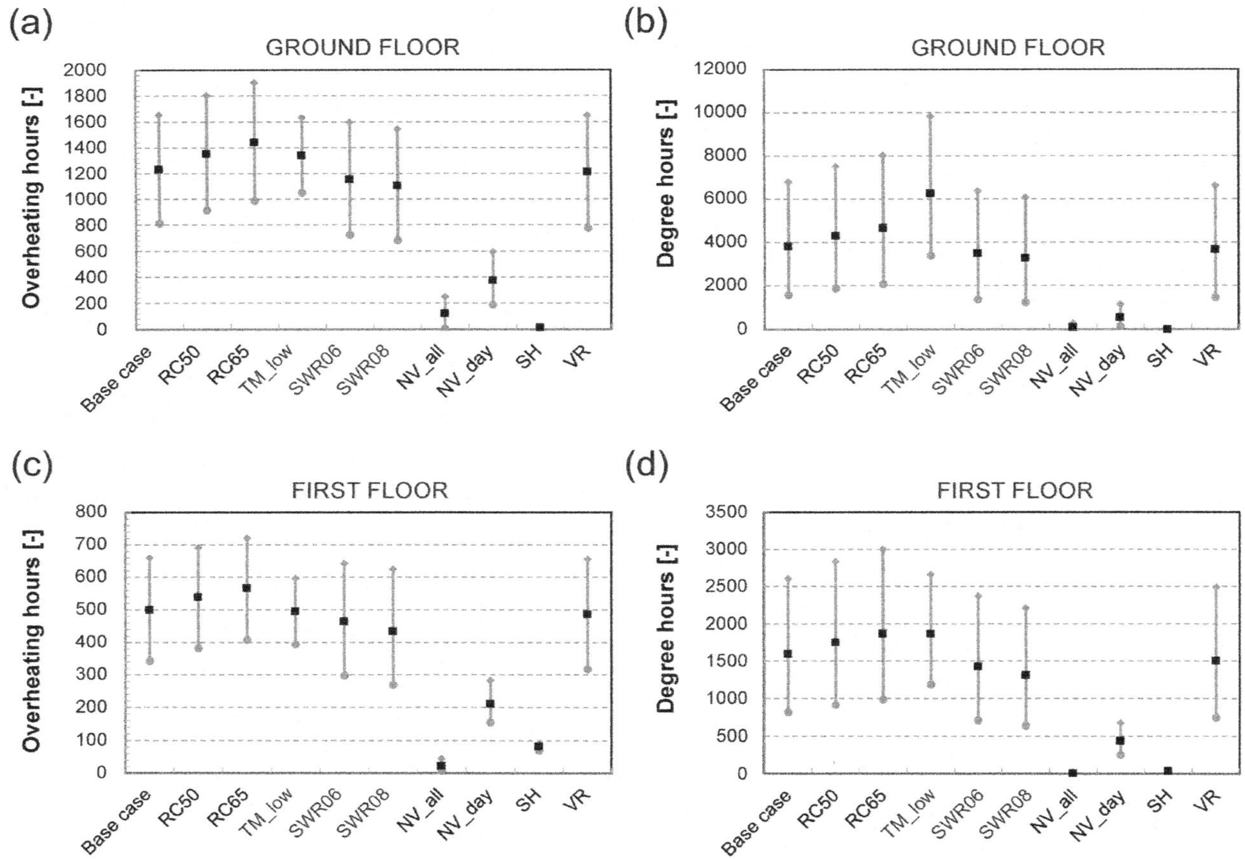

Figure 22.9 Number of overheating hours (a, c) and degree hours (b, d) for different cases of the terraced house with construction characteristics from 2012. (a, b) Ground floor. (c, d) First floor.

Legend: ■ = the average of the four orientations, ● = minimum value, ♦ = maximum value

Source: van Hooff et al. 2014, ©Elsevier, reproduced with permission

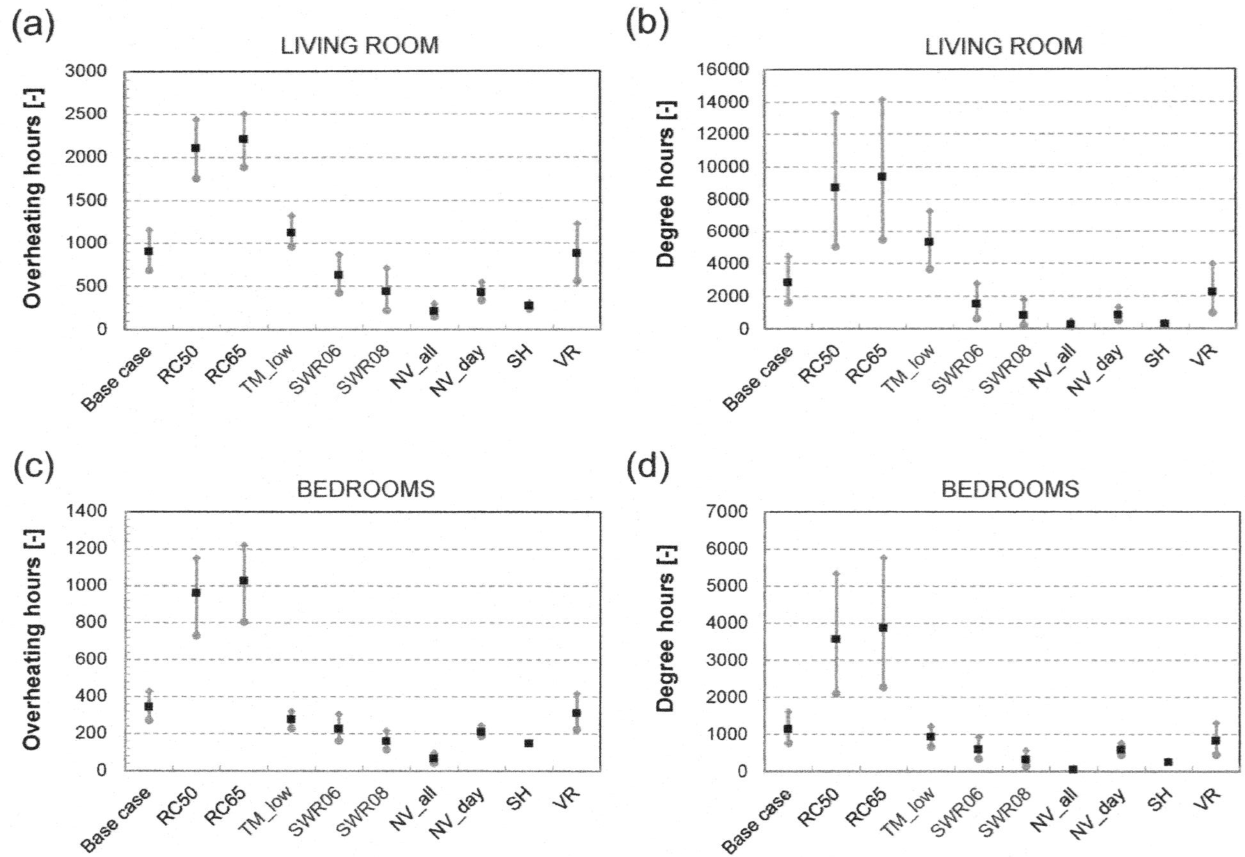

Figure 22.10 Number of overheating hours (a, c) and degree hours (b, d) for different cases of the apartment with construction characteristics from the 1970s. (a, b) Ground floor. (c, d) First floor.

Legend: ■ = the average of the four orientations, ● = minimum value, ♦ = maximum value

Source: van Hooff et al. 2014, ©Elsevier, reproduced with permission

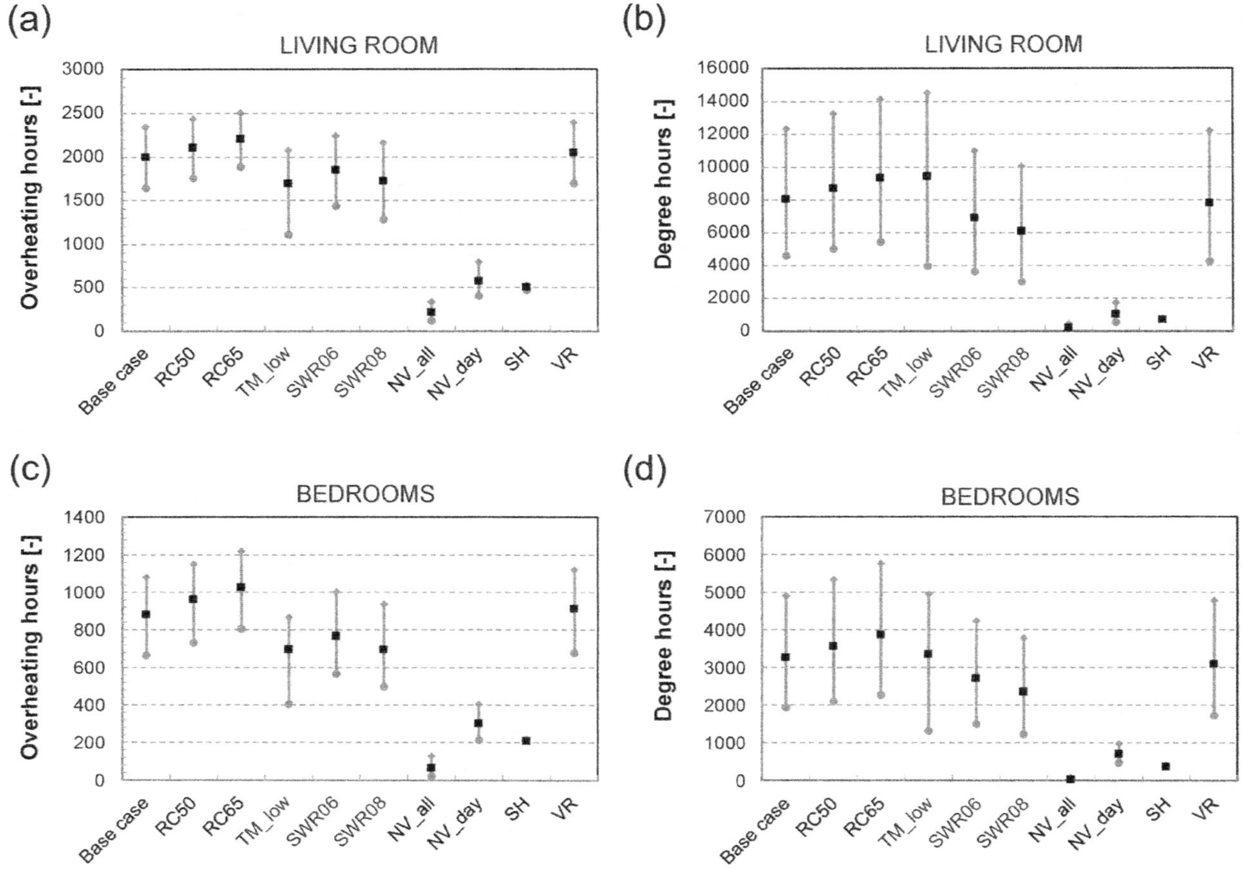

Figure 22.11 Number of overheating hours (a, c) and degree hours (b, d) for different cases of the apartment with construction characteristics from 2012. (a, b) Ground floor. (c, d) First floor.

Legend: ■ = the average of the four orientations, ● = minimum value, ♦ = maximum value

Source: van Hooff et al. 2014, ©Elsevier, reproduced with permission

In order to assess heat waves, urban heat islands and climate change adaptation measures for urban areas, different simulation tools can be employed. Due to the complexity of the related physical processes, multiple simplifications are necessary in the simulations. Validation of the results is therefore important to ensure reliable results. Simulation techniques that have shown reliable outcomes to predict the temperature distribution for urban regions are the energy balance model (or urban canopy model) and numerical simulation with computational fluid dynamics (CFD) (Mirzaei and Haghighat 2010; Toparlar et al. 2017). The energy balance model was first suggested by (Oke 1982), and it applies the law of conservation energy for a control volume. Energy balance models generally have a short calculation time, as they use a limited number of nodes. The main weakness of these models is the absence of the velocity field, which is necessary to model the effects of wind flow and related processes and to accurately determine sensible and latent heat fluxes. This shortcoming can therefore substantially affect the temperature distribution, rendering the absence of the velocity field a major limitation of energy balance models.

CFD instead couples the temperature and velocity fields inside an urban area as it solves for both these variables. This generally results in a more accurate prediction of the temperature distribution. However, careful and high-resolution modeling with a high number of control volumes or finite elements is necessary in order to obtain accurate and reliable results, resulting in longer computation times. Several guidelines are available to help achieve accurate and reliable simulations of urban physics processes with CFD by limiting errors and uncertainties (e.g. Casey and Wintergerste 2000; Franke et al. 2007, 2011; Tominaga et al. 2008; van Hooff and Blocken 2010; Blocken et al. 2007, Blocken 2015). In order to investigate the ability of CFD to accurately reproduce surface temperatures in an urban area during a heat wave, Toparlar et al. (2015) conducted a case study for the Bergpolder Zuid district in Rotterdam. This study is described in the next subsections.

Bergpolder Zuid district

The Bergpolder Zuid district is composed of both residential and office buildings with several narrow streets and surrounded by large avenues (Figure 22.12). Here, the classification of streets is made with respect to the aspect ratio between the street width and the adjacent building height. In general, most of the streets are fairly narrow with an aspect ratio between 1:1 and 2:1.

The main color of the building walls is red and the rooftops are commonly gray or dark gray. Street materials generally have lighter colors (light gray) compared to the building materials. The vegetation levels of the region are fairly low as trees and green fields are mostly located in a few small courtyards. The only urban water source is the canal located in the north part of the region, and there are no additional water facilities within the district, hence the district has low evapotranspiration levels, which is typical of dense metropolitan cities.

Computational geometry

In order to create an accurate geometrical model of the region, official drawings and documents were acquired from the Municipality of Rotterdam and from the database of AHN.[3] According to the database, the average height of the buildings is 12.6 m with the lowest and highest building having a height of 2.8 m and 51.0 m, respectively. Considering the surroundings of the Bergpolder Zuid district, south of it lies the central district of Rotterdam

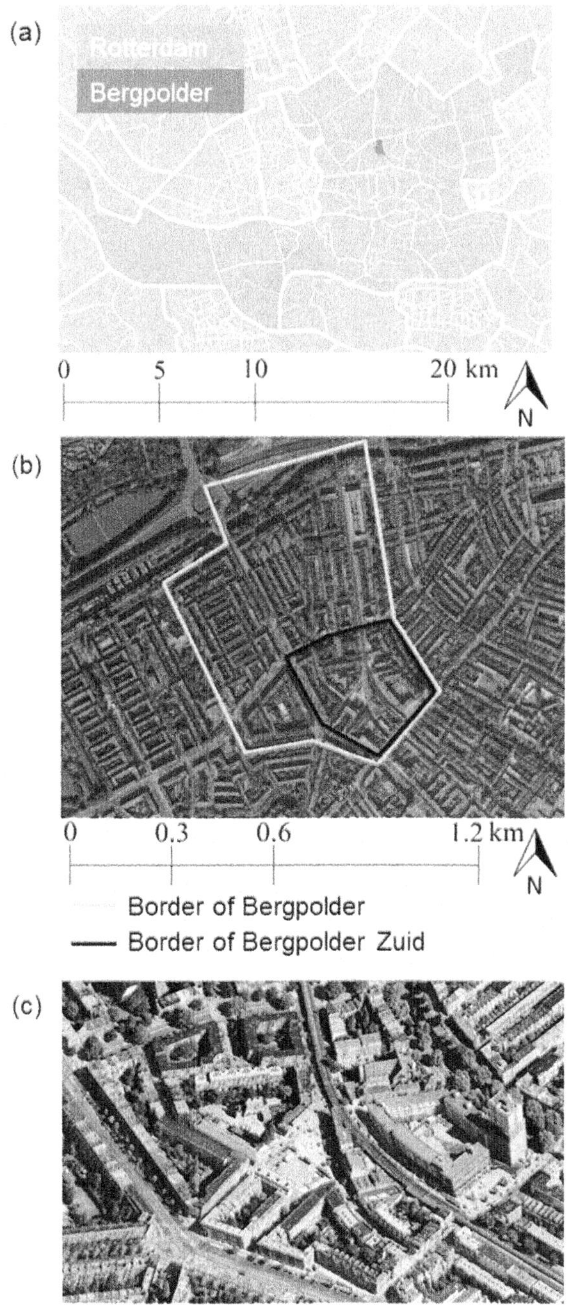

Figure 22.12 (a) Location of Bergpolder inside the city of Rotterdam (modified from Wikipedia).[1] (b) Top view with the borders of Bergpolder and Bergpolder Zuid district (modified from Google Maps[2]). (c) Aerial view of Bergpolder Zuid district (view from south) (modified from Google Maps[2]).

Source: Toparlar et al. 2015, ©Elsevier, reproduced with permission

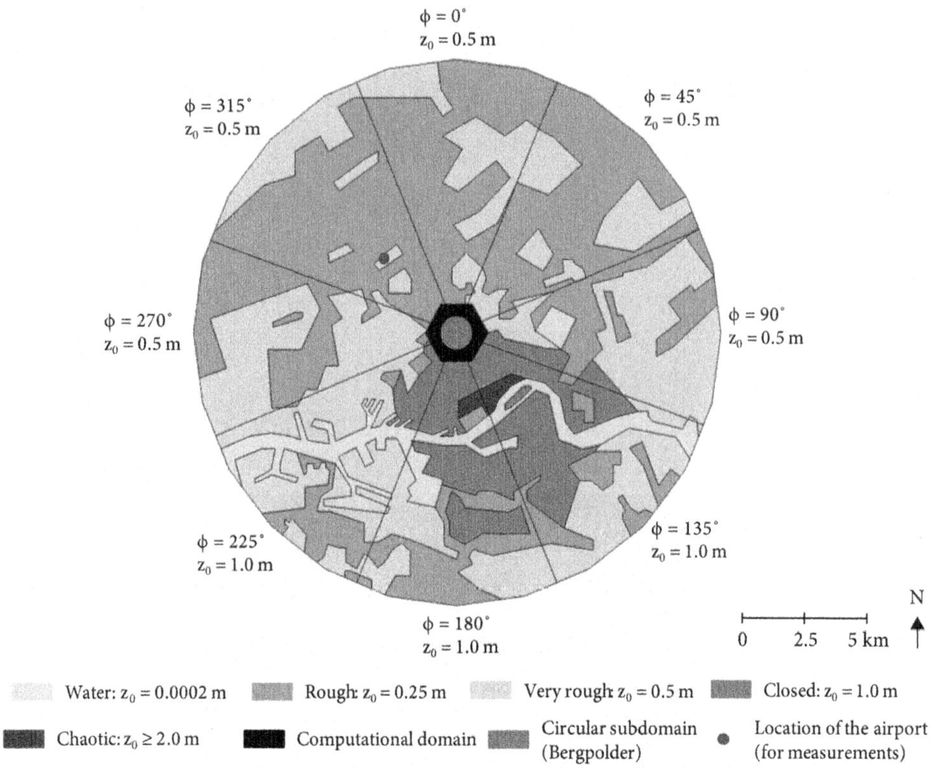

Figure 22.13 Estimated aerodynamic roughness length (z_0) for different wind direction angles and terrain types surrounding the modeled urban area

and to the north, there are mainly green fields until the city of Delft (located in northwest of the district). Based on the updated Davenport roughness classification (Wieringa 1992), the aerodynamic roughness length (z_0) of the surroundings, which is necessary as input for the CFD simulations, is determined as shown in Figure 22.13. The z_0 value is determined as a spatial average of the z_0 values of the different patches of roughness (land use) of the terrain within a 10 km radius upstream of the urban area. A similar methodology was followed in previous CFD studies investigating urban wind flow (e.g. van Hooff and Blocken 2010; Blocken et al. 2012).

Computational domain and grid

The computational domain consists of a hexagonal with edges of 1200 m surrounding a circular subdomain with a diameter of 1200 m (Figure 22.14). The height of the domain is 400 m, resulting in a blockage ratio below the recommended value of 3.0% (Franke et al. 2007; Tominaga et al. 2008; Blocken 2015). It is important to model the surrounding buildings and neighborhood around the site of interest, as they might affect the local flow field as well as the advection and turbulent diffusion of heat. The buildings inside the subdomain are modeled explicitly (i.e. with blocks representing their actual shape and size) and divided into three categories: buildings in Bergpolder Zuid with a high resolution (details of 0.5 m), buildings in the rest of the Bergpolder region (1.0 m) and additional surrounding buildings, which are modeled

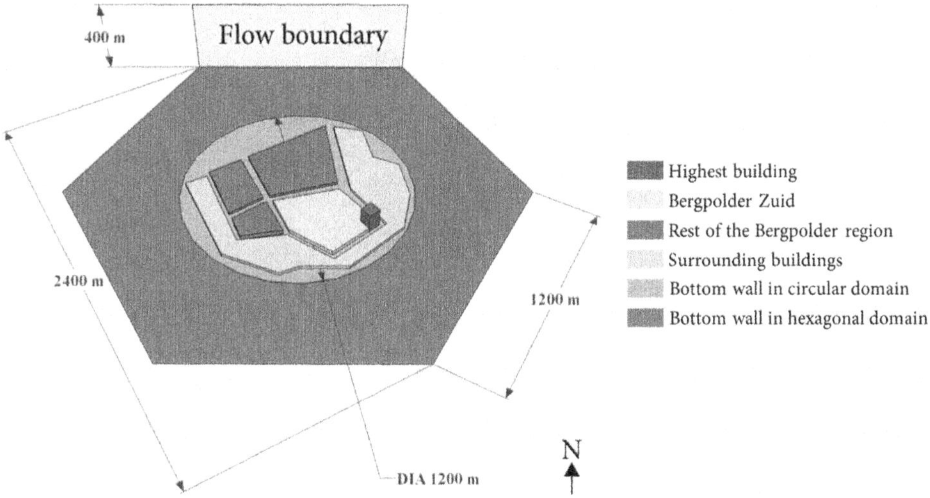

Figure 22.14 Hexagonal computational domain including the circular subdomain with different building categories

coarsely (4.0–8.0 m) and act as obstacles for the approaching flow (Figure 22.14). Elements like trees, cars, sidewalks and street poles are not modeled. The buildings and the urban terrain outside the circular domain are not modeled explicitly, but implicitly: their effect on the flow is implemented by means of appropriate values for the roughness height (k_S) and the roughness constant (C_S) in the wall functions. These roughness parameters are determined by the software dependent relationship with the aerodynamic roughness length (z_0) at the bottom of the domain (Equation 22.1) (Blocken et al. 2007).

$$k_S = \frac{9.793 \cdot z_0}{C_S} \tag{22.1}$$

Grid generation is performed with the surface-grid extrusion technique by van Hooff and Blocken (2010) with at least 10 cells along the building edges and by keeping the stretching ratio below 1.3. This results in a grid with a total of 6,610,456 hexahedral cells (Figure 22.15).

Boundary conditions

Depending on the wind direction, the wind flow is simulated with three hexagonal boundaries as velocity inlet and the other three as pressure outlet. The logarithmic mean wind speed profile (U) (m/s) is imposed with z_0 using Equation 22.2. Here u^* is the atmospheric boundary layer friction velocity (m/s), g is the von Karman constant (0.41) and z is the height coordinate (m).

$$U(z) = \frac{u^*}{\kappa} \cdot \ln\left(\frac{z + z_0}{z_0}\right) \tag{22.2}$$

The turbulence kinetic energy (k) (m^2/s^2) and turbulence dissipation rate (ε) (m^2/s^3) of the neutrally stratified atmospheric boundary layer are imposed according to (Richards and Hoxey 1993) with Equations 22.3 and 22.4. Here C_μ is a constant (0.09).

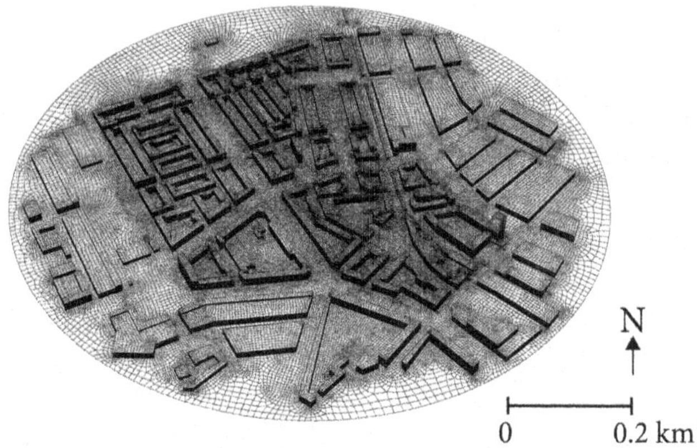

Figure 22.15 Computational grid of the circular subdomain containing the Bergpolder Zuid district

$$k = \frac{u^{*2}}{\sqrt{C_\mu}} \tag{22.3}$$

$$\varepsilon(z) = \frac{u^{*3}}{\kappa(z + z_0)} \tag{22.4}$$

In the near-wall regions, low-Re number modeling or standard wall functions (Launder and Spalding 1974) can be used. Low-Re modeling requires a very high grid resolution close to the walls in order to have control volumes in the very thin laminar sublayer of the boundary layer. Even though wall functions might perform less well for calculating convective heat transfer, they are generally used for this purpose (Blocken et al. 2009). This method is therefore used to limit the computational load.

The air temperature is set as a spatially constant value at the inlet and like the reference wind speed and total solar radiation, it is acquired by the hourly weather data from the KNMI weather station in Rotterdam. The ground plane is modeled as a 10 m thick earth layer with a temperature of 10 °C at 10 m below ground and an absorptivity of 0.6. The buildings have 0.4 m thick brick walls with an absorptivity of 0.75. The internal air temperature of the buildings is set to 24 °C. Evapotranspiration is only applied to the ground plane with a sink value set to 80 W/m^2 during the morning and afternoon and 130 W/m^2 during noon.

Solver settings

The 3-D unsteady Reynolds-averaged Navier-Stokes (URANS) equations are solved with the realizable k-ε turbulence model (Shih et al. 1995) for closure. The radiation is solved with the P-1 radiation model (ANSYS 2009a), and the Boussinesq approximation is used to include buoyancy effects. A 1-D conduction equation is used for the ground and building walls, which calculates the heat transfer through the walls. The solar calculator of ANSYS Fluent is used to estimate the sun direction vector and the diffuse part of the total radiation. The SIMPLE algorithm is applied for pressure-velocity coupling. In addition, only second-order discretization schemes ae used. A time step of 15 minutes and 60 iterations are chosen based on a time step and iteration number sensitivity analysis.

Results and validation

To evaluate the ability of the CFD simulation to reproduce some aspects of a heat wave, the results are validated/compared with experimental data from thermal infrared satellite imagery during a heat wave in July 2006 (Klok et al. 2012). The CFD simulations are performed for five consecutive days (480 time steps) to evaluate the diurnal variation of the temperature field. Ninety sampling points (based on a sensitivity analysis) are placed on the roofs and streets in the computational domain to obtain spatially average values that can be directly compared with the values from the satellite imagery. The comparison of the spatially averaged surface temperature obtained by CFD and satellite imagery is shown in Figure 22.16. Overall, there is a fairly good agreement. The CFD simulations are able to predict the surface temperatures with an average absolute temperature difference of 2.2 °C, generally due to a smaller amplitude of the diurnal variation of surface temperature. Some reasons for the surface temperature deviations are the use of a 1-D wall conduction equation instead of a 3-D one, as some heat is transferred in the planar direction. Second, the satellite data were recorded 42 times during the five days, while the simulations were based on hourly averaged meteorological data. This could cause a mismatch in average surface temperature results. In addition, the satellite data only records a single value for an entire district, of which 20% contains the Bergpolder Zuid region. Local temperature variations in the district might therefore influence the results. Finally, a simplification is made in the thermal stratification of the atmospheric boundary layer, as it was assumed to be neutral with a constant uniform temperature. In reality, the flow field might deviate from this assumption, especially during low wind speeds. Overall, the validation showed that CFD has the capability to accurately predict urban surface temperatures. However, future studies should focus on validation with data at higher spatial resolution.

Figure 22.17 shows simulation results of the wind velocity and surface temperature of the urban region at 10:00 on 15 July 2006. It can be observed that in regions with a low wind velocity (decreased convective heat transport), as in courtyards or street canyons, the surface temperatures are evidently much higher.

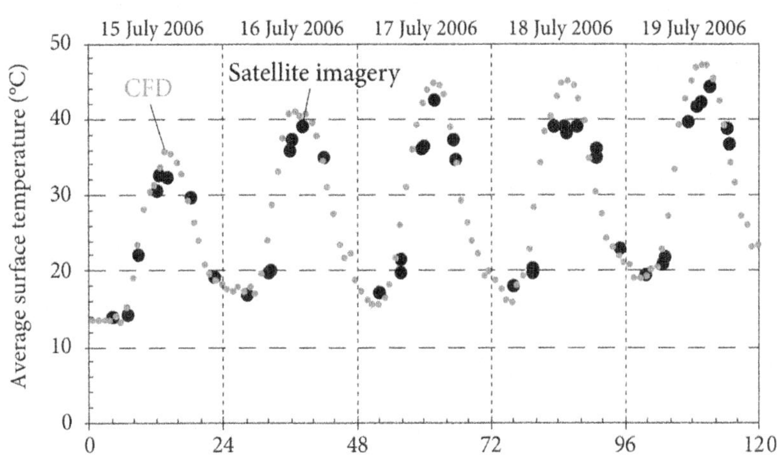

Figure 22.16 Comparison of spatially averaged surface temperature by CFD and satellite imagery for five days in July 2006

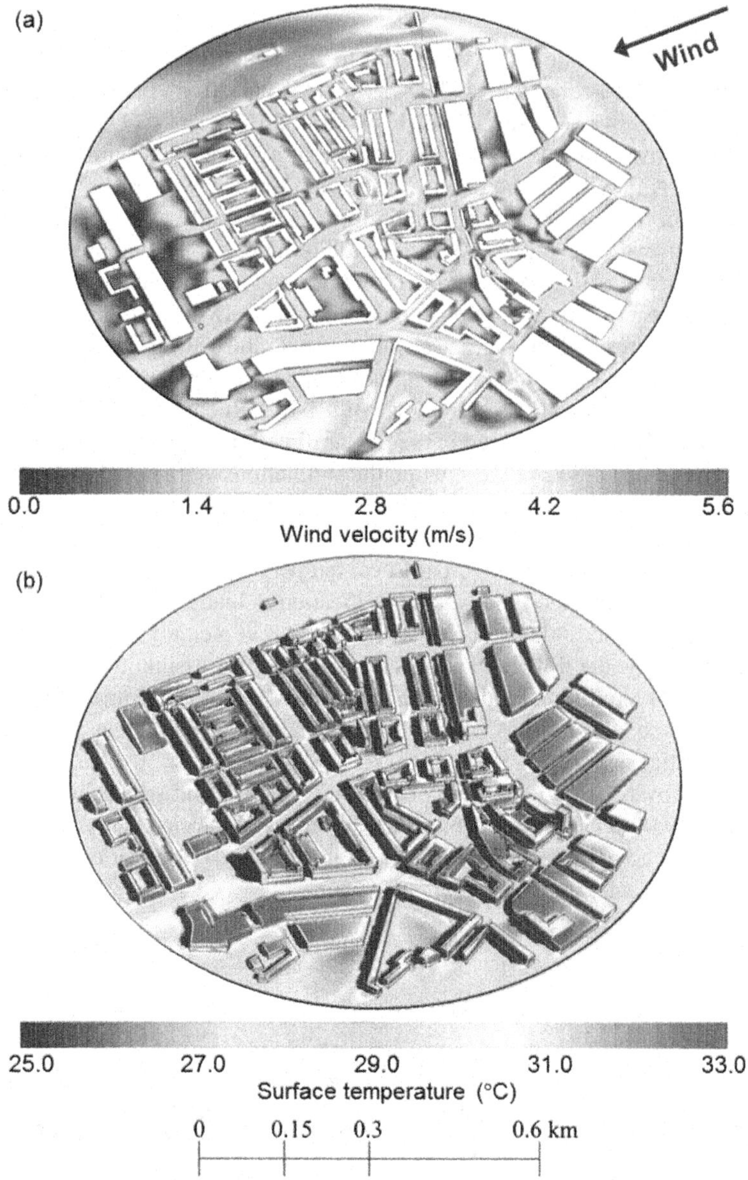

Figure 22.17 CFD simulation results of the wind velocity and surface temperature of the Bergpolder Zuid district

Source: Toparlar et al. 2015, ©Elsevier, reproduced with permission

Climate change adaptation of urban areas: effects of vegetation

Introduction

A promising climate change adaptation measure is the addition of vegetation to an urban area. The related evaporation and transpiration can help to limit high outdoor air temperatures. Vegetation can also provide shading and yield less heat storage. Examples of vegetation in urban areas are parks, green roofs and facades, urban farming and avenue trees. Many different studies have demonstrated the beneficial effects of vegetation on the temperature in urban areas (Taha 1997; Bruse and Fleer 1998; Dimoudi and Nikolopoulou 2003; Wong et al. 2003, 2010; Takahashi et al. 2004; Ali-Toudert and Mayer 2007; Alexandri and Jones 2008; Fröhlich and Matzarakis 2013).

The effects of vegetation on momentum and heat transfer can be simulated with CFD by adding source and sink terms to the momentum equations, the energy equation and the equations of the turbulence model. This approach is briefly described in the next subsection. Based on this approach, Gromke et al. (2015) provided a validation study and a case study on the effects of avenue trees and green roofs and green facades on air temperatures for a street in Arnhem, which is described in later subsections.

Vegetation effects: source and sink terms

In order to model the effects of vegetation on the mean airflow and turbulence, vegetation terms can be added to the transport equations of momentum as in Equation 22.5, turbulence kinetic energy as in Equation 22.6 and turbulence dissipation rate as in Equation 22.7 for cells that contain vegetation (Green 1992; Liu et al. 1996; Sanz 2003):

$$S_{u_i} = -\rho \cdot C_d \cdot LAD \cdot U_i \cdot \boldsymbol{U} \tag{22.5}$$

$$S_k = \rho \cdot C_d \cdot LAD \cdot \left(\beta_p \cdot \boldsymbol{U}^3 - \beta_d \cdot \boldsymbol{U} \cdot k\right) \tag{22.6}$$

$$S_\varepsilon = \rho \cdot C_d \cdot LAD \cdot \frac{\varepsilon}{k}\left(C_{\varepsilon 4} \cdot \beta_p \cdot \boldsymbol{U}^3 - C_{\varepsilon 5} \cdot \beta_d \cdot \boldsymbol{U} \cdot k\right) \tag{22.7}$$

where ρ is the density of air, C_d (= 0.2) is the leaf drag coefficient, LAD is the leaf area density, U_i is the velocity component of direction *i*, $\boldsymbol{U}$ the velocity magnitude, β_p (= 1.0) is the mean kinetic energy fraction that is converted into wake turbulence kinetic energy, β_d (= 5.1) accounts for short-circuiting of the eddy cascade, *k* is the turbulence kinetic energy, ε is the turbulence dissipation rate and $C_{\varepsilon 4}$ and $C_{\varepsilon 5}$ (= 0.9) are empirical coefficients.

Vegetation effects on momentum: validation

Validation of the vegetation effects on the mean airflow and turbulence is done by comparing the CFD simulation results with the measurement data by Amiro (1990) in a 12 m high black spruce forest. Equations 22.5–22.7 are implemented as user-defined functions (UDF) in the simulations (ANSYS 2009b). The computational domain has dimensions of 100 × 25 × 240 m (L × W × H) with coupled periodic boundary conditions for the inlet and outlet side (Figure 22.18). To obtain a streamwise wind speed of 2.6 m/s at 12 m height, an air mass flow rate of 35,000 kg/s is assigned at the inlet. For the ground plane, a no-slip condition is applied with an aerodynamic roughness length z_0 of 0.35 (k_S and C_S are calculated with Equation 22.1). Symmetry boundaries are applied for the sides and top of the domain to simulate a homogenous flow in an extended forest. The spruce forest is defined as a porous zone with the same vertical leaf area density profile (LAD(*z*)) as in the experiment by Amiro (1990). Cubical cells with a 1 m edge length are used to discretize the entire domain.

The comparison of the mean streamwise velocity (*U*) and the square of the shear stress velocity (<*u'w'*>) between the CFD results and the measurement data is shown in Figure 22.19. These results are normalized by height (*z*) at 10 m height and by the wind speed and the squared shear stress velocity at 12 m height. Overall, the simulation results exhibit a good agreement

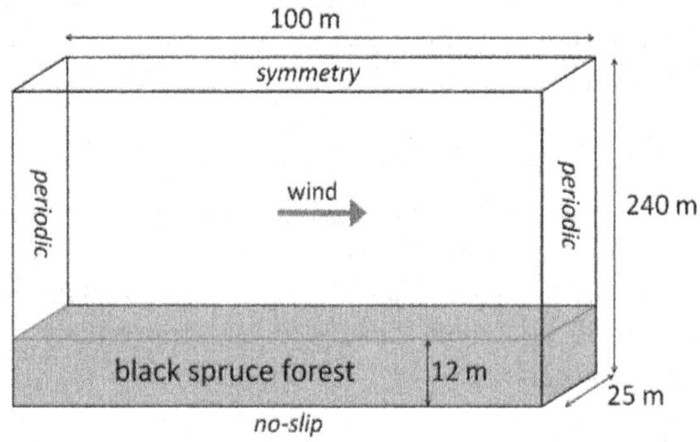

Figure 22.18 Computational domain of the black spruce forest validation study

Source: Gromke et al. 2015, ©Elsevier, reproduced with permission

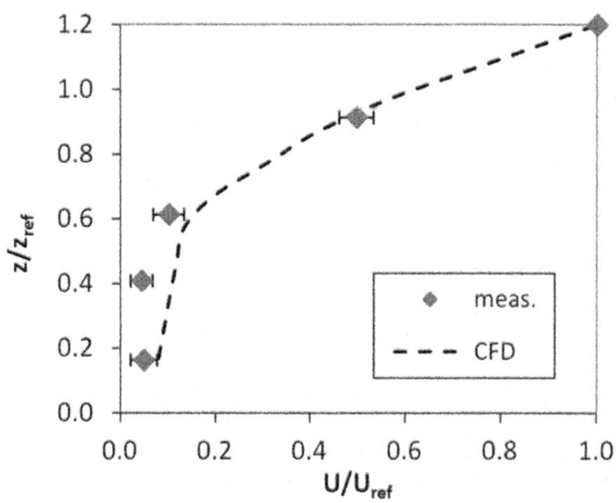

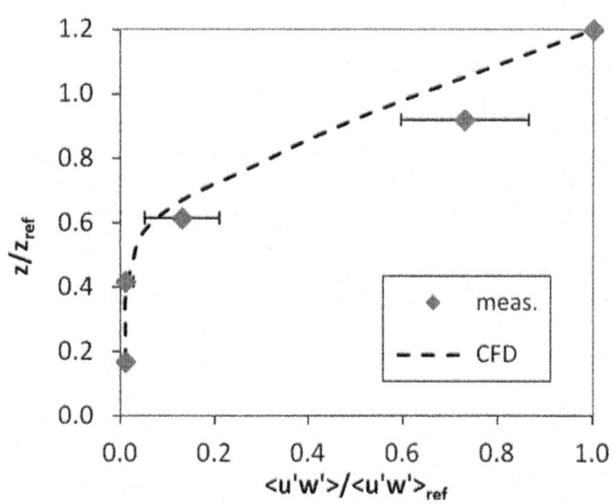

Figure 22.19 Normalized mean wind speed (top) and normalized squared shear stress velocity (bottom)

Source: Gromke et al. 2015, ©Elsevier, reproduced with permission with the measurement data. The largest deviations for the mean streamwise velocity occur at $z/z_{ref} = 0.4$. However, the wind speed at this location is still very small (0.1 m/s), the absolute difference is therefore quite small as well. For the squared shear stress, the maximum relative deviation is around 20% for the upper part of the canopy. This validation study shows that the CFD simulations with the implemented vegetation terms are capable of modeling the mean flow and shear stress velocity within a vegetation canopy.

Vegetation effects on heat transfer: validation

When air flows through vegetation, it will be cooled by transpiration mostly from the leaf surfaces. A more detailed microscale approach to model this would therefore be to consider the latent heat flux that leaves through stomata with convective heat/mass transfer coefficients (CTCs) (Defraeye et al. 2012, 2013). To reduce the associated computational costs, the volumetric cooling power of vegetation (P_C) per unit volume vegetation (V) can be used to model the vegetation effect on the change in air temperature (ΔT) by transpirational cooling with Equation 22.8. Here, $\dot{H}$ is the heat transfer rate, c_p the specific heat capacity and $\dot{m}$ is the mass flow rate.

$$\frac{H}{V} = P_c = c_p \cdot \dot{m} \cdot \Delta T \cdot \frac{1}{V} \Leftrightarrow \Delta T = P_c \cdot V \cdot \frac{1}{\dot{m}} \cdot \frac{1}{c_p} \tag{22.8}$$

In order to determine an appropriate value of P_C, a validation study is performed by steady-state 3-D RANS CFD simulations for a bare and vegetated courtyard and by comparing the simulated air temperature with the experimental data of a field study by Shashua-Bar et al. (2009, 2011). The edges of the building are discretized by cubical cells of 0.15 m length, while outside the courtyard an unstructured grid with an increasing size towards the domain boundary is created. This results in a computational domain of 670,000 hexahedral cells (Figure 22.20).

Meteorological data are used to obtain boundary conditions for three periods on 7 July 2007. The roughness length z_0 is assumed as 0.03 m for the ground and building walls. The vegetation terms in Equations 22.5–22.7 are used to model the aerodynamic effects of the trees inside the courtyard (Figure 22.20). Different P_c values are then assigned to cells that contain vegetation until an agreement in air temperature is achieved with the experimental data. The comparison between measured and simulated air temperature in a bare and vegetated courtyard at 1.5 m height for different P_c values is shown in Figure 22.21. The bare courtyards show a small deviation of only 0.4 °C on average. The average deviations for the P_c values of 250, 500 and 700 W/m^3 are 1.1, 0.5 and 0.0 °C, respectively. Since the bare courtyard has an average deviation of 0.4 °C, a P_c value of 500 W/m^3 is assumed for the courtyard trees with a LAD of 2.0 m^2/m^3. This translates to a volumetric cooling power P_c of 250 W/m^3 per unit LAD. This corresponds well to the data of Rahman et al. (2011), who reported a volumetric cooling power of 284 and 335 W/m^3 per LAD for deciduous trees during the summer. However, different values might be obtained for trees that suffer from limited water supply.

Vegetation effects in an urban area: case study

Gromke et al. (2015) investigated the transpirational cooling effect of various vegetation measures applied in an existing urban street canyon in order to reduce the air temperature during a heat wave. Three different vegetation measures are studied for the case study of the street J.P. van Muijlwijkstraat in Arnhem, the Netherlands (Figure 22.22(a)). These vegetation measures are avenue trees, green facades and green roofs. CFD simulations are carried out to model the

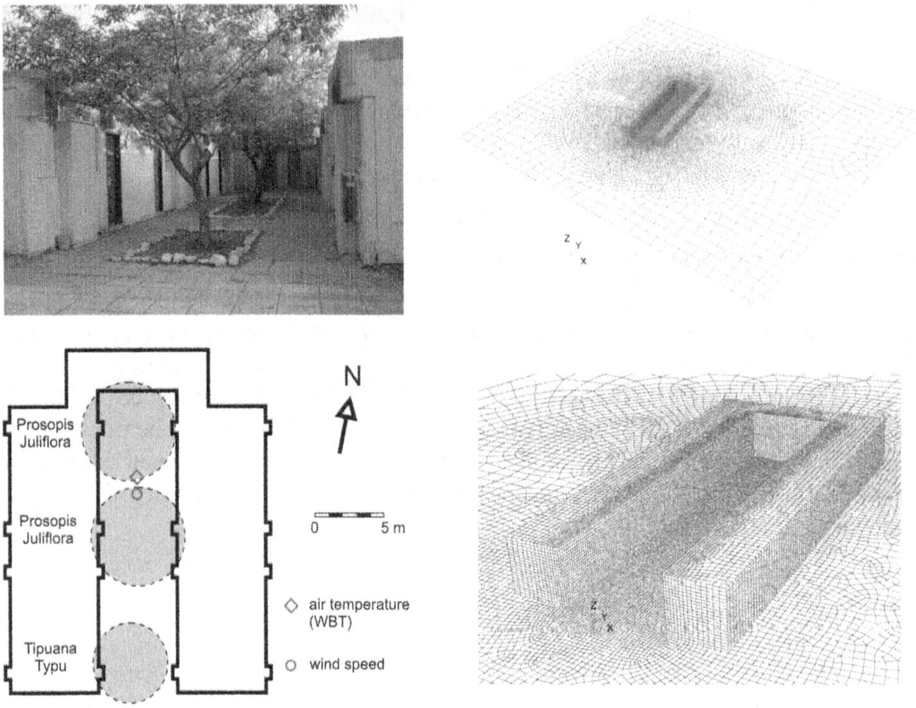

Figure 22.20 Courtyard with trees (Shashua-Bar et al. 2009, 2011); (left, top) photo of courtyard; (left, bottom) ground plan of courtyard; (right) computational grid

Source: Gromke et al. 2015, ©Elsevier, reproduced with permission

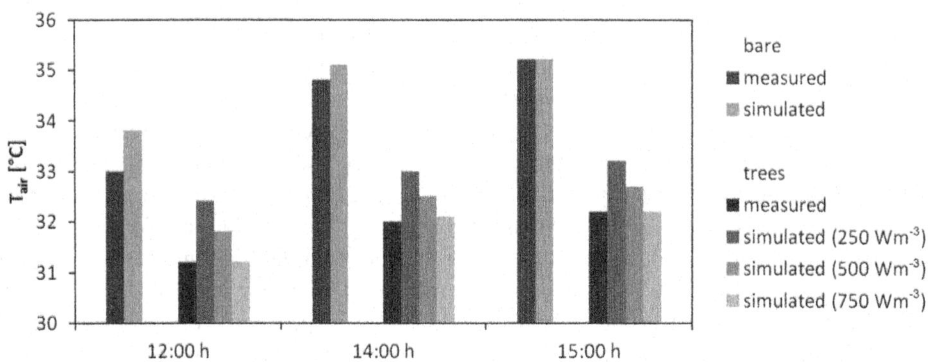

Figure 22.21 Validation results for the air temperature at 1.5 m height for a bare and vegetated courtyard

Source: Gromke et al. 2015, ©Elsevier, reproduced with permission

heat wave of 16 July 2003 with a wind direction aligned along the canyon street (east wind). The effects of vegetation on the mean flow, turbulence and transpirational cooling were validated in the foregoing sections. The simulations can therefore be considered as a reliable estimate for the flow and microclimate in an actual urban environment. However, the outcomes are dependent on the site, modeling technique and meteorological boundary conditions.

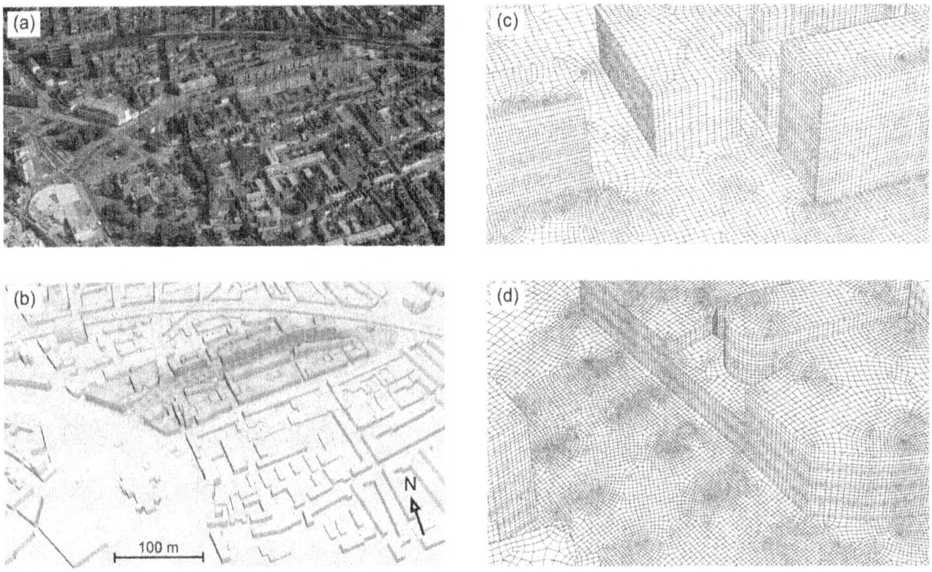

Figure 22.22 Aerial view of the J.P. van Muijlwijkstraat (a); computational grid of inner area (b); zoomed view of grid in the J.P. van Muijlwijkstraat (c, d)

Source: Gromke et al. 2015, ©Elsevier, reproduced with permission

A similar computational domain and grid and similar numerical settings and vegetation terms are used as described in foregoing sections. The computational domain has a total size of 1200 × 950 × 500 m (L × W × H), where the buildings and 350 m long canyon street in an inner area of 900 × 600 m are explicitly modeled. An aerodynamic roughness length z_0 of 0.03 m and 0 m is used for the ground and building surfaces, respectively. Outside the inner area, where no buildings are modeled, a roughness length of 0.5 m is used to represent the urban terrain (Wieringa 1992). The computational grid has a total of 35 million hexahedral cells (Figures 22.22(b)–(d)).

The current situation (reference case – status quo) and the effect of five different vegetation scenarios on the air temperature are investigated (Figure 22.23). The current situation (status quo) consists of a few small deciduous trees in the street canyon and a 600 m^2 triangle patch of trees eastwards of the canyon street. Scenario 1 is a situation with no vegetation at all. Scenario 2 applies avenue trees in the street canyon together with the trees from the current situation. Scenario 3 is a situation with the trees from the status quo together with facade greening. Scenario 4 uses the status quo together with roof greening. Finally, scenario 5 has all vegetation measures and status quo applied together.

The scenarios are simulated for 16 July 2003 at 15:00 local time with a wind speed of 5.1 m/s at 10 m height. The approach flow from the east had an air temperature of 34.5 °C. The four blue lines in Figure 22.23 indicate cross-sections where the effects of the vegetation measures on the air temperature are analyzed in more detail.

For the current situation only, the leaf area density (LAD) is set to 0.55 m^2/m^3. This corresponds to a volumetric cooling power of 137.5 W/m^3 according to the validation study. The simulation results (air temperature at 2 m height) for this scenario are shown in Figure 22.24. The average air temperature at this height is 36.11 °C. The cooling effect of the eastern patch of trees is clearly visible with cooler temperatures inside the patch of trees and downstream of it, partly inside the J.P. van Muijlwijkstraat. However, there seems to be no noticeable cooling effect of the row of small trees inside the street canyon.

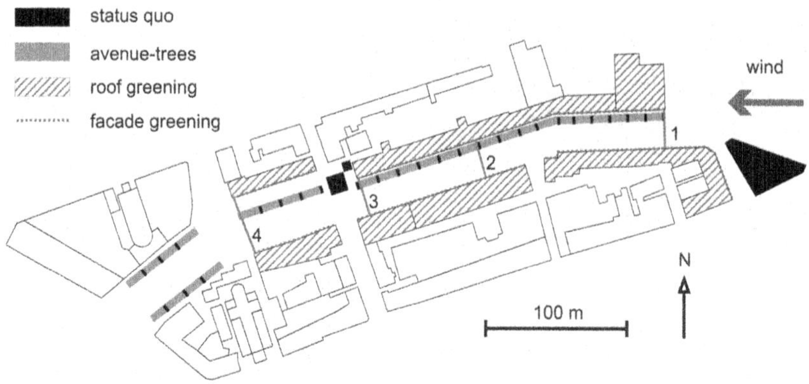

Figure 22.23 Ground plan of J.P. van Muijlwijkstraat with the current situation (status quo) and the different vegetation measures

Source: Gromke et al. 2015, ©Elsevier, reproduced with permission

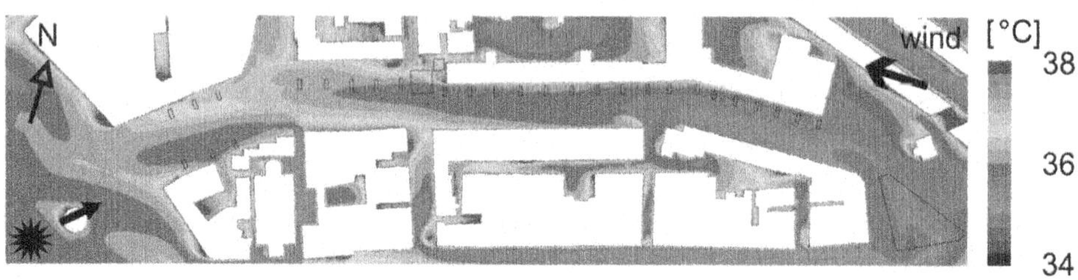

Figure 22.24 Air temperatures at 2 m height in the J.P. van Muijlwijkstraat for the current situation

Source: Gromke et al. 2015, ©Elsevier, reproduced with permission

Figure 22.25 shows the air temperature results at the four cross-sections inside the J.P. van Muijlwijkstraat. The air temperatures at ground level and at the building facades are generally higher. In addition, the average temperature slightly increases from the first to the fourth cross-section. Also there are slightly lower air temperatures at the north side of the street (right side in cross-section), except for cross-section 1. The reason for this is the presence of trees at the north side of the street canyon, which start leeward of cross-section 1. At cross-section 1, the effect of the eastern patch of trees is more pronounced.

The air temperature results of scenario 1 with no vegetation at all are shown in Figure 22.26 (2 m above ground) and Figure 22.27 (cross-sections 1 and 2) to see if the existing trees provided a noticeable cooling effect. The average temperature at 2 m height for scenario 1 is 36.28 °C (0.17 °C higher than current situation), which shows that the existing trees do not provide a significant cooling effect. The temperatures at the eastern patch of trees do however show a temperature difference of around 1 °C compared to scenario 1. This is also shown in the temperature difference in cross-section 1 as depicted in Figure 22.27, indicating a difference up to 0.5 °C at the south side close to the ground. For cross-section 2, there is a temperature difference at the north side of the street canyon, related to the absence of trees in the canyon and the flow from the eastern patch of trees.

Scenario 2 consists of the existing vegetation together with additional avenue trees with the same LAD and volumetric cooling power as the existing trees. The air temperature results are shown in Figure 22.28 (2 m above ground) and Figure 22.29 (cross-sections). The additional

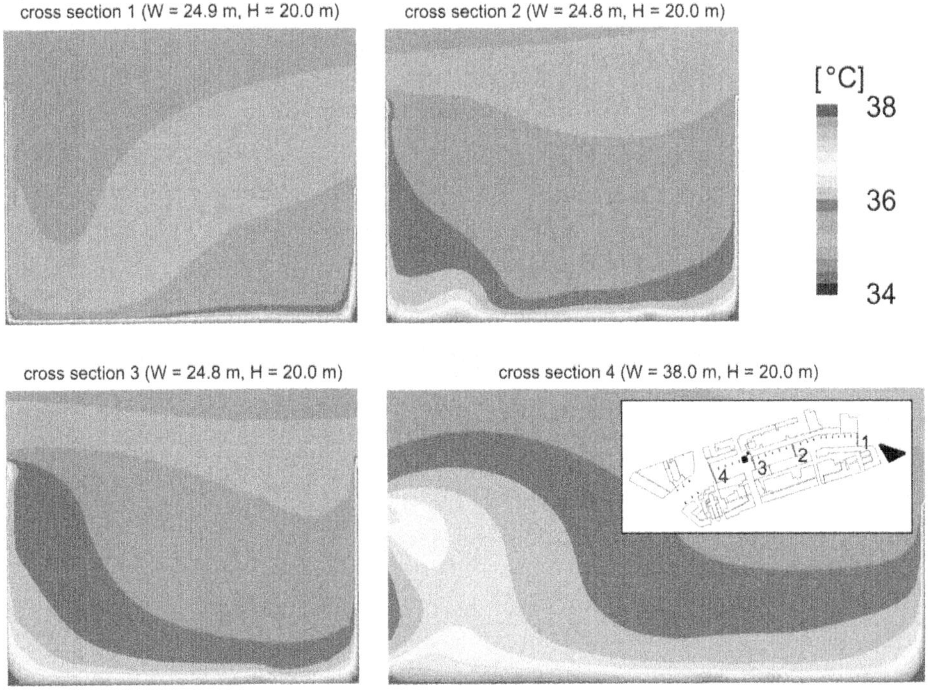

Figure 22.25 Air temperatures in the four cross-sections for the current situation

Source: Gromke et al. 2015, ©Elsevier, reproduced with permission

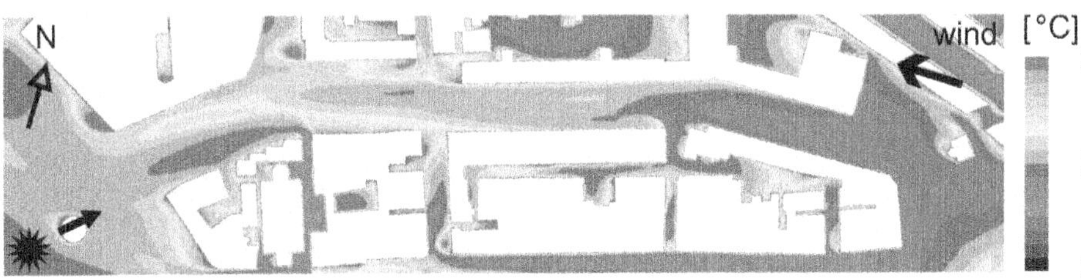

Figure 22.26 Air temperatures at 2 m height in the J.P. van Muijlwijkstraat for scenario 1 (no vegetation)

Source: Gromke et al. 2015, ©Elsevier, reproduced with permission

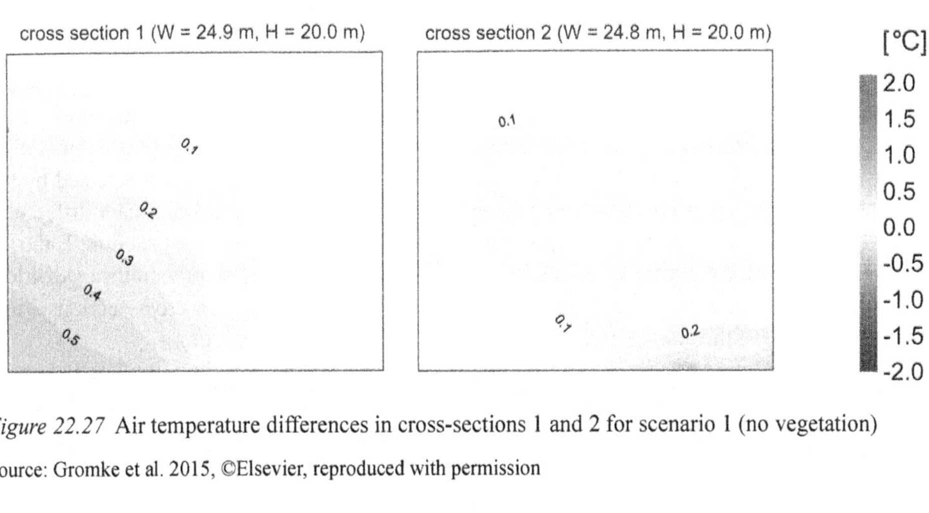

Figure 22.27 Air temperature differences in cross-sections 1 and 2 for scenario 1 (no vegetation)

Source: Gromke et al. 2015, ©Elsevier, reproduced with permission

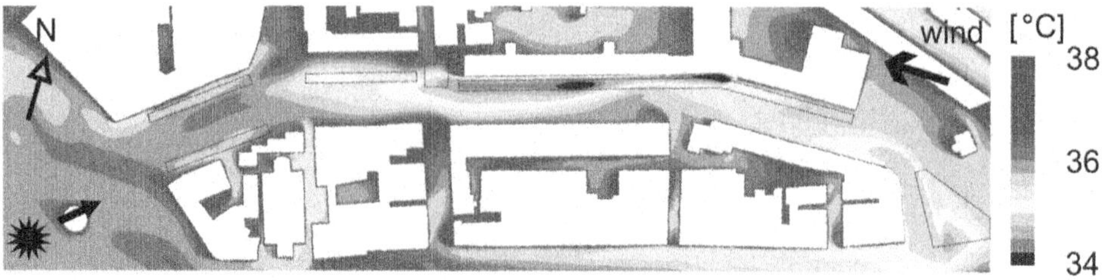

Figure 22.28 Air temperatures at 2 m height in the J.P. van Muijlwijkstraat for scenario 2 (avenue trees)

Source: Gromke et al. 2015, ©Elsevier, reproduced with permission

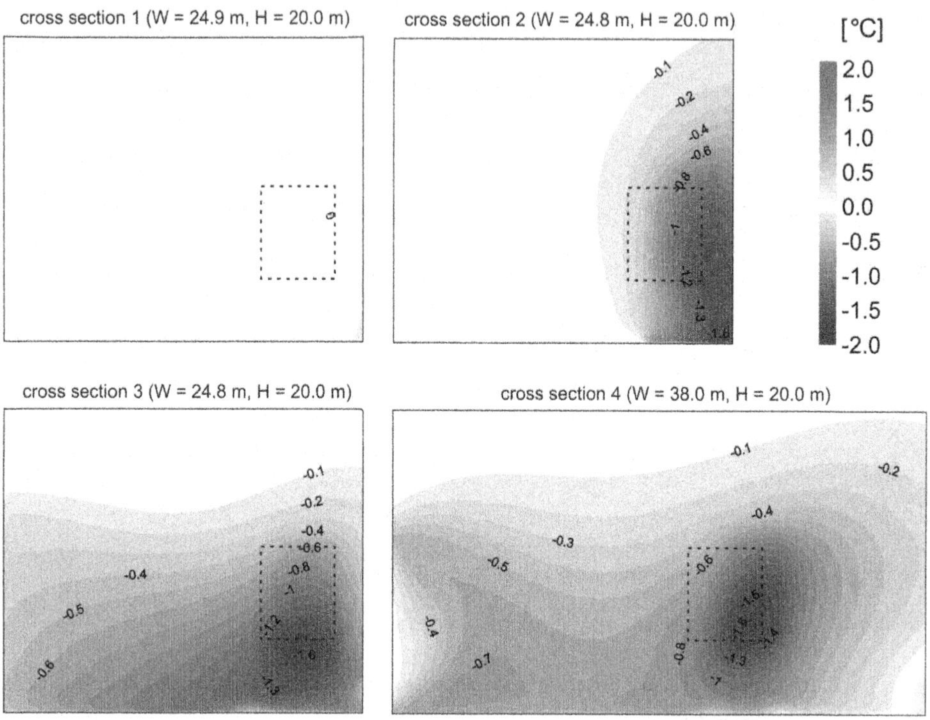

Figure 22.29 Air temperature differences in the four cross-sections for scenario 2 (avenue trees)

Source: Gromke et al. 2015, ©Elsevier, reproduced with permission

trees show a significant transpirational cooling effect for almost the whole north side of the street with temperature differences of 1–2 °C. The total average temperature reduction compared to the current situation is 0.43 °C. The air temperature at cross-section 1 is not yet affected by the avenue trees. For the other cross-sections, the air temperature does show a significant difference, especially close to the position of the avenue trees (dashed square). At cross-section 2, the air temperature between the avenue trees and building wall shows the highest temperature reduction, which suggests a flow through the trees towards the building wall here. For cross-sections 3 and 4, the whole width of the canyon street is affected by this vegetation measure.

The facade greening in scenario 3 is defined by adding a vegetation zone in all cells next to the building facades in the J.P. van Muijlwijkstraat. A LAD value of 0.75 m^2/m^3 (P_c = 187.5 W/m^3) is assigned to these cells, assuming a vegetation coverage of 50% (Susorova et al. 2013). Figure 22.30 shows the temperature difference compared with the reference case for

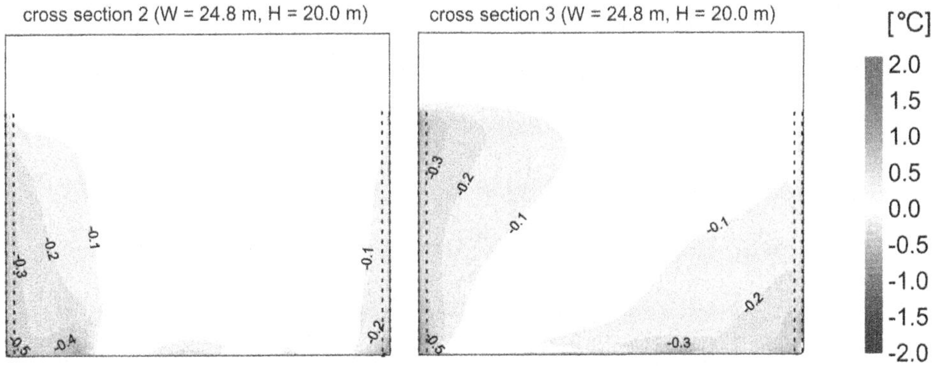

Figure 22.30 Air temperature differences in cross-sections 2 and 3 for scenario 3 (facade greening)

Source: Gromke et al. 2015, ©Elsevier, reproduced with permission

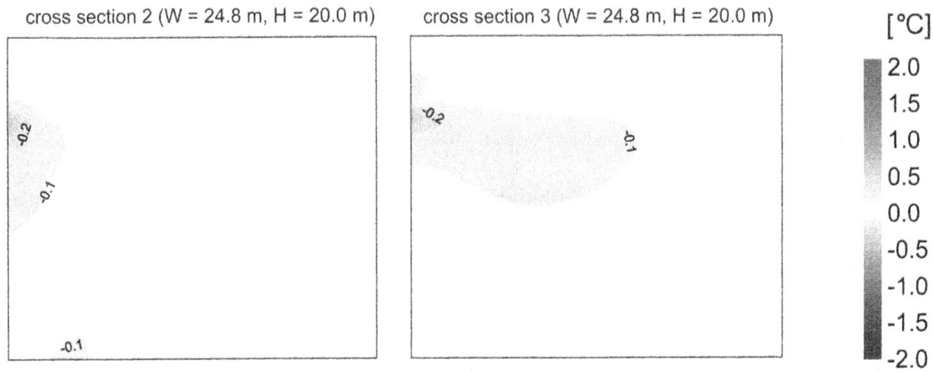

Figure 22.31 Air temperature differences in cross-sections 2 and 3 for scenario 4 (roof greening)

Source: Gromke et al. 2015, ©Elsevier, reproduced with permission

cross-sections 2 and 3. The simulation results show that the cooling effect is only present in the close vicinity of the building facades. The average temperature at 2 m above ground is also only 0.04 °C lower than the existing situation.

Roof greening (scenario 4) is defined with a LAD value of 1.5 m^2/m^3 for the cells adjacent to the roof surfaces of the buildings surrounding the street canyon. Normally higher values are present for roof greening (Barrio 1998). However, only a coverage of 25% is assumed because of obstacles such as antennas and PV cells. The air temperature differences between the existing situation and scenario 4 for cross-sections 2 and 3 is shown in Figure 22.31. Overall, the cooling effect is very small with no noticeable changes of the air temperature at 2 m height. A reason for this is that the air cooled by the roof greening is convected over the rooftops due to the eastern wind, with only a small portion entering the canyon street.

Finally, for scenario 5, all vegetation measures are applied together to investigate possible interactions. The air temperature differences between the existing situation and scenario 5 for cross-sections 2 and 3 are shown in Figure 22.32. The temperature reduction is dominated by the addition of avenue trees at the north side and green facades at the south side. The average air temperature at 2 m height is 35.59 °C (0.52 °C lower than the existing situation), which is the highest reduction in air temperature of all vegetation scenarios.

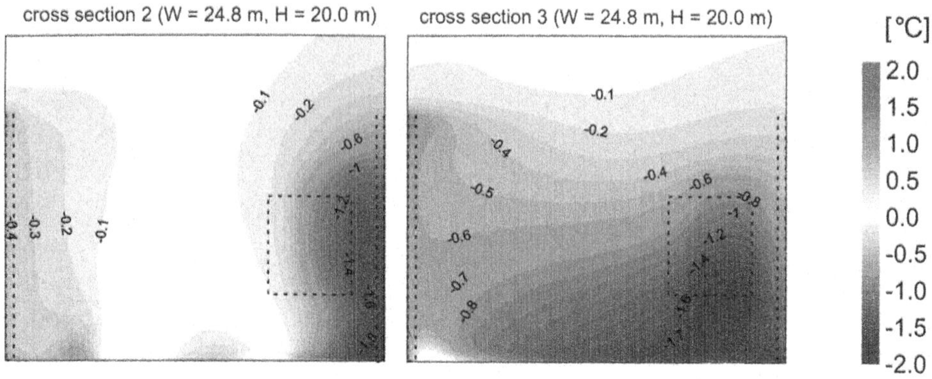

Figure 22.32 Air temperature differences in cross-sections 2 and 3 for scenario 5 (all vegetation measures together)

Source: Gromke et al. 2015, ©Elsevier, reproduced with permission

Overall, there seems to be little interaction between the different vegetation measures, when they are applied together.

Discussion

Concerning the indoor environment of buildings, building energy simulation (BES) is considered to be the most straightforward and appropriate simulation tool. In this chapter, BES was applied to three types of buildings with construction characteristics from the 1970s (low insulation levels) and 2012 (high insulation levels). Overall, applying natural ventilation throughout the day and applying exterior solar shading emerged as the most effective adaptation measures, and this was the case for all three building types with both sets of construction characteristics. Increasing the thermal resistance of the building actually had a negative effect on the overheating hours. Reducing the thermal mass in a building can have a positive effect for the indoor temperatures during the night, but only when the heat inside the building can be released either through the building envelope or by ventilation. In terms of costs (installation and maintenance), natural ventilation and exterior solar shading are considered less expensive than several other measures, such as reducing the thermal mass and providing green roofs and green facades.

Concerning the outdoor environment of buildings, energy balance models or computational fluid dynamics (CFD) can be used. For high-resolution information at the level of individual streets, CFD should be preferred over energy balance models. The validation study for the Bergpolder Zuid district showed that CFD simulations based on the 3-D unsteady Reynolds-averaged Navier-Stokes equations could reproduce the temporal evolution and the magnitude of the surface temperatures with a satisfactory accuracy.

In CFD, the effects of vegetation can be modeled with source and sink terms in the momentum and energy equations and in the equations of the turbulence model. The cooling effect by avenue trees, green roofs and green facades was simulated with CFD and with appropriate sink and source terms for a street in Arnhem under the conditions of a heat wave. Overall, the cooling effect by the avenue trees appeared most pronounced. Under the conditions of the present study, the cooling effect by green facades and roofs was very limited. By applying avenue-tree rows (with large crowns) on both sides of a street canyon, a large vegetation volume could be achieved which should lead to an even more noticeable increase in transpirational cooling. To achieve a citywide temperature reduction during the summer, a densely distributed arrangement of trees and other vegetation measures would be necessary, provided that the vegetation is regularly irrigated. However, adding avenue trees will decrease the wind speed in the street canyons, which can give rise to less ventilation and higher air pollutant concentrations due to traffic in these street canyons (Gromke and Ruck 2007).

The results found by the case study on vegetation showed some similar trends as in other studies with vegetation. Namely, the air temperature reduction of 1 °C within the first 10 m downwind of the avenue trees is in general agreement with a study by Dimoudi and Nikolopoulou (2003). Also, the air temperature reduction of the green facades up to a few meters away from the building surfaces was found by Wong et al. (2010). Finally, Alexandri and Jones (2008) also reported a similar weaker cooling effect by roof greening, compared to facade greening. Regardless of this, it is important to note that the findings in this study were obtained for a specific site and for very specific meteorological conditions. Care must therefore be taken in generalizing these results. A different wind direction might lead to less temperature reduction overall, but perhaps also to a higher cooling effect for the roof greening. Also, different wind speeds and LAD values might lead to different results for different vegetation measures. Furthermore, the vegetation was assumed to be subjected to unlimited water supply, as a limited water supply will lead to reduced volumetric cooling powers (Shashua-Bar et al. 2009, 2011). Other vegetation effects such as shading, trapping radiation and a low heat capacity were not taken into account in this study, which might have resulted in an underestimation of the air temperature reduction.

Synopsis

This chapter has presented simulations to analyze the effects of heat waves on buildings and urban areas and the effect of adaptation measures in terms of overheating hours and degree hours in buildings and surface and air temperatures in urban areas. Building energy simulation has been used to evaluate the effectiveness of a range of measures at the building scale to reduce indoor overheating. Computational fluid dynamics simulations for an urban area and for vegetative effects have been validated and subsequently applied to reproduce surface and air temperatures in two case studies.

Notes

1 Wikipedia source: http://commons.wikimedia.org/wiki/File:Rotterdamse_wijken-bergpolder.PNG
2 Google Maps source: http://goo.gl/maps/zEziy
3 AHN: Actueel Hoogtebestand Nederland (Updated Height Data of the Netherlands) www.ahn.nl/

References

Agentschap, N.L. (2013) *Referentiewoningen nieuwbouw 2013*. Sittard, The Netherlands: Agentschap NL.

Akbari, H., Bretz, S., Kurn, D.M. and Hanford, J. (1997) Peak power and cooling energy savings of high-albedo roofs. *Energy and Buildings* 25(2): 117–126.

Alexandri, E. and Jones, P. (2008) Temperature decreases in an urban canyon due to green walls and green roofs in diverse climates. *Building and Environment* 43(4): 480–493.

Ali-Toudert, F. and Mayer, H. (2007) Effects of asymmetry, galleries, overhanging facades and vegetation on thermal comfort in urban street Canyons. *Solar Energy* 81(6): 742–754.

Amiro, B.D. (1990) Comparison of turbulence statistics within three boreal forest Canopies. *Boundary-Layer Meteorology* 51(1–2): 99–121.

ANSYS. (2009a) *ANSYS FLUENT 12.0 Theory Guide*. Release 12.0 © ANSYS, Inc. 2009-01-23. ANSYS Inc, 275 Technology Drive, Canonsburg, PA 15317.

ANSYS. (2009b) *ANSYS FLUENT 12.0 UDF Manual*. Release 12.0 © ANSYS, Inc. 2009-01-23. ANSYS Inc, 275 Technology Drive, Canonsburg, PA 15317.

Awbi, H.B. (2003) *Ventilation of Buildings*. London: Taylor & Francis.

Barrio, E.P.D. (1998) Analysis of the green roofs cooling potential in buildings. *Energy and Buildings* 27(2): 179–193.

Blocken, B. (2015) Computational fluid dynamics for urban physics: Importance, scales, possibilities, limitations and ten tips and tricks towards accurate and reliable simulations. *Building and Environment* 91: 219–245.

Blocken, B., Defraeye, T., Derome, D. and Carmeliet, J. (2009) High-resolution CFD simulations for forced convective heat transfer coefficients at the facade of a low-rise building. *Building and Environment* 44(12): 2396–2412.

Blocken, B., Janssen, W.D., van Hooff, T. (2012) CFD simulation for pedestrian wind comfort and wind safety in urban areas: General decision framework and case study for the eindhoven university campus. *Environmental Modelling & Software* 30: 15–34.

Blocken, B., Stathopoulos, T. and Carmeliet, J. (2007) CFD simulation of the atmospheric boundary layer: Wall function problems. *Atmospheric Environment* 41(2): 238–252.

Bruse, M. and Fleer, H. (1998) Simulating Surface – plant – air interactions inside urban environments with a three dimensional numerical model. *Environmental Modelling & Software* 13(3–4): 373–384.

Casey, M. and Wintergerste, T. (2000) *Best Practices Guidelines: ERCOFTAC Special Interest Group on "Quality and Trust in Industrial CFD"*. Ercoftac.

Coley, D., Kershaw, T. and Eames, M. (2012) A comparison of structural and behavioural adaptations to future proofing buildings against higher temperatures. *Building and Environment* 55: 159–166.

Defraeye, T., Herremans, E., Verboven, P., Carmeliet, J. and Nicolai, B. (2012) Convective heat and mass exchange at surfaces of horticultural products: A microscale CFD modelling approach. *Agricultural and Forest Meteorology* 162: 71–84.

Defraeye, T., Verboven, P., Derome, D., Carmeliet, J. and Nicolai, B. (2013) Stomatal transpiration and droplet evaporation on leaf surfaces by a microscale modelling approach. *International Journal of Heat and Mass Transfer* 65: 180–191.

Dimoudi, A. and Nikolopoulou, M. (2003) Vegetation in the urban environment: Microclimatic analysis and benefits. *Energy and Buildings* 35(1): 69–76.

Entranze. (2010) *Share of Dwellings with Air Conditioning*. www.entranze.enerdata.eu/#/share-of-dwellings-with-air-conditioning.html.

Franke, J., Hellsten, A., Schlünzen, H. and Carissimo, B. (2007) Best practice guideline for the CFD simulation of flows in the urban environment. *COST Action*, pp. 51.

Franke, J., Hellsten, A., Schlunzen, K.H. and Carissimo, B. (2011) The COST 732 best practice guideline for CFD simulation of flows in the urban environment: A summary. *International Journal of Environment and Pollution* 44(1): 419–427.

Fröhlich, D. and Matzarakis, A. (2013) Modeling of changes in thermal bioclimate: Examples based on urban spaces in Freiburg, Germany. *Theoretical and Applied Climatology* 111(3–4): 547–558.

Green, S.R. (1992) Modelling turbulent air flow in a stand of widely-spaced trees. *Phoenics Journal Computational Fluid Dynamics and Its Applications* 5(3): 294–312.

Gromke, C., Blocken, B., Janssen, W., Merema, B., van Hooff, T. and Timmermans, H. (2015) CFD analysis of transpirational cooling by vegetation: Case study for specific meteorological conditions during a heat wave in Arnhem, Netherlands. *Building and Environment* 83: 11–26.

Gromke, C. and Ruck, B. (2007) Influence of trees on the dispersion of pollutants in an urban street canyon – experimental investgation of the flow and concentration field. *Atmospheric Environment* 41(16): 3287–3302.

Gu, L. and Crawley, D. (2009) *Overview of EnergyPlus Development: New Capabilities and Applications for Integrated Building System*. Syracuse: CHAMPS.

Henninger, R. and Witte, M. (2011) *EnergyPlus Testing with Building Thermal Envelope and Fabric Load Tests from ANSI/ASHRAE Standard 140–2007*. US Department of Energy.

IPCC. (2001) *Climate Change 2001: The Scientific Basis*. Contribution of Working Group I to the Third Assessment Report of the Intergovernmental Panel on Climate Change [Houghton, J.T., Ding, Y., Griggs, D.J., Noguer, M., van der Linden, P.J., Dai, X., Maskell, K. and Johnson, C.A. (eds)]. Cambridge, New York: Cambridge University Press, 881 pp.

Klok, L., Zwart, S., Verhagen, H. and Mauri, E. (2012) The surface heat island of rotterdam and its relationship with urban surface characteristics. *Resources, Conservation and Recycling* 64: 23–29.

Launder, B.E. and Spalding, D.B. (1974) The numerical computation of turbulent flows. *Computer Methods in Applied Mechanics and Engineering* 3(2): 269–289.

Li, D.H.W., Yang, L. and Lam, J.C. (2012) Impact of climate change on energy use in the built environment in different climate zones – a review. *Energy* 42(1): 103–112.

Liu, J., Chen, J.M., Black, T.A. and Novak, M.D. (1996) E-ε modelling of turbulent air flow downwind of a model forest edge. *Boundary-Layer Meteorology* 77(1): 21–44.

Liu, Y.S., Miao, S.G., Zhang, C.L., Cui, G.X., Zhang, Z.S. (2012) Study on micro-atmospheric environment by coupling large eddy simulation with mesoscale model. *Journal of Wind Engineering and Industrial Aerodynamics* 107–108: 106–117.

Mirzaei, P.A. and Haghighat, F. (2010) Approaches to study urban heat island – abilities and limitations. *Building and Environment* 45(10): 2192–2201.

Nishimura, N., Nomura, T., Iyota, H. and Kimoto, S. (1998) Novel water facilities for creation of comfortable urban micrometeorology. *Solar Energy* 64(4): 197–207.

Oke, T.R. (1982) The energetic basis of the urban heat island. *Quarterly Journal of the Royal Meteorological Society* 108(455): 1–24.

Pachauri, R.K. and Reisinger, A. (2007) IPCC fourth assessment report. *IPCC Fourth Assessment Report*, Geneva, Switzerland.

Peeters, L., Dear, R. de, Hensen, J. and D'haeseleer, W. (2009) Thermal comfort in residential buildings: Comfort values and scales for building energy simulation. *Applied Energy* 86(5): 772–780.

Porritt, S., Shao, L., Cropper, P. and Goodier, C. (2011) Adapting dwellings for heat waves. *Sustainable Cities and Society* 1(2): 81–90.

Porritt, S.M., Cropper, P.C., Shao, L. and Goodier, C.I. (2012) Ranking of interventions to reduce dwelling overheating during heat waves. *Energy and Buildings* 55: 16–27.

Rahman, M.A., Smith, J.G., Stringer, P. and Ennos, A.R. (2011) Effect of rooting conditions on the growth and cooling ability of Pyrus Calleryana. *Urban Forestry & Urban Greening* 10(3): 185–192.

Richards, P.J. and Hoxey, R.P. (1993) Appropriate boundary conditions for computational wind engineering models using the K-ϵ turbulence model. *Journal of Wind Engineering and Industrial Aerodynamics* 46: 145–153.

Salcedo Rahola, T.B., van Oppen, P. and Mulder, K. (2009) *Heat in the City*. Delft, The Netherlands.

Sanz, C. (2003) A note on K-ε modelling of vegetation canopy air-flows. *Boundary-Layer Meteorology* 108(1): 191–197.

Shashua-Bar, L., Pearlmutter, D. and Erell, E. (2009) The cooling efficiency of urban landscape strategies in a hot dry climate. *Landscape and Urban Planning* 92(3): 179–186.

Shashua-Bar, L., Pearlmutter, D. and Erell, E. (2011) The influence of trees and grass on outdoor thermal comfort in a hot-arid environment. *International Journal of Climatology* 31(10): 1498–1506.

Shih, T.-H., Liou, W.W., Shabbir, A., Yang, Z. and Zhu, J. (1995) A new K-ϵ Eddy viscosity model for high reynolds number turbulent flows. *Computers & Fluids* 24(3): 227–238.

Susorova, I., Angulo, M., Bahrami, P. and Stephens, B. (2013) A model of vegetated exterior facades for evaluation of wall thermal performance. *Building and Environment* 67: 1–13.

Taha, H. (1997) Urban climates and heat islands: Albedo, evapotranspiration, and anthropogenic heat. *Energy and Buildings* 25(2): 99–103.

Takahashi, K., Yoshida, H., Tanaka, Y., Aotake, N. and Wang, F. (2004) Measurement of thermal environment in Kyoto City and its prediction by CFD simulation. *Energy and Buildings* 36(8): 771–779.

Tominaga, Y., Mochida, A., Yoshie, R., Kataoka, H., Nozu, T., Yoshikawa, M. and Shirasawa, T. (2008) AIJ Guidelines for Practical Applications of CFD to Pedestrian Wind Environment around Buildings. *Journal of Wind Engineering and Industrial Aerodynamics* 96(10): 1749–1761.

Tominaga, Y., Sato, Y., Sadohara, S. (2015) CFD simulations of the effect of evaporative cooling from water bodies in a micro-scale urban environment: Validation and application studies. *Sustainable Cities and Society* 19: 259–270.

Toparlar, Y., Blocken, B., Maiheu, B. and van Heijst, G.J.F. (2017) A review on the CFD analysis of urban microclimate. *Renewable and Sustainable Energy Reviews* 80: 1613–1640.

Toparlar, Y., Blocken, B., Maiheu, B. and van Heijst, G.J.F. (2018) The effect of an urban park on the microclimate in its vicinity: A case study for Antwerp, Belgium. *International Journal of Climatology* 38: e303–e322.

Toparlar, Y., Blocken, B., Vos, P., Janssen, W., van Hooff, T., Montazeri, H. and Timmermans, H. (2015) CFD simulation and validation of urban microclimate: A case study for Bergpolder Zuid, Rotterdam. *Building and Environment* 83: 79–90.

van den Hurk, B., Klein Tank, A., Lenderink, G., van Ulden, A., Van Oldenborgh, G.J., Katsman, C., Van den Brink, H., Keller, F., Bessembinder, J. and Burgers, G. (2006) *KNMI Climate Change Scenarios 2006 for the Netherlands*. KNMI De Bilt.

Van Hooff, T. and Blocken, B. (2010) Coupled urban wind flow and indoor natural ventilation modelling on a high-resolution grid: A case study for the Amsterdam ArenA stadium. *Environmental Modelling & Software* 25(1): 51–65.

Van Hooff, T., Blocken, B., Hensen, J.L.M. and Timmermans, H.J.P. (2014) On the predicted effectiveness of climate adaptation measures for residential buildings. *Building and Environment* 82: 300–316.

van Kempen, P. (2000) Marktanalyse En-Prognose van airconditioningsystemen. *Van Kempen Onderzoek En Advies*.

Wieringa, J. (1992) Updating the Davenport roughness classification. *Journal of Wind Engineering and Industrial Aerodynamics* 41–44: 357–368.

Wong, N.H., Chen, Y., Ong, C.L. and Sia, A. (2003) Investigation of thermal benefits of rooftop garden in the tropical environment. *Building and Environment* 38(2): 261–270.

Wong, N.H., Kwang Tan, A.Y., Chen, Y., Sekar, K., Tan, P.Y., Chan, D., Chiang, K. and Wong, N.C. (2010) Thermal evaluation of vertical greenery systems for building walls. *Building and Environment* 45(3): 663–672.

Yang, A.S., Juan, Y.H., Wen, C.Y., Chang, C.J. (2017) Numerical simulation of cooling effect of vegetation enhancement in a subtropical urban park. *Applied Energy* 192: 178–200.

Zhai, Z., Johnson, M.H. and Krarti, M. (2011) Assessment of natural and hybrid ventilation models in whole-building energy simulations. *Energy and Buildings* 43(9): 2251–2261.

Recommended reading

AIAA. (1998) Guide for the verification and validation of computational fluid dynamics simulations. *American Institute of Aeronautics and Astronautics*, AIAA-G-077–1998, Reston, VA.

Allegrini, J. and Carmeliet, J. (2018) Simulations of local heat islands in Zürich with coupled CFD and building energy models. *Urban Climate* 24: 340–359.

Blocken, B. (2014) 50 Years of computational wind engineering: Past, present and future. *Journal of Wind Engineering and Industrial Aerodynamics* 129: 69–102.

Blocken, B. (2015) Computational fluid dynamics for urban physics: Importance, scales, possibilities, limitations and ten tips and tricks towards accurate and reliable simulations. *Building and Environment* 91: 219–245.

Blocken, B. (2018) LES Over RANS in building simulation for outdoor and indoor applications: A foregone conclusion?, *Building Simulation*, In press. doi.org/10.1007/s12273-018-0459-3.

Blocken, B., Stathopoulos, T., Carmeliet, J. and Hensen, J.L.M. (2011) Application of CFD in building performance simulation for the outdoor environment: An overview. *Journal of Building Performance Simulation* 4(2): 157–184.

Casey, M. and Wintergerste, T. (2000) *Best Practices Guidelines: ERCOFTAC Special Interest Group on "Quality and Trust in Industrial CFD*". Ercoftac.

Coakley, D., Raftery, P. and Kean, M. (2014) A review of methods to match building energy simulation models to measured data. *Renewable & Sustainable Energy Reviews* 37: 123–141.

Crawley, D.B., Hand, J.W., Kurnmert, M. and Griffith, B.T. (2008) Contrasting the capabilities of building energy performance simulation programs. *Building and Environment* 43: 661–673.

Ferziger, J.H. (1990) Approaches to turbulent flow computation: Applications to flow over obstacles. *Journal of Wind Engineering and Industrial Aerodynamics* 35: 1–19.

Franke, J., Hellsten, A., Schlünzen, H. and Carissimo, B. (2007) Best practice guideline for the CFD simulation of flows in the urban environment. *COST Action*, p. 51.

Gromke, C. and Blocken, B. (2015a) Influence of avenue-trees on air quality at the urban neighborhood scale. Part I: Quality assurance studies and turbulent Schmidt number analysis for RANS CFD simulations. *Environmental Pollution* 196: 214–223.

Gromke, C. and Blocken, B. (2015b) Influence of avenue-trees on air quality at the urban neighborhood scale. Part II: Traffic pollutant concentrations at pedestrian level. *Environmental Pollution* 196: 176–184.

Meroney, R.N. and Derickson, R. (2014) Virtual reality in wind engineering: The windy world within the computer. *Journal of Wind Engineering* 11(2): 11–26.

Mirzaei, P.A. and Haghighat, F. (2010) Approaches to study urban heat island – abilities and limitations. *Building and Environment* 45(10): 2192–2201.

Mochida, A. and Lun, I.Y.F. (2008) Prediction of wind environment and thermal comfort at pedestrian level in urban area. *Journal of Wind Engineering and Industrial Aerodynamics* 96(10–11): 1498–1527.

Montazeri, H., Toparlar, Y., Blocken, B. and Hensen, J.L.M. (2017) Simulating the cooling effects of water spray systems in urban landscapes: A computational fluid dynamics study in Rotterdam, the Netherlands. *Landscape and Urban Planning* 159: 85–100.

Murakami, S. (1998) Overview of turbulence models applied in CWE-1997. *Journal of Wind Engineering and Industrial Aerodynamics* 74–76: 1–24.

Murakami, S., Ooka, R., Mochida, A., Yoshida, S. and Kim, S. (1999) CFD analysis of wind climate from human scale to urban scale. *Journal of Wind Engineering and Industrial Aerodynamics* 81(1–3): 57–81.

Peng, L., Lui, J.P., Wang, Y., Chan, P.W., Lee, T.C., Peng, F., Wong, M.S. and Li, Y.G. (2018) Wind weakening in a dense high-rise city due to over nearly five decades of urbanization. *Building and Environment* 138: 207–220.

Richards, P.J. and Hoxey, R.P. (1993) Appropriate boundary conditions for computational wind engineering models using the K-ε turbulence model. *Journal of Wind Engineering and Industrial Aerodynamics* 46–47: 145–153.

Stathopoulos, T. (2002) The numerical wind tunnel for industrial aerodynamics: Real or virtual in the new millennium?, *Wind and Structures* 5(2–4): 193–208.

Toparlar, Y., Blocken, B., Maiheu, B. and van Heijst, G.J.F. (2017) A review on the CFD analysis of urban microclimate. *Renewable and Sustainable Energy Reviews* 80: 1613–1640.

Toparlar, Y., Blocken, B., Maiheu, B. and van Heijst, G.J.F. (2018) Impact of urban microclimate on summertime building cooling demand: A parametric analysis for Antwerp, Belgium. *Applied Energy* 228: 852–872.

Tominaga, Y., Mochida, A., Yoshie, R., Kataoka, H., Nozu, T., Yoshikawa, M. and Shirasawa, T. (2008) AIJ guidelines for practical applications of cfd to pedestrian wind environment around buildings. *Journal of Wind Engineering and Industrial Aerodynamics* 96(10): 1749–1761.

Tominaga, Y., Sato, Y. and Sadohara, S. (2015) CFD simulations of the effect of evaporative cooling from water bodies in a micro-scale urban environment: Validation and application studies. *Sustainable Cities and Society* 19: 259–270.

van Hooff, T., Blocken, B., Timmermans, H.J.P. and Hensen, J.L.M. (2016) Analysis of the predicted effect of passive climate adaptation measures on energy demand for cooling and heating in a residential building. *Energy* 94: 811–820.

Assignments

Assignment 1

In this exercise, the effect of passive climate change adaptation measures on the building scale is studied. For this exercise we will use EnergyPlus (version 8.7), which is a freely available building energy simulation (BES) software package; however, other BES software can be used as well.

An arbitrary simple generic building is studied, with a floor plan of 5 × 5 m^2 and a height of 3 m, resulting in an indoor volume of 75 m^3. There is one window present in the building, with a surface area of 2 × 2 m^2, which is located in the middle of the south-oriented facade (middle of window located at (2.5 m; 1.5 m); see Figure 22.33).

The construction properties are shown in Table 22.4.

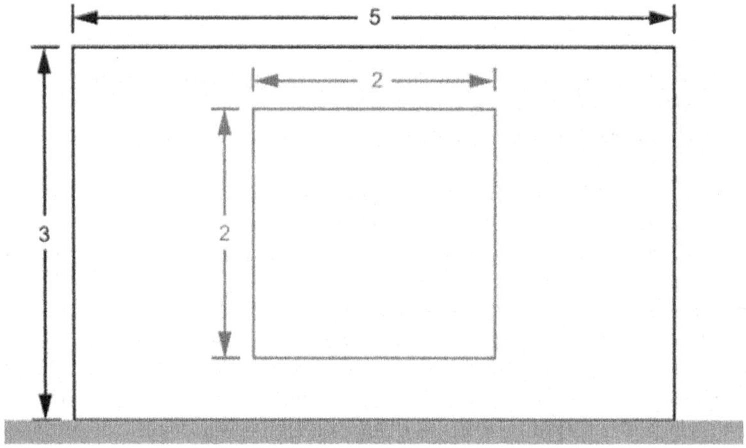

Figure 22.33 Front view of the south facade, depicting the 2 × 2 m² window located in the middle of the facade. Dimensions in m.

Table 22.4 Overview of different layers with required properties

	Thickness	*Conductivity*	*Density*	*Specific heat*	*Short-wave absorptivity*
	[m]	*[W/mK]*	*[kg/m³]*	*[J/kgK]*	*[−]*
Wall					
Outer leaf (masonry)	0.1	0.8	2000	840	0.7 (default)
Air cavity		Thermal resistance = 0.17 m²K/W			
Insulation layer	0.14	0.04	35	1470	0.7 (default)
Inner leaf (bricks)	0.1	1	2000	840	0.7 (default)
Floor (incl. soil layer)					
Soil	1	1.28	1460	800	0.7 (default)
Insulation layer	0.14	0.04	50	840	0.7 (default)
Concrete floor	0.2	1.4	2500	840	0.7 (default)
Roof					
Roof covering	0.002	0.17	1400	1470	0.7 (default)
Insulation	0.2	0.04	35	1470	0.7 (default)
Concrete roof	0.2	1.4	2500	840	0.7 (default)
	U-*factor*	*Solar heat gain coefficient*		*Visible transmittance*	
	[W/m²K]	[−]		[−]	
Window					
Simple glazing system	1.65	0.7		0.8	

An ideal loads air system is present, without active cooling and with a heating setpoint of 20 °C throughout the day. The air exchange rate is equal to ACH = 0.8, and no infiltration is assumed. There are no internal heat gains present for this case. Ground temperature is assumed to be 10 °C below the 1 m thick layer of soil.

Questions

1 Perform a simulation using weather data of Amsterdam, the Netherlands (e.g. https://energyplus.net/weather-location/europe_wmo_region_6/NLD//NLD_Amsterdam.062400_IWEC) for the reference case. Export the hourly zone air temperatures and determine the maximum value. In addition, determine the heating load.
2 Change the albedo value (short-wave reflectivity) of the outer leaf of the wall and of the roof cover to 0.8 by changing the short-wave absorptivity to 0.2 (=1 – albedo). Rerun the simulation and determine the maximum occurring zone air temperature and the heating load.
3 Change the density of the inner leaf of the wall to 100 kg/m³ to simulate the influence of a less heavy material for the inner leaf. Rerun the simulation and determine the maximum occurring zone air temperature and the heating load.
4 Perform the simulations from Question 2 for a building with less thermal insulation for the roof and walls, i.e. a 0.1 m thinner layer of thermal insulation (wall: 0.04 m instead of 0.14 m; roof: 0.1 m instead of 0.2 m, etc.). What are the maximum occurring zone air temperature and the heating load for this case? Can you explain the reason for the observed differences?

Assignment 2

The effect of transpirational cooling of air by vegetation shall be explored by CFD. Create a simple U-shaped building and equip the inner courtyard with trees analogue to Figure 22.20 (see also Gromke et al. 2015 for more information). Assign the vegetation sink and source terms provided in Equations 22.5–22.7 to cells occupied by trees. Users of ANSYS Fluent can use the user-defined function described below.

Assign a volumetric cooling power P_c to cells occupied by trees by implementing a constant value ($P_c = 200 - 400$ W/m³ per unity LAD) as a sink in the energy equation.

Perform a CFD simulation for a wind direction of your choice without taking into account the transpirational cooling effect of trees (reference scenario). Now, assign a volumetric cooling power P_c to cells occupied by trees by implementing a constant value ($P_c = 200 - 400$ W/m³ per unity LAD) as a sink in the energy equation and run the simulation. Analyze the air temperatures in the courtyard and around the building (e.g. at 1 m above ground).

Perform parameter variations:

1 double/halve the volumetric cooling power
2 double/halve the wind speed of the approach flow
3 change the direction of the approach wind
4 change value of LAD in the UDFs

Before looking at the simulation results, think about how the parameter variation will affect the air temperature. Is your expectation in qualitative/quantitative agreement with the simulation results? Explain the tendencies and observations.

User-defined function:

```
/*************************************************************/
/* UDF for specifying source terms for VEGETATION */
/*************************************************************/

#include "udf.h"

#define Cd 0.2     /* usually between 0.1 - 0.3 ; lambda = Cd * LAD */
#define LAD 2.0

real A1t = 1.0; /* Betap */
real A2t = 5.1; /* Betad */
real A3t = 0.9; /* closure constant cepsilon4 */
real A4t = 0.9; /* closure constant cepsilon5 */

/*---------------------------------------------------------*/
DEFINE_SOURCE(x_mom_source,c,t,dS,eqn)
{
real x[ND_ND];
real source;
real U   = C_U(c,t);
real V   = C_V(c,t);
real W   = C_W(c,t);
real VEL = sqrt(C_U(c,t)*C_U(c,t)+C_V(c,t)*C_V(c,t)+C_W(c,t)
           *C_W(c,t));
real k   = C_K(c,t);
real e   = C_D(c,t);
real rho = C_R(c,t);
C_CENTROID(x,c,t);

source  = -Cd*LAD*rho*VEL*U;
dS[eqn] = -Cd*LAD*rho*(U*U/VEL + VEL);

return source;
}

/*---------------------------------------------------------*/
DEFINE_SOURCE(y_mom_source,c,t,dS,eqn)
{
real x[ND_ND];
real source;
real U   = C_U(c,t);
real V   = C_V(c,t);
real W   = C_W(c,t);
real VEL = sqrt(C_U(c,t)*C_U(c,t)+C_V(c,t)*C_V(c,t)+C_W(c,t)
           *C_W(c,t));
real k   = C_K(c,t);
real e   = C_D(c,t);
real rho = C_R(c,t);
```

```
C_CENTROID(x,c,t);

source  = -Cd*LAD*rho*VEL*V;
dS[eqn] = -Cd*LAD*rho*(V*V/VEL + VEL);

return source;
}

/*---------------------------------------------------------*/
DEFINE_SOURCE(z_mom_source,c,t,dS,eqn)
{
real x[ND_ND];
real source;
real U   = C_U(c,t);
real V   = C_V(c,t);
real W   = C_W(c,t);
real VEL = sqrt(C_U(c,t)*C_U(c,t)+C_V(c,t)*C_V(c,t)+C_W(c,t)
           *C_W(c,t));
real k   = C_K(c,t);
real e   = C_D(c,t);
real rho = C_R(c,t);
C_CENTROID(x,c,t);

source  = -Cd*LAD*rho*VEL*W;
dS[eqn] = -Cd*LAD*rho*(W*W/VEL + VEL);

return source;
}

/*---------------------------------------------------------*/
DEFINE_SOURCE(tke_source,c,t,dS,eqn)
{
real x[ND_ND];
real source;
real U   = C_U(c,t);
real V   = C_V(c,t);
real W   = C_W(c,t);
real VEL = sqrt(C_U(c,t)*C_U(c,t)+C_V(c,t)*C_V(c,t)+C_W(c,t)
           *C_W(c,t));
real k   = C_K(c,t);
real e   = C_D(c,t);
real rho = C_R(c,t);
C_CENTROID(x,c,t);

source  = Cd*LAD*rho*(A1t*pow(VEL,3) - A2t*VEL*k);
dS[eqn] = Cd*LAD*rho*(- A2t*VEL);

return source;
}
```

```
/*---------------------------------------------------------------*/
DEFINE_SOURCE(eps_source,c,t,dS,eqn)
{
real x[ND_ND];
real source;
real U   = C_U(c,t);
real V   = C_V(c,t);
real W   = C_W(c,t);
real VEL = sqrt(C_U(c,t)*C_U(c,t)+C_V(c,t)*C_V(c,t)+C_W(c,t)
           *C_W(c,t));
real k   = C_K(c,t);
real e   = C_D(c,t);
real rho = C_R(c,t);
C_CENTROID(x,c,t);

source   = Cd*LAD*rho*(A3t*A1t*e/k*pow(VEL,3) - A4t*A2t*VEL*e);
dS[eqn]  = Cd*LAD*rho*(A3t*A1t/k*pow(VEL,3) - A4t*A2t*VEL);

return source;
}
```

Answers

This section provides solution keys to the projects assigned.

Assignment 1

Q1:
Maximum zone air temperature: ≈ 26.5 °C
Heating load: ≈ 2,536 kWh

Q2:
Maximum zone air temperature: ≈ 24.8 °C
Heating load: ≈ 2,817 kWh

Q3:
Maximum zone air temperature: ≈ 28.5 °C
Heating load: ≈ 2,567 kWh

Q4:
Low albedo case (see Question 2):
Maximum zone air temperature: ≈ 24.8 °C
Heating load: ≈ 2,817 kWh

Low albedo case with less thermal insulation:
Maximum zone air temperature: ≈ 23.8 °C
Heating load: ≈ 5,081 kWh

The maximum air temperature decreases when thermal insulation is decreased by 10 cm. This trend is depicted in Figure 22.34 for six days in June. The air in the building with less thermal insulation cools down faster (and to lower temperatures) than in a building with more

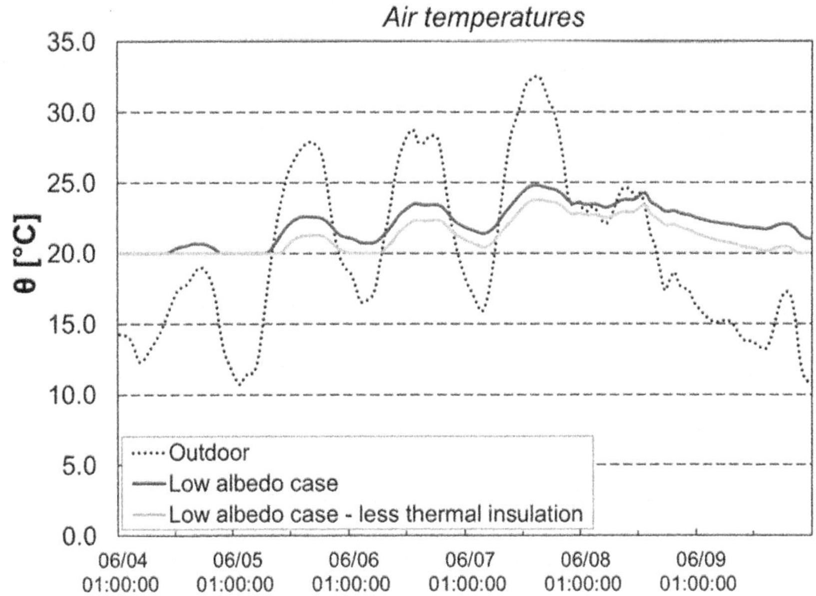

Figure 22.34 Indoor and outdoor air temperatures. Indoor air temperatures are depicted for the low albedo case with two different thicknesses of the thermal insulation layer in the building envelope.

thermal insulation. In addition, the building with more thermal insulation heats up faster when solar radiation enters the building (e.g. on 4 June), resulting in indoor air temperatures above the setpoint while the outdoor air temperatures are below the heating setpoint. Of course, the indoor air temperatures depend on many other aspects as well, e.g. ventilation and infiltration rates, solar shading, internal heat gains, etc.

The heating load obviously increases (≈ 80% increase in this case) when thermal insulation is decreased by 10 cm. This is due to the resulting lower thermal resistance of the building envelope, resulting in higher thermal transmittance and thus a higher heating demand.

Assignment 2

1 The air temperature difference relative to the reference scenario ($P_c = 0$ W/m^3) is, to a first approximation, double/halve in the cells occupied by trees if doubling/halving the volumetric cooling power and fades out with leeward distance from the trees. There should be hardly any noticeable difference in air temperature outside the courtyard.
2 The air temperature difference relative to the reference scenario ($P_c = 0$ W/m^3) is, to a first approximation, halve/double in the cells occupied by trees if doubling/halving the wind speed of the approach flow and fades out with leeward distance from the trees. There should be hardly any noticeable difference in air temperature outside the courtyard.
3 The effect of tree-induced transpirational cooling on air temperature inside the courtyard is larger the more the U-shaped building arrangement shelters the courtyard from the approach flow, i.e. the lesser the ventilation of the courtyard is. There should be hardly any noticeable difference in air temperature outside the courtyard.
4 The effect of tree-induced transpirational cooling on air temperature inside the courtyard increases with increasing LAD. There should be hardly any noticeable difference in air temperature outside the courtyard.

Index

absorption 30, 33, 117, 118, 271, 272, 274, 282, 298, 315, 319, 324, 325, 330, 331, 666
abstraction level 3, 631, 634, 635, 637, 651, 652
acoustics 118, 128, 315, 316, 317, 318, 319, 321, 322, 323, 324, 327, 328, 334, 340, 365, 391
active testing 400, 401
activity (human) 105, 117, 118, 119, 121, 150, 151, 156, 157, 159, 160, 161, 164, 166, 167, 168, 169, 180, 662, 669
adaptation measures 723, 724, 725, 726, 728, 729, 730, 733, 738, 744, 754, 755, 759
aerodynamic roughness length 617, 618, 619, 740, 741, 745, 747, 749
agent 130, 260, 350, 571, 572, 573, 574, 631, 634, 667, 671, 676, 684, 685
air changes per hour (ACH) 96, 99, 729, 761
air density 58, 78, 83, 275, 510, 530, 589, 616, 620, 745
airflow 8, 10, 28, 56, 57, 60, 76, 77, 78, 79, 80, 81, 82, 83, 84, 86, 87, 88, 92, 93, 94, 95, 96, 97, 98, 99, 100, 101, 102, 103, 104, 105, 107, 111, 110, 112, 160, 175, 180, 274, 289, 375, 401, 408, 412, 416, 417, 422, 429, 431, 432, 440, 441, 444, 445, 446, 447, 448, 455, 461, 617, 642, 643, 649, 681, 725, 745
airflow network model 76, 77, 79, 80, 81, 82, 94, 95, 98, 101, 108, 110, 111, 648, 725
air pollution 362, 724
air quality 76, 83, 95, 96, 98, 99, 105, 107, 117, 118, 131, 148, 164, 172, 289, 298
air temperature 13, 18, 21, 22, 29, 30, 35, 36, 43, 46, 47, 53, 54, 55, 56, 57, 58, 59, 60, 61, 62, 64, 72, 73, 75, 76, 83, 91, 98, 99, 101, 104, 105, 114, 121, 136, 151, 155, 157, 159, 160, 161, 166, 167, 173, 175, 180, 196, 210, 289, 293, 306, 401, 403, 411, 429, 437, 439, 440, 443, 444, 451, 454, 455, 473, 498, 499, 500, 516, 518, 519, 524, 529, 530, 543, 566, 574, 589, 591, 593, 594, 597, 598, 599, 600, 601, 602, 611, 613, 621, 681, 723, 724, 726, 729, 730, 733, 742, 744, 745, 747, 748, 749, 750, 751, 752, 753, 754, 755, 761, 764, 765
air velocity 91, 151, 155, 157, 159, 160, 167, 175, 181, 510, 527, 530
albedo value *see* shortwave reflectivity (SWR)
aspect ratio 101, 738
auralization 315, 317, 328, 329, 330, 333, 334, 336

behavioural model 683, 684
BIM ecosystem 376, 382
black box/black box model 119, 132, 361, 440, 454, 459, 460, 463, 464, 465, 466, 467, 468, 469, 477, 507, 564, 565, 566, 568, 569, 574, 634
boundary conditions 10, 20, 47, 49, 79, 80, 82, 86, 88, 90, 91, 93, 94, 102, 103, 104, 107, 112, 113, 121, 128, 151, 163, 175, 191, 192, 195, 270, 271, 275, 287, 289, 308, 315, 456, 458, 505, 511, 513, 518, 521, 522, 523, 621, 625, 636, 637, 649, 656, 664, 666, 741, 745, 747, 748
building automation system (BAS) 197, 403, 440, 443, 455, 456, 457, 465, 477, 478, 482, 487, 632, 634
building commissioning 399, 400, 402, 403, 404, 405, 422, 423, 444, 448
building delivery process 128, 140, 344, 377, 639, 645
building energy model (BEM) 8, 118, 209, 387, 455, 459, 461, 469, 472, 477, 492, 663, 671, 674, 689, 696, 697, 694, 698, 699
building information model (BIM) 9, 223, 228, 360, 362, 368, 374, 375, 376, 377, 379, 382, 383, 527, 631, 632, 639, 645, 646, 652, 656, 672, 673, 707
building operation 9, 128, 375, 399, 424, 444, 454, 455, 456, 457, 470, 479, 491, 632, 635
building policy 384, 385, 390, 391
building-to-grid integration 559, 560, 561, 562, 563, 564, 565, 569, 573, 575, 582, 584, 585
buoyancy 27, 28, 83, 88, 94, 102, 742

cavity 52, 676, 728, 760
CIBSE 8, 77, 98, 100, 206, 210, 247, 481, 678
climate change 191, 205, 206, 209, 210, 212, 213, 385, 666, 671, 701, 723, 724, 725, 726, 730, 733, 738, 744, 759
climate change adaptation 723, 724, 725, 726, 730, 733, 738, 744, 759
climate change mitigation 723, 724

climate file 207, 208, 229, 232, 233, 242, 664
clo 156, 160, 167, 169, 173, 180
cogeneration 534, 551, 552, 553, 685
combined heat and power 534, 586, 633, 697, 714
complexity 1, 3, 7, 82, 88, 93, 119, 180, 225, 239, 240, 242, 244, 287, 289, 327, 348, 357, 367, 378, 391, 455, 458, 469, 482, 485, 488, 512, 520, 587, 588, 612, 631, 635, 639, 642, 643, 652, 663, 707, 738
computational cost 120, 327, 476, 482, 563, 587, 588, 747
computational fluid dynamics (CFD) 28, 60, 76, 77, 82, 83, 84, 86, 88, 89, 91, 92, 93, 94, 95, 98, 99, 100, 102, 103, 104, 105, 107, 108, 110, 112, 146, 149, 160, 161, 162, 175, 180, 183, 195, 270, 272, 273, 287, 292, 363, 462, 631, 649, 723, 738, 740, 743, 744, 745, 747, 754, 755, 761
computational grid 742, 748, 749
conduction 13, 14, 15, 16, 17, 18, 19, 20, 21, 22, 23, 24, 26, 27, 35, 36, 43, 44, 45, 46, 48, 49, 50, 51, 60, 61, 62, 65, 70, 71, 72, 75, 150, 151, 164, 289, 495, 622, 624, 637, 648, 662, 708, 729, 742, 743
conductive resistance 51, 62
contaminant 76, 79, 81, 82, 83, 94, 95, 96, 97, 98, 103, 104, 105, 106, 107, 110, 112
convection 13, 14, 26, 27, 28, 29, 30, 32, 43, 44, 45, 46, 47, 48, 49, 50, 51, 52, 53, 54, 55, 56, 57, 58, 60, 61, 62, 63, 65, 74, 75, 87, 90, 91, 92, 105, 149, 150, 151, 157, 159, 160, 175, 177, 183, 190, 196, 275, 289, 293, 473, 474, 476, 495, 542, 545, 547, 548, 589, 594, 604, 605, 623, 637, 644, 656, 681, 688, 724, 729, 742, 743, 747
convective resistance 53, 623
cooling load 13, 18, 20, 22, 26, 29, 38, 43, 45, 46, 55, 56, 60, 61, 63, 64, 128, 439, 461, 479, 481, 484, 591, 592, 626, 646, 685, 701, 708
cooling power 13, 18, 20, 22, 26, 29, 38, 43, 45, 46, 55, 56, 60, 61, 63, 64, 128, 439, 461, 479, 481, 484, 591, 592, 626, 646, 685, 701, 708, 747, 749, 750, 755, 761, 765
cost estimating 375, 378, 379, 381, 633

data model 343, 359, 360, 640, 645, 646, 673, 680, 685
daylight autonomy 224, 250, 352, 353, 357, 716
daylight availability 143, 223, 224, 225, 239, 246, 248, 250, 251, 256, 257, 258, 259, 260, 261, 699, 716
daylight coefficient 241, 242, 253
daylight factor 224, 229, 236, 243, 244, 245, 246, 247, 248
daylight simulation 221, 222, 223, 224, 225, 226, 227, 232, 233, 235, 239, 240, 241, 242, 244, 246, 252, 257, 258, 259, 260, 261, 267, 631
decision-making 131, 343, 344, 346, 347, 355, 356, 363, 366, 367, 368, 369, 372, 373, 379, 384, 561, 588, 663, 669, 671
degree hours 171, 665, 729, 730, 731, 732, 733, 734, 735, 736, 737, 755
design day 21, 63, 194, 479
design evolution 93, 96, 98, 99, 100, 317, 330, 331, 343, 344, 346, 347, 358, 367, 369, 374, 378, 381, 382, 400, 402, 639, 656
design process 93, 96, 98, 99, 100, 317, 330, 331, 346, 347, 358, 367, 374, 378, 381, 382, 400, 402, 639, 656
design stage 118, 127, 146, 148, 180, 181, 225, 226, 316, 328, 330, 343, 358, 361, 377, 379, 382, 459, 523, 559, 561
design team 221, 223, 233, 246, 345, 346, 367, 368, 377, 378, 379, 381, 382, 632, 635, 639, 706
deterministic model 564, 565, 566, 568, 686, 712
diagnostics 131, 259, 318, 400, 401, 434, 461, 474, 632, 633, 634, 639, 645, 649, 651
diffuse daylight 223, 228, 229, 242, 247
diffuse radiation 34, 36, 37, 38, 202, 232
diffuse reflection 324, 330, 679
direct sunlight 223, 228, 229, 242, 247, 251, 255
displacement ventilation 81, 88, 94, 104, 105, 112, 113, 115, 161
domestic hot water (DHW) 535, 685, 708, 714
dry-bulb temperature 192, 194, 195, 197, 204, 612, 615, 629
duck curve 569, 575, 576, 577, 578, 713
dynamic model 454, 456, 461, 462, 470, 505, 527, 529, 586, 588, 596, 601, 676, 707

early design phase 128, 148, 181
electrical load 551, 632
electric lighting 221, 237, 240, 241, 255, 257, 259, 351
emission 2, 3, 30, 31, 119, 206, 210, 324, 326, 365, 387, 455, 471, 482, 534, 587, 616, 659, 666, 669, 671, 674, 675, 676, 696, 697, 713, 714, 715, 723, 724
empirical model 79, 87, 152, 272, 282, 467, 507, 509, 512, 526, 614
energy balance 43, 150, 158, 167, 293, 354, 495, 497, 506, 515, 524, 541, 542, 543, 548, 588, 589, 603, 664, 666, 674, 679, 738, 754
energy storage 2, 455, 456, 484, 545, 559, 561, 578, 633, 634, 644
energy use intensity (EUI) 424, 425, 699, 709, 710, 711, 712, 713, 716
equation-based 470, 504, 587, 631, 636, 637, 638, 640, 644, 645, 646, 647, 649, 650, 652, 656
evaporative cooling 307, 733, 759
evapotranspiration 664, 724, 738, 742

Fanger, P.O. 131, 147, 151, 152, 153, 155, 156, 158, 159, 160, 161, 162, 164, 165, 166, 170, 171
fault detection 400, 403, 405, 434, 437, 440, 442, 443, 444, 445, 446, 447, 456, 457, 459, 460, 463, 468, 503, 512, 523, 526, 588, 632, 633, 634, 639, 645, 649, 651

fuel cell 534, 535, 536, 537, 538, 539, 540, 541, 544, 546, 551
functional requirements 317, 346, 348, 349, 372

geographic information systems (GIS) 680, 702, 703, 707
glare 221, 226, 229, 232, 239, 248, 251, 252, 253, 254, 255, 256, 257, 259, 260
glass 33, 34, 35, 36, 51, 52, 61, 175, 409, 602, 604, 605, 611, 614, 681
glazing 2, 33, 36, 51, 177, 225, 226, 238, 240, 242, 244, 245, 246, 248, 255, 260, 409, 473, 476, 675, 728, 729, 760
gray box/gray box model 316, 454, 459, 461, 467, 468, 469, 472, 473, 474, 482, 487, 564, 565, 566
green façade 723, 745, 747, 753, 754, 755
greenhouse effect 729, 730
greenhouse gas (GHG) 2, 3, 210, 365, 366, 534, 697, 713, 714, 715, 723, 724
green roof 365, 723, 733, 745, 747, 754
grid (power/electric) 486, 559, 560, 561, 562, 563, 570, 571, 575, 576, 577, 578, 582

heat balance 13, 22, 29, 30, 34, 35, 37, 38, 43, 44, 45, 46, 47, 48, 50, 51, 52, 53, 54, 55, 56, 57, 58, 59, 60, 63, 64, 65, 70, 72, 151, 152, 153, 158, 164, 169, 171, 196, 355, 602, 611, 612, 707
heat exchanger 512, 514, 519, 524, 526, 529, 536, 539, 540, 541, 544, 545, 546, 547, 548, 553, 554, 555, 586, 592, 593, 594, 596, 598, 605, 606, 621, 622, 623, 624, 625, 626, 627, 629, 643
heating load 13, 14, 17, 18, 26, 43, 63, 64, 98, 473, 591, 701, 707, 709, 712, 716, 761, 764, 765
heat storage 18, 151, 462, 473, 637, 713, 724, 744
heat stress 150, 156, 172, 183, 190
heat transfer coefficient 35, 39, 46, 90, 91, 151, 159, 177, 412, 417, 505, 527, 540, 592, 594, 596, 598, 611, 613, 629
heat wave(s) 697, 723, 724, 725, 729, 733, 738, 743, 747, 748, 754, 755
hierarchical modeling 474, 641, 642
hybrid ventilation 76, 98, 101, 107, 108

image source method 315, 323, 325, 326, 327, 329
indoor air quality (IAQ) 76, 83, 95, 96, 98, 99, 101, 105, 107, 117, 118, 131, 148, 164, 289, 298, 470
indoor air temperature 62, 121, 136, 166, 180, 724, 726, 729, 730, 765
industry foundation classes (IFC) 376, 527, 645, 673
infiltration 13, 26, 29, 46, 55, 56, 57, 58, 59, 65, 77, 78, 79, 80, 111, 190, 196, 589, 681, 688, 695, 705, 706, 709, 729, 761, 765
integrated project delivery (IPD) 344, 374, 377, 378, 379, 381, 382
internal combustion engine 534, 535, 547, 551
interoperability 243, 374, 375, 376, 379, 585, 645, 649, 673
irradiance 196, 202, 204, 232, 240, 560, 611, 680, 682, 683

large eddy simulation (LES) 84, 281
latent heat 29, 56, 118, 151, 196, 289, 291, 293, 541, 543, 662, 685, 738, 747
leaf area density (LAD) 745, 747, 749, 750, 752, 753, 755, 761, 762, 763, 764, 765
leaf area index (LAI) 726, 729
LEED 76, 222, 223, 248, 257, 262, 353, 698
levels of detail (LOD) 146, 182, 348, 673, 704, 705
life cycle 3, 5, 6, 7, 8, 128, 362, 386, 399, 524, 632, 639, 645, 652, 656, 663, 673
life cycle analysis (LCA) 365, 375, 376
light detection and ranging (LiDAR) 674, 703, 704, 718
lighting simulation 224, 240, 257, 352, 353, 365
linear model 126, 462, 463, 464, 465, 466, 482, 504, 564, 566
load curve 697, 698, 699, 701, 710, 712, 713, 714, 717
load profile *see* load curve
low-E coating 34, 36, 53, 72
luminance 224, 227, 229, 231, 233, 234, 237, 238, 239, 241, 242, 246, 251, 252, 253, 353, 357

Markov model 122, 123, 712
mean radiant temperature 40, 41, 42, 43, 45, 53, 54, 155
mechanical ventilation 98, 101, 104, 107, 195, 477
metabolism 151, 180, 659
meteorological data 194, 200, 271, 278, 279, 281, 303, 392, 743, 747
microclimate 195, 210, 666, 699, 708, 723, 748
micro-cogeneration 534, 535, 536, 537, 538, 539, 540, 541, 542, 544, 545, 546, 547, 548, 550, 551, 553, 554, 555, 556, 714
mixed mode ventilation/building(s) 128, 166, 347, 454, 456, 472, 477, 481, 484, 594
mixing ventilation 104, 105, 106
model-based control 101, 455, 456, 461, 470, 472, 476, 634, 639
model calibration 194, 213, 329, 335, 459, 460, 471, 472, 541, 543, 544, 546, 551, 554, 678
moisture 26, 147, 151, 152, 156, 181, 191, 200, 270, 271, 272, 274, 278, 282, 289, 290, 291, 292, 293, 294, 295, 296, 297, 298, 299, 300, 301, 302, 303, 304, 305, 306, 307, 308, 487, 685
moisture damage 270, 272, 289, 292, 294, 295, 298, 307
moisture sensation 147, 152
moisture transport 26, 271, 274, 289, 300, 307
multi-agent systems/multi-agent simulation (MAS) 667, 671, 676, 684
multi-domain modeling 631, 641
multi-zone airflow network 76, 77, 79, 80, 81, 82, 94, 95, 98, 101, 107, 108, 110, 111

natural convection 28, 29
natural ventilation 13, 77, 98, 99, 100, 101, 107, 160, 190, 195, 196, 213, 463, 481, 483, 648, 725, 729, 730, 733, 754

Navier-Stokes equations 84, 87, 175, 273, 462, 742, 754
net-zero-energy building (NZEB) 13, 17
neural network 210, 464, 466, 482, 565
numerical method 20, 24, 25, 26, 70, 82, 272, 512, 610, 649, 650, 651, 656

object-oriented 631, 632, 636, 637, 638, 640, 641, 642, 644, 645, 646, 650, 652
occupancy 118, 120, 121, 122, 123, 124, 125, 126, 127, 128, 129, 131, 132, 133, 137, 138, 139, 140, 141, 167, 224, 234, 250, 255, 259, 260, 349, 388, 402, 403, 423, 425, 454, 455, 472, 479, 551, 673, 679, 683, 684, 695, 697, 711, 712, 717
occupant behavior 8, 118, 119, 123, 125, 126, 129, 131, 134, 135, 136, 137, 138, 139, 140, 143, 148, 166, 173, 221, 223, 255, 256, 260, 564, 697, 699, 705, 706, 709, 711
open-source 223, 232, 641, 650, 664, 672, 673, 680, 686
operational performance 400, 402, 405, 421
operative temperature 150, 153, 156, 157, 159, 160, 166, 167, 169, 171, 173, 176, 190, 479
optimization 9, 10, 26, 98, 194, 202, 397, 399, 400, 401, 402, 403, 404, 405, 421, 422, 423, 425, 443, 444, 448, 456, 457, 459, 460, 461, 462, 463, 467, 468, 469, 470, 471, 473, 476, 477, 478, 479, 482, 483, 489, 490, 491, 503, 509, 522, 523, 524, 525, 526, 527, 528, 559, 562, 565, 570, 587, 631, 632, 634, 651, 666, 689, 713
outdoor air temperature 18, 35, 36, 46, 53, 57, 58, 62, 72, 73, 75, 98, 99, 411, 444, 566, 574, 729, 744, 765
overcast sky 229, 231, 232, 240, 244, 246, 247, 258
overheating 148, 172, 180, 190, 194, 248, 251, 255, 548, 550, 587, 723, 725, 729, 730, 731, 732, 733, 734, 735, 736, 737, 754, 755

parametric design 377, 382
Perez sky model 231, 232, 233
performance criteria 316, 328, 343, 344, 345, 346, 347, 352, 353, 356, 360, 361, 363, 368, 369, 699
performance maps 456, 459, 465, 538
performance measure 343, 345, 346, 347, 350, 352, 353, 355, 357, 359, 368, 369, 372, 400, 433, 461, 508, 524, 526
PMV 155, 157, 158, 159, 160, 161, 162, 163, 164, 165, 166, 167, 168, 169, 171, 175, 180, 183, 190
power usage profile 559, 578, 699, 717
PPD 159, 160, 167, 171, 175, 183, 190
predicted mean vote 155, 157, 158, 159, 160, 161, 162, 163, 164, 165, 166, 167, 168, 169, 171, 180; *see also PMV*
pressure coefficient 78, 79, 99, 100
pressure difference 77, 78, 80, 98, 99
probabilistic model *see* stochastic model
productivity 2, 76, 148, 164, 645, 660, 724

quality assurance (QA) 1, 5, 8, 94, 389, 391

radiation 13, 14, 18, 20, 26, 29, 30, 31, 32, 33, 34, 36, 37, 38, 39, 40, 41, 42, 43, 44, 45, 46, 47, 48, 49, 50, 51, 53, 54, 55, 60, 61, 62, 65, 66, 72, 73, 74, 75, 92, 125, 148, 155, 156, 157, 160, 161, 162, 167, 177, 192, 193, 194, 195, 197, 200, 201, 202, 203, 206, 207, 208, 213, 220, 223, 226, 229, 232, 233, 237, 252, 289, 293, 326, 353, 354, 357, 462, 464, 473, 495, 496, 564, 602, 603, 604, 605, 662, 663, 672, 678, 679, 680, 681, 682, 683, 689, 708, 724, 726, 729, 733, 742, 755, 765
radiative resistance 51, 53
radiosity 39, 236, 237, 239, 672, 679, 682, 683
rain 194, 196, 270, 271, 272, 273, 274, 275, 276, 277, 278, 279, 280, 281, 282, 285, 288, 292, 298, 301, 306, 307, 308, 313, 314
raytracing 236, 237, 238, 239, 242
reflectance 31, 33, 66, 225, 226, 239, 245, 353, 357
regression model 125, 126, 138, 165, 424, 429, 433, 434, 697, 706
relative humidity 153, 157, 159, 160, 168, 173, 180, 196, 220, 232, 289, 293, 294, 295, 301, 408, 454
renewable energy plant 586, 587, 589, 591, 593, 595, 597, 599, 601, 603, 605, 607, 611, 613, 615, 617, 619, 621, 623, 625
renewable energy source (RES) 455, 585, 714, 724
renewable energy system 195, 389, 586, 674
resistance: conductive 51, 62; convective 53, 623; radiative 51, 53; thermal 14, 15, 16, 18, 51, 66, 120, 156, 289, 472, 498, 515, 566, 622, 624, 627, 723, 726, 728, 729, 730, 733, 754, 760, 765
response factor 20, 21, 22, 43, 60, 61, 63, 622, 624
retrofit 292, 302, 315, 316, 317, 323, 329, 333, 335, 336, 384, 400, 405, 421, 423, 461, 634, 697, 714, 715
reverberation time 319, 320, 321, 330, 331, 332, 333, 334, 340
Reynolds-averaged Navier-Stokes (RANS) equations 84, 273, 281, 742, 747, 754
robustness 3, 9, 212, 461, 472, 520, 648

sensible heat 29, 151, 156, 196, 541, 574
sensitivity analysis 9, 127, 359, 367, 368, 676, 686, 711, 742, 743
shear stress 90, 91, 745, 746, 747
shortwave radiation 32, 38, 148, 156, 293, 663, 681, 724
shortwave reflectivity (SWR) 196, 210, 355, 662, 663, 723, 725, 726, 728, 730, 733, 761, 764, 765
Simulink 469, 470, 495, 498, 599, 648, 656
sky model 196, 205, 221, 223, 224, 228, 229, 231, 232, 233, 242, 244, 246
solar heat gain coefficient (SHGC) 34, 35, 36, 760
solar photovoltaic (PV) system 196, 213, 365, 560, 561, 575, 576, 577, 578, 607, 608, 611, 612, 613, 614, 615, 616, 617, 629, 713, 714, 753
solar radiation 18, 31, 32, 33, 36, 37, 38, 39, 40, 43, 45, 46, 47, 51, 55, 60, 62, 72, 73, 74, 75, 125, 192, 193, 194, 195, 197, 200, 201, 202,

206, 207, 208, 213, 220, 232, 289, 354, 495, 496, 564, 602, 603, 604, 605, 681, 689, 729, 733, 742, 765
solar shading 707, 708, 723, 726, 729, 730, 733, 754, 765
solar thermal 196, 586, 602, 629, 632, 679, 685
sound absorption 117, 118, 319
sound pressure level 319, 330, 331, 332, 333
stack effect 77, 98, 161
stakeholder 1, 8, 9, 343, 344, 345, 346, 347, 349, 351, 352, 353, 354, 356, 361, 364, 366, 367, 368, 372, 376, 379, 381, 382, 389, 669, 673, 713, 717, 718
state space model 495, 498, 499, 568
statistical model 686, 697, 698
Stirling engine 534, 535, 551
stochastic model 122, 123, 125, 136, 137, 138, 139, 143, 200, 564, 565, 567, 679, 681, 684, 712
storage capacity 55, 559
street canyon 724, 743, 747, 749, 750, 753, 754, 755
system modeling 10, 503, 527, 559, 561, 562, 563, 564, 565, 569, 575, 578, 582, 584, 585, 631, 633, 650

target value design 378, 382
thermal bridge 14, 26, 462
thermal capacity 151, 190, 495, 496, 566, 597, 621, 726, 728
thermal comfort 29, 55, 76, 83, 92, 95, 96, 98, 99, 101, 104, 105, 128, 131, 138, 146, 148, 149, 152, 153, 154, 155, 156, 157, 160, 164, 165, 166, 167, 168, 172, 173, 177, 180, 181, 182, 183, 190, 307, 457, 481, 482, 562, 570, 643, 729
thermal conductivity 14, 20, 52, 83, 289, 460, 463, 604, 624, 625, 627, 724
thermal imaging 146, 181, 182
thermal load 13, 535, 538, 544, 551, 564
thermal mass 19, 20, 35, 51, 60, 63, 72, 456, 478, 486, 497, 544, 707, 713, 723, 725, 729, 730, 733, 754
thermal resistance 14, 15, 16, 18, 51, 66, 120, 156, 289, 472, 498, 515, 566, 622, 624, 627, 723, 726, 728, 729, 730, 733, 754, 760, 765
thermal sensation 146, 147, 148, 152, 153, 154, 155, 156, 157, 158, 161, 162, 163, 164, 166, 171, 177, 179, 180, 181, 183
thermal storage 18, 19, 29, 55, 473, 481, 527, 545, 588, 633, 685
thermoregulation 146, 150, 151, 161, 163, 180, 183
time series method 60, 70, 563, 564
TMY 197, 198, 200, 202, 206, 207, 208, 209, 212, 213, 232, 477, 574, 578, 701
transactive control 569, 570, 571, 573, 575, 576
transfer function 20, 21, 22, 23, 25, 26, 35, 38, 43, 48, 60, 61, 63, 64, 65, 70, 71, 72, 321, 328, 464, 465, 648
transient conduction 18, 20, 26, 43, 44, 48, 60, 65
transmission 18, 19, 32, 33, 226, 237, 315, 316, 319, 321, 365, 462, 478, 495, 534, 585, 614
transpiration, evapo- 664, 724, 738, 742, 744
transpirational cooling 747, 748, 752, 754, 761, 765
turbulence dissipation rate 276, 741, 745
turbulence kinetic energy 276, 741, 745
turbulence model 84, 86, 87, 88, 90, 92, 93, 103, 175, 273, 287, 742, 745, 754

uncertainty analysis 358, 359, 463, 505, 715
urban building energy models/modeling (UBEM) 659, 663, 669, 673, 674, 681, 685, 687, 694, 696, 697, 698, 699, 700, 701, 702, 703, 704, 705, 706, 707, 708, 709, 710, 711, 712, 713, 714, 715, 717, 718
urban canopy model 664, 738
urban heat island (UHI) 205, 209, 210, 212, 213, 307, 354, 355, 356, 664, 666, 701, 723, 724, 725, 738
urban modelling 5, 659, 661, 662, 663, 672, 673, 687, 688, 689, 699, 706
urban physics 723, 725, 738
useful daylight illuminance (UDI) 250, 251, 257
user behavior *see* occupant behavior
utility theory 346, 368
U-value 24, 35, 36, 46, 47, 53, 54, 55, 78, 83, 86, 87, 91, 177, 275, 276, 285, 286, 287, 288, 354, 408, 415, 416, 451, 462, 465, 474, 498, 517, 522, 562, 569, 570, 592, 593, 596, 597, 604, 605, 611, 612, 617, 618, 620, 728, 741, 745, 762, 763, 764

validation 5, 6, 8, 65, 92, 93, 94, 107, 112, 114, 118, 131, 136, 140, 141, 164, 180, 233, 239, 240, 242, 278, 317, 329, 375, 392, 458, 461, 538, 574, 651, 678, 684, 696, 699, 709, 733, 738, 743, 745, 746, 747, 748, 749, 754
validation, verification and testing (VV&T) 5, 6, 7, 8
vapour pressure 151, 155, 290
variable air volume (VAV) system/terminal 57, 58, 59, 407, 408, 411, 416, 425, 429, 431, 432, 435, 445, 452, 487, 570, 571, 572, 574, 575
vegetated roof (VR) 365, 723, 726, 729, 730, 733, 745, 747, 754, 755
ventilation performance 76, 77, 82, 83, 84, 87, 88, 91, 92, 94, 95, 103, 107, 148
verification 5, 6, 8, 92, 93, 94, 107, 347, 349, 350, 353, 356, 358, 400, 433, 458, 632, 634, 639, 647, 699
view factor 31, 32, 39, 41, 42, 47, 177, 355, 680, 682, 724
visual comfort 118, 195, 226, 228, 239, 246, 251, 256, 259, 351, 352

wall function 90, 91, 741, 742
weather data 99, 190, 191, 192, 194, 195, 196, 197, 200, 204, 205, 206, 207, 208, 209, 210, 212, 213, 292, 293, 297, 386, 390, 425, 442, 447, 456, 477, 479, 548, 578, 587, 606, 620, 695, 729, 742, 761
weather files 121, 191, 197, 202, 207, 209, 210, 477, 620, 713

white box/white box model 119, 206, 454, 458, 461, 462, 463, 467, 468, 469, 482, 564, 565, 566, 568
wind data 195, 616, 617, 620
wind flow 271, 272, 273, 275, 276, 277, 281, 285, 286, 604, 738, 740, 741
window operation 135, 136, 137, 138, 139
wind pressure 78, 79, 99, 688
wind speed 28, 46, 47, 49, 53, 60, 72, 78, 100, 194, 208, 213, 232, 274, 277, 278, 281, 283, 284, 285, 286, 287, 288, 289, 292, 293, 314, 611, 612, 616, 617, 618, 619, 620, 621, 681, 708, 741, 742, 743, 745, 746, 747, 749, 755, 761, 765

zero-energy building (ZEB) 389, 390